HANDBOOK OF
OPTICAL BIOMEDICAL DIAGNOSTICS

SECOND EDITION

Volume 1: Light–Tissue Interaction

Valery V. Tuchin
EDITOR

HANDBOOK OF OPTICAL BIOMEDICAL DIAGNOSTICS

SECOND EDITION

Volume 1: Light–Tissue Interaction

Valery V. Tuchin
EDITOR

SPIE PRESS
Bellingham, Washington USA

Library of Congress Cataloging-in-Publication Data

Names: Tuchin, V. V. (Valerii Viktorovich), editor.
Title: Handbook of optical biomedical diagnostics / Valery V. Tuchin, editor.
Other titles: Optical biomedical diagnostics
Description: Second edition. | Bellingham, Washington : SPIE Press, [2016] | Includes bibliographical references and index.
Identifiers: LCCN 2015038341| ISBN 9781628419092 (v. 1 : alk. paper) | ISBN 9781628419139 (v. 2 : alk. paper)
Subjects: | MESH: Diagnostic Imaging—methods. | Microscopy, Confocal. | Spectrum Analysis.
Classification: LCC R857.O6 | NLM WN 180 | DDC 616.07/54—dc23
LC record available at http://lccn.loc.gov/2015038341

Published by
SPIE
P.O. Box 10
Bellingham, Washington 98227-0010 USA
Phone: +1 360.676.3290
Fax: +1 360.647.1445
Email: books@spie.org
Web: http://spie.org

The content of this book reflects the work and thought of the authors and editors. Every effort has been made to publish reliable and accurate information herein, but the publisher is not responsible for the validity of the information or for any outcomes resulting from reliance thereon. All known errata will be posted on the book's page on our website.

Printed in the United States of America.
First printing.

SPIE.

Table of Contents

Editor's Introduction: Optical Methods for Biomedical Diagnosis

Valery V. Tuchin
Saratov National Research State University
National Research Tomsk State University
Institute of Precision Mechanics and Control, Russian Academy of Sciences

I.1 Historical Aspects and Brief Overview

The history of light application for monitoring tissues and cells for the purpose of disease diagnosis is presented in Refs. 1–14. In 1831, Bright had reported that sunlight or light from a candle was able to shine through the head of a patient with hydrocephalus.[10] The ability of light to transilluminate tissues was later noted by Curling in 1843, and by Cutler in 1929 for monitoring breast lesions.[11,14,15] In 1911, Hasselbalch undertook studies of ultraviolet transmission through the skin, and by the early 1930s textbooks providing good scientific data on optical transmission, absorption, and fluorescence of tissues were available. The properties of skin in the near-infrared (NIR) range were reported by Pearson and Norris in 1933 and by Hardy and Muschenheim in 1935.[4] Due to strong light scattering and autofluorescence, such early studies allowed for understanding only some of the general optical properties of tissues.

Millikan was the first to suggest the dual-wavelength optical spectroscopy method for correction of light scattering, and he was successful in metabolite analysis in humans.[1,12,16] In the 1930s, 1940s, and early 1950s, many studies of the spectroscopy of hemoglobin in tissues were undertaken.[1] *In vivo* measurements of NIR transmittance spectra of the human earlobe and cheek done by Il'ina revealed many new important details about tissue spectra, such as the presence of a water band at 980 nm[17]

The use of NIR light for deep transillumination of mammalian tissues, including the adult human head, and the diagnostic value of NIR light for the assessment of hemoglobin oxygen saturation and the cytochrome *aa3* oxidation–reduction (redox) state in thick tissues were demonstrated in 1977 by Frans Jobsis.[18,19]

For many years, Britton Chance was a pioneer in the development of tissue optics and biomedical spectroscopy.[1,2,12,20–22] He applied spectroscopy for physiological studies of bioenergetics, for trend measurements of hemoglobin oxygenation, and for investigation of cytochrome oxidation. For more precise quantification of the absorbing species in tissues and therefore of the potential utility of this approach for clinical sensing, Chance and co-workers[22] and Delpy and co-workers[23] suggested time-resolved spectroscopy using pulse transillumination and detecting—the so-called time-domain (TD) technique. Later, this approach was further developed by Patterson and co-workers[24] and Jacques[25] to be applied for reflectance measurements, and was used by many investigators for tissue studies and designing of optical diagnostic instruments.[1,2,5–14,26–28]

In 1990, Lakowicz and Berndt[29] extended the time-resolved spectroscopy of tissues by using a frequency-domain (FD) approach, which is mathematically equivalent to the time-domain approach, but allows for a more robust and sensitive measuring technique to be designed.[30] The subsequent discovery of a new type of waves—photon-density waves[8]—and their interference[31] raised the possibility that the FD approach might be able to improve significantly the spatial resolution of tissue spectroscopic analysis.[32]

Many studies on *in vitro* and *in vivo* tissue spectrophotometry using continuous-wave (CW), TD, or FD techniques are overviewed in Refs. 1–14, 26–28, and 32. The development of the cooled charge-coupled device (CCD), time- and spatial-resolved techniques, and other instruments has proceeded at an increasing pace to a wider area of NIR spectroscopy investigations and biomedical applications. At present, more than 500 NIR spectroscopy clinical instruments are available commercially for monitoring and imaging of a tissue's degree of oxygenation, concentration of oxidized cytochrome, and tissue hemodynamics.[10]

The relative simplicity of measuring the human skin reflectance and fluorescence spectra meant that these values were first obtained many years ago. Nevertheless, only in the last three decades have quantitative spectral techniques for *in vivo* monitoring and diagnosis of certain cutaneous and systematic diseases been introduced.[3,13] Historical review of these developments can be found in Refs. 3 and 33.

Various fluorescence techniques, such as those based on autofluorescence and on microscopy using fluorescent markers, time-resolved (phase and time-gated), laser scan, and multiphoton technologies, have been used to study human tissues and cells *in situ* noninvasively.[9,13,14,34–38] Fluorescence techniques are applicable to medical diagnoses of various pathologies affecting many tissues, including those involving the eye. Many robust and powerful combined optical diagnostic techniques, such as fluorescence/light scattering and fluorescence/Raman scattering, have also been designed.[35–37,39]

Raman spectroscopy, which is a great tool for studying the structure and dynamic function of biologically important molecules,[40] also has been used extensively for monitoring and diagnosis of disease *in vitro* and *in vivo* during the past three decades. Examples of its applications include cataracts, atherosclerotic lesions in coronary arteries, precancerous and cancerous lesions in human soft tissues, and bone and teeth pathologies.[13,39,41–43] The successful use of Raman spectroscopy is due to improvements in instrumentation in the NIR spectrum, where fluorescence is significantly reduced.

Among prospective noninvasive blood glucose sensing methods, optical techniques such as NIR and middle-infrared (MIR) (2.5–50 μm) spectrophotometry, fluorescence, and Raman spectroscopy are of great interest to investigators.[34,44] MIR spectroscopy—and particularly attenuated total reflectance Fourier transform infrared spectroscopy—is also important for *in vivo* monitoring of the human skin components.[13,45] MIR and Raman spectroscopy are both examples of so-called vibration spectroscopy, which is characterized by highly specific bands that are dependent on species concentration.[41–45]

Light scattering spectroscopy (LSS) is a novel technique capable of identifying and characterizing pathological changes in human tissues at the cellular and subcellular levels. It can be used to diagnose and detect disease, including through noninvasive monitoring of early cancerous changes in human epithelium.[13,46]

Quasi-elastic light scattering spectroscopy (QELSS), as applied to monitoring of dynamic systems, is based mainly on the correlation or spectral analysis of the temporal fluctuations of the scattered light intensity.[47] QELSS, which is also known as light-beating spectroscopy or correlation spectroscopy, is widely used for various biomedical applications, but especially for blood or lymph flow measurement and cataract diagnostics.[6,13,48–51] For studying optically thick tissues when multiple scattering prevails and photon migration (diffusion) within tissue is important for the character of intensity fluctuations, diffusion wave spectroscopy (DWS) is available.[8,13]

Optothermal or photothermal spectroscopy (OTS/PTS), which is based on detection of the time-dependent heat generation induced in a tissue by pulsed or intensity-modulated optical radiation, is widely used in biomedicine.[13,52–54] Among the various OTS methods, the optoacoustic (OA) and photoacoustic (PA) techniques are of great importance. They allow one to estimate the optical, thermal, and acoustic properties that depend on peculiarities of a tissue's structure.

I.2 Optical Diffuse Techniques

I.2.1 CW spectrophotometry

The specificity of optical diffuse techniques that use a CW light source and detection, applied to *in vivo* spectroscopy of thick tissues (e.g., the female breast or the newborn head) could be described by the following semi-empirical exponential equation for the collimated transmittance $T_c(\lambda)$:[13,55]

$$T_c(\lambda) = x_1 \exp[-\mu_a(\lambda)L(\lambda)x_2], \tag{I.1}$$

where $L(\lambda)$ is the total mean path length of the photons. This equation reflects the wavelength (λ) dependency on absorption, $\mu_a(\lambda)$ and reduced (transport) scattering coefficients $\mu_s'(\lambda)$; x_1 takes into account multiply scattered but nonabsorbed photons, which do not arrive at the detector, and the measurement geometry; x_2 compensates for measurement error of the slab thickness d and inaccuracies in the reduced scattering coefficient $\mu_s' = \mu(1-g)$, and μ_s and g are the tissue scattering coefficient and anisotropy factor of scattering, respectively. For a slab of thickness d, the diffusion equation can be used to calculate a total mean path length L of the photons.[24]

Equation (I.1) was successfully used for fitting of the *in vivo* measurement spectra of the female breast and estimations of the concentrations of the following absorbers: water (H_2O), fat (f), deoxyhemoglobin (Hb), and oxyhemoglobin (HbO):[55]

$$\mu_a = C_{H_2O}\,\sigma_{H_2O} + c_f\sigma_f + c_{Hb}\,\sigma_{Hb} + c_{HbO}\sigma_{HbO}, \tag{I.2}$$

where σ_i is the cross section of the absorption of the *i*th component. By varying the concentrations of the four tissue components, the measurement spectra could be fitted well using Eq. (I.2); the correlation coefficients were better than 0.99 in all cases.[55]

For many tissues, *in vivo* measurements are possible only in the geometry of the backscattering.[13,14] The corresponding relation for light reflectance R can be based on diffusion approximation. For backscattering optical spectroscopy, we have to know, in addition to the measured coefficient of reflection, the depth from which the optical signal is coming. For a spatially separated light source and detector (for example, two fibers normal to the tissue surface), that depth is defined by the photon-path distribution function for the photons migrating from a source to a detector. This spatial distribution function for a homogeneous scattering medium has a "banana" shape. The curve of the most probable direction of a photon migration of the "banana" region reaches a maximum depth, z^{max}, which depends on the source–detector separation r_{sd}:[13,56]

$$z^{\max} \approx (1/2\sqrt{2})r_{sd} \tag{I.3}$$

Instead of Eq. (I.1), which is used for *in vivo* studies in transillumination experiments, a modified form of the Beer–Lambert law is used to describe the optical attenuation in backscattering geometry:[13,56]

$$I/I_0 = \exp(-\varepsilon_{ab} \cdot c_{ab} \cdot r_{sd} \cdot \mathrm{DPF} - G_s), \tag{I.4}$$

where I is the intensity of detected light, I_0 is the intensity of the incident light, ϵ_{ab} is the absorption coefficient measured in $\mu\mathrm{mol}^{-1}\ \mathrm{cm}^{-1}$, c_{ab} is the concentration of absorber in μmol, DPF is the differential path length factor accounting for the increase of the photons' migration paths due to scattering, and G_s is the attenuation factor accounting scattering and geometry of the tissue.

When r_{sd}, DPF, and G_s are kept constant, the changes of absorbing medium concentration can be calculated using measurements of the changes of the optical density (OD), $\Delta(\mathrm{OD}) = \Delta(\log(I_0/I))$:[56]

$$\Delta c_{ab} = \Delta(\mathrm{OD})/\varepsilon_{ab} r_{sd} \mathrm{DPF}. \tag{I.5}$$

Using optical spectroscopy or imaging, the changes in the optical density are measured as follows:

$$\Delta(\mathrm{OD}) = \log(I_0/I_{\text{test}}) - \log(I_0/I_{\text{rest}}) = \log(I_{\text{rest}}) - \log(I_{\text{test}}), \tag{I.6}$$

where I_{rest} and I_{test} represent, respectively, the light scattering intensity of the object (e.g., brain tissue, skeletal muscle) detected during rest and during testing that involves induced brain activity, cold or visual testing, training, or some other experimental condition. For example, based on the OD changes at the wavelengths 760 and 850 nm, one can get either the absorption images for the two measuring wavelengths or functional images (oxygenation or blood volume) within the detection region of study:

$$\Delta(\mathrm{OD})_{oxy} = \Delta(\mathrm{OD})_{850} - \Delta(\mathrm{OD})_{760};\ \Delta(\mathrm{OD})_{total} = \Delta(\mathrm{OD})_{850} + k_{bvo}\Delta(\mathrm{OD})_{760}, \tag{I.7}$$

where $(\mathrm{OD})_{850}$ and $(\mathrm{OD})_{760}$ are the optical densities measured at the wavelengths 850 and 760 nm, and k_{bvo} is the modification factor for reducing the cross-talk between changes of blood volume and oxygenation.

The typical *in vivo* backscattering spectrum (400–700 nm) for a tissue contains the absorption bands of hemoglobin (the Soret and Q-bands).[13,57,58] It also encompasses some absorption from compounds such as flavins, beta-carotene, bilirubin, and cytochrome, among others. On the basis of measurement of the spectral differences of normal and pathological tissues, the corresponding spectral signature "identifiers" can be created. For *in vivo* medical diagnosis, the spectral "identifiers" typically use the ratios of the integrated reflection coefficients within selected spectral bands or the measurement of the spectrum slope for the selected spectral bands. As an internal standard for evaluating the absolute concentrations of the blood components in a tissue, the water band at 980 nm can be used.[57]

I.2.2 Eye tissues

Even such transparent tissues as the human cornea scatter light because the total and axial (collimated) transmissions are not identical.[13,59] Due to the low scattering, water absorption peaks are evident at 300, 980, 1180, 1450, 1900, and 2940 nm. They provide for poor transmission of light through the cornea in the ultraviolet (UV) and infrared (IR) spectral regions.

Average spectral transmittance derived from cornea transmittance measurements in the spectral range 320–700 nm on 10 subjects (14–75 years) was modeled by the following functions for the total transmittance $T_t(\lambda)$ (acceptance angle close to 180 deg) and axial transmittance $T_c(\lambda)$ (acceptance angle of about 1 deg):[60]

$$\operatorname{Log} T_t(\lambda) = -0.016 - 21 \cdot 10^8 \lambda_0^{-4}, \quad \operatorname{Log} T_c(\lambda) = -0.016 - 85 \cdot 10^8 \lambda_0^{-4}, \tag{I.8}$$

where λ_0 is the wavelength in nanometers.

The normal human eye lens is less transparent than the cornea for the visible light, because, in addition to scattering, absorption by different chromophores including 3-hydroxy-L-kynurenine-*O*-β-glucoside and age-related protein (responsible for lens yellowing in older adult subjects) is important.[13,35,61]

The sclera shows poor transparency because of strong light scattering by its structure elements (a system of polydispersive, irregularly arranged collagen cylinders immersed in the ground substance with a lower refractive index).[13] Such a fibrous structure allows for easy control of the human sclera transmittance at a refractive index matching that of collagen fibers and ground material through its impregnation by the immersion liquid.[13]

I.2.3 Time-domain method

Time-dependent radiation transfer theory (RTT) makes it possible to analyze the time response of scattering tissues.[1,2,5–14,22–26,55,62] When probing the plane-parallel layer of a scattering medium with a short laser pulse, the transmitted pulse consists of a ballistic (coherent) component, a group of photons having zigzag trajectories, and a highly intensive diffuse component. Both unscattered photons and photons undergoing forward-directed single-step scattering contribute to the intensity of the component consisting of photons traveling straight along the laser beam. This component is subject to exponential attenuation with increasing sample thickness—a factor that accounts for the limited utility of such photons for practical diagnostic purposes in medicine.

The group of snake photons with zigzag trajectories includes photons, which experience only a few collisions each. They propagate along trajectories that deviate only slightly from the direction of the incident beam and form the first-arriving part of the diffuse component. These photons carry information about the optical properties of the random medium.

The diffuse component is very broad and intense since it contains the bulk of incident photons after they have participated in many scattering acts and, in turn, migrate in different directions and have different path lengths. The diffuse component carries information about the optical properties of the scattering medium, and its deformation may reflect the presence of local inhomogeneities in the medium. However, the resolution obtained by this method at a high light-gathering power is much lower than that obtained by measuring straight-passing photons. Two principal probing schemes are conceivable—one recording transmitted photons and the other taking advantage of their backscattering.

The time-dependent reflectance is defined as[24,25]

$$R(r_{sd},t) = \frac{z_0}{(4\pi cD)^{3/2}} t^{-5/2} \exp(-\frac{r_{sd}^2 + z_0^2}{2cDt}) \exp(-\mu_a ct), \tag{I.9}$$

where t is time, $z_0 = (\mu_s')^{-1}$, and $D = 1/3(\mu_s' + \mu_a)$ is the photon diffusion coefficient in centimeters. To convert the last value to cm^2/s, it should be multiplied by c, the velocity of light in the medium (with units of cm/s).

In practice, μ_a and μ_s' are estimated by fitting Eq. (I.9) with the shape of a pulse measured by the time-resolved photon counting technique. An important advantage of the pulse method is its applicability to *in vivo* studies, in that μ_a and μ_s' can be evaluated separately using a single measurement for a definite source–detector distance r_{sd}.

I.2.4 Frequency-domain method

The frequency-domain (FD) method measures the modulation depth of scattered light intensity m_U $AC_{detector}/DC_{detector}$ and the corresponding phase shift relative to the incident light modulation phase $\Delta\Phi$ (phase lag).[1,2,5–14,29–32,62–64] Compared with the TD measurements, this method is simpler and more reliable in terms of data interpretation and noise, because it involves amplitude modulation at low peak powers, slow rise time, and hence smaller bandwidths than the TD method. Higher signal-to-noise ratios are attainable as well. Medical device FD equipment is more economic and portable.[32] However, the FD technique suffers from the simultaneous transmission and reception of signals, and it requires special efforts to avoid unwanted cross-talk between the transmitted and detected signals. The current measuring schemes are based on heterodyning of optical and transformed signals.[13,32]

The development of the theory underlying this method resulted in the discovery of a new type of waves: photon-density waves, which are progressively decaying waves of intensity. Microscopically, individual photons make random migrations in a scattering medium, but collectively they form a photon-density wave at a modulation frequency ω that moves away from a radiation source. Photon-density waves possess typical wave properties; for example, they undergo refraction, diffraction, interference, dispersion, and attenuation.[1,2,5–14,29–32,62]

In strongly scattering media with weak absorption far from the walls and from a source or a receiver of radiation, the light distribution may be regarded as a decaying diffusion process described by the time-dependent diffusion equation for photon density. For a point light source with harmonic intensity modulation at frequency $\omega = 2\pi\nu$ placed at the point $\vec{r} = 0$, an alternating component (AC) of intensity is a going-away spherical wave that has its center at the point $\vec{r} = 0$ and that oscillates at a modulation frequency with modulation depth

$$\mathrm{m}_U(\vec{r},\omega) = m_I \exp\left(\vec{r}\sqrt{\mu_a/D}\right)\exp\left(-\vec{r}\sqrt{\omega/2cD}\right), \tag{I.10}$$

and undergoes a phase shift relative to the phase value at point $\vec{r} = 0$ equal to

$$\Delta\Phi(\vec{r},\omega) = \vec{r}\sqrt{\omega/2cD}, \tag{I.11}$$

where m_I is the intensity modulation depth of the incident light.

The length of a photon-density wave, Λ_Φ, and its phase velocity, V_Φ, are defined by

$$\Lambda_\Phi^2 = 8\pi^2 cD/\omega \text{ and } V_\Phi^2 = 2cD\omega \tag{I.12}$$

.

Measuring $m_U(\vec{r},\omega)$, $\Delta\Phi(\vec{r},\omega)$ allows one to separately determine the transport scattering coefficient μ_s' and the absorption coefficient μ_a and then to evaluate the spatial distribution of these parameters.

For typical female breast tissue at 800 nm ($\mu_s' = 15$ cm^{-1}, $\mu_a = 0.035$ cm^{-1}) for $\omega/2\pi = 500$ MHz, and $c = (3 \times 10^{10}/1.33)$ cm/s, the wavelength is $\Lambda_\Phi \cong 5.0$ cm and the phase velocity is $V_\Phi \cong 1.77 \times 10^9$cm/s.

A number of FD systems demonstrating achievements in the field of optical *in vivo* diagnostics applied for clinical study have been described.[13,32] For example, to obtain quantitative measurements of the absolute optical parameters of various types of tissue, a portable, high-bandwidth

(0.3–1000 MHz), multi-wavelength (674, 811, 849, and 956 nm) frequency-domain photon migration instrument was designed.[63–66]

I.2.5 Photon-density wave interference method

The photon-density wave interference method was described first in Ref. 31 and is also known as the phase and amplitude cancellation method and as the phased-array method. It appears very promising as a means to improve the spatial resolution of the modulation technique.[13,32] This idea is based on the use of either duplicate sources and a single detector or duplicate detectors and a single source, so that the amplitude and phase characteristics can be compensated and the system becomes a differential. If equal amplitudes at 0-deg and 180-deg phases are used as sources, an appropriate positioning of the detector can lead to a null in the amplitude signal and a crossover between a 0- and 180-deg phase shift—that is, 90 deg.

In a heterogeneous medium, the apparent amplitude's null and the phase's crossover may be displaced from the geometric midline. This method is extremely sensitive to perturbation by an absorber or scatterer. A spatial resolution of approximately 1 mm for the inspection of an absorbing inhomogeneity has been achieved, and the same resolution is expected for the scattering inhomogeneity. Another good feature of the technique is that at the null condition, the measuring system is relatively insensitive to amplitude fluctuations common to both light sources. Inhomogeneities that affect a large tissue volume common to the two optical paths cannot be detected, however. The amplitude signal is less useful in imaging since the indication of position is ambiguous. Although this can be accounted by further encoding, the phase signal is more robust, and a phase noise less than 0.1 deg (signal-to-noise ratio more than 400) for a 1-Hz bandwidth can be achieved.[32]

I.2.6 Spatially modulated spectroscopy and imaging

In diffuse techniques, spatially resolved measurements have been generally limited to "multi-distance" measurements, tracking the spatial dependence of a reflected or transmitted light field generated from a point-like illumination and detection with a number of source–detector separations r_{sd}.[13] The Fourier transform equivalent to the real spatial domain is the spatial-frequency domain (SFD).[65,66] In diffractive optics, spatially structured illumination techniques are used for manipulating of optical images. Spatially modulated laser beams also have been used effectively in studies of scattering objects, including samples of tissues and blood.[13,67] This technique has mostly been applied to investigate low-scattering objects or thin tissue slices and blood layers. However, it was approved successfully for investigation of whole cataractous human eye lenses based on averaging of interferential fringes to eliminate speckle modulation.[13]

The interactions of spatially modulated light beams with diffuse media are described in Refs. 65, 66, and 68–73. Instead of laser beams, low-cost incoherent conventional white light sources are widely used with this approach. Spatially modulated imaging (SMI) provides a wide-field mapping of scattering tissues in the SFD. The spatial modulation transfer function (*s*-MTF) of a turbid tissue encodes both depth and optical property information, enabling both quantitation and tomographic imaging of the spatially varying tissue's optical properties.[71] Similarly to time-resolved methods, the SMI method can be described analytically using diffusion-based theory, or numerically using Monte Carlo simulations in the framework of an RTT-based approach. The optical properties of tissues can be recovered by analysis with the analytic diffusion model using an inversion method, such as a least-squares multifrequency fitting algorithm or a more rapid two-frequency lookup table approach.[65,66,68–73]

The spatially modulated photon density can be considered as "standing" photon-density waves. The basic principles underlying generation of spatially modulated photon-density plane waves were formulated in Ref. 71. Their properties were also described by using spatial-frequency spectral representation. As a first approximation, a diffusion theory was used to get analytical expressions valid for a relatively large transport albedo $\Lambda' = \frac{\mu_s'}{\mu_a+\mu_s'}$ and small spatial frequencies. However, based on Monte Carlo modeling of the transport equation, the results can be extended to low albedo and high spatial frequency modes.

For the time-independent form of the diffusion equation for a homogeneous medium with a semi-infinite geometry and a normally incident periodically modulated plane wave, the source function S_d can be presented in the form[71]

$$S_d = S_{do}(z)\cos(k_x x + \alpha)\cos(k_y y + \beta) \tag{I.13}$$

with spatial frequencies $f_x = (k_x/2\pi)$ and $f_y = (k_y/2\pi)$, and spatial phases α and β, extending infinitely in the tangential spatial dimensions x and y, with some arbitrary dependence on depth z.

If the medium's response is proportional to the input intensity, this sinusoidal modulation will give rise to a diffuse fluence rate U with the same frequency and phase:

$$U = U_0(z)\cos(k_x x + \alpha)\cos(k_y y + \beta). \tag{I.14}$$

A plane wave with both x and y modulation gives rise to a photon-density wave propagating with a scalar attenuation coefficient:

$$\mu'_{\text{eff}} = \sqrt{\mu_{\text{eff}}^2 + k_x^2 + k_y^2} = \frac{1}{\delta'_{\text{eff}}}, \tag{I.15}$$

where

$$\mu_{\text{eff}} = \sqrt{3\mu_a(\mu'_s + \mu_a)} \tag{I.16}$$

and δ'_{eff} is the effective light penetration depth into a scattering medium.

Although spatial anisotropy may exist in real tissues, to understand scalar photon-density wave attenuation in multiply scattering media, 1D projection can be used for simplicity; that is, $k = k_x$, with constant illumination along y ($k_y = 0$). At zero spatial frequency ($k = 0$), the effective light penetration depth into a scattering medium, δ'_{eff}, is equivalent to that of a planar (none-modulated) illumination, $\delta_{\text{eff}} = (1/\mu_{\text{eff}})$. In general, however, μ'_{eff}(and δ'_{eff}) are functions of both optical properties and the spatial frequency of illumination. Thus, at known parameters of illumination, it is possible to evaluate the optical properties of tissues.

The amplitude of the periodic wave, $U_0(z)$, is independent of the tangential spatial dimensions x and y. As a consequence, existing planar geometry solutions of the diffusion equation can be used to describe spatially modulated photon density by simply substituting μ_{eff} with the new μ'_{eff} term.

I.3 Fluorescence Spectroscopy

I.3.1 Fundamentals and methods

Fluorescence arises upon light absorption and is related to an electron's transition from the excited state to the ground state within a molecule. In the case of thin samples (e.g., biopsies that are a few micrometers in thickness), the fluorescence intensity I_F is proportional to the concentration c and the fluorescence quantum yield η of the absorbing molecules.[34,74–76] In a scattering medium, the path lengths of scattered and unscattered photons within the sample are different, and should be accounted for.[34]

At excitation of biological objects by ultraviolet light ($\lambda \leq 300$ nm), fluorescence of their components, such as proteins and nucleic acids, should be observed. Fluorescence quantum yields of all nucleic acid constituents, however, are approximately 10^{-4} to 10^{-5}, corresponding to lifetimes of the excited states in the picosecond time range. Autofluorescence (AF) of proteins is related to the amino acids tryptophan, tyrosine, and phenylalanine, which have absorption maxima at 280 nm, 275 nm, and 257 nm, respectively, and emission maxima between 280 nm (phenylalanine) and 350 nm (tryptophan).[34,74–76] Fluorescence from collagen or elastin is excited between 300 and 400 nm and shows broad emission bands between 400 and 600 nm, with maxima around 400 nm, 430 nm, and 460 nm. In particular, fluorescence of collagen and elastin can be used to distinguish various types of tissues (e.g., epithelial versus connective tissues) and their pathology.[9,13,28,58,61,74–82]

The reduced form of coenzyme nicotinamide adenine dinucleotide (NADH) is excited selectively in a wavelength range between 330 nm and 370 nm. NADH is most concentrated within mitochondria, where it is oxidized within the respiratory chain located within the inner mitochondrial membrane. Its fluorescence is an appropriate parameter for detection of ischemic or neoplastic tissues.[75] Fluorescence of free and protein-bound NADH has been shown to be sensitive to oxygen concentration. Flavin mononucleotide (FMN) and dinucleotide (FAD), which have excitation maxima around 380 nm and 450 nm, respectively, have also been reported to contribute to intrinsic cellular fluorescence.[75]

Porphyrin molecules—for example, protoporphyrin, coproporphyrin, uroporphyrin, and hematoporphyrin—occur within the pathway of biosynthesis of hemoglobin, myoglobin, and cytochromes.[75] Abnormalities in heme synthesis, such as are observed in porphyrias and some hemolytic diseases, may considerably enhance the porphyrin level within tissues. Several bacteria (e. g., *Propionibacterium acnes* and bacteria within dental caries lesions) accumulate considerable amounts of protoporphyrin. Therefore, measurements of intrinsic fluorescence appear to be a promising method for detecting acne or caries.

At present, various exogenous fluorescing dyes can be applied to probe cell anatomy and cell physiology.[75] In humans, such dyes as fluorescein and indocyanine green are already used for fluorescence angiography or blood volume determination. *In vivo* fluorescence probes rely on a family of fluorescent proteins (FP). Based their emission maxima, blue, green, yellow, and red fluorescent proteins may be distinguished. The most popular is the green fluorescent protein (GFP). After cloning of the FP gene, various FP variants with different excitation and emission properties have been produced. When genes coding for a specific cellular protein are fused with FP or its variants, functional and site-specific tracking in living cells or even whole organisms becomes possible.

Fluorescence spectra often give detailed information on fluorescent molecules, including their conformation, binding sites, and interactions within cells and tissues. Fluorescence intensity can be measured as a function of either the emission wavelength or the excitation wavelength. The fluorescence emission spectrum $I_F(\lambda)$, which is specific for any fluorophore, is commonly used in fluorescence diagnostics. Fluorescence spectrometers for *in vivo* diagnostics are commonly based on fiber-optic systems and use of an optical multichannel analyzer (OMA; a diode array or a CCD camera) as a detector of emission radiation.[58,74–78]

Various comprehensive and powerful fluorescence spectroscopies, such as microspectrofluorimetry, polarization anisotropy, time-resolved with pulse excitation and frequency-domain, time-gated, total internal reflection fluorescence spectroscopy and microscopy, fluorescence resonant energy transfer method, confocal laser scanning microscopy, and their combinations are available now.[34,74–82] These methods support the following applications (see Chapter 5 in Volume 2):[75]

1. 3D topography of specimens measured in the reflection mode for morphological studies of biological samples
2. High-resolution microscopy measured in the transmission mode
3. 3D fluorescence detection of cellular structures and fluorescence bleaching kinetics
4. Time-resolved fluorescence kinetics
5. Studies of the motions of cellular structures
6. Time-gated imaging so as to select specific fluorescent molecules or molecular interactions
7. Fluorescence lifetime imaging
8. Spectrally resolved imaging

Fluorescence is also beneficial in the practical work carried out by medical staff members. Concepts and applications of fluorescence imaging for surgeons are discussed in recent book[83] and in reviews and original papers.[84–86] Indocyanine green (ICG) is a NIR fluorescent dye that has been used in medical diagnostics for almost six decades, yet has great potential in the development of new imaging systems for several surgical specialties due to its unique molecule binding and spectral properties.[84,85] The introduction of new clinical applications has occurred especially rapidly during the last few years. ICG fluorescence imaging in the areas of plastic and reconstructive surgery, neurosurgery, and cardiac, vascular, oncological, and hepatic surgery is reviewed in Ref. 85.

The inability to identify microscopic tumors and assess surgical margins in real time during oncologic surgery, which may lead to incomplete tumor removal, prompted authors of a recent paper to develop a wearable-goggle augmented imaging and navigation system (GAINS); this system can provide accurate intraoperative imaging of tumors and sentinel lymph nodes (SLNs) in real time without disrupting normal surgical workflow.[86] The optical system projects both NIR fluorescence from the tumors and the natural-color images of tissue onto a head-mounted display without latency. Human pilot studies in breast cancer and melanoma patients using a NIR dye show that GAINS detected SLNs with 100% sensitivity. Clinical use of GAINS to guide tumor resection and SLN mapping promises to improve surgical outcomes, reduce rates of repeat surgery, and improve the accuracy of cancer staging.

I.3.2 *In vivo* human skin fluorescence

Currently, reflectance and fluorescence spectroscopies are probably the most developed among the available optical methods for investigating skin *in vivo*.

Reflectance and fluorescence from skin carry information about the structure of the epidermis and dermis, the quantity and density of blood vessels, the concentration and spatial distribution of chromophores and fluorophores in the skin, and the nature of skin metabolic processes. Typical applications include the *in vivo* quantitative analysis of skin erythema and pigmentation, determination of cutaneous color variation, monitoring of dermatological treatment effects, determination of skin photo-aging, diagnosis of skin tumors, and study of skin biophysics.[3,76,77,79–81,87,88]

The potential advantages and possible applications resulting from combined use of reflectance and fluorescence spectroscopy of the skin for the evaluation of erythema and pigmentation indices, the determination of hemoglobin oxygenation and concentration, and the investigation of the efficacy of topical sunscreens are discussed in Chapter 3 in Volume 2.[76]

Most of the biological components that are either related to the skin tissue structure or are involved in metabolic and functional processes generate fluorescence emission in the UV-visible spectral region. As a result, different morpho-functional conditions of the skin related to histological, biochemical, and physiochemical alterations can be characterized, in principle, on the basis of information available in fluorescence excitation–emission maps (EEMs).[6,76,78,80]

Among the various endogenous skin fluorophores, different forms of NAD and keratin located in the epidermis and in collagen located in the dermis can be found. The reduced (NADH) and oxidized (NAD^+) forms of NAD have roles in cellular metabolism, and the intensity of their specific fluorescence (fluorescence maxima near 460 nm and 435 nm, respectively) is used in differential diagnostics for metabolism dysfunction.[76]

Collagen and elastin are found predominantly within the papillary and reticular layers of the dermis. For these substances, both excitation light and emission light are attenuated because absorption by melanin and fluorescence intensity in the 400–480 nm range is subject to attenuation by other skin chromophores—hemoglobin, porphyrins, carotenoids, and so on.[76,81]

Recent studies of endogenous and exogenous fluorescence skin cancer diagnostics for clinical applications are overviewed in Ref. 88.

I.3.3 Advantages of multi-photon fluorescence

A new direction in laser spectroscopy of tissues and cells has emerged with the introduction of multi-photon (two-photon, three-photon) fluorescence scanning microscopy. This technique makes it possible to image functional states of an object or, in combination with autocorrelation analysis of the fluorescence signal, to determine the intercellular motility in small volumes.[13,38,89–92] Multi-photon fluorescence employs sharply focused ballistic photons at a long wavelength that provide fluorescence excitation by the second or third harmonic of the incident radiation. Fluorescence comes to a wide-aperture photodetector exactly from the focal area of the excitation beam.

A unique advantage of multi-photon microscopy is the possibility of investigating three-dimensional distributions of chromophores excited with ultraviolet radiation in relatively thick samples. Such an investigation becomes possible because chromophores can be excited (e.g., at a wavelength of 350 nm) with laser radiation whose wavelength for two-photon-excitation fluorescence (TPEF) (700 nm) falls within the range where a tissue has high transparency. Such radiation can reach deep layers and produces less damage in tissues. Fluorescent emission in this case lies in the visible range (wavelengths greater than 400 nm). It emerges comparatively easily from a tissue and reaches a photodetector, which registers only the legitimate signal from the focal volume without any extraneous background noise.

TPEF of target molecules in a tissue is a nonlinear process induced by the simultaneous absorption of two NIR photons, whose total energy is sufficient to excite the electronic state of the molecular transition. In general, photons with different wavelengths λ_1 and λ_2 can be used

$$\frac{1}{\lambda_{1f}} \cong \frac{1}{\lambda_1} + \frac{1}{\lambda_2}, \tag{I.16}$$

where λ_{1f} is the wavelength necessary to excite the fluorescence at single-photon absorption. However, the excitation by the same light source (i.e., $\lambda_1 = \lambda_2$ and $\lambda_2 \cong 2\lambda_{1f}$) is more practical.

The two-photon absorption cross section for biological molecules σ_2 is typically very small (approximately 1 GM $= 10^{-58}$ m^{4}s^{-1}), so intense photon fluxes on the order of 10^{30} photons per second per square meter (s^{-1}m^{-2}) are required. Pulsed excitation (approximately 10^{-13} s), which allows for reduction of the heat load on the tissue and selective excitation of individual electronic transitions of biological molecules, is preferable.

For a pulse laser with repetition rate $f_\mathrm{p} = 1/T$ and a duration of the rectangular pulses τ_p, the time-averaged intensity of the TPEF is expressed as[13,89]

$$\left\langle \hat{I}_{2f} \right\rangle_\mathrm{p} = \kappa\pi\sigma_2\eta \frac{P_\mathrm{ave}^2}{\tau_\mathrm{p} f_\mathrm{p}} \left[\frac{(\mathrm{NA})}{hc\lambda_\mathrm{exc}}\right]^2, \tag{I.17}$$

where κ is the coefficient taking into account the collection efficiency of the fluorescent photons, $\eta = \eta(\lambda_\mathrm{em})$ is the fluorescence quantum yield, $P_\mathrm{ave} = (\tau_p{\cdot}f_p)P_\mathrm{peak}$ is the average power, P_peak is the peak power, and NA is the numerical aperture of the microscope objective.

It follows from Eq. (I.17) that the excitation of fluorescence emission by a pulse laser with a wavelength $\lambda = 1000$ nm, an average power of 1 mW at a repetition rate of 80 MHz, and a pulse duration of 100 fs, focused by the objective with NA = 1.4 onto the tissue with a typical two-photon cross section $\sigma_2 = 10$ GM = 10^{-57} M^4c^{-1} provides a rate of fluorescence photon counting equal to 10^5 Hz.

Three-photon fluorescence microscopy of a tissue possesses the same advantages as two-photon microscopy but ensures a somewhat higher spatial resolution and provides an opportunity to excite chromophores with shorter wavelengths.

I.4 Second-Harmonic Generation (SHG)

Second-harmonic generation (SHG) is a new high-resolution nonlinear optical imaging modality for study of intact tissues and cellular structures.[13,92–94] SHG is a second-order nonlinear optical process that can arise only from media lacking a center of symmetry—for example, an anisotropic crystal or at an interface such as a membrane. This technique can be used to image highly ordered structural proteins without any exogenous labeling. Collagen is a main component of connective tissues due to its helix secondary structure, which is noncentrosymmetric; it has a dominant uniaxial second-order nonlinear susceptibility component aligned along the fiber axis and, therefore, satisfies the conditions for SHG. In the skin, SHG light is generated mostly within the dermis, rather than in cellular layers such as the epidermis or subcutaneous fat.

SHG techniques offer a number of advantages connected with the incident wavelength's division and its selectivity to tissue structure, which allow one to reject surface reflection and multiple scattering of the incident light in the upper epithelial layer without any gating technique. As in the case of multiphoton excited fluorescence, SHG arises from a very small tissue volume within a focal volume of the sharply focused NIR laser beam. As a result, it provides a high spatial resolution, in-depth probing, and separation of excitation and detection signals. In spite of the high power density in the focal spot, a very short pulse (50–200 fs) allows for generation of harmonics in the living tissue with no damage to it due to the low overall energy.[13,92–94]

In general, the nonlinear polarization for a material can be expressed as[93]

$$\mathbf{P} = \chi^{(1)}\mathbf{E} + \chi^{(2)}\mathbf{E}\mathbf{E} + \chi^{(3)}\mathbf{E}\mathbf{E}\mathbf{E} + \ldots, \tag{I.18}$$

where **P** is the induced polarization, $\chi^{(n)}$ is the nth order nonlinear susceptibility, and **E** is the electric field vector of the incident light. The first term describes normal absorption and reflection of light; the second describes SHG, sum, and difference frequency generation; and the third describes both two- and three-photon absorption, as well as third-harmonic generation and coherent anti-Stokes Raman scattering (CARS).

SHG, unlike two-photon fluorescence, does not arise from an absorptive process. Instead, an intense laser field induces a nonlinear, second-order polarization in the assembly of molecules, resulting in the production of a coherent wave at exactly twice the incident frequency (or half the wavelength).

The SHG pulse is temporally synchronous with the excitation pulse. A simplified expression for the SHG signal intensity has the form[93]

$$I(2\omega) \propto \left[\chi^{(2)} \frac{E(\omega)}{\tau_p}\right] \tau_p, \tag{I.19}$$

where $E(\omega)$ is the laser pulse energy. As in TPEF [see Eq. (I.17)], the signal is quadratic with peak power, but since SHG is an instantaneous process, a signal is generated only for the duration of the laser pulse.

I.5 Vibrational Spectroscopy

Middle-infrared (MIR) and Raman spectroscopies use light-excited vibrational energy states in molecules to obtain information about the molecular composition, structures, and interactions in a sample.[40–45,95,96] In MIR spectroscopy, infrared light from a broadband source (usually 2.5–25 μm or 4000–400 cm^{-1}) is directly absorbed to excite the molecules to higher vibrational states. In Raman scattering, event light is inelastically scattered by a molecule when a small amount of energy is transferred from the photon to the molecule (or vice versa). The energy difference between incident and scattered photons is expressed in a wavenumber shift (cm^{-1}).

The MIR and Raman spectroscopy techniques have been successfully applied to various areas addressed by clinical studies, such as cancerous tissues examination, the mineralization process of bone and teeth, tissues monitoring, glucose sensing in blood, noninvasive diagnosis of skin lesions based on benign or malignant cells, and monitoring of treatments and topically applied substances (e.g., drugs, cosmetics, moisturizers) to the skin.[41–45,95–102]

Raman spectroscopy is widely used in biological studies, ranging from studies of purified biological compounds to investigations at the level of single cells.[40,96] At present, combinations of spectroscopic techniques such as MIR and Raman spectroscopy with microscopic imaging techniques are being explored to map molecular distributions at specific vibrational frequencies on samples so as to locally characterize tissues or cells.[95–99] Chemical imaging is expected to become increasingly more important in clinical diagnosis in the future.

Because the penetration depth of MIR light in tissue extends to only a few micrometers, the attenuated total reflectance Fourier transform infrared spectroscopy (ATR-FTIR) method is well suited to study changes of the outermost cell layers of the tissue.[95]

The Raman technique exhibits certain characteristics that make it particularly suitable for studying the skin, both *in vitro* and *in vivo*.[45,95,100]

Confocal detection is particularly useful to studying the outer skin layers, such as the stratum corneum and viable epidermis. Since the dermis is much thicker than the epidermis (1–4 mm thick), it can easily be studied using a non-confocal detection scheme, given a detection volume that is large compared to the thickness of the epidermis.[95]

I.6 Coherent Anti-Stokes Raman Scattering (CARS)

Coherent anti-Stokes Raman scattering (CARS) is a third-order nonlinear optical process in which three excitation fields interact to produce a fourth field, which is detected [see Eq. (I.18)].[13,98,99] In general, two laser beams with frequencies ν_{pump} and ν_S are tuned to get their difference ($\nu_{\text{pump}} - \nu_S$) to be equal to the frequency ν_{vib} of a vibrational transition of the sampling molecules. Then the probing laser beam with frequency ν_{probe} generates resonantly a fourth enhanced field with frequency $\nu_{AS} = (\nu_{\text{pump}} + \nu_{\text{vib}})$. Typically only two laser beams are used to generate CARS signal, because a so-called frequency-degenerate optical scheme with $\nu_{\text{pump}} = \nu_{\text{probe}}$ can be applied.

The intensity of CARS signal depends quadratically on the modulus of the induced third-order polarization $\mathbf{P}^{(3)}$ in the sample [see Eq. (I.18)]

$$I_{AS} \propto |\mathbf{P}^{(3)}|^2, \tag{I.20}$$

where $\mathbf{P}^{(3)}$ depends on the third-order optical susceptibility that can be presented as a sum of the nonresonant and resonant contributions

$$\mathbf{P}^{(3)} = \left[\chi^{(3)}_{\text{nonres}} + \chi^{(3)}_{\text{res}}\right]\mathbf{E}_{\text{pump}}\mathbf{E}_{\text{prob}}\mathbf{E}_s \tag{I.21}$$

The main advantages of CARS compared to conventional Raman spectroscopy, besides the opportunity to amplify the signal by more than four orders of magnitude, are the direct signal generation, narrow band, and complete absence of the influence of autofluorescence as the signal is generated at wavelengths shorter than the wavelength of excitation. Since three nonlinear methods—CARS, TPEF, and SHG—are technically implemented using the similar experimental equipment, they are often used together as part of a multimodal approach to obtain more information about the fundamental processes in tissues and cells.[13,98,99,103,104]

I.7 Light-Scattering Spectroscopy

Based on classical measurements of light scattering, innovative techniques capable of identifying and characterizing pathological changes in human tissues at the cellular and subcellular levels have been proposed.[13,46,58,105–109] Light-scattering spectroscopy (LSS) provides structural and functional information about a tissue. This information can be used, in turn, to diagnose and monitor disease. One important application of biomedical spectroscopy is the noninvasive detection of early cancerous human epithelium.[46,106,109] The enlarging, crowding, and hyperchromaticity of the epithelium cell nuclei are common features to all types of precancerous and early cancerous conditions. LSS can be used to detect such early cancerous changes and other diseases in a variety of organs, such as the esophagus, colon, uterine cervix, oral cavity, lungs, and urinary bladder.[109]

Cells and tissues have complex structures with a very broad range of scatterer size—from a few nanometers, the size of a macromolecule; to 7–10 μm, the size of a nucleus; and to more than 20–50 μm, the size of a cell itself.[13,106] A great variety of cell organelle structures are small compared to the wavelength used in LSS. Light scattering by such particles, which is known as Rayleigh scattering, is characterized by a broad angular distribution; the scattering cross-section dependence on the particle's linear dimension a is characterized as a^6 and that on the light wavelength λ is characterized as λ^{-4}. When the particle is not small enough, coupled dipole theory or another approach such as Rayleigh–Gans approximation (RGA) can be used. RGA is particularly applicable to particles with a size comparable to the wavelength and may be useful for studying light scattering by small organelles such as mitochondria and lysosomes. With use of RGA, scattering in the forward direction prevails, and the total scattering intensity increases with the increase in the particle relative refractive index m as $(m-1)^2$ and with its size as a^6.

The scattering by a particle with dimensions much larger than the wavelength, such as a cell nucleus, can be described within the framework of the van de Hulst approximation, which enables scattering amplitudes in the near-forward direction to be obtained.[106] For large particles, the scattering intensity is highly forward directed, and the width of the first scattering lobe is approximately λ/a; the larger the particle, the stronger and narrower the first lobe. The forward scattering intensity exhibits oscillations with the wavelength. The origin of these oscillations is interference between the light ray passing through the center of the particle and a light ray not interacting with it. The frequency of these oscillations is proportional to $a(m-1)$, so it increases with the particle size and refractive index. The intensity of the scattered light also peaks in the near-backward direction, but this peak is significantly smaller than the forward-scattering peak.

These results agree well with the rigorous scattering theory developed for spherical particles (Mie theory).[110] To discriminate among the cell structure peculiarities originating from a pathology, the difference in light scattering can be used. The structures with large dimensions and high refractive index produce the scattered field that peaks in the forward and near-backward directions, whereas smaller and more optically "soft" structures scatter light more uniformly.

The photons returned after a single scattering in the backward or near-backward directions produce the so-called single-scattering component. The photons returned after multiple scattering events produce the diffuse reflectance. Although the spectra of both single-scattering and diffusive signals contain valuable information about tissue properties, the type of information they provide is different. The single-scattering component is sensitive to morphology of the upper tissue layer, which in case of any mucosal tissue almost always includes or is limited by the epithelium. Its spectroscopic features are related to the microarchitecture of the epithelial cells—that is, the sizes, shapes, and refractive indices of their organelles, inclusions, and sub-organellar components. Thus, analysis of this component might be useful in diagnosing diseases limited to the epithelium, such as preinvasive stages of epithelial cancers, dysplasias, and carcinomas *in situ* (CIS).[46,106,109]

The single-scattering component is more important in diagnosing the initial stages of epithelial precancerous lesions, whereas the diffusive component provides valuable information about more advanced stages of the disease. However, single scattering events cannot be directly observed in tissues *in vivo*, because only a small portion of the light incident on the tissue is directly backscattered.

Several methods to distinguish single scattering have been proposed. Field-based light-scattering spectroscopy[111] and optical coherence tomography (OCT)[13,112] were developed for performing cross-sectional tomographic and spectroscopic imaging. In these extensions of conventional OCT,[13] information on the spectral content of backscattered light is obtained by detection and processing of the interferometric OCT signal. These methods allow the spectrum of backscattered light to be measured either for several discrete wavelengths[111] or simultaneously over the entire available optical bandwidth from 650 to 1000 nm[112] in a single measurement.

A much simpler polarization-sensitive technique is based on the fact that initially polarized light loses its polarization when traversing a turbid tissue.[113,114] A conventional spatially resolved backscattering technique with a small source–detector separation can be used as well.[46] In that case, however, the single scattering component (2–5%) should be subtracted from the total reflectance spectra.

The promise of LSS for diagnosing dysplasia and CIS was tested in human studies in three different types of *in vivo* epithelium: columnar epithelia of the colon and Barrett's esophagus, transitional epithelium of the urinary bladder, and stratified squamous epithelium of the oral cavity.[109] The spectrum of the reflected light was analyzed to determine the nuclear size distribution. In all studied organs, a clear distinction was apparent between dysplastic and nondysplastic epithelium. Both dysplasia and CIS have a higher percentage of enlarged nuclei and, on average, a higher population density—characteristics that can be used as the basis for spectroscopic tissue diagnosis.[109]

I.8 Optical Coherence Tomography (OCT)

Optical coherence tomography (OCT) was first demonstrated in 1991.[115] Imaging was performed *in vitro* in the human retina and in atherosclerotic plaque as examples of imaging in transparent, weakly scattering media as well as highly scattering media. This is an urgent field of research, with applications attracting more and more end-users. State-of-the-art monographs, tutorials, and special issues of journals describing principles and biomedical applications of OCT are widely available.[13,51,116–131]

OCT is analogous to ultrasonic imaging that measures the intensity of reflected NIR light, rather than reflected sound waves from the sample. Time gating is employed so that the time required for the light to be reflected back, known as the echo delay time, can be used to assess the intensity of backreflection as a function of depth. Unlike in ultrasound, the echo time delay, which is on an the order of femtoseconds in optics, is measured by using an optical interferometer illuminated by a low coherent light source.

This technique is conventionally implemented with the use of a dual-beam Michelson interferometer. If the path length of light in the reference arm is changed with a constant linear speed v, then the signal arising from the interference between the light scattered in a backward direction (reflected) from a sample and the light in the reference arm is modulated at the Doppler frequency

$$f_D = \frac{2v}{\lambda}. \tag{I.22}$$

Owing to the small coherence length of a light source, the Doppler signal is produced by backscattered light only within a very small region (on the order of the coherence length l_c) corresponding to the current optical path length in the reference arm. For the light source with a Gaussian line profile

$$l_c = \frac{2\ln(2)}{\pi} \cdot \frac{\lambda^2}{\Delta\lambda}, \tag{I.23}$$

where $\Delta\lambda$ is the Gaussian line bandwidth.

If a superluminescent diode (SLD) with a bandwidth of 15–60 nm ($\lambda \approx$ 800–860 nm) is employed, the longitudinal resolution falls within the range of 5–20 μm. For a titanium–sapphire laser with a wavelength of 820 nm, the bandwidth may reach 140 nm. Correspondingly, the resolution is 2.1 μm.[116]

Transverse resolution of OCT is defined by a light beam spot, which is typically from 5 to 20 μm.

In the literature, one can find descriptions of several different OCT systems, ranging from conventional amplitude or time-domain OCT to advanced systems combining, for example, spectral-domain OCT (SD-OCT) with multi-photon tomography (MPT) for 3D multimodal *in vivo* imaging.[13,51,112,115–131] Time-domain OCT is a single-point detection technique. It can be used to generate two-dimensional OCT images up to the video rate, although such systems have a limited sensitivity or a limited space–bandwidth product (resolved pixels per dimension). For some applications, two-wavelength fiber OCT is effective. Ultrahigh-resolution fiber OCT systems are also available. Frequency- and Fourier-domain OCT techniques are based on backscattering spectral interferometry and, therefore, are also called spectral-domain OCT (SD-OCT). Such systems are widely used in biomedical studies and in clinics. Doppler OCT (DOCT) combines the Doppler principle with OCT to obtain high-resolution tomographic images of static and moving constituents in highly scattering tissues. Optical microangiography (OMAG) is an OCT technique that utilizes a constant modulation frequency to separate the signal associated with the movement in the RBS vascular bed from the backscatter signal. Correlation-map OCT (cmOCT) applies two-dimensional OCT images to reconstruct blood vessel distribution within the skin.

The specificity of conventional OCT can be improved by providing measurements of polarization properties of probing radiation when it propagates through a tissue. This approach was implemented in the polarization-sensitive OCT technique (PS-OCT). In its turn, phase-sensitive OCT (PhS-OCT) provides quantitative dispersion data that are important in predicting the propagation of light through tissues, in photorefractive surgery, and in tissue and blood refractive index measurements. PhS-OCT systems are often used in tissue elastography. Indeed, the prospective technique called optical coherence elastography (OCE) takes advantage of high-resolution OCT to provide quantitative evaluation of a tissue's mechanical properties.

Full-field or parallel OCT (FF-OCT) uses linear or two-dimensional detector arrays of, respectively, N and N^2 single detectors. The advantage of parallel OCT is that when using linear or 2-D detector arrays, the SNR can be roughly $\sqrt{N}$ and N times larger, respectively, compared to the single detector signal.

Optical coherence microscopy (OCM) is a biomedical modality for cross-sectional subsurface imaging of tissue that combines the ultimate sectioning abilities of OCT and confocal microscopy (CM). In OCM, spatial sectioning due to tight focusing of the probing beam and pinhole rejection provided by CM is enhanced by additional longitudinal sectioning provided by OCT coherence gating.

Application of the fiber-optic light-delivering and light-collecting cables allows one to build a flexible OCT system that facilitates endoscopic analysis of human tissues and organs—in particular, high-speed *in vivo* intra-arterial imaging. The feasibility of OCT ultrathin needle probes for imaging of breast cancer, dystrophic skeletal muscles, tendons, connective tissues, and air-filled lungs has already been demonstrated.

The speckle OCT method has been shown to be a viable alternative to the Doppler OCT in 2D imaging of blood flow. Flow information can be extracted using speckle fluctuations in conventional time-domain OCT. One optical coherence elastography (OCE) technique that takes advantage of the high resolution of OCT and the high sensitivity of speckles is based on speckle tracking; it has been widely studied to evaluate the skin's mechanical properties both qualitatively and quantitatively.

One important advance related to OCT systems is the combination of this technology with other optical diagnostic modalities so as to achieve a synergetic effect in diagnostic ability. Technically this combination could be realized by using miniature fiber optical probes, shared light sources, optical pathways and scanning systems, and so on. In fact, dual OCT/confocal microscopy (CM) systems have already been described. A dual imaging en face OCT/CM system was used in ophthalmology for imaging of the anterior chamber of the eye and in dentistry. The combination of dual en face OCT/CM with fluorescence imaging gives a universally applicable instrument in microscopy. All-fiber-optic-based endoscopy for simultaneous OCT and fluorescence tissue imaging provides clear visualization of structural morphologies (OCT) and fluorophore distribution (the fluorescence module). Multi-photon tomography (MPT) and SD-OCT can be used for 3D multimodal *in vivo* imaging of normal skin, nevi, scars, and pathologic skin lesions. Photothermal OCT (PT-OCT) using gold nanorods (GNRs) as contrast agents has been shown to be a potentially powerful tool for molecular imaging. Adaptive optics-assisted OCT is currently under development and holds promise for subcellular imaging in biology and medicine.

I.9 Dynamic Light-Scattering Spectroscopy and Tomography

I.9.1 Photon-correlation spectroscopy

Quasi-elastic light-scattering spectroscopy (QELSS), photon-correlation spectroscopy, spectroscopy of intensity fluctuations, and Doppler spectroscopy are synonymous terms for technologies based on the dynamic scattering of light—a capability that underlies a noninvasive method for studying the dynamics of particles on a comparatively large time scale.[13,14,47–51] The implementation of the single-scattering mode and the use of coherent light sources are of fundamental importance in this case. The spatial scale of testing for an ensemble of biological particles is determined by the inverse of the wave vector $|\bar{s}|^{-1}$:

$$|\bar{s}| = (4\pi n/\lambda_0)\sin(\theta/2), \tag{I.24}$$

where n is the refractive index and θ is the angle of scattering. With allowance for self-beating due to the photomixing of the electric components of the scattered field on a photodetector, the intensity autocorrelation function can be expressed as $g_2(\tau) = \langle I(t)I(t+\tau)\rangle$ For Gaussian statistics, this autocorrelation function is related to the first-order autocorrelation function by the Siegert formula:

$$g_2(\tau) = A\left[1 + \beta_{sb}|g_1(\tau)|^2\right], \tag{I.25}$$

where τ is the delay time; $A = \langle i\rangle^2$ is the square of the mean value of the photocurrent, or the baseline of the autocorrelation function; β_{sb} is the parameter of self-beating efficiency, $\beta_{sb} \approx 1$; and $g_1(\tau) = \exp(-\Gamma_T\tau)$ is the normalized autocorrelation function of the optical field for a monodisperse system of Brownian particles. $\Gamma_T = |\bar{s}|^2 D_T$ is the relaxation parameter and $D_T = k_B T/6\pi\eta r_h$ is the coefficient of translation diffusion, k_B is the Boltzmann constant, T is the absolute temperature, η is the absolute viscosity of the medium, and r_h is the hydrodynamic radius of a particle. Many biological systems are characterized by a bimodal distribution of diffusion coefficients, when fast diffusion (D_{Tf}) can be separated from slow diffusion (D_{Ts}) related to the aggregation of particles. The goal of QELSS is to reconstruct the distribution of scattering particles by sizes, which is necessary for the diagnosis or monitoring of a disease.

The homodyne and heterodyne photon-correlation spectrometers, the laser Doppler anemometers (LDAs), differential LDA schemes, and laser Doppler microscopes (LDMs), and laser scanning and speckle CMOS-based full-field imagers have a wide area of medical applications. In particular, they have been used to investigate eye tissues (cataract diagnosis), hemodynamics in individual vessels (vessels of eye fundus), and blood microcirculation in tissues.[13,47–51,123–126,132–139]

I.9.2 Diffusion wave spectroscopy/diffuse correlation spectroscopy

Diffusion wave spectroscopy (DWS) and diffuse correlation spectroscopy (DCS) are dynamic light scattering techniques related to the investigation of the dynamics of particles within very short time intervals.[8,13,48,49,51,140–145] A fundamental difference between these techniques and QELSS is that DWS and DCS are applicable in the case of dense media with multiple scattering, which is critical for tissues. In contrast to the case of single scattering, the autocorrelation function of the optical field $g_1(\tau)$ is sensitive to the motion of a particle on the length scale on the order of $\lambda[L/l_{tr}]^{-1/2}$, which is generally much less than λ because $L >> l_{tr}$ (L is the total mean photon path length and l_{tr} is the transport length of a photon, $l_{tr} = 1/(\mu_a + \mu_{s}\prime) \approx 1/\mu_s^\prime$). Thus, DWS/DCS autocorrelation functions decay much faster than the autocorrelation functions employed in QELSS.

Experimental implementation of DWS/DCS is very simple. A measuring system should provide irradiation of an object under study by a CW laser beam and measurement of intensity fluctuations of the scattered radiation within a single speckle with the use of a single-mode receiving fiber, photomultiplier, photon-counting system, and a fast digital correlator working in the nanosecond range.[8,13,140–145] The use of the DWS/DCS technique in medical applications has been demonstrated for blood microcirculation monitoring in the human forearm, skeletal muscle, and brain.[140–145] The autocorrelation function slope is the indicative parameter for determination of the blood flow velocity. The normalized autocorrelation function of field fluctuations can be represented in terms of two components related to the Brownian and directed motion of scatterers (erythrocytes or lymphocytes):

$$g_1(\tau) \approx \exp\{-2[\tau/\tau_B + (\tau/\tau_s)^2]L/l_{tr}\}, \tag{I.26}$$

where $\tau_B^{-1} = |\bar{s}|^2 D_B$ characterizes Brownian motion as $\tau_s^{-1} \cong 0.18 G_V|\bar{s}|l_{tr}$ (the directed flow), and G_V is the gradient of the flow rate. In Eq. (I.26), directed flow dependent on τ^2 is compared to the τ dependence for Brownian motion because particles in flows travel ballistically; also τ_B and τ_S appear separately because the different dynamical processes are uncorrelated.

I.10 Optothermal Spectroscopy and Tomography

I.10.1 Optothermal interactions

The optothermal (OT) or photothermal (PT) method detects the time-dependent heat generated in a tissue via interaction with pulsed or intensity-modulated optical radiation.[13,52–54,146–155] The thermal waves generated by the release of heat result in several effects that have given rise to various imaging techniques: optoacoustics (OA) and photoacoustics (PA); optothermal radiometry (OTR) and photothermal radiometry (PTR); and photorefractive techniques.[13,52–54,146–155] In the past, the term "optoacoustics" was used to refer primarily to the time-resolved technique utilizing pulsed lasers and measuring profiles of pressure in tissue, and the term "photoacoustics" primarily described spectroscopic experiments with CW-modulated light and a photoacoustic cell. Nowadays, the term "photoacoustics" is much more frequently used for time-resolved techniques. The informative features of the PA method allow one to estimate tissue thermal, optical, and acoustical properties, which depend on tissue structure peculiarities.

In PA techniques, microphone or piezoelectric transducers, which are in acoustic contact with the sample, are used as detectors to measure the amplitude or phase of the resultant acoustic wave. In the PTR technique, distant IR detectors and array cameras are employed for estimation and imaging of the sample surface temperature. The intensity of the signals obtained with any of the PT or PA techniques depends on the amount of energy absorbed and transformed into heat as well as on the thermo-elastic properties of the sample and its surrounding. When nonradiative relaxation is the main process in a light beam decay and extinction is not very high, $\mu_a d << 1$ (d is the length of a cylinder within the sample occupied by a pulse laser beam), the absorbed pulse energy induces the local temperature rise, which is defined by

$$\Delta T \cong E\mu_a d / c_p V\rho, \tag{I.27}$$

where c_P is the specific heat capacity for a constant pressure, $V = \pi w^2 d$ is the illuminated volume, w is the laser beam radius, and ρ is the medium density. Supposing there is an adiabatic expansion of the illuminated volume upon being heated at a constant pressure, one can calculate the change of the volume ΔV. This expansion induces a wave propagating in the radial direction at the sound speed, v_a. The corresponding change of pressure Δp is proportional to the amplitude of mechanical oscillations

$$\Delta p \approx (f_a/w)(\beta v_a/c_P)E\mu_a, \tag{I.28}$$

where β is the coefficient of volumetric expansion and f_a is the frequency of the acoustic wave.

Equations (I.27) and (I.28) present principles of various PT and PA techniques. The information about the absorption coefficient μ_a at the selected wavelength can be received from direct measurements of the temperature change ΔT (optical calorimetry), volume change ΔV (optogeometric technique), or pressure change Δp (PA technique).

I.10.2 PA technique

For a highly scattering tissue, measurement of the stress-wave profile and amplitude should be combined with measurement of the total diffuse reflectance so as to extract separately both the absorption and scattering coefficients of the sample. The absorption coefficient in a turbid medium can be estimated from the acoustic transient profile only if the subsurface irradiance is known. For the turbid media irradiated with a wide laser beam (more than 0.1 mm), backscattering causes a higher subsurface fluence rate compared with the incident laser fluence.[13] Therefore, the z-axial light distribution in tissue and the corresponding stress distribution have a complex profile with a maximum at a subsurface layer. In contrast, when the heating process is much faster than the medium expansion, then the stress amplitude adjacent to the irradiated surface $\delta p(0)$ and the stress exponential tail into the depth of tissue sample $\delta p(z)$ can be found.[146,147,151] The stress is confined temporarily during laser heat deposition when the laser pulse duration is much shorter than the time of stress propagation across the light penetration depth in the tissue sample. Such conditions of temporal pressure confinement in the volume of irradiated tissue allow for the most efficient pressure generation.

I.10.3 PTR technique

The pulse laser heating of a tissue causes perturbations of its temperature and corresponding modulation of the thermal (infrared) radiation. This pair of reactions is the basis for pulse photothermal radiometry (PTR).[13,52,54,153–155] The maximum intensity of living objects' thermal radiation approximates a wavelength of 10 μm. A detailed analysis of PTR signal formation requires knowledge of the internal temperature distribution within the tissue sample, tissue thermal diffusivity, and the absorption coefficients at the excitation μ_a and emission μ_a' (10 μm) wavelengths. And, working backward, knowledge of some of the previously mentioned parameters allows one to use a measured PTR signal to reconstruct, for example, the depth distribution of μ_a.

The pulse PTR method holds much promise for the study of the optical and thermal properties of tissues *in vitro* and *in vivo*.[13,52–54,153–155] For example, sequences (pairs) of infrared emission images recorded following pulsed laser irradiation have been used to determine the thermal diffusivity of biomaterials with high precision.[155]

I.11 Conclusion

Since publication of the first edition of the *Handbook* in 2002, optical methods for biomedical diagnostics have been further developed in many well-established, now-traditional directions, which were first summarized in the first edition. In addition, new trends have appeared. In recent years, a number of handbooks, textbooks, and special issues of journals have been published that are good companions to the coverage of topics provided in the second edition of the *Handbook*.[13,83,90,92,93,105,114,117,120,124,126,147,148,156–209]

The comprehensive *Biomedical Photonics Handbook* by Vo-Dinh covers many topics related to optical biomedical diagnostics, based on a variety of light–tissue, light–cell, and light–molecular interaction phenomena; it also includes descriptions of biosensing approaches.[159] The second edition of *Tissue Optics* by Tuchin mostly concentrates on the optics of soft and hard tissues, characterization and control of their optical properties, and light-scattering and coherent-domain methods for biomedical spectroscopy and imaging.[13]

As an introduction to the field of biomedical optics and biophotonics, three very enjoyable textbooks by Prasad,[161] Wang and Wu,[147] and Splinter and Hooper,[165] containing a number of actual examples, problems, and questions for students, might be recommended.

Optical coherence tomography is a new trend in biomedical diagnostics that is continually delivering novel modalities with better facilities in image resolution and real-time imaging.[13,117,119,120,126] The first book on OCT by Bouma and Tearney,[117] a comprehensive review paper by Fercher et al.,[119] the second volume of Tuchin's two-volume monograph,[126] the most recent exciting OCT book by Drexler and Fujimoto,[120] and the special issue of *Journal of Biomedical Optics* (*JBO*) edited by Larin et al.[131] summarize and analyze the cutting-edge OCT technologies and their biomedical applications. A brief overview of OCT fundamentals, techniques, and applications is provided by Tuchin.[13]

Another important trend in optical biomedical diagnostics is application of polarized light for tissue characterization and imaging.[114,164,178–180] The importance of the problem was underlined by the publication of a special issue of *JBO* devoted to this topic and edited by Wang et al.,[114] a monograph by Tuchin et al.,[164] and a tutorial paper by Ghosh and Vitkin.[179] Discussions of polarized light's interaction with tissues and applications for diagnostic purposes, in particular for glucose sensing, are presented in monographs by Tuchin.[13,101,177,178]

Further developments of multi-photon excitation microscopy and other methods of nonlinear microscopy are discussed in the handbook by Masters and So,[90] a book edited by Pavone,[92] and a book on SHG imaging by Pavone and Campagnola.[93]

Trends in nanobiophotonics as a novel synergetic science underlying diagnosis, prevention, and treatment of diseases including cancer, systematic conditions, and inflammatory diseases are overviewed in the special section of *JBO* edited by Tuchin et al.[171] and in a collective monograph edited by Tuchin.[177]

In vivo flow cytometry[181–185] and noninvasive blood flow imaging[123–125] in tissues are important directions for research in biomedical diagnostics. A few overview papers, book chapters, and books are available on blood flow imaging, published by Leahy and his group.[123–125]

In the previous edition of the *Handbook*, a lot of attention was paid to the characterization of optical properties of biological tissues using innovative approaches for inverse problem solution. Recent work on this front includes the collection of new experimental data and their critical analysis for different tissues in a wide spectral range.[186–191]

A comprehensive presentation of fundamentals, basic research, and medical application of biophotonics is provided in a three-volume monograph edited by Popp et al.[192–194] Some practical aspects of optical biomedical diagnostics and treatments, especially those targeting skin disease and cosmetology, are discussed in books by Wilhelm et al.,[166] Ahluwalia,[168] Baron,[172] and Querleux.[203] Photonics for solution of specific dental problems is covered in Kishen and Asundi's book.[163] All aspects of glucose noninvasive optical sensing and its impact on tissues are analyzed in the book by Tuchin.[101]

Tissue optical clearing has emerged as a hot topic in the field of optical biomedical diagnostics owing to the considerably enhanced ability of different optical methods to suppress light scattering.[13,210–240] The book on tissue and blood optical clearing by Tuchin,[212] many recent review papers,[210,211,213,216,218,219,223,240] several book chapters,[215,217,220,221,224] and a special section of *JBO*[214] are devoted to this growing area of research and the applications of temporal (reversible) control of tissue optical properties using immersion clearing. Mechanical compression and stretching are also prospective tools to improve optical imaging and diagnostics as well as the therapeutic abilities of light.[13,212,240–266]

Very recently, a brilliant textbook on *Quantitative Biomedical Optics* (which, I believe, students will accept with enthusiasm) was published by Bigio and Fantini.[267] Three more special sections of journals on urgent problems of biomedical optics and biophotonics, with a large number of papers related to optical biomedical diagnostics and specifically to polarization and optical clearing methods, were issued.[268–270]

Acknowledgments

The work was supported by Russian Presidential grant NSh-7898.2016.2, the Government of the Russian Federation grant 14.Z50.31.0004, and the Tomsk State University Academic D.I. Mendeleev Fund Program.

References

1. B. Chance, "Optical method," *Ann. Rev. Biophys. Biophys. Chem.* **20**, 1–28 (1991).
2. D. Benaron, I. Bigio, E. Sevick-Muraca, and A.G. Yodh, Eds., "Special issue honoring Professor Britton Chance," *J. Biomed. Opt.* **5**, 115–248, 269–282 (2000).
3. R.R. Anderson and J.A. Parrish, "Optical properties of human skin," *The Science of Photomedicine*, J.D. Regan and J.A. Parrish, Eds., pp. 147–194, Plenum Press, New York (1982).
4. F.A. Duck, *Physical Properties of Tissue: A Comprehensive Reference Book*, Academic, London (1990).
5. G. Müller, B. Chance, and R. Alfano, et al., Eds., *Medical Optical Tomography: Functional Imaging and Monitoring*, IS11, SPIE Press, Bellingham, Wash. (1993).
6. V.V. Tuchin, Ed., *Selected Papers on Tissue Optics Applications in Medical Diagnostics and Therapy*, **MS 102**, SPIE Press, Bellingham, Wash. (1994).
7. M.S. Patterson, "Noninvasive measurement of tissue optical properties: current status and future prospects," *Comments Mol. Cell. Biophys.* **8**, 387–417 (1995).

8. A.G. Yodh and B. Chance, "Spectroscopy and imaging with diffusing light," *Phys. Today* **48**, 34–40 (1995).
9. B.B. Das, F. Liu, and R.R. Alfano, "Time-resolved fluorescence and photon migration studies in biomedical and random media," *Rep. Prog. Phys.* **60**, 227–292 (1997).
10. M. Ferrari, D. Delpy, and D.A. Benaron, Eds., "Special section on clinical near infrared spectroscopy/imaging," *J. Biomed. Opt.* **1**, 361–434 (1996); 2, 7–41; 147–175 (1997).
11. O. Minet, G. Mueller, and J. Beuthan, Eds., Selected Papers on Optical Tomography, Fundamentals and Applications in Medicine, **MS 147**, SPIE Press, Bellingham, Wash. (1998).
12. K. Okada and T. Hamaoka, "Special section on medical near-infrared spectroscopy," *J. Biomed. Opt.* **4**, 391–428 (1999).
13. V.V. Tuchin, *Tissue Optics: Light Scattering Methods and Instruments for Medical Diagnosis*, 3rd ed., PM254, SPIE Press, Bellingham, Wash. (2015).
14. H. Wabnitz, J. Rodriguez, I. Yaroslavsky, A. Yaroslavsky, H. Battarbee, and V.V. Tuchin, "Time-resolved imaging in diffusive media," in the *Handbook of Optical Biomedical Diagnostics, Volume 1*, 2nd ed., V.V. Tuchin, Ed., Chapter 6, pp. 401–475, SPIE Press, Bellingham, Wash. (2016).
15. M. Cutler, "Transillumination as an aid in the diagnosis of breast lesions," *Surg. Gynecol. Obstet.* **48**, 721 (1929).
16. G.A. Millikan, "A simple photoelectric colorimeter," *J. Physiol.* (Lond,) **79**, 152–157 (1933).
17. A.A. Il'ina, "Transmittance of the near infrared rays by tissues of the human body," *Soviet Physiol. J.* **35**, 338–348 (1949).
18. F. Jobsis, "Non invasive monitoring of cerebral and myocardial oxygen sufficiency and circulatory parameters," *Science* **198**, 1264–1267 (1977).
19. F.F. Jobsis-vander Vliet, "Discovery of the near-infrared window into the body and the early development of near-infrared spectroscopy," *J. Biomed. Opt.* **4**, 392–396 (1999).
20. B. Chance and G.R. Williams, "A method for the localization of sites for oxidative phosphorylation," *Nature* (Lond.) **176**, 250–254 (1955).
21. B. Chance, P. Cohen, F. Jobsis, and B. Schoener, "Intracellular oxidation⊠reduction states *in vivo*," *Science* **137**, 499–508 (1962).
22. B. Chance, J.S. Leigh, H. Miyake, D.S. Smith, S. Nioka, R. Greenfeld, and G. Holtom, "Comparison of time-resolved and unresolved measurement of deoxygenation in tissue," *Proc. Natl. Acad. Sci. USA* **85**, 4971–4975 (1988).
23. D.T. Delpy, M. Cope, P. Van Zee, S. Arridge, S. Wray, and J. Wyatt, "Estimation of optical path-length through tissue from direct time of flight measurement," *Phys. Med. Biol.* **33**, 1433–1442 (1988).

24. M.S. Patterson, B. Chance, and B.C. Wilson, "Time resolved reflectance and transmittance for the non-invasive measurement of tissue optical properties," *Appl. Opt.* **28**, 2331–2336 (1989).
25. S.L. Jacques, "Time-resolved reflectance spectroscopy in turbid tissues," *IEEE Trans. Biomed. Eng.* **36**, 1155–1161 (1989).
26. W.-F. Cheong, S.A. Prahl, and A.J. Welch, "A review of the optical properties of biological tissues," *IEEE J. Quantum Electr.* **26**, 2166–2185 (1990).
27. K. Frank and M. Kessler, Eds., Quantitative Spectroscopy in Tissue, pmi Verlag, Frankfurt am Main (1992).
28. H. Moseley, Ed., "Special issue on optical radiation technique in medicine and biology," *Phys. Med. Biol.* **24**, 759–996 (1997).
29. J.R. Lakowicz and K. Berndt, "Frequency-domain measurements of photon migration in tissues," *Chem. Phys. Lett.* **166**, 246–252 (1990).
30. M.S. Patterson, J.D. Moulton, and B.C. Wilson, et al., "Frequency-domain reflectance for the determination of the scattering and absorption properties of tissue," *Appl. Opt.* **30**, 4474–4476 (1991).
31. J.M. Schmitt, A. Knüttel, and J.R. Knutson, "Interference of diffusive light waves," *J. Opt. Soc. Am. A* **9**, 1832–1843 (1992).
32. B. Chance, M. Cope, E. Gratton, N. Ramanujam, and B. Tromberg, "Phase measurement of light absorption and scatter in human tissue," *Rev. Sci. Instrum.* **698**, 3457–3481 (1998).
33. J.A. Parrish, R.R. Anderson, F. Urbach, and D. Pitts, *UV-A: Biologic Effects of Ultraviolet Radiation with Emphasis on Human Responses to Longwave Ultraviolet*, Plenum Press, New York (1978).
34. J.R. Lakowicz, *Principles of Fluorescence Spectroscopy*, 3rd ed., Springer Science + Business, New York (2006).
35. B.K. Pierscionek, Ed., "Special section on light scatter and fluorescence of the eye lens," *J. Biomed. Opt.* **1**, 241–295 (1996).
36. J.A. van Best and E.V.M.J. Kuppens, "Summary of studies on the blue-green autofluorescence and light transmission of the ocular lens," *J. Biomed. Opt.1*, 243–250 (1996).
37. N.-T. Yu, B.S. Krantz, and J.A. Eppstein, et al., "Development of noninvasive diabetes screening device using the ratio of fluorescence to Rayleigh scattered light," *J. Biomed. Opt.* **1**, 280–288 (1996).
38. W. Denk, "Two-photon excitation in functional biological imaging," *J. Biomed. Opt.* **1**, 296–304 (1996).
39. Y. Ozaki, "Medical application of Raman spectroscopy," *Appl. Spectroscopy Rev.* **24**, 259–312 (1988).
40. A.T. Tu, *Raman Spectroscopy in Biology*, John Wiley & Sons, New York (1982).
41. A. Mahadevan-Jansen and R. Richards-Kortum, "Raman spectroscopy for detection of cancers and precancers," *J. Biomed. Opt.* **1**, 31–70 (1996).

42. M.D. Morris, Ed., "Special section on biomedical applications of vibrational spectroscopic imaging," *J. Biomed. Opt.* **4**, 6–34 (1999).
43. A. Carden and M.D. Morris, "Application of vibration spectroscopy to the study of mineralized tissues (review)," *J. Biomed. Opt.* **5**, 259–268 (2000).
44. R.J. McNichols and G.L. Cote, "Optical glucose sensing in biological fluids: an overview," *J. Biomed. Opt.* **5**, 5–16 (2000).
45. G.W. Lucassen, G.N.A. van Veen, and J.A.J. Jansen, "Band analysis of hydrated human skin stratum corneum attenuated total reflectance Fourier transform infrared spectra *in vivo*," *J. Biomed. Opt.*, **3**, 267–280, (1998).
46. L.T. Perelman, V. Backman, and M. Wallace, et al., "Observation of periodic fine structure in reflectance from biological tissue: a new technique for measuring nuclear size distribution," *Phys. Rev. Lett.*, **80**, 627–630 (1998).
47. H.Z. Cummins and E.R. Pike, Eds., *Photon Correlation and Light Beating Spectroscopy*, Plenum Press, New York (1974); *Photon Correlation Spectroscopy and Velocimetry*, Plenum Press, New York (1977).
48. V.V. Tuchin, "Coherence-domain methods in tissue and cell optics," *Laser Physics*, **8**, 807–849 (1998).
49. V.V. Tuchin, "Coherent optical techniques for the analysis of tissue structure and dynamics," *J. Biomed. Opt.* **4**, 106–124 (1999).
50. A.V. Priezzhev and T. Asakura, Eds., "Special section on optical diagnostics of biological fluids," *J. Biomed. Opt.* **4**, 35–93 (1999).
51. V.V. Tuchin, H. Podbielska, and C.K. Hitzenberger, Eds., "Special section on coherence domain optical methods in biomedical science and clinics," *J. Biomed. Opt.* **4**, 94–190 (1999).
52. S.E. Braslavsky and K. Heihoff, "Photothermal methods," *Handbook of Organic Photochemistry*, J.C. Scaiano, Ed., Vol **1**, pp 327–355, CRC Press, Boca Raton, Fla. (1989).
53. V.E. Gusev and A.A. Karabutov, *Laser Optoacoustics*, AIP Press, New York (1993).
54. V.V. Tuchin, "Lasers and fiber optics in biomedicine," *Laser Phys.* **3**, 767–820, 925–950 (1993).
55. H. Heusmann, J. Kolzer, and G. Mitic, "Characterization of female breast *in vivo* by time resolved and spectroscopic measurements in near infrared spectroscopy," *J. Biomed. Opt.* **1**, 425–434 (1996).
56. H. Gong, Q. Luo, S. Zeng, S. Nioka, Y. Kuroda, and B. Chance, "Monitoring of brain activity with near infrared spectroscopy," in the *Handbook of Optical Biomedical Diagnostics, Volume 1*, 2nd ed., V.V. Tuchin, Ed., Chapter 8, pp 533–583, SPIE Press, Bellingham, Wash. (2016).
57. S.J. Matcher, "Absolute quantification in tissue near-infrared spectroscopy," in the *Handbook of Optical Biomedical Diagnostics, Volume 1*, 2nd ed., V.V. Tuchin, Ed., Chapter 9, pp 585–686, SPIE Press, Bellingham, Wash. (2016).

58. I.J. Bigio and J.R. Mourant, "Ultraviolet and visible spectroscopies for tissue diagnostics: fluorescence spectroscopy and elastic-scattering spectroscopy," *Phys. Med. Biol.* **42**, 803–814 (1997).
59. Z. S. Sacks, R. M. Kurtz, T. Juhasz, and G. A. Mourau, "High precision subsurface photodisruption in human sclera," *J. Biomed. Opt.* **7** (3), 442–450 (2002).
60. T.J.T.P. van den Berg and K.E.W.P. Tan, "Light transmittance of the human cornea from 320 to 700 nm for different ages," *Vision Res.* **33**, 1453–1456 (1994).
61. J. Dillon, "The photophysics and photobiology of the eye," *J. Photochem. Photobiol. B: Biol.* **10**, 23–40 (1991).
62. J.B. Fishkin and E. Gratton, "Propagation of photon-density waves in strongly scattering media containing an absorbing semi-infinite plane bounded by a strait edge," *J. Opt. Soc. Am. A* **10**, 127–140 (1993).
63. J.B. Fishkin, O. Coquoz, and E.R. Anderson, et al., "Frequency-domain photon migration measurements of normal and malignant tissue optical properties in a human subject," *Appl. Opt.* **36**, 10–20 (1997).
64. B. Tromberg, O. Coquoz, and J.B. Fishkin, et al., "Non-invasive measurements of breast tissue optical properties using frequency-domain photon migration," *Phil. Trans. R. Soc. Lond. B.* **352**, 661–668 (1997).
65. T.D. O'sullivan, A.E. Cerussi, D.J. Cuccia, and B.J. Tromberg, "Diffuse optical imaging using spatially and temporally modulated light," *J. Biomed. Opt.* **17**(7), 071311-1–14 (2012).
66. A. Mazhar, D. J. Cuccia, A. J. Durkin, and B. J. Tromberg, "Spatial and temporal frequency domain tissue optical imaging," *Advanced Biophotonics: Tissue Optical Sectioning*, R.K. Wang and V.V. Tuchin, Eds., pp. 109–136, CRC Press, Taylor & Francis Group, Boca Raton, Fla. (2013).
67. V.P. Ryabukho "Diffraction of interference fields on random phase objects," *Coherent-Domain Optical Methods: Biomedical Diagnostics, Environmental and Material Science*, V.V. Tuchin, Ed., pp. 235–318, Kluwer Academic Publishers, Boston (2004).
68. D.J. Cuccia, F. Bevilacqua, A.J. Durkin, and B.J. Tromberg, "Modulated imaging: quantitative analysis and tomography of turbid media in the spatial-frequency domain," *Opt. Lett.* **30**(11), 1354–1356 (2005).
69. A. Bassi, D. Cuccia, A. Durkin, and B. Tromberg, "Spatial shift of spatially modulated light projected on turbid media," *J. Opt. Soc. Am.* **25**(11), 2833–2839 (2008).
70. D. Abookasis, C. Lay, M. Mathews, M. Linskey, R. Frostig, and B. Tromberg, "Imaging cortical absorption, scattering, and hemodynamic response during ischemic stroke using spatially modulated near-infrared illumination, *J. Biomed. Opt.* **14**(2), 024033 (2009).

71. D. Cuccia, F. Bevilacqua, A. J. Durkin, F. Ayers, and B. Tromberg, "Quantitation and mapping of tissue optical properties using modulated imaging, *J. Biomed. Opt.* **14**(2), 024012-1–13 (2009).
72. A. Mazhar, D. Cuccia, S. Gioux, A. Durkin, J. Frangioni, and B. Tromberg, "Structured illumination enhances resolution and contrast in thick tissue fluorescence imaging," *J. Biomed. Opt.* **15**(1), 010506 (2010).
73. J.R. Weber, D.J. Cuccia, R. William, W.R. Johnson, G. Bearman, A.J. Durkin, M. Hsu, A. Lin, D.K. Binder, D. Wilson, and B.J. Tromberg, "Multispectral imaging of tissue absorption and scattering using spatial frequency domain imaging and a computed-tomography imaging spectrometer," *J. Biomed. Opt.* **16**, 011015 (2011).
74. H. Schneckenburger, P. Weber, T. Bruns, and M. Wagner, "Advances in fluorescence spectroscopy and imaging," *Handbook of Photonics for Medical Science*, V.V. Tuchin, Ed., pp. 119–136, CRC Press, Taylor & Francis Group, London, (2010).
75. H. Schneckenburger, W. Strauss, K. Stock, and R. Steiner, "Fluorescence technologies in biomedical diagnostics," in the *Handbook of Optical Biomedical Diagnostics, Volume 2*, 2nd ed., V.V. Tuchin, Ed., Chapter 5, pp 241–303, SPIE Press, Bellingham, Wash. (2016).
76. Y.P. Sinichkin, N. Kollias, G. Zonios, S.R. Utz, and V.V. Tuchin, "Reflectance and fluorescence spectroscopy of human skin *in vivo*," in the *Handbook of Optical Biomedical Diagnostics, Volume 2*, 2nd ed., V. V. Tuchin, Ed., Chapter 3, pp 99–190, SPIE Press, Bellingham, Wash. (2016).
77. S. Svanberg, "New developments in laser medicine," *Phys. Scripta* **T72**, 69–75 (1997).
78. R.R. Richards-Kortum, R.P. Rava, R.E. Petras, M. Fitzmaurice, M. Sivak, and M.S. Feld, "Spectroscopic diagnosis of colonic dysplasia," *Photochem. Photobiol.* **53**, 777–786 (1991).
79. H.J.C.M. Sterenborg, M. Motamedi, R.F. Wagner, J.R.M. Duvic, S. Thomsen, and S.L. Jacques, "*In vivo* fluorescence spectroscopy and imaging of human skin tumors," *Lasers Med. Sci.* **9**, 344–348 (1994).
80. H. Zeng, C. MacAulay, D.I. McLean, and B. Palcic, "Spectroscopic and microscopic characteristics of human skin autofluorescence emission," *Photochem. Photobiol.* **61**, 639–645 (1995).
81. Y.P. Sinichkin, S.R. Utz, A.H. Mavlutov, and H.A. Pilipenko, "*In vivo* fluorescence spectroscopy of the human skin: experiments and models," *J. Biomed. Opt.* **3**, 201–211 (1998).
82. Z. Malik, I. Amit, and C. Rothmann, "Subcellular localization of sulfonated tetraphenyl porphines in colon carcinoma cells by spectrally resolved imaging," *Photochem. Photobiol.* **65**, 389–396 (1997).

83. F.D. Dip, T. Ishizawa, N. Kokudo, and R. Rosenthal, Eds., *Fluorescence Imaging for Surgeons: Concepts and Applications*, Springer Science and Business Media, New York (2015).
84. J.T. Alander, I. Kaartinen, A. Laakso, T. Pätilä, T. Spillmann, V.V. Tuchin, M. Venermo, and P. Välisuo, "A review of indocyanine green fluorescent imaging in surgery," *Int. J. Biomed. Imaging*, 940585 (2012).
85. J.T. Alander, O.M. Villet, T. Pätilä, I.S. Kaartinen, M. Lehecka, T. Nakaguchi, T. Suzuki, and V. Tuchin, "Review of indocyanine green imaging in surgery," Fluorescence Imaging for Surgeons: Concepts and Applications, F.D. Dip, T. Ishizawa, N. Kokudo, and R. Rosenthal, Eds., pp 35–53, Springer Science and Business Media, New York (2015).
86. S.B. Mondal, S. Gao, and N. Zhu, et al. "Binocular goggle augmented imaging and navigation system provides real-time fluorescence image guidance for tumor resection and sentinel lymph node mapping," *Sci. Rep.* **5**, 12117; doi: 10.1038/srep12117(2015).
87. E. Borisova, P. Troyanova, P. Pavlova, and L. Avramov, "Diagnostics of pigmented skin tumors based on laser-induced autofluorescence and diffuse reflectance spectroscopy," *Quantum Electron.* **38**(6), 597–605 (2008).
88. E.G., Borisova, L.P. Angelova, and E.P. Pavlova, "Endogenous and exogenous fluorescence skin cancer diagnostics for clinical applications," *IEEE J. Sel. Top. Quant. Elect.* **20**, 7100412 (2014).
89. A. Diaspro, G. Chirico, and M. Collini, "Two-photon fluorescence excitation and related techniques in biological microscopy," *Q. Rev. Biophys.* **38**(2), 97–166 (2005).
90. B.R. Masters and P.T.C. So, Eds., *Handbook of Biomedical Nonlinear Optical Microscopy*, Oxford University Press, New York (2008).
91. W. Zheng, Y. Wu, D. Li, and J. Y. Qu, "Autofluorescence of epithelial tissue: Single-photon versus two-photon excitation," *J. Biomed. Opt.* **13** (5), 054010-1–8 (2008).
92. F.S. Pavone, Ed., *Laser Imaging and Manipulation in Cell Biology*, Wiley-VCH Verlag, Weinheim (2010).
93. F.S. Pavone and P.J. Campagnola, Eds., *Second Harmonic Generation Imaging*, CRC Press, Taylor & Francis Group, Boca Raton, Fla. (2014).
94. R. Cicchi, N. Vogler, D. Kapsokalyvas, B. Dietzek, J. Popp, and F.S. Pavone, "From molecular structure to tissue architecture: collagen organization probed by SHG microscopy," *J. Biophotonics* **6**(2), 129–142 (2013).
95. G.W. Lucassen, P.J. Caspers, G.J. Puppels, M.E. Darvin, and J. Lademann, "Infrared and Raman spectroscopy of human skin *in vivo*," in the *Handbook of Optical Biomedical Diagnostics, Volume 2*, 2nd ed., V.V. Tuchin, Ed., Chapter 4, pp 191–240, SPIE Press, Bellingham, Wash. (2016).

96. G.J. Puppels, "Confocal Raman microspectroscopy," *Fluorescent and Luminescent Probes for Biological Activity*, W. Mason, Ed., pp. 377–406, Academic Press, London (1999).
97. R. Petry, M. Schmitt, and J. Popp, "Raman spectroscopy: a prospective tool in the life sciences," *Chemphyschem.* **4**, 14–30 (2003).
98. C. Krafft and J. Popp, "Raman and CARS microscopy of cells and tissues," *Handbook of Photonics for Biomedical Science*, V.V. Tuchin, Ed., pp. 197–227, CRC Press, Taylor & Francis Group, London (2010).
99. C. Krafft, B. Dietzek, M. Schmitt, and J. Popp, "Raman and coherent anti-Stokes Raman scattering microspectroscopy for biomedical applications." *J. Biomed. Opt.* **17**(4), 040801-1–15 (2012).
100. J. Zhao, H. Lui, S. Kalia, and H. Zeng, "Real-time Raman spectroscopy for automatic *in vivo* skin cancer detection: an independent validation," Anal. Bioanal. Chem. doi: 10.1007/s00216-015-8914-9(2015).
101. V.V. Tuchin, Ed., *Handbook of Optical Sensing of Glucose in Biological Fluids and Tissues*, CRC Press, Taylor & Francis Group, London (2009).
102. W.-C. Shih, K. L. Bechtel, and M.V. Rebec, "Noninvasive glucose sensing by transcutaneous Raman spectroscopy," *J. Biomed. Opt.* **20**(5), 051036-1-5 (2015).
103. D. Li, W. Zheng, Y. Zeng, and J. Y. Qu, "*In vivo* and simultaneous multimodal imaging: Integrated multiplex coherent anti-Stokes Raman scattering and two-photon microscopy," *Appl. Phys. Lett.* **97**, 223702-1-3 (2010).
104. H.G. Breunig, R. Bückle, M. Kellner-Höfer, M. Weinigel, J. Lademann, W. Sterry, and K. König, "Combined *in vivo* multiphoton and CARS imaging of healthy and disease-affected human skin," *Microsc. Res. Tech.*, **75**, 492–498 (2012).
105. A. Wax and V. Backman, Eds., Biomedical Applications of Light Scattering, McGraw-Hill, New York (2010).
106. L.T. Perelman and V. Backman, "Light scattering spectroscopy of epithelial tissues: principles and applications," in the *Handbook of Optical Biomedical Diagnostics, Volume 2*, 2nd ed., V.V. Tuchin, Ed., Chapter 2, pp 37–98, SPIE Press, Bellingham, Wash. (2016).
107. J.R. Mourant, J. P. Freyer, and A. H. Hielscher, et al., "Mechanisms of light scattering from biological cells relevant to noninvasive optical-tissue diagnosis," *Appl. Opt.* **37**, 3586–3593 (1998).
108. R. Drezek, A. Dunn, and R. Richards-Kortum, "Light scattering from cells: finite-difference time-domain simulations and goniometric measurements," *Appl. Opt.* **38**, 3651–3661 (1999).
109. V. Backman, M. Wallace, and L.T. Perelman, et al., "Diagnosing cancers using spectroscopy," *Nature*, 405 (2000).
110. C.F. Bohren and D.R. Huffman, Absorption and Scattering of Light by Small Particles, Wiley, New York (1983).

111. C. Yang, L.T. Perelman, and A. Wax, et al., "Feasibility of field-based light scattering spectroscopy," *J. Biomed. Opt.* **5**, 138–143 (2000).
112. U. Morgner, W. Drexler, and F.X. Kartner, et al., "Spectroscopic optical coherence tomography," *Opt. Lett.* **25**, 111–113 (2000).
113. V. Backman, R. Gurjar, and K. Badizadegan, et al., "Polarized light scattering spectroscopy for quantitative measurement of epithelial cellular structures *in situ*," *IEEE J. Sel. Top. Quant. Elect.* **5**, 1019–1026 (1999).
114. L.V. Wang, G.L. Coté, and S.L. Jacques, Eds., "Special section on tissue polarimetry," *J. Biomed. Opt.* **7**(3), 278–397 (2002).
115. D. Huang, E.A. Swanson, C.P. Lin, J.S. Schuman, W.G. Stinson, W. Chang, T. Hee, M.R. Flotte, K. Gregory, C.A. Puliafito, and J.G. Fujimoto, "Optical coherence tomography," *Science* **254**, 1178–1181 (1991).
116. A.F. Fercher, "Optical coherence tomography," *J . Biomed. Opt .* **1**, 157–173 (1996).
117. B.E. Bouma and G.J. Tearney, Eds., Handbook of Optical Coherence Tomography, Marcel-Dekker, New York (2002).
118. D.A. Zimnyakov and V.V. Tuchin, "Optical tomography of tissues (overview)," *Quantum Electron.* **32**(10), 849–867 (2002).
119. A.F. Fercher, W. Drexler, C.K. Hitzenberger, and T. Lasser, "Optical coherence tomography: principles and applications," *Rep. Progr. Phys.* **66**, 239–303 (2003).
120. W. Drexler and J.G. Fujimoto, Eds., *Optical Coherence Tomography: Technology and Applications*, Springer, Berlin (2008); 2nd ed. Springer, Berlin (2015).
121. Y. Wu and X. Li, "Endomicroscopy technologies for high-resolution nonlinear optical imaging and optical coherence tomography," *Handbook of Photonics for Biomedical Science*, V.V. Tuchin, Ed., pp. 547–573, CRC Press, Taylor & Francis Group, London (2010).
122. L. An, Y. Jia, and R.K. Wang, "Label-free optical micro-angiography for functional imaging of microcirculations within tissue beds *in vivo*," *Handbook of Photonics for Biomedical Science*, V.V. Tuchin, Ed., pp. 401–422, CRC Press, Taylor & Francis Group, London (2010).
123. M. J. Leahy and G. E. Nilsson, "Biophotonic functional imaging of skin microcirculation," *Handbook of Photonics for Biomedical Science*, V.V. Tuchin, Ed., pp. 323–334, CRC Press, Taylor & Francis Group, London (2010).
124. M.J. Leahy, Ed., *Microcirculation Imaging*, Wiley-VCH Verlag, Weinheim (2012).
125. S.M. Daly and M.J Leahy, "'Go with the flow:' a review of methods and advancements in blood flow imaging," *J. Biophoton.* **6**(3), 217–255 (2013).

126. V.V. Tuchin, Ed., *Handbook of Coherent-Domain Optical Methods: Biomedical Diagnostics, Environmental Monitoring and Material Science*, Vols. 1 & 2, 2nd ed., Springer-Verlag, Berlin (2013).
127. A. Podoleanu, "Flying spot en face OCT imaging," in the *Handbook of Coherent-Domain Optical Methods: Biomedical Diagnostics, Environmental Monitoring and Material Science*, 2nd ed., V.V. Tuchin, Ed., pp. 799–856, Springer-Verlag, Berlin (2013).
128. J.F. de Boer, "Polarization sensitive optical coherence tomography: phase sensitive interferometry for multi-functional imaging," in the *Handbook of Coherent-Domain Optical Methods: Biomedical Diagnostics, Environmental Monitoring and Material Science*, 2nd ed., V.V. Tuchin, Ed., pp. 857–888, Springer-Verlag, Berlin (2013).
129. R.A. McLaughlin, D. Lorenser, and D.D. Sampson, "Needle probes in optical coherence tomography," in the *Handbook of Coherent-Domain Optical Methods: Biomedical Diagnostics, Environmental Monitoring and Material Science*, 2nd ed., V.V. Tuchin, Ed., pp. 1065–1102, Springer-Verlag, Berlin (2013).
130. L.S. Dolin, G.V. Gelikonov, V.M. Gelikonov, N.D. Gladkova, R.R. Iksanov, V.A. Kamensky, R.V. Kuranov, N.M. Shakhova, and I.V. Turchin, "OCT fundamentals and clinical applications of endoscopic OCT," in the *Handbook of Coherent-Domain Optical Methods: Biomedical Diagnostics, Environmental Monitoring and Material Science*, 2nd ed., V.V. Tuchin, Ed., pp. 999–1064, Springer-Verlag, Berlin (2013).
131. K.V. Larin, V.V. Tuchin, and A. Vitkin, Eds., "Special section on optical coherence tomography and interferometry: advanced engineering and biomedical applications," *J. Biomed. Opt.* **19**(2), 021101 (2014).
132. J.D. Briers, "Laser Doppler and time-varying speckle: a reconciliation," *J. Opt. Soc. Am. A.* **13**, 345–350 (1996).
133. A. Serov, B. Steinacher, and T. Lasser, "Full-field laser Doppler perfusion imaging and monitoring with an intelligent CMOS camera," Optics Express 13, 3681–3689 (2005).
134. K.R. Forrester, J. Tulip, C. Leonard, C. Stewart, and R.C. Bray, "A laser speckle imaging technique for measuring tissue perfusion," *IEEE Trans. Biomed. Eng.* **51**, 2074–2084 (2004).
135. Q. Liu, Z. Wang, and Q. Luo, "Temporal clustering analysis of cerebral blood flow activation maps measured by laser speckle contrast imaging," *J. Biomed. Opt.* **10**(2), 024019-1-7 (2005).
136. A.K. Dunn, "Laser speckle contrast imaging of cerebral blood flow," *Ann. Biomed. Eng.* **40**(2), 367–377 (2012).
137. D. Briers, D.D. Duncan, E. Hirst, S.J. Kirkpatrick, M. Larsson, W. Steenbergen, T. Stromberg, and O.B. Thompson, "Laser speckle contrast imaging: theoretical and practical limitations," *J. Biomed. Opt.* **18**(6), 066018 (2013).

138. S.M.S. Kazmi, E. Faraji, M.A. Davis, Y.-Y. Huang, X.J. Zhang, and A. K. Dunn, "Flux or speed? Examining speckle contrast imaging of vascular flows," *Biomed. Opt. Express* **6**(7), 2588–2608 (2015).
139. P. Zakharov and F. Scheffold, "Advances in dynamic light scattering techniques," *Light Scattering Reviews 4: Single Light Scattering and Radiative Transfer*, A.A. Kokhanovsky, Ed., pp. 433–468, Springer, Heidelberg (2009).
140. D.A. Boas, I.V. Meglinsky, and L. Zemany, et al., "Diffusion of temporal field correlation with selected applications," *SPIE CIS Sel. Papers* **2732**, 34–46 (1996).
141. G. Yu, G. Lech, C. Zhou, B. Chance, E.R. Mohler, III, and A.G. Yodh, "Time-dependent blood flow and oxygenation in human skeletal muscles measured with noninvasive near-infrared diffuse optical spectroscopies," *J. Biomed. Opt.* **10**(2), 024027-1-7 (2005).
142. T. Durduran, C. Zhou, E.M. Buckley, M.N. Kim, G. Yu, R. Choe, J.W. Gaynor, T.L. Spray, S.M. Durning, S.E. Mason, L.M. Montenegro, S. C. Nicolson, R.A. Zimmerman, M.E. Putt, J. Wang, J.H. Greenberg, J. A. Detre, A.G. Yodh, and D.J. Licht, "Optical measurement of cerebral hemodynamics and oxygen metabolism in neonates with congenital heart defects," *J. Biomed. Opt.* **15**(3), 037004-1–10 (2010).
143. I.V. Meglinski and V.V. Tuchin, "Diffusing wave spectroscopy: application for blood flow diagnostics," in the *Handbook of Coherent-Domain Optical Methods: Biomedical Diagnostics, Environmental Monitoring and Material Science*, 2nd ed., V.V. Tuchin, Ed., pp. 149–167, Springer-Verlag, Berlin (2013).
144. H.M. Varma, C.P. Valdes, A.K. Kristoffersen, J.P. Culver, and T. Durduran, "Speckle contrast optical tomography: a new method for deep tissue three dimensional tomography of blood flow," *Biomed. Opt. Express* **5**(4), 1275–1289 (2014).
145. P. Farzam and T. Durduran, "Multidistance diffuse correlation spectroscopy for simultaneous estimation of blood flow index and optical properties," *J. Biomed. Opt.* **20**(5), 055001-1–10 (2015).
146. A.A. Oraevsky, S.J. Jacques, and F.K. Tittel, "Measurement of tissue optical properties by time-resolved detection of laser-induced transient stress," *Appl. Opt.* **36**, 402–415 (1997).
147. L.V. Wang and H.-I. Wu, Biomedical Optics: Principles and Imaging, Wiley-Interscience, Hoboken, N.J. (2007).
148. L. Wang, Ed., *Photoacoustic Imaging and Spectroscopy*, CRC Press, Taylor & Francis Group, London (2009).
149. S. Hu, K. Maslov, and L.V. Wang, "Optical-resolution photoacoustic microscopy for *in vivo* volumetric microvascular imaging in intact tissues," *Handbook of Photonics for Biomedical Science*, V.V. Tuchin, Ed., pp. 361–375, CRC Press, Taylor & Francis Group, London (2010).

150. L. Song, Z. Guo, and L.V. Wang, "High-speed photoacoustic tomography," *Advanced Biophotonics: Tissue Optical Sectioning*, R.K. Wang and V.V. Tuchin, Eds., pp. 423–448, CRC Press, Taylor & Francis Group, Boca Raton, Fla. (2013).
151. I.M. Pelivanov, T.D. Khokhlova, A.A. Karabutov, and A.A. Oraevsky, "Measurement of optical fluence distribution and optical properties of tissues using time-resolved profiles of optoacoustic pressure," in the *Handbook of Optical Biomedical Diagnostics, Volume 1*, 2nd ed., V.V. Tuchin, Ed., Chapter 11, pp 735–776, SPIE Press, Bellingham, Wash. (2016).
152. V. Ntziachristos and D. Razansky, "Molecular imaging by means of multispectral optoacoustic tomography (MSOT)," *Chem. Rev.* **110**, 2783–2794 (2010).
153. V.V. Tuchin, E.I. Galanzha, and V.P. Zharov, "*In vivo* photothermal and photoacoustic flow cytometry," *Advanced Optical Cytometry: Methods and Disease Diagnoses*, V.V. Tuchin, Ed., pp. 501–571, Wiley-VCH Verlag, Weinheim (2011).
154. U.S. Sathyam and S.A. Prahl, "Limitations in measurement of subsurface temperatures using pulsed photothermal radiometry," *J. Biomed. Opt.* **2**, 251–261 (1997).
155. T.E. Milner, D.M. Goodman, and B.S. Tanenbaum, et al., "Noncontact determination of thermal diffusivity in biomaterials using infrared imaging radiometry," *J. Biomed. Opt.* **1**, 92–97 (1996).
156. D.R. Vij and K. Mahesh, Eds., Lasers in Medicine, Kluwer Academic Publishers, Boston (2002).
157. W.R. Chen, V.V. Tuchin, Q. Luo, and S.L. Jacques, "Special issue on biophotonics," *J. X-Ray Sci. Technol.* **10**(3/4), 139–243 (2002).
158. P. French and A.I. Ferguson, Eds., "Special issue on biophotonics," *J. Phys. D: Appl. Phys.* **36**(14), R207–R258, 1655–1757 (2003).
159. T. Vo-Dinh, Ed., *Biomedical Photonics Handbook*, CRC Press, Boca Raton, Fla. (2003); 2nd ed. (2014).
160. H.-P. Berlien and G.J. Müller, Eds., *Applied Laser Medicine*, Springer-Verlag, Berlin (2003).
161. P. Prasad, *Introduction to Biophotonics*, Wiley-Interscience, Hoboken, N. J. (2003).
162. R.K. Wang, J.C. Hebden, and V.V. Tuchin, Eds., "Special issue on recent developments in biomedical optics," *Phys. Med. Biol.* **49**(7), 1085–1368 (2004).
163. A. Kishen and A. Asundi, Eds., Photonics in Dentistry: Series of Biomaterials and Bioengineering, Imperial College Press, London (2006).
164. V.V. Tuchin, L. Wang, and D.A. Zimnyakov, *Optical Polarization in Biomedical Applications*, Springer-Verlag, Berlin (2006).

165. R. Splinter and B.A. Hooper, *An Introduction to Biomedical Optics*, Taylor and Francis, New York (2007).
166. K.-P. Wilhelm, P. Elsner, E. Berardesca, and H.I. Maibach, Eds., *Bioengineering of the Skin: Skin Surface Imaging and Analysis*, 2nd ed., Informa Healthcare USA, New York (2007).
167. Q. Luo, L. Wang, and V.V. Tuchin, Eds., *Advances in Biomedical Photonics and Imaging*, World Scientific, New Jersey (2008).
168. G. Ahluwalia, Ed., *Light Based Systems for Cosmetic Application*, William Andrew, Inc., Norwich, N.Y. (2008).
169. W. Bock, I. Gannot, and S. Tanev, Eds., *Optical Waveguide Sensing and Imaging,* NATO SPS Series B: Physics and Biophysics, Springer, Dordrecht (2008).
170. S. Tanev, B.C. Wilson V., V. Tuchin, and D. Matthews, Eds., "Special issue on biophotonics," *Adv. Opt. Technol.* 2008 (2008), doi:10.1155/2008/134215.
171. V.V. Tuchin, R. Drezek, S. Nie, and V.P. Zharov, Eds., "Special section on nanophotonics for diagnostics, protection and treatment of cancer and inflammatory diseases," *J. Biomed. Opt.* **14**(2), 020901, 021001⊠021017 (2009).
172. E. Baron, Ed., Light-Based Therapies for Skin of Color, Springer, N.Y. (2009).
173. K.-E. Peiponen, R. Myllylä, and A.V. Priezzhev, *Optical Measurement Techniques: Innovations for Industry and the Life Science*, Springer-Verlag, Berlin (2009).
174. V.V. Tuchin, *Lasers and Fibre Optics in Biomedical Science*, 2nd ed., Fizmatlit, Moscow (2010).
175. A.P. Dhawan, B. D'Alessandro, and X. Fu, "Optical imaging modalities for biomedical applications," *IEEE Rev. Biomed. Eng.* **3**, 69–92 (2010).
176. X.-C. Zhang and J. Xu, *Introduction to THz Wave Photonics*, Springer, New York (2010).
177. V.V. Tuchin, Ed., *Handbook of Photonics for Medical Science*, CRC Press, Taylor & Francis Group, London (2010).
178. N. Ghosh, M.F.G. Wood, and I.A. Vitkin, "Polarized light assessment of complex turbid media such as biological tissues via Mueller matrix decomposition," *Handbook of Photonics for Biomedical Science*, V.V. Tuchin, Ed., pp. 253–282, CRC Press, Taylor & Francis Group, London (2010).
179. N. Ghosh and I.A. Vitkin, "Concepts, challenges and applications of polarized light in biomedicine: a tutorial review," *J. Biomed. Opt.* **16**(11), 110801-1–29 (2010).
180. B. Kunnen, C. Macdonald, A. Doronin, S. Jacques, M. Eccles, and I. Meglinski, "Application of circularly polarized light for non-invasive

diagnosis of cancerous tissues and turbid tissue-like scattering media," *J. Biophotonics* **8**(4), 317–323 (2015).

181. E.I. Galanzha, V.V. Tuchin, and V.P. Zharov, "Advances in small animal mesentery models for *in vivo* flow cytometry, dynamic microscopy, and drug screening (invited review)," *World J. Gastroenterol.* **13**(2), 198–224 (2007).
182. V.V. Tuchin, A. Tárnok, , and V.P. Zharov, Eds., "Special issue: *in vivo* flow cytometry," *Cytometry A* **79A**(10), 737–883 (2011).
183. V.V. Tuchin, A. Tárnok, and V.P. Zharov, "*In vivo* flow cytometry: a horizon of opportunities," *Cytometry A* **79A**(10), 737–745 (2011).
184. V.V. Tuchin, Ed., *Advanced Optical Flow Cytometry: Methods and Disease Diagnoses*, Wiley-VCH Verlag, Weinheim (2011).
185. V.V. Tuchin, "*In vivo* optical flow cytometry and cell imaging," *Rivista Del Nuovo Cimento*, **37**(7), 375–416 (2014).
186. V.V. Tuchin, "Optical spectroscopy of biological materials," *Encyclopedia of Applied Spectroscopy*, D.L. Andrews, Ed., Chapter 16, Wiley-VCH Verlag, Weinheim (2009).
187. A.N. Bashkatov, E.A. Genina, and V.V. Tuchin, "Optical properties of skin, subcutaneous, and muscle tissues: a review," *J. Innov. Opt. Health Sci.* **4**(1), 9–38 (2011).
188. S.L. Jacques, "Optical properties of biological tissues: a review," *Phys. Med. Biol.* **58**(14), R37–R61 (2013).
189. S.L. Jacques, "Quick analysis of optical spectra to quantify epidermal melanin and papillary dermal blood content of skin," *J. Biophoton.* **8**, 309–316 (2015).
190. R.H. Wilson, K.P. Nadeau, F.B. Jaworski, B.J. Tromberg, and A.J. Durkin, "Review of short-wave infrared spectroscopy and imaging methods for biological tissue characterization," *J. Biomed. Opt.* **20**(3), 030901-1-10 (2015).
191. D.A. Boas, C. Pitris, and N. Ramanujam, Eds., *Handbook of Biomedical Optics*, CRC Press, Taylor & Francis Group, London (2011).
192. J. Popp, V.V. Tuchin, A. Chiou, and S.H. Heinemann, Eds., *Handbook of Biophotonics, Vol.1: Basics and Techniques*, Wiley-VCH Verlag, Weinheim (2011).
193. J. Popp, V.V. Tuchin, A. Chiou, and S.H. Heinemann, Eds., *Handbook of Biophotonics, Vol. 2: Photonics for Health Care*, Wiley-VCH Verlag, Weinheim (2012).
194. J. Popp, V.V. Tuchin, A. Chiou, and S.H. Heinemann, Eds., *Handbook of Biophotonics, Vol. 3: Photonics in Pharmaceutics, Bioanalysis and Environmental Research*, Wiley-VCH Verlag, Weinheim (2012).
195. V.V. Tuchin, *Dictionary of Biomedical Optics and Biophotonics*, SPIE Press, Bellingham, Wash. (2012).
196. A.N. Bashkatov, A.V. Priezzhev, and V.V. Tuchin, Eds., "Special issue on laser technologies for biophotonics," *Quantum Electron.* **42**(5), 379, 2012.

197. B.W. Pogue, V. Backman, S. Emelianov, C.K. Hitzenberger, P. So, and V. Tuchin, Eds., "Biomed 2012 feature issue," *Biomed. Opt. Express*, **3**(11) (2012).
198. D. Zhu, V.V. Tuchin, and Q. Luo, Eds., "Special issue on advances in biophotonics and biomedical optics," *J. Innov. Opt. Health Sci.* **6**(1⊠2) (2013).
199. R.K. Wang and V.V. Tuchin, Eds., *Advanced Biophotonics: Tissue Optical Sectioning*, CRC Press, Taylor & Francis Group, London (2013).
200. H. Jelinkova, Ed., *Lasers for Medical Applications: Diagnostics, Therapy and Surgery*, pp. 47–109, Woodhead Publishing, Cambridge, U.K. (2013).
201. A.N. Bashkatov and V.V. Tuchin, Eds., "Special issue: biophotonics," *Optics and Spectroscopy* 114 (2013).
202. F.S. Pavone, P.T.C. So, and P.M.W. French, Eds., *Proceedings of the International School of Physics* "Enrico Fermi," Course 181: Microscopy Applied to Biophotonics, Societa Italiana di Fisica, Bologna (2014).
203. B. Querleux, Ed., *Computational Biophysics of the Skin*, CRC Press, Taylor & Francis Group, London (2015).
204. I.K. Ilev, S.A. Boppart, S. Andersson-Engels, B.-M. Kim, L. Perelman, and V. Tuchin, Eds., "Biophotonics," *IEEE J. Sel. Top. Quant. Electron.* **20**(2), 6800407–7100912 (2014).
205. E.A. Genina, D. Zhu, and V.V. Tuchin, "Special issue on optical technologies in biophysics and medicine," *J. Innov. Opt. Health Sci.* **8**(3), 1502002 (2015).
206. D. Zhu, S. Zeng, and V.V. Tuchin, Eds., "Special issue on biomedical photonics," *Front. Optoelectron.* Higher Education Press and Springer-Verlag, Berlin doi: 10.1007/s12200-015-0525-8(2015).
207. A.V. Priezzhev, H. Schneckenburger, and V.V. Tuchin, Eds., "Special section on laser applications in life sciences," *J. Biomed. Opt.* **20**(5), 051001-1 (May 2015).
208. V.V. Tuchin, "Tissue optics and photonics: biological tissue structures [review]," *J. Biomed. Photon. Eng.* **1**(1), 3–21 (2015).
209. V.V. Tuchin, "Tissue optics and photonics: light-tissue interaction [review]," *J. Biomed. Photon. Eng.* **1**(2), 98–134 (2015).
210. V.V. Tuchin, "Optical clearing of tissue and blood using immersion method," *J. Phys. D: Appl. Phys.* **38**, 2497–2518 (2005).
211. V.V. Tuchin, "Optical immersion as a new tool to control optical properties of tissues and blood," *Laser Phys.* **15**(8), 1109–1136 (2005).
212. V.V. Tuchin, Optical Clearing of Tissues and Blood, **PM 154**, SPIE Press, Bellingham, Wash. (2006).
213. V.V. Tuchin, "A clear vision for laser diagnostics," *IEEE J. Select. Top. Quant. Electr.* **13**(6), 1621–1628 (2007).

214. V.V. Tuchin, R.K. Wang, and A.T. Yeh, Eds., "Special section on optical clearing of tissues and cells," *J. Biomed. Opt.* **13**, 021101–1 (2008).
215. E.A. Genina, A.N. Bashkatov, and V.V. Tuchin, "Glucose-induced optical clearing effects in tissues and blood," in the *Handbook of Optical Sensing of Glucose in Biological Fluids and Tissues*, V.V Tuchin, Ed., pp. 657–692, CRC Press, Taylor & Francis Group, Boca Raton, Fla. (2009).
216. E.A. Genina, A.N. Bashkatov, and V.V. Tuchin, "Tissue optical immersion clearing," *Expert Rev. Med. Devices* **7**(6), 825–842 (2010).
217. E.A. Genina, A.N. Bashkatov, K.V. Larin, and V.V. Tuchin, "Light–tissue interaction at optical clearing," *Laser Imaging and Manipulation in Cell Biology*, F.S. Pavone, Ed., pp. 115–164, Wiley-VCH Verlag, Weinheim (2010).
218. K.V. Larin, M.G. Ghosn, A.N. Bashkatov, E.A. Genina, N.A. Trunina, and V.V. Tuchin, "Optical clearing for OCT image enhancement and in-depth monitoring of molecular diffusion," *IEEE J. Select. Top. Quant. Electron.* **18**(3), 1244–1259 (2012).
219. D. Zhu, K.V. Larin, Q. Luo, and V.V. Tuchin, "Recent progress in tissue optical clearing," *Laser Photonics Rev.* **7**(5), 732–757 (2013).
220. R.K. Wang and V.V. Tuchin, "Optical coherence tomography: light scattering and imaging enhancement," in the *Handbook of Coherent-Domain Optical Methods: Biomedical Diagnostics, Environmental Monitoring and Material Science*, 2nd ed., V.V. Tuchin, Ed., pp. 665–742, Springer-Verlag, Berlin (2013).
221. D. Zhu, Q. Luo, and V.V. Tuchin, "Tissue optical clearing," in the *Advanced Biophotonics: Tissue Optical Sectioning*, R.K. Wang and V.V. Tuchin, Eds., pp. 621–672, CRC Press, Taylor & Francis Group, Boca Raton, Fla. (2013).
222. V. Hovhannisyan, P.-S. Hu, S.-J. Chen, C.-S. Kim, and C.-Y. Dong, "Elucidation of the mechanisms of optical clearing in collagen tissue with multiphoton imaging," *J. Biomed. Opt.* **18**(4), 046004-1–8 (2013).
223. J. Wang, Y. Zhang, P. Li, Q. Luo, and D. Zhu, "Review: tissue optical clearing window for blood flow monitoring (invited paper)," *IEEE J. Sel. Top. Quant. Electron.* **20**(2), 6801112-1–12 (2014).
224. O. Nadiarnykh and P.J. Campagnola, "SHG and optical clearing," *Second Harmonic Generation Imaging*, F.S. Pavone and P.J. Campagnola, Eds., pp. 169–189, CRC Press, Taylor & Francis Group, Boca Raton, Fla. (2014).
225. Y. Zhou, J. Yao, and L.V. Wang, "Optical clearing-aided photoacoustic microscopy with enhanced resolution and imaging depth," *Opt. Lett.* **38** (14), 2592–2595 (2013).

226. Y. Liu, X. Yang, D. Zhu, R. Shi, and Q. Luo, "Optical clearing agents improve photoacoustic imaging in the optical diffusive regime," *Opt. Lett.* **38**(20), 4236–4239 (2013).
227. Y.A. Menyaev, D.A. Nedosekin, M. Sarimollaoglu, M.A. Juratli, E.I. Galanzha, V.V. Tuchin, and V.P. Zharov, "Skin optical clearing for *in vivo* photoacoustic flow cytometry," *Biomed. Opt. Express* **4**(12), 3030–3041, 2013.
228. Y. Ding, J. Wang, Z. Fan, D. Wei, R. Shi, Q. Luo, D. Zhu, and X. Wei, "Signal and depth enhancement for *in vivo* flow cytometer measurement of ear skin by optical clearing agents," *Biomed. Opt. Express* **4**(11), 2518–2526 (2013).
229. M. Wenne, "The most transparent research," *Nat. Med.* **15**(10), 1106–1109 (2009).
230. H.U. Dodt, U. Leischner, A. Schierloh, N. Jährling, C.P. Mauch, K. Deininger, J.M. Deussing, M. Eder, W. Zieglgänsberger, and K. Becker, "Ultramicroscopy: three-dimensional visualization of neuronal networks in the whole mouse brain," *Nat. Methods* **4**(4), 331–336 (2007).
231. H. Hama, H. Kurokawa, H. Kawano, R. Ando, T. Shimogori, H. Noda, K. Fukami, A. Sakaue-Sawano, and A. Miyawaki, "Scale: a chemical approach for fluorescence imaging and reconstruction of transparent mouse brain," *Nat. Neurosci.* **14**, 1481–1488 (2011).
232. A. Ertürk, C.P. Mauch, F. Hellal, F. Förstner, T. Keck, K. Becker, N. Jährling, H. Steffens, M. Richter, M. Hübener, E. Kramer, F. Kirchhoff, H.U. Dodt, and F. Bradke, "Three-dimensional imaging of the unsectioned adult spinal cord to assess axon regeneration and glial responses after injury," *Nat. Med.* **18**, 166–171 (2012).
233. K. Becker, N. Jährling, S. Saghafi, R. Weiler, and H.U. Dodt, "Chemical clearing and dehydration of GFP expressing mouse brains," *PLoS One* **7**(3), e33916 (2012).
234. K. Becker, N. Jährling, S. Saghafi, and H.U. Dodt, "Dehydration and clearing of whole mouse brains and dissected hippocampi for ultramicroscopy," *Cold Spring Harb. Protoc*, **2013**(7), doi: 10.1101/pdb.prot075820(2013).
235. C. Leahy, H. Radhakrishnan, and V.J. Srinivasan, "Volumetric imaging and quantification of cytoarchitecture and myeloarchitecture with intrinsic scattering contrast," *Biomed. Opt. Express* **4**(10), 1978–1990 (2013).
236. K. Chung, J. Wallace, S.-Y. Kim, S. Kalyanasundaram, A.S. Andalman, T.J. Davidson, J.J. Mirzabekov, K.A. Zalocusky, J. Mattis, A.K. Denisin, S. Pak, H. Bernstein, C. Ramakrishnan, L. Grosenick, V. Gradinaru, and K. Deisseroth, "Structural and molecular interrogation of intact biological systems," *Nature* **497**, 332–337 (2013).

237. L. Silvestri, A.L. Allegra Mascaro, J. Lotti, L. Sacconi, and F.S. Pavone, "Advanced optical techniques to explore brain structure and function," *J. Innov. Opt. Health Sci.* **6**(1), 1230002 (2013).
238. B. Yang, J.B. Treweek, R.P. Kulkarni, B.E. Deverman, C.-K. Chen, E. Lubeck, S. Shah, L. Cai, and V. Gradinaru, "Single-cell phenotyping within transparent intact tissue through whole-body clearing," *Cell* **158** (4), 945–958 (2014).
239. V. Marx, "Microscopy: seeing through tissue," *Nature Methods* **11**(12), 1209–1214 (2014).
240. E.A. Genina, A.N. Bashkatov, Y.P. Sinichkin, I.Y. Yanina, and V.V. Tuchin, "Optical clearing of biological tissues: prospects of application in medical diagnostics and phototherapy [review]," *J. Biomed. Photon. Eng.* **1**(1), 22–58 (2015).
241. G.A. Askar'yan, "Enhancement of transmission of laser and other radiation by soft turbid physical and biological media," *Sov. J. Quantum Electron.* **12**(7), 877–880 (1982).
242. E.K. Chan, B. Sorg, D. Protsenko, M. O'Neil, M. Motamedi, and A.J. Welch, "Effects of compression on soft tissue optical properties," *IEEE J. Sel. Top. Quantum Electron.* **2**(4), 943–950 (1996).
243. Y.P. Sinichkin, S.R. Uts, and E.A. Pilipenko, "Spectroscopy of human skin *in vivo*: 1. Reflection spectra," *Opt. Spectrosc.* **80**(2), 228–234 (1996).
244. Y.P. Sinichkin, S.R. Uts, I.V. Meglinskii, and E.A. Pilipenko, "Spectroscopy of human skin *in vivo*: II. Fluorescence spectra," *Opt. Spectrosc.* **80**(3), 383–389 (1996).
245. H. Shangguan, S.A. Prahl, S.L. Jacques, and L.W. Casperson, "Pressure effects on soft tissues monitored by changes in tissue optical properties," *Proc. SPIE* **3254**, 366–371 (1998).
246. M.H. Khan, B. Choi, S. Chess, K.M. Kelly, J. McCullought, and J.S. Nelson, "Optical clearing of *in vivo* human skin: implications for light-based diagnostic imaging and therapeutics," *Lasers Surg. Med.* **34** (2), 83–85 (2004).
247. A. Nath, K. Rivoire, S. Chang, D. Cox, E.N. Atkinson, M. Follen, and R. Richards-Kortum, "Effect of probe pressure on cervical fluorescence spectroscopy measurements," *J. Biomed. Opt.* **9**(3), 523–533 (2004).
248. K. Rivoire, A. Nath, D. Cox, E.N. Atkinson, R. Richards-Kortum, and M. Follen, "The effects of repeated spectroscopic pressure measurements on fluorescence intensity in the cervix," *Am. J. Obstet. Gynecol.* **191**(5), 1606–1617 (2004).
249. B.W. Murphy, R.J. Webster, B.A. Turlach, C.J. Quirk, C.D. Clay, P.J. Heenan, and D.D. Sampson, "Toward the discrimination of early melanoma from common and dysplastic nevus using fiber optic diffuse reflectance spectroscopy," *J. Biomed. Opt.* **10**(6), 064020 (2005).

250. W. Chen, R. Liu, K. Xu, and R.K. Wang, "Influence of contact state on NIR diffuse reflectance spectroscopy *in vivo*," *J. Phys. D: Appl. Phys.* **38**, 2691–2695 (2005).
251. S.A. Carp, T. Kauffman, Q. Fang, E. Rafferty, R. Moore, D. Kopans, and D. Boas, "Compression-induced changes in the physiological state of the breast as observed through frequency domain photon migration measurements," *J. Biomed. Opt.* **11**(6), 064016 (2006).
252. R. Reif, M.S. Amorosino, K.W. Calabro, O. A'Amar, S.K. Singh, and I. J. Bigio, "Analysis of changes in reflectance measurements on biological tissues subjected to different probe pressures," *J. Biomed. Opt.* **13**(1), 010502 (2008).
253. Y. Ti and W.C. Lin, "Effects of probe contact pressure on *in vivo* optical spectroscopy," *Opt. Express* **16**(6), 4250–4262 (2008).
254. H. Kang, T. Son, J. Yoon, K. Kwon, J.S. Nelson, and B. Jung, "Evaluation of laser beam profile in soft tissue due to compression, glycerol, and micro-needling," *Laser Surg. Med.* **40**(8), 570–575 (2008).
255. C.G. Rylander, T.E. Milner, S.A. Baranov, and J.S. Nelson, "Mechanical tissue optical clearing devices: enhancement of light penetration in ex vivo porcine skin and adipose tissue," *Lasers Surg. Med.* **40**(10), 688–694 (2008).
256. V.V. Sapozhnikova, R.V. Kuranov, I. Cicenaite, R.O. Esenaliev, and D. S. Prough, "Effect on blood glucose monitoring of skin pressure exerted by an optical coherence tomography probe," *J. Biomed. Opt.* **13**(2), 021112 (2008).
257. A. Cerussi, S. Siavoshi, A. Durkin, C. Chen, W. Tanamai, D. Hsiang, and B.J. Tromberg, "Effect of contact force on breast tissue optical property measurements using a broadband diffuse optical spectroscopy handheld probe," *Appl. Opt.* **48**, 4270–4277 (2009).
258. J.A. Delgado Atencio, E.E. Orozco Guillén, S. Vázquezy Montiel, M. Cunill Rodríguez, J. Castro Ramos, J.L. Gutiérrez, and F. Martínez, "Influence of probe pressure on human skin diffuse reflectance spectroscopy measurements," *Optical Memory & Neural Networks* (Information Optics) **18**(1), 6–14 (2009).
259. L. Lim, B. Nichols, N. Rajaram, and J.W. Tunnell, "Probe pressure effects on human skin diffuse reflectance and fluorescence spectroscopy measurements," *J. Biomed. Opt.* **16**(1), 011012 (2011).
260. S. Ruderman, A.J. Gomes, V. Stoyneva, J.D. Rogers, A.J. Fought, B.D. Jovanovic, and V. Backman, "Analysis of pressure, angle and temporal effects on tissue optical properties from polarization-gated spectroscopic probe measurements," *Biomed. Opt. Express* **1**(2), 489–499 (2010).
261. A. Izquierdo-Roman, W.C. Vogt, L. Hyacinth, and C.G. Rylander, "Mechanical tissue optical clearing technique increases imaging resolution and contrast through ex vivo porcine skin," *Lasers Surg. Med.* **43**, 814–823 (2011).

262. M.Y. Kirillin, P.D. Agrba, and V.A. Kamensky, "*In vivo* study of the effect of mechanical compression on formation of OCT images of human skin," *J. Biophotonics* **3**(12), 752–758 (2010).
263. A.A. Gurjarpadhye, W.C. Vogt, Y. Liu, and C.G. Rylander, "Effect of localized mechanical indentation on skin water content evaluated using OCT," *Int. J. Biomed. Imag.* **2011**, 817250 (2011).
264. I.V. Ermakov and W. Gellermann, "Dermal carotenoid measurements via pressure mediated reflection spectroscopy," *J. Biophotonics* **5**(7), 559–570 (2012).
265. L.E. Dolotov and Y.P. Sinichkin, "Features of applying fiber-optic sensors in spectral measurements of biological tissues," *Opt. Spectrosc.* **115**(2), 187–192 (2013).
266. C. Li, J. Jiang, and K. Xu, "The variations of water in human tissue under certain compression: studied with diffuse reflectance spectroscopy," *J. Innov. Opt. Health Sci.* **6**(1), 1350005 (2013).
267. I. J. Bigio and and S. Fantini, *Quantitative Biomedical Optics: Theory, Methods, and Applications, Textbook*, Part of Cambridge Texts in Biomedical Engineering, Cambridge University Press, Cambridge (2016).
268. T. Novikova, I. Meglinski, J.C. Ramella-Roman, and V.V. Tuchin, Eds., "Special section on polarized light for biomedical applications," *J. Biomed. Opt.* **21**(7), 071001 (2016).
269. D. Zhu, B. Choi, E. Genina, and V.V. Tuchin, Eds., "Special section on tissue and blood optical clearing for biomedical applications," *J. Biomed. Opt.* **21**(8), 081201 (2016).
270. A.N. Bashkatov, E.A. Genina, and A.V. PriezzhevV.V. Tuchin, Eds., "Special issue on laser biophotonics," *Quantum Electronics.* **46**(6), 487 (2016).

Valery V. Tuchin is a professor and chairman of Optics and Biophotonics at National Research Saratov State University. He is also the head of laboratory at the Institute of Precision Mechanics and Control, RAS, and the supervisor of the Interdisciplinary Laboratory of Biophotonics at National Research Tomsk State University. His research interests include biophotonics, tissue optics, laser medicine, tissue optical clearing, and nanobiophotonics. He is a member of SPIE, OSA, and IEEE, guest professor of HUST (Wuhan) and Tianjin Universities of China, and adjunct professor of the Limerick University (Ireland) and National University of Ireland (Galway). He is a fellow of SPIE and OSA, and has been awarded the Honored Science Worker of the Russia, SPIE Educator Award, FiDiPro (Finland), Chime Bell Prize of Hubei Province (China), and Joseph W. Goodman Book Writing Award (OSA/SPIE).

Preface

This *Handbook* is the second edition of the monograph initially published in 2002. The first edition described some aspects of laser–cell and laser–tissue interactions that are basic for biomedical diagnostics and presented many optical and laser diagnostic technologies prospective for clinical applications. The main reason for publishing such a book was the achievements of the last millennium in light scattering and coherent light effects in tissues, and in the design of novel laser and photonics techniques for the examination of the human body. Since 2002, biomedical optics and biophotonics have had rapid and extensive development, leading to technical advances that increase the utility and market growth of optical technologies. Recent developments in the field of biophotonics are wide-ranging and include novel light sources, delivery and detection techniques that can extend the imaging range and spectroscopic probe quality, and the combination of optical techniques with other imaging modalities.

The innovative character of photonics and biophotonics is underlined by two Nobel prizes in 2014 awarded to Eric Betzig, Stefan W. Hell, and William E. Moerner "for the development of super-resolved fluorescence microscopy" and to Isamu Akasaki, Hiroshi Amano, and Shuji Nakamura "for the invention of efficient blue light-emitting diodes which has enabled bright and energy-saving white light sources." The authors of this *Handbook* have a strong input in the development of new solutions in biomedical optics and biophotonics and have conducted cutting-edge research and developments over the last 10–15 years, the results of which were used to modify and update early written chapters. Many new, world-recognized experts in the field have joined the team of authors who introduce fresh blood in the book and provide a new perspective on many aspects of optical biomedical diagnostics.

The optical medical diagnostic field covers many spectroscopic and laser technologies based on near-infrared (NIR) spectrophotometry, fluorescence and Raman spectroscopy, optical coherent tomography (OCT), confocal microscopy, optoacoustic (photoacoustic) tomography, photon-correlation spectroscopy and imaging, and Doppler and speckle monitoring of biological flows.[1–45] These topics—as well as the main trends of the modern laser diagnostic techniques, their fundamentals and corresponding basic research on laser–tissue interactions, and the most interesting clinical applications—are discussed in the framework of this Handbook. The main unique features of the book are as follows:

1. Several chapters of basic research that discuss the updated results on light scattering, speckle formation, and other nondestructive interactions of laser light with tissue; they also provide a basis for the optical and laser medical diagnostic techniques presented in the other chapters.
2. A detailed discussion of blood optics, blood and lymph flow, and blood-aggregation measurement techniques, such as the well-recognized laser Doppler method, speckle technique, and OCT method.
3. A discussion of the most-recent prospective methods of laser (coherent) tomography and spectroscopy, including OCT, optoacoustic (photoacoustic) imaging, diffusive wave spectroscopy (DWS), and diffusion frequency-domain techniques.

The intended audience of this book consists of researchers, postgraduate and undergraduate students, biomedical engineers, and physicians who are interested in the design and applications of optical and laser methods and instruments for medical science and practice. Due to the large number of fundamental concepts and basic research on laser–tissue interactions presented here, it should prove useful for a much broader audience that includes students and physicians, as well. Investigators who are deeply involved in the field will find up-to-date results for the topics discussed. Each chapter is written by representatives of the leading research groups who have presented their classic and most recent results. Physicians and biomedical engineers may be interested in the clinical applications of designed techniques and instruments, which are described in a few chapters. Indeed, laser and photonics engineers may also be interested in the book because their acquaintance with a new field of laser and photonics applications can stimulate new ideas for lasers and photonic devices design. The two volumes of this *Handbook* contain 21 chapters, divided into four parts (two per volume):

- Part I describes the fundamentals and basic research of the extinction of light in dispersive media; the structure and models of tissues, cells, and cell ensembles; blood optics; coherence phenomena and statistical properties of scattered light; and the propagation of optical pulses and photon-density waves in turbid media. Tissue phantoms as tools for tissue study and calibration of measurements are also discussed.
- Part II presents time-resolved (pulse and frequency-domain) imaging and spectroscopy methods and techniques applied to tissues, including optoacoustic (photoacoustic) methods. The absolute quantification of the main absorbers in tissue by a NIR spectroscopy method is discussed. An example biomedical application—the possibility of monitoring brain activity with NIR spectroscopy—is analyzed.

- Part III presents various spectroscopic techniques of tissues based on elastic and Raman light scattering, Fourier transform infrared (FTIR), and fluorescence spectroscopies. In particular, the principles and applications of backscattering diagnostics of red blood cell (RBC) aggregation in whole blood samples and epithelial tissues are discussed. Other topics include combined back reflectance and fluorescence, FTIR and Raman spectroscopies of the human skin *in vivo*, and fluorescence technologies for biomedical diagnostics.
- The final section, Part IV, begins with a chapter on laser Doppler microscopy, one of the representative coherent-domain methods applied to monitoring blood in motion. Methods and techniques of real-time imaging of tissue ultrastructure and blood flows using OCT is also discussed. The section also describes various speckle techniques for monitoring and imaging tissue, in particular, for studying tissue mechanics and blood and lymph flow.

Financial support from a FiDiPro grant of TEKES, Finland (40111/11) and Academic D.I. Mendeleev Fund Program of National Research Tomsk State University have helped me complete this book project. I greatly appreciate the cooperation and contribution of all of the authors and co-editors, who have done a great work on preparation of this book. I would like to express my gratitude to Eric Pepper and Tim Lamkins for their suggestion to prepare the second edition of the *Handbook* and to Scott McNeill for assistance in editing the manuscript. I am very thankful to all of my colleagues from the Chair and Research Education Institute of Optics and Biophotonics at Saratov National Research State University and the Institute of Precision Mechanics and Control of RAS for their collaboration, fruitful discussions, and valuable comments. I am very grateful to my wife and entire family for their exceptional patience and understanding.

Valery V. Tuchin
April 2016

References

1. F. A. Duck, *Physical Properties of Tissue: A Comprehensive Reference Book*, Academic, London (1990).
2. A. P. Shepherd and P. A. Oberg, *Laser Doppler Blood Flowmetry*, Kluwer, Boston (1990).
3. J. B. Pawley (Ed.), *Handbook of Biological Confocal Microscopy*, Plenum Press, New York (1990).
4. T. Wilson (Ed.), *Confocal Microscopy*, Academic Press, London (1990).
5. K. Frank and M. Kessler (Eds.), *Quantitative Spectroscopy in Tissue*, pmi Verlag, Frankfurt am Main (1992).

6. G. Müller, B. Chance, R. Alfano, et al. (Eds.), *Medical Optical Tomography: Functional Imaging and Monitoring*, **IS 11**, SPIE Press, Bellingham (1993).
7. V. V. Tuchin (Ed.), *Selected Papers on Tissue Optics Applications in Medical Diagnostics and Therapy*, Milestones Series **MS 102**, SPIE Press, Bellingham (1994).
8. B. R. Masters (Ed.), *Confocal Microscopy*, **MS 131**, SPIE Press, Bellingham (1996).
9. O. Minet, G. Mueller, and J. Beuthan (Eds.), *Selected Papers on Optical Tomography, Fundamentals and Applications in Medicine*, **MS 147**, SPIE Press, Bellingham (1998).
10. V. V. Tuchin, *Tissue Optics: Light Scattering Methods and Instruments for Medical Diagnosis*, SPIE Tutorial Texts in Optical Engineering, Tutorial Text Series, **38** SPIE Press, Bellingham (2000).
11. B. R. Masters (Ed.), *Selected Papers on Optical Low-Coherence Reflectometry and Tomography*, **MS 165**, SPIE Press, Bellingham (2001).
12. B.E. Bouma and G.J. Tearney (Eds.), *Handbook of Optical Coherence Tomography*, Marcel-Dekker, New York (2002).
13. T. Vo-Dinh (Ed.), *Biomedical Photonics Handbook*, Boca Raton, CRC Press (2003); 2nd ed. (2014).
14. H.-P. Berlien and G.J. Müller (Eds.), *Applied Laser Medicine*, Springer-Verlag, Berlin (2003).
15. P. Prasad, *Introduction to Biophotonics*, Wiley-Interscience, Hoboken, New Jersey (2003).
16. J.R. Lakowicz, *Principles of Fluorescence Spectroscopy*, 3rd ed., Springer Science + Business, New York (2006).
17. V.V. Tuchin, *Tissue Optics: Light Scattering Methods and Instruments for Medical Diagnosis*, 2nd ed., **PM 166** (2007); 3rd ed., PM254, SPIE Press, Bellingham, WA (2015).
18. L.V. Wang and H.-I. Wu, *Biomedical Optics: Principles and Imaging*, Wiley-Interscience, Hoboken, New Jersey (2007).
19. Q. Luo, L. Wang, and V.V. Tuchin (Eds.), *Advances in Biomedical Photonics and Imaging*, World Scientific, New Jersey, London, Singapore et al. (2008).
20. G. Ahluwalia (Ed.), *Light Based Systems for Cosmetic Application*, William Andrew, Inc., Norwich, New York (2008).
21. W. Bock, I. Gannot, and S. Tanev (Eds.), *Optical Waveguide Sensing and Imaging*, NATO SPS Series B: Physics and Biophysics, Springer, Dordrecht (2008).
22. W. Drexler and J.G. Fujimoto (Eds.), *Optical Coherence Tomography: Technology and Applications*, Springer, Berlin (2008); 2nd ed. Springer, Berlin (2015).

23. E. Baron (Ed.), *Light-Based Therapies for Skin of Color*, Springer, New York (2009).
24. K.-E. Peiponen, R. Myllylä, and A. V. Priezzhev, *Optical Measurement Techniques, Innovations for Industry and the Life Science*, Springer-Verlag, Berlin, Heidelberg (2009).
25. L. Wang, Ed., *Photoacoustic Imaging and Spectroscopy*, CRC Press, Taylor & Francis Group, London (2009).
26. V.V. Tuchin (Ed.), *Handbook of Optical Sensing of Glucose in Biological Fluids and Tissues*, CRC Press, Taylor & Francis Group, London (2009).
27. A. Wax and V. Backman (Eds.), *Biomedical Applications of Light Scattering*, McGraw-Hill, New York (2010).
28. V. V. Tuchin, *Lasers and Fiber Optics in Biomedical Science*, 2nd ed., Fizmatlit, Moscow (2010).
29. X.-C. Zhang and J. Xu, *Introduction to THz Wave Photonics*, Springer, New York (2010).
30. V.V. Tuchin (Ed.), *Handbook of Photonics for Medical Science*, CRC Press, Taylor & Francis Group, London (2010).
31. F. S. Pavone (Ed.), *Laser Imaging and Manipulation in Cell Biology*, Wiley-VCH Verlag GmbH & Co. KGaA, Weinheim (2010).
32. V.V. Tuchin (Ed.), *Advanced Optical Flow Cytometry: Methods and Disease Diagnoses*, Wiley-VCH Verlag GmbH & Co. KGaA, Weinheim (2011).
33. D. A. Boas, C. Pitris, and N. Ramanujam (Eds.), *Handbook of Biomedical Optics*, CRC Press, Taylor & Francis Group, London (2011).
34. J. Popp, V.V. Tuchin, A. Chiou, and S.H. Heinemann (Eds.), *Handbook of Biophotonics*, vol. 1: Basics and Techniques, WILEY-VCH Verlag GmbH & Co. KGaA, Weinheim (2011).
35. J. Popp, V.V. Tuchin, A. Chiou, and S.H. Heinemann (Eds.), *Handbook of Biophotonics*, vol. 2: Photonics for Health Care, WILEY-VCH Verlag GmbH & Co. KGaA, Weinheim (2012).
36. J. Popp, V.V. Tuchin, A. Chiou, and S.H. Heinemann (Eds.), *Handbook of Biophotonics*, vol. 3: Photonics in Pharmaceutics, Bioanalysis and Environmental Research, WILEY-VCH Verlag GmbH & Co. KGaA, Weinheim (2012).
37. V.V. Tuchin, *Dictionary of Biomedical Optics and Biophotonics*, SPIE Press, Bellingham, WA (2012).
38. M. J. Leahy (ed.), *Microcirculation Imaging*, Wiley-VCH Verlag GmbH & Co. KGaA, Weinheim (2012).
39. R.K. Wang and V.V. Tuchin (Eds.), *Advanced Biophotonics: Tissue Optical Sectioning*, CRC Press, Taylor & Francis Group, London (2013).
40. H. Jelinkova (Ed.), *Lasers for Medical Applications: Diagnostics, Therapy and Surgery*, Woodhead Publishing, Ltd., Cambridge (2013).

41. F. S. Pavone and P. J. Campagnola (Eds.), *Second Harmonic Generation Imaging*, CRC Press, Taylor & Francis Group, Boca Raton, London, New York (2014).
42. F.S. Pavone, P.T.C. So, and P.M.W. French (Eds.), *Proc. of the International School of Physics 'Enrico Fermi,'* Course 181 – Microscopy Applied to Biophotonics, Societa Italiana di Fisica, Bologna (2014).
43. B. Querleux (Ed.), *Computational Biophysics of the Skin*, CRC Press, Taylor & Francis Group, London (2015).
44. F.D. Dip, T. Ishizawa, N. Kokudo, and R. Rosenthal (Eds.), *Fluorescence Imaging for Surgeons: Concepts and Applications*, Springer Science + Business Media, New York (2015).
45. I. J. Bigio and S. Fantini, *Quantitative Biomedical Optics: Theory, Methods, and Applications*, Cambridge University Press, Cambridge (2016).

List of Contributors

Britton Chance
University of Pennsylvania, USA

Jurgen Claassen
Radboud University Medical Centre, Netherlands

Sergio Fantini
Tufts University, USA

George Filippidis
Institute of Electronic Structure and Laser, Greece

Hui Gong
Huazhong University of Science and Technology, China

Alexander A. Karabutov
Moscow State University, Russia

Nikolai G. Khlebtsov
Institute of Biochemistry and Physiology of Plants and Microorganisms of the Russian Academy of Sciences, Saratov and Saratov National Research State University, Russia

Tatiana D. Khokhlova
University of Washington, USA

Vesa Kiviniemi
Medical Research Center of Oulu, Finland

Yasufumi Kuroda
University of Pennsylvania, USA

Qingming Luo
Huazhong University of Science and Technology, China

Irina L. Maksimova
Saratov National Research State University, Russia

Stephen J. Matcher
University of Sheffield, UK

Igor Meglinski
University of Oulu, Finland

Teemu Myllylä
University of Oulu, Finland

Shoko Nioka
University of Pennsylvania, USA

Alexander A. Oraevsky
TomoWave Laboratories and the University of Houston, USA

Theodore G. Papazoglou
Institute of Electronic Structure and Laser, Greece

Ivan M. Pelivanov
Moscow State University, Russia

Alexander B. Pravdin
Saratov National Research State University, Russia

Juan Rodriguez
St. Louis College of Pharmacy, USA

Angelo Sassaroli
Tufts University, USA

Vladislav Toronov
Ryerson University, Canada

Valery V. Tuchin
Saratov National Research State University, National Research Tomsk State University, and the Institute of Precision Mechanics and Control, Russian Academy of Sciences, Russia

Heidrun Wabnitz
Physikalisch-Technische Bundesanstalt (PTB), Germany

Lihong V. Wang
Washington University in St. Louis, USA

Anna N. Yaroslavsky
University of Massachusetts–Lowell and Massachusetts General Hospital, USA

Ilya V. Yaroslavsky
IPG Photonics, Inc., USA

Giannis Zacharakis
Institute of Electronic Structure and Laser, Greece

Shaoqun Zeng
Huazhong University of Science and Technology, China

Dmitry A. Zimnyakov
Saratov State Technical State University, Russia

Part I: Light–Tissue Interaction: Diagnostic Aspects

This part, which is essential for understanding subsequent parts of the book, focuses on the fundamental physics of light propagation in turbid media such as biological tissues. It is divided into five chapters. Chapter 1 begins with a discussion of the extinction and scattering of light in disperse systems and the theoretical models related to this concept. The chapter covers single-particle scattering of light, as well as extinction and scattering by aggregated and compounded structures. The turbidity spectra of disperse systems with random and oriented particles are presented. Particular attention is paid to the extinction and scattering of light by plasmon-resonant particles such as small metal spheres, nanorods, and nanoshells and to the applications of these phenomena in biomedicine. Next, the chapter addresses aspects of tissue structure and the related optical models that involve light scattering by densely packed particles. Two approaches to the description of light propagation in turbid media are considered. The first approach involves analytical theory, which is based on a solution of the wave equation for an electromagnetic field in spatially inhomogeneous media and which allows for a rigid analysis of the interference effects resulting from the superposition of partially scattered waves. The second approach is based on a numerical simulation of radiative transfer in turbid media by means of the Monte Carlo technique. Tissue diagnostics using nephelometry and polarization methods are discussed. The potential of tissue probing with circularly polarized light is considered. The technique of tissue optical clearing with the use of biologically compatible immersion agents is discussed. Finally, the possibility to use the Poincaré sphere formalism for polarization diagnostics and visualization is presented.

Chapter 2 covers the optics of blood, a vital component of biological tissues. Blood provides an important contrast medium for optics-based diagnostics as well as therapeutics. For example, blood optics is the key to optically based oximetry and port-wine stain therapy. The chapter covers the physical properties of blood cells, the optical properties of oxyhemoglobin and deoxyhemoglobin, the absorption and scattering of light by a single erythrocyte, the optical properties of blood, and an overview of the practical relevance of blood optics.

Chapter 3 focuses on the propagation of pulses and photon-density waves in turbid media. It is shown that in the case of radiative transfer in linear turbid media, both forms of optical response of a probed medium can be related by applying a Fourier transformation. The chapter covers the time-dependent light transport theory, the Monte Carlo method, the diffusion approximation to the exact radiative transfer equation and beyond, and the role a finite scattering time serves in the formation of the temporal response of a probed scattering medium. The analytical approach provides insight into physical problems, whereas the Monte Carlo approach is able to realistically solve complex problems numerically.

Chapter 4 discusses some fundamental phenomena that accompany the propagation of electromagnetic waves in disordered multiple scattering systems and can play an important role in coherent light interaction with tissues. These phenomena include the weak localization of light in disordered and weakly ordered media and the manifestation of such as coherent backscattering; the existence of temporal and spatial correlations of multiple scattered light; the correlation transfer in disordered media that can be considered as the physical basis for diagnostics and visualization of dynamic turbid media; the effect of similarity in multiple scattering that allows one to establish the relations between various optical properties of a scattering system. Also, various methods for evaluation of the photon path distributions in scattering media as the background for light-media interaction analysis are considered. Some applications of coherent backscattering spectroscopy and full-field speckle correlometry to tissue diagnostics, analysis of microfluidics in porous systems, and monitoring of phase separation process in "liquid crystal–prepolymer" systems are discussed.

Chapter 5 covers tissue phantoms, which are important for laboratory experiments and for the calibration of instruments. The following issues are considered: a general approach to phantom development, scattering media for phantom preparation, and light-absorbing media for phantom preparation. Also, the concept of "smart phantoms" mimicking organs as well as some practical realizations of this concept are presented. Finally, phantoms containing optically active media are discussed.

Dmitry A. Zimnyakov
Lihong V. Wang
Co-editors

Chapter 1
Introduction to Light Scattering by Biological Objects

Nikolai G. Khlebtsov
Institute of Biochemistry and Physiology of Plants and Microorganisms, Russian Academy of Sciences, Saratov, Russia

Saratov National Research State University, Saratov, Russia

Irina L. Maksimova
Saratov National Research State University, Saratov, Russia

Igor Meglinski
University of Oulu, Oulu, Finland

Lihong V. Wang
Washington University in St. Louis, St. Louis, USA

Valery V. Tuchin
Saratov National Research State University, Saratov, Russia

National Research Tomsk State University, Tomsk, Russia

Institute of Precision Mechanics and Control, Russian Academy of Sciences, Saratov, Russia

1.1 Introduction

The practices of scattering and absorbing electromagnetic radiation enjoy wide use in various fields of science and technology that aim to study the structure and properties of inhomogeneous media. The theory and practice of light-scattering methods is now a fairly well-developed field, owing to the methods' profound importance for applications such as atmospheric and oceanic optics,[1–5] radio-wave propagation and radio communication,[6] physical chemistry of solutions and colloids,[7] materials science and chemical technology,[8] biophysics, and laser biomedicine.[9,10] The theoretical models, experimental measurement procedures, and data interpretation methods have been developed by experts in various disciplines (from astrophysics to laser ophthalmology); therefore, there are certain traditions and terminological barriers that hinder effective interactions among various research groups. To illustrate, for experts in atmospheric optics and astrophysics, the ideology of radiation transfer theory (RTT)[2,3] is natural, and using language related to the apparatus of correlation functions and structural scattering factor is more habitual when interpreting data by small-angle x-ray scattering or neutron scattering.[7,8] Another example is the composite medium technology whose basic concepts are effective dielectric permeability and effective refractive index.[11] In colloid optics, the model of scattering via an isolated particle is most popular; this model is described either in terms of Maxwell's rigorous electromagnetic theory or on the basis of various approximations.[1,7,12]

Despite terminological and other differences, the basis of many methods using the scattering of neutrons, x-radiation, light, or radio waves proves to be sufficiently versatile. With respect to the scattering of electromagnetic waves of various frequencies, this versatility is probably explained by the common classical basis, Maxwell's electromagnetic theory, which is applied with physical models of scattering media. Even in the case of particle scattering due to potentials associated with this or another inhomogeneity of the medium,[13] the general theoretical interpretation of the scattering (e.g., in terms of T-matrix formalism) may be exactly the same as that in the case of electromagnetics.[14]

In view of the great diversity and structural complexity of biological systems,[9,10] the development of adequate optical models of the scattering and absorption of light is often the most complex step of a study. These models include virtually all sections of dispersion media optics: (1) simple single-scattering approximation; (2) incoherent multiple scattering, described by the radiation transfer equation (RTE); and (3) multiple-wave scattering in condensed systems of electrodynamically interacting scatterers and inhomogeneities. Quite plainly, such a broad range of problems rules out the possibility of a more or less detailed treatment of technical details. Therefore, the material presented in this chapter includes only certain elements of the theoretical apparatus used in the above-mentioned sections of scattering media optics; otherwise, it includes references to the relevant literature.

1.2 Extinction and Scattering of Light in Disperse Systems: Basic Theoretical Approaches

Three major directions can be distinguished in scattering media optics. The first is associated with the solution of diffraction problems for individual scatterers and with the determination of how absorption and scattering characteristics are linked to the optical, geometrical, and structural parameters of particles.[12,15] Since the second half of the 1980s, a number of novel methods and algorithms have been developed that allow, in principle, quantitative results to be obtained for a very wide class of sizes, shapes, structures, and optical parameters of particles.[15] A brief survey of particle-scattering-theory methods is given in Section 1.3; more detailed information can be found in Ref. 16.

The importance of this domain of the scattering theory for applications to actual systems is determined by the following circumstances. First, for very rarefied and optically thin media, the scattering characteristics of an ensemble of particles (intensities, integral cross sections) are found by simple summation with no regard to the interference of scattered waves. What is known as single-scattering approximation consists therein. Of course, in some special cases, this approximation should be used with caution. For example, scattering in the forward direction is coherent at least when the time of the incident light coherence is greater than the time of its propagation in the scattering layer and is less than the characteristic time of particle configuration change. For small angles, therefore, it is the scattered fields, not intensities, that should be summed. However, for the typical sizes of the scattering sample, the domain of such angles turns out to be very small and may be neglected.[1,12] The single-scattering approximation may also be applied when scattered fields are summed with regard to geometrical phase shifts from different particles. However, the scattered fields themselves are calculated on the basis of excitation of the scatterer by the initial incident wave, i.e., with no regard to the radiation from other particles (see Section 1.7). Conditions necessary to apply the single-scattering approximation have been discussed by Mishchenko et al.[17]

Another important application of the theory of scattering by small particles is the calculation of the characteristics of the elementary scattering volume, which appears in the RTT.[2,6] Finally, the electromagnetic theory for an isolated particle determines the scattering operator, which is used in the multiple wave scattering theory (MWST)[18–20] and determines the particle's response to a specified exciting field. The RTT and MWST theories will be considered below.

The second trend of the light-scattering theory is associated with the RTE. The equation operates with photometric quantities and phenomenological characteristics of the medium such as scattering and absorption coefficients and the unit-volume-scattering indicatrix (also known as phase function or form factor).[2–4,6,21] In the RTT, multiple scattering is allowed, phenomenologically, on the basis of the energy conversation law and the ray intensity (or ray radiance) concept.[6] By now the RTE theory has been elaborated in detail (including formulas, tables,[22] and public-domain computer codes for the solution of various particular problems[23]).

The physical picture forming the basis of the RTT is based on the notion of mutual multiple reradiation of inhomogeneities. In essence, the subject of the RTT is the kinetics of photon diffusion in an inhomogeneous medium. Inclusion of the Stokes polarization parameters in the RTT leads to the matrix formulation of the RTE;[3,21] however, the physical basis of the matrix RTT remains exactly the same as that in the scalar case. The elementary volume of the scattering medium plays the role of a quadratic detector that splits the field into incoherent beams (photons), and the scattering act itself is always interpreted in the energy sense, with averaging over time and volume.[24] The notions of photon diffusion in an inhomogeneous medium naturally lead to a statistical scheme of solving the RTE in the context of Monte Carlo (MC) ideology. This method calls for considerable computer resources, but it may be used for the realistic geometries of the experiment. This question is examined at greater length in Section 1.8. The microphysical basis of RTE and its derivation from Maxwell equations and statistical properties of scattering media was discussed by Mischchenko.[25]

The third, and the most rigorous, trend of the scattering theory is in essence the electrodynamics of statistically inhomogeneous media.[20,24] This approach considers multiple wave scattering by discrete or continuous inhomogeneities, the correlation statistics of scatterers, and the vector character of the electromagnetic field.[18–20] The basic physical principles of the theory of multiple wave scattering are the same for systems with continuous and discrete inhomogeneities, but details of the formalism are somewhat different. For definiteness, we will speak below of the case of discrete scatterers.

The theory of propagation and multiple scattering of radiation in a medium of interacting scatterers belong to a fairly complex class of many-particle electrodynamic problems. For their solution, special diagram methods have been developed that lead to the Dyson equation for Green's mean function or to the Bethe-Solpiter equation for Green's function correlation.[20] Because of difficulties of a purely mathematical nature, noticeable progress in this field, including the emergence of effective numerical algorithms, has only recently been made.[26–28]

Being fairly complex mathematically, the theory of multiple wave scattering is based on simple physical principles. First, it is assumed that the spatial configuration of all particles and its statistical properties are known. Second, it is assumed that the scattering operator of an isolated particle that describes the scattered field for a given exciting field is known. Since we are dealing with a system of electrodynamically interacting multipole oscillators, the exciting field is formed from the incident field, and the multiple scattered fields from all the other particles. Hence it is clear that the main difficulty of the theory is just to find the exciting field with regard to all possible orders of scattering from all the interacting particles. The different versions of the theory differ primarily in the methods of calculation of the exciting field with regard to the statistical properties of the ensemble describing the spatial configuration and the optical properties of scatterers. Once the exciting fields have been found, further analysis consists in calculating the scattered fields of isolated particles and in adding these fields together with regard to phase shifts. Since we are dealing with random fields, one has to use the appropriate correlation analysis in order to calculate the observed photometric characteristics.

In the context of the rigorous MWST, the theory of coherent propagation of radiation in a medium with densely packed scatterers has been fairly well developed.[18,19,21] The basic result of this theory is the deduction of a *dispersion equation* for the effective wave number describing the coherent-field propagation in a medium and differing from the wave number of free space. This dispersion equation allows for the optical properties of scatterers and for the statistical properties of their spatial arrangement. As a rule, when deducing a dispersion equation, one makes certain simplifying assumptions. For example, quasi-crystalline approximation is used to break the infinite chain of multiple-scattering equations, and the Percus–Yevick approximation is used to describe pair correlations in particle positions.[29]

Significantly, there is a close relation between the theory of coherent light propagation in densely packed disperse systems and that in homogeneous molecular or crystalline media. The relationship between the refractive index n of such a medium, the concentration N, and the polarizability of molecules α is given by the well-known Lorentz–Lorenz formula[30]

$$\frac{n^2 - 1}{n^2 + 2} = \frac{4\pi}{3} N\alpha. \tag{1.1}$$

It is important for our consideration that this relation can be obtained as a result of derivation of the Ewald–Oseen extinction theorem.[30] In this derivation, the optical properties of a medium are treated on the basis of representation of the total optical field as a sum of the incident wave and all waves multiply scattered by molecules exited by an effective self-consisting field. Elimination of the incident wave and the arising of a new wave, propagating through the medium with velocity c/n, are due to the interference of the incident wave and self-consisting fields scattered by molecular dipoles in the forward direction. For other directions, the fields scattered from different, spatially fixed and physically small volumes $\delta V \ll \lambda^3$ would be totally suppressed because of the interference if no fluctuations were in the medium. However, the fluctuations violate coherence and are responsible for molecular scattering.

The problem with calculation of the effective propagation constant in a condensed medium is closely linked to the theory of effective optical constants.[11,31] The main object of this theory is to formulate a *mixing rule* where the actual dispersion medium could be replaced by a model homogeneous medium with effective optical constants. In the electrostatic case, it is required that the calculated values of the effective dielectric permeability be as close as possible to the experimental measurements of this parameter for an inhomogeneous sample. In the optical case, the calculation of absorption and scattering for a homogeneous particle with an effective ε_{eff} should be as close as possible to the results of the numerical or natural experiments for an actual inhomogeneous particle.

A fundamental peculiarity of the MWST is that the optical properties of interacting particles differ from those characteristics that are obtained in solving the scattering problem for an isolated particle. In Section 1.4, we will see that the extinction cross sections of particles in a cluster do not coincide with the usual Mie calculation. Even in the simplest case of two absolutely identical contacting spheres, the scattering cross section of each particle will depend on the the separation between particles and the bisphere orientation with respect to the incident plane wave. This type of effect is commonly called the collective scattering effect or cooperative scattering effect.[20,24]

It should be emphasized that these cooperative effects may be of different nature and hence have different experimental manifestations. First, as indicated above, the exciting local field may differ substantially from the incident-wave field and even from the mean field in a dispersion medium because of the violent local fluctuations. In this case, each particle's optical response itself will differ substantially from the isolated-scatterer case. Even if the positions of the interacting particles are absolutely random and are in no way correlated, the optical properties of such a dispersion system will differ from the case of single scattering of noninteracting particles. Another cause for the emergence of cooperative effects in densely packed systems is associated with the constructive interference of individual scattered fields, which is due to the presence of a certain order in particle arrangement. The physical mechanisms of emergence of this order are connected with the competition from attraction (repulsion) forces and the entropy factor. The correlations in particle positions may be either short range or long range, depending on the nature of these forces. In the case of noncharged particles, only the short-range order is usually observed, which may be described using the solid-sphere model, the Lennard–Jones potential, and so on.[29]

In the general case, the cooperative effects of multiple scattering include both components, and their accurate calculation is, as noted above, a challenging task. However, for biological systems, the situation is simplified a bit because the optical properties of interacting particles usually do not differ widely from the properties of their environment. Therefore, even at multiple reradiation of particles the differences of the exciting field from the incident-wave field are not too notable. This means that the major influence on the optical characteristics of the medium is exerted not by the change in the local exciting fields, but by the scattered-fields interference, which is due to correlations of the short-range positions. To a first approximation, the scattered fields themselves can be evaluated within the limits of a single approximation, i.e., assuming that the exciting field is equal to the incident-wave field. The development of that approach will be discussed below (Section 1.7).

A rigorous analysis of conditions for the applicability of this or that version of the light-scattering theory is a nontrivial problem,[20] in which one has to account for the coherence properties of the incident radiation; the size, concentration, and optical properties of particles; the characteristic time of stability of the medium microstructure (i.e., the characteristic times of the fluctuation relaxation); the geometrical parameters of the scattering sample; the photodetector characteristics; and so on.[24] For example, if the coherence time of the incident radiation is less than the time of the photon run between two consecutive scattering acts, then these scattering acts will be statistically independent. In this case—which is typical for neutron or x-ray scattering—either the RTT or the single-scattering approximation (near-transparent media) is applicable. Qualitatively, the possibility that each of the three approaches mentioned can be applied for describing radiation propagation and scattering in a dispersion medium is determined by five characteristic parameters: the wavelength λ; the size of particles (inhomogeneities) a; the mean distance between them (correlation radius) R; the geometrical thickness of the medium d; and the coherence length l_c. For concentrated dispersion media, $R \sim a$, so if the particle size is comparable to the wavelength, the scattering particles are not independent, and allowance should be made for multiple reradiation within the limits of the MWST. How the theory is formulated specifically depends on the ratio among the particle size, the wavelength, and the geometrical parameters of the scattering volume.

If the mean distance $R \gg a$, λ, l_c, (moderately concentrated and dilute media of arbitrary optical thickness), the RTT theory is applicable, and, again, its specific form will depend on the a/λ and R/d ratios (small-angle approximation, diffusion approximation, etc.[6]). From a physical standpoint, a necessary condition of RTT applicability is the smallness of the wavelength and particle size as compared with the characteristic length of the light-field extinction.[24] It should be noted, however, that sometimes the RTT is not applicable in principle. In particular, it is not applicable for describing the effect of enhanced back-scattering.[32,33]

In the case of dilute and optically thin dispersion media (when $R \gg a,\lambda$, and the optical thickness $\sigma_e d/R^3 \ll 1$, σ_e being the extinction cross section of the particle), the single-scattering approximation is applicable, in which photometric quantities, quadratic in the scattered field, are just an additive sum over the particles. It is important to keep in mind that the boundaries of applicability of single scattering differ substantially in the suspension-particle concentration (or in the optical thickness), depending on the type of effects being measured. For instance, when measuring the scattering indicatrix, one sees that the contribution of multiple scattering is manifested at considerably lower concentrations than those in the case of measurement of the extinction of a collimated light beam. The physical causes for such different sensitivities of different effects are obvious. In the case of measurement of the indicatrix, the probability of multiple scattering in a specified direction increases with an increase in particle concentration. However, when measuring the extinction of a collimated beam, one sees that the probability of multiple noncoherent scattering does not affect the result (if only the receiver's aperture is small enough). Otherwise stated, the multiplicity of noncoherent scattering does not change the collimated transmission if the multiply scattered "photon" did not get into the photodetector.

The question under discussion is intimately linked to the important question of the boundaries of applicability of Bouguer's law under conditions of multiple scattering.[34–36] As is known, Bouguer's law (also known as the Beer–Lambert law) describes the exponential extinction of the incident-light intensity in a layer of thickness d

$$I(d) = I_0 \exp(-\bar{\tau} d) = I_0 \exp(-\tau), \tag{1.2}$$

where $\bar{\tau}$ and $\tau = \bar{\tau} d$ are the turbidity and the optical thickness of a sample, which in the single-scattering approximation are proportional to the extinction cross section σ_e and to the number particle concentration N,

$$\tau = \sigma_e N d. \tag{1.3}$$

Though the particle concentration and the geometrical thickness of the layer enter into relationship (1.3) in the same way, it is important to emphasize the fundamental difference between the optically dense, concentrated systems ($R \sim a,\lambda$) and the optically thick ($\tau \gg 1$), but dilute systems ($R \gg a,\lambda$).

In the first case, the scattering characteristics of the elementary volume of a suspension differ from the dilute-system parameters because of the electrodynamic particle interaction, and the relationship between the wave number in the medium and the particle concentration is nonlinear. More specifically, Eq. (1.3) is no longer valid. Physically, this can be seen from a simple qualitative example. Let an empty glass cuvette be filled up with a water aerosol that consists of submicron-sized droplets. As the droplet concentration (or the corresponding volume fraction f) increases, the imaginary part of the effective refractive index also increases from 0 to the values determined by[1]

$$n \approx 1 + 2\pi k^{-3} N S(0) \equiv 1 + N[2\pi k^{-3} \mathrm{Re} S(0) + i\sigma_e/2\,k], \tag{1.4}$$

where $S(0)$ is the forward-scattering amplitude [we define it as the van de Hulst[1] amplitude $S(0)$ divided by the imaginary unit i]. As the scattering becomes intensive enough, the coherent wave extinction gradually deviates from the linear law (1.4), then reaches a maximum and eventually, at the volume fraction $f \to 1$ (cuvette with water!), decreases to very small values corresponding to the transmittance of clear water.

A quantitative description of the above scenario is based on the theory of multiple scattering and coherent electromagnetic wave propagation in a densely packed disperse medium.[18,19,26] For example, in the Rayleigh limit, the concentration dependence $n(f)$ for spherical particles with radius $a \ll \lambda$ can be described by a generalization of the Mossotti–Clausius or Maxwell Garnett mixture formulas[18]

$$\varepsilon \equiv n^2 = 1 + \frac{3f\alpha_1}{1 - f\alpha_1}\left\{1 + i\frac{2}{3}x^3\alpha_1 \frac{(1-f)^4}{(1-f\alpha_1)(1+2f)^2}\right\}, \tag{1.5}$$

$$\alpha_1 = \alpha/a^3 = \frac{\varepsilon_1 - 1}{\varepsilon_1 + 2}. \tag{1.6}$$

Here, $x = ka \ll 1$, ε_1 is the dielectric permittivity of spheres, the refractive index of the surrounding medium is supposed to be equal to 1, and the Percus–Yevick approximation was used to represent the pair distribution function. In the dilute case $f \ll 1$, Eq. (1.5) is equivalent to Eq. (1.4) with $S(0) \approx k^3\alpha(1 + i2x^3\alpha_1/3)$, whereas for the densely packed suspension with $f = (4\pi/3)Na^3 \to 1$, the extinction contribution due to light scattering vanishes and Eq. (1.5) reduces exactly to the Lorentz–Lorenz relation (1.1). Thus, for concentrated suspensions, the deviations from Bouguer's law are determined by the electrodynamic particle interaction and can be observed even at small geometrical thicknesses of the layer.

In the case of dilute, albeit optically dense, systems, the situation is entirely different. Here, the considerable optical thickness is ensured by the corresponding geometrical thickness of the scattering sample. The optical characteristics of the elementary volume are determined by the properties of isolated particles and are proportional to their concentration. The extinction of a direct (coherent) beam is determined by Bouguer's exponential law with optical thickness (1.3). As the geometrical path and the corresponding optical thickness increase, there occurs transformation of the primary coherent beam into multiply scattered noncoherent light. This transformation is described by the RTE, which includes Bouguer's extinction of the primary beam.[2,3,6] Thus, the difference from the single-scattering case is only that multiply scattered noncoherent light, as well as the attenuated incident light beam, also gets into the photodetector. This means that the effects of multiple light scattering during the measurement of collimated transmittance are physically manifested as effects of *small-angle scattering* or, more specifically, as effects of an increase in transmittance at the expense of single- and multiple-scattering small-angle light flux. This situation is completely analogous to the well-known influence of small-angle scattering on the measurements of transmittance of dilute suspensions.[37,38] The ratio between the attenuated incident light and the small-angle flux depends not only on the layer thickness, but also on the optical scheme of photometry.[34,35,38] For example, if the primary beam is well collimated, and the detector's receiving aperture is small, the effects of small-angle distortions due to multiple scattering will be small up to a very large optical thickness.[34–36]

1.3 Theoretical Methods for Single-Particle Light-Scattering Calculations

1.3.1 Basic parameters for single-particle light scattering

Consider the scattering by a certain particle of a monochromatic $[\exp(-i\omega t)]$ plane electromagnetic wave,

$$\mathbf{E}_0 = \mathbf{e}_0 \exp(i\mathbf{k}_0), |\mathbf{e}_0| = 1. \tag{1.7}$$

By virtue of linearity of Maxwell's equation, the scattered field $\mathbf{E}_s$ (in the direction $s = \mathbf{k}_s/k$) should be linearly related to the incident field via the *scattering affinor* $S(\mathbf{k}_s,\mathbf{k}_0)$[14,39]

$$\mathbf{E}_s = \frac{\exp(ikr)}{kr}\bar{S}(\mathbf{k}_s,\mathbf{k}_0)\mathbf{e}_0, \tag{1.8a}$$

the *vector amplitude of scattering*[14]

$$\mathbf{S}(\mathbf{k}_s,\mathbf{k}_0) = \bar{S}(\mathbf{k}_s,\mathbf{k}_0)\mathbf{e}_0, \tag{1.8b}$$

or the corresponding *amplitude scattering matrix*[15]

$$\begin{Bmatrix} E_{s1} \\ E_{s2} \end{Bmatrix} = \frac{\exp(ikr)}{kr}\begin{bmatrix} S_{11} & S_{12} \\ S_{21} & S_{22} \end{bmatrix}\begin{Bmatrix} e_{01} \\ e_{02} \end{Bmatrix}, \tag{1.9}$$

where the indices $1 = \|$ and $2 = \perp$ stand for the transverse components (i.e., $E_\| = E_\vartheta$, $E_\perp = E_\varphi$) relative to the planes containing the z-axis and the incidence direction $\mathbf{s}_0 = \mathbf{k}_0/k$ or the scattering direction $\mathbf{s} = \mathbf{k}_s/k$. As a rule, the incidence direction is chosen to be coincident with the z-axis,[1,12] so the transverse components of the fields are determined relative to the scattering plane $(\mathbf{s}, \mathbf{s}_0)$. We note that different basis vectors for the transverse components of the field and different normalization in Eq. (1.9) are chosen by different authors.[1,12,15] As a consequence, the S_{ij} elements may differ.[15] It is important to emphasize that the amplitude scattering matrix relates the field components determined in *different* coordinate systems and, therefore, *is not a tensor*.

The amplitude scattering matrix gives a full description of the scattered field, but in actual experiments photocurrents are usually measured that are proportional to quadratic field combinations. Therefore, to describe the monochromatic transverse wave one introduces four Stokes parameters, I, Q, U, V[1,12] or the corresponding column vectors $\hat{I}$. The linear relationship between the incident ($\hat{I}_0$) and scattered ($\hat{I}_s$) Stokes parameters is given by the *Mueller scattering matrix* (or simply the *scattering matrix*)[1,12]

$$\hat{I}_s = \frac{1}{k^2 r^2}\hat{F}\hat{I}_0. \tag{1.10}$$

The formulas expressing the 4×4 scattering matrix $\hat{F}$ in terms of the amplitude 2×2 matrix $\hat{S}$ can be found elsewhere.[1,12] The $\hat{F}$ matrix contains complete information about the scattering properties of the particle, but in practice one would measure only some scattered-light parameters, e.g., the scattering intensity $I_s(\theta)$ and the degree of linear polarization.

Particle scattering and absorption of light withdraws energy from the incident wave. This effect is consequently characterized by the extinction cross section σ_e having an area dimension. The product $I_0\sigma_e$ determines the total power of scattering and absorption. The scattering (σ_s) and absorption (σ_a) cross sections are determined in a similar way. According to the optical theorem,[1,12,14] the extinction cross section of linearly polarized light is proportional to the forward scattering amplitude

$$\sigma_e = \frac{4\pi}{k^2}\,\mathrm{Im}[e_0^* \bar{S}(\mathbf{k}_0,\mathbf{k}_0)\mathbf{e}_0], \tag{1.11}$$

where the asterisk denotes complex conjugation. In the more general case of the incident light with arbitrary polarization, the cross section σ_e is determined by way of the *extinction matrix.*[3,21] The scattering cross section is calculated by integration of the scattered intensity over all directions. For example, for the incident light with Stokes parameters (I_0, Q_0, U_0, V_0), we have

$$\sigma_s = \frac{1}{k^2 I_0}\int_{4\pi}[I_0 F_{11} + F_{12}Q_0 + F_{13}U_0 + F_{14}V_0]d\Omega(\mathbf{s}). \tag{1.12}$$

In the case of unpolarized light and spherically symmetrical scatterer, the relationship (1.12) is reduced to the usual integral of the indicatrix with respect to the scattering angle

$$\sigma_s = \frac{2\pi}{k^2}\int_0^{\pi} F_{11}(\theta)\,\sin\,\theta d\theta. \tag{1.13}$$

The asymmetry (anisotropy) parameter g is defined by equation

$$g = \langle\cos\,\theta\rangle = \frac{2\pi}{k^2\sigma_s}\int_0^{\pi}\cos\,\theta F_{11}(\theta)\,\sin\,\theta d\theta, \tag{1.14}$$

and its sign points to the preferential forward ($g > 0$), backward ($g < 0$), or symmetrical isotropic scattering ($g = 0$).

1.3.2 Exact analytical and numerical methods

The general statement of the single-particle scattering problem is rather simple and can be described as follows. A field $\mathbf{E}_0$ is incident on a scattering particle of volume V and excites the field $\mathbf{E}_i$ inside V. Outside V, an additional field, the diffraction field $\mathbf{E}_s$, is generated. It is required to calculate, on the basis of Maxwell's equations, the complete field $\mathbf{E}$, which equals $\mathbf{E}_i$ inside V, and $\mathbf{E}_0 + \mathbf{E}_s$ outside V, and that satisfies the boundary conditions at the particle surface as well as the radiation conditions at infinity. The general scheme being simple, a concrete solution to the problem depends essentially on the geometry of the scatterer and on the properties and structure of its substance. For example, the solution cannot be obtained in a general closed form even for a *spherical* particle with an anisotropic refractive index of a general type.[12] Therefore, in the theory of scattering by small particles various methods have been developed whose scope and effectiveness depend on the particular statement of a problem. From a current standpoint, an effective numerical algorithm realized on a personal computer is equivalent to an analytical solution, which as a rule also calls for nontrivial calculations. In this section, we will look briefly only at those methods that are most intensively used in biological applications. A more detailed discussion of the calculation methods can be found in a recent review.[16]

1.3.2.1 Separation of variables and T-matrix methods (SVM and TM)

The most important analytical solution is the theory of scattering by a homogeneous isotropic sphere, called the Mie theory[1,12] (the pioneering work on this problem is associated with a constellation of names[40–44]). To illustrate the basic results of the Mie theory, we give formulas for the extinction cross section, the scattering cross section, and the anisotropy parameter, following the designations[12]

$$\sigma_e = \frac{2\pi}{k^2} = \sum_{n=1}^{\infty}(2n+1)\mathrm{Re}(a_n + b_n), \tag{1.15}$$

$$\sigma_s = \frac{2\pi}{k^2}\sum_{n=1}^{\infty}(2n+1)(|a_n|^2 + |b_n|^2), \tag{1.16}$$

$$g = \frac{4\pi}{k^2\sigma_s}\left\{\sum_{n=1}^{\infty}\frac{2n+1}{n(n+1)}\mathrm{Re}\,(a_n b_n^*) + \sum_{n=1}^{\infty}\frac{n(n+2)}{n+1}\mathrm{Re}\,(a_n a_{n+1}^* + b_n b_{n+1}^*)\right\}, \tag{1.17}$$

where the coefficients a_n and b_n are called Mie coefficients.[12] The Mie theory is generalized for the case of concentrated spheres, particles with an inhomogeneous profile of refractive index or with the optical activity of the substance, and also for the case of illumination by a focused laser beam. The relevant literature citations and effective computer codes can be found in Refs. 16 and 23.

For an infinite circular cylinder an exact solution at perpendicular incidence was first obtained by Rayleigh,[45] who addressed this problem once again in his last work written shortly before his death.[46] Though an infinite cylinder is not a three-dimensional (3-D) scatterer, this model is helpful in understanding light scattering and extinction by an anisotropic dispersion medium (see Section 1.7 below). References to literature that generalizes a simplest model can be found in Ref. 16. The Mie solution is also generalized to an arbitrary collection of spheres, but this subject concerns the problem of scattering by particle aggregates and will be treated separately in Section 1.4.

In Helmholtz's scalar equation, the variables are separated into 11 physically interesting coordinate systems.[47] But for a vector field containing *three* scalar functions, a complete separation of variables is possible only in six systems: (1) Cartesian, (2–4) three cylindrical, (5) conical, and (6) spherical. Therefore, Möglich's[48] formal solution for spheroidal coordinates was useless for all practical purposes until the method for separation of variables was adapted by Asano and Yamamoto[49] for numerical calculations via cutoff of infinite coupled (i.e., not fully separated) equations. Farafonov[50] improved the method described in Ref. 49 by using an ingenuous scheme of splitting fields into two types with an invariant angular part. By now, a large body of factual material has been assembled on the application of SVM in the calculation of scattering by spheroidal particles.[15,51]

The T-matrix method, which is well known in quantum theory,[13] was introduced into electromagnetic scattering by Waterman.[14] Contrary to the Green function method, the T-matrix relates not the fields themselves in a coordinate representation, but the expansion coefficients of the incident and scattered fields over some complete set of vector basis functions. For example, if one expands all fields in vector spherical harmonics (VSH) $\mathbf{Y}_{mnp}^{(\tau)}$ [13,47,52]

$$\mathbf{Y}_{mn1}^{(\tau)} = \mathbf{N}_{mn}^{(\tau)}(k\mathbf{r}), \mathbf{Y}_{mn2}^{(\tau)} = \mathbf{M}_{mn}^{(\tau)}(k\mathbf{r}), \tag{1.18}$$

of the first and third kind ($\tau = 1{,}3$), then the expansion coefficients for scattered (a_{mnp}) and incident (exciting, $p_{\mu\nu q}$) fields are related by

$$a_{mnp} = \sum_{\nu=1}^{\infty} \sum_{\mu=-\nu}^{\nu} \sum_{q=1}^{2} T_{mnp,\mu\nu q} p_{\mu\nu q}. \tag{1.19}$$

After being published in Ref. 53, the T-matrix method began to be used commonly in the scattering theory.[54–56] An important strong point of the method is the possibility that the problem on the orientational averaging of the observed scattering characteristics may be solved analytically.[57,58] To illustrate, the extinction and scattering cross sections averaged over random particle orientations are expressed directly in terms of a T-matrix

$$\langle C_e \rangle = -\frac{2\pi}{k^2} \mathrm{Re}[\mathrm{Sp}(T)], \tag{1.20}$$

$$\langle C_s \rangle = \frac{2\pi}{k^2} \sum_{ab} |T_{ab}|^2, \tag{1.21}$$

where the symbol Sp in Eq. (1.20) stands for a spur over all T-matrix indices, and the symbol a or b in Eq. (1.21) signifies a combined multi-index (nmp).

1.3.2.2 Integral equation method

The integral equation method (IEM) occupies a special place in the range of theoretical methods, since it is a general approach in which, by using the affinor Green function[13]

$$G(\mathbf{r},\mathbf{r}') = (1 + k^{-2}\nabla\nabla)\frac{\exp[i\mathbf{k}(\mathbf{r}-\mathbf{r}')]}{4\pi|\mathbf{r}-\mathbf{r}'|}, \tag{1.22}$$

a boundary problem is reduced to an integral equation that includes the boundary condition and radiation condition at infinity[13,59,60]

$$\mathbf{E}(\mathbf{r}) = \mathbf{E}_0(\mathbf{r}) + k^2 \int G(\mathbf{r},\mathbf{r}')[\varepsilon(\mathbf{r}') - \hat{1}]\mathbf{E}(\mathbf{r}')d^3\mathbf{r}'. \tag{1.23}$$

Therefore, the IEM is not merely a numerical approach, but also an efficient basis for the formulation of other methods [e.g., the method of moments,[61] the T-matrix method,[53,62] or the Rayleigh–Debye–Gans (RDG) method[6,13,59,63]]. In Eq. (1.23), the scattered field is expressed by way of an unknown distribution of *bulk* sources induced by an external field. In the electromagnetic theory, another approach is also used,[60,62] in which the scattered field is calculated by way of a *surface* source distribution.

The presence of singularity in Green's function (1.22) gives rise to a "self-term" that accounts for the difference between the average Maxwell field and the local (effective) Lorenz field.[59] The various formulations of IEM are in many ways different only in the methods for allowing for the self-term and in the methods for replacing the integral equation by its discrete counterpart.[64–67] The singular equation (1.23) can be represented, by the Fourier transform, as a nonsingular Fredholm integral equation (FIE) of the second kind,[68,69] which is solved numerically after the discrete-analogue substitution. It is interesting that the FIE kernel coincides with the second Born approximation for the integral equation obtained in Ref. 63. References to papers on the application of various IEM versions in problems of single-particle scattering are available for review.[16]

1.3.2.3 Discrete dipole approximation

In the theory of scattering by small particles, this method began to be employed intensively after the publication of a paper by Pursell and Pennypacker,[70] though undoubtedly its physical basis was known and applied elsewhere previously.[30] Yurkin et al. (see, e.g., Ref. 71, and references therein) developed an alternative discrete dipole approximation code (ADA version) with enhanced efficiency. A detailed consideration of discrete dipole approximation (DDA) or ADDA capabilities and limitations can be found in Ref. 71. The basic physical approximation is in substitution of the actual scatterer for an ensemble of discrete elements with volume V_i, polarizability α_i, and dipole moments $\mathbf{d}_i = \mathbf{d}(\mathbf{r}_i)$, $i = 1 - N$. All the other calculations, e.g., those of the dipole amplitudes, the scattered field, the integral cross sections, and the scattering matrix may be done absolutely rigorously. The equations for dipole moments are not hard to write from simple considerations based on the concept of an exciting field equal to the sum of the incident wave and the fields of other dipoles at a given point

$$\mathbf{d}_i(\mathbf{r}_i) = \alpha_i\left[\mathbf{E}_0(\mathbf{r}_i) + k^3\sum_{j\neq i} G_{ij}\mathbf{d}_j(\mathbf{r}_j)\right]. \tag{1.24}$$

The tensor of dipolar scattering G_{ij} is determined by the known formulas[30]

$$G_{ij} = \exp(ikr_{ij})[G_1(kr_{ij}) + G_2(kr_{ij})\mathbf{R}_{ij}\cdot\mathbf{R}_{ij}], \tag{1.25}$$

$$G_n(z) = (-1)^n[-z^{-1} - (2n-1)(iz^{-2} - z^{-3})], \quad n = 1,2, \tag{1.26}$$

where $\mathbf{r}_{ij} = \mathbf{r}_i - \mathbf{r}_j$, $\mathbf{R}_{ij} = \mathbf{r}_{ij}/\mathbf{r}_{ij}$, and $\mathbf{a}\cdot\mathbf{b}$ is a dyadic. The solution of the system of linear equations (1.24) allows the calculation of all basic optical characteristics of an aggregate: the vector scattering-amplitude, $\mathbf{S}(\mathbf{k}_0, \mathbf{k}_s)$

$$\mathbf{S}(\mathbf{k}_0,\mathbf{k}_s) = k^3\sum_i[\mathbf{d}_i - \mathbf{s}(\mathbf{s}\mathbf{d}_i)]\exp(-i\mathbf{k}_s\mathbf{r}_i), \tag{1.27}$$

and the integral cross sections of extinction C_e, absorption C_a, and scattering $C_s = C_e - C_a$

$$C_e = 4\pi k\,\mathrm{Im}\sum_i(\mathbf{e}_0\mathbf{d}_i)\exp(-i\mathbf{k}\mathbf{r}_i), \tag{1.28}$$

$$C_e = 4\pi k\sum_i\eta_i|\mathbf{d}_i|^2, \quad \eta_i = \frac{4\pi\mathrm{Im}(\varepsilon_i)}{V_i|\varepsilon_i - 1|^2}. \tag{1.29}$$

The important question about the choice of model polarization is left beyond the limits of the phenomenological scheme just described. It is well known[1] that the simplest choice based on the Mossotti–Clausius formula does not satisfy the optical theorem. Therefore, various approximations have been proposed in the literature[64,70,72–75] that allow for the dipole-energy radiation losses and lead to complex polarizability even for the nonabsorbing dipole particle.

At first glance, it seems that the bulk IEM and DDA methods are based on different physical principles. Therefore, it is important to emphasize that the solution of Eq. (1.23) by the method of moments and the DDA solution actually lead to *identical* systems of linear equations for dipoles if equivalent assumptions have been made for polarizability in the DDA and for the self-term in the IEM.[66]

If a basis lattice has been specified and the algorithm of finding polarizability found, it remains only to fill the lattice nodes by the corresponding dipoles with regard to the inhomogeneity and shape of the scatterer. Essentially no limitations are placed on the geometry and, to a degree, on the properties of the material. This is the chief value of the method, particularly as applied to structurally complex biological objects.[76] The limitations of DDA are associated with the provision of convergence and accuracy of the results, which are impaired for optically rigid structures.

1.3.3 Approximate theories

With advances in computer facilities and new algorithms many approximate methods of the scattering theory lost their initial value. Despite this, they often give useful information for the qualitative understanding of the physics of phenomena and quantitative tendencies. In this section, we look briefly at only the basic physical principles of certain approximations. A closer look at the approximate theories and abundant references are available in Refs. 16, 77, and 78.

1.3.3.1 Rayleigh approximation

The basic ideas of all approximate methods are associated with definite regions of values of the most important diffraction parameters: the size parameter ka and the relative refractive index $m = n/n_0$. For example, if $ka \ll 1$ and $ka|m| \ll 1$, we are dealing with Rayleigh scattering, in which a particle scatters like an infinitesimal dipole $\mathbf{d} = \alpha \mathbf{e}_0$

$$\mathbf{E}_{sR} = k^3[\mathbf{d}_i - \mathbf{s}(\mathbf{s}\mathbf{d}_i)]\frac{\exp(ikr)}{kr}. \tag{1.30}$$

The dipole moment and the corresponding polarizability tensor are estimated from electrostatic equations.[1,12] The possibility of using an electrostatic approximation to calculate the dipole moment allows particles of virtually all shapes to be considered in the Rayleigh scattering theory.[79] The accuracy of the Rayleigh approximation has received in-depth treatment in Refs. 80 and 81.

Stevenson[82,83] generalized the Rayleigh theory by expanding the fields in powers of ka. This approximation was used to calculate the Mueller matrix of light scattering by random spheroids.[84] This chapter refers to other applications of the Stevenson approximation, as does Ref. 16.

1.3.3.2 Rayleigh–Debye–Gans approximation

In a large number of practically important cases, the relative refractive index of particles m is close to 1. In particular, for the overwhelming majority of biological structures, $m \sim 1$.[76] This condition is valid in an even greater number of cases for x-radiation or neutron scattering. Such particles are called "optically soft," and the corresponding approximation can be called an "approximation of optically soft particles."

The condition $|m - 1| \ll 1$ itself is not sufficient for the development of the theory, since the ratio between size and light wavelength and the phase-shift magnitude $\rho = 2ka(m - 1)$ are of major importance. The RDG approximation is applicable when two conditions are simultaneously fulfilled

$$|m - 1| \ll 1, \quad |\rho| \ll 1. \tag{1.31}$$

There are a number of names for the theory that is based on the assumptions in Eq. (1.31). The best-known name is the Rayleigh–Gans (RG) approximation.[1] We believe that the term RDG is best suited for the following reasons. The fundamental ideas of the method (including the derivation of successive approximations based on the integral relation for a scattered field) were formulated by Rayleigh in his 1881 paper,[45] and the formula for the sphere-scattering cross section was obtained by him in 1914.[85] In 1915, Debye[86] derived a general formula for the intensity of x-rays scattered by a randomly oriented particle ensemble. Later, Debye applied this approach to light scattering by polymer solutions,[87] which had a profound impact on the development of this trend.[7] In quantum mechanics, an analogue to Rayleigh iterations was developed by Born;[88] his name is rightly associated with this approximation. Thus, retaining the name of Gans in the name of the theory (RDG) makes sense only because the name RG approximation has received wide acceptance thanks to van de Hulst.[1]

The RDG approximation can be obtained by various means. For example, one may use general integral relation (1.23), assuming that the field inside the particle is the incident-wave field.[13] Another mean is based on the physical interpretation of scattering as a result of interference from the independent-dipole fields excited by an incident wave in particle volume V.[1,7] In either case, the scattered field is represented as

$$\mathbf{E}_s = \mathbf{E}_{sR} G(\mathbf{q}), \tag{1.32}$$

where $\mathbf{E}_{sR}$ is the Rayleigh scattered field (1.30), $\mathbf{q} = \mathbf{k}_s - \mathbf{k}_0$ is the scattering vector, and $G(\mathbf{q})$ is the interference function or the scattering form-factor

$$G(\mathbf{q}) = \frac{1}{V} \int \exp(-i\mathbf{q}\mathbf{r}) d^3\mathbf{r}. \tag{1.33}$$

Thus, in this approximation the polarization characteristics of scattered light do not differ from the Rayleigh case, and the angular ones differ only in the interference function. For particles of certain shapes the interference functions can be obtained as simple analytical expressions.[7] For example, for spherical particles with radius a and scattering angle θ

$$G(u) = \frac{3}{u} j_1(u), \quad u = qa = 2ka \sin\frac{\theta}{2}, \tag{1.34}$$

where $j_1(x)$ is the Bessel spherical function. Equation (1.34) is also valid for ellipsoids with semiaxes a, b, c, but in this case

$$u = \sqrt{(\mathbf{qa})^2 + (\mathbf{qb})^2 + (\mathbf{qc})^2}. \tag{1.35}$$

An important and simple generalization of RDG theory is that the dipole moment of a scatterer is calculated with regard to the possible *anisotropy of the substance*[7] or the *shape anisotropy*[84,89] of the particle. In this case, polarization effects arise that are absent in the standard version of the RDG.[1]

The usually measured quantity being the intensity of scattering from a randomly oriented particle ensemble, of practical interest is the averaged square of the interference function $<G^2(q)> = P(q)$, also called the scattering form factor.[7] In the theory of light scattering by substance-structure inhomogeneities, the analogous quantity is called the scattering structure factor.[8,90]

In the RDG approximation, the absorption cross section is proportional to volume and generally does not differ in form from the Rayleigh absorption cross section.[1] The scattering cross section should be found by integrating the intensity over all angles, since the optical theorem in the RDG approximation gives either absorption cross section or zero (nonabsorbing particles).

1.3.3.3 Anomalous diffraction and related approximations

Van de Hulst,[1] using Huygens and Babinet's principles, considered a problem on the scattering and extinction of light by a particle with size $a \gg \lambda$ and refractive index $m \sim 1$, i.e., under conditions of

$$|m - 1| \ll 1, \quad x \gg 1. \tag{1.36}$$

The formula obtained for the small-angle scattering amplitude by van de Hulst is a generalization of the Fraunhofer diffraction formula and allows the extinction cross section to be found by a simple integration

$$C_e = 2\mathrm{Re}\int[1 - \exp(-i\rho(z))]dS, \tag{1.37}$$

where $\rho(z) = 2ka(z)(m - 1)$ is the phase shift of the ray propagating along the z axis, $a(z)$ is the ray pathlength inside the particle, and the integration is performed over all rays that intersect the particle. It is not hard to see that in the anomalous-diffraction (AD) approximation, the scattering particle is treated as an ideal phase screen. In a similar way, the absorption cross section can be found as the sum of absorptions of all the rays

$$C_a = \int [1 - \exp(-2\mathrm{Im}\rho(z))] dS. \quad (1.38)$$

The applicability range of the AD theory for the scattering amplitude is restricted by small angles, but the cross-section formulas proved a very appropriate approximation for a wide range of particles. The general universal views of the dependence of the homogeneous-sphere scattering cross section on the parameter of the central-ray phase shift, $\rho = 2x(m - 1)$, is given by the well-known formula of Van de Hulst[1]

$$\sigma_s = \pi a^2 Q(\rho), \quad Q(\rho) = 2 - 4\frac{\sin\rho}{\rho} + 4\frac{1 - \cos\rho}{\rho^2}. \quad (1.39)$$

The AD approximation is easily applicable for nonspherical and inhomogeneous particles.[16] Specifically, for spheroids with the semiaxes (a,b,b) the formula for the effectiveness factor $Q(\rho)$ retains its view, but now the central-ray phase shift will depend on the symmetry-axis orientation a relative to light (direction cosine, $\cos\vartheta$)

$$\rho = 2kb(m - 1)/\sqrt{1 + (b^2/a^2 - 1)\cos^2\vartheta}. \quad (1.40)$$

The AD approximation is a version out of the wider family of high-energy approximations, including the Wentzel–Kramers–Brillouin (WKB) approximation, the eikonal approximation, etc.[13,16,91,92] The main disadvantage of all versions of high-energy approximations is that they ignore polarization effects. Meteen[93] offered an anisotropic modification of AD, in which phase shifts are allowed separately for ordinary and extraordinary waves. However, for isotropic nonspherical particles this theory reduces to a scalar version of AD.

For very large particles, the scattering can be considered as the result of interference of the rays that experienced multiple reflections and refractions in accordance with the laws of geometric optics (GO).[12,59] The diffraction at the edge of a particle, which gives a sharp peak in the small-angle region, may be accounted for separately as required. In the past few years, the GO method has been frequently applied in combination with MC simulations.[16] In the work,[94] a method called the physical optic approximation was suggested, in which the scattered field is expressed by way of the field at the particle surface according to the Huygens–Kirchhoff principle. To find the surface field, one has to use the laws of reflection and refraction. The accuracy of the GO method for nonspherical particles was studied in a recent paper[95] on the basis of a comparison with the exact T-matrix results.

1.3.4 Other methods and approximations

Apart from the rigorous methods mentioned in Section 1.3.2, other algorithms and public domain codes[23,96] have also been described in the literature. Among such methods are the finite element method (FEM),[97] the finite difference time domain method (FDTDM),[98] and the point matching method (PM) (see the corresponding references in the overview[16]). The strong point of these methods is their applicability in principle to particles of arbitrary shapes and structures. FEM can be applied to solve the Helmholtz equation while FDTDM is used for Maxwell's equations immediately. Since in their nature these methods deal with local fields, a transition to the far scattered field is a special problem.[16]

Among other approximate methods are the perturbation method[99] and certain simple approximations. The perturbation method is based on expansion of an unknown solution to the problem of scattering in terms of a small parameter in the vicinity of an exact solution. Applied to nonspherical particles, this means that the solution is sought in the form of small deviations from the Mie solution, which are caused by the small deviations of the shape from the ideal sphere.[16,99] This approach is effective for a back-of-the-envelope analysis of light scattering by particles with stochastic surface properties[100] or with a weak anisotropy of the substance.[101,102]

To describe the scattering and absorption of light by nonspherical particles, Latimer[103,104] proposed hybrid approximations based on some combinations of RDG and AD theories for spheroids and the Mie theory for spheres. In a sense, such an approach is analogous to a more pragmatic strategy in which exact formulas or numerical data are approximated by simple analytical expressions.[105,106]

1.4 Extinction and Scattering by Aggregated and Compounded Structures

In this section, we discuss methods for calculating scattering and absorption characteristics for two types of structures: (1) large clusters formed as a result of small-particle aggregation and (2) composite structures, i.e., a "host" large matrix-particle with foreign "guest" inclusions. In either case, the cluster particles may be in direct contact with the immediate neighbors or be separated by interparticle distances in accordance with the statistical properties of the pair density–density correlation function. Such structures are physical models for many biological objects or processes. More specifically, immune precipitation,[107–109] sol-particle immunoassay (SPIA),[110,111] or cell agglutination[38,112,113] are well-known examples of the formation of biological aggregates. Bacterial or eukaryotic cells are typical composite structures bounded by the cell wall and the membrane.[38,76] Methods for computing and applying light scattering in the study of aggregated and composite particles have been discussed in Refs. 114 to 116.

It should be emphasized that the cluster scattering theory can be applied to calculate light scattering by aggregated simple-in-structure biological particles or by individual complex particles. Depending on the size of cluster monomers, one can use the DDA approach (small monomers such as globular proteins, biospecific conjugates prepared from gold or silver nanoparticles,[111,113] etc.) or the more sophisticated superposition method (Section 1.4.2) if aggregates are formed from large cells like erythrocytes, leukocytes, or platelets. In spite of the fact that the present-day cluster scattering theory can consider these large aggregated cells only as homogeneous Mie spheres regardless of the complicated internal structure of actual cells, it is still very useful for describing the general futures of light scattering by the aggregated cells. In application of cluster scattering theory to complex nonspherical and nonhomogeneous biological cells or other individual structures by using DDA ideology, we can treat these complex light scattering targets as 3-D clusters built from small monomers of various sizes and optical properties.[76,115] This seems to be a powerful and versatile strategy to simulate light scattering by large, nonspherical, and nonhomogeneous individual biological particles.[117]

1.4.1 Approximate and discrete dipole approximation methods

In the case of optically soft particles, the interaction among monomers in an aggregate or a composite can be described within the limits of the approximate methods considered in the preceding sections. The simplest approximation is that of a homogeneous spherical particle with a certain effective refractive index, which is calculated using a certain mixing rule [Maxwell Garnett, Bruggeman, extended effective medium approximation (EEMA),[11] etc.]. If the size of an optically soft aggregate does not exceed the light wavelength, its scattering properties can be described within the limits of the RDG approximation, by summing the scattered fields from the monomers with regard to phase shifts. In a converse case, when the size of the aggregate is much greater than the wavelength, its extinction can be calculated by Eq. (1.37) of the AD approximation.[118]

The DDA method can be used in the optics of aggregates in two versions. If the size of monomers is much less than the light wavelength, the method is applied in its physical interpretation, when a monomer is considered as an infinitesimal point dipole and its dipole moment is calculated by electrostatics formulas.[116,119] In a more general case, the aggregated or composite structure is replaced by an ensemble of dummy dipoles on the basic lattice. In point of fact, this means that the monomers may be considered as a set of dipole subdomains. Such a finer division increases the complexity of the calculations considerably.

1.4.2 Superposition method

A rigorous method for calculating the optical characteristics using an arbitrary configuration of N spherical particles[115] can be formulated simply enough, using the *generalized* Mie theory for an isolated spherical particle[120] and the superposition principle. The essence of the generalized Mie theory is: an arbitrary electromagnetic field $\mathbf{E}^i_{inc}$ incident on the i-th particle can be represented as an expansion in terms of VSH of the first order[115,120]

$$\mathbf{E}^i_{inc} = \sum_{n=1}^{\infty} \sum_{m=-n}^{n} \sum_{p=1}^{2} E_{mn} p^i_{mnp} \mathbf{Y}^{(1)}_{mnp}(k\mathbf{r}_i), \tag{1.41}$$

where the coefficients E_{mn} depend on the choice of normalization. For a plane wave, in which the incidence direction and polarization are specified by three Euler angles (α,β,γ) in the i-th coordinate system, expansion (1.41) with coefficients

$$\bar{p}^i_{mnp} = \exp(i\mathbf{k}\mathbf{r}_i)\exp(-im\alpha)\frac{1}{n(n+1)}[\tau_{mnp}(\beta)\cos\gamma - i\tau_{mn3-p}(\beta)\sin\gamma] \tag{1.41a}$$

also holds. The functions $\tau_{mnp}(\beta)$ $(p = 1,2)$ correspond to the well-known angular functions $\tau_{mn}(\cos\beta)$, $\pi_{mn}(\cos\beta)$.[115] The radial dependence of the first-kind VSH is given by spherical Bessel functions $j_n(kr_i)$ (or Ricatti–Bessel functions[115]). An expansion analogous to Eq. (1.41) but with other coefficients also holds for the *internal* field of each particle in an aggregate. To represent a *scattered* field, one has to use third-kind VSHs, based on spherical Hankel functions. Use of the boundary conditions on the sphere surface leads to the generalized Mie theory in the form of the following simple expressions for scattered-field coefficients:[120]

$$a^i_{mnp} = \bar{a}^i_{np} p^i_{mnp}, \quad p = 1,2, \tag{1.42}$$

where $\bar{a}^i_{n1} = a^i_n$, $\bar{a}^i_{n2} = b^i_n$ are the usual Mie coefficients for an isolated i-th sphere.[12] The fundamental physical result of Eq. (1.42) is that the scattering occurs without a coupling between electromagnetic modes. In the spirit of Fresnel's analogy,[115] the Mie coefficients $\bar{a}^i_{np}$ in Eq. (1.42) can be interpreted as the *reflection* coefficients for partial modes in the plane wave decomposition. In the same way, the Mie coefficients for the *internal* field can be interpreted as Fresnel coefficients of the *transmitted* partial waves.[115]

Equation (1.42) is a key one in the theory of scattering by sphere aggregates. To find the field scattered from each particle of the aggregate, all one has to do is to find the coefficients of expansion of the exciting field, which should then be multiplied by the usual Mie coefficients. The exciting field is found by the superposition principle

$$\mathbf{E}^i_{inc} = \mathbf{E}^i_0 + \sum_{j\neq i}^{N} \mathbf{E}^i_s(j). \tag{1.43}$$

For fields $\mathbf{E}^i_s(j)$ scattered by a j-th particle, the expansion (1.41) is also applicable. However, they cannot be used directly for finding the coefficients p^i_{mnp}, included in Eq. (1.42). The point is that these expansions will be written with respect to the coordinate systems not coincident with the center of the j-th particle, whereas Eq. (1.42) is written just in the i-th basis set. This problem is solved with the help of the addition (translation) theorem for VSH.[115] Performing this translation for the VSHs describing $N - 1$ scattered fields, we can find the coefficients of expansion of the complete exciting field for the i-th particle. It is easy to see that the procedure described leads not to an explicit solution, but to a system of coupled linear equations for the coefficients p^i_{mnp}, which we will write

$$\sum_{j=1}^{N}\sum_{\nu=1}^{\infty}\sum_{\mu=-\nu}^{\nu}\sum_{p=1}^{2} H^{ij}_{mnp,\mu\nu q}\, p^j_{\mu\nu q} = \bar{p}^i_{mnp}. \tag{1.44}$$

The interaction matrix H in our formulation of the method is determined by the relationship

$$H^{ij}_{mnp,\mu\nu q} = \begin{Bmatrix} \delta_{ij}\delta_{m\mu}\delta_{n\nu} + (1-\delta_{ij})A^{ji}_{mn\mu\nu}\bar{a}^j_\nu & (1-\delta_{ij})B^{ji}_{mn\mu\nu}\bar{b}^j_\nu \\ (1-\delta_{ij})B^{ji}_{mn\mu\nu}\bar{a}^j_\nu & \delta_{ij}\delta_{m\mu}\delta_{n\nu} + (1-\delta_{ij})A^{ji}_{mn\mu\nu}\bar{b}^j_\nu \end{Bmatrix}, \tag{1.45}$$

where $A^{ji}_{mn\mu\nu}$ and $B^{ji}_{mn\mu\nu}$ are the coefficients of VSH translation, based on the spherical Hankel function of the first kind, $h^{(1)}_n(kr)$. If one multiplies Eq. (1.44) by the usual Mie coefficients a^i_{np} and uses relationship (1.42), then system (1.44) reduces to the known[115,120] system of equations directly for the scattered-field coefficients a^i_{mnp}.

From the structure of the interaction matrix it follows that an electrodynamic interaction leads to the excitement of partial modes, which might be absent in the incident radiation and the amplitude scattering matrix may become nondiagonal.

In practice, the infinite system (1.44) is cut off to a certain maximal order of the multipole expansion M that ensures the convergence of calculations of physical quantities. If Eq. (1.44) is solved and coefficients of multipole expansion (1.42) are found, one can calculate all characteristics of the cluster scattering. For example, the extinction cross section is calculated by the equation

$$C_e = \frac{4\pi}{k^2} \sum_{i=1}^{N} \sum_{n=1}^{M} \sum_{m=-n}^{n} \sum_{p=1}^{2} c_{mn} \mathrm{Re}[a^i_{mnp} (\bar{p}^i_{mnp})^*], \tag{1.46}$$

where the coefficients c_{mn} depend on particular normalization of VSH and field expansions.

To calculate the amplitude scattering matrix, one has to translate the coefficients of individual-particle decompositions to the expansion coefficients a_{mnp} with respect to the common center of the cluster. The simplest way to do this is to employ the far-field approximation[121]

$$a_{mnp} = \sum_{j=1}^{N} a^j_{mnp} \exp(-ikr_j). \tag{1.47}$$

However, this scheme is ineffective for orientational averaging (see below).

In terms of the coefficients a_{mnp}, the amplitude-scattering matrix[1] $S_{ij}(\theta,\varphi)$ is calculated in a conventional way[115,122]

$$S_{11} = \sum_{m,n,p} s_{mn} \exp(im\varphi) \tau_{mnp}(\theta) a^0_{mnp}, \tag{1.48a}$$

$$S_{12} = \sum_{m,n,p} s_{mn} \exp(im\varphi) \tau_{mnp}(\theta) a^{90}_{mnp}, \tag{1.48b}$$

$$S_{21} = \sum_{m,n,p} i s_{mn} \exp(im\varphi) \tau_{mn3-p}(\theta) a^0_{mnp}, \tag{1.48c}$$

$$S_{22} = \sum_{m,n,p} i s_{mn} \exp(im\varphi) \tau_{mn3-p}(\theta) a^{90}_{mnp}, \tag{1.48d}$$

where the coefficients s_{mn} are determined by the normalization of VSHs and field decompositions, and the superscripts 0 and 90 are related to the transversal magnetic and electric modes of the incident light. We emphasize that matrix (1.48) is written in a coordinate system associated with the incident wave $k = z$; therefore, for the averaging over cluster orientations to be performed, one has to do repeated calculations for each cluster orientation. A more effective T-matrix approach is discussed in Section 1.4.3.

Let us consider the composite structures. In order to appreciate the essence of this method of solving the problem, it will suffice to consider the simplest structure in the form of a "host" sphere with a nonconcentric spherical inclusion. We can interpret the process of scattering of such a structure in terms of the following Fresnel analogy.[115] An incident plane wave is projected onto the VSH basis, so that we have a set of incident partial modes. These modes induce in the host sphere transmitted and reflected partial waves with the reflection and transmission coefficients in accordance with the Mie theory. As has been said before, there is no mode coupling in this process. The inhomogeneity being nonconcentric with respect to the "host," each transmitted mode induces a multitude of reflection and transmission modes in the inhomogeneity. The amplitudes of the reflection and transmission are given by the Mie theory. The reflected modes of inclusion are incident *outgoing* modes with respect to the external sphere. If we reexpand these waves with respect to the "host" center, a problem arises that resembles the Mie theory but is for the outgoing waves crossing the inside spherical boundary between two media. This problem has a solution of the type (1.42). This process itself does not lead to mode coupling, but this coupling has occurred earlier, when we had to perform reexpansions of partial waves. A mathematical description of this physical picture is available in Ref. 115. If the expansion coefficients of the scattered field are found, the further calculations of the optical characteristics do not differ from the cluster case.

1.4.3 T-matrix formalism for cluster scattering

Usually, it is not just a calculation for an individual structure, but an averaged result over the statistical ensemble and orientations that is required. Formally such an averaging can be carried out by simple summation of the results calculated for various orientations and polarizations of the incident light, but this is a very ineffective way. The orientational averaging for clusters is more convenient to perform by using the T-matrix formalism,[122] as well as for nonspherical particles. Since all of the incident light properties are determined by expansion coefficients, and the T-matrix depends only on the properties of the scattering structure, intuition suggests that for an ensemble of structures with random orientations the observable physical quantities should somehow be expressed directly by way of the T-matrix. Thus, we can apply this approach to cluster structures as well; one needs only the recipe for calculating the cluster T-matrix. If we formally invert Eq. (1.44), we will get the T-matrix of an individual cluster particle[122]

$$a^i_{mnp} = \sum_j \sum_{\mu\nu q} T^{ij}_{mnp,\mu\nu q} p^j_{\mu\nu q}. \tag{1.49}$$

Using theorems of VSH addition, one can combine all of these single-particle T-matrices into a common cluster T-matrix[115,122]

$$T^0_{a,b} = \sum_{i,j} \sum_{c,d} J^{oi}_{ac} T^{ij}_{cd} J^{j0}_{db}, \tag{1.50}$$

where for simplicity we use the multi-indices a, b, c, and d to denote the order, degree, and mode (i.e., $a = mnp$, etc.). The matrices J_{ab}^{oi} describe the VSH translation to the common cluster center and are based on the spherical Bessel functions. The further calculations of the optical characteristics and their orientational averaging are performed according to the same scheme as that employed in the case of the usual nonspherical particles.[122]

1.4.4 Fractal aggregates

The structure of various biological aggregates may be described in terms of statistical (irregular) fractal clusters,[123] i.e., statistically self-similar objects with the fractal dimension $D_f < 3$ defined by power relations

$$g(r) \sim (r/R)^{D_f - 3}, \quad N \sim (R/a)^{D_f}, \tag{1.51}$$

where $g(r)$ is the binary density-density correlation function, N is the aggregate particle number, R is the average size of aggregates (r.m.s. radius R, gyration radius R_g, etc.), and a is the size of monomers. From Eq. (1.51), one can see a main property of fractal aggregates: low average density and large density fluctuations within short-range distances. A direct consequence of such a property of binary density correlations of monomers inside a cluster is the already well-known[124] power law for the angular dependence of the static structure factor (normalized intensity) of light, x-ray, or neutron scattering $S(q) \sim (qR)^{-D_f}$.

The above-presented power laws for $g(r)$ and $S(q)$ are observed in the asymptotic sense only, when the scattering vector values of the probing irradiation satisfy the strong inequality $a \ll q^{-1} \ll R$.[125] In real experiments within the visible optical region the condition $qR \gg 1$ is usually not observed rigorously, since the average size of aggregates does not, as a rule, exceed 1 μm. In these cases, the character of the decrease in density correlations while approaching the cluster boundary becomes important. This decrease is described by using the so-called cutoff function $h(r/R)$, which is included in the complete correlation function $r^{D_f-3}h(r/R)$.[126,127] Several forms of $h(x)$, including the single exponential model $h(x) \sim \exp(-bx)$, were proposed in the literature (see the discussion and relevant citations in Ref. 127). Based on experimental data[127–130] and computer simulations,[126] the following approximation seems to be most appropriate for fractal aggregates:

$$h(x = r/R) \approx \exp(-bx^{\nu}), \quad \nu \approx D_f \approx 2, \tag{1.52}$$

where $b \sim 1$ for reaction-limited aggregates (RLCA),[123,124] and $b \sim 1/2$ for diffusion-limited aggregates (DLCA).[123,124]

In fractal cluster optics, the aggregates built from small metal nanoparticles are of special interest. Large density fluctuations and strong electrodynamic interaction of metal monomers result in specific linear[119] and nonlinear[131] optical effects. The readers are referred to the relevant paper[119] and books.[116,131] Further, the monomers of an aggregate are assumed to be the optically soft particles. In this case, the optical properties of fractal aggregates can be understood in terms of the mean field theory (MFT).[132] It is assumed in MFT that all aggregate monomers have the same dipole moments directed along an average exciting field. The monomer interaction results only in renormalization of the average amplitude of dipole moments exciting by the average (mean) field. The scattered fields are calculated with taking into account the geometrical phase shifts. Therefore, the angular scattering characteristics in the RDG and MFT theories coincide.

Let us consider a monodisperse ensemble of fractal clusters each containing N small spherical particles of radius a. If N_0 is the number of primary particles per unit of suspension volume, then the scattering intensity of nonpolarized light by an ensemble of clusters may be written in the following form:[127]

$$I(q) = I_0 N_0 N |B|^2 \left[\frac{3}{16\pi}\sigma_{1s}(1+\cos^2\theta)\right] G^2(qa)S(q), \tag{1.53}$$

where σ_{1s} and $G(qa)$ are the integral scattering cross section and the scattering form factor of primary particles [Eq. (1.34)], respectively. The factor $|B|^2$ gives a mean field theory correction,[132] independent from the scattering angle θ, to the scattering structure factor that describes in the RDG approximation the effect of spatial correlations of primary particles in a cluster

$$S(q) = \frac{1}{N}\left[1 + (N-1)\int_0^\infty x^2 p(x)\frac{\sin qRx}{qRx}dx\right], \tag{1.54}$$

where $x = r/R$;

$$R^2 = 2R_g^2 = \sum_{ij} |\mathbf{r}_i - \mathbf{r}_j|^2 p(\mathbf{r}_{ij}) \tag{1.55}$$

is the r.m.s. radius of a cluster; and $p(\mathbf{r}_{ij})$ is the probability density for finding a randomly chosen pair of cluster particles separated by a vector $\mathbf{r}_{ij} = \mathbf{r}_i - \mathbf{r}_j$. Note that $p(\mathbf{r}_{ij})$ is closely related to the density–density correlation function $g(r)$.[126]

At small scattering angles, when $qR \ll 1$, the angular dependence of scattering is represented by the universal Guinier's expansion[124]

$$S(q) \approx 1 - (qR_g)^2/3, \tag{1.56}$$

while the asymptotic transition to the fractal regime $S(q) \sim (qR)^{-D_f}$ is determined by polydispersity and the form of the cut-off function.[125,126]

Detailed calculations of the scattering structure factor for various models of fractal clusters, as well as the relevant references can be found in Ref. 125.

The mean field theory factor[132] B is given by

$$B = \left[1 - \frac{3}{2}(N-1)\sin\delta\exp(i\delta)P(kR)\right]^{-1}, \quad (1.57)$$

$$P(\alpha) = \int_0^\infty x^2 p(x) P_1(kRx)dx, \quad (1.58)$$

$$P_1(z) = \frac{\exp(iz)}{z^2}\left[\sin z - \frac{z}{3}G(z) + \frac{1-iz}{z}\left(G(z) - \frac{\sin z}{z}\right)\right]q, \quad (1.59)$$

where δ is the phase shift of the p wave for a small spherical particle.[132] Analytical properties of function $P(kR)$ are discussed in Refs. 127 and 132.

Let us now consider the integral functions that describe the spectral dependence of light extinction for nonabsorbing systems. Turbidity of an ensemble of fractal clusters may be represented in the following form:[127]

$$\bar{\tau} = (N_0/N)\sigma_s = \tau_1 F_s(kR), \quad (1.60)$$

where $\tau_1 = N_0\sigma_{1s}$ is the turbidity of an ensemble of primary particles, and $F_s(kR)$ is the enhancement (coherence) factor equal to the scattering cross section of a cluster σ_s normalized on $\sigma_{1s}N$

$$F_s = \sigma_s/N\sigma_{1s}. \quad (1.61)$$

For absorbing monomers, one has to introduce the extinction cross section σ_e as well as the corresponding extinction F_e and absorption F_a factors. Then Eq. (1.60) can be rewritten as

$$\bar{\tau} = \frac{N_0}{M}\sigma_e = N_0\sigma_{1e}F_e = \tau_1 F_e, \quad (1.62)$$

$$F_e = \sigma_e/N\sigma_{1e} \equiv F_s\varpi + (1-\varpi)F_s, \quad F_a = \sigma_a/N\sigma_{1a}, \quad (1.63)$$

where $\varpi = \sigma_{1s}/\sigma_{1e}$ is the monomer albedo. In MFT, the total cluster absorption is simply equal to the sum of monomer absorptions, i.e.,

$$\sigma_a = N\sigma_{1a}, \quad F_a = 1. \quad (1.64)$$

One should remember that Eq. (1.64) holds only for weakly absorbing monomers, and it violates for metal[119] or soot[115] nanoparticles. Absorption of small clusters gives a main contribution to their total extinction; therefore, Eq. (1.64) can be used for the extinction coherence factor as well.

For small kR or for small $N \sim (R/a)^{D_f}$

$$F_s(kR) \approx N\left[1 - \frac{1}{3}(kR)^2\right], \tag{1.65}$$

so that the turbidity of an ensemble of clusters increases by the factor of N as compared to τ_1. To calculate the enhancement factor for arbitrary argument kR, the general relationship of the mean field theory can be used[127]

$$F_s(kR) = F_{1s}(kR)|B|^2, \tag{1.66}$$

where $F_{1s}(\alpha)$ is the enhancement factor in the single intracluster scattering approximation

$$F_{1s}(\alpha) = \frac{1}{N} + \frac{3}{2}\left(1 - \frac{1}{N}\right)\mathrm{Im}P(kR). \tag{1.67}$$

The behavior of the scattering enhancement factor at great N or kR values is given by the following asymptotic expression:[127]

$$F_s^{(1)} = A_1(D_f) + A(D_{f2})(kR)^{D_f-2}, \quad kR \gg 1, \tag{1.68}$$

where the constants A_1 and A_2 are determined by the cutoff correlation function (e.g., see explicit expressions in Refs. 126 and 127). The physical meaning of Eq. (1.68) is as follows: the turbidity of a cluster suspension does not depend on the cluster size R or cluster particle number N if $D_f < 2$, logarithmically depends on R if $D_f = 2$, and increases as $(kR)^{D_f-2}$ if $D_f > 2$. Note that asymptotic expression (1.67) is not applicable in the geometrical optics limit. Transition to this limit for fractal aggregates has been discussed in Ref. 133.

Aggregation of particles always leads to polydisperse systems of clusters. The evolution processes of cluster-mass distribution $f(N)$ have by now been well investigated theoretically, experimentally, and by computer simulation.[123,134] The most important result of these studies is the proof of the dynamic scaling, according to which the shape of the distribution curve is described by the universal function $\psi(N/N_n)$, which is not explicitly time dependent. The time evolution of cluster-mass distribution is described only by specific dependences of distribution moments $N_n(t)$ for DLCA and RLCA clusters.[123] Detailed consideration of polydispersity effects in fractal cluster optics can be found in Refs. 125 and 127.

1.5 Extinction and Scattering by Plasmon-Resonant Particles

1.5.1 Localized plasmon resonance of small metal spheres

Over the recent decade, gold (and other metallic) nanoparticles (NPs)[135] have attracted significant interest as a novel platform for various applications to nanobiotechnology and biomedicine[136–138] because of convenient surface bioconjugation with molecular probes[139] and remarkable plasmon-resonant optical properties.[140,141] Recently published examples include applications of NPs such as biosensorics,[142] genomics,[143] clinical chemistry,[144] immunoassays,[145] immune response enhancement,[146] detection and control of microorganisms,[147] optical imaging of biological cells (including cancer cell imaging with resonance scattering,[148] optical coherence tomography (OCT),[149] two-photon luminescence,[150] and photoacoustic[151] techniques), cancer cell photothermolysis,[152] and targeted delivery of drugs or genetic and immunological substances.[153,154] In particular, there is great interest in the development of nanoparticle-based vectors that decrease the toxicity of free drugs and ensure targeted delivery directly to tumor cells.[155–157] It should be emphasized that almost simultaneously with the beginning of the use of GNPs in biomedicine, acute questions were raised about the biodistribution and circulation of NPs in the bloodstream, their pharmacokinetics and elimination from the organism, and their possible toxicity to the organism as a whole or at the level of cyto- and genotoxicity.[158]

The absorption and scattering spectra of metal nanoparticles exhibit resonance UV–vis bands, which are absent in macroscopic samples. The nature of these bands is determined by the collective behavior of conduction electrons in the light wave field. In the electron gas of metals, collective plasma oscillations can be excited at frequency ω_p in the visible region.[12] An elementary quantum with energy $\hbar\omega_p$ is called a plasmon.[140] Because plasmons are the result of quantization of classical plasma oscillations, their properties can be derived from Maxwell's equations.

Collective fluctuations of the electron density on the boundary of a usual dielectric with the positive permittivity and a metal with the negative permittivity are called surface plasmons.[140] The excitation of surface plasmons by light is called a surface plasmon resonance (SPR) for planar structures with travelling waves and a localized surface plasmon resonance (LSPR) for metal nanoparticles.[159] These excitations are related to evanescent surface electromagnetic waves, which are not necessarily localized on the interface.

To describe the optical properties of plasmon-resonant nanostructures, one can use various computational approaches,[160] including the T-matrix method.[161] An elementary classical description of a plasmon resonance (PR) in a small metal sphere is as follows.[159] The electric field of an electromagnetic light wave displaces the cloud of free electrons and produces uncompensated charges near a particle surface and corresponding returning forces. As mentioned above, the optical resonance related to these oscillations is called the LSPR. The term "surface" is used, first, because the returning force is caused by the polarization of the particle surface. Second, the radial component of the electric field inside the particle changes as r^{n-1}, so that it is stronger localized near the particle surface[12] (for $r \to a$) with increasing the mode order $n \gg 1$. In the general case, the eigenfrequency of such a "collective" oscillator does not coincide with the wave frequency and is determined by many factors, including the concentration and effective mass of conductive electrons, the shape, structure, and size of particles, interaction between particles, and the influence of the environment. However, for the elementary description, it is sufficient to use a combination of the usual dipole (Rayleigh) approximation and the Drude theory.[12] In this case, the absorption and scattering of light by a small particle are determined by its electrostatic polarizability α, which can be calculated by using the optical permittivity $\varepsilon(\omega)$ [or $\varepsilon(\lambda)$], where ω is the angular frequency and λ is the wavelength of light in vacuum. For a small sphere of volume V and radius a in a homogeneous dielectric medium with the permittivity ε_m, we have the following expressions for the extinction, absorption, and scattering cross sections:

$$C_{\text{ext}} = C_{\text{abs}} + C_{\text{sca}} = \frac{12\pi k}{a^3}\frac{\varepsilon_m \text{Im}(\varepsilon)}{|\varepsilon - \varepsilon_m|^2}|\alpha|^2 + \frac{8\pi}{3}k^4|\alpha|^2 \simeq 4\pi k \text{Im}(\alpha), \qquad (1.69)$$

where $k = 2\pi\sqrt{\varepsilon_m}/\lambda$ is the wave number in the surrounding medium, and the polarizability α is given by[162]

$$\alpha = \frac{\alpha_0}{1 + \varphi(x)a^{-3}\alpha_0}, \qquad \alpha_0 = \frac{3V}{4\pi}\frac{\varepsilon - \varepsilon_m}{\varepsilon + 2\varepsilon_m} = a^3\frac{\varepsilon - \varepsilon_m}{\varepsilon + 2\varepsilon_m}. \qquad (1.70)$$

Here, $x = ka$, the renormalization function $\varphi(x)$ is defined as[163] $\varphi(x) = 2 + 2(ix - 1)\exp(ix) \simeq -x^2 - i(2/3)x^3$, and Eq. (1.70) reproduces Meier and Wokaun's formula,[164] which has been generalized for spheroids by Moroz.[165]

Below, we will not distinguish the electrostatic polarizability from the renormalized polarizability, which accounts for radiative damping effects. In this approximation, the extinction of a small particle is determined by its absorption $C_{abs} = C_{ext} = 4\pi k\text{Im}(\alpha)$ and scattering contribution can be neglected.

As the particle size is decreased to the value comparable with the electron mean-free path ($a \sim l_e$), deviations of the phenomenological dielectric function $\varepsilon(\omega,a)$ of the particle from the bulk values $\varepsilon(\omega) = \varepsilon(\omega,a \gg l_e)$ can be expected. A general recipe for the inclusion of macroscopic tabulated data and size effects to the size-dependent dielectric function consists of the following.[140] Let $\varepsilon_b(\omega)$ be the macroscopic dielectric function, which can be found in the literature from measurements with massive samples. Then the size-dependent dielectric function of a particle may be written as

$$\varepsilon(\omega,a) = \varepsilon_b(\omega) + \Delta\varepsilon(\omega,a), \tag{1.71}$$

where the correction term $\Delta\varepsilon(\omega,a)$ takes into account the contribution of size-dependent scattering of electrons to the Drude part of the dielectric function

$$\Delta\varepsilon(\omega,a) = \varepsilon_b^{\text{Drude}}(\omega) - \varepsilon_p^{\text{Drude}}(\omega,a) = \frac{\omega_p^2}{\omega(\omega + i\gamma_b)} - \frac{\omega_{p,a}^2}{\omega(\omega + i\gamma_p)}. \tag{1.72}$$

Here, $\gamma_b = \tau_b^{-1}$ is the volume decay constant; τ_b is the electron-free path time in a massive metal; $\omega_{p,a}$ is the plasma frequency for a particle of diameter a (we assume below that $\omega_{p,a} \simeq \omega_p$);

$$\gamma_p = \tau_p^{-1} = \gamma_b + \gamma_s = \gamma_b + A\frac{v_F}{L_{\text{eff}}} \tag{1.73}$$

is the size-dependent decay constant equal to the inverse electron mean transit time $\gamma_p = \tau_p^{-1}$ in a particle; L_{eff} is the effective electron mean-free path; γ_s is the size-dependent contribution to the decay constant; and A is a dimensionless parameter determined by the details of scattering of electrons by the particle surface. A particular value of the scattering constant A is determined by the particle size, shape, structure, etc. For details, readers are referred to papers by Coronado and Schatz[166] and Moroz.[167]

One can see from the expressions presented above that the polarizability and optical cross sections can have a strong resonance under the condition

$$\varepsilon(\omega_{\max} \equiv \omega_0) = \varepsilon(\lambda_{\max}) = -2\varepsilon_m. \tag{1.74}$$

The PR frequency can be estimated from the elementary Drude theory for the permittivity of a bulk metal

$$\varepsilon(\omega) = \varepsilon_{ib} - \frac{\omega_p^2}{\omega(\omega + i\gamma_b)}, \tag{1.75}$$

where ε_{ib} is the contribution of interband electronic transitions; ω_p is the frequency of volume plasma oscillations of free electrons; and γ_b is the volume decay constant related to the electron mean-free path l_b and the Fermi velocity v_F by the expression $\gamma_b = l_b/v_F$. By combining the above equations, one can obtain the following expressions for the resonance plasmon frequency and wavelength

$$\omega_{\max} \equiv \omega_0 = \frac{\omega_p}{\sqrt{\varepsilon_{ib} + 2\varepsilon_m}}, \quad \lambda_{\max} \equiv \lambda_0 = \lambda_p\sqrt{\varepsilon_{ib} + 2\varepsilon_m}. \tag{1.76}$$

Here, $\lambda_p = 2\pi c/\omega_p$ is the wavelength of volume oscillations of the metal electron plasma.

Equation (1.74) determines the very first ($n = 1$) dipole resonance of a spherical particle. Except this dipole resonance, higher multipoles and corresponding multipole (quadrupole, etc.) resonances can be also excited. For each multipole mode, the resonance condition exists, which is similar to Eq. (1.74) and corresponds to the resonance of the quadrupole, octupole, and so on, polarizability contributions. For spherical particles, these conditions correspond to the resonance relations for the partial Mie coefficients[12] $\omega_n = \omega_p(\varepsilon_{ib} + \varepsilon^m(n + 1)/n)^{-1/2}$, where n is the mode (resonance) number. To understand the physics of LSPRs, it is important to distinguish two possible scenarios of excitation of multipole resonances. The first case corresponds to small but nonspherical particles of irregular or uneven shape, when the distribution of induced surface charges is strongly inhomogeneous and does not correspond to the dipole distribution. This inhomogeneous distribution generates high multipoles even in the case when the system size is certainly much smaller than the wavelength of light. Prominent examples are cubic particles[168] or two contacting spheres.[169] As mentioned earlier, the field distribution near the contact point is so inhomogeneous that multipole expansions converge very slowly or diverge at all. The second scenario of high multipole excitation is realized with increasing the particle size, when the transition from the quasi-stationary to radiative regime is realized, and the contribution of higher spherical harmonics should be taken into account in the Mie series (or another multipole expansion). For example, while the extinction spectrum for a spherical silver particle with a 30-nm diameter is completely determined by the dipole contribution and has one resonance, the spectrum of a 60-nm sphere exhibits a distinct high-frequency quadrupole peak in addition to the low-frequency dipole peak.

1.5.2 Metal nanorods

Gold nanorods (NRs) possess unique optical properties because of their tunable VIS–NIR dipole plasmon resonances, as demonstrated in pioneering experiments[170] and simulations.[171] Owing to the shape- and size-controlled scattering and absorption properties, the gold NRs have found promising applications in biomedical applications, cancer diagnostics, photothermal destruction of cancer cells or murine macrophages *in vitro* (see also the citations in Ref. 172), and *in vivo* flow cytometry of circulating cells.[151] New potential biomedical applications of metal NRs can be expected because of their enhanced PR sensitivity to the dielectric environment[173] and their orientation with respect to polarized incident light,[174] including laser orientation alignment and trapping,[175] strong light scattering oscillations induced by Brownian rotation,[176] unusual depolarization of scattered light,[177] and anisotropic properties of NRs embedded in a solid matrix[178] or deposited on a substrate.[179,180] Recent advances in synthesis[181] and functionalization[182] of gold NRs will stimulate fast progress in biomedical technologies using conjugates of gold NRs with biomolecules. For a comprehensive discussion of the chemistry, optics, and biomedical applications of metal NRs, the readers are referred to recent reviews.[162,174]

The optical properties of metal NRs can be understood in terms of an approximation of the particle shape by an prolate spheroid with semiaxes (*a*,*b*, *b*) and known analytic solution for the axial polarizability tensor

$$\alpha_{\|,\perp} \equiv \alpha_{a,b} = \frac{V}{4\pi} \frac{\varepsilon - \varepsilon_m}{\varepsilon_m + (\varepsilon - \varepsilon_m) L_{\|,\perp}}, \tag{1.77}$$

where the geometrical depolarization factors[12] satisfy the relation $L_{\|} + 2L_{\perp} = 1$, and $L = 1/3$ for spheres. Unlike spherical particles, expressions of type (1.76) now predict the appearance of two resonances corresponding to oscillations of electrons across and along the symmetry axis of a particle. In particular, the supersensitive tuning of the longitudinal resonance is explained by the modification of expression (1.76)

$$\lambda_{\max,\|} = \lambda_p (\varepsilon_{ib} + (1/L_{\|} - 1)\varepsilon_m)^{1/2}. \tag{1.78}$$

It follows from Eq. (1.78) that the dipole resonances of small gold or silver spheres ($5 \leq 2a \leq 30$ nm) in water are localized near 520 or 380 nm and do not depend on the particle size. By contrast, the longitudinal LPR of silver and gold nanorods can easily be tuned across the vis–NIR band by varying the aspect ratio, as $L_{\|}^{-1} \to \infty$ for needles and $L_{\|}^{-1} \to 0$ for thin disks, which predicts the strong red shift of the resonance with increasing the axial ratio of particles (when $L_{\|} \to 0$).

Figure 1.1 shows the extinction and scattering spectra of randomly oriented gold [Fig. 1.1(a) and (c)] and silver [Fig. 1.1(b) and (d)] NRs with the equivolume diameter $D = d_{ev} = 20$ nm and the aspect ratios $1 \leq e = a/b \leq 6$. The single-particle extinction and scattering properties are expressed in terms of the corresponding efficiency factors, i.e., the corresponding cross sections normalized to the geometrical cross section of an equivolume sphere. We see that the properties of particles depend very strongly on the metal nature. First, as the aspect ratio for gold rods is increased, the resonance extinction increases approximately by a factor of five, and the Q factor also increases. For silver, vice versa: the highest Q factor is observed for spheres, and the resonance extinction for rods is lower. Second, for the same volume and axial ratio, the extinction and scattering of light by silver rods are considerably more efficient. The resonance scattering factors for silver particles are approximately five times larger than those for gold particles. Third, the relative intensity of the transverse PR of silver particles with the aspect ratio above 2 is noticeably larger than that for gold particles, where this resonance can be simply neglected. Finally, principal differences are revealed for particles with the shape factor smaller than 2 [Fig. 1.1(c) and (d)].

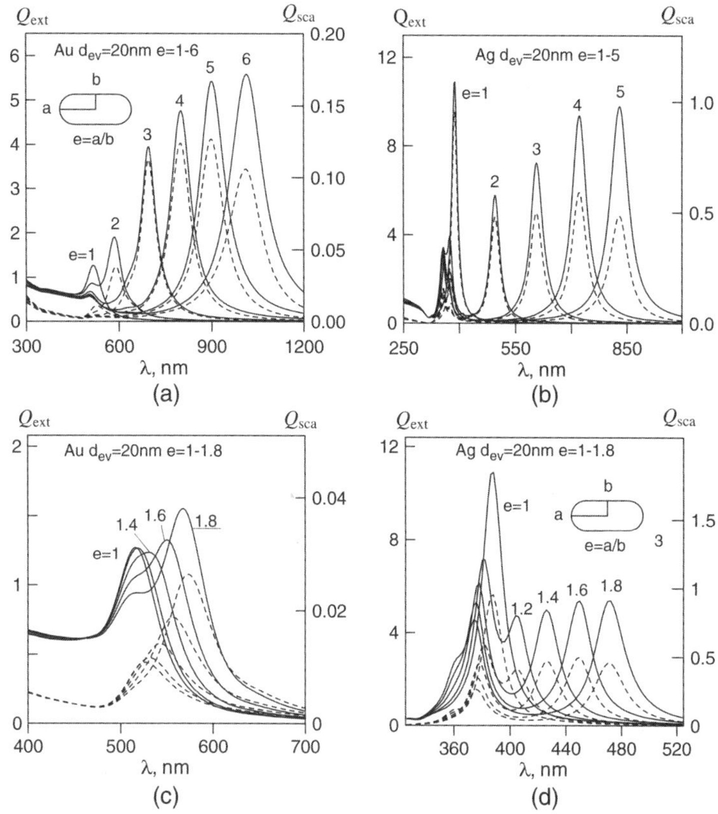

Figure 1.1 Extinction (solid curves) and scattering (dashed curves) spectra of randomly oriented (a) and (c) gold and (b) and (d) silver NRs (cylinders with semispherical ends). The equivolume diameter is 20 nm and the aspect ratio varies from 1 to 6. Panels (c) and (d) show the transformation of spectra at small deviations of the particle shape from spherical.

The resonance for gold particles shifts to the red and gradually splits into two bands with dominating absorption in the red region. The scattering band shifts to the red, and its intensity increases. For silver rods, the situation is different. The short-wavelength extinction resonance shifts to the blue, its intensity decreases, and it splits into two distinct bands. In this case, the intensity of the long-wavelength extinction band remains approximately constant; it is comparable with the short-wavelength band intensity and shifts to the red with increasing nonsphericity. The scattering and absorption spectra approximately reproduce these features.

1.5.3 Metal nanoshells

Gold nanoshells[183] are of great interest for biosensorics,[184] immunoassay, optical visualisation of biological objects, and laser photodestruction of cancer cells.[149] Note that the efficiency of nanoshells as photothermal labels can be related not only to their heating but to the formation of explosive vapor bubbles.[185] The existing protocols of synthesis of gold nanoshells with a silica core[183] allow the preparation of particles of diameter 80–400 nm with a comparatively narrow size distribution. The resonance scattering of light by such particles exceeds by more than an order of magnitude scattering from usual colloidal gold particles of diameters 10–50 nm. Very strong scattering allows one to observe individual nanoshells in the dark-field microscope and opens up the way to the development of high-contrast labels for visualization of biologically specific interactions of conjugates (particles with attached probe macromolecules) with target molecules.

The optical properties of model nanoshells in the form of a bilayer sphere with a dielectric core can be calculated easily by using various programs of the Mie solution type. The PR wavelength is determined mainly by the ratio of the shell thickness to the core diameter and by the dielectric functions of the core, shell, and environment. Although numerical analysis can be performed quite easily, it is desirable to have a simple analytic method for quantitative estimates of the polarizability of nanoshells and their dipole optical properties. Such a method has been developed in Ref. 186 to calculate an equivalent average permittivity of a multilayered sphere. The main advantage of our approach is that now we can apply usual expressions for resonances of spherical particles to multilayer spheres with the equivalent average permittivity as well.

The resonance tuning of NRs and nanoshells is related to the shape and structure of the particles, respectively. The idea of a combination of these two particles in the form of an ellipsoidal nanoshell was proposed in Ref. 187. Such a structure was later called nanorice.[188] The principle of the dipole equivalence can easily be modified for such particles as well.[161]

Consider an initial particle with known polarizability—e.g., a dielectric spheroid with the symmetry semiaxes $a_1 \geq b_1$ and the dielectric permittivity ε_1 (Fig. 1.2).

It does not matter whether the initial particle is homogeneous or layered. The goal is to find the polarizability of the same particle, which is now covered by a metallic shell with the dielectric function ε_2. To this end, the initial particle is placed in an auxiliary homogeneous dielectric medium with the permittivity ε_2 and is surrounded by an imaginary confocal spheroid (a_2, b_2), where $a_2^2 = a_1^2 + s^2$, $b_2^2 = b_1^2 + s^2$, and s is the shell thickness. Then the imaginary particle is replaced by an equivalent homogeneous particle with an average permittivity ε_{av}, which gives an equivalent dipole moment. Therefore, the following equation holds:[161]

$$\frac{V_1}{L_1}\frac{\varepsilon_1 - \varepsilon_2}{\varepsilon_1 + \varphi_1\varepsilon_2} = \frac{V_2}{L_2}\frac{\varepsilon_{av} - \varepsilon_2}{\varepsilon_{av} + \varphi_2\varepsilon_2}, \quad \varphi_i = L_i^{-1} - 1, \quad i = 1,2, \tag{1.79}$$

where V_i are the volumes and L_i are the geometrical depolarization factors of inner ($i = 2$) and outer ($i = 2$) spheroids, respectively.

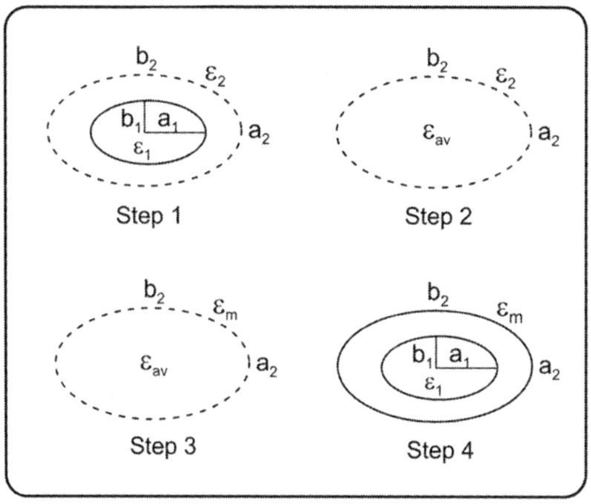

Figure 1.2 Scheme of the dipole equivalence method.[186] Step 1: The initial particle (ε_1, a_1, b_1) is embedded in an auxiliary medium with permittivity ε_2 and is surrounded by an imaginary confocal spheroid (a_2, b_2). Step 2: An equivalent particle (ε_{av}, a_2, b_2) is embedded in the auxiliary medium with permittivity ε_2. Step 3: The equivalent particle (ε_{av}, a_2, b_2) is embedded in the surrounding medium with permittivity ε_m. Step 4: The initial particle (ε_1, a_1, b_1) is covered by an outer layer (ε_2, a_2) and is embedded in the surrounding medium with permittivity ε_m. Steps 1 and 2 give identical dipoles in the auxiliary medium, whereas steps 3 and 4 give identical dipoles in the surrounding medium.

In fact, Eq. (1.79) defines the equivalent averaged permittivity, which gives identical dipoles in steps 1 and 2

$$\varepsilon_{av} = \varepsilon_2 \frac{1 + \varphi_2 f_{12}\alpha_{12}}{1 - f_{12}\alpha_{12}}, \quad \alpha_{12} = \frac{\varepsilon_1 - \varepsilon_2}{\varepsilon_1 + \varphi_1 \varepsilon_2}, \quad f_{12} = \frac{V_1 L_2}{V_2 L_1}. \tag{1.80}$$

In step 3, the equivalent particle of step 2 is placed in the surrounding medium with the dielectric permittivity ε_m. Finally, note that the dipole moment in step 3 equals that of step 4, thus giving the final analytical solution for the polarizability of the layered particle embedded in the surrounding medium

$$\alpha_{av}^{(2)} = \frac{V_2}{4\pi L_2} \frac{\varepsilon_{av} - \varepsilon_m}{\varepsilon_{av} + \varphi_2 \varepsilon_m}. \tag{1.81}$$

This solution has the same form as that for usual homogeneous spheroids. It is evident that the outlined procedure can be continued iteratively for an arbitrary number of confocal layers.

The LPR resonance condition $\mathrm{Re}(\varepsilon_{av}) \equiv \varepsilon'_{av} = -\varphi_2 \varepsilon_m$ can be recast as Eq. (1.74)

$$\varepsilon' \equiv \varepsilon'_2 = -2\varepsilon_m \varphi_\pm. \tag{1.82}$$

As the form of Eq. (1.82) is identical to that of Eq. (1.74), from Eq. (1.82), we immediately get the final solution for LPR wavelengths

$$\lambda_{\max} = \lambda_p [\varepsilon_{ib} + 2\varepsilon_m \varphi_\pm]^{1/2}. \tag{1.83}$$

In Eqs. (1.82) and (1.83), functions $\varphi_\pm$ are defined by the following expressions:

$$\varphi_\pm = \frac{p}{2} \pm \frac{1}{2}\left[p^2 - \varphi_2 \frac{\varepsilon_1}{\varepsilon_m} \frac{1 - f_{12}}{\varphi_1 - \varphi_2 f_{12}}\right]^{1/2} \tag{1.84}$$

$$p = \frac{1}{\varphi_1 - \varphi_2 f_{12}} \left(\frac{\varepsilon_1}{2\varepsilon_m}(1 + f_{12}\varphi_2) + \frac{\varphi_2}{2}(\varphi_1 + f_{12}) \right). \tag{1.85}$$

The positive and negative signs in Eq. (1.83) correspond to the LPRs of the outer and inner boundary shells, respectively. For spherical gold nanoshells, the short-wavelength inner resonance cannot be resolved in absorption or scattering spectra, whereas for silver nanoshells it is clearly seen around 350 nm.[189]

It is instructive to show that the general solution [Eqs. (1.83) through (1.85)] recovers all limiting cases of interest: metal spheroids, nanoshells, and spheres. For homogeneous spheroids, we set $f_{12} = 0$, so Eqs. (1.80) and (1.84) reduce to $\varepsilon_{av} = \varepsilon_2 \equiv \varepsilon$, $\varphi_+ = \varphi_2/2 = (L^{-1}_{\|,\perp} - 1)/2$. This yields the usual electrostatic polarizability tensor (1.77).

For spherical nanoshells, we have to set $\varphi_{1,2} = 2$ and $f_{12} = V_1/V_2 = 1 - f_s$, where f_s is the volume fraction of the metal shell. Then Eqs. (1.84) and (1.85) reduce to the previously published results

$$\varphi_\pm = \frac{1}{2}\left[p_1 \pm (p_1^2 - (\varepsilon_1/\varepsilon_m))^{1/2}\right], \quad p_1 = \frac{\varepsilon_1}{\varepsilon_m}\left(\frac{3}{4f_s} - \frac{1}{2}\right) + \frac{3}{2f_s} - \frac{1}{2}. \tag{1.86}$$

For thin shells, we get asymptotic solutions $\varepsilon = -2\varepsilon_m p_1$ and $\varepsilon = -\varepsilon_1/2p_1$. The first solution can be recast as Eq. (7.10) in Ref. 140, but the second expression differs from Eq. (7.11) in Ref. 140, which seems to be erroneous. The corresponding resonance wavelengths of a thin nanoshell are as follows:[162]

$$\lambda_{\max} \equiv \lambda_+ = \lambda_p \left(\varepsilon_{ib} + \frac{3}{f_s}\left(\varepsilon_m + \frac{\varepsilon_1}{2}\right) \right)^{1/2}, \tag{1.87}$$

$$\lambda_{\min} \equiv \lambda_- = \lambda_p \left(\varepsilon_{ib} + \frac{2\varepsilon_1 \varepsilon_m f_s}{3(\varepsilon_1 + 2\varepsilon_m)} \right)^{1/2}. \tag{1.88}$$

Again, the maximal LPR wavelength can be tuned by varying the core/shell ratio $0 < f_s \leq 1$ ($f_s = 1$ for a homogeneous sphere and $f_s \to 0$ for a thin shell). In the limit $f_s \to 0$, the first resonance wavelength $\lambda_{\max} \to \infty$ and $\lambda_{\min} \to \lambda_p \varepsilon_{ib}^{1/2}$. By contrast, Eq. (7.11) in Ref. 140 reduces to $\lambda_{\min} \to \lambda_p(\varepsilon_{ib} + \varepsilon_1/2)^{1/2}$.

Finally, the properties of thick shells are close to those for homogeneous spheres. Indeed, by setting $f_s \to 1$ in Eqs. (1.87) and (1.88), we arrive at the following expressions:

$$\lambda_{\max} = \lambda_p(\varepsilon_{ib} + 2\varepsilon_m)^{1/2}, \quad \lambda_{\min} = \lambda_p\left(\varepsilon_{ib} + \frac{\varepsilon_1}{2}\right)^{1/2}. \tag{1.89}$$

One additional important note is in order here. The above equations determine the very first ($n = 1$) dipole resonances of particles. With an increase in the particle size, higher multipoles and corresponding multipole resonances can also be excited in larger particles. For example, in the case of spherical particles, these conditions correspond to the resonance relations for the partial Mie coefficients[161] $\lambda_n = \lambda_p(\varepsilon_{ib} + \varepsilon_m(n+1)/n)^{1/2}$, where n is the mode (resonance) number. With an increase in the sphere size, the multipole wavelength also increases.[190] The multipole modes in thin spherical nanoshells can be written as follows:[161]

$$\lambda_{n,\max} = \lambda_p[\varepsilon_{ib} + 2\varepsilon_m \varphi_n]^{1/2}, \qquad \varphi_n = \frac{1}{2}\left[p_n \pm (p_n^2 - (\varepsilon_1/\varepsilon_m))^{1/2}\right], \tag{1.90}$$

with the multipole parameter p_n being defined by the following expression:

$$p_n = \frac{q_n^2 + (\varepsilon_1/\varepsilon_m) + q_n(1 - f_s)[1 + (\varepsilon_1/\varepsilon_m)]}{2q_n f_s}, \quad q_n = \frac{n+1}{n}. \tag{1.91}$$

At the lowest dipolar order $n = 1$, Eq. (1.91) reduces to Eq. (1.86).

1.5.4 Coupled plasmon resonances: bisphere and linear chain examples

Along with the optics of individual PR particles, the collective behavior of the interacting PR particles is of great interest for nanobiotechnology.[191] The analysis of its features includes the study of various structures, beginning from one-dimensional chains with unusual optical properties.[192] Another example is the optics of two-dimensional (2-D) arrays,[193] in particular, clusters of spherical particles on a substrate and 2-D planar ensembles formed by usual gold or polymer-coated spheres.[172] The unusual properties of monolayers of silver nanoparticles in a polymer film[194] and on a glass substrate[195] have been recently discovered. A review of 3-D cluster optics can be found in Refs. 140, 162, 169, and 191.

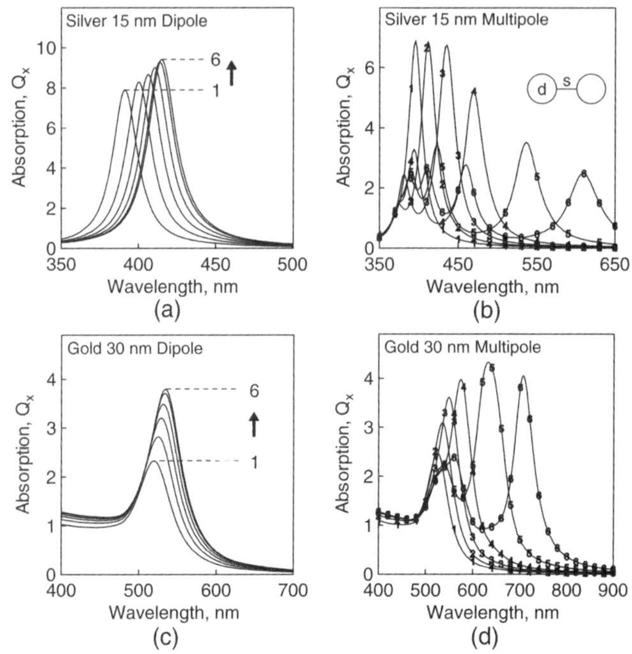

Figure 1.3 Absorption spectra at an incident light polarization parallel to the (a) and (b) silver and (c) and (d) gold bisphere axis ($x \equiv \|$). Calculations by the (a) and (b) dipole approximation and (c) and (d) GMM multipole codes for particle diameters $d = 15$ (silver) and 30 nm (gold) and the relative interparticle separations $s/d = 0.5$ (1), 0.2 (2), 0.1(3), 0.05 (4), 0.02 (5), and 0.01 (6).

In this chapter, we restrict our consideration of collective plasmons by only one instructive example: gold and silver bispheres of a diameter d separated by a variable distance s.

Figure 1.3 shows the absorption spectra calculated by the dipole and multipole approaches for two particle diameters $d = 15$ nm (silver) and 30 nm (gold) (for data for 60 nm, see Ref. 169). The numbers on the curves correspond to the relative interparticle separations s/d from 0.5 to 0.01. When the interparticle separation satisfies the condition $s/d \geq 0.5$, the absorption efficiencies approach the single-particle quantities so that the dipole and multipole calculations give identical results. However, the situation changes dramatically when the relative separation s/d is about several percent. The exact multipole approach predicts the well-known enormous theoretical[191] and experimental[196] red-shifting of spectra and their splitting[197] into two modes, whereas the dipole spectra show only a minor red shift.

In the case of silver bispheres [Fig. 1.3(a) and (b)], the resonance light scattering of 60-nm clusters exceeds the resonance absorption so that any comparison between the dipole and the multipole approaches becomes incorrect unless both scattering and absorption are taken into account for the total extinction. That is why we show only the calculated data for silver

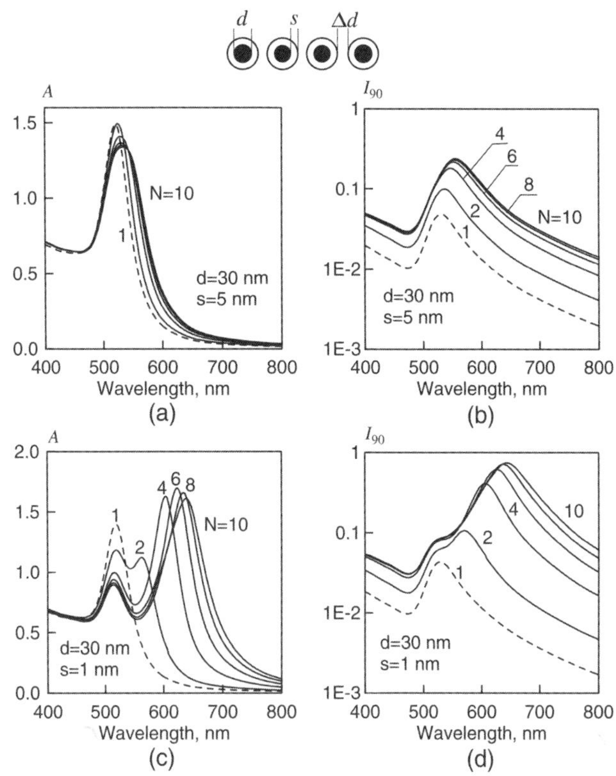

Figure 1.4 Extinction and scattering spectra of randomly oriented linear chains of two-layered gold (core)/dielectric shell particles with different particle number $N = 1$–10. Calculations for the core diameter $d = 30$ nm, (a) and (b) shell thickness $s = 5$ and (c) and (d) 1 nm, and separation distance $\Delta d = 0$.

nanospheres with $d \leq 60$ nm. At moderate separations ($s/d > 0.05$), the independent-particle spectrum splits into two modes; therefore, the data of Fig. 1.3(a) and (b) are in great part analogous to those of Fig. 1.3(c) and (d) and need no additional comment. However, at smaller separations $s/d < 0.05$, we observe the appearance of four plasmon resonances related to the quadrupole and the next high-order multipole excitations.

Strong binary coupling also is observed for linear chains of two-layer conjugates. Figure 1.4 shows the dependence of extinction $A(\lambda)$ and scattering $I_{90}(\lambda)$ spectra of randomly oriented linear chains built from two-layer particles with gold core and a dielectric shell. The number of particles is the variable parameter of curves. It is worth noting a principal difference between the spectra of densely packed [s = 1 nm, strong binary coupling, Fig. 1.4(c) and (d)] and rare chains [s = 5 nm, weak binary coupling, Fig. 1.4(a) and (b)].

The scattering intensity in Fig. 1.4(b) and Fig. 1.4(d) is expressed in the same scale and was calculated for the same gold concentration. In agreement with our previous observations,[191] the extinction spectra do not change

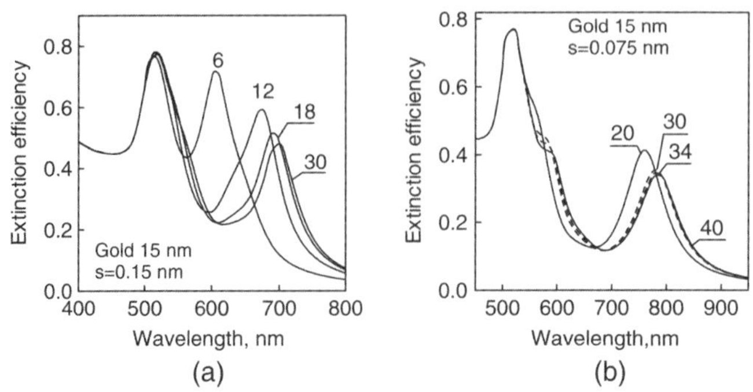

Figure 1.5 Extinction spectra of 15-nm randomly oriented gold bispheres in water, calculated by the exact T-matrix method. The sphere diameter is 15 nm, the separation distances between spheres are (a) 0.15 nm and (b) 0.075 nm, and the numbers near the curves designate the multipole orders that have been included in the single-particle field expansions of coupled equations.

essentially for small and rare aggregates [Fig. 1.4(a)], while the PR peak in the scattering spectra increases significantly due to constructive far-field interference. However, the peak position does not shift significantly.

For a thin dielectric shell, the transformation of extinction and scattering spectra are due to both the electrodynamic coupling and the far-field interference. Again, we observe a rapid saturation of particle-number effect. This can be interpreted as manifestation of an effective electrodynamic interaction between monomers, which belong to a finite conjugate group. Other examples of aggregated particles and their optical properties can be found in our works.[161,191]

The dependence of the extinction spectra on the total multipole order N_M is shown in Fig. 1.5. Remember once again that N_M means the maximal order of VSH in the coupled equations rather than the number of multipoles involved in the final calculations of optical characteristics. According to these computer experiments, one has to include extra-high single-particle multipole orders (up to 30–40) into coupled equations to calculate correctly the extinction spectra of 15-nm gold spheres separated by a 0.5–1% relative distance s/d. The extinction spectra were calculated by the exact T-matrix code for randomly oriented gold bispheres in water. It is evident from Fig. 1.4 that the convergence problems are related to calculations of the red-shifted resonance peak, which can be reproduced correctly if we retain the VSH with the order of about 30 in the case of 0.15 nm (1%) separation. For smaller separations [e.g., $s = 0.075$ nm, Fig. 1.5(b)], we note the appearance of a quadrupole resonance near 600 nm that can also be reproduced accurately only if we include multipoles of the 40th order into coupled equations.

The need to retain high multipoles for small spheres, which are themselves well within the dipole approximation, seems to be somewhat counterintuitive. It should be emphasized that the final calculations involve a rather small number of multipoles (as a rule, less than 6). However, to find these small-order contributions correctly, one needs to include many more multipoles into coupled equations such as given by Eq. (1.44). The physical origin of this unusual electrodynamic coupling was first established by Mackowski[198] for small soot bispheres. He showed that the electric-field intensity can be highly inhomogeneous in the vicinity of minimal separation points between the spheres even if the external filed is homogeneous on the scale of bisphere size. Evidently, the same physics holds in our case, as the imaginary part of the dielectric permittivity is the main parameter that determines the spatial electric-field distribution near the contact bisphere point.

1.6 Tissue Structure and Relevant Optical Models

Biological tissues are optically inhomogeneous and thus scattering processes with light propagation play a substantial role. The scattered radiation contains information about sizes and shapes of tissue structural elements, their orientation, optical constants, and other parameters. To extract this information and interpret experimental results on light scattering, one needs to develop an appropriate optical model for a tissue and on its basis to solve the inverse problem of radiation propagation and scattering in the medium.

There is no rigorous theory describing light absorption and scattering by tissues and cell structures. However, important information on their optical properties may be obtained by studying the appropriate model objects. The complicated structure of tissues, high concentration of scattering particles, variability of their sizes, shapes, and optical constants makes the problem of constructing an adequate optical model quite cumbersome. It may be assumed that models, being the basis of the scattering equation, would be best matched to real objects. The main peculiarities of scattering effects are essentially controlled by simple factors. These factors for a single particle are[35] the size and shape of the particle, the internal structure of the particle, and the inherent inhomogeneity of the particle.

1.6.1 Continuous and discrete models of tissues

Two approaches may be considered for tissue modeling, namely, tissue's representation as a medium with the continuous random spatial distribution of optical parameters or as a discrete ensemble of scatterers. The choice of one or another approach is dictated both by the features of the tissue under study and, to a considerable extent, the kind of light scattering characteristics that are to be obtained in the result of the modeling. The microstructure of biological cells and tissues is complicated and difficult to be characterized quantitatively. The microstructure of a cell or tissue is revealed in light or electron micrographs as spatial density fluctuations that are the result of the stain density variations of different structural components.[199] Most biological tissues are composed of structures with sizes that span a wide range. In the absence of a single dominant size, representation of a tissue as an ensemble of isolated and independent particles fails. Light scattering in these systems is possible to describe using the idea of a random continuum of index of refraction inhomogeneities with a varying spatial scale.[200] Similar problems have been investigated for light propagation through atmospheric turbulence[201] and for reflection by rough surfaces.[202]

The tissue representation as inhomogeneous medium with continuous spatial fluctuations of the index of refraction is, e.g., employed to examine the speckle structure of scattered radiation. The interest for the scale fluctuation properties of a speckle pattern appearing under the interaction of coherent radiation with the optically inhomogeneous objects is due to a possibility of extraction of information on the structure of scattering objects.[203]

Intriguing results for such a tissue model have been discussed by Schmitt et al.[200,204] In particular, the phase contrast microscopy was used to show the structure of the refractive index inhomogeneous in the tissues of mammals to be similar to the structure of frozen turbulence in a number of cases.[6] This result is of fundamental importance for understanding the peculiarities of radiation transfer in the tissue, and it may be a key for the solution of the inverse problem on tissue structure reconstruction. The above approach is usually exploited for tissues where there are no pronounced boundaries between elements and which are featured by significant heterogeneity. Figure 1.6 is the electronic micrograph of a rat liver represented as an example.[205]

The process of scattering at these structures may be described under certain conditions using the model of a phase screen.[206]

The second approach to tissue modeling is its representation as a system of discrete scattering particles. This chapter deals mainly with different aspects of radiation propagation in the tissues on the basis of the model of discrete scatterers. This model is to be advantageously used to describe the angular dependence of the polarization characteristics of the scattered radiation.[207]

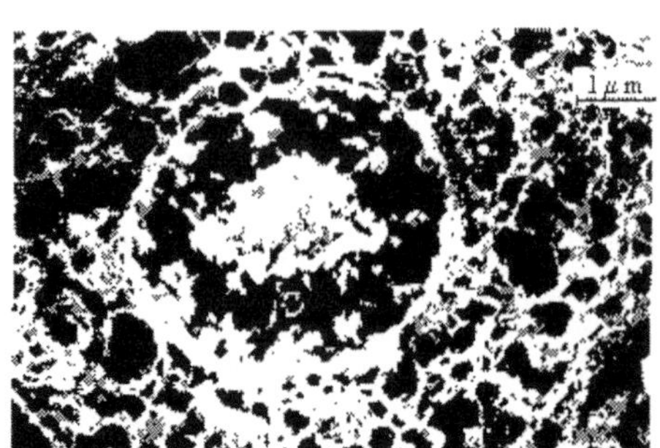

Figure 1.6 Electronic micrograph of a rat liver.[205]

Blood is an example of the most important biological disperse system that entirely corresponds to the model of discrete scatterers.

To properly choose a theoretical method and approximations for calculating the scattering characteristics of a particle system it is necessary to regard the main features of the tissue structure and namely, the shape and the size range of particles, their index of refraction and concentration. Consider some examples of discrete scatterer models for some tissue types.

1.6.2 Shape and sizes of particles in discrete tissue models

A biological medium is most often modeled by an ensemble of homogeneous spherical particles. This has a certain meaning since many cells, microorganisms and blood corpuscles are close in shape to spheres or ellipsoids. A system of noninteracting particles is the simplest version of this model. The solution of the problem of the diffraction of electromagnetic waves at a spherical particle is well known; this is the Mie solution.[12] The development of this model involves a consideration of the spherical particles structure,[15] namely, that of multilayered spheres and spheres with radial nonhomogeneity, anisotropy, and optical activity.

For tissues that have fiber structures, a system of long cylinders with different concentration is the most appropriate model. Muscular tissue, the cornea, and the sclera belong to these tissues formed essentially by collagen fibrils. The solution of the problem of diffraction at a single homogeneous or multilayered cylinder scatterer is also well known.[12]

The sizes of most biological cells and tissue structure elements are varied in a range from a few tens of nanometers to hundreds of micrometers.[9,208] Some examples are listed below. Bacteria usually have the size of a few micrometers.[208] The blood corpuscles (erythrocytes, leukocytes, and platelets) exhibit the following parameters. A normal erythrocyte in plasma has the shape of a concave–concave disc with a diameter varying from 7.1 to 9.2 μm, its thickness in its center 0.9–1.2 μm and 1.7–2.4 μm, respectively.[9] Leukocytes are formed like spheres of 8–22 μm diameter,[209] and platelets are thin discs with diameters from 2 to 4 μm. Most other human cells are in the range of 5–20 μm in diameter.[38]

The retinal nerve fiber layer comprises bundles of unmyelinated axons that run across the surface of the retina. The cylindrical organelles of the retinal nerve fiber layer (axonal membranes, microtubules, neurofilaments, and mitochondria[210]) as seen by electron microscopy were modeled as parallel cylindrical arrays in order to gain insight into their optical properties.[211] Axonal membranes, like all cell membranes, are thin (6–10 nm) phospholipid bilayers that form cylindrical shells enclosing the axonal cytoplasm. Axonal microtubules are long tubular polymers of the protein tubulin with an outer diameter of ≈25 nm, an inner diameter of ≈15 nm, and a length of 10–25 μm. Neurofilaments are stable protein polymers with a diameter of ≈10 nm.

Mitochondria are ellipsoidal organelles containing densely involved membranes of lipid and protein. They are 0.1–0.2 μm thick and 1–2 μm long. In the epidermal layer, the cells are large (average cross-sectional area ≈80 μm^2) and quite uniform in size.[200]

Hence, for most tissues the size parameter of the particle $x = 2\pi r/\lambda$ (where r is the particle radius, λ is the radiation wavelength) is varied in a range of $1 < x < 100$ in the visible region.

The size distribution of scattering particles in biological objects may be both essentially monodisperse as, e.g., in the optical eye tissues, and quite broad. There is no universal distribution function in dimension that would describe all tissues with equal adequacy. Gaussian, gamma, and power distributions are used as a function of particle size distribution in the optics of disperse systems.[9] It was particularly stated that scatterers in the epidermal layer of the skin exhibit a lognormal size distribution, whereas the spatial fluctuations in the index of refraction of dense fibrous tissues, such as the dermis, follow a power law.[200]

In cases where scattering at particles of complex shape is needed, different procedures are applied to calculate scattering at nonspherical particles, e.g., the method of T-matrices and others[15] (see above). The modeling of complexly shaped particles by spherical scatterers aggregates is possible. The light scattering peculiarities for a cell modeled by a spherical particle with inclusions are analyzed in Ref. 212. All of the above present a vast area of research; the specific nature of tissues is, however, not exhausted by these peculiarities.

1.6.3 Optical constants of tissues, heterogeneity, and optical softness

The measurement of the index of refraction for tissues and their components is one of the actual problems of tissue optics. Such studies have been performed for a comparatively long time,[213] but there is not adequate information in the literature even about an averaged index of refraction for different tissues.[9]

The matter surrounding the scatterers (intercellular liquid and cytoplasm), the so-called ground substance is composed mainly of water with salts and organic components solved in it. The ground matter index is usually taken as $n_0 = 1.35$–1.37. Scattering particles themselves (organelles, protein fibrils, membranes, protein globules) exhibit a higher density of proteins and lipids in comparison with the ground substance and thus a greater index of refraction $n_1 = 1.43$–1.47. This implies that structures with binary fluctuations of the index of refraction are the simplest model of tissue.

Absorption for most tissues in the visible region is insignificant except the absorption bands of blood hemoglobin and some other chromophores. The absorption bands of protein molecules are mainly in the near-UV region. Absorption in the IR region is essentially defined by water contained in tissues.

Many examples may be cited that confirm these regularities.[15] Brunsting and Mullaney[214] measured the indexes of refraction for cytoplasm and nucleous ovary cells of hamster and obtained the values $n_{\text{cytoplasm}} = 1.37$ and $n_{\text{nucleus}} = 1.39$. Schmitt and Kumar[215] obtained the indexes of refraction of 1.4–1.45 for the structural fibers of tissue and cell nuclei and that of 1.36 for the cell cytoplasm of different tissues. The nuclei and cell membrane of fibroblasts have an index of refraction of 1.48, the cytoplasm has an index of refraction of 1.38, and the averaged index of a cell is 1.42.[216] The index for *Bacillus subtilis* spores has a value of about 1.5, and its imaginary part is smaller than 0.01 in a wide spectral range.[217] The collagenous fibrils of cornea and sclera have a index of refraction of 1.47, and the refractive index of the ground matter is 1.35.[218]

The real part of the erythrocyte index with respect to plasma $m = 1.041 - 1.067$ ($\lambda = 600$ nm). Its imaginary part is varied within 10^{-2}–10^{-5} ($\lambda = 350$–1000 nm). The relative index for the human lymphocytes varies in the range of $1.01 < m < 1.08$.[219] The optical parameters of platelets have not been studied thoroughly; they are, however, referred to weakly absorbing soft particles (for the wavelengths bigger than 600 nm). Additional information on the refractive index of biological cells may be found in Ref. 220. A great deal of data on optical constants for various tissues is given in Ref. 9.

The above examples are evidence that tissue inhomogeneities have sizes comparable with visible or NIR wavelengths and a small relative index of refraction, hence they are to be considered as optically soft. This enables different approximation methods described in Sections 1.1–1.5 to be used for calculation.

1.6.4 Anisotropy of tissues

Many biological structures are optically anisotropic. A number of tissues contain single-axis birefringence structures.[221] For bone, muscular, and skin tissues, these are mineralized (hydroxylapatite crystals), myosinous and collagenous bundles, respectively. For the muscular tissue, the packing density of bundles in a multifractal is higher than that for the bone, and myosinous fibers in them show a smaller birefringence ($\delta n \approx 10^{-3}$). The predominant orientation of collagenous fibers in different regions of the cornea results in shape birefringence and dichroism.[222] The orientation structure of collagenous bundles of the skin dermis is extremely varied even for a small thickness of a microscopic section. Besides linear birefringence and dichroism, many biological objects show optical activity.

1.6.5 Volume fraction of the particles

An important parameter for the model of discrete particles that defines its optical properties is the packing density or, in other words, volume fraction of the particles. This characteristic may be experimentally found using electron microscopy. This is one of the straightforward approaches based on measurement of the area occupied by an element of a particular size. The selected region is viewed as a slice through a 3-D surface that is divided into subcubes for counting the number of elements with the same size. The volume fraction occupied by the elements is proportional to the number of subcubes counted. Employing this procedure and the electron microscopy data one would calculate volume fraction f_i occupied by the particles with size d_i. Unfortunately, systematic errors may take place that may lead to different values of the volume fraction evaluated for varying thicknesses of the section. The Holmes effect to Delesse's theorem has been used to eliminate these errors.[223] Estimations for a volume fraction occupied by scattering particles may also be calculated by comparing the weights of a native tissue and dry rest with the known density of separate tissue components.

As a rule, the volume fraction occupied by the scattering particles in such tissues as muscle, cornea, sclera, and eye lens covers from 20% to 40%. Conventionally the whole blood contains approximately 5×10^6 erythrocytes in 1 mm^3. Erythrocytes make up to 40% of the blood volume. The volume fraction f of erythrocytes in the blood is called the hematocrit H. For normal blood, $H = 0.4$. The remaining 60% of the blood volume is the plasma–an essentially transparent water solution of salts. The concentration of other scatterers would amount from percent fractions to a few dozens of percents.

1.6.6 Effects of spatial ordering

A substantial role in tissues is played by the spatial organization of particles forming them. As said above, with very small packing densities, we deal with incoherent single or multiple scattering at independent particles. If the volume fraction occupied by the particles is 0.1 or more, coherent concentration effects appear. The concentration of scattering particles is sufficiently high for most tissues, therefore spacing between individual scatterers is comparable with their sizes. If the particle-size distribution is close to the monodisperse one, then such dense packing entails a spatial degree of order on the arrangement of the particle. The effects of spatial ordering should be taken into account when constructing the adequate optical model of such tissues.

Spatial ordering is of utmost importance for optical eye tissues, nevertheless, aspects considered below are more or less inherent in all tissues, particularly in the skin or muscles. The spatial degree of order of densely packed systems of scattering particles ensures a high transmission of cornea and eye lens that is responsible for our seeing the surrounding world. The cornea would scatter to 90% of light incident and we would see essentially nothing if all its fibers were arranged randomly and scattered light independently. The real situation is, however, different.

The effect of multiple scattering does much to light propagation in dense disperse systems. The propagation of light in densely packed disperse biostructures shows a number of intriguing peculiarities that enable the optical eye tissues to be considered as natural photon crystals. Like artificial photon crystals presenting structures with periodical variations of the index of refraction with characteristic scales on the order of light wavelength, statistically ordered biostructures demonstrate high transmission spectral regions and a band of frequencies for which the propagation of electromagnetic waves is forbidden. Bandgaps are due to the effects of light scattering at a system of ordered particles. The positions and depth of these suppression bands highly depends on the size, refractive index, and spatial arrangement of scattering particles.

The importance of the spatial degree of order is conveniently illustrated by comparing the optical properties of the cornea with that of sclera (Fig. 1.7). Now let's consider the structure of cornea and sclera in more detail.[9,224–231]

The cornea is the frontal section of the eye's fibrous capsule, its diameter is ≈ 10 mm. The sclera is a turbid opaque tissue covering nearly 80% of the eye and serving as a protective membrane providing for, along with the cornea, counteraction against internal and external forces and thereby keeping the shape of the eye.

Both tissues are composed of collagenous fibrils immersed in a mucopolysaccharide ground substance. The collagenous fibrils have a shape

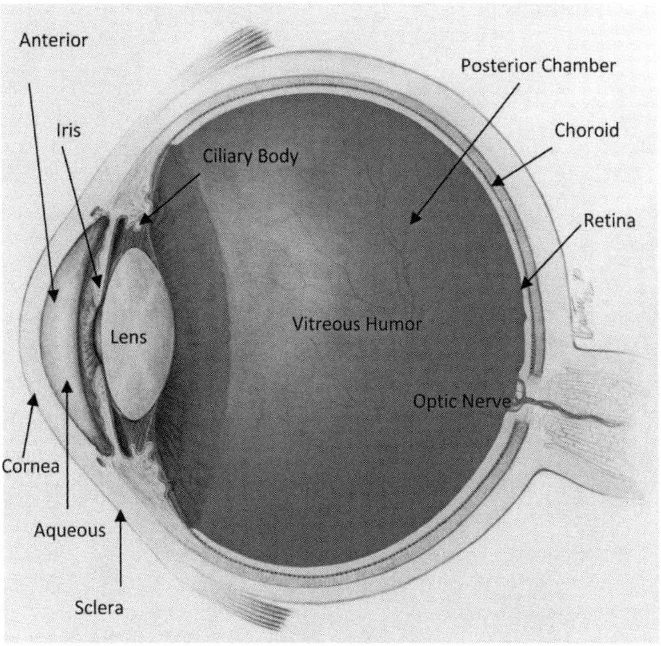

Figure 1.7 The human eye. The cornea is the transparent tissue of the anterior surface of the eye. The sclera is the white of the eye, and it is opaque.[9] National Eye Institute, NIH (Figure NEA04). See color plates.

close to that of a cylinder. They are packed in bundles like lamellae. All fibers are aligned within every lamella nearly in parallel with each other, the lamella plane and tissue surface. The indexes of refraction for the fibers and the ground substance surrounding them differ markedly from others and amount to 1.470 and 1.345, respectively.

The cornea fibril diameters varied from 20 to 31 nm for different species of animals, but their variations are less than 10% for each concrete species of animals or men, i.e., the fibrils are quite uniform in diameter (Fig. 1.8). The spacing between the fibril axes is 50–60 nm.

The collagenous fibril diameters of the sclera are widely spread from 25 to 230 nm. Spacing between the centers of adjacent fibrils in the bundle is about 100–300 nm (Fig. 1.8).[9,224,230] As has been mentioned, the collagenous fibrils are packed in bundles and lie in them nearly parallel with each other, but not so regularly as in the cornea. Furthermore, groups of fibrils in a certain bundle are separated from each other by comparatively large regions that are randomly distributed in space.

While both tissues are composed of similar molecular components, they have different microstructures and thus very different physiological functions. The sclera of the eye is opaque to light, scattering at almost all wavelengths of visible light, thus appearing white. The cornea is transparent, allowing for more than 90% of the incident light to be transmitted. The collagen fibrils in the cornea have a much more uniform size and spacing than those of the sclera,[225] resulting in a greater degree of spatial order (hexagonal quasi-crystal) in the organization of fibrils in the cornea as compared with the sclera (Fig. 1.8).[226,227]

It is believed that the spatial order and small size and spacing of the collagen fibrils are responsible for the transparency of the cornea, while the

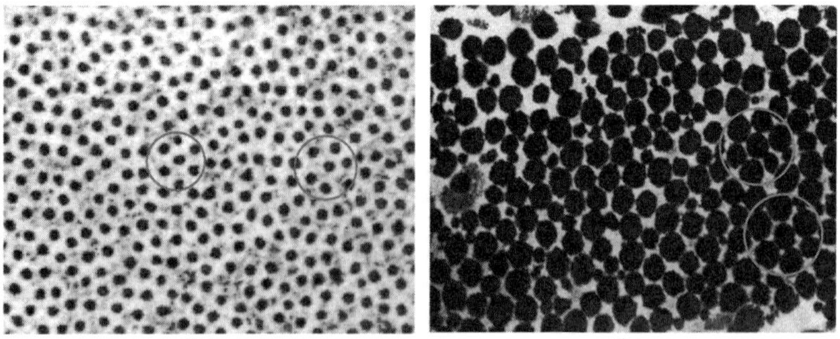

(a) (b)

Figure 1.8 The high-magnification electron micrographs of the human (a) cornea ($\times$32,000) and (b) sclera ($\times$18,000) showing a cross section of the collagen fibrils [dark circular areas embedded in a mucopolysaccharide ground substance (bright areas)]: for (a) cornea collagen fibrils have a uniform diameter and are arranged in the same direction within the lamellae and form a hexagonal quasi-crystal; (b) scleral collagen fibrils, which display various diameters, and thus more randomly packed; however, they can be locally quasi ordered.[9,230]

lack of spatial order and large size and spacing of the fibrils are responsible for the opacity of the sclera.

The transmittance of a densely packed disperse system can be calculated using the radial distribution function $g(r)$, which statistically describes the spatial arrangement of particles in this system. The function of the radial distribution of scattering centers $g(r)$ may be calculated for some tissues on evidence derived from electron microscopy.

The technique of experimental $g(r)$ determination involves the counting of the number of particles, placed at a specified spacing from an arbitrarily chosen initial particle, followed by its statistical averaging over the whole ensemble. In a 2-D case, the particle number ΔN at the spacing from r to $r + \Delta r$ is related to function $g(r)$ by the following equation:

$$\Delta N = 2\pi\rho g(r) r \Delta r, \tag{1.92}$$

where ρ is the mean number of particles for a unit area.

The radial distribution function $g(r)$ was first found for the cornea by Farrell et al.[228] using experimental data. Figure 1.9(a) depicts the typical result for one of the cornea regions, as obtained by determining the ratio of local to mean density of centers as a function of radii taken about 700 fibril centers. The results of similar studies for the eye sclera[229] are illustrated in Fig. 1.9(b); the data are calculated for several regions of sclera using electron micrographs from Ref. 230, averaging about 100 fibril centers.

Function $g(r)$ for the sclera was obtained on the basis of the spatial distribution of fibril centers neglecting discrepancy in their diameters. A noisy

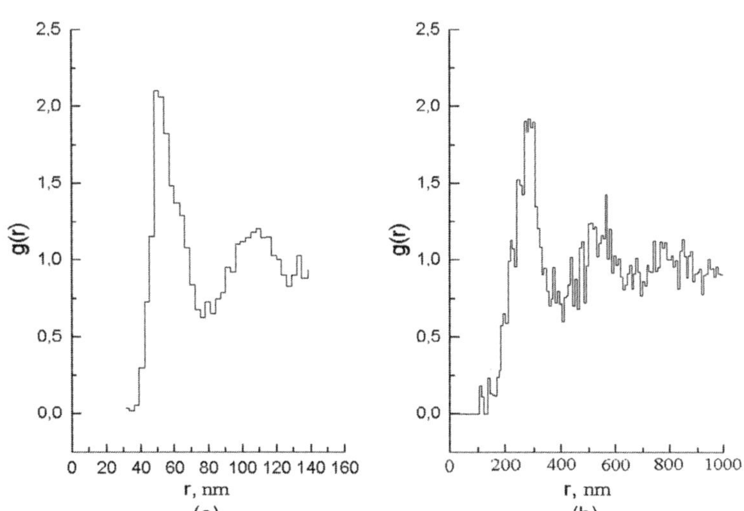

Figure 1.9 Histogram representation of the radial distribution functions $g(r)$, as obtained using the electron microscopy for (a) the rabbit cornea[228] and (b) the human sclera[229].

curve is elucidated by a small volume of statistical averaging. In spite of assumptions made earlier about a noncorrelated distribution of the sclera fibers in space,[231] the results obtained using electron microscopy attest to the presence of near-order degree in the sclera; albeit this degree of order is less pronounced as that in the cornea. The low value of interference interaction contribution is due to both a smaller degree of order in the arrangement of the sclera fibers and a strong discrepancy in the scattering amplitude at different-sized fibers within the band inherent in the sclera.

Similar results of a comparative examination of the cornea and the sclera were obtained using the method of spatial frequency analysis in Refs. 226 and 231. A linear log–log plot of the Fourier spectrum is an indication of the power-law relationship between the amplitude and the frequency of Fourier components. The 2-D Fourier spectrum is a plot of squared amplitude of the Fourier components as a function of spatial frequency. In an isotropic medium, the spatial frequency of the Fourier components is represented by the reciprocal radial distance from the center of Fourier spectrum in μm^{-1}.

The spatial density fluctuations representing the microstructure can be resolved into Fourier components.[226,231] The Fourier components of any micrograph provide a complete basis for a detailed, quantitative, and unique description of the microstructure.

The microstructural information obtained by the 2-D Fourier analysis is related to transparency and opacity of the cornea and the sclera, respectively. The distributions of the Fourier components can be determined from the profiles of the Fourier spectra (Fig. 1.10). The abscissa represents the spatial frequency of the Fourier components in units of nm^{-1}. The ordinate represents the normalized amplitude of the Fourier components in the upper-left quadrant of 2-D Fourier spectra. The vertical dashed lines represent the reference 200-nm dimension. The peaks in the profiles are due to rings in the 2-D Fourier spectra. Profile peaks represent Fourier components with higher

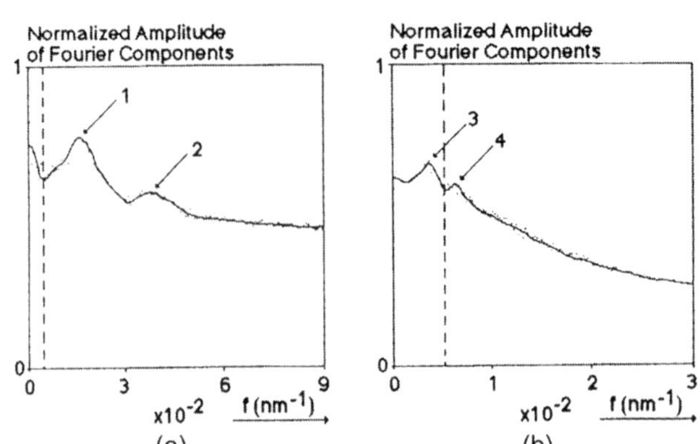

Figure 1.10 Profiles of the 2D Fourier spectra of (a) the cornea and (b) the sclera.[226]

amplitude relative to the background. These Fourier components represent the predominant spatial density fluctuations: the collagen fibril center-to-center spacing. While peaks 1 and 3 represent the fundamental Fourier components, peaks 2 and 4 represent the first harmonics. The curves in the profiles are drawn by averaging the two nearest neighbors of the pixel. The profiles showed that the major Fourier components of the cornea caused by the fibril center-to-center spacing were smaller than 200 nm. The major Fourier components of the sclera were larger than 200 nm. Also, the microstructure of the cornea revealed greater spatial order than the sclera.

The eye lens is also an example of the tissue for which the spatial degree of order is of crucial importance. Like a lens, it focuses light to form an image at the retina because of its high index of refraction and transparency. The eye lens exhibits a certain viscosity and is capable of altering its radius of curvature under the action of accommodation muscles and thus its focal length. It is gel, i.e., a cross-linked polymer plus low molecular liquid system where the polymer forms a spatial cross-linked sewn structure.[8] A similar structure is inherent in traditional marmalades and gelatin.

Brewster was the first to demonstrate regularly repeating structures in the lens tissue that is presently known as the lens fibers. The lens transparency is limited by two physical phenomena, namely, the absorption and scattering of visible light. The first phenomenon is negligible in a normal lens because the lens components contain a small number of chromophores. Light scattering proceeds at the protein conglomerates of the lens. The lens protein amounts to about 90% of its dry weight and its concentration is about 0.2 to 0.4 g/mL, i.e., it is the greatest one for tissues. The lens proteins are specific for the given tissue; they are called crystallines. This is due to the fact that in the nineteenth century they considered the lens as a crystal similar to ice. However, only the near-order degree exists in the lens and it appears to be sufficient for ensuring transparency.

The idea of the near-order degree in the organization of the lens proteins that provides for its transparency was first proposed by Benedek.[213] The main role among the ocular lens proteins is played by water-solved α crystallins whose shape is close to the spherical one with a diameter of about 17 nm. The studies of lens transparency, along with their major value, are of importance in connection with a wide propagation of such phenomenon as turbidity; the cataract. The structural bases of the cataract have remained an object of intense research for many years.[213,232,233]

The method of x-ray small-angular scattering was used to examine the tissues of the ocular lens for the bull and the frog.[234] It was shown that different structural states of crystallins may exist in the lens. A small-angular diffraction maximum is inherent in the region for Bragg distances $D = 15$–20 nm for one of the states that is explained by the near order of macromolecular complexes of crystallins (α-crystallin). No maxima were observed for other states.

The 2-D Fourier analysis has an immediate application in the study of the microstructure of the ocular lens.[235] The Fourier method can be applied to the study of microstructural change as a function of differentiation, aging, or pathogenesis in the cytoplasm or extracellular matrix.

1.6.7 Fractal properties of tissues

Almost all material, excluding monocrystals, are disordered in one or another sense. Fractal geometry[236] regards an accident quantitatively and thus it enables one to describe such random systems as polymers, colloidal aggregates, and porous materials. The qualitative feature of fractal objects is the invariance of their main geometric peculiarities, such as the magnification variation in the microscope with varying scale (see Section 1.4). The fractal properties of scattering systems strongly affect their light scattering.[124] The same mass of particles may induce small scattering in a dense cluster and significantly greater scattering in the fractal one. The fractal structure is most clearly pronounced with scattering in the case of multiple scattering.[116] The peculiarities of multiple scattering at the fractals are caused by a slowly falling correlation of the particle density. The fractal effects at the multiple scattering are observed even for fractal clusters whose sizes are shorter than the wavelength.

Since the spatial distributions of the constituents of many types of tissues appear to satisfy the conditions of statistical self-similarity,[227,237] fractal analysis may potentially provide a much simpler foundation for the analysis of tissue. Statistical self-similarity implies that the object is composed of building blocks with inherent statistical regularities that can be described by power law. The correlations of a variety of tissues in the refractive indices exhibit characteristics of a random fractal with a Hurst coefficient between 0.3 and 0.5.[200]

The tissue structure can be represented as a multifractal composed of various fractal formation types.[221,237] For the bone tissue the main fractal elements are trabeculae (formations with flatly lying mineralized fibers), and osteons (regions with the spiral-like orientation of fibers raised at an angle of 30–60 deg).[221] The above fractal types form an architectonic multifractal network. The geometric dimensions of biofractals are sufficiently great (100–1000 μm). In many cases, the fractal geometry yields a key to understanding the scattering peculiarities of these objects.

Quantitative models of the micro-optical properties of biological tissues have potential applications in several areas of biomedical optics.

1.7 Light Scattering by Densely Packed Correlated Particles

Certain regularities exist in densely packed media with characteristic dimensions on the order of the wavelength between the phases of waves scattered by adjacent particles and it is necessary to sum the amplitudes of scattered waves under the calculation of light scattered by the system with regard to phase relations. The interference interaction may result both in a varied total amount of scattered energy and an altered angular dependence of scattered light as compared with similar quantities for a system of noninteracting particles. The alteration of angular dependences can be particularly noticeable for biosystems since particle sizes are comparable with the light wavelength. In concentrated optically thick disperse systems a substantial role may also be played by the effects of multiple scattering.

The softness of the biotissue scattering particle makes it possible to restrict oneself with the single-scattering approximation in a number of cases. To correctly exploit this approximation, we need the optical thickness of the object under study to be small enough, i.e., $\tau < 0.1$.[1] For strongly scattering structures this means a necessity to restrict oneself with thin histological sections.

The theory of light scattering by densely packed correlated particles is principal for such tissues as eye lens, cornea, and sclera. This theory explains the transparency of the cornea and normal eye lens as well as their turbidity by pathology and opacity of the sclera. It was noted in Section 1.6. Besides, the theory of light scattering by densely packed correlated particles should also be important for investigations of subepithelial tissues, whole blood, and other bio-objects, which have been explored in other chapters in this book.

The characteristic relations between the scatterer sizes in tissues and the wavelength of visible light are close to similar relations between the liquid molecules and the x-ray radiation. Moreover, the near-order degree of order in the arrangement of bioparticles resembles that of the molecule degree of order in the liquid. Thus, the description of light propagation in tissues relies, to a significant extent, on the methods for describing the x-ray radiation diffraction in the liquid.[238] The light scattering in the correlated disperse systems operates with the statistical characteristics of these systems. The function of radial distribution $g(r)$ is just such a statistical characteristic of the spatial arrangement of scatterers.

1.7.1 Pair distribution function $g(r)$

Consider N spherical particles in a finite volume. To statistically describe their spatial arrangement a pair distribution function $g(r)$ may be employed.[239] The pair distribution function $g_{ij}(r)$ is proportional to the conditional probability of finding a particle of type j at the distance r from the origin given that there is a particle of type i at the origin (Fig. 1.11). For noninterpenetrable spheres, the interparticle forces are zero except for the fact that two particles cannot interpenetrate each other.

The arrangement of particles in a densely packed system is not entirely random. The near-order degree of order is observed in their arrangement; they are more ordered the greater the density of scattering centers and the narrower their distribution in size; and $g(r) = 0$ near the origin of coordinates in the

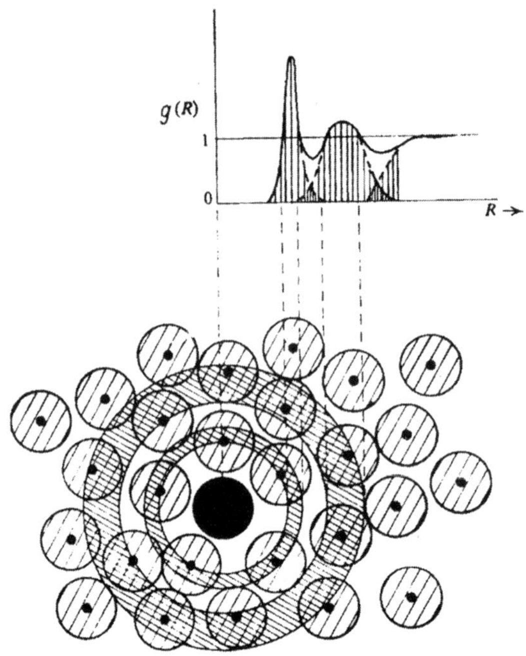

Figure 1.11 Diagram of radial distribution function $g(r)$ that is proportional to the probability of particle detection at a certain distance r from an arbitrarily fixed particle.[239]

region with a radius on the order of effective particle diameter that points to the interimpenetrability of a particle. Function $g(r)$ has a few maxima whose positions correspond to distances from the chosen particle to its first, second, etc. neighbors. Nonzero values of minima are indicative of a particle's migration between various coordination spheres. It is obvious that correlation between the pairs of particles should be degraded with growing r, hence $\lim_{r\to\infty} g(r) = 1$. Function $g(r)$ is the relation of the local density in the given sample, i.e., it is a probability to reveal a scatterer at the specified distance from the center chosen.

The medium composed of N scatterers considered here is analogous to an ensemble of mixture of L types of particles in the study of statistical mechanics, by considering the dynamics and positions of the particles with regard to the interparticle forces. Studies have been made in obtaining the pair distribution functions using various approximate theories. One of the important results is based on the Percus–Yevick (PY) approximation. As applied to a model of hard spheres distributed in a 3-D space, there exists the analytical solution of this equation. To find the function of radial distribution, the MC method is also used.

The solution of the Ornstein–Zernice equation for the case of single species has been solved by Wertheim.[240] For the case of two species, the solution can be found in the paper by Lebowitz.[241] For the case of general L species, the solution is obtained by Baxter based on a generalized

Wiener–Hopf technique.[21,242] The polydispersity of the real system is approximated by an L-step distribution function.

For monodisperse systems containing spherical particles with diameter d, $g(r)$ would be presented in the approximation of solid spheres, as follows:[243]

$$g(r) = 1 + \frac{1}{4\pi f} \int_0^\infty \frac{H_3^2(z)}{1 - H_3(z)} \frac{\sin zx}{zx} z^2 dz, \quad \text{for } x > 1, \tag{1.93}$$

where $x = r/d$,

$$H_3(z) = 24f \cdot \int_0^1 c_3(x) \frac{\sin zx}{zx} x^2 dx, \quad c_3(x) = -\alpha - \beta x - \delta x^3, \tag{1.94}$$

$$\alpha = \frac{(1+2f)^2}{(1-f)^4}, \quad \beta = -6f \frac{(1+\frac{1}{2}f)^2}{(1-f)^4}, \quad \delta = \frac{1}{2} f \cdot \frac{(1+2f)^2}{(1-f)^4}, \tag{1.95}$$

where f is the volume fraction occupied by particles.

1.7.2 Light scattering by a system of particles in the single-scattering approximation

Let us consider the light scattering by a system of N spherical particles. In general, the field affecting a particle in this system differs from the field of incident wave E_{inc} in that it also contains the total field of adjacent scatterers. Within the single-scattering approximation (Born's approximation), the field affecting the particle does not essentially differ from that of the initial wave. Regarding double scattering as a field affecting the particle, one needs to take the sum of the initial field and the single-scattered field, and so on.[20] In as much as biological particles are optically soft, the use of the single-scattering approximation often yields satisfactory results for the systems of these particles.

The field scattered by a particle with the center defined by radius-vector r_j differs from that scattered by a particle, which is placed at the origin of coordinates by a phase multiplier characterizing the phase shift of waves. This phase difference is $\frac{2\pi}{\lambda}(\mathbf{s}_0 - \mathbf{s})\mathbf{r}_j$, where $\mathbf{s}_0$ and $\mathbf{s}$ are the unit vectors of the directions of incident and scattered waves. The difference of these vectors is called the scattering vector $\mathbf{q}$, and

$$\mathbf{q} = \frac{2\pi}{\lambda}(\mathbf{s}_0 - \mathbf{s}). \tag{1.96}$$

Taking into account that the wave vector module is invariable with elastic scattering, the value of the scattering vector is found as follows:

$$q = 2\frac{2\pi}{\lambda}\sin(\theta/2), \tag{1.97}$$

where θ is the angle between directions $\mathbf{s}_0$ and $\mathbf{s}$, i.e., it is the scattering angle. The amplitude of a wave scattered by the system of N particles will be

$$E_s = \sum_{j=1}^{N} E_{sj} = \sum_{j=1}^{N} E_{0j} e^{i\mathbf{q}\mathbf{r}_j}, \tag{1.98}$$

where E_{0j} is the amplitude of scattering by an isolated particle. The intensity of single scattering for the given spatial realization of the N particle arrangement is

$$I = |E_s|^2 = \sum_{j=1}^{N} E_{oj} \sum_{i=1}^{N} E_{0i}^{*} e^{i\mathbf{q}(\mathbf{r}_j - \mathbf{r}_i)}. \tag{1.99}$$

For real systems, the mean scattering intensity by an ensemble of particles is measured due to thermal particle motion, finite measuring time, and finite area of a photodetector, as

$$<I>=<\sum_{j=1}^{N}\sum_{i=1}^{N} E_{oj} E_{0i}^{*} e^{i\mathbf{q}(\mathbf{r}_j - \mathbf{r}_i)}> . \tag{1.100}$$

The French brackets show the averaging over all possible arrangement configurations for the particles of the system. This equation presents a sum of two contributions for a noncoherent scattered intensity. One of them defines the light distribution on the assumption that there is no interference of light scattered at various particles. Another term regards the interference effect on the light field structure and depends on the degree of order in the arrangement of particles that is characterized by the radial distribution function $g(r)$. For an isotropic system of identical particles, we may write

$$<I>= N|E_0|^2 S_3(\theta), \tag{1.101}$$

$$S_3(\theta) = \left\{ 1 + 4\pi\rho \int_0^R [g(r) - 1] r^2 \frac{\sin qr}{qr} dr \right\}, \tag{1.102}$$

where $q = \frac{4\pi}{\lambda} \sin\frac{\theta}{2}$. Quantity $S_3(\theta)$ is the so-called structure factor. This structure factor describes the variation of the angular scattering intensity distribution which appears with a higher particle concentration. Within the approximation of hard spheres the structure factor is represented in the form

$$S_3(\theta) = 1/[1 - H_3(q)], \tag{1.103}$$

where $H_3(q)$ is defined by Eq. (1.94).

With small concentrations, we may make use of the approximation of excluded volume where $g(r) = 0$ with r being shorter than the particle diameter and having a unit value over long distances. In this approximation, the structure factor for a system of spherical particles takes the form that was first found by Dirac:

$$S_3(\theta) = 1 - f\Phi(qa), \tag{1.104}$$

where a is the particle radius, and $\Phi(qa)$ is the function defined by

$$\Phi(qa) = \frac{3(\sin qa - qa\cos qa)}{(qa)^3}. \tag{1.105}$$

Quantities a and λ can be on the same order for tissues, thus the first maximum of this function would be observed at angles achieving tens of degrees. Function $\Phi(qa)$ modulates the angular dependence of scattering intensity by diminishing its values at small angles and generating a diffusion ring at angles θ corresponding to the function $\Phi(qa)$ maximum.

Within a very small concentration of inhomogeneities, the structure factor is nearly a unit and the intensity of scattering by a disperse medium is essentially a sum of the contributions of independent inhomogeneities.

The structure factor changes slightly as a function of scattering angle for particles whose dimensions are much smaller than the wavelength, and the interference interaction in the systems of soft particles reveals itself mainly in a uniform decrease of scattering intensity in all directions (Fig. 1.12). For systems of large particles, the structure factor differs noticeably from a unit only in the region of small scattering angles where it is of an oscillating nature. The interference interaction of scatterers in one angular band reduces the scattering intensity and in the other band, the scattering intensity is raised as compared with that for a system of the equivalent number of independent particles (Fig. 1.13).

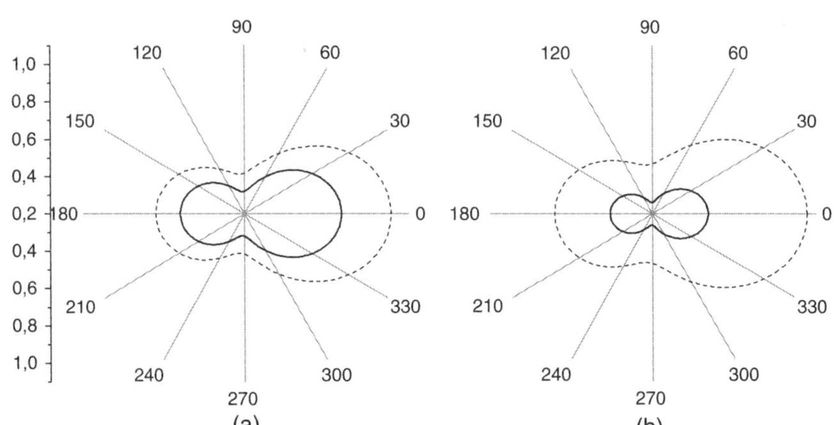

Figure 1.12 Angular scattering intensity dependences for a system of small interacting spherical particles of 50 nm in radius; $m = 1.105$, $\lambda = 633$ nm, and the volume fractions $f = 0.04$ (a) and $f = 0.1$ (b). The dotted lines indicate independent particles.

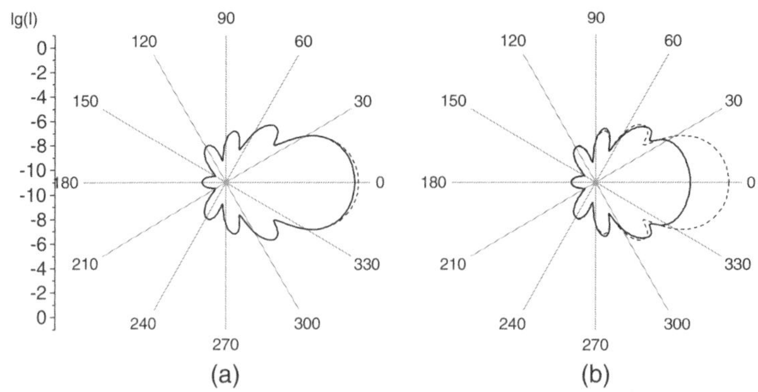

Figure 1.13 Angular scattering intensity dependences for a system of large interacting spherical particles of 500 nm in radius; $m = 1.105$, $\lambda = 633$ nm, and the volume fractions (a) $f = 0.04$ and (b) $f = 0.4$. The dotted lines indicate particles.

For the case where light is incident normally to the axes of infinitely long identically aligned cylinders with radius a, the structure factor is defined within the approximation of single scattering, as follows:

$$S_2(\theta) = \left\{1 + 8f\int_0^R [g(r) - 1]J_0\left(\frac{2\pi a}{\lambda} r \sin\theta\right) dr\right\}. \tag{1.106}$$

As the light is incident perpendicular to the cylinder axis, the scattered light propagates only in the direction perpendicular to the axis.

1.7.3 Angular characteristics for polarized light scattering

The angular dependence of the light scattering intensity of the systems of spherical and cylinder particles within the single-scattering approximation is described by Eqs. (1.101), (1.102), and (1.106). The structure factor altering the angular dependence of scattered light is defined by the statistical characteristics of the spatial particle arrangement, and it is independent of the state of light polarization. The angular dependences of all scattering matrix elements of identical particles are multiplied by the same quantity as the result of interference interaction within the single-scattering approximation. Consequently, the light scattering matrix (LSM) for this system coincides with that of an isolated particle under the commonly used normalization for the magnitude of its first element.

Unlike monodisperse systems, the normalization of matrix elements by M_{11} does not eliminate the influence of the structure factor on the angular dependences of matrix elements for differently sized densely packed particle systems.[244] As a practical matter, of particular interest are binary systems formed by a great number of small particles and a small number of coarse particles. This problem arises with the diagnostics of initial stages of protein coagulation, in particular, with the early diagnostics of the ocular lens turbidity.[244]

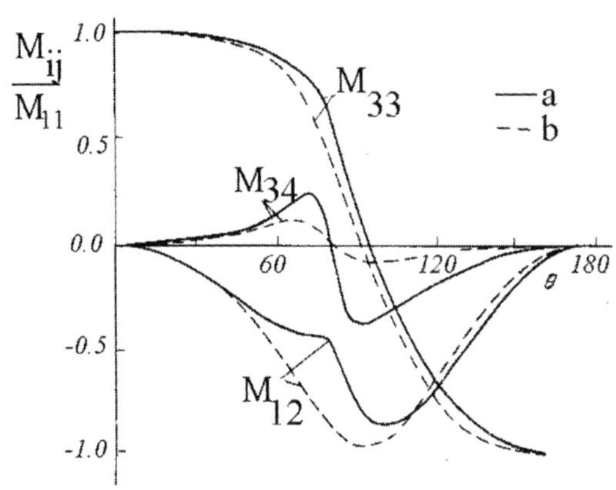

Figure 1.14 The calculated normalized LSM angular dependences of binary mixture of small and large spherical particles ($d_1 = 60$ nm, $d_2 = 500$ nm; $f_1 = 0.3$, $f_2 = 0.02$; $m = 1.107$; and $\lambda = 633$ nm): (a) interacting and (b) independent particles.

Figure 1.14 depicts the calculated results for the LSM of a mixture of spherical particles of diameters $d_1 = 60$ nm and $d_2 = 500$ nm for the volume fractions (spatial volume to total scattering system volume ratio) $f_1 = 0.3$ and $f_2 = 0.02$. For comparison, the LSM angular dependences of the same binary mixture—shown in Fig. 1.14 by a dashed line—have been calculated neglecting cooperative effects (within the assumption of an entirely arbitrary arrangement of scattering particles). The relative index of refraction $m =$ 1.07, and the light wavelength is 633 nm.

It is seen from Fig. 1.14 that the normalized LSM of a dense binary mixture is substantially altered due to the interference interaction. The high concentration of small particles is responsible for the degree of order in their arrangement followed by a lower intensity of scattering in all directions except the direction of straightforwardly propagating light. As a consequence, the results of the solution of the inverse problem for the experimental LSM of dense particle mixture that were obtained at neglecting the cooperative effects should yield an overestimated value for the relative fraction of large particles. The LSM variations due to the cooperative effects are of a more complicated nature for a binary system whose two components are sized on the order of the wavelength of incident light and they could not be interpreted so uniquely, as in the preceding case. Numerical estimates for binary systems of different compositions show[244] the considered effects to be of the most crucial importance for the LSM in the visible region for the mixtures of particles with $d_1 < 200$ nm and $d_2 > 250$ nm.

1.7.4 Spectral characteristics of scattering systems

The scattering strongly deforms the spectral biotissue characteristics since the extinction of transmitted light is defined not only by the absorption coefficient as a function of the wavelength, but also by a light fraction taken from the beam because of the scattering. The latter process complexly depends on the wavelength, structure, and size of particles.

The spectrum of collimated transmission of a disperse layer is interpreted as a spectral dependence of a weaker coherent component of light. Finding the coherent component of light, scattered at a system of inhomogeneities correlated in the space, is a complicated physical task exhibiting all of the difficulties inherent in the problem of light propagation through a system of many bodies.[243] Assuming that the intensity of the coherent light component is reduced with a longer distance by the exponent law due to scattering and absorption, the transmission of a disperse layer of thickness x would be described using the Bouguer–Beer–Lambert law

$$T = \exp(-\rho\sigma_e x), \tag{1.107}$$

where ρ is the particle density, and σ_e is the extinction cross section for a single particle of the layer. For low particle concentrations, it is equivalent to the extinction cross section of an independent particle. For greater f values, the quantity σ_e is determined not only by the properties of a single inhomogeneity but also by their volume concentration. Within the assumption that the absorption cross section is independent of the packing density, σ_e may be calculated as a sum of the absorption cross section of the independent particle and the scattering cross section obtained by taking into account correlation of the scatterers.

With the knowledge of the angular scattering intensity distribution of the system of particles, one would calculate the scattering cross section for a single particle in the system. Having integrated the scattering intensity over all directions in the space, the total amount of energy scattered by the system can be found. The scattering cross section for the system of spherical particles is obtained similar to Eq. (1.13), however, the intensity must be calculated by using the structure factor.

The scattering cross section for the system of cylinder particles, illuminated by a plane wave incident in the direction normal to the cylinder axis, is defined by numerically integrating over all possible scattering directions in a plane perpendicular to the cylinder axis

$$\sigma_s \propto \int_0^{2\pi} I(\theta)d\theta. \tag{1.108}$$

Dividing σ_s by the particle number N, one may find the scattering cross section for a single particle of the system. The result obtained may differ substantially from the scattering cross section of an independent particle.

Even the scattering cross-section for an independent particle sized on the order of a wavelength has a very strong nonmonotonous dependence on the wavelength. Effects associated with dense packing also have a substantial dependence on the wavelength. As a result, the transmission spectra for a system of identical particles can differ highly depending on the packing density and the degree of order. A clear example of such a dependence is presented by the transmission spectra of the cornea in the norm and with its turbidity caused by a disrupted spatial degree of order and by appearing regions denuded of fibrils, the so-called lakes.[245]

The extinction of a collimated incident beam due to complex angular/wavelength dependence of the scattered light in the dense systems, even for nonabsorbing ones, would result in substantial differences in the transmittance in different spectral regions. The values of the real and imaginary part of the indices of refraction depend weakly on the wavelength far from the absorption bands and may be assumed to be constant at calculations. In systems of small nonabsorbing particles, the interference interaction causes the shift of the short-wavelength transmission spectrum boundary to a smaller wavelength and a slightly greater steepness of the spectrum (Fig. 1.15). If a scattering system is formed by particles whose sizes are comparable with the wavelength, then the spectrum of this system would be nonmonotonous even with no absorption.

For particle systems with absorption, in the vicinity of the absorption bands, the real and imaginary parts of the complex index of refraction of the particles demonstrate pronounced spectral dependences. Figure 1.16 illustrates the transmission spectra calculated for two systems of differently sized particles with the refractive index described by the Lorentz contour.[12] The scattering deforms the symmetric contour of the absorption line and the spectrum appears different for systems of differently sized particles with varying packing density.

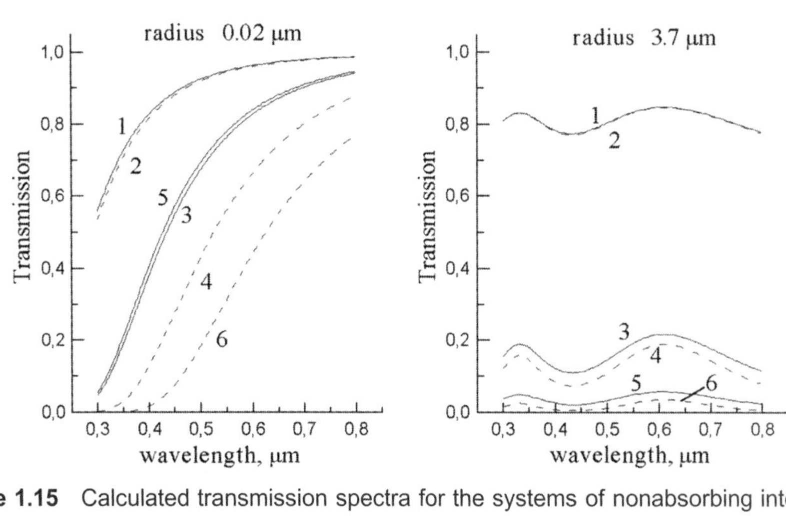

Figure 1.15 Calculated transmission spectra for the systems of nonabsorbing interacting particles (solid lines) and at approximation of noninteracting particles (dashed lines); $m = 1.1$, $f = 0.01$ (1, 2), $f = 0.1$ (3, 4), and $f = 0.2$ (5,6).

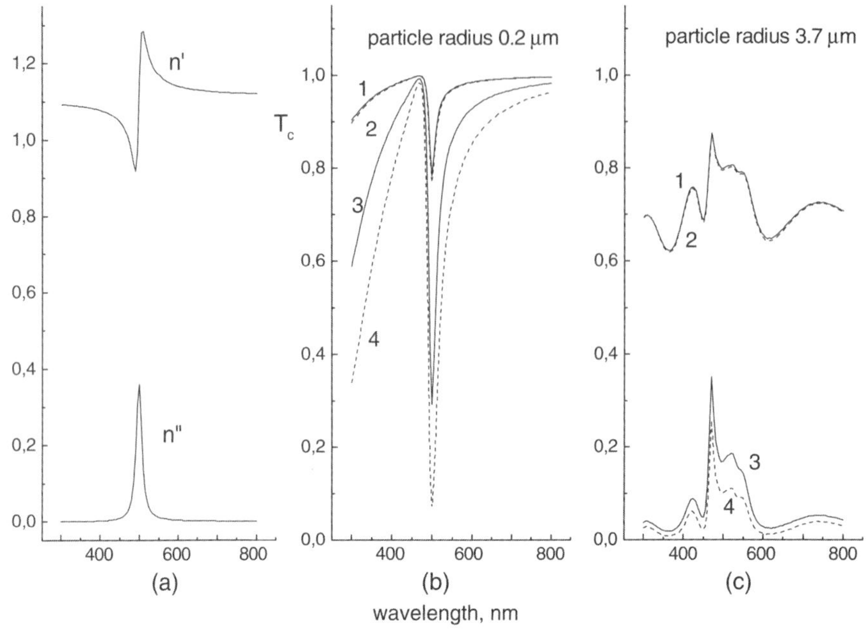

Figure 1.16 Spectral dependence of real and imaginary part of the relative complex index of refraction for (a) particles with the Lorentz contour center at $\lambda = 500$ nm; transmission spectra for the slab of thickness (b) $L = 3$ mm and (c) 0.1 mm of the disperse systems of particles (solid lines); independent particles (dotted lines); $f = 0.01$ (1,2) and $f = 0.1$ (3,4).

For example, the transmission spectrum for whole blood differs substantially from that for the hemolyzed blood because in whole blood, hemoglobin is localized in erythrocytes, which are strong light scatterers.[6] Therefore, it is much easy to estimate blood saturation with oxygen by using the absorption spectra for the hemolyzed blood.[246] A detailed review of spectral characteristics for various tissues and blood is given in Ref. 9.

For the densely packed system of large optically soft particles within the approximation of hard spheres and at neglecting of mutual particle reradiation, the following equation was obtained for the transmission of a layer with thickness L that accounts for coherent effects:[247]

$$T_c = \left[1 - \frac{2b}{(1+b)} \cdot \frac{\sigma_e}{\pi \cdot a^2} + \frac{b^2}{(1+b)^2} \cdot \frac{2\lambda}{\pi \cdot a^3}\sigma_s I_1(0)\right]^{L/2a}, \tag{1.109}$$

where $b = 1.5f \times \exp(1.5f)$, and $I_1(0)$ is the intensity of forward scattering at a single particle of diameter $2a$. This formula is transformed into the Bouguer–Beer–Lambert law for the scattering systems of noninteracting particles under the rarefaction of the scattering layer.

Not only coherent weakened light but also a portion of noncoherently scattered light is usually recorded in real experiments because of the finite angular aperture of the receiving unit. For this reason, a transmittance called the instrumental transparency found experimentally is somewhat different from the coherent transmission T_c.

For the first time, the approximation regarding the near-order degree of order has been used to describe the propagation of light in a tissue in Refs. 224, 228, and 245, with calculating the transmission spectra of the cornea. The near order in the arrangement of scattering particles and the related interference interaction of scattered light are the cause of the high transparency of the human eye optical tissues in their normal state.[248,249]

The spectral characteristics vary most strongly with high packing densities, as the volume fraction occupied by particles exceeds 50%. For these dense systems, the considered approximation of single scattering is incorrect and it is necessary to account for the effects of the reradiation of particles.

Unlike the transmission spectra, the spectral dependences of the intensity of light scattered in different directions are poorly studied. This is related, on the one hand, to experimental difficulties due to the need for standard spectral devices to be rearranged. On the other hand, a comparison of different experimental results is difficult because the form of the scattering spectrum depends substantially on the geometry of the system under study. Nevertheless, the scattering spectra are of great interest. The authors of Ref. 250 measured the absorption and scattering spectra of the chest muscle of a chicken in the visible range. The scattering spectra visually define the observed tissue color, and they can be employed for express estimating of tissue condition.

One traditional method for eye ocular lens diagnostics assumes the observation of varying color characteristics for light scattered at an angle. The qualitative analysis of composition alteration for the scattering particles of the eye lens with age by measuring the scattering spectra[251] is given in Ref. 252. The analysis of spectral scattering characteristics lies in the basis of the method for measuring the skin erythema.[253] Reference 254 describes some applications of the "scattering spectral method" along with viscosimetry and sedimentation data to the study of *Yersinia pestis* capsular protein.

1.7.5 Consideration of multiple-scattering effects in a system of densely packed particles

In real disperse systems containing a great number of particles, multiple scattering takes place. The spatial correlation of single inhomogeneities results in a necessity to consider the interference of multiple scattered waves.[20,247] In a case when effects of particle reradiation in the densely packed disperse system are not negligible, the distinction of an effective field in a medium from the incident wave field is induced. Under this condition, the approach based on the statistical theory of multiple wave scattering seems to be most promising for describing the collective interaction between an ensemble of particles and electromagnetic radiation.[20,247]

The rigorous theory of multiple wave scattering is constructed on the basis of fundamental differential equations for the fields combined with a statistical approach.[20] The resulting field in any point of the space is displayed as the sum of the incident wave fields and all possible multiple scattered waves with regard to their phases. The total field $E(r)$ at the point r is the sum of the incident field and scattered fields from all particles [Eq. (1.43)]

Twersky has derived a closed system of integral equations describing the processes of multiple scattering.[255] A rigorous solution in a general form has not been found yet for this problem. For its solution, various approximations whose efficiency is determined for every particular case should be exploited. One of the most efficient approximations to describe tissue optics is a so-called quasi-crystalline approximation proposed for densely packed systems.[256]

Averaging in Eq. (1.43) over different configurations of the system yields an infinite set of equations. The quasi-crystalline approximation corresponds to the condition when this set is limited to the second step. The obtained closed system of equations for the effective field is reduced to a system of linear equations by expanding in terms of vector spherical or cylindrical harmonics. The explicit form of coefficients is quite cumbersome for the obtained system.[21,257] Coefficients involve the radial distribution function, taking into account correlation of the particle arrangement, and the coefficients of Mie row expansion for a single particle. The equality to zero for the determinant of the system of linear equations yields the dispersion relation for the effective propagation constant k_{eff} in this medium.[258] For the systems of particles whose sizes are small compared to the wavelength, the expression for k_{eff} has the view[21]

$$k_{\text{eff}}^2 = k^2 + \frac{3fy}{D}k^2\left[1 + i\frac{2}{3}\frac{k^2a^2y}{D}S(\theta = 0)\right], \tag{1.110}$$

where

$$y = \frac{n_1^2 - n_0^2}{n_1^2 + 2n_0^2}, \quad D = 1 - fy, \quad S(\theta = 0) = \frac{1}{1 - H_3},$$
$$H_3 = -24f\left(\frac{\alpha}{3} + \frac{\beta}{4} + \frac{\delta}{6}\right),$$

f is the volume fraction of particles with the refractive index n_1 embedded into ground medium with $n = n_0$, and α, β, and δ are defined from the approximation of the hard spheres [Eq. (1.95)]. The calculated effective index of refraction $n_{\text{eff}} = n'_{\text{eff}} + in''_{\text{eff}}$ is complex, even if particles and ground medium surrounding them exhibit no intrinsic absorption. The imaginary part of the effective index of refraction n''_{eff} describes the energy loss of the incident plane wave due to scattering in all directions. The transmittance of the scattering layer with thickness x is defined as

$$T = \exp\left(-\frac{4\pi}{\lambda} n''_{\mathrm{eff}} x\right), \tag{1.111}$$

where $(4\pi/\lambda)n''_{\mathrm{eff}}$ is the extinction coefficient, and n''_{eff} decreases with a higher wavelength and nonmonotonously depends on the particle concentration. For small particle concentration, the transmittance of the layer decreases with particle concentration increase (increase of their volume fraction f); however, due to particle pair correlation starting from $f \approx 0.1$, a so-called clearing effect takes place. The real part of the effective index of refraction in this approximation is essentially independent of the wavelength and alters monotonously with growing particle concentration to approach the refractive index of the particle material. The near order appearing in the scatterers' arrangement with their greater concentration not only provides for conditions for the manifestation of the secondary scattered wave interference,[259] it also affects the character of propagation of noncoherent multiple scattered light. This may be accompanied by the so-called concentration effects of darkening and clearing.[259]

The optical softness of tissues enables calculations to employ an expansion by scattering multiplicities with restricting to low orders. In Ref. 260, an expression for the effective index of refraction of the eye cornea modeled by the system of cylinder scatterers was obtained in the framework of expansion by scattering multiplicities. The effects of polarization anisotropy were analyzed with respect to the double scattering contributions.

The problem of scattering and absorption of electromagnetic radiation by the composite medium is so complicated that the exact solution of Maxwell's equations under the appropriate boundary conditions is unpractical.[6] For this reason, it seems to be attractive to have a way to determine some averaged optical properties of heterogeneous materials that would enable us to treat the heterogeneous material in the same way that we consider homogeneous media. Based on the theory of multiple wave scattering, various expressions were derived for the effective parameters of a disperse medium.[18,261]

Using a definite set of assumptions concerning the shapes of inclusions and the topology of a mixture, one can obtain the appropriate analytical expression for an effective dielectric constant as a function of volume fraction f and dielectric constants ε_1 and ε_0 of individual components (see Section 1.2) or corresponding indices of refraction, $n_1 = \sqrt{\varepsilon_1}$ and $n_0 = \sqrt{\varepsilon_0}$. The simple mixing rules often used in various applications include volume averages of the dielectric constants or indices of refraction[11]

$$n_{\mathrm{eff}} = f n_1 + (1 - f) n_0. \tag{1.112}$$

Using the theory of multiple scattering, Twersky[262] succeeded in deriving the approximate expressions for absorption μ_a and scattering μ_s coefficients with describing the light scattering in the blood. The blood hematocrit H is related to the erythrocytes concentration ρ and to the volume of one erythrocyte V_e by the following ratio:[6]

$$\rho = H/V_e. \tag{1.113}$$

Thus, the absorption coefficient μ_a is

$$\mu_a = (H/V_e)\sigma_a. \tag{1.114}$$

For sufficiently small values of H ($H < 0.2$), the scattering coefficient is given by a similar equation

$$\mu_s = (H/V_e)\sigma_s. \tag{1.115}$$

For $H > 0.5$, the particles become densely packed, and the medium is almost homogeneous. In this case, the blood may be considered as a homogeneous medium containing hemoglobin in which scattering particles are included from plasma surrounding red blood cells (RBCs). Within the limits of $H \to 1$, "plasma particles" disappear, and the scattering coefficient should tend to zero. This results in the following approximate equation for μ_s:

$$\mu_s = \frac{H(1-H)}{V_e}\sigma_s, \tag{1.116}$$

where coefficient $(1-H)$ regards the scattering termination with $H \to 1$. Dense packing ($H = 1$) is unattainable, and the packing influence should not be described by the simple function $(1-H)$. For example, if the particles are the hard spheres, H may not exceed 0.64. Then we should take[262]

$$\mu_s = (H/V)\sigma_s F(H), \tag{1.117}$$

where function $F(H)$ must monotonously decrease from a unit with $H = 0$ to zero with a certain limit value of H. Reference 263 presents the following dependence of scattering coefficient μ_s on the hematocrit H for thin blood layers

$$\mu_s \approx H(1-H)(1.4-H). \tag{1.118}$$

1.7.6 Birefringence of a system of anisotropic particles

Besides the crystal birefringence caused by the anisotropic properties of molecules they contain, there also exists a so-called birefringence of the form.[30] The latter phenomenon may be observed in an ordered system of particles made from an optically isotropic material whose sizes are great compared to those of molecules but are small with respect to the wavelength.

The difference of refractive indices for two orthogonal states of linear polarization of an incident plane wave at light propagation in a system of thin dielectric plates is described by[30]

$$n_e^2 - n_o^2 = -\frac{f(1-f)(n_1 - n_m)}{(1-f)n_m^2 + f \cdot n_1^2}, \tag{1.119}$$

where f is the volume fraction of plates, and n_1 and n_m are the refractive indices of plates and the surrounding medium, respectively. This implies that this system behaves like a negative single-axis crystal with its optical axis being normal to the plate surface. Wiener[30] succeeded in obtaining the following expression for a system of thin cylinder rods with $f \ll 1$

$$n_e^2 - n_o^2 = \frac{f(1-f)(n_1^2 - n_m^2)^2}{(1+f)n_m^2 + (1-f)n_1^2}, \tag{1.120}$$

where f is the volume fraction of rods, and n_1 and n_m are the refractive indices of rods and the surrounding medium, respectively. This difference is always positive and, thus, the system formed by thin cylinders constitutes a positive single-axis crystal. This means that the rod system behaves like the positive single-axis crystal with its optical axis being in parallel with the rod axes. Equation (1.120) is consistent with the results obtained by Rayleigh[264] for a system of parallel cylinders, even if f is not very small compared with a unit, provided that the difference between the refractive indices n_1 and n_m is small (optically soft particles).

The author of Ref. 265 has derived a more general expression that describes the birefringence within the limit of thin cylinders for an arbitrary value of f:

$$n_e - n_o = \frac{f(1-f)(n_1 - n_m)^2}{(1-f)n_m + fn_1}. \tag{1.121}$$

This equation corresponds to the Wiener equation (1.120) with $|n_1 - n_2| \ll 1$.

The form birefringence is used in biological microscopy. The sign of the observed difference points to the particle shape being close to that of the rod or the plate and, if n_1 and n_m are known, one can assess the volume fraction of the particles. To distinguish between the form birefringence and the birefringence of the particle material, one is able to change the refractive index of the base material. The form birefringence vanishes at $n_1 = n_m$. Linear dichroism, i.e., different wave attenuation for two orthogonal polarizations, is defined by the difference between the imaginary parts of effective indices of refraction of particles and surrounding medium. Depending on the relation between the size and the optical constants of the scatterers, this difference could be positive or negative.

1.8 Application of Radiative Transfer Theory to Tissue Optics

The classical radiative transfer theory (RTT) is found on energy consideration and serves as a basis for photometry. The initial development of the RTT is usually associated with the names of Bouguer and Lambert. Photometry operates with energy characteristics that describe the light detection capabilities by using quadratic detectors. The classical phenomenological theory of light transfer makes use of the pictorial concept of ray optics supplemented by a statistical assumption of fields being entirely mutually incoherent for rays propagating in different directions. This assumption allows one to summarize the average intensities of rays coming from the different directions ignoring the phase relations. This formulation is widely used in atmosphere optics,[266] the optics of photographic layers, and tissue optics.

In the classical theory of light transfer that considers the wave field as an ensemble of incoherent rays, the main notion is the radiation intensity (or brightness) $I(r,s)$ defining the mean energy flow dP through an area da which is concentrated in a solid angle $d\Omega$ near direction s in the frequency interval $(\nu,\nu + d\nu)$,

$$dP = I(r,s)\cos\theta da d\Omega d\nu. \tag{1.122}$$

This theory is valid for an ensemble of scatterers located far from one another and has been successfully used to work out some practical aspects of tissue optics. The governing differential equation of this theory is called the Boltzmann equation used in the kinetic theory of gases and in the theory of neutron transfer.[267] The main stationary equation of the RTT for monochromatic light has the form[6,9]

$$\frac{\partial I(\bar{r},\bar{s})}{\partial \bar{s}} = -\mu_t I(\bar{r},\bar{s}) + \frac{\mu_s}{4\pi}\int_{4\pi} I(\bar{r},\bar{s}')p(\bar{s},\bar{s}')d\Omega', \tag{1.123}$$

where $I(\bar{r},\bar{s})$ is the radiation intensity at point $\bar{r}$ in direction s, (W $m^{-2}sr^{-1}$); $p(\bar{s},\bar{s}')$ is the scattering phase function; and $d\Omega'$ is the unit solid angle in direction s'. It is assumed that there are no radiation sources inside the medium. Equation (1.123) with the appropriate initial and boundary conditions defines the behavior of the radiation intensity. Equation (1.123) refers to the energy balance in an infinitely small medium volume: the rate of variation for the radiation intensity along the beam is found by scattering in the given direction s from all other directions s' (integral term) and by attenuation due to scattering and absorption [term $-\mu_t I(r,s)$].

The phase function $p(\bar{s},\bar{s}')$ describes the scattering properties of the medium and is in fact the probability density function for scattering in the direction s' of a photon traveling in the direction s; in other words, it characterizes an elementary scattering act. If scattering is symmetric relative to the direction of an incident wave, then the phase function depends only on the angle θ between directions s and s'.

The scattering indicatrix $p(\bar{s},\bar{s}')$ may be specified either as a table obtained by measuring or calculating, or as an analytical expression.

A portion of the radiation intensity that decreases due to scattering and absorption while passing through volume V containing scattering particles is called the attenuated incident intensity and satisfies the following equation:

$$\frac{\partial I_{ri}(\bar{r},\bar{s})}{\partial \bar{s}} = -\mu_t I_{ri}(\bar{r},\bar{s})\,. \tag{1.124}$$

Another portion, appearing in the medium due to scattering, is called the diffusion intensity. Since the total intensity

$$I(\bar{r},\bar{s}) = I_{ri}(\bar{r},\bar{s}) + I_d(\bar{r},\bar{s}) \tag{1.125}$$

satisfies Eq. (1.123), the diffusion intensity can be found from the equation

$$\frac{\partial I_d(\bar{r},\bar{s})}{\partial \bar{s}} = -\mu_t I_d(\bar{r},\bar{s}) + \frac{\mu_s}{4\pi}\int_{4\pi} I_d(\bar{r},\bar{s}')p(\bar{s},\bar{s}')d\Omega' + \varepsilon_{ri}(\bar{r},\bar{s}), \tag{1.126}$$

where $\varepsilon_{ri}(\vec{r},\vec{s})$ is the equivalent source function given rise by the attenuated incident intensity.[6] Equation (1.124) is the differential form of the Beer–Lambert–Bouguer law for the scattering medium with absorption. This means that the Beer–Lambert–Bouguer law is valid for any optical thickness for the attenuated incident intensity within the framework of the transfer theory. Deviations from this law, e.g., caused by dense particle packing and manifested as optical clearing or darkening at concentration change, are ignored by the RTT.

As the vector nature of the electromagnetic field is accounted for the radiation intensity should be replaced by an intensity matrix that describes not only the intensity, but also the polarization properties of radiation with μ_t and μ_s being matrix quantities as well. The order in which matrices are incorporated into the equations should be taken into account. The scalar equation (1.123) is exploited in optics to describe light radiation in cases where the polarization effects may be neglected.

The integro-differential equation (1.123) is too complex to be employed for the analysis of light propagation in scattering media directly. There are two approaches to the problem under study. With one of them, one proceeds to obtain a general solution with unknown coefficients, and then to find these coefficients by means of the appropriate boundary conditions. Another approach is in passing from the differential equations with specified boundary conditions to integral equations for the certain unknown functions. For simple geometries, the approach based on the integro-differential equation is widely used. For complex geometries, it is often more convenient to proceed with the integral equations and to obtain their approximate solutions.

The exact solutions of the radiative transfer equation (RTE) (1.123) and of the integral equation for the beam intensity were obtained only for a small number of particular cases. Planar parallel geometry and isotropic scattering are the examples for which the solutions have been found and are written in the form acceptable for calculations. In practice, one would be forced to apply to approximate solutions in most cases.

We do not aim to describe all or most methods to solve the RTE; this is referred to the literature.[6,9,20,266,267] Instead, we will consider a few approximations that are often used in tissue optics and have a clear physical meaning.

1.8.1 Approximation methods for solution of the radiation transfer equation

1.8.1.1 The first-order approximation

For weak scattering, as the scattering medium is rarefied and the scattering volume is small, iteration may be used to solve the RTE. In the first approximation, the iterative solution yields the result known as the first-order approximation of the RTT.[6] In this approximation, it is assumed that the total intensity incident on the particles is approximately equal to the attenuated incident intensity that is known. Thus, the solution within the first-order approximation is[6]

$$I(\bar{r},\bar{s}) = I_{ri}(\bar{r},\bar{s}) + I_d(\bar{r},\bar{s}), \tag{1.127}$$

$$I_d(\bar{r},\bar{s}) = \int_0^s \exp[-(\tau - \tau_1)] \cdot \left[\frac{\mu_s}{4\pi} \int_{4\pi} I_{ri}(\bar{r}_1,\bar{s}')p(\bar{s},\bar{s}')d\Omega'\right] d\bar{s}', \tag{1.128}$$

where I_{ri} is the attenuated incident intensity, I_d is the diffuse intensity, and $\tau = \int_0^s \rho\sigma_e ds$, $\tau_1 = \int_0^{s_1} \rho\sigma_e ds$ are the optical pathlengths.

The so-called first-order solution is realized for optically thin and weakly scattering media ($\tau < 1$ and single-scattering albedo $\Lambda = \sigma_s/\sigma_e < 0.5$) when the intensity of the transmitting (coherent) wave is described by Bouguer's law [Eq. (1.2)]. Given a narrow beam (e.g., a laser), this approximation may be applied to optically denser tissues ($\tau > 1$, $\Lambda < 0.9$).

1.8.1.2 Diffusion approximation

With a longer optical path, the directional diagram of scattered light is broadened and at last it becomes almost isotropic: the scattered light "forgets" the direction of initial wave propagation in the depth regime. To describe this almost isotropic radiation, a diffusion approximation would be used that would appear to be significantly simpler than the original integro-differential equation, since it is reduced to the solution of differential equations. Because of this, the diffusion approximation is widely used as an effective approach to account for multiple scattering.

As one derives the diffusion equation, one should pass from the RTE to an infinite system of equations for the coefficients of expansion of beam intensity I into a series over spherical harmonics. The reduction of this system results in the so-called P_L approximation, the simplest of which is a system of equations for four functions that are equivalent to the diffusion approximation. There exist also other means to derive the diffusion equations for the diffusion coefficient. All distinctions disappear, however, with passing to the limit of weak absorption, as it is valid for many tissues. This is easy to understand from the physics of the diffusion approximation.[20] Actually, the description of scattering as a certain diffusion process only becomes adequate if the scattering effects dominate upon the absorption ones. Also, the diffusion approximation does not allow one to describe boundary effects manifesting themselves near the boundaries where the radiation "has not yet forgotten" the boundary and initial conditions.

The diffusion theory provides a good approximation in the case of small scattering anisotropy factor $g \leq 0.1$ and large albedo $\Lambda \to 1$. For many tissues, $g \approx 0.6$ to 0.9 and can be as large as 0.990–0.999 for blood.[9,218] This significantly restricts the applicability of the diffusion approximation. It is argued that this approximation can be used at $g < 0.9$ when the optical thickness τ of an object is of the order 10–20. However, the diffusion approximation is inapplicable for an input beam near the object's surface where single- or low-step scattering prevails.

The first-order approximation is valid only when the volume density, equal to the relation between a volume occupied by the particle and the entire volume of the medium, is substantially less than 1%. If the volume density is much greater than 1%, the diffusion approximation provides for good results. For volume density on the order of 1%, neither the near-order approximation nor the diffusion approximation can be valid, and we need to solve the RTE.

1.8.1.3 Small-angular approximation

The incident wave may be represented as a narrow collimated beam in many cases of laser diagnostics of tissues. Examples are the determination of bacteria concentration in liquids using a laser beam,[268] and detection of blood oxygen saturation by means of an optical fiber catheter.[246]

For the case of large scattering particles whose diameters are comparable or greater than the wavelength, the scattering intensity in a narrow sector of angles in the forward scattering direction is high. This enables the description of collimated beams to be simplified significantly by using a small-angular approximation, which is applicable for short propagation routes where the beam remains sufficiently strongly collimated. The RTE in the small-angular approximation allows for an exact solution, however, the solution looks quite cumbersome and not friendly for numerical calculations. A simpler approximation corresponds to the consideration of the scattered light diffusion in angular variables.

1.8.1.4 Flux theory

The radiation intensity I itself is not often of interest but instead its integrals yielding the energy characteristics of the radiation field is used. If illumination is diffuse and the medium is sufficiently turbid providing the light diffusion scattering, experimental results are well described by the two-flux Kubelka–Munk theory.[9] This theory relies on a model of two light fluxes propagating in the forward and backward directions. The extension of two-flux Kubelka–Munk theory to a four-flux theory makes it possible to describe a collimated beam incidence onto the medium. The four-flux model[6,269] is actually two diffuse fluxes traveling to meet each other (Kubelka–Munk model) and two collimated laser beams, the incident beam and the beam reflected from the rear boundary of the sample.

A seven-flux model is the simplest 3-D representation of scattered radiation and an incident laser beam in a semi-infinite medium.[270] Of course, the simplicity and the possibility of expeditious calculations of the radiation dose or rapid determination of tissue optical parameters (solution of the inverse scattering problem) are achieved at the expense of accuracy.

There exist various numerical procedures to solve the RTE. A more exact solution of the RTE is possible using the discrete ordinates method (multiflux theory) in which Eq. (1.123) is converted into a matrix differential equation for illumination along many discrete directions (angles).[6] The solution approximates an exact one as the number of angles increases. Many computer programs are available to solve the RTE under different conditions.[23]

1.8.1.5 Vector radiative transfer equation

As it was already shown, the majority of tissues are turbid media showing a strong scattering and much less absorption (up to two orders of magnitude fewer than scattering in the red and NIR). Moreover, in their natural state (nonsliced), tissues are rather thick. Therefore, multiple scattering is a specific feature of a wide class of tissues.[9,271–275]

Polarization effects at light propagation through various multiple-scattering media, including tissues, are fully described by the vector RTE (VRTE).[9,268,276–291] The RTT originated as a phenomenological approach based on considering the transport of energy through a medium filled with a large number of particles and ensuring energy conservation. This medium, composed of discrete, sparsely, and randomly distributed particles, is treated as continuous and locally homogeneous. In the framework of the RTT, the scattering and absorption of the small volume element follow from the

Maxwell equations and are given by the incoherent sums of the respective characteristics of the constituent particles; the result of scattering is not the transformation of a plane incident wave into a spherical scattered wave but, rather, the transformation of the specific intensity vector (Stokes) of the incident light into the specific intensity vector of the scattered light.[56]

For macroscopically isotropic and symmetric plane-parallel scattering media, the VRTE can be substantially simplified as[56]

$$\frac{d\mathbf{S}(\bar{\mathrm{r}},\vartheta,\varphi)}{d\tau(\bar{r})} = -\mathbf{S}(\bar{\mathrm{r}},\vartheta,\varphi) + \frac{\Lambda(\bar{\mathrm{r}})}{4\pi}\int_{-1}^{+1} d(\cos\vartheta')\int_{0}^{2\pi} d\varphi'\bar{\mathbf{Z}}(\bar{\mathrm{r}},\vartheta,\vartheta',\varphi-\varphi')\mathbf{S}(\bar{\mathrm{r}},\vartheta',\varphi'), \tag{1.129}$$

where $\mathbf{S}$ is the Stokes vector; $\bar{\mathrm{r}}$ is the position vector; ϑ and φ are the angles characterizing the incident direction and the polar (zenith) and azimuth angles, respectively;

$$d\tau(\bar{\mathrm{r}}) = \rho(\bar{\mathrm{r}})\langle\sigma_e(\bar{\mathrm{r}})\rangle ds \tag{1.130}$$

is the optical pathlength element; ρ is the local particle number density; $<\sigma_e>$ is the local ensemble-averaged extinction coefficient; ds is the pathlength element measured along the unit vector of the direction of light propagation; Λ is the single-scattering albedo; ϑ' and φ' are the angles characterizing scattering direction, the polar (zenith) and the azimuth angles, respectively; $\bar{\mathbf{Z}}$ is the normalized phase matrix

$$\bar{\mathbf{Z}}(\bar{\mathrm{r}},\vartheta,\vartheta',\varphi-\varphi') = \mathbf{R}(\Phi)\mathbf{M}(\theta)\mathbf{R}(\Psi), \tag{1.131}$$

where $\mathbf{M}(\theta)$ is the single-scattering Mueller matrix; θ is the scattering angle; and $\mathbf{R}(\phi)$ is the Stokes rotation matrix for an angle ϕ:

$$\mathbf{R}(\phi) = \begin{bmatrix} 1 & 0 & 0 & 0 \\ 0 & \cos 2\phi & -\sin 2\phi & 0 \\ 0 & \sin 2\phi & \cos 2\phi & 0 \\ 0 & 0 & 0 & 1 \end{bmatrix}. \tag{1.132}$$

This phase matrix links the Stokes vectors of the incident and scattered beams, specified relative to their respective meridional planes. To compute the Stokes vector of the scattered beam with respect to its meridional plane, one must calculate the Stokes vector of the incident beam with respect to the scattering plane, multiply it by the scattering matrix (to obtain the Stokes vector of the scattered beam with respect to the scattering plane), and then compute the Stokes vector of the scattered beam with respect to its meridional plane. Such a procedure involves two rotations of the reference plane: $\Phi = -\phi$; $\Psi = \pi - \phi$ and $\Phi = \pi + \phi$; and $\Psi = \phi$. The scattering angle θ and the angles Φ and Ψ are expressed via the polar and the azimuth incident (ϑ, φ) and scattering angles (ϑ′, φ′).[9,56]

1.8.2 Monte Carlo simulation

1.8.2.1 Introduction

The MC method, being widely used for the numerical solution of the RTT equation[9,290–296] in different fields (astrophysics, atmosphere and ocean optics, etc.) appears to be especially promising for the solution of direct and inverse radiation transfer problems for media with arbitrary configurations and boundary conditions, in particular for the purposes of the medical polarization optical tomography and spectroscopy.[9,269,273,286–289,293–301] The method is based on the numerical simulation of photon transport in scattering media. Random migrations of photons inside a sample can be traced from their input until absorption or output occur.

The straightforward simulation using the MC method has the following advantages: (1) one can employ any scattering matrix, (2) there are no obstacles for the use of strongly forward-directed phase functions or experimental single-scattering matrices, (3) the polarization calculation takes only a twofold increase in computation time over that needed for the evaluation of intensity, (4) any reasonable number of detectors can be accounted for without noticeable increase of the computation time, (5) there are no difficulties in determining the radiation parameters inside the medium, and (6) it is possible to model media with complex geometry where radiance depends not only on the optical depth but also on the transverse coordinates.

The liability of the obtained results to statistical variations on the order of a few percent at an acceptable computation time is the main disadvantage of the MC technique. For a twofold increase of the accuracy, one needs a fourfold increase in the computation time. The MC method is also impractical for great optical depths ($\tau > 100$).

1.8.2.2 Simulation algorithm

A few MC codes for modeling of polarized light propagation through a scattering layer are available in the literature (see, e.g., Refs. 9, 281, and 286–301). To illustrate the MC simulation technique, the algorithm described in Ref. 300 and applied to model the angular dependencies of the scattering matrix elements is discussed. Let a flux of photons within an infinitely narrow beam be incident exactly upon the center of the spherical volume filled up by the scattering particles.[300] The path of a single photon migration in the medium is accounted for in a process of computer simulation. The photons are considered in this case as ballistic particles. Different events possible in the course of the photon migration are estimated by the appropriate probability distributions. In the model under study, the photons would either be elastically scattered or absorbed under their collisions with the medium particles. A certain outcome of every event is found by a set of uniformly distributed random numbers. The probability of scattering in the given direction is determined in accordance with scattering by a single particle. One is able to specify the cross section of scattering and values of the scattering matrix elements for every photon interaction with a scatterer.

When an incident photon enters a scattering layer, it is allowed to travel a free pathlength, l. The l value depends on the particle concentration ρ, and extinction cross section σ_e. The free pathlength l is a random quantity that takes any positive values with the probability density $p(l)$:

$$p(l) = \rho\sigma_e e^{-\rho\sigma_e l}. \tag{1.133}$$

The particular realization of the free pathlength l is dictated by the value of a random number ξ that is uniformly distributed over the interval [0, 1]:

$$\int_0^l p(l)dl = \xi. \tag{1.134}$$

Substituting Eq. (1.133) into Eq. (1.134) yields the value l of the certain realization in the form

$$l = -\frac{1}{\rho\sigma_e}\ln\xi. \tag{1.135}$$

If the distance l is larger than the thickness of the scattering system, then this photon is detected as transmitted without any scattering. If, having passed the distance l, the photon remains within the scattering volume, then the possible events of photon–particle interaction (scattering or absorption) are randomly selected.

Within the spherical system of coordinates, the probability density of photon scattering along the direction specified by the angle of scattering θ between the directions of the incident and scattered photons and by the angle ϕ between the previous and new scattering planes is given as

$$p(\theta,\phi) = \frac{I_s(\theta,\phi)\sin\theta}{\int_0^{2\pi}\int_0^{\pi} I_s(\theta,\phi)\sin\theta d\theta d\phi}, \tag{1.136}$$

where $I_s(\theta, \phi)$ is the intensity of the light scattered in the direction (θ, ϕ) with respect to the previous direction of the photon, defined by angles ϑ and φ. For spherical particles, this intensity is given by the Mie formulas with allowance for the state of polarization (SOP) of each photon. An integral $I_s(\theta, \phi)$ over all scattering directions determines the scattering cross-section

$$\sigma_s = \int_0^{2\pi}\int_0^{\pi} I_s(\theta,\phi)\sin\theta d\theta d\phi. \tag{1.137}$$

The probability density of photon scattering along the specified direction, $p(\theta, \phi)$, depends on the Mueller matrix of the scattering particle $\mathbf{M}(\theta, \phi)$ (a single-scattering matrix) and the Stokes vector $\mathbf{S}$ associated with the photon. The single-scattering Mueller matrix $\mathbf{M}(\theta, \phi)$ links the Stokes vectors of the incident $[\mathbf{S}_i(0, 0)]$ and scattered $[\mathbf{S}_s(\theta, \phi)]$ light. For spherical scatterers, the elements of this matrix may be factorized

$$\mathbf{M}(\theta,\phi) = \mathbf{M}(\theta)\mathbf{R}(\phi). \tag{1.138}$$

The elements of the single-scattering matrix $\mathbf{M}(\theta)$ of spherical particles matrix are given by the Mie formulas,[1,12] which are functions of the scattering angle θ and diffraction parameter $x = 2\pi a/\lambda$, where a is the radius of the spherical particle and λ is the wavelength in the medium.

The matrix $\mathbf{R}(\phi)$ describes the transformation of the Stokes vector under rotation of the plane of scattering through the angle ϕ, which is defined by Eq. (1.132). Thus, the intensity of the light scattered by spherical particles is determined by the expression

$$I_s(\theta,\phi) = [M_{11}(\theta)I_i + (Q_i \cos 2\phi + U_i \sin 2\phi)M_{12}(\theta)], \tag{1.139}$$

where Q_i and U_i are the components of the Stokes vector of the incident light. As it follows from this equation, the probability $p(\theta, \phi)$ [Eq. (1.136)], unlike the scattering matrix [Eq. (1.138)], cannot be factorized; it appears to be parametrized by the Stokes vector associated with the scattered photon. In this case, one should use a rejection method to evaluate $p(\theta, \phi)$.

The following method of generating pairs of random numbers with the probability density $p(\theta, \phi)$ may be used.[300] In the 3-D space, the function $p(\theta, \phi)$ specifies some surface. The values (θ, ϕ) corresponding to the distribution $p(\theta, \phi)$ are chosen using the following steps: (1) a random direction (θ_ξ, ϕ_ξ) with a uniform spatial distribution is selected, the values of the random quantities θ_ξ, and ϕ_ξ distributed over the intervals $(0, \pi)$ and $(0, 2\pi)$, respectively, are found from the equations

$$\cos\theta_\xi = 2\xi - 1, \quad \phi_\xi = 2\pi\xi, \tag{1.140}$$

where ξ is a random number uniformly distributed over the interval (0,1); (2) the surface specified by the function $p(\theta, \phi)$ is surrounded by a sphere of radius $\mathbf{R}$, equal to the maximum value of the function $p(\theta, \phi)$, and a random quantity $r_\xi = \xi\mathbf{R}$ is generated; (3) the direction (θ_ξ, ϕ_ξ) is accepted as the random direction of the photon scattering at this stage, provided the condition $r_\xi \le p(\theta_\xi, \phi_\xi)$ to be satisfied. In the opposite case, steps 1 and 2 are repeated again.

The migration of the photon in the scattering medium can be described by a sequence of transformations for the related coordinate system. Each scattering event is accompanied by a variation of the Stokes vector associated with the photon. The new Stokes vector $\mathbf{S}_{n+1}$ is a product of the preceding Stokes vector, transformed to the new scattering plane, and the Mueller matrix $\mathbf{M}_k(\theta)$ of the scattering particle

$$\mathbf{S}_{n+1} = \mathbf{M}_k(\theta)\mathbf{R}_n(\phi)\mathbf{S}_n, \tag{1.141}$$

where the matrix $\mathbf{R}_n(\phi)$ [see Eq. (1.132)] describes rotation of the Stokes vector around the axis specifying the direction of propagation of the photon before the interaction.

For the chosen scattering direction, the Stokes vector is recalculated using Eq. (1.141) and expressions for elements of the single-scattering Mueller matrix for a homogeneous sphere made of an optically inactive material.[12] The value thus obtained is renormalized so that the intensity remains equal to unity. Thus, the Stokes vector associated with the photon contains information only about the variation of the SOP of the scattered photon. Real intensity is determined by the number of detected photons in the chosen direction within the detector aperture.

The above procedure is repeated as long as the photon appears to be outside the scattering volume. In this case, if the photon propagation direction intersects the surface of the detector, the photon is detected. Upon detection, the Stokes vector is rotated from the current plane of the last scattering to the scattering plane of the laboratory coordinate system. The values obtained are accumulated in the appropriate cells of the detector whose number is defined by the photon migration direction. Furthermore, with registering, the photon is classified in accordance with the scattering multiplicity and the length of a total path. For every nonabsorbed photon, the direction and the coordinates of a point at which it escapes the scattering volume, as well as the number of scattering acts it has experienced, were also recorded. The spatial distribution of radiation scattered by the scattering volume can be obtained with regard to polarization by analyzing the above data for a sufficiently great number of photons.

To find the full LSM of an object, one has to detect the light scattering for four linearly independent states of polarization of the incident light $\mathbf{S}_{1i}$, $\mathbf{S}_{2i}$, $\mathbf{S}_{3i}$, and $\mathbf{S}_{4i}$. This allows one to construct the following system of linear equations:

$$\mathbf{C}\mathbf{M}' = \mathbf{S}', \tag{1.142}$$

where $\mathbf{M}'$ is the column matrix composed of be found matrix elements of the LSM of the object, and $\mathbf{S}'$ is the 16-element vector containing the Stokes vector elements recorded upon light scattering for the four independent states of the incident light polarization. The transformation matrix $\mathbf{C}$ is determined by the choice of the initial set of the Stokes vectors of the incident light. Having solved this system of equations for the set of Stokes vectors:

$\mathbf{S}_{1i} = (1,1,0,0)$, $\mathbf{S}_{2i} = (1,-1,0,0)$, $\mathbf{S}_{3i} = (1,0,1,0)$, and $\mathbf{S}_{4i} = (1,0,0,1)$, one finds the desired LSM of the object, $\mathbf{M}' = \mathbf{M}$

$$\mathbf{M} = \frac{1}{2}\begin{bmatrix} I_1 + I_2 & I_1 - I_2 & 2I_3 - (I_1 + I_2) & 2I_4 - (I_1 + I_2) \\ Q_1 + Q_2 & Q_1 - Q_2 & 2Q_3 - (Q_1 + Q_2) & 2Q_4 - (Q_1 + Q_2) \\ U_1 + U_2 & U_1 - U_2 & 2U_3 - (U_1 + U_2) & 2U_4 - (U_1 + U_2) \\ V_1 + V_2 & V_1 - V_2 & 2V_3 - (V_1 + V_2) & 2V_4 - (V_1 + V_2) \end{bmatrix}, \tag{1.143}$$

where the elements of the Stokes vectors of the scattered light obtained in each of these four cases are denoted as $\mathbf{S}_n = (I_n, Q_n, U_n, V_n)$, $(n = 1, 2, 3, 4)$. As a result, one may calculate the angular dependencies for all elements of LSM with allowance for the contributions of multiple scattering.

1.8.2.3 Calculation of LSM for a multiple-scattering system

The simulation was performed for the systems of spherical particles with relative index of refraction, $m = 1.2$, which are uniformly distributed within a spherical volume at volume fraction $f = 0.01$.[300] In the calculations, the illuminating beam was assumed to be infinitely narrow and incident exactly upon the center of the scattering volume in the zero angle direction, and the scattered radiation is detected at different scattering angles in the far zone by a detector with the full angular aperture of 1 deg in the scattering plane and 5 deg in a plane that is perpendicular to the scattering one.

The calculated angular distributions of the total scattering intensity for different scattering systems of spherical particles with a small radius, $a = 50$ nm, or large radius, $a = 300$ nm, are presented in Fig. 1.17. The average multiplicity of scattering of the detected radiation increases with increasing dimensions of the scattering system. For systems of small particles at illumination in the visible range (633 nm), approximation of the Rayleigh scattering is applicable. For rather small dimensions of the scattering volume of 1 mm of diameter, the contribution of single scattering is predominant. This follows from the intensity angular dependence, which is rather isotropic, Fig. 1.17(a). As the dimensions of the scattering system increase, the fraction of contributions of the higher multiplicity scattering grows as well. For a 20-mm-diameter system, the detected light contains noticeable contributions of scattering of the 10th to the 20th multiplicity. With a further increase of the system dimensions, most of the incident light is scattered in the backward direction and the scattering intensity in the forward half-plane vanishes. For this reason, beginning from a certain value, the dimensions of the scattering system hardly affect the shape of the diagram of the scattering multiplicity distribution.

Systems composed of particles with a size of the order of the wavelength [Fig. 1.17(b)] also show an increase in the contributions of higher-order

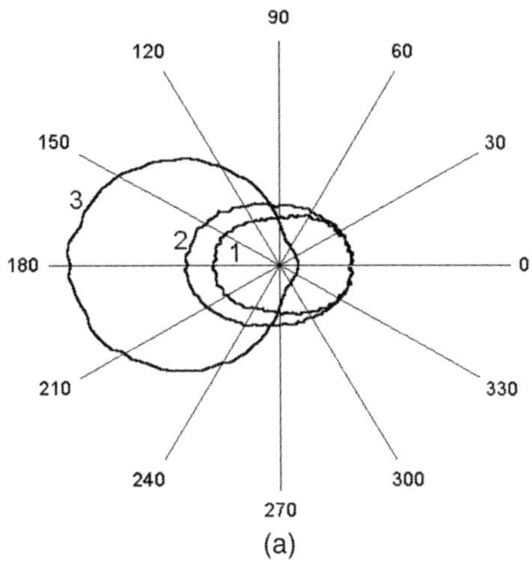

Figure 1.17 Angular distributions of the total scattering intensity for the multiple-scattering systems of spherical particles having relative refractive index $m = 1.2$ and uniformly distributed within a spherical volume at volume fraction $f = 0.01$: (a) particles with small radius, $a = 50$ nm, diameter of the system is equal to (1) 1, (2) 2, and (3) 20 mm; and (b) particles with large radius, $a = 300$ nm, diameter of the system is equal to (1) 0.002, (2) 0.2, and (3) 2 mm; the infinitely narrow unpolarized light beam is incident exactly upon the center of the scattering volume in the zero angle direction; the wavelength is 633 nm (see Refs. 9 and 300).

scattering with increasing dimensions of the scattering system. The system transforms from the forward- to backward-directed scattering mode at rather small thickness, 2 mm in diameter.

As is seen, the intensity of unpolarized light at the higher scattering multiplicity weakly depends on the scattering angle and carries almost no information about the size of the scattering particles. Note that systems of small particles at triple scattering may already be considered as nearly isotropic, while angular distributions for the large particles, strongly elongated in the forward direction at single scattering, remain anisotropic for sufficiently high scattering multiplicity [four to six scattering events for the system with a 0.2 mm diameter, Fig. 1.17(b)].

The view of the LSM elements' angular dependences under the conditions of multiple scattering differs substantially from that for the LSM of a single–scattering system. It is seen from Figs. 1.18 and 1.19 that the multiple scattering flattens the angular dependences of the LSM elements. The solid line shows the result of calculation of a normalized LSM for an isolated spherical particle with the similar radius and relative index of refraction. All elements of the LSM are normalized to the M_{11} element (total scattering intensity) along the given direction, and the element M_{11} is presented in the plot as normalized to unity in the forward direction; its actual intensity distributions are presented in Fig. 1.17.

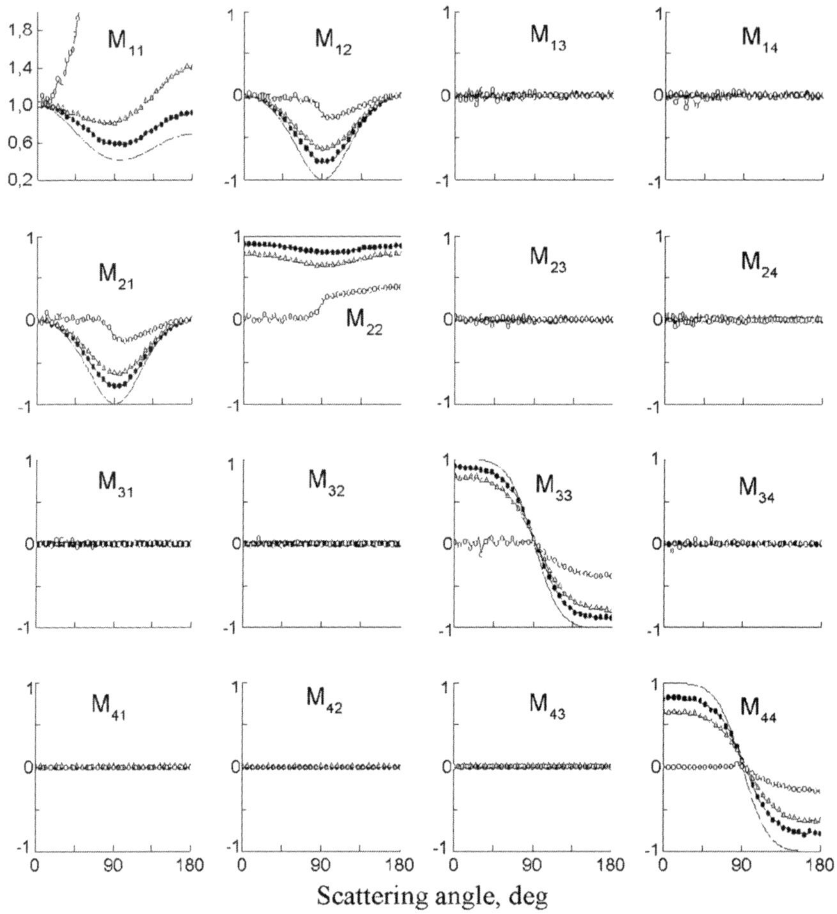

Figure 1.18 The MC simulation: the angular distributions of the LSM elements for the multiple-scattering systems of small spherical particles ($a = 50$ nm, $m = 1.2$) uniformly distributed within a spherical volume ($f = 0.01$); diameter of the system is equal to 1 mm (–•–), 2 mm (–△–), and 20 mm (–○–); the solid line shows the results of calculations in the approximation of single scattering; the infinitely narrow unpolarized light beam is incident exactly upon the center of the scattering volume in the zero angle direction; the wavelength is 633 nm (see Refs. 9 and 300).

Since the single-scattering angular distribution for particles with sizes substantially exceeding the Rayleigh limit is strongly asymmetric, the scattering intensity at large angles is very low. For this reason, one must trace the trajectories of a great number of photons to obtain good accuracy in this angular range. Therefore, to demonstrate the fine structure of the angular dependence of the matrix elements, one needs to use in the simulation at least 10^7–10^8 photons.

For the scattering by particle suspensions in a spherical volume of small diameter, almost all the detected photons are singly scattered. An increase in

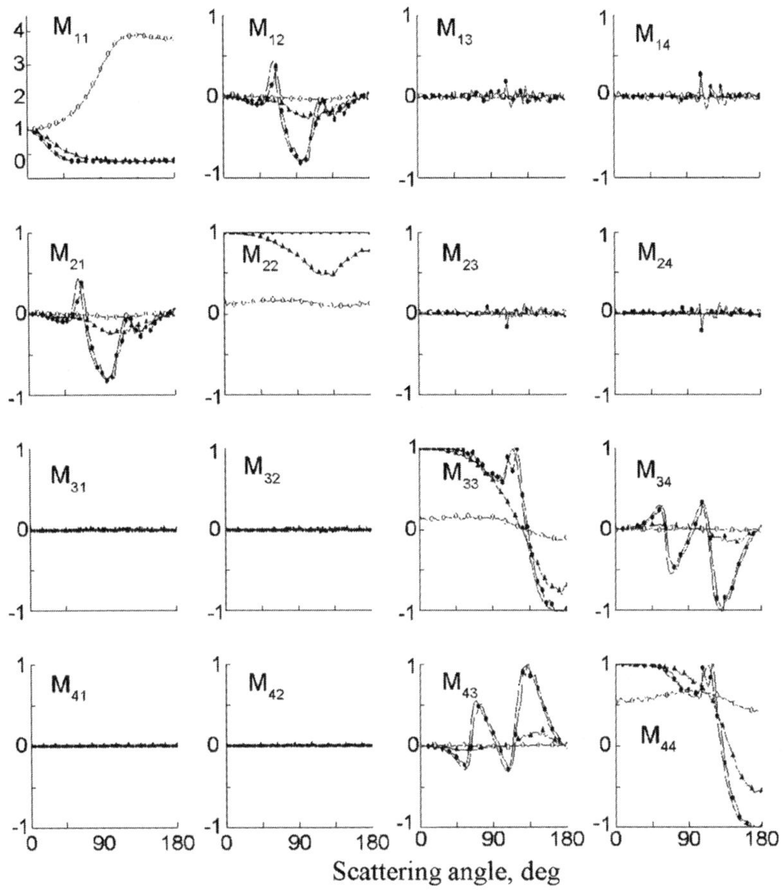

Figure 1.19 The MC simulation: the angular distributions of the LSM elements for the multiple-scattering systems of large spherical particles ($a = 300$ nm, $m = 1.2$) uniformly distributed within a spherical volume ($f = 0.01$); diameter of the system is equal to 0.002 mm (–•–), 0.2 mm (–Δ–), and 2 mm (–O–); the solid line shows the results of calculations in the approximation of single scattering; the infinitely narrow unpolarized light beam is incident exactly upon the center of the scattering volume in the zero angle direction; the wavelength is 633 nm (see Refs. 9 and 300).

the optical thickness considerably enhances the contribution of multiple scattering. The angular dependences of the LSM elements have a form close to the single-scattering LSM, provided that the optical thickness of the scattering system τ does not exceed unity for the systems of large particles considered ten or above for systems of small particles.

The multiple-scattering intensity (the element M_{11}) for a volume of large diameter decreases with increasing scattering angle slower than the single-scattering intensity. As the object diameter further increases, the backward scattering becomes predominant (see Figs. 1.17–1.19). In the systems of small particles (see Fig. 1.18), the growth of the multiple-scattering contributions is accompanied by a gradual decrease in magnitude of all of the elements except for M_{11}, i.e., the form of the LSM approaches that of the ideal depolarizer. In particular, the magnitudes of the elements M_{12} and M_{21} decrease in nearly the same way; the elements M_{33} and M_{44} also decrease in magnitude, but M_{44} decrease faster. As a result, multiple scattering gives rise to a difference in the detected values of the elements M_{33} and M_{44}, even for the systems of spherical particles. The values of the element M_{22} become smaller than unity, this decrease being more substantial in the range of scattering angles close to 90 deg. Thus, the manifestation of the effect of multiple scattering in monodisperse systems of spherical particles, which is revealed in the appearance of nonzero values of the differences $|M_{33}–M_{44}|$ and $|1–M_{22}|$, is similar to the manifestation of the effect of nonsphericity of the scatterers observed under conditions of single scattering.[207]

For large particle systems, the multiple scattering also decreases the magnitudes and smooths out the angular dependences of the normalized elements of the LSM (see Fig. 1.19). The corresponding angular dependences, as compared to the LSM of small particles, show the following specific features: the minimum value of the element M_{22} is reached not at 90 deg, but rather at large scattering angles; the fine structures of the angular dependences for all elements are smeared even in the presence of a small fraction of the multiply scattered light; and, finally, a very important result that the element M_{44}, unlike other elements, in the limit of high scattering multiplicity, tends to 0.5 rather than to zero for all scattering angles. Such a form of the LSM means that the radiation scattered by the large particles holds the preferential circular polarization at higher scattering multiplicities. This result may serve as a confirmation of preferential survival of different types of polarization under conditions of multiple scattering for different sizes of scattering particles or tissue structures.[276,279,302]

The process of multiple scattering of the photons during their migration is considered as a series of successive rotations of their coordinate systems, determined by the scattering planes and directions. Since these rotations are random, the detected photons will be randomly polarized and, hence, the detected light will be partially depolarized. The depolarization will increase with the increasing multiplicity of scattering. For the moderate optical thicknesses (object diameter of 0.2 mm, $f = 0.01$), the depolarizing ability is strongly different for different directions. The scattered light may be almost completely polarized in the region of small scattering angles and completely depolarized at large angles ($\theta = 120$ deg) and be partly polarized in the backward direction. The angular range of the strongest depolarization corresponds to the angle at which the element M_{22} acquires minimum values (see Fig. 1.19).

The simulated dependences allow one to estimate the limits of applicability of the single-scattering approximation when interpreting the results of experimental studies of disperse scattering systems. It follows from these simulations that modifications of the LSM of monodisperse systems of spherical particles due to the effects of multiple scattering have much in common with modification of the LSM of singly scattering systems upon deviation of the shape of the particles from spherical. This fact imposes serious limitations on the application of the measured LSM of tissues for the inverse problem solving to determine particle nonsphericity. The appropriate criteria to distinguish the effects of multiple scattering and particle nonsphericity have to be developed.

It is important to note that the comparison of MC simulation accounting for all orders of multiple scattering with the analytical double-scattering model had indicated no essential change in the back-scattering polarization patterns.[303] This is due to the fact that the main contribution comes from near-double-scattering trajectories in which light suffers two wide-angle scatterings and many near-forward scatterings among multiple-scattering trajectories. The contributions of such multiple but near-double scattering trajectories are obviously well approximated by the contributions of the corresponding double-scattering trajectories.

The above MC technique of photon trajectory modeling is well suited to the simulation of multiple-scattering effects in a system of randomly arranged particles. Furthermore, this scheme allows for an approximate approach to describe the interference effects caused by space particle ordering. To this end, one should include the interference of scattered fields into calculations of the single-scattering Mueller matrix and integral cross sections for a particle. In other words, at the first stage one accounts for the interference effects for simulation of the *single-scattering* properties, and then uses these properties in the MC simulation of *multiple scattering*. Such an approach is admissible if the size of a region of the local particle ordering is substantially smaller than the mean free photon pathlength.

In general, for polarized light propagated in a strongly scattering medium, the multiple scattering decreases the magnitudes and smooths out the angular dependences of the normalized LSM elements, characterizing polarized light interaction with the medium. For media composed of large particles, specified by a high degree of single-scattering anisotropy or considerable photon transport length, the scattered radiation holds the preferential circular polarization at higher scattering multiplicities. This theoretical result serves as a confirmation of preferential survival of different types of polarization under conditions of multiple scattering for different sizes of scattering particles or tissue structures.

Given the known character of the Stokes vector transformation for each scattering act, the SOP following multiple light scattering in a highly scattering medium can be found using various approximations of the multiple-scattering theory or the MC method. For small particles, the effects of multiple scattering are apparent as the broken symmetry relationship between LSM elements $M_{12}(\theta) \neq M_{21}(\theta)$, $M_{33}(\theta) \neq M_{44}(\theta)$, and a significant reduction of linear polarization of the light scattered at angles close to $\pi/2$.[304]

1.8.2.4 Degree of linear and circular polarization of light interacting with tissues

For a system of small spatially uncorrelated particles, the degree of linear ($i = L$) and circular ($i = C$) polarization in the far region of the initially polarized (linearly or circularly) light transmitted through a layer of thickness d is defined by the relation[279]

$$P_i \cong \frac{2d}{l_s} \sinh(l_s/\xi_i) \cdot \exp(-d/\xi_i), \tag{1.144}$$

where $l_s = 1/\mu_s$ is the scattering length, and

$$\xi_i = (\zeta_i \cdot l_s/3)^{0.5} \tag{1.145}$$

is the characteristic depolarization length for a layer of scatterers $d \gg \xi_i$, $\zeta_L = l_s/[\ln(10/7)]$, $\zeta_C = l_s/(\ln 2)$.

As can be seen from Eq. (1.144), the characteristic depolarization length for linearly polarized light in tissues that can be represented as ensembles of Rayleigh particles is approximately 1.4 times greater than the corresponding depolarization length for circularly polarized light. One can employ Eq. (1.144) to assess the depolarization of light propagating through an ensemble of large-scale spherical particles whose sizes are comparable with the wavelength of incident light (Mie scattering). For this purpose, one should replace l_s by the transport length $l_{tr} \cong 1/\mu_s'$ and take into account the dependence on the size of scatterers in ζ_L and ζ_C. With the growth in the size of scatterers, the ratio ζ_L/ζ_C changes. It decreases from $\sim$1.4 down to 0.5 as $2\pi a/\lambda$ increases from 0 up to $\sim$4; it remains virtually constant at the level of 0.5 when $2\pi a/\lambda$ grows from $\sim$4 to 15.

MC numerical simulations and model experiments in aqueous latex suspensions with particles of various diameters demonstrate that there are three regimes of the dependence of the ratio of the degree of linear polarization to the circular polarization for transmitted light, P_L/P_C, on d/l_{tr}.[279] In the Rayleigh range, P_L/P_C grows linearly with the increase of d/l_{tr}. In the intermediate range, this ratio remains constant. In the range of Mie scattering, this quantity decreases linearly. Such behavior of this quantity is associated with the transition of the system under study from isotropic scattering to anisotropic. Qualitatively, the physical mechanism behind the change in the depolarization is associated with the fact that a considerable probability of backward scattering in each event of light–medium interaction (isotropic scattering) does not distort linear polarization, whereas backward scattering for circular polarization is equivalent to the reversal of polarization direction (similar to reflection from a mirror), i.e., it is equivalent to depolarization. For the same reason, in the case of a strongly elongated scattering phase function, the degree of circular polarization in an individual scattering event (anisotropic scattering) for light propagating in a layer should remain nonzero for lengths greater than the degree of linear polarization.

These arguments also follow from the above MC simulation of polarized light interaction with multiple-scattering systems[300] and experimental works.[302,305] For example, at high-scattering multiplicities the radiation scattered by the large particles holds the preferential circular polarization (LSM element M_{44} is far from zero for all scattering angles) (see Fig. 1.19). At multiple scattering, the LSM for a monodisperse system of randomly distributed spherical particles is modified to be approximately identical to the single-scattering LSM of the system containing nonspherical particles, or optically active spheres.

Thus, different tissues or the same tissues in various pathological or functional states should display different responses to a probe with linearly and circularly polarized light. This effect can be employed in both optical medical tomography and for determining optical and spectroscopic parameters of tissues. As follows from Eq. (1.144), the depolarization length in tissues should be close to the mean transport pathlength l_{tr} of a photon, because this length characterizes the distance within which the direction of light propagation and consequently, the polarization plane of linearly polarized light, become totally random after many sequential scattering events.

Since the length l_{tr} is determined by the parameter g characterizing the anisotropy of scattering, the depolarization length should also substantially depend on this parameter. Indeed, the experimental data of Ref. 306 demonstrate that the depolarization length l_p of linearly polarized light, which is defined as the length within which the ratio $I_{\|}/I_{\perp}$ decreases down to 2, displays such a dependence. The ratio mentioned above varied from 300 to 1, depending on the thickness of the sample and the type of tissue. These measurements were performed within a narrow solid angle ($\sim 10^{-4}$ sr) in the direction of the incident laser beam. The values of l_p differed considerably for the white matter of brain and tissue from the cerebral cortex: 0.19 and 1.0 mm for $\lambda = 476$–514 nm and 0.23 and 1.3 mm for $\lambda = 633$ nm, respectively. Human skin dermis (bloodless) has a depolarization length of 0.43 mm ($\lambda = 476$–514 nm) and 0.46 mm ($\lambda = 633$ nm). The depolarization length at $\lambda = 476$–514 nm decreases in response to a pathological change in the tissue of aorta wall: 0.54 mm for a normal tissue, 0.39 mm for the stage of tissue calcification, and 0.33 mm for the stage of necrotic ulcer. Whole blood with a low hematocrit is characterized by a considerable depolarization length (about 4 mm) at $\lambda = 633$ nm, which is indicative of the dependence on the parameter g, whose value for blood exceeds the values of this parameter for tissues of many other types and can be estimated as 0.966–0.997[9,307] (see Chapter 2).

In contrast to depolarization, the attenuation of collimated light is determined by the total attenuation coefficient μ_t. For many tissues, μ_t is much greater than μ_s'. Therefore, in certain situations, it is impossible to detect pure ballistic photons (photons that do not experience scattering), but the forward scattered photons retain their initial polarization and can be used for imaging.[308,309] This is illustrated by Fig. 1.20, which presents the experimental data for the decay of the degree of linear polarization P_L obtained for various tissues and blood as a function of light transmission.[302]

The authors of Ref. 310 experimentally demonstrated that laser radiation retains linear polarization on the level of $P_L \leq 0.1$ within $2.5 l_{tr}$. Specifically, for skin irradiated in the red and NIR ranges, we have $\mu_a \cong 0.4\ \text{cm}^{-1}$, $\mu_s' \cong 20\ \text{cm}^{-1}$, and $l_{tr} \cong 0.48$ mm. Consequently, light propagating in skin can retain linear polarization within a length of about 1.2 mm. Such an optical path in a tissue corresponds to a time delay on the order of 5.3 ps, which provides an opportunity to produce polarization images of macro-inhomogeneities in a tissue with a spatial resolution equivalent to the spatial resolution that can be achieved by the selecting of photons using more sophisticated time-resolved techniques. In addition to the selection of diffuse-scattered photons, polarization imaging makes it possible to eliminate specular reflection from the surface of a tissue, which allows one to use this technique to image microvessels in facile skin and detect birefringence and optical activity in superficial tissue layers.[311–316]

Polarization imaging is an innovative technology in biophotonics.[9,294,308–326] The most prospective approaches for polarization tissue imaging, in particular, linear polarization degree mapping, 2-D backscattering Mueller matrix measurements, polarization-sensitive OCT, and

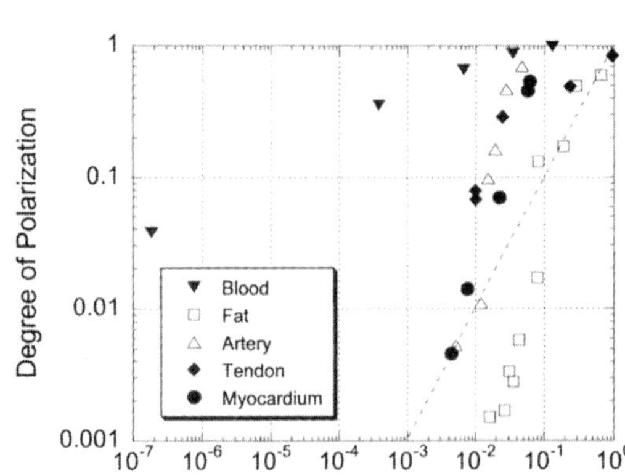

Figure 1.20 Degree of linear polarization in different tissues as a function of the sample optical transmittance, $I_{out}/I_{in} \equiv T$, on 633 nm. Each point is an average of three measurements.[302] The error bars representing standard deviation of measurements are smaller than the used symbols.

full-field polarization-speckle technique will be discussed in this and the following chapters.

The registration of 2-D polarization patterns for the backscattering of a polarized incident narrow laser beam is the basis for the polarization imaging technique. The major informative images can be received using the back-scattering Mueller matrix approach. To determine each of the 16 experimental matrix elements, a total of 16 images should be taken at various combinations of input and output polarization states.

In weakly absorbing media showing a small-angular multiple scattering, the degree of linear polarization for a Henyey–Greenstein phase function is described by[325]

$$P_L = -[(\mu_s'z)^4/2\theta^2] \cdot \left[\sqrt{1+(\theta/\mu_s'z)^2} - 1]^2 \cdot [1+(\theta/\mu_s'z)^2\right]. \tag{1.146}$$

This means that in a very small angle range ($\theta \ll \mu_s'z$) the degree of polarization does not depend on the depth (z)

$$P_L = -\theta^2/8. \tag{1.147}$$

At the wings of the scattering angle dependence ($\theta \gg \mu_s'z$), it tends to

$$P_L = -\theta^2/2, \tag{1.148}$$

which equals the degree of polarization of singly scattered light.

1.8.2.5 Simulation of two-dimensional reflection and transmission LSM

Rakovic et al.[327] present the MC-based simulations of the diffusely back-scattering intensity patterns that arise from illuminating a turbid medium with a polarized laser beam. It is rigorously shown that, because of the axial symmetry of the system, only seven elements of the effective back-scattering Mueller matrix are independent. Other studies have demonstrated that information on the properties of turbid medium can be obtained by analyzing the SOP of diffusely back-scattering polarized laser light. Applications of this technique that have been investigated include the measurements of the average particle size, the scattering coefficients and the anisotropy factor of particle suspensions,[268] and the study of biological material.[328] Hielscher et al.[268] generalized the concept of an effective Mueller matrix and measured the 2-D Mueller matrix of back-scattering light. Rakovic et al.[327] extended this work to include a theoretical, computational, and experimental verification of the use of the incoherent scattering theory to explain the patterns seen in polarized light back-scattering from the suspensions of polystyrene spheres (Fig. 1.21). Yao et al. measured depth-resolved Mueller-matrix images using polarization-sensitive optical coherence tomography.[329,330]

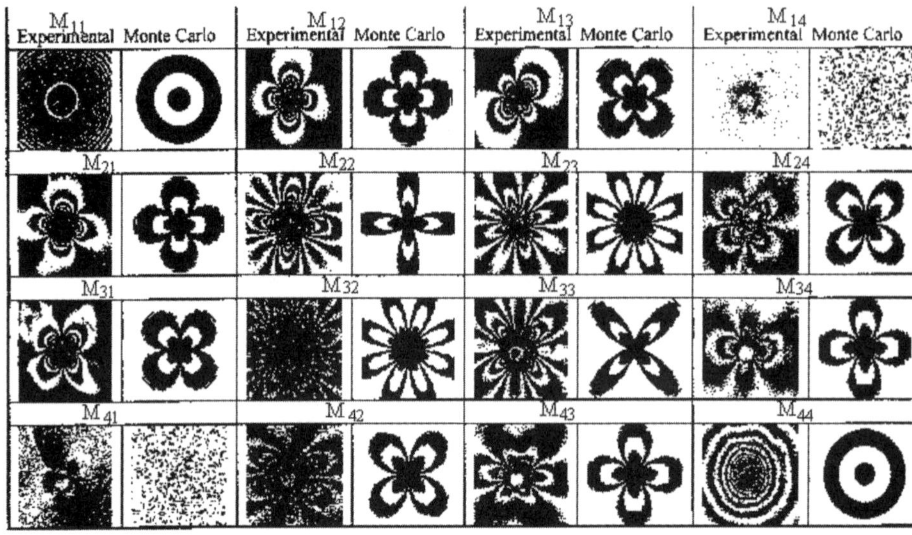

Figure 1.21 Experimental and MC counted back-scattering Mueller matrix.[327] The phantom was comprised of a 0.05-wt% suspension of polystyrene spheres (of diameter 2.02 μm) in deionized water. The approximate size of each image was 1.6 by 1.6 cm. The light wavelength was 633 nm. The smaller values correspond to contours located further from the center of each plot.

Figure 1.22 shows the reflection and the transmission Mueller matrices of a turbid medium with parameters characteristic to tissues.[294] The patterns of the reflection Mueller matrix are identical to those reported in Refs. 288 and 327 (see Fig. 1.21). The symmetries in the patterns can be explained by the symmetries in the single-scattering Mueller matrix and the turbid medium.[327] The transmission Mueller matrix has different patterns from the reflection Mueller matrix. One of the noticeable differences appears in elements M_{31} and M_{13}, which are antisymmetric in the reflection Mueller matrix but symmetric in the transmission Mueller matrix. This difference is caused by the mirror effect in the reflection process of the scattered light.

1.8.2.6 Simulation of the spectra of transmission, reflection, and scattering

The collimated transmission is understood as a detection of a relative fraction of photons running out of the scattering layer close to the forward direction that exhibit a slight lateral shift corresponding to their entrapment by the specified receiving aperture.

If light is incident onto a layer of parallel cylinder particles normal to the cylinder axes, the scattered light propagates in a plane perpendicular to these axes. Consequently, this problem has to be considered as a 2-D one. When calculating the collimated transmission or the integral characteristics of light scattering, the results calculated for linearly polarized light that is normally incident to the surface of a system of layers that are randomly

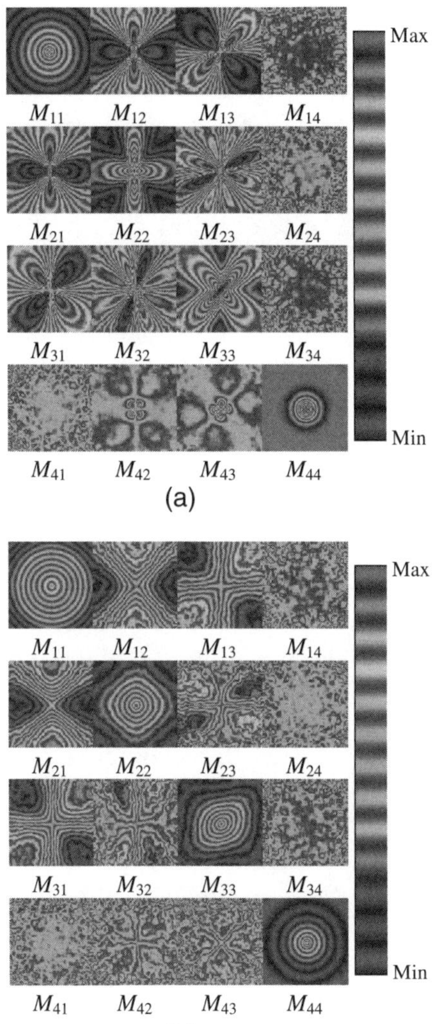

Figure 1.22 (a) Reflection and (b) transmission Mueller matrices of a slab of turbid medium with a scattering coefficient of 4 cm^{-1} and a scatterer radius of 0.102 μm.[294] The calculated Mueller-matrix elements are normalized to the M_{11} element to compensate for the radial decay of intensity. Each of the images is displayed with its own color map to enhance the image contrast. The size of each image is 4×4 cm^2. (See color plates.)

aligned around the direction of light propagation are equivalent to those obtained for light incident onto a system of uniformly oriented layers. It is, therefore, possible to restrict oneself by considering the 2-D problem if the effects of rescattering between different layers are negligible, which substantially simplifies the problem. It is, however, unfair for the angular dependence of scattering.

The simulation shows that dependence of transmission on the layer thickness is described by two exponential terms. This allows for study of the transition from predominantly single scattering to the diffusion regime of scattering as thickness increases. The MC method also makes it possible to study the kinetics of light scattering as the relative index of refraction in a disperse medium is varied. The above effects have been investigated in detail with the example of the eye sclera optical clearing (see Section 1.10 about the sclera optical clearing and temporal spectra). The MC method is also useful for simulating the spectral characteristics of scattering in different directions.[331]

1.9 Nephelometry and Polarization Methods for Diagnostics of Bioobjects

This section deals with the study of methods for the diagnostics of the parameters of biological particles and tissues with regard to the angular and polarization characteristics of elastically scattered light. These methods are simple to realize and they easily undergo computer control. Moreover, the study of the angular dependence of the scattered light intensity for measuring the size or density of particles suspended in a fluid (the so-called nephelometry) has a long history and is widely used in very different areas of biology and medicine.[332] As for the analysis of polarization effects, it is made use of very rarely with light scattering, although this analysis yields more extensive information on a scattering object. The polarization of light scattered by biological cells can reveal properties not evident from measurements of the total scattered intensity alone. Therefore, we consider in detail the problems of laser polarization nephelometry, the methods for the computer-controlled measurement of angular dependencies of the LSM elements, and potentials for diagnostics.

The most complete information on elastic light scattering by an object is contained in the LSM (see Section 1.3). The LSM elements and their dependencies on the scattering angle contain information on the structure and properties of the object under study. The solution of the appropriate inverse problems can provide the size distribution function of scattering particles, their index of refraction, shape and orientation. The rigorous solution is, however, possible only for a small class of objects. Most biological objects have a quite complex structure, hence even the solution of a straightforward problem is a quite difficult task. Along with this, in a number of cases a qualitative estimation of the object's properties would be sufficient, requiring no exact solution of the inverse problem. The general view of experimental scattering matrix and symmetry relations for its elements allow us to compare the object under study with a certain class of scattering systems. The evaluation of the state of the object under study by using the experimental LSM can be done with the help of the relations between the matrix elements. These relations can also be a criterion for estimation of the correctness of experimental results.

1.9.1 Relations between the LSM elements: depolarization criterion

The nature and mathematical expressions of relations between 16 LSM elements were considered in Refs. 333 through 335. It has been shown that there are nine equalities relating the matrix elements for a nondepolarizing system that is usually understood as a system whose Mueller matrix is obtained by coherently summing the amplitude matrices of single elements of the system.

It has been established[336,337] that for depolarizing systems six of the above mentioned equalities are transformed into the following inequalities:

$$
\begin{aligned}
&(M_{11}+M_{22})^2-(M_{12}+M_{21})^2 \geq (M_{33}+M_{44})^2+(M_{43}-M_{34})^2\\
&(M_{11}-M_{22})^2-(M_{12}-M_{21})^2 \geq (M_{33}-M_{44})^2+(M_{43}+M_{34})^2\\
&(M_{11}+M_{21})^2-(M_{12}+M_{22})^2 \geq (M_{13}+M_{23})^2+(M_{14}+M_{24})^2\\
&(M_{11}-M_{21})^2-(M_{12}-M_{22})^2 \geq (M_{13}-M_{23})^2+(M_{14}-M_{24})^2\\
&(M_{11}+M_{12})^2-(M_{21}+M_{22})^2 \geq (M_{31}+M_{32})^2+(M_{41}+M_{42})^2\\
&(M_{11}-M_{12})^2-(M_{21}-M_{22})^2 \geq (M_{31}-M_{32})^2+(M_{41}-M_{42})^2.
\end{aligned}
\tag{1.149}
$$

Another important characteristic for the scattering matrices is the number Tr(M)

$$
\mathrm{Tr}(M)=\sum_{i,j=1}^{4} M_{ij}^2, \tag{1.150}
$$

where M is the scattering matrix normalized to the first element.

It was shown that the equality Tr(M) = 4 is a necessary and sufficient condition for this matrix M to describe a nondepolarizing object.[338] For depolarizing objects,Tr(M) takes values from 1 to 4. These inequalities enable one to clearly classify various objects and in some cases to verify the experimental results. The application of the above relations to the analysis of the LSM of different biological objects has been described.[339] The symmetry relations for the LSM of an ensemble of particles are considered.[340] A collection of particles of random orientation has a scattering matrix of the form

$$
M(\theta)=\begin{pmatrix}
m_{11}(\theta) & m_{12}(\theta) & m_{13}(\theta) & m_{14}(\theta)\\
m_{12}(\theta) & m_{22}(\theta) & m_{23}(\theta) & m_{24}(\theta)\\
-m_{13}(\theta) & -m_{23}(\theta) & m_{33}(\theta) & m_{34}(\theta)\\
m_{14}(\theta) & m_{24}(\theta) & -m_{34}(\theta) & m_{44}(\theta)
\end{pmatrix}. \tag{1.151}
$$

If all particles have a plane of symmetry or, equivalently, particles and their mirror particles are presented in equal numbers, we obtain the block-diagonal structure

$$M(\theta) = \begin{pmatrix} m_{11}(\theta) & m_{12}(\theta) & 0 & 0 \\ m_{12}(\theta) & m_{22}(\theta) & 0 & 0 \\ 0 & 0 & m_{33}(\theta) & m_{34}(\theta) \\ 0 & 0 & -m_{34}(\theta) & m_{44}(\theta) \end{pmatrix}. \tag{1.152}$$

In the case of back-scattering, consequences for the linear and circular depolarization ratios have been reported by Mishchenko and Hovenier.[341]

1.9.2 Angular dependence of the scattering intensity of nondepolarized light

One of the main characteristics to be examined with light scattering is the angular dependence of the scattered intensity. The measurement is provided by illuminating an object with the light beam and recording the intensity of light scattered at different angles. Therefore, an optical nephelometer, the device measuring scattered light, is based on a source with small angular divergence and a light detector with a certain angle of view. Lasers appear to be the most suitable light sources in this case because of sufficient directness and high intensity of the laser beams.

To find particle sizes in a range of 0.02–0.2 μm, the method of the angular dependence asymmetry is used. It is based on recording of the scattering angular dependence deviation from that of Rayleigh form with greater particle size. The method of full angular dependence measurement is applied to quantify the particle sizes in a range of 0.1–10 μm, and that of small-angle light scattering is used to measure the particle sizes in a range of 1–300 μm.[342] Laser nephelometry is employed in immunology to quantitatively estimate the antigen–antibody reaction and to determine the concentration of components participating in the reaction. In particular, laser nephelometry was used to study the influence of various factors (temperature, pH of the medium, and others) on the rate of the reaction and of the formation of antigen–antibody complexes.[343]

A number of physically justified light scattering methods were developed to investigate the morphological parameters of erythrocytes (sizes, shape, concentration, and aggregation) under conditions approaching their native state.[344–346] They include the procedures of deriving the hematocrit ratio, aggregation function of erythrocytes, and concentration of hemoglobin derivatives via elastic light scattering. Particularly, a strong dependence of scattering angular dependences on the erythrocyte sizes that has been revealed for angles of 1–30 deg permits one to obtain their size-distribution function. At the same time, the surface roughness of pathological erythrocytes is determined with a high accuracy by a higher scattered laser radiation intensity within angles exceeding 90 deg.

1.9.3 Measurements of the angular dependences of the scattering matrix elements

The simplest measuring technique for the LSM elements evaluation is based on the conventional nephelometer and additional optical elements placed ahead of and after the scattering medium under study. The linear polarizers and quarter-wave plates are employed as such elements.[12] The results of such measurement are the combinations of the LSM elements obtained by multiplying the matrices of optical elements placed ahead of the scattering object, the matrices of the scattering object itself, and those of optical elements placed after the scattering object (Fig. 1.23). In general, four measurements are necessary to obtain one LSM element. Despite the fact that this technique is reasonable even if quite cumbersome, its application may be followed by significant relative errors associated with small matrix elements obtained as differences of big signals. These errors can be avoided by modulating the polarization state in the incident and/or the scattered fields.

A laser polarization nephelometer has been described with the principle of operation relying on the modulation of the polarization state of incident and scattered light by mechanically rotating polarization elements.[339,347,348] The principle of operation is described by the following matrix equation:

$$\mathbf{S} = A_a \Phi_a M \Phi_p A_p \mathbf{S_0}, \tag{1.153}$$

where $\mathbf{S}$ and $\mathbf{S_0}$ are the Stokes vectors of recorded and source radiation, respectively; A and Φ are the Mueller matrices for the linear polarizer and the phase plate, respectively. As the phase plates are rotated, the intensity recorded by a photodetector, i.e., the first element of the Stokes vector $\mathbf{S}$, would depend on time. By multiplying the matrices in Eq. (1.153) and performing the appropriate trigonometric transformations, one can show that the output intensity would be represented as a Fourier series, namely,

$$I = a_0 + \sum_{k=1}^{K} (a_{2\,k} \cos 2\,k\varphi + b_{2\,k} \sin 2\,k\varphi), \tag{1.154}$$

where φ is the angle of orientation of the phase plate axis. The coefficients of this series are defined by the values of the matrix M elements of the object

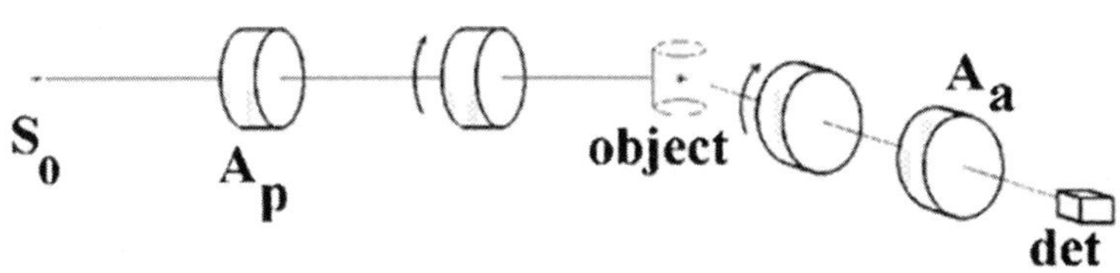

Figure 1.23 Schematics of a polarization nephelometer with the rotating phase plates.

under study, and their measurement ensures a system of linear equations to determine the matrix M. The number of equations and the degree of stipulation for this system of equations are dependent on choosing the ratio between the rotation rates of the phase plates. The proper choice of the rotation rates relation as 1:5 allows one to get an optimally stipulated system of linear equations to find the full matrix M of the object under study.

More sophisticated elecro-optical[349,350] and magnitoptical[351] modulation principles of polarization states are also used.

1.9.4 The LSM for some biological objects

A review of experimental studies of scattering polarization properties of biological tissues and cells the reader could find in the multiple literature.[9,76,207,283,294,302,306,310–321,328] The following regularities can be mentioned that allow for parameters of scattering particle structures or suspensions to be classified by analyzing their LSM. The distinction between elements M_{22} and M_{11} would serve as a measure for scattering particles to be nonspherical. These peculiarities were studied for different types of pollen[352] and marine organisms.[353,354] However, a similar distinction between elements M_{22} and M_{11} for a model of spherical particles may be caused by multiple scattering (see above).

As noted in Refs. 76, 352, and 355–358, element M_{34} is most specific for various bioparticles. This element is sensitive to small morphological alterations of scatterers. It has been shown that the element M_{34} is affected by a small surface roughness on a sphere.[356] It was also proven that M_{34} measurements may be the basis for determining the diameter of rod-shaped bacteria (*Esgherichia coli* cells) that is difficult to be measured using other techniques.[357]

In Refs. 352 and 355, the measuring results for the whole LSM of some biological particles are presented. A high specificity of element M_{34}/M_{11} was shown for every type of biological scatterer. Stable distinctions were revealed in the values of parameter M_{34}/M_{11} for spores of two mutant varieties of bacteria, which are distinguished by the variations of their specific structure and are invisible by means of traditional techniques. The distinctions of other matrix elements, however, were seen less clearly for these two types of similar scatterers. When scattering by biological particles is well described by the RG approximation, $M_{34} = 0$. Thus, M_{34} is a matrix element that suffers strongly from the deviation of the particle parameters from those that satisfy the RG approximation. Possibly, this is the reason for the M_{34}/M_{11} to be so sensitive to the characteristics of biological scatterers.[12]

The polarization characteristics for suspension of biological particles have been described in Ref. 207, where the sensitivity of different matrix elements to the variation of the scatterer shape and size is analyzed. It is noted that the magnitudes of elements M_{33} and M_{44} measured in the backward scattering direction may serve as an indicator for the particle nonsphericity.

The polarization nephelometry is also used to examine blood corpuscles. The protocol for determination of the real part of the index of refraction of blood corpuscles based on the study of the angular structure of the LSM nonzero elements is described by Korolevich et al.[346] This protocol is useful for the normal or gamma distribution of polydisperse particles and does not require data on the particle concentration; it is only necessary to obey the condition of single scattering. The technique for determination of the real part of the relative index of refraction m is reduced to finding a scattering angle at which the LSM element is zero within a range of scattering angles from 80 to 120 deg. Further, the relative index of refraction m is derived accordingly to nomograms valid in the range from 1.02 to 1.07. If the element is not zero within the angular range of 80–120 deg, then $m > 1.07$, and it is needed to determine a scattering angle at which element M_{34} is also zero. The measuring of angular dependencies for the total LSM of blood erythrocytes enables one to distinguish between disc-like and spherulated cells[345] (Fig. 1.24). As it was noted in Ref. 76, the comparison between measured signals for all types of human white blood cells allows one to distinguish between two types of granulocytes.

The results of measuring the angular dependence of the LSM elements of the sliced samples of human skin dermis, bones, and muscular tissues were presented.[359] The LSM analysis of the tissue slices shows a high level of birefringence and a random nature of orientation of the bone tissue local structures. A determined orientation of optically active bundles of the muscular tissue is revealed in a quasi-harmonic behavior of the LSM element angular dependence. The random orientation of collagen bundles in the skin dermis results in an LSM transformation to its form inherent to a multiple-scattering diffuser.

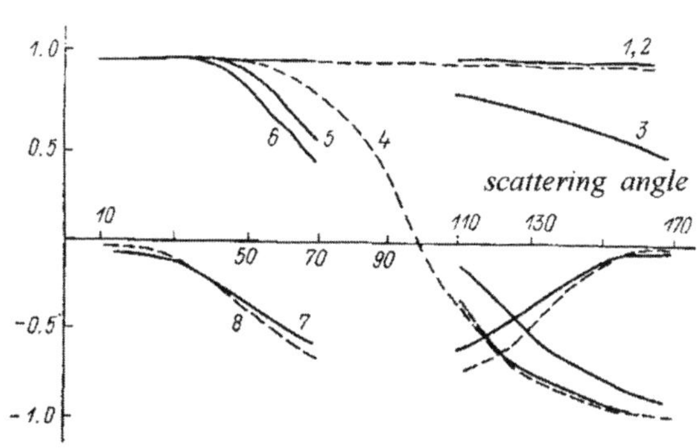

Figure 1.24 Angular distributions for the LSM elements: M_{22} (1, 2, 3) M_{33} (4, 5, 6), M_{12} (7), and M_{21} (8) for disc-like erythrocytes (3, 4, 7, 8) and spherocytes (2, 6). Dashed lines refer to theoretical analysis using the Mie formulae (from Ref. 345).

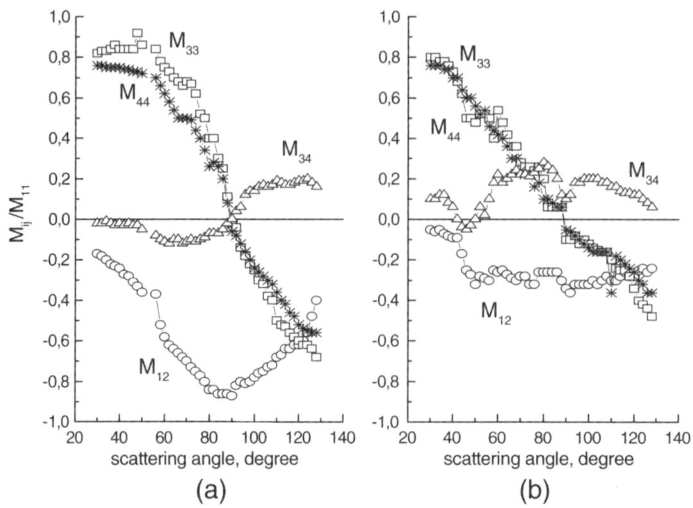

Figure 1.25 Experimental angular dependencies for the LSM elements of (a) normal (age of 56 y, 5 h after death), and (b) cataractous (age of 88 y, 5 h after death) human lenses. Measurements were performed at a wavelength of 633 nm.[360]

Study of transmission and scattering polarization properties of a tissue gives a basis that allows one to consider in what way the alteration of spatial protein organization would lead to turbidity. The measurements of the angular dependences of LSM elements for the human eye lens carried out using the laser polarization nephelometer[249] demonstrate substantial distinctions in these dependencies for normal and turbid (cataractous) lenses[360] (Fig. 1.25).

These distinctions are due to large nonspherical scattering particles appearing in the medium of a turbid lens (because of formation of high-molecular proteins). A transparent lens is featured by a monodisperse system of small-diameter scatterers. A turbid lens contains a reasonable fraction of larger scatterers. The high sensitivity of the LSM angular dependencies to the variations in the medium structure makes it possible to employ the LSM measurements for early diagnostics of the alterations in the tissue structure that are related to the cataract appearance.

The possibility of elastic light scattering utilization to quick identification of the bacteria suspensions is a highly attractive field of applications. However, the reproducibility of the light scattering measurements is often insignificant for various bacteria and other bioparticles obtained in different laboratories, unlike the scattering properties measured for suspensions of physical particles (i.e., suspension of identical polystyrene spheres serves as a standard). For example, the reproducibility and sensitivity of polarized light scattering were examined for cylinder samples of bacterial suspensions under different growing conditions.[361] The angular dependencies of certain LSM elements (predominantly M_{34}/M_{11}) were studied. The M_{34}/M_{11} angular dependencies were found to be highly correlated with the bacterial cell sizes. From the viewpoint of significant variations revealed to different growing conditions, it was concluded that a better technique is necessary to distinguish scattering patterns associated with such alterations of bacterium properties.

1.9.5 Effects of circular light probing and optical activity

Many tissues and systems of bioparticles demonstrate the effects of optical activity that are manifested in circular dichroism and circular birefringence. The optical activity of biological objects may be conditioned by the optical activity of the substance they are formed from and by their structure peculiarities. Circular intensity differential scattering (CIDS) is a difference between scattered intensities for left and right circularly polarized (RCP) incident light. The CIDS effects can be investigated by measuring the LSM element M_{14}.[12] The so-called form-CIDS is an anisotropy caused by the helical structure of a particle.[76] The CIDS interrelation with the scatterer structure has been considered by Bustamante et al.[362–364] The measurements of CIDS are used to study secondary and ternary structures of macromolecules[365] and the polymerization of hemoglobin in sickling RBC.[366]

The experimental studies of the polarization properties of laser radiation scattered by optically active tissues[359] show that laser polarization nephelometry can be employed for developing noninvasive methods for diagnostics of the bone tissue structure.

Circularly and elliptically polarized laser light and a standard optical polarimeter were used to observe the alterations of polarization properties of light back-scattered from tissue samples at cancer development.[323,367] It was shown that by tracking the Stokes vector of the detected light on the Poincaré sphere, a differentiation between normal and cancerous tissue can be assessed.

If a tissue contains a substance exhibiting optical activity, then the polarization measurements provide a noninvasive procedure for determining the concentration of this substance. For example, an important diagnostic problem is the monitoring and precise control of blood sugar elevation for patients with diabetes.[368–371]

Polarimetric quantification of glucose is based on the phenomenon of optical rotatory dispersion whereby a chiral molecule in an aqueous solution will rotate the plane of linearly polarized light passing through the solution.[6,9,294,368–371] The angle of rotation depends linearly on the concentration of the chiral species, the pathlength through the sample, and a constant for the molecule called the specific rotation. The net rotation is expressed as $\varphi = \alpha_\lambda LC$, where α_λ is the specific rotation for the species in $dm^{-1}(g/L)^{-1}$ at the wavelength λ, L is the pathlength in dm, and C is the concentration in g/L. Glucose in the body is dextrorotatory (rotates light in the right-handed direction) and has a specific rotation of $+52.6\ dm^{-1}(g/L)^{-1}$ at the sodium D-line of 589 nm.

At physiological concentrations and pathlengths of about 1 cm, optical rotations due to glucose are on the order of 0.005 deg. A number of techniques for obtaining measurements with this high degree of accuracy exist and generally fall into two categories: those that utilize crossed polarizers to measure rotation via amplitude changes, and those that measure the relative phase shift of modulated polarized light passing through the sample.[9,294,368–371]

It is difficult to measure *in vivo* sugar concentration in blood because strong light scattering by the skin causes light depolarization. For this reason, a number of investigators have suggested the anterior chamber of the eye (aqueous humor) as a site that is well suited for polarimetric measurement since scatter in the eye is generally very small compared to other tissues.

The use of optical polarimetry for the noninvasive measurement of physiological glucose concentration in the anterior chamber of the eye of New Zealand white rabbits was reported.[371] Measurements were acquired using a custom-designed laser-based optical polarimetry system (Fig. 1.26). Aqueous humor-based polarimetric measurements were obtained by coupling light through the anterior chamber of the eye. Blood glucose levels were first stabilized and then altered with intravenous dextrose and insulin administration and measured every 3–5 min with a standard glucometer and intermittently with a glucose laser analyzer. Acquired polarimetric glucose signals were calibrated to measured blood glucose concentration. Errors in glucose concentration prediction were shown to be related to gross movement of the rabbit during the procedures, incurring time-varying corneal birefringence effects that directly affect the measured polarimetric signal. These effects can be compensated for with appropriate design modifications. The technique provides a basis for the development of a noninvasive polarimetric glucose monitor for home, personal, or hospital use.

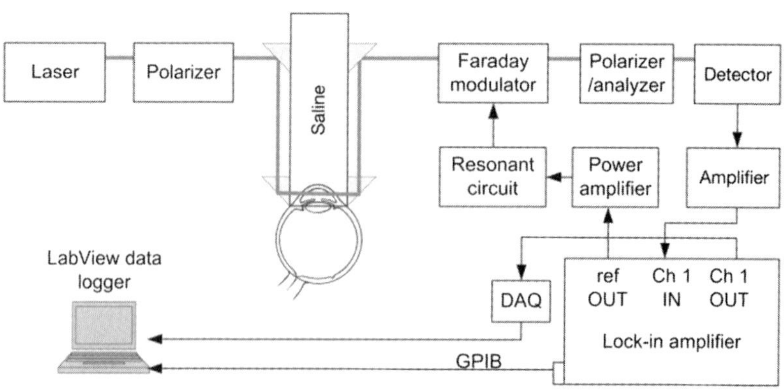

Figure 1.26 The polarimetric experimental setup employed for the sensing glucose concentration in the eye. DAQ, data acquisition; GPIB, general purpose interface bus.[371]

1.10 Controlling Optical Properties of Tissues

Reflection, absorption, and scattering in tissues and blood can be effectively controlled by different methods.[9,10,372–399] Tissue as a scattering medium shows all optical effects that are characteristic to turbid physical systems. It is well known that turbidity of a dispersive physical system can be effectively controlled using an immersion liquid by matching of refractive indices of the scatterers and the ground material. The living tissue allows one to control its optical (scattering) properties using various physical and chemical actions such as compression, stretching, dehydration, coagulation, UV irradiation, exposure to low temperature, and impregnation by chemical solutions, gels, and oils.[9,10,372–399]

Control of *in vivo* tissue optical properties is very important for many medical applications. A number of laser surgery, therapy, and diagnostic technologies include tissue compression and stretching for better transportation of the laser beam to underlying layers of tissue. The human eye compression technique allows one to perform transscleral laser coagulation of the ciliary body and retina/choroid.[375–377] The possibility of selective translucence of the upper tissue layers should be very useful for developing eye globe imaging techniques and for detecting local inhomogeneities hidden by a highly scattering medium in functional tomography. Results on control of human sclera optical properties by tissue impregnation with hyperosmotic chemicals, such as trazograph (x-ray contrast), glucose and polyethylene glycol (PEG), as well as hypaque-60 (x-ray contrast), were reported.[9,10,372–374,378,379,383–386]

In general, the reduced scattering coefficient μ_s' of tissue is dependent on refractive index mismatch between the extracellular fluid and the cellular membrane; between cytoplasm, cell nucleus, organelles, melanin, and cellular membrane; and for fibrous tissue (like sclera, dermis, and breast) between an interstitial medium and collagenous fibers. For hematous tissue like the liver, its impregnation by solutes with different osmolarity also leads to refractive index matching and reduction of the scattering coefficient, however, the effect is not so pronounced as for skin and sclera due to cells changing size as a result of osmotic stress.[372,373]

Soft tissue is composed of closely packed groups of cells entrapped in a network of fibers through which water percolates. At a microscopic scale, the tissue components have no pronounced boundaries. They appear to merge into a continuous structure with spatial variations in the refractive index. As it was discussed, to model such a complex structure, it is necessary to resort to a statistical approach.

It has already been shown that the tissue components that contribute most to the local refractive-index variations are the connective tissue fibers (bundles of elastin and collagen), cytoplasmic organelles (mitochondria, lysosomes, and peroxisomes), cell nuclei, and melanin granules.[9,204,248,378,400–403]

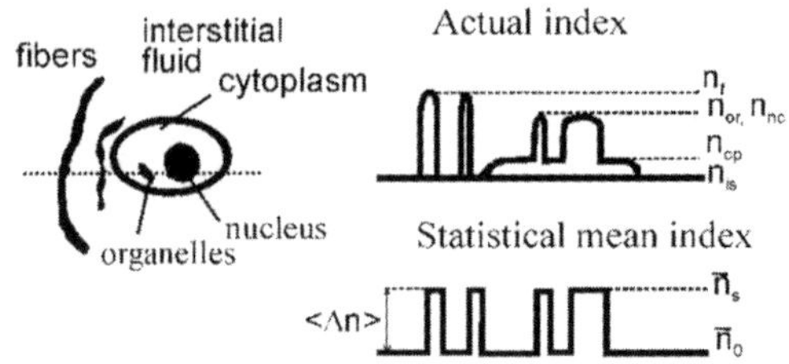

Figure 1.27 Spatial variations of the refractive index of a soft tissue.[204] A hypothetical index profile through several tissue components is shown along with the profile through a statistically equivalent volume of homogeneous particles.

Figure 1.27 shows a hypothetical index profile formed by measuring the refractive index along a line in an arbitrary direction through a volume of tissue. The widths of the peaks in the index profile are proportional to the diameters of the elements, and their heights depend on the refractive index of each element relative to that of its surroundings. In accordance with this model, the origin of the index variations will be presented by a statistically equivalent volume of discrete particles having the same index but different sizes.

The statistical mean index profile in Fig. 1.27 illustrates the nature of the approximation implied by this model. The average background index is defined as the weighted average of refractive indices of the cytoplasm and the interstitial fluid, n_{cp} and n_{is}, as[204]

$$\bar{n}_0 = f_{cp} n_{cp} + (1 - f_{cp}) n_{is}, \tag{1.155}$$

where f_{cp} is the volume fraction of the fluid in the tissue contained inside the cells, $n_{cp} = 1.367$, and $n_{is} = 1.355$.[402] Since approximately 60% of the total fluid in soft tissue is contained in the intracellular compartment, it follows from Eq. (1.155) that $\bar{n}_0 = 1.362$. The refractive index of a particle can be defined as the sum of the background index and the mean index variation

$$\bar{n}_s = \bar{n}_0 + \langle \Delta n \rangle, \tag{1.156}$$

which can be approximated by another volume–weight average

$$\langle \Delta n \rangle = f_f (n_f - n_{is}) + f_{nc}(n_{nc} - n_{cp}) + f_{or}(n_{or} - n_{cp}). \tag{1.157}$$

Here, subscripts *f*, *is*, *nc*, *cp*, and *or* refer to the fibers, interstitial fluid, nuclei, cytoplasm, and organelles, respectively, which are the major contributors to index variations. The terms in parentheses in this expression are the differences between the refractive indices of the three types of tissue components and their respective backgrounds; the multiplying factors are the volume fractions of the elements in the solid portion of the tissue. The refractive index of the connective-tissue fibers is about 1.47, which corresponds to about 55% hydration of collagen, its main component. The nucleus and the cytoplasmic organelles in mammalian cells that contain similar concentrations of proteins and nucleic acids, such as mitochondria and the ribosomes, have refractive indices that lie within a relatively narrow range (1.38–1.41).[402] Accounting for this and supposing that $n_{nc} = n_{or} = 1.40$, the mean index variation can be expressed in terms of the fibrous-tissue fraction f_f only:

$$\langle \Delta n \rangle = f_f(n_f - n_{is}) + (1 - f_f)(n_{nc} - n_{cp}). \tag{1.158}$$

Collagen and elastin fibers comprise approximately 70% of the fat-free dry weight of the dermis, 45% of the heart, and 2 to 3% of the nonmuscular internal organs.[204] Therefore, depending on tissue type, $f_f = 0.02$–0.7, which corresponds to $<\Delta n> = 0.035$–0.09.

The mean refractive index $\bar{n}$ of a tissue is defined by the refractive indices of material of the scattering centers n_s and ground matter n_0 [see Eq. (1.155)]

$$\bar{n} = f_s n_s + (1 - f_s) n_0, \tag{1.159}$$

where f_s is the volume fraction of the scatterers.

The $n_s/n_0 \equiv m$ ratio determines the scattering coefficient. For example, in a simple monodisperse model of scattering dielectric spheres (Mie theory)[404]

$$\mu_s' = 3.28\pi a^2 \rho_s \left(\frac{2\pi a}{\lambda}\right)^{0.37} \times (m - 1)^{2.09}, \tag{1.160}$$

where a is the radius of the particles and ρ_s is their volume density. Equation (1.160) is valid for noninteracting Mie scatterers $g > 0.9$; $5 < 2\pi a/\lambda < 50$; $1 < m < 1.1$.

It follows from Eq. (1.160) that even a 5% change in the refractive index of the ground matter ($n_0 = 1.35 \rightarrow 1.42$), when that of the scattering centers is $n_s = 1.47$, will cause a sevenfold decrease of μ_s'. In the limit of equal refractive indices for nonabsorbing particles, $m = 1$, and $\mu_s' \rightarrow 0$.

Light scattering and absorption of particles that compose tissue is calculated by Mie theory. The relevant parameters are the size (radius a) and shape of the particles, their wavelength-dependent complex refractive index

$$n_s(\lambda_0) = n_s'(\lambda_0) + i n_s''(\lambda_0), \tag{1.161}$$

the refractive index of the dielectric host (ground material) $n_0(\lambda_0)$, and the relative refractive index $m = n_s/n_0$. The imaginary part of the complex refractive index of scatterer material is responsible for light loss due to absorption. Mie theory yields the absorption and scattering efficiencies and the phase function from which the absorption and scattering coefficients $\mu_s = \rho\sigma_s$ and $\mu_a = \rho\sigma_a$ and the scattering anisotropy g are calculated; ρ is the scatterer (particle) density, σ_s and σ_a are the scattering and absorption cross sections, respectively [see Eqs. (1.15)–(1.17)].

The introduction of the specific scattering and absorption coefficients extrapolated to a volume fraction of 100% is useful for describing scattering and absorption properties of tissues with controlled structural properties.[405,406] In that case and when the particles are sufficiently diluted to prevent dependent scattering, the scattering, transport scattering and absorption coefficients are proportional to the dimensionless volume fraction of scatterers $c_s \equiv f_s$

$$\mu_s = c_s\bar{\sigma}_s, \mu_s' = c_s\bar{\sigma}_s[1 - g(\lambda_0, a)], \mu_a = c_s\bar{\sigma}_a, \tag{1.162}$$

where the specific scattering and absorption coefficients $\bar{\sigma}_s$ and $\bar{\sigma}_a$ are expressed in cm^{-1}. The optical parameters of broad-banded particle size distributions are averaged values over the distribution weighted by the volume fractions of particles with different diameters. The resulting specific optical coefficients are the averaged values and can be defined analogously to Eqs. (1.15)–(1.17).

Mie theory predicts that scattering introduced by spherical micrometer-sized particles is strongest if particle radius and wavelength are of the same order of magnitude. Mie theory is strictly applicable only to particles of particular regular shapes, but results are still useful if the shape is irregular. The oscillatory structure of the scattering coefficient and anisotropy factor as a function of particle size, which is observed with spherical particles (Fig. 1.28), is averaged out. The transport scattering coefficient increases

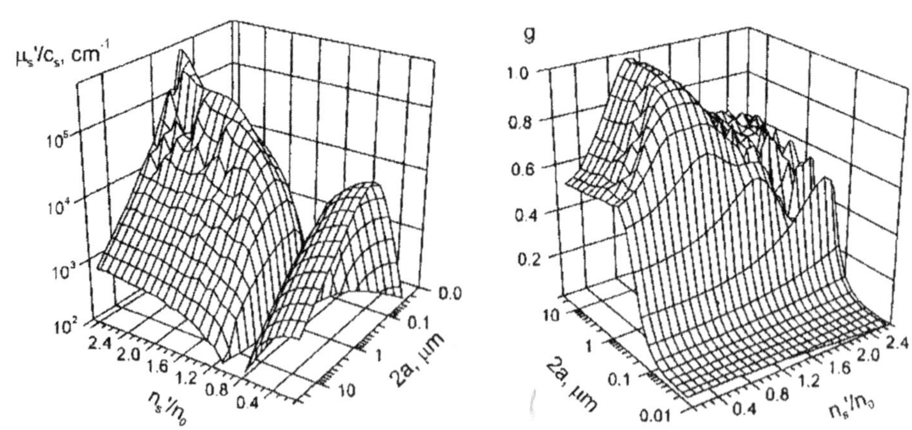

(a) (b)

Figure 1.28 Scattering properties of nonabsorbing particles at the wavelength 633 nm calculated by Mie theory. (a) Transport scattering coefficient strongly depends on both the particle size and relative refractive index. This graph is approximately symmetric. The axis of symmetry is at $n_s'/n_0 = 1$. While the transparent scattering coefficient equals zero at that point, (b) the scattering anisotropy factor is maximal. In some parts of the range shown, the functions are not monotonous but rapidly oscillating.[406]

strongly with the ratio n'_s/n_0. In its turn, the scattering anisotropy factor is maximal when this ratio approaches 1 (Fig. 1.28).

For the matched refractive indices of scatterers and background material the scattering coefficient goes to zero, that means that only absorption is responsible now for the light beam extinction. However, as it follows from Mie theory, absorbing particles suspended in an index-matched medium cause strongly forward-directed resonance scattering. Light absorption by such particles is smaller than expected from their bulk absorption coefficient.[405,406] For 1-μm diameter particles with $n_s = 1.6$ and bulk absorption coefficient of their material equal to 10^4 cm^{-1} in an index-matched medium, the particle system absorption coefficient $\mu_a = c_s \times 4120$ cm^{-1}.

If particle size and ratio of refractive indices are fixed, the wavelength dependencies are caused by the spectral variation of the ratio of particle size and wavelength. For particles with a refractive index close to that of the host (see Fig. 1.28), the scattering coefficient of the particle system with a diameter of particles smaller than the wavelength decreases with wavelength, while that of a system with a diameter of particles larger than the wavelength is almost constant. The scattering anisotropy factor depends less on the wavelength. There are plateaus if the particles are much smaller (isotropic scattering) or larger (highly anisotropic scattering) than the wavelength with a steep increase between.[406]

It is possible to increase significantly the transmission through a soft tissue by its squeezing (compressing) or stretching. The optical clearing of living tissue is due to its optical homogeneity achieved through the removal of blood and interstitial fluid from the compressed site. This results in a higher refractive index of the ground matter, whose value becomes close to that of scatterers (cell membrane, muscle, or collagen fibers).[375,376] Indeed, the absence of blood in the compressed area also contributes to altered tissue absorption and refraction properties. Certain mechanisms underlying the effects of optical clearing in tissues were proposed in Refs. 375 and 376. It should be emphasized, however, that squeezing-induced effects in tissues containing little blood, such as sclera, are characterized by a marked inertia (a few minutes) because of the relatively slow water diffusion from the compressed region.[375,378] It was suggested that compression of sclera may displace water from interspace of collagen fibrils increasing the protein and mucopolysaccharide concentrations. Since these proteins and sugars have a refractive index closer to that of the collagen fibrils, a more index-matched environment can be created. On the other hand, compression reduces specimen thickness, which might increase the effective scatterer and chromophore concentration inside the tissue.[376] Therefore, compression may also give rise to the increase in tissue scattering and absorption coefficients. Sometimes the scatterer concentration increase is likely to be more dominant than the reduction in index mismatch.[376]

It is possible to achieve a marked impairment of scattering by matching the refractive indices of scattering centers and ground matter by means of intratissue administration of appropriate chemical agents. Conspicuous experimental optical clearing in the human sclera and skin in the visible and NIR wavelength range induced by administration of x-ray contrast (verografin, trazograph, hypaque, omnipaque, glucose, propylene glycol, polyethylene glycol, glycerol and other solutions) has been described in Refs. 9, 374, and 378–399. Coordination between refractive indices in multicomponent tissues showing polarization anisotropy (e.g., cornea) leads to its decrease.[9,248] Osmotic and diffusive processes that occur in sclera treated with an optical clearing agent (OCA) are also important.[378] Osmotic phenomena appear to be involved when optical properties of cells and tissues are modulated by sugar, alcohol, and electrolyte solutions. This may interfere with the evaluation of hemoglobin saturation with oxygen or identification of such absorbers as cytochrome oxidase in tissues by optical methods.[372,373]

A marked clearing effect through the human[385] and the rat[383] skin and the rabbit sclera[385,386] occurred for an *in vivo* tissue within a few minutes of topical application or intratissue injection of glycerol, glucose, verografin, or trazograph.

Albumin, a useful protein for index matching in phase contrast microscopy experiments, can be used as the immersion medium for tissue study and imaging.[402] Proteins smaller than albumin may offer a potential alternative to the relatively high scattering of albumin. Alternatively, medical diagnosis or contrasting of a lesion image can be provided by the enhancement of tissue scattering properties by applying, for instance, the acetic acid that was used successfully as a contrast agent in optical diagnostics of cervical tissue.[402] It has been suggested that the aceto-whitening effect seen in cervical tissue is due to coagulation of nuclear proteins. Therefore, an acetic acid probe may also prove extremely significant in quantitative optical diagnosis of precancerous conditions due to its ability to selectively enhance nuclear scatter.[402]

The loss of water by tissue seriously influences its optical properties. For instance, during *in vitro* study of the human aorta the absorption coefficient increased by 20% to 50%, especially in the visible range, when an average of 46.4% of the total tissue weight was lost as a result of dehydration.[406] The weight loss was accompanied with an average thickness shrinkage of 19.5%. The loss of water decreased the sample thickness. Primarily because of shrinkage the absorption coefficient was increased in the spectral range of 400 to 1300 nm. There was only a slight increase of 2–15% of the reduced scattering coefficient in the visible range.

Natural physiological changes in cells and tissues are also responsible for their altered optical properties, which may be used as a measure of these changes, e.g., for glucose concentration monitoring.[369,380–382]

A model of the human sclera in a local region can be represented as a slab with a thickness d that is filled by thin and long dielectric cylinders (collagen fibers) with average diameter ~ 100 nm and refractive index $n_c = 1.474$. The cylinders are located in planes that are parallel to the slab surface, but within each plane their orientations are random. The interstitial space is filled by a homogeneous base substance with the refractive index $n_0 = 1.345$. This refractive index is a controlled parameter and can be changed in the range from 1.345 to 1.474. For $n_c = n_0 = 1.474$, the medium becomes totally homogeneous and optically transparent if absorption of scatterers is small.[405,406]

The described model of tissue is applicable to any fibrous tissue including skin dermis and muscle. Indeed, refractive indices and fiber diameters should be changed. The transmission of collimated light by a tissue layer of thickness d is defined as Bouguer's law [Eq. (1.2)], where $\bar{\tau} \equiv \mu_t = \mu_a + \mu_s$ is the extinction coefficient, and μ_a and μ_s are the absorption and scattering coefficients, respectively. For the human sclera at the wavelength $\lambda =$ 650 nm, the absorption coefficient $\mu_a \cong 0.08$ cm^{-1}, and the reduced scattering coefficient $\mu_s' = \mu_s(1-g) \cong 25$ cm^{-1}, where g is the scattering anisotropy factor.[408] For $g = 0.9$, $\mu_s \cong 250$ cm^{-1}.

To describe the kinetics of the refractive index and corresponding decrease of the scattering coefficient when a chemical agent diffuses within the interstitial substance of a tissue, the model of free diffusion was used.[378,409] The approximate solution of diffusion equation for a plane slab with a thickness d, at the moment $t = 0$ and concentration c_0 (the initial concentration of this substance within the slab is equal to 0; i.e., $t = 0$, $0 \leq x \leq d$, $c = 0$) has the form

$$c \cong c_0[1 - \exp(-t/\tau)], \tag{1.163}$$

where

$$\tau = \frac{d^2}{\pi^2 D} \tag{1.164}$$

is the diffusion constant, D is the diffusion coefficient, and d is the thickness of the scleral sample. This relation is very close to the relation describing diffusion through a partially permeable membrane.[378,409] These equations are valid for diffusion through a homogeneous slab when OCA is applied to both surfaces of the slab. However, due to its fibrous structure, a tissue can be presented as a porous material that leads to modification of the agent diffusion coefficient

$$D = D_0/p. \tag{1.165}$$

Here, D_0 is the chemical agent diffusion coefficient within the interstitial fluid, and p is the porosity coefficient defined as

$$p = (V - V_c)/V, \tag{1.166}$$

where V is the volume of the scleral sample and V_c is the volume of collagen fibers.

When applying an OCA, a change in pH level of the environment is very important, because swelling or shrinkage of tissue is expected. The swelling or shrinkage of a fibrous tissue is caused not only by the increase (decrease) of collagen (elastin) fibril size but also by the increase (decrease) of the sample volume due to the rise (diminution) of the mean distance between fibrils. It is well known that the change of the environmental pH to the more acidic or more alkaline side from a colloid isoelectric point increases the degree of swelling. It is explained by the appearance of a positive or negative charge of colloid particles and, therefore, increase of hydration degree. In general, the initial pH condition of the tissue under study and the acid or alkaline nature of the impregnated solution may lead to different dependencies of tissue thickness or volume on chemical agent concentration (or time of impregnation) due to changes of pH. Such behavior of a tissue sample should be taken into account when optical measurements are used for estimation of tissue properties. For example, the swelling or shrinkage was observed for different initial conditions of scleral tissue sample preparation and solutions used.[9,378,384–386]

A fiber optic photodiode array spectrometer was used for *in vitro* and *in vivo* light transmittance or reflectance measurements of sclera in the spectral range 400–800 nm.[386] In *in vitro* study, the conjunctiva and ciliary body, as well as the retina with choroid were removed. The mean thickness of samples was about 0.5 mm. A sample under study was fixed on a plastic plate with a square aperture of $5 \times 5\ \text{mm}^2$ (effective impregnation by a chemical agent via both surfaces of the sample was provided) and placed in a 5-ml cuvette filled with an OCA. Various solutions (glucose, trazograph, verografin, and propylene glycol) were used, which do not have strong absorbing bands within the used spectral range.

To understand the mechanisms of scleral tissue optical clearing, the collimated transmittance spectra and change of the scleral sample weight were measured concurrently with the administration of glucose solution.[386] Figure 1.29 illustrates the kinetics of transmittance spectra. It is easily seen that the untreated sclera is a poorly transparent media for visible light. Glucose administration makes this tissue highly transparent. As it follows from Fig. 1.30, the characteristic time response of human optical clearing is about 5 min.

Based on such measurements, the two-stage model of action of hyperosmotic liquid on a fibrous tissue was suggested.[386] At the first stage, which takes place approximately 5 min after a sample is placed in a glucose solution, besides substantial optical clearing, an increase of sample mass was observed. The latter is due to considerable swelling of collagen fibers in glucose solution with pH = 3.5 (40% glucose from a drugstore, prepared for

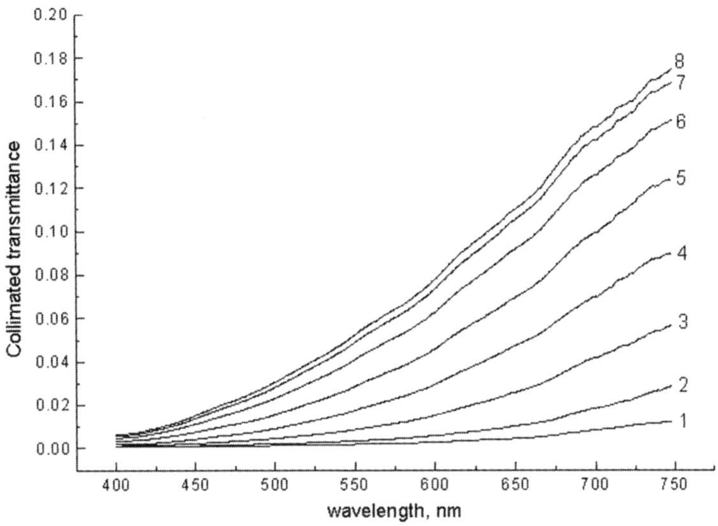

Figure 1.29 The time-dependent collimated transmittance of the human sclera sample impregnated by 40% glucose: (1) 10 sec, (2) 1 min, (3) 2 min, (4) 3 min, (5) 4 min, (6) 5 min, (7) 5 min, and (8) 8.5 min after the scleral sample was immersed in glucose.[386]

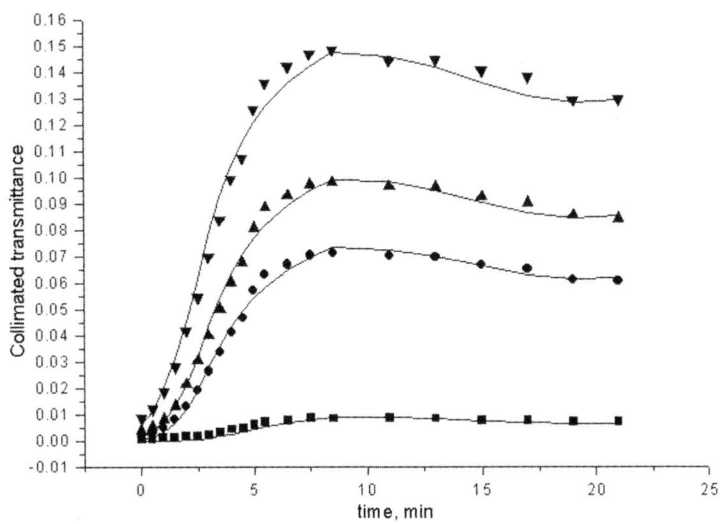

Figure 1.30 The time-dependent collimated transmittance of the human scleral sample measured at 420 nm (squares); 589 nm (circles); 630 nm (up triangles); and 700 nm (down triangles) concurrently with administration of 40% glucose.[386]

intravenous injection). In spite of tissue thickness (mass or volume) increase, the first stage is characterized by a highly increased tissue transmittance caused by a very effective refractive index matching.

During the following stage, which occurs after 5 min, the matter diffusion (and correspondingly refractive index matching) saturates, but tissue thickness increases due to the glucose solution acting in the space between fibrils. At this stage, transmittance is also saturated and even goes slightly down.

To estimate the matter diffusion coefficient, the experimental data for collimated transmittance (Figs. 1.29 and 1.30) and the time-dependent measurements of tissue sample mass (thickness) changes under immersion solution action can be used. The model of glucose transport in fibrous tissue is described in Ref. 386. Equations (1.165) and (1.166) are the basis for this model, which can be used for reconstruction of the diffusion constant. The estimated average value of the diffusion coefficient of glucose transport in the scleral sample is equal to $(3.45 \pm 0.46) \times 10^{-6}$ cm^2/sec at a temperature of 20 °C.

In vivo measurements were done for a rabbit eye (see Fig. 1.31). The surface temperature of the rabbit eye was ~38 °C. As an OCA for the optical clearing of sclera, 40% glucose was administered to the eye by drops. A significant decrease of the reflectance during the first 5 min of glucose administration is seen. Dips appearing at 420, 530, and 570 nm are caused by blood perfusion. The lower reflectance at 420 nm is caused by the strong absorption of blood. Evidently, more fast decay reflects blood perfusion kinetics due to eye conjunctiva and sclera inflammation caused by light irradiation and the osmotic action of glucose. Because blood has less influence at 630 and 700 nm, measured kinetic responses can be used for *in vivo* estimation of diffusion coefficient for glucose in scleral tissue.

The kinetic response of optical properties (modulation depth and phase shift of intensity modulation of the back-scattered light) of the human eye sclera via interval of a chemical agent (solution, gel, or oil) administration was measured using a photon-density wave (frequency-domain) technique.

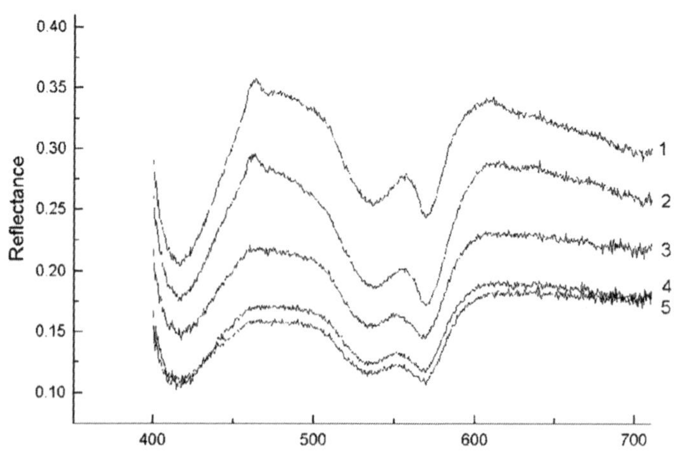

Figure 1.31 The *in vivo* time-dependent reflectance spectra of a rabbit eye sclera measured concurrently with administration of 40% glucose: (1) 1 min, (2) 4 min, (3) 21 min, (4) 25 min, and (5) 30 min after dropping the glucose solution into the rabbit eye.[386]

When intensity of the light source is modulated at a frequency ω, a photon density wave is induced in a scattering medium[9,372,380,408] (see Chapters 3 and 7)

$$A(r) = A_{dc} + A_{ac}\exp[-i(\omega t - \theta)], \tag{1.167}$$

where A_{dc}, A_{ac}, and $(\omega t - \theta)$ are the dc and ac components of the amplitude of the photon-density wave and its phase, respectively.

Photon-diffusion theory provides independent determination of the absorption and reduced scattering coefficients from measurements at a single modulation frequency. The expressions for the measured quantities as the phase delay θ and ac amplitude A_{ac} have been presented elsewhere[9] (see also Chapters 3 and 7). These expressions depend on the source–detector separation r, reduced scattering coefficient μ_s', and absorption coefficients μ_a.

The intensity and phase of photon-density waves generated by an NIR optical source were measured at several source–detector separations. The light source was a laser diode with a wavelength of 786 nm and 4-mW power at the end of a coupled multimode fiber (core diameter 62.5 μm).[374] The intensity modulation depth of approximately 80% at a frequency of 140 MHz was provided by modulation of the injection current of the laser diode. The experimental setup was designed at the University of Pennsylvania.

A multifiber detection system with small source–detector separations together with a Dicon multichannel fiber optic switcher has been used for immersion experiments on the human sclera *ex vivo*. The whole human eyeball was studied. Data shown in Fig. 1.32 are the temporal changes of ac amplitude during trazograph administration for three different source–detector separations and two different techniques of immersion solution administration—by injection and by drops. The clearing of scleral tissue was observed during the first 3 min of trazograph administration by injection. For small source–detector separations (about 1–2 mm) and a relatively large one (3.5 mm), the temporal

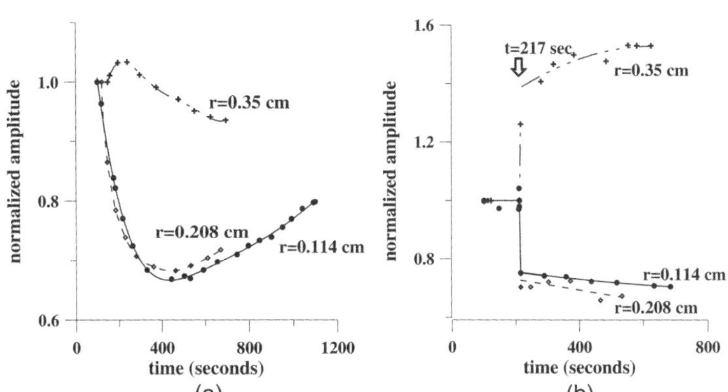

Figure 1.32 Frequency-domain measurements. The time-dependent changes in amplitude of optical signal from the human eye globe *ex vivo* after (a) trazograph injection and (b) trazograph drops in the vicinity of the detector fiber tip.[374]

dependencies are quite different. Keeping in mind that at the first 3 min after injection of the chemical agent the positive time scale corresponds to a decrease of scattering due to tissue immersion, the opposite tendencies of considered dependencies can be understood as the following. For small source–detector separation close to the back-reflectance geometry, intensity of reflected light decreases along with scattering; and for rather large separations, when lateral photon diffusion effects are important, intensity at first goes up with decreased scattering, but if scattering continues to decrease, intensity will lessen. That is why a local maximum on a curve for a separation of 3.5 mm was observed. At the third minute after chemical agent injection, due to its diffusion into neighboring tissue regions, amplitudes for all separations have a tendency to go to initial values. Another technique of chemical agent administration by drops shows the same tendencies for small and large separations as for injection, but essential changes of the amplitudes happen momentarily after chemical agent drops are applied, and then amplitudes slowly change in the opposite directions. Such behavior depends on the specific features of a chemical agent application, which are (1) superficial impregnation (underlines the importance of surface immersion effect and (2) continuous renovation of the chemical agent on the tissue surface (many drops during the measurement interval).

This study shows that the impregnation of the eye sclera by hyperosmotic chemical agents affects the reversible refractive index matching of the collagen fibrils and interstitial media that leads to dramatic reduction of the tissue scattering ability.

Inhomogeneities of the trazograph concentration in the tissue volume cause macroscopic spatial inhomogeneities of polarization structure of scattered light; such polarization structure of the sclera image can be easily observed using an optical scheme with a "white" light source and a tissue sample placed between two crossed polarizers. Figure 1.33 illustrates the evolution of such polarization images during scleral clearing.[9]

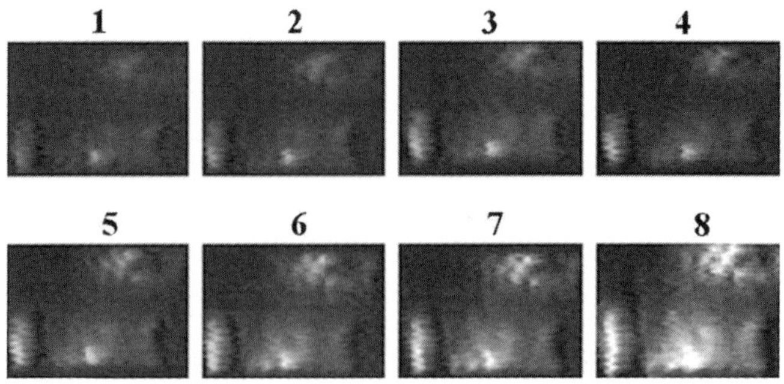

Figure 1.33 Polarization images of sclera sample (white light source, crossed polarizers) for different stages of sclera clearing. Time elapsed after trazograph administration: (1) 4 min, (2) 5 min, (3) 6 min, (4) 7 min, (5) 8 min, (6) 9 min, (7) 5 min, and (8) 10 min.[9] The tissue-supporting wires are seen for the translucent sample.

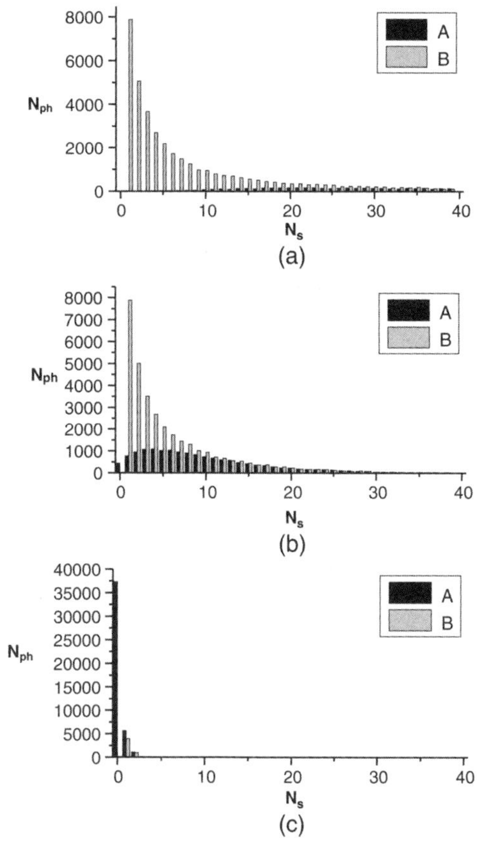

Figure 1.34 Model of the human eye sclera, refractive indices of scleral components: collagen, $n_c = 1.47$, and ground media, n_0. Calculated distributions of the number N_{ph} of (A) forward- and (B) back-scattered photons ($\lambda = 800$ nm) via number of collisions N_s (two integrating sphere geometry): (a) $n_0 = 1.35$ (normal tissue); (b) $n_0 = 1.39$ (slightly immersed tissue); and (c) $n_0 = 1.45$ (strongly immersed tissue).[9]

MC modeling of the initially collimated photons transport within the sclera tissue, represented as a fibrous structure combined of thin and long collagenous cylinders randomly distributed within the ground medium, allows one to understand the character of transition from multiple to single scattering due to refractive index matching.[9,229,378] Results of such modeling for a system of scattering cylinders with a mean diameter of 120 nm, a mean separation between fibrils centers of 285 nm, and a refractive index $n_c = 1.47$ surrounded by an interstitial medium with a changeable refractive index are shown in Fig. 1.34. Calculations were done in a geometry very close to the experimental one for 1-mm sample thickness.[9,229,378] The numbers of back- and forward-scattered photons collected by integrating spheres were calculated. It is easy to see that for partly matched refractive indices the unscattered and low-step scattered photons dominate in both directions—forward and backward. In the NIR, the optical clearing of tissue and transformation of the scattering mode from multiple to low-step or even single begins much earlier than for visible light. The validity of such calculations were proved using spectral measurements of human scleral tissue at different stages of clearing.[9,378]

It has been shown that administration of osmolytes to fibrous tissue and cell structures allows one to control effectively their optical characteristics. The kinetics of tissue optical clearing are characterized by a time response of about 5–10 min, which is defined by the dynamics of refractive index matching, which in turn depends on the diffusivity of immersion OCA in tissue and tissue thickness. Tissue swelling plays an important role in tissue clearing; it leads to saturation or even a slight reduction of tissue optical transmittance for more prolonged action of some OCAs.

Temporal optical characteristics can be used to determine the diffusion coefficient of endogenous (metabolic) and exogenous (chemical agent) fluids in the human sclera, skin, and other tissues. Obtained magnitude for the diffusion coefficient of glucose in the human sclera corresponds well to values of the diffusion coefficient for small molecules diffusing in water with some hindering caused by fibrous structure of the tissue.[409]

Particularly for blood, the refractive index mismatch between erythrocyte cytoplasm and blood plasma is important. The refractive index of erythrocyte cytoplasm is mostly defined by hemoglobin concentration (blood hematocrit).[410] The scattering properties of blood are also dependent on erythrocyte volume and shape, which are defined by blood plasma osmolarity,[113,410] and aggregation or disaggregation ability.[411–414] The feasibility of index matching as a method to overcome the limited light penetration through blood for capturing tissue images has been demonstrated.[415–417] Glucose, low- and high-molecular dextrans, x-ray contrasting, glycerol, and other agents were used to increase the refractive index of blood plasma closer to that of erythrocyte cytoplasm to improve penetration depth so that images of objects behind the blood layer could be obtained.

The reduced scattering coefficient μ_s' of blood depends on a mismatch of averaged refractive indices of blood plasma and RBC cytoplasm. The ratio $n_{RBC}/n_{bp} \equiv m_{RBC}$ determines the scattering coefficient, n_{RBC} is the mean refractive index of RBC cytoplasm, and n_{bp} is the mean refractive index of the blood plasma. For modeling the RBC ensemble as a monodisperse system of noninteracting scattering dielectric spheres of radius a irradiated at an NIR wavelength λ, when $5 < 2\pi a/\lambda < 50$, anisotropy scattering factor $g > 0.9$, and $m_{RBC} \cong 1.05$, μ_s' is well described by Eq. (1.160).

The blood plasma contains up to 91% water, 6.5%–8% (about 70 g/L) various proteins, and about 2% low-molecular compounds. Because of the low concentration and relatively low refractive index of low-molecular chemical compounds, the mean blood plasma (background) index can be estimated as the weighted average of the refractive indices of water (92%) n_w and proteins (8%) n_p

$$n_{bp} = f_w n_w + (1 - f_w) n_p, \tag{1.168}$$

where f_w is the volume fraction of water contained in plasma, $n_w = 1.329$ at 800 nm, and the index of proteins can be taken as $n_p = 1.470$.[9,218] Since approximately 92% of the total plasma is water, it follows from Eq. (1.168) that $n_{bp} = 1.340$.

The refractive index of RBC cytoplasm defined by the cell-bounded hemoglobin solution can be found from[410]

$$n_{RBC} = n_w + \beta c, \tag{1.169}$$

where c is the hemoglobin concentration in g/100 ml and $\beta = 0.001942$ at a wavelength of 589 nm. Since the scattering coefficient of blood is not significantly dependent on wavelength in the range from 580 to 800 nm, this value of β can be used for estimation of the refractive index of RBC in the NIR.[410]

As it follows from Eq. (1.160), about a tenfold reduction of the scattering coefficient μ_s' is expected when the refractive index of the blood plasma is changed from $n_{bp} = 1.340$ to 1.388 and the refractive index of RBC cytoplasm is kept constant, $n_{RBC} = 1.412$ (for hemoglobin concentration in cytoplasm of 400 g/L).[410]

A 1300-nm optical coherent tomography (OCT) system (see Chapter 7 in Volume 2) was used to take images of the reflector through circulating blood *in vitro*.[415] The total intensity of the signal off the reflector was used to represent penetration. As immersion substances, dextran (group refractive index of 1.52) and IV contrast (group refractive index of 1.46) were taken. Both dextran and IV contrast were demonstrated to increase penetration through blood: 69% ± 12% for dextran and 45% ± 4% for IV contrast. The studies of blood scattering reduction by the immersion technique using various solutions that are biocompatible with blood, such as saline, glucose, glycerol, propylene glycol, trazograph (x-ray contrasting substance for intravenous injection), and dextran, are described in Refs. 9, 416, and 417. For taking images of the reflector through a 1-mm layer of noncirculating fresh whole blood, 820- and 1310-nm OCT systems were applied. It was shown that for noncirculating blood sedimentation plays an important role in blood clearing using the immersion technique, and OCT allows for precise monitoring of blood sedimentation and aggregation. It was found that blood optical clearing is defined not only by the refractive index matching effect, but also by changes of RBC size and their aggregation ability when OCA are added. For example, when to whole blood twice diluted by a saline 6.5%-glycerol is added, the total attenuation coefficient is reduced from 42 to

20 cm^{-1}, and, correspondingly, the optical penetration at 820 nm increases to 117%. For other tested agents (glucose, dextrans, propylene glycol, and trazograph), the enhancement of penetration was from about 20% up to 150.5%.

Blood optical clearing has great perspectives for optical multimodal endoscopic imaging in clinics with using of OCT for atherosclerotic plaques and stent lesions monitoring.[418,419]

In a blood sedimentation study, the regular or irregular oscillations or jumps of the RBC/plasma boundary were observed. The 1-min time period of regular oscillations correlates well with the kinetics of the aggregation process, described by the two subsequent stages of formation of the linear and 3-D aggregates.[412,413,416,417]

Blood vessel's imaging through skin is another prospective optical clearing technology.[9,394,395,397,420–422] First demonstrated for rat skin by spectral reflectivity technique,[420] it is now widely used for different optical imaging modalities including speckle-correlation technologies.[397,422]

Blood vessels are also a pathway for OCA supply into living tissue. This was demonstrated for the first time for the animal model, when combined intravessel and intratissue injection were used for OCA delivery.[421]

1.11 Circularly Polarized Light

An alternative approach to the Mueller matrix and LSM has been developed.[367,423] The technique utilizes circularly polarized light. When circularly polarized light is multiply scattered within a turbid tissue-like scattering medium, the incident light is depolarized, and the depolarization rate depends strongly on the size and shape of scattering particles,[264,268] as well as on the number of scattering events.[424] A popular belief claims that linear polarization is better preserved in scattering tissue-like media compared to circularly polarized light. This, however, turns out to be true only when the size of scattering particles a is smaller than the wavelength of the incident light λ, $a < \lambda$. In media consisting larger scattering particle sizes ($a \geq \lambda$), the opposite situation happens. This phenomenon is known as the circular polarization *memory* of light,[425,426] and is of fundamental importance. Due to the intrinsic phase difference between its two basis linear components in the case of reflection at the medium surface, an incident state of circular polarization undergoes a flip in its helicity[426] and the same goes for the back-scattering of light. For the large particles with highly anisotropic scattering, the light becomes strongly forward scattered that preserves its initial helicity. Linear polarization possesses no such sense of the direction in which light travels.

Therefore, with a known stage of polarization of incident light, the helicity of detected optical radiation can be used to determine if it has been forward or back scattered. With the geometry of an experimental system that allows both portions of forward and back-scattered circularly polarized light to be measured, changes in the medium under observation can be resolved by looking at the contribution of each part to the final signal. When combined with the knowledge of anisotropy of scattering particles within the medium, this effect provides a method for examining relevant properties of a scattering medium, including size, shape, number density, and refractive index of scattering particles.

The experimental realization of the technique utilizing circularly polarized light is presented in Fig. 1.35.

The back-scattered light is collected at distance d from the point of light incidence (see Fig. 1.35) and then passed through a polarimeter to measure its SOP. Incident circularly polarized light is propagated through the turbid medium and undergoes multiple-scattering events before being collected by the detector. The MC-based computational studies[428] indicated that when the source and detection areas are overlapped ($d = 0$), the detected signal is likely to be overwhelmed by the cross-polarized component due to single back-scattering events at or near the surface. Separation of the source and detector increases the pathlength of the light within the sample before reaching the detector, and thus increases the contribution of the co-polarized component to the detected signal. Too large a detecting distances result in complete depolarization of the incident light. The actual value of d causing complete depolarization is related to the optical properties of the medium, specifically the transport mean-free path l_{tr} ($l_{tr} = 1/\mu_s'$).

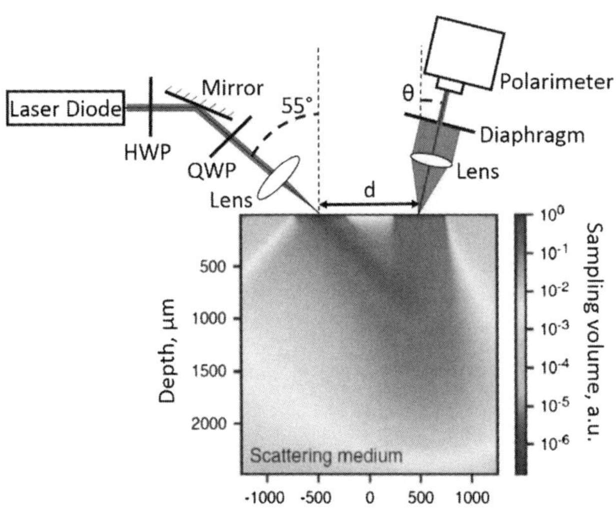

Figure 1.35 Schematic presentation of the experimental setup. Vertically polarized light from a laser diode (LQC639-30C, Newport, RI, 639 nm, 30 mW) is directed toward the sample at 55 deg from the normal. The light is then altered by a quarter wave plate into a state of RCP and/or right elliptically polarized (REP), and focused onto the surface of the tissue by a lens. The source detector separation d as well as the angle of detection θ can be varied to influence the sampling volume.[427,428] (See color plates.)

To visualize and quantify changes in light polarization associated with alterations that were built into models or changes in biological tissues, the use of the Poincaré sphere has been suggested.[367,423] The coordinates on the Poincaré sphere are defined by the Stokes parameters, defined as[1]

$$S_0 = I, \tag{1.170}$$

$$S_1 = Q = I \cdot \mathrm{DOP} \cdot \cos(2\psi)\cos(2\chi), \tag{1.171}$$

$$S_2 = U = I \cdot \mathrm{DOP} \cdot \sin(2\psi)\cos(2\chi), \tag{1.172}$$

$$S_3 = V = I \cdot \mathrm{DOP} \cdot \sin(2\chi), \tag{1.173}$$

where I is intensity, DOP is the degree of polarization and describes the portion of polarized light, χ is ellipticity angle, and ψ is azimuth angle. Using the Poincaré sphere makes it easier to visualize multiple polarization states at once for comparison purposes. The last three Stokes parameters are plotted in 3-D space, outlining a sphere with all possible polarization states (Fig. 1.36).

The Poincaré sphere has been used as a convenient graphical tool for the analysis of SOP of light scattered within biological tissue and tissue-like scattering meida.[367,423] Mapping the Stokes vector of the back-scattered light

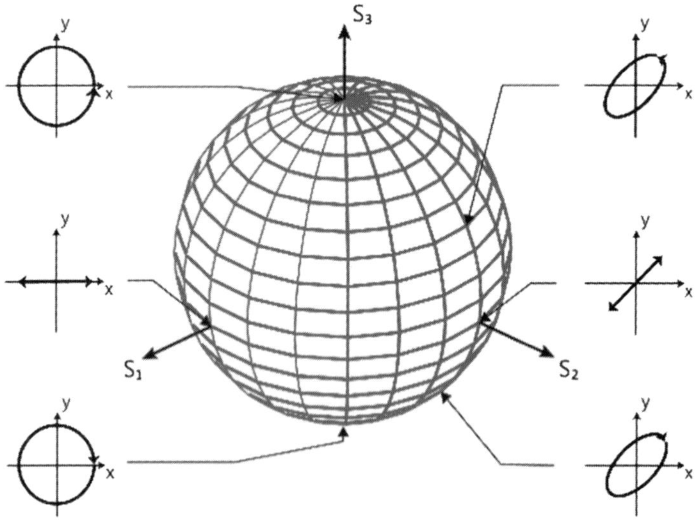

Figure 1.36 The Poincaré sphere describes all possible states of polarization. The distance from the center of the sphere indicates the degree of polarization; fully polarized states occupy a point on the surface of the sphere, whereas partially polarized states reside inside. Each point on the sphere represents one SOP so for instance a point at the northern pole $+S_3$ signifies right-handed circular polarization, and the south pole $-S_3$, left circular. States of linear polarization lie on the equator, and elliptical polarizations fill the remaining surface.

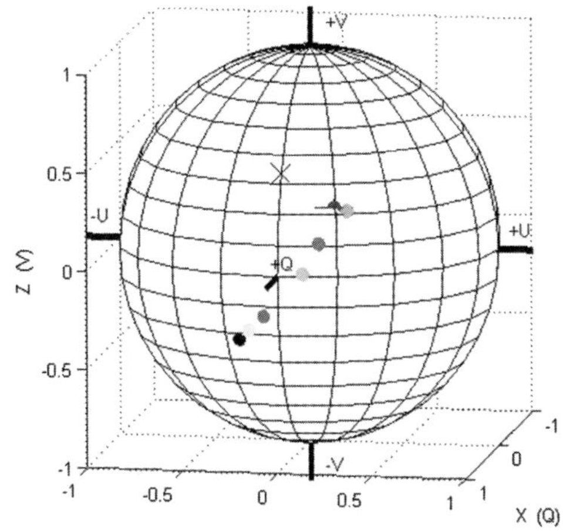

Figure 1.37 Position of the Stokes vector on the Poincaré sphere for right helically polarized light scattered in the water solutions of with dilution factor (from top): 1:1 (•– blue), 1:10 (•– green), 1:20 (•– red), 1:40 (•– cyan), 1:100 (•– magenta), 1:200 (•– yellow), and 1:500 (•– black). $d = 1$ mm and $\theta = 10$ deg. × indicates the Stokes vector of incident light.

on a Poincaré sphere demonstrated that the changes of polarization state correlates with the concentration of scattering particles and their sizes.

Figure 1.37 shows that SOP of light begins in a right handed elliptical polarization state and reverses its helicity to become left handed with the highly dilute Intralipid water solutions. The helicity of the polarization state remains right-handed with the water solutions of Intralipid up to dilution factor 1:40. After this point the polarization state changes more significantly, whereas the changes between the first and second measurement is minimal (see Fig. 1.37).

The results of measurements of the signature of the Stokes vector on the Poincaré sphere for the light scattered in the mono-disperse solutions of polystyrene microspheres (of known optical properties, see Table 1.1) are presented in Fig. 1.38.

Thus, the quantitative phantom studies utilizing polystyrene beads confirmed that the Stokes vector position on the Poincaré sphere depended on the size of the light-scattering particles, which in turn influenced the latitude of the Stokes vector mapped onto the Poincaré sphere. Based on the

Table 1.1 Optical properties of mono-disperse solutions of polystyrene microspheres used in the experiments.

	Diameter (μm)	Concentration (particles/ml)	g	l_s (cm^{-1})	μ_s' (cm^{-1})	Dilution factor
a	11 ± 0.25	3.55×10^7	0.912	68.6	6.0	1:1
b	5.9 ± 0.29	1.25×10^8	0.864	68.6	9.3	1:1.77
c	5.9 ± 0.29	8.09×10^7	0.864	44.3	6.0	1:2.74
d	5.9 ± 0.29	3.55×10^7	0.864	19.5	2.7	1:6.23

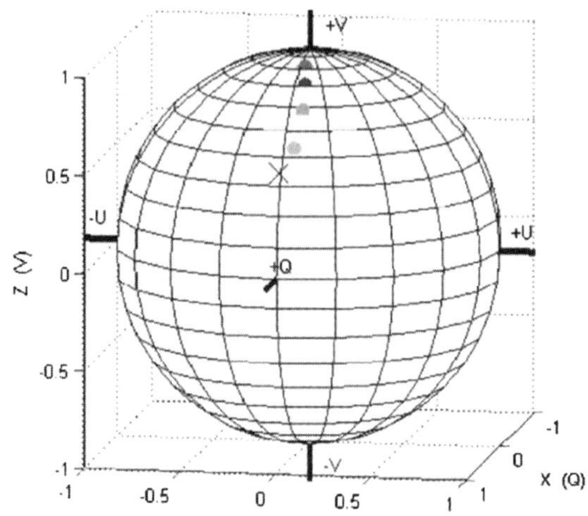

Figure 1.38 Position of Stokes vector on the Poincaré sphere for right-handed circularly/elliptically polarized light scattered in water solutions of microspheres, the scattering properties of which presented in Table 1.1: a (•– red), b (•– blue), c (•– green), and d (•– cyan), × indicates the Stokes vector of incident light. $d = 6$ mm and $\theta = 30$ deg.

Mie theory and the associated calculations of scattering pattern and polarization state, we can see that for small microspheres the scattering acts like a dipole. Figure 1.39(a) shows the S_3 component of Stokes vector (also known as V) of the light after a single-scattering event for a small (0.1 μm) microsphere as a function of the cosine of the scattering angle θ. Since the incident light is right circularly polarized the negative values for V indicate a helicity flip. We can see that back-scattering events (when cosine of scattering angle < 0) result in a helicity flip. The S-shaped curve can be explained by the attenuation of p- and s- components in terms of projections of the electric field vector. But for larger spheres this behaviour drastically changes. The S-shape is lost at a sphere size of 0.3 μm [Fig. 1.39(b)]. The angles resulting in helicity reversal become even more complicated at the sphere sizes used in the experiments described above [see Fig. 1.39(c) and (d)].

Figure 1.40 shows the results of a MC simulation[428] recording the spatial distribution of the V component of the Stokes vector on the surface of a scattering medium. The optical properties of media used in the modelling are presented in Table 1.1 (a–d), respectively. As one can see the bulk polarization state measured from the medium with the highest scattering anisotropy [see Fig. 1.40(a)] has the highest helicity preservation, or V component. The overall trend between ellipticity and changes of anisotropy of scattering is clearly seen between all the samples.

The results of phantom studies with water solutions of Intralpid and microspheres show clearly the response of backward scattered circular/elliptically polarized light. The results of MC modelling shows that phase shifts are involved in scattering by large spheres but that their helicity flips

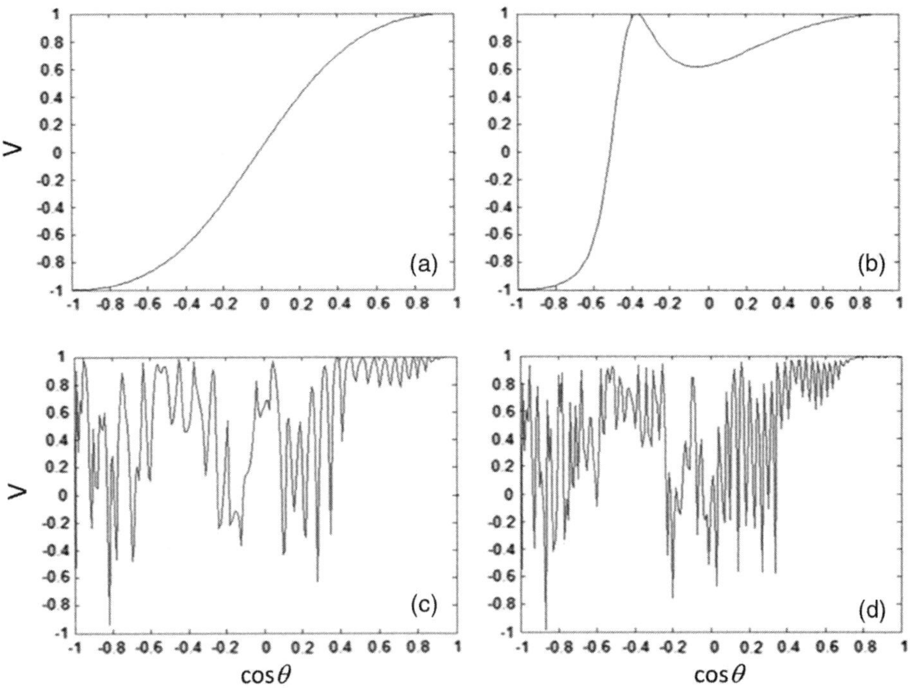

Figure 1.39 *V* component (S_3) of the outgoing Stokes vector vs. cosine of scattering angle calculated by using Mie theory[429] for the diameter of scattering particles: (a) 0.1 μm, (b) 0.3 μm, (c) 5.9 μm, and (d) 11 μm. The wavelength of light is 639 nm.

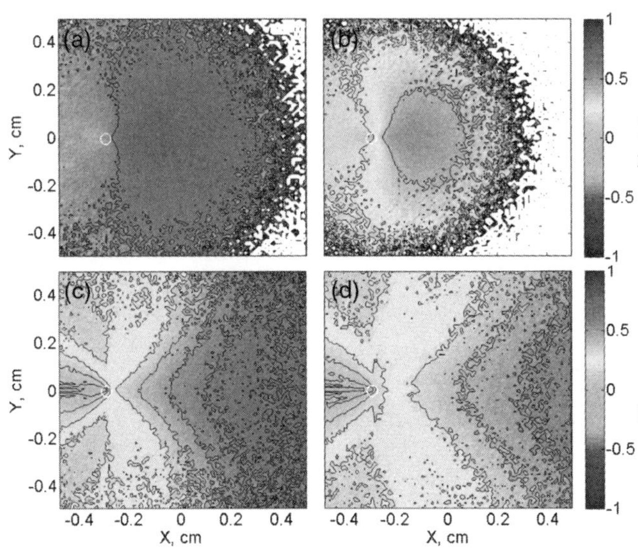

Figure 1.40 Spatial distribution of *V* component of the Stokes vector on the surface of scattering medium, obtained for circularly polarized light [1,0,0,1] of a wavelength of 639 nm incident to the surface of the medium at point (0,0) with the angle of 45 deg. The optical properties of scattering media are, respectively, presented in Table 1.1. (See color plates.)

occur at a complex range of scattering angles. The obtained results give a strong suggestion that changes in nuclear size in cells can be characterised by this technique. This provides a good foundation for further work implementing the noninvasive circularly/elliptically polarized light based diagnostic approach for early disease detection, as many forms of cancerous growths alter properties such as anisotropy and density of scattering sites.

Following the results discussed earlier, Fig. 1.41 illustrates the changes of SOP of light scattered during the tissue optical clearing by using water solutions of glycerol. As one can see in the first measurement, the initial right-handed elliptical polarization state gradually narrows and approaches linear polarization along the x axis. It is clearly seen that the incident right circularly polarized light retains its helicity over the measured time. Within the first 20 min, the SOP of the back-scattered light changed more rapidly than at later time intervals. Up to 30 min, the SOP decreases (see Fig. 1.41).

It has long been speculated that altered nuclear size and shape are representative of various underlying causes of altered nuclear morphology that occur during cancer onset. This altered nuclear morphology is thought to reflect primary changes in chromosome organization, which in turn can affect gene expression,[430] although the exact relationship between these parameters has still not been defined. It has also been proposed that the altered nuclear shape in cancer cells facilitates the formation of metastases because of reduced nuclear stiffness, which could increase the ability of transformed cells to penetrate tissue.[431] Consistent with these ideas is the notion that certain (as yet unidentified) factors in cells regulate or limit nuclear expansion.

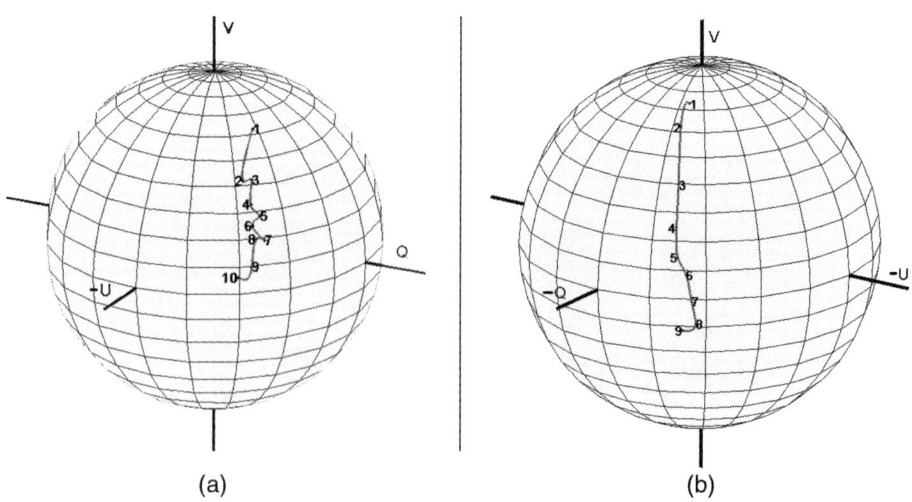

Figure 1.41 The position of the Stokes vector on the Poincaré sphere for right-handed circularly/elliptically polarized light scattered at *in vitro* tissue optical clearing by topical application of water solution of glycerol: (a) chicken skin sample and (b) bovine tendon sample. The SOP was recorded every 10 min during a time period of 100 and 90 min, respectively.

For example, several studies suggest that nuclear size is determined by cytoplasmic volume,[432] and it is well documented that the nuclear to cytoplasmic ratio decreases in cells as they move from being stem cells to being terminally differentiated.[431]

Abnormalities induced in tissues by cancerous changes include an increased nucleus-to-cytoplasm ratio and an overall increase in the volume density of cells.[433,434] These two effects, as shown with the phantom studies discussed above and confirmed by the results of theoretical studies,[435–437] impact greatly on the SOP of light propagated through the tissue. An increase in nuclear size leads to a higher forward scattering of light (Fig. 1.42).

Therefore, if the SOP of the scattered light is closer to the state of incident right circularly polarized light, i.e., to the north pole on the Poincaré sphere, then the tissue sample is either neoplastic potentially malignant or neoplastic malignant. If the state of polarization of the scattered light is close to the equator of the Poincaré sphere then the tissue sample is normal. The results of feasibility studies[439] show that the current approach can discriminate between cancerous and noncancerous tissue by mapping the Stokes vector of backscattered light on a Poincaré sphere (see Fig. 1.43).

Multiple measurements were performed in a scanning approach on a human lung metastasis of thymic carcinoma embedded in paraffin wax.[367] The samples had a variety of tissue structures present, including both healthy and cancerous tissue. A pathologist reviewed the tissue prior to taking measurements, and the region of the cancer tissue was outlined on a 5-μm tissue section via histological haematoxylin and eosin staining and light microscopy.

Figure 1.44 is shown the averaged polarization states of detected light scattered from healthy and cancerous tissues complete with standard

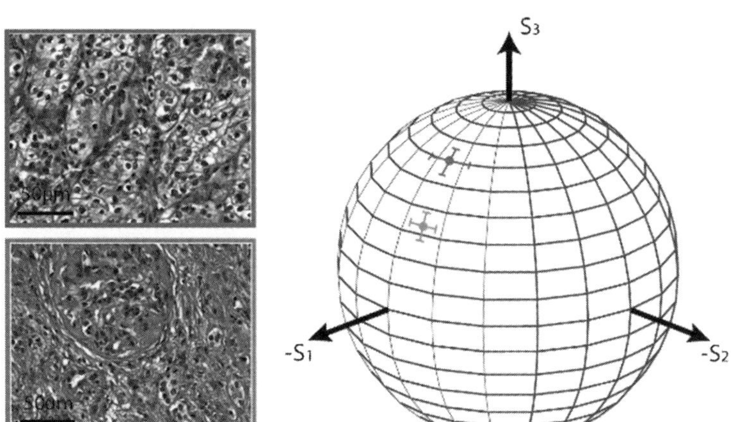

Figure 1.42 The SOP of the right-handed elliptically polarized light scattered from healthy (green) and cancerous (red) sites of kidney tissue.[438] (See color plates.)

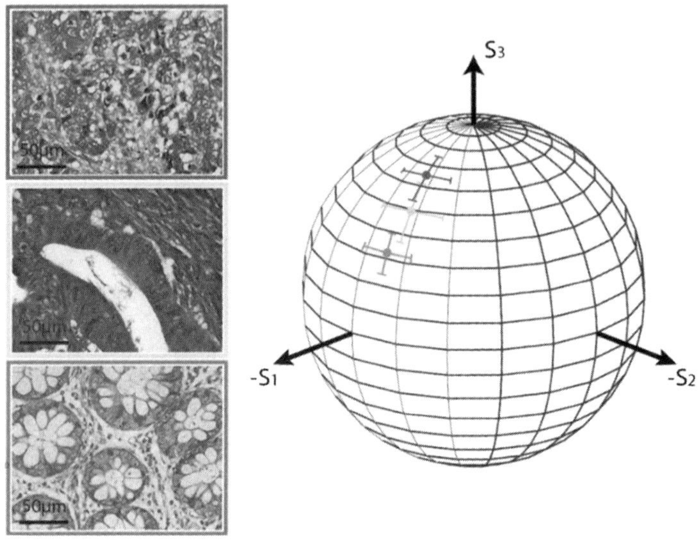

Figure 1.43 Position of the Stokes vector on the Poincaré sphere (right) correlates with the successive grades of colorectal cancer (normal colon, green; grade 2, yellow; grade 3, red) confirmed by microscopy (left three panels: normal colon, bottom; grade 2, middle; grade 3, top). (See color plates.)

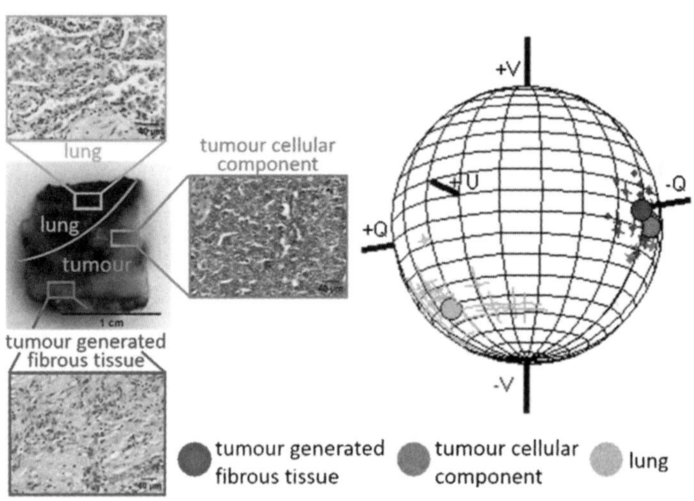

Figure 1.44 Lung tissue embedded in paraffin wax (left) shows the boundary between healthy lung and tumor tissue (red line) marked based on histological examination. Microscopy images of hematoxylin and eosin stained 5-μm tissue sections are shown in the colored squares, and their position on the block is marked by the corresponding squares. The Poincaré sphere (right) shows the mean Stokes vector of REP light back-scattered from the region of healthy lung tissue (green), tumor-generated fibrous tissue (blue) and the tumor cellular component (red); dots with error bars represent a single measurement, and the larger circles represent the mean SOP. $d = 1.5$ mm, and $\theta = 30$ deg. (See color plates.)

deviation. With right-handed elliptically polarized incident light, we can see on the figure that the ellipticity (latitude on the sphere) is much higher for the cancerous tissue. This is expected to be a result of the increased average size of lung cell nuclei (as can be seen in the corresponding microscope images). This leads to higher anisotropy and an increase in successive forward-scattering events contributing to the signal where co-polarized helicity has a stronger influence.

Thus, it has been shown that back-scattered circularly polarized light is highly sensitive to the presence of cells with enhanced plasticity. It has been found that the magnitude of the alterations defined by the Stokes vector on the Poincaré sphere is proportional to the grade of cancer. These results demonstrate a robust, sensitive, and unbiased method for identifying enhanced plasticity in tissues in real time before metastasis. In the bigger picture, this approach has the strong potential to revolutionize the current practice of cancer screening in living tissues to ensure the well-being of patients in the near future.

1.12 Summary

In this chapter, the basic theoretical approaches and methods that are widely used for calculations of angular distribution and total extinction of light scattered and absorbed by small particles or particle aggregates are discussed. At present there are a number of web sites with public domain computer codes that maintain up-to-date theoretical approaches that can be easily implemented for biomedical applications (see, e.g., Ref. 96). In the first part of the chapter, the direct and inverse spectroturbidimetric problems of disperse systems with random and ordered particle orientations have been considered in detail. We have shown how to estimate the average size, shape, and index of refraction of cells using the extinction (integral light scattering) spectra recorded at various orientations of a light beam with respect to an external orienting field.

In the second part of the chapter, it was shown that the model of discrete scatterers explains many experimental optical properties of various tissues, primarily the spectral dependences of light scattering and transmittance. It should be emphasized that a simple model of single light scattering by independent particles fails at theoretical simulation of optical properties of real tissues *in situ*, except in the case of specially prepared samples. For example, the high transparency of structurally nonhomogeneous tissues of the eye can be explained only by the concept of dependent light scattering from spatially ordered discrete scatterers. Accounting for the optical softness of structural elements of the majority of tissues, an adequate description of experimental data, including polarization and angular characteristics, can be achieved by using approximate variants of the multiple-wave scattering theory. A prospective method for controlling of tissue optical properties is discussed. It was demonstrated in *in vitro* and *in vivo* studies that an effective reduction of tissue scattering could be achieved by a refractive index matching using tissue and blood immersion by OCA.

The experimental angular dependences of an LSM, i.e., a combination of angular and polarization light scattering characteristics, creates an informative basis for solving inverse problems that are important for noninvasive optical tissue diagnostics. The actual number of examples of successful application of optical polarization techniques in tissue optics is growing, because the capabilities of this approach are rather promising, especially at using the Poincaré sphere concept for fast recognition of pathology grade.

Acknowledgments

Many results presented in this chapter on modeling and measuring of tissue optical properties were generated by Irina L. Maximova, who was a very talented scientist and a great person. She passed away in 2013 during the work on the second edition of this handbook. We will remember her forever.

VVT is thankful for support from grant No. 14-15-00186 of the Russian Science Foundation and Russian Presidential Grant NSh-7898.2016.2. The work by NGK (Sections 1.1 to 1.5) was supported by grant No. 14-13-01167 from the Russian Science Foundation.

References

1. H. C. van de Hulst, *Light Scattering by Small Particles*, Wiley, New York (1957).
2. V. V. Sobolev, *Light Scattering in Planetary Atmospheres*, Pergamon, Oxford (1974).
3. A. Z. Dolginov, Yu. N. Gnedin, and N. A. Silant'ev, *Propagation and Polarization of Radiation in Cosmic Media*, Gordon and Breach, Basel (1995).
4. K. N. Liou, *Radiation and Cloud Processes in the Atmosphere: Theory, Observation, and Modeling*, Oxford Univ. Press, New York (1992).
5. M. S. Quinby-Hunt, P. G. Hull, and A. J. Hunt, "Polarized light scattering in the marine optics," Chapter 18 in *Light Scattering by Nonspherical Particles: Theory, Measurements, and Applications*, M. I. Mishchenko, J. W. Hovenier, and L. D. Travis, Eds., Academic Press, New York, pp. 525–554 (2000).
6. A. Ishimaru, *Wave Propagation and Scattering in Random Media*, Academic Press, San Diego (1978).
7. M. Kerker, *The Scattering of Light and Other Electromagnetic Radiation*, Academic Press, San Diego (1969).

8. V. I. Klenin, *Thermodynamics of Systems with Flexible Chain Polymers*, Elsevier, Amsterdam (1999).
9. V. V. Tuchin, *Tissue Optics: Light Scattering Methods and Instruments for Medical Diagnosis*, 3rd ed., **PM 254**, SPIE Press, Bellingham, WA, 2015.
10. V. V. Tuchin, "Light scattering study of tissues," *Phys.-Uspekhi* **40**, 495–515 (1997).
11. P. Chylek, G. Videen, D. J. W. Geldart, J. S. Dobbie, and H. C. W. Tso, "Effective medium approximations for heterogeneous particles," Chapter 9 in *Light Scattering by Nonspherical Particles: Theory, Measurements, and Applications*, M. I. Mishchenko, J. W. Hovenier, and L. D. Travis, Eds., Academic Press, San Diego, pp. 273–308 (2000).
12. C. F. Bohren and D. R. Huffman, *Absorption and Scattering of Light by Small Particles*, Wiley, New York (1983).
13. R. G. Newton, *Scattering Theory of Waves and Particles*, McGraw-Hill, New York (1966).
14. P. C. Waterman, "Symmetry, unitarity, and geometry in electromagnetic scattering," *Phys. Rev. D* **3**(4), 825–839 (1971).
15. M. I. Mishchenko, J. W. Hovenier, and L. D. Travis, Eds., *Light Scattering by Nonspherical Particles: Theory, Measurements, and Applications*, Academic Press, San Diego (2000).
16. M. I. Mishchenko, W. J. Wiscombe, and L. D. Travis, "Overview of scattering by nonspherical particles," Chapter 2 in *Light Scattering by Nonspherical Particles: Theory, Measurements, and Applications*, M. I. Mishchenko, J. W. Hovenier, and L. D. Travis, Eds., Academic Press, San Diego, pp. 29–60 (2000).
17. M. I. Mishchenko, L. Liu, and G. Videen, "Conditions of applicability of the single-scattering approximation," *Opt. Express* **15**, 7522–7527 (2007).
18. K. H. Ding and L. Tsang, "Effective propagation constants in media with densely distributed dielectric particles of multiple sizes and permittivities," in *Progress in Electromagnetic Research*, PIER 1, J. A. Kong, Ed., Elsevier, New York, pp. 241–295 (1989).
19. V. K. Varadan, V. N. Bringi, and V. V. Varadan, "Coherent electromagnetic wave propagation through randomly distributed dielectric scatterers," *Phys. Rev. D* **19**(8), 2480–2489 (1979).
20. S. M. Rytov, Y. A. Kravtsov, and V. I. Tatarskii, *Principles of Statistical Radiophysics*, Vol. 4, *Wave Propagation through Random Media*, Springer-Verlag, Berlin (1989).
21. L. Tsang, J. A. Kong, and R. T. Shin, *Theory of Microwave Remote Sensing*, Wiley, New York (1985).
22. H. C. van de Hulst, *Multiple Light Scattering: Tables, Formulas and Applications*, Academic Press, San Diego (1980).

23. P. Flatau, "SCATTERLIB: light scattering codes library," http://atol.ucsd.edu/~pflatau/scatlib/.
24. G. V. Rozenberg, "Electrodynamics of statistically inhomogeneous media and radiative transfer theory," in *Theoretical and Applied Light Scattering Problems*, B. I. Stepanov and A. P. Ivanov, Eds. (in Russian), Nauka i Tekhnika, Minsk, pp. 159–170 (1971).
25. M. I. Mishchenko, "Directional radiometry and radiative transfer: a new paradigm," *J. Quant. Spectrosc. Radiat. Transfer* **112**, 2079–2094 (2011).
26. V. K. Varadan, V. N. Bringi, V. V. Varadan, and A. Ishimaru, "Multiple scattering theory for waves in discrete random media and comparison with experiments," *Radio Sci.* **18**(3), 321–327 (1983).
27. M. Kaveh, "New phenomena in the propagation of optical waves through random media," *Waves Rand. Med.* **3**(1), 121–128 (1991).
28. V. I. Kuzmin and V. P. Romanov, "Coherent phenomena in light scattering from disordered systems," *Phys.-Uspekhi* **39**(3), 231–260 (1996).
29. C. A. Croxton, *Liquid State Physics: A Statistical Mechanical Introduction*, Cambridge Univ. Press (1974).
30. M. Born and E. Wolf, *Principles of Optics*, 4th ed., Pergamon, Oxford (1968).
31. V. A. Davis and L. Schwartz, "Electromagnetic propagation in close-packed disordered suspensions," *Phys. Rev. B* **31**(8), 5155–5165 (1985).
32. K. Furutsu, "Random-volume scattering: boundary effects, cross-sections, and enhanced backscattering," *Phys. Rev. A* **43**(6), 2741–2762 (1991).
33. Yu. N. Barabanenkov, Yu. A. Kravtsov, V. D. Ozrin, and A. I. Saichev, "Enhanced backscattering in optics," in *Progress in Optics*, Vol. **29**, E. Wolf, Ed., Elsevier, Amsterdam, pp. 65–197 (1991).
34. W. G. Tam and A. Zardecki, "Multiple scattering to the Beer–Lambert law: I. open detector," *Appl. Opt.* **21**(13) 2405–2412 (1982).
35. A. Zardecki and W. G. Tam, "Multiple scattering corrections to the Beer–Lambert law: II. detector with a variable field of view," *Appl. Opt.* **21**(13), 2413–2420 (1982).
36. N. G. Khlebtsov, "Role of multiple scattering in turbidimetric investigations of dispersed systems," *J. Appl. Spectrosc.* **40**(2), 243–247 (1984); translated from *Zh. Prikl. Spektr.* **40**(2), 320–325 (1984).
37. R. O. Gumprecht and C. M. Slipcevich, "Scattering of light by large spherical particles," *J. Chem. Phys.* **57**(1), 90–95 (1953).
38. P. Latimer, "Light scattering and absorption as methods of studying cell population parameters," *Ann. Rev. Biophys. Bioeng.* **11**(1), 129–150 (1982).
39. D. S. Saxon, "Tensor scattering matrix for the electromagnetic field," *Phys. Rev.* **100**(6), 1771–1775 (1955).

40. L. Lorenz, *Oeuvres Scientifiques*, Lehman and Stage, Copenhagen (1898).
41. A. E. H. Love, "The scattering of electric waves by a dielectric sphere," *Proc. London Math. Soc.* **30**(678), 308–321 (1899).
42. G. Mie, "Beiträge zur Optik trüber Medien, speziell kolloidaler Metallösungen," *Ann. Phys.* **25**(3), 377–445 (1908).
43. P. Debye, "Der Lichtdruck auf Kugeln von beliebigem Material," *Ann. Phys.* **30**, 57–136 (1909).
44. D. W. Rayleigh, "The incidence of light upon a transparent sphere of dimensions comparable with the wavelength," *Proc. R. Soc. London Ser. A* **84**, 25–46 (1910).
45. D. W. Rayleigh, "On the electromagnetic theory of light," *Philos. Mag.* **12**(73), 81–101 (1881).
46. D. W. Rayleigh, "On the dispersal of light by a dielectric cylinder," *Philos. Mag.* **36**, 365–376 (1918).
47. P. M. Morse and H. Feshbach, *Methods of Theoretical Physics*, McGraw-Hill, New York (1953).
48. F. Möglich, "Beugungserscheinungen an Korpen von ellipsoidischer Gestalt," *Ann. Phys.* **83**(13), 609–735 (1927).
49. Sh. Asano and G. Yamamoto, "Light scattering by a spheroidal particle," *Appl. Opt.* **14**(1), 29–49 (1975).
50. V. G. Farafonov, V. B. Il'in, and Th. Henning, "A new solution of the light scattering problem for axisymmetric particles," *JQSRT* **63**(2–6), 205–215 (1999).
51. I. R. Cirik and F. R. Cooray, "Separation of variables for electromagnetic scattering by spheroidal particles," Chapter 4 in *Light Scattering by Nonspherical Particles: Theory, Measurements, and Applications*, M. I. Mishchenko, J. W. Hovenier, and L. D. Travis, Eds., Academic Press, San Diego, pp. 89–130 (2000).
52. D. A. Varshalovich, A. N. Moskalev, and V. K. Khersonskii, *Quantum Theory of Angular Momentum*, World Scientific, Singapore (1988).
53. P. W. Barber and C. Yeh, "Scattering of electromagnetic waves by arbitrarily shaped dielectric bodies," *Appl. Opt.* **14**(12), 2864–2872 (1975).
54. M. I. Mishchenko, L. D. Travis, and A. Macke, "T–matrix method and its applications," Chapter 6 in *Light Scattering by Nonspherical Particles: Theory, Measurements, and Applications*, M. I. Mishchenko, J. W. Hovenier, and L. D. Travis, Eds., Academic Press, San Diego, pp. 147–172 (2000).
55. M. I. Mishchenko, L. D. Travis, and D. W. Mackowski, "T–matrix computations of light scattering by nonspherical particles: a review," *JQSRT* **55**(5), 535–575 (1996).
56. M. I. Mishchenko, L. D. Travis, and A. A. Lacis, *Scattering, Absorption, and Emission of Light by Small Particles*, Cambridge Univ. Press, Cambridge (2002).

57. M. I. Mishchenko, "Light scattering by randomly oriented axially symmetric particles," *J. Opt. Soc. Am. A* **8**(6), 871–882 (1991).
58. N. G. Khlebtsov, "Orientational averaging of light scattering observables in the T-matrix approach," *Appl. Opt.* **31**(25), 5359–5365 (1992).
59. K. S. Shifrin, Scattering of Light in a Turbid Medium, NASA Tech. Transl., NASA TT F-477 (1968).
60. A. J. Poggio and E. K. Miller, "Integral equation solutions of three-dimensional scattering problems," in *Computer Techniques for Electromagnetics*, R. Mittra, Ed., Pergamon, Oxford, pp. 159–264 (1973).
61. J. J. H. Wang, *Generalized Moment Methods in Electromagnetics: Formulation and Computer Solution of Integral Equations*, Wiley, New York (1991).
62. S. Ström, "On the integral equations for electromagnetic scattering," *Am. J. Phys.* **43**(12), 1060–1069 (1975).
63. N. G. Khlebtsov, "Integral equation for problems of light scattering by particles of the medium," *Opt. Spectrosc.* **57**(4), 399–401 (1984).
64. G. H. Goedecke and S. G. O'Brien, "Scattering by irregular inhomogeneous particles via the digitized Green's function algorithm," *Appl. Opt.* **27**(12), 2431–2438 (1988).
65. M. F. Iskander, H. Y. Chen, and J. E. Penner, "Optical scattering and absorption by branched chains of aerosols," *Appl. Opt.* **28**(15), 3083–3091 (1989).
66. A. Lakhtakia and G. W. Mulholland, "On two numerical techniques for light scattering by dielectric agglomerated structures," *J. Res. Natl. Inst. Stand. Technol.* **98**(6), 699–716 (1993).
67. K. Lumme and J. Rahola, "Comparison of light scattering by stochastically rough spheres, best-fit spheroids and spheres," *JQSRT* **60**, 439–450 (1998).
68. N. K. Uzunoglu, B. G. Evans, and A. R. Holt, "Evaluation of the scattering of an electromagnetic wave from precipitation particles by the use of Fredholm integral equations," *Electron. Lett.* **12**(12), 312–313 (1976).
69. A. R. Holt, N. K. Uzunoglu, and B. G. Evans, "An integral equation solution to the scattering of electromagnetic radiation by dielectric spheroids and ellipsoids," *IEEE Trans. Antennas Propag.* AP **26**(5), 706–712 (1978).
70. E. M. Purcell and C. R. Pennypacker, "Scattering and absorption of light by non-spherical dielectric grains," *Astrophys. J.* **186**(2), 705–714 (1973).
71. M. A. Yurkin and A. G. Hoekstra, "The discrete-dipole-approximation code ADDA: capabilities and known limitations," *J. Quant. Spectrosc. Radiat. Transfer* **112**, 2234–2247 (2011).
72. B. T. Draine, "The discrete-dipole approximation and its application to interstellar graphite grains," *Astrophys. J.* **333**(2, Pt.1), 848–872 (1988).

73. C. E. Dungey and C. F. Bohren, "Light scattering by nonspherical particles: a refinement to the coupled-dipole method," *J. Opt. Soc. Am. A* **8**(1), 81–87 (1991).
74. B. T. Draine and J. Goodman, "Beyond Clausius–Mossotti: wave propagation on a polarizable point lattice and the discrete dipole approximation," *Astrophys. J.* **405**(2), 685–697 (1993).
75. B. T. Draine, "The discrete dipole approximation for light scattering by irregular targets," Chapter 5 in *Light Scattering by Nonspherical Particles: Theory, Measurements, and Applications*, M. I. Mishchenko, J. W. Hovenier, and L. D. Travis, Eds., Academic Press, San Diego, pp. 131–145 (2000).
76. A. G. Hoekstra and P. M. A. Sloot, "Biophysical and biomedical applications of nonspherical scattering," Chapter 20 in *Light Scattering by Nonspherical Particles: Theory, Measurements, and Applications*, M. I. Mishchenko, J. W. Hovenier, and L. D. Travis, Eds., Academic Press, San Diego, pp. 585–602 (2000).
77. A. R. Jones, "Light scattering for particle characterization," *Progr. Energy. Combust. Sci.* **25**, 1–53 (1999).
78. A. Kokhanovsky, *Optics of Light Scattering Media: Problems and Solutions*, Wiley/Praxis, Chichester (1999).
79. A. G. Ramm, "Wave scattering by small bodies," *Rep. Math. Phys.* **21**(1), 69–77 (1985).
80. J. C. Ku and J. D. Felske, "The range of validity of the Rayleigh limit for computing Mie scattering and extinction efficiencies," *JQSRT* **31**(6), 569–574 (1984).
81. R. E. Kleinman and T. B. A. Senior, "Rayleigh scattering," in *Low and High Frequency Asymptotics*, V. K. Varadan and V. V. Varadan, Eds., Elsevier, Amsterdam, pp. 1–70 (1986).
82. A. F. Stevenson, "Solution of electromagnetic scattering problems as power series in the ratio (dimension of scattering)/wavelength," *J. Appl. Phys.* **24**(9), 1134–1142 (1953).
83. A. F. Stevenson, "Electromagnetic scattering by an ellipsoid in the third approximation," *J. Appl. Phys.* **24**(9), 1143–1151 (1953).
84. N. G. Khlebtsov, "Scattering matrix of anisotropic ellipsoids comparable in size to the wavelength of light," *Opt. Spectrosc.* **46**(2), 188–190 (1979).
85. D. W. Rayleigh, "On the diffraction of light by spheres of small refractive index," *Proc. R. Soc. London Ser. A.* **90**, 219–225 (1914).
86. P. Debye, "Zerstreuung von Röntgenstrahlen," *Ann. Phys.* **46**(4), 809–823 (1915).
87. P. Debye, "Molecular-weight determination by light scattering," *J. Phys. Colloid Chem.* **51**(1), 18–31 (1947).
88. M. Born, "Quantenmechanik der Stossvorgänge," *Z. Phys.* **38**(8), 803–827 (1926).

89. J.-C. Ravey, "Light scattering in the third-order and RGD approximation: the form anisotropy in electro-optics," *J. Colloid Interface Sci.* **56**(3), 540–556 (1976).
90. R. Pecora, Ed., *Dynamic Light Scattering. Applications of Photon Correlation Spectroscopy*, Plenum, New York (1985).
91. J. D. Klett and R. A. Sutherland, "Approximate methods for modeling the scattering properties of nonspherical particles: evaluation of the Wentzel–Kramers–Brillouin method," *Appl. Opt.* **31**(3), 373–386 (1992).
92. T. W. Chen, "High energy light scattering in the generalized eikonal approximation," *Appl. Opt.* **28**(19), 4096–4102 (1989).
93. G. H. Meeten, "Induced birefringence in colloidal dispersions," *J. Chem. Soc. Faraday Trans.* **75**(10, Pt. 2), 1406–1415 (1979).
94. J.-C. Ravey and P. J. Mazeron, "Light scattering by large spheroids in the physical optics approximation: numerical comparison with other approximate and exact results," *J. Opt.* (Paris) **14**(1), 29–41 (1983).
95. A. Macke, M. I. Mishchenko, K. Muinonen, and B. E. Carlson, "Scattering of light by large nonspherical particles: ray-tracing approximation versus T-matrix method," *Opt. Lett.* **20**(19), 1934–1936 (1995).
96. T. Wriedt, "List of electromagnetic scattering codes," http://imperator.cip–iw1.uni-bremen.de.
97. P. P. Silvester and R. L. Ferrary, *Finite Elements for Electrical Engineers*, Cambridge Univ. Press, New York (1996).
98. A. Talfove, Ed., *Advances in Computational Electrodynamics: The Finite-Difference-Time-Domain Method*, Artech House, Boston (1998).
99. V. A. Erma, "Exact solution for the scattering of electromagnetic waves from bodies of arbitrary shape: III. Obstacles with arbitrary electromagnetic properties," *Phys. Rev.* **179**(5), 1238–1246 (1969).
100. K. Muinonen, "Light scattering by stochastically shaped particles," Chapter 11 in *Light Scattering by Nonspherical Particles: Theory, Measurements, and Applications*, M. I. Mishchenko, J. W. Hovenier, and L. D. Travis, Eds., Academic Press, San Diego, pp. 323–352 (2000).
101. V. N. Kuzmin and V. A. Babenko, "Light scattering by a weakly anisotropic spherical particle," *Opt. Spectrosc.* **50**(3), 269–273 (1981).
102. D. Lacoste and B. A. Van Tiggelen, "Stokes parameters for light scattering from a Faraday-active sphere," *JQSRT* **63**(2–6), 305–319 (1999).
103. P. Latimer, "Light scattering by ellipsoids," *J. Colloid Interface Sci.* **53**(1), 102–109 (1975).
104. P. Latimer, "Predicted scattering by spheroids: comparison of approximate and exact methods," *Appl. Opt.* **19**(18), 3039–3041 (1980).
105. D. Deirmendjian, *Electromagnetic Scattering on Spherical Polydispersions*, Elsevier, New York (1969).

106. G. R. Fournier and B. T. N. Evans, "Approximation to extinction efficiency for randomly oriented spheroids," *Appl. Opt.* **30**(15), 2042–2048 (1991).
107. V. A. Bloomfield, "Biological applications," in *Dynamic Light Scattering. Application of Photon Correlation Spectroscopy*, R. Pecora, Ed., Plenum, New York, pp. 363–416 (1985).
108. G. Sittampalam, "Theory of light scattering measurements as applied to immunoprecipitation reactions," *Anal. Chem.* **56**(12), 2176–2180 (1984).
109. G. K. Von Schulthess, G. B Benedek, and R. W. de Blois, "Measurement of the cluster size distributions for high functionality antigenes cross-linked by antibody," *Macromolecules* **13**(4), 939–945 (1980).
110. J. H. W. Leuvering, P. J. H. M. Thal, M. van der Waart, and A. H. W. M. Schuurs, "A sol particle agglutination assay for human chorionic gonadotrophin," *J. Immunol. Methods* **45**, 183–194 (1981).
111. J. M. C. Martin, M. Paques, T. A. van der Velden-De Groot, and E. C. Beuvery, "Characterization of antibody labeled colloidal gold particles and their applicability in a sol particle immunoassay (SPIA)," *J. Immunoassay* **11**, 31–47 (1990).
112. P. Latimer, "Blood platelet aggregometer: predicted effects of aggregation, photometer geometry, and multiple scattering," *Appl. Opt.* **22**(8), 1136–1143 (1983).
113. S. Yu. Shchyogolev, "Inverse problems of spectroturbidimetry of biological disperse systems: an overview," *J. Biomed. Opt.* **4**(4), 490–503 (1999).
114. U. Comberg and T. Wriedt, "Comparison of scattering for aggregated particles based on different models," *JQSRT* **63**(2–6), 149–162 (1999).
115. K. A. Fuller and D. W. Mackowski, "Electromagnetic scattering by compounded spherical particles," Chapter 8 in *Light Scattering by Nonspherical Particles: Theory, Measurements, and Applications*, M. I. Mishchenko, J. W. Hovenier, and L. D. Travis, Eds., Academic Press, San Diego, pp. 225–272 (2000).
116. V. A. Markel and Th. F. George, Eds., *Optics of Nanostructured Materials*, Wiley, New York (2000).
117. A. E. Moskalensky, D. I. Strokotov, A. V. Chernyshev, V. P. Maltsev, and M. A. Yurkin, "Additivity of light-scattering patterns of aggregated biological particles," *J. Biomed. Opt.* **19**, 085004 (2014).
118. P. Latimer and F. Wamble, "Light scattering by aggregates of large colloidal particles," *Appl. Opt.* **21**(12), 2447–2455 (1982).
119. V. A. Markel, V. M. Shalaev, E. B. Stechel, W. Kim, and R. L. Armstrong, "Small-particle composites: I. Linear optical properties," *Phys. Rev. B* **53**(5), 2425–2436 (1996).
120. Yu-lin Xu, "Electromagnetic scattering by an aggregate of spheres," *Appl. Opt.* **34**(21), 4573–4588 (1995).

121. Yu-lin Xu, "Electromagnetic scattering by an aggregate of spheres: far field," *Appl. Opt.* **36**(36), 9496–9508 (1997).
122. D. W. Mackowski and M. I. Mishchenko, "Calculation of the T-matrix and the scattering matrix for ensembles of spheres," *J. Opt. Soc. Am. A* **13**(11), 2266–2278 (1996).
123. P. Meakin, "Models for colloidal aggregation," *Ann. Rev. Phys. Chem.* **39**, 237–267 (1988).
124. M. Y. Lin, H. M. Lindsay, D. A. Weitz, R. C. Ball, R. Klein, and P. Meakin, "Universality of fractal aggregates as probed by light scattering," *Proc. R. Soc. London Ser. A* **423**(1864), 71–87 (1989).
125. N. G. Khlebtsov and A. G. Melnikov, "Structure factor and exponent of scattering by polydisperse fractal colloidal aggregates," *J. Colloid Interface Sci.* **163**(1), 145–151 (1994).
126. R. Botet, P. Rannou, and M. Cabane, "Sensitivity of some optical properties of fractals to the cut-off functions," *J. Phys. A: Math. Gen.* **28**, 297–316 (1995).
127. N. G. Khlebtsov, "Spectroturbidimetry of fractal clusters: test of density correlation cutoff," *Appl. Opt.* **35**(21), 4261–4270 (1996).
128. J. Cai, N. Lu, and C. M. Sorensen, "Analysis of fractal cluster morphology parameters: structural coefficient and density correlation function cutoff," *J. Colloid Interface Sci.* **171**(2), 470–473 (1995).
129. C. M. Sorensen, J. Cai, and N. Lu, "Test of static structure factors for describing light scattering from fractal soot aggregates," *Langmuir* **8**(8), 2064–2069 (1992).
130. M. Novak and T. G. Deway, Eds., *Fractal Frontiers*, World Scientific, Singapore (1999).
131. V. M. Shalaev, *Nonlinear Optics of Random Media: Fractal Composites and Metal Dielectric Films*, Springer-Verlag, Berlin (2000).
132. M. V. Berry and I. C. Percival, "Optics of fractal clusters such as smoke," *Opt. Acta.* **33**(5), 577–591 (1986).
133. N. G. Khlebtsov, "Optics of fractal clusters in the anomalous diffraction approximation," *J. Mod. Opt.* **40**(11), 2221–2235 (1993).
134. R. Jullien, "The application of fractals to investigations of colloidal aggregation and random deposition," *New. J. Chem.* **14**(3), 239–253 (1990).
135. M. C. Daniel and D. Astruc, "Gold nanoparticles: assembly, supramolecular chemistry, quantum-size-related properties, and applications toward biology, catalysis, and nanotechnology," *Chem. Rev.* **104**, 293–346 (2004).
136. T. P. Sau and D. Goia, "Biomedical application of gold nanoparticles," in *Fine Particles in Medicine and Pharmacy*, E. Matijević, Ed., Springer, New York, pp. 101–145 (2012).

137. L. A. Dykman and N. G. Khlebtsov, "Gold nanoparticles in biomedical applications: recent advances and perspectives," *Chem. Soc. Rev.* **41**, 2256–2282 (2012).
138. E. C. Dreaden, A. M. Alkilany, X. Huang, C. J. Murphy, and M. A. El-Sayed "The golden age: gold nanoparticles for biomedicine," *Chem. Soc. Rev.* **41**, 2740–2779 (2012).
139. K. E. Sapsford, W. R. Algar, L. Berti, K. B. Gemmill, B. J. Casey, E. Oh, M. H. Stewart, and I. L. Medintz, "Functionalizing nanoparticles with biological molecules: developing chemistries that facilitate nanotechnology," *Chem. Rev.* **113**, 1904–2074 (2013).
140. M. Quinten, *Optical Properties of Nanoparticle Systems*, Wiley-VCH Verlag, Weinheim (2011).
141. M. I. Stockman, "Nanoplasmonics: past, present, and glimpse into future," *Opt. Express* **19**, 22029–22106 (2011).
142. K. M. Mayer and J. H. Hafner, "Localized surface plasmon resonance sensors," *Chem. Rev.* **111**, 3828–3857 (2011).
143. D. A. Giljohann, D. S. Seferos, W. L. Daniel, M. D. Massich, P. C. Patel, and C. A. Mirkin, "Gold nanoparticles for biology and medicine," *Angew. Chem. Int. Ed.* **49**, 3280–3294 (2010).
144. P. Baptista, E. Pereira, P. Eaton, G. Doria, A. Miranda, I. Gomes, P. Quaresma, and R. Franco, "Gold nanoparticles for the development of clinical diagnosis methods," *Anal. Bioanal. Chem.* **391**, 943–950 (2008).
145. Sh. Gupta, S. Huda, P. K. Kilpatrick, and O. D. Velev, "Characterization and optimization of gold nanoparticle–based silver-enhanced immunoassays," *Anal. Chem.* **79**, 3810–3820 (2007).
146. L. A. Dykman, S. A. Staroverov, V. A. Bogatyrev, and S. Y. Shchyogolev, *Gold Nanoparticles as an Antigen Carrier and an Adjuvant*, Nova Science, New York (2010).
147. P. G. Luo and F. J. Stutzenberger, "Nanotechnology in the detection and control of microorganisms," *Adv. Appl. Microbiol.* **63**, 145–181 (2008).
148. M. K. Yu, J. Park, and S. Jon, "Targeting strategies for multifunctional nanoparticles in cancer imaging and therapy, *Theranostics* **2**, 3–34 (2012).
149. C. Loo, L. Hirsch, M. Lee, E. Chang, J. West, N. Halas, and R. Drezek, "Gold nanoshell bioconjugates for molecular imaging in living cells," *Opt. Lett.* **30**, 1012–1014 (2005).
150. J. Park, A. Estrada, K. Sharp, K. Sang, J. A. Schwartz, D. K. Smith, Ch. Coleman, J. D. Payne, B. A. Korgel, A. K. Dunn, and J. W. Tunnell, "Two-photon-induced photoluminescence imaging of tumors using near-infrared excited gold nanoshells," *Opt. Express* **16**, 1590–1599 (2008).
151. L. V. Wang, Ed., *Photoacoustic Imaging and Spectroscopy*, CRS Press, Boca Raton, FL (2009).

152. L. C. Kennedy, L. R. Bickford, N. A. Lewinski, A. J. Coughlin, Y. Hu, E. S. Day, J. L. West, and R. A. Drezek, "A new era for cancer treatment: gold-nanoparticle-mediated thermal therapies," *Small* **7**, 169–183 (2011).
153. D. Pissuwan, T. Niidome, and M. B. Cortie, "The forthcoming applications of gold nanoparticles in drug and gene delivery systems," *J. Control. Release* **149**, 65–71 (2011).
154. L. A. Dykman and N. G. Khlebtsov, "Uptake of engineered gold nanoparticles into mammalian cells, *Chem. Rev.* **114**, 1258–1288 (2014).
155. G. F. Paciotti, L. Myer, D. Weinreich, D. Goia, N. Pavel, R. E. McLaughlin, and L. Tamarkin, "Colloidal gold: a novel nanoparticle vector for tumor directed drug delivery," *Drug Deliv.* **11**, 169–183 (2004).
156. C. R. Patra, R. Bhattacharya, E. Wang, A. Katarya, J. S. Lau, S. Dutta, M. Muders, S. Wang, S. A. Buhrow, S. L. Safgren, M. J. Yaszemski, J. M. Reid, M. M. Ames, P. Mukherjee, and D. Mukhopadhyay, "Targeted delivery of gemcitabine to pancreatic adenocarcinoma using cetuximab as a targeting agent," *Cancer Res.* **68**, 1970–1978 (2008).
157. Jose L. Arias, Ed., *Nanotechnology and Drug Delivery*, CRS Press, Boca Raton, FL (2015).
158. N. G. Khlebtsov and L. A. Dykman, "Biodistribution and toxicity of engineered gold nanoparticles: a review of in vitro and in vivo studies," *Chem. Soc. Rev.* **40**, 1647–1671 (2011).
159. K. L. Kelly, E. Coronado, L. L. Zhao, and G. C. Schatz, "The optical properties of metal nanoparticles: the influence of size, shape, and dielectric environment," *J. Phys. Chem. B* **107**, 668–677 (2003).
160. M. A. Yurkin, "Computational approaches for plasmonics," in *Handbook of Molecular Plasmonics*, D. F. Sala and S. D'Agostino, Eds., Pan Stanford Publishing, Singapore, pp. 83–135 (2013).
161. N. G. Khlebtsov, "T-matrix method in plasmonics: an overview," *J. Quant. Spectrosc. Radiat. Transfer* **123**, 184–217 (2013).
162. N. G. Khlebtsov, "Optics and biophotonics of nanoparticles with a plasmon resonance," *Quant. Electron.* **38**, 504–529 (2008).
163. A. Lakhtakia, "Strong and weak forms of the method of moments and the coupled dipole method for scattering of time-harmonic electromagnetic fields," *Int. J. Mod. Phys.* **3**, 583–603 (1992)
164. M. Meier and A. Wokaun, "Enhanced fields on large metal particles: dynamic depolarization," *Opt. Lett.* **8**, 581–583 (1983).
165. A. Moroz, "Depolarization field of spheroidal particles," *J. Opt. Soc. Am. B* **26**, 517–527 (2009).
166. E. A. Coronado and G. C. Schatz, "Surface plasmon broadening for arbitrary shape nanoparticles: a geometrical probability approach," *J. Chem. Phys.* **119**, 3926–3934 (2003).

167. A. Moroz, "Electron mean-free path in a spherical shell geometry," *J. Phys. Chem. C* **112**(29), 10641–10652 (2008).
168. C. J. Noguez, "Surface plasmons on metal nanoparticles: the influence of shape and physical environment," *J. Phys. Chem. C* **111**, 3806–3819 (2007).
169. B. N. Khlebtsov, A. G. Melnikov, V. P. Zharov, and N. G. Khlebtsov, "Absorption and scattering of light by a dimer of metal nanospheres: comparison of dipole and multipole approaches," *Nanotechnology* **17**, 1437–1445 (2006).
170. Y.-Y. Yu, S.-S. Chang, C.-L. Lee, and C. R. Ch. Wang, "Gold nanorods: electrochemical synthesis and optical properties," *J. Phys. Chem. B.* **101**(34), 6661–6664 (1997).
171. S. Link, M. B. Mohamed, and M. A. El-Sayed, "Simulation of the optical absorption spectra of gold nanorods as a function of their aspect ratio and the effect of the medium dielectric constant," *J. Phys. Chem. B* **103**, 3073–3077 (1999).
172. B. N. Khlebtsov, V. P. Zharov, A. G. Melnikov, V. V. Tuchin, and N. G. Khlebtsov, "Optical amplification of photothermal therapy with gold nanoparticles and nanoclusters," *Nanotechnology* **17**(20), 5167–5179 (2006).
173. N. G. Khlebtsov, L. A. Trachuk, and A. G. Melnikov, "The effect of the size, shape, and structure of metal nanoparticles on the dependence of their optical properties on the refractive index of a disperse medium," *Opt. Spectrosc.* **98**, 83–90 (2005).
174. J. Pérez-Juste, I. Pastoriza-Santos, L. M. Liz-Marzán, and P. Mulvaney, "Gold nanorods: synthesis, characterization and applications," *Coordination Chem. Rev.* **249**, 1870–1901 (2005).
175. M. Pelton, M. Liu, H. Y. Kim, G. Smith, P. Guyot-Sionnest, and N. F. Scherer, "Optical trapping and alignment of single gold nanorods by using plasmon resonances," *Opt. Lett.* **31**, 2075–2077 (2006).
176. C. Sönnichsen and A. P. Alivisatos, "Gold nanorods as novel nonbleaching plasmon-based orientation sensors for polarized single-particle microscopy," *Nano Lett.* **5**, 301–304 (2005).
177. B. N. Khlebtsov, V. A. Khanadeev, and N. G. Khlebtsov, "Observation of extra-high depolarized light scattering spectra from gold nanorods," *J. Phys. Chem. C.* **112**(33), 12760–12768 (2008).
178. R. Atkinson, W. R. Hendren, G. A. Wurtz, W. Dickson, A. V. Zayats, P. Evans, and R. J. Pollard, "Anisotropic optical properties of arrays of gold nanorods embedded in alumina," *Phys. Rev. B* **73**, 235402–235408 (2006).
179. M. B. Cortie, X. Xu, and M. J. Ford, "Effect of composition and packing configuration on the dichroic optical properties of coinage metal nanorods," *Phys. Chem. Chem. Phys.* **8**, 3520–3527 (2006).

180. J. Grand, P.-M. Adam, A.-S. Grimault, A. Vial, M. Lamy de la Chapelle, J.-L. Bijeon, S. Kostcheev, and P. Royer, "Optical extinction spectroscopy of oblate, prolate and ellipsoid shaped goldnanoparticles: experiments and theory," *Plasmonics* **1**, 135–140 (2006).
181. Ch. J. Orendorff and C. J. Murphy, "Quantitation of metal content in the silver-assisted growth of gold nanorods," *J. Phys. Chem. B* **110**, 3990–3994 (2006).
182. H. Liao and J. H. Hafner, "Gold nanorod bioconjugates," *Chem. Mater.* **17**, 4636–4641 (2005).
183. L. R. Hirsch, A. M. Gobin, A. R. Lowery, F. Tam, R. A. Drezek, N. J. Halas, and J. L. West, "Metal nanoshells," *Ann. Biomed. Eng.* **34**(1), 15–22 (2006).
184. B. N. Khlebtsov and N. G. Khlebtsov, "Enhanced solid-phase immunoassay using gold nanoshells: effect of nanoparticle optical properties," *Nanotechnology* **19**, 435703 (2008).
185. G. Akchurin, B. Khlebtsov, G. Akchurin, V. Tuchin, V. Zharov, and N. Khlebtsov, "Gold nanoshell photomodification under single nanosecond laser pulse accompanied by color-shifting and bubble formation phenomena," *Nanotechnology* **19**, 015701 (2008).
186. B. N. Khlebtsov and N. G. Khlebtsov, "Biosensing potential of silica/gold nanoshells: sensitivity of plasmon resonance to the local dielectric environment," *J. Quant. Spectrosc. Radiat. Transfer* **106**, 154–169 (2007).
187. L. A. Trachuk, S. A. Vrublevsky, B. N. Khlebtsov, A. G. Melnikov, and N. G. Khlebtsov, "Optical properties of gold spheroidal particles and nanoshells: effect of the external dielectric medium," *Proc. SPIE* **5772**, 1–10 (2005).
188. H. Wang, D. W. Brandl, F. Le, P. Nordlander, and N. J. Halas, "Nanorice: a hybrid plasmonic nanostructure," *Nano Lett.* **6**, 827–832 (2006).
189. L. A. Trachuk, A. G. Melnikov, B. N. Khlebtsov, and N. G. Khlebtsov, "Dependence of the optical properties of metal nanoparticles on the external dielectric medium: effects of the particle size, shape, and structure," *Proc. SPIE* **5829**, 127–137 (2005).
190. K. Kolwas, S. Demianiuk, and M. Kolwas, "Optical excitation of radius–dependent plasmon resonances in large metal clusters," *J. Phys. B* **29**, 4761–4770 (1996).
191. N. G. Khlebtsov, A. G. Melnikov, V. A. Bogatyrev, and L. A. Dykman, "Optical properties and biomedical applications of nanostructures based on gold and silver bioconjugates," in *Photopolarimetry in Remote Sensing*, G. Videen, Ya. S. Yatskiv, and M. I. Mishchenko, Eds., NATO Science Series, II. Mathematics, Physics, and Chemistry, **Vol. 161**, Kluwer Academic, Dordrecht, pp. 265–308 (2004).
192. V. A. Markel, "Divergence of dipole sums and the nature of non-Lorentzian exponentially narrow resonances in one-dimensional periodic

arrays of nanospheres," *J. Phys. B: At. Mol. Opt. Phys.* **38**, L115–L121 (2005).

193. B. Lamprecht, G. Schider, R. T. Lechner, H. Ditlbacher, J. R. Krenn, A. Leitner, and F. R. Aussenegg, "Metal nanoparticle gratings: influence of dipolar particle interaction on the plasmon resonance," *Phys. Rev. Lett.* **84**, 4721–4724 (2000).
194. S. Malynych and G. Chumanov, "Light-induced coherent interactions between silver nanoparticles in two-dimensional arrays," *J. Am. Chem. Soc.* **125**, 2896–2898 (2003).
195. B. N. Khlebtsov, V. A. Khanadeyev, J. Ye, D. W. Mackowski, G. Borghs, and N. G. Khlebtsov, "Coupled plasmon resonances in monolayers of metal nanoparticles and nanoshells," *Phys. Rev. B* **77**(3), 035440 (2008).
196. K.-H. Su, Q.-H. Wei, X. Zhang, J. J. Mock, D. R. Smith, and S. Schultz, "Interparticle coupling effects on plasmon resonances of nanogold particles," *Nano Lett.* **3**, 1087–1090 (2003).
197. B. Lamprecht, A. Leitner, and F. R. Aussenegg, "SHG studies of plasmon dephasing in nanoparticles," *Appl. Phys. B* **68**, 419–423 (1999)
198. D. Mackowski, "Electrostatics analysis of radiative absorption by sphere clusters in the Rayleigh limit: application to soot particles," *Appl. Opt.* **34**, 3535–3545 (1995).
199. M. A. Hayat, *Principles and Techniques of Electron Microscopy, Biological Applications 1*, Van Nostrand Reinhold, New York (1970).
200. G. Kumar and J. M. Schmitt, "Micro-optical properties of tissue," *Proc. SPIE* **2679**, 106–116 (1996).
201. R. L. Fante, "Electromagnetic propagation in turbulent media," *Proc. IEEE* **63**, 1669–1692 (1975).
202. J. A. Ogilvy, *Theory of Wave Scattering from Random Rough Surfaces*, Hilger, Bristol (1991).
203. D. A. Zimnyakov, V. V. Tuchin, and A. A. Mishin, "Spatial speckle correlometry in applications to tissue structure monitoring," *Appl. Opt.* **36**, 5594–5607 (1997).
204. J. M. Schmitt and G. Kumar, "Optical scattering properties of soft tissue: a discrete particle model," *Appl. Opt.* **37**(13), 2788–2797 (1998).
205. O. V. Volkova, V. A. Shakhmatova, and A. A. Mironova, Eds., *Atlas of Scanning Electron Microscopy Images of Cells, Tissues, and Organs* (in Russian), Medicine, Moscow (1987).
206. J. W. Goodman, *Statistical Optics*, Wiley-Interscience, New York (1985).
207. S. Ya. Sid'ko, V. N. Lopatin, and L. E. Paramonov, *Polarization Characteristics of Suspensions of Biological Particles* (in Russian), Nauka, Novosibirsk (1990).
208. R. D. Dyson, *Cell Biology: A Molecular Approach*, Allyn and Bacon, Boston (1974).

209. A. P. Ivanov, A. Ya. Khairullina, and T. N. Kharkova, "Experimental detection of cooperative effects in a dispersing volume," *Opt. Spectrosc.* **28**(2), 380–387 (1970).
210. M. J. Hogan, J. A. Alvardo, and J. Weddel, *Histology of the Human Eye*, 638–677, W. B. Sanders, Philadelphia (1971).
211. Q. Zhou and R. W. Klinghton, "Light scattering and form birefringence of parallel cylindrical arrays that represent cellular organelles of the retinal nerve fiber layer," *Appl. Opt.* **36**(10), 2273–2285 (1997).
212. G. Videen and D. Ngo, "Light scattering multipole solution for a cell," *J. Biomed. Opt.* **3**(2), 212–220 (1998).
213. G. B. Benedek, "Theory of transparency of the eye," *Appl. Opt.* **10**(3), 459–473 (1971).
214. A. Brunsting and P. Mullaney, "Differential light scattering from spherical mammalian cells," *Biophys. J.* **14**, 439–453 (1974).
215. J. M. Schmitt and G. Kumar, "Turbulent nature of refractive-index variations in biological tissue," *Opt. Lett.* **21**, 1310–1312 (1996).
216. J. Beutham, O. Minet, J. Helfmann, M. Herrig, and G. Muller, "The spatial variation of the refractive index in biological cells," *Phys. Med. Biol.* **41**(3), 369–382 (1996).
217. P. S. Tuminello, E. T. Arakawa, B. N. Khare, J. M. Wrobel, M. R. Querry, and M. E. Milham, "Optical properties of Bacillus subtilis spores from 0.2 to 2.5 mm," *Appl. Opt.* **36**, 2818–2824 (1997).
218. F. A. Duck, *Physical Properties of Tissue: A Comprehensive Reference Book*, Academic Press, San Diego (1990).
219. K. W. Keohane and W. K. Metcalf, "The cytoplasmic refractive index of lymphocytes, its significance and its changes during active immunisation," *Quart. J. Exp. Physiol. Cognate Med. Sci.* **44**, 343–346 (1959).
220. F. P. Bolin, L. E. Preuss, R. C. Taylor, and R. J. Ference, "Refractive index of some mammalian tissues using a fiber optic cladding method," *Appl. Opt.* **28**, 2297–2303 (1989).
221. O. V. Angelski, A. G. Ushenko, A. D. Arkhelyuk, S. B. Ermolenko, and D. N. Burkovets, "Scattering of laser radiation by multifractal biological structures," *Opt. Spectrosc.* **88**(3), 444–447 (2000).
222. V. F. Izotova, I. L. Maksimova, I. S. Nefedov, and S. V. Romanov, "Investigation of Mueller matrices of anisotropic nonhomogeneous layers in application to optical model of cornea," *Appl. Opt.* **36**(1), 164–169 (1997).
223. J. D. Bancroft and A. Stevens, Eds., *Theory and Practice of Histological Techniques*, Churchill Livingstone, Edinburgh and New York (1990).
224. D. M. Maurice, *The Cornea and Sclera: The Eye*, H. Davson, Ed., Academic Press, Orlando, pp. 1–158 (1984).

225. M. S. Borcherding, L. J. Blasik, R. A. Sittig, J. W. Bizzel, M. Breen, and H. G. Weinstein, "Proteoglycans and collagen fiber organization in human corneoscleral tissue," *Exp. Eye Res.* **21**, 59–70 (1975).
226. S. Vaezy and J. I. Clark, "Quantitative analysis of the microstructure of the human cornea and sclera using 2-D Fourier methods," *J. Microsc.* **175**(Pt. 2), 93–99 (1994).
227. O. V. Angelsky and P. P. Maksimyak, "Optical correlation diagnostics of random fields and objects," *Opt. Eng.* **34**(4), 937–981 (1995).
228. R. W. Hart and R. A. Farrell, "Light scattering in the cornea," *J. Opt. Soc. Am.* **59**(6), 766–774 (1969).
229. I. L. Maksimova, D. A. Zimnyakov, and V. V. Tuchin, "Control of optical properties of biotissues: I. Spectral properties of the eye sclera," *Opt. Spectrosc.* **89**(1), 78–86 (2000).
230. Y. Kamai and T. Ushiki, "The three-dimensional organization of collagen fibrils in the human cornea and sclera," *Invest. Ophtalmol. Vis. Sci.* **32**, 2244–2258 (1991).
231. S. Vaezy and J. I. Clark, "A quantitative analysis of transparency in the human sclera and cornea using Fourier methods," *J. Microsc.* **163**, 85–94 (1991).
232. F. A. Bettelheim, "Physical basis of lens transparency," in *The Ocular Lens: Structure, Function and Pathology*, H. Maisel, Ed., Marcel Dekker, New York, pp. 265–300 (1985).
233. A. Tardieuand and M. Delaye, "Eye lens proteins and transparency from light transmission theory to solution x-ray structural analysis," *Ann. Rev. Biophys. Chem.* **17**, 47–70 (1988).
234. A. V. Krivandin, "On the supramolecular structure of eye lens crystallins: the study by small-angle x-ray scattering," *Biofizika* **46**(6), 1274–1278 (1997).
235. S. Vaezy and J. I. Clark, "Characterization of the cellular microstructures of ocular lens using 2-D power law analysis," *Ann. Biomed. Eng.* **23**, 482–490 (1995).
236. J. Feder, *Fractals*, Plenum, New York and London (1988).
237. Y. Sakurada, J. Uozumi, and T. Asakura, "Scaling properties of the Fresnel diffraction field produced by one dimensional regular fractals," *Pure Appl. Opt.* **3**(3), 374–380 (1994).
238. R. W. James, *The Optical Principles of Diffraction of X-rays: The Crystalline State II*, L. Bragg, Ed., London (1950).
239. J. M. Ziman, *Models of Disorder: The Theoretical Physics of Homogeneously Disordered Systems*, Cambridge Univ. Press, London, New York, and Melbourne (1979).
240. M. S. Wertheim, "Exact solution of the Percus–Yevick integral equation for hard spheres," *Phys. Rev. Lett.* **10**(8), 321–323 (1963).

241. J. L. Lebovitz, "Exact solution of generalized Percus–Yevick equation for a mixture of hard spheres," *Phys. Rev.* **133**(4A), 895–899 (1964).
242. R. J. Baxter, "Ornstein–Zernike relation and Percus–Yevick approximation for fluid mixtures," *J. Chem. Phys.* **52**(9), 4559–4562 (1970).
243. A. P. Ivanov, V. A. Loiko, and V. P. Dik, *Light Propagation in Densely Packed Disperse Media*, Nauka i Technika, Minsk (1988).
244. I. L. Maksimova and L. P. Shubochkin, "Light-scattering matrices for a close-packed binary system of hard spheres," *Opt. Spectrosc.* **70**(6), 745–748 (1991).
245. R. L. McCally and R. A. Farrell, "Light scattering from cornea and corneal transparency," in *Noninvasive Diagnostic Techniques in Ophtalmology*, B. R. Master, Ed., Springer-Verlag, New York (1990).
246. L. Reynolds, *Optical Diffuse Reflectance and Transmittance from an Anisotropically Scattering Finite Blood Medium*, Ph.D. thesis, Univ. of Washington, Seattle (1975).
247. V. G. Vereshchagin and A. N. Ponyavina, "Statistical characteristic and transparency of thin closely packed disperse layer," *Zh. Prikl. Spektr. (J. Appl. Spectrosc.)* **22**(3), 518–524 (1975).
248. I. L. Maksimova, V. V. Tuchin, and L. P. Shubochkin, "Polarization features of eye's cornea," *Opt. Spectrosc.* **60**(4), 801–807 (1986).
249. I. L. Maksimova, V. V. Tuchin, and L. P. Shubochkin, "Light scattering matrices of the crystalline lens," *Opt. Spectrosc.* **65**(3), 365–368 (1988).
250. G. Marguez, L. V. Wang, S.-P. Lin, J. A. Swartz, and S. L. Thomsen, "Anisotropy in the absorption and scattering spectra of the chicken breast tissue," *Appl. Opt.* **37**(4), 798–804 (1998).
251. S. Zigman, G. Sutliff, and M. Rounds, "Relationships between human cataracts and environmental radiant energy: cataract formation, light scattering and fluorescence," *Lens Eye Toxicity Res.* **8**, 259–280 (1991).
252. V. V. Tuchin and D. M. Zhestkov, "Tissue structure and eye lens transmission and scattering spectra," *Proc. SPIE* **3053**, 123–128 (1997).
253. Yu. P. Sinichkin, S. R. Utz, and H. A. Pilipenko, "Spectroscopy of human skin in vivo: I. reflection spectra," *Opt. Spectrosc.* **80**(2), 228–234 (1996).
254. N. G. Khlebtsov, V. V. Nikiforov, A. G. Melnikov, T. K. Merkulova, and L. N. Serdobintsev, "Elastic scattering spectroscopy of Yersinia pestis capsular protein solutions," *Biopolym. Cell* **6**(2), 81–87 (1990).
255. V. Twersky "Interface effects in multiple scattering by large, low refracting, absorbing particles," *J. Opt. Soc. Am.* **60**(7), 908–914 (1970).
256. M. Lax, "Multiple scattering of waves: II. The effective field in dense system," *Phys. Rev.* **85**(4), 621–629 (1952).

257. K. M. Hong, "Multiple scattering of electromagnetic waves by a crowded monolayer of spheres: application to migration imaging films," *J. Opt. Soc. Am.* **70**(7), 821–826 (1980).
258. A. Ishimary and Y. Kuga, "Attenuation constant of a coherent field in a dense distribution of particles," *J. Opt. Soc. Am.* **72**(10), 1317–1320 (1982).
259. A. N. Ponyavina, "Selection of optical radiation in scattering by partially ordered disperse media," *Zh. Prikl. Spektr. (J. Appl. Spectrosc.)* **65**(5), 721–733 (1998).
260. T. R. Smith, "Multiple scattering in the cornea," *J. Mod. Opt.* **35**(1), 93–101 (1988).
261. P. A. Zhu, A. K. Fung, and K. W. Wong, "Effective propagation constants in dense random media under effective medium approximation," *Radio Sci.* **22**, 234–250 (1987).
262. V. Twersky, "Absorption and multiple scattering by biological suspensions," *J. Opt. Soc. Am.* **60**, 1084–1093 (1970).
263. J. M. Steinke and A. P. Shepherd, "Diffusion model of the optical absorbance of whole blood," *J. Opt. Soc. Am. A* **5**, 813–822 (1988).
264. D. W. Rayleigh, "On the influence of obstacles arranged in rectangular order on the properties of a medium," *Philos. Mag.* **34**, 481–502 (1892).
265. R. P. Hemenger, "Birefringence of a medium of tenuous parallel cylinders," *Appl. Opt.* **28**(18), 4030–4034 (1989).
266. J. Lenoble, Ed., *Radiative Transfer in Scattering and Absorbing Atmospheres: Standard Computational Procedures*, A. Deepak Publishing, Hampton, VA (1985).
267. M. M. R. Williams, *Mathematical Methods in Particle Transport Theory*, Wiley, New York (1971).
268. A. H. Hielsher, J. R. Mourant, and I. J. Bigio, "Influence of particle size and concentration on the diffuse backscattering of polarized light from tissue phantoms and biological cell suspensions," *Appl. Opt.* **36**, 125–135 (1997).
269. V. V. Tuchin, S. R. Utz, and I. V. Yaroslavsky, "Tissue optics, light distribution, and spectroscopy," *Opt. Eng.* **33**(10), 3178–3188 (1994).
270. G. Yoon, A. J. Welch, M. Motamedi, and M. C. J. Van Gemert, "Development and application of three-dimensional light distribution model for laser irradiated tissue," *IEEE J. Quantum Electron.* **23**(10), 1721–1733 (1987).
271. G. Müller, B. Chance, and R. Alfano et al., Eds., *Medical Optical Tomography: Functional Imaging and Monitoring* IS11, SPIE Press, Bellingham, WA (1993).
272. G. Müller and A. Roggan, Eds., *Laser-Induced Interstitial Thermotherapy* **PM25**, SPIE Press, Bellingham, WA (1995).

273. V. V. Tuchin, Ed., *Selected Papers on Tissue Optics: Applications in Medical Diagnostics and Therapy* **MS102**, SPIE Press, Bellingham, WA (1994).
274. B. B. Das, F. Liu, and R. R. Alfano, "Time-resolved fluorescence and photon migration studies in biomedical and random media," *Rep. Prog. Phys.* **60**, 227–292 (1997).
275. Tuan Vo-Dinh, Ed., *Biomedical Photonics Handbook*, CRC Press, Boca Raton, FL (2003; 2nd ed., 2014).
276. J. M. Schmitt, A. H. Gandjbakhche, and R. F. Bonnar, "Use of polarized light to discriminate short-photons in a multiply scattering medium," *Appl. Opt.* **31**, 6535–6546 (1992).
277. P. Bruscaglioni, G. Zaccanti, and Q. Wei, "Transmission of a pulsed polarized light beam through thick turbid media: numerical results," *Appl. Opt.* **32**, 6142–6150 (1993).
278. M. I. Mishchenko, L. D. Travis, and A. A. Lacis, *Multiple Scattering of Light by Particles: Radiative Transfer and Coherent Backscattering*, Cambridge Univ. Press, Cambridge (2006).
279. D. Bicout, C. Brosseau, A. S. Martinez, and J. M. Schmitt, "Depolarization of multiply scattering waves by spherical diffusers: influence of the size parameter," *Phys. Rev. E* **49**, 1767–1770 (1994).
280. M. Dogariu and T. Asakura, "Photon pathlength distribution from polarized backscattering in random media," *Opt. Eng.* **35**, 2234–2239 (1996).
281. A. Ambirajan and D. C. Look, "A backward Monte Carlo study of the multiple scattering of a polarized laser beam," *J. Quant. Spectrosc. Radiat. Transfer* **58**, 171–192 (1997).
282. D. A. Zimnyakov, Yu. P. Sinichkin, P. V. Zakharov, and D. N. Agafonov, "Residual polarization of non-coherently backscattered linearly polarized light: the influence of the anisotropy parameter of the scattering medium," *Waves Random Media* **11**, 395–412 (2001).
283. I. A. Vitkin and R. C. N. Studinski, "Polarization preservation in diffusive scattering from in vivo turbid biological media: effects of tissue optical absorption in the exact backscattering direction," *Opt. Commun.* **190**, 37–43 (2001).
284. A. A. Kokhanovsky, "Reflection and polarization of light by semi-infinite turbid media: simple approximations," *J. Colloid Interface Sci.* **251**, 429–433 (2002).
285. L. Dagdug, G. H. Weiss, and A. H. Gandjbakhche, "Effects of anisotropic optical properties on photon migration in structured tissues," *Phys. Med. Biol.* **48**, 1361–1370 (2003).
286. H. H. Tynes, G. W. Kattawar, E. P. Zege, I. L. Katsev, A. S. Prikhach, and L. I. Chaikovskaya, "Monte Carlo and multicomponent approximation methods for vector radiative transfer by use of effective mueller matrix calculations," *Appl. Opt.* **40**(3), 400–412 (2001).

287. I. M. Stockford, S. P. Morgan, P. C. Y. Chang, and J. G. Walker, "Analysis of the spatial distribution of polarized light backscattering," *J. Biomed. Opt.* **7**(3), 313–320 (2002).
288. S. Bartel and A. H. Hielscher, "Monte Carlo simulations of the diffuse backscattering mueller matrix for highly scattering media," *Appl. Opt.* **39**, 1580–1588 (2000).
289. K. Y. Yong, S. P. Morgan, I. M. Stockford, and M. C. Pitter, "Characterization of layered scattering media using polarized light measurements and neural networks," *J. Biomed. Opt.* **8**(3), 504–511 (2003).
290. S. V. Gangnus, S. J. Matcher, and I. V. Meglinski, "Monte Carlo modeling of polarized light propagation in biological tissues," *Laser Phys.* **14**, 886–891 (2004).
291. B. C. Wilson and G. A. Adam, "Monte Carlo model for the absorption and flux distributions of light in tissue," *Med. Phys.* **10**, 824–830 (1983).
292. L.-H. Wang, S. L. Jacques, and L.-Q. Zheng, "MCML: Monte Carlo modeling of photon transport in multi-layered tissues," *Comput. Methods Prog. Biomed.* **47**, 131–146 (1995).
293. L.-H. Wang, S. L. Jacques, and L.-Q. Zheng, "CONV: convolution for responses to a finite diameter photon beam incident on multi-layered tissues," *Comput. Methods Prog. Biomed.* **54**, 141–150 (1997).
294. V. V. Tuchin, L. V. Wang, and D. A. Zimnyakov, *Optical Polarization in Biomedical Applications*, Springer-Verlag, New York (2006).
295. C. J. Hourdakis and A. Perris, "A Monte Carlo estimation of tissue optical properties for the use in laser dosimetry," *Phys. Med. Biol.* **40**, 351–364 (1995).
296. L. T. Perelman, J. Wu, I. Itzkan, and M. S. Feld, "Photon migration in turbid media using path integrals," *Phys. Rev. Lett.* **72**, 1341–1344 (1994).
297. G. Bal and M. Moscoso, "Theoretical and numerical analysis of polarization for time-dependent radiative transfer equations," *J. Quant. Spectrosc. Radiat. Transfer* **70**, 75–98 (2001).
298. X. Wang and L. V. Wang, "Propagation of polarized light in birefringent turbid media: time-resolved simulations," *Opt. Express* **9**(5), 254–259 (2001).
299. X. Wang and L. V. Wang, "Propagation of polarized light in birefringent turbid media: a Monte Carlo study," *J. Biomed. Opt.* **7**(3), 279–290 (2002).
300. I. L. Maksimova, S. V. Romanov, and V. F. Izotova, "The effect of multiple scattering in disperse media on polarization characteristics of scattered light," *Opt. Spectrosc.* **92**(6), 915–923 (2002).
301. L. V. Wang and H.-I. Wu, *Biomedical Optics: Principles and Imaging*, Wiley-Interscience, Hoboken, NJ (2007).

302. V. Shankaran, J. T. Walsh Jr, and D. J. Maitland, "Comparative study of polarized light propagation in biological tissues," *J. Biomed. Opt.* **7**(3), 300–306 (2002).
303. M. J. Rakovic and G. W. Kattawar, "Theoretical analysis of polarization patterns from incoherent backscattering of light," *Appl. Opt.* **37**(15), 3333–3338 (1998).
304. I. L. Maksimova, S. N. Tatarintsev, and L. P. Shubochkin, "Multiple scattering effects in laser diagnostics of bioobjects," *Opt. Spectrosc.* **72**, 1171–1177 (1992).
305. K. C. Hadley and I. A. Vitkin, "Optical rotation and linear and circular depolarization rates in diffusively scattered light from chiral, racemic, and achiral turbid media," *J. Biomed. Opt.* **7**(3), 291–299 (2002).
306. L. O. Svaasand and Ch. J. Gomer, "Optics of tissue," in *Dosimetry of Laser Radiation in Medicine and Biology* IS5, SPIE Press, Bellingham, WA, pp. 114–132 (1989).
307. A. N. Yaroslavsky, I. V. Yaroslavsky, T. Goldbach, and H. -J. Schwarzmaier, "Influence of the scattering phase function approximation on the optical properties of blood determined from the integrating sphere measurements," *J. Biomed. Opt.* **4**(1), 47–53 (1999).
308. H. Horinaka, K. Hashimoto, K. Wada, and Y. Cho, "Extraction of quasi-straightforward-propagating photons from diffused light transmitting through a scattering medium by polarization modulation," *Opt. Lett.* **20**, 1501–1503 (1995).
309. S. P. Morgan, M. P. Khong, and M. G. Somekh, "Effects of polarization state and scatterer concentration optical imaging through scattering media," *Appl. Opt.* **36**, 1560–1565 (1997).
310. M. R. Ostermeyer, D. V. Stephens, L. Wang, and S. L. Jacques, "Nearfield polarization effects on light propagation in random media," OSA TOPS 3, Optical Society of America, Washington, DC, pp. 20–25 (1996).
311. L. V. Wang, G. L. Coté, and S. L. Jacques, Eds., "Special section on tissue polarimetry," *J. Biomed. Opt.* **7**(3), 278–397 (2002).
312. R. R. Anderson, "Polarized light examination and photography of the skin," *Arch. Dermatol.* **127**, 1000–1005 (1991).
313. A. W. Dreher and K. Reiter, "Polarization technique measures retinal nerve fibers," *Clin. Vis. Sci.* **7**, 481–485 (1992).
314. N. Kollias, "Polarized light photography of human skin," in *Bioengineering of the Skin: Skin Surface Imaging and Analysis*, K.-P. Wilhelm, P. Elsner, E. Berardesca, and H. I. Maibach, Eds., CRC Press, Boca Raton, FL, pp. 95–106 (1997).
315. S. L. Jacques, R. J. Roman, and K. Lee, "Imaging superficial tissues with polarized light," *Lasers Surg. Med.* **26**, 119–129 (2000).
316. S. L. Jacques, J. C. Ramella-Roman, and K. Lee, "Imaging skin pathology with polarized light," *J. Biomed. Opt.* **7**(3), 329–340 (2002).

317. N. Ghosh, M. F. G. Wood, and I. A. Vitkin, "Polarized light assessment of complex turbid media such as biological tissues via Mueller matrix decomposition," in *Handbook of Photonics for Biomedical Science*, V. V. Tuchin, Ed., CRC Press, Taylor & Francis Group, London, pp. 253–282 (2010).
318. N. Ghosh and I. A. Vitkin, "Concepts, challenges and applications of polarized light in biomedicine: a tutorial review," *J. Biomed. Opt.* **16**(11), 110801-1–29 (2010).
319. S. Alali, M. Ahmad, A. J. Kim, N. Vurgun, M. F. G. Wood, and I. A. Vitkin, "Quantitative correlation between light depolarization and transport albedo of various porcine tissues," *J. Biomed. Opt.* **17**(4), 045004 (2012).
320. D. Layden, N. Ghosh, and A. Vitkin, "Quantitative polarimetry for tissue characterization and diagnosis," in *Advanced Biophotonics: Tissue Optical Sectioning*, R. K. Wang and V. V. Tuchin, Eds., CRC Press, Taylor & Francis Group, Boca Raton, FL, London, and New York, pp. 73–108 (2013).
321. A. Pierangelo, S. Manhas, A. Benali, C. Fallet, J.-L. Totobenazara, M.-R. Antonelli, T. Novikova, B. Gayet, A. De Martino, and P. Validire, "Multispectral Mueller polarimetric imaging detecting residual cancer and cancer regression after neoadjuvant treatment for colorectal carcinomas," *J. Biomed. Opt.* **18**(4), 046014-1–9 (2013).
322. E. E. Gorodnichev, S. V. Ivliev, A. I. Kuzovlev, and D. B. Rogozkin, "Transillumination of highly scattering media by polarized light," in *Light Scattering Reviews 8. Radiative Transfer and Light Scattering*, A. A. Kokhanovsky, Ed., Springer Praxis, Berlin and Heidelberg, pp. 317–361 (2013).
323. C. Macdonald, A. Doronin, A. F. Peña, M. Eccles, and I. Meglinski, "Diffusing-wave polarimetry for tissue diagnostics," *Proc. SPIE* **8940**, 894007 (2014).
324. E. Du, H. He, N. Zeng, M. Sun, Y. Guo, J. Wu, S. Liu, and H. Ma, "Mueller matrix polarimetry for differentiating characteristic features of cancerous tissues," *J. Biomed. Opt.* **19**(7), 076013 (2014).
325. E. E. Gorodnichev and D. B. Rogozkin, "Small angle multiple scattering in random inhomogeneous media," *JETP* **107**, 209–235 (1995).
326. J. Ramella-Roman, S. Prahl, and S. Jacques, "Three Monte Carlo programs of polarized light transport into scattering media: part I," *Opt. Express* **13**(12), 4420–4438 (2005); part II, *Opt. Express* **13**(25), 10392–10405 (2005).
327. M. J. Rakovic, G. W. Kattawar, M. Mehrubeoglu, B. D. Cameron, L.-H. V. Wang, S. Rastegar, and G. L. Coté, "Light backscattering polarization patterns from turbid media: theory and experiment," *Appl. Opt.* **38**, 3399–3408 (1999).

328. S. L. Jacques, M. Ostemeyer, L.-H. Wang, and D. Stephens, "Polarized light transmission through skin using video reflectometry: toward optical tomography of superficial tissue layers," in *Lasers in Surgery: Advanced Characterization, Therapeutics, and Systems VI*, R. R. Anderson, Ed., *Proc. SPIE* **2671**, 199–220 (1996).
329. G. Yao and L.-H. V. Wang, "Two-dimensional depth-resolved Mueller matrix characterization of biological tissue by optical coherence tomography," *Opt. Lett.* **24**, 537–539 (1999).
330. S. Jiao, G. Yao, and L.-H. V. Wang, "Depth-resolved two-dimensional Stokes vectors of backscattered light and Mueller matrices of biological tissue measured by optical coherence tomography," *Appl. Opt.* **39**, 6318–6324 (2000).
331. V. F. Izotova, I. L. Maksimova, and S. V. Romanov, "Color of biotissues and artificial disperse systems under diffuse scattering of light," *Proc. SPIE* **4001**, 279–285 (2000).
332. G. V. Bally, *Optics in Biomedical Sciences*, Springer-Verlag, Berlin (1982).
333. J. W. Hovenier, H. C. van de Hulst, and C. V. M. van der Mee, "Conditions for the elements of the scattering matrix," *Astron. Astrophys.* **157**, 301–310 (1986).
334. J. W. Hovenier, "Structure of a general pure Mueller matrix," *Appl. Opt.* **33**, 8318–8324 (1994).
335. E. S. Fry and G. W. Kattawar, "Relationships between elements of the Stokes matrix," *Appl. Opt.* **20**, 2811–2814 (1981).
336. R. W. Shaefer, "Inequalities between the elements of the Mueller scattering matrix," *Appl. Opt.* **20**, 2875–2881 (1981).
337. G. W. Kattawar and E. S. Fry, "Inequalities between the elements of the Mueller scattering matrix: comments," *Appl. Opt.* **21**, 18 (1982).
338. J. J. Gil and E. Bernabeu, "A depolarization criterion in Mueller matrix," *Opt. Acta* **32**, 259–261 (1985).
339. V. F. Izotova, I. L. Maksimova, and S. V. Romanov, "Utilization of relation between elements of the Mueller matrices for estimating properties of objects and the reliability of experiments," *Opt. Spectrosc.* **80**(5), 753–759 (1996).
340. J. W. Hovenier and C. V. M. van der Mee, "Relationships for matrices describing scattering by small particles," Chapter 3 in *Light Scattering by Nonspherical Particles: Theory, Measurements, and Applications*, M. I. Mishchenko, J. W. Hovenier, and L. D. Travis, Eds., Academic Press, New York, pp. 62–85 (2000).
341. J. W. Mishchenko and L. D. Hovenier, "Depolarization of light backscattered by randomly oriented nonspherical particles," *Opt. Lett.* **20**, 1356–1358 (1995).

342. A. V Priezzhev, V. V. Tuchin, and L. P. Shubochkin, *Laser Diagnostics in Biology and Medicine* (in Russian), Nauka, Moscow (1989).
343. G. J. Buffone, J. Savory, and J. Hermans, "Evaluation of kinetic light scattering as an approach to the measurement of specific proteins with the centrifugal analyzer: II. theoretical considerations," *Clin. Chem.* **21**(12), 1735–1746 (1975).
344. A. Ya. Khairullina, T. V. Oleinik, and I. L. Katsev, "Possibility of optic diagnostics of mono- and multi-layer structures simulating human organs," *Opt. Spectrosc.* **75**(1), 130–137 (1993).
345. A. N. Korolevich, A. Ya. Khairullina, and L. P. Shubochkin, "Scattering matrix for 'soft' particles," *Opt. Spectrosc.* **68**(2), 403–409 (1990).
346. A. N. Korolevich, A. Ya. Khairullina, and L. P. Shubochkin, "Effects of aggregation of large biological particles on the scattering matrix elements," *Opt. Spectrosc.* **77**(2), 278–282 (1994).
347. V. F. Izotova, I. L. Maksimova, and S. V. Romanov, "Simulation of polarization characteristics of the crystalline lens during protein aggregation with regard to multiple scattering," *Opt. Spectrosc.* **86**(6), 902–908 (1999).
348. V. F. Izotova, I. L. Maksimova, and S. V. Romanov, "Analysis of errors of a laser polarization nephelometer," *Opt. Spectrosc.* **80**(6), 905–1001 (1996).
349. A. J. Hunt and D. R. Huffman, "A new polarization-modulated light scattering instrument," *Rev. Sci. Instrum.* **44**(12), 1753–1762 (1973).
350. R. C. Thompson, J. R. Bottiger, and E. S. Fry, "Measurement of polarized light interactions via the Mueller matrix," *Appl. Opt.* **19**(8), 1323–1332 (1980).
351. R. M. A. Azzam, Polarimeter, US Patent No. 4.306.809 (1981).
352. W. S. Bickel, J. F. Davidson, D. R. Huffman, and R. Kilkson, "Application of polarization effects in light scattering: a new biophysical tool," *Proc. Natl. Acad. Sci. USA* **73**, 486–490 (1976).
353. K. J. Voss and E. S. Fry, "Measurement of the Mueller matrix for ocean water," *Appl. Opt.* **23**, 4427–4439 (1984).
354. K. D. Lofftus, M. S. Quinby-Hunt, A. J. Hunt, F. Livolant, and M. Maestre, "Light scattering by Prorocentrum micans: a new method and results," *Appl. Opt.* **31**, 2924–2931 (1992).
355. W. S. Bickel and M. E. Stafford, "Biological particles as irregularly shaped particles," in *Light Scattering by Irregularly Shaped Particles*, D. Schuerman, Ed., Plenum, New York, pp. 299–305 (1980).
356. A. G. Hoekstra and P. M. A. Sloot, "Dipolar unit size in coupled-dipole calculations of the scattering matrix elements," *Appl. Opt.* **18**, 1211–1213 (1993).

357. B. V. Bronk, S. D. Druger, J. Czege, and W. P. van de Merwe, "Measuring diameters of rod-shaped bacteria in vivo with polarized light scattering," *Biophys. J.* **69**, 1170–1177 (1995).
358. W. P. van de Merwe, Z.-Z. Li, B. V. Bronk, and J. Czege, "Polarized light scattering for rapid observation of bacterial size changes," *Biophys. J.* **73**, 500–506 (1997).
359. O. V. Angelsky, A. G. Ushenko, A. Yu. Ushenko, V. P. Pishak, and A. P. Peresunko, "Statistical, correlation, and topological approaches in diagnostics of the structure and physiological state of birefringent biological tissues," in *Handbook of Photonics for Biomedical Science*, V. V. Tuchin, Ed., CRC Press, Taylor & Francis Group, London, pp. 283–322 (2010).
360. V. V. Tuchin, I. L. Maksimova, A. N. Yaroslavskaya, T. N. Semenova, S. N. Tatarintsev, V. I. Kochubey, and V. F. Izotova, "Human eye lens spectroscopy and modeling of its transmittance," *Proc. SPIE* **2126**, 393–406 (1994).
361. W. P. van de Merwe, D. R. Huffman, and B. V. Bronk, "Reproducibility and sensitivity of polarized light scattering for identifying bacterial suspension," *Appl. Opt.* **28**(23), 5052–5057 (1989).
362. C. Bustamante, I. Tinoco, Jr., and M. F. Maestre, "Circuar intensity differential scattering of light by helical structures: III. A general polarizability tensor and anomalous scattering," *J. Chem. Phys.* **74**, 4839–4850 (1981).
363. C. Bustamante, I. Tinoco, Jr., and M. F. Maestre, "Circular intensity differential scattering of light by helical structures. IV. Randomly oriented species," *J. Chem. Phys.* **76**, 3440–3446 (1982).
364. C. Bustamante, M. F. Maestre, D. Keller, and I. Tinoco, Jr., "Differential scattering (CIDS) of circularly polarized light by dense particles," *J. Chem. Phys.* **80**, 4817–4823 (1984).
365. S. Zietz, A. Belmont, and C. Nicolini, "Differential scattering of circularly polarized light as a unique probe of polynucleosome superstructure," *Cell Biophys.* **5**, 163–187 (1983).
366. C. T. Gross, H. Salamon, A. J. Hunt, R. I. Macey, F. Orme, and A. T. Quintanilha, "Hemoglobin polymerization in sickle cells studied by circular polarized light scattering," *Biochim. Biophys. Acta* **1079**(2), 152–160 (1991).
367. B. Kunnen, C. Macdonald, A. Doronin, S. Jacques, M. Eccles, and I. Meglinski, "Application of circularly polarized light for non-invasive diagnosis of cancerous tissues and turbid tissue-like scattering media," *J. Biophotonics* **8**(4), 317–323 (2015).
368. R. J. McNichols and G. L. Cote, "Optical glucose sensing in biological fluids: an overview," *J. Biomed. Opt.* **5**(1), 5–16 (2000).
369. V. V. Tuchin, Ed., *Handbook of Optical Sensing of Glucose in Biological Fluids and Tissues*, CRC Press, Taylor & Francis Group, London (2009).

370. M. F. G. Wood, N. Ghosh, X. Guo, and I. A. Vitkin, "Towards noninvasive glucose sensing using polarization analysis of multiply scattered light," in *Handbook of Optical Sensing of Glucose in Biological Fluids and Tissues*, V. V. Tuchin, Ed., CRC Press, Taylor & Francis Group, London, pp. 469–450 (2009).
371. G. Purvinis, B. D. Cameron, and D. M. Altrogge, "Noninvasive polarimetric-based glucose monitoring: an in vivo study," *J. Diabetes Sci. Technol.* **5**(2), 380–387 (2011).
372. B. Chance, H. Liu, T. Kitai, and Y. Zhang, "Effects of solutes on optical properties of biological materials: models, cells, and tissues," *Anal. Biochem.* **227**, 351–362 (1995).
373. H. Liu, B. Beauvoit, M. Kimura, and B. Chance, "Dependence of tissue optical properties on solute-induced changes in refractive index and osmolarity," *J. Biomed. Opt.* **1**, 200–211 (1996).
374. V. V. Tuchin, J. Culver, C. Cheng, S. A. Tatarkova, M. DellaVecchia, D. A. Zimnyakov, A. Chaussky, A. G. Yodh, and B. Chance, "Refractive index matching of tissue components as a new technology for correlation and diffusing–photon spectroscopy and imaging," *Proc. SPIE* **3598**, 111–120 (1999).
375. P. O. Rol, *Optics for Transscleral Laser Applications*, Dissertation for the degree of Doctor of Natural Sciences, Zurich (1992).
376. E. K. Chan, B. Sorg, D. Protsenko, M. O'Neil, M. Motamedi, and A. J. Welch, "Effects of compression on soft tissue optical properties," *IEEE J. Sel. Topics Quantam Electron.* **2**(4), 943–950 (1996).
377. B. Nemati, H. G. Rylander, III, and A. J. Welch, "Optical properties of conjunctiva, sclera, and the ciliary body and their consequences for transscleral cyclophotocoagulation," *Appl. Opt.* **35**(19), 3321–3327 (1996).
378. V. V. Tuchin, I. L. Maksimova, D. A. Zimnyakov, I. L. Kon, A. H. Mavlutov, and A. A. Mishin, "Light propagation in tissues with controlled optical properties," *J. Biomed. Opt.* **2**(4), 401–417 (1997).
379. Z. S. Sacks, R. M. Kurtz, T. Juhasz, and G. A. Mourau, "High precision subsurface photodisruption in human sclera," *J. Biomed. Opt.* **7**(3), 442–450 (2002).
380. J. S. Maier, S. A. Walker, S. Fantini, M. A. Franceschini, and E. Gratton, "Possible correlation between blood glucose concentration and the reduced scattering coefficient of tissues in the near infrared," *Opt. Lett.* **19**, 2062–2064 (1994).
381. M. M. Cope, M. Essenpreis, and D. Bocker, "Influence of glucose concentration on light scattering in tissue-simulating phantoms," *Opt. Lett.* **19**, 2170–2172 (1994).
382. J. T. Bruulsema, J. E. Hayward, and T. J. Farrell et al., "Correlation between blood glucose concentration in diabetics and noninvasively

measured tissue optical scattering coefficient," *Opt. Lett.* **22**, 190–192 (1997).
383. G. Vargas, E. K. Chan, J. K. Barton, H. G. Rylander III, and A. J. Welch, "Use of an agent to reduce scattering in skin," *Laser. Surg. Med.* **24**, 138–141 (1999).
384. V. V. Tuchin, "Coherent optical techniques for the analysis of tissue structure and dynamics," *J. Biomed. Opt.* **4**(1), 106–124 (1999).
385. V. V. Tuchin, "Controlling of tissue optical properties," *Proc. SPIE* **4001**, 30–53 (2000).
386. A. N. Bashkatov, V. V. Tuchin, E. A. Genina, Yu. P. Sinichkin, N. A. Lakodina, and V. I. Kochubey, "The human sclera dynamic spectra: in vitro and in vivo measurements," *Proc. SPIE* **3591**, 311–319 (1999).
387. V. V. Tuchin, "Optical immersion as a new tool to control optical properties of tissues and blood," *Laser Phys.* **15**(8), 1109–1136 (2005).
388. V. V. Tuchin, "Optical clearing of tissue and blood using immersion method," *J. Phys. D: Appl. Phys.* **38**, 2497–2518 (2005).
389. V. V. Tuchin, *Optical Clearing of Tissues and Blood*, **PM 154**, SPIE Press, Bellingham, WA (2006).
390. V. V. Tuchin, "A clear vision for laser diagnostics," *IEEE J. Sel. Topics Quantam Electron.* **13**(6), 1621–1628 (2007).
391. E. A. Genina, A. N. Bashkatov, and V. V. Tuchin, "Tissue optical immersion clearing," *Expert Rev. Med. Devices* **7**(6), 825–842 (2010).
392. E. A. Genina, A. N. Bashkatov, K. V. Larin, and V. V. Tuchin, "Light–tissue interaction at optical clearing" in *Laser Imaging and Manipulation in Cell Biology*, F. S. Pavone, Ed., Wiley-VCH Verlag GmbH & Co. KGaA, Weinheim, pp. 115–164 (2010).
393. K. V. Larin, M. G. Ghosn, A. N. Bashkatov, E. A. Genina, N. A. Trunina, and V. V. Tuchin, "Optical clearing for OCT image enhancement and in-depth monitoring of molecular diffusion," *IEEE J. Sel. Topics Quantam Electron.* **18**(3) 1244–1259 (2012).
394. D. Zhu, K. V. Larin, Q. Luo, and V. V. Tuchin, "Recent progress in tissue optical clearing," *Laser Photonics Rev.* **7**(5), 732–757 (2013).
395. D. Zhu, Q. Luo, and V. V. Tuchin, "Tissue optical clearing," in *Advanced Biophotonics: Tissue Optical Sectioning*, R. K. Wang and V. V. Tuchin, Eds., CRC Press, Taylor & Francis Group, Boca Raton, FL, London, and New York, pp. 621–672 (2013).
396. R. K. Wang and V. V. Tuchin, "Optical coherence tomography: light scattering and imaging enhancement," in *Coherent-Domain Optical Methods: Biomedical Diagnostics, Environmental Monitoring and Material Science*, 2nd ed., V. V. Tuchin, Ed., Springer-Verlag, Berlin, Heidelberg, and New York, pp. 665–742 (2013).

397. J. Wang, Y. Zhang, P. Li, Q. Luo, and D. Zhu, "Review: tissue optical clearing window for blood flow monitoring (invited paper)," *IEEE J. Sel. Topics Quantam Electron.* **20**(2), 6801112-1–12 (2014).
398. O. Nadiarnykh and P. J. Campagnola, "SHG and optical clearing," in *Second Harmonic Generation Imaging*, F. S. Pavone and P. J. Campagnola, Eds., CRC Press, Taylor & Francis Group, Boca Raton, FL; London; and New York, pp. 169–189 (2014).
399. E. A. Genina, A. N. Bashkatov, Yu. P. Sinichkin, I. Yu. Yanina, and V. V. Tuchin, "Optical clearing of biological tissues: prospects of application in medical diagnostics and phototherapy [review]," *J. Biomed. Photonics Eng.* **1**(1), 22–58 (2015).
400. D. E. Freund, R. L. McCally, and R. A. Farrell, "Effects of fibril orientations on light scattering in the cornea," *JOSA A* **3**, 1970–1982 (1986).
401. R. A. Farrell, D. E. Freund, and R. L. McCally, "Research on corneal structure," *Johns Hopkins Appl. Phys. Lab. Tech. Dig.* **11**(1–2), 191–199 (1990).
402. R. Drezek, A. Dunn, and R. Richards-Kortum, "Light scattering from cells: finite-difference time-domain simulations and goniometric measurements," *Appl. Opt.* **38**(16), 3651–3661 (1999).
403. B. Beauvoit, T. Kitai, and B. Chance, "Contribution of the mitochondrial compartment to the optical properties of the rat liver: a theoretical and practical approach," *Biophys. J.* **67**, 2501–2510 (1994).
404. R. Graaff, J. G. Aarnoudse, J. R. Zijp, P. M. A. Sloot, F. F. M. de Mul, J. Greve, and M. H. Koelink, "Reduced light scattering properties for mixtures of spherical particles: a simple approximation derived from Mie calculations," *Appl. Opt.* **31**, 1370–1376 (1992).
405. G. C. Beck, N. Akgun, A. Ruck, and R. Steiner, "Design and characterization of a tissue phantom system for optical diagnostics," *Lasers Med. Sci.* **13**, 160–171 (1998).
406. G. C. Beck, N. Akgun, A. Ruck, and R. Steiner, "Developing optimized tissue phantom systems for optical biopsies," *Proc. SPIE* **3197**, 76–85 (1997).
407. W.-C. Lin, M. Motamedi, and A. J. Welch, "Dynamics of tissue optics during laser heating of turbid media," *Appl. Opt.* **35**(19), 3413–3420 (1996).
408. L. O. Svaasand, B. J. Tromberg, R. C. Haskell, T.-T. Tsay, and M. W. Berns, "Tissue characterization and imaging using photon density waves," *Opt. Eng.* **32**, 258–266 (1993).
409. A. Kotyk and K. Janacek, *Membrane Transport: An Interdisciplinary Approach*, Plenum, New York (1977).
410. A. Roggan, M. Friebel, K. Dorschel, A. Hahn, and G. Mueller, "Optical properties of circulating human blood in the wavelength range 400–2500 nm," *J. Biomed. Opt.* **4**, 36–46 (1999).
411. A. H. Gandjbakhche, P. Mills, and P. Snabre, "Light-scattering technique for the study of orientation and deformation of red

blood cells in a concentrated suspension," *Appl. Opt.* **33**, 1070–1078 (1994).

412. V. L. Voeikov, C. N. Novikov, and N. D. Vilenskaya, "Low-level chemiluminescent analysis of nondiluted human blood reveals its dynamic system properties," *J. Biomed. Opt.* **4**, 54–60 (1999).
413. A. V. Priezzhev, O. M. Ryaboshapka, N. N. Firsov, and I. V. Sirko, "Aggregation and disaggregation of erythrocytes in whole blood: study by backscattering technique," *J. Biomed. Opt.* **4**, 76–84 (1999).
414. L. D. Shvartsman and I. Fine, "Optical transmission of blood: effect of erythrocyte aggregation," *IEEE Trans. Biomed. Eng.* **50**, 1026–1033 (2003).
415. M. Brezinski, K. Saunders, C. Jesser, X. Li, and J. Fujimoto, "Index matching to improve OCT imaging through blood," *Circulation* **103**, 1999–2003 (2001).
416. X. Xu, R. K. Wang, J. B. Elder, and V. V. Tuchin, "Effect of dextran-induced changes in refractive index and aggregation on optical properties of whole blood," *Phys. Med. Biol.* **48**, 1205–1221 (2003).
417. V. V. Tuchin, X. Xu, and R. K. Wang, "Dynamic optical coherence tomography in optical clearing, sedimentation and aggregation study of immersed blood," *Appl. Opt.OT* **41**(1), 258–271 (2002).
418. Y. Ozaki, H. Kitabata, and H. Tsujioka et al., "Comparison of contrast media and low-molecular-weight dextran for frequency-domain optical coherence tomography," *Circ. J.* **76**(4), 922–927 (2012).
419. J. Li, H. Minami, E. Steward, T. Ma, D. Mohar, C. Robertson, K. Shung, Q. Zhou, P. Patel, and Z. Chen, "Optimal flushing agents for integrated optical and acoustic imaging systems," *J. Biomed. Opt.* **20**(5), 056005 (2015)
420. E. I. Galanzha, V. V. Tuchin, A. V. Solovieva, T. V. Stepanova, Q. Luo, and H. Cheng, "Skin backreflectance and microvascular system functioning at the action of osmotic agents," *J. Phys. D: Appl. Phys.* **36**, 1739–1746 (2003).
421. O. Zhernovaya, V. V. Tuchin, and M. J. Leahy, "Blood optical clearing studied by optical coherence tomography," *J. Biomed. Opt.* **18**(2), 026014-1–8 (2013).
422. R. Shi, M. Chen, V. V. Tuchin, and D. Zhu, "Accessing to arteriovenous blood flow dynamics response using combined laser speckle contrast imaging and skin optical clearing," *Biomed. Opt. Express* **6**(6), 1977–1989 (2015).
423. C. Macdonald and I. Meglinski, "Backscattering of circular polarized light from a disperse random medium influenced by optical clearing," *Laser Phys. Lett.* **8**(4), 324–328 (2011).
424. L. F. Rojas-Ochoa, D. Lacoste, R. Lenke, P. Schurtenberger, and F. Scheffold, "Depolarization of backscattered linearly polarized light," *J. Opt. Soc. Am. A* **21**, 1799–1804 (2004).

425. M. Xu and R. R. Alfano, "Circular polarization memory of light," *Phys. Rev. E* **72**, 065601(R) (2005).
426. Y. L. Kim, P. Pradhan, M. H. Kim, and V. Backman, "Circular polarization memory effect in low-coherence enhanced backscattering of light," *Opt. Lett.* **31**, 2744–2746 (2006).
427. A. Doronin and I. Meglinski, "Online object oriented Monte Carlo computational tool for the needs of biomedical optics," *Biomed. Opt. Express* **2**(9), 2461–2469 (2011).
428. A. Doronin, C. Macdonald, and I. Meglinski, "Propagation of coherent polarized light in highly scattering turbid media," *J. Biomed. Opt.* **19**(2), 025005 (2014).
429. S. Prahl, "Maetzler's MATLAB Code for Mie Theory," Oregon Medical Laser Centre: http://omlc.ogi.edu/software/mie/.
430. S. He, K. L. Dunn, P. S. Espino, B. Drobic, L. Li, J. Yu, J-M. Sun, H. Y. Chen, S. Pritchard, and J. R. Davie, "Chromatin organization and nuclear microenvironments in cancer cells," *J. Cell. Biochem.* **104**, 2004–2015 (2008).
431. K. N. Dahl, A. J. Ribeiro, and J. Lammerding, "Nuclear shape, mechanics, and mechanotransduction," *Circ. Res.* **102**(11), 1307–1318 (2008).
432. F. R. Neumann and P. Nurse, "Nuclear size control in fission yeast," *J. Cell Biol.* **179**(4), 593–600 (2007).
433. M. Webster, K. L. Witkin, and O. Cohen-Fix, "Sizing up the nucleus: nuclear shape, size and nuclear–envelope assembly," *J. Cell Sci.* **122**(10), 1477–1486 (2009).
434. Y. Masui, "Towards understanding the control of the division cycle in animal cells," *Biochem. Cell. Biol.* **70**(10–11), 920–945 (1992).
435. V. L. Kuzmin and I. V. Meglinski, "Backscattering of linearly and circularly polarized light in randomly inhomogeneous media," *Opt. Spectrosc.* **106**(2), 257–267 (2009).
436. V. L. Kuzmin and I. V. Meglinski, "Anomalous polarization phenomena of light scattered in random media," *J. Exp. Theor. Phys.* **137**(5), 742–753 (2010).
437. I. Meglinski and V. L. Kuzmin, "Coherent backscattering of circularly polarized optical radiation from a disperse random medium," *Prog. Electromagn. Res. M* **16**, 47–61 (2011).
438. I. Meglinski, C. Macdonald, A. Karl, H. Yoon, and M. Eccles, "The mapping of tissues scattering properties on the Poincaré sphere," in *Biomedical Optics, OSA Technical Digest* (Optical Society of America), paper BW3B.6 (2012).
439. I. Meglinski, C. Macdonald, A. Doronin, and M. Eccles, "Screening cancer aggressiveness by using circularly polarized light," in *Optics in the Life Sciences*, OSA Technical Digest (online) (Optical Society of America), paper BM2A.4 (2013).

Nikolai G. Khlebtsov is the head of the Nanobiotechnology Laboratory at the Russian Academy of Sciences' Institute of Biochemistry and Physiology of Plants and Microorganisms. He has published over 400 scientific works, including two monographs and seven book chapters. His current research interests are biophotonics and nanobiotechnology of plasmon-resonant and hybrid multifunctional nanoparticles, nanostructures for SERS, and biomedical applications of plasmonic and hybrid composite nanoparticles.

Irina L. Maksimova (1959–2013) was a professor and chair of Optics and Biophotonics at Saratov State University. Her research interests included light scattering, polarized light analysis, tissue optics, laser medicine, and nanobiophotonics. She authored more than 100 peer-reviewed papers. Prof. Maksimova was one of the pioneers of modeling the polarization properties of eye tissues and designing precise algorithms for their measurements. In 2013, she was named the best supervisor of Ph.D. students at SSU.

Igor Meglinski is the head of the Opto-Electronics and Measurement Techniques Laboratory at the University of Oulu, Finland. For the last 20 years, his research interests have been concerned with the interface between physics, optical and biomedical engineering, sensor technologies, and life sciences, focusing on the development of new non-invasive imaging/diagnostic techniques and their application in medicine and biology, material sciences, pharmacy, food, environmental monitoring, and health care industries. He pioneered the application of diffusing wave spectroscopy (DWS) to non-invasively monitor blood flow and blood microcirculation, the application of circularly polarized light for practical cancer diagnosis, the generalization of the Monte Carlo model to simulate the coherent effects of multiple scattering of light, and the development of online computational tools for the biophotonics and biomedical optics community. Prof. Meglinski has authored or co-authored over 200 research papers in peer-reviewed scientific journals, proceedings of international conferences, book chapters, patents, and professional magazines. He has delivered over 450 presentations at major international conferences, including over 200 invited lectures and plenary talks. He is the Node Leader for the Biophotonics4Life Worldwide Consortium (BP4L), Fellow of the Institute of Physics (London, UK), and Fellow of SPIE.

Lihong V. Wang holds the Gene K. Beare Distinguished Professorship of Biomedical Engineering at Washington University in St. Louis. His book entitled *Biomedical Optics: Principles and Imaging*, one of the first textbooks in the field, won the 2010 Joseph W. Goodman Book Writing Award. He edited the first book on photoacoustic tomography. Professor Wang has published 435 peer-reviewed journal articles and delivered 430 keynote, plenary, or invited talks. His Google Scholar *h*-index and citations have reached 100 and 40,000, respectively. His laboratory was the first to report functional photoacoustic tomography, 3D photoacoustic microscopy, photoacoustic endoscopy, photoacoustic reporter gene imaging, the photoacoustic Doppler effect, and the universal photoacoustic reconstruction algorithm. He is the Editor-in-Chief of the *Journal of Biomedical Optics*. He chairs the annual conference on Photons plus Ultrasound. He received NIH's FIRST, NSF's CAREER, NIH Director's Pioneer, and NIH Director's Transformative Research awards. He also received the OSA C. E. K. Mees Medal, IEEE Technical Achievement Award, IEEE Biomedical Engineering Award, SPIE Britton Chance Biomedical Optics Award, and Senior Prize of the International Photoacoustic and Photothermal Association for "seminal contributions to photoacoustic tomography and Monte Carlo modeling of photon transport in biological tissues." An honorary doctorate was conferred on him by Lund University, Sweden. His lab is transitioning to Caltech.

Valery V. Tuchin is a professor and chairman of Optics and Biophotonics at Saratov National Research State University. He is also the head of laboratory at the Institute of Precision Mechanics and Control, RAS, and the supervisor of Interdisciplinary Laboratory of Biophotonics at Tomsk National Research State University. His research interests include biophotonics, tissue optics, laser medicine, tissue optical clearing, and nanobiophotonics. He is a member of SPIE, OSA, and IEEE, guest professor of HUST (Wuhan) and Tianjin Universities of China, and adjunct professor of the Limerick University (Ireland) and National University of Ireland (Galway). He is a fellow of SPIE and OSA, and has been awarded Honored Science Worker of the Russia, SPIE Educator Award, FiDiPro (Finland), Chime Bell Prize of Hubei Province (China), and Joseph W. Goodman Book Writing Award (OSA/SPIE).

Chapter 2
Optics of Blood

Anna N. Yaroslavsky
Department of Physics and Applied Physics, Department of Biomedical Engineering, University of Massachusetts–Lowell, Lowell, MA, USA

Department of Dermatology, Wellman Center for Photomedicine, Massachusetts General Hospital, Boston, USA

Ilya V. Yaroslavsky
IPG Photonics, Inc., Marlborough, MA, USA

2.1 Introduction

Blood is one of the most important biological fluids. It derives its importance from the ability of red blood cells (RBCs) to reversibly bind and carry oxygen. Most living cells rely on oxidative metabolism to perform their functions, while the circulatory and respiratory systems work together to provide the necessary supply of oxygen. Nearly all circulating oxygen is bound to the hemoglobin molecule, while the rest is dissolved in the cytosol of RBC and in blood plasma. Therefore, the delivery of oxygen to different organs is determined by the blood oxygen content and blood flow at the particular location.

Optical investigations of blood properties have a long history. Slare[1] was the first to report in 1700 that blood changes color when exposed to air, and Ångström was the first, in 1855, to apply a spectrophotometric technique to study spectral characteristics of blood. Oxyhemoglobin and hemoglobin absorption bands in the visible spectral range were investigated by Hoppe-Seyler in 1862.[2] Soret described hemoglobin's optical properties in the ultraviolet in 1878.[3] In 1873, Vierordt[4,5] invented a method to determine the hemoglobin content of blood samples. Since that time, numerous investigations have been devoted to the optical properties of hemoglobin derivatives and whole blood.[6–19] For example, the differences in the absorption spectra of oxyhemoglobin (HbO_2) and deoxyhemoglobin (Hb) in the visible and near-infrared spectral ranges stimulated the development of blood oximetry in the 1930s.[20] In 1977, Jöbsis[21] discovered that near-infrared light can transilluminate a cat's head and that the optical signal can provide *in vivo* information on the biotissue HbO_2/Hb ratio and on blood volume. This investigation led to the development of tissue oximetry and, consequently, to near-infrared spectroscopy (NIRS) of different biological tissues. Besides oximetry, knowledge of blood optical properties is required for many medical diagnostic and therapeutic applications. During the past 20 years, considerable efforts were directed toward developing NIRS imaging,[22] optical biopsy,[23] photodynamic therapy,[24] and laser-induced interstitial thermo-therapy.[25] The efficacy of these procedures greatly depends on the propagation and the fluence rate distribution of optical radiation within tissue. Since most human tissues contain blood, many theoretical and experimental studies[9–12,14–18,26–37] were undertaken to predict and determine its optical properties.

Characterization of blood by optical techniques is a complex problem. Interpretation of the measured optical signal is complicated by the fact that visible and near-infrared light are both scattered and absorbed by whole blood, as well as by other biological structures. In most cases, it is much more likely that visible and near-infrared light will be scattered, rather than absorbed, by biological tissues and fluids. Therefore, to properly quantify the hemoglobin concentration and oxygen saturation of blood and other tissues, it is necessary to have an accurate model of light propagation in the biological medium. Although NIRS has demonstrated a unique ability to monitor the HbO_2 saturation of tissues,[38–41] the quantitative determination of the absolute Hb and HbO_2 concentrations is still a challenging task.[22]

The optical properties of blood are influenced by many anatomical, physiological, and biochemical parameters. Blood absorptive properties in the visible and near-infrared spectral range are determined mainly by hemoglobin oxygen saturation level and hemoglobin concentration within RBCs and in plasma. Its scattering properties depend greatly on hematocrit, hemolysis, and osmolarity. The temperature and viscosity of blood may influence the results of the measurements by altering its absorption[19,32] and flow properties.[36] The other important phenomena that should be accounted for while working with blood are sedimentation, aggregation, coagulation, orientation, and deformation of cells. It is not always feasible to control or predict the influence of each parameter on the outcome of a particular investigation; consequently, the results reported by different investigators are sometimes different or inconclusive.[30]

In this chapter, we will discuss various aspects concerning the optics of blood. In Section 2.2, we briefly describe blood content and the properties of different blood cells. In some cases, the optical properties of whole blood are assumed to be identical to the optical properties of RBCs in saline solution.[35,36] For many purposes, this is a valid assumption, but it is quite important to keep in mind that the other blood particles, such as leukocytes and platelets, can also impact optical measurements. The optical properties of oxy- and deoxyhemoglobin are summarized in Section 2.3. Since most of the hemoglobin is contained within erythrocytes, it is necessary to consider the influence of light scattering by RBCs on the transmittance and reflectance of light by whole blood. Several rigorous and empirical models used to describe single scattering of visible and near-infrared light by erythrocytes are discussed in Section 2.4. The typical volume fraction of erythrocytes in human blood (hematocrit) is about 42%, and accordingly, each cubic millimeter of blood contains approximately 5 million RBCs. This means that light is multiply scattered by whole blood. Moreover, erythrocytes form a closely packed system of scatterers, which implies that the scattering in whole blood is not independent. Therefore, to obtain the optical properties of whole blood, it is not enough to know the optical properties of a single erythrocyte. The description of light propagation in whole blood can be given in the framework of radiative transport theory by numerical solutions of the transport equation[32] and Monte Carlo (MC) technique.[31,32,36,37] Several techniques to determine optical properties of whole human blood are presented in Section 2.5. In Section 2.6, the data available from the literature on the optical properties of diluted and whole human blood are summarized. The relevance of blood optics is discussed in Section 2.7.

It is important to emphasize that in this chapter, only basic optical techniques that can be employed to determine blood optical properties are discussed, and only some issues related to this complex problem are addressed. Still, we hope that the chapter provides the reader with important information about how the optical properties of blood may be determined and how this knowledge may be applied in different fields of biomedical applications and research.

2.2 Physical Properties of Blood Cells

Whole human blood consists of plasma and blood cells, with a plasma volume of approximately 57% and a cell volume of 43%.[42] Blood plasma is about 90% water and 10% protein (hemoglobin, albumin, globulins, etc.). Hemoglobin concentration in whole blood is between 134 and 173 g/L. The volume fraction of cells within whole blood is defined as the hematocrit (Hct). Blood cells contain about 99% erythrocytes and 1% leukocytes and platelets. Blood plasma contains 10% protein. Accordingly, its refractive index is slightly higher than the refractive index of water. From the values of the volume fractions of different blood cells, and hemoglobin's absorptive properties, it is clear that the optics of whole blood under normal physiological conditions are defined principally by the optical properties of RBCs and blood plasma.

There are many important parameters that influence the results of optical experiments with flowing blood in *in vivo* and *in vitro* conditions: hematocrit, osmolarity, temperature, viscosity, oxygen saturation, deformablility, and hemolysis. To control these parameters, sophisticated equipment and elaborate techniques are required.[36] When the influence of a single parameter (e.g., temperature) on the optical properties of blood is investigated, a simpler experimental arrangement may be employed.[32] The choice of parameters to control and/or investigate is defined by the goals of a particular experiment.

When blood at rest is studied, there are at least two additional important phenomena to account for: coagulation and sedimentation. The process of blood coagulation should be inhibited during the study. Among the anticoagulants available, two types are suitable and widely used for optical experiments: heparin, and calcium chelators such as dipotassium ethylene-diamine tetraacetic acid (K_2EDTA). Heparin prevents blood clotting and does not distort red cell volumes. This anticoagulant is preferred in cases when RBC suspensions are being studied. However, it does alter the size and shape of platelets and leukocytes and is not recommended when the latter are being investigated. The optimal concentration of heparin is 15 to 20 units/mL of blood. K_2EDTA is an alternative anticoagulant. Its optimal concentration is 1 mg/mL of blood. It does not significantly distort cell volume or morphology but in excessive amounts can be hypertonic and result in cell shrinkage.[43] Sedimentation occurs because cell density is higher than the density of plasma and saline solution. The rate of erythrocyte settling depends on two forces. The downward force is a function of cell weight, and the retarding force is a function of the cell surface area exposed to the medium. The smaller the cell surface area, the higher the sedimentation rate. In addition, if a blood container is slightly tilted from the vertical position, the settling rate increases.[42] However, the sedimentation rate in normal blood is quite low (0–30 mm/h, depending on sex and age). Usually, little settling occurs within 20–30 min. Obviously, it is desirable that the optical examination of a blood sample be made before appreciable sedimentation occurs.

2.2.1 Red blood cells

An erythrocyte is one of the most highly specialized cells in the human body. It consists of a membrane surrounding a solution of protein and electrolytes. More than 95% of the protein is hemoglobin, and about 5% of the remaining proteins are the enzymes required for maintaining hemoglobin in a functional state and for producing energy. Each erythrocyte contains approximately 29 pg of hemoglobin. RBCs contain no nuclei, mitochondria, or ribosomes.[42]

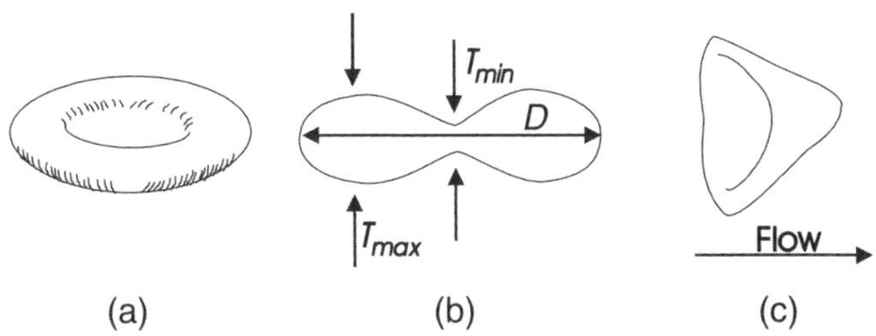

Figure 2.1 (a) Erythrocyte, (b) cross section of erythrocyte, (c) moving erythrocyte.

The lipid bilayer cell membrane has a thickness of approximately 7 nm and a refractive index of 1.46.[44] The hemoglobin concentration within an erythrocyte ranges from 299 to 357 g/L. The real part of the refractive index of the RBC is very close to 1.4 in the wavelength range from 400 to 1200 nm.[35,45,46]

The normal human erythrocyte has the shape of a biconcave disc [Fig. 2.1(a) and (b)]. The diameter D of the erythrocyte is 7.82 ± 0.62 μm, with a minimum thickness $T_{\min}$ of 0.81 ± 0.35 μm and a maximum thickness $T_{\max}$ of 2.58 ± 0.27 μm. The mean cellular volume of the red cell is 94 ± 14 μm^3, and the surface area is 135 ± 16 μm^2. The disc shape ensures nearly the maximum possible value of surface to volume ratio, which is important for facilitation of gas transfer. The dimensions of the red cell in the living state are estimated by measurements made from photomicrographs.[47,48]

The biconcave disc deforms more easily than a sphere and allows for better mobility of erythrocytes within small blood vessels such as capillaries. When the cell moves within such a vessel, the plane of the biconcave disc is oriented in the direction of the flow [Fig. 2.1(c)].[49]

In general, the shape of the erythrocyte depends on conditions within the cell and in the environment. In some cases, the RBC may become spherical. For example, osmotic swelling takes place when the cell is suspended in a hypotonic solution. In this case, the erythrocyte acquires water and swells. The volume of the cell increases, while the surface area remains nearly unchanged. When the cell becomes spherical, the cell diameter slightly decreases.

Discocyte–ehinocyte transformation occurs when the cell is exposed to high pH, stored plasma, anionic detergents, or lysolecithin, or when the cell is washed and placed between a glass and a coverslip. The same thing happens when intracellular adenosine triphosphate is depleted or intracellular calcium increases. While the discocyte–ehinocyte change occurs, the volume remains constant, and the surface area decreases. Discocyte–stomatocyte transformation takes place when the cell is exposed to low pH, cationic detergents, or phenothiazines. In this case, the cell surface area decreases, and the volume remains the same.

2.2.2 Leukocytes

The leukocytes, or white blood cells, play an important role in the defense of the organism against infection or cell damage. White blood cell concentration in normal blood is between 3.03 and 10.5×10^9/L. Among the leukocytes, there are different cell lines. Normally, the most numerous leukocytes are neutrophils, followed by lymphocytes, monocytes, eosinophils, and then basophils. Most of the cells are spheroids and range in size from 7 to 20 μm.[42] Every cell line has different properties and a unique role in defending the body. Due to the low total content in whole blood under normal physiological conditions and their negligible absorptive properties, leukocytes do not significantly influence the optical properties of blood. However, care should be taken to prevent leukocytes from clumping during optical experiments with blood.

2.2.3 Platelets

Blood platelets participate in thrombosis and in blood coagulation. Human platelets are approximately 3.6 ± 0.7 μm in diameter and 0.9 ± 0.2 μm in thickness. They are shaped like an oblate spheroid. Platelets may rapidly swell when exposed to temperatures below 37 °C.[42] A mean platelet volume of 7.1 μm^3 and a mean surface area of 22.2 μm^2 were calculated from rheo-optical sizing studies.[50] The average platelet concentration in normal venous blood is 250×10^9/L, with a range from 140 to 440×10^9/L; arterial blood has a slightly higher value. There are more platelets than leukocytes, but still the number of platelets is hardly comparable with the number of erythrocytes (approximately 5×10^{12}/L). As with the leukocytes, platelets do not contain strong chromophores.

2.3 Optical Properties of Oxyhemoglobin and Deoxyhemoglobin

Absorptive properties of human hemoglobin derivatives have been studied extensively for more than a century.[2–7,13,15,19,51,52,53] The extinction coefficients of oxy- and deoxyhemoglobin are well known for the wide spectral range, from 390 to 1000 nm (Fig. 2.2).[13] Kuenstner and Norris[52] investigated the absorptive properties of human hemoglobin derivatives in the spectral range from 1000 to 2500 nm. It is quite clear that starting from 1200 nm, the extinction coefficients of the solutions are defined by the absorption spectrum of water (saline solution).[54] Steinke and Shepherd[19] reported on the temperature dependence of the hemoglobin absorbance spectra. Zijstra et al.[53] investigated the absorption spectra of human fetal hemoglobin derivatives.

Characteristic changes in the absorptive behavior of hemoglobin, dependent on its oxygen saturation, led to the invention of clinically available blood oximeters. The principles of hemoglobin oximetry (i.e., measurements

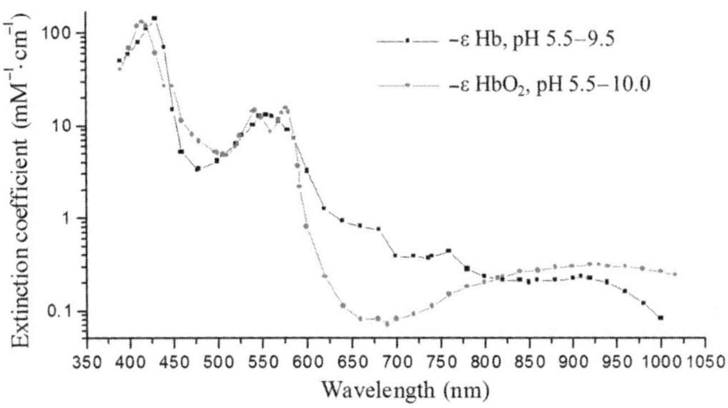

Figure 2.2 Absorption spectra of deoxy- and oxyhemoglobin in the 390-to-1000-nm-wavelength range.

of total hemoglobin concentration and its oxygen saturation) and the criteria for oximetry wavelength selection are summarized below.

Let us consider an aqueous solution that contains a mixture of deoxyhemoglobin and oxyhemoglobin. The attenuation of a monochromatic light beam, with a wavelength λ, by a hemoglobin solution may be described by the Beer–Lambert law

$$I = I_0 \exp(-\varepsilon_\lambda cd), \tag{2.1}$$

where I is a transmitted light intensity, I_0 is an incident light intensity, ε_λ is the extinction coefficient at the wavelength λ in L/(mmol · cm), c is concentration of absorbing substance in mmol/L, and d is the path length through the sample in cm.

The optical density D_λ is defined as

$$D_\lambda = \ln(I_0/I) = \varepsilon_\lambda cd. \tag{2.2}$$

To determine the total hemoglobin concentration, it is sufficient to measure optical density at the isosbestic wavelength, if ε_λ and d are known. The wavelength λ_i is called isosbestic if the extinction coefficients of deoxy- and oxyhemoglobin are equal at this wavelength. The hemoglobin oxygen saturation S can be calculated from measuring the optical density at an additional wavelength λ, where there is a large difference between the extinction coefficients of deoxy- and oxyhemoglobin. The equations, which define the total hemoglobin concentration and its oxygen saturation, are

$$D_{\lambda i} = \varepsilon_{\lambda i} cd, \tag{2.3}$$

$$D_\lambda = S\varepsilon_\lambda^o cd + (1 - S)\varepsilon_\lambda^d cd, \tag{2.4}$$

where $\varepsilon_{\lambda i}$ is the extinction coefficient of the hemoglobin at the isosbestic wavelength λ_i, ε_λ^o is the extinction coefficient of the oxyhemoglobin at the wavelength λ, and $\varepsilon_{\lambda}^{d}$ is the extinction coefficient of the deoxyhemoglobin at the wavelength λ.

In this case, the formulas to determine hemoglobin concentration and its oxygen saturation are

$$c = D_{\lambda i}/\varepsilon_{\lambda i}d, \tag{2.5}$$

$$S = [(D_{\lambda}/D_{\lambda i})\varepsilon_{\lambda i} - \varepsilon_{\lambda}^{d}]/(\varepsilon_{\lambda}^{o} - \varepsilon_{\lambda}^{d}). \tag{2.6}$$

In general, it might be preferable to use two wavelengths, λ_1 and λ_2, neither of which is isosbestic. Then both equations for determination of the hemoglobin concentration and its oxygen saturation look like Eq. (2.4), and the resulting equation for S is

$$S = \frac{\varepsilon_{\lambda 1}^{d} - \varepsilon_{\lambda 2}^{d}D_{\lambda 1}/D_{\lambda 2}}{(\varepsilon_{\lambda 2}^{o} - \varepsilon_{\lambda 2}^{d})D_{\lambda 1}/D_{\lambda 2} - (\varepsilon_{\lambda 1}^{o} - \varepsilon_{\lambda 1}^{d})}, \tag{2.7}$$

where $\varepsilon_{\lambda 1}^{o}$ and $\varepsilon_{\lambda 1}^{d}$ are the extinction coefficients at the wavelength λ_1 of the oxy- and deoxyhemoglobin, respectively; $\varepsilon_{\lambda 2}^{o}$ and $\varepsilon_{\lambda 2}^{d}$ are the extinction coefficients at the wavelength λ_2 of the oxy- and deoxyhemoglobin, respectively; and $D_{\lambda 1}$ and $D_{\lambda 2}$ are the optical densities of hemoglobin solution at the wavelengths λ_1 and λ_2, respectively.

The total hemoglobin concentration can be calculated by a substitution of Eq. (2.7) into Eqs. (2.5) or (2.6).

To analyze the specific wavelengths' sensitivity to changes in hemoglobin oxygen saturation, let us take a derivative of Eq. (2.4) with respect to S

$$\frac{dD_{\lambda}}{dS} = (\varepsilon_{\lambda}^{o} - \varepsilon_{\lambda}^{d})cd. \tag{2.8}$$

It is clear from Eq. (2.8) that the sensitivity, for given hemoglobin concentration c and pathlength d, is maximal at the wavelengths where there is maximal difference in the extinction coefficients of oxy- and deoxyhemoglobin. The sensitivity is obviously zero at all the isosbestic wavelengths. Looking at Fig. 2.2, one can notice that the sensitivity will be maximal around 430, 650–750, and 1000 nm.

Determination of oxygen saturation and total hemoglobin concentration in a hemoglobin solution is not a difficult task. In whole blood, most of the hemoglobin is contained within erythrocytes, which means that besides absorption, light scattering takes place. To solve the problem of blood oximetry *in vivo*, it is necessary to account for light absorption and scattering by an RBC, as well as for light absorption and scattering by a package of RBCs.

2.4 Absorption and Scattering of Light by a Single Erythrocyte

The scattering of visible and near-infrared light by a single RBC is a problem of great importance for *in vivo* and *in vitro* blood measurements.[9,14–16,29,31,35,49,55–57] Quantification of optical experiments with whole blood requires clear understanding of the single-scattering process.[10–12,17,18,27,29,33–37] Therefore, the process of light scattering by a single RBC has been thoroughly studied by many scientists.[14,16,28–31,33–35,37,46,55–59] Depending on the goals of the particular study, different models of single light scattering by the erythrocytes were suggested, developed, and applied. For the interpretation of flow cytometric measurements that provide important morphological and particle-sizing information, sophisticated numerical methods, such as the three-dimensional boundary-element method (BEM),[56] and theoretical methods, such as the Wentzel–Kramers–Brillouin (WKB) approximation,[55] are often employed. The researchers in Ref. 58 applied the T-matrix method[60] to model the light scattering of a single erythrocyte, assuming that it can be approximated by a volume-equivalent ellipsoid. To interpret the small-angle light scattering by the RBCs, the Fraunhofer diffraction[34,35] and anomalous diffraction[57,61] theories have been used. The researchers in Refs. 35 and 55 attempted to employ the Rayleigh–Gans approximation to calculate the scattering amplitude of a single erythrocyte. Mie theory is widely used to calculate the absorption and scattering cross sections of the RBC, approximating it by a volume-equivalent sphere.[14,29,33,35,37,62] Empirical functions, like the Henyey–Greenstein phase function (HGPF)[63] and the Gegenbauer kernel phase function (GKPF),[28] often provide a simple approximation for the erythrocyte single-scattering light distribution.[16,30–33,36,37]

2.4.1 Absorption and scattering cross sections, scattering phase function

Light interaction with a particle depends on its optical properties, such as absorption cross section, scattering cross section, scattering phase function, etc. Below we introduce the definitions of these parameters.

Let us consider an electromagnetic plane wave with unit amplitude that is incident on a particle with a relative dielectric constant $\varepsilon = \varepsilon' + i\varepsilon''$. Part of the incident radiation is scattered, and another part is absorbed by this particle (Fig. 2.3).

When the scattered electromagnetic field is observed at the distance $R > D^2/\lambda$ (D is the typical dimension of the particle, λ is the wavelength of the incident light), it is given by[45]

$$E_s(\bar{r}) = f(\bar{s},\bar{i})\exp(ikR)/R, \qquad (2.9)$$

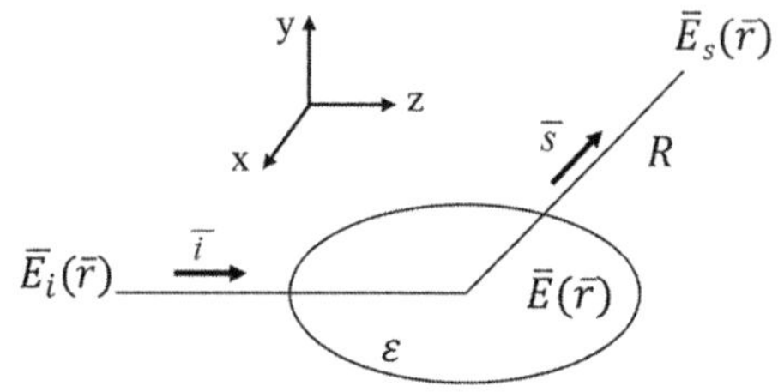

Figure 2.3 Scattering of a plane electromagnetic wave by a dielectric particle.

where $f(\bar{s},\bar{i})$ is the scattering amplitude in the far field in the direction $\bar{s}$ when the particle is illuminated by a plane wave propagating in the direction $\bar{i}$, and $k = 2\pi/\lambda$ is the wave number.

The scattering cross section of the particle is defined as

$$\sigma_s = \int_{4\pi} |f(\bar{s},\bar{i})|^2 d\omega, \tag{2.10}$$

where $d\omega$ is the differential solid angle.

The absorption cross section can be determined by integrating the product of the square of the electric field and the imaginary part of the dielectric constant of the particle over the volume V of the latter

$$\sigma_a = \int_V k\varepsilon'' |E(\bar{r}')|^2 dV'. \tag{2.11}$$

The sum of the scattering cross section and the absorption cross section is called the extinction cross section

$$\sigma_t = \sigma_a + \sigma_s. \tag{2.12}$$

If the scattering amplitude $f(\bar{i},\bar{i})$ in the forward direction is known, the extinction cross section may be determined using the forward-scattering theorem[64]

$$\sigma_t = \frac{4\pi}{k} \operatorname{Im}[f(\bar{i},\bar{i})]\hat{e}_i, \tag{2.13}$$

where $\hat{e}_i$ is a unit vector in the direction of polarization of the incident wave.

Another important characteristic of the single-scattering process is the scattering phase function $p(\bar{s},\bar{i})$. The angular distribution of the light scattered by the particle is determined by its size, shape, and relative refractive index, as well as by the wavelength of the incident light. This angular distribution of the scattered light is called the scattering phase function. It determines the amount of light scattered by the particle from the initial direction to the direction

$$p(\bar{i},\bar{s}) = \frac{1}{\sigma_s}|f(\bar{i},\bar{s})|^2. \tag{2.14}$$

There are several ways to normalize the scattering phase function. If the scattering function is treated as a probability distribution, the normalization condition should require that its integral over all angles be unity

$$\int_{4\pi} p(\bar{i},\bar{s})d\omega = 1, \tag{2.15}$$

where $d\omega$ is a differential solid angle in the direction $\bar{s}$.

It is often assumed that the probability of light scattering from the direction $\bar{i}$ into the direction $\bar{s}$ is a function only of the angle between these two directions, i.e., $p(\bar{i},\bar{s}) = p(\bar{i}\,\bar{s}) = p(\cos\theta)$. The degree of the anisotropy of the scattering phase function is characterized by the average cosine of the scattering angle $\bar{\mu}$, also referred to as the anisotropy factor

$$\bar{\mu} = \int_{4\pi} p(\bar{i}\,\bar{s})(\bar{i}\,\bar{s})d\omega. \tag{2.16}$$

2.4.2 Experimental determination of blood extinction coefficient and scattering phase function

The scattering phase function and the extinction coefficient can be directly measured by employing optical techniques and single-scattering blood samples.[10,16,30,33,35] These techniques, which are capable of measuring the properties of tissues without using any theoretical models of light propagation, are called direct methods. Several authors have reported experimental determination of blood extinction coefficients and scattering phase functions for a number of wavelengths in the visible spectral range. Scattering phase functions of the optically thin blood layers were measured by Flock et al.,[16] Yaroslavsky et al.,[33] and Hammer et al.[35] The extinction coefficients were determined by Flock et al.,[16] by Steinke and Shepherd,[18] and by Hammer et al.[35]

Direct measurements of the scattering phase function and the total attenuation coefficient require the use of the optically thin samples because it is necessary to eliminate the influence of multiple scattering and the interference of the waves scattered by neighboring erythrocytes. To ensure a single and independent scattering regime, the whole blood samples are usually diluted with an isotonic phosphate buffer solution (pH = 7.4; $n = 1.33$) to a low hematocrit of approximately 1%. To achieve complete oxygenation of blood, if desired, a sample is usually exposed to (or bubbled with) air. For the optical measurements, thin cuvettes with a thickness of 10–30 μm are used.

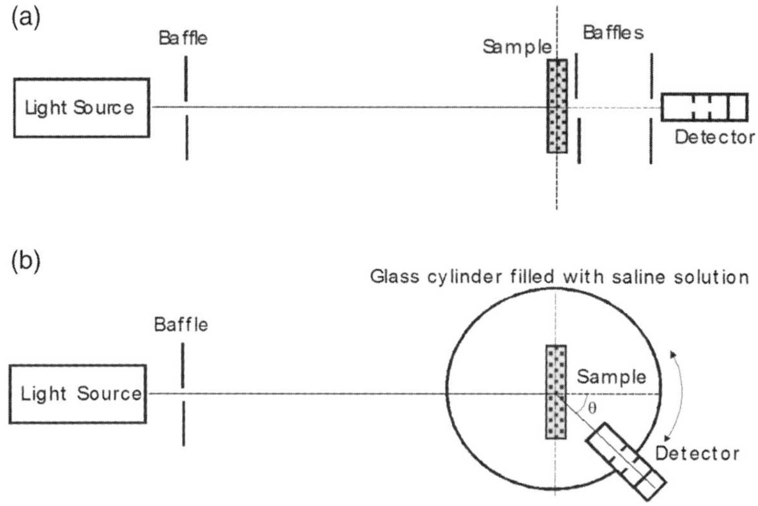

Figure 2.4 (a) Schematic for measuring the extinction coefficient and (b) schematic for measuring the scattering phase function.

Single-scattering experiments are usually performed with blood at rest, since the hematocrit of a highly diluted blood sample is not sufficiently stable within the light spot under flow conditions.[35]

Several different arrangements used to measure the extinction coefficient and the scattering phase function are described in Refs. 16, 33, and 35. A laser, a Xe lamp, and/or a Hg lamp combined with a monochromator are used as monochromatic light sources. A photomultiplier or a Si photodiode is employed as a detector. To measure the extinction coefficient, additional pinholes are used to minimize the influence of the scattered light on the registered signal [see Fig. 2.4(a)]. To measure the scattering phase function, a cuvette with blood is usually submerged into a tank filled with water or saline solution to minimize the refractive index mismatch on the boundaries of the cuvette with the surrounding medium [see Fig. 2.4(b)]. Additional corrections for the refractive index mismatch on the interfaces surrounding medium/glass and glass/blood sample are applied using the Fresnel formulas for the correct interpretation of the measured data. Possible experimental arrangements for measuring the extinction coefficient and the scattering phase function are given in Fig. 2.4(a) and (b).

Besides the simple arrangements described above, more sophisticated equipment may be used to measure the scattering phase function of a single erythrocyte in motion. For example, Shvalov et al.[55] used a scanning flow cytometer to measure the angular scattering patterns of the individual erythrocytes in the angle range from 10 to 60 deg. In this case, the interpretation of the measured angular scattering patterns is complicated by the fact that besides the stationary motion of the erythrocytes, the rotational moment affects the flow.[55] To extract the scattering phase function from the measured patterns, one has to apply additional data processing techniques, which will depend on the particular configuration and functioning of the experimental equipment.

Blood is a turbid medium with a highly anisotropic scattering phase function ($\bar{\mu} \sim 0.970$–0.999). As was shown by van de Hulst,[61] an accurate description of the scattering phase function is important for media with a high anisotropy factor. Yaroslavsky et al.[37] confirmed that the approximation of blood scattering phase function is critical for correct determination of its optical properties. However, for media with high anisotropy factors, precise measurements of the scattering phase function in the total angle range from 0 to 180 deg is a difficult technical task that demands an extremely large dynamic range of measuring equipment. Most of the scattered radiation lies in the range from 0 to 30 deg, counting from the direction of the incident beam. In addition, measurements at angles close to 90 deg are strongly affected by scattering of higher orders, even for samples of moderate optical thickness.[65] Therefore, it is highly desirable to have an appropriate analytical approximation that fits the experimental forward-scattering phase function of a single-scattering blood layer. This approximation can be used in the total angle range for the interpretation of the optical experiments with whole blood. Moreover, the scattering phase function of a single RBC cannot be adopted for blood with a physiological hematocrit ($0.37 < \text{Hct} < 0.54$) without additional verification. Reynolds et al.,[14] Roggan et al.,[36] and Hammer et al.[35] suggested that the interference of waves scattered by neighboring erythrocytes has to be taken into account in densely packed cell suspensions. Consequently, even if the scattering phase function of a single erythrocyte is known from the measurements, a valid approximation for the effective scattering phase function that adequately describes the single-scattering process in whole blood is still required.

2.4.3 Analytical and numerical methods to approximate single light scattering in blood

Several analytical, empirical, and numerical approaches are available from the literature for the theoretical treatment of the RBC single-scattering problem. The methods described below are suitable for the interpretation of the single-scattering experiments with blood and for radiative transport calculations.[45,61,64,66–68]

2.4.3.1 Mie theory

In a majority of the relevant literature, an RBC is assumed to be a homogeneous sphere with a volume equal to the volume of an average erythrocyte.[14,29,33,35,37] Several reasons suggest that Mie theory can be applied to the problem of single light scattering by erythrocytes. First, an RBC contains no cellular organelles, and its cell membrane is very thin and does not significantly affect the light-scattering process, which means that erythrocytes may be treated as a homogenous scatterers. Second, erythrocytes are easily deformable and do not preserve their biconcave discoid shape. Therefore, when randomly oriented, they may be assumed to be volume-equivalent spheres.

According to Mie theory, the scattering cross section, extinction cross section, and scattering phase functions of the RBC are defined as[66]

$$\sigma_s^{\mathrm{Mie}} = \frac{2\pi}{k^2}\sum_{n=1}^{\infty}(2n+1)(|a_n|^2 + |b_n|^2), \tag{2.17}$$

$$\sigma_t^{\mathrm{Mie}} = \frac{2\pi}{k^2}\sum_{n=1}^{\infty}(2n+1)\mathrm{Re}(a_n + b_n), \tag{2.18}$$

$$p^{\mathrm{Mie}}(\theta) = \frac{1}{k^2\sigma_s^{Mie}}(|S_1|^2 + |S_2|^2), \tag{2.19}$$

where

$$S_1(\theta) = \sum_{n=1}^{\infty}\frac{2n+1}{n(n+1)}\left\{a_n\frac{P_n^1(\cos\theta)}{\sin\theta} + b_n\frac{d}{d\theta}P_n^1(\cos\theta)\right\},$$

$$S_2(\theta) = \sum_{n=1}^{\infty}\frac{2n+1}{n(n+1)}\left\{b_n\frac{P_n^1(\cos\theta)}{\sin\theta} + a_n\frac{d}{d\theta}P_n^1(\cos\theta)\right\},$$

$P_n^1(\cos\theta)$ are the associated Legendre functions, and a_n and b_n are the coefficients of the Mie series (which depend on the wavelength λ, the size, and the relative refractive index $n = \sqrt{\varepsilon}$ of the scatterers).

The absorption cross section can be obtained from Eqs. (2.17) and (2.18) by considering that $\sigma_a = \sigma_t - \sigma_s$ [Eq. (2.12)].

Treatment of the RBC as a homogeneous Mie scatterer has proven to be a valid and fruitful approach for solving numerous problems in biomedical optics.[29,33,35] Steinke and Shepherd demonstrated that assuming a spherical shape of the erythrocyte, the value of scattering cross section of the RBC can be correctly predicted.[29] Yaroslavsky et al.[33] and Hammer et al.[35] reported that the scattering phase function of the erythrocytes could be successfully approximated in the frameworks of Mie theory. Macroscopic absorption and scattering properties, determined assuming the Mie scattering phase function for whole blood, were used to interpret light transport in it by applying diffusion approximation[14] and MC technique.[37] If randomly oriented erythrocytes or whole blood are studied, Mie theory may be successfully applied to describe single light scattering.

2.4.3.2 Wentzel–Kramers–Brillouin approximation

Single scattering of visible and near-infrared light by the RBC may be treated in the framework of the WKB approximation.[53] The WKB approximation is applicable to a particle of arbitrary shape. It is assumed that the direction and the polarization of the incident wave remain constant. This approximation is valid when the wavelength of the incident light λ is small compared to the characteristic size of the particle D and when the particle relative refractive index $n = \sqrt{\varepsilon}$ is close to unity, i.e.,

$$(\varepsilon - 1)kD \gg 1 \quad \text{and} \quad \varepsilon - 1 < 1. \tag{2.20}$$

For the erythrocyte suspended in saline solution and incident light with a wavelength of 800 nm, $\varepsilon \approx 1.1$, $k = 7.85\ \mu\text{m}^{-1}$, and $D \approx 5.5\ \mu\text{m}$. Consequently, $(\varepsilon - 1) \approx 0.1$ and $(\varepsilon - 1)kD \approx 5.18$. These values are within the range of validity of the WKB approximation, and therefore, it can be safely applied to solve the problem of RBC single light scattering.

The scattering amplitude in the far zone as given by the WKB approximation is[45]

$$f(\bar{s},\bar{i}) = \frac{k^2}{4\pi}[-\bar{s} \times (\bar{s} \times \hat{e}_i)]VS(\bar{s},\bar{i}), \tag{2.21}$$

where

$$S(\bar{s},\bar{i}) = \frac{1}{V}\int_V 2(n-1)\exp(ikz_1 + ikn(z - z_1) - ik\bar{r}' \cdot \bar{s})dV'.$$

Here, V is the volume of the particle, z_1 is z coordinate of the position where the incident wave penetrates inside the particle, and $\bar{r}'$ is the interior point of the particle.

The extinction cross section and the absorption cross section are defined by the following equations:

$$\sigma_t = k\,\text{Im}\int_V 2(n-1)\exp[-ik(n-1)z_1 + ik(n-1)z]dV', \tag{2.22}$$

$$\sigma_a = \int_V k\varepsilon'' \frac{4}{|n+1|^2}\exp[-2\,kn_i(z - z_1)]dV', \tag{2.23}$$

where $n = n_r + in_i$, the relative refractive index of a particle.

2.4.3.3 Rayleigh–Gans–Debye approximation

Several authors[35,53] have addressed the process of single light scattering by the erythrocyte using the Rayleigh–Gans–Debye (RGD) approximation. Like the WKB approximation, the RGD approximation may be applied to a particle of arbitrary shape. Within the framework of this approximation, it is assumed that the electric field inside the particle can be approximated by the incident electric field. This assumption is valid if

$$(\varepsilon - 1)kD \ll 1 \quad \text{and} \quad \varepsilon - 1 < 1. \tag{2.24}$$

The amplitude of the scattered field in the far zone according to RGD approximation is as follows:

$$f(\bar{s},\bar{i}) = \frac{k^2}{4\pi}[-\bar{s} \times (\bar{s} \times \hat{e}_i)]VS(\bar{s},\bar{i}), \tag{2.25}$$

where

$$S(\bar{s},\bar{i}) = \frac{1}{V}\int_V (n^2 - 1)\exp(ik(\bar{i} - \bar{s})\bar{r}')dV'.$$

As estimated earlier, for an erythrocyte suspended in saline solution and incident light with a wavelength of 800 nm, $(\varepsilon - 1)kD \approx 5.18$. Therefore, the RGD approximation cannot be expected to provide an accurate description of the RBC single-scattering process. Nevertheless, it was reported to provide good agreement with the RBC forward-scattering measured data at the wavelength of 577 nm.[35]

2.4.3.4 Fraunhofer and anomalous diffraction approximations

Small-angle scattering by a single erythrocyte can be considered in the framework of Fraunhofer or anomalous diffraction theories,[34,35,57,69] since for incident-visible or near-infrared light with a wavelength λ, a size parameter of RBC α is much larger than unity ($\alpha = \pi D/\lambda$). For example, for incident light with a wavelength of 800 nm, the typical size of the RBC is $D = 5.5$ μm, and the refractive index of the particle relative to saline solution is $n = 1.049$, and the size parameter of the erythrocyte equals approximately 23. In this case, the resulting small-angle distribution of the scattered light will be determined by the interference of the part of the wave that traverses the particle and another part of the wave that passes along the particle.[61] If the first part can be neglected, the simplest way to approximate the small-angle scattering by such a particle would be to apply the Fraunhofer diffraction theory. According to this theory, the scattering pattern is independent of the relative refractive index of the particle, n. The angular distribution of the scattered light is determined only by the shape and the size of the cross sectional area of the RBC. On the other hand, it was shown that the scattering properties of the biological cells depend on n.[70,71] Therefore, it is desirable to consider the interference between the light that traverses the scatterer and the light that passes along the particle. Since the erythrocytes are optically soft (i.e., $\varepsilon - 1 \ll 1$), it may be assumed that refraction does not alter the direction of the incident wave. For optically soft particles with a size much greater than the wavelength of the incident light, the anomalous diffraction theory, developed by van de Hulst,[61] is applicable. According to anomalous diffraction approximation, the small-angle scattering amplitude in the far zone ($z \to \infty$) is determined by the following equations:

$$f^{ad}(\bar{a}) = \frac{k}{2\pi i}\int \exp(-ik\bar{a}\,\bar{\rho})w(\bar{\rho})d\rho, \tag{2.26}$$

$$w(\bar{\rho}) = \{\exp[i\Phi(\bar{\rho})] - 1\}\eta(\bar{\rho}), \tag{2.27}$$

$$\Phi(\bar{\rho},\infty) = \Phi(\bar{\rho}) = k\int_{-\infty}^{\infty}[n(\bar{\rho},z') - 1]dz', \tag{2.28}$$

where $\bar{\rho} = \bar{\rho}(x,y)$ is the radial coordinate; $\bar{a}$ is the projection of the scattering direction on the plane $z = \text{const}$; $\eta(\bar{\rho})$ is the shadow function, which is equal to unity inside the particle projection and is equal to zero outside; and $n(\bar{\rho},z)$ is a complex refractive index of the scatterer.

If the first term in Eq. (2.27) is neglected, Eq. (2.26) will describe the Fraunhofer diffraction from the particle.

The scattering and the absorption cross sections of the particle are determined as the following integrals over the shadow region behind the scatterer:

$$\sigma_s = \int |w(\bar{\rho})|^2 d\bar{\rho}, \tag{2.29}$$

$$\sigma_a = \int \{1 - |\exp[i\Phi(\bar{\rho})]|^2\} d\bar{\rho}. \tag{2.30}$$

Fraunhofer and anomalous approximations are widely used for interpretation of the small-angle scattering patterns of the RBC. Ektacytometric and flow-cytometric measurements together with these theories provide important information on the cell's size, shape, deformability, and refractive index. Steekstra et al.[57] have compared the results obtained using Fraunhofer and anomalous diffraction theories with angular distributions of scattered light measured in ektacytometric experiments. The authors have come to the conclusion that anomalous diffraction approximation is in better agreement with experimental data. Hammer et al.[35] have also found that their goniophotometric measurements of the small-angle RBC single scattering agree well with predictions of the anomalous diffraction approximation. On the other hand, Borovoi et al.[69] have shown that Fraunhofer diffraction approximation can be successfully used for particle shape characterization.

2.4.3.5 Semianalytical and numerical methods

The recent advent of high-speed computers has opened the way for the application of semianalytical (T-matrix method) and numerical approaches (boundary element, finite element, finite difference, or surface and volume integral equations methods). The analytical methods described above have the disadvantage of imposing different restrictions on the size, shape, or refractive index of the scatterer. Numerical methods can solve the single-scattering problem without any significant limitations. The main disadvantage of the numerical techniques, as well as of the T-matrix formalism, is the long computational time needed to provide an accurate solution.

T-matrix formalism was developed by Waterman[72] and by Barber and Yeh.[60] This is an integral equation method, originally called the extended boundary condition method, that provides a powerful tool to solve the scattering problem with an arbitrarily shaped conducting or dielectric body illuminated by a plane electromagnetic wave. This method is, in general, suitable for particles with sizes ranging from the Rayleigh to the geometrical optics regions. A detailed description of the T-matrix technique is provided in Ref. 60. According to it, the quantities of the incident and the scattered field, as well as of the surface and the internal fields induced in the scattering object, are expanded into the infinite series using the basis functions. For a spheroidal scattering particle such as a RBC, the appropriate basis functions are spherical vector waves. The scattered field expansion coefficients are calculated from the known expansion coefficients of the incident field, employing the equivalence theorem and the integral representation of the fields internal and external with respect to the scatterer.[60,73] The matrix, which relates the expansion coefficients of the incident wave to the expansion coefficients of the scattered wave, is called the T-matrix. It is clear that the T-matrix method gives an exact solution of the light-scattering problem by a particle in a form of infinite series. Therefore, an iterative numerical method must be applied to calculate the scattering amplitude and the scattering cross section of the particle. The maximum number of terms in the expansion that are required to achieve reasonable accuracy depends on the scatterer's shape, size, and index of refraction. For spheres with a size on the order of the wavelength of the incident radiation and refractive index close to unity, the number of expansion terms is comparatively small. For the RBC, which is a biconcave disk with a size about 10 times larger than the wavelength of the incident light in the visible and near-infrared spectral range, the number of expansion terms required is much larger. For particles such as erythrocytes, large matrices should be calculated with high accuracy, and the convergence check is important. The first theoretical study of light scattering by erythrocytes using the T-matrix approach was published in 1998.[58] The authors considered the RBC to be a volume-equivalent nonabsorbing spheroid. To enable computations of the scattering amplitude for such a particle, the authors had to introduce extended precision into the program of Barber and Hill.[77] The angular scattering distributions in the far field were computed for five different RBC volume-equivalent homogeneous spheroids. The authors came to the conclusion that light scattering by the RBC volume-equivalent spheroid is highly influenced by the shape of the scattering object. The forward scattering is influenced by the optical thickness of the particle with respect to the incident field and by the shape of the spheroid. The back scattering is sensitive to the shape of the surface, which faces the incident beam.

As was mentioned before, with the development of fast modern computers, optimized numerical methods are becoming more efficient. Recently, Tsinopoulos and Polyzos[56] solved the RBC single-scattering problem using the boundary element fast Fourier transform (FFT) methodology.[75,76] The authors reported the numerical results concerning He–Ne laser light scattering by an undeformable average-sized erythrocyte. The RBC was assumed to be a nonabsorbing biconcave disc. Angular far-field distributions of the scattered light were calculated for three different orientations of the scatterer. In addition, the authors compared the results of their study with situations when the RBC is modeled as a volume-equivalent sphere and a size-volume-equivalent oblate spheroid. The authors reported that approximation of the erythrocyte by a volume-equivalent spheroid is satisfactory only for certain orientations of the cell with respect to the incident beam. The authors came to the conclusion that forward-scattered energy depends on the orientation of the RBC. They confirmed the findings of Nilsson et al.[58] that back scattering is greatly affected by the curvature of the cell surface facing the incident light.

Methods described in this section can be used to extract RBC morphological information. Precise information about the dimensions and the refractive index may be obtained by using these methods to interpret the angular scattering distributions measured on a single particle. On the other hand, it is still not feasible to use these methods to investigate whole blood, since computations for even a single erythrocyte are extremely time consuming and/or sensitive to small initial numerical errors.

2.4.3.6 Empirical phase functions

Two empirical phase functions are widely used to approximate the scattering phase function of blood. Those are the Henyey-Greenstein phase function (HGPF)[63] and the Gegenbauer kernel phase function (GKPF).[28]

The HGPF has one parameter g that coincides with the anisotropy factor $\bar{\mu}$

$$p_{hg}(\theta) = \frac{1}{4\pi}(1 - g^2)[1 + g^2 - 2g\cos(\theta)]^{-3/2},\ |g| \leq 1. \tag{2.31}$$

It may be represented as the infinite series of Legendre polynomials $P_n^1(\cos\theta)$

$$p_{hg}(\theta) = \frac{1}{4\pi}\sum_{n=0}^{\infty}(2n+1)f_n P_n^1(\cos\theta), \tag{2.32}$$

where $f_n = g^n$.

The GKPF has two variable parameters α and g

$$p_{gk}(\theta) = K[1+g^2-2g\cos(\theta)]^{-(\alpha+1)}, \tag{2.33}$$

where

$$K = \alpha g\pi^{-1}(1-g^2)^{2\alpha}[(1+g)^{2\alpha}-(1-g)^{2\alpha}]^{-1},\ \alpha > -1/2,\ |g| \leq 1.$$

This function is a generalization of the HGPF. GKPF can be reduced to HGPF by setting $\alpha = 0.5$.

GKPF may be represented as the infinite series of Gegenbauer polynomials

$$p_{gk}(\theta) = \frac{2K}{(1-g^2)}\sum_{n=0}^{\infty}\left(1+\frac{n}{\alpha}\right)C_n^{\alpha}[\cos(\theta)]g^n, \tag{2.34}$$

where C_n^{α} are Gegenbauer polynomials.[28,77]

The HGPF and the GKPF are widely employed in radiative transport calculations for whole blood because of their mathematical simplicity. Several authors reported that these phase functions are suitable for the description of the single-scattering process in whole blood.[16,33,35,36] However, it is clear that HGPF and GKPF cannot be used to accurately calculate the angular light distribution scattered by a single erythrocyte.

2.5 Optical Properties of Blood

Light propagation in turbid media such as whole blood is determined by their intrinsic optical properties. According to the radiative transfer theory, those are the absorption coefficient μ_a, the scattering coefficient μ_s, and the scattering phase function $f(\mu)$ (μ is the cosine of the scattering angle).

As mentioned earlier, direct measurements of the optical properties of blood, such as measurements of the scattering phase function and of the extinction coefficient, require use of optically thin samples.[10,16,35,37] If the scattering phase function and the extinction coefficient of the optically thin diluted blood sample are available from the goniophotometric experiments, the scattering and the absorption cross sections may be calculated using Eqs. (2.10), (2.12), and (2.14). To obtain the macroscopic absorption coefficient μ_a, and scattering coefficient μ_s, the following equations are often applied:

$$\mu_a = \frac{H}{V}\sigma_a, \tag{2.35}$$

$$\mu_s = \frac{H}{V}\sigma_s \quad \text{for} \quad H \to 0, \tag{2.36}$$

$$\mu_s = \frac{H(1-H)}{V}\sigma_s \quad \text{for} \quad H \to 1, \tag{2.37}$$

where V is the volume of an RBC and H is the volume fraction of the cells in the sample (hematocrit).

Equations (2.35) and (2.37) were obtained by Twersky[26] and were verified experimentally by Steinke and Shepherd.[78] On the other hand, Roggan et al.[36] concluded from their experiments with flowing blood that both absorption and scattering increased linearly with hematocrit for $H < 0.5$. Ishimaru[45] suggests that Eq. (2.36) is can be used for blood with sufficiently small hematocrit ($H < 0.2$), while Eq. (2.37) is valid for high hematocrits ($H > 0.5$). It is not quite clear which equation should be used for the important range of hematocrits between 0.37 and 0.47, since the results reported by different authors are controversial.

2.5.1 Integrating sphere technique

As an alternative to using the direct methods, indirect methods can be used to determine the optical properties of whole and/or diluted human blood. The method is called indirect if it requires a model of light propagation to determine the tissue's optical properties from the measured quantities. One of the indirect methods that can be used to determine optical properties of tissues *in vitro* is the integrating sphere technique. Diffuse reflectance R_d, total transmittance T_t, and collimated transmittance T_c are measured. Absorption coefficient μ_a, scattering coefficient μ_s, and anisotropy factor $\bar{\mu}$ can be obtained from these data using an inverse method based on the radiative transfer theory. When the scattering phase function $f(\mu)$ is available from goniophotometry, $\bar{\mu}$ can be readily calculated. In this case, for the determination of μ_a and μ_s, it is sufficient to measure R_d and T_t only. For the experiments with blood, a double integrating sphere configuration is preferable, since in this case, both reflectance and transmittance can be measured simultaneously (see Fig. 2.5). In the case of a double-integrating-sphere arrangement of the experiment in addition to single-integrating-sphere corrections of measured signals, multiple exchange of light between the

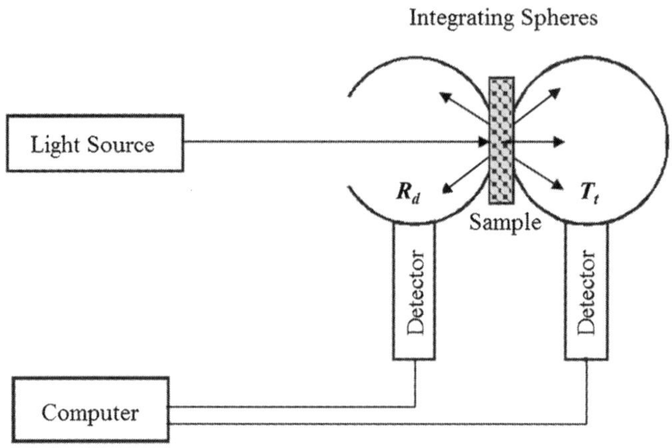

Figure 2.5 The double integrating sphere setup.

spheres should be accounted for.[79,80] The collimated transmittance measurement is usually carried out as shown in Fig. 2.4(a).

The integrating sphere technique was used by a number of investigators to determine the absorption coefficient, the scattering coefficient, the anisotropy factor, and/or the reduced scattering coefficient $\mu_s' = \mu_s(1 - \bar{\mu})$ of blood with physiological hematocrit, as well as of diluted blood.[31,32,36] Barium sulfate or Spectralon integrating spheres were used in the experiments. As in the single-scattering experiments, a laser, a Xe lamp, and a Hg lamp combined with a monochromator are used as monochromatic light sources, while a photomultiplier or a Si photodiode is employed as a detector. As mentioned above, blood has an extremely high total attenuation coefficient in the visible and near-infrared spectral range. Therefore, measuring the collimated transmittance of the undiluted blood sample with a moderate thickness ($\approx$0.1 mm) is a technically difficult task.[36] To solve this problem, a powerful light source combined with the sensitive detector must be used.[32] Alternatively, it is possible to use the third integrating sphere to collect the collimated light together with some forward-scattered light.[81] In this case, the collimated transmittance is separated from the scattered flux at the data processing stage using a MC technique or a small-angle approximation.[82,83] Different experimental arrangements to measure diffuse reflectance and transmittance of blood layers are described in detail in Refs. 31, 32, and 36.

2.5.2 Blood preparation and handling

Fresh human blood is usually obtained from healthy volunteers. If the experiments are performed with blood at rest, as in Ref. 31, blood is collected in heparinized containers. Alternatively, K_2EDTA may be added for anticoagulation,[84] as mentioned in Section 2.2. For the measurements, calibrated thin cuvettes are used (thickness 0.1–0.5 mm, slab geometry).

Standard clinical tests of blood samples are performed prior to the optical measurements. Clinical tests are necessary to determine the concentration of red and white blood cells, concentration of platelets, hematocrit, mean corpuscular volume and hemoglobin, and the other parameters of interest. To make the statistical analysis simpler, it is desirable to measure samples obtained from one donor. If the blood sample oxygenation level is of interest, it may be controlled using a conventional blood gas analyzer.[37] In most cases, the experiments are performed with either completely oxygenated or completely deoxygenated blood.[31,33,35,36,37] To obtain complete oxygen saturation, the sample is exposed to air or O_2[36,37] To completely deoxygenate blood, sodium dithionite ($Na_2S_2O_4$) is added.[37] Duration of the experiments with any particular blood sample should not exceed 20–30 min to ensure that sedimentation does not obscure the results. Light microscopy may be used prior to and after the optical experiments to check that no significant hemolysis and aggregation occurred. To be sure that neither the volume nor the surface area of the blood particles change during the experiments, the pH of the samples should be maintained at approximately 7.4.

In reality, blood is flowing through the blood vessels and, accordingly, it is preferable to study the optical properties of flowing blood. In the process of blood motion, the RBCs are subject to deformation and orientation. At lower shear rates, reversible aggregation occurs, while under higher shear rates, the erythrocytes are deformed into ellipsoids. At the same time, transversal and lateral migration of RBCs take place. On the other hand, since all these effects occur simultaneously, interpreting experiments with flowing blood is extremely difficult. Experiments with flowing undiluted and diluted blood are reported in Refs. 32, 36, and 84. Roggan et al.[36] have assembled sophisticated equipment to analyze the influence of different hematocrit, flow velocity, osmolarity, hemolysis, and oxygen saturation on the optical properties of RBC suspensions submerged into the saline solution. Nilsson et al.[32] investigated the influence of slow heating on the optical properties of flowing whole completely oxygenated blood. In the latter study, only the temperature was controlled during the experiments, while the complete clinical blood analysis was made prior to the optical measurements in a fashion similar to that described above (experiments with resting blood). The influence of shear stress on the optical properties of whole completely oxygenated blood was studied by Steenbergen et al.[83] Different experimental equipment and arrangements for the investigations of flowing blood are described in detail in Refs. 32, 36, and 84.

2.5.3 Algorithms used to determine optical properties of whole and diluted human blood from the integrating sphere measurements

Light propagation in turbid media can be described in the framework of the steady-state transport theory. If a turbid medium is irradiated by a monochromatic light with the power independent of time and there are no internal sources within the medium, the radiance $I(\bar{r},\hat{s})$ satisfies the steady-state transport equation[85]

$$\frac{\partial I(\bar{r},\hat{s})}{\partial \hat{s}} + \mu_t I(\bar{r},\hat{s}) = \frac{\mu_s}{4\pi} \iint p(\hat{s},\hat{s}') I(\bar{r},\hat{s}') d\hat{s}', \tag{2.38}$$

where $I(\bar{r},\hat{s})$ is the radiance at the point $\bar{r}$ and in the direction $\hat{s}$, $\mu_t = \mu_a + \mu_s$ is the extinction coefficient, μ_a is the absorption coefficient that defines the probability of absorption per unit pathlength, μ_s is the scattering coefficient that defines the probability of scattering per unit pathlength, $p(\hat{s},\hat{s}')$ is the scattering phase function $\bar{r} \in R^3$, and $\hat{s},\hat{s}' \in S$ is the unit sphere in R^3.

If the radiation transport is investigated in the homogeneous volume $G \subset R^3$, and ∂G is the boundary of the investigated region, the boundary conditions are given by

$$I(\bar{r},\hat{s})|_{(\hat{s}\bar{n})<0} = S(\bar{r},\hat{s}) + RI(\bar{r},\hat{s})|_{(\hat{s}\bar{n})>0}, \tag{2.39}$$

where $\bar{r} \in \partial G$, $\bar{n}$ is the outward normal to ∂G, $S(\bar{r},\hat{s})$ is the radiance of the incident light, and R is the reflection operator.

Equations (2.38) and (2.39) define the forward radiation transfer problem in the scattering and absorbing turbid medium.

Besides the radiance $I(\bar{r},\hat{s})$, the other important quantities are the fluence rate $\varphi(\bar{r})$ at the point $\bar{r}$, and the radiation flux $F(\bar{r})$ coming out of the medium, which are defined by

$$\varphi(\bar{r}) = \int_{4\pi} I(\bar{r},\hat{s}) d\omega, \tag{2.40}$$

$$F(\bar{r}) = \int_{(\hat{s}\bar{n})>0} I(\bar{r},\hat{s})(\hat{s}\bar{n}) d\omega, \quad \text{where } \bar{r} \in \partial G. \tag{2.41}$$

It should be mentioned that Eq. (2.38) has no general analytical solution in a closed form. Analytical solutions, e.g., solutions obtained in the diffusion approximation, are derived by making different assumptions. Numerical solutions of the transport equations that impose restrictions on the boundary conditions and/or the optical properties of blood are available.[27,86] Alternatively, the MC modeling technique may be used to obtain an indirect numerical solution to the transport equation without any simplifying assumptions about the boundary conditions or restrictions with respect to the optical properties.

If the optical properties, i.e., the absorption coefficient, the scattering coefficient, the anisotropy factor, and the refractive index of the blood sample, are known for a certain wavelength, then the reflected and transmitted signals can be calculated using radiative transport theory. In our case, we have an inverted situation, when the reflected and transmitted signals are measured, while the absorption coefficient, the scattering coefficient, and the anisotropy factor should be determined (the refractive index of blood in the visible and near-infrared spectral range is known from the literature[35,44–46]). For this purpose, the inverse optimization techniques are applied.[87] In the first step of the inverse algorithm, the optical properties are estimated using simple methods like the Kubelka–Munk theory[88,36] or the δ-Eddington approximation.[89,82,31] The estimated values of the optical properties are then used to calculate the reflected and transmitted signals, employing one of the models of light propagation in blood. In the next step, the calculated values are compared with the measured values. If the required accuracy is not achieved, the current optical properties are altered using one of the available optimization algorithms. The procedures of altering the optical properties and calculating the reflected and transmitted signals are repeated until the calculated values match the measured values with the required accuracy. A detailed description of the inverse algorithms used to determine the optical properties of blood from the integrating sphere measurements is given in Refs. 82 and 36.

As mentioned earlier, to use an indirect method to determine the optical properties, a model of light propagation in the tissue is required. The most widely used models are the Monte Carlo (MC) technique,[31,32,36,37,84] the diffusion approximation,[10–12,14] and the numerical solutions of the transport equation.[27,32] The diffusion approximation and the numerical solutions of the transport equation include simplifying assumptions on the geometry of the experiments and restrictions on the range where the optical properties can be determined. The MC technique is not limited with respect to geometrical configuration and the range of the optical properties; rather it requires comparatively long calculation times.

In early works, the diffusion approximation was applied to describe light transport in blood.[10,17,18] The diffusion approximation is valid and gives an accurate solution when the radiance is weakly anisotropic. This occurs far from collimated sources and boundaries, which is not the case for the reflectance and transmittance measurements of comparatively thin blood samples (0.1–3.0 mm). Another important limitation of the diffusion approximation is the requirement that $\mu_a \ll \mu_s(1 - \bar{\mu})$. This requirement is not satisfied for blood in the visible spectral range. To overcome the limitations of the diffusion approximation, more sophisticated methods such as the adding–doubling[32] and MC methods were applied to solve the direct problem in the inverse algorithm for determination of the optical properties of whole and diluted blood.

2.5.3.1 The Monte Carlo method

The MC technique was employed as a method to solve the forward problem in the inverse algorithm for the determination of the optical properties of whole and diluted human blood by Yaroslavsky et al.[31,33,37] and Roggan et al.[36] Different variations of the basic MC algorithm are described in detail by several authors.[90,91,92] The MC technique is based on the formalism of the radiative transport theory. The absorption coefficient is defined as the probability of a photon to be absorbed per unit length, and the scattering coefficient is defined as the probability of a photon to be scattered per unit length. Using these probabilities, a random sampling of photon trajectories is generated.

A basic algorithm for the generation of photon trajectories can be briefly described as follows. A photon (see Fig. 2.6), described by three spatial coordinates and two angles (x, y, z, θ, φ) is assigned its weight $W = W_0$ and placed in its initial position, depending on the source characteristics. The step size l and the direction of the photon's next movement on each step is determined using the equation

$$\eta = \int_{x_{\min}}^{x_{rnd}} p(x)dx, \tag{2.42}$$

where η is a random number uniformly distributed in the interval (0,1), $p(x)$ is a probability density function of x, $x_{\min}$ is a minimal value of x, and x_{rnd} is a random value of x.

For example, for the random stepsize l_{rnd}, the probability density function is

$$p(l) = \mu_t \exp(-\mu_t l). \tag{2.43}$$

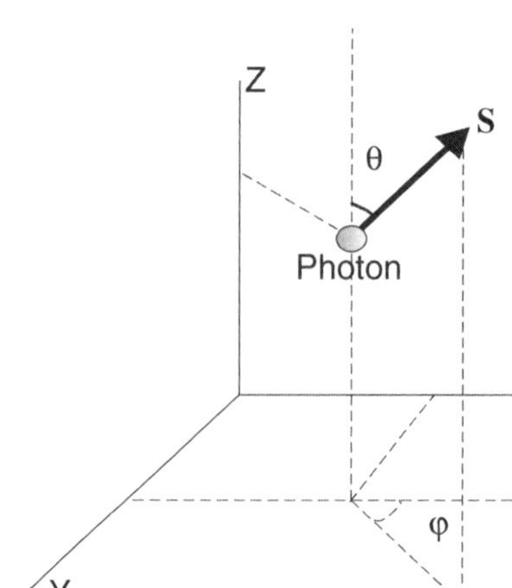

Figure 2.6 Coordinates and angles describing the movement of a photon.

Substituting Eq. (2.43) into Eq. (2.42) and performing integration, one obtains

$$l_{rnd} = -\ln(1-\eta)/\mu_t. \tag{2.44}$$

The direction of the photon's next movement is determined in the same way as described above, with the scattering phase function substituted as the probability density distribution. Several approximations for the scattering phase function of blood have been used in the MC simulations. Those are the HGPF, the GKPF, and the Mie phase function (see Sections 2.4, 2.4.3.1, and 2.4.3.6 of this chapter).

For the HGPF, the random scattering angle θ_{rnd}^{HG} is given by[90]

$$\theta_{rnd}^{HG} = \arccos\left\{\frac{1}{2g}\left[1+g^2-\left(\frac{1-g^2}{1-g+2g\eta}\right)^2\right]\right\}. \tag{2.45}$$

For the GKPF, the random scattering angle θ_{rnd}^{GK} is determined as[37]

$$\theta_{rnd}^{GK} = \arccos\left\{\left(1+g^2-1/\sqrt[\alpha]{\xi_{rnd}}\right)/2g\right\}, \tag{2.46}$$

where

$$\xi_{rnd} = 2\alpha g\eta/K + (1+g)^{-2\alpha},$$
$$K = \alpha g\pi^{-1}(1-g^2)^{2\alpha}[(1+g)^{2\alpha} - (1-g)^{2\alpha}]^{-1}.$$

For experimental phase functions, the random scattering angle can be determined in the following way.[93] If the scattering phase function is known for the discrete set of scattering angles θ_i: $f(\theta) = f(\theta_i)$, it can be determined in the total angular range using the spline-interpolation technique.[94] Then the value of the function $F_n = \int_0^{\theta_n} f(\theta)d\theta$ can be calculated numerically for any value of θ_n. It is easy to see that F_n is a nondecreasing function, which is mapping the interval (0,1). Therefore, when random value η is sampled, θ_{rnd}^{exp} is determined by setting $F_n = \eta$.

The Mie phase function can be tabulated and treated in the same way as the experimental phase function.[37]

In most cases, azimuthal symmetry is assumed. This leads to $p(\varphi) = 1/2\pi$ and, consequently, $\varphi_{rnd} = 2\pi\eta$.

On each step, the photon loses part of its weight due to absorption: $W = W(1-\alpha)$, where $\alpha = \frac{\mu_s}{\mu_t}$ is the albedo of the medium. When the photon reaches the boundary, part of its weight is transmitted according to the Fresnel equations. The amount transmitted through the boundary is added to the reflectance or transmittance. Since the refraction angle is determined by Snell's law, the angular distribution of the outcoming light can be calculated. The photon with the remaining part of the weight is specularly reflected and continues its random walk.

When the photon's weight becomes lower than a predetermined minimal value, the photon can be terminated using the "Russian roulette" procedure.[90,91] For this, the predetermined photon survival probability is compared to the generated random number. If the random number is smaller than the photon's current weight, the latter is divided by a predetermined value of survival probability, and the photon continues its random walk. If the random number is larger than the current photon weight, the photon is terminated. This procedure saves time, since it does not make sense to continue the random walk of the photon, which will not essentially contribute to the measured signal. On the other hand, it ensures that the energy balance is maintained throughout the simulation process.

The MC method has several advantages over the other methods, for it may take into account mismatched medium-glass and glass-air interfaces, losses of light at the edges of the sample, any phase function of the medium, and finite size and arbitrary angular distribution of the incident beam. If the collimated transmittance is measured, then the contribution of scattered light into the measured collimated signal can be accounted for.[82,93] The only disadvantage of this method, since it is a statistical approach, is the length of time needed to ensure good statistical convergence. The standard deviation of a quantity (e.g., diffuse reflectance or transmittance) approximated by the MC technique decreases proportionally to $1/\sqrt{N}$, where N is the total number of launched photons.

2.5.3.2 The adding–doubling method

The adding–doubling method is a numerical method based on the one-dimensional radiation transport theory. It was used by Nilsson et al.[32] as a forward model in the inverse technique to investigate temperature-induced changes in the optical properties of whole human blood.

The adding–doubling technique (ADT) is a numerical method for solving the one-dimensional transport equation in slab geometry. In addition to the slab geometry requirement, the method assumes that incident illumination is spatially uniform. Thus, finite beam size and side losses of light cannot be taken into account. ADT can be used for media with arbitrary phase function and arbitrary angular distribution of the incident radiation. The derivation and implementation of the adding–doubling method is given in detail by Prahl.[98]

ADT is based on the observation that for an arbitrary incident radiance distribution $I_{in}(\nu)$, where ν is the cosine of the polar angle, the angular distribution of the reflected radiance (normalized to an incident diffuse flux π) is given by

$$I_{ref}(\nu) = \int_0^1 I_{in}(\nu')R(\nu',\nu)2\nu' d\nu', \tag{2.47}$$

where $R(\nu', \nu)$ is the reflection redistribution function determined by the optical properties of the slab.

The distribution of the transmitted radiance can be expressed in a similar manner, with obvious substitution of the transmission redistribution function $T(\nu', \nu)$. If M quadrature points are selected to span over the interval (0,1), the respective matrices can approximate the reflection and transmission redistribution functions:

$$R(\nu_i', \nu_j) \to R_{ij}, \quad T(\nu_i', \nu_j) \to T_{ij}.$$

These matrices are referred to as the reflection and transmission operators, respectively. If a slab with boundaries indexed as 0 and 2 is comprised of two layers (01) and (12), with an internal interface 1 between the layers, the reflection and transmission operators for the whole slab (02) can be expressed as

$$\begin{aligned}
\mathbf{T}^{02} &= \mathbf{T}^{12}(\mathbf{E} - \mathbf{R}^{10}\mathbf{R}^{12})^{-1}\mathbf{T}^{01}, \\
\mathbf{R}^{20} &= \mathbf{T}^{12}(\mathbf{E} - \mathbf{R}^{10}\mathbf{R}^{12})^{-1}\mathbf{R}^{10}\mathbf{T}^{21} + \mathbf{R}^{21}, \\
\mathbf{T}^{20} &= \mathbf{T}^{10}(\mathbf{E} - \mathbf{R}^{12}\mathbf{R}^{10})^{-1}\mathbf{T}^{21}, \\
\mathbf{R}^{02} &= \mathbf{T}^{10}(\mathbf{E} - \mathbf{R}^{12}\mathbf{R}^{10})^{-1}\mathbf{R}^{12}\mathbf{T}^{01} + \mathbf{R}^{01},
\end{aligned} \tag{2.48}$$

where $\mathbf{E}$ is the identity matrix defined in this case as

$$\mathbf{E}_{ij} = \frac{1}{2\nu_i w_i}\delta_{ij}, \tag{2.49}$$

where w_i is the weight assigned to the i-th quadrature point.

The definition of the matrix multiplication also slightly differs from the standard. Specifically

$$(\mathbf{AB})_{ik} \equiv \sum_{j=1}^{M} A_{ij}2\nu_j w_j B_{jk}. \tag{2.50}$$

Equations (2.48) allow one to calculate the reflection and transmission operators of a slab when those of the comprising layers are known. The idea of ADT is to start with a thin layer for which the transfer equation can be simplified and solved with relative ease, producing the reflection and transmission operators for the thin layer, then to proceed by doubling the thickness of the layer until the thickness of the whole slab is reached. Several techniques exist for layer initialization.[96–98] The single-scattering equations for the reflection and transmission for the Henyey–Greenstein function, used in Ref. 32, are given in Refs. 99, 100, and 95. The refractive index mismatch can be taken into account by adding effective boundary layers of zero thickness and having the reflection and transmission operators determined by Fresnel's formulas. The total transmittance and reflectance of the slab are obtained by straightforward integration of Eq. (2.47). Different methods to perform the integration are discussed in Ref. 95.

The adding–doubling method provides accurate results in cases when the side losses are not significant, but it is less flexible than the MC technique.

2.6 Summary of the Optical Properties of Diluted and Whole Human Blood

In previous sections, we described several basic techniques that can be used to determine the optical properties of single RBCs and those of whole blood. Below we summarize and discuss the results provided by these techniques and those available from the literature.

2.6.1 Optical properties of blood determined using direct techniques

Optical properties of blood determined from the single-scattering experiments are summarized in Table 2.1.

Reynolds et al.[14] determined the absorption cross sections, the scattering cross sections, and the anisotropy factors of blood from the single-scattering experiments[101,102] for a number of wavelengths in the visible and near-infrared spectral range and compared the experimental values with the calculated (Mie theory) values. In Mie calculations, they used the value of 2.79 μm for the RBC radius and the value of 1.036 for the RBC refractive index (relative to blood plasma).[14,45]

Flock et al.[16] measured the total attenuation coefficient and the scattering phase function of the diluted whole blood sample at the wavelength of 632.8 nm. For the measurement, a fresh blood sample was drawn, heparinized, and diluted to a hematocrit of 1% with phosphate-buffered saline (PBS). The diluted blood sample was placed into the cuvette (< 100-μm-thick, slab geometry).

Steinke and Shepherd[29] determined the total attenuation coefficients, the scattering cross sections, and the anisotropy factors of the RBCs suspended in blood plasma and in PBS (0.9%) from the collimated transmittance and scattering phase function measurements for the wavelength of 632.8 nm and from the calculations using the Mie theory. To determine the scattering cross sections from the total attenuation coefficients, the authors used

Table 2.1. Optical properties of blood determined and approximated from the single-scattering experiments (OS = oxygen saturation).

λ (nm)	OS %	σ_a (μm^2)	σ_a (μm^2)	σ_t (μm^2)	μ_t (mm^{-1})	$\bar{\mu}$	Theory	Ref. no.
665	100	0.060	57.20		0.747	0.9951	Mie	[14]
675	100	0.060	56.14		0.744	0.9950		
685	100	0.059	55.09		0.739	0.9949		
955	100	0.191	33.47		0.666	0.9925		
960	100	0.187	33.18		0.666	0.9924		
965	100	0.185	32.90		0.665	0.9924		
665	0	0.542	56.58		0.738	0.9951		
675	0	0.535	55.53		0.735	0.9950		
685	0	0.484	54.56		0.731	0.9949		
955	0	0.090	33.54		0.668	0.9925		
960	0	0.085	33.27		0.667	0.9924		
965	0	0.080	32.98		0.666	0.9924		
630	100	0.099	56.37				Experiment	
660	100	0.066	54.20					
685	100	0.063	53.53					
800	100	0.131	42.24					
632.8	100				2.9	0.974	Experiment	[16]
							Blood (1%)	
632.8							Experiment	[29]
	100		63.82		0.709	0.9853	RBC in plasma	
	100		81.24		0.903	0.9818	RBC in PBS (0.9%)	
	100		66.62		0.740	0.9948	Mie	
	100		79.27		0.881	0.9926	RBC in plasma	
							RBC in PBS (0.9%)	
632.8	100					0.982	HGPF	[31]
						0.995	GKPF	
							(α = 1.82)	
632.8	100					0.971	HGPF	[37]
						0.997	GKPF	
							(α = 3.658)	
						0.996	Mie	
577	100			≈95		0.966	HGPF	[35]
						0.997	GKPF	
							(α = 1.5)	
						0.997	Mie	
						0.9995	RGD	

Eq. (2.36) and assumed that the absorption of light by diluted RBC samples was negligible. For the collimated transmittance measurements, the cuvettes with the thickness of 144 μm (slab geometry) were used. For the goniometrical measurements, an American Optical Hemoglobinometer cuvette (pathlength 51 μm) was employed.

Yaroslavsky et al.[31,37] measured the scattering phase functions of diluted whole blood samples; approximated the experimental scattering phase functions using Mie theory, HGPF, or GKPF; and determined the anisotropy factors for each approximation at the wavelength of 633 nm. For Mie calculations, the RBC radius was assumed to be equal to 2.995 μm and the

RBC refractive index equal to 1.04 (relative to PBS). The scattering phase functions were measured in the angle range from 2 to 18 deg. For the experiments, fresh samples of whole blood were collected into the heparinized containers and diluted with a phosphate buffer solution (pH 7.4) to a hematocrit of 0.1%. The diluted blood samples were placed into the cuvettes (10-μm-thick, slab geometry).

Hammer et al.[35] presented a comprehensive study of the RBC single-scattering behavior. The authors measured the collimated transmittance and the scattering phase functions of the RBC suspensions in isotonic phosphate buffer solution for a number of wavelengths in the visible spectral range from 458 to 660 nm. The scattering phase functions were measured for 20 scattering angles in the range between 0.75 and 14.5 deg. For the experiments, erythrocytes were separated from the blood plasma and white cell fraction and washed three times in isotonic phosphate buffer (pH 7.4). Then the RBC suspension was diluted with isotonic phosphate buffer to a hematocrit of 0.01. For the transmission measurements, a cuvette with the thickness of 100 μm (slab geometry) was used, and for the scattering phase function measurements, a cuvette with the thickness of 10 μm (slab geometry) was used. To calculate the extinction cross sections from the measurements of the extinction coefficients, the authors used the formulas (2.35) and (2.36). The measured scattering phase functions were approximated using Mie, RGD, and anomalous diffraction theories, the Henyey–Greenstein and the Gegenbauer kernel functions. Scattering and absorption cross sections calculated by Hammer et al.[35] using Mie, anomalous diffraction, and RGD theories are given in Fig. 2.7.

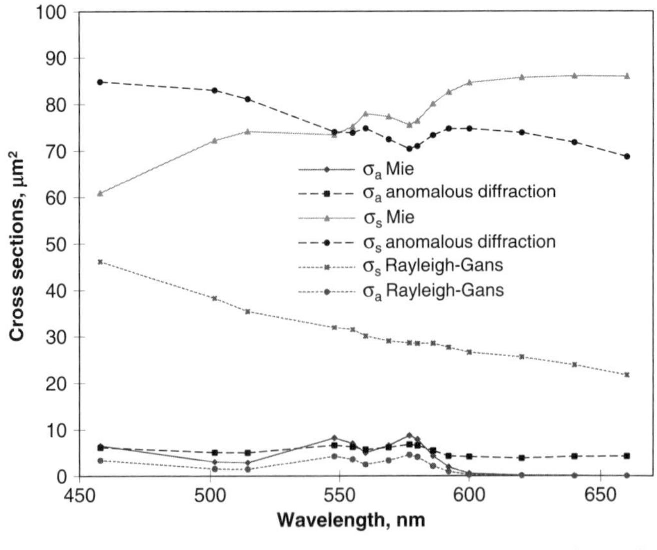

Figure 2.7 Calculated values of absorption and scattering cross sections of erythrocytes.[35]

2.6.2 Optical properties of blood determined using indirect techniques

Diluted and whole human blood optical properties determined using indirect techniques were reported by Yaroslavsky et al.,[31,37] Nilsson et al.,[32] Roggan et al.,[36] Friebel et al.,[84,82] Mienke et al.,[81] and Steenbergen et al.[83] A summary of the optical properties of diluted and whole blood determined using indirect techniques at the wavelength of 633 nm is given in Table 2.2.

Table 2.2. Overview of the optical properties of blood at 633 nm (*data taken from the graphs of the respective reference).

Ref. no.	μ_a (mm^{-1})	μ_s (mm^{-1})	μ_t (mm^{-1})	$\bar{\mu}$	μ_s' (mm^{-1})	Phase function	Conditions
[31]	1.55	64.5		0.982	1.16	HGPF	$H=45\%$,
							OS > 98%
	1.54	223.9		0.995	1.12	GKPF	
						$\alpha=1.82$	
[37]	1.52 ± 0.06	40 ± 3		0.971 ± 0.001	1.17 ± 0.12	HGPF	$H=38\%$
							OS > 98%
						GKPF	7 samples
	1.61 ± 0.06	413 ± 17		0.997 ± 0.0001	1.24 ± 0.09	$\alpha=3.658$	
						Mie	
	1.63 ± 0.05			0.9962 ± 0.0001		$r=2.995$ μm	
		239 ± 16				$n=1.04$	
[32]*						HGPF	$H=44 \pm 3\%$
							OS = 100%
							Temperatures:
	0.30				1.85		$t=25$ °C
	0.35				1.82		$t=35$ °C
	0.40				1.80		$t=42$ °C
	0.45				2.10		$t=48$ °C
	0.60				1.70		$t=54$ °C
[36]*						GKPF	$H=41\%$
						$\alpha=1.0$	
	2.0				2.0		OS = 25%
	1.6				2.0		OS = 50%
	1.2				2.0		OS = 75%
	0.7				1.8		OS = 100%
							$H=5\%$
	0.125	30				> 0.99	OS = 0%
	0.110	30				> 0.99	OS = 100%
[83]							Hct = 50 ± 0.5%
				0.950–0.963*		HGPF	OS = 100%
							Shear rates:
			≈120	0.956–0.965*			50 s^{-1}
			≈120				100 s^{-1}
			≈120	0.960–0.966*			150 s^{-1}
			≈120				200 s^{-1}
			≈120	0.962–0.967*			300 s^{-1}
							400 s^{-1}
				0.963–0.968*			500 s^{-1}
				0.963–0.970*			
				0.964–0.973*			

Yaroslavsky et al.[31] used an inverse Monte Carlo (iMC) technique to determine the optical properties of completely oxygenated whole blood samples in the near-infrared spectral range from double integrating sphere measurements. The measured values included the diffuse reflectance, the total transmittance, and the collimated transmittance. From the measured data, the absorption coefficient, the scattering coefficient, and the anisotropy factor (in the assumption of the HGPF) were derived. The spectral range investigated extended from 700 to 1200 nm. For the optical measurements, blood was collected into heparinized containers. Calibrated cuvettes (thickness 0.1 and 0.5 mm, slab geometry) were used. The hematocrit of the investigated samples was 45–46%. The optical properties of whole blood obtained in Ref. 31 are presented in Fig. 2.8(a)–(c). In addition, the effect of the scattering phase function approximation on the resulting estimates of the optical properties was analyzed. The Henyey–Greenstein and the GKPFs were considered. The calculated angular distributions of scattered light were compared with goniophotometric measurements performed at the wavelength of 633 nm. The presented data prove that the variations of the employed scattering phase function approximation can cause large discrepancies in the derived optical properties. Therefore, exact knowledge of the scattering phase function is required for the precise determination of the blood optical constants. In Ref. 37, the impact of the scattering phase function approximation on the optical properties of whole human blood determined from integrating sphere measurements using an iMC technique was investigated. The diffuse reflectance R_d and the total transmittance T_t ($\lambda = 633$ nm) of whole blood samples ($H = 38\%$; OS > 98%) were measured with double integrating sphere equipment. The scattering phase functions of highly diluted blood samples ($H = 0.1\%$; OS > 98%) were measured using a goniophotometer (see Section 2.6.1). The experimentally determined scattering phase functions were approximated with Henyey–Greenstein, Gegenbauer kernel, or Mie phase functions to preset the anisotropy factor $\bar{\mu}$ for the inverse problem. HGPF, GKPF, and MPF approximations were employed in the iMC procedure to derive the absorption coefficient μ_a and the scattering coefficient μ_s (see Table 2.2). To evaluate the obtained data, the angular distributions of scattered light for optically thick samples were calculated and results were compared with goniophotometric measurements. The results of this comparison prove that the employed approximation of the scattering phase function can have a substantial impact on the derived values of μ_s and $\bar{\mu}$, while μ_a and the reduced scattering coefficient μ_s' are much less sensitive to the exact form of the scattering phase function. In addition, it was shown that both R_d and T_t are strongly affected by the form of the phase function and that the magnitude of this influence depends on the thickness of the blood sample.

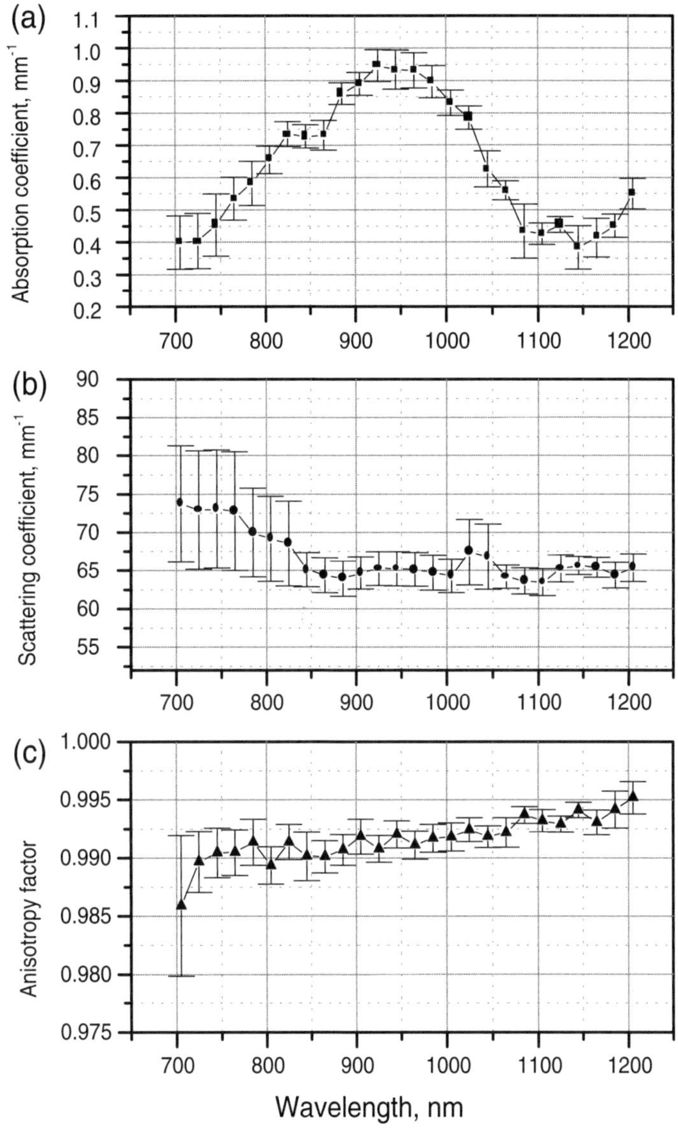

Figure 2.8 Optical properties of whole blood.[31] $H = 45$ to 46%, OS > 98%. Average of six samples. Bars are standard errors. (a) Absorption coefficient, (b) scattering coefficient, and (c) anisotropy factor.

Nilsson et al.[32] studied the influence of slow heating on the optical properties of completely oxygenated whole blood at the wavelength of 633 nm. The diffuse reflectance, the total transmittance, and the collimated transmittance were measured at different temperatures using a double integrating sphere setup. The absorption coefficient, the scattering coefficient, and the anisotropy factor (assuming the HGPF) were determined using an inverse adding–doubling method. For the measurements, whole blood was collected into the tubes, which contained EDTA to prevent coagulation. The hematocrit of the investigated samples was 44% ± 3%. During the measurements, the blood was pumped at a flow rate of 10.7 mL/min through the flow-through setup. The flow cell (length: 65 mm, height: 34 mm, total thickness: 2.5 mm) was placed between the integrating spheres. The blood sample thickness in the flow cell was 0.48 ± 0.02 mm. The temperature of blood was measured with thermocouples. The optical properties of nonheated blood were determined for the control. The blood was heated from approximately 25–55 °C at rates between 0.2 and 1.1 °C/min. While the blood was being heated, the integrating sphere measurements were continuously taken. The authors have found that changes in the optical properties of blood due to slow heating were reversible until the temperature of 44.6–46.6 °C. Coagulation of blood occurred at approximately 55 °C. The main results of this investigation are summarized in Table 2.2 and Fig. 2.9.

Roggan et al.[36] used double integrating sphere measurements and iMC simulations to investigate the optical properties of RBCs suspended in PBS under flow conditions. Reference 36 can be considered one of the most extensive studies of the macroscopic optical properties of RBC suspension,

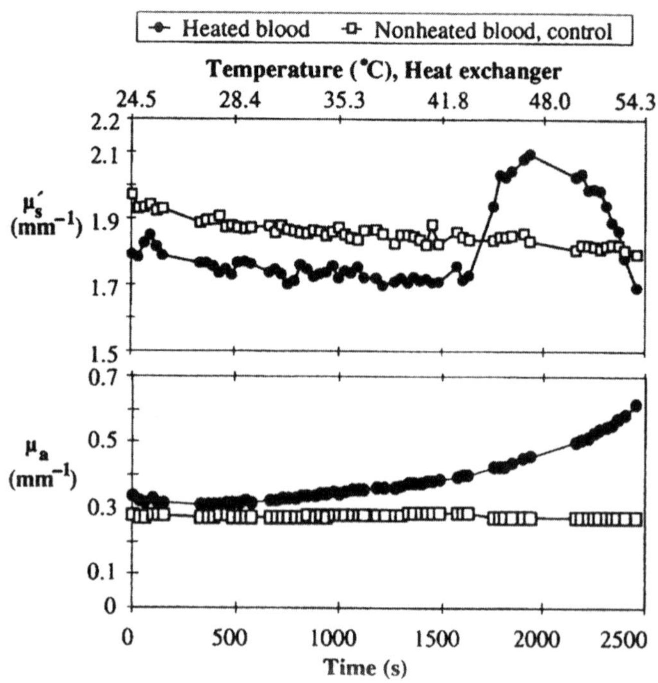

Figure 2.9 The absorption coefficient and the reduced scattering coefficient of nonheated (open squares) and heated (solid circles) whole blood ($H \approx 44\%$; OS = 100%) vs. temperature and time.[32]

determined under different physiological and biochemical conditions. The experiments at 633 nm were performed to study the influence of hematocrit, oxygen saturation, flow velocity, osmolarity, and hemolysis on the optical properties of the suspensions. In addition, the absorption coefficient, the scattering coefficient, and the anisotropy factor (assuming a GKPF with $\alpha = 1$) were determined for the oxygenated and deoxygenated RBC suspensions ($H = 5\%$) under normal physiological conditions (see Fig. 2.10). For the experiments, erythrocytes were separated from blood plasma and white cell fraction, and washed and suspended in PBS (300 mOsmol/L; pH 7.4). The hematocrit was adjusted by diluting the erythrocytes with PBS. By using PBS with different osmolarities, the osmolarity of the RBC suspension was varied. Hemolysis was induced by diluting the suspensions with distilled water. Blood oxygenation and circulation were adjusted and controlled using

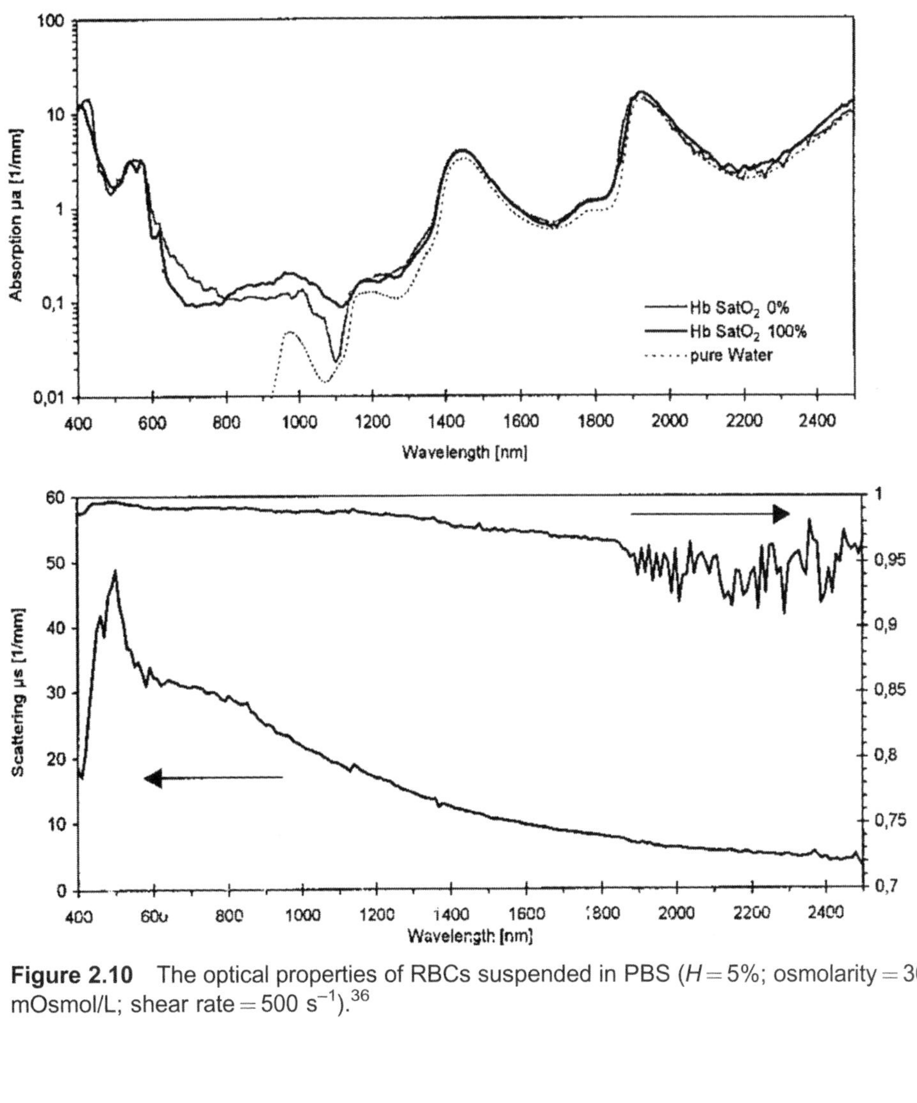

Figure 2.10 The optical properties of RBCs suspended in PBS ($H = 5\%$; osmolarity = 300 mOsmol/L; shear rate = 500 s^{-1}).[36]

an extracorporal circulation unit (Fa. Stoeckert Instrument GmgH). Blood temperature was kept constant at 20 °C. The thickness of the flow-through cuvette was 97 μm.

Analyzing the results of the study, Roggan et al.[36] came to the conclusion that an accurate approximation of the scattering phase function plays an important role in the correct determination of the optical properties of blood. It was found that absorption and scattering increased linearly with hematocrit (for $H < 50\%$). Absorption and scattering decreased slightly with increased shear rate. Among the flow parameters, axial migration was the main factor that influenced the optical properties. Erythrocyte deformation had no impact on the optical properties if RBC volume and hemoglobin content were kept constant. Hemoglobin solutions had a smaller absorption than RBC suspensions with the same concentration of hemoglobin. Obviously, the change in the oxygenation of RBC suspensions induced the expected changes in the absorption coefficient. The scattering coefficient was not affected by the change in erythrocyte oxygenation. The authors mentioned that scattering was not influenced significantly by the RBC membranes. Some of the most essential results of this work are summarized in Table 2.2. The optical properties of the RBCs suspended in PBS ($H = 5\%$; shear rate $= 500\ \text{s}^{-1}$; osmolarity = 300 mOsmol/L) are given in Fig. 2.10.

Work in this direction was continued by Friebel et al.,[84] who evaluated the Gegenbauer kernel (Reynolds–McCormick), Henyey–Greenstein, and modified HGPFs of blood diluted in PBS ($H = 0.84\%$) and undiluted blood ($H = 42.1\%$) under flowing conditions in the wavelength range from 250 to 1100 nm. The authors used a miniature blood circulation setup with a roller pump and specimen chamber constantly supplied with oxygen, nitrogen, and carbon dioxide to mimic real-life conditions. To avoid inhomogeneities, the blood was stirred within the chamber, and laminar flow was provided by a specially designed cuvette. Datasets of integrating sphere measurements and iMC simulations were used to evaluate the effective phase functions. Using the tested functions and iMC algorithms, the authors determined absorption and scattering coefficients, as well as anisotropy factors of the blood specimens. It was determined that the best results were obtained using the Gegenbauer kernel (Reynolds–McCormick) phase function with $\alpha = 1.2$ for hematocrit of 0.84% and $\alpha = 1.7$ for hematocrit of 42.1%. These results are presented in Figs. 2.11 and 2.12, respectively.

Using the same experimental setup and data processing technique, Mienke et al.[81] investigated the dependence of erythrocyte optical parameters on hematocrit. Specimens obtained from a single donor were diluted to concentrations of 0.84%, 4.0%, 5.9%, 8.6%, 17.1%, 25.6%, 33.2%, and 42.1%. The results are summarized in Fig. 2.13. In the same contribution, they evaluated the biological variability of the data, collecting specimens from 10 different donors and determining their optical properties at a fixed $H = 8.6\%$.

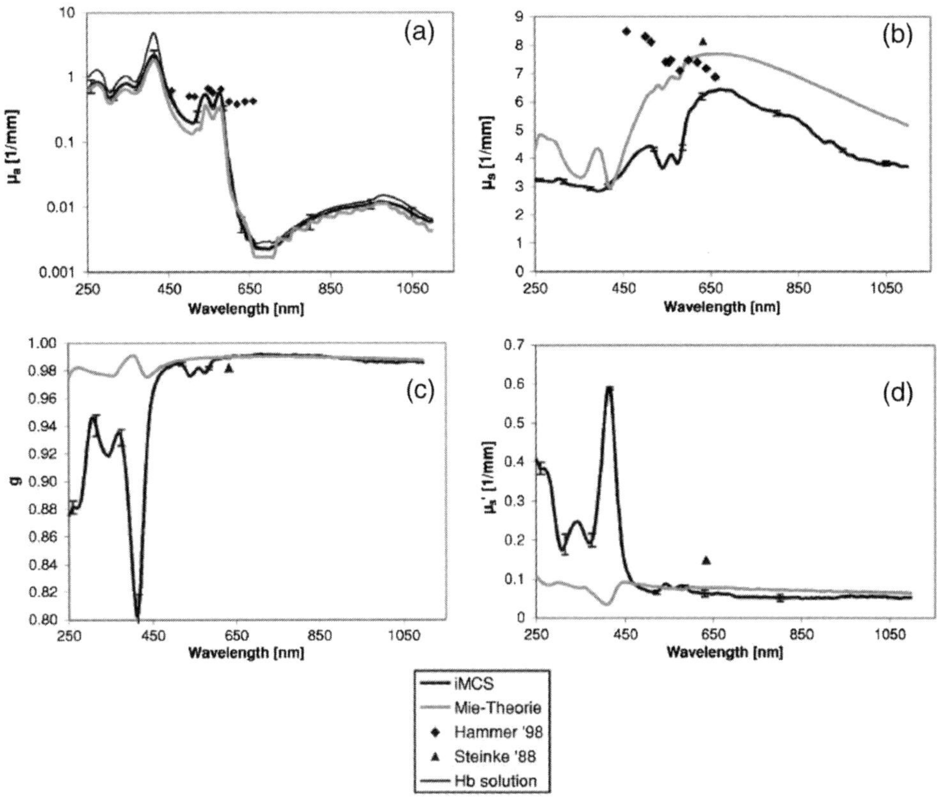

Figure 2.11 Optical properties of flowing RBC (wall shear rate 600 s^{-1}) with $H = 0.84\%$ in the wavelength range from 250 to 1100 nm, compared to absorption coefficient of a hemoglobin solution (0.27 g/L), values calculated by the Mie theory, and the data available from the literature.[84]

The obtained biologically averaged parameters were used as statistical standards to which the optical properties determined for the single specimens were related. The averaged optical properties of RBCs suspended in saline solution and in plasma are shown in Fig. 2.14. The authors showed that substituting saline solution for blood plasma decreases scattering (by 5.5% in the 250–600-nm-wavelength range and by 9.4% in the 600–1100-nm-wavelength range) and absorption (by 3.2% in the 250–600-nm-wavelength range and by 16.6% in the 600–1100-nm-wavelength range). At the same time, due to the decrease in refractive index mismatch between RBCs and the dilution medium, the anisotropy factor slightly increased (by 0.22% between 250 and 600 nm, and by 0.77% between 600 and 1100 nm).

The same research group investigated the influence of oxygen saturation on the optical properties of human RBCs ($H = 33.2\%$) under flow conditions in the wavelength range from 250 to 2000 nm.[82] Integrating sphere measurements and iMC simulations using the Gegenbauer kernel

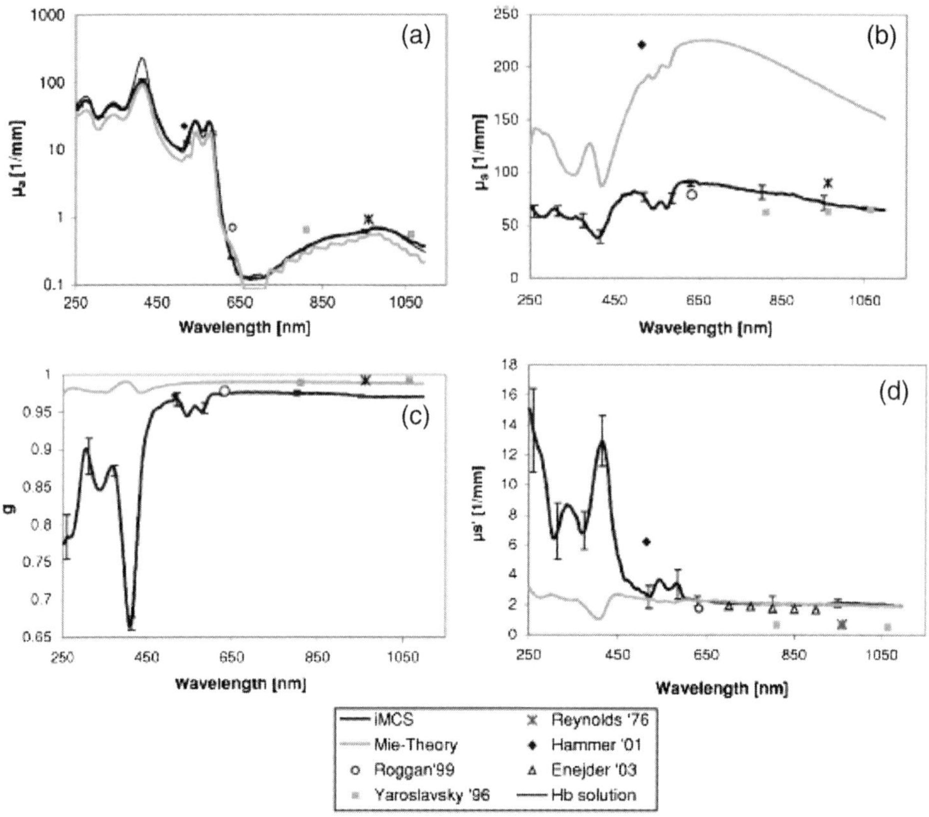

Figure 2.12 Optical properties of flowing RBC (wall shear rate 600 s^{-1}) with $H = 42.1\%$ in the wavelength range from 250 to 1100 nm, compared to the absorption coefficient of a hemoglobin solution (0.27 g/L), values calculated by the Mie theory, and the data available from the literature.[84]

(Reynolds–McCormick) phase function ($\alpha = 1.7$) were carried out for two levels of oxygen saturation (0% and 100%). Samples of human erythrocytes obtained from three donors were separated in two groups. One half of the specimens were constantly supplied with a gas mixture of O_2 (96%) and CO_2 (4%) to keep them fully oxygenated. To ensure complete deoxygenation of the second half, 0.3% $Na_2S_2O_2$ was added to the RBC suspension, and the specimen was then equilibrated with a gas mixture of 96% N_2 and 4% CO_2. The authors showed that the variation in the wavelength-dependent absorption of hemoglobin caused by oxygen saturation affects scattering and anisotropy factors as well. The anisotropy factors of fully oxygenated specimens showed decrease of 4.5% at 250 and 375 nm, and of 12% in the wavelength range from 410 to 430 nm. The oxygen-saturation-dependent decrease in the scattering coefficients reached 15% at 400 nm and was equal to 5% at 570 nm. In the wavelength range from 750 to 850 nm,

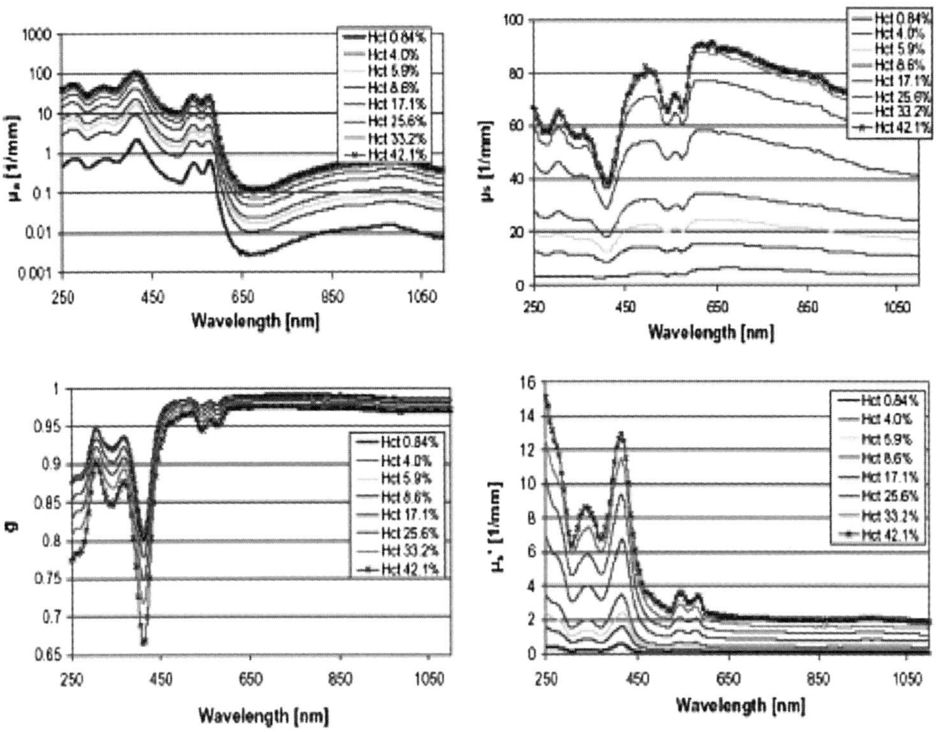

Figure 2.13 Biologically normalized optical coefficients of RBCs in saline solution of blood concentrations varying from 0.84% to 42.1% in the 250-to-1100-nm wavelength range.[81]

scattering decreased by approximately 2%. The results of the study are summarized in Figs. 2.15–2.17.

Steenbergen et al.[83] analyzed the effect of shear rate on the optical properties of the completely oxygenated whole blood. The collimated transmission and the angular distributions of the light intensity were measured at 633 nm for various shear rates (from 50 to 500 s^{-1}) and blood layer thicknesses (from 20 to 100 μm). For shear rates above 150 s^{-1}, the total attenuation coefficient was determined directly from the collimated transmission measurements. An iMC technique was used to determine the anisotropy factor from the angular light-scattering distributions. For the iMC simulations, the Henyey–Greenstein scattering phase function was adopted, the value of the total attenuation coefficient ($\mu_t = 120$ mm^{-1}) was determined from the collimated transmittance measurements, and the values of the absorption coefficient ($\mu_a = 0.7$ mm^{-1} and 1 mm^{-1}) were taken from the literature.[36]

In addition, the authors measured $\bar{\mu}$ for the blood layers with different thicknesses and determined the actual anisotropy factor by extrapolating their results to the layer thickness of zero. For the measurements, blood was

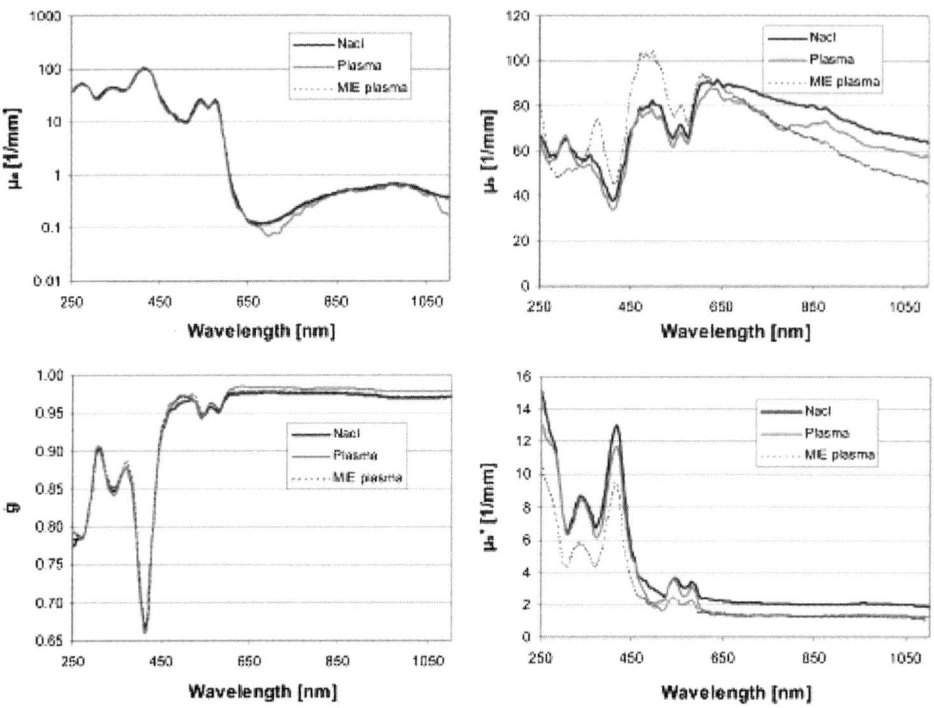

Figure 2.14 Biologically normalized optical coefficients of RBCs suspended in 0.9% PBS and in blood plasma ($H = 42.1\%$).[81]

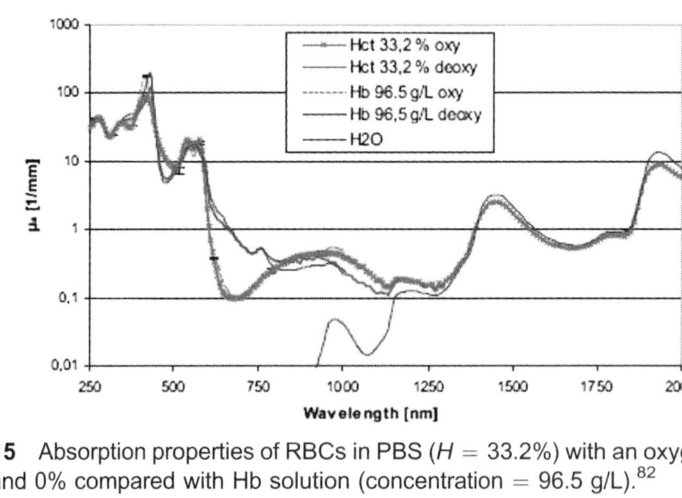

Figure 2.15 Absorption properties of RBCs in PBS ($H = 33.2\%$) with an oxygen saturation of 100% and 0% compared with Hb solution (concentration = 96.5 g/L).[82]

collected, and EDTA was added for anticoagulation. The hematocrit of the investigated blood varied between 49.5% and 50.5%. For data acquisition, the authors assembled an experimental setup that is described in detail in Ref. 83. A continuous anisotropy factor increase (from 0.95 to 0.975) with an increase

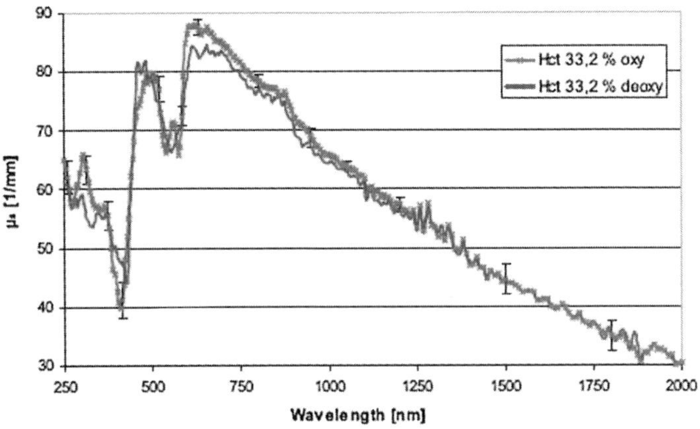

Figure 2.16 Scattering properties of RBCs in PBS ($Hct = 33.2\%$) with an oxygen saturation of 100% and 0%.[82]

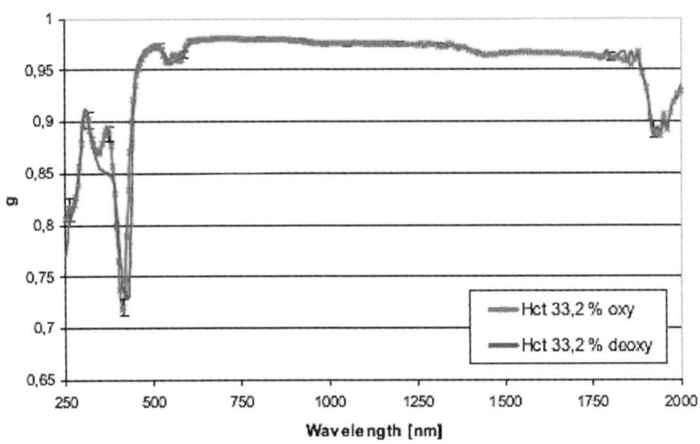

Figure 2.17 Anisotropy factor of RBCs in PBS ($H = 33.2\%$) with an oxygen saturation of 100% and 0%.[82]

in shear rate was reported. The essential results of this investigation are summarized in Table 2.2.

2.7 Practical Relevance of Blood Optics

Optical techniques provide a powerful diagnostic, therapeutic, and surgical tool for various biomedical applications. Rapid development of laser medicine has stimulated interest in the investigation of the optical properties of different biological tissues.[103–106] The efficacy of therapeutic and diagnostic procedures greatly depends on the propagation and fluence rate distribution of optical radiation within tissue. In addition, during medical procedures, the optical characteristics may change, thus altering the penetration depth of laser

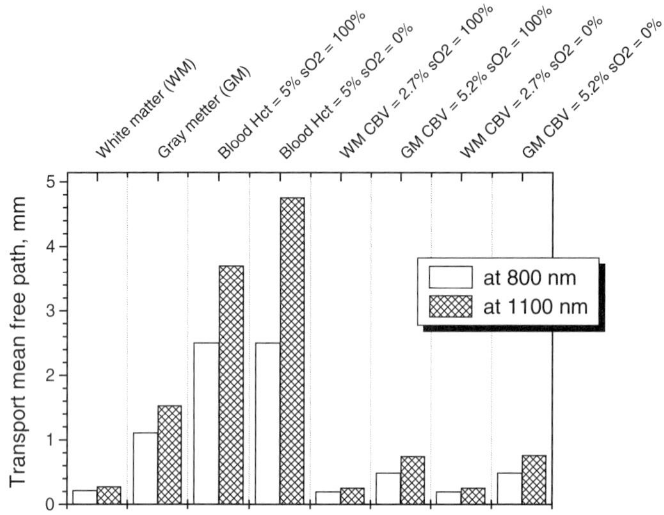

Figure 2.18 Transport mean-free path tmfp [tmfp $= (\mu_a + \mu_s')^{-1}$] at the wavelengths of 800 and 1100 nm for white brain matter (WM), gray brain matter (GM), oxygenated blood, deoxygenated blood, perfused white matter, and perfused gray matter. Based on the results of Refs. 36 and 107. Blood hematocrit of 40% was assumed in computing data for perfused tissues. CBV, cerebral blood volume.

radiation. The optical properties of blood-perfused tissues are significantly affected by tissue blood content. This is caused by two factors: First, the optical properties of the whole blood itself are substantially different from those of soft tissues, and second, whole blood is an extremely turbid medium, with a very short optical mean-free path compared to the majority of bloodless tissues. As a result, the presence of even a small amount of blood greatly changes the process of light propagation in tissues. This point is illustrated by Fig. 2.18, where the transport mean-free path has been calculated for bloodless and blood-containing brain tissues at two wavelengths.

Optical properties of other blood-containing tissues are affected in a similar manner. As a result, the optical response of tissue depends strongly on the presence of blood and on its relevant parameters, such as oxygenation state and hematocrit. Salomatina and Yaroslavsky[108] confirmed this statement in practice for skin and cartilage. They have shown that the impact of hemoglobin on the optical response of tissue significantly varies depending on the tissue state and *ex vivo* storage conditions. In this study, absorption, scattering coefficients, and anisotropy factors of *in vivo*, *ex vivo* fresh, and *ex vivo* frozen-thawed mouse ear tissue were evaluated and compared. It was discovered that spectral changes in the wavelength range dominated by blood absorption occur immediately upon transition of the tissue from the *in vivo* to the *ex vivo* state. Due to the instantaneous deoxygenation of blood, the double oxyhemoglobin peak is replaced by the single deoxyhemoglobin peak

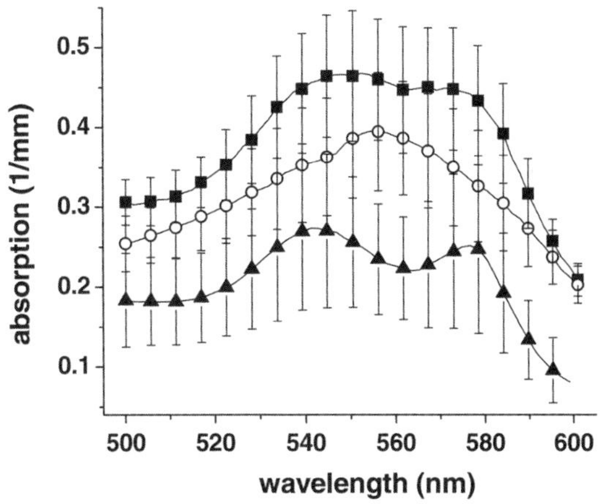

Figure 2.19 Absorption coefficients of mouse ear tissue *in vivo* (squares), *ex vivo* 0-h postmortem (open circles), and *ex vivo* frozen-thawed 73-h postmortem (triangles) in the wavelength range dominated by hemoglobin absorption. Averaged over 10 samples, the bars are standard deviations.

(Fig. 2.19). The absorption of frozen and thawed tissues exhibits a reappearance of the double peak of oxyhemoglobin. This phenomenon is associated with release of the hemoglobin from the hemolyzed erythrocytes.

Table 2.3 presents the differences in the absorption coefficients corresponding to the transition from *in vivo* to *ex vivo* states, obtained at the absorption maxima of hemoglobin (547, 555, 574 nm), lipid (1205 nm), and water (1450 nm). It reveals that the major changes in skin and cartilage optical responses occur due to changes in blood content and state. At the same time, under proper storage conditions, the changes in the optical properties at the wavelengths of the absorption maxima of the other major tissue chromophores, such as water and lipids, are less significant.

Strong hemoglobin absorption peaks in the 500–600-nm-wavelength range allow for *in vivo* noninvasive visualization of blood vessels in human papillary dermis. Since melanin, another major chromophore present in human skin, exhibits absorption that monotonously decreases with increasing wavelength, tuning the imaging wavelength to 570 nm emphasizes blood.[109] Using composite optical properties of blood containing human skin

Table 2.3. Differences in the averaged absorption coefficients obtained *in vivo* and *ex vivo*.

Wavelength (nm)	Difference in μ_a *in vivo* versus *ex vivo* (%)
547	−15
555	−13
574	−18
1205	−4
1450	+5

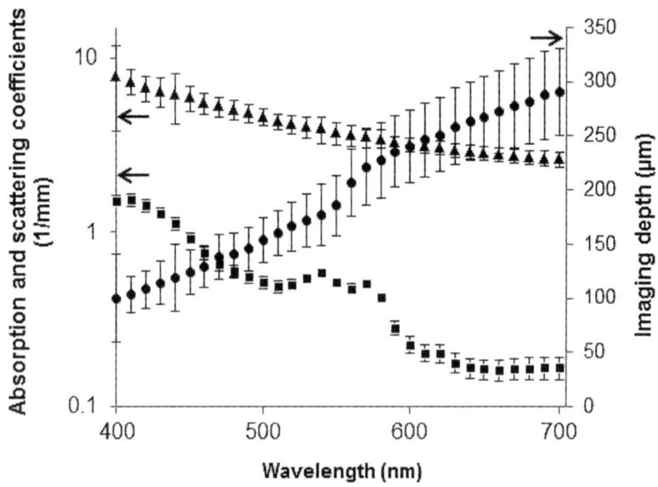

Figure 2.20 Optical properties of human dermis containing 2% oxygenated blood: triangles, scattering coefficients; squares, absorption coefficients; circles, imaging depth.[109]

(Fig. 2.20), Feng et al.[109] estimated that cross-polarized imaging of blood containing human dermis at 570 nm will predominantly register the signal from depths between 0.15 and 0.25 mm, thus making possible noninvasive visualization of the upper plexus.

Representative images of human skin, acquired at 570 nm, are shown in Fig. 2.21. In Fig. 2.21(a), a digital picture of the subjects' forehead is presented. The 570-nm cross-polarized image showing the blood vessels of the upper dermis of an 8.0×4.9 mm^2 region of interest on the subjects' forehead is shown in Fig. 2.21(b). It can be clearly appreciated that the structures overlying the upper blood plexus, with the exception of hair, are not contributing to the image. Further improvements to the clarity of the blood vessel images can be achieved by noticing that at 470 nm, absorption of hemoglobin is decreased as compared to spectral ranges of 410–420 nm and 530–590 nm, whereas the absorption of melanin is still significant. Thus, by subtracting cross-polarized 470 nm from that acquired at 570 nm, we can further improve the contrast of blood and reduce that of melanin-containing hair [Fig. 2.21(c)]. The false color image presented in Fig. 2.21(d) demonstrates that the resolution and contrast of the resulting gray scale image [Fig. 2.21(c)] is sufficient for determining the size of single blood vessels. Spectral analysis of the images acquired at 550–560 nm and 590 or 540 nm may enable determination of blood volume and oxygenation state in the single blood vessels. This capability will allow for determining the important physiological parameters of human dermis as well as for diagnosing a range of skin diseases associated with abnormal vasculature or angiogenesis, e.g., psoriasis and skin cancers.

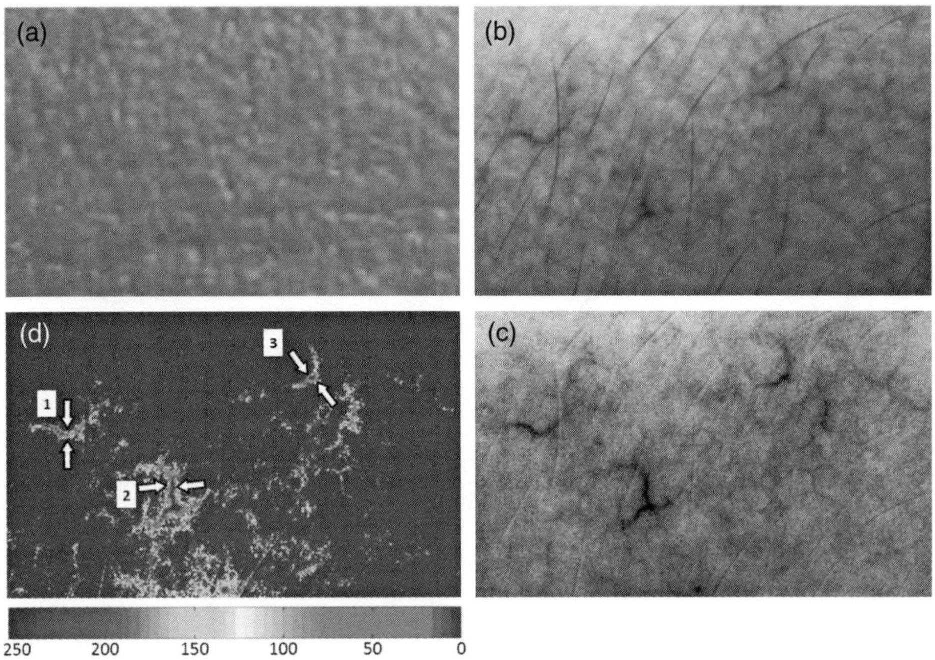

Figure 2.21 *In vivo* images of human skin. Field of view: 8.0 $\times$ 4.9 mm. (a) Digital picture; (b) cross-polarized image acquired at 570 nm emphasizes blood vessels and hair; (c) cross-polarized difference (570–470 nm) image emphasizes blood vessels; (d) cross-polarized difference (570–470 nm) image in false color enables determination of blood vessel sizes. For example, (1) is a 78-μm blood vessel, (2) is a 137-μm blood vessel, and (3) is a 98-μm blood vessel. (See color plates.)

Optical signals' sensitivity to blood content and state opens wide possibilities for optical diagnostics but also makes the dosimetry of light in tissues a more difficult task. For example, much research has been devoted to the development of optical diffuse tomography, which is a promising monitoring and imaging modality. It is known that near-infrared light in the spectroscopic transmission window between 700 and 1100 nm is predominantly absorbed by oxy- and deoxyhemoglobin. This offers a possibility of obtaining physiologically relevant information on the total hemoglobin concentration and oxygen saturation of tissues.[110] But it should be noted that quantitative information could be retrieved only if the optical properties of blood are known. Among the therapeutic procedures, laser-induced interstitial heating treatment (LITT) of deep tumors is one of the least invasive and most promising techniques. In this case, besides the knowledge of the optical properties of tumor tissue and the surrounding substances, knowledge of the blood content, its perfusion rate, and its optical properties is essential for therapy planning and for exact dosimeter. In addition, knowledge of the optical properties of tissues and blood allows for the determination of the most effective treatment wavelength, i.e., where the laser light penetration depth is maximal. Thus, an explicit account of blood content is required when modeling laser–tissue interaction, e.g., the planning of clinical procedures such as LITT or photodynamic therapy.[115] Because scattering in blood is higher than that of all biological tissues and liquids, it could be advantageous to utilize biocompatible liquids and to use optical clearing phenomena to reduce the scattering attenuation of the optical signals. This approach has been described in detail in several recent publications[111–114] (see also Chapter 1 in this volume and Chapters 3 and 10 in Volume 2). This emphasizes the need for an explicit account of blood content in the modeling of laser–tissue interaction—e.g., the planning of clinical procedures such as LITT or photodynamic therapy.[115]

The choice of wavelengths for pulse oximetry presents another example of a medical problem where knowledge of blood optical properties is of major importance. Pulse oximeters are widely used in different areas of medicine for monitoring blood oxygen saturation. Conventional pulse oximeters are used to monitor blood oxygen saturation in adults and children, whose saturation rarely falls below 70%. On the other hand, it was noticed that when the saturation level falls below 70%, conventional blood oximeters are less accurate.[116–118] It is known that typical oxygen saturation in the fetus is between 20% and 75%.[119,120] Thus, to successfully apply oximetry to monitoring fetal blood saturation during labor and delivery, it is necessary to increase the accuracy of the device for low saturation levels. There are two main criteria for choosing the wavelengths in pulse oximetry. First, for at least one of the selected wavelengths, the difference in the absorption of Hb and HbO_2 should be significant. The need to satisfy the first criterion was extensively discussed in Section 2.3. Second, the penetration depth $\delta_{eff} = 1/\sqrt{3\mu_a[\mu_a + \mu_s(1-\bar{\mu})]}$ of light into blood and the surrounding tissue (the sampling volume) should be the same for both wavelengths. When the second criterion is satisfied, differences in the signals measured at two wavelengths induced by the differences of the sampling volumes are minimized. Consequently, differences in the measured signals at two wavelengths will be defined by the difference in the absorptive behavior of oxy- and deoxyhemoglobin at these wavelengths, thus ensuring correct determination of blood oxygen saturation. Mannheimer et al.[121] addressed the problem that occurs when measuring blood with low oxygen saturation. Severinghaus and Koh[118] came to the conclusion that the wavelengths of 735 and 890 nm are more suitable for measuring low oxygen saturation, while the typical blood oximeters utilize wavelengths around 660 and 890 nm. At the wavelength of 660 nm, blood with low oxygen saturation has a higher absorption coefficient than blood with high saturation (see Fig. 2.2); accordingly, the penetration depth of light into deoxygenated blood is significantly lower. To match a penetration depth of light into blood with low oxygen saturation, it is necessary to use a longer wavelength, e.g., 735 nm, where absorption of deoxyhemoglobin is weaker, and the light penetration depths for the two wavelengths of 735 and 890 nm are similar.

Another emerging modality that relies on knowledge of blood's optical properties is optical monitoring of neuronal activity. This imaging technique makes use of the fact that during neuronal activation, local changes in absorption and scattering occur due to changes in oxygenation state and volume of hemoglobin.[122] These changes happen because neuronal activation requires increased oxygen uptake. Consequently, if tissue during neuronal activation is illuminated by visible or near-infrared light, the remitted optical signal varies in space in time. The areas where the optical variations are detected correspond to the activated cortex areas. The changes in tissue oxygenation, blood flow, and blood volume evoked by neuronal activity are collectively termed the hemodynamic response.

Optical monitoring of neuronal activity was successfully applied to solve a variety of basic science and clinical problems. In rodents, optical hemodynamic responses combined with electrophysiological techniques have been used to define the capacities and limitations of neurovascular coupling.[123–128] This information is critical to the interpretation of functional MRI (fMRI), a brain imaging technique with rapidly increasing applications. Chronic experiments have allowed investigation of brain plasticity in response to varying afferent activity.[126,127] Studies in cats have provided insight into the fine-scale functional architecture of the visual cortex.[11] In nonhuman primates, subtleties in the somatotopic organization of the primary sensory cortex have been revealed.[129–132] Intraoperative optical imaging in human patients undergoing neurosurgical procedures has shed light on the propagation of seizures[133,134] and delineated regions of eloquent brain that must be preserved during an operation.[135,136]

In conclusion, it is probably worth noting that there is hardly an area in medicine where knowledge of blood's optical properties will not bring benefits improving laser and light-based diagnostic and treatment procedures.

References

1. F. Slare, "Some effects of the air upon blood explained by an experiment of the change of colour in a clear liquid upon the admission of air," *Philadelphia Trans.* **2**, 235 (1700).
2. F. Hoppe-Seyler, "Ueber das Verhalten des Blutfarbstoffes im Spectrum des Sonnenlichtes," *Arch. Pathol. Anat. Physiol. Klin. Med.* **23**, 446 (1862).
3. J. L. Sore, "Recherches sur l'absorption des rayons ultra-violet par diverses substances," *Arch. Sci. Phys. Nat.* **10**, 429 (1883).
4. K. Vierordt, "Die Anwendung des Spectralapparates zur Photometrie der Absorptionspectren un zur Quantitativen Chemischen Analyse," *Laupp'schen Buchhandlung*, Tuebingen (1876).

5. K. Vierordt, "Die Quantitative Spectralanalyse in ihrer Anwendung auf Physiologie, Physik, Chemie und Technologie," Laupp'schen Buchhandlung, Tuebingen (1876).
6. D. L. Drabkin and J. H. Austin, "Spectrophotometric studies: I. Spectrophotometric constants for common hemoglobin derivatives in human, dog and rabbit blood," *J. Biol. Chem.* **98**, 719 (1932).
7. D. L. Drabkin and J. H. Austin, "Spectrophotometric studies: V. A technique for analysis of undiluted blood and concentrated hemoglobin solutions," *J. Biol. Chem.* **112**, 105 (1935–1936).
8. E. Gordy and L. Drabkin, "Determination of the oxygen saturation of blood by a simplified technique applicable to standard equipment," *J. Biol. Chem.* **227**, 285–299 (1957).
9. R. A. MacRae, A. McClure, and P. Latimer, "Spectral transmission and scattering properties of red blood cells," *J. Opt. Soc. Am.* **51**(12), 1366–1372 (1961).
10. R. J. Zdrojkowski and R. L. Longini, "Optical transmission through whole blood illuminated with high collimated light," *J. Opt. Soc. Am.* **59**(8), 898–903 (1969).
11. R. J. Zdrojkowski and N. R. Pisharoty, "Optical transmission and reflection by blood," *IEEE Trans. Biomed. Eng.* **BME–17**, 122–128 (1970).
12. C. C. Johnson, "Optical diffusion in blood," *IEEE Trans. Biomed. Eng.* **BME–17**, 129–133 (1970).
13. O. W. van Assendelft, "Spectrophotometry of hemoglobin derivatives," Roal Vangorcum. Assen. (1970).
14. L. Reynolds, C. Johnson, and A. Ishimaru, "Diffuse reflectance from a finite blood medium: applications to the modeling of fiber optic catheters," *Appl. Opt.* **15**, 2059–2067 (1976).
15. R. N. Pittman, "*In vivo* photometric analysis of hemoglobin," *Ann. Biomed. Eng.* **14**, 119–137 (1986).
16. S. T. Flock, B. C. Wilson, and M. S. Patterson, "Total attenuation coefficient and scattering phase function of tissues and phantom materials at 633 nm," *Med. Phys.* **14**, 835–841 (1987).
17. J. M. Steinke and A. P. Shepherd, "Diffuse reflectance of whole blood: model for a diverging light beam," *IEEE Trans. Biomed. Eng.* **BME–34**, 826–834 (1987).
18. J. M. Steinke and A. P. Shephard, "Diffusion model of the optical absorbance of whole blood," *J. Opt. Soc. Am. A* **5**, 813–822 (1988).
19. J. M. Steinke and A. P. Shepherd, "Effect of temperature on optical absorbance spectra of oxy-, carboxy, and deoxyhemoglobin," *Clin. Chem.* **38**(7), 1360–1364 (1992).
20. M. Cope, "The development of a near infrared spectroscopy system and its application for noninvasive monitoring of cerebral blood and tissue

oxygenation in the newborn infant," PhD thesis, Univ. College London (1991).
21. F. F. Jöbsis, "Noninvasive, infrared monitoring of cerebral and myocardial oxygen sufficiency and circulatory parameters," *Science* **198**, 1264–1267 (1977).
22. B. Chance and R. Alfano, "Optical tomography, photon migration and model media: theory, human studies, and instrumentation," *Proc. SPIE* **2389**, 894 (1995).
23. R. Richards-Kortum and E. Sevick-Muraca, "Quantitative optical spectroscopy for diagnostics," *Annu. Rev. Phys. Chem.* **47**, 555–606 (1996).
24. J. J. Schuitmaker, P. Baas, H. L. van Leengoed, F. W. van der Meulen, W. M. Star, and N. van Zandwijk, "Photodynamic therapy: a promising new modality for the treatment of cancer," *J. Photochem. Photobiol. B* **34**, 3–12 (1996).
25. G. Müller and A. Roggan, *Laser-Induced Interstitial Thermotherapy*, SPIE Press Vol. **PM25** (1995).
26. V. Twersky, "Absorption and multiple scattering by biological suspensions," *J. Opt. Soc. Am.* **60**, 1084–1093 (1970).
27. G. D. Pedersen, N. J. McCormick, and L. O. Reynolds, "Transport calculations for light scattering in blood," *Biophys. J.* **16**, 199–207 (1976).
28. L. O. Reynolds and N. J. McCormick, "Approximate two-parameter phase function for light scattering," *J. Opt. Soc. Am.* **70**, 1206–1212 (1980).
29. J. M. Steinke and A. P. Shepherd, "Comparison of Mie theory and the light scattering of red blood cells," *Appl. Opt.* **27**, 4027–4033 (1988).
30. W. F. Cheong, S. A. Prahl, and A. J. Welch, "A review of the optical properties of biological tissues," *IEEE J. Quantum Electron.* **26**, 2166–2185 (1990).
31. A. N. Yaroslavsky, I. V. Yaroslavsky, T. Goldbach, and H.-J. Schwarzmaier, "Optical properties of blood in the near-infrared spectral range," *Proc. SPIE* **2678**, 314–324 (1996).
32. A. M. K. Nilsson, G. W. Lucassen, W. Verkruysse, S. Andersson-Engels, and M. J. C. van Gemert, "Changes in optical properties of human whole blood in vitro due to slow heating," *Photochem. Photobiol.* **65**(2), 366–373 (1997).
33. A. N. Yaroslavsky, I. V. Yaroslavsky, T. Goldbach, and H.-J. Schwarzmaier, "Different phase function approximations to determine optical properties of blood: a comparison," *Proc. SPIE* **2982**, 324–330 (1997).
34. A. G. Borovoi and E. I. Naats, "Scattering of light by red blood cell," *J. Biomed. Opt.* **3**, 364–372 (1998).

35. M. Hammer, D. Schweitzer, B. Michel, E. Thamm, and A. Kolb, "Single scattering by red blood cells," *Appl. Opt.* **37**(31), 7410–7418 (1998).
36. A. Roggan, M. Friebel, K. Doershel, A. Hahn, and G. Mueller, "Optical properties of circulating human blood in the wavelength range 400–500 nm," *J. Biomed. Opt.* **4**(1), 36–46 (1999).
37. A. N. Yaroslavsky, I. V. Yaroslavsky, T. Goldbach, and H.-J. Schwarzmaier, "Influence of the scattering phase function approximation on the optical properties of blood determined from the integrating sphere measurements," *J. Biomed. Opt.* **4**(1), 47–53 (1999).
38. J. Wyatt, M. Cope, D. T. Delpy, S. Wray, and E. O. R. Reynolds, "Quantification of cerebral oxygenation and haemodynamics in sick newborn infants by near infrared spectrophotometry," *Lancet* **2**, 1063–1066 (1986).
39. J. S. Wyatt, M. Cope, D. T. Delpy, C. E. Richardson, A. D. Edwards, S. Wray, and E. O. R. Reynolds, "Quantification of cerebral blood volume in newborn infants by near infrared spectroscopy," *J. Appl. Physiol.* **68**, 1086–1091 (1990).
40. M. Tamura, "Non-invasive monitoring of brain oxygen metabolism during cardiopulmonary bypass by near infrared spectroscopy," *Jpn. Circ. J.* **55**, 330–335 (1991).
41. A. J. du Plessis, J. Newburger, R. A. Jones, P. Hickey, H. Naruse, M. Tsuji, A. Walsh, G. Walter, D. Wypij, and J. J. Volpe, "Cerebral oxygen supply and utilisation during infant cardiac surgery," *Ann. Neurol.* **37**, 488–497 (1995).
42. M. M. Wintrobe, *Clinical Hematology*, Lea and Febiger, Philadelphia (1981).
43. J. A. Lamasso, "Error in hematocrit value produced by excessive ethylenediamine tetra-acetate," *Am. J. Clin. Pathol.* **44**, 109 (1965).
44. R. Barer and S. Joseph, "Refractometry of living cells," *Q. J. Microsc. Sci.* **95**, 399–406, 1954.
45. A. Ishimaru, *Wave Propagation and Scattering in Random Media, Vol. 1*, Academic Press, New York, pp. 66 (1978).
46. D. H. Tycko, M. H. Metz, E. A. Epstein, and A. Grinbaum, "Flow-cytometric light scattering measurement of red blood cell volume and hemoglobin concentration," *Appl. Opt.* **24**, 1355–1365 (1985).
47. P. B. Canham and A. C. Burton, "Distribution of size and shape in populations of normal human red cells," *Circ. Res.* **22**, 405 (1968).
48. M. M. Wintrobe, "Variations in the size and hemoglobin content of erythrocytes in the blood of various vertebrates," *Folia Hematol.* **51**, 32 (1933).
49. R. Skalak and P.-I. Branemark, "Deformation of red blood cells in capillaries," *Science* **164**, 717 (1969).

50. M. M. Frojmovic and R. Panjwani, "Geometry of normal mammalian platelets by quantitative microscopic studies," *Biophys. J.* **16**, 1071 (1976).
51. S. Wray, M. Cope, D. Delpy, J. S. Wyatt, and E. O. R. Reynolds, *Biochem. Biophys. Acta* **933**, 184 (1988).
52. J. T. Kuenstner and K. H. Norris, "Spectrophotometry of human hemoglobin in the near infrared region from 100 to 2500 nm," *J. Near Infrared Spectrosc.* **2**, 59–65 (1994).
53. W. G. Zijstra, A. Buursma, and W. P. Meeuwsen-van der Roest, "Absorption spectra of human fetal and adult oxyhemoglobin, deoxyhemoglobin, carboxyhemoglobin, and methemoglobin," *Clin. Chem.* **37**(9), 1633–1638 (1991).
54. G. M. Hale and M. R. Querry, "Optical constants of water in the 200-nm to 200-μm wavelength region," *Appl. Opt.* **12**(3), 555–563 (1973).
55. A. A. Shvalov, J. T. Soini, A. V. Chernyshev, P. A. Tarasov, E. Soini, and V. P. Maltsev, "Light-scattering properties of individual erythrocytes," *Appl. Opt.* **38**(1), 230–235 (1999).
56. S. V. Tsinopoulos and D. Polyzos, "Scattering of He-Ne laser light by an average-sized red blood cell," *Appl. Opt.* **38**(25), 5499–5510 (1999).
57. G. J. Streekstra, A. G. Hoekstra, E. J. Nijhof, and R. M. Heethaar, "Light scattering by red blood cells in ektacytometry: Fraunhofer versus anomalous diffraction," *Appl. Opt.* **32**, 2266–2272 (1993).
58. A. M. K. Nilsson, P. Alsholm, A. Karlsson, and S. Andersson-Engels, "T-matrix computations of light scattering by red blood cells," *Appl. Opt.* **37**, 2735–2748 (1998).
59. P. Mazeron, S. Muller, and H. El. Azouri, "On intensity reinforcement in small-angle light scattering patterns of erythrocytes under shear," *Eur. Biophys. J.* **26**, 247–252 (1997).
60. P. Barber and C. Yeh, "Scattering of electromagnetic waves by arbitrary shaped dielectric bodies," *Appl. Opt.* **14**, 2864–2872 (1975).
61. H. C. Van de Hulst, *Light Scattering by Small Particles*, Dover, New York (1981).
62. J. M. Schmitt, J. D. Meindl, and F. G. Mihm, "An integrated circuit based optical sensor for *in vivo* measurement of blood oxygenation," *IEEE Trans. Biomed. Eng.* **33**, 98–107 (1986).
63. L. G. Henyey and J. L. Greenstein "Diffuse radiation in the galaxy," *Astrophys. J.* **93**, 70 (1941).
64. M. Born and E. Wolf, *Principles of Optics*, Macmillan, New York (1964).
65. N. G. Khlebtsov, "Role of multiple scattering in turbidimetric investigations of dispersed systems," *J. Appl. Spectrosc.* **40**, 243–247 (1984).
66. C. F. Bohren and D. R. Huffman, *Absorption and Scattering of Light by Small Particles*, Wiley, New York (1983).

67. J. A. Stratton, *Electromagnetic Theory*, McGraw-Hill, New York (1941).
68. D. S. Jones, *The Theory of Electromagnetism*, Pergamon, London (1964).
69. A. Borovoi, E. Naats, U. Oppel, and I. Grishin, "Shape characterization of a large nonspherical particle by use of its Fraunhofer diffraction pattern," *Appl. Opt.* **39**(12), 1989–1997 (2000).
70. P. F. Mullaney and P. N. Dean, "The small angle light scattering of biological cells," *Biophys. J.* **10**, 764–772 (1970).
71. W. Groner, N. Mohandas, and M. Bessis, "New optical technique for measuring erythrocyte deformability with ektacytometer," *Clin. Chem.* **26**, 1435–1442 (1980).
72. P. C. Waterman, "Symmetry, unitarity, and geometry in electromagnetic scattering," *Phys. Rev. D* **3**, 825–839 (1971).
73. S. A. Schelkunoff, *Electromagnetic Waves*, D. von Nostrand, New York (1943).
74. W. Magnus, F. Oberhettinger, and R. Soni, *Formulas and Theorems for the Special Functions of Mathematical Physics*, Springer-Verlag, New York (1966).
75. S. V. Tsinopoulos, S. E. Kattis, and D. Polyzos, "Three-dimensional boundary element analysis of electromagnetic wave scattering by penetrable bodies," *Comput. Mech.* **21**, 306–315 (1998).
76. S. V. Tsinopoulos, S. E. Kattis, and D. Polyzos, "An advanced BE/FFT methodology for solving electromagnetic wave scattering problems with axisymmetric dielectric particles," *Eng. Anal. Boundary Elem.* **23**, 155–165 (1999).
77. P. W. Barber and S. C. Hill, *Light Scattering by Particles: Computational Methods*, World Scientific, Singapore (1990).
78. J. M. Steinke and A. P. Shepherd, "Role of light scattering in whole blood oximetry," *IEEE Trans. Biomed. Eng.* **33**, 294–301 (1986).
79. J. E. Hansen and L. D. Travis, "Light scattering in planetary atmospheres," *Space Sci. Rev.* **16**, 525–610 (1974).
80. H. C. van de Hulst and M. M. Davis, *Proc. Konincl. Nederl. Akad. Wet.* **B64**, 220 (1961).
81. M. Meinke, G. Müller, J. Helfmann, and M. Friebel, "Empirical model functions to calculate hematocrit-dependent optical properties of human blood," *Appl. Opt.* **46**(10), 1742–1753 (2007).
82. M. Friebel, J. Helfmann, U. Netz, and M. Meinke, "Influence of oxygen saturation on the optical scattering properties of human red blood cells in the spectral range 250 to 2,000 nm," *J. Biomed. Opt.* **14**(3), 034001 (2009).
83. W. Steenbergen, R. Kolkman, and F. de Mul, "Light-scattering properties of undiluted human blood subjected to simple shear," *J. Opt. Soc. Am. A* **16** (12), 2959–2967 (1999).

84. M. Friebel, A. Roggan, G. Müller, and M. Meinke, "Determination of optical properties of human blood in the spectral range 250 to 1100 nm using Monte Carlo simulations with hematocrit-dependent effective scattering phase functions," *J. Biomed. Opt.* **11**(3), 34021 (2006).
85. A. N. Yaroslavsky, A. Vervoorts, A. V. Priezzhev, I. V. Yaroslavsky, J. G. Moser, and H.-J. Schwarzmaier, "Can tumor cell suspension serve as an optical model of tumor tissue in situ?" *Proc. SPIE* **3565**, 165–173 (1999).
86. I. V. Yaroslavsky, A. N. Yaroslavsky, T. Goldbach, and H.-J. Schwarzmaier, "Inverse hybrid technique for the determination of the optical properties of turbid media," *Appl. Opt.* **35**, 6797–6809 (1996).
87. A. N. Yaroslavsky, I. V. Yaroslavsky, and H.-J. Schwarzmaier, "Small-angle approximation to determine radiance distribution of a finite beam propagating through turbid medium," *Proc. SPIE* **3195**, 110–120 (1998).
88. S. Chandrasekhar, *Radiative Transfer*, Oxford Univ. Press, Oxford (1960).
89. J. Kim and J. C. Lin, "Successive order scattering transport approximation for laser light propagation in whole blood medium," *IEEE Trans. Biomed. Eng.* **45**(4), 505–510 (1998).
90. J. E. Dennis, Jr. and R. B. Schnabel, *Numerical Methods for Unconstrained Optimization and Nonlinear Equations*, Prentice-Hall, Englewood Cliffs, New Jersey (1983).
91. P. Kubelka and F. Munk, "Ein Beitrag zur Optik der Farbanstriche," *Z. Tech. Phys.* **12**, 593–601 (1931).
92. J. H. Joseph, W. J. Wiscombe, and J. A. Weinman, "The δ-Eddington approximation of radiative flux transfer," *J. Atmos. Sci.* **33**, 2452–2459 (1976).
93. M. Keijzer, S. L. Jacques, S. A. Prahl, and A. J. Welch, "Light distribution in artery tissue: Monte Carlo simulations for finite diameter laser beams," *Laser Surg. Med.* **9**, 148–154 (1989).
94. I. V. Yaroslavsky and V. V. Tuchin, "Light propagation in multilayer scattering media: modeling by the Monte Carlo method," *Opt. Spectrosc.* **72**, 505–509 (1992).
95. S. L. Jacques and L. Wang, "Monte Carlo modeling of light transport in tissue," *Optical-Thermal Response of Laser-Irradiated Tissue*, A. J. Welch and M. J. C. van Gemert, Eds., Plenum, New York (1995).
96. A. N. Yaroslavsky, "Spectroscopic investigations of biotissues and cell suspensions applied to the problems of laser diagnostic and therapy," (in Russian), PhD dissertation, Saratov State Univ., Russia (1999).
97. G. E. Forsythe, M. A. Malkolm, and C. B. Moler, *Computer Methods for Mathematical Computations*, Prentice-Hall, Englewood Cliffs, New Jersey (1977).

98. S. A. Prahl, "Light transport in tissue," PhD dissertation, Univ. of Texas at Austin (1988).
99. I. P. Grant and G. E. Hunt, "Solution of radiative transfer problems using the invariant S_n method," *Mon. Not. R. Astron. Soc.* **141**, 27–41 (1968).
100. W. M. Irvine, "Multiple scattering by large particles," *Astrophys. J.* **142**, 1463–1475 (1965).
101. H. C. van de Hulst, *Multiple Light Scattering*, Academic Press, New York (1980).
102. J. W. Pickering, C. J. M. Moes, H. J. C. M. Sterenborg, S. A. Prahl, and M. J. C. van Gemert, "Two integrating spheres with an intervening scattering sample," *J. Opt. Soc. Am. A* **9**, 621–631 (1992).
103. J. W. Pickering, S. A. Prahl, N. van Wieringen, J. F. Beek, H. J. C. M. Sterenborg, and M. J. C. van Gemert, "Double-integrating-sphere system for measuring the optical properties of tissue," *Appl. Opt.* **32**, 399–410 (1993).
104. R. Pierce, "An experimental determination of the average scattering and absorption cross sections of human red blood cells for near infrared light," MS thesis, Electrical Engineering Department, Univ. of Washington (1972).
105. L. Reynolds, J. Molcho, C. Johnson, and A. Ishimaru, "Optical cross sections of human erythrocytes," in *Proc. of 27th Annual Conf. on Eng. in Med. and Biol.* Vol. **16** (1974).
106. A. Roggan, K. Dörschel, O. Minet, D. Wolf, and G. Müller, "The optical properties of biological tissue in the near infrared spectral range: review and measurements," in *Laser-Induced Interstitial Thermotherapy*, G. Müller and A. Roggan, Eds., Vol. **PM25**, SPIE Press, pp. 10–45 (1995).
107. A. N. Yaroslavsky, P. C. Schulze, I. V. Yaroslavsky, R. Schober, F. Ulrich, and H.-J. Schwarzmaier, "Optical properties of selected native and coagulated human brain tissues *in vitro* in the visible and near infrared spectral range," *Phys. Med. Biol.* **47**, 2059–2073 (2002).
108. E. Salomatina and A. N. Yaroslavsky, "Evaluation of the *in vivo* and *ex vivo* optical properties in a mouse ear model," *Phys. Med. Biol.* **53**, 2797–2807 (2008)
109. X. Feng, R. Patel, and A. N. Yaroslavsky, "Wavelength optimized cross-polarized wide-field imaging for noninvasive and rapid evaluation of dermal structures," *J. Biophotonics* **8**(4), 324–331 (2015).
110. M. A. Franceschini, E. Gratton, and S. Fantini, "Noninvasive optical method of measuring tissue and arterial saturation: an application to absolute pulse oximetry of the brain," *Opt. Lett.* **24**(12), 829–831 (1999).
111. V. V. Tuchin, *Optical Clearing of Tissues and Blood*, Vol. **PM154**, SPIE Press, Bellingham, Washington, DC (2005).

112. V. V. Tuchin, Ed., *Handbook of Optical Sensing of Glucose in Biological Fluids and Tissues*, CRC Press, Taylor & Francis Group, London (2009).
113. D. Zhu, K. V. Larin, Q. Luo, and V. V. Tuchin, "Recent progress in tissue optical clearing," *Laser Photonics Rev.* **7**(5), 732–757 (2013).
114. V. V. Tuchin, *Chapter 9 in Tissue Optics: Light Scattering Methods and Instruments for Medical Diagnosis*, 3rd ed., Vol. **PM254**, SPIE Press, Bellingham, Washington, DC (2015).
115. J. J. Schuitmaker, P. Baas, H. L. van Leengoed, F. W. van der Meulen, W. M. Star, and N. van Zandwijk, "Photodynamic therapy: a promising new modality for the treatment of cancer," *J. Photochem. Photobiol. B* **34**, 3–12 (1996).
116. M. Sendak, A. Harris, and R. Donham, "Accuracy of pulse oximetry during arterial oxyhemoglobin desaturation in dogs," *Anesthesiology.* **68**, 111–114 (1988).
117. J. W. Severinghaus, K. Naifeh, and S. Koh, "Errors in 14 pulse oximeters during profound hypoxia," *J. Clin. Monit.* **5**, 72–81 (1989).
118. J. W. Severinghaus and S. O. Koh, "Effect of anemia on pulse oximeter accuracy at low saturation," *J. Clin. Monit.* **6**, 85–88 (1990).
119. G. Dildy, P. van den Berg, M. Katz, S. Clark, H. Jongsma, J. Nijhuis, and C. Loucks, "Intrapartum fetal pulse oximetry: fetal oxygen saturation trends during labor and relation to delivery outcome," *Am. J. OB/GYN* **171**, 679–84 (1994).
120. T. Eskes, H. Jongsma, and P. Houx, "Percentiles for gas values in human umbilical cord blood," *Eur. J. OB Reprod. Biol.* **14**, 341–346 (1983).
121. P. D. Mannheimer, J. R. Casciani, and M. E. Fein, "Wavelength selection for low-saturation blood oximetry," *IEEE Trans. Biomed. Eng.* **44**(3), 148–158 (1997).
122. L. Ostergaard, D. F. Smith, P. Vestergaard-Poulsen, S. B. Hansen, A. D. Giedde, and C. Gyldensted, "Absolute cerebral blood flow and blood volume measured by magnetic resonance imaging bolus tracking: comparison with positron emission tomography," *J. Cereb. Blood Flow Metab.* **18**, 425–432 (1998).
123. T. Bonhoeffer and A. Grinvald, "Optical imaging based on intrinsic signals: the methodology," in *Brain Mapping: The Methods*, A. Toga and J. C. Maziotta, Eds., Academic Press, San Diego, California, pp. 55–97 (1996).
124. H. S. Orbach, L. B. Cohen, and A. Grinvald, "Optical mapping of electrical activity in rat somatosensory and visual cortex," *J. Neurosci.* **5**, 1886–1195 (1985).
125. B. A. MacVicar and D. Hochman, "Imaging of synaptically evoked intrinsic optical signals in hippocampal slices," *J. Neurosci.* **11**, 1458–1469 (1991).

126. M. Jones, J. Berwick, D. Johnston, and J. Mayhew, "Concurrent optical imaging spectroscopy and laser-Doppler flowmetry: the relationship between blood flow, oxygenation, and volume in rodent barrel cortex," *Neuroimage* **13**(6), 1002–1015 (2001).
127. S. A. Sheth, M. Nemoto, M. Guiou, M. Walker, N. Pouratian, and N. Hageman et al., "Columnar specificity of microvascular oxygenation and volume responses: implications for functional brain mapping," *J. Neurosci.* **24**(3), 634–641 (2004).
128. S. A. Sheth, M. Nemoto, M. Guiou, M. Walker, and A. W. Toga, "Spatiotemporal evolution of functional hemodynamic changes and their relationship to neuronal activity," *J. Cereb. Blood. Flow Metab.* **2**, 2 (2005).
129. J. Sharma, A. Angelucci, and M. Sur, "Induction of visual orientation modules in auditory cortex," *Nature* **404**(6780), 841–847 (2000).
130. D. B. Polley, C. H. Chen-Bee, and R. D. Frostig, "Two directions of plasticity in the sensory-deprived adult cortex," *Neuron* **24**(3), 623–637 (1993).
131. T. Bonhoeffer and A. Grinvald, "Iso-orientation domains in cat visual cortex are arranged in pinwheel-like patterns," *Nature* **353**(6343), 429–431 (1991).
132. R. M. Friedman, L. M. Chen, and A. W. Roe, "Responses of areas 3b and 1 in anesthetized squirrel monkeys to single- and dual-site stimulation of the digits," *J. Neurophysiol.* **100**(6), 3185–3196 (2008).
133. H. D. Lu and A. W. Roe, "Optical imaging of contrast response in Macaque monkey V1 and V2," *Cereb. Cortex* **17**(11), 2675–2695 (2007).
134. M. Suh, S. Bahar, A. D. Mehta, and T. H. Schwartz, "Blood volume and hemoglobin oxygenation response following electrical stimulation of human cortex," *Neuroimage* **31**(1), 66–75 (2006).
135. M. Zhao, M. Suh, H. Ma, C. Perry, A. Geneslaw, and T. H. Schwartz, "Focal increases in perfusion and decreases in hemoglobin oxygenation precede seizure onset in spontaneous human epilepsy," *Epilepsia* **48**(11), 2059–2067 (2007).
136. M. M. Haglund, G. A. Ojemann, and D. W. Hochman, "Optical imaging of epileptiform and functional activity in human cerebral cortex," *Nature* **358**(6388), 668–671 (1992).

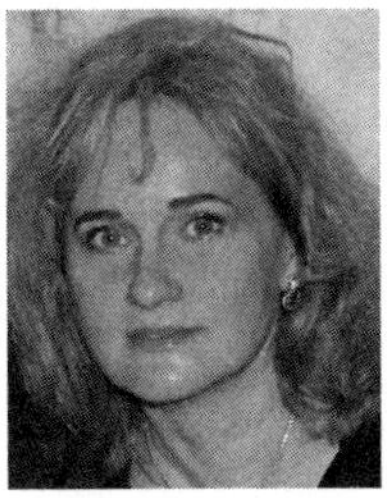

Anna N. Yaroslavsky is an associate professor of physics and director of Advanced Biophotonics Laboratory at the University of Massachusetts, Lowell, Massachusetts, USA. She obtained her PhD from the Saratov State University, Saratov, Russia, in 1999 and joined Wellman Center for Photomedicine (WCP), Boston, USA, in 2000. From 2005 to 2010, she has been an assistant professor of Dermatology at the Harvard Medical School, Boston, USA, and since 2010, she is a visiting scientist at WCP. Her expertise is in biomedical optics and medical imaging. Her research is focused on the development of optical and multimodal technologies for medical applications.

Ilya V. Yaroslavsky received his MSc degree *summa cum laude* in physics in 1990 and PhD in laser physics in 1994, both from Saratov State University, Saratov, Russia. From 1994 to 2000, he did his postdoctoral training in Heinrich Heine University, Düsseldorf, Germany, working on laser interstitial thermotherapy of brain tumors, and in Louisiana State University, Shreveport, Los Angeles, USA, developing optical diffusion techniques for stroke diagnostics. He started his industrial career at Palomar Medical Technologies, Inc., Burlington, Massachusetts, USA, in 2000 and 2012 assumed position of the vice president of Advanced Research of the company. In 2015, he joined IPG Medical Corporation as Manager for Advanced Product Development. His scientific interests include light-tissue interactions and use of lasers for biomedical applications. He has authored and coauthored more than 50 scientific papers and inventions. He served as a member of IEC TC 76 Technical Advisory Group on optical radiation safety and laser equipment.

Chapter 3
Propagation of Pulses and Photon Density Waves in Turbid Media

Ilya V. Yaroslavsky
IPG Photonics, Inc., Marlborough, MA, USA

Anna N. Yaroslavsky
Department of Physics and Applied Physics, Department of Biomedical Engineering, University of Massachusetts–Lowell, Lowell, MA, USA

Department of Dermatology, Wellman Center for Photomedicine, Massachusetts General Hospital, Boston, USA

Juan Rodriguez
St. Louis College of Pharmacy, St. Louis, USA

3.1 Introduction

The need for understanding the propagation of pulses and photon density waves in turbid media arises from the advantages that time-resolved techniques bring to the field of biomedical optical diagnostics. The most significant of these advantages is the ability of time-resolved techniques to discriminate between photons traveling different optical paths in the medium. When an adequate physical model is used to process this information, absorption and scattering contributions, as well as spatial in homogeneities in the medium, may be resolved with better accuracy than that of continuous-wave (CW) methods. Another consideration of practical importance is that a given amount of information about the medium can be collected with a smaller number of emitter–detector channels in a time-resolved experiment than in a CW one. This feature can be exemplified by a reflectance measurement done on a homogeneous semi-infinite medium at a single emitter–detector separation: the full set of optical properties can, in principle, be derived from time- or frequency-resolved data, whereas a single CW reflectance value is not sufficient for the purpose. Decreasing the number of source–detector channels helps to reduce the probability of errors related to optode/medium coupling and system calibration. Furthermore, in many measurement schemes, the intensity measurements may be eliminated completely and purely temporal characteristics, such as average arrival time (in the time domain) or phase shift (in the frequency domain), used instead. This approach may increase the signal-to-noise ratio and the overall stability of the system. These advantages come with a price: first, the time-resolved equipment is typically more complex and expensive than that of a CW technique, and second, methods used for interpretation of the data are usually more complicated. In addition, calibration of a time-domain instrument may require precise determination of the time instant when the incident pulse enters the medium. Nevertheless, it appears that for certain clinical applications, the advantages of the time-resolved techniques may outweigh their drawbacks.

The use of short laser pulses for biomedical diagnostics was first proposed in 1971 by Duguay and Mattick.[1] However, pulsed lasers had been used earlier for probing the atmosphere and ocean.[2] The similarity of the physical processes in these applications has prompted an intensive and fruitful exchange of ideas and techniques between the fields. After Jöbsis[3] put forward the idea of employing light for quantifying oxygen saturation in deep tissue regions, the time-dependent techniques have become an object of even more detailed investigation.[4] In 1988, Chance and coworkers performed for the first time a direct comparison between time-resolved and CW measurements of hemoglobin oxygenation in tissue[5] and demonstrated the potential of the time-resolved techniques. At about the same time, Delpy and coworkers initiated systematic studies of time-resolved optical responses of tissues and organs.[6] In 1990, Lakowicz and Berndt proposed the use of the frequency-domain technique, initially developed for fluorescence lifetime measurements, for elastic spectroscopy of multiple-scattering tissues.[7] Since the 1990s, the number of publications devoted to the time-resolved techniques has been growing steadily. Several excellent review papers[8–10] illuminate the advances and the milestones in the development of this rapidly progressing field.

The theoretical descriptions of propagation of pulses and photon density waves in tissue are based on the time-dependent radiative transport theory. Early systematic treatments of time-dependent radiative energy transfer in turbid media were performed by Sobolev[11] and Wing[12]. A significant role in the subsequent development of the theory was played by the ideas and methods borrowed from the atmospheric and oceanic optics.[13] Considerable efforts have been devoted to advancing the understanding of time-dependent light transport in the diffuse regime,[14] which is of great relevance for the spectroscopy and imaging of bulk tissues *in vivo*.

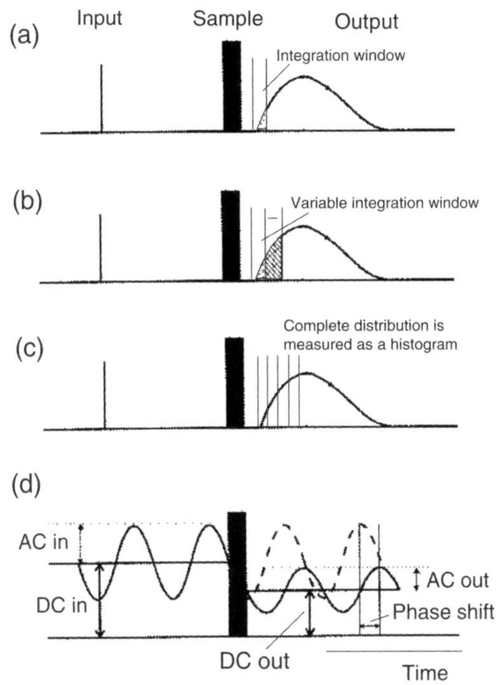

Figure 3.1 Possible implementations of the time-resolved technique. (a–c) In the time domain: (a) fixed gate width for registering ballistic photons only, (b) variable gate width for diffuse scattering corrections, and (c) registration of complete temporal profile. (d) In the frequency domain.

The time-resolved technique can be implemented in either the time domain or the frequency domain. In the former case, the medium is illuminated by a short pulse of light and the scattered or directly transmitted pulse is analyzed to extract information about the optical properties of the medium. Three main variations of the time-domain (TD) technique have been proposed. They are illustrated in Figs. 3.1(a–c). In the first scenario [see Fig. 3.1(a)], time-gating is used to measure only the earliest-arriving part of the transmitted pulse.[15] The hope is that only ballistic (unscattered) photons are measured and, therefore, Beer's law can be used to describe the attenuation of the ballistic flux. This technique is attractive because it makes data interpretation a relatively easy task. The most significant drawback of this technique is that it requires that a detectable number of ballistic photons must arrive at the detector. This is not normally the case with tissues *in vivo*. Indeed, with typical values of the total attenuation coefficient (see Table 3.1) for soft tissues in the range of 10–30 mm^{-1} (at the relevant wavelength range[16]) and geometrical distances to be probed in the order of 10 to 100 mm, attenuation of the unscattered signal is roughly between 10 and 1000 orders of magnitude. This precludes reliable measurements of the ballistic photons only. It is argued that the technique can still be valuable if it can capture the sum of

Table 3.1 Nomenclature

Quantity	Meaning	Dimension
μ_a	absorption coefficient	L^{-1}
μ_s	scattering coefficient	L^{-1}
$\mu_t = \mu_a + \mu_s$	total interaction coefficient	L^{-1}
$p(\mathbf{s},\mathbf{s}')$	scattering phase function	srad^{-1}
G	anisotropy factor	none
$d_f = 1/\mu_t$	mean free path	L
$a = \mu_s/\mu_t$	Albedo	None
$\mu'_s = \mu_s(1-g)$	reduced scattering coefficient	L^{-1}
$\mu_{tr} = \mu_a + \mu'_s$	transport coefficient	L^{-1}
D	distance	L
$\boldsymbol{X}$	set of optical properties (general)	varies
$\boldsymbol{Y}$	set of optical measurements (general)	varies
Γ	spatial area of interest	N/A
$\partial\Gamma$	boundary of Γ	N/A
$\boldsymbol{R}$	spatial coordinates	(L,L,L)
T	time	T
ε	dielectric constant	none
$\boldsymbol{S}$	unit vector of direction	none
$I(\mathbf{r},\mathbf{s},t)$	specific intensity in the time domain	$\text{W}/(L^2\text{srad})$
C	velocity of light	L/T
$t_f = 1/\mu_t c$	mean time of flight between interactions	T
t_s	mean single-scattering delay time	T
$\tilde{x}(\omega)$	Fourier transform of the function $x(t)$	varies
$a = \mu'_s/\mu_{tr}$	transport albedo	none

ballistic and so-called *snake* photons (the latter term referring to photons having undergone only a few scattering events mainly in forward direction). This sum is attenuated approximately as $\exp(-\mu_{tr}d)$ (where μ_{tr} is the transport coefficient defined below and d is the distance traveled). For typically found values of μ_{tr} in the range of 1–5 mm^{-1}, this still leads to an attenuation between 5 and 200 orders of magnitude. Simply increasing the time-gate width leads to a growing contribution of multiply scattered photons that nullifies the main original advantage of the technique. Moreover, in an imaging application, when the signal is collected at a number of spatial locations, this contribution will vary with the detector's position, resulting in further distortions in the reconstructed image. To rectify this problem, a "threshold time" technique [see Fig. 3.1(b)] has been proposed by Benaron and coworkers.[17] The idea is to select the time window individually for each image's pixel in such a way as to maintain a constant ratio of detected photons to the total number of photons in the pulse. Although this approach facilitates obtaining qualitative (pixel intensities) images of tissues, quantitative interpretation of data becomes more complex because it is no longer possible to eliminate a priori the contribution of multiply scattered light. In addition, neither of the two techniques just discussed is applicable in the reflection geometry, which is of considerable importance for many applications. In such cases, the data processing technique must rely on the information carried by multiply scattered photons. A straightforward way of collecting this information is to register the complete temporal profile of the scattered pulse [see Fig. 3.1(c)]. Having registered the profile, one may proceed in one of two ways: using the "raw" data or employing various auxiliary quantities such as average time of arrival, pulse width, different moments of the pulse shape, or slope of the trailing edge[10] In any case, a mathematical model of pulse propagation is necessary to derive the optical properties of the medium (or some functionals of them) from the measured data.

Another approach is the frequency domain (FD) technique illustrated in Fig. 3.1(d).[7,18–20] In this technique, the medium is illuminated by a modulated light source, usually with a single modulation frequency. The presence of such a source leads to periodic oscillations of the photon density inside the medium. This phenomenon is often termed *photon density waves* (PDWs). Measured quantities in this scheme are the modulation and phase shift of the scattered signal. In many instances, phase measurements alone suffice to obtain required information.

The two approaches are interconnected via Fourier transforms and theoretically have equal information content. Practical considerations dictate the choice of one technique or the other. Advantages of the FD approach are the possibility of a narrow-band detection, which can greatly increase the signal-to-noise ratio, and typically less expensive equipment. It appears that for the majority of clinical applications, the FD technique is an adequate choice. The strength of the TD technique is the possibility of obtaining the complete spectrum simultaneously and usually higher bandwidth. These features are especially advantageous when optically thick organs such as the brain are to be probed.

If a signal is collected in the time domain, data processing can often still be performed in the frequency domain after a Fourier transform. This approach is convenient and in many cases profitable in terms of computational efficiency. The opposite is not always true because usually only few modulation frequencies or even a single modulation frequency is used in FD.

This chapter discusses basic concepts and practical approaches for describing the propagation of optical pulses and photon density waves in turbid media. Section 3.2 introduces the time-dependent transport theory and its nomenclature. Section 3.3 describes several techniques useful in solving the time-dependent transport equation. Section 3.4 focuses on the Monte Carlo technique for simulating time-dependent photon migration in absorbing and scattering media. Section 3.5 is devoted to the diffusion approximation of the transport theory. In Section 3.6, several avenues for increasing the range of validity of the diffusion approximation are outlined. Finally, Section 3.7 discusses the role of a finite duration of a single scattering act on time-dependent light propagation.

3.2 Time-Dependent Transport Theory

In the field of tissue spectroscopy and imaging, one must deal with two types of problems: the *forward* and the *inverse* problems. The forward problem can be formulated in general form as follows: given the set of optical properties $\mathbf{X}$ on Γ, where Γ is the spatial domain of interest, and the set of sources $\mathbf{Q}$ on $(\Gamma + \partial\Gamma)$, where $\partial\Gamma$ is the surface of Γ, find the set of measurements $\mathbf{Y}$ at the detector locations. Naturally, the inverse problem is finding $\mathbf{X}$ from the given $\mathbf{Q}$ and $\mathbf{Y}$. This chapter is concerned mostly with the forward problem. Table 3.1 lists the basic nomenclature used in this chapter.

We will limit our attention to the monochromatic (single wavelength) case and assume that only elastic scattering takes place. Substantial literature is devoted to application of time- and frequency-domain techniques for fluorescence studies in turbid media.[21,22]

Apparently, in the time-resolved approach, both $\mathbf{Q}$ and $\mathbf{Y}$ are time- or frequency-dependent. The optical properties $\mathbf{X}$, on the other hand, are assumed not to depend on time. This assumption is clearly valid for biological tissues because the time scale of changes in the probing light intensity is always much shorter than the time scale of any possible changes in the optical properties.

What exactly is understood by the optical properties of the medium? The answer to that question depends on the model selected. The most fundamental classical way to describe propagation of light in a dielectric medium is provided by Maxwell's theory. In this case, the medium is characterized by a distribution of the complex dielectric constant $\varepsilon(\mathbf{r})$, where $\mathbf{r}$ is a spatial location inside the medium. Most natural media, including tissues, are random media. This means that the function $\varepsilon(\mathbf{r})$ is a stochastic rather than deterministic function. Treatment of the multiple-scattering regime for realistic media in the framework of the rigorous theory is a difficult, often impossible, task.[23] Therefore, a string of approximations is used to render the practical problems tractable.

One highly productive and widely recognized approach is the time-dependent transport theory (TTT). Instead of dealing with the "raw" electromagnetic field, this theory operates with the specific intensity $I(\mathbf{r},\mathbf{s},t)$, which represents the optical power passing through a unit area normal to the direction $\mathbf{s}$ at location $\mathbf{r}$. It has been shown that the specific intensity can be identified with the suitably "grained" version of the Wigner distribution function, which describes first-order coherence of the electromagnetic field.[24] This feature provides a bridge between the rigorous approach and the transport theory. Transport theory neglects the effect of long-range coherence. As a result, the differences between radiation originating from a laser and from a thermal source are lost in the formalism of the transport theory. Experimental evidence exists that for most effects of interest, the difference between coherent and non-coherent radiation is lost in a turbid medium after a distance of several tens mean free paths.[25] For soft tissues in the visible and near infrared ranges, that translates to a distance on the order of 1 mm or less. One remarkable exception is the effect of enhanced backscattering.[26] In addition, the case of highly structured tissues (such as muscle) may require special considerations.[27] Although in principle the transport equation can account for the anisotropy of the turbid medium, the bookkeeping becomes rather tedious. This chapter assumes that the medium is isotropic. Fortunately, most tissues meet this requirement.

The governing equation of the transport theory can be deduced from first principles using the relevant simplifying assumptions.[28] In general, the result of such a procedure is an equation for the specific intensity tensor, which reflects the polarization properties of light. Neglecting polarization effects further reduces this equation to a scalar equation for the specific intensity $I(\mathbf{r},\mathbf{s},t)$.

The set of the optical properties in the scalar TTT is defined as $\mathbf{X}_{TT} = \{\mu_a, \mu_s, p_T(\mathbf{s},t,\mathbf{s}',t'), n\}$. Here, μ_a is the absorption coefficient (i.e., average number of absorption events per unit pathlength); μ_s is the scattering coefficient (i.e., average number of scattering events per unit pathlength); $p_T(\mathbf{s},t,\mathbf{s}',t')$ is the time-resolved scattering phase function, which represents the probability density of a photon arriving at given location with direction s at time instant t to be scattered into a new direction $\mathbf{s}'$ at time instant t'; n is the real part of the refractive index of the medium, hereafter referred to as simply the refractive index. All the parameters of TTT can, in principle, be expressed via the microscopic distribution of the dielectric constant.[25] The exact form of this connection depends strongly on the geometry of the medium and may not be easy to establish. The interested reader is referred to.[13] This chapter treats the set of the optical properties X_{TT} as intrinsic characteristics of the medium, taking into account, however, that they represent, in fact, a result of a properly defined averaging procedure over a small but finite volume around given location.

The resulting time-dependent scalar transport equation can be written as[11]

$$\frac{\mathbf{s}\cdot\nabla I(\mathbf{r},\mathbf{s},t)}{\mu_t} + t_f\frac{\partial I(\mathbf{r},\mathbf{s},t)}{\partial t} = -I(\mathbf{r},\mathbf{s},t) + \frac{a}{4\pi}\int_{4\pi}\left[\int_{-\infty}^{t} I(\mathbf{r},\mathbf{s}',t')p_T(\mathbf{s}',t',\mathbf{s},t)dt'\right]d\Omega' + \frac{S(\mathbf{r},\mathbf{s},t)}{\mu_t}, \tag{3.1}$$

where $a = \mu_s/\mu_t$ is the single scattering albedo, $\mu_t = \mu_s + \mu_a$ is the total interaction coefficient, $t_f = 1/\mu_t c$ is the mean time of flight between interactions, c is the speed of light in the medium, and $S(\mathbf{r},\mathbf{s},t)$ is the source distribution.

Equation (3.1) is completed with the boundary conditions on $\partial\Gamma$:

$$I(\mathbf{r},\mathbf{s},t)|_{(\mathbf{sn})<0} = I_Q(\mathbf{r},\mathbf{s},t) + \Re I(\mathbf{r},\mathbf{s},t)|_{(\mathbf{sn})>0}, \quad \mathbf{r}\in\partial\Gamma, \tag{3.2}$$

where $I_Q(\mathbf{r},\mathbf{s},t)$ is the boundary distribution of the specific intensity created by external sources, $\mathbf{n}$ is the external normal to $\partial\Gamma$ at $\mathbf{r}$, and $\Re$ is the reflection operator.

We assume that the time-resolved scattering phase function $p_T(s,t,s',t')$ can be represented in factorized form as

$$p_T(s,t,s',t') = p(s,s')f(t,t'), \tag{3.3}$$

where $p(s,s')$ is the time-independent scattering phase function, and $f_s(t,t')$ is the single-scattering delay function. Furthermore, the scattering phase function is assumed to depend on the scattering angle only [i.e., $p(s,s') = p(s \cdot s')$], and the single-scattering delay function is assumed to be translationally invariant [i.e., $f_s(t,t') = f_s(t'-t)$].

Let us define single-scattering delay time t_s as the first moment of $f(t)$:

$$t_s = \int_0^\infty t f_s(t)dt. \tag{3.4}$$

The average time spent by a photon in the medium equals the product of the average number of scattering events and the sum $(t_s + t_f)$. If this time is small compared to the characteristic time scale of change in intensity of the sources, the problem can be treated as continuous wave (CW) and, accordingly, the time-independent transport equation can be used. Otherwise, one must resolve to the time-dependent theory.

Introducing the Fourier transform

$$\tilde{I}(r,s,\omega) = \int_{-\infty}^{\infty} I(r,s,t)\exp(-i\omega t)dt, \tag{3.5}$$

a corresponding transport equation in the FD can be easily obtained:

$$\frac{\mathbf{s} \cdot \nabla \tilde{I}(\mathbf{r},\mathbf{s},\omega)}{\mu_t} = -(1 + i\omega t_f)\tilde{I}(\mathbf{r},\mathbf{s},\omega) + \frac{a\tilde{f}(\omega)}{4\pi}\int_{4\pi}\tilde{I}(\mathbf{r},\mathbf{s}',\omega)p(\mathbf{s} \cdot \mathbf{s}')d\Omega' + \frac{\tilde{S}(\mathbf{r},\mathbf{s},\omega)}{\mu_t}, \tag{3.6}$$

where $\tilde{f}_s(\omega)$ is the Fourier transform of the single-scattering delay function, and $\tilde{S}(\mathbf{r},\mathbf{s},\omega)$ is the Fourier transform of the source.

In tissue optics, it is usually assumed that the single-scattering delay function may be approximated as a delta function, which is equivalent to assuming that $t_s \ll t_f$. We make this simplification for now and defer a more detailed discussion of the subject until Section 3.7.

3.3 Techniques for Solving the Time-Dependent Transport Equation

Equations (3.1) and (3.6) are integro-differential equations that allow for analytical solutions in the simplest cases only. A considerable amount of work has been devoted to developing techniques for either solving these equations directly or simplifying them using problem-specific physical conditions.[11,12,29,30]

3.3.1 Reduction to steady-state case

An interesting and potentially useful feature of Eq. (3.6) is that the equation can be interpreted as a CW transport equation for a medium with complex "optical properties:"

$$\eta_t = \mu_t(1 + i\omega t_f), \qquad b = \frac{a}{1 + i\omega t_f}. \tag{3.7}$$

Equations (3.7) are valid when $t_s = 0$ (see Section 3.7 for the more general case). The "similarity relations" enable one to use results obtained for the CW case to solve time-dependent problems.

Let us consider an example of using this technique.[31] A plane wave, harmonically modulated with a frequency ω, enters a semi-infinite medium ($z > 0$). Scattering is assumed to be isotropic, and the refractive indices of the medium and the surrounding are matched (set to unity). The incident direction of the wave coincides with the positive z-axis (see Fig. 3.2). The goal is to determine the angular distribution of the modulation and phase shift of the back-scattered radiation as well as the total outward flux at the boundary $z = 0$.

The specific intensity can be regarded as the sum of the diffuse specific intensity I_d and the reduced incident specific intensity I_{ri}:

$$I(\tau,\zeta_z,u) = I_{ri}(\tau,\zeta_z,u) + I_d(\tau,\zeta_z,u), \tag{3.8}$$

where we introduce the optical distance $\tau = \mu_t z$ and the dimensionless time $u = t/t_f$, and where the directional cosine $\zeta_z = (\mathbf{sn}_z)$, $\mathbf{n}_z$ is the unit vector

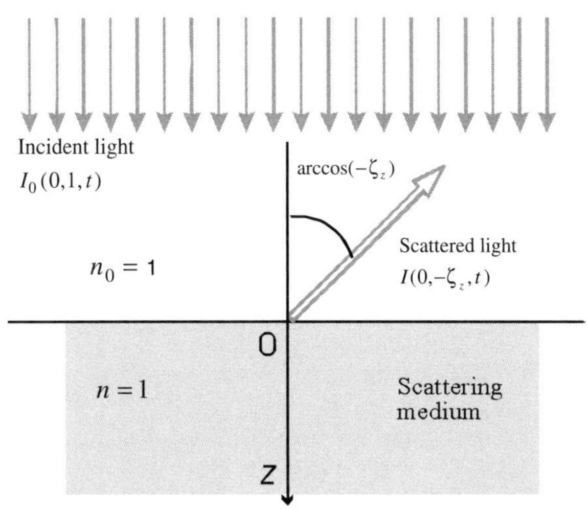

Figure 3.2 Plane-parallel wave incident on a semi-infinite medium.

in the positive direction of the z-axis. The boundary conditions (3.2) take the form

$$\begin{aligned} I_{ri}(\tau,\zeta_z,u)\big|_{\substack{\tau=0\\ \zeta_z>0}} &= Q_0\delta(\eta-1)(1+m_0\,\mathrm{Re}(\exp(i\alpha u))) \\ I_d(\tau,\zeta_z,u)\big|_{\substack{\tau=0\\ \zeta_z>0}} &= 0, \end{aligned} \tag{3.9}$$

where $\alpha = \omega t_f$ is the dimensionless modulation frequency, Q_0 is the source intensity, $\delta(\eta)$ is the Dirac delta function, and m_0 is the incident modulation.

The solution for I_{ri} is straightforward:

$$I_{ri}(\tau,\zeta_z,u) = Q_0\delta\ (\zeta_z-1)\exp(-\tau)(1+m_0\ \mathrm{Re}(\exp(i\alpha(u-\tau)))). \tag{3.10}$$

For the diffuse intensity, we have from Eq. (3.6) in the FD,

$$\zeta_z\frac{\partial\tilde{I}_d(\tau,\zeta_z,\alpha)}{\partial\tau} + (1+i\alpha)\tilde{I}_d(\tau,\zeta_z,\alpha) = \frac{a}{2}\int_{-1}^{+1}\tilde{I}_d(\tau,\zeta_z',\alpha)d\zeta_z' + \tilde{S}_d(\tau,\zeta_z,\alpha), \tag{3.11}$$

where

$$\tilde{S}_d(\tau,\zeta_z,\alpha) = \frac{a}{2}\int_{-1}^{+1} m_0Q_0\delta(\zeta_z'-1)\exp(-\tau(1+i\alpha))d\zeta_z'. \tag{3.12}$$

The solution of the respective CW problem is known:[32]

$$\tilde{I}_d(0,-\zeta_z,0)|_{\zeta_z>0} = \frac{a}{4}Q_0\frac{\varphi(\zeta_z,a)\varphi(1,a)}{\zeta_z+1}, \tag{3.13}$$

where $\varphi(\xi,a)$ is Ambartsumian's function, which can be found from the functional equation

$$\varphi(\zeta,a) = 1 + \frac{a}{2}\zeta\varphi(\zeta,a)\int_0^1\frac{\varphi(\varsigma',a)}{\zeta+\varsigma'}d\varsigma'. \tag{3.14}$$

Therefore, in our case, the solution to Eq. (3.11) for $\tau = 0$ can be written as

$$\tilde{I}_d(0,-\zeta_z,\alpha)|_{\zeta_z>0} = \frac{a}{4(1+i\alpha)}\frac{\varphi^*(\zeta_z,a)\varphi^*(1,a)}{1+\zeta_z}, \tag{3.15}$$

where $\varphi^*(\xi,a)$ is the complex-valued modified Ambartsumian's function, satisfying the equation

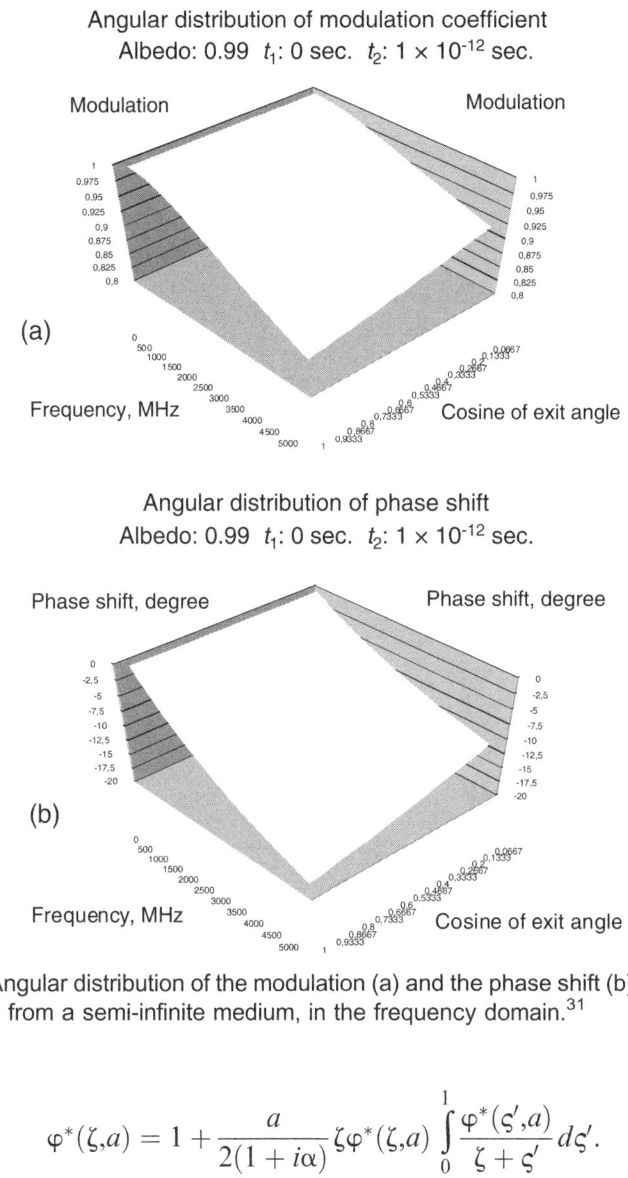

Figure 3.3 Angular distribution of the modulation (a) and the phase shift (b) of the radiation backscattered from a semi-infinite medium, in the frequency domain.[31]

$$\varphi^*(\zeta,a) = 1 + \frac{a}{2(1+i\alpha)}\zeta\varphi^*(\zeta,a)\int_0^1 \frac{\varphi^*(\varsigma',a)}{\zeta+\varsigma'}d\varsigma'. \tag{3.16}$$

Mod(I_d) and Arg(I_d) represent the modulation and phase shift of the scattered radiation, respectively. The diffusely scattered flux $\mathbf{F}_d$ at the boundary can be readily calculated:

$$\tilde{F}_d(0,\alpha) = 2\pi\int_0^1 \zeta_z \tilde{I}_d(0,-\zeta_z,\alpha)d\zeta_z. \tag{3.17}$$

The results of the calculations of the back-scattered diffuse intensity $\tilde{I}_d(0,-\zeta_z,\alpha)$ for a model medium are illustrated in Figs. 3.3(a,b). Equation (3.16) has been solved numerically, and the outcome has been

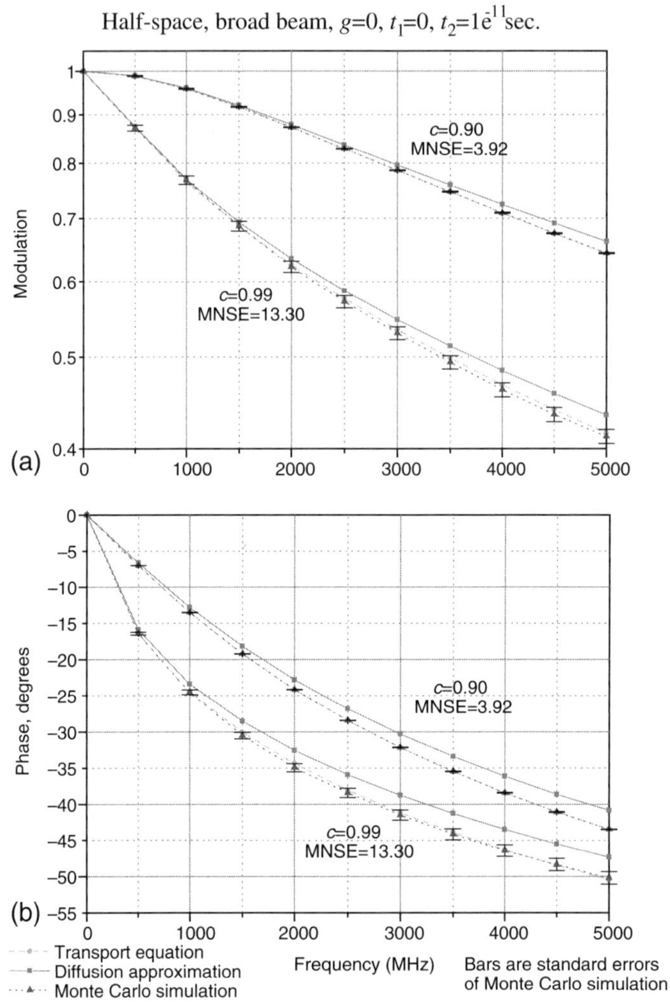

Figure 3.4 Diffuse flux backscattered from a semi-infinite medium, calculated with Eqs. (3.15–17).[46] Also shown for comparison are results of Monte Carlo simulations (Section 3.4) and the diffusion approximation (Section 3.5).

substituted into Eq. (3.15). The albedo a of the medium was set to 0.99, and the mean time of flight was set to 10^{-12} sec., which corresponds to the total interaction coefficient μ_t of 3.345 mm^{-1}. An interesting feature of the distribution is a pronounced anisotropy of both the modulation and the phase shift. Figures 3.4(a,b) represent the results of calculating the back-scattered diffuse flux $\tilde{F}_d(0,\alpha)$ as a function of the modulation frequency using Eq. (3.17). Effects of varying albedo and mean time of flight are demonstrated (see Table 3.2 for the parameters used). Also shown in Fig. 3.4 for comparison are the results of calculating the back-scattered diffuse flux with the diffusion approximation (see Section 3.5 for a detailed discussion of the diffusion approximation).

Table 3.2 Optical properties used in calculations

Case	Albedo	Mean time of flight, sec.
I	0.99	1e-12
II	0.99	1e-11
III	0.90	1e-12
IV	0.90	1e-11

A "quasi steady-state" version of the transport equation, similar to Eq. (3.6), can be obtained by using the Laplace transform instead of the Fourier transform given in Eq. (3.5). The interested reader is referred to Ref. 29 for details of this method.

Obviously, the approach discussed in this section works only if a corresponding CW solution is known or if a substantial computational economy can be achieved by solving the CW problem instead of the original time-dependent one. If this is not the case, direct numerical techniques must be used. The two most popular of these techniques are discussed in Sections 3.3.2 and 3.3.3. Reference 30 compares the results obtained with these techniques and with the diffusion approximation for the problem of the pulse propagation in a planar medium.

3.3.2 Spherical harmonics method

The basic idea of the method is the expansion of the specific intensity into a series of spherical harmonics. In the frequency domain,

$$\tilde{I}(\mathbf{r},\mathbf{s},\omega) = \sum_{l=0}^{\infty}\sum_{m=-l}^{l}\left(\frac{2l+1}{4\pi}\right)^{1/2}\tilde{I}_{lm}(\mathrm{r},\omega)\,Y_{lm}(\mathbf{s}), \tag{3.18}$$

where $Y_{lm}(\mathrm{s})$ are spherical harmonics. Similarly, for the source term,

$$\tilde{S}(\mathbf{r},\mathbf{s},\omega) = \sum_{l=0}^{\infty}\sum_{m=-l}^{l}\left(\frac{2l+1}{4\pi}\right)^{1/2}\tilde{S}_{lm}(\mathbf{r},\omega)\,Y_{lm}(\mathbf{s}). \tag{3.19}$$

The phase function is expanded in terms of a series of Legendre polynomials P_l:

$$p(\mathbf{s}'\cdot\mathbf{s}) = \sum_{l=0}^{\infty}\frac{2l+1}{4\pi}p_l P_l(\mathbf{s}'\cdot\mathbf{s}). \tag{3.20}$$

Expansion coefficients for many frequently used phase functions (in particular, Henyey–Greenstein and Gegenbauer kernel functions) can be found in Ref. 33. Substituting Eqs. (3.18)–(3.20) into Eq. (3.6), multiplying the resulting equation by $Y^{*}_{\lambda\mu}$, and integrating over 4π, one obtains

$$\left(\frac{4\pi}{2\lambda+1}\right)^{1/2}\sum_{l=0}^{\infty}\sum_{m=-l}^{l}\left(\frac{2l+1}{4\pi}\right)^{1/2}\nabla\tilde{I}_{lm}(\mathbf{r},\omega)\cdot\int_{4\pi}\mathbf{s}Y^{*}_{\lambda\mu}(\mathbf{s})Y_{lm}(\mathbf{s})d\Omega$$
$$+(1+i\omega t_f)\mu_t\tilde{I}_{\lambda\mu}(\mathbf{r},\omega)=a\mu_t p_\lambda\tilde{I}_{\lambda\mu}(\mathbf{r},\omega)+\tilde{S}_{\lambda\mu}(\mathbf{r},\omega). \tag{3.21}$$

The integration on the left-hand side of Eq. (3.21) may be carried out using a recursion relation for spherical harmonics.[29] This yields an infinite set of equations for the specific intensity expansion coefficients $\tilde{I}_{\lambda\mu}(\mathbf{r},\omega)$. Setting all coefficients with index λ exceeding a certain number L to zero truncates the number of equations to $(L+1)^2$. This procedure yields the P_L approximation of the transport equation. The resulting system of equations can be solved with a suitable numerical technique, such as finite-difference or finite-element methods. An example of using this technique for multidimensional media can be found in Ref. 34. In many cases, spatial symmetry significantly reduces the number of equations in the P_L approximation. Most common cases are planar and spherical geometries, where the number of equations is reduced to $(L+1)$.[29] Obviously, the higher the order L of the approximation, the more accurate the results are expected.

3.3.3 Discrete ordinate method

The discrete ordinate method uses a finite set of directions $\mathbf{s}_k$ ($k=1,\ldots,K$) to approximate the angular distribution of the specific intensity. The integral in the right-hand side of Eq. (3.6) is represented as

$$\int_{4\pi}\tilde{I}(\mathbf{r},\mathbf{s}',\omega)p(\mathbf{s}_l\cdot\mathbf{s}')d\Omega'\approx\sum_{k=1}^{K}\tilde{I}(\mathbf{r},\mathbf{s}_k,\omega)p_{kl}w_k, \tag{3.22}$$

where $p_{kl}=p(\mathbf{s}_k\cdot\mathbf{s}_l)$, w_k is the weight assigned to the k-th angular point. The set of the directions is determined by the selected quadrature method. Several quadrature techniques are available, including Gaussian, Lobatto, Fiveland, and Chebyshev methods.[35] Substitution of Eq. (3.22) into Eq. (3.6) leads to

$$\frac{\mathbf{s}_k\cdot\nabla\tilde{I}^k(\mathbf{r},\omega)}{\mu_t}=-(1+i\omega t_f)\tilde{I}^k(\mathbf{r},\omega)+\frac{a\tilde{f}(\omega)}{4\pi}\sum_{k'=1}^{K}\tilde{I}^{k'}(\mathbf{r},\omega)p_{k'k}w_{k'}+\frac{\tilde{S}^k(\mathbf{r},\omega)}{\mu_t}, \tag{3.23}$$

where $\tilde{I}^k(\mathbf{r},\omega)=\tilde{I}(\mathbf{r},s_k,\omega)$, $\tilde{S}^k(\mathbf{r},\omega)=\tilde{S}(\mathbf{r},s_k,\omega)$. The system of Eqs. (3.23) can be solved for the K unknowns using a variety of techniques developed for partial differential equations. As with the spherical harmonics technique, significant simplifications can be achieved when the problem possesses a symmetry. If, for instance, the specific intensity depends on a single angle only (i.e., a rotational symmetry is present), the set of directions $\mathbf{s}_k$ is reduced to a set of cosines ζ_k. Then the so-called S_K method[29] can be utilized, which assumes that the specific intensity is a linear function of angle in each of the intervals $\zeta_{k-1} < \zeta < \zeta_k$. After substitution of the corresponding expression into Eq. (3.6), the latter can be integrated on each of the angular intervals. This procedure generally leads to an improvement in accuracy with the same number of angular nodes.

A recent example of using the discrete ordinate method for image reconstruction in optical diffuse tomography can be found in Ref. 36.

3.3.4 Distributed-source approach

The distributed-source approach (DSA)[37] is not by itself a technique for solving the transport equation, but a method that allows, under certain circumstances, a simplification in the process of estimating the measurable quantities.

Suppose we are interested in calculating detector readings for a set of detectors **D**. The signal $\tilde{J}_D(\omega)$ collected by a detector D with a small area ΔS_D located at $\boldsymbol{r}_D \in (\Gamma + \partial\Gamma)$ (see Fig. 3.5) is determined in the FD by

$$\tilde{J}_D(\omega) = \Delta S_D \int\limits_{(\mathbf{n}_D\mathbf{s})>0} \tilde{I}(\mathbf{r}_D,\mathbf{s},\omega)(\mathbf{n}_D \cdot \mathbf{s})h_D(\mathbf{s},\omega)d\Omega, \tag{3.24}$$

where $\mathbf{n}_D$ is the detector's internal normal, and $h_D(\mathbf{s},\omega)$ is the detector's sensitivity function, which also accounts for signal losses at the detector-medium interface.

For an arbitrary point $\mathbf{r}_G \in \partial\Gamma$ and direction $\mathbf{s}$, Eq. (3.6) can be transformed into the following integral form:

$$\tilde{I}(\mathbf{r}_G + u\mathbf{s},\mathbf{s},\omega) = \tilde{I}(\mathbf{r}_G,\mathbf{s},\omega)\exp\left(-\int_0^u \eta_t(\mathbf{r}_G + u'\mathbf{s})du'\right) + \frac{1}{4\pi}\int_0^u \mu_s(\mathbf{r}_G + u'\mathbf{s})$$
$$\times \left[\int_{4\pi} \tilde{I}(\mathbf{r}_G + u\mathbf{s},\mathbf{s}',\omega)p(\mathbf{r}_G + u\mathbf{s},\mathbf{s},\mathbf{s}')d\Omega'\right] \times \exp\left(-\int_{u'}^u \eta_t(\mathbf{r}_G + u''\mathbf{s})du''\right)du', \tag{3.25}$$

where u is the distance counted from the point $\boldsymbol{r}_G$ in the direction $\mathbf{s}$, and the dependence of the phase function on the coordinates is explicitly shown. The absence of internal sources is assumed in Eq. (3.25).

Substituting Eq. (3.25) into Eq. (3.24) and replacing the integration over the solid angle with the integration over $(\Gamma + \partial\Gamma)$, one obtains

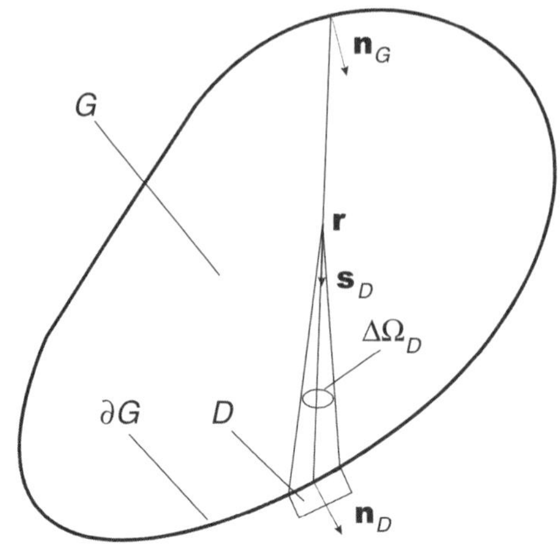

Figure 3.5 Concept of the distributed-source approach.

$$\tilde{J}_D(\omega) = \int_{\partial\Gamma} \tilde{I}(\mathbf{r},\mathbf{s}_D,\omega)\exp(-\sigma_D(\mathbf{r}))H_D(\mathbf{r},\omega)\Delta\Omega_D(\mathbf{n}_G \cdot \mathbf{s}_D)dS + \int_{\Gamma} \frac{\mu_s(\mathbf{r})}{4\pi}\left[\int_{4\pi} \tilde{I}(\mathbf{r},\mathbf{s},\omega)p(\mathbf{r},\mathbf{s}_D,\mathbf{s})d\Omega\right]\exp(-\sigma_D(\mathbf{r}))H_D(\mathbf{r},\omega)\Delta\Omega_D dV, \tag{3.26}$$

where

$$\mathbf{s}_D = \mathbf{s}_D(\mathbf{r}) = (\mathbf{r}_D - \mathbf{r})/|\mathbf{r}_D - \mathbf{r}|, \quad \Delta\Omega_D = \Delta\Omega_D(\mathbf{r}) = \Delta S_D(\mathbf{s}_D \cdot \mathbf{n}_D)/|\mathbf{r}_D - \mathbf{r}|^2,$$

$$\sigma_D(\mathbf{r}) = \int_0^{|\mathbf{r}_D-\mathbf{r}|} \eta_t(\mathbf{r} + u\mathbf{s}_D)du, \quad H_D(\mathbf{r},\omega) = \begin{cases} h_D(\mathbf{s}_D,\omega), & (\mathbf{s}_D \cdot \mathbf{n}_D) > 0 \\ 0, & \text{otherwise,} \end{cases}$$

$\mathbf{n}_G(\mathbf{r})$ is the inward normal to $\partial\Gamma$ at point $\mathbf{r}$. The signal at an extended detector with area S_D can be calculated by straightforward integration of Eq. (3.26) over S_D.

Thus, all detector readings can be calculated from the known distribution of the specific intensity with Eq. (3.26). This approach, as we will see in Section 3.4, is especially useful in the framework of the Monte Carlo technique. In addition, DSA is beneficial for solving inverse problems, such as image reconstruction in diffuse optical tomography, as mentioned in Chapter 6 and discussed in more detail in Ref. 37.

3.4 Monte Carlo Method

One drawback of the numerical techniques briefly described earlier is their tendency to grow rapidly in complexity when the geometry of the problem becomes more complicated. Complexity is an inherent feature of almost any biological system; therefore, this consideration may be an important factor when the solution technique is selected. This "complexity factor" is significantly less severe for the method known as the Monte Carlo (MC) technique.

The application of the MC technique for solving problems of the time-dependent transfer theory is based on the interpretation of the underlying physical model as a set of stochastic processes. Assuming that the probability distributions of the elementary processes are known, certain moments of more complex distributions can be calculated by sampling random variables from the elementary distributions. Examples of the elementary processes involved are photon movements without interactions and single scattering from one direction to another. Characteristics of the complex processes, in which we are interested, may include, for instance, the probability of a photon arriving at the detector at a given time interval or the average number of photons registered by a detector. Extensive literature is available on MC methods, including their use for simulating photon propagation in turbid media.[38–42,90] Of special note may be the capability of MC methods to account accurately for the transient polarization state of the propagating light[89] that is useful in situations when the incident light is highly polarized and dimensions of the area of interest are smaller than characteristic length of depolarization. Another situation when this feature is handy occurs when one must deal with a birefringent turbid medium (e.g., collagen fibers).

In the recent years, MC methods have received new impetus due to the MC technique being uniquely suitable for parallel implementation on multi-core systems (including graphic processors), which have become widely available (see, for example, Refs. 88 and 93). At the same time, the so-called White Monte Carlo (WMC) technique was expanded into a time-resolved field,[92,95] enabling simulation of photon transport for media with a certain set of the optical properties, with subsequent scaling to obtain results for a different set of the optical properties without the need to repeat simulations. These two developments have made use of the MC technique feasible as a forward method in iterative solvers of inverse problems suitable for realistic geometries as complex as the human head.[96]

3.4.1 Sampling of random variables

Selecting a random value from a known probability distribution law is of primary importance for any realization of the MC technique. This question is addressed thoroughly in a number of textbooks (see above), so it will be touched on only briefly here. Let us assume that a routine generating a random number α from a uniform distribution on the interval (0,1) is available. This is the case for almost any modern software (programming language or modeling package) that can be used for MC work. To generate a random value x_α from a probability density distribution $p(x)$, the following relation must be solved for x_α:

$$\alpha = \int_{x_{\min}}^{x_\alpha} p(x)dx, \tag{3.27}$$

where $x_{\min}$ is the minimal allowed value of x. In many cases, Eq. (3.27) can be solved analytically. For example, the Henyey–Greenstein function is a good approximation of the scattering phase function for many tissues:

$$p_{HG}(\zeta) = (4\pi)^{-1}(1 - g^2)(1 + g^2 - 2g\zeta)^{-3/2}, \tag{3.28}$$

where parameter $|g| \leq 1$ is the mean value of the scattering angle (anisotropy factor), and ζ is the cosine of the scattering angle. Substituting Eq. (3.28) into Eq. (3.27) and solving for ζ_α yields the following formula for a random value of the cosine of the scattering angle:[43]

$$\zeta_\alpha = [1 + g^2 - (1 - g^2 - \{(1 - g^2)/(1 - g + 2g\alpha)\}^2)]/2\,g. \tag{3.29}$$

Similarly, for the distribution of the free path of a photon in a homogeneous medium

$$p_{FP}(l) = \mu_t \exp(-\mu_t l), \tag{3.30}$$

a random value of the free path can be generated using the relation

$$l_\alpha = -\frac{\ln \alpha}{\mu_t}. \tag{3.31}$$

Equation (3.31) can be generalized for an inhomogeneous medium using the following technique.[44] Let $\mu_{t\mathrm{MAX}}$ be the maximal attenuation coefficient in the volume of interest Γ: $\mu_t(\boldsymbol{r}) \leq \mu_{t\mathrm{MAX}}\ \forall \boldsymbol{r} \in \Gamma$. Two sequences of random numbers are generated at each step of the simulation: $d_1,..,d_n$ according to the distribution

$$p(d) = \mu_{t\mathrm{MAX}} \exp(-\mu_{t\mathrm{MAX}} d) \tag{3.32}$$

and $\alpha_1,..,\alpha_n$ according to the uniform distribution in the interval (0,1). The third sequence $l_1,..,l_n$ is calculated using the formula

$$l_n = \sum_{k=1}^{n} d_k. \tag{3.33}$$

A number N_s is determined according to the criterion

$$N_s = \min\left\{ n; \alpha_n \leq \frac{\mu_t(\boldsymbol{r} + l_n \mathbf{s})}{\mu_{t\mathrm{MAX}}} \right\}, \tag{3.34}$$

where $\mathbf{s}$ is the scattering direction. Finally, we accept

$$l_\alpha = l_{N_s}, \tag{3.35}$$

where l_α is the random pathlength.

If an analytical solution of Eq. (3.27) is impossible or too cumbersome, a numerical technique must be used. The simplest approach[45] is to tabulate the cumulative density function

$$P(x) = \int_{x_{\min}}^{x} p(x)dx$$

for a number of values x_i, $i = 1,..,N$, $x_1 = x_{\min}$. Having generated the random number α, the corresponding value x_α can be found by computing the number n such that $P_n < \alpha < P_{n+1}$ and interpolating between x_n and x_{n+1}.

3.4.2 Generic time-resolved Monte Carlo algorithm

A typical problem in the tissue optics is finding a time-resolved signal produced on a detector D by a light pulse originating from a source Q. If the source is sufficiently small (such as a fiber tip or a focused laser beam), the time-resolved response of the detector $J^*_D(t)$ for any given profile of the incident pulse $I^*_Q(t)$ can be found by convolution

$$J^*_D(t) = \int_0^t J_D(t-\tau)I^*_Q(\tau)d\tau, \tag{3.37}$$

where $J_D(t)$ is the response of the detector D to a δ-pulse originating at the source Q. Obviously, the ω-component of the Fourier image $\tilde{J}_D(\omega)$ of $J_D(t)$ represents the response of the detector D to a harmonically modulated (at the angular frequency ω) signal at Q.

Let us consider a generic (also referred to as "elementary" or "analogous") Monte Carlo algorithm for simulating propagation of a δ-pulse of light in a turbid medium. It is assumed that the refractive index does not change over the volume of the medium. The photon is characterized by its coordinate vector **r** and its direction **s**. As a preliminary step, the time interval of interest $(0, t_{\max})$ is divided into L time "bins." For each time bin l, a counter n_l is set and initiated at zero. The total photon counter N_{ph} is also initiated at zero. The following sequence is then executed:

1. Generate a random launch position and direction of a photon from the source Q. Often, the source may be approximated by a point source, in which case the launch position of the photon is predetermined. If the source is unidirectional, the launch direction is also predetermined. Set the total travel time t_t to zero. Increment the photon counter by one.
2. Generate a random pathlength l_α using Eq. (3.31) or the "maximal cross-section" technique described in Section 3.4.1.

3. Calculate the trial time $t_\alpha = l_\alpha / c$ and the trial position $\mathbf{r}_\alpha = \mathbf{r} + l_\alpha \mathbf{s}$. If the vector $\boldsymbol{r}$ is stored as its coordinates (x, y, z) and the direction $\mathbf{s}$ is stored as its directing cosines $(\zeta_x, \zeta_y, \zeta_z)$, the coordinates of the trial position are determined by the formulas

$$v_\alpha = v + \zeta_v l_\alpha, v = x,y,z. \tag{3.38}$$

4. Check for the intersection of the segment $(\mathbf{r}_\alpha,\mathbf{r})$ with the sensitive area of the detector D. If such an event happens, calculate the intersection coordinates $\mathbf{r}^*$ and time to intersection $t^* = (\mathbf{r}^* - \mathbf{r})/c$. Add t^* to the total travel time. Calculate the number L^* of the time bin corresponding to the total travel time. Increment the respective counter n_{L^*} by one, terminate the photon, and go to step 1. If there is a nonzero probability of reflection of the photon from the detector's surface, it can be taken into account in the same way as reflections from the boundaries (see next step).
5. Check for intersection of the segment $(\mathbf{r}_\alpha,\mathbf{r})$ with the boundaries. If such an event happens, calculate the probability of reflection R^*. Generate a random number α from the uniform distribution on (0,1). If $\alpha < R^*$, recalculate position and direction of the photon according to the boundary reflection law. Usually, Fresnel's reflection is assumed.[43] If $\alpha > R^*$, terminate the photon and go to step 1.
6. Move the photon to the new position (i.e., set $\mathbf{r} = \mathbf{r}_\alpha$) and increment the total travel time by t_α.
7. Select the interaction type (absorption or scattering). To this end, generate a random number α from the uniform distribution on (0,1) and compare it to the albedo a at the interaction point. If $\alpha > a$ (absorption occurs), terminate the photon and go to step 1.
8. Generate a random scattering direction from the scattering phase function $p(\mathbf{s},\mathbf{s}')$. Usually, the random cosine ζ_α of the scattering angle is generated. Eq. (3.29) represents the generating formula for the Henyey–Greenstein phase function. The azimuth scattering angle φ_α is generated from a uniform distribution on the interval $(0,2\pi)$.
9. Calculate new direction of the photon $\boldsymbol{s}'$. The following formulas can be used:

$$\zeta'_x = \zeta_x \zeta_\alpha - (\zeta_y \sin\varphi_\alpha + \zeta_x \zeta_z \cos\varphi_\alpha)\sqrt{\frac{1-\zeta_\alpha^2}{1-\zeta_z^2}},$$

$$\zeta'_y = \zeta_y \zeta_\alpha + (\zeta_x \sin\varphi_\alpha - \zeta_y \zeta_z \cos\varphi_\alpha)\sqrt{\frac{1-\zeta_\alpha^2}{1-\zeta_z^2}},$$

$$\zeta'_z = \zeta_z \zeta_\alpha + (1-\zeta_z^2)\cos\varphi_\alpha \sqrt{\frac{1-\zeta_\alpha^2}{1-\zeta_z^2}}. \tag{3.39}$$

10. If the total photon counter N_{ph} exceeds a predetermined maximal number, exit. Go to step 2.

The histogram n_l, normalized by the total number of photons N_{ph}, is accepted as an estimate of the detector's response to the incident δ-pulse. We will use the acronym GMCA (Generic Monte Carlo Algorithm) to refer to this algorithm hereafter.

The most significant drawback of GMCA for applications in tissue optics is the fact that usually a large number of photons is required to achieve an estimate of $J_D(t)$ with a reasonable accuracy. The following are several techniques ("tricks"), commonly referred to as variance reduction techniques, that can be used to reduce this number.

3.4.3 Photon weighting

One simple and common way to increase the efficiency of MC simulations is photon weighting.[43] The principle of this technique is to assign to each photon a weight W_k, where k is the current interaction's number and, instead of randomly drawing the type of interaction on step 7 of the GMCA, simply decrease the weight according to

$$W_k = a(\boldsymbol{r}_k) W_{k-1}, \tag{3.40}$$

where $\boldsymbol{r}_k$ is the site of k-th interaction. The initial weight W_0 is set to 1. Accordingly, when the photon arrives at the detector (step 4 of GMCA), the counter n_{L^*} should be incremented by the current weight of the arriving photon. Similarly, when the interaction with the boundaries is processed (step 5 of the GMCA), the weight is multiplied by the reflection coefficient.

If the weight of the photon drops after multiple interactions to such a low value that it would not contribute significantly to the detector reading, the photon should be terminated. To maintain the energy balance, the following procedure, known as "Russian roulette,"[43] can be used. The minimal acceptable photon weight W_c is predetermined. When the current weight drops below W_c, a random number α is generated. If $\alpha < R_c$, where R_c is the predetermined survival probability, the photon weight is multiplied by $1/R_c$; otherwise, the photon is terminated.

3.4.4 Shortcut technique in the frequency domain

An obvious way to obtain frequency-resolved detector responses $\tilde{J}_D(\omega)$ is to calculate the TD response $J_D(t)$ and use a fast Fourier transform according to Eq. (3.5). If, however, only a single or a few modulation frequencies are of interest, another approach, known as the shortcut technique,[46–48] can be used.

The idea of the shortcut technique is to simulate the photon transport directly in the frequency domain as dictated by Eq. (3.6). Surprisingly, only a few modifications to the GMCA suffice to accomplish this task. The quantization of the time scale is no longer required. Instead, the signal at the detector D is estimated as a *complex* sum

$$\tilde{J}_D(\omega*) = \frac{1}{N_{ph}} \sum_{i=1}^{N_D} W^i_{K_t} \exp(i\omega^* t^i_t), \tag{3.41}$$

where N_D is the total number of photons registered by the detector D, ω^* is the frequency of interest, and K^i_t is the total number of interactions for the i-th photon. If estimates of detector readings at several modulation frequencies are required, they can be obtained from the same set of photon trajectories.

The use of the shortcut MC technique is illustrated in Figs. 3.4(a,b) for the test problem, which was discussed in Section 3.3.1.

3.4.5 Local estimate technique

This technique has been proposed and used by many groups under different names. It was referred to as the "local estimate" in Ref. 44, as the "semianalytical model" in Ref. 49, and as the "intensity reference" scheme in Ref. 50. These works were concerned with CW MC simulations. References 51 and 47 have placed the technique into the context of TD and FD simulations, respectively.

The development of the technique has been stimulated by the following observation: if the sensitive area of the detector D is small and optical distance between the source and the detector is large, only a small fraction of the photons launched in GMCA reach the detector and therefore contribute to the signal of interest. For simulations of realistic tissues and phantoms, this fraction may, in fact, be unacceptably small to produce any meaningful estimate of the detector reading. Is there a way to increase the efficiency of the simulations by somehow using the photons that never reach the detector in the analogous simulations? The integral form of the transport equation (3.25) (or its TD equivalent) suggests a positive answer to that question. The local estimate technique is a way of achieving such an increase. We consider the implementation of the local estimate technique in the frequency domain. The basis of the technique is the distributed-source approach, discussed previously in Section 3.3.4. Specifically, the expressions of Eq. (3.26) can be rewritten as

$$\tilde{J}_D(\omega) = \int_\Gamma \left[\int_{4\pi} \tilde{I}(\mathbf{r},\mathbf{s},\omega)\mu_t(\mathbf{r})P_s(\mathbf{r},\mathbf{s}_D,\mathbf{s})d\Omega \right] P_D(\mathbf{r},\omega)dV, \tag{3.42}$$

where

$$P_D(\mathbf{r},\omega) = \exp(-\sigma_D(\mathbf{r}))H_D(\mathbf{r},\omega), \quad P_s(\mathbf{r},\mathbf{s}_D,\mathbf{s}) = \frac{a(\mathbf{r})}{4\pi} p(\mathbf{r},\mathbf{s}_D,\mathbf{s})\Delta\Omega_D. \tag{3.43}$$

The first term of the right side of Eq. (3.26) (direct contribution of the photons reflected from the boundaries) has been neglected in Eq. (3.42). This is a reasonable assumption when the detector is oriented inward toward the medium and the medium is optically thick. The function P_s represents the probability of scattering of a photon at given location into the solid angle spanned over the sensitive area of the detector. The modulus of the complex-valued function P_D represents the probability of the photon reaching and being registered by the detector without having further interactions with the medium. Because the probability of a photon visiting a given location is proportional to the product of the total interaction coefficient and the specific intensity, the signal at the detector D can be estimated with MC simulations using the formula

$$\tilde{J}_D(\omega) = \frac{1}{N_{ph}} \sum_{i=1}^{N_{ph}} \sum_{k=1}^{K_t^i} W_k^i \exp(-i\omega(t_t^i)_k) P_s(\mathbf{r}_k, \mathbf{s}_D, \mathbf{s}_k) P_D(\mathbf{r}_k, \omega). \tag{3.44}$$

The expressions of Eq. (3.43) for P_D and P_s are valid, strictly speaking, only for an infinitesimally small solid angle $\Delta\Omega_D$. If the solid angle is large, the integration over the surface of the detector is necessary. Equation (3.43) should be replaced with

$$P_s(\mathbf{r}, \mathbf{s}_D, \mathbf{s}) P_D(\mathbf{r}, \omega) = \frac{a(\mathbf{r})}{4\pi} \int_{\Delta S_D} \exp(-\sigma^*(r)) H^*(\mathbf{r}, \omega) p(\mathbf{r}, \mathbf{s}^*, \mathbf{s}) \frac{(\mathbf{s}^* \cdot \mathbf{n}_D)}{|\mathbf{r}^* - \mathbf{r}|^2} dS^*, \tag{3.45}$$

where

$$\mathbf{s}^* = \mathbf{s}^*(\mathbf{r}) = (\mathbf{r}^* - \mathbf{r})/|\mathbf{r}^* - \mathbf{r}|,$$

$$\sigma^*(\mathbf{r}) = \int_0^{|\mathbf{r}^* - \mathbf{r}|} \eta_t(\mathbf{r} + u\mathbf{s}^*) du,$$

$$H^*(\mathbf{r}, \omega) = \begin{cases} h_D(\mathbf{s}^*, \omega), & (\mathbf{s}^* \cdot \mathbf{n}_D) > 0 \\ 0, & \text{otherwise,} \end{cases}$$

and integration is performed over the starred coordinates.

Such integration, if performed at each interaction site, would be very time-consuming and could nullify the time gains afforded by the local estimate technique. Instead, the point r_D is *randomly* selected at the surface of the detector, and the quantities P_s and P_D are estimated using Eqs. (3.43). This procedure still requires knowledge of the solid angle $\Delta\Omega_D$. Although it is possible to calculate $\Delta\Omega_D$ precisely, it is impractical due to high computational costs. The following approximation can be used, first suggested in Ref. 52 and valid for a circular detector,

$$\Delta\Omega_D = \pi \frac{l_D}{\rho} \frac{1}{h(1-h)} [(\tau - h)g(h) + h(1-\tau)g(1)], \tag{3.46}$$

where ρ and l_D are the coordinates of the current point with respect to the center of the detector aperture (measured in the radii of the detector), τ and h are adjustable parameters, and

$$g(h) = \ln\left\{ \frac{[\rho h + (\rho^2 + l_D^2)^{1/2}]^2}{[\rho h - 1 + (1 - 2\rho h + \rho^2 - l_D^2)^{1/2}][\rho h + 1 + (1 + 2\rho h + \rho^2 + l_D^2)^{1/2}]} \right\}. \tag{3.47}$$

The authors of Ref. 52 suggest the following values of the parameters:

$$h = 0.93, \qquad \tau = 0.684.$$

If the medium is inhomogeneous, the integral in the definition of σ_D can be estimated using a similar approach [i.e., selecting a random point on the ray ($\mathbf{r}$, $\mathbf{r}_D$) and evaluating the total interaction coefficient at that point]. Alternatively, if the detector is sufficiently small, the probabilities P_D can be calculated beforehand for a number of spatial nodes and stored in an array. During the simulations, the required values of probabilities are obtained by interpolating between the spatial nodes. This approach provides, in addition, a means for optimizing the Russian roulette procedure. Specifically, the efficiency of the Russian roulette can be increased by noticing that the photons that traveled far enough toward the detector represent a greater "value" for the simulations than the photons that wandered away from the detector.[47] Therefore, the critical weight in the Russian roulette can be adjusted as a function of the current photon position to reflect this fact. The following formulas can be used:

$$\begin{aligned} W_c(\mathbf{r}) &= W_{c\max} - \frac{P_D(\mathbf{r}) - P_{D\min}}{P_{D\max} - P_{D\min}}(W_{c\max} - W_{c\min}), \quad P_{D\min} \le P_D(\mathbf{r}) \le P_{D\max}, \\ W_c &= W_{c\max}, \quad P_D(\mathbf{r}) < P_{D\times\min}, \\ W_c &= W_{c\min}, \quad P_D(\mathbf{r}) < P_{D\times\max}, \end{aligned} \tag{3.48}$$

where the maximal and minimal probabilities $P_{D\min}$ and $P_{D\max}$ are preselected or estimated on a "learning stage" (typically, first several thousand photons). The survival probability in the Russian roulette can be adjusted in a similar manner.

Figures 3.6(a,b) illustrate the time gain afforded by the local estimate technique compared to GMCA for a test problem (see figure caption for the parameters of the test). Figure 3.7 represents an example of using the local estimate technique to calculate the TD response of a cylindrical phantom.

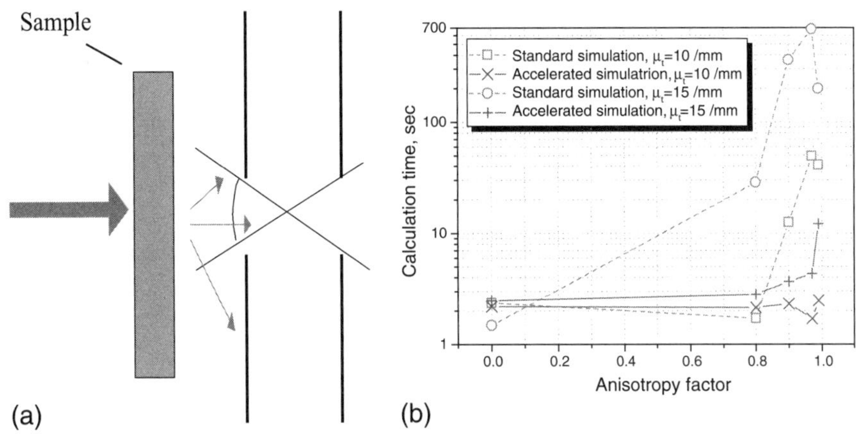

Figure 3.6 Comparison between the standard and "local estimate" Monte Carlo simulations for a test CW transmittance problem.[47] (a) Geometry of the test problem. Geometrical/optical properties: beam diameter: 3 mm; beam divergence: 0.22 deg; sample thickness: 0.4 mm; sample albedo: 0.99; phase function: Henyey–Greenstein; first aperture: 1 mm; second aperture: 1 mm; distance to the first aperture: 0 mm; distance between the apertures: 100 mm. (b) Computation time as a function of the anisotropy factor of the test medium for the two techniques.

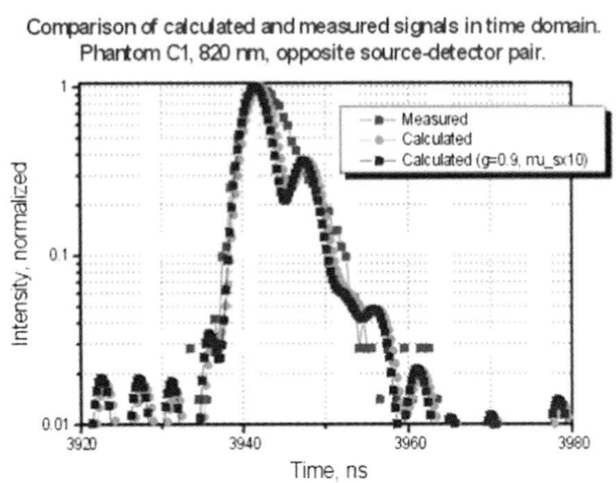

Figure 3.7 Comparison of measured data and Monte Carlo–simulated optical response of a cylindrical phantom. Monte Carlo results are convolved with the modulation transfer function of the setup.

Although the local estimate technique reduces the variance of MC simulations, it also increases the processing time for each interaction. In particular, the geometrical factors should be carefully considered when the technique is implemented in an MC algorithm. The authors of Ref. 53 have found that the optimal solution could be to use the local estimate technique in an area adjacent to the detector only.

3.4.6 Hybrid technique

Monte Carlo simulations yield a solution that converges to the exact solution of the time-dependent transport equation when the number of photons increases infinitely. It is known, however, that under certain conditions, the transport equation can be reduced to a simpler mathematical model. One of these simplifications is the diffusion approximation, which is discussed in more detail in Section 3.5, which follows. For most soft tissues, the diffusion approximation becomes valid in the areas far from the sources and boundaries. Moreover, it has been demonstrated numerically and experimentally that even for the detectors located *at* the boundaries, the diffusion approximation still can deliver an acceptable solution, provided that the source-detector separation is sufficiently large. Solving the diffusion equation usually requires less computational resources (in particular, CPU time) than performing MC simulations. Hence, one can think of a hybrid model, which would combine the accuracy of MC simulations in the vicinity of the source and the speed of a diffusion equation solver in the remote areas. Such model was first proposed for CW problems by Wang and Jacques.[54] It has been extended to the FD case by Alexandrakis et al.[55]

The idea of the hybrid technique is straightforward. Let us assume that the Green's function of the diffusion equation in the frequency domain $\tilde{G}_{dif}(\mathbf{r},\mathbf{r}',\omega)$ for a given medium is known. The Green's function describes the response of the detector located at the point $\mathbf{r}$ to a δ-like source located at the point $\mathbf{r}'$ and modulated with the angular frequency ω. For a nonlocalized source $\tilde{S}_{dif}(\mathbf{r},\omega)$, the signal can be obtained by integration

$$\tilde{J}_{dif}(\mathbf{r},\omega) = \int_{\Gamma} \tilde{G}_{dif}(\mathbf{r},\mathbf{r}',\omega)\tilde{S}_{dif}(\mathbf{r}',\omega)d\mathbf{V}'. \tag{3.49}$$

The volume Γ is divided into two subvolumes: the volume Γ_{MC}, where the MC simulations are performed, and the volume Γ_{dif}, where the diffusion approximation is used. The volume Γ_{dif} is divided into a number of voxels. Photons are launched from the source according to the standard MC technique. Once a photon crosses from the volume Γ_{MC} into the volume Γ_{dif}, its weight is recorded in the corresponding voxel and its trajectory is terminated. Such a photon is considered to have established an isotropic source. If a photon reaches a detector before leaving the volume Γ_{MC}, it is considered to have contributed to the MC part of the signal $\tilde{J}_{MC}(\mathbf{r},\omega)$. After a sufficient number of photons have been simulated, the source distribution for the diffusion approximation is calculated from the voxel distribution of the recorded photon weights using normalization. The source distribution is then substituted into Eq. (3.49) to obtain the diffusion part of the signal. Finally, the signal is computed by simple addition:

$$\tilde{J}(\mathbf{r},\omega) = \tilde{J}_{dif}(\mathbf{r},\omega) + \tilde{J}_{MC}(\mathbf{r},\omega). \tag{3.50}$$

This scheme has been implemented by the authors of Ref. 55 for the case of a two-layered planar medium.

3.5 Diffusion Approximation

The diffusion approximation (DA) is the lowest-order spherical harmonics approximation (P_1). We consider it in a separate section due to its importance in biomedical optics: many practical problems can be treated with acceptable accuracy within the framework of the DA.

3.5.1 Time-dependent diffusion equation

Let us consider the derivation of the diffusion equation in the frequency domain. It is possible to follow the general spherical harmonics procedure discussed in Section 3.3.2 and obtain four equations for the four quantities $\tilde{I}_{\lambda\mu}(\mathbf{r},\omega),\lambda,\mu = 0,1$. However, the structure of the coefficients allows for a significantly simpler approach, adopted in the majority of the textbooks for the CW case.[13,29]

The initial observation is that in the P_1-approximation, the diffusion component of the specific intensity can be represented as

$$\tilde{I}_d(\mathbf{r},\mathbf{s},\omega) \approx \tilde{U}_d(r,\omega) + \frac{3}{4\pi}\tilde{\mathbf{F}}_d(\mathbf{r},\omega)\cdot\mathbf{s}, \tag{3.51}$$

where $\tilde{U}_d(r,\omega)$ and $\tilde{\mathbf{F}}_d(\mathbf{r},\omega)$ are the frequency-domain diffuse fluence rate and diffuse flux, respectively. The formal similarity of Eq. (3.6) to the steady-state transport equation can be used to derive the respective diffusion equation following the steady-state procedure.[13] The following relation for $\tilde{U}_d(\mathbf{r},\omega)$ is obtained (refractive index is assumed constant over Γ):

$$\eta_{tr}\nabla\frac{1}{\eta_{tr}}\nabla\tilde{U}_d(\mathbf{r},\omega) - \gamma(\omega)\tilde{U}_d(\mathbf{r},\omega) = -\tilde{S}_d(\mathbf{r},\omega), \tag{3.52}$$

where

$$\tilde{S}_d(\mathbf{r},\omega) = 3\eta_{tr}\mu_s\tilde{U}_{ri}(\mathbf{r},\omega) - 3\left(\nabla + \frac{\nabla\eta_{tr}}{\eta_{tr}}\right)\cdot\int_{4\pi}\tilde{K}_{ri}(\mathbf{r},\mathbf{s},\omega)\mathbf{s}d\Omega,$$

$$\tilde{K}_{ri}(\mathbf{r},\mathbf{s},\omega) = \frac{\mu_s}{4\pi}\int_{4\pi}\tilde{I}_{ri}(\mathbf{r},\mathbf{s},\omega)p(\mathbf{s}\cdot\mathbf{s}')d\Omega', \qquad \gamma(\omega) = 3\eta_a\eta_{tr},$$

$\tilde{U}_{ri}(\mathbf{r},\omega)$ is the reduced incident fluence rate, η_a is the complex "quasi-absorption" coefficient, and η_{tr} is the complex "quasi-transport" coefficient. The latter two quantities are given by

$$\eta_a = \mu_a + i\frac{\omega}{c}, \quad \eta_{tr} = \mu_s' + \eta_a, \tag{3.53}$$

where $\mu_s' = (1-g)\mu_s$ is the reduced scattering coefficient.

Eq. (3.52) can be simplified for the case of low modulation frequencies. In particular, the following condition must be satisfied:

$$\frac{\omega^2}{\mu_a \mu_{tr} c^2} \ll 1. \tag{3.54}$$

The use of Eq. (3.54) and the condition $\mu_a \ll \mu_{tr}$ reduces the expression for γ to

$$\gamma = 3\mu_a \mu_{tr} + \frac{3i\omega\mu_{tr}}{c}. \tag{3.55}$$

Under these assumptions, Eq. (3.52) can be rewritten as

$$\nabla \frac{1}{3\mu_{tr}} \nabla \tilde{U}_d(\mathbf{r},\omega) - \mu_a \tilde{U}_d(\mathbf{r},\omega) - \frac{i\omega}{c} \tilde{U}_d(\mathbf{r},\omega) = -\frac{\tilde{S}_d(\mathbf{r},\omega)}{3\mu_{tr}}. \tag{3.56}$$

Equation (3.56) is the standard diffusion equation in the frequency domain. Its time-domain counterpart can be written as

$$\nabla \frac{1}{3\mu_{tr}} \nabla U_d(\mathbf{r},t) - \mu_a U_d(\mathbf{r},t) - \frac{1}{c}\frac{\partial U_d(\mathbf{r},t)}{\partial t} = -\frac{S_d(\mathbf{r},t)}{3\mu_{tr}}. \tag{3.57}$$

Equation (3.56) should be completed with the boundary conditions, and Eq. (3.57) requires initial and boundary conditions. One problem with the boundary conditions in the DA is that the fixed angular structure of the radiance, dictated by Eq. (3.51), may not be compatible with the general boundary conditions, given in Eq. (3.2). Therefore, approximate boundary conditions must be used. The interested reader is referred to Refs. 13, 56, and 57.

In addition, there is an on-going discussion in the literature on how the accuracy of DA may be improved by alternative choices of the coefficients in Eqs. (3.56) and (3.57). See Refs. 58 to 61 for more details on this subject.

3.5.2 Solutions for simple geometries

For many simple geometries, solutions of the time-dependent diffusion equation can be obtained in a closed analytical form. This is of special advantage when solving the inverse problems, where economy of computational resources often becomes a crucial factor. An excellent review of the solutions for a number of relevant geometries, both in time and in frequency domain, can be found in Ref. 14. Here we consider two important examples in the frequency domain.

3.5.2.1 Infinite medium

For an infinite homogeneous medium and a delta-like source term

$$\tilde{S}_d(\mathbf{r},\omega) = 3\tilde{P}(\omega)\mu_{tr}\delta(\mathbf{r}), \tag{3.58}$$

the solution of Eq. (3.56) is given by

$$\tilde{U}_d^{(\mathrm{inf})}(\mathbf{r},\omega) = 3\tilde{P}(\omega)\mu_{tr}\frac{\exp(-\sqrt{\gamma(\omega)}r)}{4\pi r}, \tag{3.59}$$

where $\tilde{P}(\omega)$ is the Fourier transformed source power. Explicit expressions for the modulus and the phase of $\tilde{U}_d^{(\mathrm{inf})}(\mathbf{r},\omega)$ can be written as

$$\begin{aligned} \mathrm{Mod}(\tilde{U}_d^{(\mathrm{inf})}) &= \frac{3\mathrm{Mod}(\tilde{P})\mu_{tr}}{4\pi r}\exp\left(-\left[9\mu_{tr}^2\left\{\mu_a^2+\frac{\omega^2}{c^2}\right\}\right]^{1/4}\cos\left(\frac{\varphi}{2}\right)r\right), \\ \mathrm{Arg}(\tilde{U}_d^{(\mathrm{inf})}) &= \mathrm{Arg}(\tilde{P}) - \left[9\mu_{tr}^2\left\{\mu_a^2+\frac{\omega^2}{c^2}\right\}\right]^{1/4}\sin\left(\frac{\varphi}{2}\right)r, \end{aligned} \tag{3.60}$$

where

$$\varphi = \arctan\left[\frac{\omega}{\mu_a c}\right].$$

3.5.2.2 Semi-infinite medium

One practically important case is a semi-infinite homogeneous medium with mismatched boundary conditions. A commonly used approximation for this problem is the replacement of the distributed source in Eq. (3.56) by a point source located at the depth $z_0 = 1/\mu_{tr}$ below the surface (so-called effective source approximation). The following relations for the outward flux are obtained:[62]

$$\begin{aligned} \mathrm{Mod}(\tilde{F}_d^{(\text{s-inf})}) &= \frac{6\tilde{P}\mu_{tr}\exp(-\rho Y_1)}{(4\pi)^2\rho^3}(1+\rho Y_2+\rho^2 Y_4^2)^{1/2}(z_0+z_b) \\ &\times\left(z+\frac{1}{\mu_{tr}}\left[1-\frac{(z_b+z_0)^2+3z^2}{2\rho^2}\left\{2+\frac{1+\rho Y_1}{1+\rho Y_2+\rho^2 Y_4^2}+\rho Y_1\right\}\right]\right), \\ \mathrm{Arg}(\tilde{F}_d^{(\text{s-inf})}) &= \arctan\left(\frac{\rho Y_3}{1+\rho Y_1}\right)-\rho Y_3, \end{aligned} \tag{3.61}$$

where

$$Y_1 = k_d V^+/\sqrt{2}, \qquad Y_2 = k_d V^+\sqrt{2}, \qquad Y_3 = k_d V^-/\sqrt{2},$$
$$Y_4 = k_d(1+x^2)^{1/4}, \qquad V^+ = (\sqrt{1+x^2}+1)^{1/2},$$
$$V^- = (\sqrt{1+x^2}-1)^{1/2}, \qquad x = \frac{\omega}{c\mu_a}, \qquad k_d = \sqrt{3\mu_a\mu_{tr}}, \tag{3.62}$$

$z_b = 2A/3\mu_{tr}$ is the distance between the real and extrapolated boundaries, the constant A is determined by the refractive index of the medium,[63] ρ is the distance from the source in the lateral plane, and the z axis is directed inward toward the medium. Equation (3.61) is valid for sufficiently large lateral distances, so that $\rho^2 \gg (z_b + z_0 \pm z)^2$.

The semi-infinite case is of special interest for diffuse spectroscopy and imaging of tissues because it represents the simplest possible model of bulk tissue *in vivo* when the sources and detectors are placed at the surface. In addition, this geometry corresponds to the experimental arrangement illustrated in Fig. 3.8, which is a logical choice of source-detector configuration when both sources and detectors are to be combined in a single probe and the measurements are to be taken noninvasively. For these reasons, many groups have carried out detailed studies of this geometry.[57,62,64] Figures 3.9 and 3.10 demonstrate comparisons of DA predictions of Eq. (3.61) with MC simulations for this geometry for matched and mismatched boundary conditions, respectively.

3.5.3 Numerical techniques

A variety of numerical techniques for solving the diffusion equation for complex geometries and inhomogeneous media are available. The two most

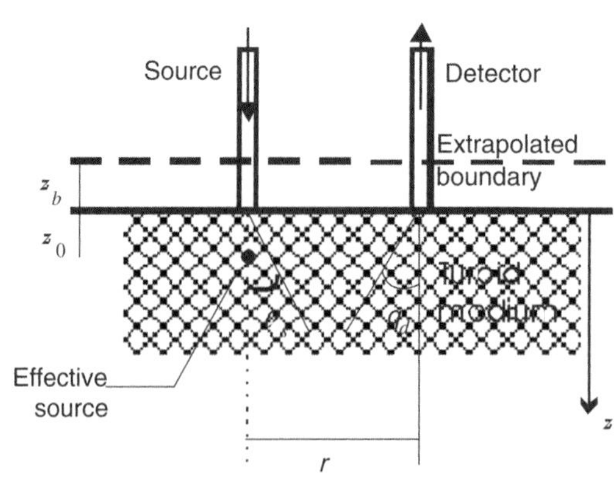

Figure 3.8 Semi-infinite spatially resolved geometry[82]. The extrapolated boundary and effective source used in the diffusion approximation are shown. z_b is the distance between the extrapolated boundary and the real boundary; z_0 is the depth of the effective source; ρ is the source-detector separation; θ_s and θ_d are the half-aperture angles of the source and the detector, respectively.

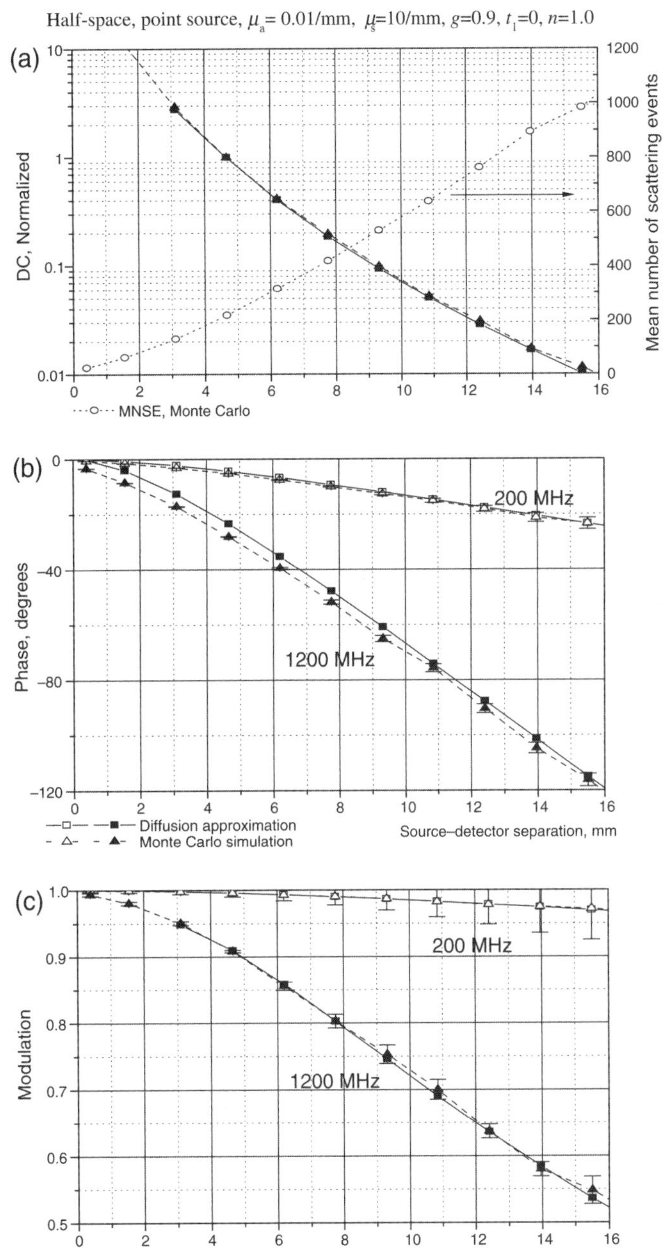

Figure 3.9 Normalized DC and mean number of scattering events (MNSE) (a), modulation (b), and phase (c) of the reflected signal for matched boundary conditions.[46]

popular ones are the finite-difference method (FDM) (see, e.g., Ref. 65) and the finite-element method (FEM) (see, e.g., Ref. 66). Commercial and public domain solvers are available that implement those techniques.[67,68] Recently, highly optimized methods for semi-infinite inhomogeneous geometries were

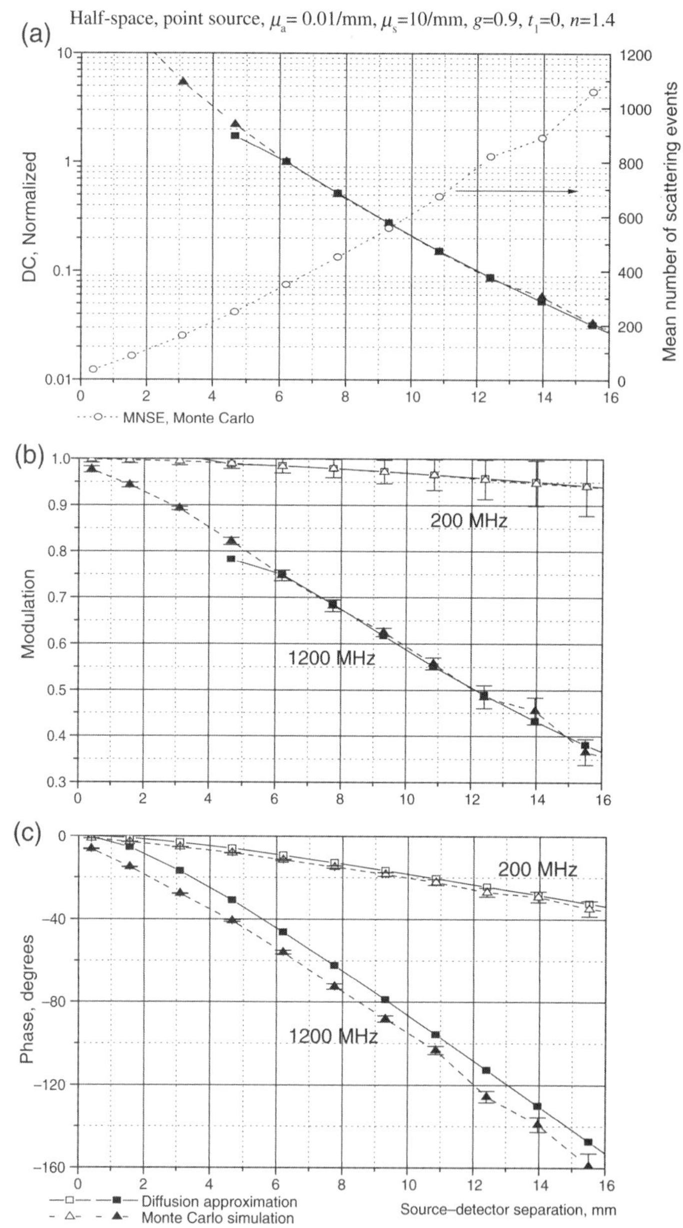

Figure 3.10 The same as Fig. 3.9 but for mismatched boundary conditions (n = 1.4).[46]

developed (e.g., Ref. 94) in the context of providing a fast forward solver for inverse problems.

The distributed-source approach (see Section 3.3.4) can also be used in the framework of DA.

In this case, the similarity relations can be used, and Eq. (3.26) is simplified to

$$\tilde{J}_D(\omega) = \int_\Gamma \tilde{K}_D(\mathbf{r},\omega) Z_D(\mathbf{r}) dV, \tag{3.63}$$

where

$$\begin{aligned} \tilde{K}_D(\mathbf{r},\omega) &= \mu_{tr}(\mathbf{r}) \tilde{U}(\mathbf{r},\omega) \exp(-i\omega t_D) \tilde{H}_D(\mathbf{r},\omega) \frac{\Delta\Omega_D}{4\pi}, \\ Z_D(\mathbf{r}) &= a\prime(\mathbf{r}) \exp(-\tau\prime_D(\mathbf{r})), \end{aligned} \tag{3.64}$$

$a'(\mathbf{r}) = \mu_s'(\mathbf{r})/\mu_{tr}(\mathbf{r})$ is the transport albedo, μ_{tr} is the transport coefficient, μ'_s is the reduced scattering coefficient, and $\tau'_D(\mathbf{r})$ and $t_D(\mathbf{r})$ are the reduced optical distance and time of flight from point $\boldsymbol{r}$ to the detector D, respectively.

Figures 3.11(a–f) illustrate this approach. The FD diffusion equation has been solved for a two-dimensional inhomogeneous medium with the optical properties shown in Figs. 3.11(a,b), using the MUDPACK package.[67,69] Figures 3.11(c,d) show the calculated modulus and phase of the kernel $\tilde{K}_D(\boldsymbol{r},\omega)$. Figures 3.11(e,f) demonstrate the modulus and the phase of the detected signal calculated with Eq. (3.63).

3.6 Beyond Diffusion Approximation

Although the DA is an extremely useful approach combining accuracy and relative simplicity, its validity deteriorates sharply when either the optical properties of the medium or the time/frequency ranges fall beyond certain limits. Two situations are of considerable importance for tissue optics. The first arises when the basic condition of the DA ($\mu_a \ll \mu_{tr}$) is not valid. For example, this condition is easily violated in whole blood, where the anisotropy factor g can reach extremely high values (see Chapter 2). Another example is provided by body regions filled with low scattering fluids,[61,70–72] such as cerebrospinal fluid and transparent fluid in cysts. The second situation arises when one is interested in early-arriving (i.e., quasi-ballistic) photons. As discussed in Section 3.1, the quasi-ballistic photons can be valuable for imaging because they retain more information on their trajectories than the diffuse photons. Yet, the DA is not valid for short arrival times, as many authors have discussed.[59] Both of the preceding cases are handled correctly by the full time-dependent transport equation, but it would be highly desirable to have simplified models capable of dealing with these situations.

Let us obtain the approximate time-dependent transport equation for strongly forward-scattering media. This approach is based on the expansion of the specific intensity in a Taylor series around the scattering direction. The idea was first proposed in Ref. 73 for the CW case and extended to the FD case in Ref. 74.

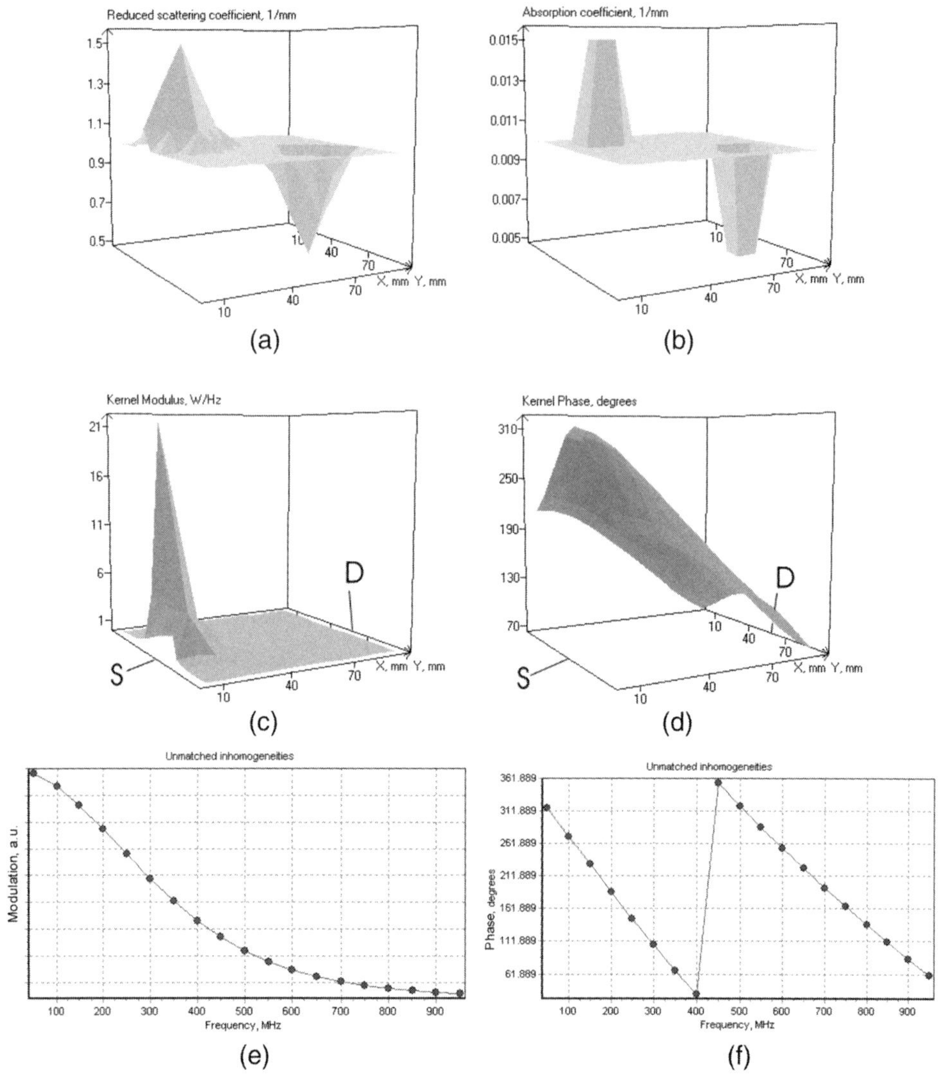

Figure 3.11 Frequency-domain response of an inhomogeneous medium, calculated with the diffusion approximation and the distributed-source approach (DSA). (a,b) Optical properties of the medium: absorption and reduced scattering coefficients, respectively. (c,d) Modulus modulus and phase of the DSA kernel inside the medium. Also shown are positions of the source (S) (50,0) mm and of the detector (D) (60,100) mm. (e,f) Modulation and phase of the registered signal.

We start with the FD transport equation (3.6). Let us introduce the coordinates ξ and η as the projections of the vector $\mathbf{s}'$ on the plane perpendicular to the vector $\mathbf{s}$. The specific intensity $\tilde{I}(\mathbf{r},\mathbf{s},\omega)$ for a fixed $\mathbf{r}$ can then be represented as a function of ξ and η:

$$H(\xi,\eta) = \tilde{I}[\mathbf{r},\mathbf{s}'(\xi,\eta),\omega]. \tag{3.65}$$

The function H can be expanded in a Taylor series (around $\xi = 0$ and $\eta = 0$):

$$H(\xi,\eta) = H(0,0) + \xi \left.\frac{\partial H}{\partial \xi}\right|_{\xi,\eta=0} + \eta \left.\frac{\partial H}{\partial \eta}\right|_{\xi,\eta=0} + \left.\frac{\xi^2}{2}\frac{\partial^2 H}{\partial \xi^2}\right|_{\xi,\eta=0} + \left.\frac{\eta^2}{2}\frac{\partial^2 H}{\partial \eta^2}\right|_{\xi,\eta=0} + \xi\eta \left.\frac{\partial^2 H}{\partial \xi \partial \eta}\right|_{\xi,\eta=0} + \ldots. \tag{3.66}$$

Substitution of Eq. (3.66) into the scattering integral on the right-hand side of Eq. (3.6) yields

$$\frac{1}{4\pi}\int_{4\pi} \tilde{I}(\mathbf{r},\mathbf{s}\prime,\omega)p(\mathbf{s}\cdot\mathbf{s}\prime)d\Omega\prime = \frac{1}{4\pi}H(0,0) + K\left.\left[\left(\frac{\partial^2}{\partial \xi^2} + \frac{\partial^2}{\partial \eta^2}\right)H\right]\right|_{\xi,\eta=0} + K\prime \left.\left[\left(\frac{\partial^2}{\partial \xi^2} + \frac{\partial^2}{\partial \eta^2}\right)^2 H\right]\right|_{\xi,\eta=0} + \ldots. \tag{3.67}$$

where

$$K = \frac{1}{8}\int_{-1}^{1}(1-\mu\prime^2)p(\mu\prime)d\mu\prime, \quad K\prime = \frac{1}{128}\int_{-1}^{1}[(1-\mu\prime^2)]^2 p(\mu\prime)d\mu\prime, \tag{3.68}$$

$\mu' = \cos(\theta')$, and θ' is the scattering angle.

Substituting Eq. (3.67) into Eq. (3.6) and neglecting all but the first two terms, one obtains the following forward-scattering approximation of the transport equation in the frequency domain:

$$s\cdot\nabla\tilde{I}(\mathbf{r},\mathbf{s},\omega) = -\mu_t(1+i\alpha\beta_2)\tilde{I}(\mathbf{r},\mathbf{s},\omega) + a\mu_t(\tilde{I}(\mathbf{r},\mathbf{s},\omega) + \Delta_s\tilde{I}(\mathbf{r},\mathbf{s},\omega)) + \tilde{S}(\mathbf{r},\mathbf{s},\omega), \tag{3.69}$$

where Δ_s is the 2-D Laplace operator acting on the angular coordinates. Conventional representation of $\mathbf{s}$ as

$$\mathbf{s} = (\sin\theta\cos\varphi, \sin\theta\sin\varphi, \cos\vartheta) \tag{3.70}$$

gives the following expression for Δ_s:

$$\Delta_s = \frac{\partial}{\partial \mu}(1-\mu^2)\frac{\partial}{\partial \mu} + \frac{1}{1-\mu^2}\frac{\partial^2}{\partial \phi^2}, \tag{3.71}$$

where $\mu = \cos(\theta)$.

Equation (3.69) is a good approximation of Eq. (3.6) when the following conditions are satisfied:

$$K \ll 1, \quad K' = O(K^2). \tag{3.72}$$

In addition, the incident light distribution should have a larger angular width compared to K.

For the Henyey–Greenstein scattering phase function, which adequately describes the single scattering law in many tissues, the parameter K can be expressed as

$$K_{HG} = \frac{1}{6}(1 - g^2). \tag{3.73}$$

Let us consider as an example the problem of finding the distribution of the frequency-domain total intensity $\tilde{U}_{pt}(r,\omega)$ in an infinite homogeneous medium for an isotropic point source located at the coordinate origin. The boundary conditions in this case demand that the intensity should approach zero when the distance from the source increases. Technically, it is easier to solve the problem for a *plane* isotropic source. The total intensity in the latter case $\tilde{U}_{pl}(r,\omega)$ completely determines the solution of the original problem:

$$\tilde{U}_{pt}(r,\omega) = -\frac{1}{2\pi r}\frac{\partial}{\partial r}\tilde{U}_{pl}(r,\omega). \tag{3.74}$$

Let the z axis be perpendicular to the source plane and the angle θ counted from the positive direction of the z axis. Then Eq. (3.69) takes the form

$$\mu\frac{\partial}{\partial z}\tilde{I}(z,\theta,\omega) + A_1\tilde{I}(z,\theta,\omega) = A_2\frac{\partial}{\partial\mu}(1 - \mu^2)\frac{\partial}{\partial\mu}\tilde{I}(z,\theta,\omega) + \tilde{S}_0\delta(z), \tag{3.75}$$

where

$$A_1 = a\mu_t K, \quad A_2 = \mu_t(1 - a + i\alpha\beta_2). \tag{3.76}$$

Because the specific intensity $\tilde{I}(z,\theta,\omega)$ can be considered as an even and periodic function of θ, it can be expanded into the following trigonometric series:

$$\tilde{I}(z,\theta,\omega) = \sum_{m=0}^{\infty}\tilde{L}_m(z,\omega)\cos(m\theta). \tag{3.77}$$

Substitution of Eq. (3.77) into Eq. (3.76) yields the following system of differential equations for the coefficients $\tilde{L}_m(z,\omega)$:

$$\frac{1}{2}\left(\frac{\partial\tilde{L}_{m-2}}{\partial z}-\frac{\partial\tilde{L}_{m+2}}{\partial z}\right)+B_{m-1}\tilde{L}_{m-1}+C_{m+1}\tilde{L}_{m+1}=\tilde{M}_{m-1}-\tilde{M}_{m+1}, \quad (3.78)$$

where

$$B_m = A_1 + A_2 m(m+1), \quad C_m = -A_1 + A_2 m(1-m),$$

and $\tilde{M}_m$ are the coefficients in the corresponding trigonometric expansion of the source term.

Taking a Fourier transform with respect to the coordinate z,

$$\tilde{N}_m(k,\omega) = \int_{-\infty}^{\infty} \tilde{L}_m(z,\omega)\exp(-ikz)dz, \quad (3.79)$$

one obtains a system of algebraic equations for the $\tilde{N}_m$'s:

$$\frac{ik}{2}(\tilde{N}_{m-2}-\tilde{N}_{m+2})+B_{m-1}\tilde{N}_{m-1}+C_{m+1}\tilde{N}_{m+1}=\tilde{S}_0\delta_{1\,m}, \quad (3.80)$$

where $\delta_{lm} = 1$ for $m = 1$ and 0 otherwise.

The sought total intensity $\tilde{U}_{pt}(r,\omega)$ can be expressed as

$$\tilde{U}_{pt}(r,\omega) = -\frac{i}{2\pi r}\times\int_{-\infty}^{\infty}\left[\sum_{m=0}^{\infty}\tilde{N}_m(k,\omega)\gamma_m\right]\times k\exp(ikr)dk, \quad (3.81)$$

where

$$\gamma_m = \begin{cases} 2, & \text{for } m=0 \\ 0, & \text{for } m \text{ odd} \\ \frac{-2}{m^2-1}, & \text{for } m \text{ even.} \end{cases} \quad (3.82)$$

Thus, the total intensity depends only on the even harmonics in the expansion of Eq. (3.77).

The meaningful truncations of the system of Eq. (3.80) are obtained if one considers a number of equations M_{tr}, which is a multiple of 4.

An example of using this technique is illustrated in Figs. 3.12(a–c), which also show a comparison with MC simulations and DA predictions.

Equation (3.69) accurately describes high-frequency PDWs, provided that the conditions of Eq. (3.72) are satisfied. Correspondingly, the TD counterpart of Eq. (3.69) should accurately describe the intensity of the early-arriving portion of the scattered light.

An alternative approach is to use higher-order approximations of the spherical harmonics techniques, such as P_3-approximation.[75,76]

Another avenue to a better description of the early-arriving photons is to use the path integral formulation, which has been developed in Refs. 77 to 79.

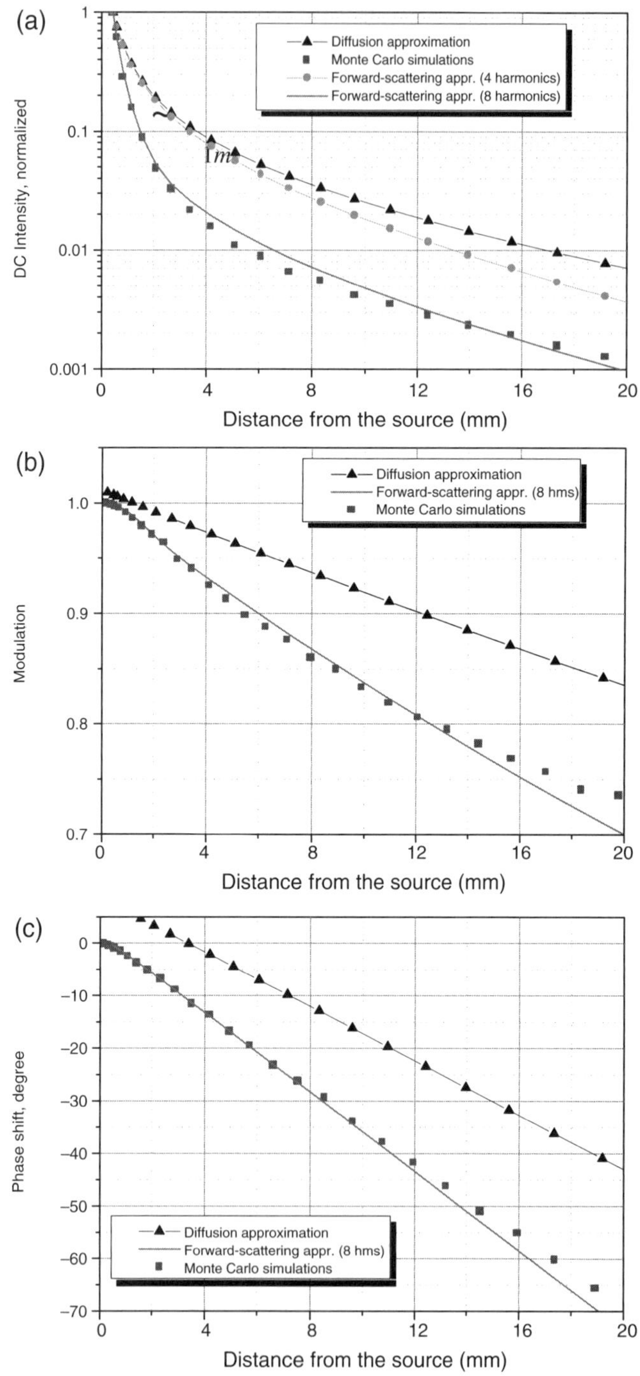

Figure 3.12 DC (a), modulation (b), and phase shift (c) of the radiation from a point source in a strongly anisotropically scattering medium (at 800 MHz), calculated with different models.

This approach makes explicit the probability of a particular path taken by a photon. The validity of this technique has been experimentally verified in Ref. 80. A similar approach, combining a single large-angle scattering with multiple small-angle scatterings and designed specifically for the back-scattering case, has been proposed in Ref. 81.

3.7 Role of the Single-Scattering Delay Time

In the preceding sections of this chapter, the single-scattering delay time t_s was assumed to be negligible compared to the mean time of flight t_f. In this section, we will consider how the time-dependent transport theory can be modified to account for a finite scattering delay time.

Following Ref. 11, we assume that the scattering delay function can be approximated by the exponential decay function:

$$f_s(t-t') = \frac{1}{t_s}\exp\left(-\frac{t-t'}{t_s}\right). \tag{3.83}$$

For the FD formulation, the Eqs. (3.7) must be replaced with

$$\eta_t = \mu_t(1+i\omega t_f), \quad b = \frac{a}{(1+i\omega t_f)(1+i\omega t_s)}. \tag{3.84}$$

It can be shown[82] that in the diffusion approximation, for low modulation frequencies, a nonzero-scattering delay time leads to the following modification of the FD diffusion equation (for homogeneous medium):

$$\nabla^2\tilde{U}_d(\mathbf{r},\omega) - 3\mu_a\mu_{tr}\tilde{U}_d(\mathbf{r},\omega) - 3i\omega\mu_{tr}\left(\frac{1}{c}+t_s\mu_s\right)\tilde{U}_d(\mathbf{r},\omega) = -\tilde{S}_d(\mathbf{r},\omega). \tag{3.85}$$

It can be further observed that Eq. (3.85) is equivalent to the standard diffusion equation (3.56) if the velocity of light c in the medium is replaced with the "transport"[83] velocity $c_{eff} = c_0/n_{eff}$, where the effective refractive index n_{eff} is given by

$$n_{eff} = n + t_s\mu_s c_o. \tag{3.86}$$

One interesting implication of Eq. (3.86) is the effect of a nonzero single-scattering delay time on the optical properties determined from time- or frequency-resolved measurements. As can be seen from Eqs. (3.58) and (3.86), the FD response of an infinite medium with a nonzero time t_s is undistinguishable from that of an effective medium with a zero time t_s and the apparent optical properties, given by

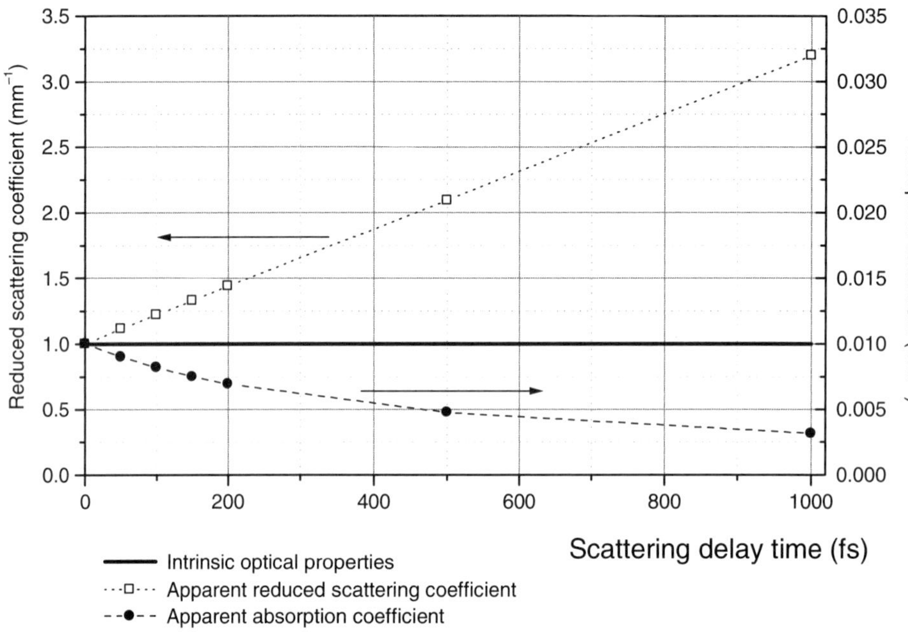

Figure 3.13 Apparent absorption coefficient μ_a^{app} (closed circles) and apparent reduced scattering coefficient $\mu_s'^{app}$ (open squares) obtained using the standard diffusion approximation from the data generated by the modified diffusion approximation. The infinite geometry was assumed. The data were generated for the following intrinsic optical properties: $\mu_a = 0.01$ mm^{-1}, $\mu_s = 10$ mm^{-1}, $g = 0.9$, and $n = 1.4$.[82]

$$\mu_a^{app} = \frac{n}{n_{eff}}\mu_a, \quad \mu_{tr}^{app} = \frac{n_{eff}}{n}\mu_{tr}. \tag{3.87}$$

As a result, the standard diffusion theory in the infinite geometry would yield apparent rather than true optical properties of the medium (with a perfect fit to experimental data). The magnitude of this deviation increases with the time t_s. This behavior is illustrated in Fig. 3.13.

The relevance of the single-scattering delay effect for biomedical optics depends, of course, on the ratio t_s/t_f in tissues. No experimental data on the single-scattering delay time in tissue are currently available. One may expect, however, that the role of the single-scattering delay is most pronounced for optically dense tissues such as blood. The single-scattering delay time can be calculated from the knowledge of the geometrical structure and the dielectric constant distribution of the scattering centers. One approach to doing so is referred to as the "energy delay" model. The principal idea of this model is to associate the scattering-delay time with the time required to store potential energy inside the scatterer,[84] that is,

$$t_s(\lambda) = \frac{W(\lambda)n}{\sigma_{sca}(\lambda)U_0 c}, \tag{3.88}$$

where λ is the vacuum wavelength of light, W is the electromagnetic energy inside the scatterer, σ_{sca} is the scattering cross-section, and U_0 is the energy density of the incident wave. For spherical dielectric scattering particles, the energy W can be calculated from Mie theory[85] as

$$W(\lambda) = U_0 \pi r^3 \sum_{n=1}^{\infty} \frac{2n+1}{y^2 - y^{*2}} \left\{ \alpha_n \left[\frac{A_n^*(y)}{y} - \frac{A_n(y)}{y^*} \right] + \beta_n \left[\frac{A_n^*(y)}{y^*} - \frac{A_n(y)}{y} \right] \right\}, \tag{3.89}$$

where

$$\begin{aligned} A_n(y) &= \frac{\psi'_n(y)}{\psi_n(y)}, \\ \alpha_n &= \frac{m^2 + m^{*2}}{2} [md_n\psi_n(y)][md_n\psi_n(y)]^* + mm^*[mc_n\psi_n(y)][mc_n\psi_n(y)]^*, \\ \beta_n &= \frac{m^2 + m^{*2}}{2} [mc_n\psi_n(y)][mc_n\psi_n(y)]^* + mm^*[md_n\psi_n(y)][md_n\psi_n(y)]^*, \\ y &= mx, \end{aligned} \tag{3.90}$$

r is the radius of the spherical scatterer, $m = m(\lambda)$ is the relative complex refractive index of the particle, ψ_n are the Ricatti–Bessel functions, c_n and d_n are coefficients of Mie series, x is the size parameter, and * stands for complex conjugate. The scattering delay time predicted by the model demonstrates a complicated dependence on the optical and geometrical properties of the medium. Figure 3.14 demonstrates predictions of the "energy delay" model for an erythrocyte, when the latter is approximated by a sphere with a radius of 3 μm. From these data, one can obtain an idea of the possible impact of a single-scattering delay on the time-dependent photon migration in blood (and blood-perfused soft tissues). For instance, the model yields the time t_s of 19.7 fs for blood at the wavelength of 765 nm. Integrating-sphere measurements yield the following optical properties of whole human blood at 765 nm (blood sample from a healthy volunteer; hematocrit 41%; oxygen saturation >98%; Mie scattering phase function and an an average index of refraction of 1.38 assumed in data processing): a scattering coefficient of 197.5 mm^{-1} and an absorption coefficient of 0.485 mm^{-1}. These numbers translate into the average time of flight between scattering events of 23.2 fs. Obviously, single-scattering delay effect should have a significant influence on the formation of the time-dependent optical response of such a medium.

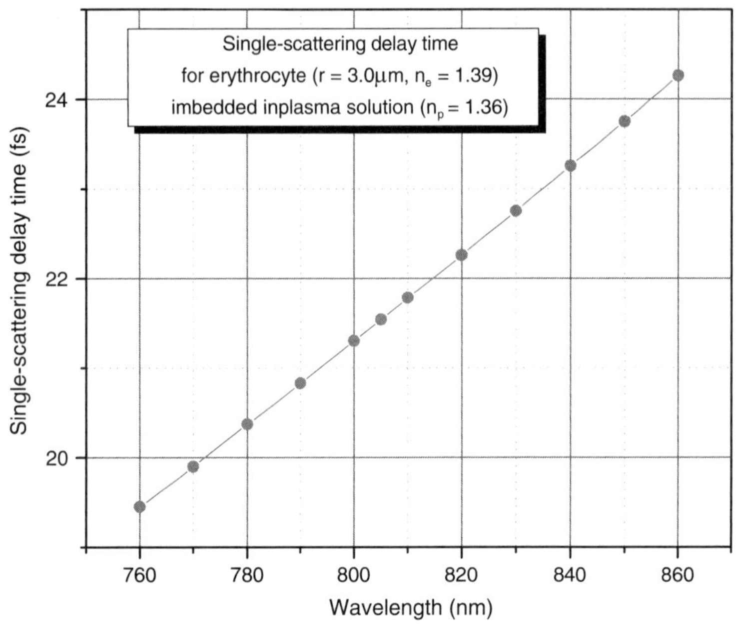

Figure 3.14 Single-scattering delay of a red blood cell as a function of wavelength. Calculated with the energy-delay model.

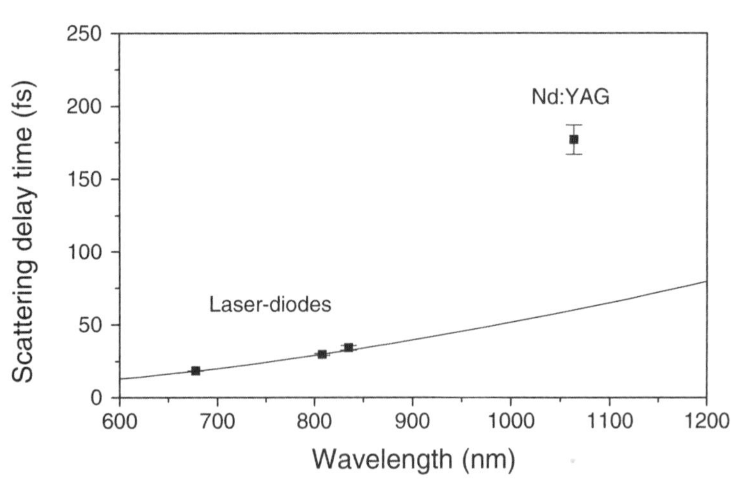

Figure 3.15 Comparison of the experimentally determined scattering delay times (symbols, mean +/– standard error) with the predictions of the Mie model (line). Phantom: monodisperse quartz spheres (radius 255 nm) embedded in an optically clear polyester resin.[86]

Equation (3.87) suggests how the single-scattering delay time can be determined experimentally. Two sets of data are required: one from a CW and one from a time-resolved experiment. Assuming that CW data are "true" and the time-resolved data are "apparent," one can immediately infer the value of the single-scattering delay time. Such an experiment on a tissue phantom (quartz spheres embedded in polyester resin) has been reported in Ref. 86. Figure 3.15 shows the measured scattering-delay times as well as predictions of the "energy delay" model. Furthermore, authors of Ref. 87 have compared optical properties of tissue phantoms measured using spatially and temporally resolved techniques and have found a trend compatible with the hypothesis of scattering delay time contribution.

3.8 Concluding Remarks

This chapter represents a brief glimpse of the time-dependent transport theory, which plays an ever-increasing role in the development of novel instruments for biomedical optical diagnostics. Among techniques and methods described in this chapter, it seems unlikely that just one can prevail in a foreseeable future as an "ideal" tool for solving all problems in time-resolved data interpretation. Rather, theoretical methods and approximations should be selected carefully according to the requirements of a particular diagnostic technique.

This work has been supported in part by the Biomedical Research Foundation of Northwest Louisiana.

References

1. M. A. Duguay and A. T. Mattick, "Ultrahigh speed photography of picosecond light pulses and echoes," *Appl. Opt.* **10**, 2162–2170 (1971).
2. C. G. Bachman, *Laser Radar Systems and Techniques*, Artech House, Dedham, MA (1979).
3. F. F. Jöbsis, "Noninvasive infrared monitoring of cerebral and myocardial oxygen sufficiency and circulatory parameters," *Science* **198**, 1264–1267 (1977).
4. K. Shimizu, A. Ishimaru, and L. Reynolds, "Backscattering of a picosecond pulse from densely distributed scatterers," *Appl. Opt.* **18**, 3484–3488 (1979).
5. B. Chance, J. Leigh, H. Miyake, D. S. Smith, S. Nioka, R. Greenfeld, M. Finander, K. Kaufmann, W. Levy, M. Young, P. Cohen, H. Yoshioka, and R. Boretsky, "Comparison of time-resolved and -unresolved measurements of deoxyhemoglobin in brain," *Proc. Nat. Acad. Sci. USA* **85**, 4971–4975 (1988).
6. D. T. Delpy, S. R. Arridge, M. Cope, D. Edwards, E. O. R. Reynolds, C. E. Richardson, S. Wray, J. S. Wyatt, and P. van der Zee, "Quantitation of pathlength in optical spectroscopy," *Adv. Exp. Med. Biol.* **248**, 41–46 (1989).
7. J. R. Lakowicz and K. Berndt, "Frequency-domain measurements of photon migration in tissues," *Chem. Phys. Lett.* **166**, 246–252 (1990).

8. J. C. Hebden, S. Arridge, and D. T. Delpy, "Optical imaging in medicine: I. Experimental techniques," *Phys. Med. Biol.* **42**, 825–840 (1997).
9. B. Chance, M. Cope, E. Gratton, N. Ramanujam, and B. Tromberg, "Phase measurement of light absorption and scatter in human tissue," *Review of Scientific Instruments* **69**, 3457–3481 (1998).
10. S. R. Arridge, "Optical tomography in medical imaging," *Inverse Problems* **15**, R41–R93 (1999).
11. V. V. Sobolev, *A Treatise on Radiative Transfer*, Van Nostrand Reinhold, Princeton, NJ (1963).
12. G. M. Wing, *An Introduction to Transport Theory*, Wiley, New York(1962).
13. A. Ishimaru, *Wave Propagation and Scattering in Random Media*, Academic Press, New York (1978).
14. S. R. Arridge, M. Cope, and D. T. Delpy, "The theoretical basis for the determination of optical pathlengths in tissue: temporal and frequency analysis," *Phys. Med. Biol.* **37**, 1531–1560 (1992).
15. L. Wang, P. P. Ho, X. Liang, H. Dai, and R. R. Alfano, "Ballistic 2-D imaging through scattering wall using ultrafast Kerr gate," *Science* **253**, 769–771 (1991).
16. V. V. Tuchin, *Tissue Optics*, SPIE Press, Bellingham, VA (2000).
17. D. A. Benaron and D. K. Stevenson, "Optical time-of-flight and absorbance imaging of biologic media," *Science* **259**, 1463–1466 (1993).
18. B. Tromberg, L. O. Svaasand, T. T. Tsay, and R. C. Haskell, "Properties of photon density waves in multiple-scattering media," *Appl. Opt.* **32**, 607–616 (1993).
19. A. H. Hielscher, F. K. Tittel, and S. L. Jacques, "Photon density wave diffraction tomography," *OSA Proceedings on Advances in Optical Imaging and Photon Migration* **21**. R. R. Alfano, Ed., pp. 78–82 (1994).
20. D. A. Boas, M. A. O'Leary, B. Chance, and A. Yodh, "Scattering of diffuse photon density waves by spherical inhomogeneities within turbid media: Analytic solution and applications," *Proc. Nat. Acad. Sci. USA* **91**, 4887–4891 (1994).
21. M.A O'Leary, D. A. Boas, B. Chance, and A. Yodh, "Fluorescence lifetime imaging in turbid media," *Opt. Lett.* **21**, 158–160 (1996).
22. T. L. Troy and E. M. Sevick-Muraca, "Fluorescence lifetime imaging and spectroscopy in random media," *Applied Fluorescence in Chemistry, Biology, and Medicine*, W. Rettig, B. Strehmel, S. Shrader, and H. Seifert. Springer Berlin Heidelberg, 3–36(1999).
23. R. L. Fante, "Relationship between radiative-transport theory and Maxwell's equations in dielectric media," *J. Opt. Soc. Am.* **71**, 460–468 (1981).
24. J. Sajeev, P. Gendi, and Y. Yumin, "Optical coherence propagation and imaging in a multiple scattering medium," *J. Biomed. Opt.* **1**, 180–191 (1996).

25. A. J. Welch, M. J. C. van Gemert, W. M. Star, and B. C. Wilson, "Definitions and overview of tissue optics." *Optical-thermal response of laser-irradiated tissue*. Plenum Press, New York, 15–46 (1995).
26. K. M. Yoo, F. Liu, and R. R. Alfano, "Biological materials probed by the temporal and angular profiles of the backscattered ultrafast laser pulses," *J. Opt. Soc. Am. B* **7**, 1685–1693 (1990).
27. V. Twersky, "Absorption and multiple scattering by biological suspensions," *J. Opt. Soc. Am.* **60**, 1084–1093 (1970).
28. Yu. N. Barabanenkov and M. Yu. Barabanenkov, "Radiative transfer theory with time delay for the effect of pulse entrapping in a resonant random medium: general transfer equation and point-like scatterer model," *Waves in Random Media* **7**, 607–633 (1997).
29. K. M. Case and P. F. Zweifel, *Linear Transport Theory*, Addison-Wesley, Reading, MA (1967).
30. K. Mitra and S. Kumar, "Development and comparison of models for light-pulse transport through scattering-absorbing media," *Appl. Opt.* **38**, 188–196 (1999).
31. I. V. Yaroslavsky, H.-J. Schwarzmaier, A. N. Yaroslavsky, and V. V. Tuchin, "The radiative transfer equation and its diffusion approximation in the frequency-domain technique: a comparison," *Proc. SPIE* **2326**, 465–474 (1994).
32. I. V. Busbridge, *The Mathematics of Radiative Transfer*, Cambridge University Press, London and New York (1960).
33. H. C. van de Hulst, *Multiple Light Scattering (Tables, Formulas, and Applications)*, Academic Press, New York (1980).
34. C. R. de Oliveria and K. Tahir, "Higher-order transport approximations for optical tomography applications," *Proc. SPIE* **3194**, 212–218 (1997).
35. S. Kumar, A. Majumdar, and C. L. Tien, "The differential discrete-ordinate method for solutions of the equation of radiative transfer," *Transactions of ASME* **112**, 424–429 (1990).
36. A. D. Klose and A. H. Hielscher, "Iterative reconstruction scheme for optical tomography based on the equation of the radiative transfer," *Med. Phys.* **26**, 1698–1707 (1999).
37. I. V. Yaroslavsky, A. N. Yaroslavsky, H. D. Battarbee, C. Sisson, and J. G. Rodriguez, "Distributed-source approach to image reconstruction in diffuse optical tomography," *Proc. SPIE* **3917**, 219–224 (2000).
38. J. M. Hammersley and D. C. Handscomb, *Monte Carlo Methods*, Wiley, New York (1964).
39. G. Fishman, *Monte Carlo: Concepts, Algorithms, and Applications*, Springer, New York (1996).
40. V. S. Antyufeev, *Monte Carlo Method for Solving Inverse Problems of Radiation Transfer*, De Gruyter, Berlin (1999).

41. S. L. Jacques and L. Wang, "Monte Carlo Modeling of Light Transport in Tissue," *Optical-Thermal Response of Laser-irradiated Tissue.* A. J. Welch and M. J. C. van Gemert, Eds., pp. 73–100, Plenum Press, New York (1995).
42. L. Wang and S. L. Jacques, "Optimized radial and angular positions in Monte Carlo modeling," *Med. Phys.* **21**, 1081–1083 (1994).
43. M. Keijzer, S. L. Jacques, S. A. Prahl, and A. J. Welch, "Light distributions in artery tissue: Monte Carlo simulations for finite-diameter laser beams," *Lasers in Surgery and Medicine* **9**, 148–154 (1989).
44. S. M. Ermakov and G. A. Mikhailov, *Course of Statistical Simulations*, Nauka, Moscow (1976).
45. J. R. Zijp and J. J. ten Bosch, "Use of tabulated cumulative density functions to generate pseudorandom numbers obeying specific distributions for Monte Carlo simulations," *Appl. Opt.* **33**, 533–534 (1994).
46. I. V. Yaroslavsky, A. N. Yaroslavsky, H.-J. Schwarzmaier, G. G. Akchurin, and V. V. Tuchin, "New approach to Monte Carlo simulation of photon transport in the frequency-domain," *Proc. SPIE* **26**, pp. 45–55 (1995).
47. I. V. Yaroslavsky, A. Terenji, S. Willmann, A. N. Yaroslavsky, H. Busse, and H.-J. Schwarzmaier, "Small-volume tissue spectroscopy using photon-density waves: apparatus and technique," *Proc. SPIE* **3597**, 465–473 (1999).
48. M. Testorf, U. Osterberger, B. Pogue, and K. Paulsen, "Sampling of time- and frequency-domain signals in Monte Carlo simulations of photon migration," *Appl. Opt.* **38**, 236–245 (1999).
49. L. R. Poole, D. D. Venable, and J. W. Campbell, "Semianalytic Monte Carlo radiative transfer model for oceanographic lidar systems," *Appl. Opt.* **20**, 3653–3656 (1981).
50. R. R. Meier, J.-S. Lee, and D. E. Anderson, "Atmospheric scattering of middle uv radiation from an internal source," *Appl. Opt.* **17**, 3216–3225 (1978).
51. E. Tinet, S. Avrillier, and J. M. Tualle, "Fast semianalytical Monte Carlo simulation for time-resolved light propagation in turbid media," *J. Opt. Soc. Am. A* **13**, 1903–1915 (1996).
52. P. Oliver and D. Gagnon, "Mathematical modeling of solid angle function, part I: approximation in homogeneous medium," *Optical Engineering* **32**, 2261–2270 (1993).
53. S. Chatigny, M. Morin, D. Asselin, Y. Painchaud, and P. Beaudry, "Hybrid Monte Carlo for photon transport through optically thick scattering media," *Appl. Opt.* **38**, 6075–6086 (1999).
54. L. Wang and S. L. Jacques, "Hybrid model of Monte Carlo simulation and diffusion theory for light reflectance by turbid media," *J. Opt. Soc. Am. A* **10**, 1746–1752 (1993).

55. G. Alexandrakis, T. J. Farell, and M. S. Patterson, "Monte Carlo diffusion hybrid model for photon migration in a two-layer turbid medium in the frequency domain," *Appl. Opt.* **39**, 2235–2244 (2000).
56. R. C. Haskell, L. O. Svaasand, T. T. Tsay, T. C. Feng, M. McAdams, and B. Tromberg, "Boundary conditions for the diffusion equation in radiative transfer," *J. Opt. Soc. Am. A* **11**, 2727–2741 (1994).
57. A. Kienle and M. S. Patterson, "Improved solutions of the steady-state and time-resolved diffusion equations for reflectance from a semi-infinite turbid medium," *J. Opt. Soc. Am. A* **14**, 246–254 (1997).
58. A. Ya. Polishchuk, S. Gutman, M. Lax, and R. R. Alfano, "Photon-density modes beyond the diffusion approximation: scalar wave-diffusion equation," *J. Opt. Soc. Am. A* **14**, 230–234 (1997).
59. D. J. Durian and J. Rudnick, "Photon migration at short times and distances and in cases of strong absorption," *J. Opt. Soc. Am. A*, **14**, 235–245 (1997).
60. A. D. Kim and A. Ishimaru, "Optical diffusion of continuous-wave, pulsed, and density waves in scattering media and comparisons with radiative transfer," *Appl. Opt.* **37**, 5313–5319 (1998).
61. A. H. Hielscher, R. E. Alcouffe, and R. L. Barbour, "Comparison of finite-difference transport and diffusion calculations for photon migration in homogeneous and heterogeneous tissues," *Phys. Med. Biol.* **43**, 1285–1302 (1998).
62. S. Fantini, M. A. Franceschini, and E. Gratton, "Semi-infinite geometry boundary problem for light migration in highly scattering media: a frequency-domain study in the diffusion approximation," *J. Opt. Soc. Am. B*, **11**, 2128–2138 (1994).
63. M. Keijzer, W. M. Star, and P. R. M. Storchi, "Optical diffusion in layered media," *Appl. Opt.* **27**, 1820–1824 (1988).
64. M. S. Patterson, B. Chance, and B. C. Wilson, "Time-resolved reflectance and transmittance for the noninvasive measurement of tissue optical properties," *Appl. Opt.* **28**, 2331–2336 (1989).
65. B. Pogue, M. S. Patterson, H. Jiang, and K. Paulsen, "Initial assessment of a simple system for frequency-domain diffuse optical tomography," *Phys. Med. Biol.* **40**, 1709–1729 (1995).
66. M. Schweiger and S. Arridge, "The finite element model for the propagation of light in scattering media: frequency-domain case," *Med. Phys.* **34**, 2683–2687 (1997).
67. MUDPACK: Multigrid Software for Elliptic Partial Differential Equations, www.scd.ucar.edu/css/software/mudpack/.
68. Time-Resolved Optical Absorption and Scatter Tomography, www.ucl.ac.uk/medphys/.

69. J. Adams, "Multigrid Fortran software for the efficient solution of linear elliptic partial differential equations," *Appl. Math . Comp.* **34**, 113–146 (1989).
70. M. Firbank, S. Arridge, M. Schweiger, and D. T. Delpy, "An investigation of light transport through scattering bodies with non-scattering regions," *Phys. Med. Biol.* **41**, 767–783 (1996).
71. J. Ripoll, M. Nieto-Vesperinas, S. Arridge, and H. Dehghami, "Boundary conditions for light propagation in diffusive media with nonscattering regions," *J. Opt. Soc. Am. A* **17**, 1671–1682 (2000).
72. S. Arridge, H. Dehghami, M. Schweiger, and E. Okada, "The finite element model for the propagation of light in scattering media: a direct method for domains with nonscattering regions," *Med. Phys.* **27**, 252–265 (2000).
73. R. P. Heminger, "Solutions of a modified transport equation for multiple scattering in suspensions of highly anisotropic scatterers," *J. Opt. Soc. Am.* **64**, 503–509 (1974).
74. I. V. Yaroslavsky, A. N. Yaroslavsky, and H.-J. Schwarzmaier, "Approximate time-dependent equation of radiative transfer for strongly forward-scattering media," *Proc. SPIE* **3194**, 285–294 (1997).
75. D. A. Boas, H. Liu, M. A. O'Leary, B. Chance, and A. Yodh, "Photon migration within P3 approximation," *Proc. SPIE* **2389**, 240–247 (1995).
76. H. Jiang, "Optical image reconstruction based on the third-order diffusion equation," *Optics Express* **4**, 241–246 (1999).
77. L. T. Perelman, J. Wu, I. Itzkan, and M. S. Feld, "Photon migration in turbid media using path integrals," *Phys. Rev. Lett.* **72**, 1341–1344 (1994).
78. L. T. Perelman, J. Wu, I. Itzkan, Y. Wang, R. R. Dasari, and M. S. Feld, "Time-dependent photon migration using path integrals," *Phys. Rev. E* **51**, 6134–6141 (1995).
79. L. T. Perelman, J. N. Winn, J. Wu, R. R. Dasari, and M. S. Feld, "Photon migration of near-diffusive photons in turbid media: a Lagrangian-based approach," *J. Opt. Soc. Am. A* **14**, 224–229 (1997).
80. J. N. Winn, L. T. Perelman, K. Chen, J. Wu, R. R. Dasari, and M. S. Feld, "Distribution of the paths of early-arriving photons traversing a turbid medium," *Appl. Opt.* **37**, 8085–8091 (1998).
81. W. Cai, B. Luo, M. Lax, and R. R. Alfano, "Time-resolved optical backscattering model in highly scattering media," *Opt. Lett.* **23**, 983–985 (1998).
82. I. V. Yaroslavsky, A. N. Yaroslavsky, V. V. Tuchin, and H.-J. Schwarzmaier, "Effect of the scattering delay on time-dependent photon migration in turbid media," *Appl. Opt.* **36**, 6529–6538 (1997).

83. D. Livdan and A. A. Lisyansky, "Diffusion of classical waves in random media with microstructure resonances," *J. Opt. Soc. Am. A* **13**, 844–850 (1996).
84. A. Lagendijk and B. A. van Tiggelen, "Resonant multiple scattering of light," *Phys. Rep.* **270**, 143–216 (1996).
85. A. Bott and W. Zdunkowski, "Electromagnetic energy within dielectric spheres," *J. Opt. Soc. Am. A* **4**, 1361–1365 (1987).
86. S. Willmann, A. Terenji, H. Busse, I. V. Yaroslavsky, A. N. Yaroslavsky, H.-J. Schwarzmaier, and P. Hering, "Scattering delay time of Mie scatterers determined from steady-state and time-resolved optical spectroscopy," *J. Opt. Soc. Am. A* **17**, 745–749 (2000).
87. J. Swartling, J. S. Dam, and S. Andersson-Engels, "Comparison of spatially and temporally resolved diffuse-reflectance measurement systems for determination of biomedical optical properties," *Appl. Opt.* **42**, 4612–4620 (2003).
88. Q. Fang and D. A. Boas, "Monte Carlo simulation of photon migration in 3D turbid media accelerated by graphics processing units," *Opt. Express* **17**, 20178–20190 (2009).
89. X. Wang, L. V. Wang, C. Sun, and C. Yang, "Polarized light propagation through scattering media: time-resolved Monte Carlo simulations and experiments," *J. Biomed. Opt.* **8**, 608–617 (2003).
90. C. Zhu and Q. Liu, "Review of Monte Carlo modeling of light transport in tissues," *J. Biomed. Opt.* **18**, 050902–050902 (2013).
91. X. Wang and L. V. Wang," Propagation of polarized light in birefringent turbid media: a Monte Carlo study," *J. Biomed. Opt.* **7**, 279–290 (2002).
92. E. Alerstam, S. Andersson-Engels, and T. Svensson, "White Monte Carlo for time-resolved photon migration," *J. Biomed. Opt.* **13**, 041304–041304-10 (2008).
93. E. Alerstam, W. C. Yip Lo, T. D. Han, J. Rose, S. Andersson-Engels, and L. Lilge, "Next-generation acceleration and code optimization for light transport in turbid media using GPUs," *Biomed. Opt. Express* **1**, 658–675 (2010).
94. A. H. Barnett, "A fast numerical method for time-resolved photon diffusion in general stratified turbid media," *J. Computational Physics* **201**, 771–797 (2004).
95. Q. Liu and N. Ramanujam, "Scaling method for fast Monte Carlo simulation of diffuse reflectance spectra from multilayered turbid media," *J. Opt. Soc. Am. A* **24**, 1011–1025 (2007).
96. D. Boas, J. Culver, J. Stott, and A. Dunn, "Three dimensional Monte Carlo code for photon migration through complex heterogeneous media including the adult human head," *Opt. Express* **10**, 159–170 (2002).

Ilya V. Yaroslavsky received his MSc degree *summa cum laude* in physics in 1990 and PhD in laser physics in 1994, both from Saratov State University, Saratov, Russia. From 1994 to 2000, he did his postdoctoral training in Heinrich Heine University, Düsseldorf, Germany, working on laser interstitial thermotherapy of brain tumors, and in Louisiana State University, Shreveport, Los Angeles, USA, developing optical diffusion techniques for stroke diagnostics. He started his industrial carrier at Palomar Medical Technologies, Inc., Burlington, Massachusetts, USA, in 2000 and 2012 assumed position of the vice president of Advanced Research of the company. In 2015, he joined IPG Medical Corporation as Manager for Advanced Product Development. His scientific interests include light-tissue interactions and use of lasers for biomedical applications. He has authored and coauthored more than 50 scientific papers and inventions. He served as a member of IEC TC 76 Technical Advisory Group on optical radiation safety and laser equipment.

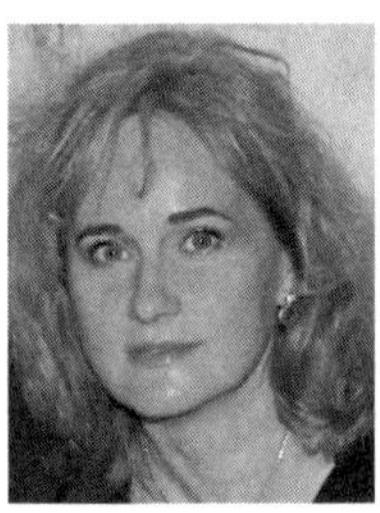

Anna N. Yaroslavsky is an associate professor of physics and director of Advanced Biophotonics Laboratory at the University of Massachusetts, Amherst, Massachusetts, USA. She obtained her PhD from the Saratov State University, Saratov, Russia, in 1999 and joined Wellman Center for Photomedicine (WCP), Boston, USA, in 2000. From 2005 to 2010, she has been an assistant professor of Dermatology at the Harvard Medical School, Boston, USA, and since 2010, she is a visiting scientist at WCP. Her expertise is in biomedical optics and medical imaging. Her research is focused on the development of optical and multimodal technologies for medical applications.

Chapter 4
Coherence Phenomena and Statistical Properties of Multiply Scattered Light

Dmitry A. Zimnyakov
Yuri Gagarin State Technical University of Saratov, Saratov, Russia

4.1 Introduction

This chapter considers important phenomena appearing as a result of the interaction of coherent light with optically dense disordered media. These media are characterized by an extreme level of randomness. Light is typically scattered numerous times and loses information about its initial propagating direction but nevertheless preserves its coherence in certain conditions. The situation of coherent light propagating significant distances in scattering systems is not obvious. However, there are several classical examples of coherence persisting despite multiple scattering by the random media. Weak localization of light and related effects such as coherent backscattering is one example. Another illustration is the existence of temporal and spatial correlations of multiply scattered light, which reveals information on the microscopic dynamic and structure properties of the scattering system.

Most of the theoretical and experimental studies related to different manifestations of coherence in multiple scattering have been published during the last two decades, beginning with the classical works of Golubentsev,[1] Stephen,[2] and MacKintosh and John.[3] This chapter briefly reviews the common aspects of coherence of light in multiple scattering that are important in gaining a better understanding of the optics of condensed media and of practical applications in industrial and medical diagnostics.

We begin our consideration with the effect of weak localization of light and its consequences on coherent backscattering (Section 4.2). This is one of the most important fundamental phenomena of classical optics of condensed matter and disordered media. We illustrate the basic ideas of the classical physics of localization using several examples borrowed from the lecture "Localization of light in disordered and periodic dielectrics" given by John at Erice Summer School (July 1993) (see also Ref. 4). Statistical (correlation) properties of the multiply scattered light and methods of studying the optically dense disordered and weakly ordered systems by means of correlation spectroscopy are considered in Section 4.3. Different approaches for the analysis of distributions of effective optical paths of partial components of scattered optical fields are discussed in Section 4.4 as the basis for quantification of the statistical properties of the multiply scattered light. Section 4.5 analyzes some fundamental relations between correlation and polarization characteristics of the multiply scattered coherent light that can be interpreted as the existence of self-similarity in multiple scattering. Finally, Section 4.6 discusses applications of the coherence phenomena in multiple scattering for characterization of scattering systems in biomedicine, material science, and other areas.

4.2 Weak Localization of Light in Disordered and Weakly Ordered Media

The existence of localized (bound) states is usually associated with massive particles that are forced by the periodic or random coordinate-dependent potential distributions. This effect was first pointed out in 1958 by Anderson[5] for the case of electrons in three-dimensional disordered conductors. This phenomenon has a quantum-mechanical nature and can be analyzed in terms of a solution of the Schrödinger equation that describes the behavior of an electron with energy E and effective mass m^* in the presence of a random potential $V(x)$ related to the structure disorder of conductor

$$\left[-\frac{\hbar^2}{2m^*}\nabla^2 + V(x)\right]\psi(x) = E\psi(x). \tag{4.1}$$

For sufficiently negative energy values, the localized electron states correspond to deep potential fluctuations that can trap an electron. With an increase of energy or a decrease of the root mean square (rms) value of the potential fluctuations V_{rms}, the probability for an electron to tunnel between two neighboring bound states increases. Introducing the correlation length of potential fluctuations a, we can evaluate the corresponding energy scale as $E_a \equiv \frac{\hbar^2}{2m^* a^2}$. It can be shown that if the potential fluctuations are small enough (the so-called weak disorder limit, $V_{\text{rms}} \ll E_a$), then for an energy value equal to $-V_{\text{rms}}^2/E_a$, the electron can propagate through the whole disordered solid due to the successive tunneling between the localized states. This process is like a slow diffusion that gives rise to a certain level of conductivity. Further increase of the energy allows the electron to traverse the solid with relative ease. The critical energy value, $-V_{\mathrm{rms}}^2/E_a$, can be termed as the mobility edge (lecture "Localization of light in disordered and periodic dielectrics" given by John at Erice Summer School, July 1993; see also Ref. 4).

In the strong disorder limit when $V_{\mathrm{rms}} \gg E_a$, the mobility edge is located in the conduction band continuum; thus, the corresponding energy states become localized. This is the so-called Anderson transition.

In the case of classical waves, certain analogies with quantum-mechanical localization can be considered. In particular, in 1983, John and Stephen[6] showed that in a three-dimensional solid, some but not all of the normal modes of vibrations (phonon states) were localized. The possibility of the existence of localized photon states in the case of a three-dimensional random dielectric medium with real positive refractive index was first proposed in 1984 by John.[7]

The possibility of localization of the electromagnetic waves in the disordered dielectric medium can be analyzed by considering the classical wave equation for the electric field amplitude, which is written in a form resembling the Schrödinger equation:

$$-\nabla^2\bar{E} + \bar{\nabla}(\bar{\nabla}\cdot\bar{E}) - \frac{\omega^2}{c^2}\varepsilon_{\mathrm{fluct}}(x)\bar{E} \equiv \varepsilon_0\frac{\omega^2}{c^2}\bar{E}. \tag{4.2}$$

Here, ω and c are correspondingly the frequency and the speed of the electromagnetic wave; the total dielectric constant is separated into the average value ε_0 and spatially fluctuating component $\varepsilon_{\mathrm{fluct}}(x)$. This fluctuating part plays a role similar to the random potential $V(x)$ associated with disordered structure in the Schrödinger equation [Eq. (4.1)] and causes scattering of the monochromatic electromagnetic wave. In the case of lossless dielectric material (with a real and positive dielectric constant everywhere) the quantity $\varepsilon_0\omega^2/c^2$, similar to an energy eigenvalue in the case of the Schrödinger equation, is always positive. Thus, in the case of electromagnetic waves, localization cannot be achieved by the usual diminishing of the photon energy as happens for Anderson localization of electrons. On the other hand, at higher photon energies, the ray optics analysis becomes valid and interference effects play an insignificant role in the light transport through the dielectric medium. Hence, for both cases (the high-frequency as well as the low-frequency limit), we can obtain not localized, but extended states of the electromagnetic field, and we should seek the localization conditions in the intermediate region between high-energy and low-energy limits (localization window).

Localization of light in the highly scattering disordered media is manifested as the existence of microscopic resonances of the electromagnetic field at distances comparable with the wavelength of light due to the constructive interference of the multiply scattered partial components of the total electromagnetic field into the scattering medium. Such resonances can be thought of as existing due to the presence of a microcavity or microscopic resonator with a certain geometry and large photon-scattering cross section. These microcavities appear due to the random configurations of the scattering sites that induce constructive interference of the scattered components of the electromagnetic field. Thus, photons within such "virtual" resonators will have a long dwell time before leaking. The product of the optical frequency and the leakage time is the quality factor of such microscopic resonators. This factor is inversely proportional to the frequency width of the microscopic scattering resonance.

The localization condition in the case of electromagnetic waves in disordered media can be obtained from analysis of the wavelength dependency of the scattering system parameters in the intermediate region between the photon high-energy and low-energy limits. The most important and universal parameter controlling the transport of classical waves in the disordered system is the elastic mean free path that can be determined as the characteristic scale of the full randomization of the wave propagation direction in the disordered medium. For distances larger than l^*, wave transport in the medium can be considered a diffusion process. Bearing in mind that interference effects cause significant spatial fluctuations of the wave intensity in the disordered medium, we can conclude that the elastic mean free path plays the central role in the physics of localization of classical waves. When l^* becomes comparable with the wavelength, the interference of the multiply scattered partial components of the propagating wave significantly influences the propagation conditions and causes the transition from extended to localized states of the classical wave. For qualitative analysis, to characterize this transition in its most simple and nonrigid form, one can use the simplified version of the Ioffe–Regel condition for free photon localization:[7,8] $2\pi l^*/\lambda \approx 1$. Thus, all of the normal modes of the electromagnetic field for which $l^* \leq \lambda/2\pi$ can be considered the localized ones.

Let us illustrate the possibility of the transition from the extended modes of the electromagnetic wave to the localized states in the disordered medium using a simple example. For discrete scattering systems consisting of spherical dielectric particles, the elastic mean free path is a function of the particle concentration, their size parameter ka (k is the wave number of light, and a is the particle radius), and the refractive indices of the particle and surrounding background medium. For small-sized particles with $a \ll \lambda$, the Rayleigh scattering regime takes place and hence $l^* \sim \lambda^4$. [This follows from the analysis of the wavelength dependence of the scattering cross section σ and the relation between l^*, σ, concentration of the particles n, and the mean scattering length l in the case of Rayleigh scattering limit: $l^* \approx l = (n\sigma)^{-1}$.] Thus, in the small energy limit, l^* monotonically rises as the fourth power of the wavelength. On the other hand, the transition into the high-energy region, $\lambda \ll a$, results in

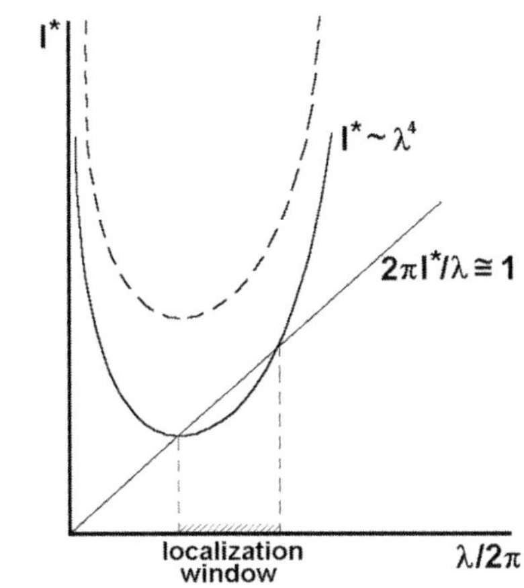

Figure 4.1 Scheme illustrating the transition from the extended states of the electromagnetic field in the disordered system of Rayleigh scatterers to the localized states (Ref. 4). If the scattering microstucture is characterized by a single spatial scale associated with the scatterer radius, then $\lambda \gg a$ and $\lambda \ll a$ are weakly scattering regimes of Rayleigh scattering and classical geometric optics, respectively. For low values of the refractive indices ratio n_a/n_b (subscripts *a* and *b* correspond to scatterer and background medium, respectively), all states of the electromagnetic field are extended (upper curve). For values of n_a/n_b not less than 2.0, localized states of electromagnetic field appear (lower curve).

reduced interference effects; in this case, we can use the following formal relation: $\lambda \rightarrow 0$; $l^* \rightarrow \infty$. Thus, the dependence of the elastic mean free path on the wavelength is characterized by the presence of a minimum in the intermediate region $0 < \lambda < \infty$. This minimum plays an important role in the transition from extended modes of the electromagnetic field to localized modes and back again with the change of the photon energy. As illustrated in Fig. 4.1, using the Ioffe–Regel condition for localization in the preceding form $2\pi l^* \cong \lambda$, we can find the position of the localization window for disordered media consisting of Rayleigh scatterers in the region of the crossover between the high-energy and low-energy modes of the electromagnetic field. The necessary condition for this is an appropriate volume density of the scattering sites and/or appropriate values of the refractive indices of the scatterers and surrounding background medium. In further analysis, considering the more real physical picture and taking into account the existence of the Mie scattering regime in the intermediate region $\lambda \sim a$, we can expect some peculiarities in the position of localization window to appear due to the presence of Mie resonances.

If the first Mie resonance is characterized by a large enough value of the scattering cross section (due to large values of the n_a/n_b ratio) and the volume fraction of the scattering particles is large, then the transport mean free path would be adequately small and the Ioffe–Regel condition for localization would be satisfied. We also would expect that for higher-volume densities of the scattering sites, the localization window would appear in the vicinity of the first Mie resonance. But for such high-volume densities of the scattering particles, the average distance between spheres will be less than the characteristic size of the "region of influence" related to the scattering cross section and all the scatterers will be optically connected. Such "optical connections" result in the elastic mean free path that will increase, not decrease, with the rise of volume fraction of scattering particles. Thus, from the viewpoint of the model of independent scatterers or microscopic resonances, free photon criterion of localization (which corresponds to the applied form of the Ioffe–Regel condition) is very sensitive to the changes of the optical properties of the scattering medium and, as mentioned by John, "is a very delicate one [criterion] to achieve."

Expanding these ideas to systems of partially correlated scatterers (i.e., weakly ordered media), we find a new quality to the phenomenon of light localization in highly scattering media. The effects of spatial correlation of a great number of "microcavities," or microscopic "resonators," will lead to the interference interactions between different microresonances. As a result, stronger macroscopic resonances will appear and "in this case the scale of the localized state of the electromagnetic field is large compared to the optical wavelength and spans many individual microcavities" (S. John, "Localization of light in disordered and periodic dielectrics," lecture, Erice Summer School, July 1993). In short, changing from disordered to strongly correlated media causes the appearance of the photonic bandgap as the manifestation of the localization of light.

Coherent backscattering is one of the most familiar examples of the manifestation of weak localization of coherent light propagating in multiple-scattering disordered systems. The first experimental study of this phenomenon—giving evidence of the existence of the classical analogy of the quantum-mechanical Anderson localization in the optical domain—was carried out by M. Van Albada and A. Lagendijk[9] and P. E. Wolf and G. Maret.[10] To illustrate the physical picture of coherent backscattering in bulk scatterers illuminated by a plane monochromatic wave, let us consider the following simple qualitative model (Fig. 4.2).

Taking into consideration the multiple scattering of a plane monochromatic wave characterized by the sequence of wave vectors $\bar{k}_0, \bar{k}_1, \bar{k}_2, \ldots, \bar{k}_m$, one can see that for fixed directions of incidence ($\bar{k}_0$) and of observation ($\bar{k}_m$), each sequence (a) has a reverse counterpart (b) given by $\bar{k}_0, -\bar{k}_{b-11}, -\bar{k}_2, \ldots -\bar{k}_{-1}, \bar{k}_m$. In the specific case of backscattered light, when $\bar{k}_m = -\bar{k}_0$, the waves associated with direct and inverse paths (a and b) have equal complex amplitudes ($A_a = A_b$) after medium. Thus, these scattered waves interfere constructively and the corresponding intensity, $4|A_a|^2$, is twice the incoherent background intensity estimated without taking into consideration interference effects. The effect of constructive interference for backscattered light is illustrated in Fig. 4.2.

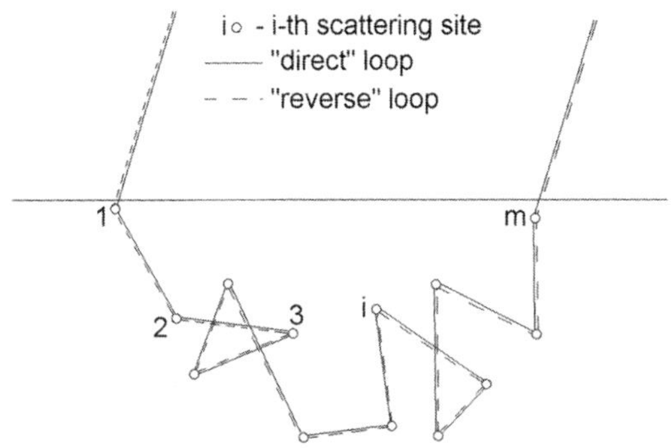

Figure 4.2 Simple geometrical interpretation of coherent backscattering as the manifestation of weak localization of light in disordered scattering media.

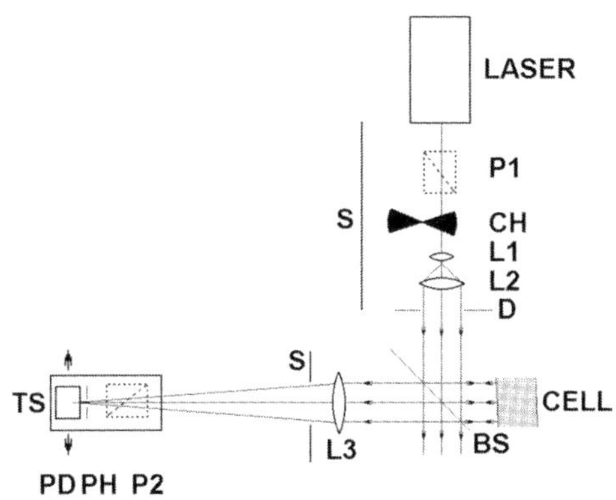

Figure 4.3 Typical scheme for measuring coherent backscattering in disordered media (Ref. 9). P1, P2: polarizers; CH: chopper; L1, L2, L3: lenses; D: diaphragm; BS: beam splitter; S: screen; TS: translation stage; PD: photodetector; and PH: pinhole.

Further, if $\bar{k}_m$ differs from $\bar{k}_0$, the phase shift between the partial contribution characterized by "direct" and "inverse" propagation through the scattering medium (a and b, respectively) can be estimated as $(\bar{k}_0 + \bar{k}_m) \cdot (\bar{r}_1 - \bar{r}_m)$, where $\bar{r}_1$ and $\bar{r}_m$ are the positions of the first and last scattering centers, respectively, in the considered sequence of the scattering events. In this case, the average interference term due to all contributions with m scattering events will be nonzero and positive for $|\bar{k}_m + \bar{k}_0| < L_m^{-1}$, where L_m is the average diameter of the loops corresponding to trajectories of scattered waves in the scattering medium (these loops are not necessarily closed). It is obvious that the minimum value of L_m is the average distance between two successive scattering events [i.e., the elastic mean free path l (in the case of isotropic scattering when $l \approx l^*$)]. Hence, we should expect an increase of the scattered intensity from the incoherent background value by up to a factor of 2 inside a cone of angular width of order λ/l centered at the backscattering direction. The overall effective cross section, including the effect of coherent multiple scattering, is thus larger than the classical one (σ_0) by a quantity of order $(\lambda/l)^2\sigma_0$ (in three dimensions) and the transport coefficient associated with the wave has a relative correction proportional to $-(\lambda/l)^2$. This is a result confirmed by more sophisticated diagrammatic expansions.[11] More rigidly, this result can be obtained using diagrammatic calculations of the intensity of light backscattered from a semi-infinite medium. These calculations give the following form for the angular dependence of the backscattered light:

$$I(\theta) \sim \frac{1}{(l^{-1} + k\theta)^2}\left\{1 + \frac{1}{lk\theta}[1 - \exp(-4k\theta z_{ext}/3)]\right\}, \tag{4.3}$$

where z_{ext} is the so-called extrapolation length that can be obtained from the boundary conditions for the diffuse light propagation in the semi-infinite medium and is of the order of l. In the case of anisotropic scattering, the mean scattering length l should be replaced with the transport mean free path l^*. For small scattering angles, the angular dependence of the intensity of coherently backscattered light in the case of anisotropic scattering has the asymptotic form

$$I(\theta) \sim 1 - \frac{5}{3}k\theta\, l^*, \tag{4.4}$$

which predicts the triangular form of the coherent backscattering peak observed in experiments.[11] Thus, we can see that the coherent backscattering of light from the dimensionless medium is controlled by the dimensionless parameter kl^* and damping of coherent backscattering occurs when $kl^* \sim 1$, which is similar to the Ioffe–Regel condition for free photon localization presented previously.

One of the typical schemes used to observe the coherent backscattering effect[9] is shown in Fig. 4.3. A He-Ne laser is used as an illumination source. Divergence of the illuminating beam is reduced to less than 0.1 mrad using a telescopic system consisting of the lenses L1 and L2. The expanded beam is reflected onto the sample from a beam splitter. The angular dependence of the backscattered intensity is measured by the movable detector in the dependence on the angle of scattering using the detector with the pinhole placed in a focal plane of the lens placed immediately behind the beam splitter. The state of polarization of the backscattered light is studied by means of a polarizer (Nicol prism) that is fitted directly to the detector housing. The cell with a multiple-scattering medium (such as aqueous suspensions of polystyrene beads) is usually titled with respect to the optical axis to keep its window reflections away from the detector. An optical chopper is used for modulation of the incident light to allow lock-in detection to improve the signal-to-noise ratio in the backscattering intensity measurements.

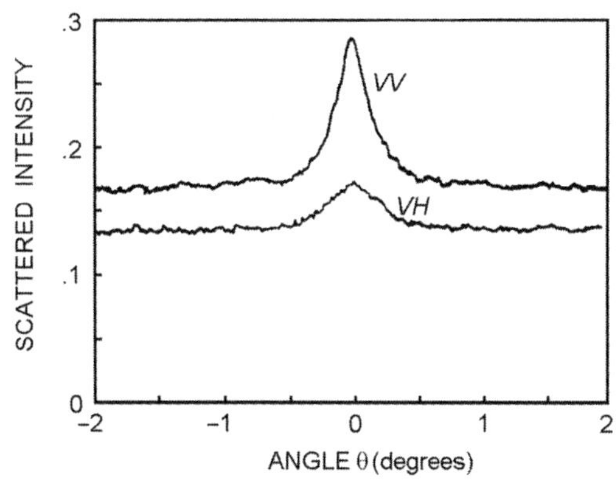

Figure 4.4 Angular dependencies of the backscattered light intensity (in arbitrary units) for copolarized (VV) and cross-polarized components of the scattered light (VH) (Ref. 10).

Experiments with a phantom scattering media, such as aqueous suspensions of polystyrene beads, show that the height of the coherent backscattering peak is roughly constant, while its width $\Delta\theta$ increases with increasing volume fraction f of the scattering particles; this is consistent with the theoretical prediction $\Delta\theta \sim \lambda/l$ [i.e., $\Delta\theta \sim f$ (for scattering media consisting of noninteracting scattering centers)].

One of the intrinsic properties of coherent backscattering in disordered media is its high polarization sensitivity; experiments with aqueous suspensions of polystyrene beads show that the height of the copolarized backscattering peak relative to the incoherent background intensity is larger than that for the cross-polarized component (see Fig. 4.4). Following Wolf and Maret,[10] we can qualitatively explain the similar polarization sensitivity due to the angular dependence of the light scattered from each scattering center. The scattered amplitude in direction $\bar{k}'$ is proportional to the projection of the incident polarization P onto the plane normal to $\bar{k}'$ [i.e., proportional to $\bar{k}' \times (\bar{P} \times \bar{k}')$]; hence, the amplitudes corresponding to the loop $(\bar{k}_0, \bar{k}_1, \ldots, \bar{k}_m, \bar{k}_0)$ and its inverse are proportional to

$$\bar{k}_0 \times [\bar{k}_0 \times [\bar{k}_m \times [\bar{k}_m \cdots\cdots [\bar{k}_1 \times [\bar{k}_1 \times \bar{P}]] \cdots]]]$$

and

$$\bar{k}_0 \times [\bar{k}_0 \times [\bar{k}_1 \times [\bar{k}_1 \cdots\cdots [\bar{k}_m \times [\bar{k}_m \times \bar{P}]] \cdots]]],$$

respectively; these two expressions are generally not equal.

Considering the approach suggested by Van Albada and Lagendijk,[9] we can describe the evolution of the two-dimensional polarization vector using a set of 2×2 matrices (which are similar to Jones matrices), resulting in the relation between the incoming and outgoing light $\bar{P}_{\text{out}} = M\bar{P}_{\text{in}}$, where matrix

M is the product of a set of matrices each of which corresponds to one of the scattering events. In principle, the resulting light intensity should be obtained from the square of the sum over all possible light paths. Van Albada and Lagendijk have found experimentally that the multiple-scattering background (i.e., outside the enhanced backscattering cone) is totally depolarized; thus, outside the cone, it is allowed to provide an incoherent summation to obtain the scattered light intensity. From this, one can conclude the equality of $\langle M_{11}^2\rangle$, $\langle M_{22}^2\rangle$, $\langle M_{12}^2\rangle$, $\langle M_{21}^2\rangle$:

$$\langle M_{11}^2\rangle = \langle M_{22}^2\rangle = \langle M_{12}^2\rangle = \langle M_{21}^2\rangle, \tag{4.5}$$

where the brackets denote a summation over all possible light paths. Inside the critical cone, one should coherently add to each light path its inverse, which can be obtained from a given path by reversing all of its momenta. To calculate the effect of the reverse path, Van Albada and Lagendijk have used the result that the matrix for the reverse path $\tilde{M}$ has given by $\tilde{M}_{ij} = M_{ji}$ and symmetry requirements are manifested in the vanishing of all cross products: $\langle M_{ij}M_{kl}\rangle \sim \delta_{ik}\delta_{jl}$. Applying these relations, they have found that the enhancement factor (which describes the height of the backscattering compared with incoherent background) is equal to 2 for the backscattered light polarized parallel to the incident light (copolarized component) and 1 for the perpendicular (cross-polarized) component. In practice, the experimentally observed enhancement factor is usually less than 2 for the copolarized component and larger than 1 for the cross-polarized one.[9]

4.3 Correlation Properties of Multiply Scattered Coherent Light: Basic Principles and Methods

4.3.1 Theoretical background for correlation analysis of multiply scattered dynamic speckles

The existence of finite spatial and temporal correlation scales for amplitude and intensity fluctuations of coherent light propagating in optically dense random media is another manifestation of coherence effects of multiply scattered light by disordered and weakly ordered media.[12] If coherent light is scattered by a nonstationary disordered medium, then intensity fluctuations detected at a fixed observation point can be characterized by the temporal correlation function of the intensity fluctuations as follows:

$$\tilde{G}_2(\tau) = \langle \delta I(t+\tau)\delta I(t)\rangle = \langle I(t+\tau)I(i)\rangle - \langle I(t)\rangle^2. \tag{4.6}$$

This expression is actually the fourth-order correlation function of the complex amplitude of the detected electromagnetic wave $E(t)$ and can be factorized as

$$\langle \delta I(t+\tau)\delta I(t)\rangle = |\langle \delta E(t+\tau)\delta E^{*}(t)\rangle|^{2}, \tag{4.7}$$

where symbol * denotes the complex conjugation. For certain scattering conditions (zero mean value of the complex amplitude of the scattered field and ergodicity of the scattering system), the relation between the correlation function of the scattered light intensity $G_2(\tau) = \langle I(t+\tau)I(t)\rangle$ and the correlation function of the complex amplitude $G_1(\tau) = \langle E(t+\tau)E^{*}(t)\rangle$ is determined by the Siegert formula

$$g_2(\tau) = G_2(\tau)/G_2(0) = 1 + \beta|g_1(\tau)|^2 = 1 + \beta|G_1(\tau)/G_1(0)|^2, \tag{4.8}$$

where the factor β depends on the detection conditions and is equal to 1 under ideal circumstances. For simplicity, the scalar wave approach is often used to describe the statistics of multiply scattered coherent light. Despite the obvious physical restrictions of this approach, it provides adequately valid results for the vast majority of scattering systems, provided that appropriate scattering and detection conditions are chosen. Moreover, the scalar wave formalism can be modified appropriately to describe the propagation of polarized light in disordered media.[3]

Propagation of coherent electromagnetic waves in random media can be considered as a sequence of statistically independent scattering events, taking place in the moment t at positions $\bar{r}_1(t), \bar{r}_2(t), \ldots$. Each scattering event is characterized by the wave vector $\bar{k}_i$. Thus, the coordinate-dependent part of the complex amplitude of the scattered field can be presented as

$$\delta E(t) \sim \exp \sum_i j\bar{k}_i \bar{r}_i(t). \tag{4.9}$$

We now follow the physical picture first presented by Maret and Wolf.[13] The scattered field [Eq. (4.9)] interferes with itself but at time $t+\tau$. In this analysis, we neglect the time delay of light propagation; correspondingly, we neglect displacements of scatterers during this propagation time. Thus, the temporal correlation function of the scattered field fluctuations can be expressed as

$$G_1(\tau) \sim \left\langle \exp\left\{ j\sum_i \bar{k}_i[\bar{r}_i(t+\tau) - \bar{r}_i(t)] \right\} \right\rangle. \tag{4.10}$$

For the sequence of statistically independent scattering events, considering the Gaussian distribution of the phase differences $\bar{k}_i[\bar{r}_i(t+\tau) - \bar{r}_i(t)]$ for each scattering event, we obtain

$$G_1(\tau) \sim \sum_n \rho(n) \exp\left(-\frac{n}{6}\langle q^2\rangle\langle \Delta r^2(\tau)\rangle\right), \tag{4.11}$$

where averaging over the ensemble of scattering sites is carried out by introducing weighting factors $\rho(n)$ that describe the contributions of each sequence of n scattering events. Thus, the factor $\exp(-n\langle q^2\rangle\langle \Delta r^2(\tau)\rangle/6)$ can be interpreted as the *single-path field correlation function.* The term $\langle \Delta \bar{r}^2(\tau)\rangle$ determines the variance of the displacements of the scattering particles per observation time τ, and the value $\langle q^2\rangle$ determines the variance of the momentum transfer per scattering event. In the case of anisotropic scattering, this magnitude can be determined by introducing the transport mean free path l^* and the anisotropy factor or mean cosine of the scattering angle g:

$$\langle q^2\rangle = 2k^2(1-g) = 2k^2\frac{l}{l^*}. \tag{4.12}$$

In further consideration, replacing the discrete scattering model with the continuous distribution of effective optical paths and carrying out the integration over the ensemble of effective optical paths, we obtain

$$G_1(\tau) \sim \int_0^\infty \exp\left(-\frac{k^2\langle \Delta \bar{r}^2(\tau)\rangle s}{3l^*}\right)\rho(s)ds. \tag{4.13}$$

Here the value of the effective optical path for each scattering sequence is $s \approx nl$ and the probability density function of the effective optical paths corresponds to the set of weighting factors $\{\rho(n)\}$ in the case of the discrete scattering model.

Equation (4.13) plays a central role in the theory of the dynamic scattering of coherent light by optically dense dynamic media. It is easy to see that the temporal correlation function of the complex amplitude can be interpreted as the Laplace transform of the probability density function independently of the dynamic properties of the scattering medium. These properties are determined by the dependence of $\langle \Delta \bar{r}^2(\tau)\rangle$ on time lag τ. The most well-known and universally adopted type of the phantom scattering media used to study the statistical properties of multiple scattered light is the Brownian scattering medium. This medium is usually prepared as an aqueous suspension of polystyrene beads (monodisperse multiple-scattering medium) or aqueous solution of Intralipid (polydisperse multiple-scattering medium). In this case, the argument of the exponential kernel [Eq. (4.13)] can be transformed by substituting $\langle \Delta \bar{r}^2(\tau)\rangle = 6D_B\tau$ (D_B is the self-diffusion coefficient of the scattering particles) for $\frac{2\tau s}{\tau_0 l^*}$. The parameter $\tau_0 = (k^2 D_B)^{-1}$ can be interpreted as the characteristic time for the displacement of a scatterer to a distance equal to the wavelength. Thus, it is necessary to have the analytical form of the probability density function of the effective optical paths to calculate $G_1(\tau)$. The function $\rho(s)$ depends on the illumination and detection conditions as well as the optical properties and geometry of the scattering medium and can be obtained by solving the radiative transfer equation. For simple cases, analytical expression for $\rho(s)$ can be found. For example, in the case of coherent light propagation between an isotropic point-like source and point-like detector, which are embedded in an infinite isotropic, highly scattering, nonabsorbing medium, $\rho(s)$ is given by

$$\rho(s) = K\left(\frac{3}{4\pi l^* s}\right)^{3/2} \exp\left[-\frac{3|\bar{r} - \bar{r}'|^2}{4l^* s}\right], \tag{4.14}$$

where K is a normalization constant. The source–detector separation $|\bar{r} - \bar{r}'|$ must be significantly larger than the transport mean free path for Eq. (4.14) to hold. This case is also referred to as the diffusion regime.

One of the essential properties of the pathlength ensemble described by Eq. (4.14) is the lack of statistical moments of the optical paths such as the mean value and the variance. This results from the presence of the slow-decaying "tail" of the s distribution, which decreases as $s^{-1.5}$ for large, effective optical paths. In other words, the mean value of the effective optical path of light diffusely propagating between an isotropic light source and a detector through an infinite nonabsorbing and highly scattering media is infinite. To choose the characteristic spatial scale for such a scattering system, one can use the modal, or the most probable value of the effective optical path: $s_{\mathrm{mod}} = |\bar{r} - \bar{r}'|^2/2l^*$. It is easy to see that this magnitude sharply increases with an increase of the source–detector distance. For example, for an infinite nonabsorbing medium with $l^* = 50$ μm and a separation between source and detector equal to 5 cm, the most probable value of the effective optical path will be equal to 25 m. For real, spatially bounded scattering systems with finite absorption, the contribution of the long photon trajectories is significantly suppressed and the ensemble of effective optical paths will be characterized by finite values of the low-order statistical moments.

Substituting Eq. (4.14) for Eq. (4.13) and carrying out the integration, we obtain

$$G_1(\tau) \sim \frac{\exp(-\sqrt{6\tau/\tau_0}|\bar{r} - \bar{r}'|/l^*)}{|\bar{r} - \bar{r}'|}. \tag{4.15}$$

The nonanalytical form of $G_1(\tau)$, related to the diverging behavior of its first- and higher-order derivatives for small values of τ [i.e., $G_1'(\tau) \to \infty$ as $\tau \to 0$], is caused by the previously mentioned slow decay of the pathlength. The appearance of absorbance, characterized by the value of the absorption coefficient μ_a and described by the additional Bougier factor $\exp(-\mu_a s)$ in Eq. (4.13) will lead to dramatic changes in $G_1(\tau)$ for small time scales; in this case, it can be expressed as

$$G_1(\tau) \sim \frac{\exp(-\sqrt{6\tau/\tau_0 + 3\mu_a l^*}|\bar{r} - \bar{r}'|/l^*)}{|\bar{r} - \bar{r}'|}. \tag{4.16}$$

More generally, with an arbitrary type of the particle dynamics, the temporal correlation function has the following form:

$$G_1(\tau) \sim \frac{\exp(-\sqrt{k^2\langle\Delta^2 r(\tau)\rangle + 3\mu_a l^*}|\bar{r} - \bar{r}'|/l^*)}{|\bar{r} - \bar{r}'|}. \tag{4.17}$$

Thus, for scattering particles that exhibit fractal-like motions with a power-law increase of the time-dependent variance $\langle\Delta^2 r(\tau)\rangle \sim \tau^\nu$ and in the range of time and space scales for which $k^2\langle\Delta^2 r(\tau)\rangle \gg 3\mu_a l^*$ and simultaneously $k^2\langle\Delta^2 r(\tau)\rangle \ll |\bar{r} - \bar{r}'|^2/l^{*2}$ (which could be in the case of weakly absorbing media), the detected speckle intensity fluctuations can be considered as a one-dimensional fractal-like process with an asymptotic power-law decay of the temporal correlation function for small time scales: $G_1(\tau) \sim 1 - K\tau^{\nu/2}$ (Ref. 14). As a rule, for real multiple-scattering systems, the region of fractality of the speckle intensity fluctuations spans no more than one to two decades of the time scale.[15]

Thus, one of the central points of each dynamic light-scattering problem in the case of optically dense random media is the correct evaluation of the probability density function of the effective optical paths for the analyzed scattering system. More detailed consideration of the basic principles of such evaluation for multiple-scattering systems with various geometries are discussed in Section 4.3.2.

4.3.2 Diffusing-wave spectroscopies and related techniques

The basis for *correlation spectroscopy*, or *diffusing-wave spectroscopy* (DWS), is the correlation analysis of temporal fluctuations of light propagating in optically dense, weakly absorbing, nonstationary media. This analysis is carried out to study the dynamic properties of scattering systems. Similar information about the scattering media can be obtained using spectral analysis of the intensity fluctuations of multiply scattered dynamic speckles. However, the DWS methods are preferable in the case of optically dense media that are characterized by a strong extinction of the probe light and very broad spectra of the speckle intensity fluctuations. For these media, more advanced instrumentation (e.g., photon counting and digital correlators) for the analysis of the detected intensity fluctuations exist.

A typical scheme for a DWS experiment is illustrated in Fig. 4.5. Light emitted by a single-mode laser propagates through a multiple-scattering dynamic medium (sample). As a result of superposition of the partial components of the scattered light, a random nonstationary interference pattern appears that is associated with the spatial-temporal fluctuations of the scattered optical field. This interference pattern, or *dynamic speckle pattern*,

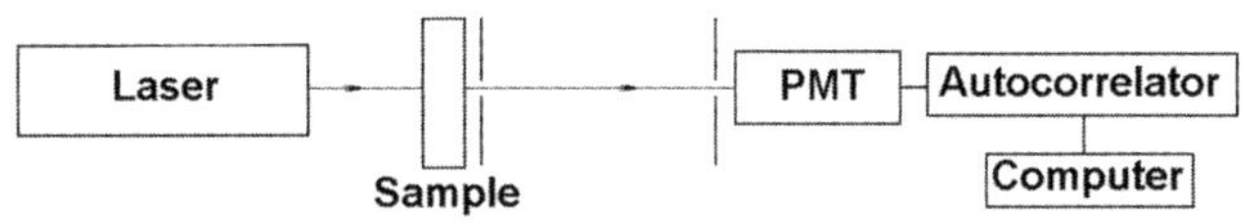

Figure 4.5 A schematic sketch of a typical DWS experiment.

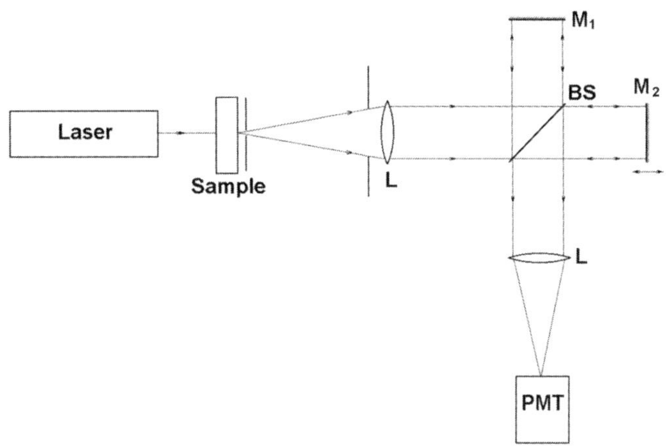

Figure 4.6 Optical scheme of the diffusing-wave interferometer for correlation analysis of scattered light with small time scales (Ref. 16).

contains information about the dynamic properties of the scattering system. Part of the scattered light is selected by the collimator (collimating system can consist of two pinhole diaphragms, as shown in Fig. 4.5) and falls into the photosensitive area of the photodetector.

The photomultiplier tube (PMT) operating in photon counting mode is usually necessary to obtain sufficient sensitivity to optically dense media. In this case, the output PMT signal is a random sequence of the amplified electron pulses that are then processed by a digital correlator. Commercially available digital correlators allow analysis of temporal fluctuations of the scattered light with a bandwidth on the order of 100 MHz, or, correspondingly, with a temporal resolution of 10 ns.

Different approaches are necessary to perform the analysis at shorter time scales. A good example of such an approach is the application of interferometers to induce the light beatings by mixing the two identical optical signals with a controlled phase delay in one channel of the interferometer (see, for example, Ref. 16). The principles of such *diffusing wave interferometry* are illustrated in Fig. 4.6. This scheme, as applied to the analysis of the dynamics of multiple-scattering media such as aqueous suspensions of polystyrene beads, enables a temporal resolution on the order of 1 ns. This is comparable with the time scale of hydrodynamic interactions for the scattering particles.[17] In this case, the time lag τ is determined by the optical path difference between two arms of the optical interferometer. The sample is illuminated by the single-mode laser beam as in the case of conventional DWS experiments. The major difference is that the output dynamic speckle field, $E(t)$, is collected, collimated, and directed into a Michelson interferometer. If the lengths of arms are equal, respectively, to L_1 and L_2, then after the recombination, the beams will be delayed by $\tau = 2(L_1 - L_2)/c$. If the intensities of both beams are equal, then the intensity of light detected by a PMT can be written as

$$I(\tau) = \frac{1}{4}\langle|E(t)|^2\rangle\{1 + [\langle E^*(t+\tau)E(t)\rangle/\langle|E(t)|^2\rangle]\cos\omega\tau\}. \quad (4.18)$$

where ω is the angular frequency of the probe light. The envelope multiplying the oscillating term is usually considered the visibility of the observed interference pattern. At the same time, this is the temporal correlation function of the scattered light evaluated for the given value of τ. Thus, analyzing the decay of the envelope with an increase of τ due to changing the position of mirror M_2, one can reconstruct the form of the normalized field correlation function for the required range of time scales.

Boas, Bizheva, and Siegel[18] developed a modification of the original DWS technique with a selection of partial contributions of the scattered optical field, characterized by the given value of the pathlength s. This method is based on the use of a low-coherence interferometer to discriminate the short-path and long-path components of the scattered field. The corresponding instrumentation is shown in Fig. 4.7. In this case, the cut-on and cut-off values of the effective optical paths are determined by the optical path difference between two arms of the low-coherence interferometer as well as by the spectral bandwidth and the central wavelength of the light source. In this system, the single-mode fiber-optic interferometer is illuminated with an 850-nm superluminescent diode (SLD). Interferometer adjustment is provided by changing the position of the retroreflector in the reference arm. Focusing lenses in the sample and reference arms of the interferometer are mounted on

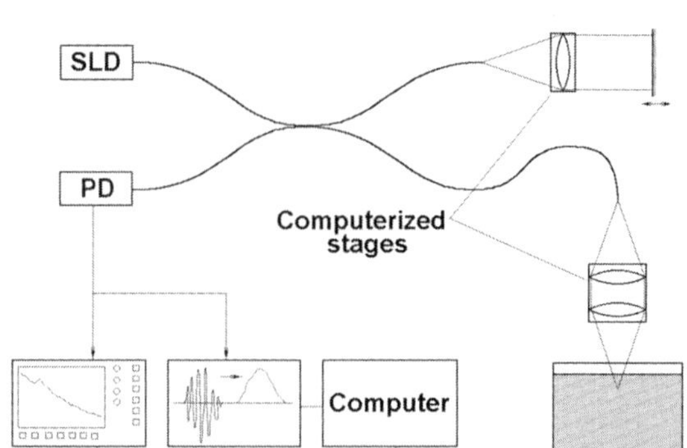

Figure 4.7 Schematic of the dynamic low-coherence interferometer system (Ref. 18).

computer-controlled stages to provide the possibility of exact matching of the optical paths in both arms.

It can be concluded from the analysis of Eq. (4.13) that in the case of probing Brownian multiple-scattering media, the detected AC signal is characterized by the single-path temporal correlation function that has the typical exponential form

$$G_1^s(\tau) \sim \exp(-K\tau) \tag{4.19}$$

and the corresponding spectral density can be described by a Lorentzian. The value of K in the discussed case should depend strongly on scattering conditions: in the case of small sample depths (i.e., small pathlengths) that are of the order of transport mean free paths for the scattering medium, only single- and low-order-scattered components of the diffuse retroreflected light will be selected by the low-coherence interferometer. In this case, the single-path temporal correlation function has the typical "single-scattering" form and the parameter K, which is related to the Lorentzian linewidth of the detected light beatings, does not depend on the sample depth. But with increasing pathlength, when it becomes significantly larger than l^*, light beatings will be induced by the contributions that are scattered many times. In this case, the single-path temporal correlation function [Eq. (4.19)] has the typical "multiple-scattering" form

$$G_1^s(\tau) \sim \exp\left[-\frac{2\tau s_{\mathrm{sel}}(\Delta z)}{\tau_0 l^*}\right]. \tag{4.20}$$

Thus, dependence of the selected value s_{sel} on the sample depth Δz leads to an increase of the spectral width of the detected signal of Δz. Such a physical picture is in qualitative agreement with experimental results obtained with a two-layer scattering system.[18] The first 580-μm-thick layer consists of 1.20-μm polystyrene (PS) beads (0.5% volume fraction) separated from the second layer (4% suspension of 0.22-μm PS beads) by a 150-μm-thick glass cover. Volume fractions for both layers were chosen to provide a mean scattering length equal to ≈100 μm. Due to anisotropic scattering in the first layer (mean cosine of the scattering angle is approximately equal to 0.89 at 830 nm), its thickness is less than the transport mean free path (≈1000 μm); consequently, single-scattered contributions strongly dominate in the formation of the detected light beating signal. On the other hand, the second layer, consisting of smaller particles, is characterized by a significantly smaller value of the anisotropy parameter ($g \approx 0.267$) and $l^* \approx 140$ μm for this scattering medium. Thus, if the sample depth exceeds this value, transition from the single- to the multiple-scattering mode of the formation for the detected optical signal takes place and the spectral width of the observed light beatings becomes dependent on the sample depth. In this experiment, such a transition

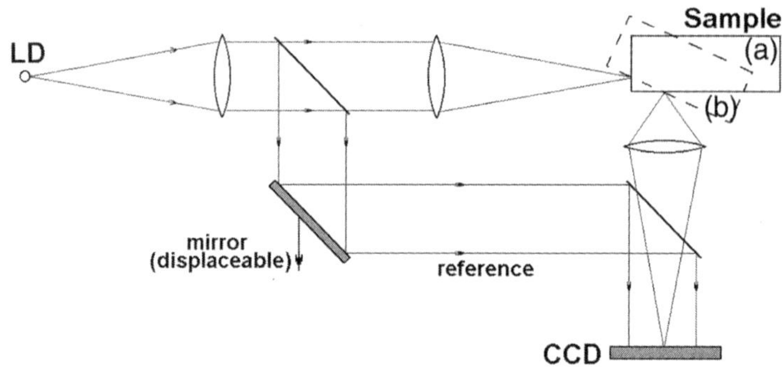

(a) observation of the transversal cross-section
(b) observation of the surface

Figure 4.8 Imaging system with a low-coherence interferometer for the study of path distributions in multiple-scattering media by means of speckle contrast analysis in the image plane (Ref. 19).

was observed when the beam waist in the sample was embedded in the second layer at depths larger than 300 μm.

A closely related technique allowing observation of partially coherent light propagation in multiple-scattering media with a very high temporal resolution was reported by Hausler et al.[19] This method is based on the application of a low-coherence source (laser diode at 670 nm operating below threshold) in combination with a Mach–Zehnder interferometer. A multiple-scattering sample (solid polyester with embedded monodispersive SiO_2 spheres; reduced scattering coefficient of 92 cm^{-1}; anisotropy parameter of 0.84) was placed in one of the arms of the interferometer (Fig. 4.8). Light scattered out of the surface of the scattering sample, and the reference wave was superimposed on the CCD (charge-coupled device sensor). The fraction of components of the scattered optical field with effective optical paths matched the propagation path of the reference beam inducing speckle modulation in the corresponding regions of the image of the sample surface; other fractions caused the incoherent summation of intensities with the intensity of the reference beam. The image, as a result, was a superposition of the reference and the object wave and was recorded by the CCD sensor. To improve the quality of the analyzed images, the subtraction technique was used: after exposure, the reference phase was shifted by π. The incoherent fractions of the signal were the same in both exposures. Therefore, the difference between these two images contained only the coherent fraction of the signal. In the processed image (Fig. 4.9), dark spots, or speckle modulation of the surface image, will be caused by the partial components of the scattered field that have run the same pathlength but along different individual paths. A particular feature of the recorded image in Fig. 4.9(c) is the sharply curved edge of the impulse response, the "photon horizon." This

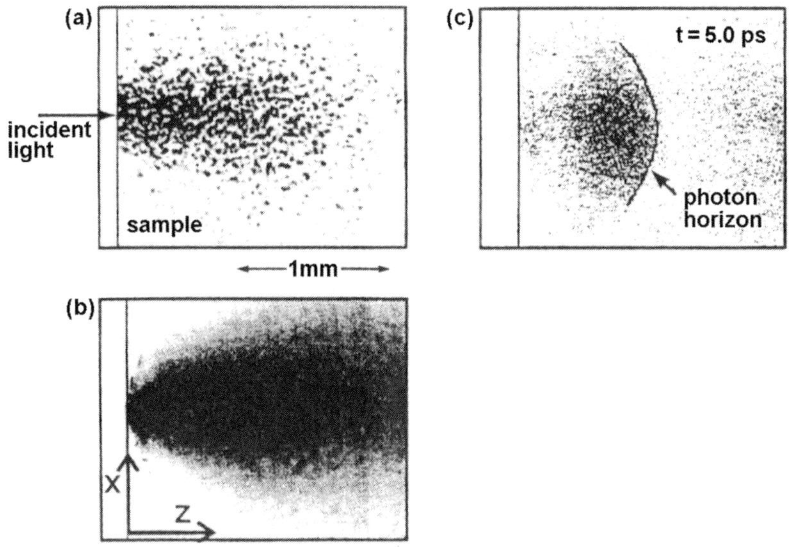

Figure 4.9 Images of the scattering process with different coherence lengths and different kinds of observation (Ref. 19). (a) fully coherent illumination, direct observation; (b) broad-band illumination, direct observation; (c) broad-band illumination, extraction of the coherent part.

curve defines the maximum penetration depth in the analyzed scattering system for each reference pathlength. Thus, different penetration depths can be visualized through suitable adjustment of the reference pathlength. The pathlength resolution for this system is determined by the coherence length l_C of the light source ($l_C = 30$ μm with a resulting time resolution of about 100 fs[19]).

One analytical approach in the diffusing-wave spectroscopy of nonstationary turbid media is the analysis of correlation transport. This transport can be viewed as a propagation of a correlation "wave" outward from sources, and its scattering by macroscopic inhomogeneities can be associated with spatial variations of dynamical or optical properties. Evolution of the spatial–temporal correlation function of the optical field fluctuations due to light propagation in free space has been analyzed in early works of Born and Wolf.[20] Later, Ackerson et al.[21] showed that certain analogies exist between the transport of correlation in disordered scattering media and transport of photons, which can be described by the well-known radiative transport equation. The main feature of correlation transport relates to the accumulation of the decay of the correlation function caused by each scattering event during the propagation of the correlation wave in the scattering system. In this case, considering the "stationary correlation transport" through the scattering medium (steady state) probed by a continuous-wave source, one can modify the radiative transfer equation in its usual form to obtain the corresponding correlation transport equation:

$$\nabla G_1(\bar{r},\tilde{\Omega},\tau) + \mu_t G_1(\bar{r},\tilde{\Omega},\tau) = \mu_s \int G_1(\bar{r},\tilde{\Omega}',\tau) g_1^s f(\tilde{\Omega},\tilde{\Omega}') d\tilde{\Omega}' + S(\bar{r},\tilde{\Omega}). \quad (4.21)$$

The temporal correlation function of the scattered field fluctuations $G_1(\bar{r},\tilde{\Omega},\tau)$ depends on the detection point position ($\bar{r}$) and direction in turbid medium ($\tilde{\Omega}$) chosen for correlation analysis; $\mu_t = \mu_s + \mu_a$. The term $g_1'(\tau) = \exp(-\frac{1}{6}q^2\langle\Delta\bar{r}^2(\tau)\rangle)$, which corresponds to single scattering, describes the accumulation of correlation decay due to sequences of scattering events; $f(\tilde{\Omega},\tilde{\Omega}')$ is the phase function of the scattering medium; and $S(\bar{r},\tilde{\Omega})$ is the light source distribution.

In the case of validity of the standard diffusion approximation, the stationary correlation transport equation [Eq. (4.21)] can be rewritten in the following form:[22]

$$(D_\gamma \nabla^2 - c\mu_a - \frac{1}{3} c\mu_s' k_0^2 \langle\Delta\bar{r}^2(\tau)\rangle) G_1(\bar{r},\tau) = -cS(\bar{r}),$$

where $D_\gamma = cl^*/3$ is the photon diffusion coefficient, c is the light speed in the scattering medium, and μ_s' is the reduced scattering coefficient. The term $\frac{1}{3}v\mu_s' k_0^2\langle\Delta\bar{r}^2(\tau)\rangle$ describes additional losses of correlation due to dynamic scattering in disordered media and can be interpreted as "correlation absorption" caused by the dynamic processes.

The presence of any kind of scattering medium dynamics is manifested as the appearance of the additional "absorbance" term in the correlation diffusion equation. Thus, numerical solution of this equation for a given source and detector positions with respect to embedded dynamic inhomogeneities can be used as the basis for an inverse problem solution (reconstruction of the inhomogeneity "image"). This technique has been applied by Boas, Campbell, and Yodh[23] and was verified in experiments with a multiple-scattering "static" object (titanium dioxide-resin cylinder) containing a "dynamic" inhomogeneity (spherical space filled with a water solution of Intralipid). The sample was illuminated by an Ar-ion laser through a fiber-optic light-guiding system; scattered light was collected by a single-mode fiber-optic light collector and detected by a photon-counting system. The scattered light intensity fluctuations as random sequences of photo-count pulses were processed by a digital autocorrelator to obtain $G_2(\tau)$ for given illumination and detection conditions. Apart from the geometry of the scattering system, angular scanning of the object was carried out; measurements were made every 30 deg at the surface of the cylinder with source–detector angular separations of 30 deg and 170 deg. Results of the inhomogeneity image reconstruction are shown in Fig. 4.10.

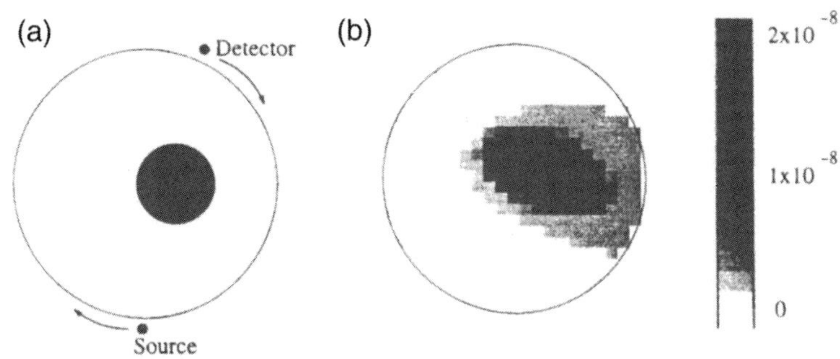

Figure 4.10 Imaging of the dynamic multiple-scattering inhomogeneity embedded in the static scatterer by means of the correlation diffusion analysis (Ref. 23). The static scatterer is a 4.6-cm-diameter cylinder with $l^* = 0.25$ cm and $\mu_a = 0.002\ \text{cm}^{-1}$. The dynamic scatterer is a 1.3-cm-diameter spherical cavity filled with a colloid with $l^* = 0.25$ cm, $\mu_a = 0.002\ \text{cm}^{-1}$, and $D_B = 1.5 \times 10^{-8}\ \text{cm}^2\text{s}^{-1}$. A slice of the image is presented in (b). The values of the reconstructed particle diffusion coefficient are given by the gray-level scale.

4.4 Evaluation of the Pathlength Density: Basic Approaches

4.4.1 The concept of the pathlength density for description of light propagation in disordered media

Evaluation of the probability density function of optical paths $\rho(s)$ for given illumination and detection conditions is the basis for quantitative analysis of various coherence and polarization phenomena in the case of multiple scattering of light by disordered media. Many modern methods of the incoherent diagnostics and visualization of optically dense media include pathlength distribution analysis as part of the image reconstruction problem. These methods are the frequency-domain[24,25] and time-domain techniques[26,27] as well as different modifications of the transillumination and conventional back-projection reconstruction techniques.[28,29] In this way, evaluation of the probability density function $\rho(s)$ is one of the most generalized approaches to the description of light propagation in disordered media.

The rigorous approach to such evaluation is to find a Green's function for the nonstationary radiative transfer problem. This function describes the "optical impulse response" of the studied scattering medium for given illumination and detection conditions. Unfortunately, Green's functions can be obtained in the rigorous analytical form only for a small number of scattering systems with simple geometry. Application of numerical methods or Monte-Carlo simulations provides evaluation of $\rho(s)$ with required accuracy but does not allow one to obtain the pathlength density in the compact analytical form. But in some cases, the radiative transfer equation can be reduced to approximations that have simpler forms compared with the initial equation. These approximations that provide the adequate accuracy of analytical evaluation of $\rho(s)$ for certain conditions are considered next.

4.4.2 Diffusion approximation

This approach is valid in case of weak absorption and strong isotropic scattering (i.e., when $\mu_a^{-1} = l_a \gg l^*$ and $|\bar{r} - \bar{r}'| \gg l^*$) and allows one to describe spatial distributions of the effective optical paths with an acceptable accuracy. Another condition is that the observation point be far from a light source or boundaries. The criterion of applicability of the diffusion approximation depends on specific scattering and observation conditions (whether a localized or extended light source is used; value of the mean cosine of the scattering medium is small); in any case, separation between light source and observation point should be several times larger than the transport mean free path l^* for a scattering medium. In particular, for multiple scattering, Brownian media with slab geometry illuminated by a collimated plane wave application of the diffusion approximation for analysis of transmitted light characteristics such as temporal correlation function and normalized intensity is justified when the slab thickness is not less than $(4 \div 10)$ l^*. This "critical" value of the slab thickness depends on the scattering anisotropy.[30] Also, a well-known example of the "diffusion approximation breakdown" is the significant discrepancy between calculated and experimentally obtained temporal correlation functions of the light backscattered from the nonstationary bounded medium. Such discrepancy is manifested in the case of small distances between the source and detector located on the boundary or in the case of usage of the extended light sources. This "breakdown" is caused by existence of the nonphysical modes of the diffusion equation solution for the scattered optical field. These modes are characterized by propagation times smaller than the ballistic time estimated as $|\bar{r} - \bar{r}'|/c$. Besides, this approximation does not take into account the low-order-scattered contributions of the scattered optical field.

Using the diffusion approximation, light transport in the isotropic scattering medium with negligible small absorption is described by the solution of the corresponding lossless diffusion equation (see, for example, Ref. 31)

$$(\gamma^2\nabla^2 - \partial/\partial t)\Phi(\bar{r},t) = -q(\bar{r},t), \tag{4.22}$$

where $\Phi(\bar{r},t)$ is the photon density, γ^2 is the light diffusion coefficient determined as $\gamma^2 = cl^*/3$, and $q(\bar{r},t)$ is the isotropic source term caused by output intensity distribution. In the case of absorbing media, the additional term $-\mu_a c$ appears on the left-hand side of Eq. (4.22) (this is a so-called lossy diffusion approximation). Solution $\Phi(\bar{r},t)$ is derived in a volume Ξ bounded by a surface $\partial\Xi$—in particular, the output flux on the boundary, which is evaluated as

$$\Gamma(\varsigma,t) = -\gamma^2(\partial/\partial n)\Phi(\bar{r},t)|_{\text{boundary}} = -\gamma^2\vec{n}\cdot\vec{\nabla}\Phi(\bar{r},t)|_{\text{boundary}}, \tag{4.23}$$

where ς is a point on the bounding surface and $\vec{n}(\varsigma)$ is the outward interface normal. This expression determines the time-dependent intensity detected by the point-like detector located in ς. The probability density function of optical paths in the case of the point-like continuous-wave source is similar to the impulse response of the scattering medium in the detection point or temporal profile of the detected intensity or Green's function of the nonstationary diffusion equation. This temporal profile can be obtained if a short light pulse $I_0\delta(t-t_0)$ from the point-like source is used to probe a scattering medium. Further, the Green's function of the diffusion equation [Eq. (4.22)] can be used to obtain the pathlength statistics in the case of the extended surface or bulk light sources by integration over the region of the source location.

The most simple and basic case is the previously mentioned probability density function [Eq. (4.14)], which corresponds to the Green's function of the diffusion equation. One of the convenient ways to obtain the pathlength density for geometrically simple but practical applications cases (such as semi-infinite space and infinite slab) is to use the method of images. In such a way, the corresponding Green's function is constructed as the finite or infinite sum of the Green's functions of the diffusion equation [Eq. (4.22)] for appropriately located virtual sources. The necessary condition is the equality of the diffusing photon density $\Phi(\bar{r},t)$ to zero on the boundaries $\partial\Xi$.

In particular, analysis of light diffusion in semi-infinite space can be provided by assuming that light is coming from an instantaneous point source at a depth z_0 below the surface.[31,32] The parameter z_0 is of the order of the transport mean free path and corresponds to the depth at which angular distribution of initially collimated light becomes isotropic. The method of images puts a negative source of the same strength at a position outside the slab. This negative source is the mirror image of the positive source. The resulting sum of the photon densities, which are induced by both sources, is the solution of the diffusion equation and satisfies the zero boundary condition. Thus, the corresponding photon density can be written as[31]

$$\phi(\bar{r},\bar{r}',t,t_0) = \frac{\phi_0}{[4\pi^2(t-t_0)]^{3/2}} \times \exp\left\{-\left[\mu_a c\{t-t_0\} + \frac{x^2+y^2}{4\gamma^2(t-t_0)}\right]\right\} \times\{\exp[-(z-z_0)^2/4\gamma^2(t-t_0)] - \exp[-(z+z_0)^2/4\gamma^2(t-t_0)]\}. \tag{4.24}$$

Here the "real" instantaneous point source is located at $\bar{r}' = (0,0,-z_0)$ and the detector is located at $\bar{r} = (x,y,z)$. It is easy to see that $\phi(\bar{r},\bar{r}',t,t_0) = 0$ if $z=0$. Pathlength density $\rho(s)$ in case of scattered light detection on the boundary can be obtained by replacing $(t-t_0)$ with s/c and, after this, by substituting Eq. (4.24) for Eq. (4.23).

Figure 4.11 illustrates application of the method of images to analysis of light diffusion in the case of infinite scattering slab illumination by the point-like source. For this scattering geometry, the pathlength density can be

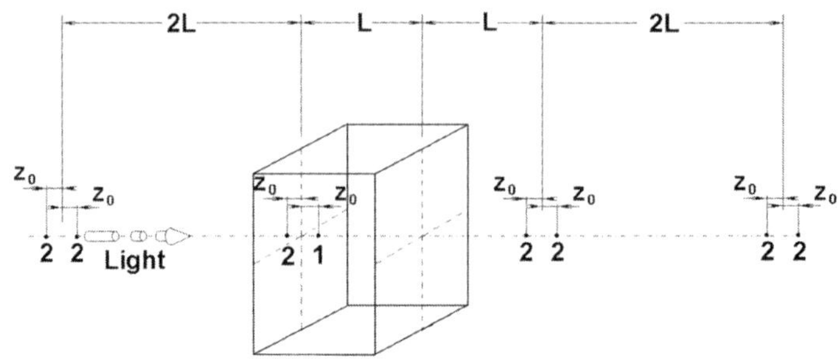

Figure 4.11 Application of method of images in the case of infinite scattering slab. 1: real source; 2: image sources. In this example "imaging boundary" coincides with interface (i.e., $z_b = 0$).

evaluated as an infinite sum of terms. Each of them is similar to terms in Eq. 4.24.

The models shown previously correspond to the case of the same optical properties (in particular, the refractive indices) of the scattering medium and free space, where the diffusing light is absent. Thus, the "imaging boundary" that "mirrors" the real source to provide the image source is considered to be coinciding with the interface between the scattering medium and free space. More rigorous analysis[33,34] taking into account the boundary conditions leads to separation of the imaging plane and interface by the distance z_b called the extrapolation length. Thus, the diffusing light intensity vanishes at the distance z_b from the interface, and distance between interface and image source should be considered equal to $2z_b + z_0$. Extrapolation length in units of the transport mean free path can be evaluated as $z_b = 2(1 + R_2) / 3(1 - R_2)$, where R_1, R_2 are calculated from the angle-dependent reflectivity $R_W(\upsilon)$ of scattering medium as

$$R_n = \int_0^1 (n+1)\theta^n R_W(\upsilon) d\upsilon,$$

and the angle $\cos^{-1}\upsilon$ is measured with respect to the interior normal.

The "diffusion" probability density function $\rho(s)$ for two important cases is listed in Table 4.1. In these expressions, K is the corresponding normalization constant, estimated using the relation $\int_0^\infty \rho(s)ds = 1$, where I is the source intensity.

The pathlength density $\rho(s)$ in the case of extended light sources (e.g., if the scattering medium is illuminated by a plane wave) can be obtained by integrating the corresponding relation over the coordinates (x,y). The temporal correlation function $G_1(\tau)$ for given scattering geometry and detection conditions will be obtained by substituting a corresponding expression for $\rho(s)$ in Eq. (4.23) and, after this, in Eq. (4.13) and integration over s.

Table 4.1 Pathlength densities $\rho(s)$ for semi-infinite space and infinite slab illuminated by the point-like source.

Case	Probability density function $\rho(s)$
Half-space	$K s^{-3/2}\exp\{-\mu_a s-3(x^2+y^2)/4 l^* s\} \times [\exp\{-3(z-z_0)^2/4 l^* s\}$ $-\exp\{-3(z+z_0+2z_b)^2/4 l^* s\}]$
Infinite slab	$K\{\exp[-\mu_a s-3(x^2+y^2)/4l^* s]/s^{3/2}\}\times\sum_{n=-\infty}^{\infty}[\exp[-3\{z-2n(L+2z_b)-z_0\}^2/4l^* s]$ $-\exp[-3\{z-2n(L+2z_b)+z_0+2z_b\}^2/4l^* s]]$

4.4.3 Other approaches

The limitations of the diffusion theory of light transport just mentioned do not allow one to provide correct evaluation of the pathlength density in the vicinity of the light sources or the boundaries for propagation distances of the order of the transport mean free path. Thus, in the case of anisotropic scattering, when the scattering medium is characterized by large values of g, traveling light should undergo many scattering events to randomize its propagation direction. In this case, higher-order approximations of the radiative transfer equation can be used to describe such scattering regimes. One of the examples is the so-called P_N approximation. To simplify analysis of the radiative transfer equation, it may be transformed using an expansion in spherical harmonics.[35] As a result, a set of $(N+1)^2$ coupled partial differential equations can be obtained. For odd N, such equations may be reduced to a single $(N+1)$th-order differential equation. The diffusion approximation considered previously is the partial case of the P_N approximation with $N=1$. A similar approach has been suggested by John and MacKintosh.[3] This method allows one to take into account contributions of single-scattered and low-order-scattered components in formation of a scattered field in disordered media with anisotropic scattering.[3]

In recent years, Durian et al.[36,37] considered light transport through disordered scattering media using the telegrapher equation,

$$\gamma_0^2\nabla^2\Phi(\bar{r},t)=\gamma_0^2\frac{\partial^2\Phi(\bar{r},t)}{\partial t^2}+(1+2\gamma_0^2\mu_a)\frac{\partial\Phi(\bar{r},t)}{\partial t}+\mu_a(1+\gamma_0^2\mu_a)\Phi(\bar{r},t). \tag{4.25}$$

Equation (4.25) is written in dimensionless form by applying the corresponding dimensionless temporal and spatial variables: the time is measured in units of l^*/c, and the lengths are measured in terms of the transport mean free path l^*; the dimensionless light diffusion coefficient γ_0^2 is equal to 1/3; and the dimensionless absorption coefficient μ_a is introduced as the ratio l^*/l_a (l_a is the absorption length). The light source can be introduced in Eq. (4.25) as it was made in the case of the light diffusion equation [Eq. (4.22)], or it can be introduced by the Green's function techniques. In this case, the boundary condition should be taken as

$$0 = \left[(1 + \gamma_0^2 \mu_a) + z_e \bar{n}\, \bar{\nabla} + \gamma_0^2 \frac{\partial}{\partial t} \right] \Phi(\bar{r},t) \Big|_{\text{boundary}}$$

and the emerging flux should be evaluated as

$$\Gamma(\varsigma,t) = \frac{\gamma_0^2}{z_e} \Phi(\bar{r},t) \Big|_{\text{boundary}} .$$

The usual diffusion equation, extrapolation boundary conditions, and Fick's law are recovered by the previously written relations in the limit of long times and weak absorption.

Physically, the second time derivative of the right-hand side of Eq. (4.25) accounts for the ballistic nature of short-time photon migration, and in this case, the wave equation is recovered. On the contrary, at long times, when the photon density is slowly varying and when there is negligibly small absorption, the second time derivative is negligibly small and Eq. (4.25) recovers the usual diffusion equation.

The Green's function of the telegrapher Eq. (4.25) in dimensionless form can be written as[37]

$$\phi(\bar{r},t) = -\phi_0 \frac{\exp[-\{\mu_a + (2\gamma_0^2)^{-1}\}t]}{4\pi r} \left\{ [\mu_a + (2\gamma_0^2)^{-1}] + \frac{\partial}{\partial t} \right\} \times \frac{\partial}{\partial r} [I_0\{(2\gamma_0^2)^{-1} \sqrt{t^2 - r^2}\}\, \Theta\,(t - r)], \tag{4.26}$$

where the instantaneous point source is located in the origin and the light pulse is emitted in the moment $t' = 0$, $I_0(x)$ is the modified Bessel function of order zero, and $\Theta(x)$ is the Heaviside function that is equal to 0 for negative arguments and to 1 for positive arguments.[37] This term provides the zero values of the photon density for $r > t$ in accordance with causality related to finite speed of light. For $r < t$, photon density can by presented as

$$\phi(\bar{r},t) = \phi_0 \frac{\exp[-\{\mu_a + (2\gamma_0^2)^{-1}\}t]}{4\pi(2\gamma_0^2)\sqrt{t^2 - r^2}} \left[(\mu_a + (2\gamma_0^2)^{-1}) \times I_1((2\gamma_0^2)^{-1} \sqrt{t^2 - r^2}) + \frac{(2\gamma_0^2)^{-1} t}{\sqrt{t^2 - r^2}} I_2((2\gamma_0^2)^{-1} \sqrt{t^2 - r^2}) \right]. \tag{4.27}$$

In the long-time limit, $t \gg 1$ and $r \ll 1$, the "telegrapher" Green function [Eq. (4.27)] asymptotically changes to the expression that is similar to the "diffusion" Green function in the weak absorption limit:

$$\phi(\bar{r},t) \approx \phi_0 \frac{1+\gamma_0^2\mu_a}{(4\pi\gamma_0^2 t)^{3/2}} \exp(-\mu_a t) \exp\left(-\frac{r^2}{4\gamma_0^2 t}\right). \tag{4.28}$$

On the contrary, in the short-time limit, Eq. (4.27) predicts the ballistic behavior of the path density that can be considered as the solution of the wave equation in the case of $\mu_a = 0$:

$$\phi(\bar{r},t) \approx \phi_o \left[\frac{\{\mu_a + (2\gamma_0^2)^{-1}\}(2\gamma_0^2)^{-2}\Theta(t-r)}{8\pi} + \frac{\{\mu_a + (2\gamma_0^2)^{-1}\}\delta(t-r) + \delta'(t-r)}{4\pi r}\right] \exp(-\mu_a t). \tag{4.29}$$

Equations (4.27)–(4.29) can be used to express the probability density function of the dimensionless pathlength s/l^* by substituting this ratio as the dimensionless time. Figure 4.12 illustrates the normalized dependencies of $\phi(\bar{r},t)$ on the dimensionless distance between source and detector. These graphs were obtained[37] for the case of light propagation from the instantaneous point source in a strongly absorbing medium (absorption length is equal to the transport mean free path) using the Monte-Carlo simulation (a), solution of the telegrapher equation [Eq. (4.27)] (b), and using the diffusion approximation (c). It is easy to see the "noncausal" behavior of $\phi(\bar{r},t)$ for short distances in the case of diffusion approximation (e.g., nonzero magnitude for values of the dimensionless distance $r > 1$). In the large-path limit, when $s/l^* \geq 20.0$, both values of $\phi(\bar{r},t)$ (the "telegrapher" value as well as the "diffusion" value) demonstrate practically the same behavior.

4.5 Manifestations of Self-Similarity in Multiple Scattering of Coherent Light by Disordered Media

Certain dimensionless relations between the wavelength of the propagating light, the spatial scales that characterize the optical properties of the scattering medium, and its size can be introduced for the description of spatial-temporal distributions of the statistical moments of scattered optical fields. In particular, the form of the enhanced backscattering peak for small scattering angles can be presented by the following relation (Section 4.2):

$$I(\theta) \sim 1 - \frac{10\pi}{3}\eta\theta,$$

where the dimensionless parameter η is equal to the ratio of the transport mean free path l^* to the wavelength λ of light.

Thus, it should be expected that in the case of coherent light probing of multiple-scattering media with differing optical properties (μ_s and g) but equal

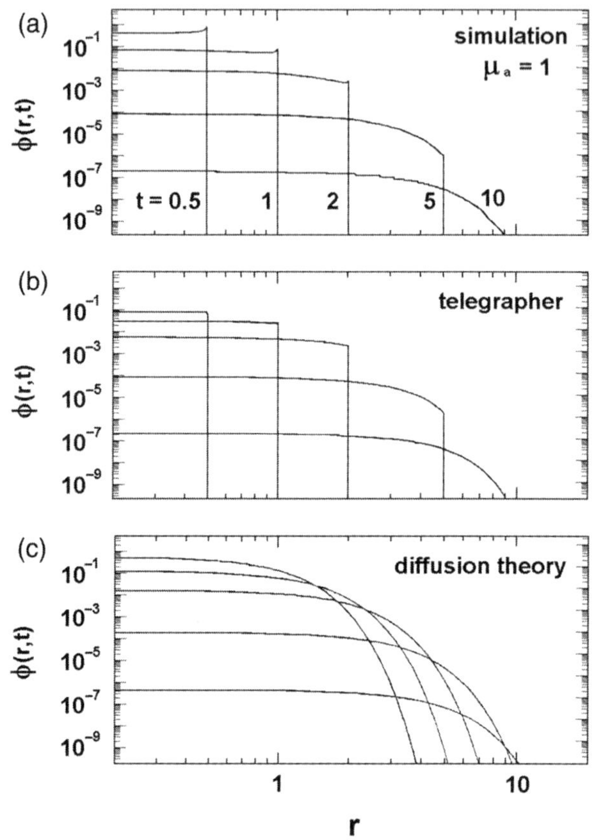

Figure 4.12 The photon density distributions in the dimensionless coordinates (*r*,*t*) for instantaneous point source in infinite scattering medium with strong absorption (Ref. 37). (a) the Monte Carlo simulation, (b) the "telegrapher" theory, and (c) the diffusion theory.

values of the ratio η, the same forms of the enhanced backscattering peak will be observed; this sameness can be interpreted as the property of similarity of angular distributions of the backscattered intensity.

In addition, in the case of the multiple scattering of coherent light by disordered systems, a group of "relaxation" phenomena can also be considered as manifestations of the similarity, or its likelihood, in multiple scattering. This manifestation is related to the same forms of dependencies of certain statistical moments of scattered optical fields on the specific spatial scales. These spatial scales characterize the decay of the corresponding moments during coherent light propagation in the disordered media. The following "relaxation" phenomena will be considered here:

- the existence of temporal correlations of amplitude and intensity fluctuations of scattered optical fields in a fixed detection point for nonstationary systems of scattering particles.

- the evolution of the polarization state of light propagating in the disordered scattering media.
- the manifestation of Bougier's law in the case of multiple scattering.

This relaxation behavior of the statistical moments of the scattered optical fields will be considered in terms of a phenomenological approach based on the concept of an ensemble of optical paths of partial contributions or components of a scattered optical field.

For the diffusion scattering mode, each partial component of the multiply scattered optical field can be interpreted as a sequence of a great number N of statistically independent scattering events. Such a sequence is characterized by the value of the effective optical path s. Introducing the probability density function of optical paths $\rho(s)$ as the generalized characteristics of the ensemble of partial components, we can express the statistical moments of the scattered optical field as the integral transforms of $\rho(s)$. In the weak scattering limit, when $l, l^* \gg \lambda$, second-order statistical moments such as the mean intensity of the scattered field, the temporal correlation function of the field fluctuations, and the degree of polarization at the fixed detection point can be expressed in the case of $N = s/l \gg 1$ as the Laplace transforms of $\rho(s)$ (Section 4.3). In particular, the mean intensity of the scattered optical field for a multiple-scattering medium with nonzero absorption can be written using the modified Bougier's law,

$$\langle I \rangle \cong \int_0^\infty \exp(-\mu_a s)\rho(s)ds = \int_0^\infty \exp(-s/l_a)\rho(s)ds, \tag{4.30}$$

where the averaging is carried out over all possible configurations of the scattering centers. The following normalization condition takes place for this approach:

$$\int_0^\infty \rho(s)ds = \langle I \rangle_0, \tag{4.31}$$

where $\langle I \rangle_0$ is the mean intensity in the absence of absorption.

For nonstationary disordered media consisting of moving scattering particles, the normalized temporal autocorrelation function of the scattered field fluctuations can be expressed as [Eq. (4.13)]

$$g_1(\tau) \approx \frac{\int_0^\infty \exp[-B(\tau)s/l]\rho(s)ds}{\int_0^\infty \rho(s)ds}, \tag{4.32}$$

where $B(\tau)$ is determined by the mean square of displacement of the scattering particles for the observation time τ. For Brownian multiple-scattering systems

(Section 4.2), the exponential kernel of the integral transform [Eq. (4.32)] equals $\exp(-2\tau s/\tau_0 l^*)$.

From the viewpoint of the discussed phenomenological approach, the relaxation of the initial polarization state of the coherent light propagating in the disordered multiple-scattering medium is caused by the intensity flux interchange between partial components of the scattered field with different polarization states. In particular, for initial linear polarization of the propagating light linearly "copolarized" and "cross-polarized," partial components of the scattered field can be considered. The first of them is characterized by the same direction of polarization azimuth as the incident illuminating beam; the other one, by the orthogonal direction. In a similar way, the interrelation between the left and right circularly polarized components can be analyzed if the illuminating light with initial circular polarization is used.[38] In terms of the approach used to describe the polarization relaxation in the disordered media, the expressions for intensities of light propagating in two different polarization channels can be written as

$$\langle I\rangle_{\text{same}} \approx K_2 \int_0^\infty [1 + K_1 \exp(-s/\xi_P)]\rho(s)ds,$$
$$\langle I\rangle_{\text{opposite}} \approx K_4 \int_0^\infty [1 - K_3 \exp(-s/\xi_P)]\rho(s)ds, \tag{4.33}$$

where the subscript "same" corresponds to the polarization channel with the polarization state identical to the initial polarization state of the propagating light, and the subscript "opposite" denotes the alternative polarization state. The parameters $K_1 \div K_4$ and ξ_P characterize the decay rate of the polarization state of light propagating in the disordered medium; the parameter ξ_P can be interpreted as the "depolarization length," which depends on the optical properties of the scattering medium. The following normalization conditions can be obtained using the obvious relation[38]

$$<I>=<I>_{same} + <I>_{opposite} \;:\; K_1 \equiv K_3 \equiv 1; \quad K_2 \equiv K_4 \equiv 0.5.$$

If a linearly polarized probe light is used, the following simple expression for the degree of polarization can be written:

$$P = \frac{<I_x> - <I_y>}{<I_x> + <I_y>}, \tag{4.34}$$

where the mean intensities of the "copolarized" ($<I_x>$) and "cross-polarized" ($<I_y>$) components of scattered light are obtained by averaging the ensemble of all possible configurations of the scattering structure. Thus, the degree of polarization in the case of damping of the initial linear polarization during the light propagation in the disordered medium can be expressed as

$$P = \frac{\int_0^\infty \exp(-s/\xi_P)\rho(s)ds}{\int_0^\infty \rho(s)ds}. \tag{4.35}$$

Comparing Eqs. (4.30), (4.32), and (4.35), we can introduce the relaxation scales $\xi_{<I>} = l_a = \mu_a^{-1}$, $\xi_g = \tau_0 l^*/2\tau$, and $\xi_P = ml^*$, which characterize the decay of the corresponding statistical moment of the scattered field (or a combination of moments such as the degree of polarization) when the probe light propagates through the disordered medium. In the latter case, the value of the ratio $m = \xi_P/l^*$ is determined by the optical properties of the elementary scatterers forming the scattering medium. In the case of temporal correlation of the scattered field fluctuations, the relaxation scale depends on the time delay τ used for estimating the value of the correlation function. Also, m depends on the initial polarization state of the propagating light (linear or circular) as well as on the size parameter ka of the elementary scatterers; for small-sized scatterers ($ka \ll 1$) $m_L > m_C$. That is, the linear polarization (L) "survives" better than the circular one (C); in contrast, if $ka \gg 1$ ("anisotropic" scattering), then $m_L < m_C$ (Ref. 39).

In the following analysis, an infinite macroscopically homogeneous slab illuminated by a point-like collimated light source will be considered as the model of a multiple-scattering medium. The illuminating source is located in the origin ($x = 0$; $y = 0$; $z = 0$), and the distribution of the optical paths is analyzed for the transmission geometry on the output boundary of the slab ($z = L$, where L is the slab thickness and the z axis is directed normally to the slab boundaries). For such scattering conditions and diffusion light propagation in the scattering slab, the probability density function of optical paths can be presented using the method of images as the infinite sum (Section 4.4),

$$\rho(s,x,y,z) = \sum_{n=-\infty}^{+\infty} [\rho'(s,x,y,z,n) - \rho''(s,x,y,z,n)], \tag{4.36}$$

where each term is related to the Green function of the light diffusion equation for the case of infinite homogeneous isotropic scattering medium, which describes light propagation from the point-like source $I_0\delta(t - t_0)\delta(\bar{r} - \bar{r}_0)$ [the optical path for a partial component is estimated as $s = c(t - t_0)$, where c is the speed of light in the scattering medium]:

$$\rho' = \left[\frac{3}{4\pi l^* s}\right]^{1.5} \exp\left\{-\frac{3(x^2+y^2)}{4l^* s}\right\} \exp\left\{-\frac{3(z-2nL-z_0)^2}{4l^* s}\right\};$$
$$\rho'' = \left[\frac{3}{4\pi l^* s}\right]^{1.5} \exp\left\{-\frac{3(x^2+y^2)}{4l^* s}\right\} \exp\left\{-\frac{3(z-2nL+z_0)^2}{4l^* s}\right\}. \tag{4.37}$$

Thus, the simplified method of images (extrapolation length is equal to zero and imaging boundary coincides with interface) used in this case. As shown below, such an approach provides the adequate accuracy used to interpret the experimental results.

The probability density function determined by Eqs. (4.36) and (4.37) satisfies the zero boundary conditions for both interfaces ($z = 0$ and $z = L$); Fick's law should be used to find $\tilde{\rho}(s)$, which characterizes the diffusing light received by a point-like detector placed on the output interface ($z = L$):

$$\tilde{\rho}(s,0,0,L) = -D\bar{n}_z \mathrm{grad}\rho(s,0,0,z)|_{z=L}, \tag{4.38}$$

where D is the light diffusion coefficient equal to $cl^*/3$ and $\bar{n}_z$ is the interface normal.

For a macroscopically homogeneous scattering slab illuminated by the point-like collimated source and probed in transmission mode by the point-like detector, we can introduce the Laplace transform of the probability density $\tilde{\rho}(s)$ in the following form:

$$\Lambda(\alpha,0,0,L) = \Lambda(\alpha) = \int_0^\infty \exp(-\alpha s)\tilde{\rho}(s,0,0,L)ds. \tag{4.39}$$

Substituting Eqs. (4.36) and (4.37) for Eq. (4.38) and carrying out the Laplace transform of $\tilde{\rho}(s)$, we can obtain $\Lambda(\alpha)$ after some calculations as

$$\Lambda(\alpha) \sim \sum_{n=-\infty}^{n=+\infty} \left(\left\{ \frac{1}{|L(2n+1)-z_0|} + \sqrt{\frac{3\alpha}{l^*}} \right\} \frac{1}{|L(2n+1)-z_0|} \right.$$
$$\times \exp\left\{ -\sqrt{\frac{3\alpha}{l^*}} [|L(2n+1)-z_0|] \right\} - \left\{ \frac{1}{|L(2n+1)+z_0|} + \sqrt{\frac{3\alpha}{l^*}} \right\} \tag{4.40}$$
$$\left. \times \frac{1}{|L(2n+1)+z_0|} \exp\left\{ -\sqrt{\frac{3\alpha}{l^*}} [|L(2n+1)+z_0|] \right\} \right).$$

Thus, the normalized values of statistical moments of the scattered field such as $\langle I \rangle, P, g_1(\tau)$ can be obtained for the on-axis detector position from Eq. (4.40) by substituting the corresponding relaxation parameters $\xi_{<I>}, \xi_P$ and ξ_g for α^{-1}. In particular, the degree of polarization of the scattered light propagated in the nonabsorbing multiple-scattering medium is equal to

$$P = \Lambda(\xi_P^{-1})/\Lambda(0). \tag{4.41}$$

The intensities of copolarized and cross-polarized components of the scattered field can be determined as $<I>_{\parallel} \sim 0.5[\Lambda(0) + \Lambda(\xi_P^{-1})]$ and $<I>_{\perp} \sim 0.5\,[\Lambda(0) - \Lambda(\xi_P^{-1})]$, respectively. The normalized value of mean intensity in the case of light propagation in the scattering medium with nonzero absorption will be expressed as

$$\langle \tilde{I} \rangle = \Lambda(\mu_a)/\Lambda(0), \tag{4.42}$$

where the corresponding value of the mean intensity in the case of light propagation in the nonabsorbing medium is used for normalization.

Furthermore, the normalized temporal autocorrelation function of the cross-polarized component has the following form:

$$g_{1\perp}(\tau) = \frac{[\Lambda(2\tau/\tau_0 l^*) - \Lambda(2\tau/\tau_0 l^* + \xi_P^{-1})]}{[\Lambda(0) - \Lambda(\xi_P^{-1})]}. \tag{4.43}$$

Similarly, the degree of polarization of light scattered by the absorbing medium is equal to

$$P = \Lambda(\xi_P^{-1} + \mu_a)/\Lambda(\mu_a). \tag{4.44}$$

Thus, the similarity of the statistical properties of multiply scattered optical fields can be interpreted as the interrelation between relaxation scales for different statistical moments of the scattered field, which leads to the equality of these scales for equal values of normalized statistical moments. To verify the discussed relations between different statistical moments of the scattered field as the manifestation of similarity in the case of multiple scattering, an experimental study of the dependencies of $< \tilde{I} >$, $g_1(\tau)$ and P on transport mean free path l^* and absorption length l_a of the scattering medium has been carried out. Both optical parameters were varied during the experiments by changing the volume concentration of the scattering fraction and added absorber. Water suspensions of polystyrene spheres with diameters 0.261, 0.460, 0.605, 1.07, and 2.19 μm were applied as phantom scattering media, and the cylindrical glass cuvettes with flat front and back walls (with thickness in the direction of light propagation equal to 10 mm or 20 mm and diameter equal to 60 mm) were used as containers for polyball suspensions.

CW (continuous wave) lasers with a linearly polarized output such as the single-mode Ar-ion laser operating at 514 nm and diode-pumped Nd laser operating at 532 nm (second harmonics) were used as an illumination source. An illuminating laser beam was focused on the central zone of the front wall of the cuvette using a lens with a 200-mm focal length; thus, the diameter of the light spot on the input surface of the scattering medium was approximately equal to 50 μm and was smaller compared with the characteristic values of estimated mean free path for polyball suspensions used in the experiment. Estimations of the path density values in the case of the slab geometry and collimated point-like source have shown that for the range of l^* studied in our experiments, we can exclude from consideration the effect of the cuvette side wall on the intensity distribution inside the slab due to the negligibly small values of path density $\rho(s)$, for the distance from the light source axis that is equal to or exceeds the cuvette radius.

Forward-scattered light was collected in the on-axis detection position by the single-mode optical fiber placed behind the back wall of the cuvette at a very small distance from the cuvette surface (less than 0.5 mm); to provide the measurements of intensity of the copolarized and cross-polarized components of the scattered light, a film-like polarization filter was fixed directly on the back wall of the cuvette, and to choose the required type of polarization of the detected scattered light, the cuvette with the polarization filter was turned with respect to the polarization direction of the probe beam.

A photon-counting photomultiplier tube was used as the detector of the collected light; a digital correlator BIC-9000 was used for processing the detected signal. The mean values of the intensity of the cross-polarized and copolarized components were determined by measuring the mean count rates for the corresponding positions of the polarization filter.

An added absorber such as a water solution of black India ink was applied to analyze the influence of absorption of the scattering medium on the normalized statistical moments of the scattered light; this substance, with 0.1% volume concentration, is characterized by a mean value of the absorption coefficient equal to 6.7 cm^{-1} in the spectral range $510 \div 540$ nm.

The phenomenological approach discussed previously in the case of diffusion propagation of the forward-scattered light through the slab predicts the decay of polarization degree with the increase of the slab thickness that is very close to exponential damping beginning from certain values of dimensionless slab thickness L/l^*. The position of the crossover between the region of exponential decay and the region of slow saturation of P with increasing mean transport free path for theoretical curves $P = f(L/l^*)$ depends on the dimensionless depolarization length m. Figure 4.13 illustrates these curves obtained using Eqs. (4.40) and (4.41) for different values of m.

Experimental dependencies of the degree of polarization on the dimensionless scattering coefficient $L\mu_s$ for polystyrene spheres in water in the case of the transmission geometry, presented in Fig. 4.14, also show the approximately exponential decay of P in the region of values of degrees of polarization less than $0.85 \div 0.90$ with increasing scattering coefficient. In estimates of dimensionless scattering coefficients, the value of L was taken to be equal to the cuvette thickness, and the scattering coefficient μ_s was calculated for given values of scatterer diameter and concentration using the Mie theory.

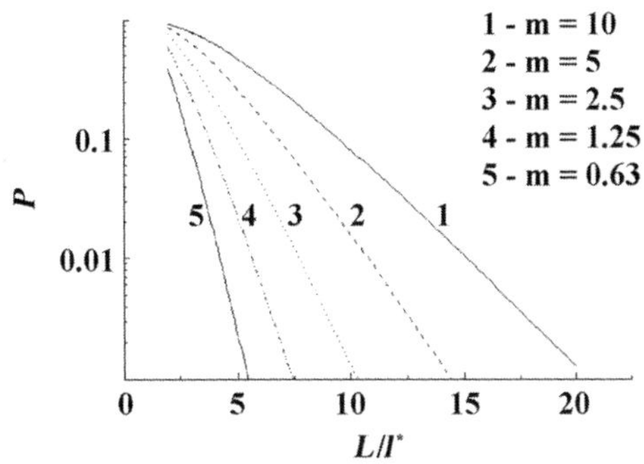

Figure 4.13 The theoretical dependencies of the degree of polarization in the transmission mode on the dimensionless slab thickness (the diffusion approximation).

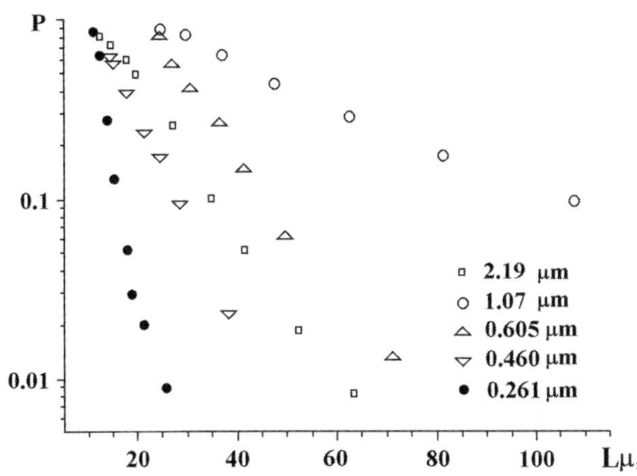

Figure 4.14 Measured values of the degree of polarization in the case of linear polarization of probe light against the dimensionless scattering coefficient. The following wavelength and cuvette thickness values were used: 514 nm (Ar-ion laser) and 10 mm for 0.460 μm and 1.07 μm particles; 532 nm (diode-pumped Nd laser) and 20 mm for 0.261 μm, 0.605 μm, and 2.19 μm particles.

The expected behavior of the dimensionless depolarization length m in the dependence on the size parameter of scatterers ka (see the previous discussion) is manifested in differing values of the slope of dependencies of P on $L\mu_s$ plotted in semi-logarithmic coordinates (Fig. 4.14). For "small-sized" polyballs (0.261, 0.460, and 0.605 μm), the slope value decreases with increasing ka; the minimal slope is observed for 1.07-μm polystyrene spheres; for the light source ($\lambda = 514$ nm) used, it corresponds to the size parameter ka equal to ≈ 8.78. But for 2.19-μm polyballs, value of the slope $\Delta(lgP)/\Delta(L\mu_s)$ sharply increases.

The value of $m = \xi/l^*$ for linearly polarized light can be estimated by comparing the experimentally obtained slope magnitudes $\Delta(\lg P)/\Delta(L\mu_s)$ with the corresponding calculated values of $\Delta(\lg P)/\{\Delta(L/l^*)\}$ (Fig. 4.13); in this

Table 4.2 The values of the dimensionless depolarization length m for polystyrene spheres in water.

Particle diameter, μm	$\frac{\Delta(\lg P)}{\Delta(L/l)}$, experimental	Anisotropy parameter g (Mie theory)	$\frac{\Delta(\lg P)}{\Delta(L/l^*)}$, calculated	m, diffusion approximation
0.261	≈0.135	≈0.68	≈0.42	≈2.00
0.460	$\approx 5.60 \times 10^{-2}$	≈0.83	≈0.33	≈3.10
0.605	$\approx 3.48 \times 10^{-2}$	≈0.88	≈0.29	≈3.65
1.07	$\approx 1.25 \times 10^{-2}$	≈0.93	≈0.18	≈9.20
2.19	$\approx 3.85 \times 10^{-2}$	≈0.88*	≈0.32	≈3.20

*In accordance with Mie theory, g diminishes with increasing $2\pi a/\lambda$ for this region of values of the size parameter.

case, the above-mentioned relations between the transport mean free path l^* and the scattering length l as well as between l and the scattering coefficient ($l^* = l/(1-g)$; $\mu_s = l^{-1}$) should be taken into account. The results of such estimates for polyball suspensions are presented in Table 4.2. The values of g were also calculated from Mie theory. The presence of Mie resonances for theoretical dependencies of l and g on size parameter ka will cause significant oscillations of estimated values of the normalized depolarization length for large-sized scattering particles ($ka \gg 1$). Taking into account the common physical assumptions, we should expect the slow saturation of oscillating value of m with the increasing size parameter for the diffusion mode of light propagation in the weak scattering limit ($l^*, l \gg \lambda, a$).

Similar tendencies in polarization decay in the case of coherent light propagation in the disordered slabs were emphasized earlier by Bicout et al.[39] (scattering medium, water suspensions of 0.22-μm, 0.48-μm, and 1.05-μm polyballs; light source, laser diode operating at 0.67 nm) and Sankaran et al.[40] (objects, water suspensions of 1.072-μm polyballs, Intralipid 20%, porcine adipose tissue and porcine miocardium, light source, He-Ne laser operating at 632.8 nm). Bicout et al. also mentioned the presence of a maximum characteristic scale of depolarization for both polarization states of the probe beam (linear as well as circular) for the value of a size parameter approximately equal to 8.

Another method of estimating the depolarization length for the disordered nonstationary multiple-scattering media is the comparison of characteristic scales of depolarization and decorrelation of the scattered light.[41,42] The quantity τ_{cd} can be introduced as the value of correlation time at which the values of the normalized autocorrelation function of the field fluctuations and the degree of polarization of the scattered field are equal. The quantity $\tau_{cd} = l^*/2\xi_i D_s k_0^2$, which can be loosely designated as the characteristic correlation time, is a universal parameter of a multiple-scattering Brownian medium. This parameter does not depend on the concentration of the scattering particles (as $l^* \sim c^{-1}$ and $\xi \sim l^*$) and is determined only by the optical properties of the scatterers. As a consequence of the hypothesis that the kernels of the integral transforms in expressions (4.32) and (4.35) have a

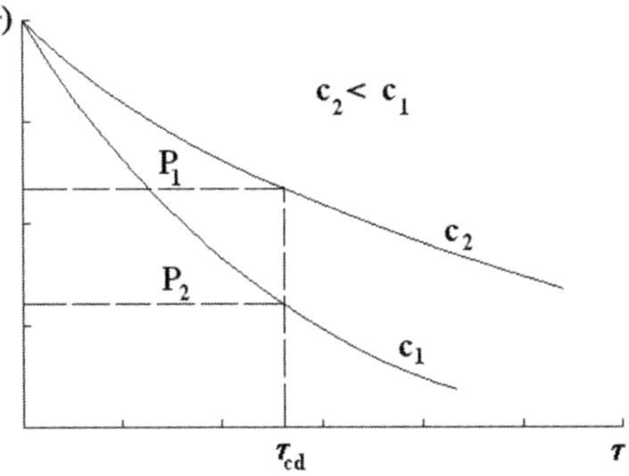

Figure 4.15 Method for determining the characteristic correlation time for multiple-scattering Brownian medium.

similar form, the universality of the parameter τ_{cd} should also be manifested in its independence, within wide limits, on the conditions of the correlation-polarization experiment. These conditions are, for example, the geometry of the scattering medium, the angle of incidence of the probe beam, and detection of the scattered radiation, which influence only the form of $\rho(s)$. Figure 4.15 illustrates the method of determining τ_{cd} for multiple-scattering Brownian media with different concentrations of scattering particles.

The absence of a concentration dependence of the characteristic correlation time for multiple-scattering Brownian ensembles of noninteracting particles was verified in experiments with aqueous suspensions of polystyrene beads with diameters of 0.46 and 1.07 μm.[41,42]

An Ar-ion laser was used as the illumination source in these experiments. The values of τ_{cd} were determined by the method illustrated in Fig. 4.15. Normalized values of module of the field correlation functions were obtained from experimentally measured intensity correlation functions using the Siegert relation. Moreover, measurements of the "conventional" correlation time as the half-width of the normalized field correlation functions were performed. Figure 4.16 shows a logarithmic plot of the experimentally measured concentration dependencies of τ_{cd} and $\Delta\tau_{0.5}$ ("conventional" correlation time estimated as the half-width of the correlation peak).

Analysis of the experimental data shows that in the experimental range of concentrations of the aqueous suspensions of polystyrene beads, the concentration dependencies $\Delta\tau_{0.5} = \varphi(c)$ are close to power-law functions $\Delta\tau_{0.5} \sim c^{-\alpha}$. The exponents α in the power-law functions approximating the experimental values of $\Delta\tau_{0.5}$ in Fig. 4.16 are $\approx$2.21 and $\approx$1.96 for polystyrene beads of diameters 0.46 and 1.07 μm, respectively. These values are in satisfactory agreement with the value $\alpha = 2$ obtained in the diffusion approach. Specifically, as mentioned in Ref. 3, for an optically thick layer of thickness L consisting of Brownian scattering particles and for the isotropic point-like source and detector, the normalized autocorrelation function of the

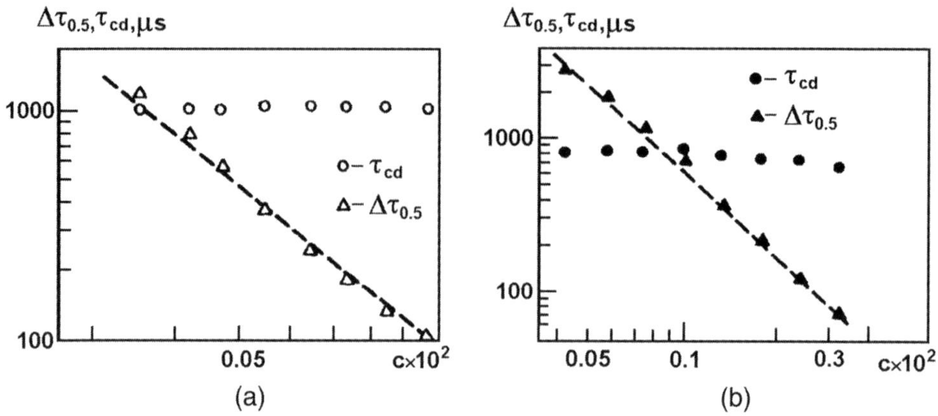

Figure 4.16 Concentration dependencies of the characteristic correlation time and half-width of the autocorrelation function of intensity fluctuations for aqueous suspensions of polystyrene beads (left: bead diameter 0.46 μm; right: bead diameter 1.07 μm).

amplitude fluctuations of the scattered coherent radiation has the form $g_1(\tau) \approx \exp(-\sqrt{6\tau/\tau_0}\frac{L}{l^*})$.

Thus, the values of dimensionless depolarization length for water suspensions of 0.46-μm and 1.07-μm polyballs estimated from corresponding values of characteristic correlation time ($m_{0.46} \approx 3.44$ and $m_{1.07} \approx 8.73$) are in satisfactory agreement with the values presented in Table 4.2.

Figure 4.17 depicts the dependencies of normalized intensity of light transmitted through the cuvette with 0.605-μm and 2.19-μm polyball suspensions on the absorption coefficient varied by adding an absorbing substance (water solution of black India ink). Semilogarithmic coordinates are used for the analysis of the decay tendencies of $<\tilde{I}>$ with diminishing of the relaxation parameter ξ_a associated with the absorption length. In the case of the application of the diffusion approximation, the decay of normalized intensity with increasing μ_a can be described by Eq. (4.42), where $\Lambda(0)$ is evaluated for given values of the transport mean free path and slab thickness in the absence of added absorption.

One of the important questions about the radiative transfer theory is the dependence of the effective value of the light diffusion coefficient D (see, for example, Ref. 31) and, correspondingly, the effective value of the transport mean free path, on the absorption coefficient of the scattering medium. Thus, the application of an appropriate model describing this dependence allows one to obtain the best fit of experimental dependencies of $<\tilde{I}>$ on μ_a.

The classical diffusion approximation leads to the well-known relationship between D and μ_a,

$$D = c/\{3[\mu_a + (1-g)\mu_s]\}. \tag{4.45}$$

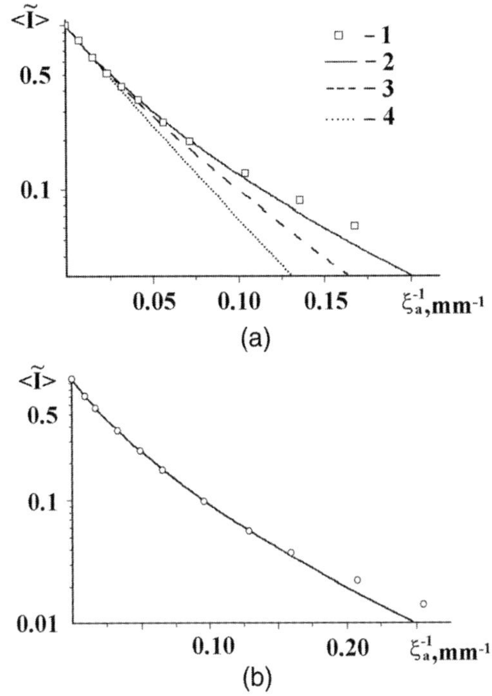

Figure 4.17 The normalized intensity of the transmitted light vs. the absorption coefficient of the scattering medium. (a) 0.605-μm particles, volume concentration 0.05%; (b) 2.19 μm particles, volume concentration 0.11%. Cuvette thickness: 20 mm; wavelength used: 532 nm. Measured values: open squares for 0.605-μm particles; open circles for 2.19-μm particles. Fitting curves, diffusion approximation with different values of *p*; (a) 0, full curve, 2; 1/3, broken curve, 3; 1, dotted curve, 4; (b) 0, solid line.

In accordance with this formula, increasing absorption causes the decrease of the effective value of the transport mean free path and, as a result, leads to changes in the statistics of the effective optical paths. (In particular, the modal value of the optical path in the case of an infinite scattering medium increases as the inverse of l^*.)

However, in recent years, the validity of this assumption has been disputed by several research groups. The question "Does the diffusion coefficient depend on absorption?" was addressed by Durduran et al.[43] They found that the diffusion approximation gives better agreement with the results of the Monte-Carlo simulation provided that the light diffusion coefficient was taken independently of the absorption coefficient of the scattering medium. The concept of the independence of the light diffusion coefficient from the absorbing properties of the light-transferring medium has also been discussed by Furutsu and Yamada,[44] Furutsu,[45] and Bassani et al.[46] In contrast, another approach, which was considered by Durian et al.[47] and is based on the application of the telegrapher equation for the description of nondiffuse regimes of light propagation in the disordered media, leads to a "weaker" dependence of D on μ_a compared with the classical diffusion approach. It was proposed that the light diffusion coefficient can be presented as[47]

$$D = c/\{3[p\mu_a + (1 - g)\mu_s]\} \tag{4.46}$$

and was found that the value $p = 1/3$ provides the best agreement of calculation data obtained using the "telegrapher approach" with the results of the Monte-Carlo simulation of light transport in the case of a semi-infinite scattering medium with nonzero absorption. To verify the validity of the above-mentioned approaches in the particular case of the description of decay of the mean intensity of light transmitted through a scattering slab with nonzero absorption, the fitting of the experimental dependency of $<\tilde{I}> (\mu_a)$ for 0.605-μm polyballs was carried out by means of the diffusion approach [Eqs. (4.40) and (4.42) with a substituted effective value of the transport mean free path obtained from Eq. (4.46) for different values of p]. The results of such a verification are shown in Fig. 4.17(a); it shows that the minimal discrepancy between the plotted experimental values of normalized intensity and the fitting curves corresponds to $p = 0$ (i.e., the concept of independence of light diffusion coefficient on μ_a provides the better fit with the experimental data). The corresponding fitting curve for 2.19-μm spheres obtained using the diffusion approximation with $p = 0$ is presented in Fig. 4.17(b). In both cases (0.605- and 2.19-μm polyballs), the transport mean free path for the medium with no added absorber has been used as a free parameter in the fitting procedure; values of l^* that provide the best fit with experimental data using Eqs. (4.40) and (4.42) have been found to be equal to $l^*_{0.605} \approx 4.10$ mm and $l^*_{2.19} \approx 3.85$ mm, respectively.

One of the manifestations of the increasing absorption is the increase in the degree of polarization of the transmitted light due to the suppression of the long-path components of the scattered field; in accordance with Eqs. (4.30), (4.33), and (4.34), the degree of polarization in the case of light propagation in scattering media with nonzero absorption can be expressed as

$$P = \int_0^\infty \exp[-(\mu_a + \xi_P^{-1})s]\rho(s)ds \Big/ \int_0^\infty \exp(-\mu_a s)\rho(s)ds. \tag{4.47}$$

Figure 4.18 shows the dependencies of the degree of polarization of transmitted light on the absorption coefficient for 0.605- and 2.19-μm polyball suspensions with an added absorber in the case of linear polarization of the probe beam. Thus, the cutoff of diffusing partial components with optical paths much larger than $l_a = \mu_a^{-1}$ causes the monotonic increase of P with a slow saturation as the absorption length diminishes. Full and dotted curves in Fig. 4.18 are obtained by the calculation of values of the degree of polarization using the modification of diffusion approximation with the light

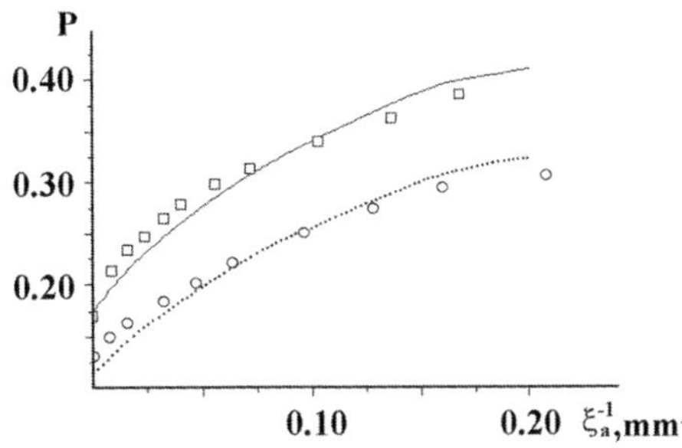

Figure 4.18 The degree of polarization of the transmitted light (linear polarization) vs. the absorption coefficient of the scattering medium; 0.605-μm and 2.19-μm particles, cuvette thickness, 20 mm; wavelength used, 532 nm. Measured values: open squares for 0.605-μm particles; open circles for 2.19-μm particles; fitting curves, diffusion approximation with $p = 0$.

diffusion coefficient independent of the absorption coefficient of the light-transferring medium ($p = 0$). Fitting curves have been obtained with the previously presented values of the transport mean free path: $l_{0.605} \approx 4.1$ mm and $l_{2.19} \approx 3.85$ mm; the value of dimensionless depolarization length m has been used as a free parameter in the fitting procedure; the values that correspond to the best fit equal $m_{0.605} \approx 3.45$ and $m_{2.19} \approx 2.95$ and are in satisfactory agreement with the results presented in Table 4.2.

One of the distinctive examples of the similarity manifestation in the decay tendencies of the statistical moments of the scattered field is the practically identical forms of the experimentally obtained dependencies of the transmitted light normalized intensity (for suspensions of 0.605-μm and 2.19-μm particles in a 20-mm cuvette) and the module of the normalized autocorrelation functions of amplitude fluctuations $|g_1(\xi_G^{-1})|$ (for a suspension of 0.46-μm particles in a 10-mm cuvette) on the generalized dimensionless relaxation parameter $\alpha L^2/l^*$ (Fig. 4.19). Values of α are the inverse of the absorption or, correspondingly, the correlation relaxation length. The absolute values of the field correlation functions were evaluated from the measured intensity correlation functions $G_2(\tau)$ using the Siegert relation. The values of the correlation relaxation, or decay parameter ξ_G, were estimated from the corresponding values of the time delay τ for a given diameter and different concentrations of polyballs such as $\xi_G = l^*\tau_0/2\tau$. (The value $\tau_0 \approx 3.9 \times 10^{-3}$ s for 0.46-μm particles in water was estimated using values of k for Ar-ion laser light and $D_B \approx 9.25 \times 10^{-13}$ m^2/s obtained by the interpolation of data presented in Ref. 22.).

Intensity correlation functions have been obtained for different volume concentrations of 0.46 μm polyballs in water that are varied approximately four times. Thus, the identity of the decay behavior of $g_1(L^2/l^*\xi_G^{-1})$ and $<\tilde{I}>(L^2/l^*\xi_\alpha^{-1})$ takes place even for relatively small values of relaxation parameters ξ_G and ξ_α, where the contribution of the short-path part of the

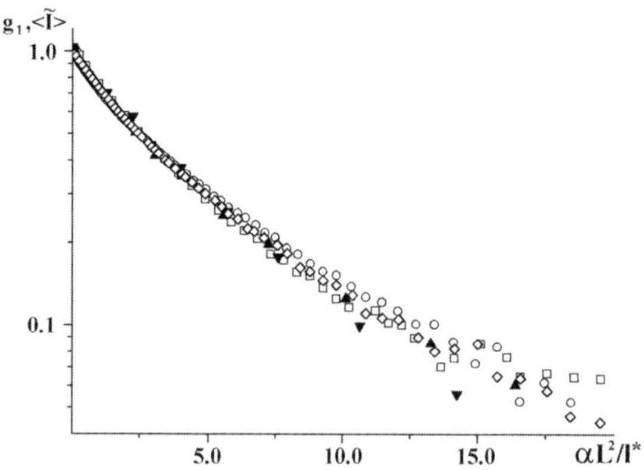

Figure 4.19 The manifestation of the similarity of decay tendencies of the normalized intensity and the temporal correlation function of the amplitude fluctuation with increase of the dimensionless relaxation parameter $\alpha L^2/l^*$. Open symbols, $|g_1(\xi_G^{-1})|$ for 0.46-μm particles; wavelength used, 514 nm; cuvette thickness, 10 mm; solid phase concentrations: 0.25% (squares), 0.11% (circles), and 0.064% (diamonds). Full symbols, $<\tilde{I}>$ for 0.605-μm (up triangles) and 2.19-μm particles (down triangles); wavelength used, 532 nm; plotted values correspond to Fig. 4.17.

s distribution to the corresponding statistical moment is significant and the diffusion approximation is not adequately valid ($\xi_a^{-1} \geq 0.2\ \text{mm}^{-1}$; see Fig. 4.17).

Analyzing expression (4.40) for $\Lambda(\alpha)$, we can see that in the case of strong scattering $L \gg l^*$, the behavior of decaying statistical moments is controlled by the one dimensionless parameter $\alpha L^2/l^*$; the ratio L^2/l^* is associated with the characteristic value of the pathlength distribution. For example, in the case of an infinite scattering medium with an embedded isotropic point-like source and a detector separated by the distance L, the modal value of the optical paths can be estimated as $0.5L^2/l^*$. With the increase of the transport mean free path, another dimensionless criterion z_0/L becomes significant; this ratio characterizes the influence of the slab boundaries on the distribution of effective optical paths.

Thus, despite the apparent simplicity of the exponential relaxation model, it gives an adequate description of the decay tendencies of the scattered field moments even for sufficiently nondiffuse modes of coherent light propagation in the disordered media. This is largely due to the decay of the corresponding parameter (the temporal correlation function of the scattered field fluctuations or the degree of polarization) being controlled by the number of statistically independent scattering events (i.e., the value of scattering length l) but not the characteristic spatial scale, controlling the light diffusion (the transport mean free path l^*). Thus, for anisotropic scattering, when l^* exceeds the scattering length significantly (see values of g presented in Table 4.2), we should expect only insignificant deviations from exponential relaxation even for such scattering conditions when diffusion approximation is close to failure (when the L/l^* ratio is of the order of 4–6). It is obvious that in the case of the Rayleigh scattering mode, when $g \approx 0$ and l^* is comparable to the scattering length, the failure of exponential scattering model should be observed for larger values of L/l^* when compared to the anisotropic scattering.

Comparing the transport mean free path for phantom media such as polyball suspensions, obtained by fitting the experimental dependencies as optimal values of the free parameter, with similar magnitudes calculated from Mie theory shows insignificant discrepancies of these values, which typically do not exceed 20%. Thus, the initial values of l^* for 0.605- and 2.19-μm polyball suspensions used in the experiments with added absorber [Fig. 4.17(a–b), respectively] have been evaluated using Mie theory as equal to ≈4.6 mm (0.605 μm; volume concentration 0.05%) and to ≈4.3 mm (2.19 μm; volume concentration 0.11%). The differences between these values and the previously presented results of the fitting procedure using the diffusion approximation ($l^*_{0.605} \approx 4.1$ mm and $l^*_{2.19} \approx 3.85$ mm) are caused largely by the errors of evaluation of the scattering particle concentrations and their optical parameters. Also, these discrepancies can be due to the application of nonrigid boundary conditions in the form $z_0 = l^*$ for Eq. (4.40); this relation can lead to underestimated or overestimated values of the path density $\rho(s)$ in the dependence on actual values of the diffusion reflectance coefficient for the slab boundaries. Increasing divergence between the experimental plots and the fitting curves (diffusion approximation with $p = 0$, Fig. 4.17) with increasing absorption coefficient is an obvious manifestation of the "breakdown" of diffusion theory for short distances and in the presence of significant absorption.

The recent studies of the influence of absorption in random media on the coherence effects in multiple scattering are related to consideration of the absorption effect on DWS measurements of the particle mobility.[48] The basic equation used in this work for the temporal autocorrelation function of the scattered field has the following form:

$$g_1(\tau) = \int_0^\infty \rho(s) \exp\left[-\left(\frac{2\tau}{\tau_c} + \frac{l^*}{l_a}\right)\frac{s}{l^*}\right] ds, \tag{4.48}$$

where $2\tau/\tau_c$ is equal to $k^2\langle\Delta r^2\rangle/3$ (as usual, k is the wave number of propagating light and $\langle\Delta r^2\rangle$ is the mean square displacement of scattering particles over the observation time τ. It is easy to see that the proposed expression corresponds to the renormalization of the pathlength probability density $\rho(s)$ by the Bougier factor $\exp(-s/l_a)$ [for comparison, see Eq. (4.47)]. Figure 4.20 illustrates the effect of absorption on the dependencies of $\langle\Delta r^2\rangle$ on the time lapse. (These dependencies were retrieved using the conventional approach to DWS data processing without taking into account the influence of absorption.) The probed systems were water suspensions of polystyrene

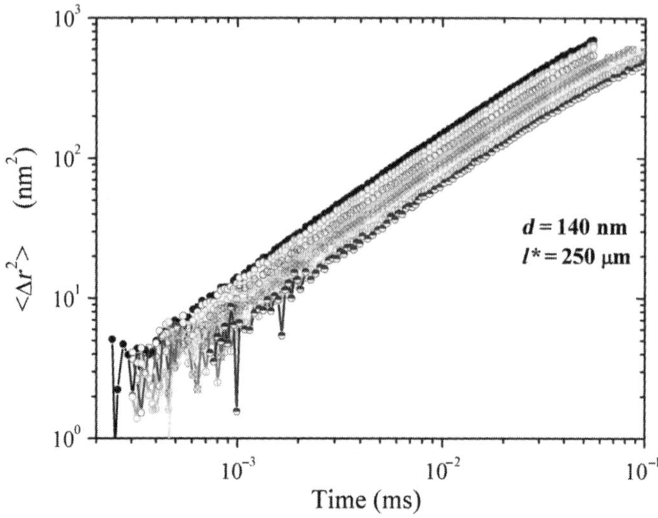

Figure 4.20 The uncorrected dependencies of the mean square displacement of scattering particles on the time lapse. Reprinted from Ref. 48. The legend is presented in Fig. 4.21.

microspheres of 140 nm; specific amounts of a dilute India ink solution (1 wt. %) were added to the microsphere sample suspensions as a light-absorbing agent. Before DWS experiments, the reflectance and transmittance measurements were performed on samples using an integrating sphere. The integrating sphere employs a photomultiplier tube attached at the wall sphere, and laser light at a wavelength of 514.5 nm was employed. Scattering and absorption coefficients of the samples were recovered using the inverse adding doubling method. Figure 4.21 shows the results of the absorption correction on the dependencies of the mean square displacement of scattering particles on the time lapse (designations are the same for both curves; the legend is presented in Fig. 4.21). However, the obtained results do not clarify the issue regarding the dependence (or independence) of the light diffusion coefficient on absorption. The issue is that the experimentally used values of the absorption length l_a significantly exceed the values of the transport mean free path l^* (see Figs. 4.20 and 4.21).

4.6 Diagnostic Applications of Light Coherence Phenomena in Multiple Scattering: Recent Applications in Biomedicine and Material Science

The relationships of structural and dynamic properties of random media with correlation characteristics of multiple scattered coherent light allow the consideration of angular or temporal correlation analysis of a scattered light field as a basis for medium characterization. In particular, attempts to apply

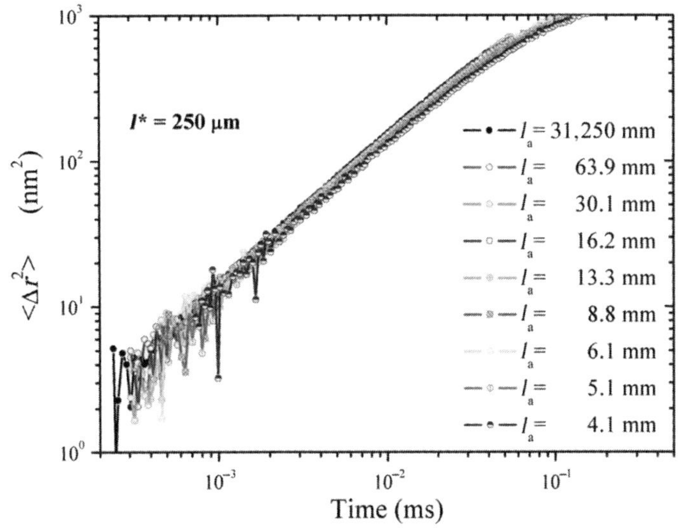

Figure 4.21 The above-presented dependencies after correction. Reprinted from Ref. 48.

the coherent backscattering effect in biomedical diagnosis were reported[49–51] in the beginning of the current century. The root principle of the proposed approach is based on the interrelation between the half-width of enhanced or coherent backscattering (CBS) peak and the transport mean free path l^* of light propagation in a probed random medium [Eqs. (4.3) and (4.4)]. Application of a broadband illumination of examined samples by a source with low spatial coherence in combination with a spectrally selected detection of backscattered light gives the possibility for retrieval of the wavelength dependence $l^*(\lambda)$. This approach also has a potential for deep-resolved measurements of l^*. The optical scheme of the bench-top prototype of a *coherent backscattering spectrometer* is presented in Fig. 4.22. A Xe lamp is used as a broadband illumination source; the lens system L1, L2 with the aperture A1 positioned in the Fourier plane is applied to vary the spatial coherence length of the illuminating beam. In the experiments (Refs. 49, 50), the spatial coherence length was varied from 110 μm to 180 μm. A double-slit interferometer was used to measure the actual value of the coherence length. The additional light-delivery arm with a high-coherence source (He-Ne laser) was used for comparison and adjustment of the instrument. Light backscattered by a sample SS (studied sample) fell on the entrance slit of the imaging spectrometer SP coupled with the CCD camera. The raw spectral imaging data for rat colon tissue, which were obtained using the instrument prototype, are presented in Fig. 4.23. The wavelength dependence of l^* can be retrieved from the wavelength dependence of the CBS peak half-width using the following approximate expression, which is valid in the case of diffusion approximation:

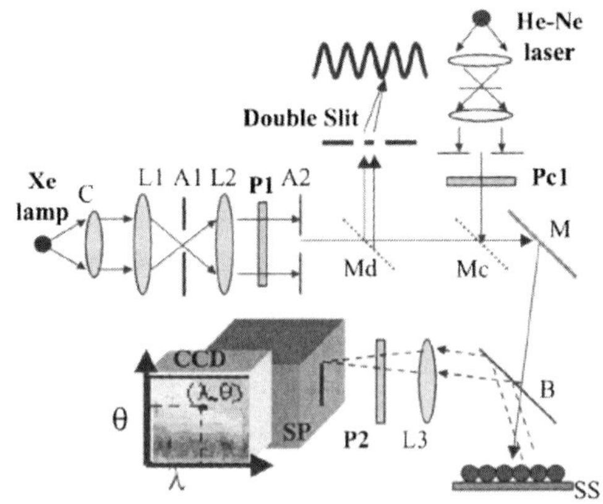

Figure 4.22 The scheme of coherent backscattering spectrometer (the bench-top prototype). Reprinted from Ref. 49.

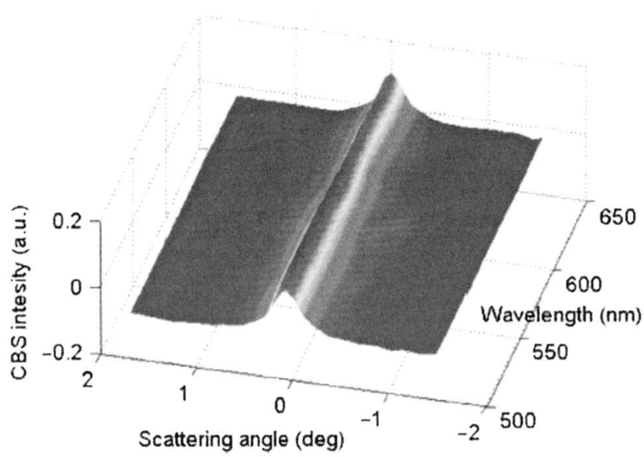

Figure 4.23 The raw spectral imaging data for rat colon tissue obtained in the low-coherent backscattering experiment. Reprinted from Ref. 50.

$$\theta_{0.5} \approx \frac{\lambda}{3\pi l^*}.$$

Diagnostic potential of the proposed approach in detection of early precancer is illustrated in Fig. 4.24. Here the conventional spectrum of normalized diffuse intensity obtained from rat colon tissue treated with azoxymethane (AOM) is compared with the coherent backscattering spectrum from the same site.

AOM-treated rat colon tissue was used as one of the most popular and robust animal cancer models. The coherent backscattering spectrum was obtained with the value of spatial coherence length equal to 140 μm with the use of integration over a scattering angle within the CBS peak. The diffuse backscattering intensity spectrum exhibits the significant contribution of

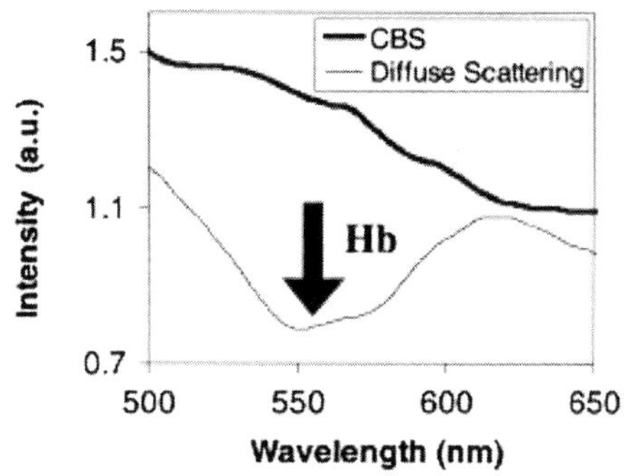

Figure 4.24 The comparison of diffuse backscattering and coherent backscattering spectra for the precancer model (rat colon tissue treated with AOM). Reprinted from Ref. 49.

hemoglobin in the tissue absorption near 550 nm (marked by the arrow). Note that the CBS spectrum is free of this peculiarity. This circumstance supports the conclusion that the CBS peaks arise from photons propagating predominantly in the epithelium.

The slope of CBS spectrum in the spectral region between 520 nm and 650 nm can be considered the diagnostic parameter; comparison of the slopes for control and precancerous rat colon tissues shows a remarkable difference between them.

Another example is the result of coherent backscattering spectroscopy of two-layered phantoms.[50] These phantoms consisted of an optically thick turbid base layer containing polystyrene microspheres and red blood cells. It was covered by superficial layers of different thicknesses without any absorbers. The CBS spectra for the different values of superficial layer thickness are presented in Fig. 4.25. Gradual disappearance of hemoglobin absorption bands with the increasing thickness of a superficial layer clearly demonstrates the effect of CBS peak formation due to photons backscattered in superficial layers.

Use of broadband light sources with the small coherence length in coherent backscattering probing of random media such as biological tissues provides some benefits compared with traditional coherent sources. The benefits are as follows:[50]

- The spectrally resolved CBS measurements are possible in the wide range of the wavelengths.
- In the case of broadband illumination and spectrally resolved detection, the CBS peaks are not masked by speckles.
- The proposed approach may provide the possibility of measuring very narrow CBS peaks, which are typical for biological tissues with large values of the transport mean free path.
- The coherent backscattering spectroscopy carried out with the changeable coherence length of a broadband illuminating source makes it possible to provide depth-resolved probes of random media.

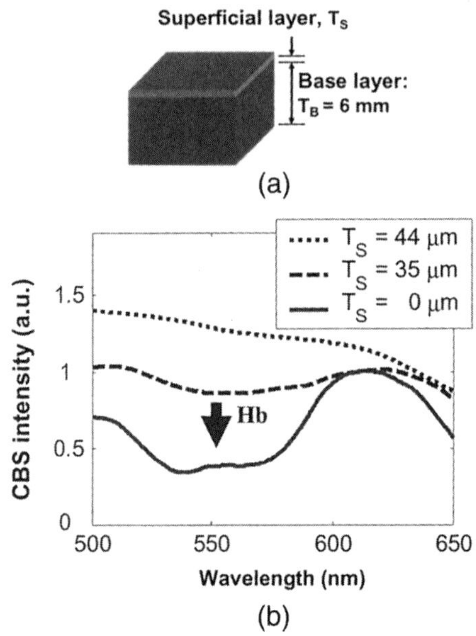

Figure 4.25 The results of coherent backscattering spectroscopy of two-layered phantoms. Reprinted from Ref. 50.

The previously mentioned peculiarities of the low-coherent backscattering spectroscopy [also defined as the low-coherence enhanced backscattering (LEBS) spectroscopy,[49–51]] such as the broadening of CBS peaks and speckle suppression following from broadband illumination and used restrictions for the spatial coherence length of illuminating light, provide the potential for depth-resolved probes of biological tissues. This is illustrated by LEBS experiments with two-layered tissue models (Fig. 4.26). In this figure, the set of CBS peaks corresponding to the different values of superficial layer thickness in a two-layered model is displayed. The angular distributions of intensity in CBS peaks are directly related to the probability distributions for photons to emerge from the surface at a given radial distance from a point of incidence. Thus, the analysis of angular distributions of enhanced backscattered intensity can be redirected to the problem of depth penetration evaluation in the dependence of the photon source—photon detector separation.

Analysis of the temporal correlations of multiply scattered light in applications to characterization of the structural and dynamical properties of random media is the more popular diagnostic tool compared with measurements of the angular correlations of scattered light (such as, in particular, the coherent backscattering measurements). The key principles of the *diffusing wave spectroscopy* of random media were established in the late eighties of the past century (see Refs. 13, 52). Early works on light correlation analysis dealt with relatively simple, well-defined, and ergodic

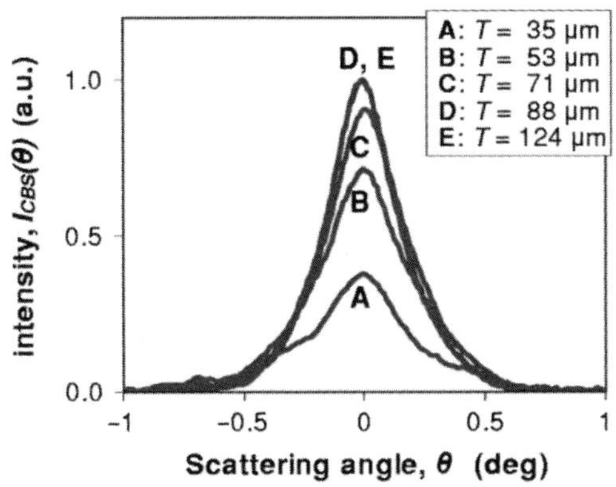

Figure 4.26 The shapes of coherent backscattering peaks for two-layered tissue phantoms with the different thickness of the superficial layer. Reprinted from Ref. 51.

multiple-scattering dynamic systems as Brownian ensembles or regular flows of spherical particles (see, for example, Refs. 13, 52, 53). The next stage in the development of this approach is related to the probing of various soft media with complex structure and dynamics (including systems with the expressed nonergodicity). Typical examples of these media are biological tissues treated by various physical and chemical factors, aggregated colloidal systems, glassy systems, foams, turbulent flows of particles, and porous media interacting with liquids and supercritical fluids. The problems of light correlation diagnostics of complex soft media are still relevant today.

The most common biomedical applications of the light correlation techniques in the past decade were the probes of deep tissue blood flow (see, for example, Refs. 54–63). Due to high sensitivity of the DWS method to variations in the scatterer mobility (such high sensitivity is caused by the accumulation of random phase shift in sequences of scattering events), it allows the characterization of complex motions of scatterers at microscopic level (such as complex dynamics of red blood cells in vessels). The recent results in the DWS probing of red blood cells dynamics[64] in living tissues have shown the possibility of identifying relative red blood cell motion in blood flow. This relative motion is associated with the diffusion-like cell motion manifestative at large time scales while ballistic cell motion gives the contribution to detected intensity fluctuations at short time scales. These results imply that in addition to tissue perfusion measurements, diffuse correlation spectroscopy may be a useful tool for blood rheology studies.

The advantages given by light correlation methods in the diagnostics of blood dynamics in living tissues have stimulated further progress in the instrumentation and software arrangement for photon correlation measurements. The implementation of fast software correlators in laboratory and clinical practice is the typical example; in particular, the fast-Fourier-transform-based software correlator considered in Ref. 65 has advantages such as flexibility in raw photon count data preprocessing and low cost. Besides, it offers smoother starting and ending plateaus when compared to hardware correlators, which could directly benefit the fitting results without too much sacrifice in the processing speed.

Another wide area of application of the DWS and other light correlation techniques is the analysis of microrheology, complex granular flows at the microscopic level, and processes in jammed colloidal systems in material science, the food and drug industry, and chemical technologies.[66–77]

In particular, the correlation analysis of the intensity fluctuations of multiply scattered light can also be used for imaging and quantitative characterization of local instabilities appearing in the course of liquid-gas interface growth in porous media. In this case, local boundaries between liquid and gaseous phases in single pores act like discrete dynamic scatterers.[78] At the final stages of the formation of stochastically shaped global boundaries between liquid and gaseous phases in porous matrices due to the gravity and capillary forces–driven imbibition, the boundary motion is controlled by the random process of initiation, development, and disappearance of "avalanches" (local instabilities of the interface). Typically, a stable, or "sleeping," boundary fragment suddenly (due to thermal fluctuations at the microscopic level) begins to move over a distance of few pores, and then this movement decays. Consequently, the final stage of interface formation, when the capillary forces are about to be compensated by the gravity, can be considered as a sequence of discrete jump-like motions. The probability of appearance of such avalanches slowly decreases in time but never falls to zero. We can consider certain analogies between this process and other stochastic physical processes (e.g., domain boundaries formation in ferromagnetic materials, percolation and conductivity processes in complex nonlinear systems, and phase transitions and other critical phenomena). That is why the analysis and visualization of avalanche-like formation of liquid-gas interfaces in the course of porous media imbibition can be very useful for better understanding of the fundamental features of various stochastic physical processes.

The potential of temporal correlation analysis of intensity fluctuations for avalanche-like interface motions in liquid-saturated porous layers was demonstrated in Ref. 79. Vertically placed porous layers (filter paper samples of various porosities) in the course of water imbibition were transilluminated by a broad collimated beam of a He-Ne laser. Forward-scattered light was detected by a CCD camera with a lens adjusted to obtain the speckle-modulated image of the rear surface of a porous layer at the CCD sensor. Image sequences were captured in the course of interface growth and then were processed; the temporal correlation analysis of pixel brightness fluctuations was performed for each pixel in the captured images. The first cumulant of temporal correlation function of pixel brightness fluctuations was applied as the imaging parameter for displaying the interface shape at various stages of its growth (see

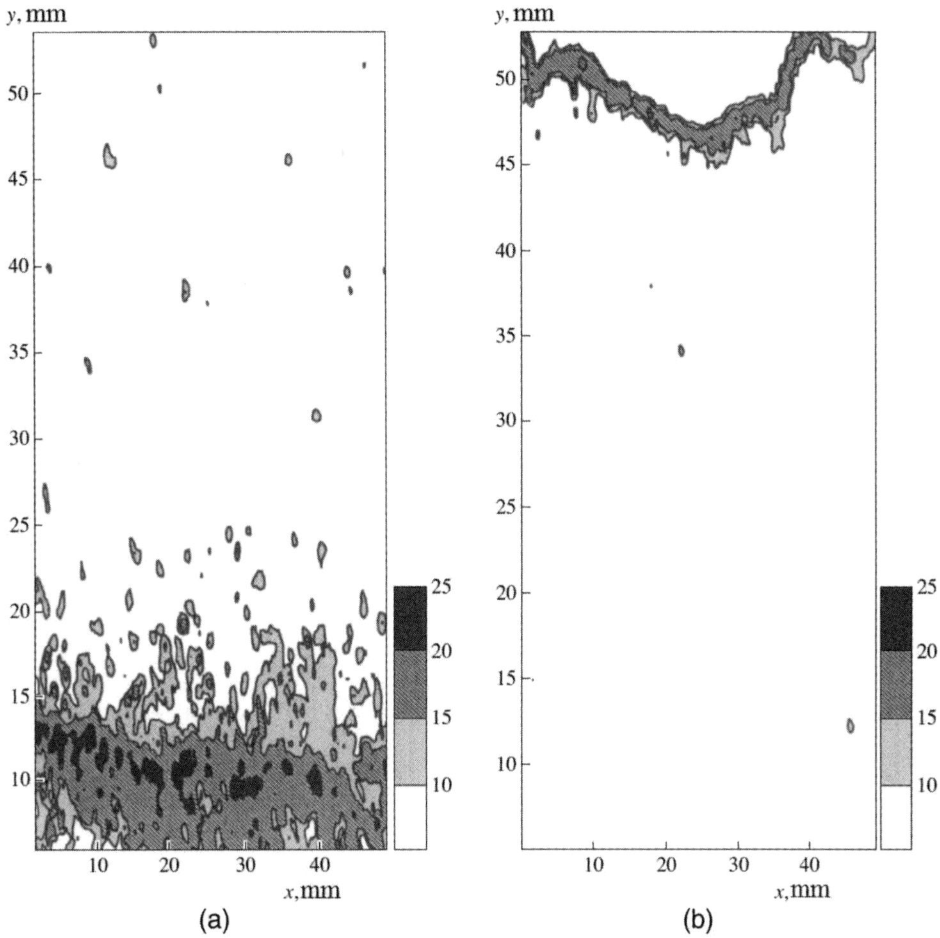

Figure 4.27 Visualization of growing interfaces in porous layers with the use of the first cumulant of intensity autocorrelation function as the imaging parameter (reprinted from Ref. 79). Filter paper with the thickness of 100 μm was used as the studied sample; a: the initial stage of interface growth; b: the final stage (close to the pinned state).

Fig. 4.27). The gradual decrease in the rate of avalanche appearance in the course of the interface approaching its stable state can be characterized by Fig. 4.28, where the dependence of the normalized value of the first cumulant on the characteristic parameter (the interface height normalized by its expected supreme value at the pinned state) is displayed. The results of cumulant analysis in combination with the results of noncoherent (white light) probes of growing interfaces allowed the establishment of fundamental relationships between the critical indices characterizing the scaling properties of interface growth at microscopic and macroscopic levels.

The conventional application of temporal correlation analysis of light intensity fluctuations (e.g., the diffusing light spectroscopy) to

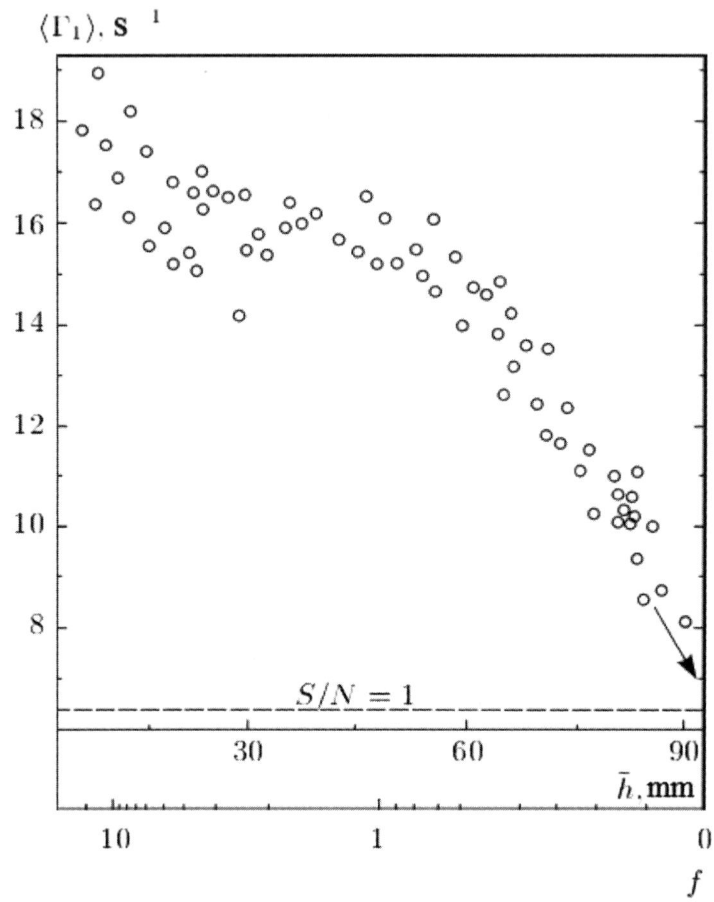

Figure 4.28 The dependence of the first cumulant averaged over the area of active growth on the normalized interface height (the pinned state height was used for normalization) (reprinted from Ref. 79). Filter paper with a thickness of 100 μm was used as the studied sample.

characterization of discrete scattering systems is related to quantification of any kind of translational or rotational motions of scattering particles. For example, such translational motions can occur in the form of classical or generalized Brownian motion, laminar or turbulent flow, a combination of flow and Brownian motions, and motions of scatterers through the complex three-dimensional network of channels (such as erythrocytes through the microcapillary net). Examples of light correlation probes of such dynamic multiple-scattering systems were presented previously. In addition, another physical mechanism of dynamic multiple scattering of partial waves propagating in random media consisting of motionless scatterers can be considered. In this case, the basic cause of stochastic modulation of propagating waves is the variations in the scattering efficiency for scatterers in the course of their growth or degeneration. Compared with the case of translational motion of scatterers, when the stochastic phase modulation of propagating waves occurs, the systems of growing or degenerating particles predominantly cause the amplitude modulation of probe light. Stochastisity of modulation in this case is caused by the natural spread in values of scatterer size.

This kind of dynamic multiple scattering of light should manifest itself in the course of two- or multiphase systems formation due to the process of phase separation, when one phase or several phases are segregated as microscopic domains fixed in a matrix with high viscosity. A typical example of such a process is the formation of random ensembles of submicrometer- and micrometer-sized liquid-crystal (LC) domains in polymer matrix due to UV irradiation of prepolymer/liquid crystal mixture.[80] Dynamic light scattering of these systems in the backscattering mode exhibits existence of three stages associated with formation of the LC droplet ("domain") structures in polymer matrices.[81] The first latent stage with a duration of several minutes or longer (depending on the volume fraction of a prepolymer in the mixture and the intensity of UV-irradiating light) is characterized by the absence of remarkable backward scattering (probed layers of initial material are almost transparent). At the second stage, the initial formation of the LC domain structure occurs, and the intensity of a backscattered signal rises abruptly.

Correspondingly, the fluctuation component of the backscattered signal also increases rapidly. The final stage, or the stage of structure stabilization, is featured by slow changes in the average intensity of backscattered light and the intensity of the fluctuation component. Figure 4.29 shows the typical scheme of a light correlation experiment with UV-cured LC-prepolymer mixtures; the principle of multispeckle diffusing-wave spectroscopy[82] is used in this setup. Because of the relatively large correlation times of backscattered intensity fluctuations typical for these scattering systems, it is not necessary to use CCD or CMOS cameras with high frame rates. Figure 4.30 shows the typical dependencies of the backscattered intensity for layers of LC-prepolymer mixtures on the time lapse in the course of UV irradiation. Comparison of the results of Monte Carlo simulation of dynamic light scattering by ensembles of growing motionless scatterers with the experimentally obtained data for LC-prepolymer mixtures containing various volume fractions of liquid crystal confirmed validity of the developed model and its high potential for the monitoring of synthesis of LC-polymer composite materials.

One of the interesting examples of intensity correlation analysis in studying the complex dynamic of nonconsolidated and partially consolidated random media is related to the use of multispeckle correlation spectroscopy of relaxation dynamics in soft porous media saturated by subcritical liquids and supercritical fluids (SCFs).[83] Interest in such systems is due to spreading implementation of supercritical fluidic technologies in various areas of modern industry.

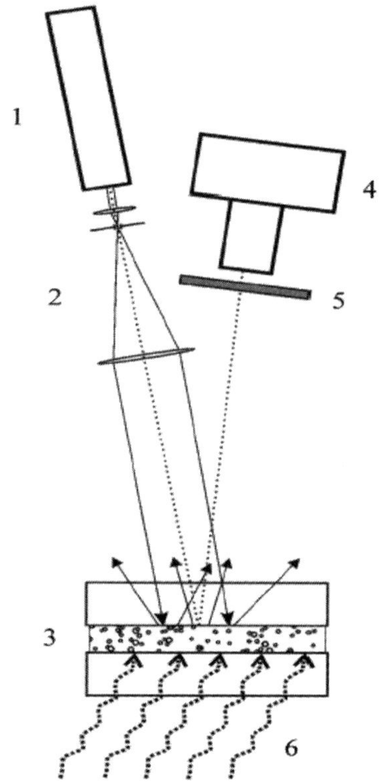

Figure 4.29 Schematic of the experimental setup. 1: He-Ne laser; 2: beam expander with the pinhole diaphragm; 3: sample under study between two quartz substrates; 4: CMOS camera with the lens; 5: interference filter; 6: wavy arrows indicating irradiation of the sample for activation of photo-polymerization (reprinted from Ref. 81).

The structure of deformable porous media saturated by a near-critical liquid or SCF is very sensitive to the thermodynamic state of the system "porous medium–saturating substance;" this causes the liability in the local positions of scatterers in such systems due to any changes in thermodynamic parameters (typically, the pressure or the temperature). The possible approach to characterization of these systems (and to analysis of interactions between porous matrices and saturating substances) is the study of relaxation of local deformation in the porous matrix under the condition of jump-like changes in the pressure of saturating substance. In particular, the similar approach applied when studying mechanical properties of constructional materials with the use of saturating subcritical liquids was defined by G. W. Scherer as the "dynamic depressurization."[84]

Light correlation experiments with fibrous porous layers saturated by near-critical carbon dioxide below and above the critical point ($T_C \approx$ 304.1282 K) have shown the strong dependence of temporal parameters of the

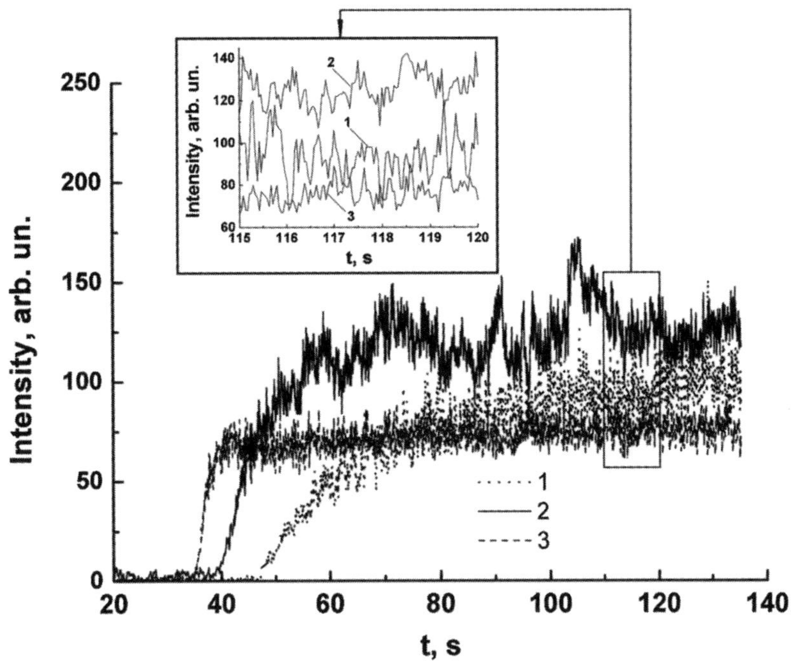

Figure 4.30 Values of intensity of detected signal (in arbitrary units) versus the time interval after the beginning of photo-polymerization; sample 1: the weight fractions of LC807 and NOA65 are 40% and 60%, respectively; sample 2: the weight fractions of LC807 and NOA65 are 50% and 50%, respectively; sample 3: the weight fractions of LC807 and NOA65 are 60% and 40%, respectively. From the start of the polymerization process, the stabilization stage begins for sample 1 after ≈60 s, for sample 2 after ≈50 s, and for sample 3 after ≈40 s. Inset shows the forms of detected fluctuating signals at small time scales in the course of the stabilization stage (reprinted from Ref. 81).

deformation relaxation in the course of dynamic depressurization on detuning of the system temperature relative to T_C. A multispeckle version of scattered light correlometer was used in the forward scattering mode of sample probing (see Fig. 4.31).

Deformable porous layers consisting of closely packed cellulose or PTFE (Teflon films with fibrillar porous) fibrils exhibit the decaying relaxation dynamic after jump-like changes in the pressure. This decaying dynamics is characterized by two specific time scales differing strongly in values: the correlation time of intensity fluctuations and the relaxation time of dynamic scattering. (The latter parameter characterizes the time interval of system transition to a new stable state.) Because the significant difference between these time scales (few decades in the magnitude) the different data processing algorithms were used for their evaluation. The correlation time was estimated using the conventional correlation analysis of pixel brightness fluctuations with the appropriate spatial-temporal averaging. To evaluate the relaxation time, the normalized power of the fluctuating component of

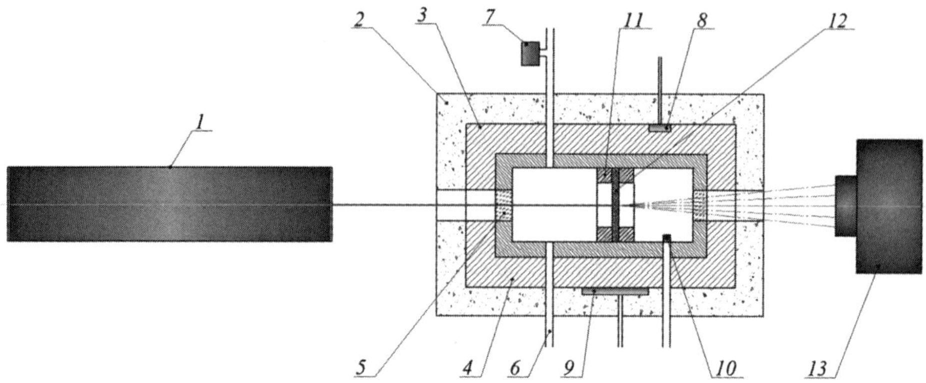

Figure 4.31 Scheme of the experimental setup. 1: He-Ne laser; 2: foamed plastic jacket; 3: aluminum housing; 4: high-pressure cell; 5: sapphire window; 6: high-pressure capillary; 7: pressure sensor; 8: quartz oscillator as a temperature sensor; 9: heater; 10: thermocouple; 11: sample holder; 12: sample under study; 13: CMOS camera (reprinted from Ref. 83).

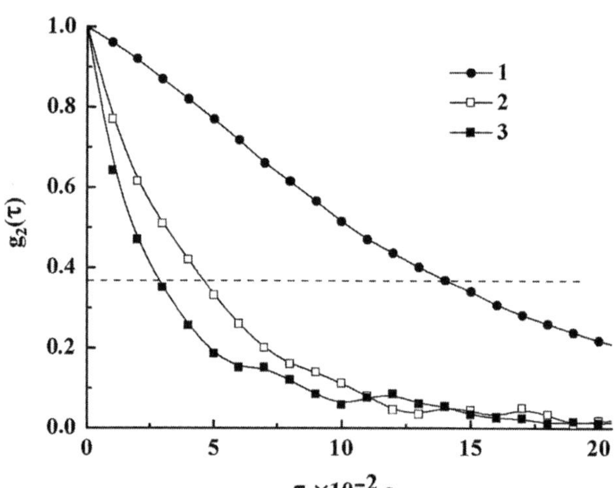

Figure 4.32 Typical normalized correlation functions of intensity fluctuations at the initial stage of relaxation in the system's "porous layer–subcritical liquid/supercritical fluid." Dashed line corresponds to $1/e$ level. Samples: paper layers; 1—T = 298.16 °K; 2—T = 300.66 °K; 3—T = 303.66 °K (reprinted from Ref. 83).

the detected optical signal was estimated in the dependence on the time lapse also using averaging in space and time domains. The relaxation time was estimated as the time interval of $1/e$ decay of the normalized power with respect to its initial value. Figures 4.32 and 4.33 show the typical dependencies of both parameters on the system temperature.

Analysis of the dependencies of both time scales (the correlation time and the relaxation time) on the system temperature showed the possibility of DWS

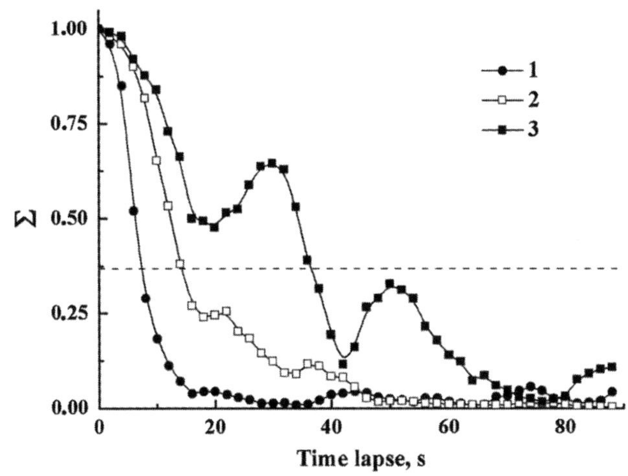

Figure 4.33 Evolution of the normalized power of intensity fluctuations in the course of relaxation of the porous matrices with subcritical liquid/supercritical fluid. Dashed line corresponds to 1/*e* level. Samples: PTFE fibrillar layers. The notations of the graphs are the same as in Fig. 4.32 (reprinted from Ref. 83).

characterization of the viscoelastic properties of porous matrices and peculiarities of the matrix-fluid interaction at the microscopic level.

4.7 Conclusion

The existence of coherence phenomena accompanying the laser light propagation in random multiple-scattering media gives the opportunity to develop novel approaches and techniques for optical diagnostics of macroscopically inhomogeneous weakly ordered scattering systems (e.g., biotissues). In particular, presence of the spatial and temporal correlations of speckle intensity fluctuations induced by the multiple scattering of probe light in nonstationary objects is the physical basis for different DWS techniques. These techniques have become a classic example of high efficiency of laser light scattering technologies for diagnostic applications in biology and medicine. Further development of the light scattering technologies can be based on the statistical analysis of multiply scattered speckles induced by the nondiffuse propagation of coherent light in the weakly ordered media such as tissues. Also, analysis of the vector statistics of multiply scattered optical fields can be considered another approach for the functional diagnostics of tissue structure.

Acknowledgments

The work on this chapter was supported in part by Russian Foundation for Basic Research (grant # 16-02-00458a) and by the Russian Science

Foundation (RSF, grant #16-19-10455). The author is grateful for the support in the framework of government tasks for higher education institutions and scientific organizations in the field of scientific activity.

References

1. A.A. Golubentsev, "On the suppression of the interference effects under multiple scattering of light" (in Russian), *Zh. Eksp. Teor. Fiz.* **86**, 47–59 (1984).
2. M.J. Stephen, "Temporal fluctuations in wave propagation in random media," *Phys. Rev. B.* **37**, 1–5 (1988).
3. F.C. MacKintosh and S. John, "Diffusing-wave spectroscopy and multiple scattering of light in correlated random media," *Phys. Rev. B.* **40**, 2382–2406 (1989).
4. S. John, "Localization of light," *Physics Today* **32**, (May, 1991).
5. P.W. Anderson, "Absence of diffusion in certain random lattices," *Phys. Rev.* **109**, 1492–1505 (1958).
6. S. John and M. Stephen, "Wave propagation and localization in a long-range correlated random potential," *Phys. Rev. B.* **28**, 6358 (1983).
7. S. John, "Electromagnetic absorption in a disordered medium near a photon mobility edge," *Phys. Rev. Lett.* **53**, 2169–2172 (1984).
8. A.F. Ioffe and A.R. Regel, *Prog. Semicond.* **4**, 237 (1960).
9. M.P. Van Albada and A. Lagendijk, "Observation of weak localization of light in a random medium," *Phys. Rev. Lett.* **55**, 2692–2695 (1985).
10. P.E. Wolf and G. Maret, "Weak localization and coherent backscattering of photons in disordered media," *Phys. Rev. Lett.* **55**, 2696–2699 (1985).
11. E. Akkermans, P.E. Wolf, R. Maynard, and G. Maret, "Theoretical study of the coherent backscattering of light by disordered media," *J. Phys. France* **49**, 77–98 (1988).
12. V.L. Kuz'min and V.P. Romanov, "Coherent phenomena in light scattering from disordered systems," *Soviet Phys. Usp.* **166**, 247–277 (1996).
13. G. Maret and P.E. Wolf, "Multiple light scattering from disordered media. The effect of Brownian motions of scattterers," *Z. Phys. B* Condensed Matter, **65**, 409–413 (1987).
14. J. Feder, *Fractals*, Plenum Press, New York (1988).
15. D. Avnir, O. Biham, D. Lidar (Hamburger), and O. Malcai, "On the abundance of fractals," *Fractal Frontiers'97*, M. M. Novak and T. G. Dewey, Ed., pp. 199–234, World Scientific, Singapore (1997).
16. A.G. Yodh, N. Georgiades, and D.J. Pine, "Diffusing-wave interferometry," *Opt. Communications* **83**, 56–59 (1991).
17. M.H. Kao, A.G. Yodh, and D.J. Pine, "Observation of Brownian motion on the time scale of hydrodynamic interactions," *Phys. Rev. Lett.* **70**, 242–245 (1993).

18. D.A. Boas, K.K. Bizheva, and A.M. Siegel, "Using dynamic low-coherence interferometry to image Brownian motion within highly scattering media," *Optics Letters* **23**, 319–321 (1998).
19. G. Hausler, J.M. Herrmann, R. Kummer, and M.W. Lindner, "Observation of light propagation in volume scatterers with 10^{11}-fold slow motion," *Optics Letters* **21**, 1087–1089 (1996).
20. M. Born and E. Wolf, *Principles of Optics*, Pergamon Press, London (1964).
21. B.J. Ackerson, R.L. Dougherty, N.M. Reguigui, and U. Nobbman, "Correlation transfer: application of radiative transfer solution methods to photon correlation problems," *J. Thermophys. and Heat Trans.* **6**, 577–588 (1992).
22. D.A. Boas and A.G. Yodh, "Spatially varying dynamical properties of turbid media probed with diffusing temporal light correlation," *JOSA A* **14**, 192–215 (1997).
23. D.A. Boas, L.E. Campbell, and A.G. Yodh, "Scattering and imaging with diffusing temporal field correlations," *Phys. Rev. Lett.* **75**, 1855–1858 (1995).
24. B. Chance, M. Maris, J. Sorge, and M.Z. Zhang, "A phase modulation system for dual wavelength difference spectroscopy of haemoglobin deoxygenation in tissue," *Proc. SPIE* **1204**, 481–491 (1990).
25. K.W. Berndt and J.R. Lakowicz, "Detection and localization of absorbers in scattering media using frequency domain principles," *Proc. SPIE* **1431**, 149–158 (1991).
26. S.R. Arridge, P. Van der Zee, M. Cope, and D.T. Delpy, "New results for the development of infra-red absorption imaging," *Proc. SPIE* **1245**, 91–103 (1990).
27. L. Wang, Y. Liu, P.P. Ho, and R.R. Alfano, "Ballistic imaging of biomedical samples using picosecond optical Kerr gate," *Proc. SPIE* **1431**, 97–101 (1991).
28. G. Jarry, S. Ghesquerre, J.M. Maarek, S. Debray, Bui-Mong-Hung, and D. Laurent, "Imaging mammalian tissues and organs using laser collimated transillumination," *J. Biomed. Eng.* **6**, 70–74 (1984).
29. P.C. Jackson, P.H. Stevens, J.H. Smith, D. Kear, H. Key, and P.N.T. Wells, "The development of a system for transillumination computed tomography," *Br. J. Radiol.* **60**, p. 375–380 (1987).
30. P.D. Kaplan, M.H. Kao, A.G. Yodh, and D.J. Pine, "Geometric constraints for the design of diffusing-wave spectroscopy experiments," *Applied Optics* **32**, 3828–3836 (1993).
31. S.R. Arridge, M. Cope, and D.T. Delpy, "The theoretical basis for the determination of optical pathlengths in tissue: temporal and frequency analysis," *Phys. Med. Biol.* **37**, 1531–1560 (1992).
32. M.S. Patterson, B. Chance, and B.C. Wilson, "Time-resolved reflectance and transmittance for the non-invasive measurement of the tissue optical properties," *Applied Optics* **28**, 2331–2336 (1989).

33. J.X. Zhu, D.J. Pine, and D.A. Weitz, "Internal reflection of diffuse light in random media," *Phys. Rev. A* **44**, 3948–3959 (1991).
34. R.C. Haskell, L.O. Svaasand, T.-T. Tsay, T.-C. Feng, M.S. McAdams, and B.J. Tromberg, "Boundary conditions for the diffusion equation in radiative transfer," *JOSA A* **11**, 2727–2741 (1994).
35. H. Bremmer, "Random volume scattering," *Radio Sci. J. Res.* **680**, 967–981 (1964).
36. P.-A. Lemieux, M.U. Vera, and D.J. Durian, "Diffusing-light spectroscopies beyond the diffusion limit: the role of ballistic transport and anisotropic scattering," *Phys. Rev. E* **57**, 4498–4515 (1998).
37. D.J. Durian and J. Rudnick, "Photon migration at short times and distances and in cases of strong absorption," *JOSA A* **14**, 235–245 (1997).
38. F.C. MacKintosh, J.X. Zhu, D.J. Pine, and D.A. Weitz, "Polarization memory of multiply scattered light," *Phys. Rev. B* **40**, 9342–9345 (1989).
39. D. Bicout, C. Brosseau, A.S. Martinez, and J.M. Schmitt, "Depolarization of multiply scattering waves by spherical diffusers: influence of size parameter," *Phys. Rev. E* **49**, 1767–1770 (1994).
40. V. Sankaran, M.J. Everett, D.J. Maitland, and J.T. Walsh, "Comparison of polarized light propagation in biologic tissue and phantoms," *Opt. Lett.* **24**, 1044–1046 (1999).
41. D.A. Zimnyakov and V.V. Tuchin, "About interrelations of distinctive scales of depolarization and decorrelation of optical fields in multiple scattering," *JETP Letters* **67**, 455–460 (1998).
42. D.A. Zimnyakov, V.V. Tuchin, and A.G. Yodh, "Characteristic scales of optical field depolarization and decorrelation for multiple scattering media and tissues," *Journal of Biomedical Optics* **4**, 157–163 (1999).
43. T. Durduran, A.G. Yodh, B. Chance, and D.A. Boas, "Does the photon-diffusion coefficient depend on absorption?" *JOSA A* **14**, 3358–3365 (1997).
44. K. Furutsu and Y. Yamada, "Diffusion approximation for a dissipative random medium and the applications," *Phys. Rev. E.* **50**, 3634–3640 (1994).
45. K. Furutsu, "Pulse wave scattering by an absorber and integrated attenuation in the diffusion approximation," *JOSA A* **14**, 267–274 (1997).
46. M. Bassani, F. Martelli, G. Zaccanti, and D. Contini, "Independence of the diffusion coefficient from absorption: experimental and numerical evidence," *Opt. Lett.* **22**, 853–855 (1997).
47. D.J. Durian, "The diffusion coefficient depends on absorption," *Opt. Lett.* **23**, 1502–1505 (1998).
48. E. Sarmiento-Gomez, B. Morales-Cruzado, and R. Castillo, "Absorption effects in diffusing wave spectroscopy," *Applied Optics* **53**, 4675–4682 (2014).

49. Y.L. Kim, Y. Liu, V.M. Turzhitsky, H.K. Roy, R.K. Wali, and V. Backman, "Coherent backscattering spectroscopy," *Optics Letters* **29**, 1906–1908 (2004).
50. Y.L. Kim, Y. Liu, R.K. Wali, H.K. Roy, and V. Backman, "Low-coherent backscattering spectroscopy for tissue characterization," *Applied Optics* **44**, 366–377 (2005).
51. Y.L. Kim, Y. Liu, V.M. Turzhitsky, R.K. Wali, H.K. Roy, and V. Backman, "Depth-resolved low-coherence enhanced backscattering," *Optics Letters* **30**, p. 741–743 (2005).
52. D.J. Pine, D.A. Weitz, P.M. Chaikin, and E. Herbolzheimer, "Diffusing wave spectroscopy," *Phys. Rev. Lett.* **60**, 11341137 (1988).
53. X-L. Wu, D.J. Pine, P.M. Chaikin, J.S. Huang, and D.A. Weitz, "Diffusing-wave spectroscopy in a shear flow," *JOSA A* **7**, 15–20 (1990).
54. C. Cheung, J.P. Culver, K. Takahashi, J.H. Greenberg, and A.G. Yodh, "In vivo cerebrovascular measurement combining diffuse near-infrared absorption and correlation spectroscopies," *Phys. Med. Biol.* **46**, 2053–2065 (2001).
55. J.P. Culver, T. Durduran, D. Furuya, C. Cheung, J.H. Greenberg, and A. G. Yodh, "Diffuse optical tomography of cerebral blood flow, oxygenation, and metabolism in rat during focal ischemia," *J. Cereb. Blood Flow Metab.* **23**, 911–924 (2003).
56. C. Zhou, G. Yu, D. Furuya, J.H. Greenberg, A.G. Yodh, and T. Durduran, "Diffuse optical correlation tomography of cerebral blood flow during cortical spreading depression in rat brain," *Opt. Express* **14**, 1125–1144 (2006).
57. J. Li, M. Ninck, L. Koban, T. Elbert, J. Kissler, and T. Gisler, "Transient functional blood flow change in the human brain measured noninvasively by diffusing-wave spectroscopy," *Opt. Lett.* **33**, 2233–2235 (2008).
58. S.A. Carp, G.P. Dai, and D.A. Boas, "Validation of diffuse correlation spectroscopy measurements of rodent cerebral blood flow with simultaneous arterial spin labeling MRI; towards MRI-optical continuous cerebral metabolic monitoring," *Biomed. Opt. Express* **1**, 553–565 (2010).
59. G. Yu, T.F. Floyd, T. Durduran, C. Zhou, J. Wang, J.A. Detre, and A.G. Yodh, "Validation of diffuse correlation spectroscopy for muscle blood flow with concurrent arterial spin labeled perfusion MRI," *Opt. Express* **15**, 1064–1075 (2007).
60. M. Belau, M. Ninck, G. Hering, L. Spinelli, D. Contini, A. Torichelli, and T. Gisler, "Noninvasive observation of skeletal muscle contraction using near-infrared time-resolved reflectance and diffusing-wave spectroscopy," *J. Biomed. Opt.* **15**, 057007 (2010).
61. Y. Shang, T.B. Symons, and T. Durduran, "Effects of muscle fiber motion on diffuse correlation spectroscopy blood flow measurements during exercise," *Biomed. Opt. Express* **1**, 500–511 (2010).

62. G. Dietsche, M. Ninck, C. Ortolf, J. Li, F. Jaillon, and T. Gisler, "Fiber-based multispeckle detection for time-resolved diffusing-wave spectroscopy: characterization and application to blood flow detection in deep tissue," *Appl. Opt.* **46**, 8506–8514 (2007).
63. M. Ninck, M. Untenberger, and T. Gisler, "Diffusing-wave spectroscopy with dynamic contrast variation: disentangling the effects of blood flow and extravascular tissue shearing on signals from deep tissue," *Biomed. Opt. Express* **1**, 1502–1513 (2010).
64. S.A. Carp, N. Roche-Labarbe, M.-A. Franceschini, V.J. Srinivasan, S. Sakadžić, and D.A. Boas, "Due to intravascular multiple sequential scattering, Diffuse Correlation Spectroscopy of tissue primarily measures relative red blood cell motion within vessels," *Biomed. Opt. Express* **2**, 2047–2054 (2011).
65. J. Dong, R. Bi, J.H. Ho, P.S.P. Thong, K.-C. Soo, and K. Lee, "Diffuse correlation spectroscopy with a fast Fourier transform-based software autocorrelator," *Journal of Biomedical Optics* **17**, 097004 (2012).
66. F. Scheffold, S. Romer, F. Cardinaux, H. Bissig, A. Stradner, L.F. Rojas-Ochoa, V. Trappe, C. Urban, S.E. Skipetrov, L. Cipelletti, and P. Schurtenberger, "New trends in optical microrheology of complex fluids and gels," *Progr. Colloid. Polym. Sci.* **123**, 141–146 (2004).
67. A. Duri, D.A. Sessoms, V. Trappe, and L. Cipelletti, "Resolving long-range spatial correlations in jammed colloidal systems using photon correlation imaging," *Phys. Rev. Lett.* **102**, 085702-1–085702-4 (2009).
68. L. Brunel, A. Brun, P. Snabre, and L. Cipelletti, "Adaptive speckle imaging interferometry: a new technique for the analysis of microstructure dynamics, drying processes and coating formation," *Optics Express* **15**, 15250–15259 (2007).
69. F. Croccolo, H. Bataller, and F. Scheffold, "A light scattering study of non-equilibrium fluctuations in liquid mixtures to measure the Soret and mass diffusion coefficient," *J. Chem. Phys.* **137**, 234202-1–234202-7 (2012).
70. M. Alexander and D.G. Dalgleish, "Application of transmission diffusing wave spectroscopy to the study of gelation of milk by acidification and rennet," *Colloids Surf B Biointerfaces* **38**, 83–90 (2004).
71. K. Kim and H.K. Pak, "Diffusing-wave spectroscopy study of microscopic dynamics of three-dimensional granular systems," *Soft Matter* **6**, 2894–2900 (2010).
72. C. Hanotin, Ph. Marchal, L.J. Michot, C. Baravian, and S. Kiesgen de Richter, "Dynamics of vibrated granular suspensions probed by mechanical spectroscopy and diffusing wave spectroscopy measurements," *Soft Matter* **9**, 9352–9360 (2013).

73. A. Le Bouil, A. Amon, J.-C. Sangleboeuf, H. Orain, P. Bésuelle, G. Viggiani, P. Chasle, and J. Crassous, "A biaxial apparatus for the study of heterogeneous and intermittent strains in granular materials," *Granular Matter* **16**, 1–8 (2014).
74. C. Oelschlaeger, M. Cota Pinto Coelho, and N. Willenbacher, "Flexibility and structure of polysaccharide hyaluronan solutions: a high frequency rheology and diffusing wave spectroscopy study" *Biomacromolecules* **14**, 3689–3696 (2013).
75. M. Erpelding, B. Dollet, A. Faisant, J.J. Crassous, and A. Amon, "Diffusing-wave spectroscopy contribution to strain analysis," *Strain* **49**, 167–174 (2013).
76. A. Raudsepp, A.J. Sutherland-Smith, and M.A.K. Williams, "Rotating angled plate diffusing wave spectroscopy," *International Journal of Nanotechnology* **5**, 573–582 (2014).
77. J. Liu, V. Boyko, Z. Yi, and Y. Men, "Temperature-dependent gelation process in colloidal dispersions by diffusing wave spectroscopy," *Langmuir* **29**, 14044–14049 (2013).
78. D.A. Zimnyakov, P.V. Zakharov, V.A. Trifonov, and O.I. Chanilov, "Dynamic light scattering study of the interface evolution in porous media," *JETP Letters* **74**, 216–221 (2001).
79. D.A. Zimnyakov, A.V. Sadovoy, M.A. Vilenskii, P.V. Zakharov, and R. Myllyla, "Critical behavior of phase interfaces in porous media: analysis of scaling properties with the use of noncoherent and coherent light," *JETP* **108**, 311–325 (2009).
80. F. Roussel, C. Canlet, and B.M. Fung, "Morphology and orientation order of nematic liquid crystal droplets confined in a polimer matrix," *Phys. Rev. E* **65**, 021701-1–021701-9 (2002).
81. D.A. Zimnyakov, A.A. Isaeva, E.A. Isaeva, O.V. Ushakova, S.P. Chekmasov, and S.A. Yuvchenko, "Analysis of the scatter growth in dispersive media with the use of dynamic light scattering," *Applied Optics* **51**, C62–C69 (2012).
82. P. Snabre and J. Crassous, "Multispeckle diffusing wave spectroscopy colloidal particles suspended in a random packing of glass spheres," *Eur. Phys. J. E* **29**, 149–155 (2009).
83. D.A. Zimnyakov, S.P. Chekmasov, O.V. Ushakova, E.A. Isaeva, V.N. Bagratashvili, and S.B. Yermolenko, "Laser speckle probes of relaxation dynamics in soft porous media saturated by near-critical fluids," *Applied Optics* **53**, B12–B21 (2014).
84. G.W. Scherer, "Dynamic pressurization method for measuring permeability and modulus: I. Theory," *Mat. Constr.* **39**, 1041–1057 (2006).

Dmitry A. Zimnyakov is a professor and chairman of the Physics Department at the Yuri Gagarin Saratov State Technical University. He is also the chief researcher of the laboratory, Institute of Precision Mechanics and Control, RAS. He is a member of international scientific societies SPIE and OSA, and a Soros Professor in 2000–2001. His research interests include biophotonics, statistical optics, laser physics, and biomedical optics. He is the author, editor, or coauthor of over 20 books, chapters in books, special issues of journals, conference proceedings, textbooks, lecture manuals, brochures, and more than 200 scientific articles and reviews.

Chapter 5
Tissue Phantoms

Alexander B. Pravdin
Saratov National Research State University, Saratov, Russia

George Filippidis, Giannis Zacharakis, and Theodore G. Papazoglou
Institute of Electronic Structure and Laser, Foundation for Research and Technology, Hellas, Greece

Valery V. Tuchin
Saratov National Research State University, Saratov, Russia
National Research Tomsk State University, Tomsk, Russia
Institute of Precision Mechanics and Control, Russian Academy of Sciences, Saratov, Russia

5.1 Introduction

We are currently witnessing a steady growth of interest in optical methods for medical diagnostics and treatment. The reason for this growth is that optical methods have the advantage of being inherently noninvasive. Many research groups are working on the theoretical foundation and measuring techniques that allow for the reconstruction of tissue's intrinsic optical properties from optical (spectral) signals that are measured on the surface of biological objects. The detection and localization of optical inhomogeneities, such as tumors and hematomas, deep within tissue is one example. Another example is the development of light-based therapeutic methods [selective tissue ablation, PUVA irradiation, and photodynamic therapy (PDT), for instance], which needs the assessment of intensity (fluence) fields within the tissue or organ that is treated.

However, the evaluation of measuring techniques and validation of theoretical predictions on light propagation in tissues is hardly possible in direct experiments on actual bio-objects. One encounters wide variations of morphological and biochemical parameters that are beyond the control of the experimenter. If diagnostic equipment is to be used on a daily basis, then stable and reproducible calibration methods need to be developed. For this purpose, stable and reproducible test objects that mimic tissue optical characteristics are needed.

The very development of optical medical techniques and technology at all stages, from elaborating the concept to obtaining the necessary operating parameters, requires calibration and verification of design tools and methods. An optical medical apparatus should come complete with tissue phantoms for testing and optimization of the device hardware and software with a variety of applications; for training operating, attending, and maintenance personnel; and for providing comparability of measurement data obtained with different hardware in different laboratories.[1,2]

Since the late 1980s, various models that mimic the optical and structural properties of tissue have been developed for many different areas of optical diagnostics. Tissue phantoms with predetermined optical and geometrical parameters were used in experimental testing of theoretical predictions (simulation results) for light propagation within tissue.[3–9] Tissue models have been developed for tissue fluorescence spectroscopy and quantitative detection of fluorophores (including photosensitizers) in tissues;[10–20] time-, frequency-, and spatial-domain tomography and spectroscopy; [1,2,21–26] and noninvasive glucose monitoring,[27–38] oxygenation monitoring and oxymetry,[1,39–42] optoacoustics,[29,32–34,42–46] pulsed photothermal measurements,[47–49] Doppler flowmetry,[50–52] and measurement of polarization degree decay.[53,54] In addition, tissue phantoms are used in spatially resolved reflectance measurements[1,55–59] and hyperspectral imaging,[60,61] including the use of multicolor phantoms.[62] On-phantom diffuse reflectance measurements may be performed in the steady state,[6,41,55,63–66] time-domain,[63,67–70] and frequency-domain.[71,72] Tissue phantoms were also used successfully in studies concerned with polarization gating,[73] time-resolved fluorescence,[74] and CW optical tomography.[75]

Calibration of a reflective confocal microscope[76] and optical coherence tomography systems[76–81] was performed using phantoms. Tissue phantoms also have been developed for calibration of the Raman signals from different types of tissue.[76,82,83] Liquid tissue-like phantoms may be used in assessment of effectiveness of optical clearing agents.[84,85]

Finally, tissue-like phantoms have been considered in research connected with therapeutic implementation of optical radiation, including light dosimetry,[86,87] laser ablation,[45,88] and PDT.[12,19,87,89,90]

5.2 General Approaches to Phantom Development

5.2.1 Basic concept

Light propagation in optically inhomogeneous turbid media such as biological tissues, which are predominantly light scattering in the visible range, is commonly described by the Boltzmann transport equation. This equation accounts for optical properties of tissue in the form of absorption coefficient μ_a, scattering coefficient μ_s, and scattering phase function. These quantities form the basis of phantom modeling. The light propagation in an artificial medium is similar to that in actual tissue, when μ_s, μ_a, and the anisotropy factor (as an approximation for phase function), g, of the phantom are similar to the values found in a natural object. In soft tissues, typical values for the coefficients are $\mu_a \approx 0.5$ to $5.0\ \text{cm}^{-1}$, $\mu_s \approx 0.2$ to $400\ \text{cm}^{-1}$, and $0.8 < g < 0.99$ in the visible and NIR ranges.

When the diffusion approximation to the Bolzmann equation is used for the problem under study and the shape of the phase function is not essential, modeling of only two parameters—μ_a and the transport scattering coefficient, $\mu_s{}' = \mu_s(1 - g)$—provides the necessary light transport similarity. For some purposes, the similarity between the optical properties of a phantom and tissue may be on an equal level of values of the diffusion coefficient, $D = \mu_a/\mu_{\text{eff}}^2$, where μ_{eff} is the effective attenuation coefficient, and $\mu_{\text{eff}}^2 = 3\mu_a[\mu_a + \mu_s(1 - g)]$. On this basis, scaled phantom models can be developed in which scattering and absorption of light proceed in the same manner as in tissue but over larger dimensions.[7]

When the task is to model a tissue with a complex architecture (or even a whole organ) or to prepare the test object for an evaluation of imaging techniques, the "macroscopic" geometry of the natural object should be reproduced in a phantom. As the steps in modeling realistic complex geometries, *i* layered samples, *ii* inserted 2D and 3D inhomogeneities, and *iii* phantoms mimicking whole organs may be considered.[3,5,7,10,11,21,37,55,56,69,91–94] Some of the offered phantom systems exhibit realistic optical properties over a wide wavelength range.[3–5,93,94]

One of the most commonly encountered features is a layered tissue structure. Multilayer phantoms have been developed to mimic, for example, skin,[47,65] the human head,[41,63] and *cervix uteri.*[13,14]

When phantoms of complex structures are developed, the refractive indices of layers (inclusions) should be accounted for, as a mismatch at the interfaces may influence the light propagation within tissue. The value of refractive index also plays an important role when the tissue-air interface is modeled. The accepted typical values for soft tissues are -1.33 to 1.50.[95]

A straightforward approach to modeling optical properties of tissue is to reproduce scattering and absorption coefficients independently by mixing relevant proportions of purely scattering (no absorption in wavelength region of interest) and purely absorbing media. The latter includes predominantly dyes that have negligible scattering because they are dispersed on a molecular level (no mesoscopic inhomogeneities of refractive index).

Phantoms developed for use in fluorescence studies also contain fluorescent dyes, which may be photosensitizers or natural fluorophores. The criteria of quantification of fluorophores to be added are based on their absorption coefficients at the excitation wavelength and spectral fluorescence yields.[10,11,17]

Both scatterers and absorbers/fluorophores are suspended in an appropriate transparent host medium (ground material), which forms the volume of the phantom. The demand on the optical properties of the host material, apart from being nonscattering and nonabsorbing, is that the value of refractive index is close to the value found in actual tissue.

When choosing a host for a particular phantom, the issue of component compatibility should be taken into consideration. This is important because some commonly used scatterers are colloidal systems and may aggregate in an inappropriate diluent, and some particles (e.g., polystyrene microspheres) may dissolve in organic solvents. Besides chemical stability, stable spectroscopic properties should be considered because the absorbers and fluorophores may be influenced markedly by solvent/matrix effects.

According to the mechanical properties of the host used, tissue phantoms can be divided in two classes—liquid and solid phantoms. Liquid phantoms, in which scatterers and absorbers are mixed in correct diluent, are rather easy to prepare. However, samples of realistic complexity are not possible.[10,11] Such phantoms have been commonly used in modeling of infinite or semi-infinite media with absorption, scattering, and (optionally) fluorescence. The advantage of a liquid phantom is that fluence or radiance detectors can move easily within the phantom volume.

In solid phantoms, the host material serves as a mechanical basis. Typical host materials are polymers[10,11,26,47,50,68,89] and aqueous gels.[10,11,13,14,21,43,48,67,69,72]

In Ref. 96, a comprehensive review on tissue phantom studies conducted prior to 2006, gelatin, agar, and polyvinyl alcohol (cryogel) gels and polyester, epoxy, polyurethane, and room-temperature vulcanizing (RTV) silicone resins are presented as typical host materials in the preparation of solid tissue-like phantoms. From this list, hydrogel materials provide a semisolid medium that is chemically compatible with the addition of polar organic molecules and are optimal for scientific laboratory studies. Polyester, polyurethane, and silicone host materials provide essentially stable phantoms that are suitable for routine calibration and testing of measuring and imaging systems.

An overview of research on the development of solid phantoms for OCT, simulating the optical, mechanical, and structural properties of a number of tissues, is given in Ref. 77. The authors focused on the analysis of phantoms based on silicone, fibrin, and cryogels of polyvinyl alcohol (PVA-C) as the most promising materials for solid phantoms, with possibly the most accurate reproduction of the properties of biological tissues.

The ability of these materials to hold a desired shape provides opportunity for manufacturing geometrically complex inhomogeneous phantoms. For example, whole organs can be mimicked by cutting and stacking slabs or using molding techniques. One of the advantages of such solid media is that the refractive index matching at interfaces is nearly perfect due to the absence of walls or partitions. Furthermore, the aggregation and sedimentation of scattering particles, a common problem in liquid phantoms, is not a problem in solid phantoms; the hardness of the host material does not allow the scattering particles to move.

However, dyes that are dissolved in a polymer or gel matrix may diffuse unobstructed from one slab to another, thus reducing the stability of the phantom optical parameters. To circumvent this problem, along with the problem of spectral shifts caused by interactions of absorbers and fluorophores with other components of a phantom, it has been suggested that absorbing and fluorescing particles suspended in host material are substituted for dyes.[10,11]

Optical phantoms can be manufactured on the basis of purely artificial materials. In addition, when specific problems are solved, *i* engineered tissues, *ii ex vivo* samples of tissues,[96] and *iii in vivo* animal tissues after a special surgery simulating a particular pathological state may also serve as phantoms.[61]

In the systematic design of tissue phantoms, it is important that the optical parameters of the sample be predictable from the individual characteristics of its constituents and from the system composition. However, the instruments currently being developed for traceable measurement of the diffuse optical properties of tissue mimicking phantom materials (see, for example, Ref. 97) can allow us to waive this requirement.

The practical implementation of the on-phantom measurements gives rise to additional constraints. For example, physical parameters of the phantom system should be stable temporarily—not only the optical parameters but also the mechanical stability (evaporation of solvents, aging of polymers, and degradation of constituents by bacteria). Furthermore, the preparation process should be as simple, quick, and safe as possible.[10,11]

5.2.2 Mie theory predictions for scattering and absorption properties of particle suspensions

Calculations based on Mie theory are conventionally performed to predict scattering and absorption properties of particle suspensions that mimic the actual light scattering and absorption in tissues. Although Mie theory is applicable only in the case of some regular-shaped particles, its predictions are often considered acceptable for the particles of irregular shapes. The theory yields the scattering cross section, σ_s (cm^2), scattering anisotropy, g, and absorption cross section, σ_a (cm^2), of the particles. For nonabsorbing particles, the first two (scattering) parameters are given by[98]

$$\sigma_s = (\lambda_0^2/2\pi n_0^2)\sum_{n=1}^{\infty}(2n+1)(|a_n|^2+|b_n|^2), \tag{5.1}$$

$$g = \frac{\lambda_0^2}{\pi n_0^2 \sigma_s}\left\{\sum_{n=1}^{\infty}\frac{2n+1}{n(n+1)}\mathrm{Re}(a_n b_n^*) + \sum_{n=1}^{\infty}\frac{n(n+2)}{n+1}\mathrm{Re}(a_n a_{n+1}^* + b_n b_{n+1}^*)\right\}, \tag{5.2}$$

where

$$a_n = \frac{\psi_n(\alpha)\psi_n'(m\alpha) - m\psi_n(m\alpha)\psi_n'(\alpha)}{\xi(\alpha)\psi_n'(m\alpha) - m\psi_n(m\alpha)\xi_n'(\alpha)}, \tag{5.3}$$

$$b_n = \frac{m\psi'(m\alpha)\psi_n(\alpha) - \psi_n(m\alpha)\psi_n'(\alpha)}{m\psi_n'(m\alpha)\xi_n(\alpha) - \psi_n(m\alpha)\xi_n'(\alpha)}, \tag{5.4}$$

$$m = \frac{n_p}{n_0}; \qquad \alpha = \frac{2\pi r n_0}{\lambda_0}, \tag{5.5}$$

r = radius of spherical scattering particles (cm),
λ_0 = light wavelength in vacuum (cm),
ψ_n, ξ_n, ψ'_n, ξ'_n = Ricatti-Bessel functions of the first or second kind,
n_0 = refractive index of ground (host) material,
n_p = refractive index of scattering particle material, and
* denotes the complex conjugate.

When all scattering events can be treated independently (the volume fraction of scatterers should not exceed 1%–10%), the scattering coefficient (cm^{-1}) can be expressed with σ_s as $\mu_s = \sigma_s \cdot \rho$, where ρ is the scattering particle density (cm^{-3}). For absorbing particles, a complex refractive index must be considered, and the relevant Mie theory expressions yield σ_s, σ_a (absorption cross section, which gives $\mu_a = \rho \cdot \sigma_a$), and g.

From the preceding expressions, it can be easily seen that the particle characteristics governing their scattering properties are radius, r, and refractive index, n_p. The controlling parameters are the ratios r/λ_0 and n_p/n_0, which implies that μ_s and g for given scatterers depend on the radiation wavelength and host material used.

Calculations based on Mie theory[10,11] predict that the transport scattering coefficient rises steeply when n_p/n_0 departs from unity [see Fig. 5.1(a)]. The symmetry about $n_p/n_0 = 1$ suggests that dispersions with $n_p/n_0 < 1$ may be used as scattering media in phantoms. For example, such media may be foams, provided that the size of bubbles can be controlled during preparation. If n_p is close to n_0, sufficiently large particles ($r > 5\lambda$) scatter light mostly in the forward direction and the anisotropy factor g approaches 1 [see Fig. 5.1(b)]. Figure 5.1(c) shows that the transport scattering coefficient of suspensions with the constant volume fraction of particles maximizes when the particle dimension is comparable (or of the same order) with the light wavelength.

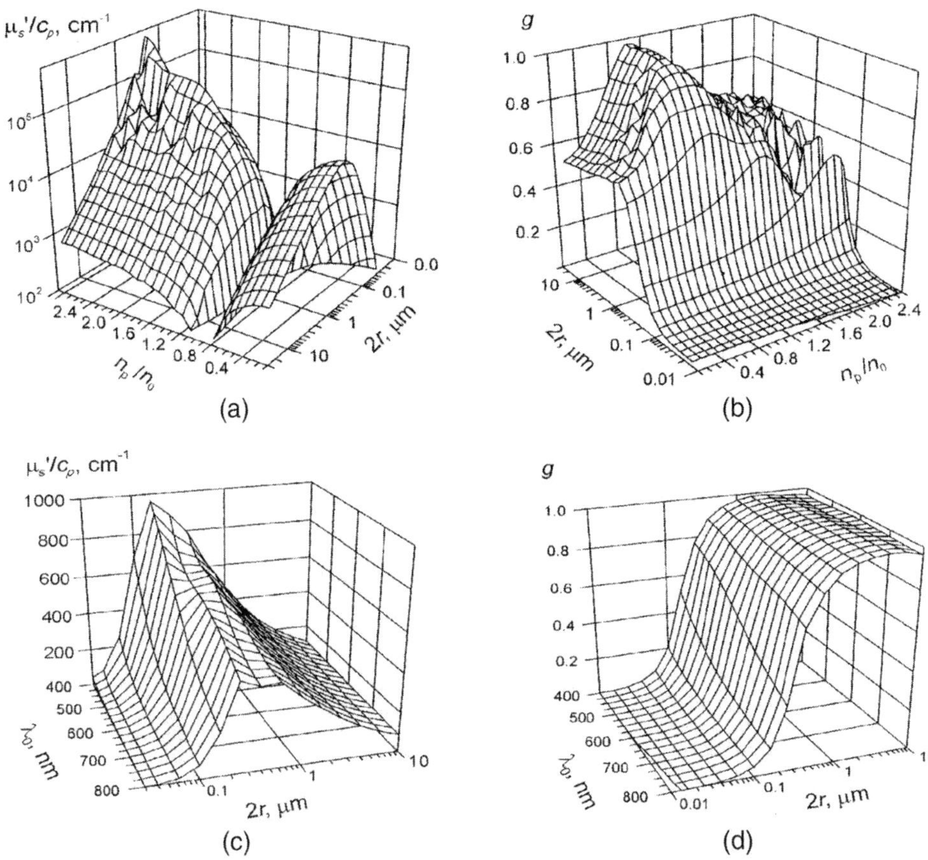

Figure 5.1 Scattering properties calculated using Mie theory for nonabsorbing spherical particles: (a),(b) at $\lambda_0 = 633$ nm; (c),(d) at $n_p/n_0 = 1.07$. c_p is the dimensionless volume fraction of scattering particles in suspension. (Reprinted from Ref. 11 with minor changes.)

Sections of the characteristic surfaces by $r = const$ planes in Figs. 5.1(c–d) reveal the wavelength dependence of scattering parameters. In calculations, n_o and n_p were assumed to be independent of λ_0; thus, spectral variations of μ_s' and g are connected only with the influence of parameter α in the Mie formulas.

Scattering by particles that are smaller or comparable in size with the wavelength is spectrally dependent; the transport scattering coefficient decreases with λ_0. In the case of large-diameter scatterers, μ_s' is nearly constant. The spectral dependence of the scattering anisotropy parameter for different sized particles is similar. The scattering characteristic of large particles show almost no dependence on wavelength, whereas the scattering characteristics of particles with diameters comparable to the wavelength or smaller show a strong dependence on wavelength.

Actual biological tissues usually show high values of g (typical for large particles) as well as a steep increase of the transport scattering coefficient toward a shorter wavelength (typical for small-size scatterers). This observation can not be modeled using a monodisperse particle suspension. Therefore, several authors have suggested that a mixture of large and small particles may be better suited to describe real tissue.[10,11,99]

When a mixture of particles or a suspension with broadband size distribution is employed, effective transport parameters of the scattering medium may be expressed[10,11] in terms of average values weighted by volume fractions of particles of different size:

$$\mu_s' = \frac{3c_p}{4\pi}\sum_i \frac{\sigma_{si}\upsilon_i(1-g_i)}{r_i^3}, \tag{5.6}$$

$$g = \frac{\sum_i \frac{\sigma_{si}\upsilon_i g_i}{r_i^3}}{\sum_i \frac{\sigma_{si}\upsilon_i}{r_i^3}}, \tag{5.7}$$

where c_p is the fraction of suspension volume occupied by particles;

υ_i is the volume fraction of particles of radius r_i, calculated as $\frac{r_i^3\nu_i}{\sum_i r_i^3\nu_i}$;

ν_i is the relative number of particles of radius r_i in suspension;

σ_{si} is the scattering cross section for particles of radius r_i;

g_i is the scattering anisotropy factor for particles of radius r_i.

The relative number of particles of a given radius in suspension may be determined from images made with an electron microscope.[85]

The optical properties of the suspension of absorbing particles also may be predicted using calculations based on Mie theory. The scattering cross section, σ_s (cm^2), and scattering anisotropy, g, of the particles are given by Eqs. (5.1) and (5.2), respectively, taking into account that in the case of absorbing particles, the refractive index n_p is a complex quantity. The absorption cross section of the particles, σ_a (cm^2), is given by[98]

$$\sigma_a = (\lambda_0^2/2\pi n_0^2)\sum_{n=1}^{\infty}(2n+1)[\mathrm{Re}(a_n+b_n) - (|a_n|^2+|b_n|^2)]. \tag{5.8}$$

The definitions of the incorporated quantities are given after Eqs. (5.1) and (5.2); n_p has a complex value, the imaginary part of which can be expressed from the absorption coefficient of particle material, μ:

$$\mathrm{Im}[n_p(\lambda_0)] = \frac{\mu(\lambda_0)\lambda_0}{4\pi\mathrm{Re}[n_p(\lambda_0)]}. \tag{5.9}$$

Following Beck et al.,[10,11] the absorption coefficient of a suspension of particles with broadband size distribution can be expressed in terms of average values weighted by volume fractions of particles of different size:

$$\mu_a = \frac{3C_p}{4\pi}\sum_i \frac{\sigma_{a_i}\vartheta_i}{r_i^3}, \tag{5.10}$$

where σ_{a_i} is the absorption cross section for the particles of radius r_i. The definitions of other quantities are given after Eqs. (5.6) and (5.7).

Mie theory predicts that the scattering by absorbing particles dispersed in a nonabsorbing medium with a refractive index n_0 will become a strongly forward-directed resonance scattering when the real part of the refractive index of the particle material is equal to n_0. For such particles, the apparent absorption, μ_a, is smaller than expected from the absorption coefficient of bulk material, μ. For example, 1-μm-diameter particles with $\mathrm{Re}(n_p) = n_0 = 1.6$ and $\mu = 10^4$ cm^{-1} yield only $\mu_a = 4.12 \cdot 10^3$ cm$^{-1} \times C_p$.[10,11]

The Mie theory formulas cited are applicable strictly to spherical particles; nevertheless, the results are often considered acceptable for the systems of particles of irregular shape.

5.3 Scattering Media for Phantom Preparation

5.3.1 Fat emulsions as scattering media in tissue phantoms

The most common scattering media used in tissue phantom manufacturing, especially in the preparation of liquid systems, are fat emulsions (Intralipid, Nutralipid, and Liposyn) for feeding patients intravenously.[6,7,10,11,15,16,19–21,41,47,55,63–65,67,71,75,100–103] These products contain soybean oil, egg phospholipids, and glycerol and are the suspensions of roughly spherical fat droplets dispersed in water. The formulation of 10% Intralipid (Kafi Pharmacia) is reported in Ref. 100 as 10 g of purified soybean oil, 1.2 g of purified egg phospholipids, and 2.2 g of anhydrous glycerol per 100 ml of water solution (dispersion).

Because these emulsions are produced for medical application and not as optical standards, their properties may vary with different batches, and the measured optical properties may vary from one research group to another. For example, the mean size of scattering oil droplets in Intralipid 10% has been estimated by Choukeife and L'Huillier[104] as 402–426 nm, but Driver et al.[105] found that more than 75% of droplets in Intralipid had diameters equal to or less than 125 nm.

In Ref. 85 a typical electron-microscopy image of an Intralipid sample along with the size distribution of 2901 particles in the size range from 50 nm to 400 nm (the mean particle diameter being 225.7 nm) is presented. In experiments, the authors used 20% Intralipid supplied by Sichuan Kelun Pharmaceutical Co., Ltd, China. It contains 3 g lecithin and 50 g soybean oil in 250 ml of solution (dispersion).

As a systematical construction of a phantom requires that the individual optical properties of its components are known beforehand, a number of measurements using different methods have been performed to determine the parameters of scattering media. Okada et al.[63] listed for 1% v/v concentration of Intralipid (Liposyn 20%) at 633 nm the following values: $\mu_s' = 14.0 \pm 0.5\ \text{cm}^{-1}$ (measured by frequency-domain diffuse reflectance technique); $\mu_s = 71\ \text{cm}^{-1}$ (determined with collimated transmission measurements), $g = 0.8$ thus being adopted; and $\mu_a = 0.005\ \text{cm}^{-1}$ (determined using a video reflectometry apparatus).

Based on a comparison between experimental wavelength dependencies of absorption length, transport length, and anisotropy factor for 2% v/v solution of Intralipid 10% and Mie theory, the use of the following approximation[106] was suggested:

$$\mu_s'(l)(\text{cm}^{-1}) \approx 1.6 \cdot 10^3\lambda^{-1};$$

$$g(\lambda) \approx 1.1 - 0.58 \cdot 10^{-3}\lambda$$

for $400\ \text{nm} < \lambda < 1000\ \text{nm}$.

Optical properties for different concentrations of Intralipid 10% at 1064 nm were obtained by integrating sphere and collimated transmission measurements followed by inverse adding-doubling calculations:[58]

$$\mu_s/\%(\text{cm}^{-1}/\%) = 1.30 \pm 0.047$$

$$\mu_a/\%(\text{cm}^{-1}/\%) = 0.054 \pm 0.02$$

$$g = 0.5 \pm 0.02,$$

where the absorption coefficient is corrected for water absorption (see Fig. 5.2).

Wolf et al.[41] accepted the approximate value of 5–8 cm^{-1} for μ_s' of 0.5% Intralipid in the NIR range and supposed that μ_s' increased proportionately with Intralipid concentration (the concentration range of 0.5%–2.0% was used).

Optical properties of pure Intralipid (10% solids; Kibavitrum, Inc.) were determined by Jacques and Flock[102] using steady-state optical measurements of collimated transmittance and total diffuse reflectance (for a semi-infinite layer) of both pure Intralipid and Intralipid with a known amount of India ink added. The results are depicted in Table 5.1.

Figure 5.3 shows the spectral dependencies of absorption (μ_a) and scattering (μ_s') coefficients for 1% Lyposyn.[71] The spectra were obtained using a frequency-domain *in vivo* spectrometer. The increase of absorption at the wavelengths exceeding 700 nm most likely results from water absorption. Wilson et al.[71] note that the reduced scattering coefficient is in 10%–15%

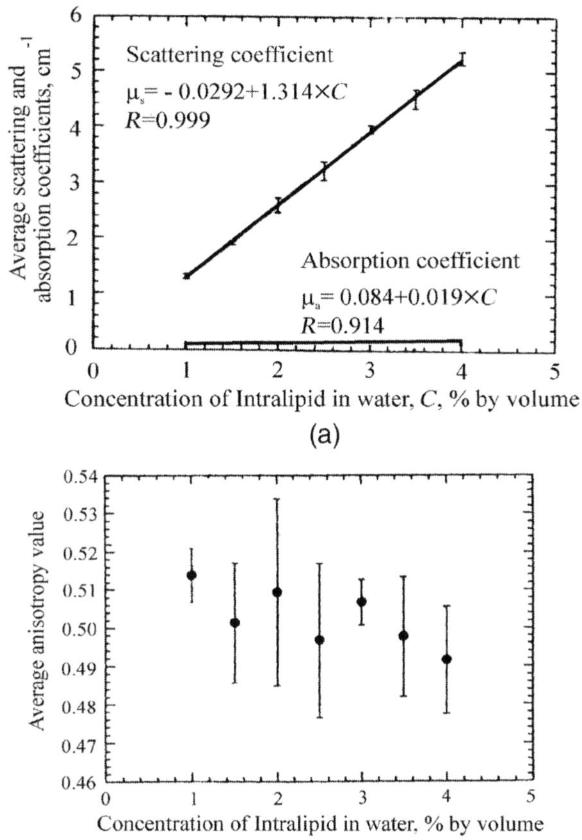

Figure 5.2 (a) The absorption and scattering coefficients and (b) scattering anisotropy factor at 1064 nm for varying concentrations of Intralipid 10% suspended in water. (From Ref. 7 with minor changes.)

Table 5.1 Optical properties of Intralipid 10%.[102]

Wavelength, λ (nm)	Absorption coefficient, μ_a (cm^{-1})	Scattering coefficient, μ_s (cm^{-1})	Anisotropy factor, g
488	0.07	617	0.80
633	0.02	313	0.71
1064	0.10	78	0.68

agreement with the published results obtained using CW spectroscopy. These figures are within the batch-to-batch variability of Lyposyn.

In constructing a liquid tissue phantom for fluorescence measurements,[15] the scattering coefficients of 79 and 47 cm^{-1} were adopted for 1% v/v Intralipid at 514 and 635 nm, respectively. For the analogous Intralipid solution containing one percent solids (this term is often used in relation to Intralipid, despite the disperse phase in this case being liquid oil), the optical

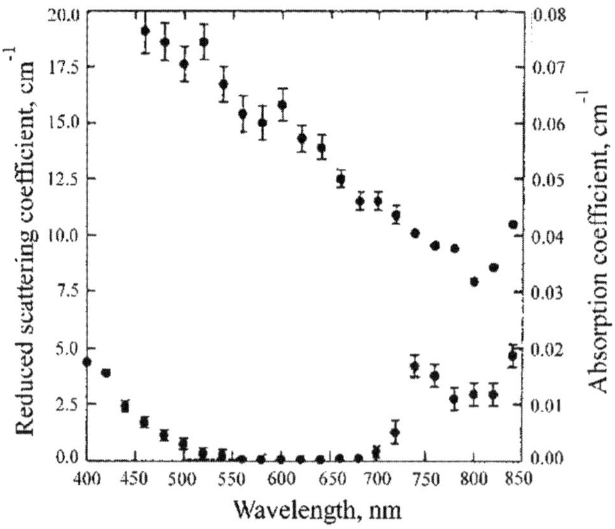

Figure 5.3 Wavelength dependencies of absorption and scattering coefficients for 1% Lyposyn, obtained using a frequency-domain spectrometer. (Adapted from Ref. 71.)

properties at 633 nm were determined.[6] The collimated transmission method determined the value of 34.4 ± 0.1 cm^{-1} for the total attenuation coefficient, μ_t. The added absorber technique gave $\mu_a = 0.011 \pm 0.001$ cm^{-1}. The scattering coefficient was assumed to be equal to μ_t because of negligibly small μ_a. The direct measurement of scattering phase function by goniometry gave $g = 0.68$.

This low value of g, as compared with most actual tissues, illustrates the general feature of Intralipid: while it reproduces the scattering coefficients of tissue (μ_s) well, it underestimates at the same time the anisotropy factor. Thus, using Intralipid, one must select the μ_s value that gives a transport scattering coefficient, $\mu_s' = \mu_s(1 - g)$, equivalent to that of the tissue; thus, we obtain the phantom that models the fluence field within the tissue only on the level of diffusion approximation or the principle of similarity.[107] In other words, we consider the fluence distributions in the tissue and phantom to be similar when these two scattering and absorbing media have the same μ_a and μ_s' but not the complete set of optical parameters (μ_a, μ_s, and g). To make the value of the scattering anisotropy parameter of Intralipid-based phantoms closer to the values for actual tissue, a fraction of large particles (e.g., 1–10-μm-diameter SiO_2 particles[12]) may be added. This scarcely affects the wavelength dependence of anisotropy factor, but enhances the value of g.

Although liquid Intralipid phantoms are generally stable during a short-term measurement, Intralipids are colloidal systems, and considerable changes in the pH or ion content of dispersive medium (diluent) may provoke an aggregation of scatterers over time.

In the preparation of solid Intralipid–based phantoms,[10,11] it was observed that transport scattering coefficients of Intralipid in gel depended strongly on the amount of gelling agent. For example, 2% w/w of agarose caused μ_s' to reduce by 30%, as compared to a water suspension. Beck et al.[10,11] suggested that the effect may be caused by the increased temperature during gel preparation or by the structural changes of the scatterers induced by interactions with agarose.

5.3.2 Milk in phantoms

Cow milk has also been applied as a light-scattering medium in a number of phantom measurements.[53,73,108–110] The system of scattering milk particles comprises two subsystems: (1) emulsion of fat spheres and (2) colloidal protein particles. The size of the spheres falls into the range of 0.5–20 μm, most of the particles having a diameter of 2–4 μm. The dimensions of the protein fraction are distributed within the range of 0.02–1 μm as follows: 0.02–0.04 μm = 30%; 0.04–0.08 μm = 30%; 0.08–0.15 μm = 30%; 0.15–1 μm = 10%.[53]

Both whole and skim milk are used in phantom preparations. Their scattering properties should differ because in the latter case, the fraction of large particles, which impart a higher g value and a weaker wavelength dependence, is removed. Optical properties of homogenized whole milk (3.5% fat) and milk diluted with water were estimated in visible and NIR regions using a time-resolved transmittance method by Wabnitz et al.[109] Here the refractive index of the suspensions was taken to be $n = 1.33$, and the sample thickness was 2 cm (see Table 5.2).

Based on the measurements of total transmittance, diffuse reflectance, and collimated transmittance, the optical properties of undiluted milk (3.5% fat) were obtained over the wavelength range of 700–1100 nm.[110] The refractive index of milk at 650 nm was found to be $n = 1.36$. Experimental data were processed using both the inverse δ-Eddington method (IDE) and the inverse Monte Carlo method combined with the small-angle approximation of radiative transfer theory (IMC + SAA). For 880 nm, the averaging of the results obtained in Ref. 73 for the milk samples of three different thicknesses gives μ_a (IDE) = 4.83 cm^{-1}, μ_a (IMC + SAA) = 0.86 cm^{-1}; μ_s(IDE) = 131.9 cm^{-1}, μ_s(IMC + SAA) = 133.0 cm^{-1}; g(IDE) = 0.741, g(IMC + SAA) = 0.799.

At a 700 nm, the scattering properties of milk obtained by Yaroslavsky et al.[110] are close to the estimations given by Wabnitz et al.[109]

Table 5.2 Estimations of optical properties of whole milk dilutions. (Data from Ref. 109.)

Volume fraction of milk, %	Estimation for μ_s' (cm^{-1})	Estimation for μ_a (cm^{-1})
6	5	<0.015
12	10	<0.015
25	18	<0.015
50	30	<0.015
100	46	<0.015

5.3.3 Polymer latex spheres in construction of tissue-like phantoms

Other scatterers frequently used in construction of tissue-like phantoms are micron-sized polystyrene latex spheres.[17,18,27,28,43–45,48,51,52,54,68,90,93,111,112] Compared with Intralipid, these products provide a much better optical standard. In general, commercially available latexes have a narrow size distribution, and the manufacturer can provide the needed distribution information. Mie theory formulas can be used to calculate the scattering properties of sphere suspensions, provided that the condition of independent scattering is observed.

Kaplan et al.[112] obtained from Mie calculations the scattering properties, at 514.5 nm, of 0.205 μm-diameter and 0.460 μm-diameter polystyrene spheres suspended in water. For the smaller particles (with the volume fraction in suspension $C_p = 0.01798 \pm 0.00013$), the calculated value of the reduced scattering coefficient, μ_s', was 83 cm^{-1} and the anisotropy factor $g =$ 0.875. The figures for 0.460 μm particles ($C_p = 0.01665 \pm 0.00013$) were 73 cm^{-1} and 0.962, respectively. In calculations, the refractive index value $n_p = 1.583$ was used for the particles and $n_0 = 1.336$ was used for water.

Table 5.3 shows the results of the calculations for particles embedded in gelatin gel (the assumed value of host material refractive index was $n_0 = 1.35$) for the wavelength of 632.8 nm performed by Steenbergen et al.[52]

Scattering parameters in the ultraviolet region ($\lambda = 335$ nm) for 0.9-μm spheres, taken as 10% suspension, are reported by Oraevsky et al.[44,45] as $\mu_s =$ 6090 cm^{-1}, and g is 0.918, which corresponds to the reduced scattering coefficient, μ_s', of 500 cm^{-1}.

When a phantom for use with different wavelengths is developed, the spectral dependencies of refractive index should be known. For polystyrene and water, data for the refractive indices in the visible were extrapolated into the NIR region using an approximation by the Cauchy fit of the form[27,28]

$$n_{\text{polystyrene}}(\lambda) = 1.5626 + 1.169 \cdot 10^3/\lambda^2 – 1.125 \times 10^9/\lambda^4 + 1.72 \cdot 10^{14}/\lambda^6$$

$$n_{\text{water}}(\lambda) = 1.3199 + 6.876 \cdot 10^3/\lambda^2 – 1.132 \cdot 10^9/\lambda^4 + 1.11 \cdot 10^{14}/\lambda^6,$$

where λ should be substituted in nanometers.

Being colloidal systems, polystyrene sphere suspensions may undergo aggregation, which will affect their scattering properties. Dense suspensions

Table 5.3 Scattering properties (at 632.8 nm) of polystyrene particles embedded in gelatin gel.[52]

Particle diameter (μm)	Scattering cross section (cm^2)	Anisotropy factor
0.21	$3.3 \cdot 10^{-11}$	0.36
0.48	$1.35 \cdot 10^{-9}$	0.81
1.05	$2.412 \cdot 10^{-5}$	0.92

require sonication before phantom preparation, and liquid phantoms should be stirred before each measurement. When solid phantoms are prepared, sonicating warm gels with scatterers before casting are also advisable. The addition of a surfactant (0.1% by weight of sodium dodecyl sulfate) to ensure long-term stability of the suspension was reported in the literature.[112] Polystyrene particles proved to be stable in a hydrophilic enviroment.[10,11] Therefore, these particles are suitable for use as constituents of solid phantoms based on hydrophilic gels (gelatin, agarose, agar-agar), whereas organic solvents and monomers that are employed to elaborate polymer-based solid phantoms may dissolve the spheres.

5.3.4 Mineral particles as scatterers in solid phantoms

Mineral particles are expected to be stable in both hydrophilic and hydrophobic environments, and they are used as a scattering medium in various solid phantoms (TiO_2;[13,14,41,63,69,70,72,87,89,93] SiO_2;[21,113] talc[88]). The drawback that constrains the employment of these scatterers in liquid phantoms is their high specific weight, which causes rapid sedimentation of particles in a nonviscous host medium.

Table 5.4 presents the relative refractive indices for nonabsorbing mineral particles embedded in solid hosts.

Hielscher et al.[69] used layered tissue phantoms to experimentally test simulation results obtained with time-resolved Monte Carlo code. Tissue layers with defined scattering properties were generated by adding TiO_2 powder to gelatin gel. The particles exhibited Gaussian distributed sizes with the mean value of 280 μm and full width at half maximum of 80 μm. In an epoxy resin host, Simpson et al.[93] suspended TiO_2 particles that were three orders of magnitude smaller in diameter ($d = 0.3$ μm). To model the optical properties of atheromatous plaque at 1064 nm and 1300 nm, ground talc, with average particle size of 30–40 μm, dispersed in 2% agar-agar gel was used.[88]

Table 5.4 Relative refractive indices of massive mineral particles in an aqueous gel ($n = 1.33$) and polyorganosiloxane rubber ($n = 1.40$).[11]

Substance	In aqueous gel	In rubber
SiO_2	1.10	1.04
γ-Al_2O_3	1.20	1.14
$BaSO_4$	1.23	1.17
MgO*	1.31	1.24
α-Al_2O_3	1.33	1.26
TiO_2**	1.95	1.86

*MgO may be unstable in hydrophilic media as it reacts with water, yielding magnesium hydroxide.

**Nanoparticles of TiO_2 are known to catalyze photodestruction of some dyes and organic molecules.

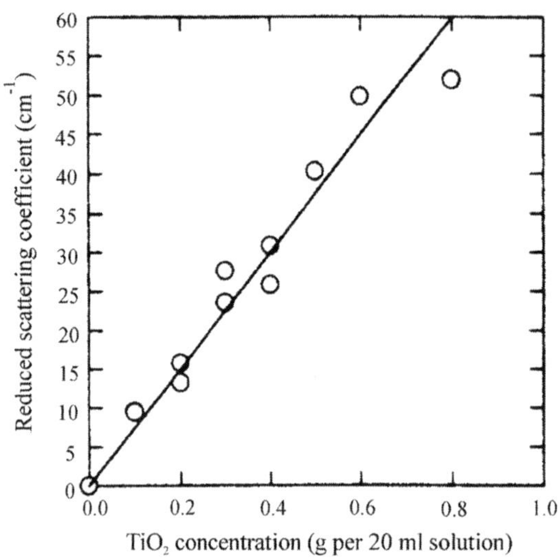

Figure 5.4 Reduced scattering coefficient at 690 nm vs. TiO_2 content. TiO_2 powder was added to the mixture of 7% gelatin, and 0.0002% India ink solution was added to 20 ml of phosphate buffered saline. (From (with minor changes) B. Pogue, L. Lilge, M. Patterson, B. Wilson, and T. Hasan, "Absorbed photodynamic dose from pulsed versus continuous wave light examined with tissue-simulating dosimeters," *Appl. Opt.*, **36**(28), 7257–7269 (1997). ©1997 OSA. Reprinted with permission from The Optical Society.)

The reduced scattering coefficient of TiO_2 powder (Fisher Chemical, Boston, Massachusetts) suspended in 7% gelatin gel was measured at 690 nm using a diffuse reflectance instrument.[87] For the sample that contained 0.2 g of TiO_2 per 20 ml of gel, the μ_s' value of 14 ± 2 cm^{-1} was obtained. A linear increase of the reduced scattering coefficient coinciding with the rise of TiO_2 concentration was observed in the series of samples (see Fig. 5.4).

The samples were prepared by mixing the components in boiling phosphate buffer, then casting after stirring for 2 minutes. In separate measurements, the samples were checked for spatial homogeneity of the reduced scattering coefficient. The variation of μ_s' appeared to be less than $\pm 10\%$ within one sample; the intersample variation was also $\pm 10\%$.[87]

For a suspension of nonabsorbing spherical particles, its scattering properties, μ_s' and g, are uniquely determined, according to Mie theory, by n_p/n_0, particle size and concentration. But the effective values of μ_s' and g and their spectral dependencies for mixtures of particles, when scatterers are made out of different substances and have different dimensions, may be adjusted to the desired figures by variation of mixture composition. It has been demonstrated[11] that even a binary mixture allows construction of scattering media with distinct μ_s' wavelength dependencies and nearly coincident scattering anisotropy factors. Figure 5.5 demonstrates the results of Mie theory calculations for two binary mixtures of monodisperse mineral particles.

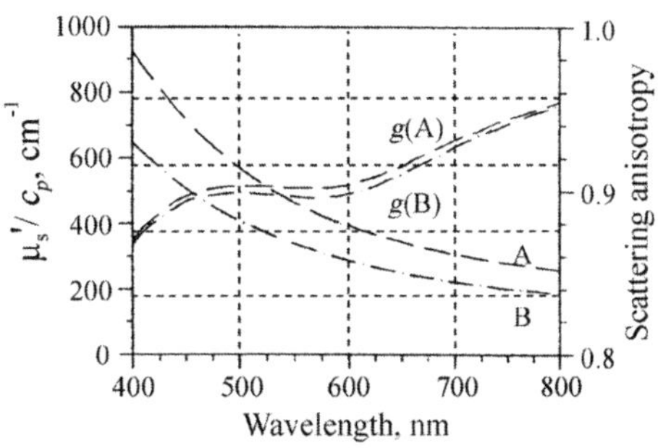

Figure 5.5 Spectral dependencies of scattering parameters of binary mixtures of nonabsorbing particles: (A) 1:19 mixture of 0.1-μm particles, $n_p/n_0 = 1.85$ (corresponds to TiO_2 in polyorganosiboxane), and 5-μm particles, $n_p/n_0 = 1.25$ (corresponds to Al_2O_3 in polyorganosiboxane); (B) 3:7 mixture of 0.1-μm particles, $n_p/n_0 = 1.25$, and 5-μm particles, $n_p/n_0 = 1.25$. (Reprinted from Ref. 11 with minor changes.)

Along with massive mineral scatterers, porous spherical γ-Al_2O_3 particles were applied in the preparation of a tissue-like phantom.[10,11] The particles had a diameter of 5.3 μm ± 1.0 μm, with nearly symmetrical size distribution. Pores of 10-nm diameter occupied 65% of the particle volume and were assumed to be filled with polyorganosiloxane. The effective refractive index, n_{eff}, of such "impregnated" particles was calculated as follows using Bruggeman theory:

$$c_{\text{por}}\frac{n_p^2 - n_{\text{eff}}^2}{n_p^2 + 2n_{\text{eff}}^2} + (1 - c_{\text{por}})\frac{n_0^2 - n_{\text{eff}}}{n_0^2 + 2n_{\text{eff}}^2} = 0, \tag{5.11}$$

where

n_p is the refractive index of bulk γ-Al_2O_3;
n_0 is the refractive index of the host rubber;
$c_{\text{por}} = 0.65$ is the volume fraction of pores in the particles.

The value of $n_{\text{eff}} = 1.52$ at 546 nm was obtained and used for phantom characterization.

W. Steenbergen and F. de Mul[50] provided an example of the scattering system with $n_p/n_0 < 1$. Hollow microspheres with an external diameter of 1 μm were embedded in polyvinylalcohol films. The cavities of the spheres in dehydrated state were filled with gas ($n_{\text{gas}} \approx 1$). Calculations based on Mie

Table 5.5 Scattering properties of gas-filled microspheres embedded in a solid host matrix calculated using Mie theory. (Data from Ref. 50.)

	λ = 633 nm	λ = 780 nm
σ_s (cm^2)	1.116×10^{-8}	0.997×10^{-8}
g	0.8447	0.8359

theory were performed for 0.8-μm-diameter spheres, with $n_p = 1$, in the host material with $n_0 = 1.53$ (see Table 5.5).

5.4 Light-Absorbing Media for Phantom Preparation

5.4.1 Common microscopy stains in liquid and solid phantoms

Light absorption by tissue is modeled in phantoms by reproducing the value of the absorption coefficient, μ_a. In most cases when the phantoms are employed, it is sufficient to give an account of light propagation at one wavelength. This does increase the variety of dyes that may be considered as candidates for use in phantoms. Such dyes should have conspicuous absorbance at the wavelength of interest and be soluble in the host material. In addition, the dyes should meet the general requirements of stability, safety, and phantom component compatibility.

Common microscopy stains, which are easily available and widely used in laboratories, absorb in the visible and NIR spectral regions, and they are water-soluble. Because of this, these dyes are often used in the preparation of water-based liquid phantoms and solid phantoms based on aqueous gels.

Semi-infinite liquid phantoms with methylene blue as an absorber were used in measurements of angular radiance in turbid absorbing media at 630 nm.[99] To vary the absorption coefficient of the medium, the stock solution of dye in distilled water at the concentration of 1 mg · ml^{-1} was added to water + 10% Intralipid mixtures at different proportions.

The radiance dependency on viewing angle of detector measured within the phantom that contained 3.2% of 10% Intralipid, 0.5% of methylene blue stock solution, and 96.3% of distilled water overlapped the radiance dependency obtained by Monte Carlo simulations using the following values of optical parameters of the medium: $\mu_a = 0.95$ cm^{-1}, $\mu_s = 150$ cm^{-1}, g = 0.975.[100]

In Ref. 64, to check the validity of spatially resolved steady-state diffuse reflectance spectroscopy, a series of measurements was performed on liquid phantoms (Intralipid in phosphate-buffered saline) with an increasing amount of Evans blue as an absorber. The absorption properties of a pure solution of Evans blue were measured using a spectrophotometer.

Spectrophotometry is a conventional technique for the isolated characterization of dyes that are used in phantoms. The specific absorption of the dyes dissolved in a host medium is measured at chosen wavelengths. These data are then used for the prediction of light absorption properties of turbid phantoms loaded with the dyes.

Indocyanine green has been employed as an NIR absorber in the preparation of gelatin gel-based layered solid phantoms. These layered phantoms were used to test reconstruction techniques for time-resolved measurements. The phantom incorporated two layers of gelatin gel. Scattering and absorption properties of the layers were controlled by the amounts of

Intralipid and indocyanine green introduced into the gel. The reduced scattering coefficient, μ_s', was 10 cm^{-1}, and the values of the absorption coefficient at 752 nm were varied from $\mu_a = 0.12$ cm^{-1} to 0.36 cm^{-1}.[67]

In the case of aqueous gel-based phantoms, the measurements of optical properties of absorbing dyes should not be conducted on their stock solutions, but on scatterer-free gel+absorber samples because the binding of the absorber to the gelling agent (usually polypeptide or polysaccharide) macromolecules may result in a pronounced spectral shift of absorption.

A liquid semi-infinite phantom with a dye as absorber was used in experimental validation of the forward-adjoint fluorescence model.[18] The optical properties of the aqueous phantom medium were controlled by the contents of a fluorophore (endogenous fluorophore NADH), a scatterer (suspended polystyrene microspheres), and an absorber (ethyl orange). The principle of fluorescence measurements places additional constraints on the absorber. The absorber should not fluoresce under the excitation wavelength used in the experiment. Ethyl orange was tested with 351-nm argon-ion laser radiation used for NADH fluorescence excitation and was found to have negligible fluorescence. The optical mean-free path was measured on a stock solution of each component with a spectrometer modified to a narrow-beam geometry. Fluorescent emission of NADH at 470 ± 5 nm (but not at the 540-nm emission spectrum maximum) was recorded because the absorption of both NADH and ethyl orange does not depend on the wavelength over this spectral interval.[18]

Several groups used the microscopy stain trypan blue in liquid and solid tissue phantoms. In a series of papers,[19,111] polarization properties of diffuse reflectance were studied on semi-infinite liquid turbid samples. The feasibility of video imaging technique using polarized light reflectometry has been demonstrated by Jacques et al.[111] on human skin scaled phantom (3.5 mm of phantom corresponded to $\approx$1 mm of skin) prepared on a water base, polystyrene spheres (0.9-μm diameter), and trypan blue. The optical properties of the phantom were determined at 633 nm and 792 nm using video reflectometry measurements and diffusion theory for analysis. (see Table 5.6)

Doornbos et al.[64] employed 13 liquid tissue-simulating phantoms to evaluate the accuracy of spatially resolved absolute diffuse reflectometry. To prepare the phantoms, different amounts of trypan blue were successively added to Intralipid solutions of different concentrations. The "true" optical

Table 5.6 Optical properties of human skin scaled phantom.[111]

	633 nm	792 nm
μ_a	0.20 cm^{-1}	0.02 cm^{-1}
μ_s	69.8 cm^{-1}	2518 cm^{-1}
g	0.913	0.939
μ_s'	6.073 cm^{-1}	6.86 cm^{-1}

properties of the samples were measured independently by the frequency-domain technique. The transport scattering coefficients were between 5 and 10 cm^{-1}, and the absorption coefficients were between 0.02 and 1 cm^{-1} at 633 nm. For each solution, the absorption coefficient of trypan blue was determined by means of conventional spectrophotometry.

Solid semi-infinite scattering and absorbing phantoms with known optical properties were used to test the algorithms for obtaining optical properties of a turbid material from pulsed photothermal radiometry data.[48] In those studies, aqueous gel was chosen as the host material because of its thermal diffusivity being practically equal to the thermal diffusivity of most soft biological tissues. Varying amounts of polystyrene microspheric colloid and trypan blue dye were added to the melted collagen gel, yielding the samples with absorption coefficients ranging from 20 to 144 cm^{-1} and a reduced scattering coefficient from 0 to 150 cm^{-1} at 627 nm. The absorption of the dye stock solution was measured with a conventional spectrophotometer. The scattering properties of the polystyrene microspheres were calculated using Mie theory.

5.4.2 Dyes as light-absorbing components of tissue-simulating phantoms

There are particular problems (e.g., measurements in the infrared spectral region, studies of fluorescence excitation and emission spectra, and manufacturing of the phantoms based on a hydrophobic host) that impose additional requirements for the properties of the absorber used in the phantom. This leads to the necessity of an extended list of dyes beyond the common microscopy stains.

The validity of combining integrating sphere measurements and the Monte Carlo inverse technique to determine near-infrared optical properties of *ex vivo* human skin and subcutaneous tissues was estimated using liquid and solid phantoms containing infrared dyes as absorbers.[93] One series of experiments in the spectral interval of 650–1000 nm was performed on the liquid samples made of 1.27-μm polystyrene spheres with an infrared dye (S109564, Zeneca) in aqueous suspension. The absorption coefficient of the solution was estimated from the known absorption spectra of the dye, and the reduced scattering coefficient was calculated from Mie theory.

In the liquid samples placed between glass coverings, the edge effects may distort the homogeneity of the optical coefficients. To eliminate this hindrance, the solid phantoms [on "Araldite" epoxy resin base with 0.3-μm TiO_2 particles as scatterers and Project 900 NP dye (Zeneca specialties, Manchester, U.K.) as an NIR absorber] were substituted for the liquid phantoms.[93] Steenbergen and de Mul[50] used the water-soluble infrared absorber IRAWS1 (provided by Zeneca specialties) to provide absorption from $\mu_a = 0.5$ cm^{-1} to 1 cm^{-1} in the solid polyvinyl alcohol-based phantom at the frequently used wavelength of 780 nm.

To demonstrate the validity of the method using dual-wavelength time-gated spectroscopy for determination of the concentration of absorber in turbid media, liquid phantoms for transmission-mode measurements at 782 nm and 831 nm were prepared.[26] Greenish-brown ink (Chugai Kasei) used as a pure absorber was added to 420 ml of 1% aqueous Intralipid 20% suspension in 0.07-ml steps up to 0.56 ml total volume of ink. The absorption coefficients of ink solutions of various concentrations were premeasured using a spectrophotometer. The measurements gave absorption coefficients for a series of phantoms in the ranges from $2.45 \cdot 10^{-2}$ to $2.94 \cdot 10^{-1}$ cm^{-1} at 782 nm and from $3.10 \cdot 10^{-2}$ to $1.85 \cdot 10^{-1}$ cm^{-1} at 831 nm. The determined values for water absorption, $2.45 \cdot 10^{-2}$ cm^{-1} at 782 nm and $3.10 \cdot 10^{-2}$ cm^{-1} at 831 nm, were accounted for.[48]

The multilayered 2% agar gel phantoms with Intralipid 10% as a scattering component and a soluble dye (safranin) as an absorber were used to study the total reflection of light from neonatal skin and the pattern of light distribution within layered tissue.[65] The concentration of the absorber and scatterer was scaled down, and this allowed modeling of light distribution in thin layers of actual skin on the large-scale gel phantoms.

As the optical properties of a layer can be expressed *via* the dimensionless quantities of albedo, $\mu_s/(\mu_a + \mu_a)$, optical thickness, $(\mu_a + \mu_s)d$, and anisotropy, g, the light distribution pattern at various wavelengths of interest can be reproduced using the same experimental wavelength. Figure 5.6 depicts the observed distribution of 633-nm radiation within the layered gel structure. To the right is the scale of corresponding dimensions in the skin for 460-nm illumination. Different layers of the phantom correspond to the actual skin layers, whose optical characteristics are summarized[65] in Table 5.7.

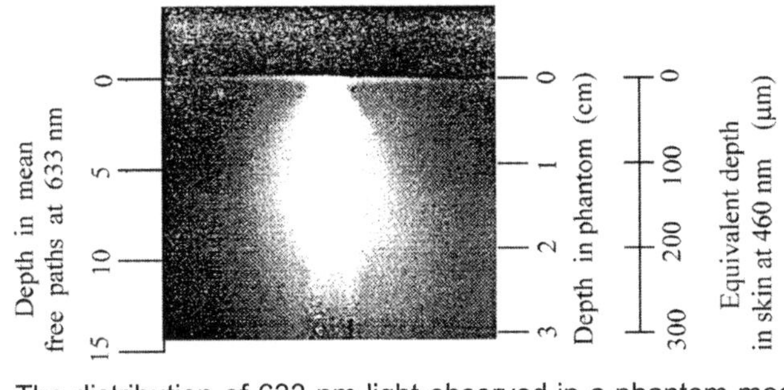

Figure 5.6 The distribution of 633-nm light observed in a phantom model of skin.[65]

Table 5.7 Optical properties of skin layers at 460 nm.[65]

Layer	μ_a (cm^{-1})	μ_s (cm^{-1})	g	thickness (μm)
Pigmented epidermis	14	370	0.9	~65
Papillary dermis	5	370	0.9	~55
Venous plexus	18	333	0.9	~50
Reticular dermis	5	440	0.9	~1000

Makropoulou et al.[103] performed a series of experiments to demonstrate the ability of laser-induced spectroscopic techniques for quantitative estimation of tissue chromophores concentration. The measurements were carried out on liquid and agar gel-based scattering and absorbing phantoms. Liquid phantoms were prepared by diluting the stock Intralipid solutions with distilled water to the concentrations of 1%–20% of Intralipid. Commercially available red dye Carmine–E120 and green dye Chlorophyll–E140 were used as light absorbers (i.e., chromophores). Optical parameters, μ_a, μ_s, g, and μ_s', of the phantoms at 543 and 633 nm were calculated, with Kubelka-Munk theory, from the integrating sphere reflectance/transmittance measurements.

To develop a phantom suitable for testing the method of hematoporphyrin derivative quantification in tissue by fluorescence measurement using dual-wavelength excitation and dual-wavelength detection, an absorber that absorbs excitation light much better than fluorescence was required.[20] As the 405- and 435-nm wavelengths were used for excitation and fluorescence was recorded over the region of 550–725 nm, the food colorant European Sunset (E110) was chosen as an absorbing component. The 1-mg/ml dye solution has absorption coefficients of 28 cm^{-1} and 34 cm^{-1} at 405 nm and 435 nm, respectively, and only ~0.1 cm^{-1} at 560–700 nm, as determined using a standard spectrophotometer. Because the laser dye DCM, which was introduced into the phantom to imitate tissue autofluorescence, also absorbs light, the concentration of European Sunset was made lower.

When manufacturing the tissue-simulating phantoms intended for the study of the effect of photosensitizer (Photofrin) photobleaching on surface fluorescence signal, Farrell et al.[89] used a dye (Lewiscraft, # 88204) as an absorbing component of the styrene resin phantom, which absorbs light in the spectral region of Photofrin's fluorescence emission. At 630 nm (one of the photosensitizer's fluorescence peaks), the absorption coefficient of the phantom, including absorption by both the dye and sensitizer, was measured as 0.19 ± 0.04 cm^{-1}. Contrary to this approach, in the development of standard samples for the noninvasive fluorimetric evaluation of sulphanated aluminum phthalocianine concentration in tissue, only the scattering coefficient of the samples was adjusted to the value pertained to the tissue, and tissue absorption over the 630–700-nm region was considered negligible.[19]

A method of correction of fluorescence emission and excitation spectra for the effect of internal absorption was tested in the measurements on liquid scattering phantoms containing the fluorophore and nonfluorescent dye, the latter imparting absorption to the phantoms both in fluorophore excitation and emission spectral regions.[16] In experiments connected with the correction of emission spectra, proflavine (3,6-diamino acridine) hemisulfate dihydrate was used as a fluorophore. A triphenylmethane dye, basic fuchsine (pararosaniline chloride), was used as a nonfluorescent absorber. The absorption spectrum of basic fuchsine (maximum at ≈540 nm) overlaps both absorption (maximum at ≈460 nm) and emission (maximum at 503 nm) spectra of proflavine+DNA complex. The complex of fluorescent dye (proflavine) with DNA was used to prevent fluorophore from interacting with the absorber and scatterer, which might lead to the resonant energy transfer and formation of nonfluorescing complexes. Positively charged planar molecules of 3,6-diaminoacridines are known to intercalate between the base pairs and bind strongly to DNA, while positively charged but nonplanar molecules of triphenylmethane dyes cannot totally intercalate into DNA macromolecules. This results in a space separation of two dyes. All solutions containing DNA (and corresponding control solutions without fluorophore) were prepared in the 0.15M phosphate buffered saline. In the experiments on excitation spectra correction, a solution of Kiton Red 620 (Sulforhodamine 620, laser-grade) in deionized water was used as a fluorophore. In those experiments, basic fuchsine was employed again because its absorption spectrum overlaps the absorption (excitation) spectrum of Kiton Red. In the preparation of all phantoms, Intralipid 10% aqueous solution was diluted in proper proportion to obtain the phantom medium with necessary photon mean free path. Basic fuchsine was added to the phantoms in the concentrations ranging from 0.09 mM to 0.6 mM.

The molecular diffusion coefficient of 2′7′-dichlorofluorescin diacetate (DCFDA), whose oxidized form absorbs (peaking at 485 nm) and fluoresces (peaking at 528 nm) in the visible, in 7% gelatin gel prepared on phosphate-buffered saline has been measured at room temperature using a CCD camera coupled to a fluorescence microscope.[87] On a 3-mm-thick layer of gelatin gel containing DCFDA and a benzoporphyrin derivative monoacid used as a photosensitizer of DCFDA oxidation, CW 690-nm laser radiation was focused on a 2-mm-diameter spot, producing an initial region of oxidized, and thus fluorescent, DCFDA localization. The fluorescent image of the spot obtained by the fluorescence microscope was then captured by a CCD camera for a series of time points after the irradiation. The full width at a half-magnitude (FWHM) border of the spot expanded from a 2-mm to 3-mm diameter during approximately one hour. The FMHM vs. time dependency was then fitted to a diffusion equation to give $2.5 \cdot 10^{-5}$ mm^2/s for the molecular diffusion coefficient of the dye. It was concluded[87] that the diffusion of the molecules would not result in a large redistribution of the fluorophore within one hour. Refrigerating the sample also could minimize the effects of diffusion.

5.4.3 Inorganic ions as absorbers in solid and liquid tissue phantoms

Inorganic ions may be considered as an individual group of light-absorbing components of phantoms. An aqueous solution of potassium chromate (K_2CrO_4) was used in the study of laser-induced transient stress in nonscattering solutions and turbid aqueous gels and laser ablation of clear homogeneously absorbing liquid.[44,45]

According to Oraevsky et al.,[44,45] K_2CrO_4 offers the following advantages over solutions of organic dyes: (1) potassium chromate solution absorbs over visible and near-UV regions; (2) the solution practically does not fluoresce, so the total absorbed light energy is converted into heat; and (3) being photochemically stable, this solution retains its optical properties at high laser irradiance. The water solution of K_2CrO_4 at the concentration of 0.035 g/cm^3 has the absorption coefficient of approximately 1000 cm^{-1} at 355 nm.[44,45] A series of solution concentrations were obtained by diluting the starting solution with distilled water. Before experiments were performed, the solutions were filtered through a 0.22-μm filter to avoid any effects of dust and microbubbles.

To obtain a medium that more closely reproduces optical and thermomechanical properties of soft tissues, aqueous gel phantoms were developed.[44,45] The samples were prepared with gelatin gel (5% w/w gelatin in water for clear phantoms and 10% w/w gelatin in water for turbid phantoms) colored with potassium chromate. The absorption coefficient of such gels was assumed to be the same as in the potassium chromate solutions used in gel preparation. Polystyrene microspheres were added to the gels at concentrations of approximately 2%, which gave $\mu_s' = 99$ cm^{-1}. In the samples, the ratio of transport scattering coefficient to absorption coefficient was 81.5. This is a typical figure for biological tissues, for which values from 70 to 100 over the spectral interval of 600–1000 nm have been reported.[44,45]

For the complex systems that incorporate K_2CrO_4, the issue of chemical compatibility and stability should be considered. Chromate ion is a strong oxidizing agent that may be reduced (e.g., by alcohol), thereby changing its color from yellow to bluish-green.

Water solutions of another inorganic salt, $CuSO_4$, also were employed as an absorbing component in liquid phantoms. Steady-state measurements of transmission under infinite-boundary conditions were performed using these phantoms. The optical diffusion coefficient in a homogenous, highly-absorbing turbid medium was determined from the measurement results.[101] Measurements were performed at 809 nm, near the maximum of $CuSO_4$ absorption in NIR. $CuSO_4$ was chosen as a phantom absorber because $CuSO_4$ is nonfluorescent (as is K_2CrO_4) and does not alter the scattering coefficient of Intralipid during measurement. Using conventional spectrophotometry, the value of 27.6 $cm^{-1}M^{-1}$ (where M stands for molarity in mol/liter) was obtained for the absorption coefficient of $CuSO_4$ aqueous solutions at 809 nm.[101]

5.4.4 from the dyes to pigments and absorbing particles in phantoms

We have already mentioned that the diffusion of dye within the gel media is a shortcoming of macroscopically inhomogeneous phantoms that reproduce actual morphological structures. In such phantoms, partitions between the volumes (more often the layers) of different optical properties are avoided to provide refractive index matching. The problem of dye diffusion within gel can be alleviated to some extent using pigments instead. Pigments are present in a system as insoluble particles of micro- and submicrometer dimensions, contrary to dyes, which are dispersed on a molecular level. On the other hand, because the particles are the inhomogenities of mesoscopic dimensions, they contribute to the light scattering occurring within the phantom; thus, strictly speaking, the pigments cannot be considered as independent light absorbers.

One of the readily available and widely used pigments is India ink, which consists of a suspension of lampblack particles in aqueous medium. According to Madsen et al.,[114] the suspension is a mixture of small (0.1 μm) and large (1.0 μm) particles.

As there did not appear to be a water-soluble, nontoxic molecular absorber for 1064 nm, India ink was used to develop liquid tissue phantoms for measurements at this wavelength.[7] Optical properties of the ink were determined from a series of ink suspensions in deionized water, with concentrations ranging from 0.05%–0.5% by volume. For each suspension, the measurements of collimated (total) and diffuse transmission and diffuse reflectance were made. Optical properties of the samples were then calculated from the experimental data using an inverse adding-doubling method. The results are shown in Fig. 5.7, a linear fit being applied to the data. The average values of optical properties per concentration of India ink (Pelikan, Germany) in water at 1064 nm were calculated in Ref. 7 as $\mu_s/\%_{\text{per volume}} = 4.64 \pm 2.07\ \text{cm}^{-1}/\%$, $\mu_a/\%_{\text{per volume}} = 35.99 \pm 4.28\ \text{cm}^{-1}/\%$, and $g = 0.30 \pm 0.18$.

The value for anisotropy is also obtained by averaging the samples with different ink concentrations. In general, the anisotropy of the scatterer suspension should not reveal a dependency on concentration. That was

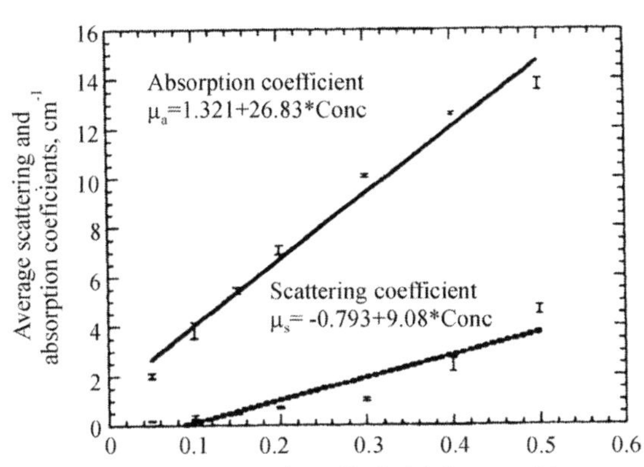

Figure 5.7 The absorption and scattering coefficients at 1064 nm for varying concentrations of water suspensions of India ink.[7]

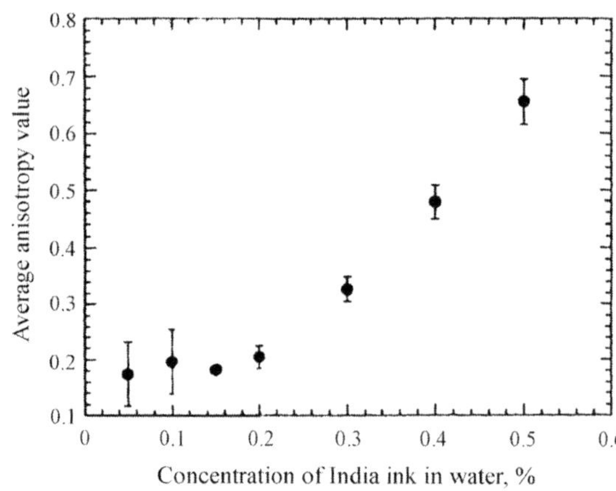

Figure 5.8 The anisotropy at 1064 nm for varying concentrations of India ink water suspension.[7]

the case for Intralipid suspensions [see Fig. 5.2(b)], but for the ink samples, the growth of the anisotropy factor with concentration is conspicuous (see Fig. 5.8). Two reasons for such a behavior were suggested.[7] The first reason is connected with the change of the ink's particle size distribution with the rise of concentration. The small particles have a much lower anisotropy factor than do the larger ones. The albedo of ink suspension increases with the increase of the concentration of large particles. Figure 5.7 shows that μ_s increases with the rise of ink concentration. At higher (0.3%–0.5%) concentrations of ink, larger particles (~1.0 μm) may strongly influence the suspension's anisotropy. At these increased concentrations, there may also be a tendency for the particles to coalesce. Royston et al.[7] suggested that a surfactant be added to ink suspensions to reduce coalescence of the ink particles.

The second reason for the nonlinearity of g versus ink concentration dependency may be purely computational. For ink concentrations exceeding 0.2%, the diffuse transmission was very weak and the insufficient measured signal might not allow the inverse adding-doubling algorithm to arrive at an accurate solution for the anisotropy factor value.[7]

India ink suspensions at substantially lower concentrations were used to impart tissue-like absorption properties to gelatin-based optical dosimeters.[87] The calibration of absorption coefficients of phantoms at 690 nm vs. the concentration of India ink in 7% gelatin gel with a fixed TiO_2 (used as scatter) concentration of 0.3 g per 20 ml gel was performed using an inverse algorithm that matches the data of diffuse reflectance measurements with Monte Carlo predictions. The results are plotted in Fig. 5.9. The best fit of the obtained points gave $\mu_a = 35 \pm 2\ \text{cm}^{-1}$/% per volume of India ink.

Liquid phantoms made of diluted (usually up to 1%) Intralipid or Lyposin suspensions with India ink added as an absorber were employed in various

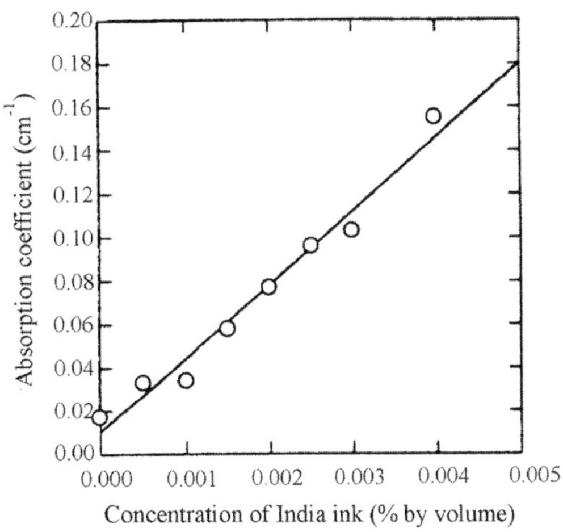

Figure 5.9 Absorption coefficient measured in dosimeters with tissue-like optical properties vs. India ink content. Ink was added to a mixture of 20 ml of 7% gelatin in phosphate-buffered saline and 0.3 g TiO_2. (From (with minor changes) B. Pogue, L. Lilge, M. Patterson, B. Wilson, and T. Hasan, "Absorbed photodynamic dose from pulsed versus continuous wave light examined with tissue-simulating dosimeters," *Appl. Opt.*, **36**(28), 7257–7269 (1997). © 1997 OSA. Reprinted with permission from The Optical Society.)

fields of optical measurements.[6,15,71,75] India ink is known to have a relatively constant absorbance over the visible region.[114] It was used as an inert absorber to simulate the absorption of hemoglobin and melanin (endogenous chromophores present in tissue) in phantoms developed for evaluating the fiber-optic bundle for quantitative fluorescence measurement from tissue. Preliminary measurements gave the value of 50 cm^{-1} as the absorption coefficient of 1% (by volume solution) of India ink at 633 nm, and the wavelength variation over the spectral region of 514–635 nm was less than 5%.[15]

The adequacy of Monte Carlo simulations of light transport in tissue has been tested on liquid Intralipid+India ink phantoms at 633 nm. But here, India ink was considered as an effectively pure (independent) absorber. The absorption coefficient of the ink at known dilutions was measured with a conventional spectrophotometer. In phantoms, the absorption coefficient was varied in the range of $0.01 < \mu_a < 2$ cm^{-1}, which corresponded to the optical properties of typical normal tissue.[6]

Solid phantoms incorporating India ink as an absorber were used in the studies of light propagation in layered tissues.[47,69,72] The influence of layered tissue structure on time-resolved reflectance data was studied at 780 nm on the phantoms mimicking a structure such as the skull encapsulating the brain. The phantom layers were made of gelatin gel with TiO_2 as the scatterer and India ink in different concentrations as the absorber. The optical properties of different layers were measured by time-resolved reflectance on homogeneous bulk samples, and in two-layer phantoms, the gels with $\mu_a = 0.50$ cm^{-1},

0.04 cm^{-1}, 0.045 cm^{-1}, and 0.87 cm^{-1} were used.[69] The same components (gelatin gel, titania, and India ink) were employed in the preparation of the layers of two-layer phantoms, with the optical properties in NIR similar to those of soft tissues, for the experimental study of the role of superficial layer in frequency-domain spectroscopic measurements of strongly scattering media.[72]

A method of measuring subsurface temperature using pulsed photothermal radiometry was evaluated in experiments on actual skin and scattering or scattering+absorbing multilayered gel-based phantoms.[47] Two types of hydrophilic host media were used: 5% gelatin gel (170-μm-thick layers) and polyacrylamide gel (70-μm-thick layers). In both types of gel, Intralipid served as the scattering component and India ink as the absorber. Absorption coefficients were measured with a conventional spectrophotometer before Intralipid was added and gelling occurred. In pulsed photothermal radiometry experiments, the phantom samples were irradiated by 1-μs laser pulses at 506 nm. Layers with the following absorption coefficients imparted by India ink were used in the phantoms:[47]

$\mu_a = 400$ cm^{-1} modeled blood absorption (this value reproduced the blood absorption at 577 nm, the wavelength chosen for treatment of port wine stains);
$\mu_a = 300$ cm^{-1} reproduced absorption by port wine stain;
$\mu_a = 30$ cm^{-1} corresponded to absorption within epidermis;
$\mu_a = 10$ cm^{-1} mimicked absorption by dermis.

Graphite powder, which may be considered a black pigment, was used as an absorbing material in silicone-based solid two-layer phantoms with polystyrene spheres as scatterers. Theoretical results concerning light propagation in two-layer turbid media were confirmed by the results of measurement of time-resolved reflection at 528 nm performed on these phantoms. For the second layers of two phantoms used, the absorption coefficients were determined as $\mu_a = 0.074 \pm 0.005$ cm^{-1} and $\mu_a = 0.19 \pm 0.01$ cm^{-1}.[68] Measurement of absolute steady-state spatially resolved reflectance at 543 nm[55] for those media gave: $\mu_a = 0.09$ cm^{-1} and $\mu_a = 0.24$ cm^{-1}, respectively.

The problem of rapid diffusion of dyes in gels and plastics, as well as the problem of chemical instability of dyes during polymerization, and the problem of spectral shifts caused by the interaction of dyes with host materials and scatterers may be overcome by the use of absorbing particles instead of dyes.[10,11]

It is evident that the absorbing particles suspended in ground material, unlike the dyes dissolved in the host medium, cannot be considered independent absorbers when one calculates the optical properties of a phantom. However, Mie theory enables the evaluation of their contribution to the total scattering. In most cases, the increments in the absorption coefficient

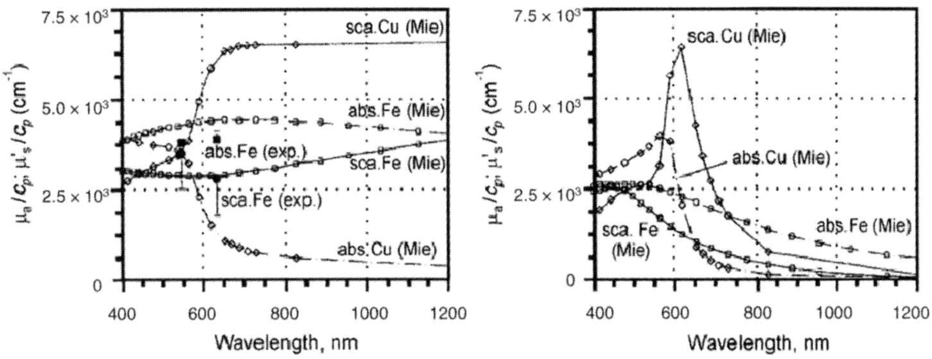

Figure 5.10 Specific transport scattering and absorption coefficients (C_p is the fraction of suspension volume occupied by particles) of the metallic particles, calculated using Mie theory. (The left-hand graph is for particles with a diameter of 2.5 μm, and the right-hand graph is for particles with a diameter of 0.1 μm). (From Ref. 11 with minor changes.)

and the reduced scattering coefficient imparted by the addition of highly-absorbing particles are on the same order of magnitude.[10,11]

In practice, metal particles and ground colored glass of optical filters have been suggested as absorbing particles without fluorescence. It has been demonstrated that the filter glass (NG-1, Schott) can be ground with a ball grinder into irregularly shaped particles with typical diameters of 1 μm, and even ruby can be ground to particles of 3-μm mean size.

Scattering and absorption of metal particles in polyorganosiloxane (POS, $n_0 = 1.40$ at 589 nm) were calculated[10,11] using Mie theory based on complex refractive index values obtained from the literature (see Fig. 5.10).

For calculations, the refractive index of the host medium was assumed to be independent of the wavelength. The experimental points for the iron particles were obtained using the integrating sphere technique. The scattering anisotropy factor of the large particles is positive and varies from 0.5 to 0.7; for the small particles, $|g| < 0.1$.[11]

During phantom preparation,[10,11] the iron particles as well as the scattering particles were dispersed in a mixture of POS components. Polymerization in castling molds typically took less than 30 minutes at 80° C. A short polymerization period at high temperature and favorable orientation of the molds during solidification favors homogeneous particle distribution within the samples. The sedimentation rate in POS can be estimated as less than 0.1 mm/h, even for heavy iron particles. Although the metal microparticles have a high specific surface and iron is a chemically reactive element, no evidence has been found for chemical reactions (other than POS polymerization itself) during POS solidification.

It has been found that scattering properties are modified when iron particles are introduced to the polymer in combination with Al_2O_3 particles. The addition of iron particles causes a 5% decrease of μ_s'/C_p for aluminum oxide particles. The alteration was presumed to be related to the slight aggregation between different kinds of particles; however, the inspection by light microscopy showed no obvious aggregates.[10,11] The iron particles used as an absorbing component in the polymer-based tissue phantoms were proven to keep their optical properties during POS polymerization and several months afterwards.[10,11]

Optical properties of quantum dots in the form of individual particles and clusters in the application of the development of new types of optical molecular imaging phantoms, including the phantoms meant for the calibration of the luminescence lifetime, are described in detail in Ref. 38.

5.4.5 Phantoms containing hemoglobin

Most of the phantoms mentioned were designed for use in the measurements at one particular wavelength. But when the aim is to reproduce in a material model the light-induced fluorescence of either endogenous (in the case of autofluorescence) or exogenous (often a photosensitizer) fluorophore in actual tissue, the relevant propagation conditions for the radiation of both the excitation and emission wavelengths should be modeled in the phantom. Thus, the phantom for fluorescence measurements should have—over a broad spectral range—the same absorption coefficient and scattering coefficient as the tissue under study. The main chromophore responsible for tissue absorption in the visible region is hemoglobin, and it has been introduced into liquid and solid phantoms as an aqueous solution, a suspension of red blood cells (erythrocytes), as well as whole or diluted blood.

Gel-based optical phantoms that reproduce absorption and scattering characteristics of biological tissues in the broad spectral window (between 400 and 650 nm) were designed and fabricated for use in PDT and fluorescence spectroscopy.[12] Such phantoms can be used simultaneously at different wavelengths, providing realistic propagation differences that are characteristic of actual tissue. In those phantoms, the ink and blood served as optical absorbers. While the absorption coefficient of the ink was assumed to be constant between 400 and 650 nm, the erythrocyte suspension provided tissue-like wavelength dependence of the absorption coefficient of the phantom. The effect of erythrocyte suspension absorption on the absorption coefficient of the phantom, $\mu_a = \mu_a$ (ink) + μ_a (blood), may be evaluated as μ_a (blood at 500 nm) $= 2.3 \times C$ cm^{-1}, where C is the erythrocyte concentrate content in the phantom, expressed in %% per volume.[12]

μ_a (blood) wavelength dependence is shown as a blood absorption spectrum in Fig. 5.11. The contribution of the known concentration of erythrocytes to the phantom absorption coefficient over a region of 300–800 nm can be predicted using the right-hand vertical axis. Scattering by the blood cells in the phantom was assumed to be negligible between 400 and 650 nm. Because different batches of erythrocyte concentrate might differ in their optical properties, the absorption of concentrate was checked before the phantoms were prepared; at 500 nm, for the light pathlength of 1 cm, the

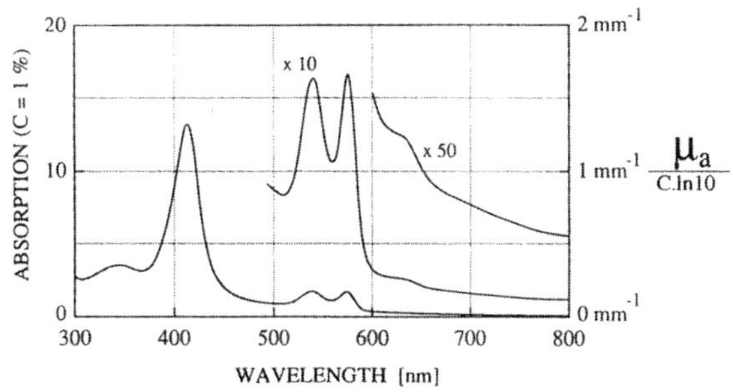

Figure 5.11 The absorption spectrum of diluted human blood. μ_a is the absorption coefficient of the phantom due to the blood only. The concentration of erythrocyte concentrate, *C*, is expressed in %% per volume. (From G.Wagnieres, S.Cheng, M. Zellweger, N.Utke, D.Braichotte, J.Ballini, and H.van den Bergh, "An optical phantom with tissue-like properties in the visible for use in PDT and fluorescence spectroscopy," *Phys. Med. Biol.*, **42**(7), 1415–1426 (1997); http://dx.doi.org/10.1088/0031-9155/42/7/014 © Institute of Physics and Engineering in Medicine. Reproduced by permission of IOP Publishing. All rights reserved.)

absorption should be equal to 0.9 for a concentration of 1% in physiological saline (0.9% NaCl). When deviations were revealed, the concentrate concentration was corrected with physiological saline.[12]

In the course of phantom preparation, human erythrocyte concentrate was never added to agarose gel, which was already loaded with scatterers (silica powder and Intralipid) and ink solution, until the mixture was cooled to 40°C. This temperature is low enough not to induce a thermal damage of the erythrocytes and hemoglobin, but sufficient for the gel to remain liquid.

To verify experimentally the theoretic expression relating measured fluorescence to the intrinsic fluorescence coefficient, water-based (phosphate-buffered saline, pH = 7.2) semi-infinite phantoms with identical fluorophore concentration (0.5 μM) and varying absorption and scattering were used (see Table 5.8).[17] Hemoglobin extracted from fresh human blood has been employed as an absorbing component. The absorption of hemoglobin covered both excitation and emission bands of Rhodamine 6G, which was used as a sole fluorophore in the phantom. The 514-nm line of argon laser was used for fluorescence excitation.

The set of phantoms with varying scattering and absorption coefficients imitated tissue optical properties over an extensive range. The diffuse reflection measured on the phantoms varied from 0.04 to 0.4 at the excitation wavelength and from 0.02 to 0.8 within the R6G emission band. That allowed the evaluation of the intrinsic fluorescence recovery method for absorption-dominant, comparable absorption and scattering, or scattering-dominant optical conditions in tissue.

However, the experimental results indicative of Rodamine 6G fluorescence quenching by hemoglobin in solution were obtained. Nonradiative energy transfer was suggested as a possible mechanism of quenching.[17] Thus, for fluorescence measurement in phantoms containing fluorophore,

Table 5.8 Tissue phantom composition. (From C.Gardner, S.Jacques, and A.Walch, "Fluorescence spectroscopy of tissue: recovery of intrinsic fluorescence from measured fluorescence," *Appl. Opt.*, **35**(10), 1780–1792 (1996). ©1996 OSA. Reprinted with permission from The Optical Society.)

R6G, μm	Polystyrene microspheres (vol. %)	Hb/HbO_2 (blood vol. %)	μ_s' at 550 nm (cm^{-1})	μ_a at 514 nm, 550 nm (cm^{-1})
0.5	0.16	8	7.5	12; 17
0.5	0.16	4	7.5	6.0; 8.4
0.5	0.16	1	7.5	1.5; 2.1
0.5	0.48	8	25	12; 17
0.5	0.48	4	25	6.0; 8.4
0.5	0.48	1	25	1.5; 2.1

hemoglobin cannot be considered a purely absorbing component and should therefore be used with caution.

5.5 Smart Phantoms

5.5.1 Multifunctional phantoms

Bovine hemoglobin was used as a contrasting absorbing agent in the preparation of human breast phantoms for evaluating the capability of laser optoacoustic imaging for tumor visualization.[43] Absorption by water in 10% gelatin gel, $\mu_a = 0.11$ cm^{-1} at 1064 nm, and scattering by polystyrene microspheres or milk, $\mu_s = 2.92$ cm^{-1} at 1064 nm, resulted in the effective attenuation coefficient of 1.0 cm^{-1} at 1064 nm for the bulk body of the phantom (see Fig. 5.12). Spherical phantoms of tumors were made of the

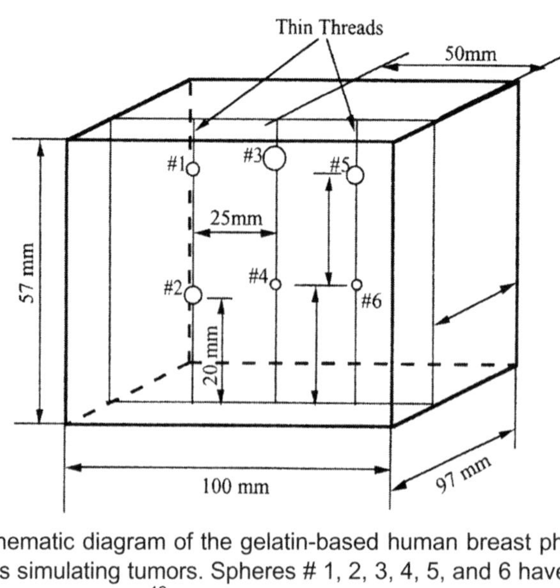

Figure 5.12 Schematic diagram of the gelatin-based human breast phantom with highly-absorbing spheres simulating tumors. Spheres # 1, 2, 3, 4, 5, and 6 have diameters of 3, 4, 6, 2, 4, and 2 mm, respectively.[43]

same turbid gel, but they were colored with a solution of bovine hemoglobin with an absorption coefficient of 0.75 cm^{-1} at 1064 nm. That value corresponded to 16% blood content in tissue and was sevenfold higher than the μ_a of the surrounding gel. Breast phantoms were obtained through sequential molding. Prefabricated hemoglobin-colored spheres were positioned with white nylon threads inside the Plexiglas mold, where colorless liquid gel close to solidification temperature was poured.[43] In the phantoms obtained, there is no partition between volumes with different absorption levels. On the one hand, this realistically imitates the interface between a tumor and surrounding tissue, and on the other hand, it allows diffusion of hemoglobin into the body of the phantom, which distorts the optical characteristics of the structure.

5.5.2 Phantoms mimicking vascular systems

Liquid but inhomogeneous phantoms of human tissue were employed in the quantitative analysis of the bluish appearance of blood vessels. These vessels contain red or dark red blood and are located under skin.[66] Measurements of spatially resolved diffuse reflectance at different wavelengths have been performed on a model blood vessel submerged in a highly scattering and weakly absorbing liquid medium. Scattering was imparted to the medium by 20% Lyposin diluted with water to give $\mu_s' = 10$ cm^1 at 633 nm, and the absorption of human dermis associated with blood in a capillary network was simulated by adding 0.34% (per volume) oxygenated blood to the medium.

A cylindrical glass tube filled with whole blood modeled the blood vessel. The distance from the top of the cylinder to the surface of the surrounding medium was 1.4 mm. The inner diameter of the tube was 1.2 mm, and the thickness of the glass wall was ~0.2 mm. Deoxygenated blood was drawn directly from the vein into the heparinized tube, whereas the sample representing arterial blood was heparinized and shaken to yield the oxygenated state.

The reduced scattering coefficients and absorption coefficients of the phantom medium are shown in Table 5.9. The coefficients were measured using frequency domain diffuse reflectometry. The value of scattering anisotropy parameter was set to 0.8.

Results of Monte Carlo simulations using known optical coefficients of tissue phantoms and human blood were compared to the data obtained from on-phantom measurements to check the experimental and Monte Carlo techniques that address the problem of the bluish color of veins.[66]

Measurements essential toward understanding light absorption by cerebral blood in various blood vessels were undertaken on heterogeneous tissue-vessel models.[70] The phantom used in the study (see Fig. 5.13) had the structure akin to the tissue model (see below) described by Mendelson and Kent,[40] but the measurements were performed at a reflection geometry.

Table 5.9 Optical parameters for a highly scattering and weakly absorbing liquid medium of tissue phantom. (From (with minor changes) Alwin Kienle, Lothar Lilge, I. Alex Vitkin, Michael S. Patterson, Brian C. Wilson, Raimund Hibst, and Rudolf Steiner, "Why do veins appear blue? A new look at an old question," *Appl. Opt.*, **35**(7), pp. 1151–1160 (1996). © 1996 OSA. Reprinted with permission from The Optical Society.)[66]

	Lyposyn plus blood	
Wavelength λ, nm	μ_a, cm^{-1}	μ_s', cm^{-1}
450	0.85	14.2
500	0.39	12.5
550	0.67	11.4
633	0.022	10.0
700	0.015	8.2

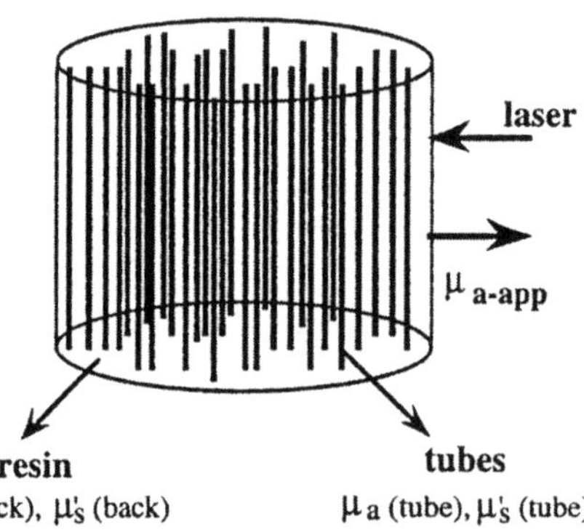

Figure 5.13 The heterogeneous tissue-vessel phantom for NIR time-resolved reflectance measurements.[70]

The body of the phantom imitated a low-absorbance tissue and was made of polyester resin with TiO_2 added to give the reduced scattering coefficient, μ_s' (back), from 6 to 9.2 cm^{-1}, whereas the absorption coefficient of the medium, μ_a (back), was from 0.02 to 0.06 cm^{-1}. The cylindrical holes (tubes) that simulate blood vessels made up 18% and 20% of the entire sample volume. Samples with either 3.2-mm-diameter holes or 6.4-mm-diameter holes were used in the experiments. Absorbant turbid solutions that filled the tubes imitated the blood in vessels. 0.5% or 1.0% Intralipid was used to ensure light scattering by the "blood." Aqueous solutions of either methemoglobin or ink served as absorbers.[70]

Methemoglobin [the form of hemoglobin that contains Fe(III)] in alkaline medium has absorption peaks at 417, 540, and 578 nm, which are close in terms of wavelength to the peaks of oxyhemoglobin but are considerably less intensive. In some cases of phantom preparation, the lack of ability to bind oxygen may give methemoglobin an advantage over hemoglobin, as the absorption spectrum of the latter depends on the oxygen saturation of phantom medium. Liu et al.[70] made the methemoglobin solution from hemoglobin by adding potassium ferricyanide.

The absorption coefficients, μ_a (tube), of the methemoglobin or ink turbid solutions at different absorber concentrations were determined separately using a time-resolved reflection technique. In time-resolved reflection measurements on the tissue-vessel phantoms at 670 nm, the methemoglobin solutions with absorption coefficient values up to ≈ 0.32 cm^{-1} were used. Estimation of the absorption coefficient of human blood at 780 nm gives $\mu_a^{780} = 8$ mM $\times$ 0.46 cm^{-1}mM^{-1} = 3.68 cm^{-1}, where 8 mM is the assumed hemoglobin concentration in the blood and 0.46 cm^{-1} mM^{-1} is the specific absorption of hemoglobin at 780 nm. Based on this estimation, the measurements at 780 nm were performed with high μ_a (tube) values that were provided in the phantoms by ink solutions.[70]

A class of phantoms, which received the name "dynamic phantoms," generally combines a complex phantom structure that mimics the vascular system of tissue with the implementation of pumping of a blood-simulating liquid or whole blood through this artificial vascular network.[81,76,115,116] Dynamic phantoms are often built on the basis of microfluidic device technology[115,116] and used in the calibration of instruments for blood flow velocity measurement and imaging, including speckle and Doppler systems.[81,76,115,116] Another type of dynamic phantom accounts for differences in the absorption by hemoglobin in the superficial and deep layers of a tissue and can simulate brain activity and other systemic changes.[59]

A blood-like liquid component was employed in a phantom with varied local oxygenation states. Using an optical coherence tomography system, the μ_a images at the wavelengths of 761 nm, 791 nm, and 835 nm were obtained on this phantom; then based on 761 nm and 835 nm data, the images of local distribution of oxy- and deoxyhemoglobin were calculated.[39]

The 80-mm-diameter body of the phantom was made of silicone resin with scatterer added in a concentration to obtain $\mu_s' = 10$ cm^{-1}. Through two blood tubes (10-mm-diameter holes running in the body of the phantom along its axis), a solution with variable oxygenation was circulated by a peristaltic pump. The solution consisted of erythrocyte suspension (1.5% or 3.0%) in a phosphate buffered saline with milk added to give a μ_s' value of about 10 cm^{-1}.[21] Here and in some optical phantoms described next, the choice of erythrocytes or hemoglobin has been determined by the unique alterations of hemoglobin absorption spectra that attend the change of oxygenation state. Oda et al.[39] adjusted the oxygenation state of solution employed in the phantom using a nitrogen-oxygen gas system.

Circulation of blood with a variable oxygenation state has been employed in a more complicated tissue-like phantom developed for the testing of noninvasive pulse oximeters.[40] Besides the similarity of optical properties, some other criteria were met in the development of this tissue phantom: the material used to construct the body of the phantom was flexible enough to simulate arterial pulsations similar in size and shape to actual photoplethysmograms; the material was biocompatable; and the ratio of whole blood volume to bloodless tissue volume was approximately 2.0% in the phantom, based on the average values inherent in real tissue.

The phantom was made of semitransparent medical-grade silicone elastomer with flesh-colored pigment (Fe–Mg–TiO_2) added to simulate lightly pigmented skin. Blood flowed through fifteen parallel 0.5mm-diameter channels in the silicone body of the phantom. Circulating blood was prepared by suspending packed red blood cells in plasma up to a hematocrit of ~41%. The blood thus obtained was purified from blood cell aggregates by filtration using a transfusion filter. pH of the blood was adjusted to 7.4 ± 0.1 by the isotonic solution of $NaHCO_3$.

The blood was pumped through the phantom using a pulsatile blood pump. A photoplethysmographic waveform measured from an average-size human finger was reproduced in the phantom by adjusting the stroke volume of the pump operating at a stroke rate of 70 strokes/minute, the fixed pulsative phase being 35% systole and 65% diastole.

For the measurements, the blood was first deoxygenated (to the saturation level of ~20%) by passing a mixture of 95% N_2 and 5% CO_2 gas though the disk oxygenator connected in series with the phantom. The blood in the oxygenator was then exposed to room air for short intervals, thus it gradually became oxygenated. In measurements, blood temperature was maintained at $37° \pm 0.5°C$.

It has been shown that the tissue phantom gives a possibility of controlled and reproducible measurement of oxygen saturation of arterial hemoglobin with the use of a pulse oxymeter. The suggested phantom has the potential to assess the effect of blood hematocrit, methemoglobin concentration, arterial blood pulsations, and blood temperature that affect the accuracy of pulse oximetry.[40]

5.5.3 Phantoms of organs

Optical phantoms of whole organs are described in the literature.[14,117,118] For example, human skin was modeled by a film of hydrated type I collagen.[49,119] This film contained variable amounts of subsurface absorbers positioned at a given depth and simulated discrete chromophores buried in multilayered human skin. The absorbers were obtained by staining a collagen film with triphenylmethane dye, which absorbs optimally at 585 nm. Discrete chromophores were prepared by cutting the 125-μm-thick stained collagen film with known optical absorption ($\mu_a = 400$ cm^{-1}) into several thin strips (100 to 300 μm wide). A skin phantom was constructed by positioning variably spaced (50 to 700 μm) absorbing thin strips underneath the 110-μm thick nonabsorbing collagen film. The absorbing thin strips and nonabsorbing

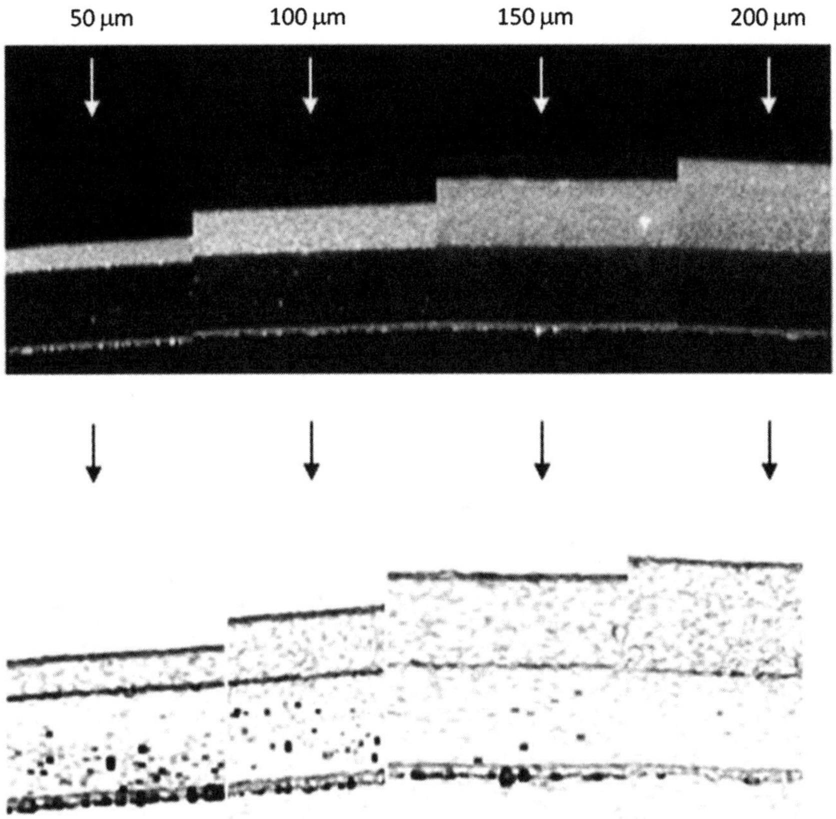

Figure 5.14 OCT images of four two-layer phantoms with gold nanoparticles in the bottom layer and TiO_2 particles in the top layers. The figures above the image stand for the thickness of TiO_2 containing layer in the phantom. The lower, more contrast, image is a version of the original top image processed to enhance visibility of gold nanoparticles.[78]

films were mounted on a 10-mm-thick collagen sponge to simulate an infinite half-space as in actual skin and subjacent tissues.

In Ref. 78, the design and characteristics of solid phantoms with silicone elastomer as a host material are described. Using thin (50 μm) layers of phantom material, the authors managed to construct complex phantoms with flow channels, phantoms with wavy interfaces (model of dermal *papillae*), and layered and curved structures simulating human retina. In the phantom layer material, the appropriate absorption was achieved by the addition of ABS 551 (Exciton, Dayton, Ohio) dye, and the scattering was provided by TiO_2 nanoparticles or micron-size SiO_2 particles. In the absence of absorbers, OCT measurements showed a linear dependence of the attenuation coefficient on the concentration of the scatterers (nanoparticles). Phantom characterization using confocal microscopy revealed a uniform distribution of scatterers in silicone elastomer, although with some clustering. In the absence of the scatterers, optical transmission spectroscopy showed the linear dependence of the absorption coefficient on the dye concentration. The phantom material samples appeared to be stable for six months. The authors also demonstrated the possibility of inclusion of gold nanoparticles in the phantoms. This practice may be useful, for example, when calibrating optical devices and techniques for diagnosis of premalignant condition. Figure 5.14 shows compounded OCT image of four two-layer phantoms with gold nanoparticles (10^9 ml^{-1}) in bottom layer and TiO_2 scatterers ($\mu_t = 4$ mm^{-1}) in top layers of various thickness. The nanoparticles can be seen in OCT images, but their visibility decreases with the increase of upper layer thickness.[78]

A solid phantom with embedded nanoparticles was designed, manufactured, and tested.[80] The phantom fabricated on the basis of transparent epoxy resin ($n = 1.47$) with dispersed therein rare, a mean distance between particles being 46 nm, silica-gold nanoshells (213 nm-diameter silica core with 19 nm-thick gold shell) was built into a commercially available artificial model of the eye and intended for the characterization, under the conditions close to actual *in vivo* measurements, of the point spread function of OCT devices used in retinal imaging.

In Ref. 14, phantoms reproducing the autofluorescence of cervix uteri in the broad spectral range were designed and developed based on a combination of natural fluorophores and chromophores and simulation of the architecture of the biological object and lightscattering and absorption in it. The histologically observed laminated structure of the organ was reproduced by a three-layer structure. Gelatin (Sigma, G2500) gel was used as a host material in the preparation of all three layers of the phantom. In the course of the work, three types of neutral scatterers were tried—$CaCO_3$, SiO_2, and TiO_2—but finally only TiO_2 (Du Pont, Ti-Pure R-900) with the average particle diameter of 410 nm was used in the phantom layers preparation for being chemically inert and having a high relative refractive index in gelatine gel.

The outer (upper) 300 μm layer of the phantom structure reproduced squamous cervical epithelium. Native UV-induced (presumably keratin) fluorescence of epithelium was imitated by the fluorescence of 10% w/w gelatine gel used as a host material in the "epithelial" layer preparation. The introduction of NADH (nicotinamide adenine dinucleotide, reduced form) to the concentration of $2 \cdot 10^{-6}$ mol/L in the "epithelium" made it possible to attain a good agreement of fluorescence (365-nm excitation) spectra of a cervix sample (*ex vivo*) and three-layer phantom over the spectral interval of 400–500 nm; FAD (flavin adenine dinucleotide) added to the concentration of $4 \cdot 10^{-6}$ mol/L gave the similarity of the spectra over 500–650 nm. To imitate the scattering properties of epithelium, the TiO_2 powder was added to the concentration that gave in visible the μ_s value identical to that the actual epithelium exhibited *ex vivo*. Mie theory–based evaluation of the scattering anisotropy factor for the TiO_2 particles embedded in gelatin gel gave the average over the visible range value of 0.6 that is lower than the figures for actual human soft tissue, so the reduced scattering coefficient for the "epithelial" layer may exceed the value for actual tissue.[14]

The second layer reproduced the thin dense basal membrane on the epithelium-connective tissue boundary. The 20-μm layer was made out of clean (neither scatterers nor absorbers in visible) dry gelatin.

The lowest (the third) layer in the phantom reproduced the stroma of cervix uteri. The thickness of the layer, 2.2 mm, was sufficiently large, and under the conditions of fluorescence measurements, the layer behaved as semi-infinite. The stromal capillary network was not modeled in the phantom; instead, the light absorption by the stromal blood was reproduced by the addition of human hemoglobin to the "stroma" in the concentration that corresponds to the blood content of the actual tissue. To imitate the scattering properties of the stroma, the TiO_2 powder was added to the hemoglobin-loaded gel to the concentration that gave in visible the μ_t value identical to that the actual stroma exhibited *ex vivo*. Fifteen percent w/w gelatin gel served as a host material (mechanical base) of the third layer and provided the imitation of collagen and elastin fluorescence of actual stroma.[14]

In Ref. 14, the structure and composition of the phantoms is consistently substantiated and the detailed technique of manufacturing is presented. The phantoms described provide under the UV (360 ± 20 nm) excitation the reproduction (over a wide spectral interval) of fluorescence spectra of normal cervix uteri and the areas of extinct squamous epithelium. The results of on-phantom measurements, allowed the authors to make assumption about the nature of the weakly expressed maximum ($\lambda_{em} \approx 510$nm) observed *ex vivo* in the fluorescence spectra of actual cervix uteri.[14]

One more example of the whole organ modeling comprises phantoms of a human head.[41,92]

While pulse oximeters operate at transmission geometry, quantification of tissue hemoglobin concentration and oxygen saturation in massive organs, such as a skull with brain tissue, requires the implementation of near-infrared spectrophotometric instruments with a reflection geometry. To validate an instrument of that type and the algorithms that evaluate the cerebral concentration of hemoglobin and oxygen saturation, a silicon rubber-liquid phantom mimicking the dimensions, shape, and layered structure of a neonatal head was constructed.[41]

Skin and skull were represented in the phantom by a 3.5-mm-thick and 11-cm-diameter hemispherical shell whose size corresponded to the head of a newborn infant of 40 weeks gestational age. The shell was made of transparent silicon rubber with TiO_2 added as scatterer and the Zeneca dye as absorber. These constituents provided $\mu_s' \approx 17\ \text{cm}^{-1}$ and $\mu_a \approx 0.5\ \text{cm}^{-1}$. The instrument's optical sensor (NIR spectrophotometer) was placed at the outside bottom of the hemisphere. Cerebrospinal fluid (CSF) surrounding the brain was mimicked by the 0.5-mm-thick layer of polypropylene attached to the surface of the shell above the sensor.

The hemisphere served as a container for the liquid component of the phantom. This rubber shell filled with aqueous solution imitated the brain. It had a volume of approximately 200 ml and contained (1) 0.5%, 1%, 1.5%, or 2% Intralipid that provided different levels of scattering; (2) physiological saline (0.9% NaCl); (3) 0.5% yeast suspension; (4) 0.15% (maximum) glucose; and (5) 60-μmol l^{-1} hemoglobin from a packed erythrocyte concentrate. For 0.5% Intralipid, the μ_s' value of 5–8 cm^{-1} was adopted, which is close to $\approx$5 cm^{-1} for a neonatal brain. During the measurements, the solution was constantly stirred by a mixer to prevent sedimentation. Two parameters that might influence the oxygen-binding properties of hemoglobin, temperature and pH, were kept within a physiological range. Temperature was maintained at 37 °C by a small heater; pH was adjusted by adding the phosphate buffer to the solution. In the measurements, the predeoxygenated solution was reoxygenated using oxygen until the pO_2 was >15 kPa; then it was gradually deoxygenated through oxygen consumption by yeast, for which purpose glucose was added in the increments of 0.05% until the total deoxygenation was achieved.

Current oxygen concentration in the solution was measured independently with a pO_2 electrode. Readings from the pO_2 monitor were recorded simultaneously with the optical raw data measured at 776.5 nm, 819.9 nm, 871.4 nm, and 908.7 nm and the calculated concentrations.

The performed on-phantom measurements[41] showed that when applied to the data obtained on a neonatal head phantom with a spherical geometry and layered structure, the algorithms used to determine the cerebral concentrations of hemoglobin and the oxygen saturation considerably underestimate the changes in hemoglobin concentration.

Phantoms reproducing the structure and optical properties of an adult human head were employed in the investigation (with the use of time-of-flight measurements) of the effect of different tissue layers on the propagation of NIR radiation through brain and surrounding tissues.[92] All of the phantoms were the layered structures; the simplest three-layered model consisted of plain layers mimicking scalp and skull (12 mm), gray matter (4 mm), and white matter (gray matter and white matter block in the phantom was 80-mm thick). In the plane four-layered model, the surface level was 10-mm thick, and between it and the gray matter, a 2-mm-thick layer that imitated cerebrospinal fluid (CSF) was placed. In the simple and sophisticated brain models, the boundary between gray matter and white matter was uneven: the gray matter had rectangular 10 mm in depth and 9 mm in width bulges placed every 15 mm. The sophisticated brain model also incorporated 10-mm deep, 1-mm wide, and clear (medium imitating CFS) slots in the bulges that imitate sulci on the brain surface. The geometry and dimensions of the models was taken from magnetic resonance imaging data for an adult head. The scalp and skull, gray matter, and white matter layers of the phantom were made of epoxy resin. TiO_2 and IR dyes were added to adjust the transport scattering coefficient and absorption coefficient of the phantom media to the values characteristic of actual tissues. Clean CFS was imitated by glycerol. On-phantom measurements of the temporal point spread function were performed using a picosecond pulsed laser and streak camera.

5.6 Phantoms with Optically Active Media

5.6.1 Introduction

The importance of phantoms in biomedical imaging has been demonstrated thoroughly in the previous sections; in this section, phantoms with some unique properties will be described. These phantoms are produced using a specific category of optically active material, which under the correct irradiation conditions produce a coherent laser-like emission. This emission is the product of multiple scattering and increased pathlength of photons propagating inside an active medium.

In the past twenty-five years, there has been an increasing interest in media that are characterized by strong scattering and absorption and in strong amplification of the light traveling through them. These media have an unexpected behavior when strong scattering is involved, and even though scattering has been considered detrimental to laser action when certain conditions are met, they produce what has been called random laser action. It was the first observation by Lawandy et al.,[120] who demonstrated isotropic laser-like emission from an optically pumped solution of Rhodamine 640 perchlorate with TiO_2 scatterers dispersed in a methanol solution that gave the boost this phenomenon has seen recent years and provided a starting point for the investigation of the properties of such materials. Nevertheless, it was Letokhov[121] who while investigating the optical properties of a random medium, which simultaneously amplified and scattered light, theoretically proposed this effect in the mid-1960s. Since then, many experiments have been carried out to examine the origins and various features of the narrow-linewidth emission they observed.[122–125] All of these experiments confirmed that the scatterers play the role of mirrors in a conventional laser cavity to provide the necessary feedback for laser action. The observed response can be interpreted in terms of the photons' increased pathlength inside the medium as they undergo multiple scattering events (schematically represented in Figure 5.15). Consequently, the contribution of stimulated emission is increased, and when the amplification of the fluorescence light overcomes a critical value, the observed fluorescence signal is dramatically narrowed in both its temporal and spectral width, and its intensity is increased significantly. Simultaneously, several theoretical studies[126–129] have been conducted to provide models that will describe the behavior of these materials efficiently and help in

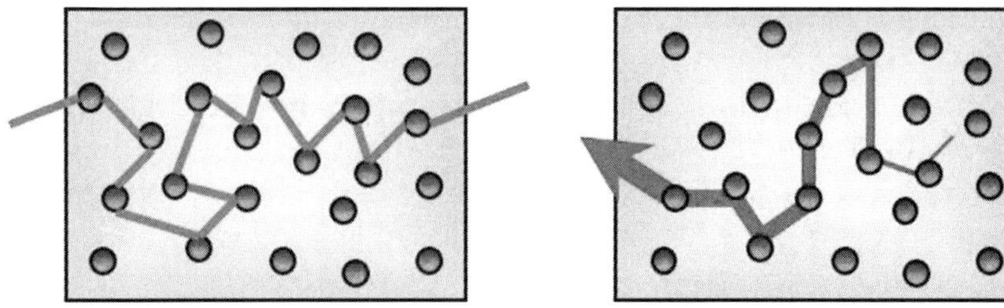

Figure 5.15 Schematic illustration of the amplification of fluorescence emission as photons experience multiple scattering inside an active medium [G. Zacharakis, unpublished data] (see color plates).

understanding the underlying mechanisms that are responsible for their laser-like characteristics.

5.6.2 Optically active phantoms

These materials can be made by grinding laser crystals or by embedding fluorophores and scattering particles in a solid matrix such as polymer or gelatin or by suspending the scattering particles and the absorbing molecules inside a solution.[130–132] The latter two arrangements, with the proper choice of absorption and scattering properties, can be thought of as simulating the behavior of biological tissue. Another type of random lasing materials are produced by embedding ZnO nanoparticles in a polymer matrix.[122] In this case, scattering and absorption are offered by the same particles. In all types of phantoms, the combination of amplifying and multiple-scattering properties results in the emission of laser-like radiation.

The liquid samples presented here were solutions of dyes in methanol placed inside a 1-cm × 1-cm × 3-cm glass cuvette. The scattering particles were polystyrene microspheres with concentrations of $3–10 \times 10^{10}$ cm^{-3}. The dyes used were Rhodamine 101 (R101), Rhodamine 6G (R6G), and DCM [4-dicyanomethylene-2methyl-6-(p-dimethylaminostyryl)-4H-pyran]. Samples of different dye (3×10^{-3} M to 1×10^{-2} M) and scatterers' ($10^{12}–10^{14}$ cm^{-3}) concentrations were prepared and irradiated with different excitation energies that varied from 0.5–30 μJ/pulse. The particle concentration corresponded to a scattering mean free path of the order of a micrometer, calculated as $l_s = 4\pi/n\sigma$.

Solid polymer samples were prepared by mixing poly(methyl methacrylate) and dye solutions in a 1-cm × 1-cm × 2-cm Teflon cuvette together with the scattering particles. The scatterers were TiO_2 nanoparticles (~400-nm diameter, ~4 kg/l density). Evaporation of a solvent (dichloromethane CH_2Cl_2) resulted in creation of a 1-mm-thick polymer sheet. The dye used was Rhodamine 6G in a concentration of 10^{-2} M, and the concentration of the scatterers was 1.2×10^{12} cm^{-3} (15-mg Ti-Pure in 0.1 cm^3 of the final solid phantom).

Solid gelatin samples containing Coumarin 307 were made in the form of 1-cm × 1-cm × 1-cm gelatin cubes. Two Coumarin 307 stock solutions (4×10^{-3} M and 1.33×10^{-3} M) in ethanol and a series of stock suspensions of TiO_2 in water (containing from 1–42 mg/ml TiO_2) were prepared. The substrates exhibited scattering mean free paths from ~10 to ~10^3 μm at 800 nm (as calculated by Mie theory) assuming negligible absorption. The corresponding absorption length at 400 nm was of the order of 1 mm (measured in methanol solution). To obtain the gel base of the sample, 0.45 ml of appropriate stock suspension was added to 0.1 g of gelatin. The mixture was then melted in a water bath at 50–60°C under intense stirring. 0.45 ml of appropriate Coumarin solution was added just before the mixture was cast in a Teflon mold.

The phantoms are used with nanosecond or femtosecond pulsed laser systems, and an ultrafast time-domain detector (e.g., streak camera) is needed for detection of the emission. If combined with a spectrograph, a simultaneous spectro-temporal analysis can be achieved as in the example shown in Fig. 5.16. In this case, the beam was focused on the samples, creating a spot size of 0.15 mm^2 while the energy/pulse was controlled by neutral density filters and varied from 50–150 μJ/pulse. The resulting fluorescence signal was spectrally resolved using a 0.10-m spectrograph and was temporally analyzed using a Hamamatsu C5680 Streak Camera (temporal resolution of 2 ps). The same setup provided the ability to perform statistics on the photon-number distributions for different time delays and at different wavelengths, reaching resolutions down to a single-photon event. This enabled investigation of the temporal coherence of the emitted light.

The emission of such optically active phantoms presents a unique and remarkable feature: the width both in time and wavelength domains is dramatically reduced when the excitation light energy is higher than a threshold value. This effect is attributed to stimulated emission caused by the

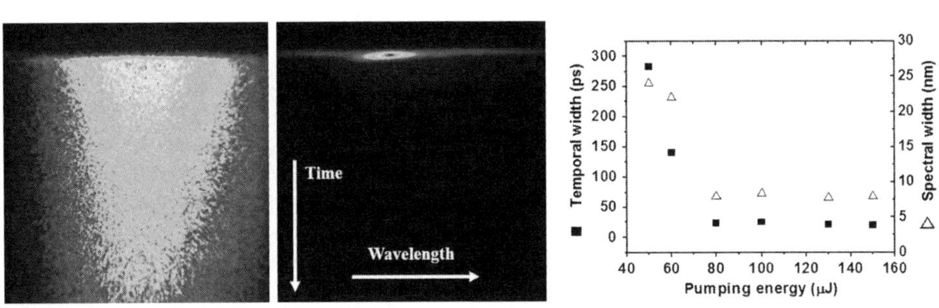

Figure 5.16 Spectro-temporal narrowing of the emission from random lasing materials when excited with energy above threshold. Reproduced with permission from G. Zacharakis, N. A. Papadogiannis and T. G. Papazoglou, "Random lasing following two-photon excitation of highly scattering gain media," *Appl. Phys. Lett.* **81**(14), 2511–2513 (2002); http://dx.doi.org/10.1063/1.1511284 Copyright 2002, AIP Publishing LLC (see color plates).

increased pathlength of the fluorescence photons inside the amplifying multiple scattering medium. This increased pathlength leads to laser-like emission when the gain exceeds the losses of the scattering cavities. The coherence properties can then be investigated by obtaining photon statistics for the emitted light above and below threshold.[130–132]

The photon-number distributions from the sample for an image obtained while excited above threshold are shown in Fig. 5.17. The different graphs correspond to distributions measured for different time delays after the excitation has ended and for a time interval of 30 ps.

The distribution of the short time delays is characteristic of coherent light. The divergence from the perfect Poisson distribution suggests that the light consists of a superposition of more than one pattern of incoherent as well as coherent light–a scheme that could be explained as a competition of localized (coherent) and extended (incoherent) modes.

Furthermore, one can see that as the time delay increases, the Poisson-shaped distribution is lost and is replaced by a Bose–Einstein distribution that is characteristic of incoherent light. This is due to the scattering the fluorescence light undergoes before exiting the sample, which is detrimental to the characteristics of the emitted coherent light. The temporal coherence is lost completely 100 ps after the end of the excitation, as shown in the inset of Fig. 5.17, in which the photon count distribution calculated for a time delay of

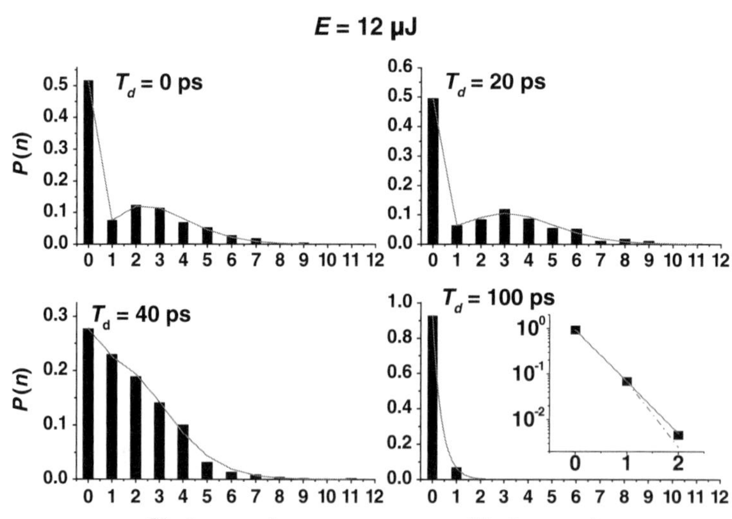

Figure 5.17 Photon-count distributions of the laser-like emission (above threshold $E = 12$ μJ) in increasing time delays after the excitation has finished. The total number of photon counts is kept constant. (From G. Zacharakis, N. Papadogiannis, G. Filippidis, T.G. Papazoglou, "Photon statistics of the laserlike emission from polymeric scattering gain media," *Opt. Lett.*, **25**(12), 923–925 (2000). © 2000 OSA. Reprinted with permission from The Optical Society.)

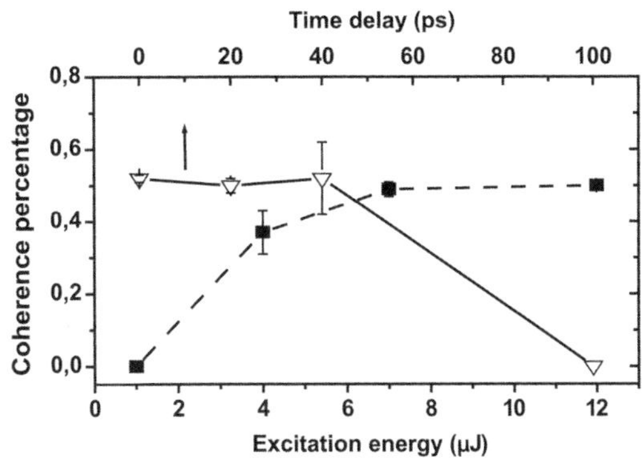

Figure 5.18 Coherent percentages of the photon-count distributions for different time delays (open triangles) and different excitation energies (filled squares) as calculated from the weight of the Poisson part of the fitting formula. (From G. Zacharakis, N. Papadogiannis, G. Filippidis, T.G. Papazoglou, "Photon statistics of the laserlike emission from polymeric scattering gain media," *Opt. Lett.*, **25**(12), 923–925 (2000). © 2000 OSA. Reprinted with permission from The Optical Society.)

100 ps plotted on a logarithmic scale is presented. The two curves correspond to fittings by use of Poisson (dashed-dotted line) and Bose–Einstein (solid line) distributions.

After the curves are fitted with a linear superposition of a Poisson and a Bose–Einstein function, the percentages of the coherent and incoherent components of the emitted light can be calculated. Figure 5.18 depicts these coherence percentages for different time delays with respect to laser excitation and different laser-excitation energies. For an excitation energy of 12 μJ and for zero delay time ($T_d = 0$ ps), the coherent component corresponds to 51% of the total light, whereas after 100 ps, the light is fully incoherent. This is an indication of the time needed for the total loss of coherence after laser excitation (dephasing time). The coherence percentage increases quickly as a function of laser-excitation energy and reaches a plateau after 7 μJ.

The coherence of the random lasing light is destroyed by the intense scattering the photons undergo during their propagation outward from the pumping volume and from the competition of different localized and extended propagating modes.

5.6.3 Conclusion

The possible applications of such active tissue phantoms are diverse and can cover a wide range of fields—from biomedical imaging to material science, new light sources, optoelectronics, and pure fundamental physics studies.

Biomedical research could enhance imaging of cancer tumors especially when they reside superficially or are found through endoscopy, taking advantage of the different emission characteristics in healthy and malignant tissue.[133,134] In therapeutic applications, such media can assist photodynamic

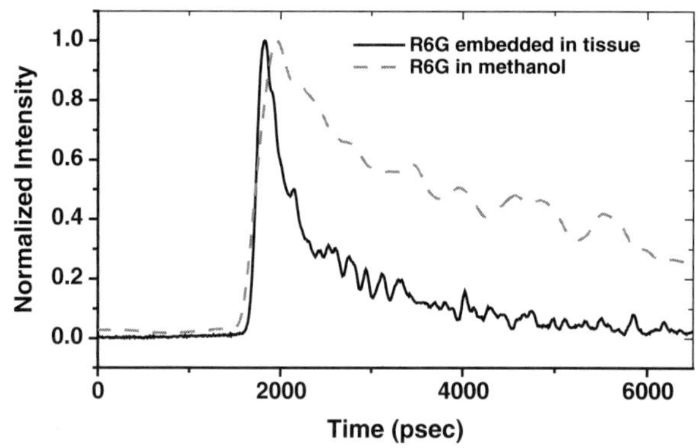

Figure 5.19 Temporal narrowing of Rhodamine 6G fluorescence when embedded inside chicken breast tissue. [G. Zacharakis, unpublished data.]

therapy by customizing the optical characteristics to the specific target. When combined with appropriate photosensitizers such as photofrin and benzophoryn derivatives, the high energy density and the narrow spectral bandwidth of the emission make them ideal for treating skin lesions.[135]

In a preliminary study of random lasing emanating from biological tissue, R6G was embedded in chicken breast tissue and irradiated with energies below and above the lasing threshold to reveal a temporal behavior shown in Fig. 5.19, where the fluorescence of the fluorophore is narrowed by the scattering of tissue. This initial observation opens the way for studying the behavior of the recent smart fluorescent biological markers in highly scattering environments such as biological tissue with the potential of advancing the performance of modern optical imaging systems.[133,134]

Exploiting and manipulating the coherence characteristics of the emission of active phantoms by adjusting excitation energy and scattering properties can open the way for new light sources in specific biomedical applications. When nonresonant feedback is exploited, the emission is uniform, with low coherence and no speckles. This can significantly increase the performance of optical coherence tomography and imaging endoscopy where high-intensity light needs to be transmitted through fibers.[136,137]

Another interesting application of microscopic active phantoms is monitoring the flow of liquids by adding small amounts of the media and detecting the enhanced emission over large flow distances.[138,139]

5.7 Summary

Tissue phantoms are beneficial for testing theoretical predictions; calibrating instruments; and verifying techniques for various methods of optical biomedical diagnostics, including time-, frequency-, and spatial-domain diffuse tissue tomography and spectroscopy, spatially resolved reflectance measurements, hyperspectral imaging, and fluorescence spectroscopy and imaging.

Tissue-like phantoms were also developed for use in noninvasive glucose monitoring, oxygenation monitoring and oxymetry, optoacoustic and pulsed photothermal measurements, Doppler flowmetry, and measurement of polarization degree decay and for calibration of reflective confocal microscopes and optical coherence tomography (OCT) systems and calibration of the Raman signals from different types of tissue. Liquid and porosity tissue phantoms may be used to assess the effectiveness of optical clearing agents.

Phantoms are also beneficial for the research connected with therapeutic implementation of optical radiation (e.g., light dosimetry, laser ablation, and PDT).

In the phantom design for particular biomedical applications, the issues of predictability of phantom optical properties from the characteristics of constituents and phantom composition, component compatibility, simplicity and low cost of fabrication technique, acceptable service life of the phantom, and the safety of preparation and use are to be considered.

Depending on the phantom host medium, two types of phantoms are used—liquid and solid. In the liquid phantom, components are dissolved/suspended in a proper diluent (in most cases, water). Such phantoms are easy to prepare and are used in modeling optically infinite or semi-infinite media. The bio-objects of complex geometry (tissue inhomogeneities, organs) are imitated by solid phantoms. Gelatine, agar, or polyvinyl alcohol gels, fibrin, polyester, epoxy resin, polyurethane, or silicone are successfully used as host material in such phantoms, with polyester, epoxy resin, polyurethane, and silicone giving essentially stable samples that are suitable for routine calibration of optical biomedical systems.

The most common materials used to impart scattering to the phantom are lipid-based emulsions, TiO_2, SiO_2, or aluminum oxide nano/micropowders, as well as polymer microspheres. To achieve appropriate absorption, the wide spectrum of absorbers is used—from common microscopy stains, commercially available dyes (including infrared), and inorganic ions to pigments and absorbing particles. Various biological materials, hemoglobin, erythrocytes, and whole and diluted blood are also used for this purpose. Even the modern products of nanotechnology, quantum dots, and nanoshells have found an application in tissue phantom design.

Of particular interest are the complex multifunctional phantoms, phantoms mimicking the vascular system, and phantoms of organs. So-called dynamic phantoms combine to form a complex structure often presented using the model of the tissue vascular system and the blood-simulating liquid or whole blood with varied oxygenation states pumped through these artificial vessels. Such phantoms can be used for calibrating instruments measuring/imaging blood flow velocity and for mimicking brain activity and other systemic alterations associated with the blood flow and blood oxygenation.

In biomedical optics research, on-phantom simulation of the creation of an active medium within the tissue could help solve the problem of enhancing tumor imaging, especially when the tumor resides superficially or is imaged via endoscopy, taking advantage of the different emission characteristics in healthy and malignant tissue. In therapeutic applications, such media can assist photodynamic therapy by customizing the optical characteristics to the specific target. The high energy density and the narrow spectral bandwidth of the emission make them ideal for treating skin lesions when combined with appropriate photosensitizers.

It is likely that in years to come, we will witness further progress in the development and implementation of new optical biomedical diagnostics methods and techniques, applying the intensive design and development of new tissue-like phantoms with renewed parameters and enhanced performance.

Acknowledgments

ABP and VVT are thankful for support from the Russian Presidential grant NSh-7898.2016.2. VVT is also thankful to the Government of Russian Federation (grant No. 14.Z50.31.0004) and The Tomsk State University Academic D.I. Mendeleev Fund Program.

References

1. A. Pifferi, A. Torricelli, A. Bassi, P. Taroni, R. Cubeddu, H. Wabnitz, D. Grosenick, M. Möller, R. Macdonald, J. Swartling, T. Svensson, S. Andersson-Engels, R. L. P. van Veen, H. J. C. M. Sterenborg, J.-M. Tualle, H. L. Nghiem, S. Avrillier, M. Whelan, and M. Stamm, "Performance assessment of photon migration instruments: the MED-PHOT protocol," *Appl. Opt.* **44**, 2104–2114 (2005).
2. A. E. Cerussi, R. Warren, B. Hill, D. Roblyer, A. Leproux, A. F. Durkin, T. D. O'sullivan, S. Keene, H. Haghany, T. Quang, W. M. Mantulin, and B. J. Tromberg, "Tissue phantoms in multicenter clinical trials for diffuse optical technologies," *Biomed. Opt. Express* **3**), 966–971 (2012).
3. J. R. Mourant, I. J. Bigio, J. Boyer, T. M. Johnson, J. Lacey, A. G. Bohorfoush, and M. Mellow, "Elastic scattering spectroscopy as a diagnostic tool for differentiating pathologies in the gastrointestinal tract: preliminary testing," *J. Biomed. Opt.* **1**, 192–199 (1996).
4. J. R. Mourant, J. Boyer, A. H. Hielscher, and I. J. Bigio, "Influence of the scattering phase function on light transport measurements in turbid media performed with small source-detector separations," *Opt. Lett.* **21**, 546–548 (1996).
5. J. R. Mourant, J. P. Freyer, A. H. Hielscher, A. A. Eick, D. Shen, and T. M. Johnson, "Mechanisms of light scattering from biological cells relevant to noninvasive optical-tissue diagnostics," *Appl. Opt.* **37**, 3586–3593 (1998).

6. S. Flock, M. Patterson, and B. Wilson, "Monte Carlo modeling of light propagation in highly scattering tissues: II. Comparison with measurements in phantoms," *IEEE Trans. Biomed. Eng.* **36**, 1169–1173 (1989).
7. D. Royston, R. Poston, and S. Prahl, "Optical properties of scattering and absorbing materials used in the development of optical phantoms at 1064 nm," *J. Biomed. Opt.* **1**, 110–116 (1996).
8. J. Fu, G. Quan, and H. Gong, "A simple method for prediction of the reduced scattering coefficient in tissue-simulating phantoms," *J. Innov. Opt. Health Sci.* **3**, 53–59 (2010).
9. S. C. Kanick, V. Krishnaswamy, U. A. Gamm, H. J. C. M. Sterenborg, D. J. Robinson, A. Amelink, and B. W. Pogue, "Scattering phase function spectrum makes reflectance spectrum measured from Intralipid phantoms and tissue sensitive to the device detection geometry," *Biomed. Opt. Express* **3**, 1086–1100 (2012).
10. G. Beck, N. Akgun, A. Ruck, and R. Stainer, "Design and characterization of a tissue phantom system for optical diagnostics," *Lasers Med. Sci.* **13**, 160–171 (1998).
11. G. Beck, N. Akgun, A. Ruck, and R. Steiner, "Developing optimized tissue phantom systems for optical biopsies," *Proc. SPIE* **3197**, 76–85 (1997).
12. G. Wagnieres, S. Cheng, M. Zellweger, N. Utke, D. Braichotte, J. Ballini, and H. van den Bergh, "An optical phantom with tissue-like properties in the visible for use in PDT and fluorescence spectroscopy," *Phys. Med. Biol.* **42**, 1415–1426 (1997).
13. S. Chernova, O. Kasimov, L. Kuznetsova, T. Moskalenko, and A. B. Pravdin, "Ex vivo and phantom fluorescence spectra of human cervical tissue," *Proc. SPIE* **4001**, 290–298 (2000).
14. S. P. Chernova and A. B. Pravdin, "Development of the multilayer phantoms for spectrofluorimetric studies," *Asian Journal of Physics* **15**, 119–129 (2006).
15. B. Pogue and G. Burke, "Fiber-optic bundle design for quantitative fluorescence measurement from tissue," *Appl. Opt.* **37**, 7429–7436 (1998).
16. N. Zhadin and R. Alfano, "Correction of the internal absorption effect in fluorescence emission and excitation spectra from absorbing and highly scattering media: theory and experiment," *J. Biomed. Opt.* **3**, 171–186 (1998).
17. C. Gardner, S. Jacques, and A. Walch, "Fluorescence spectroscopy of tissue: recovery of intrinsic fluorescence from measured fluorescence," *Appl. Opt.* **35**, 1780–1792 (1996).
18. R. Crilly, W.-F. Cheong, B. Wilson, and J. Spears, "Forward-adjoint fluorescence model: Monte Carlo integration and experimental validation," *Appl. Opt.* **36**, 6513–6519 (1997).
19. V. Loschenov, E. Luckjanetz, A. Stratonnikov, N. Torshina, and A. Volkova, "The noninvasive evaluation of absolute fluorochrom

concentration on various tissues in vivo by means of standard samples with modeled optical properties," *Proc. SPIE* **2326**, 415–419 (1995).
20. M. Sinaasappel and H. Sterenborg, "Quantification of the hematoporphyrin derivative by fluorescence measurement using dual-wavelength excitation and dual-wavelength detection," *Appl. Opt.* **32**, 541–548 (1993).
21. R. Cubeddu, A. Pifferi, P. Taroni, A. Torricelli, and G. Valentini, "Time-resolved imaging on a realistic tissue phantom: $\mu s'$ and μa images versus time-integrated images," *Appl. Opt.* **35**, 4533–4540 (1996).
22. D. Chursin, V. Shuvalov, and I. Shutov, "Optical tomograph with photon counting and projective reconstruction of the parameters of absorbing 'phantoms' in extended scattering media," *Quantum Electronics* **29**, 921–926 (1999).
23. L. Spinelli, F. Martelli, A. Farina, A. Pifferi, A. Torricelli, R. Cubeddu, and G. Zaccanti, "Calibration of scattering and absorption properties of a liquid diffusive medium at NIR wavelengths. Time-resolved method," *Opt. Express* **15**, 6589–6604 (2007).
24. H. Xu and M. Patterson, "Determination of the optical properties of tissuesimulating phantoms from interstitial frequency domain measurements of relative fluence and phase difference," *Opt. Express* **14**, 6485–6501 (2006).
25. T. T. A. Nguyen, H. N. D. Le, M. Vo, Z. Wang, L. Luu, and J. C. Ramella-Roman, "Three-dimensional phantoms for curvature correction in spatial frequency domain imaging," *Biomed. Opt. Express* **3**, 1200–1214 (2012).
26. Y. Tsuchiya, Y. Ueda, H. Zhang, Y. Yamashita, M. Oda, and T. Urakami, "Analytical expressions for determining the concentrations of absorber in turbid media by time-gating measurements," *OSA TOPS* **21**, 67–72 (1998).
27. M. Kohl, M. Cope, M. Essenpreis, and D. Bocker, "Influence of glucose concentration on light scattering in tissue-simulating phantoms," *Opt. Lett.* **19**, 2170–2172 (1994).
28. M. Kohl, M. Essenpreis, and M. Cope, "The influence of glucose concentration upon the transport of light in tissue-simulating phantoms," *Phys. Med. Biol.* **40**, 1267–1287 (1995).
29. K. Quan, G. Christison, H. MacKenzie, and P. Hodgson, "Glucose determination by a pulsed photoacoustic technique: an experimental study using a gelatin-based tissue phantom," Phys. Med. Biol., 1911–1922 (1993).
30. J. S. Maier, S. A. Walker, S. Fantini, M. S. Franceschini, and E. Gratton, "Possible correlation between blood glucose concentration and the reduced scattering coefficient of tissues in the near infrared," *Opt. Lett.* **19**, 2062–2064 (1994).
31. J. T. Bruulsema, J. E. Hayward, T. J. Farrell, M. S. Patterson, L. Heinemann, M. Berger, J. Koschinsky, T. Sandahal-Christiansen,

H. Orskov, M. Essenpreis, G. Schmelzeisen-Redeker, and D. Böcker, "Correlation between blood glucose concentration in diabetics and noninvasively measured tissue optical scattering coefficient," *Opt. Lett.* **22**, 190–192 (1997).
32. H. A. MacKenzie, H. S. Ashton, S. Spiers, Y. Shen, S. S. Freeborn, J. Hannigan, J. Lindberg, and P. Rae, "Advances in photoacoustic noninvasive glucose testing," *Clin. Chem.* **45**, 1587–1595 (1999).
33. M. Kinnunen and R. Mylylä, "Effect of glucose on photoacoustic signals at the wavelength of 1064 and 532 nm in pig blood and Intralipid," *J. Phys. D: Appl. Phys.* **38**, 2654–2661 (2005).
34. Z. Zhao, "Pulsed Photoacoustic Techniques and Glucose Determination in Human Blood and Tissue," Ph.D. diss., University of Oulu, 2002.
35. R. Myllylä, Z. Zhao, and M. Kinnunen, "Pulsed photoacoustic techniques and glucose determination in human blood and tissue," *Handbook of Optical Sensing of Glucose in Biological Fluids and Tissues*. V. V. Tuchin, Ed., pp. 419–455, CRC Press, Boca Raton, FL (2009).
36. J. Qu and B. C. Wilson, "Monte Carlo modeling studies of the effect of physiological factors and other analytes on the determination of glucose concentration in vivo by near infrared optical absorption and scattering measurements," *J. Biomed. Opt.* **2**, 319–325 (1997).
37. J. J. Burmeister, H. Chung, and M. A. Arnold, "Phantoms for noninvasive blood glucose sensing with near infrared transmission spectroscopy," *Photochem. Photobiol.* **67**, 50–55 (1998).
38. H. G. Kang, M. L. Clarke, S. H. De Paoli Lacerda, A. Karim, L. F. Pease III, and J. Hwang, "Multimodal optical studies of single and clustered colloidal quantum dots for the long-term optical property evaluation of quantum dot-based molecular imaging phantoms," *Biomed. Opt. Express* **3**, 1314–1325 (2012).
39. I. Oda, H. Eda, Y. Ito, Y. Oikawa, Y. Wada, Y. Tsunazawa, M. Oda, K. Ohta, Y. Yamashita, and Y. Tsuchiya, "Reconstructed images for oxy- and deoxyhemoglobin for a phantom with varies local oxygenation using optical CT," *Advances in Optical Imaging and Photon Migration, Technical Digest, Optical Society of America*, 153–155 (1998).
40. Y. Mendelson and J. Kent, "An in vitro tissue model for evaluating the effect of carboxyhemoglobin concentration on pulse oximetry," *IEEE Trans. Biomed. Eng.* **36**, 625–627 (1989).
41. M. Wolf, M. Keel, V. Dietz, K. Siebenthal, H. Bucher, and O. Baenziger, "The influence of a clear layer on near-infrared spectrophotometry measurements using a liquid neonatal head phantom," *Phys. Med. Biol.* **44**, 1743–1753 (1999).
42. R. O. Esenaliev, I. V. Larina, K. V. Larin, D. J. Deyo, M. Motamedi, and D. S. Prough, "Optoacoustic technique for noninvasive monitoring of blood oxygenation: a feasibility study," *Appl. Opt.* **41**, 4722–4731 (2002).

43. R. Esenaliev, A. Karabutov, F. Tittel, B. Fornage, S. Thomsen, C. Stelling, and A. Oraevsky, "Laser optoacoustic imaging for breast cancer diagnostics: limit of detection and comparison with X-ray and ultrasound imaging," *Proc. SPIE* **2979**, 71–82 (1997).
44. A. Oraevsky, S. Jacques, and F. Tittel, "Measurement of tissue optical properties by time-resolved detection of laser-induced transient stress," *Appl. Opt.* **36**, 402–415 (1997).
45. A. Oraevsky, S. Jacques, R. Esenaliev, and F. Tittel, "Pulsed laser ablation of soft tissues, gels and aqueous solutions at temperatures below 1000C," *Lasers Surg. Med.* **18**, 231–240 (1995).
46. M. Kinnunen and R. Myllylä, "Application of optical coherence tomography, pulsed photoacoustic technique, and time-of-flight technique to detect changes in the scattering properties of a tissue-simulating phantom," *J. Biomed. Opt.* **13**, 024005-1–024005-9 (2008).
47. U. Sathyam and S. Prahl, "Limitations in measurement of subsurface temperatures using pulsed photothermal radiometry," *J. Biomed. Opt.* **2**, 251–261 (1997).
48. S. Prahl, I. Vitkin, U. Bruggemann, B. Wilson, and R. Anderson, "Determination of optical properties of turbid media using pulsed photothermal radiometry," *Phys. Med. Biol.* **37** 1203–1217 (1992).
49. T. E. Milner, D. M. Goodman, B. S. Tanenbaum, B. Anvari, and J. S. Nelson, "Noncontact determination of thermal diffusivity in biomaterials using infrared imaging radiometry," *J. Biomed. Opt.* **1**, 92–97 (1996).
50. W. Steenbergen and F. de Mul, "Application of a novel laser Doppler tester including a sustainable tissue phantom," *Proc. SPIE* **3252**, 14–25 (1998).
51. F. de Mul, M. Koelink, M. Kok, and J. Greve, "Photon migration in upscaled tissue models: measurements and Monte Carlo simulations." *Proc. SPIE* **1888**, 129–140 (1993).
52. W. Steenbergen, M. Stratum, F. de Mul, and J. Greve, "Coherence effects in laser Doppler blood flowmetry," *Proc. SPIE* **2982**, 6–17 (1997).
53. S. Chernova, A. Pravdin, and V. Tuchin, "On polarized collimated transmittance of tissue-like phantom," *Proc. SPIE* **2981**, 230–234 (1997).
54. V. Sankaran, J. Walsh, and D. Maitland, "Polarized light propagation in biological tissue and tissue phantoms," *Proc. SPIE* **4001**, 54–62 (2000).
55. A. Kienle, L. Lilge, M. Patterson, R. Hibst, R. Steiner, and B. Wilson, "Spatially resolved absolute diffuse reflectance measurements for noninvasive determination of the optical scattering and absorption coefficients of biological tissue," *Appl. Opt.* **35**, 2304–2314 (1996).
56. A. Kienle, M. S. Patterson, N. Dognitz, R. Bays, G. Wagnieres, and H. van de Bergh, "Noninvasive determination of the optical properties of twolayered turbid medium," *Appl. Opt.* **37**, 779–791 (1998).
57. R. A. J. Groenhuis, J. J. ten Bosch, and H. H. Ferwerda, "Scattering and absorption of turbid materials determined from reflection measurements. 2. Measuring method and calibration," *Appl. Opt.* **22**, 2463–2467 (1983).

58. F. Martelli and G. Zaccanti, "Calibration of scattering and absorption properties of a liquid diffusive medium at NIR wavelengths. CW method," *Opt. Express* **15**, 486–500 (2007).
59. T. Funane, H. Atsumori, M. Kiguchi, Y. Tanikawa, and E. Okadac, "Dynamic phantom with two stage-driven absorbers for mimicking hemoglobin changes in superficial and deep tissues," *J. Biomed. Opt.* **17**, 047001-1–04700-11 (2012).
60. M. L. Clarke, J. Y. Lee, D. V. Samarov, D. W. Allen, M. Litorja, R. Nossal, and J. Hwang, "Designing microarray phantoms for hyperspectral imaging validation," *Biomed. Opt. Express* **3**, 1291–1299 (2012).
61. R. X. Xu, D. W. Allen, J. Huang, S. Gnyawali, J. Melvin, H. Elgharably, G. Gordillo, K. Huang, V. Bergdall, M. Litorja, J. P. Rice, J. Hwang, and C. K. Sen, "Developing digital tissue phantoms for hyperspectral imaging of ischemic wounds," *Biomed. Opt. Express* **3**, 1433–1445 (2012).
62. D. V. Samarov, M. L. Clarke, J. Y. Lee, D. W. Allen, M. Litorja, and J. Hwang, "Algorithm validation using multicolor phantoms," *Biomed. Opt. Express* **3**, 1300–1311 (2012).
63. E. Okada, Y. Tanigawa, Y. Yamada, M. Firbank, and D. Delpy, "Investigation of the direct and indirect signal contributions of brain haematoma in near infrared spectroscopy," *OSA TOPS* **21**, 7–10 (1998).
64. R. Doornbos, R. Lang, M. Aalders, F. Cross, and H. Sterenborg, "The determination of in vivo human tissue optical properties and absolute chromophore concentrations using spatially resolved steady-state diffuse reflectance spectroscopy," *Phys. Med. Biol.* **44**, 967–981 (1999).
65. I. Saidi, S. Jacques, and F. Tittel, "Monitoring neonatal bilirubinemia using an optical patch," *Proc. SPIE* **1201**, 569–578 (1990).
66. A. Kienle, L. Lilge, I. Vitkin, M. Patterson, B. Wilson, R. Hibst, and R. Steiner, "Why do veins appear blue? A new look at an old question," *Appl. Opt.* **35**, 1151–1160 (1996).
67. A. Awata, Y. Kato, and K. Shimizu, "Imaging of absorption distribution in diffuse medium using backscattered light," *OSA TOPS* **21**, 133–137 (1998).
68. A. Kienle, T. Glanzmann, G. Wagnieres, and H. Bergh, "Investigation of two-layered turbid media with time-resolved reflectance," *Appl. Opt.* **37**, 6852–6862 (1998).
69. A. Hielscher, H. Liu, B. Chance, F. Tittel, and S. Jacques, "Time-resolved photon emission from layered turbid media," *Appl. Opt.* **35**, 719–728 (1996).
70. H. Liu, A. Hielscher, B. Beauvoit, L. Wang, S. Jacques, F. Tittel, and B. Chance, "Near infrared spectroscopy of a heterogeneous turbid system containing distributed absorbers," *Proc. SPIE* **2326**, 164–172 (1995).
71. B. Wilson, M. Patterson, and B. Pogue, "Instrumentation for in vivo spectroscopy and imaging," *Proc. SPIE* **1892**, 132–147 (1993).

72. M. Franceschini, S. Fantini, L. Paunescu, J. Maier, and E. Gratton, "Influence of a superficial layer in the quantitative spectroscopic study of strongly scattering media," *Appl. Opt.* **37**, 7447–7458 (1998).
73. S. Schilders, X. Gan, and M. Gu, "Resolution improvement in microscopic imaging through turbid media based on differential polarization gating," *Appl. Opt.* **37**, 4300–4302 (1998).
74. R. Manoharan, K. Shafer, L. Perelman, J. Wu, K. Chen, G. Deinum, M. Fitzmaurice, J. Myles, J. Crowe, R. Dasari, and M. Feld, "Raman spectroscopy and fluorescence photon migration for breast cancer diagnosis and imaging," *Photochem. Photobiol.* **67**, 15–22 (1998).
75. H. Jiang, K. Paulsen, U. Osterberg, and M. Patterson, "Improved continuous light diffusion imaging in single- and multitarget tissue-like phantoms," *Phys. Med. Biol.* **43**, 675–693 (1998).
76. V. V. Tuchin, A. N. Bashkatov, E. A. Genina, V. I. Kochubey, V. V. Lychagov, S. A. Portnov, N. A. Trunina, D. R. Miller, S. Cho, H. Oh, B. Shim, M. Kim, J. Oh, H. Eum, Y. Ku, D. Kim, and Y. Yang, "Finger tissue model and blood perfused skin tissue phantom," *Proc. SPIE* **7898**, 78980Z-1-78980Z-11 (2011).
77. G. Lamouche, B. F. Kennedy, K. M. Kennedy, C.-E. Bisaillon, A. Curatolo, G. Campbell, V. Pazos, and D. D. Sampson, "Review of tissue simulating phantoms with controllable optical, mechanical and structural properties for use in optical coherence tomography," *Biomed. Opt. Express* **3**, 1381–1398 (2012).
78. D. M. de Bruin, R. H. Bremmer, V. M. Kodach, R. de Kinkelder, J. van Marle, T. G. van Leeuwen, and D. J. Faber, "Optical phantoms of varying geometry based on thin building blocks with controlled optical properties," *J. Biomed. Opt.* **15**, 025001-1–025001-10 (2010).
79. R. C. Chang, P. Johnson, C. M. Stafford, and J. Hwang, "Fabrication and characterization of a multilayered optical tissue model with embedded scattering microspheres in polymeric materials," *Biomed. Opt. Express* **3**, 1326–1339.
80. A. Agrawal, M. Connors, A. Beylin, C.-P. Liang, D. Barton, Y. Chen, R. A. Drezek, and T. J. Pfefer, "Characterizing the point spread function of retinal OCT devices with a model eye-based phantom," *Biomed. Opt. Express* **3**, 1116–1126 (2012).
81. A. V. Bykov, A. P. Popov, A. V. Priezzhev, and R. Myllylä, "Multilayer tissue phantoms with embedded capillary system for OCT and DOCT imaging," *Proc. SPIE* **8091**, 80911R-1-80911R-6 (2011).
82. J.-L. H. Demers, S. C. Davis, B. W. Pogue, and M. D. Morris, "Multichannel diffuse optical Raman tomography for bone characterization in vivo: a phantom study," *Biomed. Opt. Express* **3**, 2299–2305 (2012).
83. I. E., Iping Petterson, F. W. Esmonde-White, W. de Wilde, M. D. Morris, and F. Ariese, "Tissue phantoms to compare spatial and

temporal
offset modes of deep Raman spectroscopy," *Analyst* **140**, 2504–2512 (2015).

84. J. A. Viator, B. Choi, G. M. Peavy, S. Kimel, and J. S. Nelson, "Spectra from 2.5-15 μm of tissue phantom materials, optical clearing agents and ex vivo human skin: implications for depth profiling of human skin," *Phys. Med. Biol.* **48**, N15–N24 (2003).
85. X. Wen, V. V. Tuchin, Q. Luo, and D. Zhu, "Controling the scattering of Intralipid by using optical clearing agents," *Phys. Med. Biol.* **54**, 6917–6930 (2009).
86. P. Greguss, "Applied dosimetric measurements," *SPIE* **IS5**, 203–215 (1989).
87. B. Pogue, L. Lilge, M. Patterson, B. Wilson, and T. Hasan, "Absorbed photodynamic dose from pulsed versus continuous wave light examined with tissue-simulating dosimeters," *Appl. Opt.* **36**, 7257–7269 (1997).
88. V. Vinduska, "Exogenous chromophores in Nd:YAG laser selective ablation of the model tissue," *Proc. SPIE* **1878**, 214–223 (1993).
89. T. Farrell, R. Hawkes, M. Patterson, and B. Wilson, "Modeling of photosensitizer fluorescence emission and photobleaching for photodynamic therapy dosimetry," *Appl. Opt.* **37**, 7168–7183 (1998).
90. A. Nilsson, R. Berg, and S. Andersson-Engels, "Measurements of the optical properties of tissue in conjunction with photodynamic therapy," *Appl. Opt.* **34**, 4609–4619 (1995).
91. V. V. Tuchin, *Tissue Optics: Light Scattering Methods and Instruments for Medical Diagnosis*, 3rd ed., SPIE Press, Bellingham, WA (2015).
92. E. Okada, M. Firbank, M. Schweiger, S. R. Arridge, M. Cope, and D. T. Delpy, "Theoretical and experimental investigation of near-infrared light propagation in a model of the adult head," *Appl. Opt.* **36**, 21–31 (1997).
93. R. Simpson, M. Kohl, M. Essenpreis, and M. Cope, "Near-infrared optical properties of ex vivo human skin and subcutaneous tissues measured using the Monte Carlo inversion technique," *Phys. Med. Biol.* **43**, 2465–2478 (1998).
94. J. Laufer, C. R. Simpson, M. Kohl, M. Essenpreis, and M. Cope, "Effect of temperature on the optical properties of ex vivo human dermis and subdermis," *Phys. Med. Biol.* **43**, 2479–2489 (1998).
95. F. P. Bolin, L. E. Preuss, R. C. Taylor, and R. J. Ference, "Refractive index of some mammalian tissues using a fiber optic cladding method," *Appl. Opt.* **28**, 2297–2301 (1989).
96. B. W. Pogue and M. S. Patterson, "Review of tissue simulating phantoms for optical spectroscopy, imaging and dosimetry," *J. Biomed. Opt.* **11**, 041102-1–041102-16 (2006).

97. P. Lemaillet, J.-P. Bouchard, and D. W. Allen, "Development of traceable measurement of the diffuse optical properties of solid reference standards for biomedical optics at National Institute of Standards and Technology," *Applied Optics* **54**, 6118–6127 (2015).
98. R. Bays, G. Wagnières, D. Robert, J. F. Theumann, A. Vitkin, J. F. Savary, P. Monnier, and H. van den Bergh, "Three-dimensional optical phantom and its application in photodynamic therapy," *Lasers Surg. Med.* **21**, 227–234 (1997).
99. R. Graaff, J. G. Aarnoudse, J. R. Zijp, P. M. A. Sloot, F. F. M. de Mul, J. Greve, and M. H. Koelink, "Reduced light scattering properties for mixtures of spherical particles: a simple approximation derived from Mie calculations," *Appl. Opt.* **31**, 1370–1376 (1992).
100. O. Barajas, A. Ballangrud, G. Miller, R. Moore, and J. Tulip, "Monte Carlo modelling of angular radiance in tissue phantoms and human prostate: PDT light dosimetry," *Phys. Med. Biol.* **42**, 1675–1687 (1997).
101. T. Nakai, G. Nishimura, K. Yamamoto, and M. Tamura, "Expression of optical diffusion coefficient in high-absorption turbid media," *Phys. Med. Biol.* **42**, 2541–2549 (1997).
102. S. Jacques and S. Flock, "Time-resolved reflectance spectroscopy," *Proc. SPIE* **1525**, 35–40 (1991).
103. M. Makropoulou, E. Drakaki, G. Stamatakos, and A. Serafetinides, "Quantitative estimation of absorbing chromophores in tissue simulators based on laser induced spectroscopy and scattering measurements," *Proc. SPIE* **4162**, 76–58 (2000).
104. J. E. Choukeife and J. P. L'Huillier, "Measurements of scattering effects within tissue-like media at two wavelength of 632.8 and 680 nm," *Lasers Med. Sci.* **14**, 286–296 (1999).
105. I. Driver, J. W. Feather, P. R. King, and J. B. Dawson, "The optical properties of aqueous suspensions of Intralipid, a fat emulsion," *Phys. Med. Biol.* **34**, 1927–1930 (1989).
106. B. B. Das, L. Liu, and R. R. Alfano, "Time-resolved fluorescence and photon migration studies in biomedical and random media," *Rep. Prog. Phys.* **60**, 227–292 (1997).
107. H. C. van de Hulst, *Multiple Light Scattering Tables, Formulas, and Applications*, Academic, New York (1980).
108. J. Linford, S. Shalev, J. Bews, R. Brown, and H. Schipper, "Development of a tissue-equivalent phantom for diaphanography," *Med. Phys.* **13**, 869–875 (1986).
109. H. Wabnitz, R. Willenbrock, J. Neukammer, U. Sukowski, and H. Rinneberg, "Spatial resolution in photon diffusion imaging from measurements of time-resolved transmittance," *Proc. SPIE* **1888**, 48–61 (1993).

110. I. Yaroslavsky, A. Yaroslavsky, T. Goldbach, and H.-J. Schwarzmaier, "Inverse hybrid technique for determining the optical properties of turbid media from integrating-sphere measurements," *Appl. Opt.* **35**, 6797–6809 (1996).
111. S. Jacques, M. Ostermeyer, L. Wang, and D. Stephens, "Polarized light transmission through skin using video reflectometry: toward optical tomography of superficical tissue layers," *Proc. SPIE* **2671**, 199–210 (1996).
112. P. Kaplan, M. Kao, A. Yodh, and D. Pine, "Geometric constraints for the design of diffusing-wave spectroscopy experiments," *Appl. Opt.* **32**, 3828–3836 (1993).
113. S. Willmann, A. Terenji, I. V. Yaroslavsky, T. Kahn, P. Hering, and H.-J. Schwarzmaier, "Determination of the optical properties of a human brain tumor using a new microspectrophotometric technique," *Proc. SPIE* **3598**, 233–239 (1999).
114. S. J. Madsen, M. S. Patterson, and B. C. Wilson, "The use of India ink as an optical absorber in tissue-simulating phantoms," *Phys. Med. Biol.* **37**, 985–993 (1992).
115. A. B. Parthasarathy, W. G. Shin, X. J. Zhang, and A. K. Dunn, "Laser speckle contrast imaging of flow in a microfluidic device," *Proc. SPIE* **6446**, 644604 (2007).
116. L. Luu, P. A. Roman, S. A. Mathews, and J. C. Ramella-Roman, "Microfluidics based phantoms of superficial vascular network," *Biomed. Opt. Express* **3**, 1350–1364 (2012).
117. M. Wolf, M. Keel, V. Dietz, K. Siebenthal, H. Bucher, and O. Baenziger, "The influence of a clear layer on near-infrared spectrophotometry measurements using a liquid neonatal head phantom," *Phys. Med. Biol.* **44**, 1743–1753 (1999).
118. C. Chen, F. Klämpfl, C. Knipfer, M. Riemann, R. Kanawade, F. Stelzle, and M. Schmidt, "Preparation of a skin equivalent phantom with interior micron-scale vessel structures for optical imaging experiments," *Biomed. Opt. Express* **5**, 3140–3149 (2011).
119. T. E. Milner, D. M. Goodman, B. S. Tanenbaum, and J. S. Nelson, "Depth profiling of laser-heated chromophores in biological tissues by pulsed photothermal radiometry," *J. Opt. Soc. Am. A* **12**, 1479–1488 (1995).
120. N. M. Lawandy, R. M. Balachandran, A. S. L. Gomes, and E. Sauvian, "Laser action in strongly scattering media," *Nature* **368**, 436–438 (1994).
121. V. S. Letokhov, "Generation of light by a scattering medium with negative resonance absorption," *Sov. Phys. JEPT* **26**, 835–840 (1968).
122. D. Anglos, A. Stassinopoulos, R. N. Das, G. Zacharakis, M. Psylaki, R. Jakubiak, R. A. Vaia, E. P. Giannelis, and S. H. Anastasiadis, "Random laser action in organic-inorganic nanocomposites," *J. Opt. Soc. Am. B* **21**, 208–213 (2004).

123. H. Cao, "Review on latest developments in random lasers with coherent feedback," *J. Phys. A: Math. Gen.* **38**, 10497–10535 (2005).
124. K. L. van der Molen, A. P. Mosk, and A. Lagendijk, "Quantitative analysis of several random lasers," *Optics Communications* **278**, 110–113 (2007).
125. D. S. Wiersma, "The physics and applications of random lasers," *Nature Physics* **4**, 359–367 (2008).
126. S. John and G. Pang, "Theory of lasing in a multiple-scattering medium," *Phys. Rev. A* **54**, 3642–3652 (1996).
127. R. M. Balachandran, N. M. Lawandy, and J. A. Moon, "Theory of laser action in scattering gain media," *Opt. Lett.* **22**, 319–321 (1997).
128. D. S. Wiersma and A. Lagendijk, "Light diffusion with gain and random lasers," *Phys. Rev. E* **54**, 4256–4265 (1996).
129. L. Florescu and S. John, "Photon Statistics and Coherence in Light Emission from a Random Laser," *Phys. Rev. Lett.* **93**, 013602-1–013602-4 (2004).
130. G. Zacharakis, G. Heliotis, G. Filippidis, D. Anglos, and T. G. Papazoglou, "Investigation of the laserlike behavior of polymeric scattering gain media under subpicosecond laser excitation," *Appl. Opt.* **38**, 6087–6092 (1999).
131. G. Zacharakis, N. Papadogiannis, G. Filippidis, and T. G. Papazoglou, "Photon statistics of the laser like emission from polymeric scattering gain media" Opt. Lett. **25**(12), 923–925 (2000).
132. G. Zacharakis, N. A. Papadogiannis, and T. G. Papazoglou, "Random lasing following two-photon excitation of highly scattering gain media," *Appl. Phys. Lett.* **81**(14), 2511–2513 (2002).
133. R. C. Polson and Z. V. Vardeny, "Random lasing in human tissues," *Appl. Phys. Lett.* **85**, 1289–1291 (2004).
134. M. Siddique, L. Yang, Q. Z. Wang, and R. R. Alfano, "Mirrorless action from optically pumped dye-treated animal tissue," *Opt. Commun.* **117**, 475–479 (1995).
135. R. M. Balachandran, D. P. Pacheco, and N. M Lawandy, "Laser action in polymeric gain media containing scattering particles," *Appl. Opt.* **35**, 640–643 (1996).
136. F. Auzel and P. Goldner, "Coherent light sources with powder: stimulated amplification versus super-radiance, *J. Alloys Compounds* **300**, 11–17 (2000).
137. B. Redding, M. A. Choma, and H. Cao, "Speckle-free laser imaging using random laser illumination" *Nat. Photonics* **6**, 355–359 (2012).
138. D. S. Wiersma, "The smallest random laser," *Nature* **406**, 132–133 (2000).
139. H. Cao, J. Y. Xu, D. Z. Zhang, and R. P. Chang, "Microlaser made of disordered media," *Appl. Phys. Lett.* **76**, 2997–2999 (2000).

Alexander B. Pravdin graduated from the Chemistry Department of Moscow State University in 1979. After graduation for five years, he worked at the institutes of USSR Academy of Sciences. From 1984 to 1987, he was a PhD student at the Chemistry Department of MSU from which he received a candidate of sciences (PhD) degree in physical chemistry in 1990. From 1988, he works at the Optics Department (later—Optics and Biophotonics Department) of Saratov State University, Saratov, Russia, where he is currently an associated professor. He lectures to students specializing in the field of physics of living systems and biological and medical photonics and guides some of them during their diploma project. He supervises PhD students; under his supervision/cosupervision two Candidate of Sciences (in biophysics) theses have been defended. His research interests are focused on biophotonics and biomedical optics including autofluorescence of highly scattering biological objects, fluorescence of globular protein/dye systems, photobiological action of UV-A and visible radiation, polarization properties of tissues and scattering systems, and alterations in tissue optical properties during optical clearing. He has published more than 50 articles in peer-reviewed journals and peer-reviewed proceedings and more than 30 papers in conference proceedings. He is a coauthor of four educational books. He was also a coeditor of a book of international conference proceedings. He is a laureate of the award (for 2010) of International Academic Publishing Company "Nauka" for the best publication in its journals.

George Filippidis received his BSc degree from the Physics Department of the University of Crete in 1995 and his doctorate degree from the Medical School of the same University in 2000. He is a senior research scientist at the Institute of Electronic Structure and Laser—Foundation for Research and Technology (IESL-FORTH) Crete, Greece, in-charge of the nonlinear microscopy group. He performed his postdoctoral studies at IESL-FORTH. His research group activities include the *in vivo* elucidation of molecular mechanisms and biological processes by employing various spectroscopic techniques with emphasis on advanced microscopic modalities, optical characterization of tissue, nanosurgery, and the implementation of nonlinear image contrast modalities (MPEF, SHG, THG) as novel diagnostic tools for cultural heritage studies. He has published 48 articles in international refereed journals (h index 17), 26 full-length articles in edited and refereed proceedings volumes, 17 conference proceedings and 2 book chapters. He was cosupervisor in one PhD, four masters and seven diploma theses from the Physics Department of the

University of Crete. He is coauthor in one patent. He was the lead scientist in two large European and one Greek research grants (1.3 M€).

Giannis Zacharakis received his BSc in physics in 1997 and his PhD in 2002 from the University of Crete. He is currently a researcher at the Institute of Electronic Structure and Laser, Foundation for Research and Technology-Hellas (FORTH) where he is leading the *In Vivo* Imaging Lab. He is an elected member of the Council of the European Society for Molecular Imaging. He is also a visiting lecturer at the Department of Biology and the scientific coordinator of the European Molecular Imaging Doctorate School Network for the University of Crete. During 2003 and 2004, he has been a research fellow in radiology at the Harvard University, Boston. Between 2005 and 2009, he has been a research associate at the FORTH-IESL and a research scientist since then. His main fields of interest are biophotonics, biomedical, and optical imaging with the aim of designing and realizing of key enabling technologies and tools for end users in biology and in clinical practice and advancing our knowledge on fundamental light—biological–tissue interactions. In particular, he is interested in the noninvasive visualization of specific molecular targets and pathways by exploiting light-tissue interactions, e.g., the fluorescence signal emitted by contrast agents attached to cells or molecules. He has 32 publications in international peer-reviewed journals, 39 articles in peer-reviewed conference proceedings, over 1000 citations, 1 chapter, and 1 patent, Patent No: WO2005089637. He has more than 50 presentation in international conferences, where he has been awarded three first prizes and has been invited more than 20 times in international conferences, schools, and meetings.

Theodore G. Papazoglou graduated with honors from the Physics Department of the University of Crete in 1985 and followed his postgraduate studies at the University of Southern California (PhD in biomedical engineering in 1989). After his postdoctoral training (1989–1990, Laser Research Laboratory, Cedars Sinai Medical Center, Los Angeles, USA), he worked as independent researcher at the Institute of Electronic Structure and Lasers of the Foundation for Research and Technology-Hellas (FORTH-IESL). He lectured in parallel at the University of Crete (teaching undergraduate laser and medical physics courses, supervising six doctoral, four master's students in their respective theses and guiding several undergraduate students during their diploma project). In 1994, he was elected at a tenure-track position at FORTH-IESL (assistant researcher) and in 1999, he attained his tenure (associate researcher). During the same period, he served as technical manager of the UV Laser Facility at FORTH (activity funded and maintained on a competitive basis by the European Commission as a Large Scale research facility). The outcome of his research work in that period has been published in some 70 articles in international peer-reviewed journals (410 citations by others in 365 citing articles in early 2015), 2 patents and 3 chapters in books. He has also coedited four books of proceedings of international conferences (SPIE, OSA, etc.) and had over 60 presentations in international conferences, including 13 invited scientific talks. His research work attracted ca. €1.5M in competitive research grants over a period of 11 years (1992–2003).

Since 2004, he is working at the European Commission and in 2009, he joined the European Research Council. His coauthorship of this chapter is based on the work performed at FORTH-IESL.

Valery V. Tuchin is a professor and chairman of Optics and Biophotonics at Saratov National Research State University. He is also the head of laboratory at the Institute of Precision Mechanics and Control, RAS, and the supervisor of Interdisciplinary Laboratory of Biophotonics at National Research Tomsk State University. His research interests include biophotonics, tissue optics, laser medicine, tissue optical clearing, and nanobiophotonics. He is a member of SPIE, OSA, and IEEE, guest professor of HUST (Wuhan) and Tianjin Universities of China, and adjunct professor of the Limerick University (Ireland) and National University of Ireland (Galway). He is a fellow of SPIE and OSA, and has been awarded the Honored Science Worker of the Russia, SPIE Educator Award, FiDiPro (Finland), Chime Bell Prize of Hubei Province (China), and Joseph W. Goodman Book Writing Award (OSA/SPIE).

Part II: Tissue Near-Infrared Spectroscopy and Imaging

Probing biological tissue with near-infrared light can provide a wealth of physiological and functional information related, for example, to hemodynamics and oxygenation parameters. Spatially resolved studies afford the intriguing opportunity to investigate the spatial organization of the interrogated tissue and perform functional imaging. These optical imaging capabilities are complementary to existing medical imaging modalities such as radiography, computed tomography (CT), magnetic resonance imaging (MRI), and positron emission tomography (PET). Optical techniques are used extensively in biomedical research. Over the past 30 years, advances in instrumentation, theoretical modeling, and data processing techniques have led to the development of novel exciting approaches to spectroscopy and imaging of living tissue. The broad objective of these developments is the capability of reconstructing the spatial distribution of local optical characteristics (such as, e.g., absorption coefficient) within tissue *in vivo*. These characteristics are then correlated to biophysically meaningful parameters such as chromophore concentrations and tissue morphology. This information, in turn, is used to make clinically relevant observations. A number of significant milestones toward this objective have been achieved. Nevertheless, a widespread application of optical spectroscopy and imaging of human tissue into general clinical practice is just beginning to happen. A good indication of the expending mainstream presence is ever-growing list of optical imaging procedures covered by CPT (current procedure terminology) codes. The main complication in the optical study of tissue stems from the fact that near-infrared light in tissue is much more strongly scattered than absorbed. In fact, most tissues in the human body are *optically turbid*, thus leading to a much more complex (in comparison to traditional spectroscopy of clear samples) data interpretation. Not only does scattering dominate absorption, but, to make matters worse, most tissues are optically thick, resulting in an attenuation to the optical signal of about one order of magnitude per centimeter. Therefore, reliable detection of the optical signal and subsequent extraction of useful information from that signal becomes a formidable task indeed.

The use of time-resolved techniques is one of the ways to deal with the problem of light scattering in tissue. In this approach, the emission of the light source is not constant, as in continuous-wave (CW) methods, but is time dependent. One inherent quality of this approach is its ability to discriminate among photons having traveled different distances within a turbid medium such as tissue. This is a very attractive feature that leads to measurements of both absorption and scattering properties of tissues, and to more effective spatially resolved measurements. Two faces of this approach are termed *time domain* and *frequency domain*. Chapter 6 describes the principles of the time-domain approach, while Chapter 7 is dedicated to the frequency domain.

CW techniques, while usually unable to achieve absolute readings, can be very effective in measuring temporal changes in the tissue optical properties. This is demonstrated in Chapter 8, which reports CW applications of near-infrared functional imaging of the brain. The noninvasive monitoring of brain activity is among the most fascinating and promising applications of near-infrared tissue imaging. Since it usually investigates an evoked cerebral response, it can be effectively performed by monitoring changes in the optical signals used to probe the brain cortex. Chapter 9 analyzes in detail how to perform absolute measurements or quantified trend measurements using various approaches to near-infrared spectroscopy. In particular, the chapter discusses the measurements of a number of hemodynamic and oxygenation parameters by near-infrared methods. These parameters measured with optical methods can be effectively complemented by the information content of data and images generated by other techniques such as electro-encephalography (EEG), magneto-encephalography (MEG), and functional magnetic resonance imaging (fMRI). Such multi-modality approaches in brain research are presented in Chapter 10.

The pulsed optoacoustic method is another elegant approach to deal with the optical turbidity of tissues. This approach is based on the generation of acoustic waves inside the tissue as a result of the localized absorption of pulsed laser radiation. The general idea behind this method is to join the strengths of optical and acoustic methods by taking advantage of the functional sensitivity featured by near-infrared light and the relatively high spatial resolution afforded by ultrasound imaging. Optoacoustics and the method based on measuring profiles of laser-induced pressure (LIP) are presented in Chapter 11.

In summary, this section combines an introduction to the near-infrared techniques for the optical study of tissues with a snapshot of current applications and developments in this dynamic and exciting field. We trust that these chapters will be of interest to the interdisciplinary community of physicists, engineers, and medical doctors looking for innovative approaches to noninvasive medical diagnostics.

Sergio Fantini
Ilya V. Yaroslavsky
Co-editors

Color Plate Section

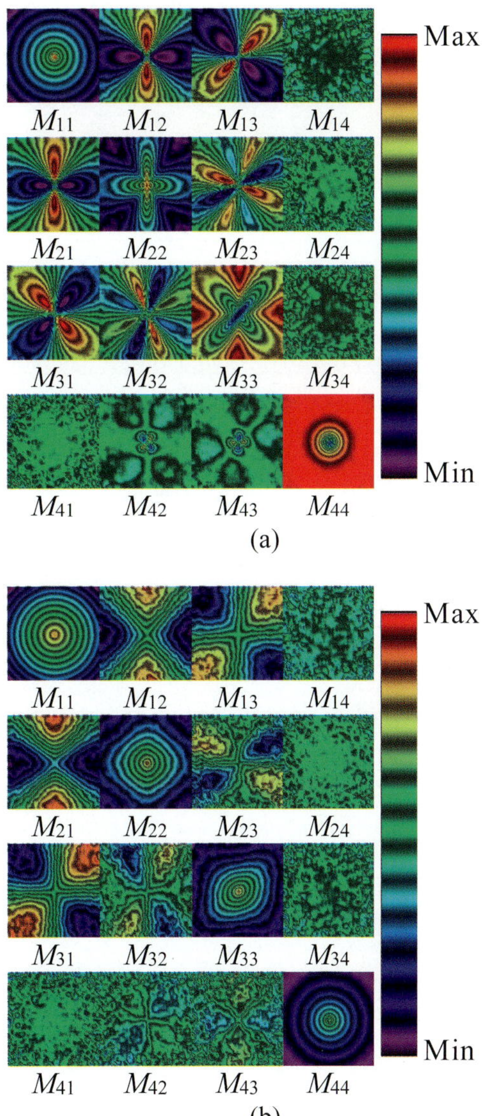

Figure 1.22 (a) Reflection and (b) transmission Mueller matrices of a slab of turbid medium with a scattering coefficient of 4 cm^{-1} and a scatterer radius of 0.102 mm.[294] The calculated Mueller-matrix elements are normalized to the M11 element to compensate for the radial decay of intensity. Each of the images is displayed with its own color map to enhance the image contrast. The size of each image is 4×4 cm^2.

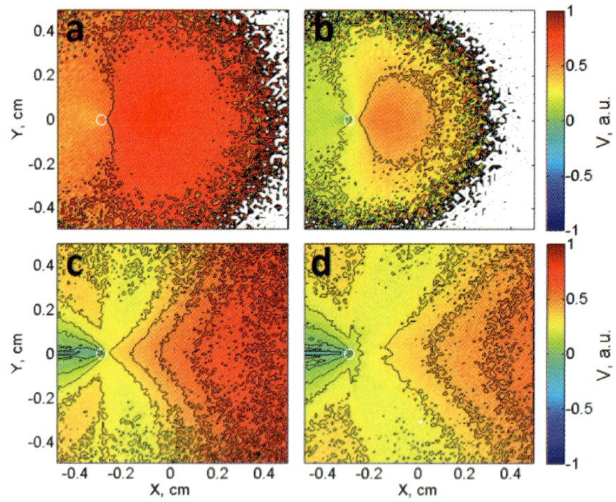

Figure 1.40 Spatial distribution of the *V* component of the Stokes vector on the surface of scattering medium, obtained for circularly polarized light [1,0,0,1] of a wavelength of 639 nm incident to the surface of the medium at point (0,0) with the angle of 45 deg. The optical properties of scattering media are, respectively, presented in Table 1.1.

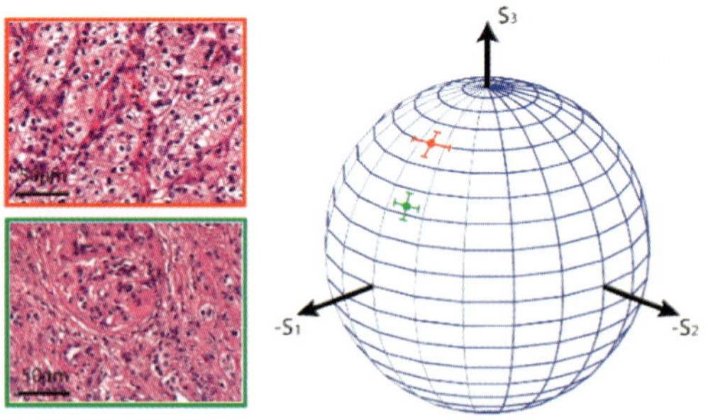

Figure 1.42 The SOP of the right-handed elliptically polarized light scattered from healthy (green) and cancerous (red) sites of kidney tissue.438

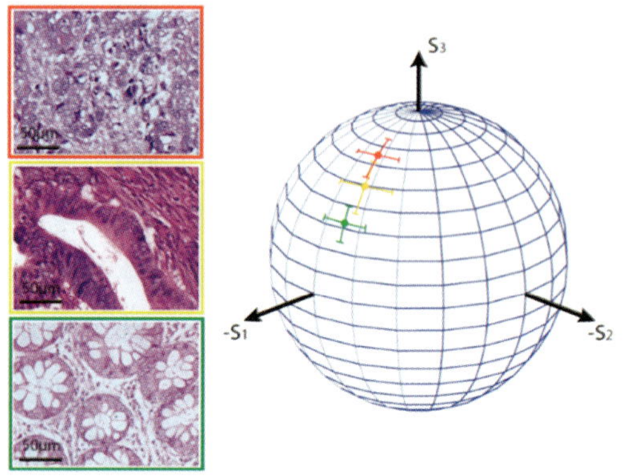

Figure 1.43 Position of the Stokes vector on the Poincaré sphere (right) correlates with the successive grades of colorectal cancer (normal colon, green; grade 2, yellow; grade 3, red) confirmed by microscopy (left three panels: normal colon, bottom; grade 2, middle; grade 3, top).

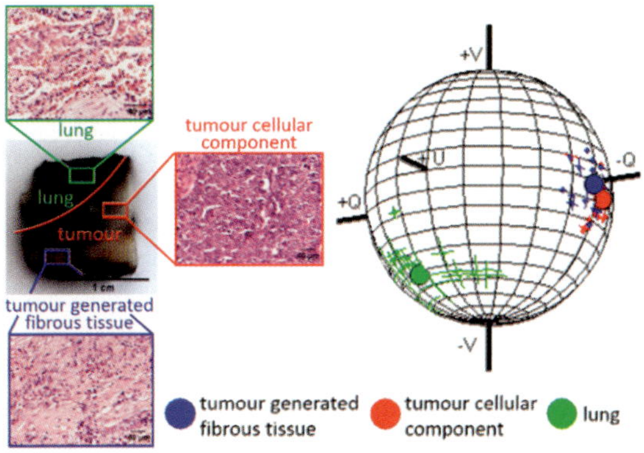

Figure 1.44 Lung tissue embedded in paraffin wax (left) shows the boundary between healthy lung and tumor tissue (red line) marked based on histological examination. Microscopy images of hematoxylin and eosin stained 5-mm tissue sections are shown in the colored squares, and their position on the block is marked by the corresponding squares. The Poincaré sphere (right) shows the mean Stokes vector of REP light back-scattered from the region of healthy lung tissue (green), tumor-generated fibrous tissue (blue) and the tumor cellular component (red); dots with error bars represent a single measurement, and the larger circles represent the mean SOP. d = 1.5 mm, and θ = 30 deg.

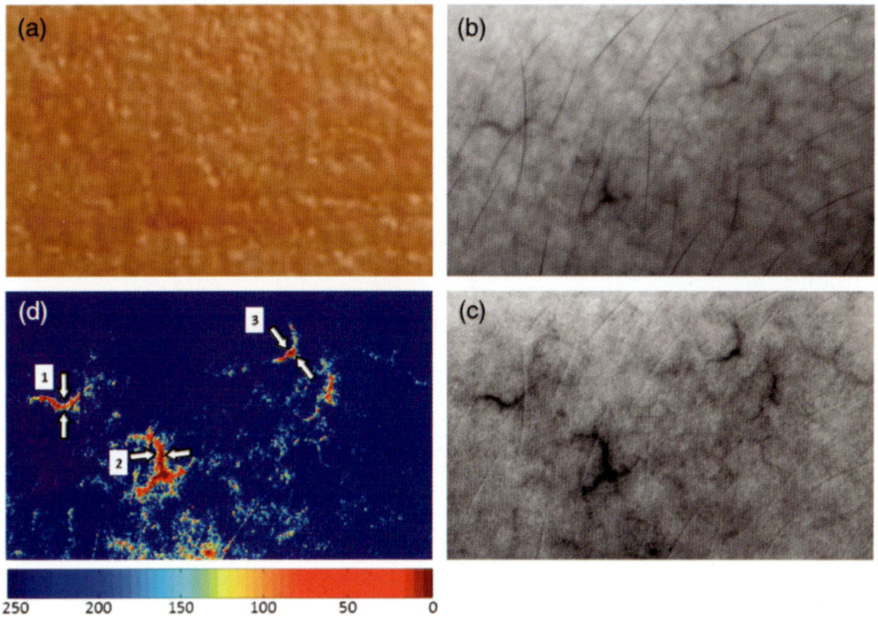

Figure 2.21 *In vivo* images of human skin. Field of view: 8.0 × 4.9 mm. (a) Digital picture; (b) cross-polarized image acquired at 570 nm emphasizes blood vessels and hair; (c) cross-polarized difference (570–470 nm) image emphasizes blood vessels; (d) cross-polarized difference (570–470 nm) image in false color enables determination of blood vessel sizes. For example, (1) is a 78-μm blood vessel, (2) is a 137-mm blood vessel, and (3) is a 98-mm blood vessel.

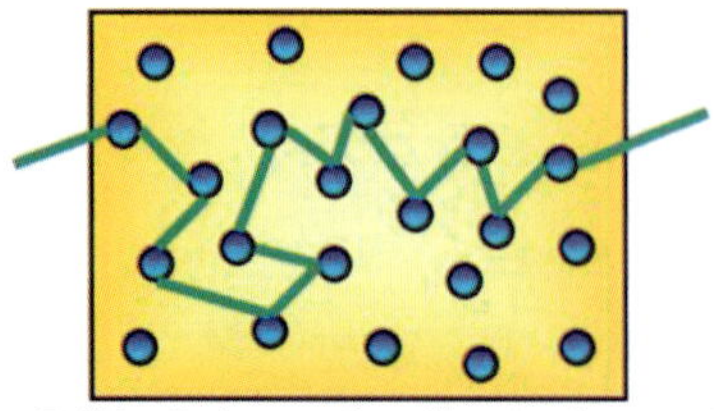

Figure 5.15 Schematic illustration of the amplification of fluorescence emission as photons experience multiple scattering inside an active medium. [G. Zacharakis, unpublished data].

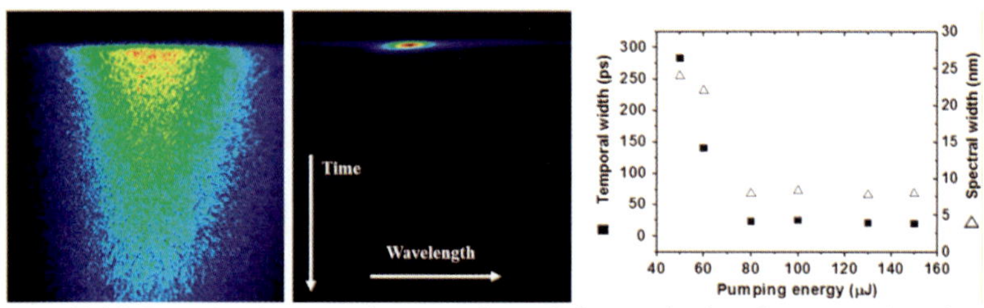

Figure 5.16 Spectro-temporal narrowing of the emission from random lasing materials when excited with energy above threshold. Reproduced with permission from G. Zacharakis, N. A. Papadogiannis and T. G. Papazoglou, “Random lasing following twophoton excitation of highly scattering gain media,” *Appl. Phys. Lett.* **81**(14), 2511–2513 (2002); http://dx.doi.org/10.1063/1.1511284 Copyright 2002, AIP Publishing LLC.

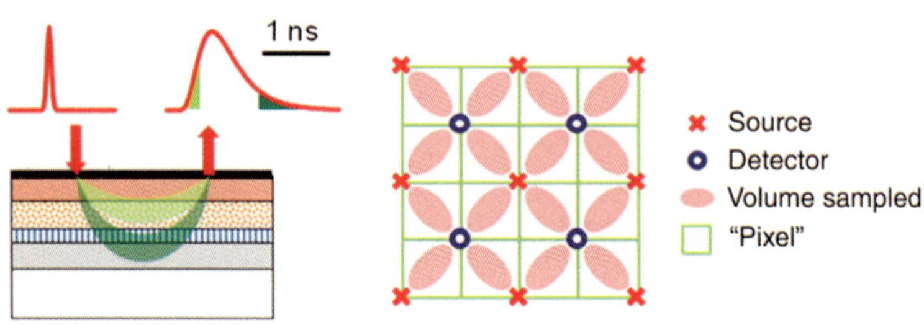

Figure 6.9 Depth selectivity (left) and lateral resolution (right) in time-domain optical brain imaging.

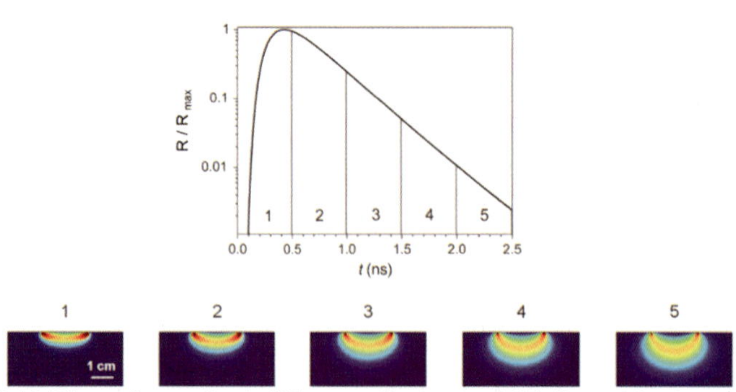

Figure 6.10 Time-dependent diffuse reflectance $R(t)$ (logarithmic scale) (top) and sensitivity profiles for photon counts in five consecutive time windows, integrated over y (bottom). Calculations of sensitivities were based on a diffusion model for perturbation by a point-like absorber according to Ref. 53. Source at $(x, y, z) = (0, 0, -1\text{ mm})$, detector at (20 mm, 0, 0), $\mu_s' = 1\text{ mm}^{-1}$, $\mu_a = 0.01\text{ mm}^{-1}$, $n = 1.4$.

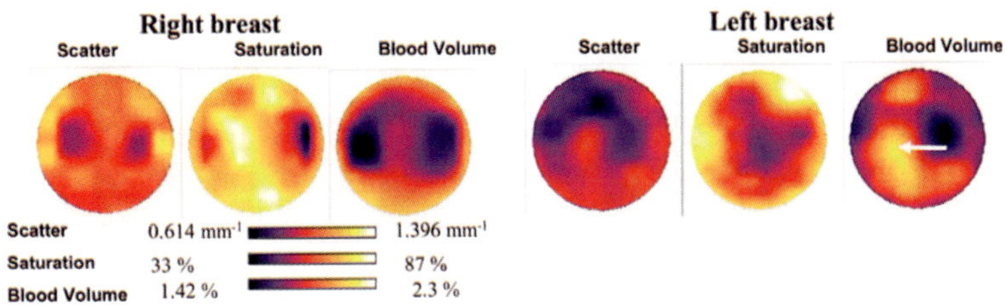

Figure 6.28 Tomographic images of reduced scattering coefficients, oxygen saturation, and blood volume from a 45-year-old woman with a carcinoma in the left breast (indicated by the arrow). From Ref. 104.

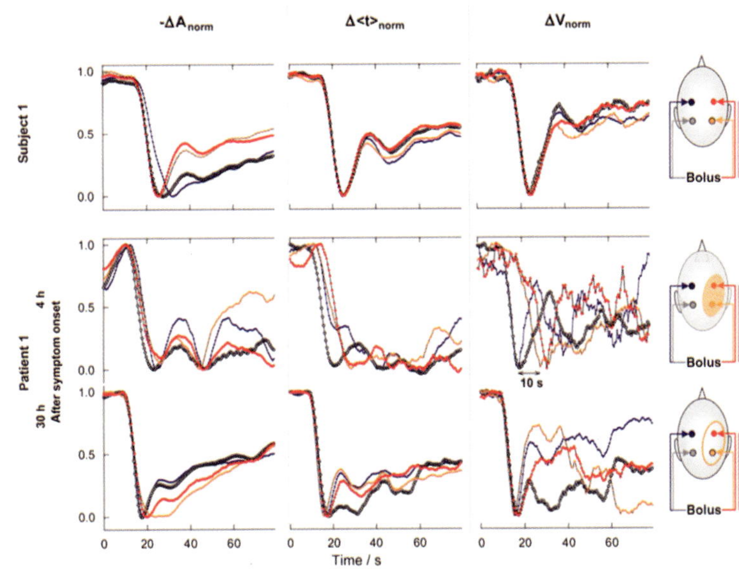

Figure 6.30 Changes in attenuation $-\Delta A$, mean time of flight $\Delta\langle t\rangle$, and variance ΔV recorded at two sites on each hemisphere after administration of an ICG bolus (intravenous injection at time t = 0), for a healthy subject and a patient. All signals were rescaled to a range from 0 to 1. Reprinted from Ref. 155 with permission from Elsevier (see color plates).

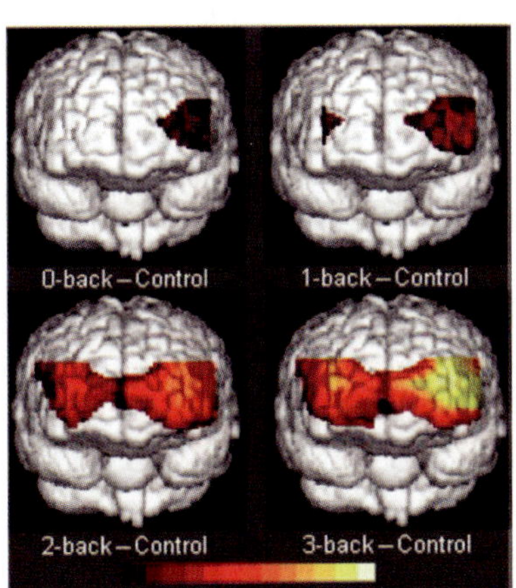

Figure 8.9 The significant activation distribution in the prefrontal lobe averaged across all 23 subjects in 0-back to approximately 3-back memory tasks.

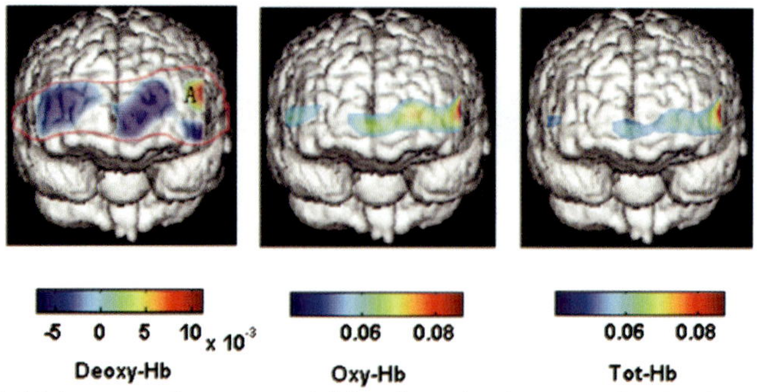

Figure 8.17 Image of averaged blood activation amplitude for an ocular attention task, including changes in the concentrations of deoxyhemoglobin and oxy-hemoglobin, as well as the blood volume.

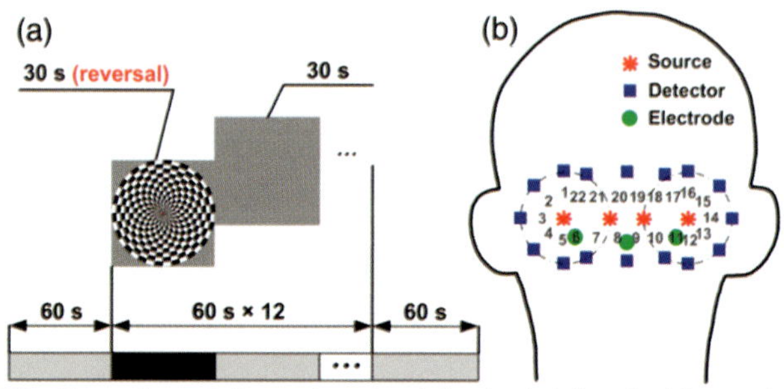

Figure 8.31 Experimental protocol. (a) A total of 12 cycles of 30-s checkerboard reversal and 30-s blank screen periods were presented between 60-s baselines. (b) Configuration of the optical probe. The probe comprised 4 sources (red asterisks) and 16 detectors (blue squares), yielding a total of 22 channels (gray numbers). The source–detector distance was 3 cm. From left to right, the green circles represent electrodes O_1, O_z, and O_2, respectively, and electrodes O_1 and O_2 were placed at the midpoints of channels 6 and 11, respectively.

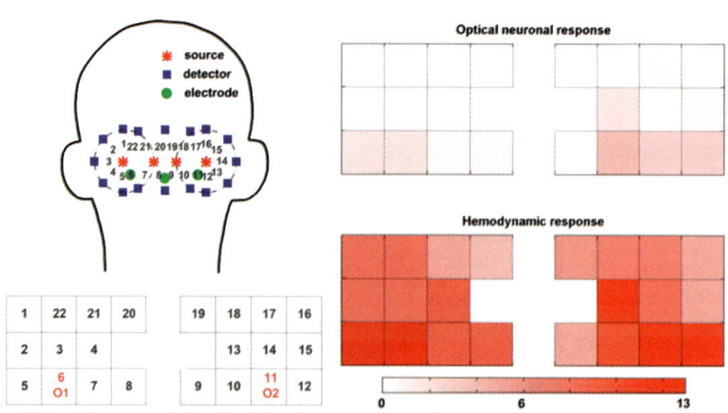

Figure 8.32 The spatial distribution maps of the optical neuronal signal (top right) and the hemodynamic response (bottom right). The top left graph shows the configuration of the optical probe on the head. The bottom left graph shows the distribution of the channels corresponding to the spatial distribution maps; the numbers marked in red indicate that the channel corresponded to the electrode positions of O_1 (channel 6) and O_2 (channel 11). The color indicates the number of the subjects who showed a significant response in the corresponding channel.

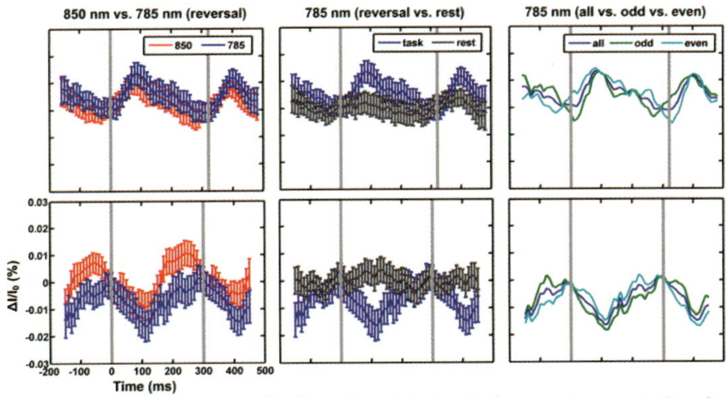

Figure 8.33 Optical neuronal signals obtained from channel 6 of subject number 3 (the top row) and channel 10 of subject number 5 (the bottom row). The error bars represent the standard errors.[78]

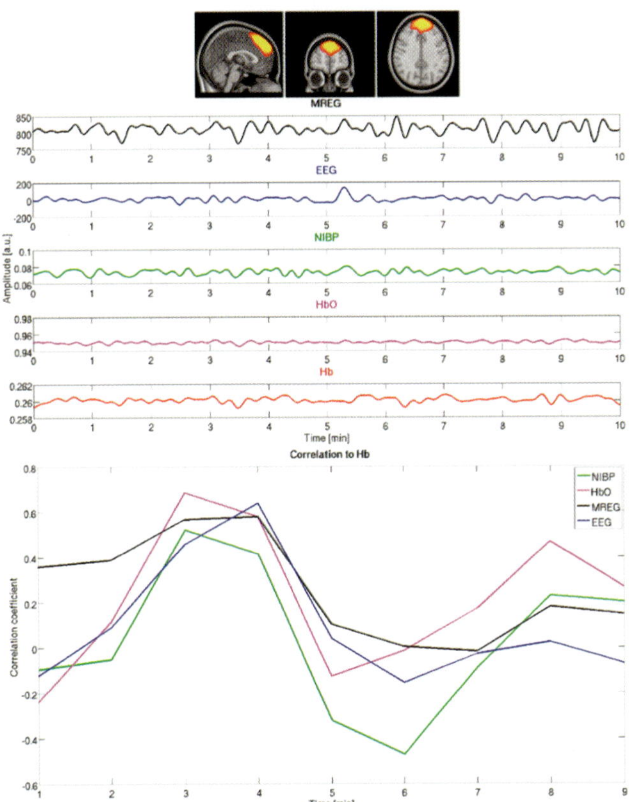

Figure 10.11 Synchronous human multimodal hepta-scan data measured during 3T MRI scanning. (top) Hot spots (bright colors) in a functional brain network belonging to the upmost anterior part of the default mode brain network (DMN) in MNI space. (middle) Time-domain activity of fMRI MREG data of the DMN in black sampled in 10 Hz. As seen below the MREG graph, the best-fit independent component is EEG, followed by NIBP, HbO2, and HHb (here marked as Hb) measured over the DMN in the forehead using fNIRS. (bottom) Correlation of the HbO2 signal to all other signals in sliding windows over 9 min, showing synchronous modulation of the measured signal correlation and indicating the strong additive power of multimodal imaging.

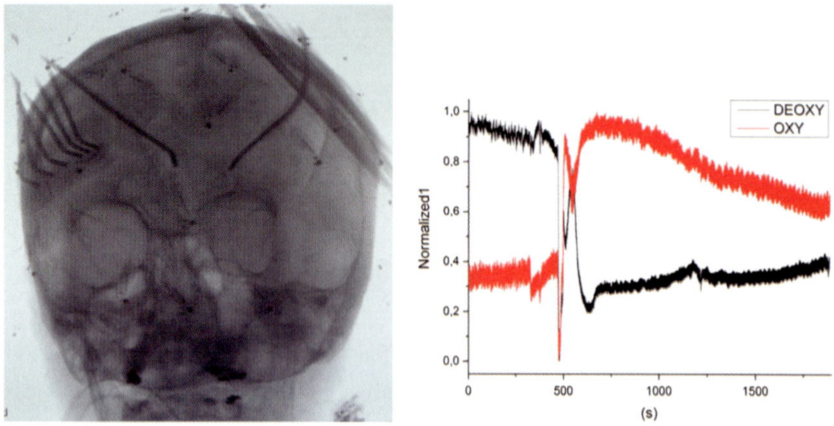

Figure 10.12 (left) A plain x-ray image from a digital subtraction angiography laboratory illustrating sensor positions during BBBD treatment. Clearly visible in black are optical fibers and EEG electrodes attached to the head, illustrating the positions at which the sensors are measuring. (right) The NIRS signal shows a robust change in amplitude after infusion of mannitol (time point of approximately 500 s) to disrupt the BBB.

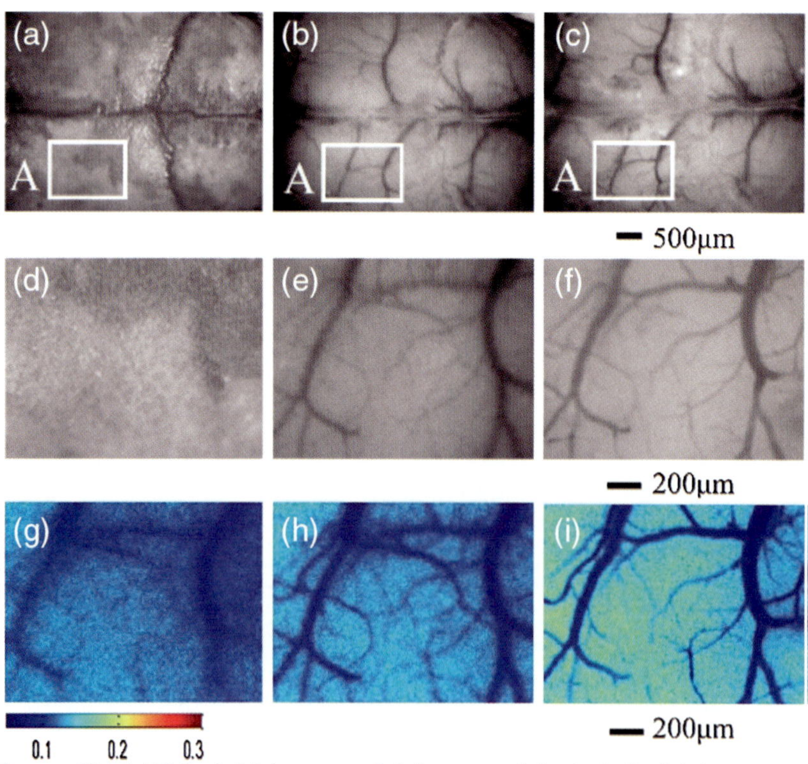

Figure 10.22 White-light images of (a) mouse intact skull, (b) transparent skull after SOCS treatment for 25 min, and (c) with removed rectangle area A. The corresponding magnified white images (d–f) and speckle contrast images (g–i) are within the rectangle area A shown in (a–c).

Chapter 6
Time-Resolved Imaging in Diffusive Media

Heidrun Wabnitz
Physikalisch-Technische Bundesanstalt (PTB), Berlin, Germany

Juan Rodriguez
St. Louis College of Pharmacy, St. Louis, USA

Ilya Yaroslavsky
IPG Photonics, Inc., Marlborough, USA

Anna Yaroslavsky
University of Massachusetts, Lowell; Massachusetts General Hospital, Boston, MA, USA

Valery V. Tuchin
Saratov National Research State University, Saratov, Russia

National Research Tomsk State University, Tomsk, Russia

Institute of Precision Mechanics and Control, Russian Academy of Sciences, Saratov, Russia

6.1 Introduction

6.1.1 Looking through turbid tissues with conventional imaging techniques

The medical use of light to probe deep tissues dates back to the 19th century, beginning with Bright's 1831 report on hydrocephalus.[1] Bright noted that sunlight or light from a candle was able to shine through the head of a patient who suffered from hydrocephalus, a disorder characterized by excess cerebrospinal fluid in and around the brain. The ability of light to transilluminate tissues was later noted by Curling in 1843 in conjunction with problems of the scrotum,[2] and by Cutler in 1929 in relation to breast lesions.[3]

The year 1977 represented another pivotal point for the field of optical biomedical diagnostics. It was then that Frans Jöbsis demonstrated the deep transillumination of mammalian tissues using near-infrared (NIR) light, including the chest of a dog and an adult human head from temple to temple.[4] Furthermore, by combining the penetrating properties of NIR light with spectroscopy, he demonstrated the diagnostic value of NIR light for the assessment of hemoglobin oxygen saturation and the cytochrome a-a3 redox state.

Given the cellular metabolic information afforded by NIR spectroscopy, coupled with its deeply penetrating and noninvasive properties and the intrinsic low cost of optical technology, it is apparent that this new biomedical tool holds great promise for diagnosing a variety of pathologies.

One fundamental limitation of the use of NIRS for diagnostic purposes is the problem of localizing the origin of the signals. One can easily demonstrate the extent of this limitation with simple experiments, which most of us have done at one time or another. For example, in a darkened room, take a bright penlight or a red laser pointer and place it under your thumb. As you activate the light source, you will readily observe your fingernail and the surrounding tissue glowing with deep red light, confirming the ability of red/NIR radiation to transilluminate thick mammalian tissue. A typical pattern of light emerging from the finger is shown in Fig. 6.1. What is notably missing from this light pattern is imaging information about the internal structure of the thumb. There is no evidence, for example, that the bone in the finger is casting a shadow. To the unaided eye, the inside of the thumb appears to be filled with a homogeneous medium.

The failure to image internal structure in such a simple experiment results from the turbid nature of biological tissues. Light propagating through such

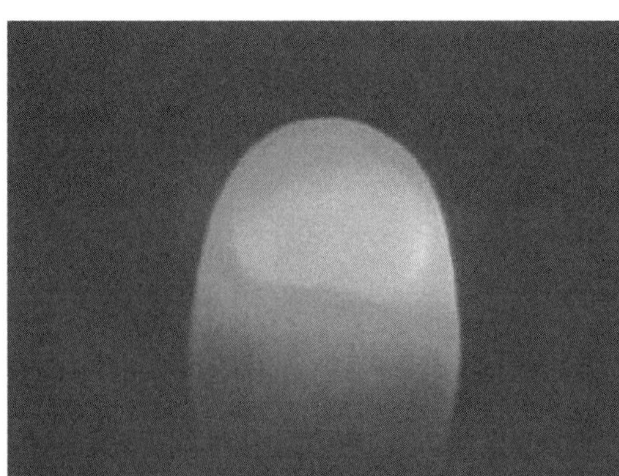

Figure 6.1 A simple transillumination experiment performed with a penlight placed behind a finger. The image clearly shows that the transilluminated light is diffused to the point where no internal structure from within the finger is noticeable.

media is strongly scattered by cellular constituents such as cell organelles and membranes. We now know that light scatters after propagating only a few tens of microns inside tissues, and that after approximately a millimeter of penetration the scattered light assumes a completely random direction. Furthermore, as was demonstrated by Frans Jöbsis, bone tissue is permeable to red/NIR light. When both of these effects are combined, it is not surprising that the internal structure will never be seen in thick tissues with conventional imaging techniques.

6.1.2 Sharpening images in diffusive media: the early history of the time-resolved method

The difficulty with conventional optical imaging through living systems resides in the lateral migration of photons caused by tissue scattering. Little can be done to minimize this scattering process without significantly affecting tissue structure and function. A more desirable approach is to examine carefully the properties of the photons that emerge from the tissue, in hopes that some of their characteristics, such as time of flight, will provide additional information that aids in the localization of internal structures.

In 1971, Duguay and Mattick first demonstrated an image enhancement technique for scattering media with potential applicability for biological tissues.[5] In their paper—which actually became famous for the first-ever photograph of a laser pulse in mid-flight—the authors described a way to see an object that is covered by a piece of tissue paper. Their idea was to illuminate the object through the tissue paper by using a picosecond pulse of light and to record the reflected light with their newly developed ultrafast camera. Since the tissue paper and the object were located in two different planes, the reflected light had to contain two components—namely, that of the tissue and to a lesser amount that of the object. More importantly, these reflected components had to arrive at the camera at slightly different times. If the camera shutter opened only during the interval of time (time gating) when the reflection from the object arrived, its image could be isolated from the stronger diffused reflection from the tissue that arrived earlier. This time-gating scheme is illustrated in Fig. 6.2.

The success of the Duguay and Mattick's technique is demonstrated in Fig. 6.3. The upper-left panel shows a photograph of the uncovered object. The upper-right panel shows the same view after the tissue paper is placed in front of the object. The bottom panels shows two photographs obtained with the ultrafast camera, which was gated to the precise time when the light scattered from the object arrives at the camera.

The work of Duguay and Mattick represented a leap forward in imaging technology—one that the authors noted could also enable imaging of biological structure obscured by scattering tissues. The proposed imaging technique was an optical equivalent of radar or sonar, and hence became known as optical ranging. Nearly a decade later, Martin et al. demonstrated the use of the same technique to perform optical ranging in the interior of a

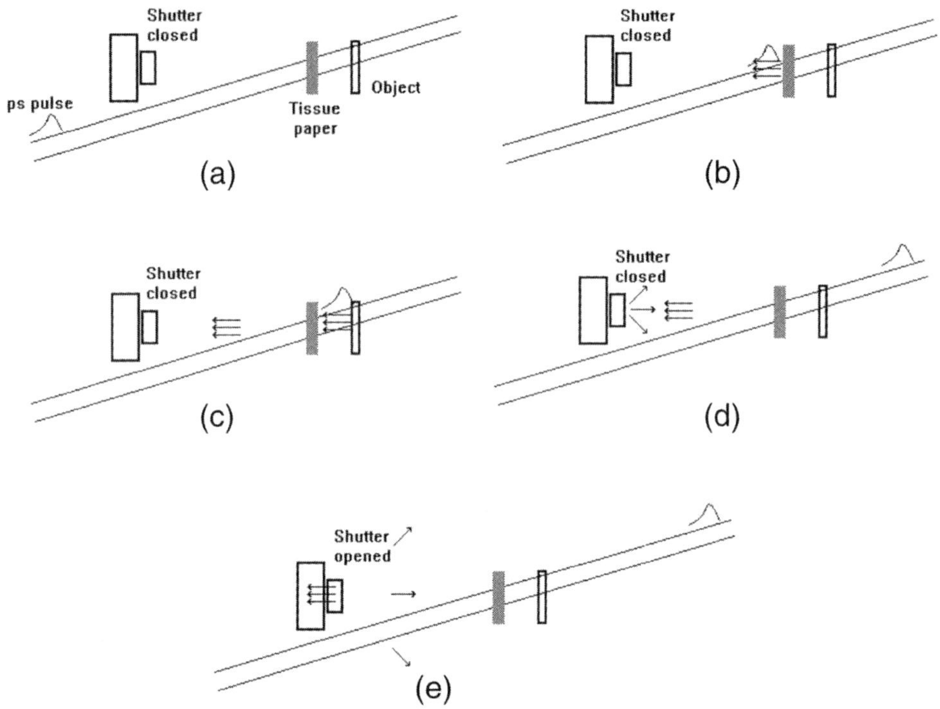

Figure 6.2 Time-gating scheme proposed by Duguay and Mattick for seeing through diffusive media.[5] (a) A pulse of light lasting a few picoseconds is directed at the sample. (b) Light from the pulse is scattered from the tissue in the direction of the camera. (c) Some light passes through the tissue; a few picoseconds later, it scatters from the object in the direction of the camera. (d) The light scattered from the tissue reaches the camera but is rejected by the shutter that is still closed at this time. (e) Just before the light scattered from the object reaches the camera, the shutter opens, thereby capturing only the image emanating from the object.

saline-perfused calf heart, with the aid of an optical fiber inserted through the heart wall.[6] Later, Fujimoto et al. applied optical ranging, using femtosecond pulses, to measure the thickness of the cornea of a rabbit eye and of skin structures (e.g., stratum corneum, epidermis) *in vivo*.[7]

Although optical ranging potentially affords micron-scale resolution of hidden structures, it is limited to depths of a millimeter or so due to its inability to account for multiply scattered light—a phenomenon that dominates the signals obtained from thicker biological tissues. This limitation is discussed in greater detail in Section 6.2.3. Thus, despite the early work of Duguay and Mattick, Martin et al., and Fujimoto et al., the use of time-resolved techniques for imaging through diffusive media did not flourish until the early 1990s. This delay in development was due in part to the absence of a basic understanding of how photons propagate in diffusive media, including analytical models and experimental results. Such ground-breaking work eventually occurred in the 1980s.[8–12]

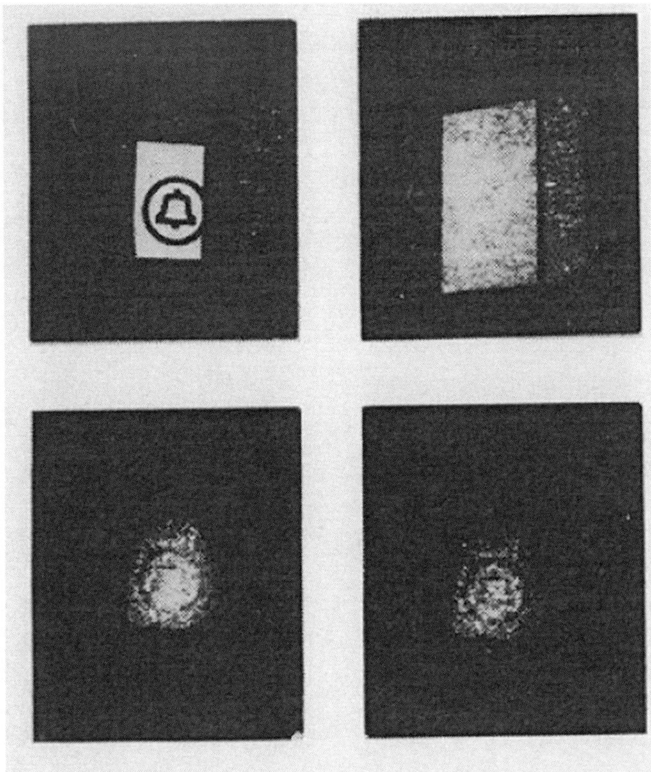

Figure 6.3 The shape of the object shown in the upper-left panel, which is hidden by the tissue shown in the upper-right panel, is successfully recovered by time-gated imaging, as shown in the lower panels. These results were reproduced from Ref. 5.

An example of the experimental work conducted during that era is shown in Fig. 6.4. The curves shown in the figure were taken from the work of Chance et al.[12] Curves such as these provide a foundation for our discussion of time-resolved imaging. The curve labeled "Excitation" shows the temporal profile of a light pulse propagating ballistically between the source and a detector located 4 cm away. In the presence of an intervening biological medium, photons experience a series of random collisions that cause them to follow different paths and to emerge from the medium over a wide range of times. The curve labeled "Response" shows the temporal profiles obtained when photons diffuse through 4 cm of human muscle. The response curve shows clearly that the emerging photons are widely spread in time. The curve also reveals that a very small fraction of the photons propagates in a time comparable to those photons in the excitation curve. These photons must propagate in a quasi-ballistic trajectory between the emitter and the detector. Other time points in the response curve contain more complicated trajectories through the tissue.

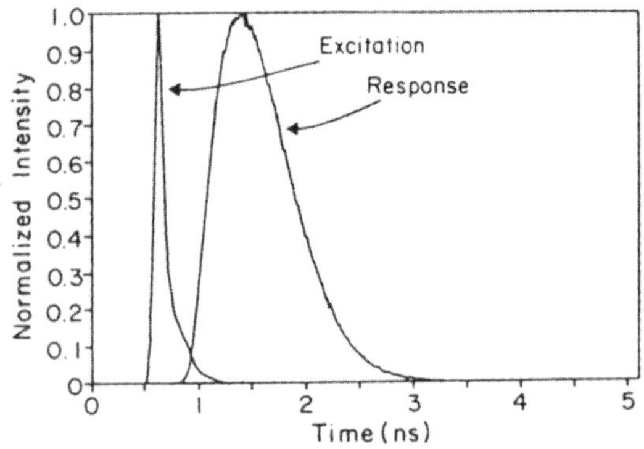

Figure 6.4 Many of the advances made in time-resolved imaging through diffusive media were founded on the ground-breaking work of Refs. 8–12. The graphs reprinted from the work of Chance et al. demonstrate the temporal effects observed when a light pulse propagates through thick biological tissues.[12]

On the theoretical side, major steps were taken in modeling the propagation of photons in biological tissue. These include the Monte Carlo method as well as the analytical equations that describe photon propagation in random media. The Monte Carlo method is a computer simulation method in which photons "play" roulette every time they encounter a scattering site in a way that simulates their random scattering through tissues. Such a technique was implemented initially for the study of light absorption and flux in biological tissues, which were quantities of interest for photodynamic therapy.[8] The development of analytical equations includes the adaptation of the "transport equation" to biological media, along with the development of its approximate form known as the "diffusion equation." The transport equation describes the photon traffic moving in and out of each volume element of a medium as a function of time, taking into account the directionality of the photons. If the volume elements are large enough, then the photons can be multiply scattered and lose all sense of directionality. In this limit, the volume element can be assumed to scatter photons isotropically. Under this assumption, the net flow of photons from one volume element to another will depend mainly on the difference in the number of photons in those volume elements. This is the diffusion limit. An estimate of the photon density in the sample can be obtained using the diffusion approximation. The diffusion approximation for light transport is mathematically simple and has been shown to be a reasonably good approximation in thick human tissue. This equation is usually expressed in the form

$$\frac{1}{c}\frac{\partial\Phi(r,t)}{\partial t} - D\nabla^2\Phi(r,t) + \mu_a\Phi(r,t) = S(r,t), \tag{6.1}$$

where

$$D = [3(\mu_a + \mu_s')]^{-1} \tag{6.2}$$

is the diffusion coefficient in the sample, μ_s' is the reduced scattering coefficient of the medium, μ_a is the medium's absorption coefficient, c is the speed of light at the location r, $\Phi(r, t)$ is the number of photons per unit volume at time t, and $S(r, t)$ is the photon source at location r and time t.[9] Equation (6.1) is valid for media satisfying the condition $\mu_a \ll \mu_s'$, and far from boundaries and sources. It should be noted that there are more recent discussions to prefer $D = 1/(3\mu_s')$ to Eq. (6.2) (see Sections 3.7 and 3.8 of Ref. 13).

A clear advantage of the diffusion approximation is its ability to yield analytic solutions in simple cases. Of particular interest to this chapter is a solution of the diffusion equation that gives the time-resolved transmittance for a homogeneous diffusing slab of thickness d, which is given in Ref. 9 as

$$\begin{aligned} T(\rho,d,t) = {} & (4\pi Dc)^{-3/2} t^{-5/2} e^{-\mu_a ct} e^{-\rho^2/(4Dct)} \\ & \times \sum_{n=1}^{\infty} \{[(2n-1)d - z_0] e^{-[(2n-1)d - z_0]^2/(4Dct)} \\ & - [(2n-1)d + z_0] e^{-[(2n-1)d + z_0]^2/(4Dct)}\}, \end{aligned} \tag{6.3}$$

where n is a positive integer, and

$$z_0 = 1/\mu_s'. \tag{6.4}$$

A comprehensive summary of the theory of light propagation in diffusive media, solutions, and software implementations can be found in Martelli et al.[13]

The aim of this chapter is to review the advances in time-resolved diffuse optical imaging that have occurred in the last 25 years and their applicability to clinical diagnostics. We begin with a discussion in Section 6.2 of general considerations that affect the spatial resolution of time-resolved methods when applied to biological tissues. This important issue directly impacts the clinical applicability of this technology. In Section 6.3, we review the technology currently available for the production and detection of fast optical signals, which affects issues of performance versus cost trade-offs. In Section 6.4, we describe some of the approaches developed over the years and the technical challenges they have addressed. These approaches have led to clinical applications that are discussed in Section 6.5. Finally, in Section 6.6, we conclude with some general remarks regarding the successes of time-resolved imaging as well as its prospects for the future.

6.2 General Concepts in Time-Resolved Imaging through Highly Diffusive Media

6.2.1 Transmittance methods

6.2.1.1 Time-gated shadowgraphs

The first theoretical analysis of time-resolved imaging through diffusive media was done by Maarek et al.[14] Their earlier work relied on the projection of collimated light through thin tissue samples to produce shadows of the internal structure.[15] The resulting images looked like x-ray pictures—that is, they were two-dimensional projections sometimes known as shadowgraphs. To extend this technique to thick, multiple-scattering tissues, time-gated selection of unscattered photons was proposed. Maarek et al. used the Monte Carlo method to simulate the transmission of continuous and pulsed light through rectangular slabs of tissue, measuring 4 and 8 cm in thickness.[14] The tissue slabs contained a cylindrical inclusion located in the mid-plane and possessing optical properties similar to blood. In these simulations the detector was placed directly across the sample from the emitter, and the sample was scanned between them. As the inclusion crossed the path between the emitter and the detector, a drop in transmittance was recorded. The depth of the shadow cast by the inclusion was assessed with a contrast function.

The authors found that the contrast profiles obtained with continuous light clearly revealed the location of the inclusion. The results also showed that the contrast was greatly enhanced when pulsed light and time gating were used. Thus, time-resolved transillumination appeared to be a technique capable of producing a significant enhancement in imaging inside thick biological media.

The early work of Maarek et al. was greatly extended with additional simulations by Hebden and Kruger.[16] Their work focused on the lateral migration of photons that propagate from an emitter to a detector located across a slab. This lateral spread is fundamentally what limits the spatial resolution for imaging structures embedded inside diffusive media by the method of transillumination. The authors considered several factors affecting the lateral spread, including depth, sample thickness, and distance traveled by photons. Their results confirmed that the spread increases with depth, reaching a maximum near the center of the sample. They also found that the spread increases linearly with increasing sample thickness. The use of time-of-flight information was explored as well, and it was found that early-arriving photons at the detector stray less from their path from the emitter to the detector than do late-arriving photons. A significant enhancement in spatial resolution was therefore seen when early-arriving photons were used for imaging. At the same time, the authors' results showed that this gain in resolution came at the expense of the rejection of most of the photons, therefore resulting in very weak signals. This trade-off led the authors to suggest that arbitrarily high resolutions may not be practical with the aid of time-resolved methods.

Hebden later examined this question in greater detail with experimental comparisons.[17] He concluded that spatial resolutions of about half a centimeter might be ultimately possible with temporal gates of 10 picoseconds (ps). From time-resolved measurements of the edge spread function for transillumination of turbid slabs of thickness d, Wabnitz and Rinneberg concluded that the spatial resolution can be improved to 0.3 d at best when taking into account the first 1% of the transmitted time-of-flight distribution.[18]

Gandjbakhche et al. analyzed the transverse spread of photons in the mid-plane of the sample, based on random walk theory on a lattice.[19] In this model, photons propagate randomly in a cubic, discrete lattice of sites separated by a distance equivalent to the mean-free path between scattering sites in a diffusive medium. One advantage of this model is its ability to quantify the probability that a photon will be found at some position in the sample after a given number of steps. Using these probability distributions, the authors derived an analytical expression for the probability that a photon, detected at a given time delay Δt following the ballistic photons, would cross the mid-plane of the sample at a given distance from the emitter–detector axis. The derived expressions, although complex in appearance, yielded simpler Gaussian-like distributions, particularly for short time delays less than the mean transit time of photons. The full width at half maximum (FWHM) of the distributions was found to fit quite well to the expression

$$\mathrm{FWHM} = 0.94\left(\frac{c\Delta t}{\mu_s'}\right)^{1/2}, \tag{6.5}$$

where c is the speed of light in the medium and μ_s' is the reduced scattering coefficient. For a more accurate theoretical result that includes anisotropic scattering, see Ref. 20. Therefore, for time gates on the order of a picosecond, and in tissues for which μ_s' is approximately 1 mm^{-1} and $c = 2.2 \times 10^8$ m/s, the expected time-resolved imaging resolution predicted by Eq. (6.5) is 1.5 mm. Such a resolution approaches that achieved by x-rays, but it is obtained at the expense of the amount of light collected. Under these circumstances, according to the estimates given by Gandjbakhche et al., the time-gated signal would experience an attenuation exceeding 20 orders of magnitude for a medium like breast tissue 51 mm in thickness and for near-infrared radiation (which mammalian tissue absorbs to the least extent).[19] Such attenuation levels are several orders of magnitude higher than is technically possible to measure even under the most idealized conditions. A balance should be reached between the desired spatial resolution and the measurability of the transillumination signal. In this instance it could mean relaxing the resolution requirements to 10 mm [corresponding to a 50-ps gate time according to Eq. (6.5)], for which the attenuation levels fall into a range of 10^{-12} according to the estimates published by Gandjbakhche et al.[19]

Hebden et al. approached the problem of extreme signal attenuation observed in early-gate shadowgraphs by estimating the number of early-arriving photons through extrapolation from the temporal spread of the late-arriving photons.[21,22] This technique, which is referred to as *temporal extrapolation*, was achieved by first curve-fitting the temporal profile of the transilluminated signals with an analytical solution that is applicable to the transmission through a homogeneous slab. Once the fitted curves were obtained, the profile was integrated over the desired early time window to obtain an estimate of the desired time-gated signal. The investigators found that temporal extrapolation is effective at estimating the amplitude of early time gates down to 200 ps. This gate duration enhanced the resolution of the images by a factor of approximately 2 over the resolution obtained with conventional transillumination imaging.

Gandjbakhche et al. also made an important observation concerning the ability of early-gate signals to resolve absorptive and scattering inclusions.[19] Their findings suggested that the contrast for absorptive inclusions was relatively small, given that the absorption coefficients for mammalian tissues are very small in the near-infrared spectrum. This finding can be understood by referring to general considerations of propagation of light in diffusive media. The early-gate technique rests on the selection of less diffuse photon paths to enhance the imaging contrast. Photons following such paths exhibit a transmittance T that approaches the one described by the Beer–Lambert law

$$T = e^{-(\mu_s+\mu_a)l} \tag{6.6}$$

where μ_a and μ_s are the absorption and scattering coefficients, respectively, and l is the thickness of the medium. The coefficient μ_a is typically three orders of magnitude smaller than μ_s in the near-infrared range. Therefore, even large changes in μ_a will produce relatively small changes in the attenuation compared to changes in μ_s. Early-gate shadowgraphs should therefore emphasize scattering contrast profiles.

6.2.1.2 Diffuse transmittance imaging

While early work focused on improving spatial resolution in shadowgraph-like images, this aspect of the technology did not turn out to have a major impact on clinical optical breast imaging. Instead, in this area, the focus changed to making full use of the information content of the whole temporal profile of transmitted photons. The applicability of the Beer–Lambert law in Eq. (6.6) is strictly true only for the earliest ballistic (or unscattered) photons. If the time gates are set to capture late-arriving photons, which are the ones that are multiply scattered in the medium, then the effective path length l of the photons can increase dramatically. These photons, therefore, have an increased likelihood for absorption by the tissue. Consequently, the contrast profiles obtained with late gates should be more sensitive to absorption changes than those realized with early gates.[23] In fact, the complementary information provided by early and late gates can be utilized to obtain information about an inclusion inside a diffusive medium.[24] This possibility was vividly demonstrated experimentally by Cubeddu et al.[25,26]

The optical properties μ_s' and μ_a of the bulk medium (healthy tissue) can be quantified by a nonlinear fit based on the solution of the diffusion equation for a homogeneous slab. The same approach can be utilized to obtain qualitative images of scattering and absorbing inhomogeneities. Cubeddu et al. found that the scattering maps produced sharp images of the scattering inclusions alone.[25,26] However, maps of the absorption coefficients obtained from the fits revealed broader features that did not satisfactorily decouple the absorption from the scattering components. The authors observed that a better way to produce absorption-sensitive maps was to display transmittance in a time gate placed on the trailing edge of the signal.

The quantitative characterization of an inclusion is much more complex. The inverse problem is ill posed, and the transillumination image does not allow the size, position in depth, and absorption and scattering coefficients to be retrieved simultaneously. If, however, prior information is available regarding the size, shape, and location of the inclusion, its optical properties can be reconstructed employing an appropriate forward model. This approach shares some similarities with the analysis related to tomographic methods discussed in the next section.

6.2.2 Time-resolved optical tomography

6.2.2.1 The back-projection technique

The development of imaging modalities such as computed tomography (CT) and magnetic resonance imaging (MRI) has revolutionized the way medicine probes the body without undertaking invasive surgery. With the discovery of the diagnostic value and the deep tissue penetration of NIR light, it eventually became apparent that a novel imaging modality was possible. This new form of tomography lent itself to localizing and quantifying key functional parameters, including hemoglobin, myoglobin, and cytochromes. Furthermore, the optical technology needed to realize this new concept in medical imaging was relatively inexpensive and portable, in contrast to other imaging modalities.

One of the first optical tomography systems that utilized the time-resolved method to image the interior of diffusive media was demonstrated by Hebden and Wong.[27] Their approach was similar to the one developed for computed tomography, in which a rotating array of parallel beams of x-rays probes the sample at evenly spaced angles and creates a sequence of shadowgraphs

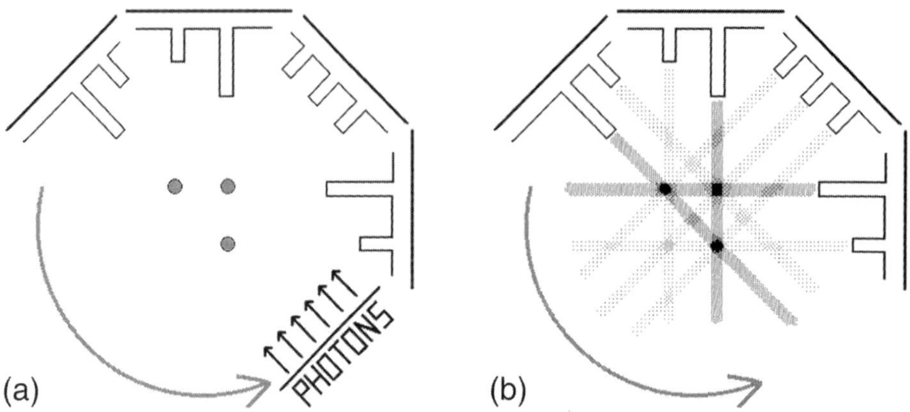

Figure 6.5 (a) An emitter array is rotated to a given position where it illuminates the sample. The shadow observed on the opposite side of the sample, known as the projection, contains information about the attenuation and the position of inclusions inside the sample. The emitter array is subsequently rotated to a new position and its projection is measured again. (b) The interior of the sample is reconstructed from all measured projections by "projecting" the attenuation back into the sample. This reconstruction technique is known as a *back projection*.

known as "projections" that are measured at the detectors. This process is shown diagrammatically in Figs. 6.5(a). The projections are then used to obtain a reconstruction of the interior of the sample, by an inverse process known as a back projection, which is illustrated in Fig. 6.5(b).

Hebden and Wong followed the same approach using an early-gate optical detection system to ensure that only quasi-ballistic, x-ray-like photons were registered across their sample. They also employed filtered back projections, which is a modified back-projection technique commonly used in CT that nullifies the spatial broadening inherent to back projections, thereby successfully reconstructing slice images of the interior of the phantom. However, these images also revealed unusual features related to the index of refraction variations at the boundaries, and to the assumption that the total attenuation through the sample equals the sum of the attenuations of all of the volume elements added along the path of the radiation. Unlike in x-ray computed tomography, where every pixel in the image corresponds to the attenuation coefficient at that location, the information derived from back projections in time-resolved imaging is far more complex, particularly for longer time-gate durations that capture a significant number of multiply scattered photons.

The complexity of the information derived from back projections in diffusive media was addressed by Yamada et al.[28] and by Oda et al.,[29] using the Temporally Extrapolated Absorbance Method (TEAM) concept. The idea behind the TEAM concept was to use absorption information from multiply scattered photons to better estimate the absorbance of the ballistic photons. To accomplish this, the method requires that a time-resolved signal be obtained from a reference sample possessing a scattering coefficient and a geometry similar to the test sample. When the time-dependent signal from the reference is divided by the signal from the test sample, for every point in time, then a relative transmission curve is obtained as a function of the photon time of flight (or photon path length) through the sample. Upon taking the log of this ratio, the curve should grow or decay linearly as predicted by the Beer–Lambert law. A linear curve fit is made to this curve, and the extrapolated intercepts at zero time are thus obtained. These intercepts represent the added absorption $\Delta\mu_a$ experienced by ballistic photons in the sample, relative to those in a test sample. These absorption data, which yield transmission estimates for unscattered photons traversing the test sample, can then be used to obtain $\Delta\mu_a$ maps of the interior of the sample by the back-projection method.

6.2.2.2 Diffuse tomography methods

The tomographic approach discussed previously relied solely on information obtained from early gates and on their back projections for the generation of tomographic images. From a conceptual standpoint, this is the simplest approach to optical imaging in biological media since it takes advantage of reconstruction techniques already established for x-ray computed tomography and ignores the problems associated with multiply scattered photons. However, this approach also ignores additional imaging information that can be extracted from multiply scattered photons utilizing their time-of-flight through the sample. Benaron and Stevenson vividly demonstrated this relationship in a relatively simple way, illustrated in Fig. 6.6.[30]

These experimental results demonstrated not only that the early-arriving photons are sensitive to the inhomogeneities present *along* the axis connecting the emitter and the detector, but also that late-arriving photons are affected mostly by inhomogeneities located *off* the axis. Therefore, each time slice contains complementary information that could aid in the localization of inhomogeneities and in the determination of their optical properties, simultaneously.

Many approaches have since been established that take advantage of the full temporal response of the signals to generate tomographic maps that separate the contributions of absorption and scattering.[31–37] These approaches aim at recovering such maps from measured signals, a process referred to as the "inverse problem." It usually involves the following steps:

1. Divide the test object into volume elements, or voxels, and assign starting absorption and scattering coefficients to each of them.
2. Distribute light emitters and detectors around the test object in a way that probes all the voxels.
3. Measure the photon time-of-flight distribution for each emitter–detector combination.

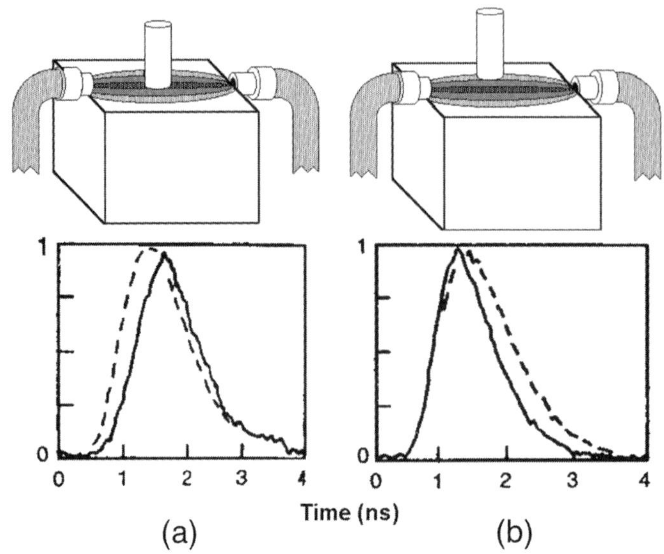

Figure 6.6 The temporal–spatial relationship associated with time-of-flight curves demonstrated by Benaron and Stevenson.[30] Benaron's experiments demonstrated that the time of arrival of photons at the detector is related to their lateral migration. In (a), a comparison is made between the time of flight and the normalized transmission curve acquired with an axial inclusion (solid curve) and without it (dashed). The difference clearly shows that the early-arriving photons, corresponding to those that travel close to the emitter–detector axis, are absorbed preferentially. Conversely, the late-arriving photons are more sensitive to the presence of inclusions located away from the emitter–detector axis, as shown in (b).

4. Simulate the photon time-of-flight distribution for each emitter–detector pair, using the assumed values of the absorption and scattering coefficients at every voxel.
5. Compare all of the measured and simulated distributions. If they match, then the assumed values for the coefficients are the correct ones and the reconstruction is concluded. If not, then adjust the coefficients in a way that minimizes the error between simulated and measured data, and return to step (4).

Step (4), which is the converse to the inverse problem, is known as the "forward problem." In principle, this step can be carried out using a variety of simulation approaches to photon propagation in diffusive media. These include the Monte Carlo method and analytical or numerical solutions to the transport equation or the diffusion equation.

A limiting factor in implementing steps (4) and (5) in a clinical setting is the limited time available to reconstruct an image, particularly in areas of critical care. In such cases, the diagnostic procedures demand that images be available within minutes of the scanning process. Traditionally, such quickness of response seemed unlikely to be attained with Monte Carlo methods for imaging through thick tissues, given today's computer speeds. For this reason, the tomographic reconstruction techniques that attempt to image arbitrary distributions of optical properties have utilized numerical or analytical solutions to the diffusion equation, both of which can be estimated relatively rapidly. However, with the tremendous increase in computer speed and, in particular, with the implementation of Monte Carlo codes for time-resolved photon migration on fast and highly parallelized graphic processing units (GPUs), the application of this method for forward simulations in fit routines is no longer out of reach.[38] Indeed, in 2009, Fang and Boas reported an acceleration ratio more than 300 times faster than conventional CPU computation.[39]

One forward solver that is widely used to obtain numerical solutions of the diffusion equation is the finite-element method (FEM). A fast GPU implementation was presented recently.[40] The TOAST++ open-source software environment for solving the forward and inverse problems in diffuse optical tomography is based on FEM.[41]

An alternative approach to the forward problem involves an approximation that greatly simplifies the computational steps. In many cases, such as in mammography, the aim is to image an abnormality, such as cancer, in a background of otherwise healthy tissue. If the assumption is made that the abnormality will behave as a relatively small perturbation in the background properties of the tissue, then a perturbation model can be employed.[42–45] The signals transmitted through the medium can then be thought of as the addition of the background solution and the contribution from the inhomogeneity. The contribution from the inhomogeneity contribution will vary in direct proportion to the difference in its optical properties relative to the background properties. Higher-order perturbation methods were developed to extend the applicability of this approach beyond small changes in absorption.[46,47]

Time-domain solutions of the forward problem can also make use of analytical solutions of the diffusion equation that have been found in the frequency domain—in particular, those related to the diffraction of photon density waves on spherical inclusions.[48] By calculating the solution for multiple frequencies and performing a Fourier transformation, the time-domain solution can be obtained. Unlike the first-order perturbation models, this approach is not limited to small changes in the optical properties. It has been applied to the characterization of tumor optical properties for time-domain optical mammography in the slab geometry by Grosenick et al.[49]

After step (4) in the resolution of the inverse problem, in which the responses for each emitter–detector combination are simulated by one of the methods discussed earlier, they are compared with the experimental responses, as outlined in step (5). If the difference in the two responses is greater than desired, then the absorption and scattering coefficients chosen for the voxels are adjusted in a way that minimizes the difference. This error minimization process is very similar to the ones commonly used by nonlinear curve-fitting routines. In the case of optical tomographic imaging in diffusive media, the error minimization involves fitting all of the emitter–detector signals to a large number of parameters that include the absorption and scattering parameters of every voxel. The task can be enormously computationally intense, even for relatively coarse tomographic maps. For instance, a $20 \times 20 \times 20$ map (8000 voxels) will require an error minimization with 8000 absorption fitting parameters and 8000 scattering fitting parameters. The minimization procedure includes mathematical calculations of the derivatives with respect to all the fitting parameters that are needed to determine the next minimizing combination. Therefore error minimization, even with simple forward models such as the diffusion equation, can be computationally intensive.

In addition to the challenges posed by error minimization, the reconstruction process is further complicated by poor convergence to noise-free solutions. This problem arises from the poor sensitivity of the transmitted signals to random fluctuations in the optical properties, which in turn results in a high sensitivity of the reconstructed images to fluctuations in the experimental signals. In other words, the addition of experimental noise can still permit solutions that minimize the error but that do not satisfactorily reconstruct the image. The affected reconstructions typically display noisy variations in the absorption and scattering properties. However, if it is known *a priori* that such rapid fluctuations are physically unreasonable, then it is possible to add this information during the reconstruction, through the process known as *regularization.* Regularization is implemented during the error calculation step by adding terms to the error that grow in value with increasing spatial fluctuations.

For a comprehensive overview of methods for solving the forward and inverse problems in optical tomography, see the reviews performed by Arridge.[35,36]

6.2.3 Depth-resolved imaging

6.2.3.1 Coherent back-scattering

The early experimental works of Duguay and Mattick,[5] as extended by Martin et al.[6] and Fujimoto et al.,[7] relied on the use of ultrafast time gating to capture the photons that propagate unscattered (or ballistically) into the sample, scatter off some interface, and then propagate back out ballistically. This technique, which was referred to earlier as optical ranging, is also known as depth-resolved imaging. The earlier work on depth-resolved imaging was very limited in terms of the depth resolution by blurring due to multiply scattered photons. The origin of this difficulty is illustrated in Fig. 6.7.

Figure 6.7 demonstrates the circumstances in which multiply scattered photons from shallower regions can arrive at the detector simultaneously with singly scattered photons from deeper regions. Time-gating alone cannot differentiate between the two cases.

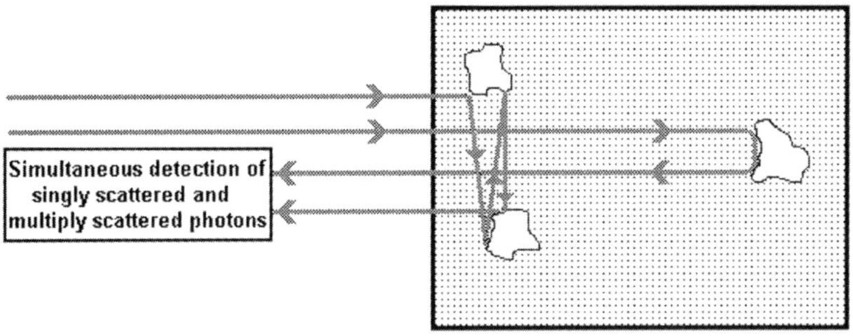

Figure 6.7 A depiction of the limitation encountered in depth-resolved imaging using a time-gated approach. The synchronous arrival of singly scattered and multiply scattered photons prevents temporal measurements from resolving the depth of the scatterers.

Work performed in the early 1990s by Huang et al. demonstrated that the coherent properties of singly scattered photons could be utilized to differentiate them from the incoherent, multiply scattered photons with the aid of interferometry.[50] This idea was extended later to time-resolved imaging by several groups, including Hyde et al.[51] and Hauger et al.[52] Their approach relied on superimposing the back-scattered light from the sample with a reference light pulse inside a nonlinear crystal—that is, a crystal whose response to light depends on that light's intensity. The essence of their experimental setups is captured in Fig. 6.8. In this figure, a coherent, pulsed light source is split into two parts, one of which is directed at the sample and the other to a nonlinear crystal. The back-scattered light from the sample is

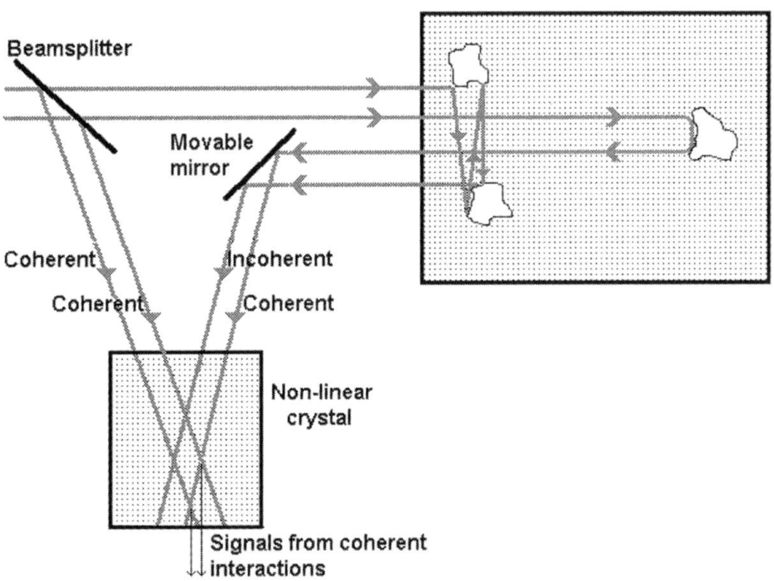

Figure 6.8 A diagram that captures the essence of how depth-resolved experiments discriminate against incoherent, and consequently diffuse, back-scattering.

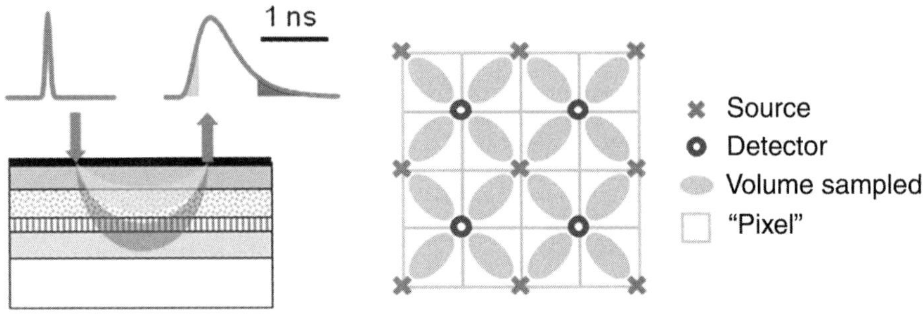

Figure 6.9 Depth selectivity (left) and lateral resolution (right) in time-domain optical brain imaging (see color plates).

collected and directed to the crystal by means of a movable mirror that varies the path length between the sample and the crystal. When the back-scattering overlaps temporally with the split portion of the source, and both signals are mutually coherent, then a strong nonlinear response is elicited from the crystal that can be separated spatially from the other signals. The technique therefore provides an effective way to reject the incoherent component of the signal, and to measure the strength of scattering at a well-defined depth in the sample.

The reliance on coherent interactions with back-scattered radiation limits this technique to depths where the radiation retains a high degree of coherence. This is expected to be the case for depths of up to $1/\mu_s'$, which is typically on the order of 1 mm for most tissues.

6.2.3.2 Diffuse reflectance imaging

Some depth resolution can also be achieved in time-resolved diffuse reflectance imaging of biological tissues. In this case the diffusely scattered light is collected up to a few centimeters apart from the point of illumination, on the same side of the tissue. In a way, this approach can be considered a kind of incomplete tomography. As a rule of thumb, which is valid for the integral intensity, the penetration depth is limited to approximately half the source–detector separation. As a consequence, thick tissues cannot be sampled completely. However, the reflectance geometry is the only way to perform optical brain imaging in adults.

In this geometry the advantage of time-resolved measurements is striking: The penetration depth is time-dependent, as illustrated in Fig. 6.9(a). Measuring the time of flight gives access to the total path length over which a photon is propagated inside the highly scattering medium. Although the concrete course of the random trajectory remains unknown, it is obvious that trajectories with short path lengths are more confined to superficial regions, whereas photons with longer path lengths have a higher probability of visiting deep regions. In reality, the sensitivity regions (sometimes called photon bananas) related to early- and late-arriving photons are not that well

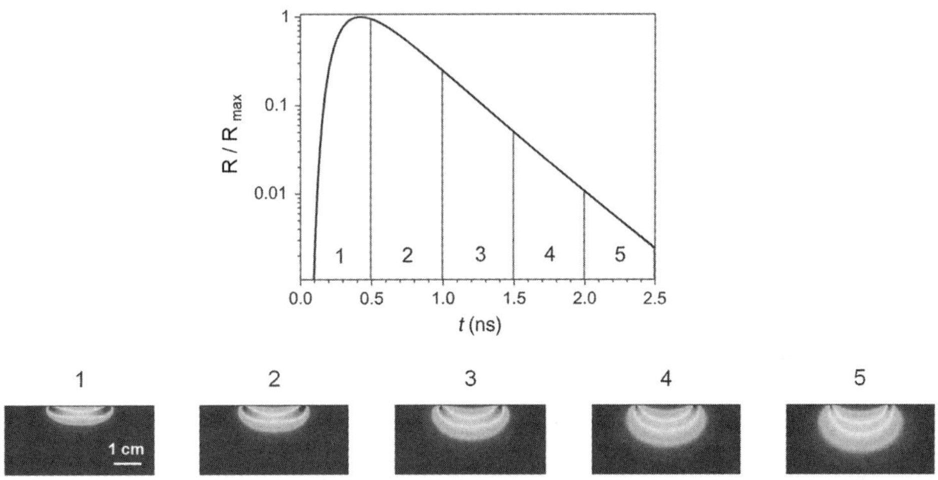

Figure 6.10 Time-dependent diffuse reflectance $R(t)$ (logarithmic scale) (top) and sensitivity profiles for photon counts in five consecutive time windows, integrated over y (bottom). Calculations of sensitivities were based on a diffusion model for perturbation by a point-like absorber according to Ref. 53. Source at $(x, y, z) = (0, 0, -1$ mm), detector at (20 mm, 0, 0), $\mu_s' = 1$ mm^{-1}, $\mu_a = 0.01$ mm^{-1}, $n = 1.4$.

separated, but rather are blurry and overlap. This situation can be seen in Fig. 6.10. Nevertheless, time-resolved measurements allow hemodynamic changes in deep and in superficial tissues to be separated.

There is no doubt that information from the cerebral cortex can indeed be obtained when measured at a source–detector separation of typically 3 cm. However, there are two problems. First, the relative sensitivity to absorption changes in the cortical region is low, it is given by the ratio of the path length that a photon propagates in the cortex to the total path length from source to detector. Hoshi et al. estimated this ratio to be between 5% and 22% for the integrated reflectance [continuous wave (CW) method] measured at a source–detector separation of 3 cm on the adult head.[54] Second, the photons that reach the detector have also propagated through superficial tissue layers—in particular, the skin—and this compartment adds unwanted signal contributions. Time-resolved methods offer the best way to cope with these problems, making use of the time dependence of the partial path length in the different compartments.

Lateral imaging is achieved by covering an area of interest on the surface with multiple source–detector pairs arranged in a grid, as depicted in Fig. 6.9(b). This approach is often referred to as optical topography, meaning that the sampled tissue is confined to a limited range beneath the surface. The tissue volume interrogated by each source–detector combination is given by the banana-shaped sensitivity region, as shown in Fig. 6.9(a). Therefore the lateral spatial resolution is mainly limited by the source–detector separation, and the number of "pixels" is rather small.

For an alternative imaging approach, the time-domain approach makes it possible to decrease the source–detector separation to zero while retaining the depth sensitivity,[55] by applying the null source–detector separation approach.[56] Provided that photons with short times of flight can be eliminated by gated detection, the performance for late photons is at least as good as that with source–detector separations of a few centimeters. The sensitivity region becomes more confined and rotationally symmetric. In this case, the source–detector grid can be more densely spaced, and the lateral resolution is limited only by the extension of the sensitivity regions. Note, however, that such approach is not feasible with CW technology.

Major applications of time-domain diffuse reflectance imaging are functional brain imaging and bolus tracking of an optical contrast agent. In both cases, absorption changes in the brain with respect to a baseline state are the quantities of interest, but changes also occur in superficial tissues. If the absorption changes are small, linear models can be applied for reconstruction. That is, the changes in a set of measurands M_i and the absorption changes $\Delta\mu_{a,j}$ in the jth compartment (e.g., layer, voxel, tissue type) are linked by

$$\Delta M_i = \sum_j S_{i,j}\Delta\mu_{a,j} \tag{6.7}$$

where $S_{i,j}$ are sensitivity factors that depend on the geometry and on the optical properties of the various compartments.

Steinbrink et al. reconstructed the absorption changes in layered turbid media from the changes in the ratio of the full temporal profile of distribution of times of flight of photons (DTOFs) measured in the activated and baseline states, for a single source–detector separation.[57] The relevant sensitivity factors are time-dependent mean partial path lengths (TMPP), illustrated in Fig. 6.11. They were obtained from Monte Carlo (MC)

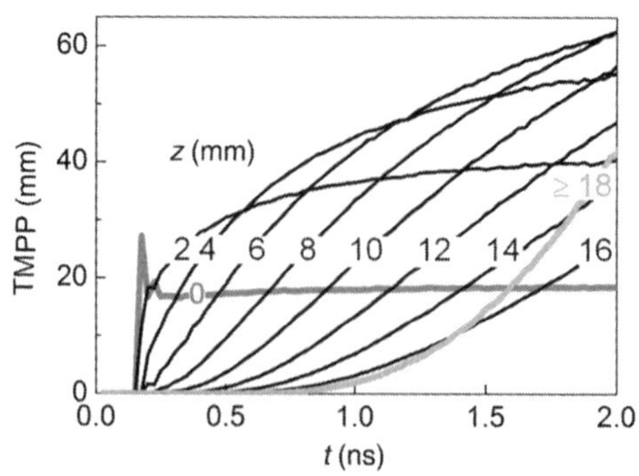

Figure 6.11 Time-dependent mean partial path length (TMPP) in layers extending in depth from z to $z + 2$ mm for a homogeneous semi-infinite medium ($\mu_a = 0.01$ mm^{-1}, $\mu_s' =$ 1.0 mm^{-1}, $\rho = 30$ mm, $n = 1.33$, isotropic scattering and no Fresnel reflection at the boundaries was assumed). After Steinbrink et al.[57]

simulations with known (or assumed) optical properties. Depending on the total time of flight *t*, individual photons spend different times (and travel related optical path lengths) in the various layers. All photons, irrespective of their total time of flight, take, on average, approximately the same amount of time to traverse a thin superficial layer. For deeper layers, the TMPP increases with total time *t*. The deepest layers can be reached only by photons with a long time of flight, so the TMPP increases for very late-arriving photons only. The shape of the dependence of TMPP on *t* is a kind of fingerprint of the related layer. The solution of the system of linear equations as given in Eq. (6.7) can be regarded as finding that linear combination of the functions TMPP(*t*) that best fits the measured ratio of DTOFs. The amount of detail in the resulting depth-resolved reconstruction is limited due to the diffuse nature of photon propagation. In general, the maximum number of compartments for which independent absorption changes can be retrieved is restricted to two or three.

Liebert et al. proposed an algorithm based on statistical moments of DTOFs $N(t)$, meaning integral N_{tot} (equivalent to a CW NIRS measurement), mean time of flight $<t>$ (first moment) and variance V (second central moment), and measurements at four distances.[58] To retrieve absorption changes in two layers with known thickness of the upper layer, the knowledge of the changes in these moments at a single source–detector separation is sufficient.[59,60] An advantage of using moments is that changes in both mean time of flight and variance are independent of the instrument response function (IRF) and deconvolution is not necessary. Selb et al. developed a method to perform a three-dimensional linear reconstruction of localized absorption changes, based on time-gated diffuse reflectance measurements with multiple sources and detectors.[61]

The quantification of absorption changes by the methods discussed so far relies on sensitivity factors that depend on the geometry and baseline optical properties of the tissue. These values are not exactly known and exhibit substantial interindividual variability. The problem can be tackled in a different way by identifying quantities (also called data types) that exhibit a selective sensitivity to deep (cerebral) absorption changes.[62] This approach allows one to distinguish between cerebral and extracerebral changes and to analyze their time courses without the need of quantitative reconstruction. Figure 6.12 compares the depth sensitivities - for the case of layer-wide absorption changes - that are achieved by using the variance and the integral of the DTOF. The integral N_{tot} is sensitive to both deep and superficial changes, whereas V is almost exclusively sensitive to deep changes. With the time window approach, selectivity for deep changes can be obtained on the basis of the ratio of the signals in the late and early time windows, whereas early time windows alone feature depend on superficial changes only.[63]

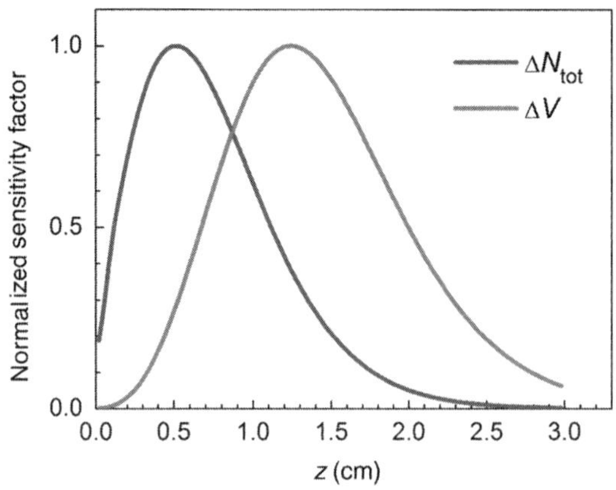

Figure 6.12 Depth-dependent sensitivity profiles for small absorption changes, for integral N_{tot} and variance V (same geometry and parameters as in Fig. 6.11).

6.3 Experimental Tools for Time-Resolved Imaging

6.3.1 General considerations

Since Jöbsis's early work on the diagnostic value of near-infrared light for mammalian tissues,[4] our understanding of the optical properties of tissues has grown significantly.[64] Our current understanding suggests that the spectral range of deep penetrability lies in the range of 600 to 1200 nm, throughout which many spectral markers, associated with a variety of biomolecules, can be found. This spectrum is also known as the "diagnostic window."

We know now that for most mammalian tissues the photons travel freely for approximately 0.1 mm before they collide with a scatterer, and that after 10 or so collisions they assume a totally random orientation. The mean-free path before scattering is therefore 0.1 mm, and the *reduced* mean-free path (the distance after which photons become totally randomized) is approximately 1 mm. It has also been ascertained experimentally that the average trajectory length for photon absorption is much longer, usually on the order of 100 mm. The path lengths given previously correspond to a scattering coefficient of $\mu_s = 10\ \mathrm{mm}^{-1}$ and a reduced (or transport) scattering coefficient of $\mu_s' = 1\ \mathrm{mm}^{-1}$. The absorption coefficient is typically on the order of $\mu_a = 0.01\ \mathrm{mm}^{-1}$. Finally, the average index of refraction is approximately 1.4 for most soft tissues. These numbers have some important implications for the design of diagnostic medical tools based on light propagation through tissues, which are addressed next.

(a) Ballistic photons must be sampled on a picosecond time scale

The average time Δt elapsed between scattering events is related to the mean free path $\Delta x = 1/\mu_s$ by

$$\Delta t = n\Delta x/c, \tag{6.8}$$

where n is the index of refraction and c is the speed of light in a vacuum. From this equation one estimates Δt to be approximately 0.5 ps for the scattering length given earlier. The time gates that are used to capture ballistic photons cannot exceed a few Δt, since otherwise photons would have enough time to scatter out and back into the path of the ballistic photons and to be captured by the detector. Therefore the time gate needed to capture true ballistic photons must not exceed a few mean-free scattering times—that is, a picosecond or so.

(b) Ballistic photons are attenuated by four orders of magnitude for every millimeter traveled

The scattering length assumed earlier implies that only a fraction 1/e of the incident photons is expected to traverse a depth of 0.1 mm without scattering. Consequently, ballistic photons are attenuated by a factor of approximately $(1/e)^{10}$ for every millimeter of tissue traveled, corresponding to an attenuation factor of more than four orders of magnitude per millimeter.

(c) The diffuse signal rises and falls on a time scale of hundreds of picoseconds

The diffuse signal refers to the large and broad temporal response typically observed from thick tissues (several centimeters thick), as exemplified in Fig. 6.4. The diffuse signal is generated by multiply scattered photons, which can be envisioned as undergoing a random walk in the tissue. If a tissue has a thickness equivalent to N reduced mean-free paths, then the number of reduced mean-free paths taken to diffuse over that distance should be approximately N^2, according to random walk theory.[65] Therefore, if a tissue is 1 cm thick, corresponding to approximately 10 reduced mean-free paths, then the average number of steps taken to traverse the tissue should be on the order of 100, for a total trajectory length of 100 mm. The time required to cover this distance, according to Eq. (6.8), is roughly 500 ps. Consequently, the signal is expected to rise during this time. One of the factors competing with this growth is the decay of the photons inside the tissue due to photon absorption. From the absorption value quoted earlier, the average distance traveled before encountering absorption is on the order of 100 mm, for a corresponding travel time of 500 ps. From these hand-waving arguments one concludes that the temporal distribution of photons arriving at a detector exhibits a rise and fall that takes place on a time scale of hundreds of picoseconds, as can be seen in Fig. 6.4.

(d) Diffuse signals are attenuated by one order of magnitude for every centimeter traveled

The diffuse signal is not expected to be attenuated as rapidly as the ballistic signals since it contains some of the photons that scattered out of the ballistic beam and rescattered back into the original direction of propagation. A rough estimate of the attenuation experienced by multiply scattered photons can be obtained with Eq. (6.3) when using $\mu_s' = 1$ mm^{-1}, $\mu_a = 0.01$ mm^{-1}, $n = 1.4$ and integrating over time. For tissue thicknesses d of 2, 3, and 4 cm, one obtains diffuse transmittance values opposite to the source ($\rho = 0$) of 2.6×10^{-5} mm^{-2}, 2.6×10^{-6} mm^{-2}, and 3.1×10^{-7} mm^{-2}, respectively. This estimate is consistent with the rule of thumb often used in the field, which states that the diffuse signal is attenuated by about one order of magnitude for every centimeter of tissue traveled.

(e) Unlike x-ray CT, the radiation employed in optical scanners often cannot travel in open air spaces

The design of conventional CT scanners is based on the ability of x-rays to propagate mostly in ballistic trajectories through biological tissue. The emitter and detectors are placed at a fixed distance from the body, in a large ring known as the *gantry*. This ring is larger than the width of a human body, which necessarily places some air space between the body and the emitter and detectors. Such "open-air" geometry works well for CT, but often presents a challenge for optical imaging in thick tissues, where the transmitted radiation can vary by many orders of magnitude across the sample. The problem in such a case is associated with stray light, which is inevitably generated from and collected with open-air light-guiding elements such as mirrors and lenses. Stray light can easily mask faint signals associated with light transmission from the thicker portions of a sample.

With these considerations in mind, one can then narrow significantly the choices of experimental tools available today for time-resolved optical imaging.

6.3.2 Pulsed light sources

The need to quantify the distribution of photon time of flight is at the basis of time-resolved optical imaging in diffusive media. For mammalian tissues, the required time resolutions of hundreds of picoseconds for diffuse optical tomography and single picoseconds for ballistic imaging can be satisfied only with equally short-duration pulsed light sources. Examples of these pulsed systems are discussed next.

6.3.2.1 Mode-locked lasers

The development of the laser in the early 1960s permitted the exploration of coherent optical effects in ways never before possible. One laser technique is known as *mode locking*.[66] This effect takes advantage of the laser's ability to synchronize all the light frequencies (or modes) it generates internally to create a train of short pulses. The effect is akin to the sound produced by mechanical clocks that operate at slightly different but evenly spaced speeds. Periodically, the clocks will tick simultaneously, producing a loud sound in the process.

After a while the clocks dephase and consequently the loudness is dampened. The duration of the loudness is determined by the dephasing time, which is related to differences in the speeds of the clocks, or their internal frequencies. The same principle operates inside mode-locked lasers: The pulse duration is related to the spread of frequencies (or the bandwidth) available to the lasing medium.

Mode locking can be achieved inside a laser in a variety of ways, which can be classified as either active or passive mode locking. *Active* mode locking involves an external agent that modulates the light amplification process inside the laser cavity. This forces the laser to lock its modes in a way that synchronizes its pulse generation with the window of time for best lasing imposed by the external modulator. The external modulation must be tightly synchronized with the round-trip time of the pulses inside the laser. Active mode locking is achieved principally by acousto-optical modulation or synchronous pumping. In acousto-optical modulation, an ultrasound periodically modulates the transmission of a crystal placed inside the cavity. For lasers that use another laser as the energy source (or pump), a way to modulate the lasing action is simply to modulate the pump laser—a technique known as synchronous pumping.

In contrast, *passive* mode locking is achieved internally in the laser when a nonlinear substance, such as a saturable dye, is present inside the cavity. Saturable dyes are substances that absorb light at the laser wavelength, to the point where they would normally prevent laser action. However, these dyes can also be bleached when the intensity reaches a certain level. With an appropriate choice of dye and concentration, it is possible to set the threshold of bleaching such that only the powerful mode-locked pulses can be transmitted and amplified in the laser cavity.

Recently, another form of passive mode locking known as self-locking has come to dominate the field of ultrafast lasers. Self-locking relies on an effect in which very intense light distorts the index of refraction of a medium inside the laser cavity, causing the medium to focus the light. This phenomenon can occur inside a cavity when some laser modes become locked. It results in light inside the cavity focusing more tightly on the lasing medium. If the increased focusing improves the efficiency of the laser, which is the case if the pump beam is narrower than the laser beam, then more modes become locked and additional focusing takes place. This positive feedback loop can continue until the laser is fully mode locked.

The neodymium glass laser was the first mode-locked laser to become widely available commercially, which happened beginning in the 1970s. It used a saturable absorber to achieve a pulse duration of approximately 10 ps and a pulse energy of a few millijoules at 1054 nm. Duguay and Mattick used this laser in the experiment discussed in Section 6.1.2. The mode-locked dye laser then became very popular in the 1980s and early 1990s. Its broader bandwidth permitted much shorter pulses, from 1 ps for a relatively simple design to 6 fs for the most sophisticated designs.

All mode-locked dye lasers require a companion pump laser, which may be continuous wave or mode locked. CW-pumped dye lasers use a saturable dye to achieve mode locking. Pumping with a mode-locked laser will induce mode locking in the dye laser, provided that the pulse rate from the pump matches the pulse rate inside the dye laser cavity. This technique is known as synchronous pumping. Mode-locked dye lasers operate very effectively in the visible light spectrum and adequately up to wavelengths of approximately 820 nm.

The lasers discussed in this section were originally constructed for scientific investigation in research laboratories. These systems required frequent adjustments to maintain acceptable performance. Necessary adjustments included changes in the dye concentrations, flow patterns, mechanical alignment of the mirrors, and frequency of the mode-locking system due to thermal expansion of the laser cavity during operation. Although such technical procedures are acceptable in research laboratories staffed with technically trained personnel and with research planning that accounts for these delays, they are not suited for medical diagnostic purposes that demand reliable and immediate performance.

A significant advancement in laser technology emerged in the 1990s with the development of the mode-locked titanium sapphire laser. The Ti:sapphire laser is configured optically as a dye laser in which the dye flow through the cavity has been replaced with a solid that is highly resistant to optical damage. This arrangement frees the lasing medium from mechanical fluctuations. This innovation turned the Ti:sapphire system into a "turnkey" mode-locked laser. Using a self-locking mode-locking arrangement, this laser routinely achieves pulse widths of less than 100 fs, with pulse energies of approximately 10 nJ, and a repetition rate in the tens of megahertz. Such performance makes Ti:sapphire lasers ideally suited for time-correlated single photon counting, a technique that is described later in this chapter. Furthermore, the laser medium allows for tunable output ranging from 690 to 980 nm, thus covering a large portion of the diagnostic window. A drawback of these systems is their cost, which ranks as very expensive compared to other laser systems.

6.3.2.2 Pulsed semiconductor lasers

The history of the semiconductor laser, also known as the diode laser, is rich in struggle and triumph.[66] The first semiconductor lasers were announced in 1962 and are historically among the earliest laser systems developed. Despite its early start, the diode laser did not achieve practical levels of operation until the mid-1970s. The laser was based on the semiconductor gallium arsenide (GaAs) and emitted light at a wavelength of 905 nm. By the early 1980s, diode lasers began to emerge from a variety of gallium–aluminum–arsenide (GaAlAs) alloys that permitted operation at other wavelengths, such as

780 nm. The GaAlAs laser became enormously popular due to its widespread use in CD players. The late 1980s saw the introduction of the gallium indium phosphide (GaInP) laser, which operated at 670 nm and was widely adopted in scanners and laser pointers. During the 1990s, laser diodes appeared commercially in a variety of wavelengths covering the 635- to 1500-nm spectrum. Around 2000, laser diodes operating between 370 and 520 nm also became available.

Laser diodes can be pulsed by injecting a current through the device for the desired period of time. Driver circuits have been constructed commercially that pulse laser diodes for times as short as 50 ps for low-current lasers. Pulsed operation was traditionally achieved for single-mode laser diodes that have a typical CW output power of 200 mW in the NIR range. When pulsing these lasers at 50 MHz, for example, an average output power of a few milliwatts can be obtained. The output power can be adjusted, but with increasing power the pulse width also increases into the range of a few hundred picoseconds. An advantage of these lasers is that their repetition rate is not fixed by their resonator length, as in the case of a Ti:sapphire laser, but rather is given by the injected current pulse train. Some lasers offer a switchable repetition rate.

Picosecond pulse operation with superior parameters was recently achieved for transversal multimode laser diodes that have a CW power of up to a few watts. For example, 150-ps pulses were obtained at 20-mW average power for a repetition rate of 50 MHz, corresponding to a single pulse energy of 0.4 nJ and a peak power of more than 2.5 W. The output of these lasers cannot be coupled into single-mode lasers, but this is not a limitation for instruments in diffuse optical imaging that use multimode fibers.

Picosecond diode lasers offer the advantage of much lower cost and smaller size than traditional picosecond laser systems and are well suited for compact, portable clinical instruments. The ongoing development focusing on higher output power at short pulse widths will contribute to improved performance of such instruments.

6.3.2.3 Other laser systems

Recent developments in fiber lasers providing picosecond or even femtosecond pulse trains have yielded promising candidates for clinical applications—in particular, supercontinuum lasers. These devices are robust, maintenance-free, rather compact units. The oscillator and amplifier are fiber based, and a broad continuum of wavelengths, ranging from visible wavelengths up to far beyond 1000 nm, can be generated by photonic crystal fibers. The pulse width is less than 100 ps. Their repetition rate of tens of megahertz makes such devices suitable for time-correlated single photon counting (TCSPC). Through spectral filtering, a tunable source is obtained. Specifically, acousto-optic tunable filters (AOTF) are very flexible in terms of their ability to select and tune wavelengths. They also allow for fast switching between wavelengths within a few microseconds.

6.3.3 Detection systems based on time-correlated single photon counting

6.3.3.1 TCSPC principle

Time-correlated single photon counting is based on the detection of single photons of a periodic light signal, the measurement of the related detection times, and the accumulation of a histogram of these individual time measurements that represents the original waveform.[67,68] Because of its excellent sensitivity, linearity, and dynamic range, TCSPC is well suited for recording the temporal profile of the light emerging from tissue. In turn, it is employed in the vast majority of clinical applications of time-domain diffuse optical imaging. In diffuse optics, this waveform is often termed distribution of times of flight of photons (DTOF), referring to the result of multiple measurements of the time of flight of individual photons through the turbid medium; another frequently used term is temporal point spread function (TPSF). TCSPC has its technical predecessor in nuclear physics, where the "delayed coincidence" method was used to determine the lifetime of unstable nuclei. The first use of this approach to record the temporal profile of light pulses was published in 1961.[69] In the following decades, TCSPC was mainly developed as a tool to measure fluorescence decay profiles. In the early 1990s, it found its first applications in diffuse optics.

TCSPC is the method of choice when low light levels are to be detected, which is the case in diffuse optical imaging of tissues. The detector output consists of isolated single-photon pulses. The probability of detecting more than one photon in a signal period is negligible, as illustrated in Fig. 6.13. In Fig. 6.13, with a periodic pulse sequence (a) impinging on the sample, a periodic sequence of DTOFs (b) is the expected output. For low light levels, a quasi-random sequence of single-photon pulses (c) is obtained. While their amplitude and shape are irrelevant, their temporal position within the respective individual signal period is measured. The counts of single-photon events as a function of time, accumulated over many pulse periods, yield the measured DTOF.

In its classical implementation, TCSPC was a rather slow method. It measured the time interval between the excitation pulse provided by a flash lamp (start) and each detected single photon (stop). For a typical repetition rate of the excitation of 10 kHz, the maximum photon count rate was on the order of only 100 s^{-1}, leading to collection times of hours. However, with the availability of pulsed laser sources with repetition rates of several tens of megahertz, together with faster TCSPC electronics, maximum count rates increased by several orders of magnitude.

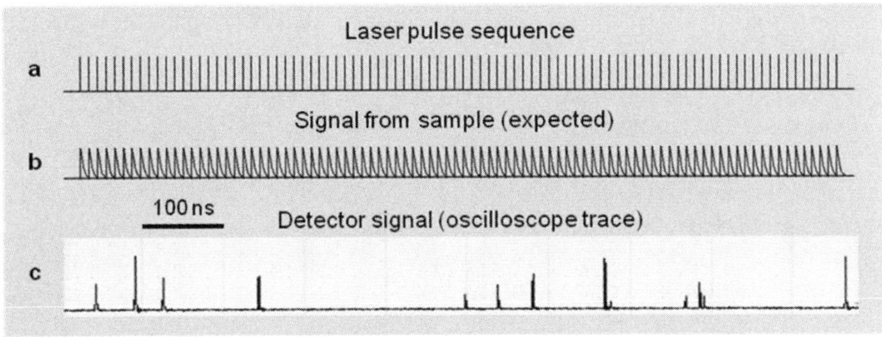

Figure 6.13 Typical single-photon detector signal at a pulse repetition rate of 80 MHz. Adapted from Ref. 70, with permission.

For light sources with a high and stable repetition rate, the timing scheme can be reversed; that is, the time measurement can be started by the single-photon pulse and stopped by the following excitation (sync) pulse. The maximum count rate is limited by the dead time of the detector and the electronics, following the processing of a single-photon event. State-of-the-art implementations have dead times of approximately 100 ns, resulting in a maximum count rate of several 10^6 s^{-1}. Such high count rates make it possible to record DTOFs with an acceptable signal-to-noise ratio and collection times in the millisecond range, and to follow dynamic processes on a sub-second time scale. Note that it is often not easy to harvest such high photon flux from tissues several centimeters apart from the point of light injection, bearing in mind the limitations due to laser safety requirements—for example, considerations related to the intensity impinging on the skin.

Figure 6.14 shows the basic architecture of TCSPC electronics. The constant-fraction discriminator (CFD) ensures that the determination of the

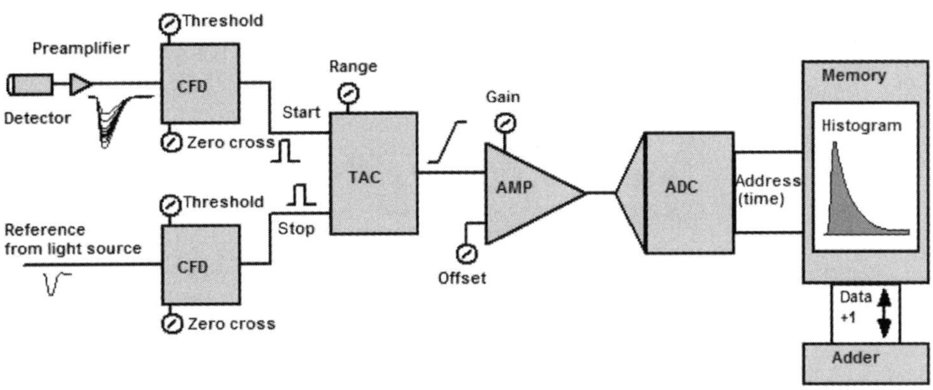

Figure 6.14 Main building blocks of TCSPC electronics operating in the reverse start–stop regime. CFD, constant-fraction discriminator; TAC, time-to-amplitude converter; ADC, analog-to-digital converter. Adjustable parameters are indicated. From Ref. 70, with permission.

time of arrival of the incoming pulses is independent of their amplitude. The time-to-amplitude converter (TAC) acts like a stopwatch and measures the interval between the standardized start and stop pulses delivered by the CFDs by means of a linear voltage ramp. The analog-to-digital converter (ADC) converts the voltage to a memory address in which the event is added. As an alternative to the TAC-ADC combination, a time-to-digital converter (TDC) can be applied. Both approaches have advantages and disadvantages,[68] but at present they can achieve similar performances. Modern TCSPC implementations have multidimensional capabilities.[71] That is, a multitude of optical waveforms that depend on parameters—such as wavelength, time during the experiment, or spatial coordinates—can be recorded in parallel.

To be suitable for TCSPC, photodetectors have to fulfill several requirements. Their gain must be sufficient to deliver a useful electronic output pulse for a single photon. They must be fast, and their output pulses must be short enough to resolve the individual single-photon pulses even at a high count rate. It is important to note that the time resolution achievable by TCSPC recording is considerably better than the width of these single-photon pulses since only their timing matters. The limiting factor for time resolution is the transit time spread (TTS), meaning the uncertainty of the time interval between the photon's arrival at the detector and the delivery of the corresponding output pulse. This situation is illustrated in Fig. 6.15(a), which shows a typical set of output pulses for a fast photomultiplier. Their pulse height is rather variable, but their width is nearly constant—in this case, approximately 1 ns. When plotting normalized single-photon pulses as a function of time after photon arrival at the detector, as shown in Fig. 6.15(b), middle, the TTS becomes visible. It determines the width of the instrument response function in the ideal case of infinitely short light pulses hitting the detector. As a rule of thumb, the TTS is by approximately an order of magnitude better than the width of the (analog) single-photon pulse. The timing uncertainty of TCSPC electronics itself is negligible compared to the TTS. A comprehensive review of detectors for TCSPC, their principles, and their properties, as well as the specifics of a number of individual types, can be found in Ref. 70.

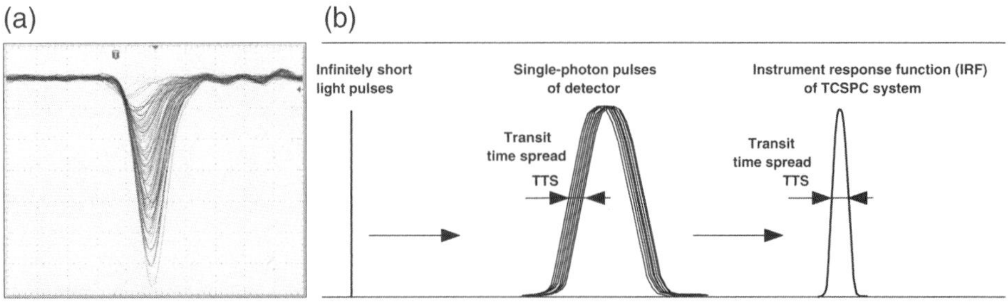

Figure 6.15 Time resolution in TCSPC. (a) Single-photon pulses from a photomultiplier as an oscilloscope recording (1 ns / div). (b) TTS and time resolution.

The IRF of a complete TCSPC-based system, including light source and optical system, depends not only on the TTS, but also on the laser pulse shape and the pulse broadening due to temporal dispersion in the fiber optics. It can be considered as a convolution of these individual effects.

6.3.3.2 Detectors for TCSPC

6.3.3.2.1 Photomultiplier tubes

The development of photomultiplier tube (PMT) detectors can be traced to 1936, when the principal contemplated application was sound-on-film pickup.[72] The PMT has become the device of choice for spectroscopic and biomedical applications that demand the detection of exceedingly faint light signals. Their sensitivity, coupled with continued improvements, has made the PMT a tough act to follow technologically, even today.

PMTs come in a variety of packages designed for optimal performance at different wavelengths and time regimens. These devices are now offered in a variety of sizes and shapes. Regardless of the packaging, the fundamental operating principle, illustrated in Fig. 6.16, remains the same: electron multiplication.

All PMTs contain a photosensitive surface, called the photocathode, that responds to the absorption of a photon by releasing an electron. In traditional PMTs [see Fig. 6.16(a)], the released electron is then pulled by an electric field into a nearby plate, called a dynode. The electric field is generated by a voltage difference between the dynode and the photocathode, which is

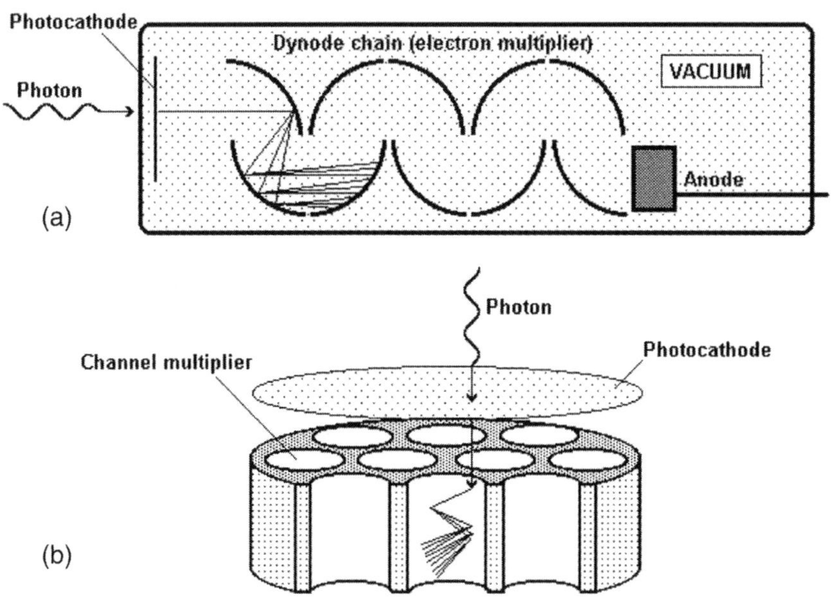

Figure 6.16 Design of photomultiplier tubes. (a) Dynode-based design. (b) Microchannel-plate PMT.

typically on the order of 100 V. By the time the electron reaches the dynode, it has acquired enough kinetic energy to free many electrons from the surface of the dynode. These electrons are, in turn, accelerated by another voltage drop across to the second dynode, where each frees many more electrons. In this way, electrons are multiplied as they cascade from dynode to dynode. By the time the cascade reaches the anode, it can pack more than 10^6 electrons, which can easily be registered by an external electronic circuit. In fact, this signal can be readily observed on an oscilloscope without further processing.

As explained earlier, the time resolution in TCSPC is given by the TTS. This spread is related to the overall transit time in the dynode system, which itself depends on the distances and voltages present. This reason explains why the compact "metal channel" PMTs in a TO8 housing that are often used in systems for diffuse optics provide a better time resolution than the fastest classical PMTs based on glass tubes. The active diameter of the former PMTs' photocathodes is typically between 5 and 8 mm. Their TTS is between 140 and 200 ps.

Another type of PMT, known as the microchannel-plate PMT or simply MCP, offers a much shorter time response. The design of the MCP is illustrated in Fig. 6.16(b). In the MCP, the electron multiplication takes place inside a capillary channel with a high voltage maintained across it, which behaves as a continuous dynode chain. The length of the channel is kept short—a factor that is the key to the short time response offered by these devices. The TTS in recent MCP-PMT devices is as low as 25 to 30 ps. Another advantage of the MCP is its imaging amplification capability when combined with a spatially resolving anode structure. NIR MCP-PMTs have a comparably large photocathode (diameter 10 mm or larger), which is advantageous when collecting diffuse light. Nevertheless, it is important not to concentrate the light to a too-small portion of this area, but rather to engage many microchannels because their individual recovery times after an amplification process are long (approximately 1 μs). A major disadvantage of this type of detector is its cost, which exceeds that of the traditional PMT by more than an order of magnitude.

6.3.3.2.2 Single-photon avalanche diodes

Photodiodes are semiconductor detectors that have garnered widespread use because of their sensitivity, speed, compactness, and low cost. Photodiodes make use of the internal photo effect and do not need a photocathode. Standard photodiodes are not suitable for single photon counting because of their low gain. Avalanche photodiodes rely on the "avalanche effect." which is a suitable internal gain mechanism; however, the achievable stable gain of 10^2 to 10^3 is not yet sufficient for accurate single-photon detection. Single-photon avalanche diodes (SPAD) were developed in the 1980s that operate in "Geiger mode,"[73] with a TTS as small as 20 ps. In these devices, the reverse voltage of

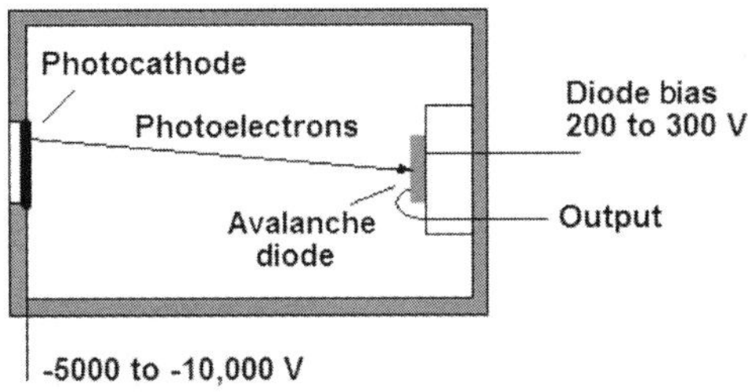

Figure 6.17 Principle of a hybrid PMT. From Ref. 70, with permission.

the diode is set several volts above the breakdown voltage. To avoid the avalanche destroying the diode, a passive—or, even better, an active—quenching circuit restores normal operation after detection of a photon.[74] SPADs are compact and robust devices with a high quantum yield. Their major disadvantage is their small active area (diameter typically less than 200 μm), which is required to achieve fast operation.

6.3.3.2.3 Hybrid photomultiplier tubes

Hybrid photodetectors, which have recently become available, combine photoelectron emission from a photocathode with the amplification mechanism of an avalanche diode, as shown in Fig. 6.17. A photoelectron emitted by the photocathode is accelerated by a strong electrical field and injected directly into a silicon avalanche diode. There, it generates a large number of electron–hole pairs in the silicon. These carriers are further amplified by the linear gain of the avalanche diode that is operated below the breakdown threshold.

The principles utilized in the hybrid PMT offer a number of advantages. The TTS can be as small as 50 ps. Moreover, the IRF of a hybrid PMT employed in TCSPC has a clean shape, without extended tails or secondary peaks. The most striking advantage is the virtual absence of any signal-dependent background. Such afterpulsing background is present in conventional PMTs (due to ionization of residual gas molecules in the vacuum) as well as in SPADs (due to trapped carriers of the previous avalanche breakdown). It is advantageous to use compact hybrid detector modules that incorporate the high-voltage power supply as well as a low-noise wide-band amplifier.[75]

6.3.4 Other high-speed detection systems

6.3.4.1 Streak cameras

In many ways, the streak camera is one of the most sophisticated detectors available for time-resolved spectroscopy. The extraordinary performance of this camera is achieved with the design illustrated in Fig. 6.18. For the purpose of simplicity, the device shown in the figure does not implement spectral capabilities.

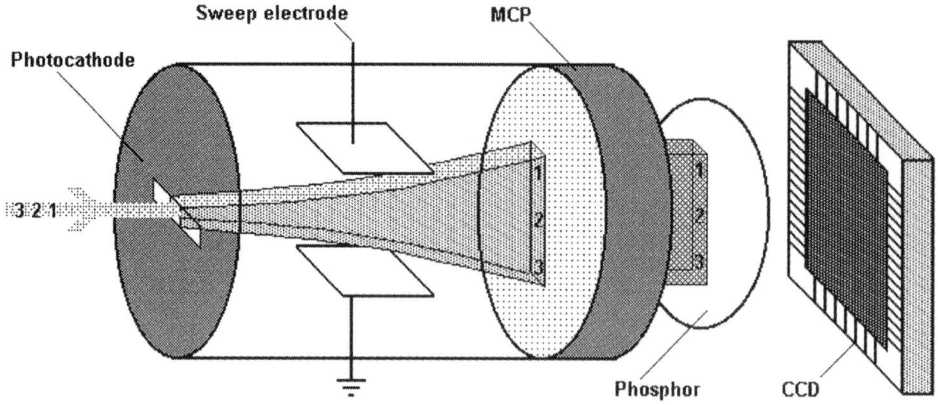

Figure 6.18 Working principle behind the streak camera.

Light that falls on the photocathode causes electrons to stream out of its surface. These electrons are accelerated toward a set of plates that are subjected to a fast voltage change. The electrons are deflected by the plates at different angles, depending on the voltage present across the plates at that particular time. As they fall on the MCP surface, they are multiplied by a factor of 10^4 and exposed to a phosphor screen, where they are converted into an optical signal again. In this way, a luminous streak is produced across the surface with a brightness at any point that is proportional to the number of photons incident at that particular time. This luminous streak can be recorded with a sensitive imaging camera, such as charge-coupled device (CCD) camera. The resulting overall sensitivity of the streak camera can in this way approach that of a photomultiplier tube, but with an unmatched temporal resolution.

For added functionality, the streak camera can monitor independent signals across the length of its entrance slit, such as the signals originating from the dispersed output of a monochromator. This arrangement allows simultaneous light detection at multiple wavelengths.

In most respects, the streak camera seems ideally suited for time-resolved imaging through diffuse media. It allows the measurement of temporal profiles with an unsurpassed time resolution of 1 ps, at multiple wavelengths or points in space simultaneously. However. its performance is reportedly affected by several distortions, including the nonlinearity of the time axis and the nonuniform background noise, which may change from one experiment to another.[76] Another disadvantage is the cost, which exceeds that of photomultiplier tubes by more than two orders of magnitude.

6.3.4.2 Gated cameras

Complete two-dimensional (2D) gated images can be obtained by combining a CCD camera with an ultrafast shutter. A gated image intensifier allows a gate width in the picosecond range to be achieved. The gating is performed by means of a grid behind the photocathode. As long as the grid voltage is negative relative to the photocathode, the photoelectrons cannot pass the grid. When a positive gate pulse is applied, the photoelectrons can pass and are accelerated toward one or more microchannel plates, in which they are amplified. The amplified electron pulses hit a fluorescent screen that is monitored by a CCD camera. The spatial resolution is better than 100 μm, based on a 18-mm diameter of the intensifier.

It is challenging to achieve very short gate widths since, for example, the electric field between the cathode and the gate has to follow a very short gating pulse with the required speed. For this purpose, low-conductivity photocathodes are mandatory. The shortest gate widths realized in commercial instruments are less than 100 ps at a repetition rate of 10 kHz, and less than 300 ps at a repetition rate up to 110 MHz. These parameters make fast-gated cameras useful tools in diffuse optics, and they have been applied in a number of experiments.

Nevertheless, gated cameras also have several drawbacks. While a 2D image is obtained at once, the recording of a full temporal profile requires sequential acquisitions at stepwise-increased gate delays. The gate cuts out a small fraction of the temporal profile of incoming light, limiting the recording efficiency. The photocathode is always exposed to the total light intensity. Time resolution, sensitivity, and dynamic range are also generally lower with these devices compared to TCSPC.

6.3.5 Light guides

Optical fibers are often the device of choice for light delivery to, and light collection from, a diffusive medium. These flexible fibers possess the ability to deliver light directly to the medium easily, economically, and precisely. Their confinement and jacketed insulation cut down on the stray light that often plagues alternative light guiding elements, such as mirrors and lenses. The fibers are also effective light collectors when placed in the immediate proximity of the medium.

Optical fibers can be constructed from a variety of transparent materials, but by far the most widely used is the glass fiber. Glass offers very low losses in the near-infrared spectrum and, therefore, is ideally suited for applications in the field of biomedical optics. Glass fibers can be made into various diameters, ranging from a few microns to approximately 1 mm. Beyond this size, the fiber becomes very stiff and no longer serves the purpose of a flexible waveguide. However, by bundling numerous small fibers, it is possible to construct a large-diameter light guide that is still relatively flexible.

When the source of light is a highly coherent source such as a laser, which is usually the case for time-resolved imaging methods, then the light can be easily focused into a single fiber and in this way delivered to the sample. However, some restrictions do apply. One important consideration is the coupling

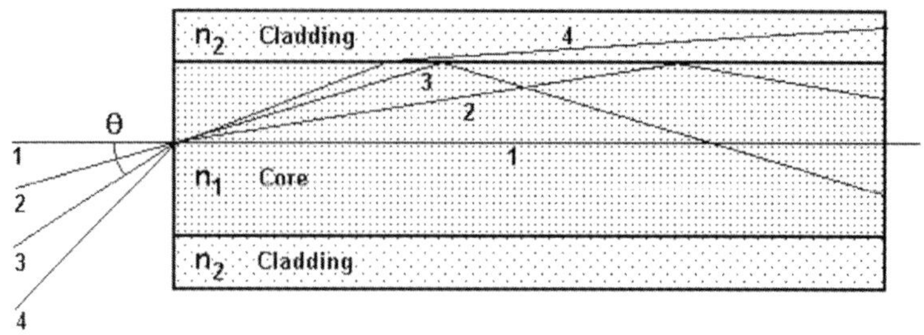

Figure 6.19 Cross section of a step-index optical fiber.

efficiency into the fiber. This restriction is related to the mechanism that confines light for propagation down the fiber, known as total internal reflection, and is illustrated in Fig. 6.19 for a type of fiber known as step index.

Step-index fibers have a uniform cylindrical glass core and are surrounded by a cladding made of a glass having a lower refractive index. The total internal reflection that takes place between the core and the cladding in the fiber depends on two factors—namely, the relative difference in the indices of refraction n_1 and n_2, and the angle of incidence on the cladding. The latter characteristic is controlled by the initial coupling angle θ into the fiber and the external index of refraction n. As Fig. 6.19 depicts, when the angle θ is large enough, the condition for total internal reflection is violated and the ray escapes through the cladding. This is what happens to ray 4 in Fig. 6.19. One can show from the law of refraction that the largest angle of incidence that allows propagation down the fiber is given by $n\sin(\theta) = \sqrt{n_1^2 - n_2^2}$.[77] The factor $n\sin(\theta)$ is also known as the numerical aperture (NA). A common NA for commercial fibers is 0.22, but higher NAs are also available. Instruments for diffuse optical imaging often use fiber bundles of a few millimeters diameter made of thin step-index fibers of higher NA (e.g., 0.54) so as to collect as much as possible light exiting the tissue. Unfortunately, this advantage comes at the expense of decreased bandwidth (in other words, temporal broadening), as discussed later.

Short light pulses that propagate through optical fibers are broadened by two effects. The first is known as modal dispersion. This effect arises when light is coupled into different paths (or modes) inside the fiber, as illustrated by rays 1–3 in Fig. 6.19 for a step-index fiber. Since their path lengths through the fiber are different, the signals from each of these modes will exit the fiber at different times, causing the pulse to broaden. One can show from simple geometrical arguments that the maximum broadening Δt for a fiber of length l is given by the expression

$$\Delta t = \frac{n_1 l}{c}\left(1\Big/\sqrt{1 - \frac{\mathrm{NA}^2}{n_1^2}} - 1\right) \approx \frac{l}{2n_1 c}\mathrm{NA}^2 \tag{6.9}$$

in which the approximation holds for small NA/n_1.[78] If NA = 0.22 and n_1 = 1.50, then $\Delta t/l$ is 55 ps/m, while NA = 0.54 gives 180 ps/m. Such effects are of significance for time-resolved optical imaging where fibers or fiber bundles of 2-m length or greater are often used. It is important to adequately consider these effects when measuring the instrument response function.[78] In this measurement, care has to be taken to fill the modes of the fibers in the same way as with light coming from tissue—that is, with a wide angular distribution. This can be achieved by, for example, inserting a piece of paper or another thin scatterer in the beam path. Sometimes fibers are also utilized to delay optical signals. With n_1 = 1.5, the delay is 5 ns per meter, similar to the delay in electrical cables. In time-resolved measurements with high dynamic range, reflections at the end faces of optical fibers back into the fiber and toward the detector again are often disturbing. The time at which they occur can be estimated by considering the delay when passing through the fiber back and forth.

Modal dispersion can be largely overcome with the use of single-mode fibers or graded-index (GRIN) fibers. Single-mode fibers are optical fibers with cores on the order of a few microns in diameter—small enough to restrict light to only one mode of propagation. Unfortunately, it is not always easy to couple the output of a laser into a single-mode fiber, and these fibers are susceptible to damage due to the high optical fields. In GRIN fibers, the refractive index gradually decreases from the center of the core to the outer radius. GRIN fibers possess larger core diameters, typically ranging from 60 to 100 μm. In these fibers, the light rays that travel in the modes with longer path lengths also propagate to a greater extent in the part of the fiber with the lower index of refraction. The longer path length of the higher modes is therefore compensated for by their faster speed, leading to a significant reduction in modal dispersion. A typical range of modal dispersion in GRIN fibers is 1 to 10 ps/m.

The second form of broadening in optical fibers is known as material broadening. This mechanism is directly dependent on the spectral bandwidth of the laser pulse and the index dispersion in the fiber. The broadening arises when the wavelength components fall out of step as they travel with varying speeds. The resulting dispersion per unit length of fiber is obtained with $-(\lambda/c)(d^2n/d\lambda^2)\Delta\lambda$. A typical value of material dispersion in the near-infrared spectrum (800 nm) is 0.1 ps/nm/km, which is negligible compared to modal dispersion for multimode fibers.

6.4 Technical Designs for Time-Resolved Imaging

6.4.1 Transmittance imaging

6.4.1.1 Time-gated 2D projections

In the early days of time-resolved imaging through thick tissues, a number of different approaches were investigated. An overview of these technologies is given in this section. Only a few of them made their way into clinics; those approaches are discussed in more detail in the next section.

The inability to see through diffusive media under normal lighting conditions was underscored by the example shown at the beginning of this chapter in Fig. 6.1, where the bone structure in the finger was noticeably absent. The use of time gating to resolve the bones in human fingers was, in fact, one of the first demonstrations of an optical shadowgraph made in 1990 by Andersson-Engels et al.[79] In this experiment, a mode-locked dye laser and an MCP-based time-correlated single photon counting system were used to probe turbid samples. These samples included a 3.5-cm breastlike tissue phantom containing inclusions and human fingers. The time gate was set to detect photons arriving during the first 100 ps. Line scans [one-dimensional (1D) imaging] were achieved by scanning the sample across the emitter–detector axis. The results, which are reproduced in Fig. 6.20, demonstrated a

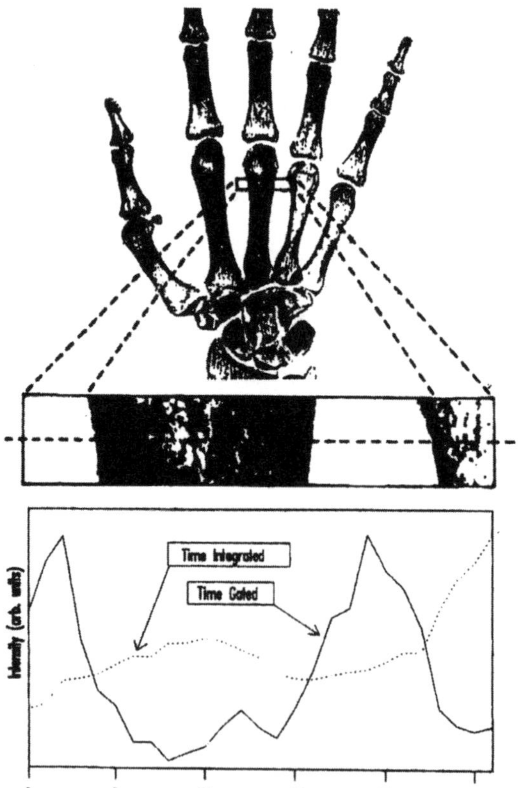

Figure 6.20 Demonstration of an optical shadowgraph of human fingers, made in 1990 by Andersson-Engels et al.[79] The transmission profiles clearly show an enhancement in contrast with the "time-gated" method over the "time-integrated" method.

significant contrast enhancement with the time-gated method, particularly in the finger-transmission experiment.

Within a few short months, another report of time-gated transillumination was published by Hebden et al.[80] In this case, a mode-locked dye laser and a streak camera were used to probe a 4-cm-thick sample containing a diffusive solution and several inclusions at varying depths. Line scans and 2D time-gated shadowgraphs were generated with a single emitter–detector pair. The image was reconstructed one point at a time, translating the emitter–detector axis after each measurement was made. Varying time-gate widths were investigated, ranging from 37.5 to 375 ps. As expected, sharper images of the inclusions, including the ones located at greater depths, were obtained with the use of shorter time gates. However, the results also showed that images cannot be sharpened indefinitely due to increasingly poorer signal-to-noise ratios at the shortest time gates, caused by the scarcity of photons exiting at those early times.

The image acquisition in a shadowgraph, when performed one pixel at a time as proposed earlier, requires an extended period of time that may not be appropriate for *in vivo* samples or clinical applications where subject movement is inevitable. In such a case, it is instead desirable to use an imaging technique capable of recording a whole 2D projection at once. This can be accomplished with the ultrafast camera developed by Duguay and Mattick, based on the optical Kerr effect.[5] In the normal Kerr effect, a strong electric field modifies the index of refraction of a medium in the direction of the field, causing light that passes through it to rotate its polarization. If the medium is placed between crossed polarizers, then the combined transmission will be nulled when the field is switched off, but will increase when the field is activated. The same effect can be accomplished with the electric field associated with electromagnetic waves, particularly with intense laser pulses. In this case, the Kerr gate will transmit only when an intense pulse is incident on the medium.

Alfano et al. pioneered the adaptation of the optical Kerr effect to meet the needs of biomedical applications.[81–86] They developed Kerr–Fourier imaging, a technique that enhances the temporal rejection of the optical Kerr shutter. The basis for this technique is illustrated in Fig. 6.21. The output of a mode-locked dye laser is first split into strong and weak components. The weak component is directed to the diffusive medium from which it emerges with a distribution of time delays and directions. The portion that scatters in the forward direction is normally extinguished as it passes through two crossed polarizers (labeled P in Fig. 6.21). The strong component is directed at the Kerr material, where it induces a temporary index of refraction change along the direction of polarization of the beam. The induced birefringence causes the scattered light to rotate its polarization, thereby allowing it to pass through the second polarizer. A delay line is incorporated to adjust the delay of the time gate. The lens arrangement shown in Fig. 6.21, traditionally

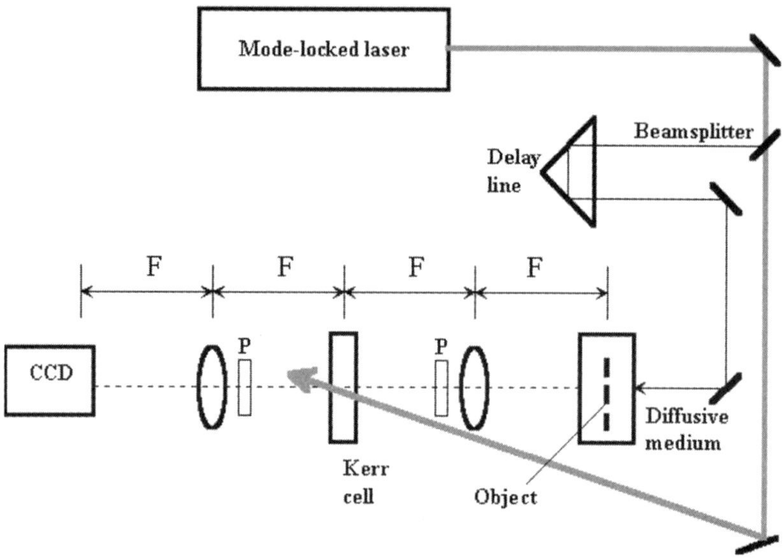

Figure 6.21 The Kerr–Fourier shutter. (Reproduced with permission from *Current Science.*)

known as a Fourier filter, is designed to filter the collimated transmission from the more intense diffuse one. The positioning of the first lens is such that the collimated rays that traverse the sample are focused at the Kerr cell and traverse it unimpeded, whereas the scattered rays are focused outside the cell and eliminated from the path of the camera. The second lens is positioned to recollimate the rays before they enter the camera.

In most investigations of time-resolved imaging, the time-gating process is essentially accomplished by extinction of all the photons outside a desired time window. An alternative approach to time gating is to amplify the photons inside the time window.[87,88] This has been achieved with a Raman amplifier, with a gain of 10^6, which was placed at the output of the diffusive media. Raman amplification is a process in which a weak light pulse gains strength from an intense one by a coupling mechanism that involves the vibrational modes of the medium that light passes through. Since the weak pulse must overlap temporally in the medium with the intense pulse to achieve amplification, this method offers a time-gating window that is set by the pulse duration of the intense (or pump) pulse, which in the case of the previously mentioned studies was 30 ps.[87,88] This method also offers limited imaging (2D) capability within the field of view of the pump pulse in the amplifier.

A related approach has been pursued by Lantz et al. by using another amplification process known as parametric amplification.[89] This type of amplification is generated in crystals irradiated by intense optical fields that drive the electrons in the material to respond nonlinearly. When this happens, the crystal can convert a high-energy photon into two lower-energy photons.

The conversion is stimulated by existing lower-energy photons. In Lantz's design, a potassium titanyl phosphate (KTP) nonlinear crystal was pumped with 532-nm light. The medium was illuminated with 1064-nm photons. Every 1064-nm photon that transilluminated the sample then interacted with a 532-nm photon in the crystal, which resulted in the creation of two additional photons at 1064 nm. Such a multiplicative effect results in the amplification of the image only at a time when it overlaps with the pump pulse.

6.4.1.2 Diffuse transmittance imaging

Since breast tissue has a comparably low blood content, it is rather transparent in the near-infrared spectral range, and the diffuse transmittance through up to 8 or 10 cm of tissue can be recorded. Optical mammography can therefore be performed in several basic geometries—the parallel slab geometry with a slightly compressed breast, the tomographic geometry with an uncompressed breast, and the reflectance geometry. The slab geometry has several advantages. The image is obtained by scanning the source and detector fibers across the area of the breast with a large number (typically approximately 1000) of scan positions, providing a high lateral resolution for inhomogeneities close to the surfaces. The signal does not change substantially across the whole scan. 2D diffuse projection images can be immediately obtained, without spatial reconstruction, by plotting any parameter derived from the measured time-of-flight distributions as a function of the x-y scan position. These images can be directly compared with x-ray images obtained in the same projection (compression direction). While numerous phantom and laboratory-based studies have been performed in the context of optical mammography with a variety of systems, this section focuses on the instruments that have been used in clinical studies.

The first clinical time-domain optical mammograph was developed at the Physikalisch-Technische Bundesanstalt (PTB) in Berlin.[91,92] It is shown schematically in Fig. 6.22 in its final configuration. The breast was slightly compressed between two glass plates, and the source fiber and the detection fiber bundle were moved in tandem in steps of 2.5 mm in a meander-like scan across the area of the breast. The repetition rate of the picosecond diode lasers was 24 MHz. The three pulse trains were interleaved on the nanosecond time scale by delaying them with respect to one another, so as to perform a quasi-parallel acquisition at the three wavelengths. The DTOFs recorded could be assigned to the respective wavelengths by their delay. A fiber bundle with a diameter of 4 mm and a numerical aperture of 0.54 was used. With a collection time at each scan position of 150 ms, a breast scan was completed within 3 to 10 min, depending on size. For a typical total photon count rate of $6 \times 10^5\ s^{-1}$, the DTOF at each wavelength contained approximately 3×10^4 photons. The photon count rate strongly increased while approaching the edges of the breast due to the decreasing breast thickness. The scan direction

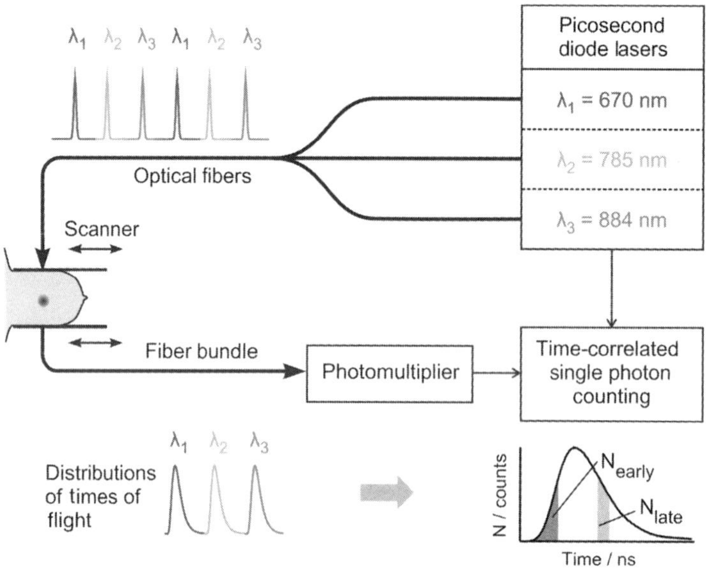

Figure 6.22 Block diagram of the first-generation PTB optical mammograph.

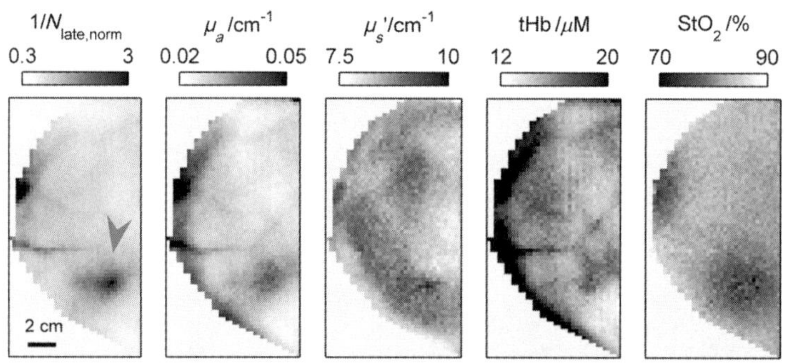

Figure 6.23 Optical mammograms (craniocaudal view) of a patient with an invasive ductal carcinoma indicated by the arrow. From left to right: Reciprocal normalized photon counts in a late time window of the DTOFs (670 nm), absorption coefficient μ_a (670 nm) reduced scattering coefficient μ_s' (670 nm) total hemoglobin concentration tHb and blood oxygen saturation StO_2 in percent.

was turned as soon as the count rate reached a predefined threshold. During analysis, a correction for the varying thickness was applied, exploiting the mean time of flight as a relative measure for breast thickness.[91]

The initial analysis of the measured arrays of DTOFs included plotting photon counts in selected time windows. Images of μ_s' and μ_a were derived by fitting the solution of the diffusion equation for a homogeneous slab to each DTOF. From the absorption coefficients at the various wavelengths, maps of total hemoglobin concentration and blood oxygen saturation in the tissue were calculated. Figure 6.23 shows an example of these mammograms. Since the volume sampled by the photons on their way from source to detector extends over the full thickness of the slab and is larger than a localized lesion, this homogeneous approach underestimates the changes between the lesion and the surrounding tissue. To create a more realistic estimate of the tumor's optical properties, an inhomogeneous model was applied by solving the diffusion equation for a homogeneous slab with a spherical inhomogeneity,[49] based on a model for the diffraction of photon density waves.[48] This model is not limited to small absorption changes. The required prior knowledge of the size of the lesions was derived from histopathology or other imaging modalities. An advanced multichannel time-domain mammograph developed by PTB included the option of fluorescence detection and simultaneous scans with additional detection fiber bundles offset from the central position.[93] The latter approach enabled the depth positions of lesions to be estimated.

An optical mammograph based on the same principles as the one in Ref. 91 was built at Politecnico di Milano.[94] It included four wavelengths between 680 nm and 975 nm, to investigate not only the concentrations of oxyhemoglobin and deoxyhemoglobin, but also the concentrations of lipids and water in the breast. An upgraded version of the instrument contained seven wavelengths of light from 630 nm up to 1060 nm, which enabled the additional quantification of collagen as another relevant tissue constituent.[95] Light at the four visible wavelengths and the three infrared wavelengths was collected by a bifurcated fiber bundle, spectrally filtered, and detected by two different photomultipliers.

6.4.2 Time-resolved optical tomography

The tomographic approach is limited to tissues that still have a sufficient transmittance when the source and the detector are placed on opposite sides of the medium. This is the case for the uncompressed breast and the neonatal head. A challenge in optical tomography is the large difference in the amount of light to be detected near and far from the source.

The first time-resolved optical tomographic images obtained through diffuse media were reported by Hebden and Wong in 1993.[96] The investigators employed time gating to transilluminate a cylindrical diffusive phantom, containing a series of inclusions, with quasi-ballistic photons. Picosecond pulses of light generated from a synchronously mode-locked dye laser were focused on the phantom. Light collected on the opposite side of the phantom was captured with a streak camera. A series of projections, each obtained by translating the sample across the emitter–detector axis, were performed subsequent to rotating the phantom in equal angular intervals of 12 degrees. Tomographic images were finally reconstructed with a standard back-projection method used in CT scanners.

The images obtained with the early approach of Hebden and Wong were noticeably fuzzy; in this sense, they provided an early glimpse into the challenges involved in attaining higher-resolution tomographic images with light. Furthermore, the sample to be imaged had to be mechanically translated and rotated, an arrangement that is unlikely to be practical for *in vivo* imaging.

One of the first optical tomography systems designed to meet the needs of *in vivo* imaging was presented by Hintz et al.[97] This system was designed to provide continuous monitoring of the brain tomographically and was targeted to the neonatal population. The apparatus consisted of an array of optical fibers, arranged in a stationary band that was strapped around the patient's head or a head phantom. Half of the fibers delivered light to the sample from pulsed laser diodes, and the other half guided light scattered from the sample to a detection system based on a picosecond-resolution photon counting system.

The tomography design reported by Hintz et al. used three laser diodes that operated at different wavelengths, along with a single detector. The acquisition of temporal curves, from all emitter–detector positions at all three wavelengths, was achieved sequentially with the aid of fiber-optic network switches. The process required an extensive amount of time to complete and, as noted by the authors, represented one of the greatest challenges facing optical tomography in diffusive media. Another related challenge was the wide variability in light intensity as a function of emitter–detector distances. On the one hand, this variation forced the investigators to use attenuation at short emitter–detector distances. On the other hand, it led to significantly longer acquisition times at larger emitter–detector distances.

The data obtained from the tomography system were analyzed with the diffusion equation, so as to obtain an average scattering coefficient through the sample. Tomographic images were obtained with simple back projections. The investigators reported obtaining images with a resolution that was depth dependent—for example, 1 to 2 mm at 2 cm versus 1 cm at 5 cm. A sample of their images is shown in Fig. 6.29 in the next section describing clinical applications. Images acquired in a clinical setting were reported to take two hours.

The challenges identified by Hintz et al. have been partially addressed by other research groups. Eda et al. described in detail a system developed in Japan that utilized parallel signal detection from detector sites around a diffuse medium.[98] In this system, the emitter distribution system was similar to the one described by Hintz et al.[97] Specifically, the output of three picosecond diode lasers was distributed *sequentially* to 64 sites equally spaced around the sample via mechanical network switches. However, the light scattered from the sample was collected at the 64 sites *simultaneously* by a network of detectors and signal processors operating in parallel. Each detection channel began with a custom-designed PMT mount containing an array of mechanically activated attenuators that balanced the collected signal with those from other channels. This was followed by parallel time-correlated single-photon counting circuits, one for each of the 64 detection sites. As with the design described by Hintz et al., the prototype created by Eda et al. utilized custom-designed fiber assemblies that permitted the delivery of signals to, and collection of signals from, the same site on the sample. In this case, the delivery fiber, consisting of a single 400-μm fiber, was embedded in the middle of a fiber bundle that was 3 mm in diameter. Acquisition time for a set of 64 × 63 emitter–detector temporal curves was reported by the investigators to be about 30 min for a phantom.

Rodriguez et al. developed a prototype for time-resolved optical tomography at increased speed, avoiding slow mechanical switching between sources.[99] In their design, which was based on nitrogen-pumped dye lasers, the emitter fibers were bundled together at the source, where they were illuminated simultaneously with every laser pulse. Each of these fibers was made increasingly longer so that its delivered pulses arrived at each of the emitting sites at an incrementally later time on the nanosecond time scale, thus enabling a quasi-parallel acquisition of signals from several sources.

Schmidt et al. described in great detail a prototype that was developed at University College London (UCL).[100] A schematic of this system is shown in Fig. 6.24. The system shared many of the features of the prototypes described previously, such as sequential laser pulse delivery via a fiber switch and simultaneous collection of signals by a parallel network of detector circuits based on time-correlated single photon counting. Unlike the prototypes discussed previously, the UCL system[100] used a Ti:sapphire system that delivered pulses significantly more energetic than those achieved by

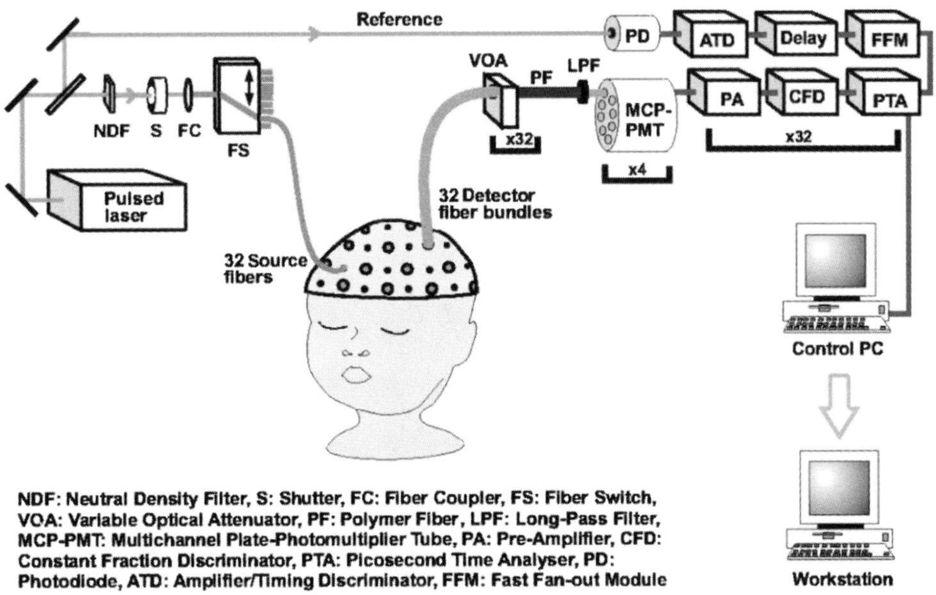

Figure 6.24 A schematic representation of the UCL prototype. (Reprinted with permission from Ref. 100. Copyright 1999, American Institute of Physics.).

picosecond laser diodes, while providing tunability over a wide spectral range. The prototype also included special multianode MCPs that are capable of detecting eight separate optical signals independently. Data acquisition times of 10 to 20 min were reported for their prototype on a phantom, which included a 30-s detection time for each illumination site and a 3.5-s switching time from one illumination site to the next.

The authors also identified numerous sources of problems that are inherent to these tomographic systems. These included the need to ascertain the temporal delays in each of the fibers (also pointed out by Hintz et al.[97]), the presence of temporal shifts of approximately 5 ps/h (even after a 10-h warm-up period), cross-talk between channels from the multianode detectors and from the fiber switch, and reflections from the ends of the fibers.

The photon time-of-flight information obtained with this system was processed with FEM-based reconstruction software developed by the same group, known as TOAST (Temporal Optical Absorption and Scattering Tomography).[101,102] The method and software were steadily improved over the years, resulting in the TOAST++ open-source software environment for solving the forward and inverse problems in diffuse optical tomography.[41]

Figure 6.25 demonstrates the imaging capability of the prototype at this stage of development. The image was acquired with a cylindrical phantom and with the optodes aligned in a single plane. This system was used for clinical studies of the brain of neonates and later for breast studies. The breast investigations were first performed with rings holding the fibers in air and then replaced with cups filled with scattering liquid.[103] In this way the problems associated with calibrating the coupling between fibers and tissue were avoided. In addition, a differential reconstruction could be performed using a measurement with the liquid-filled cup without the breast as reference.

Recently, the UCL group presented an advanced prototype for optical tomography called MONSTIR II and its performance characterization.[104] This system's new features include a supercontinuum laser source with two AOTFs performing rapid switching between four selected wavelengths. Light

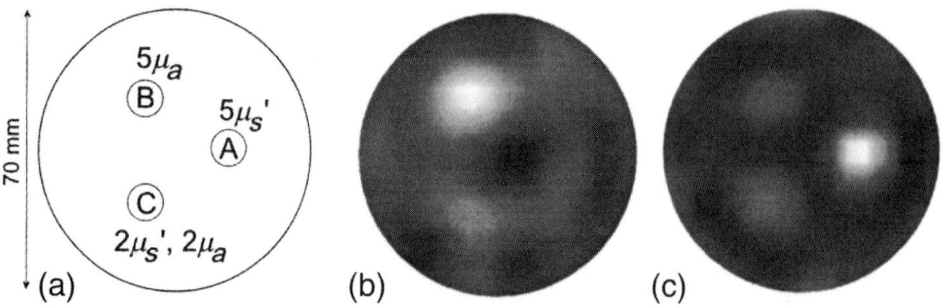

Figure 6.25 Tomographic maps obtained with the UCL prototype, depicting (b) absorption and (c) scattering cross sections of phantom (a). Reproduced with permission from Ref. 101.

is detected by 32 PMTs in parallel. By using eight-channel routing electronics, groups of eight of the PMTs share one high-performance TCSPC module.

A group at Hamamatsu Photonics K.K. and University School of Medicine in Hamamatsu City, Japan, also developed a prototype of a time-domain optical mammograph based on the tomography approach.[105] The breast was suspended in a liquid-filled hemispherical cup of 128-mm diameter, holding 48 fiber bundles, each of which combined a source fiber and a detection bundle. The matching fluid, which comprised Intralipid, water, and carbon ink, was adapted to each patient's breast and kept at 37 °C. A mode-locked Ti:sapphire laser was used as a light source and tuned to 765 nm, 800 nm, and 835 nm. A 1×48 fiber switch was utilized to select the illumination point. The detection was performed with 48 channels, each consisting of a fiber bundle, variable attenuator, PMT, and TCSPC electronics. Image reconstruction of local changes in μ_a was based on the method of the time-resolved photon path distribution.[106]

A multimodal approach for quantitative diffuse optical tomography was pursued by Ntziachristos et al.[107] Concurrent examination enabled magnetic resonance (MR) and diffuse optical imaging under identical physiological conditions and geometry. This approach offers the possibility to employ MR images as *a priori* information to improve the optical tomographic reconstruction. The optical setup included laser diodes and TCSPC detection. The breast was softly compressed between two plates holding MR coils and optical fibers. A total of 24 source fibers were mounted on one plate in a 3×8 grid with 1.25-cm adjacent fiber separations, and 8 detector fiber bundles in a 2×4 grid (separation 2.5 cm) on the other plate. The laser was coupled to the source fibers sequentially by an optical switch while all 8 detection channels were recording in parallel. The authors reported the results of a study on 14 patients with malignant and benign breast lesions, analyzing hemoglobin concentration and saturation in the lesions.[108] Moreover, indocyanine green (ICG) was used as optical contrast agent to provide differential absorption contrast.[107]

6.4.3 Reflectance imaging

6.4.3.1 Depth-resolved coherence imaging

One of the first demonstrations of depth-resolved imaging through diffusive media was achieved with the photorefractive crystal barium titanate, BaTiO3. This crystal possesses an index of refraction that changes in the presence of light. This property has been exploited in the past to store holographic patterns; these patterns are created when an image pattern and a reference light beam interfere inside the crystal. Hyde et al. exploited this property to store images collected from well-defined depths within a diffuse medium.[51]

Their experimental setup was similar to the schematic shown in Fig. 6.8. The output from a frequency-doubled mode-locked Ti:sapphire laser

(70-fs pulses, at a wavelength of 450 nm) was split into two beams—one that was used to illuminate the diffusive medium and one that was used to provide a reference inside a $BaTiO_3$ crystal. When the path length of the reference coincided with that of the light back-scattered from a plane inside the medium, light interference patterns were generated in the crystal containing imaging information about the plane. With sufficient exposure, the interference pattern eventually built up an index of refraction grating that contained spatial variation information about the imaging plane. This grating remained stored in the crystal and could be "read" subsequently by the reference beam after the scattered light was blocked. The image read contained a sharp image of the imaging plane since the grating was created by the coherent (or ballistic) radiation originating from the image plane. Hyde et al. reported a spatial resolution of about 50 μm, transverse and in depth, to an imaging depth of 5 scattering mean-free paths.[51]

Another approach for depth-resolved imaging has been advanced by Zinth et al.[52,109,110] In their approach, a femtosecond laser beam was split into two beams, one of which was tightly focused in the sample, while the other was directed to a beta barium borate (BBO) nonlinear crystal where it serves as a reference beam. Again, this geometry was very similar to the one depicted in Fig. 6.8. Inside the sample, the light scattered back along the path of the beam was reflected back out. The back-scattered light was collected and focused into the nonlinear crystal, where it interfered coherently with the reference beam, if their path lengths were equal. As a result of the nonlinear interaction with BBO, a new beam was generated that bisected the signal and reference beams; this new beam's strength was proportional to the scattered signal at a particular depth. As the path length of the reference beam increased, it overlapped with light that scattered from progressively deeper regions of the sample. In this way, the system was able to obtain a depth scan of the sample from a particular point on the surface. With additional transverse scanning of the probe beam, the cross sections of the sample were obtained.

Baigar et al. have reported several results obtained with this technique on test objects consisting of multilayered scattering media and in scattering media containing a pinlike inclusion.[110] Their results indicate that a depth resolution of 10 μm can be achieved with this approach. The investigators also reported depth-resolved imaging from porcine skin to a depth of 1 mm. A cross section of the skin obtained is shown in Fig. 6.26. A skin layer 0.7 to 0.8 mm in thickness is readily apparent in the image.

6.4.3.2 Diffuse reflectance imaging

Functional near-infrared spectroscopy (fNIRS) of the brain has become one of the major targets in time-domain diffuse optical imaging. Its major advantage over the CW method is the ability to discriminate between cerebral and extracranial absorption changes. In their recent review of time-domain

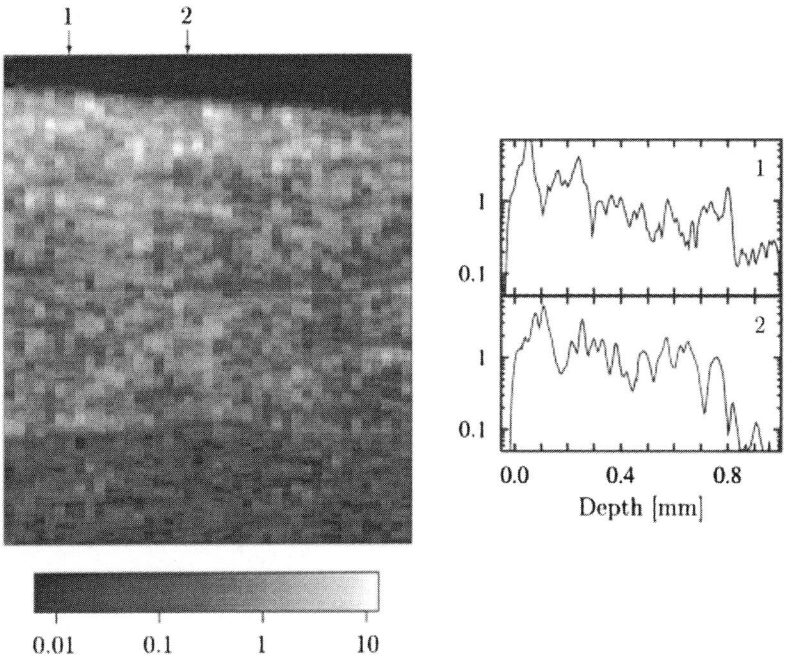

Figure 6.26 An optical biopsy of porcine skin obtained with the depth-resolved method described in Ref. 110. (Reprinted with permission.)

fNIRS imaging, Torricelli et al. gave a comprehensive overview of various generations of time-domain fNIRS systems and their components as well as their performance.[111] This section introduces design features of a few recent systems, with a focus on the imaging performance.

The compact, portable prototype time-domain optical brain imagers developed for clinical studies by groups in Berlin,[112] Milan,[113] and Warsaw[114] were based on picosecond diode lasers, fast compact photomultipliers and multiboard TCSPC systems, optical fibers for light delivery and fiber bundles for light collection. With these systems, topographic mapping of brain activation was achieved by grids of source–detector pairs. Piezomechanical fiber switches were employed to switch between various source fibers, while 4 or 8 TCSPC boards were used for parallel detection. A specific challenge in time-domain functional brain imaging is the small magnitude of changes to be detected, which are typically in the range of a few percent in amplitude and approximately 1 ps in mean time of flight. The detection of such small changes with a good signal-to-noise ratio requires sufficient input power (within safety limits), high responsivity of the detection system, and the acquisition at high photon count rates of at least approximately 10^6 s^{-1}. Another challenge is the requirement to perform dynamic measurements—that is, recordings from multiple source–detector pairs and wavelengths to be completed within 1 s or less. One of the related solutions included cascades of fiber switches to combine multiplexing of wavelengths and source positions (16 sources and 8 detectors).[115] Another advanced solution was based on rapid multiplexing of wavelengths by burst-mode operation of picosecond diode lasers on the μs time scale in combination with encoding of the source position on the nanosecond time scale by fibers of different length (5 sources and 4 detectors per hemisphere).[116]

Selb et al. found a different way to implement multiple parallel detection channels, by using a gated intensified CCD (ICCD) camera as a multichannel detector, in combination with detection fiber optics.[117] The need for a sequential scan of the electronic gate was avoided by combining 7 fibers of different lengths (in 10-cm increments corresponding to a 500-ps delay time) in each of 25 detection bundles, thereby covering the complete DTOF with a single electronic delay. The beam of a Ti:sapphire laser was sequentially coupled into 42 source fibers. Lateral imaging performance and depth discrimination were demonstrated by phantom and *in vivo* tests.

Mazurenka et al. described the first implementation of non-contact scanning for time-domain optical brain imaging, adopting the null source–detector separation approach.[56,118] By using a galvanometer scanner and a short source–detector separation of a few millimeters, a dense, flexible grid of measurement positions was realized (e.g., 32×32 pixels in a 4×4 cm^2 area) and scanned with a frame rate of 1/s. A supercontinuum laser with AOTF allowed for fast switching between two wavelengths (line by line). The key element in this implementation was a fast-gated SPAD detector that was employed to eliminate the intense early part of the DTOFs.[119] The remaining late part was recorded by imaging TCSPC.[71] A step width on the order of 1 mm definitely leads to oversampling when imaging features at the depth of the cerebral cortex. However, such spatial resolution can be useful to study superficial hemodynamic changes, especially in the vasculature. While brain activation—in particular, by motor stimulation—was successfully detected and localized by this approach, substantial challenges remain. The small size of the detector (100 μm) prevented an efficient detection of light emerging from the tissue. Another disadvantage compared to fiber-based systems was the strong decrease in the recorded count rate in the presence of hair.

Commercial time-domain NIRS devices were developed by Hamamatsu Photonics K.K. in Japan—namely, the single-channel TRS-10 device[120] and later the two-channel TRS-20 device[121] that operates at 760, 800, and 830 nm. Data analysis relied on a fit of the DTOF with a homogeneous diffusion model and provided oxyhemoglobin and deoxyhemoglobin concentrations. Interestingly, it was shown recently that this approach yields a higher sensitivity to deep absorption changes and a lower sensitivity to superficial absorption changes compared to two devices based on the CW and frequency-domain technologies.[122]

An important element in assessing and improving technical designs is a quantitative performance characterization. The consortium of a European project developed two protocols for assessing the performance of time-domain optical brain imagers and applied them in multiple-laboratory comparisons. The Basic Instrumental Performance protocol addressed the direct measurement of relevant device characteristics,[123] while the nEUROPt protocol was based on dedicated inhomogeneous phantoms and included measurements to characterize sensitivity, spatial resolution, and quantification of localized absorption changes.[124]

6.5 Toward Clinical Applications

6.5.1 Time-domain optical mammography

X-ray mammography is used routinely today as a diagnostic tool for the detection and location of breast tumors. Although its diagnostic value is undeniable, it suffers from several drawbacks. The procedure, which requires breast compression, can be highly uncomfortable. The exposure to x-rays, which is a form of ionizing radiation, always raises safety concerns. Finally, x-ray mammographs often offer ambiguous results that can be confirmed only with much more invasive procedures. A need exists for additional weapons against this disease that are comfortable, safe, and more accurate when making a diagnosis.

Much of the research in near-infrared diffuse optical imaging has been drawn into providing an alternative imaging modality for the diagnosis of breast cancer. Several reviews provide a comprehensive overview of the advances in optical mammography.[125,126] In 2008, Leff et al. published a systematic review of clinically oriented studies based on CW, frequency-domain, and time-domain techniques.[127] In relation to time-domain diffuse optics, a recent overview specifically addressed various applications to assess breast physiology and pathology.[128]

Optical mammography has naturally attracted the interest of industry. Continuous-wave methods have been pursued, in particular, by several companies that attempted to develop commercial optical mammography systems, including Philips (continuous-wave approach) and Siemens (frequency-domain approach). Imaging Diagnostic Systems Inc. (IDSI) of Fort Lauderdale, Florida, surprised the scientific community when it announced the development of a commercial prototype as early as 1995. In the IDSI system, the patient lay prone on a table with a breast pendant from an aperture in the bed, into a chamber where the scanning took place. The early prototype contained a mode-locked Ti:sapphire laser that delivered subpicosecond pulses to the breast, and a large array of avalanche photodiodes collected light transmitted through the breast. Later IDSI successfully changed its strategy to produce CW optical mammographs. Another attempt with time-domain mammography was made by the Canadian company Advanced Research Technologies (ART). The ART system known as Softscan contained a mode-locked Ti:sapphire laser and a streak camera. The pendulant breast was softly compressed and scanned in transmittance. Several prototypes were built and successfully applied in clinical studies,[125,129] but the company could not continue this activity.

In the research community, the first substantial clinical studies based on time-domain optical mammography were performed within the European project OPTIMAMM (2000–2004) in Berlin and Milan. The group in Berlin investigated 154 patients suspected of having breast cancer.[91] Optical mammograms, as shown in Fig. 6.23, were recorded in two projections (craniocaudal and mediolateral); they were compared with the corresponding x-ray mammograms and MR mammograms, if available, and a visibility score was assigned. Out of 102 histologically confirmed tumors, 72 (70%) were detected retrospectively in both optical projection mammograms; in an additional 20 cases (20%), the tumors were detected in one projection only. The group in Milan tested their instrument on a total of 194 patients with 225 malignant and benign lesions.[130] Out of 52 cancers, 41 (79%) were detected in two views and 9 more (17%) in one view only. In addition, a significant number of cysts (82) were analyzed. The detection rate was 83% in both projections; it was 90% if a detection in one view was regarded as sufficient. Generally, the intensity in a late time window at wavelengths of less than 700 nm turned out to exhibit the highest tumor contrast and contrast-to-noise ratio. This finding can be attributed to the typically higher blood content due to higher vascularization in malignant lesions. It was confirmed by the more detailed assessment based on tumor optical properties.[49] In Fig. 6.27, the resulting total hemoglobin concentration (tHb) and oxygen saturation are

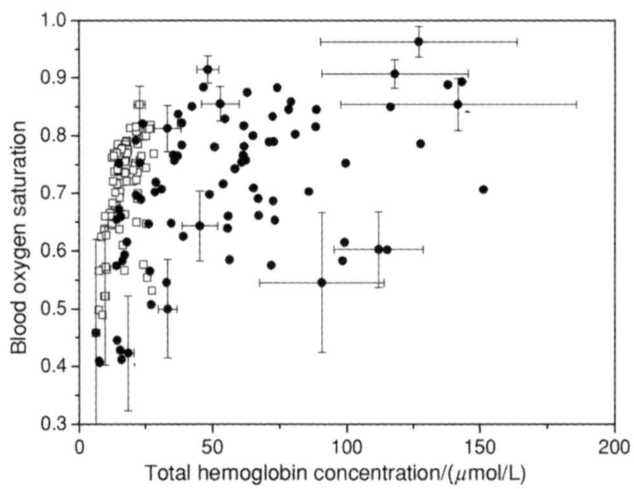

Figure 6.27 Blood oxygen saturation versus total hemoglobin concentration of tumors (solid circles) and corresponding healthy breast tissue (open squares) for 87 patients. If available, data are averaged over both projection mammograms. From Ref. 49, with permission.

plotted for 87 carcinomas. While tHb in the carcinomas was increased with respect to the tHb in the surrounding healthy tissue by about a factor of 3 on average, oxygen saturation turned out to be a poor discriminator.

In another part of the OPTIMAMM project, a group at University College London tested the time-domain optical tomography approach on the uncompressed breast in a study involving 21 patients.[131] Based on 3D absorption and scattering images, 17 out of 19 cases of lesions of different nature were detected. A second study with an improved system and adopting a liquid-filled cup arrangement was performed on a cohort of 38 patients with various lesions.[104] An example of the tomograms obtained in this study is shown in Fig. 6.28. The results were consistent with previous studies. For breast cancer, in most cases an increased absorption rate and increased regional blood volume were found. Only half of the fibroadenomas could be observed.

While their sensitivity for cancer detection may be acceptable, specificity is problematic for mammograms based on intrinsic optical contrast. This challenge is not specific to the time-domain implementation of optical mammography. Benign lesions, notably mastopathies, also exhibit an elevated blood content. Such conditions may frequently lead to false-positive findings, thereby rendering optical mammography inadequate for screening. In turn, this situation inspired studies with optical contrast agents.

So far, no tumor-specific optical contrast agents have become available for application in humans. Studies in this area are presently limited to indocyanine green (ICG), a dye that is routinely used in medical practice—for example, in liver function diagnostics and cardiac output assessment. Hagen et al. exploited the enhanced permeability and retention effect of macromolecules in tumors to distinguish malignant from benign lesions.[132] Due to the strong binding of ICG to plasma proteins (macromolecules), this complex can extravasate only through the leaky capillaries in carcinomas, where it remains for a longer time because of the impaired lymphatic system. With a combination of a bolus and a continuous infusion over 20 min, a prolonged

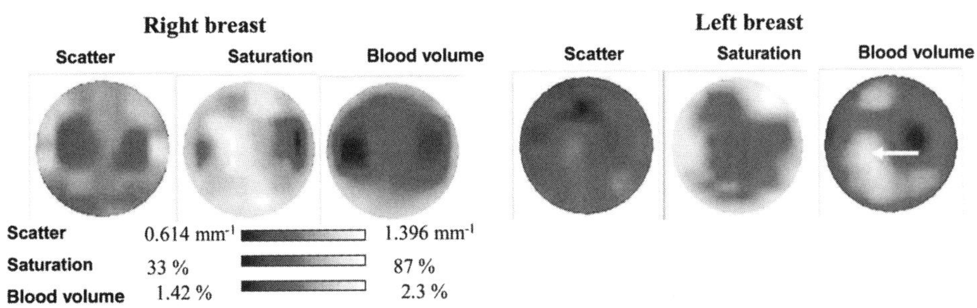

Figure 6.28 Tomographic images of reduced scattering coefficients, oxygen saturation and blood volume from a 45-year-old woman with a carcinoma in the left breast (indicated by the arrow). Reprinted from Ref. 104 with permission from the Optical Society of America. (See color plates.)

interaction time between ICG and tissue was achieved. A measure of the extravasated ICG concentration in the tissue was obtained by recording the ICG fluorescence about 25 min after the end of the dye infusion, when the ICG was mostly cleared from the blood (extravascular phase). In a clinical study that utilized on the PTB fluorescence mammography technique, 20 patients with suspicious lesions were investigated.[93] Ratio images obtained by dividing fluorescence by transmittance images in the extravascular phase clearly showed carcinomas, whereas fibroadenomas did not exhibit extravasation. These results suggest that imaging the extravasation of ICG-labeled macromolecules is a promising way to differentiate malignant from benign breast lesions.[133]

Most of the work in the field of therapy monitoring has relied on the CW or frequency-domain approaches. Monitoring during neoadjuvant therapy has been a significant focus of research in optical mammography since 2004. Chemotherapy is widely used to treat breast cancer, especially to reduce the size of large carcinomas before surgery. Estimation of the individual response to the treatment is relevant to avoid exposure to ineffective drugs with severe side effects or to change the treatment. In this context, the ability of diffuse optical imaging to quantify local hemoglobin concentrations can be an advantage,[134] as can its noninvasive nature.

The group at University College London was the first to test the benefit of time-domain optical tomography on the uncompressed breast in this application in two pilot studies,[135,136] using the instrument mentioned earlier.[104] The first study (4 cases) addressed the neoadjuvant chemotherapy applied prior to surgery in women with locally advanced breast cancer.[135] Measurements were taken before, during, and after treatment over a period lasting up to 6 months. Total hemoglobin concentration and oxygen saturation of the healthy tissue were shown to be reproducible. Both parameters changed during therapy. The second report presented 3 case studies from postmenopausal women with estrogen receptor–positive tumors treated with neoadjuvant hormone therapy.[136] The results of the optical measurements—in particular, total hemoglobin concentration in the tumors—correlated with the treatment outcomes.

Ueda et al. reported on monitoring of the early optical response to single-agent bevacizumab in 7 patients (4 responders and 3 non-responders).[137] The authors used a handheld probe based on the TRS-20 system (Hamamatsu Photonics K.K., Japan),[138] in combination with an ultrasound system; they performed a set of time-resolved reflectance measurements at a 3-cm source–detector separation. The results showed a low tumor oxygenation level to be characteristic of the nonresponders. Although the number of cases was far too small to reach definitive conclusions about the applicability of this approach, it is obvious that time-domain optical mammography has the potential to provide valuable physiological information for early prediction of response to therapy.

Important diagnostic information about the breast can be derived from a quantification of other chromophores than oxyhemoglobin and deoxyhemoglobin. Breast density, for example, is a recognized independent risk factor for developing breast cancer. Its noninvasive assessment is possible by optical methods. Taroni et al. performed time-domain optical mammography employing their seven-wavelength instrument[130] on 147 subjects to characterize the breast tissue composition for each subject based on its oxyhemoglobin and deoxyhemoglobin, water, lipids, and collagen content.[139] Moreover, the scattering parameters, i.e., the amplitude and the slope characterizing the wavelength dependence of the reduced scattering coefficient, were determined. In a thorough statistical analysis, collagen (content and scattering parameters) was shown to be the key parameter for identifying women at high risk of developing breast cancer—a factor that may be connected with high breast density. The quantification of the collagen content may also be beneficial to distinguish malignant from benign lesions, as was shown by the characterization of 32 malignant and 29 benign lesions based on a perturbation model analysis of 7-wavelength mammograms.[140]

During the last 15 years of intense research, the focus in optical mammography research has shifted from development of a screening tool for initial detection of lesions toward therapy monitoring and detailed characterization of lesions as well as risk assessment. The question is still open as to whether time-domain optical mammography can become a valuable, routinely used adjunct to existing diagnostic modalities. Several promising approaches have been identified but require validation in larger clinical studies and support from advanced technological developments before they are widely adopted.

6.5.2 Time-domain optical brain imaging

6.5.2.1 Optical tomography of the infant brain

Another potential application of time-resolved imaging is for detection and monitoring of cerebral hemorrhage and oxygenation in neonates. The small size of the infant's head, combined with the thin skull at this stage of development, makes it possible to transilluminate the head without elevated optical intensities. As a consequence, optical spectroscopy can be used to probe every portion of an infant's brain and, in turn, to obtain tomographic images of key physiological parameters, including blood volume, blood flow, and tissue oxygenation. Furthermore, the portability of NIR technology and the safety of the radiation it uses enable optical scanners to be brought to the intensive care unit, where they can monitor continuously the condition of critically ill patients.

The first clinical trials of this technology in the neonatal population were reported in Refs. 97 and 141. The results demonstrated that time-resolved optical tomography successfully detected cerebral hemorrhages, which were confirmed by other imaging modalities. In some cases the optical scanner

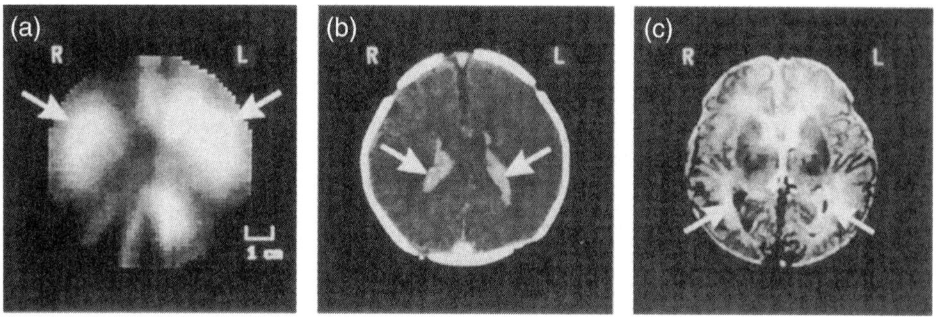

Figure 6.29 A comparison between (a) an optical scan, (b) an MRI scan, and (c) a CT scan. The neonate had bilateral intraventricular hemorrhage, in the locations indicated by the arrows (from Ref. 141).

yielded false-negative results, but these could be accounted for by factors such as positioning of the optodes, aging blood clots that were blanched, and the small size of the bleed. The investigators suggested that they could overcome these factors with improved technology. Figure 6.29 shows an example of an optical scan performed on a neonate with intraventricular hemorrhage, in areas indicated by arrows.

Later, the group at UCL applied its 32-channel tomographic instrument, MONSTIR,[100] to image blood volume and oxygenation in the brain of preterm infants in the clinical setting.[142] Successful 3D reconstructions were obtained in several cases of healthy preterm infants, showing regional variations in cerebral blood volume and tissue oxygen saturation.[142] In two cases with intraventricular hemorrhage, related increased blood volume and decreased oxygenation were reported.[142,143] Functional images were obtained from responses to passive arm movements in 4 out of 6 neonates examined.[144] Six of 8 images showed the largest change in total hemoglobin in the region of the contralateral motor cortex. Although the optical method can provide valuable information, a number of challenges related to the practicality of the measurement and the accuracy of the reconstruction procedure remain.

6.5.2.2 Functional optical brain imaging and cerebral oximetry in adults

The field of fNIRS has been growing tremendously since 1992, as vividly shown in the historical review by Ferrari and Quaresima,[145] in parallel with the availability of a number of commercial devices based on CW technology and a few frequency-domain devices. While brain activity monitoring is discussed in more detail in Chapter 8, this section focuses on applications of the time-domain methodology. Torricelli et al. provided a thorough review of the various traditional and state-of-the-art systems developed by several research groups, including their *in vivo* applications.[111] This section will focus on clinical studies involving patients as an indicator of attempts to pave the way to clinical diagnostic applications of the method.

Initial *in vivo* applications demonstrated the depth-resolving capability of the time-domain method.[57,58] Among the various studies conducted in healthy adult subjects, several were aimed at the validation of the separation between cerebral and superficial hemodynamic responses and an in-depth investigation of the confounding influence of hemodynamic changes in the skin, in particular in the prefrontal area,[146,147] highlighting the value of time-resolved measurements.

One of the first applications of time-domain NIRS to patients was published by Hoshi et al. in 2006; these researchers employed the TRS-10 device to investigate resting hypofrontality in 14 schizophrenic patients, by measuring the total hemoglobin concentration on both sides of the forehead.[148] Ohmae et al. monitored 23 patients who underwent coronary-artery bypass surgery by deriving the (absolute) concentrations of oxyhemoglobin and deoxyhemoglobin as well as oxygen saturation of blood in brain tissue from TRS-10 measurements.[149] With the same device, Sato et al. examined 8 patients during carotid endarterectomy.[150] They compared the concentration changes of oxyhemoglobin and deoxyhemoglobin obtained from fitting the DTOFs with a solution of the diffusion equation with those obtained by means of the modified Beer–Lambert law as used for CW fNIRS measurements.

The brain imagers developed by the groups in Berlin, Milan, and Warsaw were also employed in a number of clinical fNIRS studies. A combined setting with direct current (DC)–magnetoencephalography was applied to assess neurovascular coupling in healthy subjects[151] and stroke patients[152] performing a motor task, by comparing the time courses of hemodynamic and electrophysiological responses. Visani et al. studied the neuronal and hemodynamic responses to motor stimulation in 10 epileptic patients with movement disorders by using a multimodal approach including time-domain fNIRS, EEG, and fMRI.[153] Kacprzak et al. monitored 16 patients with atherosclerotic disease during carotid endarterectomy.[154] Time courses of changes in the cerebral oxyhemoglobin and deoxyhemoglobin concentrations derived from a depth-selective layered reconstruction were recorded during cross-clamping of the carotid arteries.

6.5.2.3 Perfusion assessment by ICG bolus tracking

The kinetics of a bolus of the optical contrast agent indocyanine green (ICG) provide information on the regional tissue perfusion, similarly to perfusion-weighted MRI. A bolus is injected intravenously, and its shape when passing through the tissue of interest is analyzed. The optical method has the potential to become a neuromonitoring tool in such cases that is applicable at the bedside. The need to distinguish between cerebral and superficial bolus signals is particularly relevant since the dye bolus is present in the whole circulation and can mask the bolus in the cortex.

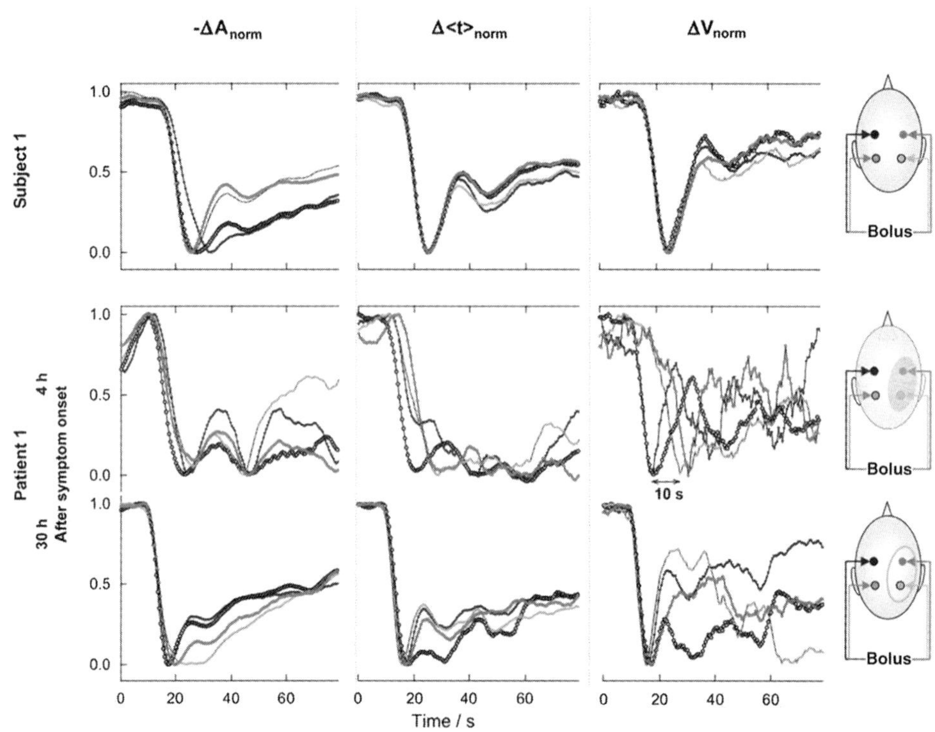

Figure 6.30 Changes in attenuation $-\Delta A$, mean time of flight $\Delta <t>$, and variance ΔV recorded at two sites on each hemisphere after administration of an ICG bolus (intravenous injection at time $t = 0$), for a healthy subject and a patient. All signals were rescaled to a range from 0 to 1. Reprinted from Ref. 155 with permission from Elsevier (see color plates).

In 2005, Liebert et al. reported the first application of time-domain optical brain imaging to the assessment of cerebral perfusion in two patients with cerebral perfusion deficits.[155] Figure 6.30 illustrates the kind of signals obtained, as well as the advantage of selecting the signal from the cortex by using variance of the DTOFs. For the healthy subject, the changes in the moments representing the bolus appear virtually simultaneously in both hemispheres. For the stroke patient suffering from an ischemic stroke, the bolus is delayed in the impaired hemisphere. The delay is most pronounced in the variance signal, while the attenuation signals (equivalent to a CW measurement) are not sufficient to distinguish between healthy and diseased tissue. A thrombolytic therapy resulted in normal perfusion, as indicated by the results obtained after 30 hours.

Steinkellner et al. reported the results of a study on 10 patients who had experienced acute unilateral ischemic stroke. In all patients, a delay (between 0.2 and 5.2 s) of the bolus over the affected hemisphere as compared to the unaffected hemisphere was found.[156] Weigl et al. applied the ICG bolus tracking method in a study on patients with post-traumatic brain injury—specifically,

11 patients with brain hematoma and 15 patients with brain edema.[157] Apart from detecting absorption changes, they also analyzed the fluorescence from the ICG bolus. Diop et al. combined the ICG bolus technique with another promising optical technique to continuously monitor cerebral perfusion—namely, diffuse correlation spectroscopy (DCS)—in a study on piglets.[158] They showed that the time-domain bolus measurement can be employed to calibrate the DCS measurement to yield an absolute quantification of cerebral blood flow.

6.6 Conclusions

From the early days of Bright's report on hydrocephalic transillumination[1] to the clinical applications of today, the field of optical imaging through diffusive media has undergone tremendous advances. Some were made possible by technological advances such as picosecond pulse generation and detection. Others were fueled by theoretical modeling of experimentally observed signals, which allowed, for example, the separation of the contributions of absorption and scattering and the reconstruction of tomographic images in diffusive media.

Two major directions of clinical application of time-domain diffuse optical imaging have been targeted by a number of research groups. In optical mammography, the time-domain method allows for quantification of optical properties of both bulk tissue and lesions. The clinical studies performed so far in this area have included hundreds of patients, yielding a rich pool of experience. The initial idea to employ the optical method for screening turned out to be unrealistic, mainly because of insufficient specificity. However, optical mammography seems to be a promising tool to monitor the response to neoadjuvant chemotherapy by assessing total hemoglobin concentration and oxygen saturation. In addition, the quantification of chromophores other than oxyhemoglobin and deoxyhemoglobin and of scattering parameters is expected to become a way to identify women at high risk of developing breast cancer.

Optical brain imaging is a rapidly growing field in which the time-domain method so far has gained a rather limited share. However, its advantages are obvious: Discrimination between cerebral and superficial signal contributions and their separate quantification is feasible, even at a single source–detector separation. There are a multitude of diverse applications for the versatile tool of optical brain imaging, ranging from diagnostics in neurology and psychiatry, psychology and development, to basic research. The spread of the new technology into these fields was clearly facilitated by the availability of a number of commercial systems based on the CW technology, with hundreds of units being sold. A number of clinical feasibility studies have demonstrated the ability of the time-domain technology to yield superior results. Nevertheless, in their competition with CW systems, the time-domain systems still have disadvantages in terms of compactness and costs. This situation is likely to change in the near future. There are currently ongoing formidable developments in picosecond laser technology, fast detectors based on novel principles, as well as faster and more compact multichannel single-photon counting electronics that are being taken up rapidly by research groups. The availability of commercial devices based on such novel technologies will be crucial for the further spread of the time-domain methodology and its broader adoption for routine use in clinics.

In summary, time-domain diffuse optical imaging has made substantial progress in the last two decades and promises to become a relevant tool in routine clinical diagnostics in several areas. We therefore anticipate that this field will experience a further growth in the coming years.

Acknowledgments

The authors wish to thank Harold Battarbee (formerly with Louisiana State University Health Sciences Center, USA, now retired) for his substantial contribution to the first edition of this chapter.

HW is grateful for research funding received from the European Community's Seventh Framework Programme (FP7/2007-2013) under grant agreement No. FP7-HEALTH-F5-2008-201076. VVT is thankful for support from the Government of Russian Federation (grant No. 14.Z50.31.0004), Russian Presidential grant NSh-7898.2016.2, and the Tomsk State University Academic D.I. Mendeleev Fund Program.

References

1. O. Minet, G. J. Mueller, and J. Beuthan, Eds., "Selected papers on optical tomography: fundamentals and applications in medicine," *SPIE Milestone Series* **MS147**, 6–12 (1998).
2. O. Minet, G. J. Mueller, and J. Beuthan, Eds., "Selected papers on optical tomography: fundamentals and applications in medicine," *SPIE Milestone Series* **MS147**, 13–25 (1998).
3. M. Cutler, "Transillumination as an aid in the diagnosis of breast lesions," *Surg. Gynecol. Obstet.* **48**, 721 (1929).
4. F. Jöbsis, "Non invasive monitoring of cerebral and myocardial oxygen sufficiency and circulatory parameters," *Science* **198**, 1264–1267 (1977).
5. M. A. Duguay and A. T. Mattick, "Ultrahigh speed photography of picosecond light pulses and echoes," *Appl. Opt.* **10**, 2162–2170 (1971).
6. J. L. Martin, Y. LeCarpentier, A. Antonetti, and G. Grillon, "Picosecond laser stereometry light scattering measurements on biological material," *Med. Biol. Eng. Comput.* **18**, 250–252 (1980).

7. J. G. Fujimoto, D. De Silvestry, E. P Ippen, C. A. Puliafito, R. Margolis, and A. Oseroff, "Femtosecond optical ranging in biological systems," *Opt. Lett.* **11**, 150–152 (1986).
8. B. C. Wilson and G. Adam, "A Monte-Carlo model for the absorption and flux distribution of light in tissue," *Med. Phys.* **10**, 824–830 (1983).
9. M. Patterson, B. Chance, and B. C. Wilson, "Time-resolved reflectance and transmittance for the non-invasive measurement of tissue optical properties," *Appl. Opt.* **28**, 2331–2336 (1989).
10. K. M. Yoo and R. R. Alfano, "Time-resolved coherent and incoherent components of forward light scattering in random media," *Opt. Lett.* **15**, 320–322 (1990).
11. D. T. Delpy, M. Cope, P. van der Zee, S. Arridge, S. Wray, and J. Wyatt, "Estimation of optical pathlength through tissue direct time of flight measurement," *Phys. Med. Biol.* **33**, 1433–1442 (1988).
12. B. Chance, S. Nioka, J. Kent, K. McCully, M. Fountain, R. Greenfeld, and G. Holtom, "Time-resolved spectroscopy of hemoglobin and myoglobin in resting and ischemic muscle," *Anal. Biochem.* **174**, 698–707 (1988).
13. F. Martelli, S. Del Bianco, A. Ismaelli, and G. Zaccanti, *Light Propagation through Biological Tissue and Other Diffusive Media*, *SPIE*, Bellingham, Wash. (2010).
14. J. M. Maarek, G. Jarry, J. Crowe, Bui-Mong-Hung, and D. Laurent, "Simulation of laser tomoscopy in a heterogeneous biological medium," *Med. Biol. Eng. Comput.* **24**, 407–414 (1986).
15. G. Jarry, S. Ghesquiere, J. M. Maarek, F. Fraysse, S. Debray, Bui-Mong-Hung, and D. Laurent, "Imaging mammalian tissues and organs using collimated transillumination," *J. Biomed. Eng.* **6**, 70–74 (1984).
16. J. C. Hebden and R. A. Kruger, "Transillumination imaging performance: spatial resolution simulation studies," *Med. Phys.* **17**, 41–47 (1990).
17. J. C. Hebden, "Evaluating the spatial resolution performance of a time-resolved optical imaging system," *Med. Phys*, **19**, 1081–1087 (1992).
18. H. Wabnitz and H. Rinneberg, "Imaging in turbid media by photon density waves: spatial resolution and scaling relations," *Appl. Opt.* **36**(1), 64–74 (1997).
19. A. H. Gandjbakhche, R. Nossal, and R. F. Bonner, "Resolution limits for optical transillumination of abnormalities deeply embedded in tissues," *Med. Phys.* **21**, 185–191 (1994).
20. V. Chernomodik, R. Nossal, and A. H. Gandjbakhche, "Point spread functions of photons in time-resolved transillumination experiments using simple scaling arguments," *Med. Phys.* **23**, 1857–1861 (1996).
21. J. C. Hebden, D. J. Hall, and D. T. Delpy, "The spatial resolution performance of a time-resolved optical imaging system using temporal extrapolation," *Med. Phys.* **22**, 201–208 (1995).

22. J. C. Hebden, D. J. Hall, M. Firbank, and D. T. Delpy, "Time-resolved optical imaging of a solid tissue-equivalent phantom," *Appl. Opt.* **34**, 8038–8047, (1995).
23. A. H. Gandjbakhche, R. F. Bonner, R. Nossal, and G. H. Weiss, "Absorptivity contrast in transillumination imaging of tissue abnormalities," *Appl. Opt.* **35**, 1767–1774 (1996).
24. A. H. Gandjbakhche, V. Chernomordik, J. C. Hebden, and R. Nossal, "Time-dependent contrast functions for quantitative imaging in time-resolved transillumination experiments," *Appl. Opt.* **37**, 1973–1981 (1998).
25. R. Cubeddu, A. Pifferi, P. Taroni, A. Torricelli, and G. Valentini, "Imaging of optical inhomogeneities in highly diffusive media: discrimination between scattering and absorption contributions," *Appl. Phys. Lett.* **69**, 4162–4164 (1996).
26. R. Cubeddu, A. Pifferi, P. Taroni, A Torricelli, and G. Valentini, "Imaging with diffusing light: an experimental study of the effect of background optical properties," *Appl. Opt.* **37**, 3564–3573 (1998).
27. J. C. Hebden and K. B. Wong, "Time-resolved optical tomography," *Appl. Opt.* **32**, 372–380 (1993).
28. Y. Yamada, Y. Hasegawa, and H. Maki, "Simulation of time-resolved optical computer tomography imaging," *Opt. Eng.* **32**, 634–641 (1993).
29. I. Oda, H. Eda, Y. Tsunazawa, M. Takada, Y. Yamada, G. Nishimura, and M. Tamura, "Optical tomography by the temporally extrapolated absorbance method," *Appl. Opt.* **35**, 169–175 (1996).
30. D. A. Benaron and D. K. Stevenson, "Optical time-of-flight and absorbance imaging of biologic media," *Science* **259**, 1463–1466 (1993).
31. S. R. Arridge, M. Schweiger, M. Hiraoka, and D. T. Delpy, "A finite element approach to modelling photon transport in tissue," *Med. Phys.* **20**, 299–309 (1993).
32. J. Chang, W. Wu, Y. Wang, H. L. Graber, and R. L. Barbour, "Regularized progressive expansion algorithm for recovery of scattering media from time-resolved data," *J. Opt. Soc. Am. A* **14**, 306–312 (1997).
33. R. Model, O. M. Orlt, M. Walzel, and R. Hünlich, "Optical imaging: three-dimensional approximation and perturbation approaches for time-domain data," *Appl. Opt.* **37**, 7968–7976 (1998).
34. F. Gao, H. Niu, H. Zhao, and H. Zhang, "The forward and inverse models in time-resolved optical tomography imaging and their finite-element method solutions," *Image Vis. Comp.* **16**, 703–712 (1998).
35. S. R. Arridge, "Optical tomography in medical imaging," *Inverse Probl.* **15**(2), R41 (1999) [doi:10.1088/0266-5611/15/2/022].
36. S. R. Arridge, "Methods in diffuse optical imaging," *Philos. Transact. A Math. Phys. Eng. Sci.* **369**(1955), 4558–4576 (2011) [doi: 10.1098/rsta.2011.0311].

37. V. V. Tuchin, *Tissue Optics: Light Scattering Methods and Instruments for Medical Diagnosis*, 3rd ed., PM254, SPIE Press, Bellingham, Wash. (2015).
38. E. Alerstam, T. Svensson, and S. Andersson-Engels, "Parallel computing with graphics processing units for high-speed Monte Carlo simulation of photon migration," *J. Biomed. Opt.* **13**(6), 060504 (2008) [doi: 10.1117/1.3041496].
39. Q. Fang and D. A. Boas, "Monte Carlo simulation of photon migration in 3D turbid media accelerated by graphics processing units," *Opt. Express* **17**(22), 20178–20190 (2009).
40. M. Schweiger, "GPU-accelerated finite element method for modelling light transport in diffuse optical tomography," *Int. J. Biomed. Imaging* **2011**, e403892 (2011) [doi:10.1155/2011/403892].
41. M. Schweiger and S. Arridge, "The Toast++ software suite for forward and inverse modeling in optical tomography," *J. Biomed. Opt.* **19**(4), 040801 (2014) [doi: 10.1117/1.JBO.19.4.040801].
42. S. R. Arridge, "Photon-measurement density functions. Part I: analytical forms," *Appl. Opt.* **34**(31), 7395–7409 (1995).
43. J. C. Hebden and S. R. Arridge, "Imaging through scattering media by the use of an analytical model of perturbation amplitudes in the time domain," Appl. Opt. **35**(34), 6788–6796 (1996).
44. S. Carraresi, T. S. Shatir, F. Martelli, and G. Zaccanti, "Accuracy of a perturbation model to predict the effect of scattering and absorbing inhomogeneities on photon migration," *Appl. Opt.* **40**(25), 4622–4632 (2001).
45. M. Morin, S. Verreault, A. Mailloux, J. Fréchette, S. Chatigny, Y. Painchaud, and P. Beaudry, "Inclusion characterization in a scattering slab with time-resolved transmittance measurements: perturbation analysis," *Appl. Opt.* **39**(16), 2840–2852 (2000).
46. B. Wassermann, "Limits of high-order perturbation theory in time-domain optical mammography," *Phys. Rev. E Stat. Nonlin. Soft Matter Phys.* **74**(3 Pt 1), 031908 (2006).
47. A. Sassaroli, A. Pifferi, D. Contini, A. Torricelli, L. Spinelli, H. Wabnitz, P. Di Ninni, G. Zaccanti, and F. Martelli, "Forward solvers for photon migration in the presence of highly and totally absorbing objects embedded inside diffusive media," *J. Opt. Soc. Am. A Opt. Image Sci. Vis.* **31**(3), 460–469 (2014).
48. D. A. Boas, M. A. O'Leary, B. Chance, and A. G. Yodh, "Scattering of diffuse photon density waves by spherical inhomogeneities within turbid media: analytic solutions and applications," *Proc. Nat. Acad. Sci. USA* **91**(11), 4887–4891 (1994).
49. D. Grosenick, H. Wabnitz, K. T. Moesta, J. Mucke, P. M. Schlag, and H. Rinneberg, "Time-domain scanning optical mammography: II. Optical properties and tissue parameters of 87 carcinomas," *Phys. Med. Biol.* **50**(11), 2451–2468 (2005) [doi:10.1088/0031-9155/50/11/002].

50. D. Huang, E. A. Swanson, C. P. Lin, J. S. Shuman, W. G. Stinton, W. Chang, M. R. Hee, T. Flotte, K. Gregory, C. A. Puliafito, and J. G. Fujimoto, "Optical coherence tomography," *Science* **254**, 1178–1181 (1991).
51. S. C. W. Hyde, N. P. Barry, R. Jones, J. C. Dainty, and P. M. W. French, "High resolution depth resolved imaging through scattering media using time resolved holography," *Opt. Comm.* **122**, 111–116 (1996).
52. C. Hauger, E. Baigar, T. Wilhelm, and W. Zinth, "Time-resolved backscattering of femtosecond pulses from scattering media: an experimental and numerical investigation," *Opt. Comm.* **131**, 351–358 (1996).
53. S. Carraresi, T. S. Shatir, F. Martelli, and G. Zaccanti, "Accuracy of a perturbation model to predict the effect of scattering and absorbing inhomogeneities on photon migration," *Appl. Opt.* **40**(25), 4622–4632 (2001).
54. Y. Hoshi, M. Shimada, C. Sato, and Y. Iguchi, "Reevaluation of near-infrared light propagation in the adult human head: implications for functional near-infrared spectroscopy," *J. Biomed. Opt.* **10**(6), 064032 (2005) [doi: 10.1117/1.2142325].
55. S. Del Bianco, F. Martelli, and G. Zaccanti, "Penetration depth of light re-emitted by a diffusive medium: theoretical and experimental investigation," *Phys. Med. Biol.* **47**(23), 4131–4144 (2002).
56. A. Torricelli, A. Pifferi, L. Spinelli, R. Cubeddu, F. Martelli, S. Del Bianco, and G. Zaccanti, "Time-resolved reflectance at null source–detector separation: improving contrast and resolution in diffuse optical imaging," *Phys. Rev. Lett.* **95**(7), 078101 (2005).
57. J. Steinbrink, H. Wabnitz, H. Obrig, A. Villringer, and H. Rinneberg, "Determining changes in NIR absorption using a layered model of the human head," *Phys. Med. Biol.* **46**(3), 879–896 (2001) [doi:10.1088/0031-9155/46/3/320].
58. A. Liebert, H. Wabnitz, J. Steinbrink, H. Obrig, M. Moeller, R. Macdonald, A. Villringer, and H. Rinneberg, "Time-resolved multidistance near-infrared spectroscopy of the adult head: intracerebral and extracerebral absorption changes from moments of distribution of times of flight of photons," *Appl. Opt.* **43**(15), 3037–3047 (2004) [doi: 10.1364/AO.43.003037].
59. M. Kacprzak, A. Liebert, P. Sawosz, N. Żolek, and R. Maniewski, "Time-resolved optical imager for assessment of cerebral oxygenation," *J. Biomed. Opt.* **12**(3), 034019 (2007).
60. A. Jelzow, H. Wabnitz, I. Tachtsidis, E. Kirilina, R. Brühl, and R. Macdonald, "Separation of superficial and cerebral hemodynamics using a single distance time-domain NIRS measurement," *Biomed. Opt. Express* **5**(5), 1465–1482 (2014) [doi: 10.1364/BOE.5.001465].

61. J. Selb, A. Dale, and D. A. Boas, "Linear 3D reconstruction of time-domain diffuse optical imaging differential data: improved depth localization and lateral resolution," *Opt. Express* **15**(25), 16400–16412 (2007).
62. H. Wabnitz, M. Moeller, A. Liebert, H. Obrig, J. Steinbrink, and R. Macdonald, "Time-resolved near-infrared spectroscopy and imaging of the adult human brain," *Adv. Exp. Med. Biol.* **662**, 143–148 (2010) [doi:10.1007/978-1-4419-1241-1_20].
63. D. Contini, L. Spinelli, A. Torricelli, A. Pifferi, and R. Cubeddu, "Novel method for depth-resolved brain functional imaging by time-domain NIRS," *Proc. SPIE* **6629**, 662908 (2007) [doi:10.1117/12.728104].
64. V. V. Tuchin, "Light scattering study of tissues," *Physics-Uspekhi* **40**, 495–515 (1997).
65. V. G. Kolinko, F. F. M. de Mul, J. Greve, and A. V. Priezzhev, "Probabilistic model of multiple light scattering based on rigorous computation of the first and second moments of photon coordinates," *Appl. Opt.* **35**(22), 4541–4550 (1996).
66. P. W. Milonni and J. H. Eberly, *Lasers*, Wiley, New York (1988).
67. D. V. O'Connor and D. Phillips, *Time-Correlated Single Photon Counting*, Academic Press, London (1984).
68. W. Becker, *Advanced Time-Correlated Single Photon Counting Techniques* (Springer Series in Chemical Physics, Vol. **81**), Springer-Verlag Berlin Heidelberg (2005).
69. L. M. Bollinger and G. E. Thomas, "Measurement of the time dependence of scintillation intensity by a delayed coincidence method," *Rev. Sci. Instrum.* **32**(9), 1044–1050 (1961) [doi: 10.1063/1.1717610].
70. W. Becker, *The BH TCSPC handbook*, 6th ed. Becker & Hickl GmbH, Berlin (2014).
71. W. Becker, Ed., *Advanced Time-Correlated Single Photon Counting Applications* (Springer Series in Chemical Physics, Vol. **111**), Springer (2015).
72. R. W. Engstrom, "*Photomultiplier handbook*," Pub. PMT-62 (RCA Corp., Princeton, N.J., 1980); available from Burle Industries Inc., Lancaster, Pa., as Pub. **TP-136**, (1989).
73. S. Cova, A. Lacaita, M. Ghioni, G. Ripamonti, and T. A. Louis, "20-ps timing resolution with single-photon avalanche diodes," *Rev. Sci. Instrum.* **60**(6), 1104–1110 (1989) [doi: 10.1063/1.1140324].
74. S. Cova, M. Ghioni, A. Lacaita, C. Samori, and F. Zappa, "Avalanche photodiodes and quenching circuits for single-photon detection," *Appl. Opt.* **35**(12), 1956–1976 (1996).
75. W. Becker, B. Su, and A. Bergmann, "Better FLIM and FCS data by GaAsP hybrid detectors," *Proc. SPIE* **7569**, 75690S (2010) [doi:10.1117/12.843649].

76. D. J. Hall, J. C. Hebden, and D. T. Delpy, "Imaging very-low-contrast objects in breastlike scattering media with a time-resolved method," *Appl. Opt.* **36**, 7270–7276 (1997).
77. F. L. Pedrotti and L. S. Pedrotti, *Introduction to Optics*, 2nd ed., Prentice Hall, New Jersey, p. 506, (1993).
78. A. Liebert, H. Wabnitz, D. Grosenick, and R. Macdonald, "Fiber dispersion in time domain measurements compromising the accuracy of determination of optical properties of strongly scattering media," *J. Biomed. Opt.* **8**(3), 512–516 (2003) [doi: 10.1117/1.1578088].
79. S. Andersson-Engels, R. Berg, S. Svanberg, and O. Jarlman, "Time-resolved transillumination for medical diagnostics," *Opt. Lett.* **15**, 1179–1181 (1990).
80. J. C. Hebden, R. A. Kruger, and K. S. Wong, "Time-resolved imaging through a highly scattering medium," *Appl. Opt.* **30**, 788–794 (1991).
81. L. Wang, P. P. Ho, C. Liu, G. Zhang, and R. R. Alfano, "Ballistic 2-D imaging through scattering walls using an ultrafast Kerr gate," *Science* **253**, 769–771 (1991).
82. L. Wang, P. P. Ho, X. Liang, H. Dai, and R. R. Alfano, "Kerr–Fourier imaging of hidden objects in thick turbid media," *Opt. Lett.* **18**, 241–243 (1993).
83. L. M. Wang, P. P. Ho, and R. R. Alfano, "Double-stage picosecond Kerr gate for ballistic time-gated optical imaging in turbid media," *Appl. Opt.* **32**, 535–540 (1993).
84. B. B. Das and R. R. Alfano, "Ultrafast time-gated approach in optical biomedical imaging," *Curr. Sci.* **77**, 885–893 (1999).
85. L. M. Wang, P. P. Ho, and R. R. Alfano, "Time-resolved Fourier spectrum and imaging in highly scattering media," *Appl. Opt.* **32**, 5043–5048 (1993).
86. R. R. Alfano, X. Liang, L. M. Wang, and P. P. Ho, "Time-resolved imaging of translucent droplets in highly scattering turbid media," *Science* **264**, 1913–1915 (1993).
87. M. D. Duncan, R. Mahon, L. L. Tankersley, and J. Reintjes, "Time-gated imaging through scattering media using stimulated Raman amplification," *Opt. Lett.* **16**, 1868–1870 (1991).
88. R. Mahon, M. D. Duncan, L. L. Tankersley, and J. Reintjes, "Time-gated imaging through dense scatterers with a Raman amplifier," *Appl. Opt.* **18**, 7425–7433 (1993).
89. G. Le Tolguenec, F. Devaus, and E. Lantz, "Two-dimensional time-resolved direct imaging through thick biological tissues: a new step toward non-invasive medical imaging," *Opt. Lett.* **24**, 1047–1049 (1999).
90. O. Jarlman, R. Berg, S. Andersson-Engels, S. Svanberg, and H. Pettersson, "Time-resolved white light transillumination for medical imaging," *Acta Radiol.* **38**, 185–189 (1997).

91. D. Grosenick, H. Wabnitz, H. H. Rinneberg, K. T. Moesta, and P. M. Schlag, "Development of a time-domain optical mammograph and first in vivo applications," *Appl. Opt.* **38**(13), 2927–2943 (1999).
92. D. Grosenick, "Photon counting in diffuse optical imaging." *Advanced Photon Counting*, P. Kapusta, M. Wahl, and R. Erdmann, Eds., pp. 343–365, Springer (2015).
93. D. Grosenick, A. Hagen, O. Steinkellner, A. Poellinger, S. Burock, P. M. Schlag, H. Rinneberg, and R. Macdonald, "A multichannel time-domain scanning fluorescence mammograph: performance assessment and first in vivo results," *Rev. Sci. Instrum.* **82**(2), 024302 (2011) [doi: 10.1063/1.3543820].
94. A. Pifferi, P. Taroni, A. Torricelli, F. Messina, R. Cubeddu, and G. Danesini, "Four-wavelength time-resolved optical mammography in the 680–980-nm range," *Opt. Lett.* **28**(13), 1138–1140 (2003).
95. P. Taroni, A. Pifferi, E. Salvagnini, L. Spinelli, A. Torricelli, and R. Cubeddu, "Seven-wavelength time-resolved optical mammography extending beyond 1000 nm for breast collagen quantification," *Opt. Express* **17**(18), 15932–15946 (2009).
96. J. C. Hebden and K. S. Wong, "Time-resolved optical tomography," *Appl. Opt.* **32**, 372–380, (1993).
97. S. Hintz, D. A. Benaron, J. P. Van Houten, J. L. Duckworth, F. W. H. Liu, S. D. Spilman, D. K. Stevenson, and W. Cheong, "Stationary headband for clinical time-of-flight optical imaging at the bedside," *Photochem. Photobiol.* **68**, 361–369 (1998).
98. H. Eda, I. Oda, Y. Ito, Y. Yada, Y. Oikawa, Y. Tsunazawa, Y. Tsuchiya, Y. Yamashita, M. Oda, A. Sassaroli, Y. Yamada, and M. Tamura, "Multichannel time-resolved optical tomographic imaging system," *Rev. Sci. Instrum.* **70**, 3595–3602 (1999).
99. J. Rodriguez, C. Quarles, C. Sisson, and H. D. Battarbee, "High bandwidth optical tomography." *Proc. SPIE* **3657**, 375–383 (1999).
100. F. E. W. Schmidt, M. E. Fry, E. M. C. Hillman, J. C. Hebden, and D. T. Delpy, "A 32-channel time-resolved optical tomographic imaging system," *Rev. Sci. Instrum.* **71**, 256–265 (2000).
101. M. Schweiger and S. R. Arridge, "A system for solving forward and inverse problems in optical spectroscopy and imaging," *OSA TOPS Adv. Opt. Imaging Photon Migration* **2**, 263–268 (1996).
102. S. R. Arridge, J. C. Hebden, M. Schweiger, F. E. W. Schmidt, M. E. Fry, E. M. C. Hillman, H. Dehghani, and D. T. Delpy, "A method for three-dimensional time-resolved optical tomography," *Int. J. Imaging Syst. Technol.* **11**(1), 2–11 (2000).
103. L. C. Enfield, A. P. Gibson, N. L. Everdell, D. T. Delpy, M. Schweiger, S. R. Arridge, C. Richardson, M. Keshtgar, M. Douek, and J. C. Hebden, "Three-dimensional time-resolved optical mammography of the

uncompressed breast," *Appl. Opt.* **46**(17), 3628 (2007) [doi: 10.1364/AO.46.003628]

104. R. J. Cooper, E. Magee, N. Everdell, S. Magazov, M. Varela, D. Airantzis, A. P. Gibson, and J. C. Hebden, "MONSTIR II: a 32-channel, multispectral, time-resolved optical tomography system for neonatal brain imaging," *Rev. Sci. Instrum.* **85**(5), 053105 (2014) [doi: 10.1063/1.4875593].
105. Y. Ueda, K. Yoshimoto, E. Ohmae, T. Suzuki, T. Yamanaka, D. Yamashita, H. Ogura, C. Teruya, H. Nasu, E. Ima, H. Sakahara, M. Oda, and Y. Yamashita, "Time-resolved optical mammography and its preliminary clinical results," *Technol. Cancer Res. Treat.* **10**(5), 393–401 (2011).
106. Y. Tsuchiya, "Photon path distribution in inhomogeneous turbid media: theoretical analysis and a method of calculation," *J. Opt. Soc. Am. A Opt. Image Sci. Vis.* **19**(7), 1383–1389 (2002).
107. V. Ntziachristos, A. G. Yodh, M. Schnall, and B. Chance, "Concurrent MRI and diffuse optical tomography of breast after indocyanine green enhancement," *Proc. Natl. Acad. Sci. U.S.A.* **97**(6), 2767–2772 (2000) [doi: 10.1073/pnas.040570597].
108. V. Ntziachristos, A. G. Yodh, M. D. Schnall, and B. Chance, "MRI-guided diffuse optical spectroscopy of malignant and benign breast lesions," *Neoplasia* **4**(4), 347–354 (2002) [doi: 10.1038/sj.neo.7900244].
109. C. Hauger, E. Baigar, and W. Zinth, "Induced backscattering due to reflecting surfaces in highly scattering media," *Opt. Comm.* **133**, 72–76 (1997).
110. E. Baigar, C. Hauger, and W. Zinth, "Imaging within highly scattering media using time-resolved backscattering of femtosecond pulses," *Appl. Phys. B* **67**, 257–261 (1998).
111. A. Torricelli, D. Contini, A. Pifferi, M. Caffini, R. Re, L. Zucchelli, and L. Spinelli, "Time domain functional NIRS imaging for human brain mapping," *NeuroImage* **85**(Pt 1), 28–50 (2014) [doi: 10.1016/j.neuroimage.2013.05.106].
112. H. Wabnitz, M. Moeller, A. Liebert, A. Walter, R. Erdmann, O. Raitza, C. Drenckhahn, J. P. Dreier, H. Obrig, J. Steinbrink, and R. Macdonald, "A time-domain NIR brain imager applied in functional stimulation experiments," *Proc. SPIE* **5859**, 58590H (2005) [doi:10.1117/12.632837].
113. D. Contini, A. Torricelli, A. Pifferi, L. Spinelli, F. Paglia, and R. Cubeddu, "Multi-channel time-resolved system for functional near infrared spectroscopy," *Opt. Express* **14**(12), 5418–5432 (2006).
114. M. Kacprzak, A. Liebert, P. Sawosz, N. Zolek, and R. Maniewski, "Time-resolved optical imager for assessment of cerebral oxygenation," *J. Biomed. Opt.* **12**(3), 034019 (2007) [doi: 10.1117/1.2743964].

115. R. Re, D. Contini, M. Turola, L. Spinelli, L. Zucchelli, M. Caffini, R. Cubeddu, and A. Torricelli, "Multi-channel medical device for time domain functional near infrared spectroscopy based on wavelength space multiplexing," *Biomed. Opt. Express* **4**(10), 2231–2246 (2013) [doi: 10.1364/BOE.4.002231].
116. O. Steinkellner, H. Wabnitz, A. Walter, and R. Macdonald, "Multiple source positions in time-domain optical brain imaging: a novel approach," *Proc. OSA-SPIE* **8799**, 87990K (2013) [doi:10.1117/12.2032518].
117. J. Selb, D. K. Joseph, and D. A. Boas, "Time-gated optical system for depth-resolved functional brain imaging," *J. Biomed. Opt.* **11**(4), 044008 (2006) [doi: 10.1117/1.2337320].
118. M. Mazurenka, L. Di Sieno, G. Boso, D. Contini, A. Pifferi, A. D. Mora, A. Tosi, H. Wabnitz, and R. Macdonald, "Non-contact in vivo diffuse optical imaging using a time-gated scanning system," *Biomed. Opt. Express* **4**(10), 2257–2268 (2013) [doi: 10.1364/BOE.4.002257].
119. A. Pifferi, A. Torricelli, L. Spinelli, D. Contini, R. Cubeddu, F. Martelli, G. Zaccanti, A. Tosi, A. Dalla Mora, F. Zappa, and S. Cova, "Time-resolved diffuse reflectance using small source–detector separation and fast single-photon gating," *Phys. Rev. Lett.* **100**(13), 138101 (2008).
120. M. Oda, Y. Yamashita, T. Nakano, A. Suzuki, K. Shimizu, I. Hirano, F. Shimomura, E. Ohmae, T. Suzuki, and Y. Tsuchiya, "Near-infrared time-resolved spectroscopy system for tissue oxygenation monitor," Proc. SPIE **4160**, 204–210 (2000) [doi:10.1117/12.407622].
121. Y. Yamashita, "S8-1: principle of time-resolved spectroscopy for quantitative tissue oxygenation monitoring," *Clin. Neurophysiol.* **124** (8), e22 (2013) [doi: 10.1016/j.clinph.2013.02.041].
122. S. Gunadi, T. S. Leung, C. E. Elwell, and I. Tachtsidis, "Spatial sensitivity and penetration depth of three cerebral oxygenation monitors," *Biomed. Opt. Express* **5**(9), 2896–2912 (2014) [doi: 10.1364/BOE.5.002896].
123. H. Wabnitz, D. R. Taubert, M. Mazurenka, O. Steinkellner, A. Jelzow, R. Macdonald, D. Milej, P. Sawosz, M. Kacprzak, A. Liebert, R. Cooper, J. Hebden, A. Pifferi, A. Farina, I. Bargigia, D. Contini, M. Caffini, L. Zucchelli, L. Spinelli, R. Cubeddu, and A. Torricelli, "Performance assessment of time-domain optical brain imagers, part 1: basic instrumental performance protocol," *J. Biomed. Opt.* **19**(8), 086010 (2014) [doi: 10.1117/1.JBO.19.8.086010].
124. H. Wabnitz, A. Jelzow, M. Mazurenka, O. Steinkellner, R. Macdonald, D. Milej, N. Żołek, M. Kacprzak, P. Sawosz, R. Maniewski, A. Liebert, S. Magazov, J. Hebden, F. Martelli, P. Di Ninni, G. Zaccanti, A. Torricelli, D. Contini, R. Re, L. Zucchelli, L. Spinelli, R. Cubeddu, and A. Pifferi, "Performance assessment of time-domain optical brain

imagers, part 2: nEUROPt protocol," *J. Biomed. Opt.* **19**(8), 086012 (2014) [doi: 10.1117/1.JBO.19.8.086012].

125. V. Venugopal and X. Intes, "Recent advances in optical mammography," *Curr. Med. Imaging Rev.* **8**(3), 244–259 (2012).
126. P. Taroni, A. Pifferi, G. Quarto, L. Spinelli, A. Torricelli, and R. Cubeddu, "Diffuse optical imaging: application to breast imaging." *Emerging Imaging Technologies in Medicine*, M. A. Anastasio and P. La Riviere, Eds., Boca Raton, London, New York, pp. 139–154, Taylor & Francis (2012).
127. D. R. Leff, O. J. Warren, L. C. Enfield, A. Gibson, T. Athanasiou, D. K. Patten, J. Hebden, G. Z. Yang, and A. Darzi, "Diffuse optical imaging of the healthy and diseased breast: a systematic review," *Breast Cancer Res. Treat.* **108**(1), 9–22 (2008) [doi: 10.1007/s10549-007-9582-z].
128. G. Quarto, A. Torricelli, L. Spinelli, A. Pifferi, R. Cubeddu, and P. Taroni, "Breast monitoring by time-resolved diffuse optical imaging." *Advanced Time-Correlated Single Photon Counting Applications*, W. Becker, Ed., pp. 587–611, Springer (2015).
129. H. Soliman, A. Gunasekara, M. Rycroft, J. Zubovits, R. Dent, J. Spayne, M. J. Yaffe, and G. J. Czarnota, "Functional imaging using diffuse optical spectroscopy of neoadjuvant chemotherapy response in women with locally advanced breast cancer," *Clin. Cancer Res. Off. J. Am. Assoc. Cancer Res.* **16**(9), 2605–2614 (2010) [doi:10.1158/1078-0432.CCR-09-1510].
130. P. Taroni, A. Torricelli, L. Spinelli, A. Pifferi, F. Arpaia, G. Danesini, and R. Cubeddu, "Time-resolved optical mammography between 637 and 985 nm: clinical study on the detection and identification of breast lesions," *Phys. Med. Biol.* **50**(11), 2469–2488 (2005) [doi:10.1088/0031-9155/50/11/003].
131. T. Yates, J. C. Hebden, A. Gibson, N. Everdell, S. R. Arridge, and M. Douek, "Optical tomography of the breast using a multi-channel time-resolved imager," *Phys. Med. Biol.* **50**(11), 2503–2517 (2005) [doi:10.1088/0031-9155/50/11/005].
132. A. Hagen, D. Grosenick, R. Macdonald, H. Rinneberg, S. Burock, P. Warnick, A. Poellinger, and P. M. Schlag, "Late-fluorescence mammography assesses tumor capillary permeability and differentiates malignant from benign lesions," *Opt. Express* **17**(19), 17016–17033 (2009) [doi: 10.1364/OE.17.017016].
133. A. Poellinger, S. Burock, D. Grosenick, A. Hagen, L. Lüdemann, F. Diekmann, F. Engelken, R. Macdonald, H. Rinneberg, and P.-M. Schlag, "Breast cancer: early- and late-fluorescence near-infrared imaging with indocyanine green: a preliminary study," *Radiology* **258**(2), 409–416 (2011) [doi: 10.1148/radiol.10100258].
134. A. Cerussi, D. Hsiang, N. Shah, R. Mehta, A. Durkin, J. Butler, and B. J. Tromberg, "Predicting response to breast cancer neoadjuvant

chemotherapy using diffuse optical spectroscopy," *Proc. Natl. Acad. Sci. U.S.A.* **104**(10), 4014–4019 (2007) [doi: 10.1073/pnas.0611058104].

135. L. C. Enfield, G. Cantanhede, D. Westbroek, M. Douek, A. D. Purushotham, J. C. Hebden, and A. P. Gibson, "Monitoring the response to primary medical therapy for breast cancer using three-dimensional time-resolved optical mammography," *Technol. Cancer Res. Treat.* **10**(6), 533–547 (2011).
136. L. Enfield, G. Cantanhede, M. Douek, V. Ramalingam, A. Purushotham, J. Hebden, and A. Gibson, "Monitoring the response to neoadjuvant hormone therapy for locally advanced breast cancer using three-dimensional time-resolved optical mammography," *J. Biomed. Opt.* **18**(5), 56012 (2013) [doi: 10.1117/1.JBO.18.5.056012].
137. S. Ueda, I. Kuji, T. Shigekawa, H. Takeuchi, H. Sano, E. Hirokawa, H. Shimada, H. Suzuki, M. Oda, A. Osaki, and T. Saeki, "Optical imaging for monitoring tumor oxygenation response after initiation of single-agent bevacizumab followed by cytotoxic chemotherapy in breast cancer patients," *PLoS One* **9**(6) (2014) [doi: 10.1371/journal.pone.0098715].
138. S. Ueda, N. Nakamiya, K. Matsuura, T. Shigekawa, H. Sano, E. Hirokawa, H. Shimada, H. Suzuki, M. Oda, Y. Yamashita, O. Kishino, I. Kuji, A. Osaki, and T. Saeki, "Optical imaging of tumor vascularity associated with proliferation and glucose metabolism in early breast cancer: clinical application of total hemoglobin measurements in the breast," *BMC Cancer* **13**, 514 (2013) [doi:10.1186/1471-2407-13-514].
139. P. Taroni, G. Quarto, A. Pifferi, F. Ieva, A. M. Paganoni, F. Abbate, N. Balestreri, S. Menna, E. Cassano, and R. Cubeddu, "Optical identification of subjects at high risk for developing breast cancer," *J. Biomed. Opt.* **18**(6), 060507 (2013) [doi: 10.1117/1.JBO.18.6.060507].
140. G. Quarto, L. Spinelli, A. Pifferi, A. Torricelli, R. Cubeddu, F. Abbate, N. Balestreri, S. Menna, E. Cassano, and P. Taroni, "Estimate of tissue composition in malignant and benign breast lesions by time-domain optical mammography," *Biomed. Opt. Express* **5**(10), 3684–3698 (2014) [doi: 10.1364/BOE.5.003684].
141. S. R. W. Cheong, J. P. Van Houten, D. K. Stevenson, and D. A. Benaron, "Bedside imaging of intracranial hemorrhage in the neonate using light: comparison with ultrasound, computed tomography, and magnetic resonance," *Pediatr. Res.* **45**, 54–59 (1999).
142. T. Austin, A. P. Gibson, G. Branco, R. M. Yusof, S. R. Arridge, J. H. Meek, J. S. Wyatt, D. T. Delpy, and J. C. Hebden, "Three dimensional optical imaging of blood volume and oxygenation in the neonatal brain," *NeuroImage* **31**(4), 1426–1433 (2006) [doi: 10.1016/j.neuroimage.2006.02.038].

143. J. C. Hebden, A. Gibson, R. M. Yusof, N. Everdell, E. M. C. Hillman, D. T. Delpy, S. R. Arridge, T. Austin, J. H. Meek, and J. S. Wyatt, "Three-dimensional optical tomography of the premature infant brain," *Phys. Med. Biol.* **47**(23), 4155–4166 (2002).
144. A. P. Gibson, T. Austin, N. L. Everdell, M. Schweiger, S. R. Arridge, J. H. Meek, J. S. Wyatt, D. T. Delpy, and J. C. Hebden, "Three-dimensional whole-head optical tomography of passive motor evoked responses in the neonate," *NeuroImage* **30**(2), 521–528 (2006) [doi: 10.1016/j.neuroimage.2005.08.059].
145. M. Ferrari and V. Quaresima, "A brief review on the history of human functional near-infrared spectroscopy (fNIRS) development and fields of application," *NeuroImage* **63**(2), 921–935 (2012) [doi: 10.1016/j.neuroimage.2012.03.049].
146. F. Aletti, R. Re, V. Pace, D. Contini, E. Molteni, S. Cerutti, A. Maria Bianchi, A. Torricelli, L. Spinelli, R. Cubeddu, and G. Baselli, "Deep and surface hemodynamic signal from functional time resolved transcranial near infrared spectroscopy compared to skin flowmotion," *Comput. Biol. Med.* **42**(3), 282–289 (2012) [doi: 10.1016/j.compbiomed.2011.06.001].
147. E. Kirilina, A. Jelzow, A. Heine, M. Niessing, H. Wabnitz, R. Brühl, B. Ittermann, A. M. Jacobs, and I. Tachtsidis, "The physiological origin of task-evoked systemic artefacts in functional near infrared spectroscopy," *NeuroImage* **61**(1), 70–81 (2012) [doi: 10.1016/j.neuroimage.2012.02.074].
148. Y. Hoshi, T. Shinba, C. Sato, and N. Doi, "Resting hypofrontality in schizophrenia: a study using near-infrared time-resolved spectroscopy," *Schizophr. Res.* **84**(2–3), 411–420 (2006) [doi: 10.1016/j.schres.2006.03.010].
149. E. Ohmae, M. Oda, T. Suzuki, Y. Yamashita, Y. Kakihana, A. Matsunaga, Y. Kanmura, and M. Tamura, "Clinical evaluation of time-resolved spectroscopy by measuring cerebral hemodynamics during cardiopulmonary bypass surgery," *J. Biomed. Opt.* **12**(6), 062112 (2007) [doi: 10.1117/1.2804931].
150. C. Sato, T. Yamaguchi, M. Seida, Y. Ota, I. Yu, Y. Iguchi, M. Nemoto, and Y. Hoshi, "Intraoperative monitoring of depth-dependent hemoglobin concentration changes during carotid endarterectomy by time-resolved spectroscopy," *Appl. Opt.* **46**(14), 2785–2792 (2007).
151. T. H. Sander, A. Liebert, B. M. Mackert, H. Wabnitz, S. Leistner, G. Curio, M. Burghoff, R. Macdonald, and L. Trahms, "DC-magnetoencephalography and time-resolved near-infrared spectroscopy combined to study neuronal and vascular brain responses," *Physiol. Meas.* **28**(6), 651–664 (2007) [doi:10.1088/0967-3334/28/6/004].
152. S. Leistner, T. Sander-Thoemmes, H. Wabnitz, M. Moeller, M. Wachs, G. Curio, R. Macdonald, L. Trahms, and B.-M. Mackert, "Non-

invasive simultaneous recording of neuronal and vascular signals in subacute ischemic stroke," *Biomed. Tech. (Berl)* **56**(2), 85–90 (2011) [doi: 10.1515/BMT.2011.002].
153. E. Visani, L. Canafoglia, I. Gilioli, D. R. Sebastiano, V. E. Contarino, D. Duran, F. Panzica, R. Cubeddu, D. Contini, L. Zucchelli, L. Spinelli, M. Caffini, E. Molteni, A. M. Bianchi, S. Cerutti, S. Franceschetti, and A. Torricelli, "Hemodynamic and EEG time-courses during unilateral hand movement in patients with cortical myoclonus: an EEG-fMRI and EEG-TD-fNIRS study," *Brain Topogr.* (2014) [doi: 10.1007/s10548-014-0402-6].
154. M. Kacprzak, A. Liebert, W. Staszkiewicz, A. Gabrusiewicz, P. Sawosz, G. Madycki, and R. Maniewski, "Application of a time-resolved optical brain imager for monitoring cerebral oxygenation during carotid surgery," *J. Biomed. Opt.* **17**(1), 016002 (2012) [doi: 10.1117/1.JBO.17.1.016002].
155. A. Liebert, H. Wabnitz, J. Steinbrink, M. Möller, R. Macdonald, H. Rinneberg, A. Villringer, and H. Obrig, "Bed-side assessment of cerebral perfusion in stroke patients based on optical monitoring of a dye bolus by time-resolved diffuse reflectance," *NeuroImage* **24**(2), 426–435 (2005) [doi: 10.1016/j.neuroimage.2004.08.046].
156. O. Steinkellner, C. Gruber, H. Wabnitz, A. Jelzow, J. Steinbrink, J. B. Fiebach, R. Macdonald, and H. Obrig, "Optical bedside monitoring of cerebral perfusion: technological and methodological advances applied in a study on acute ischemic stroke," *J. Biomed. Opt.* **15**(6), 061708 (2010) [doi: 10.1117/1.3505009].
157. W. Weigl, D. Milej, A. Gerega, B. Toczylowska, M. Kacprzak, P. Sawosz, M. Botwicz, R. Maniewski, E. Mayzner-Zawadzka, and A. Liebert, "Assessment of cerebral perfusion in post-traumatic brain injury patients with the use of ICG-bolus tracking method," *NeuroImage* **85**(Pt 1), 555–565 (2014) [doi: 10.1016/j.neuroimage.2013.06.065].
158. M. Diop, K. Verdecchia, T.-Y. Lee, and K. St Lawrence, "Calibration of diffuse correlation spectroscopy with a time-resolved near-infrared technique to yield absolute cerebral blood flow measurements," *Biomed. Opt. Express* **2**(7), 2068–2081 (2011) [doi: 10.1364/BOE.2.002068].

Heidrun Wabnitz is a senior scientist at the Department of Biomedical Optics at Physikalisch-Technische Bundesanstalt (PTB) in Berlin. She received a diploma in physics as well as a Dr. rer. nat. degree from Friedrich Schiller University in Jena. Her fields of work in Jena included picosecond spectroscopy of molecules and time-resolved fluorescence laser scanning microscopy. She joined PTB in 1991 where she focused on diffuse optical imaging and spectroscopy of tissues with picosecond time resolution, in particular on projects related to optical mammography and optical brain imaging. She is active in the development of time-domain instrumentation, modeling and data analysis, performance characterization of instruments, standardization, and clinical applications.

Juan Rodriguez earned his PhD in physics from the University of Arkansas, where he received training in optics and laser physics. He then served as a postdoctoral fellow at Washington University in St. Louis where he investigated ultrafast phenomena related to photosynthesis. From 1990 to 2014, he assumed a faculty position at Centenary College of Louisiana where he taught a variety of physics courses including optics, electromagnetism, and biophysics. During that time he collaborated with several faculty members at the Louisiana State University Medical Center on projects related to biomedical optics and nitric oxide biology. Currently, he is at the St. Louis College of Pharmacy where his research focuses on areas related to vision and physics education.

Ilya V. Yaroslavsky received his MSc degree *summa cum laude* in physics in 1990 and PhD in laser physics in 1994, both from Saratov State University, Saratov, Russia. From 1994 to 2000, he did his postdoctoral training in Heinrich Heine University, Düsseldorf, Germany, working on laser interstitial thermotherapy of brain tumors, and in Louisiana State University, Shreveport, Los Angeles, USA, developing optical diffusion techniques for stroke diagnostics. He started his industrial career at Palomar Medical Technologies, Inc., Burlington, Massachusetts in 2000 and 2012 assumed position of the Vice President of Advanced Research of the company. In 2015, he joined IPG Medical Corporation as manager for Advanced Product Development. His scientific interests include light-tissue interactions and use of lasers for biomedical applications. He has authored and coauthored more than 50 scientific papers and inventions. He served as a member of IEC TC 76 Technical Advisory Group on optical radiation safety and laser equipment.

Anna N. Yaroslavsky is an associate professor of physics and director of the Advanced Biophotonics Laboratory at the University of Massachusetts, Lowell, USA. She obtained her PhD from the Saratov State University, Saratov, Russia, in 1999 and joined Wellman Center for Photomedicine (WCP), Boston, USA, in 2000. From 2005 to 2010, she has been an assistant professor of dermatology at the Harvard Medical School, Boston, USA, and since 2010, she is a visiting scientist at WCP. Her expertise is in biomedical optics and medical imaging.

Her research is focused on the development of optical and multimodal technologies for medical applications.

Valery V. Tuchin is a professor and chairman of optics and biophotonics at the Saratov State University. He is also the head of laboratory, Institute of Precision Mechanics and Control, RAS, and the supervisor of Interdisciplinary laboratory of Biophotonics at the Tomsk State University. His research interests include biophotonics, tissue optics, laser medicine, tissue optical clearing, and nanobiophotonics. He is a member of SPIE, OSA, and IEEE, guest professor of HUST (Wuhan) and Tianjin Universities of China, and adjunct professor of the Limerick University (Ireland) and National University of Ireland (Galway). He is a fellow of SPIE and has been awarded Honored Science Worker of the Russia, SPIE Educator Award, FiDiPro (Finland), and Chime Bell Prize of Hubei Province (China).

Chapter 7
Frequency-Domain Techniques for Tissue Spectroscopy and Imaging

Sergio Fantini and Angelo Sassaroli
Department of Biomedical Engineering, Tufts University,
Medford, MA, USA

7.1 Introduction

In frequency-domain spectroscopy, the intensity of the light source is sinusoidally modulated at a frequency f. One can fully describe the modulated intensity using three parameters—namely, the average intensity (or DC, from the electrical terminology of direct current), the amplitude of the intensity oscillations (or AC, from the electrical equivalent of alternating current), and their phase (Φ). The modulation depth is defined as the ratio AC/DC. Because the phase and the modulation depth are key features of frequency-domain spectroscopy, the term *phase modulation* is sometimes used interchangeably with *frequency domain*. The phase of the measured intensity wave is related to the optical propagation time through the medium. If τ is a typical propagation time—say, the mean photon time of flight between the light source and the optical detector—then the phase of the detected signal (relative to the light source) is on the order of $\omega\tau$, where $\omega = 2\pi f$ is the angular modulation frequency. To obtain phase measurements with a good signal-to-noise ratio, it is required that $\omega\tau$ be approximately equal to 1 (corresponding to a relatively large phase angle of 1 rad, or approximately 57 deg), which is the condition that guides the choice of the modulation frequency, f. In the case of near-infrared spectroscopy and imaging of tissues, the mean photon time of flight is on the order of 1 ns for source–detector separations of a few centimeters.[1,2] Consequently, the condition $\omega\tau \approx 1$ results in $f \approx 100$ MHz, which falls within the radio-frequency range. Frequency-domain optical studies of biological tissues typically use intensity-modulation at frequencies on the order of 100 MHz.

In this chapter, we describe frequency-domain instrumentation (Section 7.2), diffusion theory (Section 7.3), and representative *in vivo* applications of frequency-domain spectroscopy (Section 7.4) and imaging (Section 7.5). We have aimed to make this chapter self-contained, while providing extensive references to the literature as a guide for additional reading and for in-depth coverage of topics that are merely mentioned here due to space considerations.

7.2 Instrumentation, Modulation Methods, and Signal Detections

Frequency-domain spectroscopy can be implemented using several instrumental and data processing approaches.[3] In Sections 7.2.1–7.2.4, we describe various light sources and modulation methods, as well as principles of homodyne and heterodyne detection schemes. In Section 7.2.5, we describe a specific frequency-domain instrument for near-infrared tissue spectroscopy.

7.2.1 Light sources and modulation techniques

Semiconductor lasers, or laser diodes (LDs), and light-emitting diodes (LEDs) can be intensity modulated by driving them with an oscillating current. As a result of the relatively fast response time of laser diodes, they can be modulated at frequencies up to the gigahertz range. The LEDs' modulation frequency bandwidth is typically limited to 150 MHz, and they emit light over a spectral wavelength bandwidth of approximately 50–80 nm. Consequently, LEDs can be used to measure continuous spectra.[4] One can find a number of LDs and LEDs emitting in the wavelength region of interest for noninvasive tissue spectroscopy, which extends from 600 to 1100 nm. Laser diodes are the most commonly used light sources in frequency-domain optical studies of tissue because of their cost-effectiveness, ease of modulation, and effective coupling to fiber optics, in addition to the fact that tissue spectroscopy can be effectively performed using a few discrete wavelengths (see Sections 7.4.2 and 7.4.3). For a thorough description of the state-of-the-art technology of semiconductor lasers (as well as detectors), we refer readers to Razeghi.[5]

The emission of continuous-wave lasers and arc lamps can be modulated by using devices based on the electro-optical effect (Pockels cells)[6] or on the acousto-optical effect.[7,8] A *Pockels cell* is a birefringent crystal whose indices of refraction can be varied by applying an electric field. The application of a time-varying voltage to the Pockels cell modulates the relative phase delay of the light components polarized along the two principal axes of the cell. If this relative phase delay, or retardation, oscillates between 0 and π, one achieves an intensity modulator by placing the modulated Pockels cell between two crossed linear polarizers, each at an angle of 45 deg with respect to the principal axes of the cell. In fact, no light is transmitted when the retardation is 0, while all light is transmitted when the retardation is π. An *acousto-optic modulator* is a material that uses the piezoelectric and photo-elastic effects to convert an oscillating electric field into mechanical vibrations, which in turn induce a spatially dependent index of refraction in the material. When a standing acoustic wave is established, the acousto-optic crystal behaves as an oscillating refractive index grating that modulates the transmitted light by time-varying diffraction. Both electro-optic and acousto-optic devices require the light beam to be collimated. In the case of arc lamps, appropriate collimation optics are required. Pockels cells provide effective modulations up to approximately 500 MHz, and acousto-optic modulators up to approximately 300 MHz.

The laser wavelength is chosen on the basis of the requirements of the particular application (e.g., absorption band of a chromophore, optimal penetration depth in tissues). Examples of externally modulated continuous-wave (CW) lasers suitable for optical studies of tissues include the krypton ion (647 nm) and He-Ne (633 nm) lasers. Dye lasers pumped by either argon or krypton lasers afford continuous tunability over a wide spectral range that covers the whole visible band. Arc lamps (e.g, Xe, Xe-Hg) provide continuous spectral emission from the UV range (230 nm) to the near-infrared range (1100 nm). Therefore, they are ideal sources for spectroscopic studies when a broadband and continuous spectral range is required.

7.2.2 Pulsed sources

It is possible to achieve a large modulation bandwidth by exploiting the harmonic content of pulsed sources with high repetition rates. These sources can be either mode-locked pulsed lasers (e.g, Nd:YAG, Ti:sapphire, dye lasers),[9] photonic crystal fibers excited by femtosecond pulses,[10] or synchrotron radiation.[11,12] The repetition rate of the pulses corresponds to the fundamental frequency, whereas the temporal pulse width is inversely related to the frequency bandwidth of the power spectrum. The power spectrum of mode-locked lasers extends well above 10 GHz—an upper limit in frequency-domain spectroscopy imposed by the optical detectors (and by the strong attenuation in tissues of light that is intensity modulated at high frequencies) rather than by the light sources. The wavelengths of the previously mentioned lasers are 1064 nm for the Nd:YAG laser, 660–1180 nm (tunable) for the Ti:sapphire laser, and 625–780 nm (tunable) for dye lasers using DCM or oxanine 1 dyes. The excitation of highly nonlinear photonic crystal fibers with ultrafast (sub-picosecond) laser pulses yields supercontinuum generation over the visible to near-infrared spectral region. This light source provides a powerful tool for time-resolved, broadband tissue spectroscopy.[13] A unique pulsed source is provided by synchrotron radiation, which continuously covers the UV/visible/near-infrared spectrum.

7.2.3 Optical detectors

Optical detectors employed in frequency-domain spectroscopy include photomultiplier tubes (PMTs),[14–17] microchannel plate photomultipliers (MCP-PMT),[18,19] avalanche photodiodes (APDs),[20,21] and charge-coupled device (CCD) cameras in conjunction with a gated image intensifier.[22–26] In all cases, frequency-domain data can be obtained at the modulation frequency f by homodyne detection, or at an offset frequency Δf (the so-called cross-correlation frequency; see Section 7.2.4) by heterodyne detection. In the latter case, the down-conversion from the source modulation frequency f to the cross-correlation frequency Δf can occur either internally to the detector, by modulating the detector gain at a frequency $f+\Delta f$, or externally, by electronically mixing the detector output at frequency f with the down-conversion signal at frequency $f+\Delta f$.

Photomultiplier tubes are highly sensitive detectors. The cathode sensitivity is typically 50 mA/W, and a current amplification of approximately 10^6 determines an anode sensitivity as high as 0.1 A/μW. PMTs can operate in the visible and near-infrared ranges at wavelengths up to approximately 1000 nm. For internal down-conversion, their gain is modulated by a signal applied to the second dynode of the amplification chain. The typical rise time of a PMT, which is in the nanosecond range, allows for a modulation bandwidth of several hundred megahertz. The faster response of micro-channel-plate PMTs makes them suitable devices for modulation frequencies of up to several gigahertz. The frequency down-conversion can be performed by modulating the grid of the MCP-PMT[19] or by feeding the MCP-PMT output into an electronic mixer.[18]

CCD cameras with modulated image intensifiers are powerful tools for frequency-domain imaging,[26] fluorescence spectroscopy,[24] and fluorescence lifetime microscopy.[27] (Elder *et al*, 2006). Finally, avalanche photodiodes can work at higher modulation frequencies and/or longer wavelengths than those allowed by PMTs. The APD output is usually directed to an electronic mixer for frequency down-conversion.

7.2.4 Homodyne and heterodyne detection

Homodyne detection systems are the precursors to heterodyne systems, and are typically used for simplicity of implementation. Homodyne systems detect amplitudes and phase shifts at the original modulation frequency without any down-conversion, using in-phase/quadrature (I/Q) demodulators. Homodyne detection is used in a variety of frequency-domain systems.[28–31]

Heterodyne detection consists of down-converting the frequency f of the detected signal to a lower frequency Δf, which is termed the cross-correlation frequency. In the digital acquisition method,[32] after current-to-voltage conversion and low-pass filtering, the low-frequency signal is digitized.

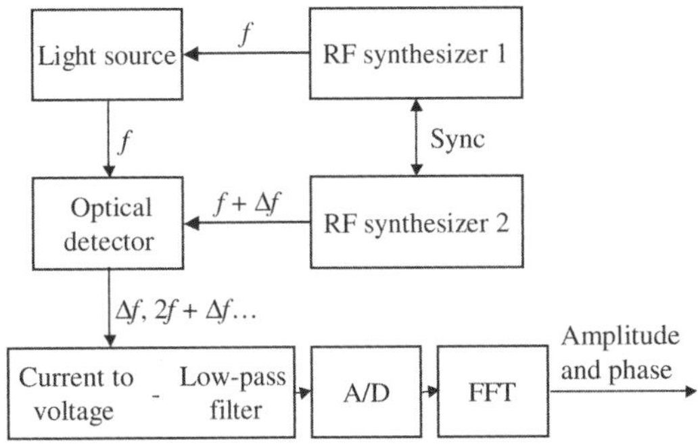

Figure 7.1 Heterodyne detection scheme and digital signal processing with a Fourier filter. The intensity of the light source is modulated at a radio-frequency (RF) f, typically on the order of 100 MHz. The beating between the detected signal at frequency f and the down-conversion signal at frequency $f+\Delta f$ determines a low-frequency component Δf that is filtered, digitized by analog-to-digital (A/D) conversion, and fast Fourier transformed (FFT) to yield the amplitude and phase readings. The two radio-frequency synthesizers are synchronized (sync).

The A/D sampling rate is an integer multiple of Δf ($n\Delta f$), so that the digital sampling is synchronous with the cross-correlation signal. The sampling theorem states that n must be at least 2 for the complete determination of the harmonic at frequency Δf.[33] A higher sampling rate allows for the discrimination of higher harmonics. Typical values of n range from 4 to 256.

The digital signal processing consists of a preliminary average of the data sampled over a number of cross-correlation periods, followed by the discrimination of the frequency harmonic at Δf by discrete Fourier transform. The first filter (averaging) causes destructive interference of all the frequencies that are not multiples of Δf, whereas the discrete Fourier transform resolves each harmonic of Δf. The rejection of higher harmonics may be accomplished to within a factor greater than 2000.[32]

Figure 7.1 shows a schematic diagram of the heterodyne detection and digital acquisition. Heterodyne systems have been used in a variety of frequency-domain systems.[34–36]

7.2.5 A frequency-domain tissue spectrometer

We now describe a specific frequency-domain instrument for near-infrared spectroscopy and imaging of tissues that was originally conceived in Enrico Gratton's group at the University of Illinois at Urbana–Champaign,[37] and subsequently developed into a commercial system by ISS, Inc., of Champaign, Illinois. A schematic diagram of the instrument is shown in Fig. 7.2. The light sources are laser diodes that are DC-biased near threshold for laser operation. A superimposed 110 MHz radio-frequency current signal modulates the optical output of the laser diodes with a modulation depth close to 100%. The optical detectors are photomultiplier tubes (Hamamatsu Photonics R928)

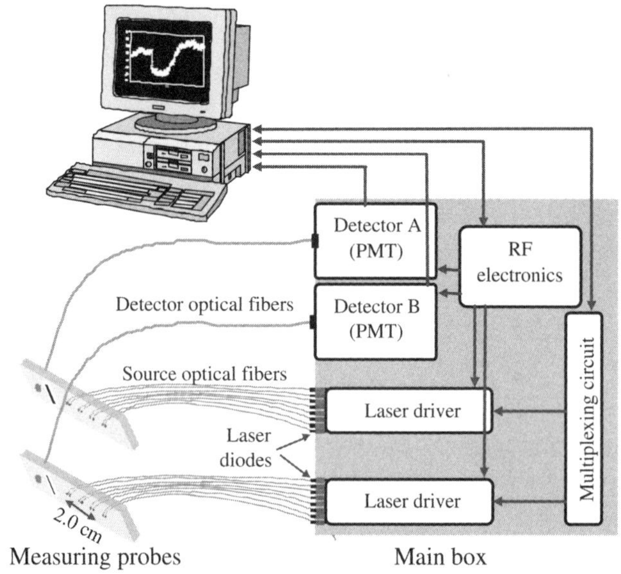

Figure 7.2 Schematic diagram of a frequency-domain tissue spectrometer (OxiplexTS, ISS, Inc., Champaign, IL). In the configuration for tissue oximetry (shown here), each optical probe contains eight illumination optical fibers and one collection optical fiber. These fibers are coupled to eight laser diodes (four emitting at 690 nm and four at 830 nm) and to a photomultiplier tube (PMT) detector, respectively. RF electronics modulate the output of the light sources at a frequency of 110 MHz, and the PMT gain at a frequency of 110.005 MHz. A multiplexing circuit turns the lasers on and off in sequence at an adjustable rate controlled by the software (a typical multiplexing rate of 100 Hz corresponds to a 10-ms on-time per laser). The two optical probes can operate in parallel at two different tissue locations.

whose sensitivity is modulated by a 110.005-MHz signal (amplitude ≈32 dBm) applied to the second dynode of the PMTs. This heterodyne detection scheme down-converts the 110-MHz signal to the cross-correlation frequency of 5 kHz. The signal processing proceeds as shown in Fig. 7.1 to yield the average value, the amplitude, and the phase of the detected modulated signal.

The choice of the laser diodes for tissue oximetry (the case shown in Fig. 7.2) fulfills the requirement of having two wavelengths on the opposite sides of the near-infrared isosbestic point of hemoglobin at 800 nm for maximal sensitivity (see Section 7.4.3). The choice of laser diodes emitting at 758 and 830 nm is ideal, as these wavelengths correspond to relatively flat portions of the absorption spectra of oxyhemoglobin and deoxyhemoglobin (see Fig. 7.5 later in this chapter); in particular, these wavelengths closely match a local maximum and a local minimum, respectively, in the deoxyhemoglobin spectrum. However, the particular values of these wavelengths are not critical, so that the 758-nm lasers may be replaced by 690-nm lasers.

Furthermore, the laser diodes are mounted on interchangeable boards, so that one can change the laser wavelength according to the specific application.

For instance, one can use boards containing lasers emitting at eight different wavelengths for tissue spectroscopy. For example, Fig. 7.6 (which appears later in this chapter) reports data collected at 633, 670, 751, 776, 786, 814, 830, and 841 nm.[38] Alternatively, one can use a laser board containing laser diodes emitting at the same wavelength to maximize the number of source locations in single-wavelength optical imaging. In general, the instrument provides 2 parallel detectors and 16 sources, resulting in 32 source–detector channels. The optical probes shown in Fig. 7.2 are designed to perform quantitative tissue oximetry. They implement the multi-distance measurement scheme that was found to be robust and accurate for *in vivo* applications.[4,39] This scheme consists of placing multiple (in this case, four) illumination optical fibers at different distances from the single collection optical fiber. In this fashion, one measures the spatial dependencies of the DC intensity, AC amplitude, and phase data, which are related to the tissue optical coefficients by diffusion theory (see Section 7.3.5). If two wavelengths are used, each illumination location consists of two illumination fibers (one per wavelength).

One of the advantages of the multi-distance scheme is its insensitivity to motion artifacts, which allows for measurement while the subject moves or exercises [see Fig. 7.7(c) later in this chapter]. Multiple laser diodes time-share the optical detector by means of a multiplexing circuit that turns the laser diodes on and off in sequence, so that only one laser is on at any given time. The multiplexing rate, which determines the on-time of each laser diode, can be adjusted through the software associated with the device.

A typical multiplexing rate of 100 Hz corresponds to a 10 ms on-time per diode. The laser rise time, which is determined by the protection circuitry, is approximately 1 ms. Therefore, the data collected during the first 1 ms of the on-time period of each laser are discarded. During a typical 10-mslaser on-time, 45 cross-correlation periods are processed (the 5-kHz cross-correlation frequency corresponds to a period of 0.2 ms). As few as 25 cross-correlation periods (5 ms) can be processed per laser on-time, leading to a maximum multiplexing rate of approximately 170 Hz. Faster computers and higher cross-correlation frequencies may further increase the data acquisition rate. A number of multiplexing cycles (selected by software) can be averaged to increase the signal-to-noise ratio when fast measurements are not required. The fluctuations of DC intensity, AC amplitude, and phase data due to instrumental noise, which depends on the acquisition time, are typically much smaller than those due to physiological dynamics observed *in vivo*. The noise can be estimated by the standard deviation of the temporal fluctuations of the measured data.

Table 7.1 reports the instrumental noise measured for a tissue-simulating optical phantom that has optical coefficients typical of blood-perfused tissues in the near-infrared wavelength range. In particular, the noise-to-signal ratio (defined as the standard deviation divided by the mean value) is reported for

Table 7.1 Instrumental noise estimated by the temporal standard deviation of the raw data (DC intensity, AC amplitude, and phase) and the optical coefficients (absorption and reduced scattering) measured on a solid tissue-like phantom with the frequency-domain tissue spectrometer of Fig. 7.2. The acquisition time for the raw data (average intensity, amplitude, and phase) is 160 ms. Because the measurement of the optical coefficients at each wavelength requires the readings from four source locations (four source–detector separations), the acquisition time for μ_a and μ_s' is 160 ms $\times$ 4 = 640 ms).

		Instrumental noise	
Parameter	**Acquisition time (ms)**	**At 690 nm**	**At 830 nm**
Average intensity (DC)	160	0.05%	0.05%
Amplitude (AC)	160	0.2%	0.1%
Phase (Φ)	160	0.1°	0.1°
Absorption coefficient (μ_a)	640	0.0007 cm^{-1}	0.0006 cm^{-1}
Reduced scattering coefficient (μ_s')	640	0.05 cm^{-1}	0.04 cm^{-1}

both DC intensity and AC amplitude. For the phase data, the relevant parameter characterizing the noise level is the standard deviation (reported in degrees). The noise affecting these three parameters is reported at both wavelengths of 690 nm and 830 nm. The data in Table 7.1 refer to an acquisition time of 160 ms (achieved by averaging sixteen 10-ms on-time cycles) for DC, AC, and phase data, and an acquisition time of 640 ms for the optical coefficients. The instrumental errors decrease at longer acquisition times.

7.3 Frequency-Domain Diffusion Theory for Quantitative Tissue Spectroscopy

7.3.1 The Boltzmann transport equation

The Boltzmann transport equation (BTE) is a balance relationship that describes the flow of particles in scattering and absorbing media. The propagation of light in optically turbid media can be modeled by the transport equation, where the photons act as the transported particles. If we denote the angular photon density as $u(\mathbf{r},\hat{\Omega},t)$, which is defined as the number of photons per unit volume per unit solid angle traveling in direction $\hat{\Omega}$ at position $\mathbf{r}$ and time t, we can write the BTE as

$$\frac{\partial u(\mathbf{r},\hat{\Omega},t)}{\partial t} = -\hat{\Omega}\cdot\nabla u(\mathbf{r},\hat{\Omega},t) - v(\mu_a+\mu_s)_u(\mathbf{r},\hat{\Omega},t) + v\mu_s\int_{4\pi} u(\mathbf{r},\hat{\Omega}',t)f(\hat{\Omega}',\hat{\Omega})d\hat{\Omega}' + q(\mathbf{r},\hat{\Omega},t), \tag{7.1}$$

where v is the speed of light in the medium, μ_a is the absorption coefficient (units of cm^{-1}), μ_s is the scattering coefficient (units of cm^{-1}), $f(\hat{\Omega}',\hat{\Omega})$ is the phase function or the probability density that a photon that travels along direction $\hat{\Omega}'$ is scattered into direction $\hat{\Omega}$, and $q(\mathbf{r},\hat{\Omega},t)$ is the source term.[40] $q(\mathbf{r},\hat{\Omega},t)$has units of $s^{-1}m^{-3}sr^{-1}$ and represents the number of photons injected by the light source per unit volume, per unit time, per unit solid angle at position **r**, time *t*, and direction $\hat{\Omega}$. The time derivative on the left-hand side of Eq. (7.1) represents the temporal variation of the angular photon density. Each one of the terms on the right-hand side represents a specific contribution to this variation. The first term is the net gain of photons at position **r** and direction $\hat{\Omega}$ due to the flow of photons. The second term is the loss of photons at **r** and $\hat{\Omega}$ as a result of collisions (absorption and scattering). The third term is the gain of photons at **r** and $\hat{\Omega}$ due to scattering. Finally, the fourth term is the gain of photons due to the light sources. Let us now define some of the quantities used to describe photon transport.

Angular photon density: $u(\mathbf{r},\hat{\Omega},t)$
$u(\mathbf{r},\hat{\Omega},t)$ is defined such that $u(\mathbf{r},\hat{\Omega},t)d\mathbf{r}d\hat{\Omega}$ represents the number of photons in $d\mathbf{r}$ that travel in a direction within $d\hat{\Omega}$ around $\hat{\Omega}$. The units of $u(\mathbf{r},\hat{\Omega},t)$ are $m^{-3}sr^{-1}$.

Photon radiance: $L(\mathbf{r},\hat{\Omega},t)$
$L(\mathbf{r},\hat{\Omega},t) = vu(\mathbf{r},\hat{\Omega},t)$. $L(\mathbf{r},\hat{\Omega},t)d\hat{\Omega}$ represents the number of photons traveling per unit time per unit area (perpendicular to $\hat{\Omega}$) in a range of directions within $d\hat{\Omega}$ around $\hat{\Omega}$. The units of $L(\mathbf{r},\hat{\Omega},t)$ are $s^{-1}m^{-2}sr^{-1}$.

Photon density: $U(\mathbf{r}, t)$
$U(\mathbf{r},t) = \int_{4\pi} u(\mathbf{r},\hat{\Omega},t)d\hat{\Omega}$. The photon density is the number of photons per unit volume. The units are m^{-3}.

Photon fluence rate: $E_0(\mathbf{r}, t)$
$E_0(\mathbf{r},t) = vU(\mathbf{r},t) = \int_{4\pi} L(\mathbf{r},\hat{\Omega},t)d\hat{\Omega}$ The photon fluence rate is defined as the number of photons traveling per unit time per unit area (perpendicular to the direction of propagation) over all directions. The units are $s^{-1}m^{-2}$.

Photon current density, or photon flux: $\mathbf{J}(\mathbf{r}, t)$
$J(\mathbf{r},t) = \int_{4\pi} L(\mathbf{r},\hat{\Omega},t)\hat{\Omega}d\hat{\Omega}$. The photon flux is a vector that represents the net flow of photons. Its direction points in the direction of the net flux, while its amplitude gives the net number of photons transmitted per unit time per unit area in that direction. The units of $\mathbf{J}(\mathbf{r}, t)$ are $s^{-1}m^{-2}$.

These definitions can be extended to describe radiant energy (instead of photon number) at any given wavelength λ, or frequency ν, by replacing the word "photon" with "energy," and by introducing a factor $h\nu$ in all definitions ($h\nu$ is the photon energy, where h is Planck's constant). A complete nomenclature for quantities used in medical optics can be found in Hetzel et al.[41]

7.3.2 Derivation of the diffusion equation from the BTE

To model light propagation in highly scattering media, which include most biological tissues, it is useful to derive the diffusion equation (DE) from the

BTE, since the DE is often an excellent approximation and is a much simpler equation to solve. For this purpose, we expand the angular photon density $u(\mathbf{r},\hat{\Omega},t)$, the source term $q(\mathbf{r},\hat{\Omega},t)$, and the phase function $f(\hat{\Omega}',\hat{\Omega})$ into spherical harmonics $Y_l^m(\hat{\Omega})$.[42–44] The so-called P_N approximation to the Boltzmann equation is based on such an expansion. As a result of the completeness property of the spherical harmonics, any function $h(\theta, \varphi)$ (where the angles θ and φ identify a direction $\hat{\Omega}$in the space) can be expanded in the Laplace series

$$h(\theta,\varphi) = \sum_{l=0}^{\infty} \sum_{m=-l}^{l} h_{lm} Y_l^m(\hat{\Omega}), \tag{7.2}$$

where h_{lm} are coefficients independent of θ and φ, and $\hat{\Omega}$ is the unit directional vector defined by $\hat{\Omega} = \sin\theta\cos\varphi\hat{\mathbf{x}} + \sin\theta\sin\varphi\hat{\mathbf{y}} + \cos\theta\hat{\mathbf{z}}$.[45] Accordingly, we expand $u(\mathbf{r},\hat{\Omega},t)$, $q(\mathbf{r},\hat{\Omega},t)$, and $f(\hat{\Omega}',\hat{\Omega})$ into spherical harmonics as follows:

$$u(\mathbf{r},\hat{\Omega},t) = \sum_{l=0}^{\infty} \sum_{m=-l}^{l} u_{lm}(\mathbf{r},t)\, Y_l^m(\hat{\Omega}), \tag{7.3}$$

$$q(\mathbf{r},\hat{\Omega},t) = \sum_{l=0}^{\infty} \sum_{m=-l}^{l} q_{lm}(\mathbf{r},t)\, Y_l^m(\hat{\Omega}), \tag{7.4}$$

$$f(\hat{\Omega}' \cdot \hat{\Omega}) = \sum_{l=0}^{\infty} \frac{2l+1}{4\pi} f_l P_l(\hat{\Omega}' \cdot \hat{\Omega}) = \sum_{l=0}^{\infty} \sum_{m=-l}^{l} f_l Y_l^{m*}(\hat{\Omega}')\, Y_l^m(\hat{\Omega}). \tag{7.5}$$

In Eq. (7.5), we have assumed that the phase function $f(\hat{\Omega}',\hat{\Omega})$ depends only on the dot product $\hat{\Omega}' \cdot \hat{\Omega}$ (i.e., on the cosine of the scattering angle γ). Equation (7.5) is based on a Legendre series representation and on the addition theorem for Legendre polynomials.[45] Note that the coefficients q_{lm} and f_l in Eqs. (7.4) and (7.5) are known, while the coefficients u_{lm} in Eq. (7.3) should be solved for.

By substituting Eqs. (7.3)–(7.5) into Eq. (7.1), and using the orthogonality property of the spherical harmonics, we obtain an infinite set of coupled partial differential equations for the unknowns u_{lm}. For fixed integer values $l = L$ and $m = M$, we find that the partial differential equation for u_{LM} contains u_{lm} with indices l ranging only from $L-1$ to $L+1$, and m ranging from $M-1$ to $M+1$, while the only coefficients of the source and phase function expansion are q_{LM} and f_L, respectively. Therefore, the expansion of the BTE into spherical harmonics leads to an infinite set of equations with indices L (ranging from 0 to ∞) and M (ranging from $-L$ to L). Truncation of the Laplace series at $L = N$ leads to the so-called P_N approximation. In particular, the P_1 approximation consists of retaining the terms of the Laplace series with $L \leq 1$. The four equations for u_{LM} in the P_1 approximation are those for indices $(L,M) = (0, 0), (1, -1), (1, 0), (1, 1)$. It can be shown that these four equations can be consolidated into the following two equations:

$$\frac{\partial}{\partial t} U(\mathbf{r},t) + v\mu_a U(\mathbf{r},t) + \nabla \cdot \mathbf{J}(\mathbf{r},t) = S_0(\mathbf{r},t), \tag{7.6}$$

$$\frac{1}{v}\frac{\partial}{\partial t}\mathbf{J}(\mathbf{r},t) + (\mu_s(1 - <\cos\gamma>) + \mu_a)\mathbf{J}(\mathbf{r},t) + \frac{1}{3}v\nabla U(\mathbf{r},t) = \mathbf{S}_1(\mathbf{r},t), \tag{7.7}$$

which are coupled partial differential equations for the photon density ($U(\mathbf{r}, t)$) and the photon flux ($\mathbf{J}(\mathbf{r}, t)$). The source terms in Eq. (7.6) and in Eq. (7.7)—$S_0(\mathbf{r}, t)$ and $\mathbf{S}_1(\mathbf{r}, t)$, respectively—are related to the source coefficients $q_{0,0}$, $q_{1,-1}$, $q_{1,0}$, and $q_{1,1}$. In particular, we observe that $\mathbf{S}_1(\mathbf{r}, t) = 0$ for an isotropic source term. One can define the quantity $\mu_s' = \mu_s(1 - <\cos\gamma>)$, which is termed the reduced scattering coefficient,[46] from the scattering coefficient (μ_s) and the average cosine of the scattering angle ($<\cos\gamma>$), which in turn depends on the phase function. The coupled Eqs. (7.6) and (7.7) of the P_1 approximation can be written as a single equation, which for homogeneous media (i.e., for spatially independent μ_s', μ_a) reduces to the following equation:

$$\begin{aligned}\nabla^2 U(\mathbf{r},t) = {} & \frac{3}{v^2}\frac{\partial^2 U(\mathbf{r},t)}{\partial t^2} + \frac{1}{D}\left(1 + \frac{3D}{v}\mu a\right)\frac{\partial U(\mathbf{r},t)}{\partial t} + \frac{v\mu_a}{D}U(\mathbf{r},t) \\ & - \frac{3}{v^2}\frac{\partial S_0(\mathbf{r},t)}{\partial t} - \frac{1}{D}S_0(\mathbf{r},t) + \frac{3}{v}\nabla \cdot \mathbf{S}_1(\mathbf{r},t),\end{aligned} \tag{7.8}$$

where we have defined the diffusion coefficient $D = v/[3(\mu_s' + \mu_a)]$.

By making a few assumptions, which are often satisfied in the case of near-infrared light (approximately 600–1000 nm) propagation in biological tissue, Eq. (7.8) reduces to the diffusion equation (DE). The assumptions are the following:

1. Strongly scattering regime, or $\mu_a \ll \mu_s'$. This condition means that a photon, on average, will undergo many effectively isotropic scattering events before being absorbed. In this case, $3D\mu_a/v \equiv \mu_a/(\mu_a + \mu_s') \ll 1$, and the second term on the right-hand side of Eq. (7.8) reduces to $(1/D)\partial U(\mathbf{r},t)/\partial t$.
2. The time scale of the variations of $U(\mathbf{r},t)$ and $S_0(\mathbf{r},t)$ is much longer than the average time between collisions $1/[v(\mu_a + \mu_s')]$. This condition can be expressed by the formal operators inequality: $\partial/\partial t \ll v(\mu_a + \mu_s') \equiv$

$v^2/(3D)$ (where the factor $v^2/(3D)$ is the "multiplication operator"). Consequently,

$$\frac{3}{v^2}\frac{\partial^2 U(\mathbf{r},t)}{\partial t^2} \ll \frac{1}{D}\frac{\partial U(\mathbf{r},t)}{\partial t}, \tag{7.9}$$

$$\frac{3}{v^2}\frac{\partial S_0(\mathbf{r},t)}{\partial t} \ll \frac{1}{D}S_0(\mathbf{r},t). \tag{7.10}$$

In the frequency domain, where the harmonic time dependence is given by a factor $\exp(-i\omega t)$, the time derivative operator becomes a multiplication operator (given by the factor "$-i\omega$"). Here, ω is the angular modulation frequency of the intensity modulation (which should not be confused with the frequency of light). Consequently, the condition $\partial/\partial t \ll v(\mu_a + \mu_s') \equiv v^2/(3D)$ becomes $\omega \ll v^2/(3D)$, which poses an upper limit to the modulation frequency. In the case of biological tissues, the DE usually breaks down at modulation frequencies on the order of 1 GHz.[18]

3. The source term is isotropic; that is, $\mathbf{S}_1(\mathbf{r},t) = 0$.

With these assumptions, the P_1 approximation of Eq. (7.8) reduces to the diffusion equation for homogeneous media:

$$\frac{\partial U(\mathbf{r},t)}{\partial t} = D\nabla^2 U(\mathbf{r},t) - v\mu_a U(\mathbf{r},t) + S_0(\mathbf{r},t), \tag{7.11}$$

and the photon flux $\mathbf{J}(\mathbf{r},t)$ is related to the photon density $U(\mathbf{r},t)$ by Fick's law:

$$\mathbf{J}(\mathbf{r},t) = -D\nabla U(\mathbf{r},t). \tag{7.12}$$

7.3.3 The diffusion equation in the frequency domain

The frequency-domain diffusion equation is a particular case of Eq. (7.11) for the harmonically modulated source term

$$S_0(\mathbf{r},t) = g(\mathbf{r})[P_{DC} + P_{AC}(\omega)e^{-i\omega t}], \tag{7.13}$$

where $g(\mathbf{r})$ is a spatially dependent source factor (with units m^{-3}), whereas P_{DC} and $P_{AC}(\omega)$ express the DC and AC source power (with units s^{-1}), respectively. We note that the source term given by Eq. (7.13) represents a physical source term only if $P_{DC} > P_{AC}(\omega)$. The source term of Eq. (7.13) can be broken into the *DC* and *AC* source terms:

$$\begin{aligned} S_0(\mathbf{r},t) &= S_{DC}(\mathbf{r}) + S_{AC}(\mathbf{r},t) \\ S_{DC}(\mathbf{r}) &= g(\mathbf{r})P_{DC} \\ S_{AC}(\mathbf{r},t) &= g(\mathbf{r})P_{AC}(\omega)e^{-i\omega t}. \end{aligned} \tag{7.14}$$

Even though the DE associated with the source term of Eq. (7.13) is the correct frequency-domain DE (containing a physically meaningful source term), one may also write the frequency domain diffusion equation in terms of the sole AC source term, with the convention that the DC term corresponds to the special case $\omega = 0$. A formal solution of the equation associated to the AC source term has the form $U(\mathbf{r},\omega,t) = U(\mathbf{r},\omega)e^{-i\omega t}$. If we use the formal substitution of operators $\partial/\partial t \rightarrow -i\omega$ in Eq. (7.11), we are left with the Helmholtz equation for $U(\mathbf{r}, \omega)$ (i.e., the frequency-domain DE)

$$(\nabla^2 + k^2)U(\mathbf{r},\omega) = -\frac{g(\mathbf{r})P_{AC}(\omega)}{D}, \tag{7.15}$$

where $k^2 = (i\omega - v\mu_a)/D$. We observe that for $\omega = 0$, Eq. (7.15) reduces to the DC diffusion equation. We note that since the DC-domain and the frequency-domain (FD) diffusion equations are formally equivalent, we can obtain the solutions in one domain from those found in the other domain. The only difference between the solutions in the two domains lies in the definition of k^2, which in FD contains $i\omega$ (and therefore is a complex number), but in DC does not (since $\omega = 0$) such that k becomes a pure imaginary number.

7.3.4 Solutions to the frequency-domain diffusion equation

7.3.4.1 Infinite geometry

The solution to the FD diffusion equation [Eq. (7.15)] for a homogeneous, infinite medium containing a harmonically modulated point source at $\mathbf{r} = 0$ [i.e., $g(\mathbf{r}) = \delta(\mathbf{r})$, where $\delta(\mathbf{r})$ is the Dirac delta] emitting a power $P_{AC}(\omega)$ is given by

$$U(\mathbf{r},\omega) = \frac{P_{AC}(\omega)}{4\pi D}\frac{e^{\mathrm{i}kr}}{r}, \tag{7.16}$$

where r is the distance from the point source.[47] Since the wave number k is complex, $U(r, \omega)$ is a complex function of the real variables r and ω. The explicit expressions for the amplitude $\mathrm{AC} = |U(r, \omega)|$ and phase $\Phi(r, \omega) = \mathrm{Arg}(U(r, \omega))$ of the photon-density wave $U(r, \omega)$ are

$$\mathrm{AC} = \frac{P_{AC}(\omega)}{4\pi D}\frac{e - r(v\mu_a/2D)^{1/2}\left[\left(1 + \frac{\omega^2}{v^2\mu_a^2}\right)^{1/2} + 1\right]^{1/2}}{r}, \tag{7.17}$$

$$\Phi(r,\omega) = r(v\mu_a/2D)^{1/2}\left[\left(1 + \frac{\omega^2}{v^2\mu_a^2}\right)^{1/2} - 1\right]^{1/2} + \Phi_s, \tag{7.18}$$

where Φ_s is the source phase in radians.[39,48,49] Note that Eq. (7.16) is also a solution to the DC diffusion equation obtained for $\omega = 0$. In this case, $k = iv\mu_a/D$ and Eq. (7.16) becomes a real function of the real variable r.

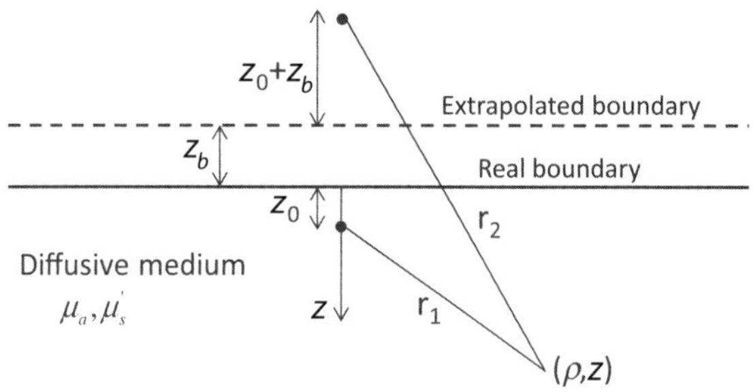

Figure 7.3 Schematic representation of the semi-infinite medium geometry. The diffusive medium is bounded by the plane $z = 0$. The extrapolated boundary is identified by the equation $z = -z_b$. The coordinates of the real and imaginary point sources are $(\rho, z) = (0, z_0)$, and $(\rho, z) = (0, -z_0 - 2z_b)$, respectively. Given an arbitrary field point (ρ, z), r_1 and r_2 are the distances from the field point to the real and imaginary point sources, respectively.

7.3.4.2 Semi-infinite geometry

Analytical solutions to the diffusion equation in the frequency-domain have been reported for a semi-infinite medium,[39,49,50] infinite slab, cylindrical, and spherical geometries.[51] Here, we report the solution in the semi-infinite medium geometry with extrapolated boundary conditions (EBCs). For the EBCs, we refer to Contini et al.[52] One method to obtain this solution is to make use of the solution in the infinite medium geometry and to apply the method of images for realizing the boundary conditions.[53]

Figure 7.3 shows a schematic diagram of the semi-infinite medium geometry with the diffusive medium found in the lower half space ($z > 0$). The extrapolated boundary, which is defined by those points where the photon density is zero ($U(\mathbf{r}_b, \omega) = 0$), is identified by the plane $z = -z_b$, where $z_b = 2A(n)D/v$, and v is the speed of light in the diffusive medium. The expression of the coefficient $A(n)$, which depends on the refractive index mismatch between medium and the surroundings (n), is given by Contini et al.[52] The positions of the real and imaginary sources are $(\rho, z) = (0, 0, z_0)$ and $(\rho, z) = (0, 0, -z_0 - 2z_b)$, respectively. By applying the method of images, the solution of Eq. (7.15) for $g(\mathbf{r}) = \delta(\mathbf{r} - \mathbf{r}_0)$ (with $\mathbf{r}_0 = (0, 0, z_0)$ and $z_0 = 1/\mu_s')$ is

$$U(r,\omega) = \frac{P_{AC}(\omega)}{4\pi D}\left(\frac{e^{ikr_1}}{r_1}\frac{e^{ikr_2}}{r_2}\right), \tag{7.19}$$

where r_1 and r_2 are defined by

$$r_1 = \sqrt{(z - z_0)^2 + \rho^2}$$

$$r_2 = \sqrt{(z + z_0 + 2z_b)^2 + \rho^2} \tag{7.20}$$

The output flux, also called single-point reflectance, is defined by

$$R(\rho,\omega) = \mathbf{J}(\rho,\omega) \cdot (-\hat{\mathbf{z}})_{|z=0}, \tag{7.21}$$

which, after using Fick's law [as given by Eq. (7.12)], becomes

$$R(\rho,\omega) = D \frac{\partial U(r,\omega)}{\partial z}\bigg|_{z=0}. \tag{7.22}$$

By carrying out the derivative in Eq. (7.22), we obtain

$$R(\rho,\omega) = \frac{P_{AC}(\omega)}{4}\left[\frac{z_0}{r_{10}^2} e^{ikr_1}\left(\frac{1}{r_{10}} - ik\right) + \frac{z_0 + 2z_b}{r_{20}^2} e^{ikr_2}\left(\frac{1}{r_{20}} - ik\right)\right] \tag{7.23}$$

where r_{10} and r_{20} are given by Eq. (7.20) with $z = 0$. Again, the AC(ρ, ω) and phase Φ(ρ, ω) are given by AC(ρ, ω) = |*R*(ρ, ω)|, Φ(ρ, ω) = Arg[*R*(ρ, ω)].

7.3.4.3 Two-layered geometry

A two-layered medium is a special case of a non-homogeneous medium. Here, we rewrite the frequency-domain DE for a non-homogenous medium in a slightly different form:

$$\nabla \cdot [D_0(\mathbf{r})\nabla E_0(\mathbf{r},\omega)] - \left[\mu_a(\mathbf{r}) + i\frac{\omega}{v}\right] E_0(\mathbf{r},\omega) = -P_{AC}(\omega)\delta(\mathbf{r}). \tag{7.24}$$

Equation (7.24) is written for the unknown fluence rate (E_0) and is derived from the time-domain DE by assuming that the temporal dependence of the source is "exp($i\omega t$);" also, we have redefined the diffusion coefficient, indicated here as D_0, which is related to the previous one (D) by the relationship $D_0 = D/v$. Both forms of the optical diffusion coefficient are commonly found in the literature.

Layered diffusive media are characterized by plane-bounded layers having different thicknesses, optical properties, and refractive indices. In the literature, FD solutions of the DE have been reported for a two-layered, laterally unbounded medium, with the second layer having infinite thickness,[54] and also for a stack of *n* layers.[55] In the second case, the solution of the DE was derived for a cylindrical stack of layers, which simulates a laterally infinite stack of layers when the radius of the cylinder is much larger than the source–detector separation. By using a cylindrical stack of layers (instead of a laterally unbounded stack of layers), the computational problem becomes less challenging.

Figure 7.4 shows a schematic diagram of a two-layered stack in cylindrical geometry, where l_i ($i = 1, 2$) and μ_{ai}, μ_{si}' ($i = 1, 2$) are the thicknesses and optical properties of the two layers, respectively. For a

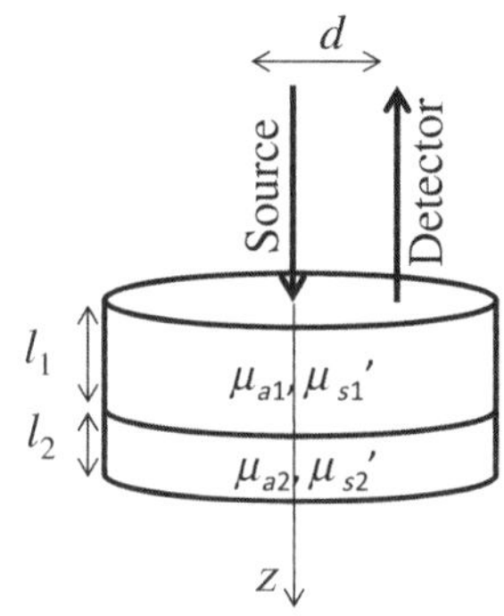

Figure 7.4 Schematic illustration of a two-layered stack in cylindrical geometry (extrapolated boundaries are omitted). The solution provided in the text is obtained when $l_1 = L$ and $l_2 \rightarrow \infty$. The origin of the cylindrical coordinate system is the point where the collimated light source is incident on the medium.

point source incident at the center of a layered cylindrical medium, the general solution of the two-layer DE in cylindrical coordinates ($\mathbf{r} = (\rho, \theta, z)$) is given by

$$E_{0\,k}(\mathbf{r},\omega) = \frac{P_{AC}(\omega)}{\pi a'^2} \sum_{n=1}^{\infty} G_k(s_n,z,\omega) J_0(s_n\rho) J_1^{-2}(a's_n), \tag{7.25}$$

where E_{0k} is the photon fluence rate in the kth layer of the medium (i.e., $k = 1$, 2), s_n are the positive roots of the 0th-order Bessel function of the first kind divided by $a' = a + z_b$ (where a is the radius of the cylinder), and J_m is the Bessel function of the first kind of order m.[55] Also, z_b is the distance between the extrapolated and the real boundary, $z_b = 2D_{01}(1 + R_{\text{eff}})/(1 - R_{\text{eff}})$, and R_{eff} is the fraction of photons that are internally diffusely reflected at the cylinder boundary.[49] Here, we report the solution for G_k only for the first layer ($k = 1$), since it is the layer where the reflectance is calculated. More details on the derivation of Eq. (7.26) and the expressions of G_k for arbitrary k can be found in Liemert and Kienle.[55]

For the first layer (the one illuminated by the light source), G_1 is given by the following expression:

$$\begin{aligned} G_1(s_n,z,\omega) = {} & \frac{\exp(-\alpha_1|z - z_0|) - exp[\alpha_1(z + z_0 + 2z_b)]}{2D_{01}\alpha_1} \\ & + \frac{\sinh[\alpha_1(z_0 + z_b)]\sinh[\alpha_1(z + z_b)]}{D_{01}\alpha_1 \exp[\alpha_1(L + z_b)]} \\ & \times \frac{D_{01}\alpha_1 - D_{02}\alpha_2}{D_{01}\alpha_1 \cosh[\alpha_1(L + z_b)] + D_{02}\alpha_2 \sinh[\alpha_1(L + z_b)]} \end{aligned} \tag{7.26}$$

where L is the thickness of the first layer, and α_k is given by

$$\alpha_k = \sqrt{\frac{\mu_{ak}}{D_{0\,k}} + s_n^2 + \frac{i\omega}{D_{0\,k^v}}} k = 1,2. \tag{7.27}$$

Equation (7.26) is obtained from Eq. (7.15) for the limiting case when the second layer is infinite in the z direction (see Fig. 7.4) and for the case of two layers having the same refractive indices.[55] In Eq. (7.27), D_{0k} is the diffusion coefficient of the kth layer. For the calculation of the reflectance (R), one may apply Fick's law:

$$R(\rho,\omega) = D_{01}\frac{\partial}{\partial z}E_{01}(\rho,z)|_{z=0} \tag{7.28}$$

From the expression of the reflectance (which is a complex function), we can calculate the AC(ρ, ω) and phase $\Phi(\rho, \omega)$ as follows: AC(ρ, ω) = $|R(\rho, \omega)|$, $\Phi(\rho, \omega) = \mathrm{Arg}[R(\rho, \omega)]$.

7.3.5 Multi-distance tissue spectroscopy

Multi-distance tissue spectroscopy is a method to measure the optical properties of highly scattering media and biological tissue from frequency-domain measurements at multiple source–detector distances. From Eq. (7.17), we can see that

$$\ln(r\mathrm{AC}) = \ln\left(\frac{P_{\mathrm{AC}}(\omega)}{4\pi D}\right) - r(v\mu_a/2D)^{1/2}\left[\left(1+\frac{\omega^2}{v^2\mu_a^2}\right)^{1/2}+1\right]^{1/2} \tag{7.29}$$

Therefore, in the infinite medium geometry, both phase [Eq. (7.18)] and ln (rAC) are linear functions of the source–detector separation (r). This linear dependence leads to an analytical inversion procedure to evaluate μ_a and μ'_s using, for example, the slopes of ln(rAC) and phase (S_{AC} and S_{PH}, respectively) as a function of source–detector distance:[4]

$$\mu_a = \frac{\omega}{2v}\left(\frac{S_{\mathrm{PH}}}{S_{\mathrm{AC}}} - \frac{S_{\mathrm{AC}}}{S_{\mathrm{PH}}}\right) \tag{7.30}$$

$$\mu_s' = \frac{S_{\mathrm{AC}}^2 - S_{\mathrm{PH}}^2}{3\mu_a} - \mu_a. \tag{7.31}$$

Equations (7.30) and (7.31) can be used also in the semi-infinite medium geometry if we replace, to a first approximation, the slope of ln(rAC) with the slope of ln(r^2AC).[39,56]

7.3.6 Multi-frequency tissue spectroscopy

It is also possible to measure the optical properties of highly scattering media or tissue by using a different approach that relies on data collected at multiple modulation frequencies. Through inspection of Eq. (7.17) (for the AC) and Eq. (7.18) (for the phase) in infinite media, and similar equations derived from the complex reflectance [Eq. (7.23)] in semi-infinite media, one can see that by fixing the source–detector separation, the AC and phase data become nonlinear functions of the angular modulation frequency (ω). Multi-frequency instruments have been devised on the basis of a sweep of modulation frequencies in a wide range (300 kHz–3 GHz),[57] even though the frequency range 100–700 MHz has been most commonly used for measurements in tissues.[58] The fitting procedures to retrieve the optical properties are nonlinear[49] and, therefore, are more complicated than those used for multi-distance tissue spectroscopy. Despite this complication, the advantage of using multi-frequency methods is the reduced sensitivity to the heterogeneity of tissues, since the source–detector distance is fixed and the collected photons consequently probe a more confined region of tissue compared to the case of multi-distance methods.

7.4 Tissue Spectroscopy and Oximetry

7.4.1 Optical properties of biological tissue

As discussed in Section 7.3.1, the propagation of light inside biological tissues can be described in terms of a flow of photons. From this perspective, the light source injects a given number of photons per unit time, per unit volume, per unit solid angle into a specific tissue location; these photons then travel inside the tissue along certain trajectories. The collective motion of photons along these trajectories is called *photon migration*. While traveling inside tissues, photons can undergo a number of different processes, such as absorption, elastic scattering, inelastic scattering (for instance, dynamic light scattering and Raman scattering), and fluorescence. In this chapter, we consider only absorption (Section 7.4.1.1) and elastic scattering (Section 7.4.1.2) events. The scattering process from non-stationary scattering centers in tissues (e.g., cells, cellular organelles), which strictly speaking is an inelastic process, involves such small wavelength changes that it is also referred to as quasi-elastic scattering. Quasi-elastic scattering is the basis for dynamic light scattering,[59] diffusing-wave spectroscopy,[60] diffuse correlation spectroscopy,[61,62] and laser Doppler flowmetry.[63,64] When a photon is absorbed, it disappears and transfers its energy to the absorbing center. When a photon is elastically (or quasi-elastically) scattered, its direction of propagation changes, while its wavelength remains essentially unchanged. By contrast, Raman scattering

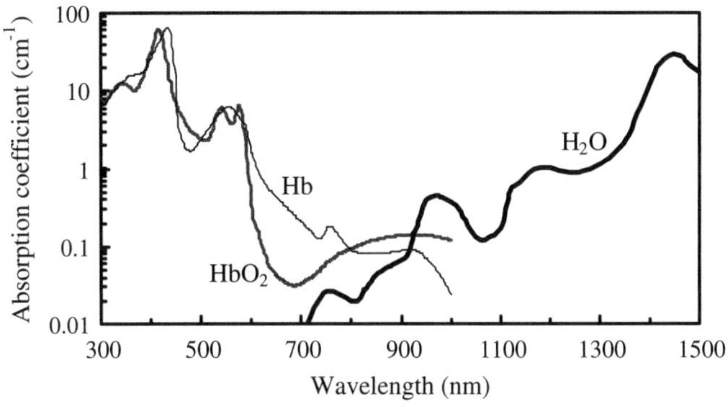

Figure 7.5 Absorption spectra of the three dominant near-infrared chromophores in tissues: oxyhemoglobin (HbO_2), deoxyhemoglobin (Hb), and water (H_2O). The absorption coefficient is defined to base *e*. The concentrations of Hb and HbO_2 are both assumed to be 50 μM, a typical value for blood-perfused tissues. These spectra are obtained from compiled absorption data for water[65] and hemoglobin.[66]

and fluorescence processes cause larger photon wavelength changes related to the involvement of vibrational energy levels.

7.4.1.1 Absorption

The main absorbers of near-infrared (NIR) light in blood–perfused tissues are oxyhemoglobin, deoxyhemoglobin, and water. Their absorption spectra between 300 and 1500 nm are shown in Fig. 7.5, which is obtained from compiled absorption data for water[65] and hemoglobin.[66] In Fig. 7.5, the concentrations of oxyhemoglobin and deoxyhemoglobin are assumed to be 50 μM, which is a typical value in blood-perfused tissues. We observe that the absorption coefficients reported in Fig. 7.5 are defined in terms of base *e* logarithms, and those for hemoglobin refer to a complete molecule (four heme groups). The "medical spectral window" extends from approximately 700 to 900 nm, where the absorption of light is relatively small (see Fig. 7.5). As a result, light in this spectral window penetrates deeply into tissues, allowing for noninvasive investigations. The optical penetration depth into tissues is limited at shorter wavelengths by the hemoglobin absorption, and at longer wavelengths by the water absorption. Other NIR absorbers in tissues that may be important in particular cases include myoglobin, lipids, cytochrome *c* oxidase, melanin, bilirubin, and collagen.

The absorption properties of tissue are described by the absorption coefficient (μ_a), which is defined as the inverse of the average photon pathlength before absorption. From this definition, it follows that $1/\mu_a$ is the average distance traveled by a photon before being absorbed. In the NIR spectrum, typical values of μ_a in tissue range from 0.02 to 0.30 cm^{-1}. Consequently, the photon mean-free path for absorption ranges between approximately 3 and 50 cm.

7.4.1.2 Scattering

The scattering properties are mainly determined by the size of the scattering particles relative to the wavelength of light, and by the refractive index mismatch between the scattering particles and the surrounding medium. In biological tissues, the scattering centers are mostly cells and cellular organelles. In the medical spectral window (700–900 nm), cellular organelles have dimensions comparable to the wavelength, and their index of refraction is relatively close to that of the cytosol and extracellular fluid. As a result, light scattering in tissue is mainly forward directed (i.e., the scattering angle γ introduced in Section 7.3.2 is much less than 90 deg) and shows a weak wavelength dependence.

As discussed in Section 7.3.2, the scattering properties of tissues are described by two parameters: (1) the scattering coefficient (μ_s), defined as the inverse of the average photon pathlength between successive scattering events, and (2) the average cosine of the scattering angle ($f_1 = <\cos\gamma>$). From the definition of μ_s, it follows that $1/\mu_s$ is the average distance traveled by a photon between successive scattering events. Even though each scattering event is mainly forward directed, after a number of collisions a photon loses the memory of its original direction of propagation. Under these conditions, the photon has experienced an effectively isotropic scattering event. As discussed by Zaccanti et al., the reduced scattering coefficient [$\mu_s' = (1 - f_1)\,\mu_s$] represents the inverse of the average distance over which the direction of propagation of a photon is randomized.[46] In other words, we can say that $1/\mu_s'$ is the average distance between successive, effectively isotropic scattering events. Note that μ_s' coincides with μ_s in the case of isotropic scattering ($f_1 = 0$). Typical values of μ_s' in biological tissue range from 2 to 20 cm^{-1}, while f_1 is typically approximately 0.9 (so that μ_s' is about one order of magnitude smaller than μ_s).[67,68] The average distance traveled by a photon in tissue before losing memory of its initial direction of propagation is typically a few millimeters or less.

7.4.2 Absorption spectroscopy of tissue

Since μ_s' is typically much larger than μ_a, NIR light propagation in tissue is dominated by scattering. This is one of the conditions discussed in Section 7.3.2 for the validity of the diffusion equation [Eq. (7.11)]. The frequency-domain solution given by Eq. (7.16) provides a quantitative description of photon migration in an infinite medium with uniform optical properties. However, biological tissues are neither infinite nor homogeneous. With regard to tissue inhomogeneity, Eq. (7.16) is a reasonable approximation as long as the characteristic size of the tissue inhomogeneities (e.g., muscle fibers, blood vessels) is small compared to the source–detector separation. More precisely, Monte Carlo simulations and diffusion theory calculations have shown that the homogeneity of the distribution of hemoglobin (the main tissue absorber in the NIR wavelength range) in tissue is a valid hypothesis for the case of "smaller" blood vessels, which are characterized by the following relation: $d < 0.06/\mu_a^{vessel}$ where d is the diameter and μ_a^{vessel} is the absorption coefficient of the blood vessel.[69]

A second problem is determined by the actual, irregular geometries of finite volumes of investigated tissues. For example, when collecting reflectance data, where the source and the detector are located on the same side of the tissue, one typically applies the simplified approach of semi-infinite boundary conditions. This approximation implies that the tissue acts as an effective half-space, which is a reasonable assumption if the tissue depth is greater than the optical penetration depth (typically on the order of 2–3 cm or less). This assumption is not valid in a transmission geometry (where the source and the detector are located on opposite sides of the tissue) or in the case of a smaller tissue volume. In these cases, the application of more appropriate geometries (such as a slab, cylinder, or sphere) may be needed.[51]

Tissue spectroscopy aims at determining certain properties of the investigated tissue volume (e.g., the oxygen saturation or the concentration of hemoglobin in skeletal muscle) on the basis of measurement of the optical properties (absorption and scattering) of the tissue. Since the tissue absorption coefficient is influenced by a number of chromophores (e.g., oxyhemoglobin, deoxyhemoglobin, water, cytochrome *c* oxidase, melanin, bilirubin, lipids), multi–wavelength measurements have been employed to determine the relative contributions of each chromophore. The basic idea is that the contribution to μ_a from the ith chromophore can be written as the product of the extinction coefficient (ε_i) times the concentration (C_i) of that chromophore. As a result, in the presence of N chromophores, the absorption coefficient μ_a at wavelength λ_j is given by

$$\mu_a(\lambda_j) = \sum_{i=1}^{N} \varepsilon_i(\lambda_j) C_i. \tag{7.32}$$

If the extinction spectra $\varepsilon_i(\lambda)$ of all N species are known (and are all different from each other), the concentrations C_i can be determined by measuring μ_a at N or more wavelengths, so that the linear system of Eq. (7.32) is fully determined. This approach requires that μ_a be measured independently of μ_s'. In the frequency domain, Eqs. (7.16)–(7.18) for the infinite medium geometry, or the corresponding equations for semi-infinite or other geometries, can be used to quantitatively measure μ_a and μ_s'.[50] The spatial dependence[4] or the frequency dependence[21,49,58] of the AC amplitude and phase can also be effectively used for *in vivo* tissue spectroscopy. It is worth noting that this approach to tissue spectroscopy yields quantitative, absolute values of the concentration of chromophores in tissue.

7.4.3 Quantification of hemoglobin concentration and saturation in tissue

In many cases, the absorption spectra of tissues can be well described by considering only three chromophores—namely, oxyhemoglobin, deoxyhemoglobin, and water. For example, Fig. 7.6 shows the absorption coefficients of human muscle (forearm) and brain (prefrontal cortex accessed through the forehead) measured at eight NIR wavelengths (squares) under resting conditions. The lines in Fig. 7.6 are the best-fit absorption spectra corresponding to a linear combination of the water, oxyhemoglobin, and deoxyhemoglobin extinction spectra. In the fits, the water concentration (by volume) is assumed to be 75% for muscle tissue and 80% for brain tissue,[70] while the concentrations of oxyhemoglobin, [HbO_2], and deoxyhemoglobin, [Hb], are the fitting parameters. The best-fit spectra of Fig. 7.6 yield values for [HbO_2] and [Hb] of 58 μM and 26 μM, respectively, in the muscle, and 30 μM and 10 μM, respectively, in the brain. Consequently, the measured hemoglobin saturation in the muscle is 69%;[38] in the brain, it is 75%.[71] The good fit between the experimental data and the three-component absorption spectra indicates that oxyhemoglobin and deoxyhemoglobin are indeed the dominant absorbing species over the wavelength range considered (630–840 nm) (water absorption introduces only a relatively small correction at these wavelengths). This result allows for measurements of the oxygen saturation of hemoglobin in tissues using just two wavelengths.

The possibility of using dual-wavelength optical measurements for blood oximetry has been recognized for a long time.[72] In recent decades, it has been exploited by pulse oximeters to measure the arterial saturation.[73] The novelty introduced by frequency-domain and time-domain spectroscopy is the capability of performing absolute tissue oximetry.[37,74]

The two wavelengths λ_1 and λ_2 for near-infrared oximetry are usually chosen such that $\lambda_1 < \lambda_{iso} \leq \lambda_2$, where λ_{iso} is the NIR isosbestic wavelength at

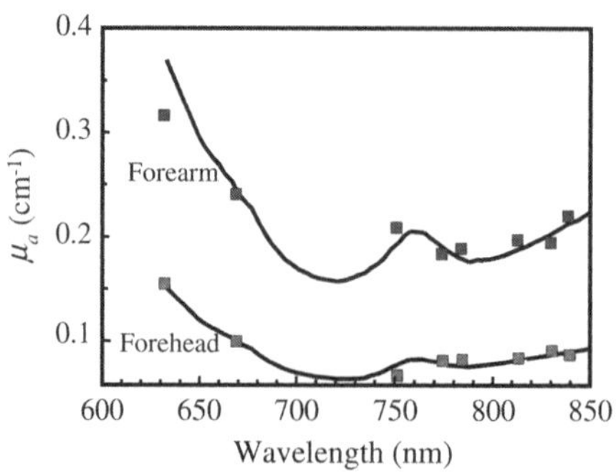

Figure 7.6 Quantitative absorption spectroscopy of tissues *in vivo*.[38,71] The squares represent the eight-wavelength spectra measured with frequency-domain spectroscopy in the skeletal muscle and in the forehead (as labeled) of human subjects. The lines are best fits using a linear combination of the extinction spectra of oxyhemoglobin, deoxyhemoglobin, and water.

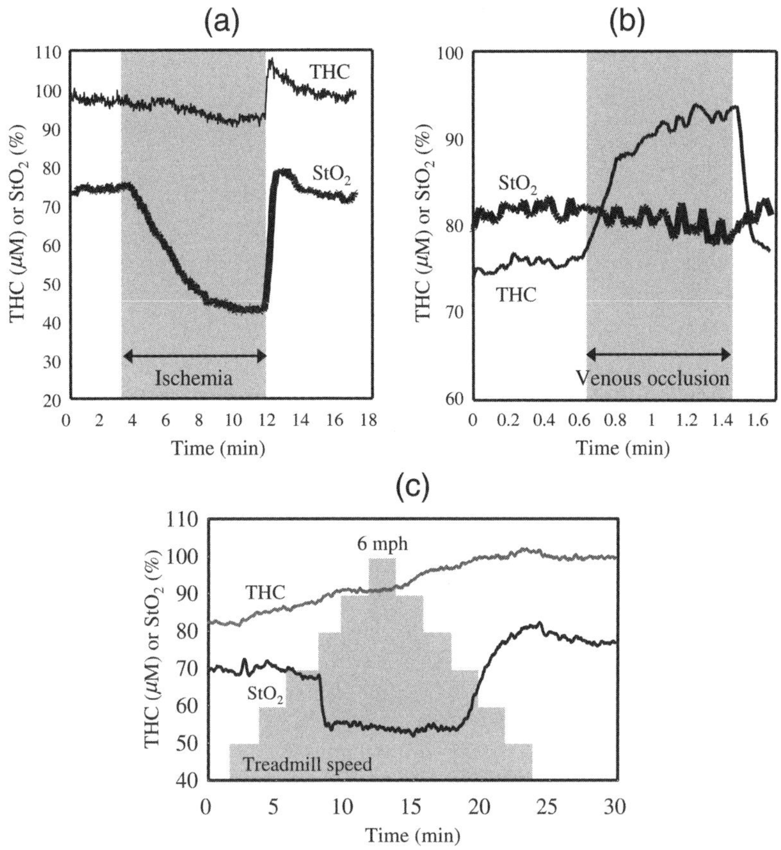

Figure 7.7 Temporal traces of optically measured total hemoglobin concentration (THC = $[HbO_2] + [Hb]$) and tissue saturation (StO_2) in human subjects. The measurements were conducted on a forearm muscle during (a) arterial occlusion or (b) venous occlusion, and (c) on a calf muscle during treadmill exercise. In (c), the step function represents the treadmill speed, which was changed by increments of 1 mph every 2 min.[75] Running started at a speed of 4 mph. The frequency-domain measurements allow for absolute *y*–axis labeling.

which the extinction coefficients of oxyhemoglobin and deoxyhemoglobin have the same value (λ_{iso} is approximately 800 nm, as can be seen in Fig. 7.4). This choice maximizes the sensitivity of the optical measurement to changes in the tissue oxygenation. The measurement of μ_a at two wavelengths translates Eq. (7.32) into a linear system of two equations (one for each wavelength) and two unknowns (the tissue concentrations of oxyhemoglobin and deoxyhemoglobin). Its solution gives the oxyhemoglobin and deoxyhemoglobin concentrations, which in turn yield the total hemoglobin concentration, $THC = [HbO_2] + [Hb]$, and the tissue saturation, $StO_2 = [HbO_2]/THC$.

Figures 7.7(a), (b), and (c) show typical traces of total hemoglobin concentration and tissue saturation measured with the frequency-domain tissue oximeter described in Section 7.2.5 and illustrated in Fig. 7.2. Panels (a) and (b) of Fig. 7.7 refer to measurements on the brachioradialis muscle (forearm) of a human subject during arterial and venous occlusion, respectively; panel (c) reports a measurement on the vastus lateralis muscle (thigh) during treadmill exercise.[75]

The two wavelengths employed in the particular case of Fig. 7.7 are 758 and 830 nm, and the acquisition time per data point is 1.28 s. The traces appear in real time on the computer screen during the examination. We stress again that the *y*–axes for hemoglobin concentration and saturation are quantitative and absolute. Figure 7.7(a) shows that the tissue desaturation during ischemia results from a rate of decrease in [HbO_2] that is equal to the rate of increase in [Hb] (because the total hemoglobin concentration remains approximately constant during this ischemia protocol). The rate of conversion of HbO_2 to Hb can be used to quantify the muscle oxygen consumption.[76,77] The main effect of the venous occlusion [Fig. 7.7(b)] is to increase the hemoglobin concentration, as a result of blood accumulation. In fact, while the arterial inflow is unaffected (at least initially) by the venous occlusion, the venous outflow is blocked. The initial rate of increase of THC during venous saturation can be used to measure the muscle blood flow.[78–81] Figure 7.7(c) provides an example of an oxygenation and hemodynamics study *in vivo* during muscle exercise, with potential applications in the area of vascular disease diagnosis and sports medicine.[75]

As shown in Fig. 7.6, near-infrared spectroscopy lends itself to measurements on the human head as well. Figure 7.8(a) shows the arrangement of optical fibers on the forehead of a human subject. The cerebral oxygenation may be varied by changing the fraction of oxygen inspired by the subject. The resulting changes in the cerebral tissue oxygenation (StO_2) are reported in the lower trace of Fig. 7.8(b). Figure 7.8(b) also shows a comparison between two measurements of arterial saturation (SaO_2); one with pulse oximetry, and the other with frequency-domain tissue oximetry.[71] Both approaches make use of the oscillatory components of the optical signals at the frequency of the heartbeat (which are directly associated with the arterial pulsation, and therefore with arterial blood volume). The key difference is that while the pulse oximeter applies an empirical calibration to translate the pulsatile components of the optical densities into a reading of arterial saturation, the frequency-domain oximeter directly measures the pulsatile components of the absorption coefficients and quantifies the arterial saturation by applying Eq. (7.32).[71]

The lack of discomfort associated with near-infrared measurements *in vivo*, along with their noninvasive character, renders near-infrared tissue oximetry a suitable technique for newborn infants. For instance, Fig. 7.9 shows a comparison between the systemic reading of arterial saturation

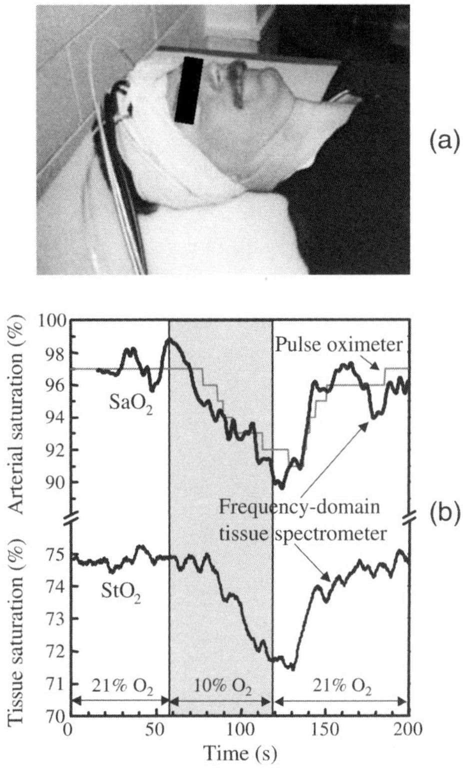

Figure 7.8 (a) Application of optical fibers to the forehead of a human subject for frequency-domain brain oximetry. (b) Temporal traces of arterial saturation (SaO_2) measured systemically (on a finger) by a pulse oximeter and locally (on the forehead) by a frequency-domain tissue oximeter that also measured the cerebral tissue saturation (StO_2).[71] The fraction of inspired oxygen was varied between 21% and 10% (by volume) as indicated in the figure.

(with a pulse oximeter attached to a toe) and the cerebral oxygenation measured locally with a frequency-domain tissue oximeter. In this case, the decrease in the measured oxygenation was induced by a controlled decrease in the fraction of inspired oxygen. The arterial saturation and the cerebral saturation show a qualitatively similar trend. However, a local cerebral deoxygenation would be detected only by the NIR cerebral measurement. The manual application of the optical probe shown in Fig. 7.9(a) can be replaced by a flexible headband[82,83] or by an infant cap for continuous long-term monitoring.

Questions about the safe applicability of frequency-domain instrumentation and methods to newborn infants have inspired a number of studies. Some of them are basic studies of the baseline optical properties and cerebral architecture,[84,85] whereas others focus on the development of brain functions and impairments[86,87] or on pathological conditions.[88] It is worth noting the

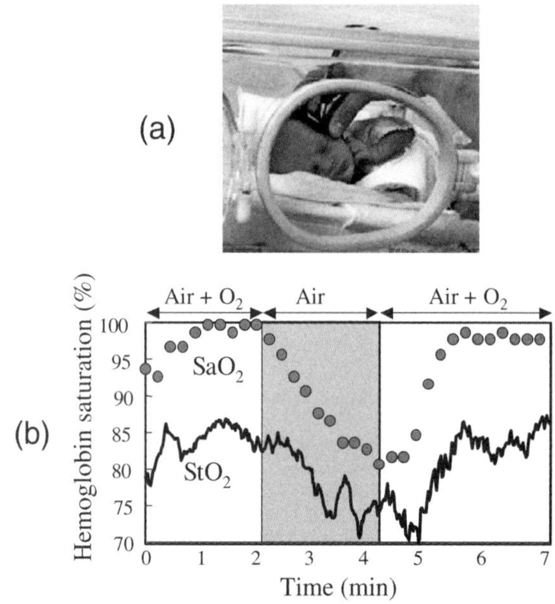

Figure 7.9 Frequency-domain, near-infrared oximetry of the newborn infant brain. (a) Photograph showing the application of a flexible optical probe to the infant's head. (b) Temporal traces of cerebral tissue saturation (StO_2) measured with frequency-domain optical oximetry, and systemic arterial saturation (SaO_2) measured with a pulse oximeter at an infant's toe. Deoxygenation is achieved by a controlled reduction in the fraction of inspired oxygen. (Data collected in collaboration with Gianpaolo Donzelli and Simone Pratesi, University of Florence, Italy.)

development of hybrid systems that combine frequency-domain spectroscopy and diffuse correlation spectroscopy (DCS) to concurrently measure blood volume and an index proportional to blood flow velocity.[87,88]

7.4.4 Absolute brain measurements with semi-infinite and two-layer models

We have recently studied the difference in the baseline values of the optical properties, translated into baseline values of absolute oxyhemoglobin, deoxyhemoglobin, and total hemoglobin concentrations, in two groups of individuals.[89] The first group consisted of 36 elderly individuals (mean age, 85 ± 6 y), and the second group consisted of 19 healthy young adults (mean age, 28 ± 4 y). Measurements were performed using the multichannel, dual-wavelength (690 and 830 nm), frequency-domain tissue oximeter described in Section 7.2.5 (manufactured by ISS Inc., Champaign, Illinois). We used multi-distance tissue spectroscopy together with the solution of the diffusion equation in the semi-infinite medium geometry to retrieve the absolute concentrations of the hemoglobin species (see Sections 7.3.4.2 and 7.3.5). To test the reproducibility of our findings, the measurements on the elderly group

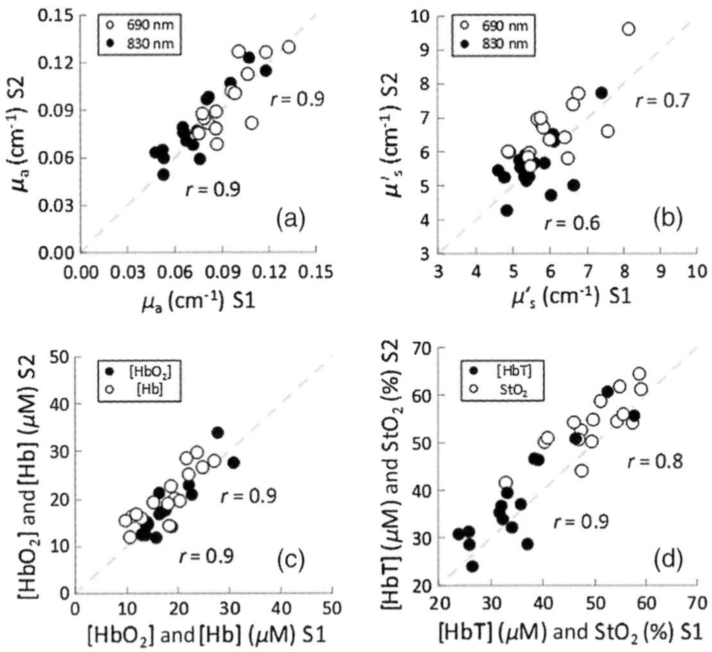

Figure 7.10 Frequency-domain measurements of the optical properties (top panels) and hemoglobin-related parameters (bottom panels) of the prefrontal cortex of a group of elderly subjects (85 ± 6 years old).[89] We performed measurements in two sessions separated by 5 months. Measured parameters in session 2 (S2) versus the corresponding parameters measured in session 1 (S1) illustrate the reproducibility of the absolute frequency-domain measurements. Dashed lines represent the ideal identity line.

were repeated in two different sessions, which were separated in time by 5 months. Source–detector separations for the probe used in session 1 ranged from 2 to 3.5 cm in 0.5-cm increments. Source–detector separations for the probe used in session 2 ranged from 0.8 to 3.8 cm in 0.5-cm increments.

Figure 7.10 shows the results of the reproducibility test. As we can see, there is a good reproducibility of the optical properties (top panels) and of the hemoglobin concentrations and oxygen saturation (bottom panels), validating the robustness of the frequency-domain, multi-distance method for absolute brain measurements.

We also tested the validity of the assumption of tissue homogeneity. Because different sets of source–detector separations probe different tissue regions (which extend to deeper volumes for longer source–detector distances), only under conditions of tissue homogeneity would one retrieve the same optical properties (and hemoglobin concentrations) regardless of the set of source–detector separations used. Figure 7.11 shows the absolute optical properties (top and middle panels) and the absolute total hemoglobin concentration and tissue saturations (bottom panel) retrieved by the multi-distance method when different sets of source–detector separations were used

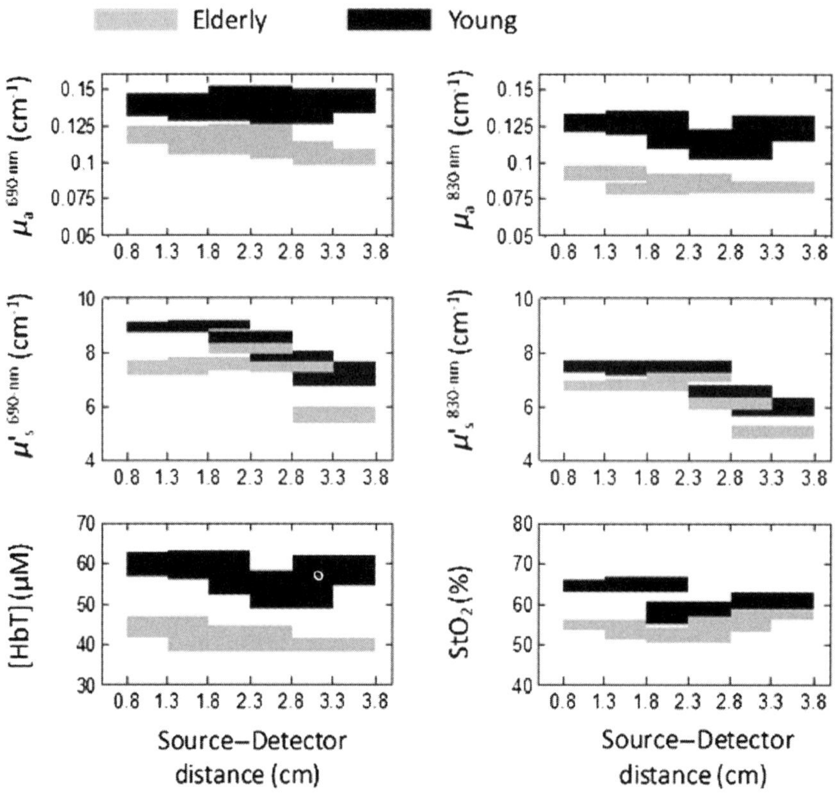

Figure 7.11 Dependence of measured parameters on source–detector separation in elderly (gray bars) and young (black bars) subjects.[89] The top panels report the measured μ_a at 690 and 830 nm. The middle panels report the measured μ_s' at 690 and 830 nm. The bottom panels report measured [HbT] and StO_2. The gray (elderly) and black (young) data blocks represent the mean (block center) and standard error (block thickness) of the measurements at the range of source–detector distances corresponding to the horizontal extent of the data blocks.

(represented in the figures by the horizontal bars).[89] For both elderly (gray boxes) and younger people (black boxes), there is a clear dependence of the retrieved parameters on the set of distances used, which confirms the effects of the inhomogeneous tissue geometry on the absolute optical measurements on the subject's forehead.

To summarize the differences in the measured parameters in the younger and older groups, we have used a set of source–detector separations that were common to the measurements on both groups. The results are summarized in Table 7.2. We note that the results found in the group of young subjects compare well with other studies reported in the literature based on time-domain methods[90] and frequency-domain methods.[91]

The results of Fig. 7.11, which reveal the effect of the inhomogeneous tissue architecture on the optical measurements, motivates research based on

Table 7.2 Summary of absolute measurements in elderly and young subjects.[89] Values in parentheses represent the standard error on the last significant digit of the mean.

	$\mu_a^{690\,nm}$ (cm^{-1})	$\mu_a^{830\,nm}$ (cm^{-1})	$\mu_s^{\prime 690\,nm}$ (cm^{-1})	$\mu_s^{\prime 830\,nm}$ (cm^{-1})	[HbO_2] (uM)	[Hb] (uM)	[HbT] (uM)	StO_2 (%)2
Elderly	0.11(1)	0.08(1)	6.9(2)	6.0(2)	20(1)	18(1)	38(2)	52(2)
Young	0.13(1)	0.11(1)	7.7(4)	6.6(3)	30(3)	22(2)	52(3)	58(3)

Table 7.3 Optical coefficients of the first layer (subscript 1) and second layer (subscript 2), and thickness of the first layer (*L*) retrieved by analyzing with a two-layer model the optical data collected on the forehead of three human subjects.[92]

	μ_{a1} (mm^{-1})		$\mu'_{s1}(mm^{-1})$		*L* (mm)		μ_{a1} (mm^{-1})		$\mu'_{s2}(mm^{-1})$	
Subject #	**690 nm**	**830 nm**	**690 nm**	**830 nm**	**690 nm**	**830 nm**	**690 nm**	**830 nm**	**690 nm**	**830 nm**
1	0.007	0.006	1.2	1.1	11.8	11.4	0.019	0.016	0.2	0.2
2	0.009	0.009	1.3	1.2	13.0	12.3	0.020	0.020	0.4	0.3
3	0.008	0.008	1.3	1.1	14.5	14.0	0.022	0.023	0.4	0.2
mean	*0.008*	*0.008*	*1.3*	*1.1*	*13.1*	*12.6*	*0.020*	*0.020*	*0.3*	*0.2*
stdev	*0.001*	*0.002*	*0.1*	*0.1*	*1.3*	*1.3*	*0.002*	*0.003*	*0.1*	*0.1*

more complex tissue models to better represent geometrical and inhomogeneous tissue features. For example, it is possible to carry out an inversion procedure on *in vivo* optical data collected on adult human heads, based on the two-layer model described in Section 7.3.4.3.[92] This approach relies on an inversion procedure with six unknown parameters (the absorption and reduced scattering coefficients of the two layers, the thickness of the first layer, and an amplitude parameter) on *in vivo* optical data. The procedure has been shown to be robust in terms of the initial guess (a variety of parameter initializations lead to the same results) and to retrieve optical properties in a range of values that compares well with previously published results that also used the two-layer geometry model. The data were collected with multiple source–detector separations in the range 8–48 mm.

Tables 7.3 and 7.4 report the results of the two-layer inversion procedure on three healthy human subjects (all males; age: 29 ± 2 y) for the optical coefficients and first-layer thickness (Table 7.3) and for the hemoglobin-related quantities (Table 7.4). In Table 7.3, notice how the inversion procedure recovers a thickness of the "first layer" that is consistent at the two wavelengths and within measured anatomical ranges.[93] Also, the reduced scattering coefficient at 690 nm is never smaller than that at 830 nm in both layers—a result that confirms the reliability of the method (the reduced scattering coefficient should decrease with wavelength). The optical properties and hemoglobin concentrations found by our methods are within the range reported by previous studies,[90,94] and no *a priori* anatomical information (such as the thickness of the superficial layer) was required. AC and phase

Table 7.4 Concentration and saturation of hemoglobin from the optical coefficients reported in Table 7.3, retrieved from optical data measured on the forehead of three human subjects using a two-layer model.[92] The first layer is labeled "superficial" and the second layer is labeled "cerebral." Group mean and standard deviation (stdev) are reported in the bottom two rows.

	[HbO_2] (μM)		[Hb] (μM)		[HbT] (μM)		Sat (%)	
Subject #	**Superficial**	**Cerebral**	**Superficial**	**Cerebral**	**Superficial**	**Cerebral**	**Superficial**	**Cerebral**
1	8.0	38.0	9.9	30.3	17.9	68.3	44.8	55.6
2	19.8	51.8	12.4	31.2	32.2	83.0	61.5	62.4
3	17.1	63.7	10.5	31.9	27.6	95.6	61.9	66.6
mean	*15.0*	*51.2*	*10.9*	*31.1*	*25.9*	*82.3*	*56.05*	*61.6*
stdev	*6.2*	*12.9*	*1.3*	*0.8*	*7.3*	*13.6*	*9.78*	*5.6*

data from the multi-distance measurements were also inverted by using a semi-infinite medium geometry (results not shown), and the retrieved oxyhemoglobin, deoxyhemoglobin, and total hemoglobin concentrations were in between those measured for the superficial layer and for the brain (see Table 7.4). This result supports the straightforward interpretation that the accuracy of brain measurements based on the assumption of tissue homogeneity is impacted by the optical properties of the superficial, extracerebral tissue layer. Finally, we note that the relatively lower values of the reduced scattering coefficients found in the second layer are also consistent with reported values in the literature.[90] These lower values may be due to the effect of the cerebrospinal fluid (CSF) layer, which is relatively optically clear.

7.4.5 Measurements of optical scattering in tissue

Traditionally, frequency-domain optical studies have targeted the absorption properties of tissues to extract quantitative information on the concentration and oxygen saturation of hemoglobin. The strong scattering of light in tissue constitutes a complication in quantitative absorption spectroscopy. However, the scattering properties of tissue may themselves provide information about physiologically relevant parameters. For instance, it has been shown that mitochondria are the main source of light scattering in the liver, and possibly in other tissues as well.[95] Since a number of metabolic processes related to cellular respiration occur in the mitochondria, the reduced scattering coefficient may be related to the cellular activity and viability. Scattering properties of tissue may also provide information on the size, morphology, and density of cellular organelles. For example, it was observed that smaller cellular organelles, such as mitochondria (0.3–0.7 μm in cross-sectional diameter), and cell nuclei (3–10 μm in diameter) feature scattering with distinct angular properties,[96] as predicted by Mie theory. Based on these principles, a variety of light scattering techniques have been developed in an effort to develop a new class of biomedical sensing and diagnostic tools.[97]

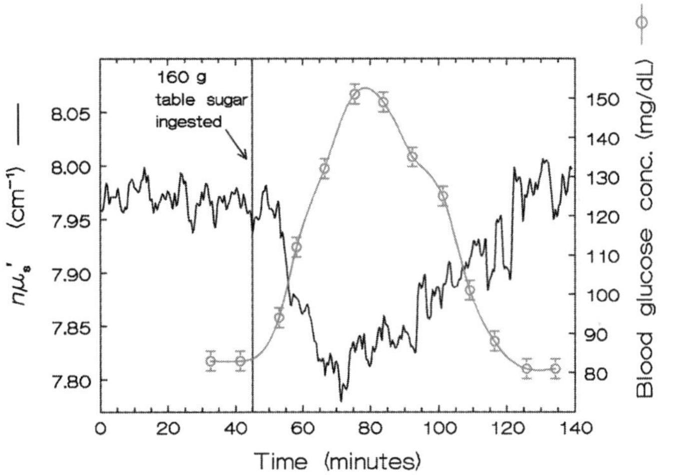

Figure 7.12 Comparison between the time-varying blood glucose concentration and the temporal evolution of the reduced scattering coefficient (μ_s') measured on the thigh muscle of a human subject by frequency-domain spectroscopy during a glucose tolerance test.[100] The diffusion model is not able to separate μ_s' from the refractive index n of the tissue [in Eq. (7.16), n and μ_s' are coupled in the expression for $D = v/[3(\mu_s' + \mu_a)] \cong c/(3n\,\mu_s')$, where c is the speed of light in vacuum], so that the y–axis reports the product $n\mu_s'$.

In an effort to exploit the diagnostic value of scattering properties of tissue, some studies have investigated the correlation between blood glucose concentration and the reduced scattering coefficient of tissue.[98–100] This correlation is suggested by the fact that the scattering properties of tissues depend on the mismatch between the refractive indices of the scattering centers (cells, cellular organelles) and the host medium (extracellular fluid, cytosol). Since the index of refraction of the extracellular fluid is sensitive to glucose concentration, it is a plausible hypothesis that μ_s' would be affected by the blood glucose concentration. *In vitro* studies have predicted that the sensitivity of μ_s' to the blood glucose level should be adequate to monitor glucose concentrations within the physiological range (80–120 mg/dL in blood).[100] This prediction is confirmed by Fig. 7.12, which shows a trace of μ_s' measured by frequency-domain NIR spectroscopy on the thigh muscle of a healthy 26-year old male subject during a glucose tolerance test.[100] The reduced scattering coefficient decreases as the blood glucose increases, because of the glucose-induced reduction in the refractive index mismatch between the scattering centers and the host fluid. However, this kind of measurement yields only a relative determination of glucose concentration. Furthermore, a number of additional factors, such as changes in temperature and in pH, may lead to variations in μ_s' that compete with the glucose-induced effects. Theoretical and experimental studies of the optical properties of the human sclera (the white outer coat enclosing the eyeball) have also shown a significant scattering change induced by glucose and by other osmotically active species.[101] While the monitoring of glucose concentration by NIR spectroscopy is not ready for clinical application, it does nevertheless point out the diversity of potential diagnostic applications of medical optics.

7.5 Optical Imaging of Tissues

7.5.1 General concepts

Imaging applications rely on the sensitivity of optical measurements to the optical properties of tissues—namely, absorption and scattering. Consequently, the contrast in NIR imaging originates from spatial variations in the optical absorption and scattering properties of tissue. These spatial variations can be due to a local change in hemoglobin concentration or saturation (for instance, a hematoma or heterogeneous vascularization, or a local perturbation to blood flow or oxygen demand), in the tissue architecture (for example, microcalcifications or structural rearrangement of muscle fibers), or in the size or concentration of cellular organelles. In any case, NIR imaging is sensitive to physical properties of tissues that are different than those probed by current diagnostic imaging modalities such as x-ray tomography (mass density, atomic number), ultrasound imaging (gradients or discontinuities in acoustic impedance), magnetic resonance imaging (proton density, nuclear relaxation times), and positron emission tomography (accumulation of a radioactive tracer). In this sense, NIR imaging is not necessarily an alternative to these imaging modalities, but rather might constitute an adjunct technique providing an effective complement to existing methods. We point out here that the promise of diffuse optical imaging lies not in the ability to achieve a high spatial resolution (which is intrinsically limited by the diffusive nature of light propagation in tissues), but rather in the possibility of obtaining a high contrast and specificity to abnormal tissue conditions.

NIR tissue imaging aims to generate spatial maps that display either structural or functional properties of tissues. Since optical imaging deals with the spatial distribution of optical properties, Eq. (7.11) must be replaced with the diffusion equation for inhomogeneous media:

$$\frac{\partial U(\mathbf{r},t)}{\partial t} - \nabla \cdot D(\mathbf{r})\nabla U(\mathbf{r},t) + v\mu_a(\mathbf{r})U(\mathbf{r},t) = q(\mathbf{r},t). \tag{7.33}$$

Equation (7.33) (or the corresponding equation in the frequency domain) should be used to solve the forward problem, which consists of finding the photon density $U(\mathbf{r}, t)$ corresponding to a given spatial distribution of the optical properties and of the photon sources. The output power per unit detector area at an arbitrary point of the medium boundary ($\mathbf{r}_b$), at time t, which we define as the detected intensity $I(\mathbf{r}_b, t)$, is calculated by using $I(\mathbf{r}_b, t) = \mathbf{J}(\mathbf{r}_b, t)\ \ \mathbf{n}$ (where $\mathbf{n}$ is the unit vector perpendicular to the boundary and outwardly oriented) and Fick's law [Eq. (7.12)]. Analytical solutions of

Eq. (7.33) are available for several regularly bounded (or unbounded) homogeneous media, including single regularly shaped defects,[47,102–104] and also for layered media.[105–108] For arbitrary inhomogeneous cases, Eq. (7.33) can be solved using numerical methods such as the finite difference method (FDM)[109,110] or the finite element method (FEM).[30,111–114] Alternative approaches for solving Eq. (7.33) include perturbation theory[115–120] or lattice random walk models.[121,122] However, perturbation and random walk methods are used mainly for regularly bounded geometries and focal defects.

Another important method is the Monte Carlo (MC) method, which is able to solve the more general BTE [Eq. (7.1)]; for this reason, it is suitable for general random media and for all those cases when diffusion conditions are not fulfilled (e.g., low-scattering tissue regions, small tissue volumes). In the last 10 years, MC methods have been proliferating since the advent of graphic processor units (GPU) that enabled the speed of calculations to increase by approximately 2–3 orders of magnitude. A comprehensive review of MC methods used in tissue optics is provided by Zhu and Liu.[123]

The goal of diffuse optical tomography (DOT),[124–126] which is a branch of optical imaging aimed at three-dimensional imaging of thick tissues, is to solve the inverse imaging problem. This problem consists of determining the spatial distribution of μ_a and D (or μ_a and μ_s') within a volume V, from optical measurements at the boundary of V.[42,127] One of the simplest way to solve the inverse problem for inhomogeneous media is to use perturbation methods that keep only the linear term of a series expansion of the detected intensity as a function of the optical properties.[128] The inverse problem reduces to inverting the Jacobian, $\mathbf{J}$ (i.e., the first derivative matrix), of the measured quantities with respect to the optical properties (or, since $\mathbf{J}$ is not generally a square matrix, inverting the square matrix $\mathbf{J}^T\mathbf{J}$), under some regularization conditions. The Jacobian matrix can be estimated using MC simulations,[129] FEM,[130] or analytical models for regular geometries.[128] However, prior knowledge of the background optical properties is required. Usually, linear perturbation inversion procedures rely on some a priori knowledge and are used for simple reconstruction problems, such as the case of a single defect embedded in an otherwise homogeneous medium.[116] The main practical disadvantage of the perturbation method is that it requires a pair of measurements with and without the perturbation, and is therefore better suited for dynamic—rather than baseline—optical properties reconstruction. For *in vivo* applications, this is sometimes impractical, except when dynamic perturbations are induced by evoked cerebral activity or by the injection of optical contrast agents.[131]

Nonlinear approaches are based on minimization of the distance (in a measurement space) between the experimental data and the projection (according to the forward problem) of the distribution of optical properties onto the measurement space. The minimization process involves repeated iterations, so the computational speed of the forward model is a relevant factor. Regularization methods are also used to tackle the ill-posed nature of the inverse imaging problem.[132,133] One example of a nonlinear approach is the Newton–Raphson scheme for minimization, which has been used in conjunction with an FEM forward model in optical tomography.[134–136] The previously mentioned methods usually require relatively long computation times that render them unsuitable for real-time imaging. For applications where real-time information is important, simplified algorithms based on back projection[137–140] or on direct projection for 2D mapping[141] have been proposed.

Finally, there is some interest in using the BTE [Eq. (7.1)] as a forward solver in inversion procedures. The BTE is more accurate than the diffusion equation, especially for data collected in the presence of clear layers (such as the cerebrospinal fluid around the brain) or at short source–detector distances, as required to probe small tissue volumes.[142–145]

7.5.2 The phase information in frequency-domain optical imaging

The key difference between continuous-wave (CW) and frequency-domain spectroscopy and imaging is the phase measurement capability afforded by the frequency domain. For optical studies of tissues, the phase information can be exploited in at least two ways.

First, by combining the phase measurement with the DC intensity, AC amplitude, or modulation depth measurements, one can separately measure the absorption and the reduced scattering coefficients of tissues. This capability is of paramount importance for quantitative tissue spectroscopy, imaging, and oximetry. In this chapter, we have illustrated the ability of frequency-domain spectroscopy to quantify the absorption and the reduced scattering coefficients (Figs. 7.6 and 7.11), and to perform absolute tissue oximetry (Figs. 7.7–7.11).

The second relevant aspect of phase measurements is that the phase features a different spatial sensitivity with respect to the DC intensity and the AC amplitude. The region of sensitivity can be evaluated by the effect on the measured quantities of a small perfectly absorbing object located at $\mathbf{r}$. In particular, the DC, AC, and phase sensitivities can be expressed by the parameters $\varepsilon_{DC}(\mathbf{r}) = DC(\mathbf{r})/DC_0$, $\varepsilon_{AC}(\mathbf{r}) = AC(\mathbf{r})/AC_0$, and $\varepsilon_{\Phi}(\mathbf{r}) = \Phi(\mathbf{r}) - \Phi_0$, respectively, where DC_0, AC_0, and Φ_0, are the measured values in the absence of the absorbing object, and $DC(\mathbf{r})$, $AC(\mathbf{r})$, and $\Phi(\mathbf{r})$ are the measured values in the presence of the absorbing object at point $\mathbf{r}$. The theoretical and experimental regions of sensitivities for the DC, AC, and phase are illustrated in Fig. 7.13 for an infinite medium (other conditions are listed in the figure caption).[146] Figure 7.13 shows that the DC and AC measurements are most sensitive to the region along the line joining the source and the detector, which, by contrast, is not the highest region of sensitivity for the phase. In a noninvasive tissue measurement, which is better modeled by a semi-infinite

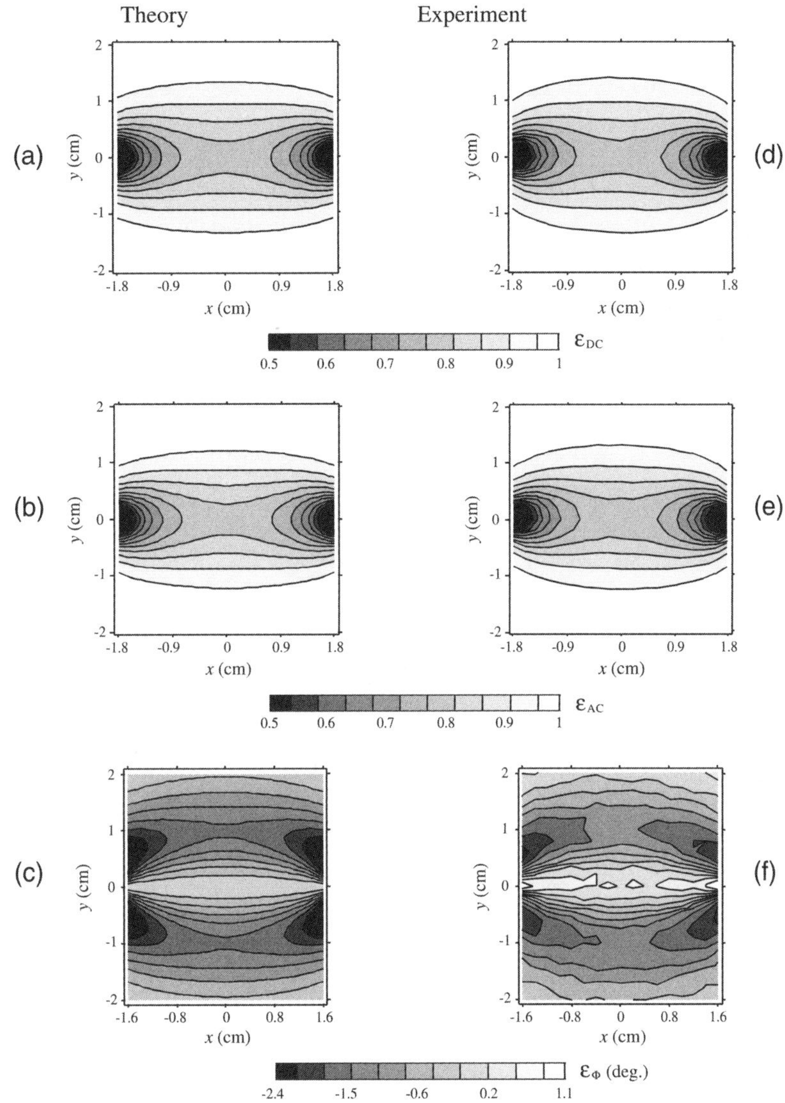

Figure 7.13 Regions of sensitivity for the DC [(a) and (d)], AC [(b) and (e]), and phase [(c) and (f)] in an infinite medium. The source and the detector (not shown) are located respectively at (−2, 0) cm and (2, 0) cm. The optical properties of the medium are $\mu_a = 0.03$ cm^{-1} and $\mu_s' = 19$ cm^{-1}, and the source modulation frequency is 120 MHz. The gray scales represent the parameters ε_{DC}, ε_{AC}, and ε_Φ, which are defined in the text. The left panels are theoretical predictions based on diffusion theory, whereas the right panels are experimental results.[146]

geometry, where the illuminating and collecting optical fibers are located on the tissue/air interface, the phase measurement is typically less sensitive to the superficial tissue layer with respect to the DC and AC measurements. As a result, the phase information may help to achieve depth sensitivity or some level of spatial discrimination in optical imaging of tissues. The drawback of phase data is that typical perturbations induced either by a focal "defect" in tissue (like a tumor) or by hemodynamic changes experienced by tissue during functional activation in the brain are typically detected with a poorer signal-to-noise ratio in phase measurements than intensity measurements.

Applications of frequency-domain techniques to the optical study of biological tissues exploit these two features, with the objective of developing innovative approaches for medical imaging and diagnosis.

7.5.3 Optical mammography

One of the major applications of near-infrared imaging and DOT is optical mammography.[147,148] From a practical point of view, the human breast readily lends itself to near-infrared tissue spectroscopy and imaging. In fact, it is easily accessible, and its relatively low hemoglobin concentration (10–20 μM)[149,150] leads to near-infrared absorption coefficients (approximately 0.02–0.04 cm^{-1}) that allow for a deep optical penetration depth and adequate optical transmission through the entire breast. Furthermore, there is much appeal in the possibility of using non-ionizing radiation to detect breast cancer or to monitor the efficacy of therapeutic regimens.

The optical detection of breast cancer relies on the perturbations induced by cancer on the host tissue, rather than on a direct detection of optical signatures associated with cancer cells.[151] These perturbations include angiogenesis,[151,152] alterations to the blood flow and oxygenation,[152] and fibroblast proliferation,[151] all of which may yield modifications to the optical absorption and scattering properties of breast tissue. However, it is still unclear whether these cancer-induced optical perturbations can lead to an effective optical approach to breast cancer detection over the wide range of cases observed in the clinical and screening practice (e.g., different types and stages of cancer, various benign lesions, different levels of tissue heterogeneity). *In vitro* optical studies of normal and diseased breast tissues have not found significant differences between their optical properties.[153,154] This result may suggest that the optical contrast induced by cancer alone (the *intrinsic contrast*) may not allow for a sufficient sensitivity and specificity for breast cancer detection by optical mammography. Nevertheless, one should consider the fact that optical contrast *in vivo* may benefit from additional sources of contrast related to blood flow and metabolic rate of oxygen, as well as some aspects of the tissue architecture. In fact, early *in vivo* studies indicated that cancerous breast tissue has a significantly larger absorption coefficient and a marginally greater reduced scattering coefficient with respect to normal breast tissue.[155–157] Further studies—in some cases, extending the spectral information by considering multiple wavelengths or broadband spectroscopy—have reproducibly found that breast cancer features an intrinsic optical contrast associated with greater hemoglobin concentration, greater water content, and lower lipids content, with respect to healthy tissue; less reproducible results include a potential tumor contrast associated with a lower hemoglobin saturation and a greater scattering coefficient. For a review of optical mammography based on intrinsic contrast, see Fantini and Sassaroli.[158]

Because it is yet to be demonstrated that optical mammography can rely solely on the intrinsic optical contrast associated with cancer to effectively detect breast cancer, it has also been proposed the use of exogenous dyes as optical contrast agents.[131,159–162] Another important application of optical mammography is to monitor the effectiveness of neoadjuvant chemotherapy or hormonal therapy.[163–170] In this case, the issue is not the detection of breast cancer but rather the characterization of its evolution during therapeutic regimens, with the overall goal of identifying responders and nonresponders early in the treatment period.

Frequency-domain optical methods have been widely used for breast imaging because of their capability to separate the absorption and the reduced scattering coefficients and to provide data with a richer information content with respect to CW data.[20,150,171–174] The absolute measurement of optical properties leads to the characterization of healthy and malignant tissues in terms of oxyhemoglobin, deoxyhemoglobin, and total hemoglobin concentrations, as well as lipids and water content. In an early FD prototype for breast imaging developed by Siemens, the breast was imaged in two projections, leading to craniocaudal (cc) and an oblique (ob) views similar to those obtained with x-ray mammography.[175] A pilot clinical study conducted with this instrument led to a sensitivity (fraction of cancerous lesions successfully detected) of 72% and a specificity (fraction of noncancerous lesions or normal tissue inhomogeneities correctly evaluated as negative) of 52% within a clinical population of 131 patients.[176] This result was consistent with the sensitivity of 73% obtained in a study of 69 patients[177] with a similar frequency-domain prototype developed independently by the research laboratories of Carl Zeiss (Oberkochen, Germany).[178]

Frequency-domain instrumentation has developed in two different directions:

1. Systems that are more user-friendly and are devised to be used mainly in a clinical setting. These systems include imaging setups with a handheld sensor probe that is placed on the breast to collect optical data in a reflection geometry.[179,180] The probe can be manually scanned on the breast, so no breast compression is required and the data acquisition can be conducted at the bedside. The data analysis is usually carried out by using simpler forward and inverse models.
2. More complex systems that collect data in a transmission geometry[28,181] or according to an arrangement of the illumination and collection optical fibers that lends itself to full tomographic reconstructions.[182] While these instruments have also been used in clinical settings, they are more

sophisticated and require complex inversion procedures to retrieve the optical images and the clinically relevant parameters. These inversion procedures are devised for realistic geometries of the breast and arbitrary distributions of the optical properties.[113] The data acquisition cannot be conducted at the bedside, and the breast is slightly compressed either between parallel plates[28,181] or surrounded by source–detector fibers arrays.[182]

In the first group of FD instruments,[179,180] the capability of separating the absorption and reduced scattering coefficient is obtained by multi-frequency tissue spectroscopy (see section 7.3.6). The FD instrument used by Cerussi et al., described in detail in previous work,[57,183] is a hybrid CW/FD instrument that allows for the reconstruction of absolute oxyhemoglobin, deoxyhemoglobin, and total hemoglobin concentrations and also for water and lipids content in breast tissue.[179] The reconstruction of these parameters is obtained by using a diffusion model for a semi-infinite medium geometry. The instrument, through manual scanning of the breast, also has imaging capabilities.[166] In a study of 58 malignant breast tumors, it was found that intrinsic physiological properties were significantly different in tumors and normal tissues.[179] Specifically, it was found that malignant lesions were characterized by a decreased lipid content, increased water content, and increased oxyhemoglobin and deoxyhemoglobin concentrations.

The FD instrument used by Zhu et al. is a hybrid ultrasound (US)/FD instrument that includes a handheld probe with a US transducer at the center and optical fibers for delivering and collecting light at the periphery.[180] The optical data are collected in a reflectance mode. A modified Born approximation is used for recovering the tumor optical properties, while the US system provides information on the size and location of the lesion.

The system used by Choe et al. is also a hybrid CW/FD instrument.[28] The FD capability is used to collect reflectance data so to provide bulk tissue concentrations of hemoglobin, by using a multi-spectral method in the homogeneous, semi-infinite medium geometry (four lasers are modulated at 70 MHz). After the bulk properties are estimated, the CW system, which utilizes approximately 4×10^4 source–detector pairs for each wavelength in a transmittance geometry (the detector being a CCD camera), is used to reconstruct the optical properties with a 3D nonlinear inversion procedure. The data acquisition is carried out with the subject lying prone on a bed, with the breast immersed in a cell containing a liquid suspension of intralipid, water, and ink. This approach is used to improve the optical contact between optodes and tissue and to decrease the measured dynamic range. More details about the instrument can be found in the report by Culver et al.[29] In a study involving 51 lesions (from 47 subjects), the malignant tumors (41 cases) showed statistically significant increases in oxyhemoglobin, total hemoglobin, and scattering coefficient.[28] In contrast, benign tumors (10 cases) did not show any parameters that were statistically significantly different than normal tissue. The authors concluded that it is possible to discriminate benign and malignant lesions with sole intrinsic contrast.

The system used by Fang et al. is an integrated optical/x-ray system that allows for the coregistration of both types of images.[181] The x-ray system identifies different tissue types within the breast, such as fibroglandular, adipose, and muscular tissues. Absolute estimation of oxyhemoglobin, deoxyhemoglobin, and total hemoglobin concentrations can be carried out by applying the optical system to each tissue region identified by x-rays.[184] The optical component of the system is a hybrid CW/FD system, where both data types are collected in a transmission geometry. The FD capability allows for initial assessment of the bulk properties of the tissue, while full 3D reconstruction is achieved by using the CW data. The homogeneous properties (found with FD data) are used as the initial guess in the 3D reconstruction. With this system, the subject is in a standing position, as when x-ray mammography is used, with the breast compressed between parallel plates.

In a relatively large clinical study on 189 breasts from 125 patients, including 138 healthy breasts and 51 breasts with lesions, it was found that the malignant tumors (26 cases) had a higher total hemoglobin concentration than the fibroglandular tissue in the same breast, while benign lesions (17 cases) and cysts (8 cases) had lower contrast than that found in malignant tissue.[184] Similarly to Choe et al.,[28] the authors concluded that it is possible to differentiate benign and malignant lesions with the intrinsic optical contrast associated with the total hemoglobin concentration in tissue.

The system used by McBride et al. uses a tomographic arrangement of sources and detectors (placed around the subject's breast), with the subject lying on a bed in a prone position.[182] The system is a pure FD system (modulation frequency: 100 MHz) and uses 16 optical fibers placed around the breast. When one of them is used to deliver light in the tissue, the other 15 optical fibers collect the light from different locations. The system has been upgraded by expanding the number of source–detector fibers to feature three rings of 16 fibers each.[185] This instrument has been used in a number of *in vivo* studies to image the distribution of NIR chromophores in breasts tissue (oxyhemoglobin, deoxyhemoglobin, total hemoglobin, water, and lipids)[185–187] and to monitor the effectiveness of neoadjuvant chemotherapy.[165]

7.5.4 Imaging of finger joints

Frequency-domain methods have also been used for imaging joints and bones—for example, in the early detection of arthritis.[30,188] In the study performed by Xu et al., the FD data were collected by using a circular arrangement of illumination and collection optical fibers placed around a cylindrical scattering phantom (5 cm in diameter), with a finger holder in the center.[188] With this arrangement, the finger can be considered as a perturbation to the uniform optical properties of the phantom. A 3D nonlinear inversion procedure (the regularized Newton's method), using the solution of the DE as a forward solver, was used to retrieve the optical properties. In the study performed by Hielscher et al., a FD system where the modulation frequency is swept up to 1 GHz was used.[30] The modulated laser light ($\lambda = 670$ nm) was shone on the back of a finger and scanned across the proximal interphalangeal joints. An ultrafast gated ICCD camera was used as the optical detector. Surface coordinates of the finger (used to generate surface mesh for image reconstruction) were generated by a laser scanner unit. The reconstruction of the optical properties was carried out by using a 3D nonlinear inversion procedure using the FD solution of BTE as a forward solver.

7.6 Prospects for Frequency-Domain Spectroscopy and Imaging of Tissue

Two major advances, one theoretical and one experimental, that occurred in the late 1980s have significantly impacted the field of diffuse optical spectroscopy and imaging of tissue. The first advance was the introduction of quantitative theoretical models to describe light propagation in biological tissues. The second advance was the introduction of time-resolved techniques that feature time-varying source emissions and time-sensitive detection. In this chapter, we have touched on both of these developments by describing diffusion theory and frequency-domain instrumentation, and we have described some representative applications that were enabled by these theoretical and experimental approaches. Even though CW methods, with their less complex instrumentation and more straightforward implementation into broadband spectral system, do play an important role in diffuse optical measurements of tissue, there is no question that the richer information content of time-resolved spectroscopy (in the frequency domain, as presented here, or in its even more powerful time-domain version) offers important opportunities for quantitative optical characterization of tissues. Such quantitative optical characterization of tissue has empowered diffuse optical studies of tissues, and has contributed to moving diffuse optical techniques closer to achieving clinical applicability.

Acknowledgments

We acknowledge support from the U.S. National Institutes of Health, Grant No. R01 CA154774.

References

1. D. T. Delpy, M. Cope, P. van der Zee, S. Arridge, S. Wray, and J. Wyatt, "Estimation of optical pathlength through tissue from direct time of flight measurement," *Phys. Med. Biol.* **33**, 1433–1442 (1988).
2. M. S. Patterson, B. Chance, and B. C. Wilson, "Time resolved reflectance and transmittance for the non-invasive measurement of optical properties," *Appl. Opt.* **28**, 2331–2336 (1989).
3. B. Chance, M. Cope, E. Gratton, N. Ramanujam, and B. J. Tromberg, "Phase measurement of light absorption and scatter in human tissue," *Rev. Sci. Instrum.* **69**, 3457–3481 (1998).
4. S. Fantini, M. A. Franceschini, J. B. Fishkin, B. Barbieri, and E. Gratton, "Quantitative determination of the absorption spectra of chromophores in strongly scattering media: a light-emitting-diode based technique," *Appl. Opt.* **33**, 5204–5213 (1994).
5. M. Razeghi, *Technology of Quantum Devices*, Springer, New York (2010).
6. E. Gratton and M. Limkeman, "A continuously variable frequency-domain cross-correlation phase fluorometer with picosecond resolution," *Biophys. J.* **44**, 315–324 (1983).
7. D. W. Piston, G. Marriott, T. Radivoyevich, R. Clegg, T. M. Jovin, and E. Gratton, "Wide-band acousto-optic light modulator for frequency-domain fluorometry and phosphorimetry," *Rev. Sci. Instrum.* **60**, 2596–2600 (1989).
8. R.D. Spencer and G. Weber, "Measurement of subnanosecond fluorescence lifetimes with a cross-correlation phase fluorometer," *Ann. N.Y. Acad. Sci.* **158**, 361–376 (1969).
9. J. R. Alcala, E. Gratton, and D. M. Jameson, "A multifrequency phase fluorometer using the harmonic content of a mode-locked laser," *Anal. Instrum.* **14**, 225–250 (1985).
10. A. Bassi, J. Swartling, C. D'Andrea, A. Pifferi, A. Torricelli, and R. Cubeddu, "Time-resolved spectrophotometer for turbid media based on supercontinuum generation in a photonic crystal fiber," *Opt. Lett.* **29**, 2405–2407 (2004).
11. G. De Stasio, N. Zema, F. Antonangeli, A. Savoia, T. Parasassi, and N. Rosato, "Plastique: a synchrotron radiation beamline for time resolved fluorescence in the frequency domain," *Rev. Sci. Instrum.* **62**, 1670–1671 (1991).
12. E. Gratton, D. M. Jameson, N. Rosato, and G. Weber, "Multifrequency cross-correlation phase fluorometer using synchrotron radiation," *Rev. Sci. Instrum.* **55**, 486–493 (1984).
13. A. Bassi, A. Farina, C. D'Andrea, A. Pifferi, G. Valentini, and R. Cubeddu, "Portable, large-bandwidth time-resolved system for diffuse optical spectroscopy," *Opt. Express* **15**, 14482–14487 (2007).

14. E. Gratton, W. W. Mantulin, M. J. van de Ven, and J. B. Fishkin, "A novel approach to laser tomography," *Bioimaging* **1**, 40–46 (1993).
15. B. Guan, Y. Zhang, and B. Chance, "A new multi-wavelength phase modulation system for photon diffusion studies," *Proc. SPIE* **2979**, 838–846 (1997).
16. A. Knüttel, J. M. Schmitt, and J. R. Knutson, "Spatial localization of absorbing bodies by interfering diffusive photon-density waves," *Appl. Opt.* **32**, 381–389 (1993).
17. E. M. Sevick, J. K. Frisoli, C. L. Burch, and J. R. Lakowicz, "Localization of absorbers in scattering media by use of frequency-domain measurements of time-dependent photon migration," *Appl. Opt.* **33**, 3562–3570 (1994).
18. J. B. Fishkin, S. Fantini, M. J. van de Ven, and E. Gratton, "Gigahertz photon density waves in a turbid medium: Theory and experiments," *Phys. Rev. E* **53**, 2307–2319 (1996).
19. J. R. Lakowicz and K. Berndt, "Frequency-domain measurements of photon migration in tissues," *Chem. Phys. Lett.* **166**, 246–252 (1990).
20. X. D. Li, T. Durduran, A. G. Yodh, B. Chance, and D. N. Pattanayak, "Diffraction tomography for biochemical imaging with diffuse-photon density waves," *Opt. Lett.* **22**, 573–575 (1997).
21. S. J. Madsen, E. R. Anderson, R. C. Haskell, and B. J. Tromberg, "Portable, high-bandwidth frequency-domain photon migration instrument for tissue spectroscopy," *Opt. Lett.* **19**, 1934–1936 (1994).
22. K. W. Berndt and J. R. Lakowicz, "Detection and localization of absorbers in scattering media using frequency-domain principles," *Proc. SPIE* **1431**, 149–160 (1991).
23. T. E. French, "The development of fluorescence lifetime imaging and an application in immunology," Ph.D. Thesis, Dept. of Physics, University of Illinois at Urbana–Champaign (1996).
24. A. Godavarty, M. J. Eppstein, C. Zhang, S. Theru, A. B. Thompson, M. Gurfinkel, and E. M. Sevick-Muraca, "Fluorescence-enhanced optical imaging in large tissue volumes using a gain-modulated ICCD camera," *Phys. Med. Biol.* **48**, 1701–1720 (2003).
25. M. Gurfinkel, T. S. Pan, and E. M. Sevick-Muraca, "Determination of optical properties in semi-infinite turbid media using imaging measurements of frequency-domain photon migration obtained with an intensified charge coupled device," *J. Biomed. Opt.* **9**, 1336–1346 (2004).
26. U. J. Netz, J. Beuthan, and A. H. Hielscher, "Multipixel system for gigahertz frequency-domain optical imaging of finger joints," *Rev. Sci. Instrum.* **79**, 034301 (2008).
27. A. D. Elder, J. H. Frank, J. Swartling, X. Dai, and C. F. Kaminski, "Calibration of a wide-field frequency-domain fluorescence lifetime

microscopy system using light emitting diodes as light sources," *J. Microsc.* **224**, 166–180 (2006).
28. R. Choe, S. D. Konecky, K. Lee, T. Durduran, D. R. Bush, S. Pathak, B. J. Czerniecki, J. Tchou, D. L. Fraker, A. DeMichele, B. Chance, S. R. Arridge, M. Schweiger, J. P. Culver, M. D. Schnall, M. E. Putt, M. A. Rosen, and A. G. Yodh, "Differentiation of benign and malignant breast tumors by in-vivo three-dimensional parallel-plate diffuse optical tomography," *J. Biomed. Opt.* **14**, 024020 (2009).
29. J. P. Culver, R. Choe, M. J. Holboke, L. Zubkov, T. Durduran, A. Slemp, V. Ntziachristos, B. Chance, and A. G. Yodh, "Three-dimensional diffuse optical tomography in the parallel plane transmission geometry: evaluation of a hybrid frequency domain/continuous wave clinical system for breast imaging," *Med. Phys.* **30**, 235–247 (2003).
30. A. Hielscher, H. K. Kim, L. D. Montejo, S. Blaschke, U. J. Netz, P. A. Zwaka, G. Illing, G. A. Müller, and J. Beuthan, "Frequency-domain optical tomographic imaging of arthritic finger joints," *IEEE Trans. Med. Imaging* **30**, 1725–1736 (2011).
31. G. Yu, T. Durduran, D. Furuya, J. H. Greenberg, and A. G. Yodh, "Frequency-domain multiplexing system for in vivo diffuse light measurements of rapid cerebral hemodynamics," *Appl. Opt.* **42**, 2931–2939 (2003).
32. B. A. Feddersen, D. W. Piston, and E. Gratton, "Digital parallel acquisition in frequency domain fluorometry," *Rev. Sci. Instrum.* **60**, 2929–2936 (1989).
33. R. N. Bracewell, *The Fourier Transform and Its Applications*, 3rd ed. McGraw-Hill, New York (2000).
34. N. G. Chen, M. Huang, H. Xia, D. Piao, E. Cronin, and Q. Zhu, "Portable near-infrared diffusive light imager for breast cancer detection," *J. Biomed. Opt.* **9**, 504–510 (2004).
35. K.-S. No, R. Kwong, P. H. Chou, and A. Cerussi, "Design and testing of a miniature broadband frequency domain photon migration instrument," *J. Biomed. Opt.* 050509 (2008).
36. A. G. Orlova, I. V. Turchin, V. I. Plehanov, N. M. Shakhova, I. I. Fiks, M. I. Kleshnin, N. Y. Konuchencko, and V. A. Kamenshy, "Frequency-domain diffuse optical tomography with single source–detector pair for breast cancer detection," *Laser Phys. Lett.* **5**, 321–327 (2008).
37. A. Fantini, M. A. Franceschini, J. S. Maier, S. A. Walker, B. Barbieri, and E. Gratton, "Frequency-domain multichannel optical detector for non-invasive tissue spectroscopy and oximetry," *Opt. Eng.* **34**, 32–42 (1995).
38. M. A. Franceschini, E. Gratton, D. Hueber, and S. Fantini, "Near-infrared absorption and scattering spectra of tissues in vivo," *Proc. SPIE* **3597**, 526–531 (1999).

39. A. Fantini, M. A. Franceschini, and E. Gratton, "Semi-infinite-geometry boundary problem for light migration in highly scattering media: a frequency-domain study in the diffusion approximation," *J. Opt. Soc. Am. B* **11**, 2128–2138 (1994).
40. J. J. Duderstadt and L. J. Hamilton, *Nuclear Reactor Analysis*, Wiley, New York (1976).
41. F. Hetzel, M. S. Patterson, L. Preuss, and B. Wilson, *Recommended Nomenclature for Physical Quantities in Medical Applications of Light*, AAPM Report No. 57, American Institute of Physics, Woodbury, N.Y. (1996), pp. 1–6.
42. S. R. Arridge, "Optical tomography in medical imaging," *Inverse Problems* **15**, R41–R93 (1999).
43. D. A. Boas, "Diffuse photon probes of structural and dynamical properties of turbid media: Theory and biomedical applications," Ph.D. Thesis, Dept. of Physics, University of Pennsylvania (1996).
44. J.-M. Kaltenbach and M. Kaschke, "Frequency- and time-domain modelling of light transport in random media," *in Medical Optical Tomography: Functional Imaging and Monitoring*,G. J. Muller et al., Eds. SPIE, Bellingham, Wash. (1993), pp. 65–86.
45. H. W. Wyld, *Mathematical Methods for Physics*, Addison-Wesley, Reading, Mass. (1994), Chapter 3.
46. G. Zaccanti, E. Battistelli, P. Bruscaglioni, and Q. Wei, "Analytic relationships for the statistical moments of scattering point coordinates for photon migration in a scattering medium," *Pure Appl. Opt.* **3**, 897–905 (1994).
47. D. A. Boas, M. A. O'Leary, B. Chance, and A. G. Yodh, "Scattering of diffuse photon density waves by spherical inhomogeneities within turbid media: analytic solution and applications," *Proc. Natl. Acad. Sci. USA* **91**, 4887–4891 (1994).
48. J. B. Fishkin and E. Gratton, "Propagation of photon-density waves in strongly scattering media containing an absorbing semi-infinite plane bounded by a straight edge," *J. Opt. Soc. Am. A* **10**, 127–140 (1993).
49. R. C. Haskell, L. O. Svaasand, T. T. Tsay, T. C. Feng, M. S. McAdams, and B. J. Tromberg, "Boundary conditions for the diffusion equation in radiative transfer," *J. Opt. Soc. Am. A* **11**, 2727–2741 (1994).
50. M. S. Patterson, J. D. Moulton, B. C. Wilson, K. W. Berndt, and J. R. Lakowicz, "Frequency-domain reflectance for the determination of the scattering and absorption properties of tissue," *Appl. Opt.* **30**, 4474–4476 (1991).
51. S. R. Arridge, M. Cope, and D. T. Delpy, "The theoretical basis for the determination of optical pathlengths in tissue: temporal and frequency analysis," *Phys. Med. Biol.* **37**, 1531–1560 (1992).

52. D. Contini, F. Martelli, and G. Zaccanti, "Photon migration through a turbid slab described by a model based on diffusion approximation. I. Theory," *Appl. Opt.* **36**, 4587–4599 (1997).
53. E. Zauderer, *Partial Differential Equations of Applied Mathematics*, 2nd ed. Wiley, New York (1989), p. 484.
54. A. Kienle, M. S. Patterson, N. Dognitz, R. Bays, G. Wagnieres, and H. van den Bergh, "Noninvasive determination of the optical properties of two-layered turbid media," *Appl. Opt.* **37**, 779–791 (1998).
55. A. Liemert and A. Kienle, "Light diffusion in a turbid cylinder. II. Layered case," *Opt. Express* **18**, 9266–9279 (2010).
56. S. Fantini, D. Hueber, M. A. Franceschini, E. Gratton, W. Rosenfeld, P. G. Stubblefield, D. Maulik, and M. R. Stankovic, "Non-invasive optical monitoring of the newborn piglet brain using continuous-wave and frequency-domain spectroscopy," *Phys. Med. Biol.* **44**, 1543–1563 (1999).
57. T. H. Pham, O. Coquoz, J. B. Fishkin, E. Anderson, and B. J. Tromberg, "Broadband bandwidth frequency domain instrument for quantitative tissue optical spectroscopy," *Rev. Sci. Ins.* **71**, 2500–2513 (2000).
58. A. E. Cerussi, D. Jakubowski, N. Shah, F. Bevilacqua, R. Lanning, A. J. Berger, D. Hsiang, J. Butler, R. F. Holcombe, and B. J. Tromberg, "Spectroscopy enhances the information content of optical mammography," *J. Biomed. Opt.* **7**, 60–71 (2002).
59. J. Ricka, "Dynamic light scattering with single-mode and multimode receivers," *Appl. Opt.* **32**, 2860–2875 (1993).
60. D. J. Pine, D. A. Weitz, P. M. Chaikin, and E. Herbolzheimer, "Diffusing-wave spectroscopy," *Phys. Rev. Lett.* **60**, 1134–1137 (1988).
61. D. A. Boas and A. G. Yodh, "Spatially varying dynamical properties of turbid media probed with diffusing temporal light correlation," *J. Opt. Soc. Am. A* **14**, 192–215 (1997).
62. T. Durduran and A. G. Yodh, "Diffuse correlation spectroscopy for non-invasive, micro-vascular cerebral blood flow measurements," *NeuroImage* **85**, 51–63 (2014).
63. I. M. Braveman, "Anatomy and physiology of the cutaneous microcirculation," *Bioengineering of the Skin: Cutaneous Blood Flow and Erythema*,E. Berardesca, P. Elsner, and H. I. Maibach, Eds. CRC Press, Boca Raton, FL (1995).
64. V. Rajan, B. Varghese, T. G. van Leeuwen, and W. Steenbergen, "Review of methodological developments in laser Doppler flowmetry," *Laser Med. Sci.* **24**, 269–283 (2009).
65. G. M. Hale and M. R. Querry, "Optical constants of water in the 200 nm to 200 m wavelength region," *Appl. Opt.* **12**, 555–563 (1973).
66. S. Prahl, online tabulated molar extinction coefficients of oxy-hemoglobin and deoxy-hemoglobin: http://omlc.ogi.edu/spectra/hemoglobin/summary.html Last accessed May 11, 2016.

67. W. F. Cheong, S. A. Prahl, and A. J. Welch, "A review of the optical properties of biological tissues," *IEEE J. Quantum Electron.* **26**, 2166–2185 (1990).
68. S. L. Jacques, "Optical properties of biological tissue: a review," *Phys. Med. Biol.* **58**, R37--R61 (2013).
69. M. Firbank, E. Okada, and D. T. Delpy, "Investigation of the effect of discrete absorbers upon the measurements of blood volume with near-infrared spectroscopy," *Phys. Med. Biol.* **42**, 465–477 (1997).
70. F. A. Duck, *Physical Properties of Tissues: A Comprehensive Reference Book*, Academic Press, San Diego, Calif. (1990).
71. M. A. Franceschini, E. Gratton, and S. Fantini, "Non-invasive optical method to measure tissue and arterial saturation: an application to absolute pulse oximetry of the brain," *Opt. Lett.* **24**, 829–831 (1999).
72. G. A. Millikan, "The oximeter, an instrument for measuring continuously the oxygen saturation of arterial blood in man," *Rev. Sci. Instr.* **13**, 434–444 (1942).
73. Y. Mendelson, "Pulse oximetry: theory and applications for noninvasive monitoring," *Clin. Chem.* **38**, 1601–1607 (1992).
74. M. Miwa, Y. Ueda, and B. Chance, "Development of time resolved spectroscopy system for quantitative non-invasive tissue measurement," *Proc. SPIE* **2389**, 142–149 (1995).
75. V. Quaresima, M. A. Franceschini, S. Fantini, E. Gratton, and M. Ferrari, "Difference in leg muscles oxygenation during treadmill exercise by a new near infrared frequency-domain oximeter," *Proc. SPIE* **3194**, 116–120 (1998a).
76. T. R. Cheatle, L. A. Potter, M. Cope, D. T. Delpy, P. D. Coleridge Smith, and J. H. Scurr, "Near-infrared spectroscopy in peripheral vascular disease," *Br. J. Surg.* **78**, 405–408 (1991).
77. R. A. de Blasi, M. Cope, and M. Ferrari, "Oxygen consumption of human skeletal muscle by near-infrared spectroscopy during tourniquet-induced ischemia in maximal voluntary contraction," *Adv. Exp. Med. Biol.* **317**, 771–777 (1992).
78. C. Casavola, L. A. Paunescu, S. Fantini, and E. Gratton, "Blood flow and oxygen consumption with near-infrared spectroscopy and venous occlusion: spatial maps and the effect of time and pressure of inflation," *J. Biomed. Opt.* **5**, 269–276 (2000).
79. R. A. de Blasi, M. Ferrari, A. Natali, G. Conti, A. Mega, and A. Gasparetto, "Noninvasive measurement of forearm blood flow and oxygen consumption by near-infrared spectroscopy," *J. Appl. Physiol.* **76**, 1388–1393 (1994).
80. S. Homma, H. Eda, S. Ogasawara, and A. Kagaya, "Near-infrared estimation of O2 supply and consumption in forearm muscles working at varying intensity," *J. Appl. Physiol.* **80**, 1279–1284 (1996).

81. M. C. P. van Beekvelt, W. N. J. M. Colier, B. G. M. van Engelen, M. T. E. Hopman, R. A. Wevers, and B. Oeseburg, "Validation of measurement protocols to assess oxygen consumption and blood flow in the human forearm by near-infrared spectroscopy," *Proc. SPIE* **3194**, 133–144 (1998).
82. S. R. Hintz, D. A. Benaron, J. P. van Houten, J. L. Duckworth, F. W. H. Liu, S. D. Spilman, D. K. Stevenson, and W.-F. Cheong, "Stationary headband for clinical time-of-flight optical imaging at the bedside," *Photochem. Photobiol.* **68**, 361–369 (1998).
83. S. R. Hintz, W.-F. Cheong, J. P. van Houten, D. K. Stevenson, and D. A. Benaron, "Bedside imaging of intracranial hemorrhage in the neonate using light: comparison with ultrasound, computed tomography, and magnetic resonance imaging," *Pediatr. Res.* **45**, 54–59 (1999).
84. P. Y. Lin, N. Roche-Labarbe, M. Dehaes, A. Fenoglio, P. E. Grant, and M. A. Franceschini, "Regional and hemispheric asymmetries of cerebral hemodynamic and oxygen metabolism in newborns," *Cereb. Cortex* **23** 339–48. (2013).
85. J. Zhao, H. S. Ding, X. L. Hou, C. L. Zhou, and B. Chance, "In vivo determination of the optical properties of infant brain using frequency-domain near-infrared spectroscopy," *J. Biomed. Opt.* **10**, 024028 (2005).
86. M. A. Franceschini, S. Thaker, G. Themelis, K. K. Krishnamoorthy, H. Bortfeld, S. G. Diamond, D. A. Boas, K. Arvin, and P. E. Grant, "Assessment of infant brain development with frequency-domain near-infrared spectroscopy," *Pediatr. Res.* **61**, 546–551 (2007).
87. N. Roche-Labarbe, A. Fenoglio, A. Aggarwal, M. Dehaes, S. A. Carp, M. A. Franceschini, and P. E. Grant, "Near-infrared spectroscopy assessment of cerebral oxygen metabolism in the developing premature brain," *J. Cereb. Blood Flow Metab.* **32**, 481–488 (2012).
88. M. Dehaes, A. Aggarwal, P. Y. Linn, C. Rosa Fortuno, A. Fenoglio, N. Roche-Labarbe, J. S. Soul, M. A. Franceschini, and P. E. Grant, "Cerebral oxygen metabolism in neonatal hypoxic ischemic encephalopathy during and after therapeutic hypothermia," *J. Cereb. Blood Flow Metab.* **34**, 87–94 (2014).
89. B. Hallacoglu, A. Sassaroli, M. Wysocki, E. Guerrero-Berroa, M. Schnaider Beeri, V. Haroutunian, M. Shaul, I. H. Rosenberg, A. M. Troen, and S. Fantini, "Absolute measurement of cerebral optical coefficients, hemoglobin concentration and oxygen saturation in old and young adults with near-infrared spectroscopy," *J. Biomed. Opt.* **17**, 081406 (2012).
90. L. Gagnon, C. Gauthier, R. D. Hoge, F. Lesage, J. Selb, and D. A. Boas, "Double layer estimation of intra and extracerebral hemoglobin concentration with a time resolved system," *J. Biomed. Opt.* **13**, 054019 (2008).

91. R. Gatto, W. E. Hoffman, M. Mueller, C. Paisansathan, and F. Charbel, "Age effects on brain oxygenation during hypercapnia," *J. Biomed. Opt.* **12** 062113 (2007).
92. B. Hallacoglu, A. Sassaroli, and S. Fantini, "Optical characterization of two-layered turbid media for non-invasive, absolute oximetry in cerebral and extracerebral tissue," *PLoS One* **8**, e64095 (2013).
93. F. B. Haeussinger, S. Heinzel, T. Hahn, M. Schecklmann, A.-C. Ehlis, and A. J. Fallgatter, "Simulation of near-infrared light absorption considering individual head and prefrontal cortex anatomy: implications for optical neuroimaging," *PLoS One* **6**, e26377 (2011).
94. J. Choi, M. Wolf, V. Toronov, U. Wolf, C. Polzonetti, D. Hueber, L. P. Safonova, R. Gupta, A. Michalos, W. Mantulin, and E. Gratton, "Noninvasive determination of the optical properties of adult brain: near-infrared spectroscopy approach," *J. Biomed. Opt.* **9**, 221–229 (2004).
95. B. Beauvoit, T. Kitai, and B. Chance, "Contribution of the mitochondrial compartment to the optical properties of the rat liver: a theoretical and practical approach," *Biophys. J.* **67**, 2501–2510 (1994).
96. J. R. Mourant, J. P. Freyer, A. H. Hielscher, A. A. Eick, D. Shen, and T. M. Johnson, "Mechanism of light scattering from biological cells relevant to noninvasive optical-tissue diagnostics," *Appl. Opt.* **37**, 3586–3593 (1998).
97. A. Wax and V. Backman, *Biomedical Applications of Light Scattering*, McGraw-Hill, New York (2010).
98. J. T. Bruulsema, J. E. Hayward, T. J. Farrel, M. S. Patterson, L. Heinemann, M. Berger, T. Koschinsky, J. Sandahl–Christiansen, H. Orskov, M. Essenpreis, G. Schmelzeisen-Redeker, and D. Böcker, "Correlation between blood glucose concentration in diabetics and noninvasively measured tissue optical scattering coefficient," *Opt. Lett.* **22**, 190–192 (1997).
99. M. Kohl, M. Cope, M. Essenpreis, and D. Böcker, "Influence of glucose concentration on light scattering in tissue-simulating phantoms," *Opt. Lett.* **19**, 2170–2172 (1994).
100. J. S. Maier, S. A. Walker, S. Fantini, M. A. Franceschini, and E. Gratton, "Possible correlation between blood glucose concentration and the reduced scattering coefficient of tissues in the near infrared," *Opt. Lett.* **19**, 2062–2064 (1994).
101. V. Tuchin, I. L. Maksimova, D. A. Zimnyakov, I. L. Kon, A. H. Mavlutov, and A. A. Mishin, "Light propagation in tissues with controlled optical properties," *J. Biomed. Opt.* **2**, 304–321 (1997).
102. D. A. Boas, M. A. O'Leary, B. Chance, and A. G. Yodh, "Detection and characterization of optical inhomogeneities with diffuse photon density waves: a signal to noise analysis," *Appl. Opt.* **36**, 75–92 (1997b).

103. P. N. den Outer, Th. M. Nieuwenhuizen, and A. Lagendijk, "Location of objects in multiple-scattering media," *J. Opt. Soc. Am. A* **10**, 1209–1218 (1993).
104. S. A. Walker, D. A. Boas, and E. Gratton, "Photon density waves scattered from cylindrical inhomogeneities: theory and experiments," *Appl. Opt.* **37**, 1935–1944 (1998).
105. A. Kienle, T. Glanzmann, G. Wagnieres, and H. V. Bergh, "Investigation of two-layered turbid medium with time-resolved reflectance," *Appl. Opt.* **37**, 6852–6862 (1998b).
106. F. Martelli, A. Sassaroli, S. Del Bianco, Y. Yamada, and G. Zaccanti, "Solution of the time-dependent diffusion equation for layered random media by the eigenfunction method," *Phys. Rev. E* **67**, 056623 (2003).
107. J. Ripoll, V. Ntziachristos, J. P. Culver, D. N. Pattanayak, A. G. Yodh, and M. Nieto-Vesperinas, "Recovery of optical parameters in multiple-layered diffusive media: theory and experiments," *J. Opt. Soc. Am. A* **18**, 821–830 (2001).
108. J. Sikora, A. Zacharopoulos, A. Douiri, M. Schweiger, L. Horesh, A. R. Arridge, and J. Ripoll, "Diffuse photon propagation in multilayered geometries," *Phys. Med. Biol.* **51**, 497–516 (2006).
109. B. W. Pogue, M. S. Patterson, H. Jiang, and K. D. Paulsen, "Initial assessment of a simple system for frequency domain diffuse optical tomography," *Phys. Med. Biol.* **40**, 1709–1729 (1995).
110. Y. Yao, Y. Wang, Y. Pei, W. Zhu, and R. L. Barbour, "Frequency domain optical imaging of absorption and scattering distributions by a Born iterative method," *J. Opt. Soc. Am. A* **14**, 325–342 (1997).
111. S. R. Arridge, M. Schweiger, M. Hiraoka, and D. T. Delpy, "A finite element approach to modelling photon transport in tissue," *Med. Phys.* **20**, 299–309 (1993a).
112. R. L. Barbour, H. L. Graber, Y. Pei, S. Zhong, and C. H. Schmitz, "Optical tomographic imaging of dynamic features of dense-scattering media," *J. Opt. Soc. Am.* **18**, 3018–3036 (2001).
113. H. Dehghani, M. E. Eames, P. K. Yalavarthy, S. C. Davis, S. Srinivasan, C. M. Carpenter, B. W. Pogue, and K. D. Paulsen, "Near infrared optical tomography using NIRFAST: algorithm for numerical model and image reconstruction," *Commun. Numer. Methods Eng.* **25**, 711–732 (2008).
114. H. Jiang, K. D. Paulsen, U. L. Osterberg, B. W. Pogue, and M. S. Patterson, "Optical image reconstruction using frequency-domain data: simulations and experiments," *J. Opt. Soc. Am.* 13, 253–266 (1996).
115. D. A. Boas, "A fundamental limitation of linearized algorithms for diffuse optical tomography," *Opt. Express* 1, 404–413 (1997).
116. D. Grosenick, A. Kummrow, R. Macdonald, P. M. Schlag, and H. Rinneberg, "Evaluation of higher-order time domain perturbation

theory of photon diffusion on breast-equivalent phantoms and optical mammograms," *Phys. Rev. E* **76**, 061908 (2007).

117. M. R. Ostermeyer and S. L. Jacques, "Perturbation theory for diffuse light transport in complex biological tissues," *J. Opt. Soc. Am. A* **14**, 255–261 (1997).
118. A. Sassaroli, F. Martelli, and S. Fantini, "Perturbation theory for the diffusion equation by use of the moments of the generalized temporal point-spread function: I. Theory," *J. Opt. Soc. Am. A* **23**, 2105–2118 (2006).
119. A. Sassaroli, F. Martelli, and S. Fantini, "Perturbation theory for the diffusion equation by use of the moments of the generalized temporal point-spread function: II. Continuous wave results," *J. Opt. Soc. Am. A* **23**, 2119–2131 (2006).
120. B. Wassermann, "Limits of high-order perturbation in time-domain optical mammography," *Phys. Rev. E* **74**, 031908 (2006).
121. V. Chernomordik, D. Hattery, A. Gandjbakhche, A. Pifferi, P. Taroni, A. Torricelli, G. Valentini, and R. Cubeddu, "Quantification by random walk of the optical parameters of nonlocalized abnormalities embedded within tissuelike phantoms," *Opt. Lett.* **25**, 951–953 (2000).
122. A. H. Gandjbakhche, R. F. Bonner, R. Nossal, and G. H. Weiss, "Absorptivity contrast in transillumination imaging of tissue abnormalities," *Appl. Opt.* **35**, 1767–1774 (1996).
123. C. Zhu and Q. Liu, "Review of Monte Carlo modelling of light transport in tissues," *J. Biomed. Opt.* **18**, 050902 (2013).
124. H. Dehghani, S. Srinivasan, B. W. Pogue, and A. Gibson, "Numerical modelling and image reconstruction in diffuse optical tomography," *Phil. Trans. R. Soc. A* **367**, 3073–3093 (2009).
125. T. Durduran, R. Choe, W. B. Baker, and A. G. Yodh, "Diffuse optics for tissue monitoring and tomography," *Rep. Prog. Phys.* **73**, 076701 (2010).
126. A. H. Hielscher, A. Y. Bluestone, G. S. Abdoulaev, A. D. Klose, J. Lasker, M. Stewart, U. Netz, and J. Beuthan, "Near-infrared diffuse optical tomography," *Dis. Markers* **18**, 313–337 (2002).
127. S. R. Arridge and J. C. Schotland, "Optical tomography: forward and inverse problems," *Inv. Prob.* **25**, 123010 (2009).
128. J. C. Hebden and S. R. Arridge, "Imaging through scattering by use of an analytical model of perturbation amplitudes in the time domain," *Appl. Opt.* **35**, 6788–6795 (1996).
129. H. L. Graber, J. Chang, J. Lubowsky, R. Aronson, and R. L. Barbour, "Near-infrared absorption imaging of dense scattering media by steady-state diffusion tomography," *Proc. SPIE* **1888**, 372–386 (1993).
130. S. R. Arridge and M. Schweiger, "Photon-measurement density functions. Part 2: Finite-element-method calculations," *Appl. Opt.* **34**, 8026–8037 (1995).

131. V. Ntziachristos, A. G. Yodh, M. Schnall, and B. Chance, "Concurrent MRI and diffuse optical tomography of breast after indocyanine green enhancement," *Proc. Natl. Acad. Sci. USA* **97**, 2767–2772 (2000).
132. M. Belge, M. E. Kilmer, and E. L. Miller, "Efficient determination of multiple regularization parameters in a generalized L-curve framework," *Inv. Prob.* **18**, 1161–1183 (2002).
133. S. H. Katamreddy and P. K. Yalavarthy, "Model-resolution based regularization improves near infrared diffuse optical tomography," *J. Opt. Soc. Am. A* **29**, 649–656 (2012).
134. S. R. Arridge, M. Schweiger, M. Hiraoka, and D. T. Delpy, "Performance of an iterative reconstruction algorithm for near-infrared absorption imaging," *Proc. SPIE* **1888**, 360–371 (1993b).
135. H. Jiang, K. D. Paulsen, U. L. Osterberg, B. W. Pogue, and M. S. Patterson, "Simultaneous reconstruction of optical absorption and scattering maps in turbid media from near-infrared frequency-domain data," *Opt. Lett.* **20**, 2128–2130 (1995).
136. B. W. Pogue, T. O. McBride, J. Prewitt, U. L. Österberg, and K. D. Paulsen, "Spatially variant regularization improves diffuse optical tomography," *Appl. Opt.* **38**, 2950–2961 (1999).
137. D. A. Benaron, D. C. Ho, S. Spilman, J. P. Van Houten, and K. D. Stevenson, "Non-recursive linear algorithms for optical imaging in diffusive media," *Adv. Exp. Med. Biol.* **361**, 215–222 (1994).
138. S. B. Colak, D. G. Papaioannou, G. W.'t Hooft, M. B. van der Mark, H. Schomberg, J. C. J. Paasschens, J. B. M. Melissen, and N. A. A. J. van Asten, "Tomographic image reconstruction from optical projections in light-diffusing media," *Appl. Opt.* **36**, 180–213 (1997).
139. V. V. Lyubimov, A. G. Kalintsev, A. B. Konovalov, O. V. Lyamtsev, O. V. Kravtsenyuk, A. G. Murzin, O. V. Golubkina, G. B. Mordvinov, L. N. Soms, and L. M. Yavorskaya, "Application of the photon average trajectories method to real-time reconstruction of tissue inhomogeneities in diffuse optical tomography of strongly scattering media," *Phys. Med. Biol.* **47**, 2109–2128 (2002).
140. S. A. Walker, S. Fantini, and E. Gratton, "Image reconstruction using back-projection from frequency-domain optical measurements in highly scattering media," *Appl. Opt.* **36**, 170–179 (1997).
141. M. A. Franceschini, V. Toronov, M. E. Filiaci, E. Gratton, and S. Fantini, "On-line optical imaging of the human brain with 160–ms temporal resolution," *Opt. Express* **6**, 49–57 (2000).
142. P. Gonzalez-Rodriguez and A. D. Kim, "Diffuse optical tomography using the one-way radiative transfer equation," *Biomed. Opt. Express* **6**, 2006–2021 (2006).
143. A. D. Klose, U. Netz, J. Beuthan, and A. Hielscher, "Optical tomography using the time-independent equation of radiative transfer.

Part 1: forward model," *J. Quant. Spectrosc. Radiat. Transfer* **72**, 691–713 (2002).
144. K. Ren, G. S. Abdoulaev, G. Bal, and A. H. Hielscher, "Algorithm for solving the equation of radiative transfer in the frequency domain," *Opt. Lett.* **29** 578–580 (2004).
145. P. Wang, S. R. Arridge, and M. Jiang, "Radiative transfer equation for media with spatially varying refractive index," *Phys. Rev. A* **90**, 023803 (2014).
146. S. Fantini, M. A. Franceschini, S. A. Walker, J. S. Maier, and E. Gratton, "Photon path distributions in turbid media: applications for imaging," *Proc. SPIE* **2389**, 340–349 (1995b).
147. S. G. Demos, A. J. Vogel, and A. H. Gandjbakhche, "Advances in optical spectroscopy and imaging of breast lesions," *J. Mammary Gland Biol. Neoplasia* **11**, 165–181 (2006).
148. D. R. Leff, O. J. Warren, L. C. Enfield, A. Gibson, T. Athanasiou, D. K. Patten, J. Hebden, G. Z. Yang, and A. Darzi, "Diffuse optical imaging of the healthy and diseased breast: a systematic review," *Breast Cancer Res. Treat.* **108**, 9–22 (2008).
149. V. Quaresima, S. J. Matcher, and M. Ferrari, "Identification and quantification of intrinsic optical contrast for near-infrared mammography," *Photochem. Photobiol.* **67**, 4–14 (1998b).
150. B. J. Tromberg, O. Coquoz, J. B. Fishkin, T. Pham, E. R. Anderson, J. Butler, M. Cahn, J. D. Gross, V. Venugopalan, and D. Pham, "Non-invasive measurements of breast tissue optical properties using frequency-domain photon migration," *Phil. Trans. R. Soc. Lond. B* **352**, 661–668 (1997).
151. S. Thomsen and D. Tatman, "Physiological and pathological factors of human breast disease that can influence optical diagnosis," *Ann. NY Acad. Sci.* **838**, 171–193 (1998).
152. P. Vaupel, F. Kallinowski, and P. Okunieff, "Blood flow, oxygen and nutrient supply, and metabolic microenvironment of human tumors: a review," *Cancer Res.* **49**, 6449–6465 (1989).
153. V. G. Peters, D. R. Wyman, M. S Patterson, and G. L. Frank, "Optical properties of normal and diseased human breast tissues in the visible and near-infrared," *Phys. Med. Biol.* **35**, 1317–1334 (1990).
154. T. L. Troy, D. L. Page, and E. Sevick-Muraca, "Optical properties of normal and diseased breast tissues: prognosis for optical mammography," *J. Biomed. Opt.* **1**, 342–355 (1996).
155. S. Fantini, S. A. Walker, M. A. Franceschini, M. Kaschke, P. M. Schlag, and K. T. Moesta, "Assessment of the size, position, and optical properties of breast tumors in vivo by non-invasive optical methods," *Appl. Opt.* **37**, 1982–1989 (1998).

156. D. Grosenick, H. Wabnitz, H. H. Rinneberg, K. T. Moesta, and P. M. Schlag, "Development of a time-domain optical mammograph and first in vivo applications," *Appl. Opt.* **38**, 2927–2943 (1999).
157. B. J. Tromberg, N. Shah, R. Lanning, A. Cerussi, J. Espinoza, T. Pham, L. Svaasand, and J. Butler, "Non-invasive in vivo characterization of breast tumors using photon migration spectroscopy," *Neoplasia* **2**, 26–40 (2000).
158. S. Fantini and A. Sassaroli, "Near-infrared optical mammography for breast cancer detection with intrinsic contrast," *Ann. Biomed. Eng.* **40**, 398–407 (2011).
159. A. Corlu, R. Choe, T. Durduran, M. A. Rosen, M. Schweiger, S. R. Arridge, M. D. Schnall, and A. G. Yodh, "Three-dimensional in vivo fluorescence diffuse optical tomography of breast cancer in humans," *Opt. Express* **15**, 6696–6716 (2007).
160. S. Nioka, S. B. Colak, X. Li, Y. Yang, and B. Chance, "Breast tumor images of hemodynamic information using a contrast agent with back projection and FFT enhancement," *OSA Trends in Optics and Photonics Vol. 21: Advances in Optical Imaging and Photon Migration*, J. G. Fujimoto and M. S. Patterson, Eds. Optical Society of America, Washington, D.C. (1998), pp. 266–270.
161. B. Riefke, K. Licha, and W. Semmler, "Contrast media for optical mammography," *Radiologe* **37**, 749–755 (1997).
162. S. van de Ven, S. Elias, A. Wiethoff, M. Van der Voort, A. Leproux, T. Nielsen, B. Brendel, L. Bakker, M. van der Mark, W. Mali, and P. Luijten, "A novel fluorescent imaging agent for diffuse optical tomography of the breast: first clinical experience in patients," *Mol. Imaging Biol.* **12**, 343–348 (2010).
163. A. E. Cerussi, D. Hsiang, N. Shah, R. Mehta, A. Durkin, J. Butler, and B. J. Tromberg, "Predicting response to breast cancer neoadjuvant chemotherapy using diffuse optical spectroscopy," *PNAS* **104**, 4014–4019 (2007).
164. A. E. Cerussi, V. W. Tanamai, D. Hsiang, J. Butler, R. S. Mehta, and B. J. Tromberg, "Diffuse optical spectroscopic imaging correlates with final pathological response in breast cancer neoadjuvant chemotherapy," *Phil. Trans. R. Soc. A* **369**, 4512–4530 (2011).
165. S. Jiang, B. W. Pogue, C. M. Carpenter, S. P. Poplack, W. A. Wells, C. A. Kogel, J. A. Forero, L. S. Muffly, G. N. Schwartz, K. D. Paulsen, and P. A. Kaufman, "Evaluation of breast tumor response to neoadjuvant chemotherapy with tomographic diffuse optical spectroscopy: case studies of tumor region-of-interest changes," *Radiology* **252**, 551–560 (2009).
166. D. Roblyer, S. Ueda, A. Cerussi, W. Tanamai, A. Durkin, R. Mehta, D. Hsiang, J. A. Butler, C. McLaren, W. Chen, and B. J. Tromberg,

"Optical imaging of breast cancer oxyhemoglobin flare correlates with neoadjuvant chemotherapy response one day after starting treatment," *Proc. Natl. Acad. Sci. USA* **30**, 14626–14631 (2011).

167. H. Soliman, A. Gunasekara, M. Rycroft, J. Zubovits, R. Dent, J. Spayne, M. J. Yaffe, and G. J. Czarnota, "Functional imaging using diffuse optical spectroscopy of neoadjuvant chemotherapy response in women with locally advanced breast cancer," *Clin. Cancer Res.* **16**, 2605–2614 (2010).
168. S. Ueda, D. Roblyer, A. Cerussi, A. Durkin, A. Leproux, Y. Santoro, S. Xu, T. D. O' sullivan, D. Hsiang, R. Mehta, J. Butler, and B. J. Tromberg, "Baseline tumor oxygen saturation correlates with a pathologic complete response in breast cancer patients undergoing neoadjuvant chemotherapy," *Cancer Res.* **72**, 4318–28 (2012).
169. Q. Zhu, P. A. DeFusco, A. Ricci, Jr., E. B. Cronin, P. U. Hegde, M. Kane, B. Tavakoli, Y. Xu, J. Hart, and S. H. Tannenbaum, "Breast cancer: assessing response to neoadjuvant chemotherapy by using US-guided near-infrared tomography," *Radiology* **266**, 433–442 (2013).
170. Q. Zhu, S. Tannenbaum, P. Hegde, M. Kane, C. Xu, and S. H. Kurtzman, "Noninvasive monitoring of breast cancer during neoadjuvant chemotherapy using optical tomography with ultrasound localization," *Neoplasia* **10**, 1028–1040 (2008).
171. M. A. Franceschini, K. T. Moesta, S. Fantini, G. Gaida, E. Gratton, H. Jess, W. W. Mantulin, M. Seeber, P. M. Schlag, and M. Kaschke, "Frequency-domain techniques enhance optical mammography: initial clinical results," *Proc. Natl. Acad. Sci. USA* **94**, 6468–6473 (1997).
172. K. T. Moesta, S. Fantini, H. Jess, S. Totkas, M. A. Franceschini, M. Kaschke, and P. M. Schlag, "Contrast features of breast cancer in frequency-domain laser scanning mammography," *J. Biomed. Opt.* **3**, 129–136 (1998).
173. B. W. Pogue, M. Testorf, T. McBride, U. Osterberg, and K. Paulsen, "Instrumentation and design of a frequency-domain diffuse optical tomography imager for breast cancer detection," *Opt. Express* **1**, 391–403 (1997).
174. S. Zhou, C. Xie, S. Nioka, H. Liu, Y. Zhang, and B. Chance, "Phased array instrumentation appropriate to high precision detection and localization of breast tumor," *Proc. SPIE* **2979**, 98–106 (1997).
175. L. Götz, S. H. Heywang-Köbrunner, O. Schütz, and H. Siebold, "Optische mammographie an präoperativen patientinnen," *Akt. Radiol.* **8**, 31–33 (1998).
176. S. Fantini, E. L. Heffer, M. A. Franceschini, L. Götz, A. Heinig, S. Heywang-Köbrunner, O. Schütz, and H. Siebold, "Optical mammography with intensity-modulated light," *Proc. of In Vivo Optical Imaging*

Workshop, Sept. 16–17, 1999, National Institutes of Health, Bethesda, MD,A. Gandjbakhche, Ed. Optical Society of America (2000), pp. 111–117.

177. H. Jess, H. Erdl, K. T. Moesta, S. Fantini, M. A. Franceschini, E. Gratton, and M. Kaschke, "Intensity-modulated breast imaging: technology and clinical pilot study results," *OSA Trends in Optics and Photonics on Advances in Optical Imaging and Photon Migration*, Vol. **2**., R. R. Alfano and J. G. Fujimoto, Eds. Optical Society of America, Washington, D.C. (1996), pp. 126–129.
178. M. Kaschke, H. Jess, G. Gaida, J. M. Kaltenbach, and W. Wrobel, "Transillumination imaging of tissue by phase modulation techniques," *Advances in Optical Imaging and Photon Migration*,R. R. Alfano and J. G. Fujimoto, Ed., *Proc. OSA* **21**, 88–92 (1994).
179. A. E. Cerussi, N. Shah, D. Hsiang, A. Durkin, J. Butler, and B. J. Tromberg, "In vivo absorption, scattering, and physiologic properties of 58 malignant breast tumors determined by broadband diffuse optical spectroscopy," *J. Biomed. Opt.* **11**, 044005 (2006).
180. Q. Zhu, C. Xu, P. Guo, A. Aguirre, B. Yuan, F. Huang, D. Castilo, J. Gamelin, S. Tannenbaum, M. Kane, P. Hedge, and S. Kurtzman, "Optimal probing of optical contrast of breast lesions of different size located at different depths by US localization," *Technol. Cancer Res. Treat.* **5**, 365–380 (2006).
181. Q. Fang, S. A. Carp, J. Selb, G. Boverman, Q. Zhang, D. B. Kopans, R. H. Moore, E. L. Miller, D. H. Brooks, and D. A. Boas, "Combined optical imaging and mammography of the healthy breast: optical contrast derived from breast structure and compression," *IEEE Trans. Med. Imaging* **28**, 30–42 (2009).
182. T. O. McBride, B. W. Pogue, S. Jiang, U. L. Österberg, and K. D. Paulsen, "A parallel-detection frequency-domain near-infrared tomography system for hemoglobin imaging of the breast in vivo," *Rev. Sci. Inst.* **72**, 1817–1824 (2001).
183. D. B. Jakubowski, A. E. Cerussi, F. Bevilacqua, N. Shah, D. Hsiang, J. Butler, and B. J. Tromberg, "Monitoring neoadjuvant chemotherapy in breast cancer using quantitative diffuse optical spectroscopy: a case study," *J. Biomed. Opt.* **9**, 230–238 (2004).
184. Q. Fang, J. Selb, S. A. Carp, G. Boverman, E. L. Miller, D. H. Brooks, R. H. Moore, D. B. Kopans, and D. A. Boas, "Combined optical and X-ray tomosynthesis breast imaging," *Radiology* **258**, 89–97 (2011).
185. B. W. Pogue, S. Jiang, H. Dehghani, C. Kogel, S. Soho, S. Srinivasan, X. Song, T. D. Tosteson, S. P. Poplack, and K. D. Paulsen, "Characterization of hemoglobin, water, and NIR scattering in breast tissue: analysis of intersubject variability and menstrual cycle changes," *J. Biomed. Opt.* **9**, 541–552 (2004).

186. H. Dehghani, B. W. Pogue, S. P. Poplack, and K. D. Paulsen, "Multiwavelength three-dimensional near-infrared tomography of the breast: initial simulation, phantom, and clinical results," *Appl. Opt.* **42**, 135–145 (2003).
187. S. Srinivasan, B. W. Pogue, S. Jiang, H. Dehghani, C. Kogel, S. Soho, J. J. Gibson, T. D. Tosteson, S. P. Poplack, and K. D. Paulsen, "Interpreting hemoglobin and water concentration, oxygen saturation, and scattering measured in vivo by near-infrared breast tomography," *Proc. Natl. Acad. Sci. USA* **100**, 12349–12354 (2003).
188. Y. Xu, N. Iftimia, H. Jiang, L. L. Key, and M. B. Bolster, "Three-dimensional diffuse optical tomography of bones and joints," *J. Biomed. Opt.* **7**, 88–92 (2002).

Sergio Fantini is a professor of biomedical engineering at Tufts University. He contributed to the early development of frequency-domain spectroscopy techniques for quantitative optical measurements on biological tissues. His research interests are in the area of diffuse optical spectroscopy and imaging of tissue, with applications to optical mammography, tissue oximetry, functional brain imaging, and the quantitative assessment of cerebral hemodynamics. He is an inventor on 11 patents, a coauthor in about 200 scientific publications, and has cowritten (with Prof. Irving Bigio, Boston University) a textbook on quantitative biomedical optics.

Angelo Sassaroli is a research assistant professor at Tufts University. He has been actively involved in research on diffuse optics since his graduate studies at the University of Florence, Italy, and the University of Electro-Communications, Tokyo, Japan, where he received his PhD in physics in 2002. His research activities have focused on theoretical, computational, and data analysis aspects of continuous-wave and time-resolved diffuse optics, including *in vivo* applications of near-infrared spectroscopy with potential clinical impact. He has authored or coauthored more than 60 peer-reviewed publications.

Chapter 8
Monitoring of Brain Activity with Near-Infrared Spectroscopy

Hui Gong, Qingming Luo, and Shaoqun Zeng
Britton Chance Center for Biomedical Photonics, Wuhan National Laboratory for Optoelectronics, Huazhong University of Science and Technology, Wuhan, China

Shoko Nioka, Yasufumi Kuroda, and Britton Chance
University of Pennsylvania, Philadelphia, PA, USA

8.1 Introduction

Biological tissue is relatively transparent to light in the near-infrared range between 700 and 1000 nm. This is due to the fact that water absorption and hemoglobin absorption are relatively small within this wavelength region. Several kinds of methods have been used to investigate human brain activity such as positron emission tomography (PET) and functional magnetic resonance imaging (fMRI). Recent research has shown that brain activity is associated with changes in the optical properties of brain tissue.[1] The activity of nerve and brain cells has been reported to be associated with changes in the optical properties of the nervous system,[2] cell cultures,[3] and bloodless brain slices,[4] as well as in intact cortical tissues.[5–7] Optical signals have been used to map brain function after surgical exposure of cortical tissue in animal[5] and human subjects.[8] Chance et al. measured the brain responses to functional activation with near infrared spectroscopy in 1993.[9] Tamura et al. analyzed the spatio-temporal changes in the oxygenation states through simultaneous near-infrared spectroscopy (NIRS) measured at five typical locations on the brain.[10] Maki et al. mapped the NIR topography on the human brain's motor cortex by measuring the changes in oxygenation caused by motor activity induced by finger stimulation.[11]

Brain activity is accompanied by certain physiological events that can be subdivided into cellular physiological events and intravascular events. Cellular physiological events occur intracellularly or at the cell membrane, whereas intravascular events are mediated by neurovascular coupling and occur within the vascular space. The activity of neurons is characterized by ion and water fluxes across the neurons' membranes, which induce changes in the membrane potential as well as changes in the electrical and magnetic fields. The main ions involved are Na^+, K^+, Cl^-, and Ca^{2+}, with the ion shifts inducing changes in their intracellular and extracellular concentrations. Furthermore, second messenger systems, such as cAMP (Ca^{2+} itself also serves as a second messenger), are activated by such ion shifts. Research has also showed that brain cells' activity is accompanied by glucose consumption,[12] oxygen consumption,[13] and oxidation of intracellular NADH, flavoproteins, and cytochrome *c* oxidase (CO).[14] In addition to the cellular events, local brain activity induces an increase in local cerebral blood volume (CBV) and blood flow (CBF), termed neurovascular coupling.[15]

Functional optical imaging measures the interactions of photons with brain tissue. As mentioned earlier, brain activity is accompanied by physiological events. Therefore, to understand how functional optical imaging maps the brain activity, we must clarify the relationship between optical and physiological parameters. Villringer and Chance classified optical measurements of brain activity as either extrinsic (using exogenous contrast agents) or intrinsic (without exogenous contrast agents).[1] For example, the brain activity associated with cerebral blood volume, cerebral blood oxygenation, and intracellular energy status influences the values of certain intrinsic parameters—for example, the concentration of total hemoglobin, deoxyhemoglobin and oxyhemoglobin, and cytochrome *c* oxidase—that in turn can be measured through their influence on optical parameters—for example, light absorption. Light scattering occurs at the borders of media with different refractive indices. Activity-related light scattering changes have been measured in isolated axons, neuronal cell cultures, brain slices, and the neurohypophysis *in vitro*,[16] as well as in intact animals.[17] Recently, even noninvasive measurements in human adults have been reported.[18] Two types of light scattering signals associated with brain cell activity have been reported: (1) A fast signal, which may be related to changes in the index of refraction of neuronal membranes, has a response time on the order of probably less than a millisecond; and (2) a slower scattering signal, which may be related to high [K^+]-associated glial swelling, has a response time on the order of a few seconds. These signals have been detected in hippocampal brain slices as well as in intact animals. Fluorescence techniques may reveal NADH oxidation and flavoprotein oxidation. As an intracellular marker of brain activity, the CO oxidation–reduction (redox) state can be measured with visible light and near-infrared light since CO has a characteristic absorption pattern that is influenced by its redox state.

Functional cortical imaging is a fundamental technique in the monitoring of brain activity and detection of abnormalities. Several techniques, such as electroencephalography (EEG), fMRI, and PET, can resolve physiological function in the cerebral cortex.

The methods of near-infrared spectroscopy and imaging are still in an early phase of application in neuroscience. Compared to other functional neuroimaging methods, such as PET and fMRI, NIRS lacks, in particular, spatial resolution and depth penetration, which limits its use in most current studies to the cortical gray matter. Nevertheless, the NIR methods do have some unique properties that may offer advantages over other methods. In contrast to other techniques (in particular, fMRI), NIRS offers enhanced biochemical specificity through its measurement of concentrations of biochemically well-defined substances such as oxy-Hb, deoxy-Hb, and CO redox state. Whereas other functional neuroimaging methods, such as fMRI, PET, EEG, and magnetoencephalography (MEG), measure either correlates of neuronal activity (EEG, MEG) or the vascular response to it, optical methods, in principle, can assess not only aspects of the vascular response, such as oxy-Hb and deoxy-Hb, but also intracellular events, such as CO redox state and light scattering. Optical methods have intrinsically good temporal resolution. Even in transcranial examinations in human subjects, a temporal resolution in the millisecond range seems feasible. Since optical systems are usually flexible and portable, bedside examinations are easily feasible, enabling patients to be examined repetitively or monitored continuously for extensive time periods. Finally, the cost of even highly sophisticated time-resolved NIR imaging systems of the future will be far less than the costs of other functional neuroimaging devices.

Based on the current status of the developments in optical imaging, and considering the relative advantages and disadvantages as compared to other functional neuroimaging methods, a number of applications are envisioned for optical methods. In basic neuroscience, optical methods represent a unique approach to study neurovascular coupling, since both ends of this process (neuronal and vascular) can be measured simultaneously. In cognitive neuroscience, optical methods may be useful in localizing brain activity, particularly in situations in which other methods are not applicable. For example, optical measurements could be performed in individuals who are walking or under other natural conditions that are not easily accommodated by other functional methods. Whether the detection of fast optical signals corresponding to neuronal activity will enable the detection of signals in cases where the electrical or magnetic dipoles cannot be picked up on the head's surface remains to be elucidated in future studies. For developmental studies, it should be possible to examine young children who would not lie quietly in an MRI or PET scanner. Interesting applications may also be found in clinical neuroscience. Children with attention deficits, who are also not easily studied with PET or fMRI, may represent a group of patients who are candidates for optical studies. Monitoring cerebral oxygenation over several hours in patients at the bedside is probably a prerequisite for the detection of peri-infarct depolarization (PID) in patients with stroke or spreading depression (SD) equivalents during migraine attacks. The ability to detect PID could eventually lead to an application for optical methods in clinical neurology, as the recent animal data suggest that pharmacologic inhibition of PID is an efficient approach for stroke treatment.[19] Another potential clinical application is the detection of hematoma.[20]

NIRS opens up a new avenue for the resolution of cortical activity noninvasively. Several NIRS studies in recent years have demonstrated that changes in brain activity can be assessed noninvasively in adult human subjects. NIRS has been used to assess brain activities such as motor activity, visual activation, auditory stimulation, and performance of cognitive tasks. Based on absorption and scattering measurements, four types of activity-related signals have been recorded noninvasively: (1) changes in hemoglobin oxygenation, (2) changes in blood volume, (3) changes in CO oxidation, and (4) fast optical signals presumably related to changes in light scattering.

The prefrontal lobe is important to human cognition, and it is related to many kinds of brain functions, such as learning and memory, planning, strategy, thinking, attention, emotion, and so on. Using functional imaging techniques, such as PET and fMRI, many researchers have studied the prefrontal lobe's role in forming episodic memories, including the encoding and retrieval stages of this process. The results indicate that the left prefrontal lobe makes greater contributions to semantic encoding, whereas the right prefrontal lobe has a stronger effect on the retrieval process. This hemispheric encoding/retrieval asymmetry (HERA) model was originally put forward by E. Tulving in 1992.[21]

The prefrontal cortex (PFC) has previously been identified by neuroimaging studies as a region likely to participate in memory encoding. The left PFC supports cognitive operations, including analysis of the meaning of items and comparison of items. Functional NIR imaging has the advantage of monitoring the PFC's responses during mental work and pattern recognition.

Both Wagner et al.[22] and Brewer et al.[23] reported fMRI results about building memories. However, this imaging technology cannot fully address more complicated questions, such as clarifying the nature of the relationship between the prefrontal and parahippocampal cortex during encoding. Meanwhile, this technique has not been widely used in psychological testing because of its high cost and its equipment setup, which can invoke claustrophobia attacks in some patients. Compared with fMRI, fNIRI is a good choice for this kind of study.

Three major experimental techniques exist in the field of NIR spectroscopy. The first is continuous-wave spectroscopy, which measures changes in transmitted intensity due to tissue absorption and scattering between the detector and source.[24] The second is the intensity modulation method, in which the light source (typically a current-modulated diode laser) is modulated at single or multiple radio frequencies.[25] The detection system in this case consists of a demodulator, which measures the detected intensity (*I*), phase shift (Φ), and modulation depth (*M*) relative to the input signal, as a function of the source–detector separation. The third is the time-resolved approach, which uses short pulses of light injected into the tissue. This detection process requires either a time-correlated single photon counting (TCSPC) system[26] or a synchroscan streak camera.[27] The advantages of TCSPC relative to streak camera systems include the former's larger detector area and a significantly better temporal linearity and dynamic range.

We developed a continuous-wave (CW) NIR imager that could localize brain activity by functional imaging.[28] The functional near-infrared imager (fNIRI) includes four light sources and 10 detectors. Each three-wavelength laser-emitting diode can emit light at 735 nm, 805 nm, and 850 nm. Using this fNIRI system, the motor function of the motor cortex area, the visual function of the occipital area, and the cognitive activity of the frontal forehead area of the human brain were measured when the subjects were stimulated by moving fingers, viewing a flashing light, and performing a cognitive test, respectively. The experimental results showed that fNIRI can be used for functional images of brain activity, based on changes in oxygenation and blood volume caused by the stimulus.[29–31] For measuring fast neuronal signals, we have developed a new CW-NIR system that benefits from a simultaneous sampling method with a separate high-speed ADC for each input channel.[32]

In this chapter, we largely focus on continuous-wave light fNIRI, moving from principles to instrumentation in Section 8.2. Optical imaging results of brain activity with the CW functional NIRS imager are described in Section 8.3.

8.1.1 Brain mapping by time-resolved and frequency-domain imaging systems

The time-resolved system (TRS) for diffuse tomography was first developed in the early 1990s for breast cancer detection.[33] The TRS has an advantage in quantifying the absorption and scattering coefficients, thereby ensuring accurate imaging of hemoglobin oyygenation.[34] TRS commercial instruments are available (e.g., ART, Hamamatsu) and, in theory, can be used for brain studies. TRS brain imaging systems have been created in a few institutes.[35–37] Challenges for TRS specifically in brain measurement applications have related to the requirement for faster temporal resolution than is needed in the systems used for breast cancer detection. Researchers have compensated for this requirement by using faster repetition times of 2–3 MHz to increase the signal-to-noise ratio of the time-of-flight measurement [often referred to as temporal point spread function (TPSF)].[38,39] The resulting systems measure time of flight and its variances. Moreover, because they are independent of light intensity, the systems are resistant to noises and changing coupling coefficients. Another benefit is that they assess selected photons from deeper tissue, especially when measuring variance in the time of flight, as these data come from considerably deeper (approximately 1.5 cm in depth in SD 3-cm separation) in the brain cortical area. Another option is to use ratios of photon counts in time windows.[40] With this approach, one can make direct use of the fact that, on average, later photons have penetrated deeper parts of the tissue.

Frequency-domain (FD) brain imaging systems have also been implemented, with many studies on their use having been published.[41] Major benefits of the FD systems are that, through phase and amplitude information, they can reduce the scattering coefficient data, and that quantitative data on absorption of hemoglobin are available similar to the case for the TRS. A commercially available (e.g., ISS Inc.) development of interest is that through scattering measurements, optical neuronal imaging in human is possible,[42] with the peak signal being only 20 to 60 ms from the stimulation.

8.1.2 The concepts of NIRS signals as a measure of neuronal activities

Since we measure hemoglobin in the blood in the brain, it is important to know and understand how neuronal activity is related to vascular responses. This section introduces a few issues related to neurovascular coupling, metabolism of neurons, and astrocytes.

8.1.2.1 Time delay between neuronal activation and blood signals

Time between neuronal depolarization and mitochondrial activation is 5 ms, according to measurement of membrane potentials and NADH oxidation.[43] The onset of blood oxygenation and increased blood flow signals and neuronal activation occurs within 1–4 s,[44] and the BOLD signal increase is delayed by 2–3 s, followed by a ramp of 6–12 s to a plateau or peak value. The signal returns to the baseline with a similar ramp.[45–47] Data from monkey studies report a time course of the hemodynamic response that is similar to a low-pass-filtered expression of the total neural activity.[48] Major differences in those two modalities—that is, neuronal signals and blood-related signals—are the peak time and the time of deactivation in the blood-related trends in a short neuronal activation, as neuronal deactivation occurs immediately after the activation. The mean onset of the BOLD signal is 2.16 s after neural activation, reaching a plateau approximately 7 s later.[48] In rats, deoxy-Hb has been shown to peak at 3.25–3.75 s, whereas the profiles of oxy-Hb started at 1.25 s and peaked at 3.0–3.75 s.[49] However, when neuronal stimulation is longer, there is no difference in the trends between neuronal signals and blood signals.[44] Instant and local vascular responses to the neuronal action are due to nitrous oxide release, local chemical release (adenosine), and calcium sensing of astrocyte activity.[50]

8.1.2.2 Problems surrounding astrocytes and high oxygen tension during activation

Astrocytes surround the capillaries forming the blood–brain barrier, and play key roles in clearing the neurotransmitters and ions that are produced by neuronal activation. Astrocytes are also activated by the neuronal activation that occurs with high glucose consumption, with the end product of lactate being shuttled to neurons to be used as their fuel. The neurons use primarily oxidative ATP synthesis, whereas astrocytes rely on aerobic glycolysis to produce the lactate is sent to neurons.[51] Neurons need oxygen, but the longer diffusion distance may require higher oxygen pressure in the capillaries to meet this demand. In contrast, there is no competition for oxygen when neurons are activated. Several explanations for the high oxygenation have been proposed, including the longer diffusion path of oxygen to reach the neurons[46] and imprecision of the vascular response (watering the garden for the sake of one flower).[52]

8.1.2.3 What is the smallest activation unit in the brain?

As yet, not enough information is available to answer the question of what the smallest activation unit is. No transferable studies exist that might link cellular and tissue culture studies of neurons and astrocytes and brain functional studies *in vivo*. However, the facts that MRI and NIRS show quite large areas of vascular responses and that the waveform of the vascular hemodynamics in a single vascular site is very similar to averaged waveforms in a millimeter or more cortical area[53] suggest there are large areas of neuronal and astrocytic activation in at least one brain function. More studies are needed to clarify to verify the findings from the noninvasive MRI and NIRS technologies. Several models and observations provide more detailed analysis for addressing the pertinent questions.[54–56]

8.1.2.4 Problems of adapting NIR to brain functional imaging

Diffuse optical tomography—indeed, photon migration tomography in general—has two shortcomings that are largely attributable to the scattering property of the tissues (approximately 10 per centimeter), even though NIR imaging has opened up new opportunities for mapping the human brain. The first problem is the depth penetration; it is 2 cm, which allows the researcher to look only at the cortical layer. The second challenge relates to the spatial resolution, which is also determined by the scattering function; it is on the order of centimeters in the brain. Another problem is that the scattering coefficient changes with blood flow, such that alterations in blood volume hinder the quantitative reduction of oxygenation data.[57] Further, diffuse optical technology is very sensitive to the surface signals when CW amplitude is used, such that the optode interface coupling to human subjects produces a considerable amount of noise. At this point, the TRS seems to solve this problem by enabling researchers to choose a deeper light path.[35] Moreover, many publications have confirmed aspects of the validity and feasibility of the NIR techniques, such as their reproducibility.[58]

8.2 Continuous-Wave Functional Near-Infrared Imaging

8.2.1 Photon migration

8.2.1.1 Light absorption changes with oxygenation and blood volume and algorithm for the instrumentation

Photon migration through biological and other turbid (multiple-scattering) media can be described by the well-known diffusion theory. In this approach, the photon fluence density function$\varphi(\vec{r},t)$, which has units of number of photons per unit area per unit time, satisfies the equation

$$\frac{1}{c}\frac{\partial\phi(\vec{r},t)}{\partial t} - D\nabla\phi(\vec{r},t) + \mu_a\phi(\vec{r},t) = S(\vec{r},t), \tag{8.1}$$

where μ_a is the absorption coefficient measured in units of inverse length and $D = 1/3[\mu_a + (1 - g)\mu_s]$ is the diffusion coefficient measured in units of length. Here, $g = \langle\cos\theta\rangle$ (θ is the scattering angle), and μ_s is the scattering coefficient, which is also measured in units of inverse length; $\mu_s' = (1 - g)\mu_s$ is the transport or reduced scattering coefficient, which describes isotropic diffusion processes at large distances. $c = c_0/n$ is the speed of light inside the diffusive (multiple scattering) medium with a refractive index n. $S(\vec{r},t)$ is the source function for the diffusive light.

For semi-infinite geometry, which is the most relevant case for noninvasive medical applications, we may take

$$S(\vec{r},t) = S_0[\delta(\vec{r} - z_0\hat{z}) - \delta(\vec{r} + z_0\hat{z})], \tag{8.2}$$

for the source, as shown in Fig. 8.1(a). When the light source is CW, S_0 is simply a constant. Feng et al. gave the analytical derivation of the photon-path distribution function for the single-source and single-detector configuration.[58] For semi-infinite geometry, the desired photon-path distribution function at position $\vec{r}$ [for photons injected at $\vec{r}_0 = 0$ and detected at $\vec{r}_d = (d,0,0)$] can be written as

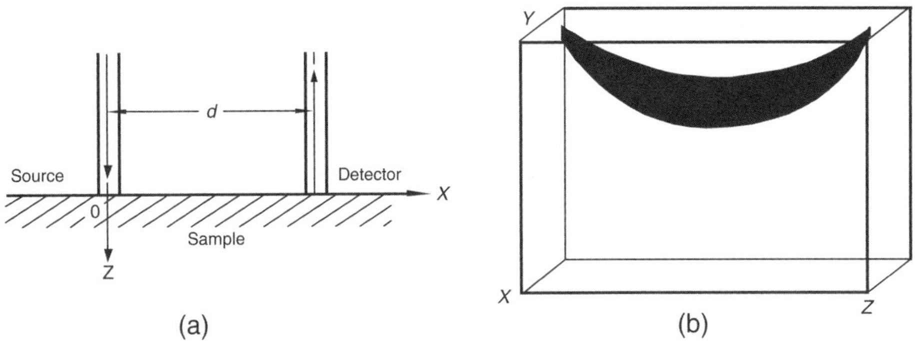

Figure 8.1 (a) Semi-infinite geometry and (b) banana-shaped region of photon-path distribution.[29]

$$P(x,y,z) = \frac{z^2 \exp(-k\{(x^2+y^2+z^2)^{1/2} + [(d-x)^2+y^2+z^2]^{1/2}\})}{(x^2+y^2+z^2)^{3/2}[(d-x)^2+y^2+z^2]^{3/2}} \times [k(x^2+y^2+z^2)^{1/2}+1]\{k[(d-x)^2+y^2+z^2]^{1/2}+1\}, \quad (8.3)$$

where k is defined as $k = 1/L_a = (\mu_a/D)^{1/2} = (3\mu_a\mu_s')^{1/2}$. In the multiple-scattering regime, L_a determines the range of diffusive light migration. (In a transparent medium, μ_a^{-1} determines the decay of light.) From Eq. (8.3), the peak position of the photon-path distribution function P (x, y, z) within a cross section in the y-z plane for a fixed $0 < x < d$ can be determined. For symmetry, it has $y = 0$. If we take its z coordinate as $z_0(x)$, this function represents the modal line of the banana region. In the weak absorption limit, $kd \ll 1$, $z_0(x)$ is given by

$$z_0(x) \approx [(1/8)(\{[x^2+(d-x)^2]^2 + 32x^2(d-x)^2\}^{1/2} - x^2 - (d-x)^2)]^{1/2} \quad (8.4)$$

At $x = d/2$, the modal line of the banana region reaches a maximum depth:

$$z_0^{\max} \approx (\sqrt{2}/4)d \quad (8.5)$$

As in the infinite geometry case, we can define the boundary of the banana-shaped region in which detected photon paths are concentrated by

$$r(x,y_b,z_b) = P(x,y_b,z_b)/P(x,0,z_0) = B, \quad (8.6)$$

where $B < 1$ is a previously described constant and $(y_b,\ z_b)$ gives the boundaries of the banana region as a function of x $(0 < x < d)$. Figure 8.1(b) shows a boundary of the banana-shaped region with some specific μ_a, μ_s, and g.

Delpy et al. indicated that the mean optical path length of photon migration was approximately 5 times longer than the distance between light input and output couplers.[60] Monte Carlo simulations, which also have a high probability of transit from input to output, show only those photons described by the banana shape.

8.2.1.2 A Monte Carlo model in multi-voxeled tissues

A Monte Carlo model of steady-state and time-resolved photon migration in multi-voxeled tissues (MCMV) has been designed for three-dimensional (3D) tissues with arbitrary boundaries or complex structures. Comparisons of MCMV with classical MCML (steady-state Monte Carlo simulation for multilayered tissue) and diffusion approximation (DA) demonstrate that MCMV provides an alternative to MCML and DA for the existing simulations. Furthermore, MCMV can serve as a resource for characterizing photon migration in real biological tissues with subtle anatomical structure described by voxel data set.

The implementation of MCMV maintains the style of MCML, and also preserves some functions such as the presentation of the remaining running time and the allocation of variables dynamically. To trace the photon's path in the multi-voxeled tissue, an optimized and concise frame is created, voxel-related variables are added, continuous absorption weighting is employed, photon reaction on the voxel-based three-dimensional boundary is modeled, and the search method for hitting a point on the voxeled boundary is established in MCMV. A time-spatial resolved method to record absorption and flux for multi-voxeled tissue is also established in MCMV to perform both steady-state and time-resolved simulations. Additionally, the photon injection is modeled at an arbitrary angle and position to expand the scope of application.

A five-layer brain model may be chosen to test the steady-state MCMV. Figure 8.2(a) shows the fluence field with contours drawn every 30 dB; Fig. 8.2(b) shows the depth-resolved absorption determined by MCMV, MCML, and tMCimg. The medium in this simulation is a semi-infinite brain model with skin, skull, CSF, gray matter, and white matter. The optical properties of these tissues[61–64] are defined in Table 8.1. In this model, the

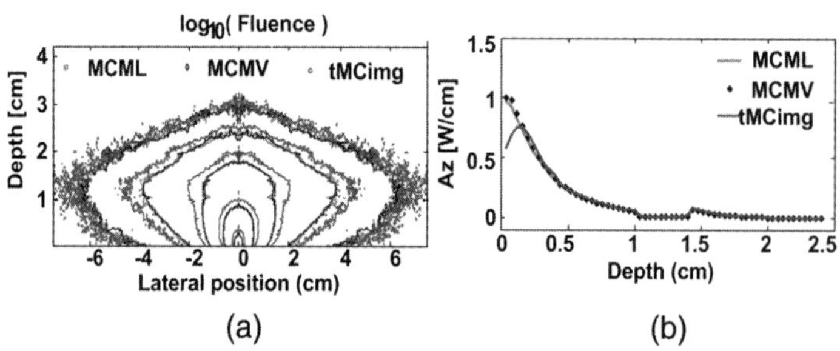

Figure 8.2 Comparison of MCMV, MCML and tMCimg for (a) the photon fluence field and (b) the depth-resolved absorption within a semi-infinite five-layer brain medium.

Table 8.1 Optical properties.

Tissue Type	μ_a (cm^{-1})	μ_s (cm^{-1})	g	n
Skin	0.18	190	0.9	1.37
Skull	0.16	160	0.9	1.43
Muscle	1.40	500	0.9	1.4
CSF	0.04	28	0.9	1.33
Artery	2.33	500	0.99	1.4
Vein	2.38	522	0.99	1.4
Gray matter	0.36	220	0.9	1.37
White matter	0.14	910	0.9	1.37
Air	0.003	0.8	1.0	1.00

tissue thicknesses are 0.4, 0.6, 0.4, 0.52, and 2.08 cm, respectively.[62,65] The radius and thickness of each grid in the MCML simulation are both 0.04 cm. The size of each voxel in the MCMV and tMCimg simulations is $0.04 \times 0.04 \times 0.04$ cm^3.

Figure 8.2(a) shows that the result of MCMV matches the result of MCML over four orders of magnitude, but there is a significant distance between the curves from tMCimg and MCML, especially for the middle three contours. The fluence field calculated by tMCimg is spread approximately 0.5 cm away from MCML in the same scale. Thus, significant errors could arise in tissue tomography predictions made by tMCimg simulation. The depth-resolved absorption A_d calculated by tMCimg in Fig. 8.2(b) is notably smaller than that found with MCML when the depth is smaller than 0.5 cm. Conversely, MCMV agrees well with MCML in terms of A_d. Additionally, Fig. 8.2(a) shows that the variations in the fluence field found with MCMV and MCML are approximately the same quantity, whereas tMCimg shows a larger variation, especially in the region far from the light source. The oddly and nearly lowermost A_d within the depth range of (1, 1.4) cm in Fig. 8.2(b) is caused by the low scattering and low absorption region for CSF, which is rightly located in the same depth range.

Ten computations consisting of 10^5 photons each were used to calculate the averages and standard errors of R_d and A as described in Table 8.2. The means of R_d and A calculated by MCMV agree with those found with MCML in three orders of magnitude, while those calculated by tMCimg agree with MCML in only two orders of magnitude. The standard errors from MCMV and MCML are approximately the same quantity.

Table 8.2 Comparison of the results from MCMV and MCML for a five-layer brain tissue model.

Source	R_d	A
MCMV	0.62066 ± 0.00033	0.35496 ± 0.00033
MCML	0.62105 ± 0.00029	0.35454 ± 0.00029
tMCimg	0.63901 ± 0.00038	0.33659 ± 0.00038

Values are mean ± standard error.

In this model, light transport in the prefrontal lobe of the Digital Chinese Human is simulated with MCMV. The optical properties of each involved tissue type are shown in Table 8.1. One of the simulation results is given on the website for our lab, which quantitatively supports the experimental and theoretical investigation based on simplified brain tissue models.[66,67]

8.2.2 Instrumentation and performance

8.2.2.1 Light absorption changes with oxygenation and blood volume and algorithm for the instrumentation

In spectral range 700–900 nm, which is also known as the optical window, scattering—rather than absorption—dominates photon transport in the human brain, and water shows a much lower absorption in this spectral range than in others. The fNIRI device was designed based on the optical window in which oxygenated hemoglobin (HbO_2) and deoxygenated hemoglobin (Hb) have significantly different absorption coefficients at 735 nm and 850 nm. Utilizing these two wavelengths, we can determine the relative concentrations of the two states of hemoglobin, oxygenated and deoxygenated.

As mentioned in Section 8.2.1, the true path length of photon migration in tissues from source to detector is much longer than the geometrical distance. A modified form of the Beer–Lambert law is used to describe the optical attenuation:

$$I/I_0 = e^{-\varepsilon \cdot c \cdot d \cdot DPF - G}, \tag{8.7}$$

where I_0 represents the incident light intensity, I is the detected light intensity, ε is the absorption coefficient of the absorbing medium, d is the distance between the points where light enters and leaves the tissue, c is the concentration of the medium, DPF is the differential path length factor, and G is a constant attenuation factor related to the optical properties and geometry of the tissue.

We may assume d, DPF, and G to be constant if we calculate only changes in the absorbing medium concentration

$$\Delta c = \frac{\Delta OD}{\varepsilon \cdot d \cdot DPF}, \tag{8.8}$$

where

$$OD = \log(I_0/I). \tag{8.9}$$

The imager measures the optical density (OD) change

$$\Delta OD = \log(I_0/I_{test}) - \log(I_0/I_{rest}) = \log(I_{rest}) - \log(I_{test}), \quad (8.10)$$

where I_{rest} and I_{test} represent the light intensity detected by the detector during rest and test, respectively.

All outputs from the fNIRI are converted to optical density. In this way, all data are referenced directly to the modified Beer–Lambert law. Based on the OD changes at wavelengths 735 nm and 850 nm, we can get either the absorption images in the two measuring wavelengths or the functional images (oxygenation or blood volume) in the detection region of study.

Based on the fact that hemoglobin is the main chromophore at the specified wavelengths, the following equations can be seen as the fundamental equations:

$$\Delta OD_{850} = k\Delta[\mathrm{HbO_2}] + k'\Delta[\mathrm{Hb}] \quad \Delta OD_{735} = t\Delta[\mathrm{HbO_2}] + t'\Delta[\mathrm{Hb}] \quad (8.11)$$

where ΔOD_{850} and ΔOD_{850} indicate optical density changes at 850 nm and 735 nm, respectively; and $\Delta[HbO_2]$ and $\Delta[Hb]$ denote the changes in the concentrations of HbO_2 and Hb, respectively. The following equations were deduced from earlier expressions:

$$\begin{aligned} \Delta[\mathrm{HbO_2}] &= R[(\Delta OD_{850} - (k'/t')\Delta OD_{735}] \\ \Delta[\mathrm{Hb}] &= R(t/t')[(k/t)\Delta OD_{735} - \Delta OD_{850}] \end{aligned} \quad (8.12)$$

If R is assumed to be equal to $\frac{1/t}{k/t-k'/t'}$, this value is considered to be 1 in this case.

The ratios k/t, k'/t' and t/t' can be determined by calculating the ratios of the optical density changes in a fully oxygenated state (oxy) and a fully deoxygenated state (deoxy):

$$\begin{aligned} &k/t = \Delta OD_{850oxy}/\Delta OD_{735oxy}, \quad k'/t' = \Delta OD_{850deoxy}/\Delta OD_{735deoxy}, \\ &t/t' = \Delta OD_{850oxy}/\Delta OD_{735deoxy}. \end{aligned} \quad (8.13)$$

From the results obtained from the test of the blood model, the optical density changes associated with three wavelengths in the fully oxygenated and deoxygenated states were plotted against the concentration of red blood cells (RBC). The experimental curve illustrated a linear relationship between ΔOD and RBC concentration. Based on the result, the ratios k/t, k'/t' and t/t' can be obtained by calculating the slope of these curves. We ultimately adopted 1.74, 0.56, and 0.45, respectively, for these values in our algorithm. Therefore, the changes of HbO_2 and Hb concentrations can be obtained by the following equations:

$$\Delta[\text{HbO}_2] = \Delta OD_{850} - 0.56\Delta OD_{735} \quad \Delta[\text{Hb}] = 0.783\Delta OD_{735} - 0.45\Delta OD_{850}. \tag{8.14}$$

A summation calculation can be performed to determine the total change in hemoglobin concentration (corresponding to the blood volume) within the detection area:

$$\Delta BV = \Delta[\text{HbO}_2] + \Delta[\text{Hb}].$$

8.2.2.2 Imager description

Our device includes 4 three-wavelength laser-emitting diodes (LEDs) and 10 preamplifier photodiodes used for the imager probe. Each three-wavelength LED can emit near-infrared light at 735 nm, 805 nm, and 850 nm. Each preamplifier photodiodes is used as a detector and has a 2.29 mm $\times$ 2.29 mm detection area. The imager is depicted in Fig. 8.3. The LEDs are controlled by a computer and turned on and off sequentially. After amplification, the detector outputs are sampled by an A/D board and recorded by the computer in accordance with the detection sections shown in Fig. 8.3. There are 16 sections, and the total detection area is 4.4 cm $\times$ 15 cm. In each detection section, the distance between the LED and the detector is 2.88 cm, which is sufficient to pick up the photons from 1.33-cm-deep tissue, as demonstrated by the penetration test.

The imager housing is made of silicone rubber, which means it is flexible and can measure either brain or muscle activities. The LEDs and detectors are blocked by the rubber, so no light is transmitted along the surface of the tissue. During the measurement, the imager probe must be fixed and shielded perfectly.

A monolithic driver (TB62705AN from TOSHIBA) supplies a highly stable, constant DC current for the LEDs. The output current value can be set

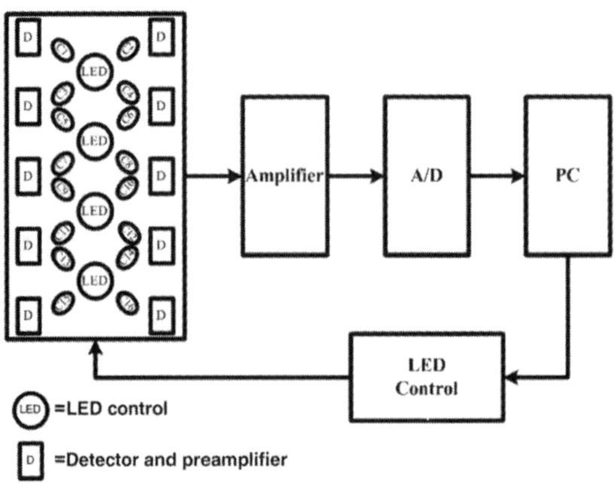

Figure 8.3 Diagram of the imager and detection sections (1,2, ..., 16).[29]

using an external resistor. As a result, all 10 outputs have almost the same current levels, and the maximum current capacity is 90 mA. The signals detected by all of the 10 preamplifier photodiodes are fed into post-amplifiers together with low-pass filters, which further magnify and purify these signals. The data acquisition board (DAQ) in the instrument hosts the electronic systems that turn the light sources on and collect the back-reflected light from the probe. The analog inputs of DAQ are converted to 12-bit digital signals that are transmitted to a notebook PC through a USB port.

The system has an adjustable sampling rate, resulting in temporal resolutions ranging from 100 ms to 1000 ms. Users can adjust the resolution from the software control panel.

8.2.2.3 Measuring and imaging algorithms

The measuring and imaging algorithms consist of three parts. The first part deals with control of the imager and data acquisition. Raw data are acquired from the detector after amplification. A calibration of the instrument is performed before data acquisition on each test, so as to eliminate the error caused by the offset. The calibration program measures the offset, and then stores the offset data for subtractive cancellation.

The second part of the software comprises the data processing program. Interpolation, FFT, and polynomial algorithms are used to multiply the data from 4×4 points to 64×64 points. This step prepares the data for intensity or functional images. If the goal is to produce an intensity image, this program processes the data and outputs only the OD results related to light intensity of each wavelength. For a functional image, the program calculates the OD_{oxy} and OD_{bv} values according to Eqs. (8.11) and (8.12).

The third part is the display program. It is combined with the data processing program and has a user-friendly interface. The user may view data and images in real time. A chart recorder is emulated, and the data are dynamically drawn along with the successive measurements. Four kinds of images (735 nm, 850 nm, oxygenation, and blood concentration) can be shown on the monitor sequentially. These images are scaled with the specific density (256 scales). The user is able to change the scale, choose the detection period, and/or average the data from the same or different data files.

8.2.2.4 Noise and drift test

To evaluate the system performance of the imager, we measured the dark current from the detectors and the amplifiers. In this test, we displayed the signal output from the output of the amplifier when all the light sources were turned off. The black curve in Fig. 8.4 shows the result from one of the 16 channels. The noise level from the detector and amplifiers was approximately 0.2 mV (RMS), which is compatible with A/D resolution. We also found that the offset caused by the dark current was much bigger than the noise level and

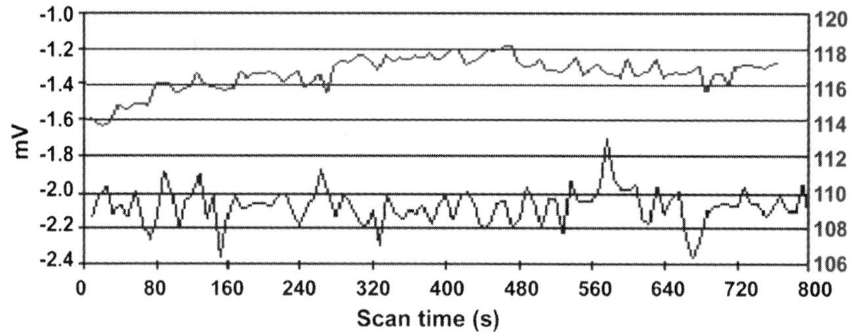

Figure 8.4 Noise output after the detector preamplifier with no incident light on the detector (black curve) and during a stationary test when the imager probe was put on a brain model (blue curve) ($\mu_a = 0.14\ cm^{-1}$, $\mu_s' = 8.5\ cm^{-1}$). The optical density change due to the drift and noise was less than 0.007.[29]

was different for different channels. However, the range of the variation was 2.4 mV.

The blue curve in Fig. 8.4 shows the noise and drift levels during a stationary test. In this test, the imager was put on a solid brain model and fixed in place by tape. Measurements were taken over periods as long as 800 seconds. Based on the curve, the absolute noise level was approximately 2 mV, and the drift level was smaller than 4 mV. Compared with the results expressed by the black curve, it appears that the main noise source in this system was the light sources—the LEDs. The *OD* change of the noise and drift in a stationary test was less than 0.007, which is within one scale level of the image.

In summary, the resolution of our data acquisition system is good enough for the present detection system. It is can be changed with the software, from 0.305 mV to 2.44 mV. The corresponding dynamic range of detectable signal levels is 0.305 mV to 5.0 mV.

8.2.2.5 High signal-to-noise ratio CW-NIRS system

We can only infer neuronal activity from hemodynamic changes that occur consequent to neuronal activity indirectly, via neurovascular coupling by fMRI or traditional CW-NIRS. Previous *in vitro* and animal studies have shown that neuronal activity (such as the depolarization or hyperpolarization of neuronal cells) can directly cause variations in optical scattering—namely, the fast neuronal signal. Therefore, the fast neuronal signal may reflect the functional activity of nerves directly, instead of requiring that we measure the neuronal activity with hemodynamic parameters indirectly.

The fast neuronal signal exhibits an intensity change of less than 0.1%. The light intensity loses 7–9 orders of magnitude when the photons penetrate the tissue with an inter-optode distance of 3–4 cm—a distance that is commonly chosen to measure the response from the brain cortex through an intact head. Therefore, if the power of the light source is on the order of 10 mW (i.e., within safe exposure limits), the amplitude of a fast neuronal signal should be an order lower than picowatts. To detect the fast neuronal signal, the detection sensitivity of the CW-NIRI system should be at least on the order of 0.1 pW. The fast neuronal signal usually shows a latency of approximately 50–150 ms. Therefore, the temporal resolution of the CW-NIRI system should be more than 20 Hz.

We have developed a CW-NIRS imager for detecting fast neuronal signals *in vivo* and noninvasively.[32] To do so, we adopted a frequency-multiplexed scheme for the light sources and used a simultaneous sampling method with a separate high-speed ADC for each input channel to achieve high temporal resolution. We implemented system control and signal processing using digital technology as much as possible to reduce noise due to analog circuits, making the system less complex and offering better flexibility. This system, which is depicted in Fig. 8.5, can achieve a high detection sensitivity through digital lock-in detection with large point samples.

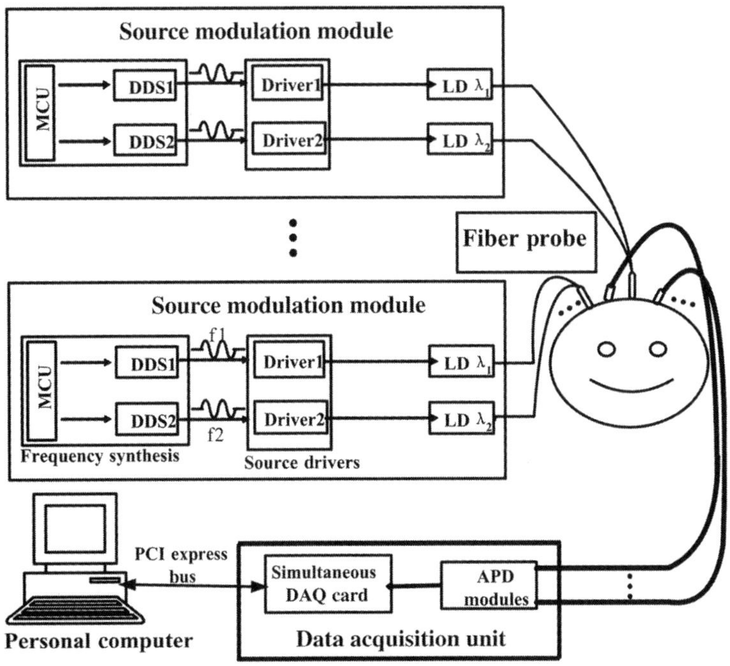

Figure 8.5 Block diagram of the high signal-to-noise ratio CW-NIRS system. MCU: microcontrol unit; DDS: direct digital synthesis; f1 and f2: modulation frequencies; LD: laser diode; APD: avalanche photodiode; DAQ: data acquisition.[32]

8.3 Monitoring of Human Brain Activity with CW Functional Optical Imager

The history of NIR human brain imaging follows Ogawa's 1993 discovery of the BOLD signal by using MRI. The application of this imaging technology to brain functional activity prompted the optical communities to do the same. NIR technologies, for their part, were first used to study hemoglobin concentrations and oxygen saturation in human tissues more than 20 years before the MRI application was developed. Thus, it is fair to say that the two modalities, fMRI and fNIR, were first applied to human functional brain mapping at almost the same time. However, the development of fNIR has lagged behind that of fMRI, as evidenced by the predominance of publications focused on fMRI up to 2008 (20 times fewer publications were devoted to fNIR). The fNIR literature has been not grown much over the last 10 years. Several factors related to the technology account for this relatively low level of interest. Because it cannot penetrate to depths of more than 2 cm, NIR imaging is limited to only brain cortical studies, especially those focusing on the control and executive functional areas of the brain. In addition, the spatial resolution of NIR imaging is not better than the centimeter order. We still have more theoretical verifications and feasibility studies of this technology than studies of human brain mapping utilizing fNIR. Nevertheless, many studies addressing the potential of NIR human brain imaging have been published. This section highlights some examples of published and unpublished data based on our CW system.

8.3.1 Motor cortex in finger tapping

With the fNIR imager described in Fig. 8.3, absorption changes in the contralateral primary motor cortex area due to motor stimulation with voluntary finger tapping were monitored. The probe has nine sources and four detector pairs placed as shown in Fig. 8.6(a). With this configuration, absorption in 16 subareas were measured and calculated as shown in Fig. 8.6(a) with dotted lines. Fig. 8.6(b) shows an axial slice obtained from a layer 2 cm under the surface where the optical probe was placed. To coregister with the image from a functional magnetic resonance imaging (fMRI) system, filled tubes with water and 1 mM copper sulfate were placed at the corners of the optical probe to identify the optical position on the parietal region in fMRI. With a 4-T MRI, the fMRI image of the parietal region activated by stroking fingers was obtained. The image required four 30-s on/off intervals. One of these markers can still be seen near the lower-right corner of the rectangle marking the position of the optical probe on the fMRI image.

A separate finger-stroking study with the CW optical functional imager pad fitted into the copper sulfate–defined rectangle was recorded and imaged for a similar interval. The resulting image is shown as in Fig. 8.6(c).

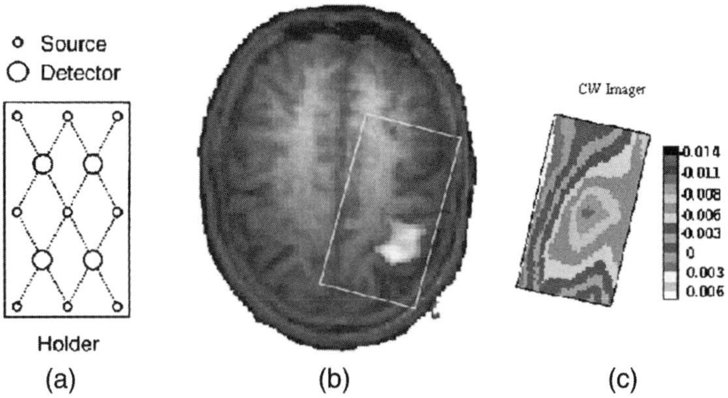

Figure 8.6 Application of a cw functional optical imager to monitoring motor cortex activity. (a) Probe configuration. (b) fMRI BOLD image of motor cortex activation during finger tapping. (c) fNIRI image of the relative change in the blood concentration during the same activation protocol and on the same volunteer as in (b). Both images show blood concentration increases.[30]

Compared with Fig. 8.6(b), the fMRI image is projected at the surface of the brain and can be compared to the displaced optical image in Fig. 8.6(c). Figures 8.6(b) and (c) show that there is rather good congruence of the two images, although fMRI measures larger vessels and deoxyhemoglobin, and the optical method measures total oxyhemoglobin plus deoxyhemoglobin in the capillary bed. The difference in the shape of the activation is explained by the NIR low-resolution mapping.

Since the CW functional optical imager was developed in 1996,[28] a multichannel phased array imager[68] and a time-resolved imager[69] have been developed and also used to monitor brain activity. Figure 8.7 shows optical

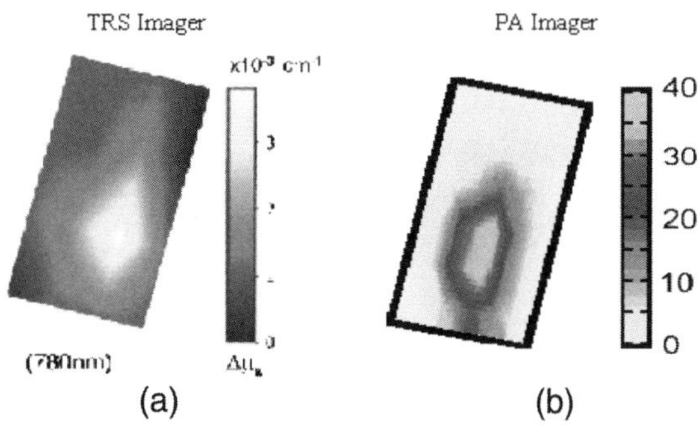

Figure 8.7 Optical imaging of motor cortex activity with NIR (a) time-resolved optical imager[70] and (b) phased array imager.[69] Both of the images are absorption changes maps obtained using the same activation protocol, on the same volunteer as in the fMRI image shown in Fig. 8.6(b). A total of 4×4 spatially resolved spectroscopic measurements were considered, which were indicated in Fig. 8.6(a) with dotted lines. The maximum absorption change coregisters well with the fMRI image.

images of motor cortex activity measured by the TRS imager and the PA imager. Both images are absorption changes maps obtained through the same activation protocol as with the CW imager described previously, on the same volunteer as in the fMRI image shown in Fig. 8.6(b). A total of 4 × 4 spatially resolved spectroscopic measurements were considered, which are indicated in Fig. 8.6(a) with dotted lines. Both images show that the maximum absorption change coregisters well with the fMRI image.

Time-resolved spectroscopy imaging has the advantage of being able to quantify absorption changes. Under the assumption of an invariable scattering background, the TRS system has been shown to attain sensitivity on the order of 10^{-4} cm^{-1} and accuracy on the order of 10^{-3} cm^{-1}. The limitations of this system are mainly due to laser instabilities.

A phased array imager or intensity-modulated system can affect a higher resolution by increasing the number of sources and detectors and by employing the amplitude and phase cancellation method. These changes permit a large increase in sensitivity and resolution. The capability of multi-wavelength operation also gives enhanced sensitivity for oxyhemoglobin and deoxyhemoglobin concentrations and for light scattering changes.

8.3.2 *n*-back test

The definition of working memory has evolved from the concept of a unitary short-term memory system. Compared with short-term memory, working memory provides temporary storage and manipulation of the information necessary for complex cognitive tasks. The *n*-back test is one kind of experimental pattern for working memory.

8.3.2.1 Subjects

Thirty-four undergraduate students (18 males and 16 females, age range 18–24 years, right-handed, with normal or corrected-to-normal vision, healthy) volunteered to participate in this study as potential subjects. Each participant was given two blocks of trials: one block for a two-back task and the other block for a three-back task. Twenty-seven subjects passed the prescreening criterion of at least 60% accuracy on both matches and mismatches, and could participate in the formal experiment, including a session of practice and a session of functional NIRS testing. Another 5 subjects were excluded because of some problems during the experiments. The other 23 subjects were retained for further analysis.

8.3.2.2 Materials and procedure

Subjects were presented with a sequence of 45 English letters. The presentation time of each letter was 500 ms, and the interval time between successive letters was 2.5 s during which the screen was blank. When each letter was presented, subjects were required to judge whether it matched the

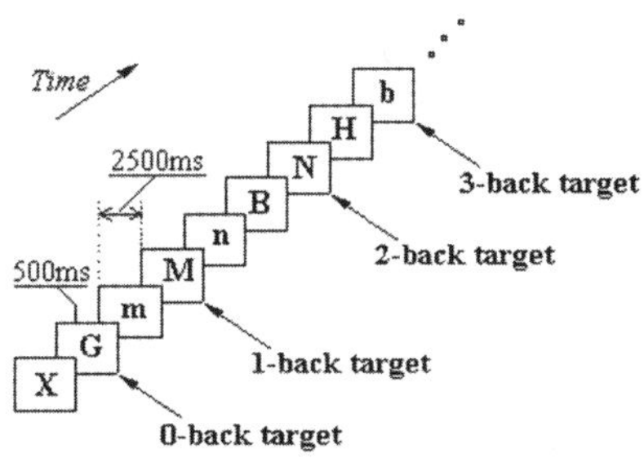

Figure 8.8 Schematic diagram for *n*-back task in present experiment. The tasks differ in working memory load, varying from zero- to three-back.

letter presented *n* items back in the series (homophony, not case sensitive; for instance, "m" matching "M"). Subjects gave their responses by pressing the computer mouse once to indicate a match or twice to indicate a mismatch. There were five tasks—four memory tasks and one control task—in the experiment, as shown in Fig. 8.8. John Jonides et al. have described the details in their report.[71]

8.3.2.3 Results and discussions

Performance accuracies and response times from the practice and NIRS test sessions were analyzed respectively. Repeated measures analysis of variance (ANOVA) revealed that the memory load had a significant effect on response times and accuracies in both test sessions.

The significant activation distribution in the prefrontal lobe is shown in Fig. 8.9, averaged across all 23 subjects in 0-back to approximately 3-back

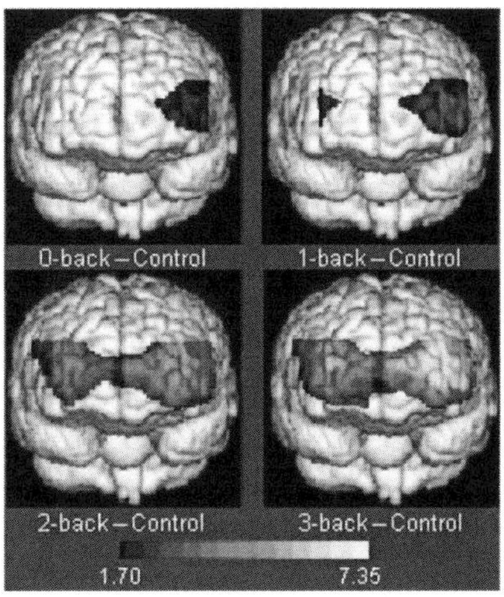

Figure 8.9 The significant activation distribution in the prefrontal lobe averaged across all 23 subjects in 0-back to approximately 3-back memory tasks (see color plates).

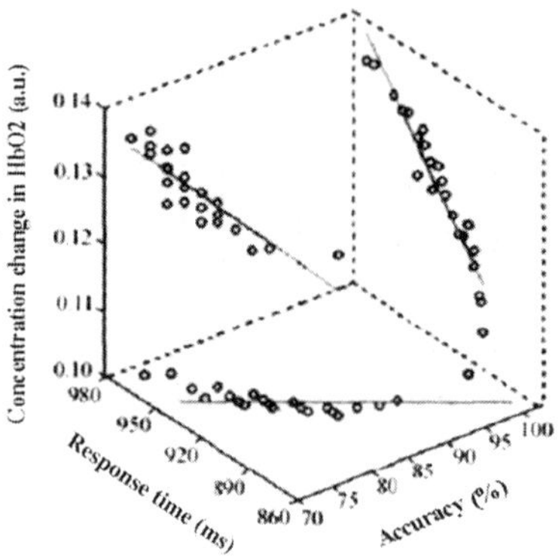

Figure 8.10 The relationship between activation of PFC and the behavioral parameters in the 3-back task.

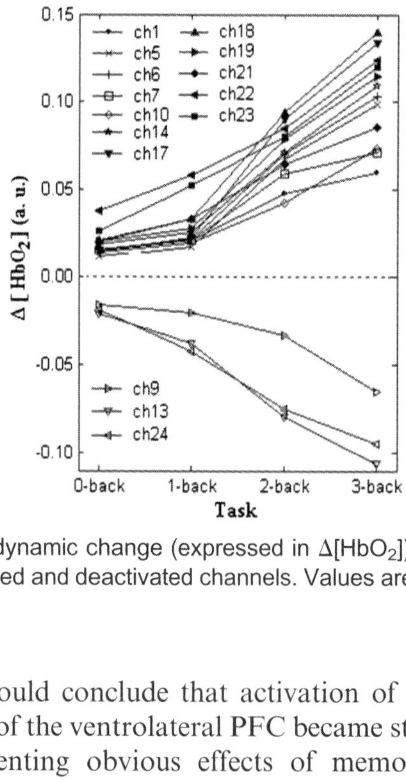

Figure 8.11 The hemodynamic change (expressed in $\Delta[HbO_2]$) as a function of memory load in each of the activated and deactivated channels. Values are the mean change across 23 subjects.

memory tasks. We could conclude that activation of the prefrontal, dorsal lateral PFC and part of the ventrolateral PFC became stronger as the memory load increased, presenting obvious effects of memory load. The *n*-back memory shows obvious functional laterality in the ventrolateral PFC.

Under the higher memory load, there was stronger activation in PFC for the subjects with low correct rate or long response time had. In 3-back tasks, Δ[oxy-Hb] had a significantly positive correlation with response time and a negative correlation with accuracy, as shown in Fig. 8.10. Figure 8.11 depicts the hemodynamic change with the memory load.

8.3.3 The study of children with developmental dyslexia

Children with developmental dyslexia have low reading ability relative to their intelligence quotient. The prevalent performance seen with dyslexia is inadequate phonologic cognition.

8.3.3.1 Subjects

Twenty dyslexic elementary school students (right-handed, age range 10–12 years, 7 females and 13 males) participated in this experiment with parental consent. Twenty control children (right-handed, age range 10–12 years, 7 females and 13 males) were also identified. All of the children had normal intelligence and no clinical signs of brain disorders.

8.3.3.2 Materials and procedure

The Chinese characters were chosen because they had only two pronounced syllables. Each subject performed two tasks. The first task was phonological awareness: The subject was asked to judge whether the two characters had an identical initial consonant. The second task was phonological decoding: The subject was asked to replace the initial consonant of the right character with the one of the left character. The first task had 25 pairs of Chinese characters and the second task had 20 pairs. The probe covering the left PFC had 8 channels (Fig. 8.12).

8.3.3.3 Results and discussions

During the phonological awareness processing (Fig. 8.13), the blood parameters showed a slight change in normal subjects in region 5, compared with the subjects with dyslexia. In addition, the blood parameters' average change in the children with dyslexia in region 3 was distinct from that in the normal subjects (deoxy-Hb: $t = 9.406$, $p < 0.001$, oxy-Hb: $t = 7.949$, $p < 0.001$, tot-Hb: $t = 11.929$, $p < 0.001$). During the phonological decoding processing (Fig. 8.14), there were significant differences between normal and dyslexic subjects in regions 4 and 5. Therefore, the ventrolateral and middorsal PFC presented uncommon patterns of hemodynamic parameters in dyslexic children.

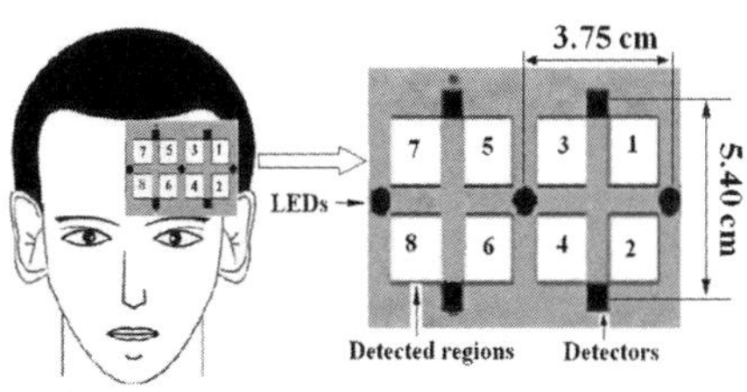

Figure 8.12 Probe placement on left PFC.

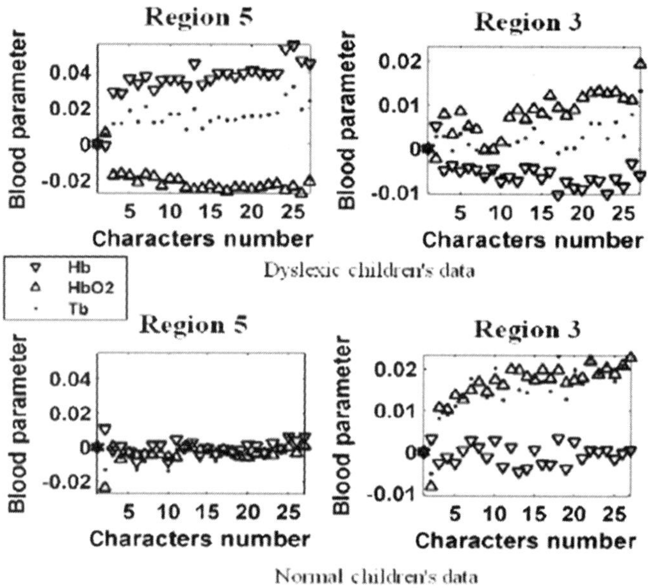

Figure 8.13 The average changes in hemodynamic parameters for 18 dyslexic and 20 normal subjects for each pair of characters during the phonological awareness processing.

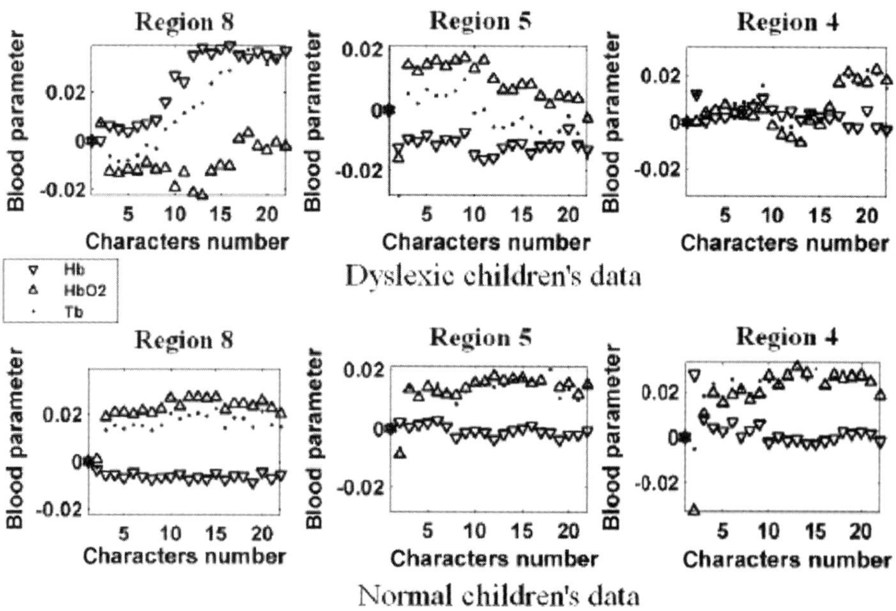

Figure 8.14 The average changes in homodynamic parameters for 18 dyslexic and 20 normal subjects for each pair of characters during the phonological decoding processing.

In the phonological processing, the left PFC became activated. In turn, different changes in the blood parameters were observed in normal and dyslexic children.

8.3.4 Stem recognition performance measurement

8.3.4.1 Subjects

Eighteen Chinese college students (right-handed, age range 18–23 years, 9 males and 9 females) participated in this experiment. All subjects had normal or corrected-to-normal vision. No subject had a history of neurological, major medical, or psychiatric disorders, and none was taking medication during the period of measurement.

8.3.4.2 Materials and procedure

The experiment utilized a block design, including verbal control, verbal memory, spatial control, and spatial memory (Fig. 8.15). There was a 1-minute rest period before the first block, after the last block, and between two blocks. In the verbal recognition task, English letters were the target and probing stimuli. In the spatial recognition task, black dots were the targets and circles were the probing stimuli. In each block, matched and unmatched cases were pseudorandom and presented with 50% frequency.

8.3.4.3 Results and discussions

In the spatial memory task, the middle and right PFC had strong activation, as in the studies using other techniques. Activation of the right inferior PFC was also observed in the verbal memory task. A possible reason for this finding was that the subjects were all native Chinese speakers, so English was not the primary language for any of them. Possibly, the native Chinese speakers use spatial-related structure to code English words. The oxygenation activation of the left prefrontal cortex for verbal memory was much stronger

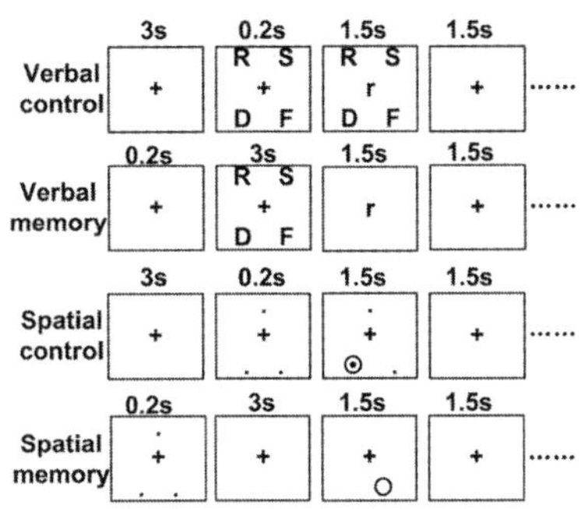

Figure 8.15 A schematic paradigm of verbal control, verbal memory, spatial control, and spatial memory in the stem recognition task.

than that for spatial memory. This result supports the notion that lateralization in the left hemisphere occurs for verbal memory.

8.3.5 Pinpoint source location for ocular nonselective attention

8.3.5.1 Subjects

Fifteen native Chinese college students (8 males and 7 females, mean age 22.6 ± 1.8 years, educated age 16.8 ± 1.2 years, right-handed) participated in this study with oral consent. Participants had normal or corrected-to-normal vision, had no personal or family history of neurological disorders, and were all healthy. During the experiment, blood oxygenation change of the prefrontal lobe was monitored with fNIRI and electrocortical potential waves were monitored by an ERP system. All data obtained with the fNIRI and ERP were effective. The accuracy of the behavior parameters from each subject was nearly 100%.

8.3.5.2 Materials and procedure

The stimulus was a regular triangle (untargeted stimulation), an inverted regular triangle (targeted stimulation), or patterns with rambling lines (novel stimulation: different from each other) presented as white lines against a black background. There were 140 untargeted stimuli, 30 target stimuli, and 30 unfamiliar stimuli presented pseudorandomly. The presentation time for each stimulus was 100 ms. The interval between stimuli while waiting for the subject's response (SOA) was randomly within the range of 900 to1100 ms. The unfamiliar stimulus and target stimulus did not appear more than two successive times. Participants were asked to press a button in response to the inverted triangle as soon as possible, but instructed not to react to other stimuli. Half of the subjects responded with the right hand, and half with the left hand. The design of the experiment is depicted in Fig. 8.16.

This study used the portable fNIRI instrument. The probe was fixed on the skin of the prefrontal lobe with stretch fabric. The bottom side of the probe was just up to the brow, and the midline of the probe was placed in the sagittal plane of the brain. A system consisting of 72-channel Neuroscan SynAmps electrocortical potential collection and a Ag/AgCl electrode cap was

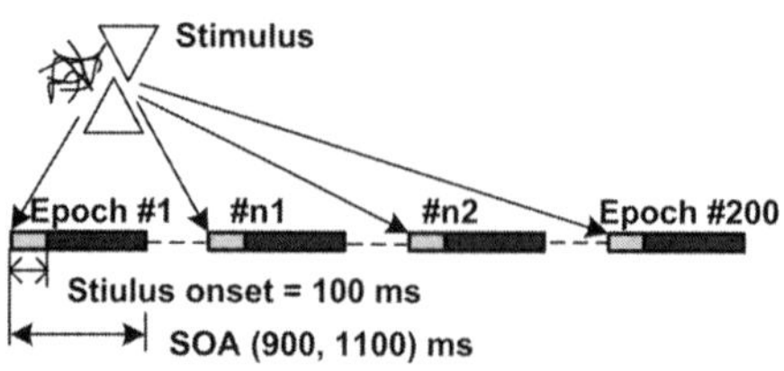

Figure 8.16 The design of ocular nonselective attention experiment.

utilized to record EEG data. The EEG records and behavior data were recorded simultaneously.

8.3.5.3 Results and discussions

ERP data showed that in the middle cortex, two types of attention activated similar patterns of P3a waves, but the nonselective attention showed higher activation with a higher peak and a longer time interval for the P3a component (300–500 ms). Thus, the middle cortex was not the special area distinguishing the reaction to nonselective attention from the reaction to common attention. Prior studies have shown that the P3a component is related to the reaction to nonselective attention. The peak of the P3a component appeared approximately 330 ms after the onset of the novel stimuli, which is why the potential P330 was analyzed.

The typical concentration changes curve for most brain functional activities measured by the fNIRI was obtained. It demonstrated stronger positive activation of Δ[oxy-Hb], weak positive activation of Δ[tot-Hb], and weak negative activation of Δ[deoxy-Hb], as shown in Fig. 8.17. The Δ[deoxy-Hb] map showed the smallest region with positive activation—region A. Because the region between FP1 and F7 near region A was activated by nonselective attention, and because of the coupling between the vascular and neuron changes, we can conclude that region A was the source of the response to ocular nonselective attention, as shown in Fig. 8.18.

Although fNIRI is still a developing technique, the accuracy and advantages of its application in functional source locations were well demonstrated in this study. The combination of two technologies can compensate for the limitations of each technology when applied separately, thereby identifying the precise location of the source of some brain functional activity. Such combinations hold promise for revealing more information about the workings of various brain mechanisms.

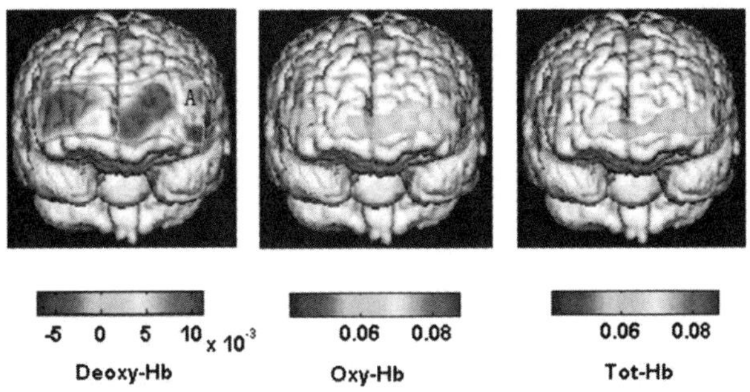

Figure 8.17 Image of averaged blood activation amplitude for an ocular attention task, including changes in the concentrations of deoxyhemoglobin and oxy-hemoglobin, as well as the blood volume (see color plates).

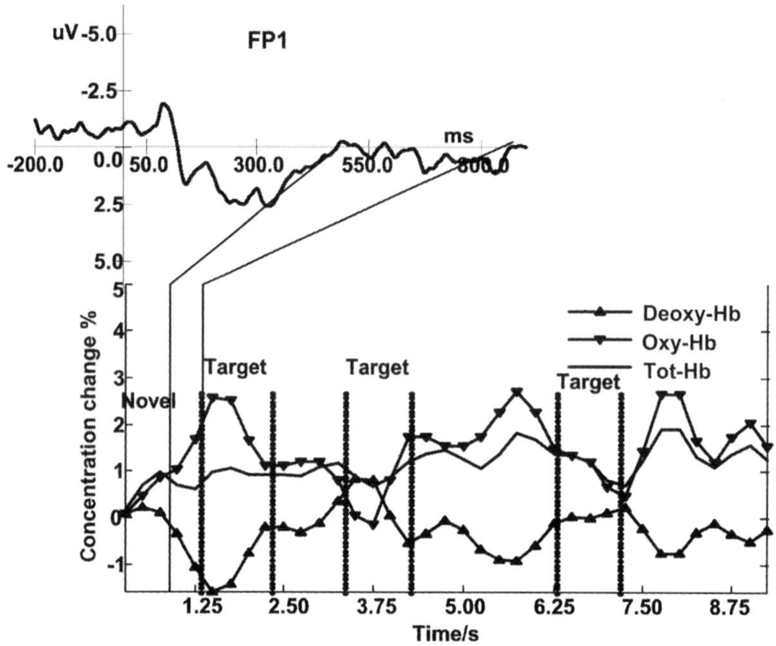

Figure 8.18 Comparisons between blood activation changes and brain waves for novel stimuli in the sites monitored by an FP1 electrode.

8.3.6 Cognitive Conflict Control

Cognitive conflict control provides a flexible solution with complicated situation. This ability is a necessary part of daily life. The Stroop paradigm has traditionally been used to investigate cognitive conflict.[72,73] When the stimulus of the Stroop paradigm carries interference or facilitation information, the brain is activated to handle the conflict process.

8.3.6.1 Subjects

Sixteen healthy native Chinese university students (right-handed, age range 21–23 years, 6 females and 10 males) participated in this experiment with oral consent.[74] All had normal or corrected-to-normal vision and normal color vision. No subject had a history of neurological, major medical, or psychiatric disorders, and none was taking medication during the period of measurement.

8.3.6.2 Materials and Procedure

The stimulus picture had two Chinese characters. The upper one was colorized (blue, red, yellow, or green) and the lower one was black (Fig. 8.19). The subjects were instructed to decide whether the upper character's color and

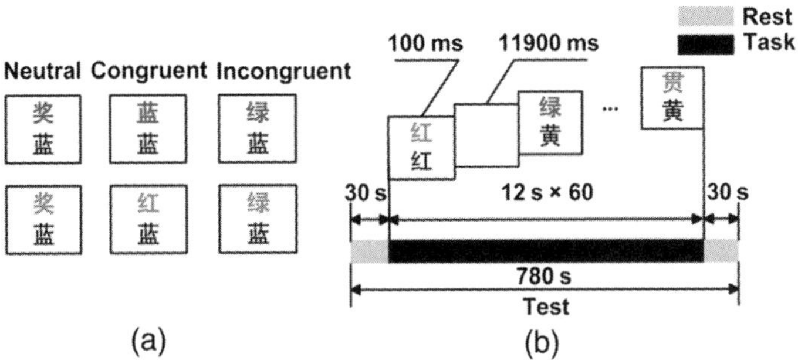

Figure 8.19 (a) Three kinds of stimuli: neutral, congruent, incongruent. (b) The procedure for presenting the stimuli.

the lower character's meaning were consistent and to press button F (consistent) or button J (inconsistent) as their response.

There were three kinds of stimuli: congruent, incongruent, and neutral condition. The discrimination factor was the upper character. In the congruent condition, the upper character's meaning and color were congruent; in the incongruent condition, the meaning and color were incongruent; in the neutral condition, the upper character was a no-color-meaning one (贯, 奖, 球, 涂). Between the rest times, which lasted 30 s each, 60 trials (20 neutral stimuli, 20 congruent stimuli, 20 incongruent stimuli) were displayed in pseudorandom order.

The fNIRI probe consisted of separate optodes (two sources and four detectors) defining four channels in the PFC (Fig. 8.20). Otherwise, the electrophysiological signals were recorded by Neuroscan's 64-channel system (ERP). Only FP1, FP2, AF3, and AF4 were used symmetrical to fNIRI channels.

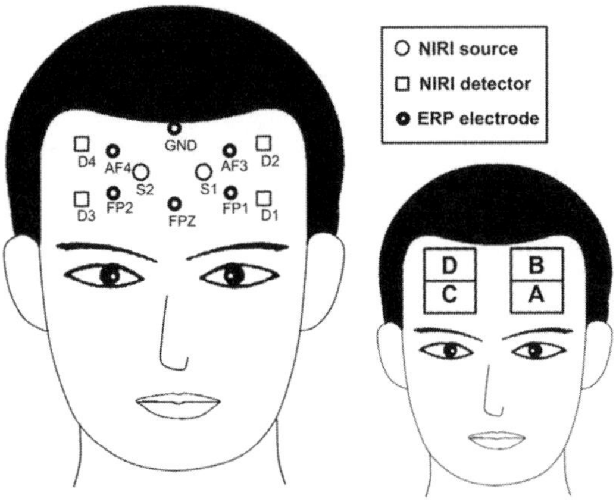

Figure 8.20 The locations of the NIRS optodes and ERP electrodes (left) and the four channels (right).

8.3.6.3 Results and discussions

The mean reaction time was longer in the incongruent condition (1319.65 ± 67.14 ms) than in the congruent condition (1236.91 ± 74.08 ms) or the neutral condition (1297.47 ± 70.34 ms). This difference was statistically significant ($F = 3.658$, $p < 0.05$).

For fNIRI, the signal's average amplitude was highest in channel A. The incongruent condition had the largest amplitude and shortest latency time. For the ERP device, the last two obvious feature waveforms (N500: negative waveform in the 450–550 ms time window; P600: positive waveform in the 600–700 ms time window) were chosen. The amplitude of P600 was largest in the incongruent condition, but the amplitude of N500 was largest in the congruent condition. The latencies of P600 and N500 were consistent. The incongruent condition had the shortest latency.

To evaluate the relationship between the fNIRI and ERP signals, the latencies and amplitudes were compared. As Table 8.3 shows, P600 had more significant correlations with hemodynamic change amplitudes than N500. However, both P600 and N500 had strong correlations with oxy-Hb in terms of latency (Fig. 8.21).

Considering that the electrophysiological signal is transmitted quickly, the amplitude is sensitive to the distance from the generator of the waveform. P600's generator was located in the PFC, whereas N500's generator was not.

Table 8.3 Correlations of NIRS and ERP signals amplitudes.

	Oxy-Hb Peak	Deoxy-Hb 1	Deoxy-Hb 2	Oxy-Hb Mean	Deoxy-Hb Mean
P600	*	—	*	*	**
N500	—	—	—	—	—

*Correlation is significant at the 0.05 level (2-tailed).
**Correlation is significant at the 0.01 level (2-tailed).
⁻Correlation is not significant.

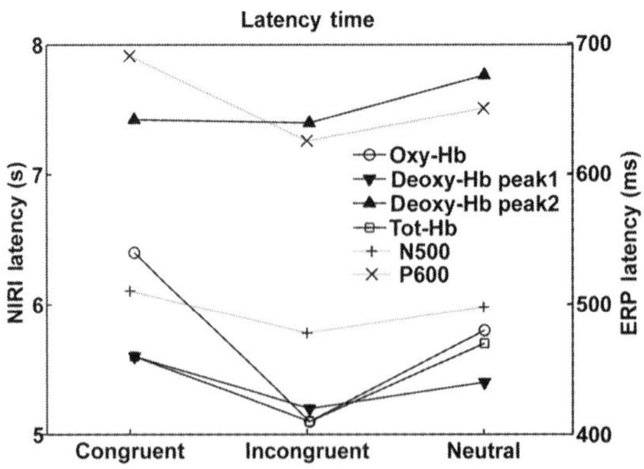

Figure 8.21 Latency time of NIRS and ERP averaged across all subjects and channels. Only deoxy-Hb had two peak values in each channel, so the NIRS signals had four parameters (oxy-Hb peak, tot-Hb peak, deoxy-Hb peak 1, and deoxy-Hb peak2). The ERP device had N500 and P600 as its two parameters.[74]

The results from the combined fNIRI and ERP testing match those from the previous studies that measured outcomes using one of these techniques separately.[75,76] In summary, the approach combining fNIRI and ERP appears feasible and the correlations of hemodynamic and electrophysiological parameters are reliable. These correlations provide the effective cognitive neutral information, and offer proof for neurovascular coupling.

8.3.7 Motor skill learning

8.3.7.1 "Tie a knot"

The "tie a knot" protocol has 12 high school student subjects learn how to tie the sailor's knots known as a "clove hitch" and a "bowline." Since these knots are not common and new to the subjects, they learn how to tie them by viewing a video, memorizing the procedure, and then trying to tie the knots on their own. We also asked the subjects to think about the knot-tying procedure ("imagery practice"), which is a practice technique often used in sports competitions. NIR images were obtained during four periods: (1) after resting, (2) after looking at the video demonstration (15 seconds), (3) after 10 seconds of imagery practice (controls were ask to wait 10 seconds but were not assigned any task during this period), and (4) after an attempt to the knots for 15 seconds. The subjects were asked to repeat the sequence if they could not tie the knots within 15 seconds, and to repeat these steps until they could finish within 15 seconds.

The behavioral results showed that the imagery practice did not shorten the learning time; instead, subjects learned more slowly at first, needed to repeat the steps 5 times on the first day. However, both groups (learners and controls) successfully completed the test on the fourth day (i.e., they tied the knot on the first attempt on day 4). The imagery practice seemed to cause subjects to take a longer time to learn and was associated with lower oxygenation; oxygenation increased when learning imagery practice was combined with better memory experience. During the execution of the tying attempts, the PFC oxygenation decreased by 30% with motor activation (Fig. 8.22 left).

8.3.7.2 Driving skill test and learning

We tested whether the PFC became activated while learning a driving skill and whether this activity had any relation to the level of driving skill achieved ($n = 8$). Further, we tested whether the PFC activation changed when driving was distracted by eating and use of a phone ($n = 20$) with 4 levels of difficulty of driving. Since the high school students were learning driving at the age of 16–17 years, no subjects were skilled drivers, and many described their abilities as quite "novice" in relation to making left and right turns. We used a

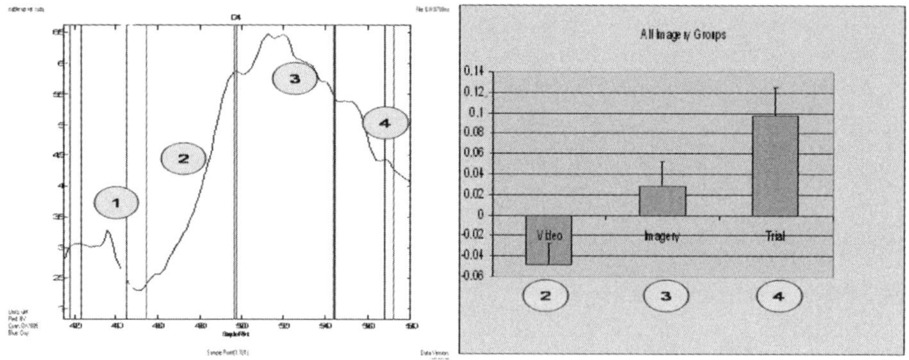

Figure 8.22 The left graph shows a learner subject's time course of PFC oxygenation profile. The four periods represent (1) resting baseline, (2) video watching, (3) imagery, and (4) tying knot trial. The right graph depicts the overall averages for oxygenation during (2) video watching, (3) imagery, and (4) knot trials when the subjects were learning.

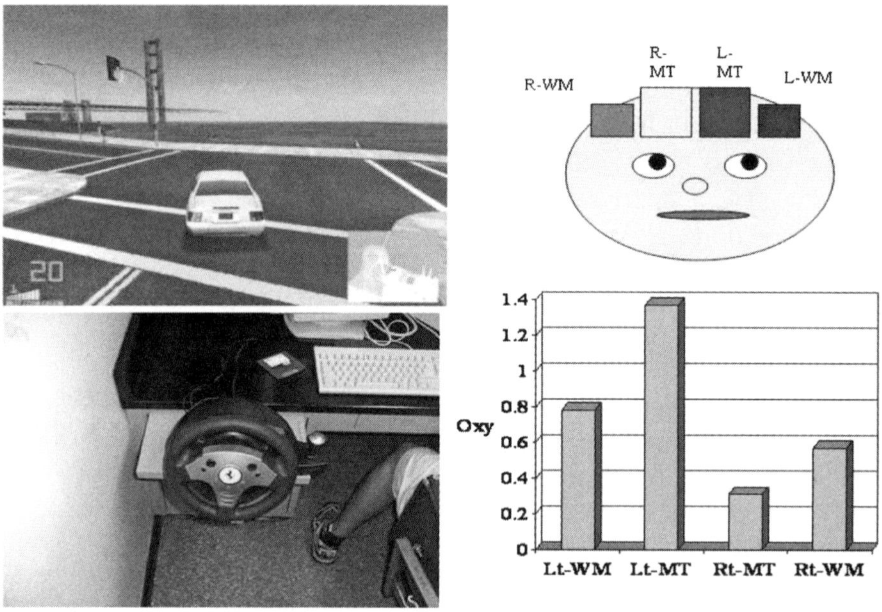

Figure 8.23 (left) A subject drives on San Francisco streets using a PC monitor and a wheel. (right) The CW-NIRS monitors the student's prefrontal area, including the medial anterior (MT) and dorsolateral area (WM). In this study, the left brain was found to be more oxygenated during the simulated driving.

PC driving model of San Francisco streets with a regular car handle (Fig. 8.23) to mimic the real driving.

Unexpectedly, we found more left-side oxygenation in the anterior medial area (Broadman 9 and 10) and the working memory area of the dorsolateral PFC, suggesting that not only did the subjects watch the constantly changing environments, but they also analyzed them logically by applying semantic information that is instantly used for judgments and motor reactions. A negative linear relation was found between skill development and oxygenation in these areas of the PFC.

The distractions associated with eating and phone conversations, as expected, caused subject to make more errors in each of the four difficulty levels of driving. During the experience of driving with the distractions, different responses in the PFC were likewise noted. Future studies are expected to elucidate how the brain deals with many different types of distractions.

8.3.8 Thinking process and learning: "insight signal" through verbal stimuli

8.3.8.1 Word association

"Insightful thinking" during problem solving is a thinking process that does not predict when a solution can be reached, but rather is characterized by the sudden arrival of a solution. Many problem-solving processes are based on the "sudden insight" processes associated with creative thinking, mathematical concept discovery, and so on. The word association test, for example, is known to produce the insightful thinking process. We applied this protocol by asking subjects to find a word related to three presented words (Fig. 8.24 left); after the solution was found, the subject pressed a key. In addition, we asked subjects how suddenly the answers came in each problem, while a brain mapping device monitored their prefrontal cortical activity continuously. Figure 8.24 presents a typical sudden insight signal taken from the left medial

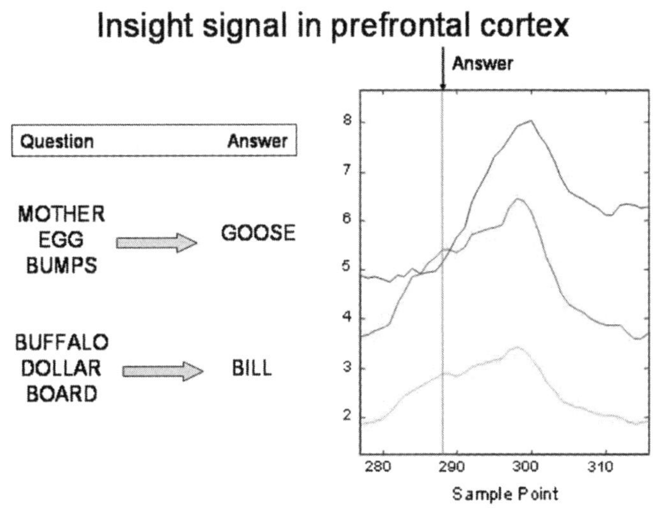

Figure 8.24 (left) Word association protocol. (right) Blue line = oxygenation; red line = blood volume before answering a question (black bar).

PFC (right). The blood volume increased 2 seconds before the answer in this typical case, followed by the oxygenation. Since oxygenation is expressed as [HbO_2-Hb], deoxyhemoglobin is the component whose concentration increases at the time of the "sudden insight signal," matching the time of mitochondrial use of oxygen during ATP synthesis.

8.3.8.2 Anagram test and learning

An anagram is a game in which the player tries to form a word from scrambled letters. Solving an anagram is also known to trigger "sudden insight."[77] We used the anagram game to look at the cognitive process associated with performance behavior and learning. The protocol used a block design for the anagram difficulty task, and 3 letter-anagram to 4, to 5 and again 4 to 3-letter anagram was chronologically tested each for 1 min (Fig. 8.25). The average oxygenation for the subject during the entire 1-min test was recorded to determine brain activation.

Protocol 1 consisted of 10 novice-learning subjects. A significant difference was noted with their 5-letter anagram performance: 6 subjects demonstrated a gradual decline in performance with increasing numbers of letters in the anagram (low performers), whereas the other 4 subjects did not show this trend with 4- and 5-letter anagrams (high performers). Brain activity analysis was then focused on the differences between the low performers and high performers, in an effort to determine the responsible area and the intensity of brain activation related to their performance differences.

Bilateral deactivations were noted with 3-letter anagram solving (Fig. 8.26, bottom) and bilateral activations with 5-letter anagram solving (Fig. 8.26, top) tasks, in all groups of subjects.

In contrast, in the 4-letter anagram task, brain activation differed between the low and high performers; high performers exhibited high oxygenation to a greater extent in the left PFC than in the right PFC. This finding correlates with the performance data for the high performers. The data suggest that the high activity of the left PFC most likely accounts for the high performance level in 4-letter anagram solving. This difference is shown in Fig. 8.27(c).

The differences between the high and low performers in the 4-letter anagram solving are depicted in Fig. 8.27(c). Association of the anterior PFC

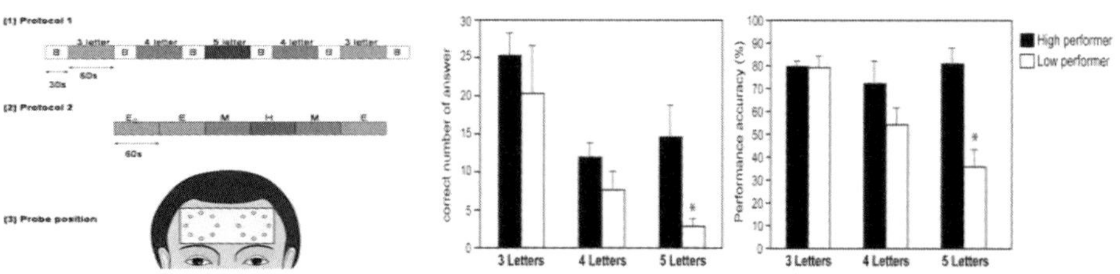

Figure 8.25 (left) Anagram protocol and behavior outcome. (middle) Correct number of answers. (right) Performance accuracy.

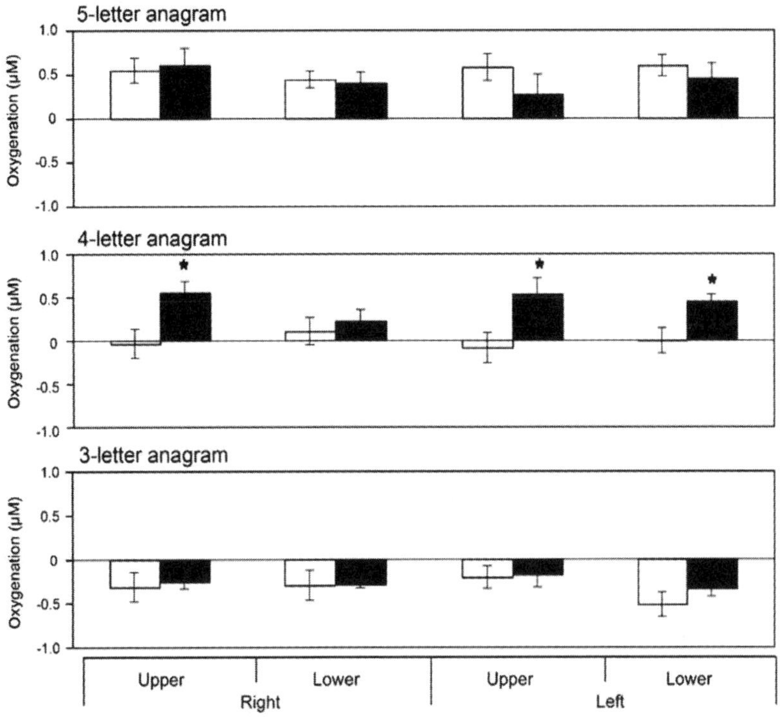

Figure 8.26 Anterior PFC (aPFC) responses to 3-, 4- and 5-letter anagram solving tasks. In the 4-letter anagram solving task (middle), the low (white bar) and high (black bar) performers differed in the degree of activation.[77]

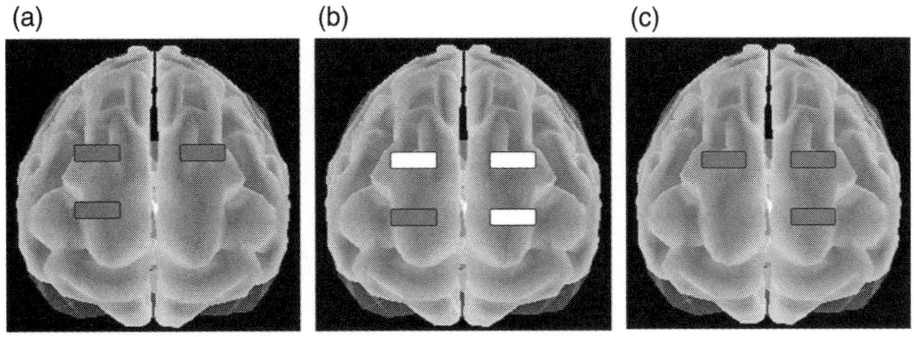

Figure 8.27 aPFC activation during the 4-letter anagram task (compared to the 3-letter anagram task): (a) low performers; (b) high performers; and (c) differences in aPFC activity between high and low performers (b–a). The white and gray bars indicate approximately 0.8 μM and 0.4 μM oxygenation, respectively.

(aPFC) activation with the difficulty of the complex task suggests that the aPFC is part of a circuit for execution of task performance. In addition, the greater use of the aPFC by untrained high performers suggests that the role of the aPFC may be to increase efficiency during a problem-solving task.

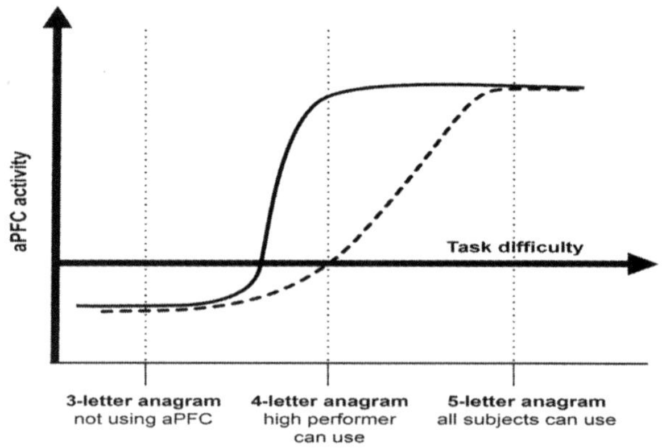

Figure 8.28 Proposed neuronal circuit intensity changes related to the task difficulty during a novice-learning task. The better performance is associated with high cognitive activity in the middle degree of difficulty, where the difficulty intersects with the challenge effort required to conquer the task.

We hypothesize that high performers can engage the scaffolding function for problems of a much wider range of difficulty than can low performers, as shown in Fig. 8.28. In this way, good performers can control (orchestrate) the complex task to make anagram solving easier by searching and retrieving words more vigorously for better success.

In a second protocol, another 6 subjects underwent 6 weeks of training in solving anagrams. The training reduced the level of oxygenation in their aPFC while solving these problems. In particular, subjects with lower baseline skill in anagram production showed a larger reduction in oxygenation where true performance gains occurred (medium difficulty) and a smaller reduction where the performance gains were limited (hard anagrams).

8.3.9 PFC responses to emotional stresses

8.3.9.1 Emotional face recognition

We studied face emotion recognition and prefrontal (BA10) brain activity using a CW imager. Five emotional faces were presented in a monitor, and 13 healthy high school students were asked to recognize the five types of emotions (happy, angry, fear, sad, and no emotion) that the faces presented. The PFC was activated more when subjects were presented with happy, fearful, and angry faces than when they saw sad faces. There were also brain activity differences depending on the face's gender (Fig. 8.29), and the faces of young adults and adults stimulated the PFC of the subjects to a greater extent than did older adult faces in general. These results suggest that the BA 10 is used for processing of facial recognition, and we can observe these functions by applying the NIR cognometer.

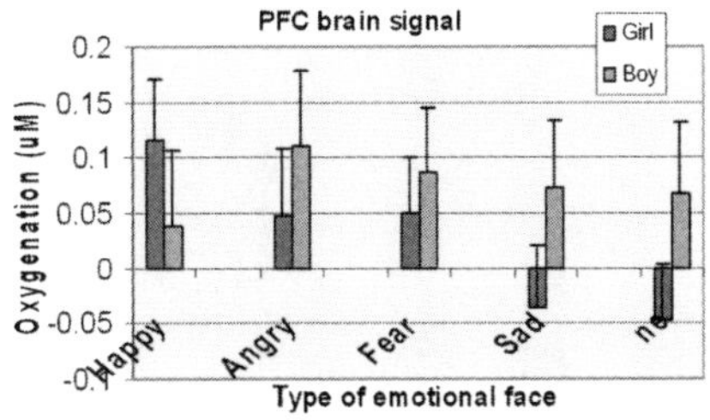

Figure 8.29 Emotional face recognition protocol. There were gender differences in Broadman 10 oxygenation with the happy face (female dominant) and with negative-emotion faces (male dominant).

8.3.9.2 Deception and social inhibition

Social norms influence human behaviors and responses to others' expectations, especially in young people. When we ask embarrassing questions related to social norms in young people, such as those dealing with use of drugs, stealing, and lying, there must be activation of the PFC. Indeed, we know that the medial and anterior PFC has a direct connection to the emotional circuit consisting of the orbito-frontal cortex, amygdala, and hippocampus. Further, a "lie" signal originates close to the PFC, in the anterior cingulate cortex (langlaben). Thus we expect some signals to be related to the decision making that occurs during lying.

The protocol in this study sought to determine whether following the social norm of telling the truth enables the PFC to work with less oxygenation compared with not following the social norm. Twelve students (mixed genders) participated in the study. First, we asked the participants questions about their social background; these questions were followed by questions addressing socially unacceptable behavior that had been mentioned previously. Natural and nontargeted questions were used for controls. We asked participants to answer the questions with "yes" (imposed yes) and "no" (imposed no). The results, which are shown in Fig. 8.30, showed remarkable oxygenation levels generally occurred when subjects were responding to embarrassing questions (against social norms), albeit with some gender differences.

As seen in the results, large differences in oxygenation between answering "yes" and answering "no" were observed with significant between-subject effects; we call this finding the "deception effect'" Moreover, similarities in oxygenation between "yes" and "no" answers were observed frequently; we call this finding a signal of social "excitation" or embarrassment. Interestingly, in many cases, the embarrassment signals were stronger than the deception signals, resulting in negative deception signals; that is, participants showed negative oxygenation during their false answers. Nevertheless, on average, false answers show positive responses in oxygenation level.

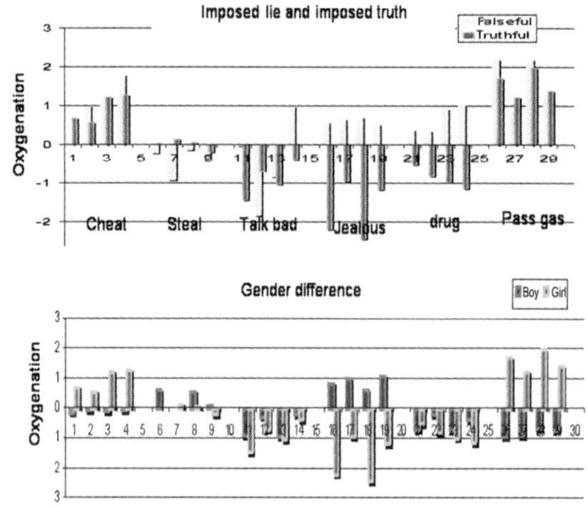

Figure 8.30 Six boys and girls were tested by asking them questions about behaviors outside social norms—in particular, questions containing the keywords "cheating," "stealing," "talk bad," "jealous," "drugs," and "passing gas." The four bars for each question indicate oxygenation changes in the left lateral/center and right center/lateral areas of the PFC. The upper graph shows the oxygenation changes when a subject answered truthfully or falsely. The bottom graph shows gender differences when each gender answered truthfully. Data were normalized based on control data, which were obtained during nontargeted questions.

8.3.10 Optical neuronal signals in the visual cortex

We investigated the feasibility of performing noninvasive intensity measurements of optical neuronal signals within the visual cortex. In this study, we used a homemade CW-NIRS system with high detection sensitivity (on the order of 0.1 pW).[32]

8.3.10.1 Materials and procedure

A full-field circular black-and-white reversing checkerboard stimulus was presented to maximally stimulate the visual cortex. A custom-made "U"-shaped support was used to restrain the head of the participant. Each measurement included a 1-min baseline measurement and measurement during a 720-s sequence of alternating periods of 30-s stimulation and 30-s rest, followed by an additional 1-min baseline measurement (Fig. 8.31). The experiment lasted for 14 min to avoid tiring the participants with a prolonged experimental time.

8.3.10.2 Results

We obtained slow hemodynamic signals in all participants. Figure 8.32 presents the averaged the spatial distribution map of the hemodynamic responses across all participants. Figure 8.33 shows the representative optical neuronal signals from two subjects.

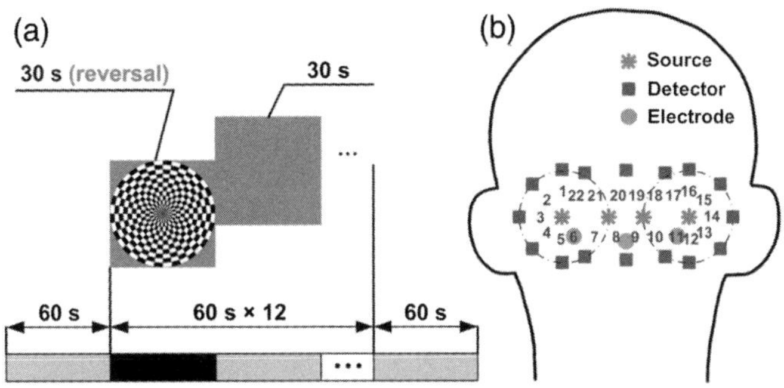

Figure 8.31 Experimental protocol. (a) A total of 12 cycles of 30-s checkerboard reversal and 30-s blank screen periods were presented between 60-s baselines. (b) Configuration of the optical probe. The probe comprised 4 sources (red asterisks) and 16 detectors (blue squares), yielding a total of 22 channels (gray numbers). The source–detector distance was 3 cm. From left to right, the green circles represent electrodes O_1, O_z, and O_2, respectively, and electrodes O_1 and O_2 were placed at the midpoints of channels 6 and 11, respectively.[78] (See color plates.)

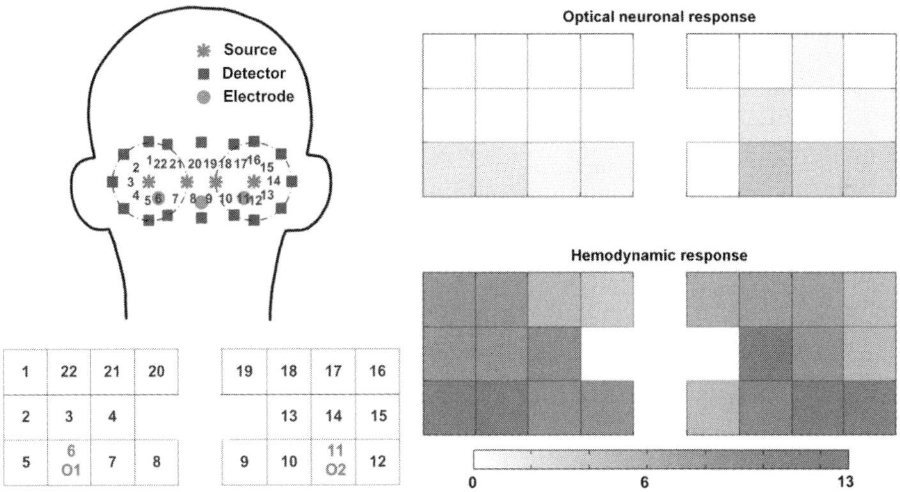

Figure 8.32 The spatial distribution maps of the optical neuronal signal (top right) and the hemodynamic response (bottom right). The top left graph shows the configuration of the optical probe on the head. The bottom left graph shows the distribution of the channels corresponding to the spatial distribution maps; the numbers marked in red indicate that the channel corresponded to the electrode positions of O_1 (channel 6) and O_2 (channel 11). The color indicates the number of the subjects who showed a significant response in the corresponding channel. (See color plates.)

8.4 Future Prospects

In this chapter, we have described the principles, algorithm, and performance of a CW functional near-infrared imager. This device is a simple and low-cost instrument compared with EEG, fMRI, and PET. Even so, it has several advantages in monitoring brain activity, especially in temporally and spatially

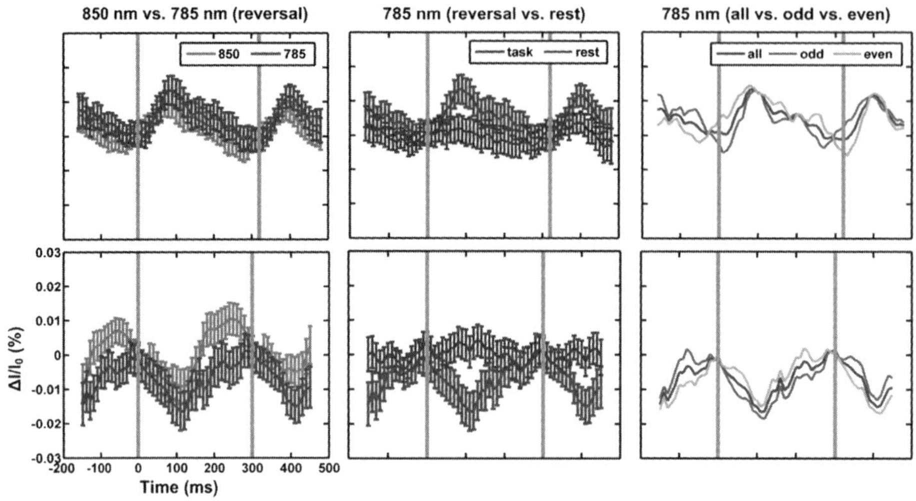

Figure 8.33 Optical neuronal signals obtained from channel 6 of subject number 3 (the top row) and channel 10 of subject number 5 (the bottom row). The error bars represent the standard errors.[78] (See color plates.)

resolving cerebral blood flow and determining relative oxygenation of the cortex. As discussed in Section 8.2, the penetration depth of the fNIRI depends on the source–detector spacing, which also determines the spatial resolution. Recent work on simulation and visualization of light propagation in the Visible Chinese Human Head for fNIRI may help to optimize the imager system.[79,80] To develop better spatial resolution and localization precision without losing any depth information, one may keep the same source–detector spacing and use more detectors and light sources in the same total detection area by overlapping the detection sections. Temporal resolution is dominated by the light sources, as discussed in Section 8.2. It could be improved by using better light sources such as laser diodes.

NIR techniques have the capability to realize multi-wavelength operation, thereby providing enhanced sensitivity for oxyhemoglobin and deoxyhemoglobin concentrations and for light scattering changes, although these technologies lack the resolution of MRI or the chemical specificity of PET. The most important thing, however, is that the NIR optical method opens up new fields of study relative to the human population. Investigations of adults under conditions of simulated or real stress may yield data indicating the areas of the brain that have important effects upon functional performance. Importantly, the study of optical neuronal signals within the visual cortex in the intact human brain using a CW-NIRS system has been demonstrated to offer a high detection sensitivity.[78,79]

NIRS cannot avoid interference from information outside the brain, such as that produced by the scalp, the bone, and so on (the subtraction may decrease the interference). Despite this shortcoming, NIRS has distinctive advantages over other functional brain imaging techniques. For example, it could be operated noninvasively and conveniently. These characteristics make it practical and affordable for large populations, thereby opening up new areas for using the complex technology of measuring brain function. NIRS is portable and low cost. In cognitive tasks, subjects can report their answers orally. Furthermore, the temporal resolution could be a millisecond or less, at least in theory.

Given these factors, NIRS—though developed only in recent years—has good prospects as a means to explore the human brain's secrets. Time-resolved and intensity-modulated optical imaging systems provide even more choices for the monitoring of human brain activity when combined with optical methods. The rapid technological progress being made in application of fNIRS will greatly advance the understanding of the functioning human brain.[81]

References

1. A. Villringer and B. Chance, "Noninvasive optical spectroscopy and imaging of human brain function," *Trends Neurosci.* **20**, 435–442 (1997).
2. D.K. Hill and R.D. Keynes, "Optical changes in stimulated nerve," *J. Physiol.* **108**, 278–281 (1949).
3. R.A. Stepnoski, A. LaPorta, F. Raccuia-Behling, G.E. Blonder, R.E. Slusher, and D. Kleinfeld, "Noninvasive detection of changes in membrane potential in cultured neurons by light scattering," *Proc. Natl. Acad. Sci. USA* **88**, 9382–9386 (1991).
4. B.A. MacVicar and D. Hochman, "Imaging of synaptically evoked intrinsic optical signals in hippocampal slices," *J. Neurosci.* **11**, 1458–1469 (1991).
5. A. Grinvald, E. Lieke, R.D. Frostig, C.D. Gilbert, and T. N. Wiesel, "Functional architecture of cortex revealed by optical imaging of intrinsic signals," *Nature* **324**, 361–364 (1986).
6. F.F. Jobsis, "Intracellular metabolism of oxygen," *Am. Rev. Respir. Dis.* **110**, 58–63 (1974).
7. S.I. Harik, J.C. LaManna, A.I. Light, and M. Rosenthal, "Cerebral norepinephrine: influence on cortical oxidative metabolism in situ," *Science* **206**, 69–71 (1979).
8. M.M. Haglund, G.A. Ojemann, and D.W. Hochman, "Optical imaging of epileptiforme and functional activity in human cerebral cortex," *Nature* **358**, 668–671 (1992).
9. B. Chance, Z. Zhuang, C. UnAh, C. Alter, and L. Lipton, "Cognition activated low frequency modulation of light absorption in human brain," *Proc. Natl. Acad. Sci. USA* **90**, 3770–3774 (1993).
10. Y. Hoshi, H. Onoe, Y. Watanabe, J. Andersson, M. Bergstrom, A. Lilja, B. Langstom, and M. Tamura, "Nonsynchronous behavior of neuronal

activity, oxidative metabolism and blood supply during mental tasks in man," *Neurosci. Lett.* **172**, 129–133 (1994).
11. A. Maki, Y. Yamashita, Y. Ito, E. Watanabe, Y. Mayanagi, and H. Koizumi, "Spatial and temporal analysis of human motor activity using noninvasive NIR topography," *Med. Phys.* **22**(12), 1997–2005 (1995).
12. W. Chen, E.J. Novotny, X. Zhu, D.L. Rothman, and R.G. Shulman, "Localized 1H NMR measurement of glucose consumption in the human brain during visual stimulation," *Proc. Natl. Acad. Sci. USA* **90**, 9896–9900 (1993).
13. P.T. Fox, M.E. Raichle, M.A. Mintun, and C. Dence, "Nonoxidative glucose consumption during focal physiologic activity," *Science* **241**, 462–464 (1988).
14. J.C. LaManna, T.J. Sick, S.M. Pikarsky, and M. Rosenthal, "Detection of an oxidizable fraction of cytochrome oxidase in intact rat brain," *Am. J. Physiol.* **253**, C477–C483 (1987).
15. A. Villringer and U. Dirnagl, "Coupling of brain activity and cerebral blood flow: basis of functional neuroimaging," *Cerebrovasc. Brain Metab. Rev.* **7**, 240–276 (1995).
16. B.M. Salzberg and A.L. Obaid, "Optical studies of the secretory event at vertebrate nerve terminals," *J. Exp. Biol.* **139**, 195–231 (1988).
17. D. Malonek and A. Grinvald, "Interactions between electrical activity and cortical microcirculation revealed by imaging spectroscopy: implication for functional brain mapping," *Science* **272**, 551–554 (1996).
18. G. Gratton, P.M. Corballis, E. Cho, M. Fabiani, and D.C. Hood, "Shades of gray matter: noninvasive optical images of human brain responses during visual stimulation," *Psychophysiology* **32**, 505–509 (1995).
19. K.A. Hossmann, "Periinfarct depolarizations," *Cerebrovasc. Brain Metab. Rev.* **8**, 195–208 (1996).
20. C.S. Robertson, S.P. Gopinath, and B. Chance, "A new application for near-infrared spectroscopy: detection of delayed intracranial hematomas after head injury," *J. Neurotrauma* **12**(suppl), 591–600 (1995).
21. E. Tulving, S. Kapur, F.I.M. Craik, and R. Habib, "Hemispheric encoding/retrieval asymmetry in episodic memory: positron emission tomography," *Proc. Natl. Acad. Sci. USA* **91**, 2016–2020 (1994).
22. A.D. Wagner, D.L. Schacter, M. Rotte, W. Koutstaal, A. Maril, A.M. Dale, B.R. Rosen, and R.L. Buckner, "Building memories: remembering and forgetting of verbal experiences as predicted by brain activity," *Science* **281**, 1188–1190 (1998).
23. J.B. Brewer, Z. Zhao, J.E. Desmond, G.H. Glover, and J.D.E. Gabrieli, "Making memories: brain activity that predicts how well visual experience will be remembered," *Science* **281**, 1185–1197 (1998).

24. B. Chance, J.S. Leigh, H. Miyake, D.S. Smith, S. Nioka, R. Greenfeld, M. Finander, K. Kaufmann, W. Levy, M. Young, P. Cohen, H. Yoshioka, and R. Boretsky, "Comparison of time-resolved and -unresolved measurement of deoxyhemoglobin in brain," *Proc. Natl. Acad. Sci. USA* **85**, 4971–4975 (1988).
25. E.M. Sevick, B. Chance, J.C. Leigh, S. Nioka, and M. Maris, "Quantitation of time- and frequency-resolved optical spectra for the determination of tissue oxygenation," *Anal. Biochem.* **195**, 330–351 (1991).
26. J.P. Van Houten, D.A. Benaron, S. Spilman, and D.K. Stevenson, "Imaging brain injury using time-resolved near infrared light scanning," *Pediatr. Res.* **39**, 470–476 (1996).
27. J.C. Hebden, D.J. Hall, M.F. Firbank, and D.T. Delpy, "Time-resolved optical imaging of a solid tissue-equivalent phantom," *Appl. Opt.* **34**, 8038 (1995).
28. Q. Luo, S. Nioka, and B. Chance, "Imaging on brain model by a novel optical probe: fiber hairbrush." *OSA TOPS on Advances in Optical Imaging and Photon Migration*, R.R. Alfano and J.G. Fujimoto, Eds., II, 155–159 (1996).
29. Q. Luo, S. Nioka, and B. Chance, *Functional Near-Infrared Imager, in Optical Tomography and Spectroscopy of Tissue: Theory, Instrumentation, Model, and Human Studies II*, B. Chance and R. Alfano, Eds., *Proc. SPIE* **2979**, 84–93 (1997).
30. B. Chance, Q. Luo, S. Nioka, D.C. Alsop, and J.A. Detre, "Optical investigations of physiology: a study of intrinsic and extrinsic biomedical contrast," *Phil. Trans. R. Soc. Lond. B* **352**, 707–716 (1997).
31. H.R. Heekeren, R. Wenzel, H. Obrig, J. Ruben, J.P. Ndayisaba, Q. Luo, A. Dale, S. Nioka, M. Kohl, U. Dirnagl, A. Villringer, and B. Chance, "Towards noninvasive optical human brain mapping: improvements of the spectral, temporal, and spatial resolution of near infrared spectroscopy." *Optical Tomography and Spectroscopy of Tissue: Theory, Instrumentation, Model, and Human Studies II*, B. Chance and R. Alfano, Eds., *Proc. SPIE* **2979**, 847–857 (1997).
32. Z. Zhang, B. Sun, H. Gong, L. Zhang, J. Sun, B. Wang, and Q. Luo, "A fast neuronal signal-sensitive continuous-wave near-infrared imaging system," *Rev. Sci. Instrum.* **83**(9), 094301-1-9 (2012),
33. K. Suzuki, K. Yamashita, K. Ohta, and B. Chance, "Quantitative measurement of optical parameters in the breast using time-resolved spectroscopy: phantom and preliminary in vivo results," *Invest. Radiol.* **29**(4), 410–414 (1994).
34. E.M. Sevick, B. Chance, J. Leigh, S. Nioka, and M. Maris, "Quantitation of time- and frequency-resolved optical spectra for the determination of tissue oxygenation," *Anal. Biochem.* **195**(2), 330–351 (1991).

35. H. Wabnitz, M. Moeller, A. Liebert, A. Walter, R. Erdmann, O. Raitza, C. Drenckhahn, J. P. Dreier, H. Obrig, J. Steinbrink, and R. Macdonald, "A time-domain NIR brain imager applied in functional stimulation experiments." *Photon Migration and Diffuse-Light Imaging II*, K. Licha and R. Cubeddu, Eds., Proc. OSA-SPIE Biomed Opt. **5859**, 58590H (2005).
36. E. Ohmae, Y. Ouchi, M. Oda, T. Suzuki, S. Nobesawa, T. Kanno, E. Yoshikawa, M. Futatsubashi, Y. Ueda, H. Okada, and Y. Yamashita, "Cerebral hemodynamics evaluation by near-infrared time-resolved spectroscopy: correlation with simultaneous positron emission tomography measurements," *Neuroimage* **29**(3), 697–705 (2006).
37. Y. Ueda, T. Yamanaka, D. Yamashita, T. Suzuki, E. Ohmae, M. Oda, and Y. Yamashita, "Reflectance diffuse optical tomography: its application to human brain mapping," *Japan. J. Appl. Physics* **44**, 1203–1206 (2005).
38. D. Contini, A. Torricelli, A. Pifferi, L. Spinelli, F. Paglia, and R. Cubeddu, "A multi-channel time-resolved system for functional near infrared spectroscopy," *Opt. Express.* **14**, 5418–5432 (2006).
39. M. Kacprzak, A. Liebert, P. Sawosz, N. Zolek, and R. Maniewski, "Time-resolved optical imager for assessment of cerebral oxygenation," *J. Biomed. Op.t* **12**, 034019 (2007).
40. J.J. Selb, J.J. Stott, M.A. Franceschini, A.G. Sorensen, and D.A. Boas, "Improved sensitivity to cerebral hemodynamics during brain activation with a time-gated optical system: analytical model and experimental validation," *J. Biomed. Opt.* **10**, 11013 (2005).
41. M.A. Franceschini, S. Thaker, G. Themelis, K.K. Krishnamoorthy, H. Bortfeld, S.G. Diamond, D.A. Boas, K. Arvin, and P.E. Grant, "Assessment of infant brain development with frequency-domain near-infrared spectroscopy," *Pediatr. Res.* **61**, 546–551 (2007).
42. G. Gratton, C.R. Brumback, B.A. Gordon, M.A. Pearson, K.A. Low, and M. Fabiani, "Effects of measurement method, wavelength, and source–detector distance on the fast optical signal," *Neuroimage* **32**(4), 1576–1579 (2006).
43. J. Talbot, J.N. Barrett, E.F. Barrett, and G. David, "Stimulation-induced changes in NADH fluorescence and mitochondrial membrane potential in lizard motor nerve terminals," *J. Physiol.* **15**(579 Pt 3), 783–798 (2007).
44. J. Berwick, D. Johnston, M. Jones, J. Martindale, C. Martin, A.J. Kennerley, P. Redgrave, and J.E.W. Mayhew, "Fine detail of neurovascular coupling revealed by spatiotemporal analysis of the hemodynamic response to single whisker stimulation in rat barrel cortex," *J. Neurophysiol.* **99**, 787–798 (2008).
45. N.K. Logothetis, H. Guggenberger, S. Peled, and J. Pauls, "Functional imaging of the monkey brain," *Nat. Neurosci.* **2**, 555–562 (1999).

46. R.B. Buxton, E.C. Wong, and L.R. Frank, "Dynamics of blood flow and oxygenation changes during brain activation: the balloon model," *Magn. Reson. Med.* **39**, 855–864 (1998).
47. J. Frahm, G. Kruger, K.D. Merboldt, and A. Kleinschmidt, "Dynamic uncoupling and recoupling of perfusion and oxidative metabolism during focal brain activation in man," *Magn. Reson. Med.* **35**, 143–148 (1996).
48. N.K. Logothetis, J. Pauls, M. Augath, T. Trinath, and A. Oeltermann, "Neurophysiological investigation of the basis of the fMRI signal," *Nature* **412**, 150–157 (2001).
49. S.A. Sheth, M. Nemoto, M. Guiou, M. Walker, N. Pouratian, and A.W. Toga, "Linear and nonlinear relationships between neuronal activity, oxygen metabolism, and hemodynamic responses," *Neuron* **42**, 347–355 (2004).
50. T. Takano, G.F. Tian, W. Peng, N. Lou, W. Libionka, X. Han, and M. Nedergaard, "Astrocyte-mediated control of cerebral blood flow," *Nat. Neurosci.* **9**, 260–267 (2006).
51. P.J. Magistretti, "Cellular bases of functional brain imaging: insights from neuron–glia metabolic coupling," *Brain Res.* **886**(1 – 2), 108–112 (2000).
52. R. Turner, "How much cortex can a vein drain? Downstream dilution of activation-related cerebral blood oxygenation changes," *Neuroimage* **16** (4), 1062–1067 (2002).
53. J. Mayhew, D. Johnston, J. Berwick, M. Jones, P. Coffey, and Y. Zheng, "Spectroscopic analysis of neural activity in brain: increased oxygen consumption following activation of barrel cortex," *Neuroimage* **12**(6), 664–675 (2000).
54. D.A. Boas, S.R. Jones, A. Devor, T.J. Huppert, and A.M. Dale, "A vascular anatomical network model of the spatio-temporal response to brain activation," *Neuroimage* **40**(3), 1116–1129 (2008).
55. T.J. Huppert, M.S. Allen, H. Benav, P.B. Jones, and D.A. Boas, "A multicompartment vascular model for inferring baseline and functional changes in cerebral oxygen metabolism and arterial dilation," *J. Cereb. Blood Flow Metab.* **27**(6), 1262–1279 (2007).
56. A. Devor, P. Tian, N. Nishimura, I.C. Teng, E.M. Hillman, S.N. Narayanan, I. Ulbert, D.A. Boas, D. Kleinfeld, and A.M. Dale, "Suppressed neuronal activity and concurrent arteriolar vasoconstriction may explain negative blood oxygenation level-dependent signal," *J. Neurosci.* **27**(16), 4452–4459 (2007).
57. M. Tomita, M. Ohtomo, and N. Suzuki. "Contribution of the flow effect caused by shear-dependent RBC aggregation to NIR spectroscopic signals," *Neuroimage* **33**(1), 1–10 (2006).
58. H. Sato, M. Kiguchi, A. Maki, Y. Fuchino, A. Obata, T. Yoro, and H. Koizumi. "Within-subject reproducibility of near-infrared spectroscopy

signals in sensorimotor activation after 6 months," *J. Biomed. Opt.* **11**(1), 014021 (2006).
59. S. Feng, F. Zeng, and B. Chance, "Photon migration in the presence of a single defect: a perturbation analysis," *Appl. Opt.* **34**(19), 3826–3837 (1995).
60. D.T. Deply, S.R. Arridge, M. Cope, D. Edwards, E.O. Reynolds, C.E. Richardson, S. Wray, J. Wyatt, and P. van der Zee, "Quantitation of pathlength in optical spectroscopy," *Adv. Exp. Med. Biol.* **247**, 41–46 (1989).
61. E. Okada and D.T. Delpy, "Near-infrared light propagation in an adult head model. I. Modeling of low-level scattering in the cerebrospinal fluid layer," *Appl. Opt.* **42**(16), 2906–2914 (2003).
62. E. Okada and D.T. Delpy, "Near-infrared light propagation in an adult head model. II. Effect of superficial tissue thickness on the sensitivity of the near-infrared spectroscopy signal," *Appl. Opt.* **42**(16), 2915–2922 (2003).
63. W.F. Cheong, S.A. Prahl, and A.J. Welch, "A review of the optical properties of biological tissues," *IEEE J. Quantum Electron.* **26**(2), 2166–2185 (1990).
64. A.N. Yaroslavsky, P.C. Schulze, and I.V. Yaroslavsky, "Optical properties of selected native and coagulated human brain tissues in vitro in the visible and near infrared spectral range," *Phys. Med. Biol.* **47**, 2059–2073 (2002).
65. S.Q. Luo, "Digital Chinese Human dataset and its applications," *Conf. Proc. IEEE Eng. Med. Biol. Soc.* **3**, 1850–1852 (2004).
66. E. Okada, M. Firbank, M. Schweiger, S.R. Arridge, M. Cope, and D.T. Delpy, "Theoretical and experimental investigation of near-infrared light propagation in a model of the adult head," *Appl. Opt.* **36**(1), 21–31 (1997).
67. Y. Fukui, Y. Ajichi, and E. Okada, "Monte Carlo prediction of near-infrared light propagation in realistic adult and neonatal head models," *Appl. Opt.* **42**(16), 2881–2887 (2003).
68. P. Li, S. Zeng, Q. Luo, J. Yang, and L. Guan, "Left prefrontal cortex activation during semantic encoding accessed with functional near infrared imaging," *Space Med. Medical Eng.* **13**(2), 79–83 (2000).
69. B. Chance, E. Anday, S. Nioka, S. Zhou, L. Hong, K. Worden, C. Li, T. Murray, Y. Ovetsky, D. Pidikiti, and R. Thomas, "A novel method for fast imaging of brain function, noninvasively, with light," *Opt. Express* **2**(10), 411–423 (1998).
70. V. Ntziachristos, X. Ma, A. G. Yodh, and B. Chance, "Multichannel photon counting instrument for spatially resolved near infrared spectroscopy," *Rev. Sci. Instrum.* **70**, 193–201 (1999).

71. P. Rama, J.B. Sala, J.S. Gillen, J.J. Pekar, and S.M. Courtney, "Dissociation of the neural systems for working memory maintenance of verbal and nonspatial visual information." *Cogn. Affect. Behav. Neurosci.* **1**(2), 161–171 (2001).
72. J. Jonides, E.H. Schumacher, E.E. Smith, E.J. Lauber, E. Awh, S. Minoshima, and R.A. Koeppe, "Verbal working memory load affects regional brain activation as measured by PET," *J. Cogn. Neurosci.* **9**(4), 462–475 (1997).
73. R.C. Kadosh, K.C. Kadosh, A. Henik, and D.E.J. Linden, "Processing conflicting information: facilitation, interference, and functional connectivity," *Neuropsychologia* **46**(12), 2872–2879 (2008).
74. S. Zysset, K. Muller, G. Lohmann, and D.Y. von Cramon, "Color–word matching stroop task: separating interference and response conflict," *Neuroimage* **13**(1), 29–36 (2001).
75. J. Zhai, T. Li, Z. Zhang, and H. Gong, "Hemodynamic and electrophysiological signals of conflict processing in the Chinese-character Stroop task: a simultaneous near-infrared spectroscopy and event-related potential study," *J. Biomed. Opt.* **14**(5), 054022-1-7 (2009).
76. J. Markela-Lerenca, N. Illeb, S. Kaisera, P. Fiedlera, C. Mundta, and M. Weisbroda, "Prefrontal-cingulate activation during executive control: which comes first?," *Cogn. Brain Res.* **18**, 278–287 (2004).
77. M.L. Schroeter, S. Zysset, T. Kupka, F. Kruggel, and D.Y. von Cramon, "Near-infrared spectroscopy can detect brain activity during a color – word matching Stroop task in an event-related design," *Human Brain Mapping* **17**, 61–71 (2002).
78. S. Nioka, JI. Tracy, S. Rains, S. Bunce, and B. Chance, "Anagram problem-solving and learning in anterior prefrontal cortex," *J. Innovative Opt. Health Sci.* **1**, 157–175 (2008).
79. B. Sun, L. Zhang, H. Gong, J. Sun, and Q. Luo, "Detection of optical neuronal signals in the visual cortex using continuous wave near-infrared spectroscopy," *Neuroimage* **87**, 190–198 (2014).
80. T. Li, H. Gong, and Q. Luo, "Visualization of light propagation in Visible Chinese Human Head for functional near-infrared spectroscopy," *J. Biomed. Opt.* **16**(4), 045001-1-6 (2011).
81. T. Li, Q. Luo, and H. Gong, "MCVM: Monte Carlo modeling of photon migration in voxelized media," *J. Innovative Opt. Health Sci.* **3**, 91–102 (2010).
82. M. Ferrari and V. Quaresima, "A brief review on the history of human functional near-infrared spectroscopy (fNIRS) development and fields of application," *Neuroimage* **63**(2), 921–935 (2012).

Hui Gong is one of the founding members of Britton Chance Center for Biomedical Photonics, Huazhong University of Science and Technology and Wuhan National Laboratory for Optoelectronics. Her research interests focus on optical neuroimaging and visible brainwide networks. She has made many fundamental contributions to the development and applications of the Micro-Optical Sectioning Tomography (MOST). Her team made creative contributions to whole mouse brain staining, MOST system design and implementation, high-resolution mouse brain atlas and visualization, which won the China's Top Ten Major Scientific Progresses in 2011. She is devoted to optimize a portable near-infrared spectroscopy system and made novel applications on brain activities such as attention, working memory, emotion, and functional connectivity. As first or corresponding author, she has published papers in science, nature methods, nature communications and neuroimage, and so on.

Qingming Luo is the vice-president of Huazhong University of Science and Technology and executive deputy director of Wuhan National Laboratory for Optoelectronics. He is an elected fellow of the International Society for Optics and Photonics, The Institution of Engineering and Technology, and The Optical Society. His research interests focus primarily on multiscale optical bioimaging and cross-level information integration. Since 1996, he has been devoted to new techniques and novel applications in life sciences, including laser speckle imaging and combination with optical intrinsic signal imaging, small animal imaging of fluorescence diffusion optical tomography coregistered with micro-CT, micro-optical sectioning tomography, and functional near-infrared imaging. He is currently leading the project "Visible Brainwide Networks" at single-neuron resolution and the chief scientist of National Major Scientific Instruments and Equipment's Development Project "Instrument Development and Application Demonstration of the Micro-Optical Sectioning Tomography System". He created "the most detailed three-dimensional map of all the connections between the neurons in a complete mouse brain" and "demonstrated the first long-range tracing of individual axons in the mouse brain". He is also the chief scientist of the project "Novel Technologies and Methods of Optical Molecular Imaging for Protein Function *in vivo*" supported by National Basic Research Program of China, which aims to develop a series of novel technologies and methods of optical molecular imaging, such as the high-resolution fast microscopy for simultaneously monitoring the molecular events of multiple proteins, multiscale optical imaging for protein function, cross-level integration and visualization of the related information of the protein function. He holds 60 patents and has coauthored more than 200 papers in peer-reviewed journal, including *Science, Nature Cell Biology, Nature Communication, PNAS, Optics Letters, Optics Express*, and *Journal of Biomedical Optics*, with an *h*-index of 42 according to Google Scholars. He won the Cheung Kong Professorship of Ministry of Education of China in 1999, the National Science Fund for Distinguished Young Scholars in 2000, the second place prize in State Natural Sciences Award in 2010, China's Top Ten Major Scientific Progress, and the second place prize in State Technological Invention Award in 2014.

Shaoqun Zeng Biomedical Photonics distinguished professor of "Chang Jiang Scholars Program" of the Ministry of Education of China (2007), is a professor and doctoral supervisor of Huazhong University of Science and Technology, deputy director of Britton Chance Research Center for Biomedical Photonics, leader of preparatory team of the Key Laboratory of Biomedical Photonics of Ministry of Education, China, and the associate dean of School of Life Science and Technology, Huazhong University of Science and Technology.

He received his BS and PhD degrees in physical electronics from Huazhong University of Science and Technology in 1990 and 1996, respectively, stayed on in the university as a teacher in April, 1996. He conducted the cooperative research in University of Pennsylvania from October, 1997 to February, 1998 and visited Yale University and researched there many times from July, 2002 to June, 2005. Since working, he mainly studied technologies such as femtosecond laser biomedical imaging, new technologies and new methods in high resolution optical functional imaging of protein and biotechnology information. In recent years, he has successfully presided over and accomplished two projects of National Natural Science Foundation of China and one National "863" Program. His achievement of the femtosecond laser biomedical imaging was elected as one of "Major Achievements in Optics of China 2007," the femtosecond laser random scan microscopic imaging system invented by him has improved the observation rate by two orders of magnitude applied by such well-known academic institutions as Yale University and his Research and Survey Report has been published by the first class periodical of neurosciences Neuron. He has been authorized with eight Chinese Patents and additionally has nine Chinese Patents, one US Patent and one international PCT patent that have been reviewed. He won one first prize of Prize of Hubei Province for Natural Science in 2007 (ranking the third), one first prize of the Science and Technology Progress Award of Hubei Province in 2006 (ranking the fifth), and one second prize of Prize of

Hubei Province for Natural Science in 2003 (ranking the second) and has published more than 90 academic papers about optical microscopic imaging and the new methods and instruments of protein functional imagining, among which more than 60 academic papers have been collected by SCI, including the first class periodicals of optics, instruments and neuroscience, such as Optics Letters, Review of Scientific Instruments and Neuron.

He was selected as one of the key members of the Outstanding Young Teachers (2002), one of the new century excellent talents in University by the Ministry of Education (2006) and one of the Excellent Science and Technology Workers of Wuhan City (2007) successively. He is also the member of Society of Photo-Optical Instrumentation Engineers, American Academy of Neurology, and Chinese Optical Society, a senior member of the Chinese Institute of Electronics and an editorial board member of ACTA LASER BIOLOGY SINICA. He was once the Chairmen of Organizing Committee of the fifth to seventh International Symposiums on Biomedical Optics (BiOS) (2006–2008, in Wuhan), served as a member of International Program Committee of the Special Session on BIOS-Optical Diagnostics and Sensing of SPE Photonics West for many times and serves as reviewers of international journals, such as Optics Letters, Optics Express, Journal of Biomedical Optics, Applied Optics, IEEE Sensors and Journal of X-ray Science and Technology.

Shoko Nioka started her carrier as an anesthesiologist in Kyoto Prefectural University of Medicine for 10 years. After being a visiting professor in UCLA Sports Medicine and Muscle Physiology, she worked with Britton Chance in Respiratory Physiology and Biochemistry and Biophysics at the University of Pennsylvania for 27 years. Her research fields include NMR and optical spectroscopy and imaging methods to study the blood hemodynamics and mitochondrial bioenergetics in muscle under exercises and in functioning brain, resulting in a number of publications. She joined several small business companies for development and commercialization of the portable devices for studying muscle exercise and brain functional activity and the fetal pulse oximeter, NIRS breast cancer detectors. To date, she has been working on fluorescence biomarkers in Radiology UPENN as well as for fluorescence oral cancer detection in National Cheng Kung University in Taiwan.

Yasufumi Kuroda received his bachelor of pedagogy and master of pedagogy degrees in mathematics education from Osaka Kyoiku University, Japan, in 1988 and 1990, respectively. He received a PhD in human sciences by Osaka University, Japan, in 2005. He is a professor at the Kyoto University of Education, Japan, since 2014. His research interests include mathematics education, neuroscience, and educational neuroscience as interdisciplinary study.

Chapter 9
Signal Quantification and Localization in Tissue Near-Infrared Spectroscopy

Stephen J. Matcher
University of Sheffield, Sheffield, UK

9.1 Introduction

Tissue near-infrared (NIR) spectroscopy is now more than 30 years old. The early promise of the technique as a unique tool for the noninvasive measurement of brain and muscle oxygenation is even now only beginning to be fulfilled. Despite the efforts of many hundreds of researchers, the basic problems associated with quantification and localization of the detected signal remain severe hindrances to the widespread clinical adoption of this technique. In contrast to its ubiquitous relative the pulse oximeter, the clinical cerebral and muscle oximeter is still rarely used outside the specialist research laboratory. A number of factors give rise to this situation:

- To be of immediate clinical utility, a cerebral oximeter should provide an absolute measurement of the hemoglobin oxygenation in the tissue of interest. Such information is readily available if blood samples are drawn from a patient and analyzed in a co-oximeter. Similarly, the pulse oximeter can provide this information for hemoglobin located in the arterial vascular bed. Unfortunately, providing a measurement of the average hemoglobin oxygenation in a thick tissue section such as the brain is fraught with complexities—so many, in fact, that early commercial cerebral oximeters did not attempt to measure this quantity at all. Instead, these devices measured only the change in tissue oxygenation, with reference to an unknown starting point. Such measurements can be of use to the physiologist, but are of relatively little use to, say, the neurologist who is running a head injury unit.

Although some absolute hemodynamic variables such as absolute cerebral blood volume and flow can be estimated using such "trend" measurements, the procedure is time-consuming, is cumbersome, and requires both skilled operators and relatively sophisticated data interpretation.

- The head and other organs are physically heterogeneous, and it is not obvious which of several physiologically distinct regions contributes the majority of the measured signal. The most obvious example is the adult head, where the tissue of interest, the brain, must be interrogated through overlying tissues (the scalp, skull, and meninges) that are of considerable thickness and whose blood supplies are largely independent from the brain's own blood supply. While modern instrumentation can successfully transilluminate the neonatal head, this is generally not possible with the much larger adult head. Thus, measurements must be made in a "reflection" mode, with source–detector spacing not exceeding approximately 5.0 cm. A naïve analysis then suggests that the interrogating light has sampled these overlying tissues to a similar or greater extent than the cerebral tissues. In consequence, there has been a long-standing controversy about whether the oxygenation changes detected by NIR oximeters are truly reflective of changes occurring within the cerebral gray and white matter, particularly when used on adults.

It is the goal of this chapter to review the subject of quantification and localization of oxygenation measurements in tissue near-infrared spectroscopy. Considerable progress has been made in the last 10 years in instrumentation, data analysis, and theoretical modeling of light propagation in biological tissues. A number of instruments are now available that are capable of measuring absolute hemoglobin oxygenation, several of them commercially. New theoretical models have been developed that, given accurate knowledge of the optical properties of various tissues, can predict which regions of a heterogeneous organ contribute to the measured signal. New application areas have emerged—in particular, the exciting new field of functional near-infrared studies of the brain. Finally, the ultimate goal of localized spectroscopy—namely, the generation of oxygenation images—now appears to be a less remote possibility than before, with encouraging results having recently been obtained on tissue phantoms.

9.2 Oximetry

The continued availability of oxygen is vital for all animal life. Oxygen is a necessary substrate in the oxidative phosphorylation pathway, in which respiratory cells synthesize the energy-rich phosphorus compound adenosine triphosphate (ATP). Without oxygen, these cells must resort to anaerobic glycolysis to generate the energy needed to maintain their vitality. This reaction is much less efficient than oxidative phosphorylation and cannot meet the energy requirements of the cells for more than a brief period. The measurement of tissue oxygen levels is thus of great interest, and it can be accomplished in a number of ways.

The most direct form of measurement is to measure the partial pressure of oxygen dissolved within the interstitial fluid or blood. This can be accomplished invasively by inserting a polarographic needle electrode into the tissue, or noninvasively by using a transcutaneous oxygen electrode. In the latter technique, a heater placed on the skin causes sufficient vasodilation such that considerable amounts of oxygen diffuse across the epidermis and can be measured by a conventional noble-metal electrode placed on the skin surface.[1] The technique can also be used to measure CO_2 levels,[2] but it has the disadvantage that normal adult epidermis is frequently so thick that insufficient oxygen can diffuse across it to the surface to yield an accurate measurement. Transcutaneous electrodes are routinely used in neonatal patients in intensive care, however, in whom the epidermis is much thinner. A further disadvantage is that only systemic changes in blood oxygenation will produce changes in transcutaneous O_2; local changes in, for example, cerebral oxygen levels caused by birth asphyxia[3] will not be directly detectable.

Such problems have motivated the development of truly noninvasive techniques for measuring tissue oxygenation, and optical spectroscopy has become established as the technique of choice for routine clinical monitoring.

9.2.1 Optical spectroscopy

Optical spectroscopy is a well-established technique for determining the chemical composition of dilute samples of a substance. Consider a cuvette containing a nonscattering clear liquid, into which is dissolved an absorbing compound, presents at an overall concentrationc_{ab}. The conventional units of c_{ab} are moles per liter (molar, M). Further, consider that a collimated beam of radiation of intensity I_0 illuminates the cuvette and that this beam traverses a physical path d, before emerging with a reduced intensity I. The Beer–Lambert law then states that I and I_0 are related by

$$\frac{I}{I_0} = \exp(-\mu_a d), \tag{9.1}$$

where μ_a is the absorption coefficient of the medium (conventional units cm^{-1}). A common convention is to express the overall reduction of intensity in terms of a logarithmic unit, the optical density (OD), where an attenuation of 1 OD corresponds to a 10-fold reduction in intensity. Hence, attenuation A is given by

$$A = \log_{10}\left(\frac{I_0}{I}\right) = \frac{\mu_a d}{\ln 10}. \tag{9.2}$$

If the absorbing substance is dissolved uniformly throughout the host liquid, then the overall absorption coefficient μ_a can be expressed in terms of the absorber concentration c_{ab}:

$$\mu_a = c_{ab}\varepsilon_{ab},$$

where ε_{ab} is known as the "specific absorption coefficient." The conventional units of μ_a and ε_{ab} are somewhat different. μ_a is defined in terms of natural logarithms and has units of cm^{-1}. ε_{ab} is defined as the number of ODs of attenuation produced by the absorber, at a concentration of 1 millimolar and over a physical path of 1 cm. Hence, to convert between ε_{ab} in units of OD cm^{-1} mM^{-1} and μ_a in units of cm^{-1}, one must multiply by a scaling factor

$$\mu_a = c_{ab} \cdot \varepsilon_{ab} \cdot \ln(10) = 2.3c_{ab} \cdot \varepsilon_{ab}. \tag{9.3}$$

Now consider that the cuvette contains not one, but N distinct absorbing compounds with concentrations $c_{ab_i}(i = 1 \ldots N)$. If these compounds do not chemically interact, then the overall absorption coefficient is simply obtained by adding the contributions of each individual absorber:

$$\mu_a = 2.3\sum_{i=1}^{N} c_{ab_i}\varepsilon_{ab_i}. \tag{9.4}$$

These equations form the basis of spectroscopic chemical assay techniques. The Beer–Lambert law shows that μ_a can be determined readily from a measurement of overall light attenuation, provided the physical pathlength d is known. If μ_a is then determined at M wavelengths (where $M \geq N$), then the concentrations c_{ab_i} can be determined by solving the resulting system of M linear simultaneous equations (see Section 9.6).

The utility of optical spectroscopy to perform oximetry becomes apparent when one considers the specific absorption coefficient spectrum of the metalloprotein hemoglobin ε_{ab} (λ). Hemoglobin, which is found within the red blood cells, binds to oxygen molecules in regions of high pO_2 and unbinds from oxygen in regions of low pO_2, thereby transporting oxygen from the lungs to other areas of the body. Hemoglobin was one of the first compounds to be studied by spectroscopy, with its optical absorption spectrum being reported by Hartridge and Hill in 1914. From this and subsequent work, it has become well established that hemoglobin shows distinctive absorption bands in the UV (the Soret band), visible, and near-infrared ranges, all of which change form depending on whether the molecule is bound to oxygen. Figure 9.1 shows the visible and near-infrared absorption spectra of oxyhemoglobin (HbO_2) and deoxyhemoglobin (Hb).

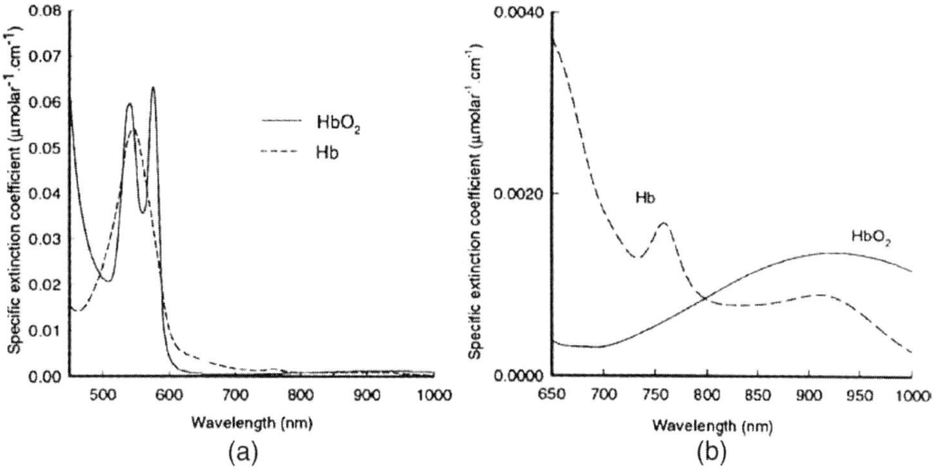

Figure 9.1 The specific absorption coefficient spectra of oxyhemoglobin and deoxyhemoglobin in the visible and near-infrared spectral range.[4]

It is apparent that a spectroscopic analysis of blood can, in principle, determine the concentrations of the oxygenated and deoxygenated forms of hemoglobin. This simple observation has led to a number of technical applications in the healthcare field, perhaps the most important and widespread being the optical co-oximeter. This device performs a rapid spectral analysis of a blood sample drawn from a patient and measures the absolute concentrations of Hb and HbO_2. Many co-oximeters also determine levels of carboxyhemoglobin (HBCO) and hemiglobin (Hi). The measurement of absolute hemoglobin saturation (SO_2),

$$SO_2(\%) = \frac{HbO_2}{Hb + HbO_2} \times 100, \tag{9.5}$$

is of vital importance clinically because it is related indirectly (via the "oxygen dissociation curve") to the partial pressure of oxygen within tissue (pO_2), which in turn gives a direct measure of the availability of oxygen to the cell mitochondria (the site of oxidative phosphorylation).

The measurement of SO_2 by co-oximetry requires drawing blood samples, which has clear drawbacks in the clinical setting, particularly for neonatal monitoring. The co-oximeter typically requires a blood sample of 0.3 mL, which poses difficulties if continuous monitoring is required, given that preterm infants may have a total blood supply of just a few deciliters. Regular spectroscopic readings can be obtained by inserting fiber-optic reflectance probes into the arterial or venous vessels,[5] but these devices themselves have problems associated with blood cell aggregation on the sensing tip, and regular calibration against conventionally sampled blood is generally required.

9.2.2 Noninvasive hemoglobin spectroscopy

The field of noninvasive hemoglobin spectroscopy received significant attention during World War II, as Allied bomber crews were by then reliant on an artificially maintained oxygen supply due to the high operational altitude of their aircraft. A failure of the oxygen supply would clearly be catastrophic, so a noninvasive means of determining SO_2 was sought. In 1942, Millikan developed the first practical device, which used optical radiation to transilluminate the pinna of the ear.[6] This device was refined by others, but it never offered the stability and reproducibility required for clinical use until Hewlett-Packard introduced an ear oximeter that, by measuring at eight wavelengths and using a heater to promote maximal skin vasodilation, finally offered usable performance. This device, while being widely regarded as the "gold standard" for noninvasive hemoglobin saturation measurements, proved too bulky and expensive to gain widespread application in the clinic.

A decisive breakthrough occurred in 1975 when Nakajima introduced the pulse oximeter.[7] This device relies on the fact that the concentration of hemoglobin in the arterial vascular bed displays a pulsatile variation in synchrony with the heartbeat, whereas the capillary and venous beds do not exhibit this effect. In consequence, if a thin tissue section such as the ear is transilluminated, the optical attenuation displays a pulsatile component, whose variation is due solely to arterial blood. By subtracting attenuation spectra at systole and diastole, a difference spectrum is obtained that effectively represents the attenuation spectrum of arterial blood alone. It is then relatively straightforward to extract from this the relative concentrations of Hb and HbO_2, in much the same manner as in an *in vitro* co-oximeter. The pulse oximeter is, at the time of writing, essentially the only noninvasive optical oximeter that can be found in widespread clinical use.

9.2.3 Near-infrared spectroscopy (NIRS)

Both the ear oximeter and the pulse oximeter are restricted to measuring optical attenuation across comparatively thin tissue sections. In 1977, Jöbsis pointed out that modern near-infrared light sources and detectors had reached the point at which attenuation as high as 8–10 ODs could be measured using integration times of a few seconds. Furthermore, in the 600–1000 nm spectral region, a unique transparency "window" exists, in which tissue displays attenuation per unit distance as low as 1 OD cm^{-1}. Jöbsis effectively invented the field of tissue near-infrared spectroscopy (NIRS) by successfully collecting NIR transillumination spectra across the intact cat head and thereby monitoring the oxidation–reduction (redox) state of respiratory chain enzymes within the cell mitochondria.[8] In 1985, Ferrari et al. reported the first successful NIR measurements on the human brain.[9] In 1988, Cope and Delpy reported a system capable of transilluminating the intact head of a human neonate.[10] The field of tissue near-infrared spectroscopy has subsequently expanded enormously. To date, 1000 or so refereed papers have appeared that describe both the methodology and clinical applications.

9.3 Tissue Near-Infrared Spectroscopy

Within biological tissues, a number of substances ("chromophores") display significant absorption in the wavelength range 600 to 1000 nm. These substances can be categorized into those that exhibit oxygen-dependent absorption and those whose absorption is essentially fixed during the course of a clinical measurement.

9.3.1 Oxygen-dependent chromophores

Oxygen-dependent chromophores are the compounds of primary interest in tissue NIR spectroscopy, as their absorption spectra are directly influenced by the presence of oxygen.

9.3.1.1 Hemoglobin

Hemoglobin is a complex metalloprotein (molecular weight = 64,500) that consists of four heme groups. Each of these heme groups can bind one molecule of O_2 by undergoing a conformal change that causes the O_2 molecule to become physically (rather than chemically) bound. This conformal change shifts the absorption band exhibited by the Fe atom within each heme group, producing the observed spectral difference between Hb and HbO_2.

9.3.1.2 Cytochrome aa_3 (cytochrome-oxidase)

Cytochrome aa_3 is an enzyme located within the cell mitochondria that catalyzes the final step of the respiratory chain in which oxygen is reduced to water. The enzyme contains two copper centers, denoted as CuA and CuB. The CuA center is responsible for the distinctive broad absorption spectrum centered at approximately 830 nm when the enzyme is oxidized. This band disappears when the enzyme is reduced, so measuring the amplitude of this band can, in principle, allow one to determine the proportion of the enzyme that is oxidized relative to the proportion that is reduced (the "redox" state). This potential application has received enormous attention because it would allow for tracking oxygen utilization directly to the site of ATP synthesis. Since the concentration of this enzyme is effectively a constant during a single measurement, only the difference spectrum need be considered. Figure 9.2 shows the difference spectrum, measured on a sample of the purified enzyme *in vitro*.[11]

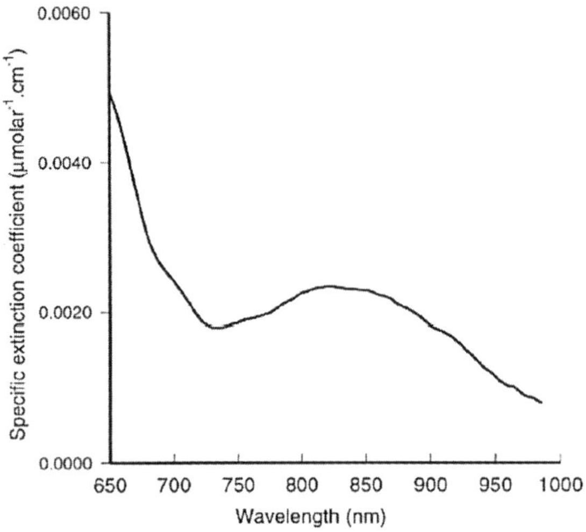

Figure 9.2 The specific absorption coefficient difference spectrum of cytochrome oxidase.[11]

9.3.1.3 Myoglobin

Myoglobin, a metalloprotein, is closely related to hemoglobin and also fulfills an oxygen-delivery role. It is found in muscle but not in brain tissue. Its high affinity for oxygen means that it does not offload its oxygen until very low tissue pO_2 values are reached. In muscle, it serves as a reservoir to store and then slowly release the oxygen delivered by blood hemoglobin. Its absence in brain tissue helps to explain that organ's extreme sensitivity to oxygen deficiency. The NIR absorption spectrum of myoglobin is essentially identical to that of hemoglobin, so that it is not possible to monitor myoglobin independently without resorting to a supplementary technique such as magnetic resonance spectroscopy (MRS).

9.3.2 Oxygen-independent chromophores

Oxygen-independent chromophores are compounds that display significant absorption in the NIR, but their absorption spectra do not vary significantly as the blood/tissue oxygen level varies over a physiologically reasonable range. Their presence can be ignored when using a trend or quantified trend monitor and when deriving hemodynamic variables such as cerebral blood flow (CBF) from such trend measurements. However, their presence must be accounted for when extracting measurements of tissue oxygenation from absolute measurements of μ_a at various wavelengths (see Section 9.5).

9.3.2.1 Water

Water accounts for more than 80% of the volume of the human body and is a uniquely versatile solvent in which a host of biochemical reactions can take

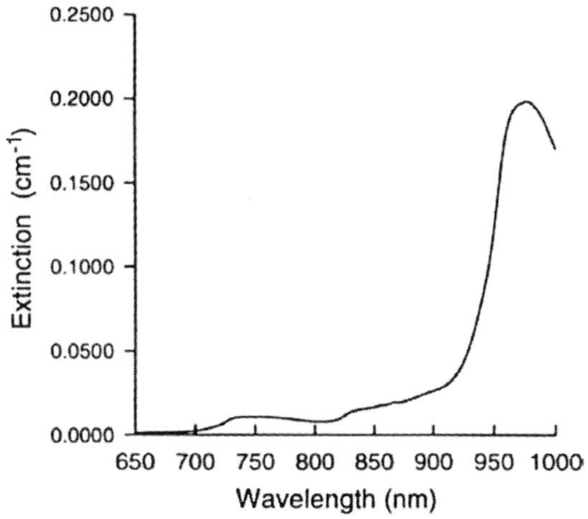

Figure 9.3 The specific absorption coefficient of pure water.[14]

place. The O—H bond of H_2O displays a characteristic vibrational spectrum, with absorption peaks at (in decreasing order of intensity) 10.6 μm, 1.95 μm, 1.4 μm, and 0.98 μm. Still-weaker overtones can be observed at 840 nm and 760 nm. At wavelengths less than 760 nm, water is effectively transparent until the UV range. Figure 9.3 shows the NIR absorption spectrum of 100% water (an effective concentration of 56 *M*). Water can exist both in the free state and bound to macromolecules. While this differing state produces large changes in, for example, the magnetic resonance precessional frequency, there is little evidence that the absorption spectrum of water is greatly affected by it. Temperature and salinity, in contrast, produce measurable changes in the water absorption spectrum.[12] The shift due to temperature has been proposed as a means of measuring body temperature *in vivo*.[13]

9.3.2.2 Lipids

Lipids are found throughout the body, but particularly in adipose tissue. These compounds also constitute a large component of the cell membrane and the myelin sheath surrounding nerve axons. In postmenopausal women, adipose tissue comes to dominate the composition of the breast. The C—H bond produces a vibrational spectrum reminiscent of water, with a strong NIR band at 930 nm (Fig. 9.4) and a weaker overtone at 760 nm.

9.3.2.3 Other cytochromes

The respiratory chain contains four other cytochrome enzymes: cytochrome *b* (cyt *b*), cytochrome c_1 (cyt c_1), cytochrome *c* (cyt *c*), and cytochrome *a* (cyt *a*). However, the high specific absorption coefficient of cyt aa_3, coupled with its high concentration and possession of a distinctive absorption band in the

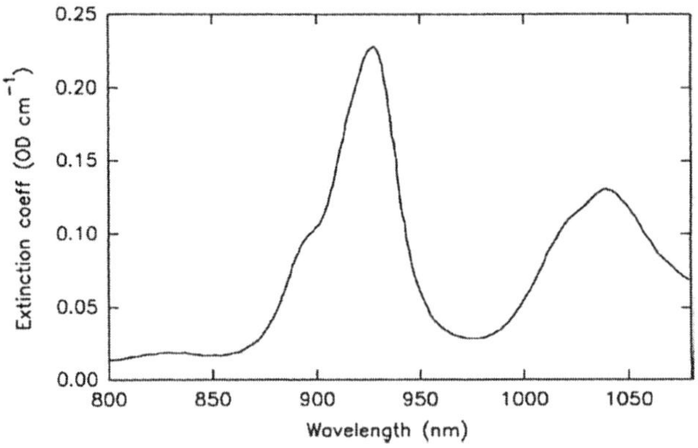

Figure 9.4 The specific absorption coefficient of lipids (pork fat).[15]

NIR, means that cyt aa_3 is the enzyme of most interest in tissue NIR spectroscopy.

9.4 Spectroscopy in a Highly Scattering Medium

The ultimate aim of tissue near-infrared spectroscopy is to determine the absolute concentrations of various compounds in an organ of interest from a spectrum of light attenuation measured across that organ. Unfortunately, while it is technically fairly simple to obtain the attenuation spectrum, it is much more challenging to extract from these data the absolute concentrations of the chromophores in the tissue. The primary reason for this difficulty is simple: NIR photons interact with tissue via two main mechanisms, only one of which, absorption, is directly determined by the chromophore concentrations. NIR photons are also subjected to intense elastic scattering, which itself causes light attenuation independent of the concentration of chromophore. This situation stands in contrast to the situation that normally pertains in analytical spectroscopy, where liquid samples are deliberately prepared to possess very low levels of light scattering. This is clearly not possible to achieve *in vivo*, because the elastic light scattering is produced by the structure of intact cells. The effect of light scattering on tissue spectroscopy is profound; the remainder of this chapter will consider how elastic light scattering complicates the task of measuring both the absolute concentration of absorbers within tissue and changes in those concentrations. Some of the attempts that have been developed to overcome the problem will be discussed as well.

In Section 9.2.1, we described the steps needed to determine absorber concentrations in a nonscattering medium. Recall that, in the absence of scattering, the absorption coefficient μ_a can be directly obtained from the measured attenuation A via the Beer–Lambert law. Hence, one can determine the concentration of N absorbers by making measurements of A at M ($\geq N$) wavelengths and solving the resulting set of M linear simultaneous equations.

Unfortunately, this simple methodology cannot be applied to biological tissues, because such tissues are multiple scattering in the NIR spectral region. Elastic scattering is the dominant scattering process in biological tissues; it arises due to the refractive index mismatch between intracellular and extracellular fluids (see Chapter 1). A variety of mathematical models have been developed to describe both single and multiple light scattering, but for now we will be content to describe the scattering process using a quantity analogous to μ_a—namely, the scattering coefficient μ_s.

Consider again the cuvette mentioned earlier. If this cuvette now contains a liquid that does not absorb light but merely scatters it, then the outgoing intensity will still be reduced—not because light energy has been converted to another form, but simply because photons have been deflected away from their original direction. If both the incident beam divergence and the detector acceptance angle are small, then the incident and detected intensities are related by

$$\frac{I}{I_0} = \exp(-\mu_s d). \tag{9.6}$$

In biological tissue and in the NIR, μ_a is typically 0.4 cm^{-1}, whereas μ_s is almost three orders of magnitude higher, at 100 cm^{-1}. For NIR light propagating through tissue, scattering events are thus much more probable than absorption events, which invalidates the simple Beer–Lambert law.

A number of theoretical models exist for describing the propagation of light in such a multiple-scattering medium (see Section 9.8.1). For now, let us consider plotting graphically the result of performing a simple experiment in which the absorption coefficient of both the nonscattering and scattering liquids are increased from zero, leading to an increase in measured light attenuation A. The results of this hypothetical experiment are shown graphically in Fig. 9.5.

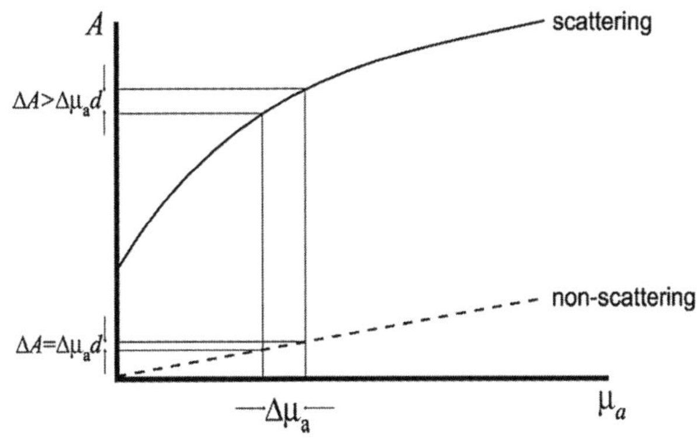

Figure 9.5 Attenuation versus absorption coefficient, plotted for both scattering and nonscattering samples.

When plotted against μ_a, the nonscattering liquid will clearly yield a straight line of gradient d that passes through the origin. The results for the scattering solution differ in two ways:

1. Since scattering produces attenuation in the absence of absorption, the intercept of the curve is no longer zero.
2. The curve is no longer linear, and has a gradient that exceeds d at low μ_a, and asymptotically approaches d at large μ_a. This effect is a direct result of the increased path length traversed by photons in propagating across a distance d. Photons now describe a random walk and propagate along a convoluted path whose length considerably exceeds d. As a consequence, a given increase in μ_a produces a larger increase in A than would result in the absence of scattering. This effect is most pronounced when μ_a is small, as photons can then perform a random walk over considerable distances before being absorbed. As μ_a increases, one clearly tends toward the situation in which the normal Beer–Lambert law applies, and hence the slope of $A(\mu_a)$ approaches d.

The ultimate goal of analytical spectroscopy is to infer μ_a at several wavelengths and hence determine the concentrations of the absorbers. This is comparatively simple when the Beer–Lambert law applies, as the measurable value (A) is straightforwardly related to the desired quantity (μ_a). In a multiple-scattering medium, however, this is not a trivial task, as the precise form of $A(\mu_a)$ is generally not known. One can identify two main problems:

- The nonzero intercept makes it extremely difficult to infer an absolute value of μ_a given an absolute measurement of A. Accurately calculating this intercept requires detailed knowledge of both the scattering properties of the tissue and its geometry.
- Making accurate absolute measurements of A in a clinical context is not a trivial task. Apart from experimental difficulties, factors such as skin pigmentation are major sources of interference in the measurement.

As a consequence, the goal of absolute, quantified tissue oximetry using NIR light is a major research objective. A less ambitious goal is quantified trend oximetry, in which only changes in oxygenation (relative to an unknown starting point) are measured. Clearly, this approach eliminates the need to know the intercept of the $A(\mu_a)$ curve, but it does require knowledge of the local gradient of the curve at the operating point of the instrument (see Section 9.6). The original NIR oximeters were less ambitious still, providing only unquantified trend measurements (i.e., they could tell whether oxygen levels were rising or falling but not by how much). Figure 9.6 illustrates these three levels of sophistication by showing the measurements that each type of instrument would produce during a hypothetical arteriovenous occlusion experiment on the human forearm.

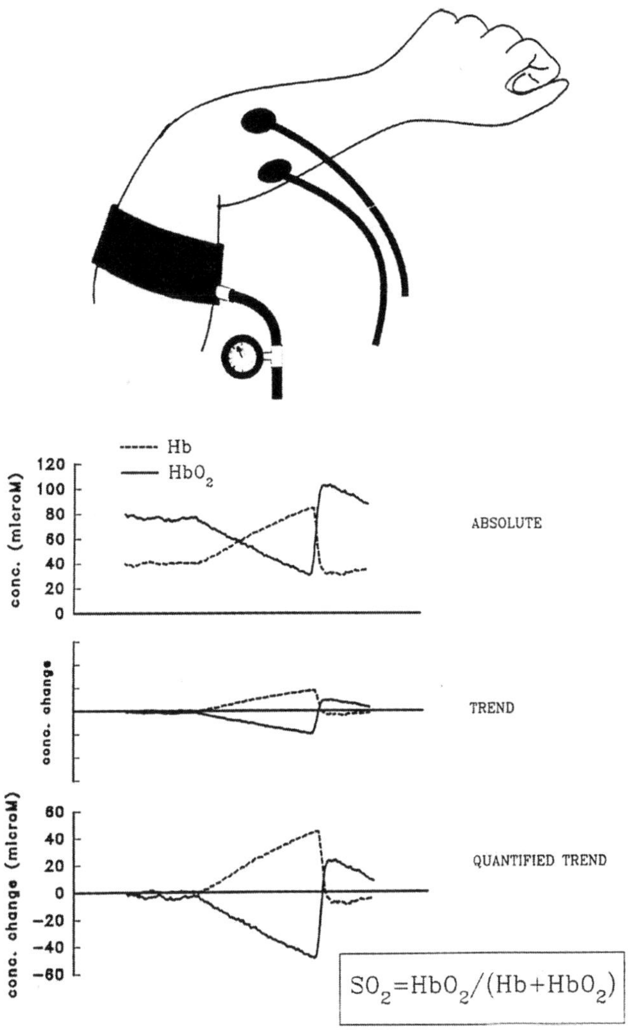

Figure 9.6 The output of three NIR oximeters of varying sophistication during a hypothetical ischemic deoxygenation of the forearm muscle. (top) Absolute quantified oximeter. (middle) Unquantified trend monitor. (bottom) Quantified trend monitor.

We will now go on to consider the theory and the practical implementation of both quantified trend and absolute oximeters.

9.5 Absolute Measurements

The most direct route to an absolute measurement of chromophore concentration is, of course, to directly measure μ_a. If μ_a can be estimated at M wavelengths $\lambda_1, \ldots, \lambda_M$, then N concentration values $c_{ab_1}, \ldots, c_{ab_N} (N \leq M)$ can be estimated using multilinear regression (see Section 9.6) In this section we will describe various methods that have been proposed to accomplish this. Broadly speaking, we will discriminate between methods that employ a rigorous model of light transport to relate measured properties of the exiting light distribution to μ_a and μ_s from "chemometric" methods that develop an empirical calibration algorithm given a set of training data. Among methods using a model of light transport, we will further discriminate between methods that rely on the diffusion approximation and those adopting a more general approach.

In the clinical context, most interest revolves around the measurement of μ_a, as this allows direct measurement of hemoglobin oxygenation and cytochrome oxidase redox state. However, μ_s can also contain useful information and has been suggested as a means of monitoring glucose concentration,[16] neuronal activation,[17] and muscular dystrophy.[18]

9.5.1 Use of a "forward model" of light transport

These methods employ a mathematical relationship (a "forward model") to predict various properties of the emerging light distribution in terms of μ_a and μ_s. Given a function f predicting at least two independent properties of the exit light $\wp_{\text{out}}$ (e.g., intensity or mean photon time of flight of the exiting photons)

$$\wp_{out} = f(\mu_a, \mu_s),$$

then the measured values $\wp_{\text{out}_i}^{(m)}$ $(i = 1, 2)$ can be used to extract estimates for μ_a and μ_s by solving, for example, the nonlinear system of equations

$$\wp_{\text{out}_i}^{(m)} - f_i(\mu_a, \mu_s) = 0 \tag{9.7}$$

using an iterative scheme such as the Newton–Raphson method.[19] In some circumstances, a large number M $(M > 2)$ of measurements may be available, in which case a nonlinear least squares approach in which

$$\chi^2 = \sum_{i=1}^{M} [\wp_{\text{out}_i}^{(m)} - f_i(\mu_a, \mu_s)]^2 \tag{9.8}$$

is minimized with regard to μ_a and μ_s can be used.

The most accurate forward model for light transport in a multiple-scattering medium (excepting wave-theory formalisms) is the Boltzmann transport equation,[20] which can account for all features of light transport apart from wave effects such as interference. In general, however, this model is mathematically intractable, so it is necessary to simplify it. The most popular method is to expand the photon radiance at every spatial point into a spherical harmonics power series. If the expansion is truncated at the Nth term, the resulting approximation is termed the P_N approximation[21] and the Boltzmann transport equation reduces to a set of N coupled differential equations.

For $N = 1$ (in which the radiance is described by an isotropic term plus a small directional flux) the spatio-temporal variation of the photon density Φ (number of photons per unit volume) is described by a single equation, the P_1 (or telegraphers) equation:

$$-D\nabla^2\Phi(\bar{r},t) + \left[\mu_a c + \left(1 + \frac{3D\mu_a}{c}\right)\frac{\partial}{\partial t} + \frac{3D}{c^2}\frac{\partial^2}{\partial t^2}\right]\Phi(\bar{r},t) = \left(1 + \frac{3D}{c^2}\frac{\partial}{\partial t}\right)q_0(\bar{r},t). \quad (9.9)$$

The term D is the photon diffusion coefficient $\frac{c}{3(\mu_a+\mu_s')}$, where $\mu_s' = \mu_s(1-g)$ is the "reduced scattering coefficient," and g is the "scattering anisotropy factor" $(-1 \leq g \leq 1)$.[20] Typical values of g for biological tissue are in the range 0.7–0.98.[22]

If $\Phi(\bar{r},t)$ and $q_0(\bar{r},t)$ are assumed to vary only slowly with time, then the second-order time-derivative on the LHS and the first-order derivative on the RHS can be neglected, yielding the familiar diffusion equation

$$-D\nabla^2\Phi(\bar{r},t) + \left[\mu_a c + \frac{\partial}{\partial t}\right]\Phi(\bar{r},t) = q_0(\bar{r},t). \quad (9.10)$$

The diffusion equation is by far the most popular forward model in use at the moment.

The value of the diffusion equation in this context is that, for simple geometries, it yields comparatively straightforward closed-form expressions for the intensity of light at the tissue boundary as functions of position and time. A variety of simple geometries can be solved, such as the infinite medium, semi-infinite half-space,[23] infinite slab, infinite cylinder, finite cylinder, and sphere.[24] A Fourier transform procedure can be used to obtain relations between intensity and source modulation frequency instead of intensity and time of flight,[24] which allows frequency-domain instrumentation to be used. We will now describe three principal methods by which μ_a and μ_s' may be estimated using diffusion theory.

9.5.1.1 Spatially resolved spectroscopy (SRS)

Consider the semi-infinite half-space geometry below, with a pencil beam of photons incident at $r_{sd} = z = 0$.

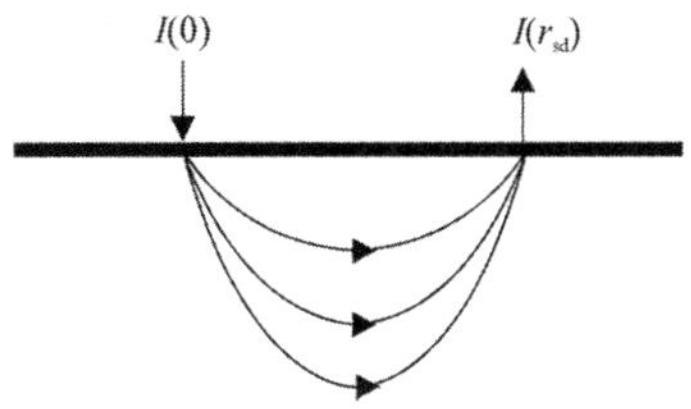

If the boundary is refractive index matched, then the popular Dirichlet boundary condition requires that $\Phi = 0$ at the boundary.[25] In this case, the method of images can be used to derive the diffuse reflectance as a function of r_{sd}, $R_d(r_{sd})$:

$$R_d(r_{sd}) = \frac{z_0}{2\pi} \frac{\exp[-\mu_d(z_0^2 + r_{sd}^2)^{1/2}]}{z_0^2 + r_{sd}^2} \left[\mu_d + \frac{1}{(r_{sd}^2 + z_0^2)^{1/2}} \right]$$

Here, $z_0(\approx 1/\mu_s)$ is the depth of the equivalent "isotropic" point source that is considered to arise from the directional randomization of the incident photons. μ_d is the "inverse diffusion length" and is given by the formula

$$\mu_d = \sqrt{3\mu_a(\mu_a + \mu_s')}$$

If the reflectance is measured at large r_{sd} ($r_{sd} >> z_0$), then the preceding expression simplifies to

$$R_d(r_{sd}) = z_0 \frac{\exp(-\mu_d r_{sd})}{2\pi r_{sd}^2} \left(\mu_d + \frac{1}{r_{sd}} \right),$$

Multiplying by r_{sd}^2 and taking the natural logarithm yields

$$\ln[r_{sd}^2 R(r_{sd})] = -\mu_d r_{sd} + \ln\left(\mu_d + \frac{1}{r_{sd}} \right) + \text{const}, \tag{9.11}$$

which suggests that provided $r_{sd} \gg 1/\mu_d$, a plot of $\ln[r_{sd}^2 R_d(r_{sd})]$ versus r_{sd} should be linear with a slope of $-\mu_d$. This provides a way of estimating μ_d without having to make measurements of $R_d(r_{sd})$, which are calibrated absolutely.

To obtain μ_a and μ_s' separately, some additional measurement is necessary. Patterson et al.'s original suggestion was to measure the absolute overall diffuse reflectance of the sample [i.e., $R_d(r_{sd})$ integrated over the entire surface], which then yields the transport albedo $\mu_s'/(\mu_s'+\mu_a)$.

Alternatively, the form of $R_d(r_{sd})$ at small r_{sd} has been exploited to yield an estimate for μ_s'.[26] This technique has been applied to the noninvasive measurement of the optical properties of the esophagus, using an endoscopic spatially resolved probe.[27]

Originally suggested as a method for determining photosensitizer concentrations in photodynamic therapy (PDT), the spatially resolved method has obvious applications in tissue oxygenation monitoring and has been implemented in variety of ways. Liu et al. have presented an analysis similar to that given previously, but using a diffusion model incorporating the more realistic case of a refractive index-mismatched boundary.[38] This does not affect the slope of $r_{sd}^2 R_d(r_{sd})$ substantially, yielding the following approximation for $\ln[r_{sd}^2 R_d(r_{sd})]$:

$$\ln[r_{sd}{}^2 R_d(r_{sd})] = -\mu_d r_{sd} - \ln(a/l_t) + \ln\left(\mu_d + \frac{1}{r_{sd_0}}\right) \tag{9.12}$$

where r_{sd_0} is the mean separation between source and detector, $\langle r_{sd} \rangle$. The constant a is defined by $a = 2\pi/(1 + 2A/3)$, where A is an internal reflection parameter, depending on the relative refractive index mismatch at the boundary. Liu et al. rely on absolute measurements of $R_d(r_{sd})$ to provide an accurate value for the intercept of $r_{sd}^2 R_d(r_{sd})$ versus r_{sd} in addition to the slope, which clearly yields an estimate for μ_s', given that a and r_{sd_0} are known.

The authors have implemented this system using a simple arrangement of light bulbs placed at distances of 3, 5, and 7 cm from an optical receiver. In ink and Intralipid phantoms, a good agreement between expected and derived μ_a is obtained in the range 0.2 to 0.4 cm^{-1}, although the derived value then begins to systematically under-read the expected value.

The SRS principle has been implemented by at least three companies. One of the earliest examples was a two-detector device, the INVOS-3100 (Somanetics, Troy, Michigan).[28] Hamamatsu Photonics developed a prototype system in the mid-1990s that was subsequently marketed as the NIRO-300.[29] This system uses three detectors and four wavelengths, and has two independent measuring heads. Unlike the method of Liu et al.,[38] this system does not attempt to measure μ_s' separately, but assumes values for the wavelength dependence of μ_s' based on measurements using time-resolved spectroscopy (Section 9.5.1.2).[30] This allows μ_d at the measurement wavelengths to be converted to μ_a, whereupon the absolute oxyhemoglobin and deoxyhemoglobin concentrations (and hence hemoglobin saturation) can be derived using standard multilinear regression. The accuracy of the device has been assessed in both hemoglobin phantoms and *in vivo* on forearm muscle during arterial and venous occlusions.[31] In a phantom containing blood, Intralipid, and yeast (to consume oxygen), excellent agreement with hemoglobin saturation as measured by co-oximetry was achieved. In the forearm ischemia studies, the agreement with results from a time-resolved spectroscopy system (see next section) was similarly encouraging (Fig. 9.7).

When used to make measurements *in vivo* on the adult head, however, reports differ as to the accuracy of the SRS device. Al-Rawi et al. have used a prototype version of the NIRO-300 on patients undergoing cardiopulmonary bypass (both those undergoing hypothermia during bypass and those remaining normothermic).[5] In this surgical procedure, blood samples can be drawn at regular intervals via a bulb inserted into the jugular vein and venous saturation (S_vO_2) measured by co-oximetry. Since the onset of bypass is expected to produce a global cerebral desaturation, a correlation plot of mixed hemoglobin saturation (S_mO_2) as measured by the NIRO-300 against S_vO_2 should represent a powerful validation of the instrument. Al-Rawi et al. reported a significant ($p < 0.05$) correlation between these saturation

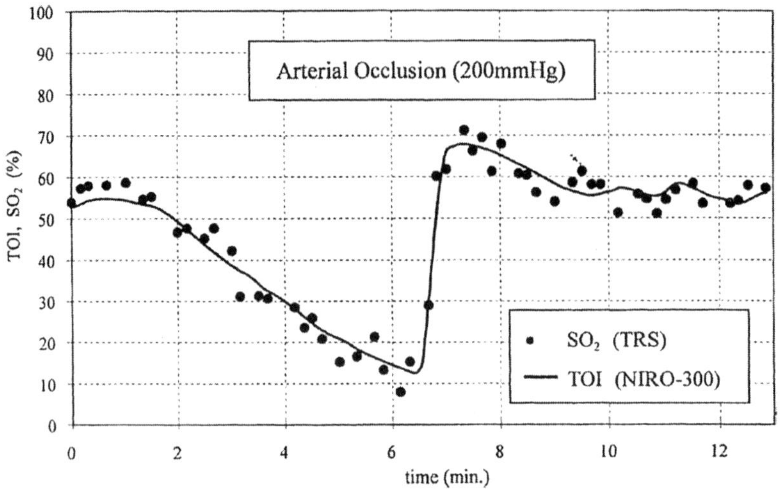

Figure 9.7 Changes in the absolute hemoglobin saturation of forearm muscle during arterio-venous occlusion. A comparison is made between two devices: a time-resolved oximeter (•) and a spatially resolved oximeter (—).[31]

measurements in only 12 of 24 patients studied; thus, they conclude that the device in its present form is not a suitable replacement for jugular venous sampling.

Elwell et al. compared S_mO_2 measurements obtained with a four-detector predecessor of the NIRO-300 against noninvasive measurements of S_vO_2 obtained using the venous occlusion method of Yoxall (Section 9.7.1).[32] A quantified trend monitor (the NIRO-500) was used for these latter measurements, so that the two sets of measurements were made on opposite sides of the head. Although the S_mO_2 tracked S_vO_2 in terms of trend, there was a consistent tendency for S_mO_2 to underread S_vO_2. This result is puzzling, as S_mO_2, which represents a mixed arterial, capillary, and venous signal, should overread S_vO_2.

Quaresima et al. performed a similar study on a larger group (16 adults), but used the NIRO-300 to measure both S_mO_2 and—via its trend monitoring capability—S_vO_2.[33] This ensured that the same volume of brain was studied by both methods. Here a reasonable correlation ($r^2 = 0.56$) was obtained between the two measurements (Fig. 9.8). More studies are clearly needed to resolve the debate concerning these and similar instruments.

The NIRO-300 has been used for muscle studies as well. Boushel et al. have studied the absolute oxygenation of the calf muscle and Achilles tendon during dynamic plantar flexion. The evidence from S_mO_2 measurement, combined with measurements of calf blood flow, cardiac output, and arterio-venous O_2 difference (from blood samples), suggests that the tendon undergoes an increase in perfusion that parallels that in the calf muscle. This finding argues against earlier theories in which connective tissues such as the

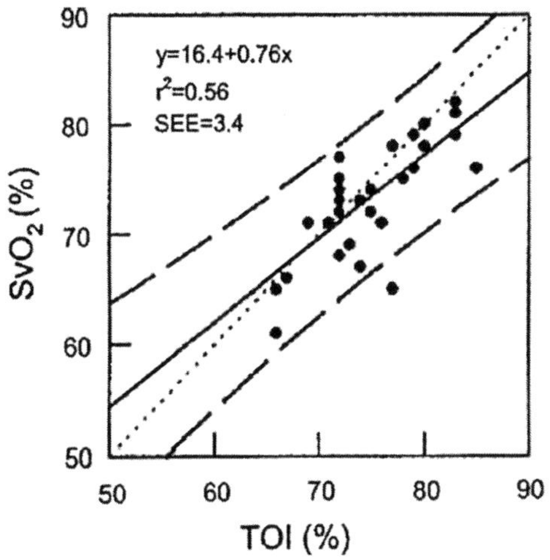

Figure 9.8 Comparison between venous saturation as measured by occlusion plethysmography (S_vO_2) and mean cerebral saturation as measured by spatially resolved spectroscopy—that is, tissue oxygenation index (TOI). Subjects were normoxic adults.[33]

tendon were considered to act as circulatory shunts during exercise, decreasing their perfusion so as to divert blood to the muscles.

Both Shimadzu Corporation and Critikon have released absolute oximeters based on the SRS principle.

9.5.1.2 Time-resolved spectroscopy (TRS)

An alternative approach for determining μ_a and μ_s' emerges when we consider illuminating the tissue sample shown in the previous section with an ultrashort (typically 10 ps) pulse of light. In this case, the temporal evolution of the light remitted at $z = 0$ is described by a solution to the time-dependent diffusion equation

$$-D\nabla^2\Phi + \mu_a c\Phi + \frac{\partial\Phi}{\partial t} = q_0(t),$$

while photon flux (the number of photons per unit area per unit time) at the surface results from Fick's law:

$$J(r_{sd},t) = (-D|\nabla\Phi|)|_{\partial G}.$$

Again applying the Dirichlet boundary condition and the method of images yields the solution[23]

$$J(r_{sd},t) = (4\pi D)^{-3/2} z_0 t^{-5/2} \exp(-\mu_a ct) \cdot \exp\left(-\frac{r_{sd}^2 + z_0^2}{4Dt}\right). \tag{9.13}$$

Plotting this function for representative values of μ_a, μ_s', r_{sd}, and z_0 leads to a curve with the appearance shown in Fig. 9.9.

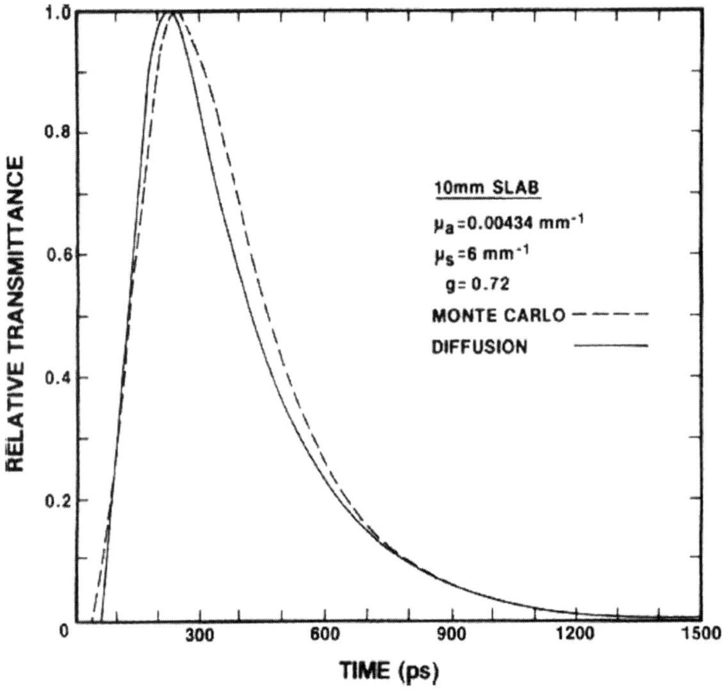

Figure 9.9 The form of $J(r_{sd},t)$ predicted by Eq. (9.13), compared with the predictions of a Monte Carlo simulation.[23] $J(r_{sd},t)$ is in general merely proportional to the ordinate values in this figure, as time-resolved measurements of flux are generally not obtained in photometrically calibrated units.

Inspection of the formula in Eq. (9.13) suggests that the decay of this curve is dominated by the term $\exp(-\mu_a ct)$, while the rise is more sensitive to D (and hence μ_s'). Thus, the curve contains enough information to extract independent estimates of μ_a and μ_s'—a task that can be performed in a variety of ways.

The most direct way is to perform a least-squares fit to the measured data using the previously describe model, with μ_a, μ_s', and an arbitrary scaling factor (whose final value is discarded) as the free parameters. This particular model can clearly be linearized with respect to the fitting parameters μ_a and μ_s' by taking the natural logarithm. However, for more complicated geometries (e.g., the slab) this is not possible, so a nonlinear (e.g., Levenburg–Marquardt) optimization algorithm must be employed. Such algorithms can prove problematic (1) because of the time required for convergence and (2) because of problems of convergence to a true global minimum of the χ^2 error norm.

An alternative suggestion[34] is to note that

$$\lim_{t\to\infty} \frac{d}{dt} \ln J(r_{sd},t) = -\mu_a c, \tag{9.14}$$

which implies that fitting the exponentially falling tail of $J(r_{sd},t)$ can yield an estimate of μ_a. With this, μ_s' can be extracted by noting that $\ln[J(r_{sd}, t)]$ reaches its maximum value at t_{max}, where

$$\mu_s' = \frac{1}{3r_{sd}^2}(4\mu_a c^2 t_{max}^2 + 10ct_{max}) - \mu_a. \tag{9.15}$$

Unfortunately, this suggestion proves problematic, chiefly because by the time $J(r_{sd}, t)$ has decayed sufficiently to make the approximation accurate, the data have become subject to a rather poor signal-to-noise ratio (SNR).

As an alternative, Matcher has proposed a simple, approximate step to extract μ_a and μ_s' from the mean time of $J(t)$ in addition to t_{max}.[35] Integrating $tJ(t)$ yields the following expression for the photon mean time of flight $\langle t\rangle = L/c$:

$$\langle t\rangle = \frac{r_{sd}^2}{2[D + r_{sd}(\mu_a cD)^{1/2}]}.$$

By making some approximations, consistent with the usual values for tissue optical properties and measurement geometry, this simplifies to

$$\mu_a = \frac{2.5t_{max} - \langle t\rangle}{c(\langle t\rangle^2 - t_{max}^2)}. \tag{9.16}$$

When the geometry is more complicated, the coefficients change slightly. This technique represents a rapid, closed form method for obtaining μ_a, which does not rely exclusively on low-SNR data. One restriction is that it systematically misreads μ_a by several percent due to the various approximations. Another restriction is that it is not robust when the signal is convolved with an instrumental impulse response of significant temporal width that is not symmetrical in time.

The problem of accounting for an instrumental response function can also be addressed using an interesting spatio-temporally resolved technique.[36] Here the basic idea is to rely completely on measurements of t_{max} and its variation with r_{sd}. The motivation for this is that t_{max} is comparatively unaffected when the signal is convolved with an instrumental response function, even of substantial width. Re-expressing Eq. (9.15) yields

$$r_{sd}^2 \approx 4D\mu_a ct_{max}^2 + 10Dt_{max}.$$

Hence, fitting r_{sd}^2 versus t_{max} with a quadratic would, in principle, yield estimates for μ_a and D. In practice, the term in t_{max}^2 is small compared to that in t_{max}, so that this approach yields a poor signal-to-noise ratio. Instead, a linear fit is used to obtain an estimate for D, which is then used to remove the

r_{sd}-dependence from the map $J(r_{sd})$ by multiplying each curve by $t^{5/2} \cdot \exp[-r_{sd}^2/(4Dt)]$. Since the modified map $J'(r_{sd}, t)$ has the form

$$J'(r_{sd},t) = \text{const} \cdot (4\pi D)^{-3/2} z_0 \exp(-\mu_a ct), \tag{9.17}$$

the dependency on r_{sd} is removed. This then allows one to average the curves over all values of r_{sd} and obtain an estimate for μ_a of high SNR. The estimate for μ_a can then be used to improve the estimate for D obtained using the previous equation. The process is iterated until μ_a and D converge to stable values.

The previous methods have all assumed that the measurement geometry is semi-infinite. Madsen et al. have systematically evaluated the effect on the derived transport coefficients when the geometry differs substantially from this ideal case.[37] For a large phantom, transport coefficients were obtainable from TRS data that differ by at most 10% from the theoretical values. However, for spherical and cylindrical phantoms of small size (e.g., 2.5 cm), μ_a can be overestimated by as much as 50% and μ_s' underestimated by a similar amount. These results can be understood intuitively, as the presence of boundaries at which photons are lost causes a preferential loss of long-path-length photons. The suppression of long-pathlength photons relative to short ones is analogous to the effect of an increased μ_a or reduced μ_s'.

For cases where the sample is very small, an interesting variant of the time-resolved technique is the "sample substitution" method.[38] The sample of interest is immersed into a larger "host" medium (typically a scattering liquid medium such as Intralipid) whose transport properties can be altered. Provided the refractive indices of the sample and host media are matched, then the insertion of the sample into the host medium should not cause a perturbation to $J(t)$ measured across a line of sight through the sample when the optical properties of sample and host are matched in μ_a and μ_s'. The procedure is thus to endow the liquid phantom with a range of μ_a and μ_s' values and measure these using a time-resolved system. The sample is then placed in a new liquid phantom, and the μ_a/μ_s' values varied over the same range as before. When the μ_a/μ_s' values derived for the sample/host combination match those derived for the host alone, the sample optical properties must equal the host optical properties.

An alternative to the use of the diffusion equation is to use the Monte Carlo technique (Chapter 3) to compute $J(t)$ for a range of μ_a/μ_s' values. Barilli et al.[47] have used this technique to extract the optical properties of human skeletal forearm muscle using time-resolved measurements. By approximating the forearm as an infinite cylinder of radius 4 cm, launching photons into the medium at a point on the surface, and detecting photons emerging at various azimuthal angles between 0 and 180 deg, a set of $J(t)$ curves can be derived that accurately reflects the true geometry. Such curves can be used in a variety of ways. Barilli et al. characterize them using functions of a form motivated by the result from diffusion theory:

$$J(t) = \text{const} \cdot t^{-\alpha} \exp(-\beta\mu_s'/t) \cdot \exp(-\mu_a ct), \tag{9.18}$$

but with the parameters α and β derived by least-squares fitting the Monte Carlo results. These curves are, in turn, used to fit the experimental $J(t)$ profiles and extract estimates for the tissue μ_a and μ_s.

9.5.1.3 The Microscopic Beer–Lambert Law

An alternative means of extracting an estimate of μ_a from a TPSF that is based on the "microscopic Beer–Lambert law" (MBLL) has been described.[39] This law is a generalization of the Beer–Lambert law that is valid for a multiple-scattering medium. Consider first a multiple-scattering medium with zero absorption. Then, define a function $s(\mu_s, t)$ that represents the probability that an injected photon emerges from the medium having propagated for a time t (i.e., has traversed a zigzag path of length ct) in the medium. Now consider that the medium acquires a homogeneously distributed absorption coefficient μ_a. The probability of a photon propagating for a time t in the medium is modified and is now represented by the function $P(\mu_s, \mu_a, t)$. The MBLL can then be simply stated as

$$P(\mu_s,\mu_a,t) = s(\mu_s,t) \cdot \exp(-\mu_a ct). \tag{9.19}$$

Consider that two measurement types are made: the overall time-integrated flux $J(\mu_s, \mu_a)$ and the mean path length traversed by photons $L(\mu_s, \mu_a)$. These quantities are related to $s(\mu_s, \mu_a, t)$ by

$$J(\mu_s,\mu_a) = \text{const} \cdot \int_0^\infty s(\mu_s,t) \cdot \exp(-\mu_a ct) \cdot dt$$

$$L(\mu_s,\mu_a) = \frac{\int_0^\infty ct \cdot s(\mu_s,t) \cdot \exp(-\mu_a ct) \cdot dt}{\int_0^\infty s(\mu_s,t) \cdot \exp(-\mu_a ct) \cdot dt}.$$

As is discussed more fully in Section 9.6, a natural interpretation for L results from considering $\partial \ln J/\partial\mu_a$:

$$\frac{\partial \ln J}{\partial \mu_a} = \frac{1}{J}\frac{\partial J}{\partial \mu_a} = \frac{-1}{J}\int_0^\infty ct \cdot s(\mu_s,t) \cdot \exp(-\mu_a ct)dt = -L(\mu_s,\mu_a).$$

Integrating both sides with regard to μ_a, we obtain

$$\ln J = -\int_0^{\mu_a} L(\mu_s,\mu_a) \cdot d\mu_a + \ln \int_0^\infty s(\mu_s,t)dt.$$

Also, if J is now defined as the detected flux (rather than the emerging flux), then an additional constant b must be introduced to account for detector efficiency:

$$\ln J = -\int_0^{\mu_a} L(\mu_s,\mu_a)\cdot d\mu_a + \ln\int_0^{\infty} s(\mu_s,t)dt + \ln b. \tag{9.20}$$

A number of schemes can be developed to use this relation to derive absolute chromophore concentration. To illustrate the basic idea, we will consider the simplest scheme in which measurements of J and L are made at two wavelengths λ_1 and λ_2 to determine the concentration c_{ab} of a single absorber. Denoting the specific absorption coefficients at the two wavelengths as ε_{ab_1} and ε_{ab_2}, respectively, we can write

$$c_{ab} = \frac{\mu_{a_2} - \mu_{a_1}}{\varepsilon_{ab_2} - \varepsilon_{ab_1}}.$$

From the previous equation, we can express the ratio of intensity detected at these two wavelengths as

$$\ln\frac{J_2}{J_1} = -\int_{\mu_{a_1}}^{\mu_{a_2}} L(\mu_s,\mu_a)d\mu_a + \ln\left[\int_0^{\infty} s_2(\mu_s,t)dt/\int_0^{\infty} s_1(\mu_s,t)dt\right] + \ln(b_2/b_1).$$

To simplify this equation still further, an approximation must be made in which it is implicitly assumed that L varies linearly with regard to μ_a. In this case, the first integral on the RHS can be evaluated using the trapezoidal rule to yield

$$\ln\frac{J_2}{J_1} \approx -\frac{(\mu_{a_2}-\mu_{a_1})[L(\mu_{s_2},\mu_{a_2}) + L(\mu_{s_1},\mu_{a_1})]}{2} + \ln\left[\int_0^{\infty} s(\mu_{s_2},t)dt/\int_0^{\infty} s(\mu_{s_1},t)dt\right] + \ln(b_2/b_1),$$

from which the concentration c_{ab} can be derived as

$$c_{ab} \approx \frac{2}{(\varepsilon_{ab_2}-\varepsilon_{ab_1})[L(\mu_{s_2},\mu_{a_2}) + L(\mu_{s_1},\mu_{a_1})]}\left(\ln\frac{J_2}{J_1} - \ln\frac{\int_0^{\infty} s(\mu_{s_2},t)}{\int_0^{\infty} s(\mu_{s_1},t)} - \ln\frac{b_2}{b_1}\right). \tag{9.21}$$

Finally, an approximation for $\int_0^{\infty} s(\mu_{s_2},t)dt/\int_0^{\infty} s(\mu_{s_1},t)dt$ is required. Zhang et al. have used diffusion theory applied to both a semi-infinite half-space and a slab geometry to show that

$$\int_0^{\infty} s(\mu_{s_2},t)dt/\int_0^{\infty} s(\mu_{s_1},t)dt \approx \mu_{s_2}/\mu_{s_1}. \tag{9.22}$$

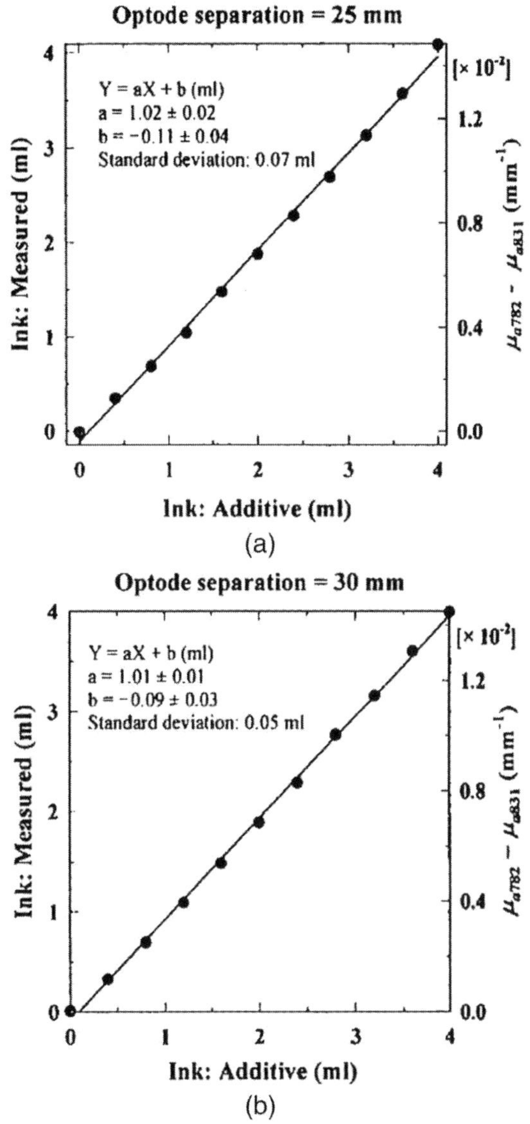

Figure 9.10 The absolute concentration of a dye in a scattering solution as measured using the MBLL approach. Plots show the correlation between measured and actual dye concentrations for two different values of the source–detector spacing.[39]

Note that Eq. (9.22) does not require knowledge of the source–detector spacing r_{sd}. In addition, apart from the last approximation, no reference to any particular medium geometry has been made. The MBLL approach is therefore expected to be robust when used in a variety of situations. In a large, homogeneous phantom the algorithm yields impressive results, as Fig. 9.10 shows.

9.5.1.4 Practical TRS systems and their applications

A variety of systems have been assembled to implement time-resolved spectroscopy. Basically the requirements can be summarized as follows:

- The generation of pulses of near-infrared ($\lambda = 700$–1000 nm) light, of typical duration less than 100 ps (i.e., short compared with the temporal dispersion experienced by photons traveling over several centimeters of tissue); and
- The detection of the transmitted light pulse, which will typically have been attenuated to a level of a few hundred picowatts, with a temporal resolution better than 100 ps.

The most popular light source is the mode-locked laser.[14,40] Originally, dye lasers were common, but nowadays the device of choice is the titanium–sapphire laser, which can generate pulses of duration less than 100 fs, with repetition rates of 80 MHz, mean powers as high as 1 W, and—a characteristic important for spectroscopic applications—a wavelength continuously tunable from 700 to 1000 nm.[41] Semiconductor lasers are also capable of generating short-duration pulses (typically 50 ps), given suitable driver circuits, and have been successfully incorporated into portable TRS systems.[42]

For detection, the synchroscan streak camera is the benchmark instrument, since it offers better than 10-ps temporal "jitter" combined with photon-counting detection across most of the near-infrared window. Another popular system is the photon-counting photomultiplier coupled to a time-to-amplitude converter (TAC) and pulse height analyzer (PHA).[42],[43] While less expensive than the streak camera, the TAC system has a fundamentally lower dynamic range, as only one photon can be recorded per conversion cycle. Other, more exotic possibilities have been explored for measuring the signal, chiefly for imaging applications. These include Kerr gating,[44] stimulated Raman scattering,[45] and optical parametric amplification.[46] None of these technologies has found widespread application in TRS systems.

TRS has been employed in a number of studies. Suzuki et al. have determined the optical properties of female breast tissue *in vivo*, using a pulsed semiconductor laser and TAC system.[31] For a sample of 30 Japanese women, age 23 to 74, μ_a and μ_s' values were determined as functions of age, body mass index (BMI), breast thickness, number of pregnancies, and menstrual status (i.e., premenopausal versus postmenopausal). Both transport parameters were found to decrease with age, BMI, and menstrual status, consistent with known changes in breast composition (i.e., the replacement of glandular with adipose tissue). No significant correlation was found between the transport properties and menstrual cycle, breast thickness, or number of pregnancies. Heussmann et al. have similarly used a Ti:sapphire and streak camera system to obtain breast optical properties *in vivo* at 800 nm. μ_a was found to vary between 0.017 cm^{-1} and 0.045 cm^{-1}, while μ_s' varied between 7.2 cm^{-1} and 13.5 cm^{-1}.

Matcher et al. have used the wavelength tunability of the Ti:sapphire laser to measure μ_a and μ_s' from 760 to 1000 nm on the adult head, forearm, and calf.[30] These data are important in that they are used by the Hamamatsu NIRO-300 SRS machine to convert its measurements of μ_d into estimates of μ_a.

Barilli et al. determined μ_a and μ_s' for forearm skeletal muscle, using probe spacing values between 2.5 and 4.5 cm. Reasonably consistent values were obtained, suggesting that the probed tissue volume was roughly homogeneous.[47] Mean values of 0.17 cm^{-1} and 10 cm^{-1} for μ_a and μ_s', respectively, were obtained at 800 nm. These are significantly different from the results obtained by Matcher et al. at this same wavelength (0.23 cm^{-1} and 6.8 cm^{-1} for μ_a and μ_s', respectively).

9.5.1.5 Frequency-domain spectroscopy

Rather than acquire the data in the time domain, it is equally feasible to acquire the data in the frequency domain.[48] In this technique, which is well established in the field of fluorescence lifetime measurements,[49] the sample is illuminated with a continuous—rather than pulsed—laser beam whose intensity is modulated at (typically) radio-frequencies on the order of 100 MHz. Temporal dispersion in the time domain is then equivalent to frequency-dependent attenuation and phase shift of the Fourier components of the transmitted light signal. In essence, then, the frequency-domain method amounts to measuring each of these Fourier components independently. The measurements of interest are the AC modulation depth $m_U(r_{sd}, f)$ and the absolute phase shift $\Delta\Phi$ (r_{sd}, f) of the intensity at frequency f and source–detector spacing r_{sd}. A simple Fourier transform of Eq. (9.13) shows that for a semi-infinite, homogeneous half-space, these quantities are given by

$$\begin{aligned} m_U(r_{sd},f) &= \frac{(1+\psi_0^2+2\psi_i)^{1/2}}{(1+\psi_\infty)}\exp(\psi_\infty-\psi_i), \\ \Delta\Phi(r_{sd},f) &= \psi_r - \tan^{-1}\left(\frac{\psi_r}{1+\psi_i}\right), \end{aligned} \tag{9.23}$$

where

$$\begin{aligned} \psi_0 &= \{3[\mu_a+(1-g)\mu_s](r_{sd}^2+z_0^2)[(\mu_a c)^2+(2\pi f)^2]^{1/2}c^{-1}\}^{1/2}, \\ \psi_r &= -\psi_0 \sin\left(\frac{\theta}{2}\right), \psi_\mathrm{i} = \psi_0 \cos\left(\frac{\theta}{2}\right), \theta = \arctan\left(\frac{2\pi f}{\mu_a c}\right), \\ \psi_\infty &= \{3\mu_a[\mu_a+(1-g)\mu_s](r_{sd}^2+z_0^2)\}^{1/2}. \end{aligned}$$

It is possible in principle to determine μ_a and μ_s' given measurements of m_U and $\Delta\Phi$ at a single frequency. Figure 9.11 plots the contours of modulation depth and phase shift (measured at 400-MHz modulation

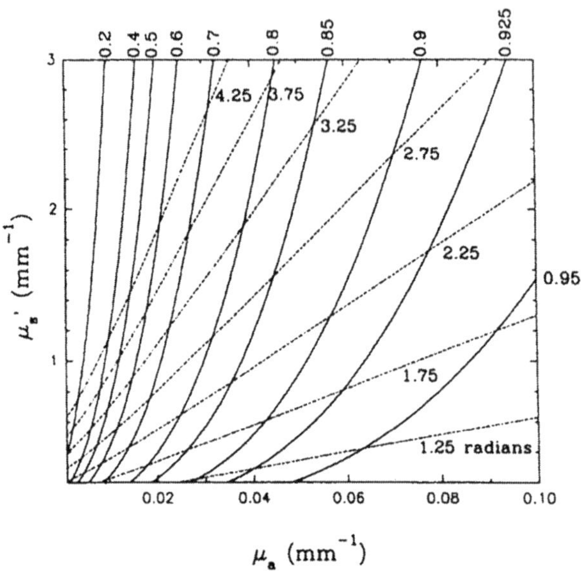

Figure 9.11 Lines of constant $m_U(r_{sd}, f)$ and $\Delta\Phi(r_{sd}, f)$ as functions of μ_a and μ'_s.[48] The results are calculated using Eq. (9.23) and are for $f = 400$ MHz and $r_{sd} = 5$ cm. The intersection point of such curves allows one to estimate μ_a and μ'_s from $m_U(r_{sd}, f)$ and $\Delta\Phi$ (r_{sd}, f).

frequency and 50-mm source–detector separation) versus μ_a and μ'_s. The intersection points of the solid and dotted contours define the solutions to the nonlinear equations defined earlier and can be located, for instance, using a Newton–Raphson root-finding algorithm.

In practice, this approach can be problematic chiefly because at modulation frequencies less than 100 MHz the modulation depth is very close to unity and, consequently, difficult to measure accurately. Small calibration errors induced, for example, when determining the initial modulation depth of the applied light beam, can render the measurement of $m_U(r_{sd}, f)$ highly unreliable.

One possible solution to this dilemma is to combine an absolute measurement of phase shift obtained at a low frequency with a measurement of μ_d obtained using the SRS technique (Section 9.5.1.1). However, this approach can still be problematic, as it requires an absolute calibration of the phase shift.

Fantini et al. have described a device that combines frequency-domain and spatially resolved measurements to eliminate the need for an absolute calibration of the modulation depth or phase shift.[50,51] Their device, the "OMNIA," performs frequency-resolved measurements at four spatial locations and two wavelengths (715 nm and 850 nm) by using eight separate, intensity-modulated light-emitting-diode sources. Gratton and Fishkin have used a solution of the diffusion equation for an infinite, homogeneous medium to obtain the following expressions for the phase shift $\Delta\Phi$ and the DC and AC components of the photon density $DC_{detector}$ and $AC_{detector}$:

$$\begin{aligned}\Delta\Phi &= r_{sd}\left(\frac{c^2\mu_a^2 + 4\pi^2 f^2}{D^2}\right)^{1/4} \sin\left[\frac{1}{2}\arctan\left(\frac{2\pi f}{c\mu_a}\right)\right],\\ \ln(r_{sd}DC_{detector}) &= -r_{sd}\sqrt{\frac{\mu_a c}{D}} + \ln\left(\frac{S}{4\pi D}\right),\\ \ln(r_{sd}AC_{detector}) &= -r_{sd}\left(\frac{c^2\mu_a^2 + 4\pi^2 f^2}{D^2}\right)^{1/4} \cos\left[\frac{1}{2}\arctan\left(\frac{2\pi f}{c\mu_a}\right)\right] + \ln\left(\frac{Sm_I}{4\pi D}\right),\end{aligned} \tag{9.24}$$

which show that $\Delta\Phi$, $\ln(r_{sd}DC_{detector})$ and $\ln(r_{sd}AC_{detector})$ vary linearly with r_{sd}, with constants of proportionality dependent on μ_a and D. S is the source strength (photons s^{-1}), and $c = c_0/n$, where n is the medium refractive index. This spatial and frequency-resolved technique thus allows one to determine μ_a and μ_s' using only relative measurements of the amplitude and phase of the remitted signal. Since three independent measurements are available (i.e., $DC_{detector}$, $AC_{detector}$, and $\Delta\Phi$), any of three possible pairs of measurements can be used to obtain the transport properties. This technique has been shown, in a large homogeneous phantom, to produce some of the most accurate measurements of μ_a and μ_s' available in the literature. By paying careful attention to the design of the optical probe, Gerken and Faris have measured the transport properties of a dye/microsphere/water phantom, measuring μ_a to within 1% and μ_s' to within 3% of their true values.[52]

An alternative scheme to increase the robustness of the estimates of μ_a and μ_s' is to sweep the modulation frequency over a wide band. Tromberg et al. describe a system that uses modulation frequencies from 300 kHz to 1 GHz.[53] A continuous sweep of this frequency range is carried out in 1 second. Once again, models extracted using diffusion theory are used to fit the model data. Unlike the multidistance method described earlier, this technique still requires an absolute calibration of the phase and amplitude, which can be accomplished, for example, by constructing a solid plastic phantom[54] of known optical properties and making a reference measurement of amplitude and phase across it.

A further variant on the frequency-domain approach has been suggested by Kohl et al.[55] Rather than consider how the slope of AC intensity and phase vary with source–detector distance r_{sd}, Kohl et al. have considered how these quantities vary with μ_a. In particular, consider a frequency-domain spectrometer operating at a single source–detector separation and measuring remitted flux a distance r_{sd} from the source. Also consider that μ_a then increases by a small amount $\Delta\mu_a$. From the earlier discussion, it is clear that this will produce an increase in attenuation ΔA and a decrease in phase shift—that is, path length $c\Delta<t>$. In the clinical context, such a change in μ_a might be brought about by a small change in blood volume or oxygenation. In a phantom, it could be produced by tuning the emission wavelength of the laser source by a small amount. Kohl et al. note that a regression of ΔA against $\Delta<t>$ is roughly linear for small $\Delta\mu_a$, with a slope proportional to the absolute value of μ_a. This can be seen most easily by considering the dominant terms in the expressions for A and $<t>$ for a semi-infinite half-space, as an example:

$$R_d(r) = z_0\left(\frac{1}{r} + \mu_d\right)\frac{\exp(-\mu_d r)}{2\pi r^2}, \quad \langle t\rangle(r) = \frac{r^2}{2[D + r\cdot(\mu_a cD)^{1/2}]},$$

where $r = (r_{sd}^2 + z_0^2)^{1/2}$. Defining the attenuation A as the base 10 logarithm of the remittance R_d, and differentiating both expressions with respect to μ_a yields

$$\frac{\partial A}{\partial\mu_a} = \frac{3}{2\ln 10}\frac{r}{1/r + \mu_d}(2\mu_a + \mu_s'),$$
$$\frac{\partial\langle t\rangle}{\partial\mu_a} = \frac{-3}{2(1/r + \mu_d)^2}\left[\frac{r}{2}\frac{\mu_s'}{(\mu_a D/c)^{1/2}} - 1\right],$$

Hence,

$$\left(\frac{\partial A}{\partial\mu_a}\right)\bigg/\left(\frac{\partial\langle t\rangle}{\partial\mu_a}\right) = \frac{-(1 + r\mu_d)c}{\ln 10\left(\frac{r\mu_d}{2} - \frac{\mu_a}{\mu_s'}\right)}\mu_a.$$

With typical source detector separations of 2 to 4 cm, μ_a typically 0.05 to 0.5 cm^{-1}, and μ_s' typically 10 to 20 cm^{-1}, it is evident that the term $r\mu_d/2 \gg \mu_a/\mu_s'$ and $r\mu_d \gg 1$, so that the preceding expression simplifies to

$$\left(\frac{\partial A}{\partial\mu_a}\right)\bigg/\left(\frac{\partial\langle t\rangle}{\partial\mu_a}\right) \approx \frac{-2c}{\ln 10}\mu_a. \tag{9.25}$$

Figure 9.12 shows results obtained on a liquid phantom. The solid line shows the theoretical μ_a versus λ, and the symbols show measurements obtained at four discrete wavelengths using a frequency-domain spectrometer and the previously described method. As can be seen, theory and experiment are in excellent agreement.

Numerous systems have been constructed to perform frequency-domain measurements. As mentioned, one of the chief attractions of frequency-domain measurements is the comparative simplicity of the instrumentation. The essential requirements of such a system are

1. A means of generating a continuous-wave light beam whose intensity is modulated at radio-frequencies
2. A light detector sensitive in the NIR range (700–1000 nm) whose response time is sufficient to detect this modulation frequency

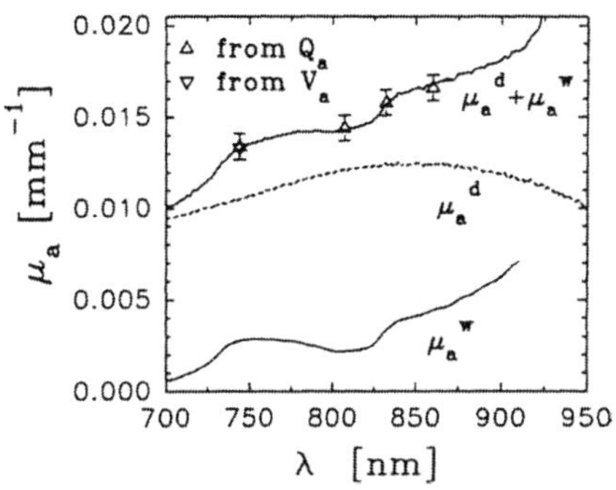

Figure 9.12 The absolute value of μ_a determined from changes in attenuation and phase as functions of small changes in μ_a. Open symbols = experimentally measured values of μ_a; solid line = expected values of μ_a based on the known dye concentration.[55]

3. A means of measuring the relative demodulation and phase shift of the remitted light relative to the incident light

Fantini et al. describe a system in which light is generated by broadband LEDs.[50] Modulation of the intensity is achieved via modulation of the LED drive current, which is performed at a frequency of 120 MHz. The light detection system is a high-speed photomultiplier tube (PMT), the Hamamatsu R928. This device is equipped with a multialkali extended red-response photocathode providing roughly 0.3% detective quantum efficiency (DQE) throughout the NIR window. The tube possesses a 2-ns rise time, equivalent to a 500-MHz bandwidth. A heterodyne technique is used to measure the amplitude and phase of the detected system, which is achieved not by heterodyning the electrical output from the PMT, but rather by modulating the gain of the PMT directly. The PMT accelerating voltage is modulated at 120.0004 MHz, producing a 400-Hz beat frequency signal that can be acquired directly into a host computer using a 12-bit ADC card. Each period of the 400-Hz wave is digitized at 16 points and averaged between 8 and 50 of these periods. A discrete Fourier transform (DFT) is then used to extract the ac modulus and phase information. This system has been used to measure absolute hemoglobin saturation in exercising muscle.[56]

9.5.2 Chemometric methods

It is possible to avoid the use of an analytical model of light transport in tissue completely if measurements can somehow be made in which a set of "independent" data (e.g., an attenuation spectrum) is measured and simultaneously the corresponding "dependent" data, such as S_mO_2, is determined by some independent means. It is then possible to use a class of techniques known collectively as "chemometrics" to generate an entirely empirical algorithm that relates the independent data to the dependent data.[57] This is achieved by first collecting a large set of independent and dependent data (the "calibration" set) in which the dependent data is varied so as to span the range that will be encountered in the actual measurements. A least-squares procedure [conventionally, partial least squares (PLS)] is then used to generate a set of near-orthogonal "basis functions" that, when linearly combined, optimally span the calibration data set. If these same basis functions are then used to fit an unknown data set (using conventional multilinear regression), the resulting weighting coefficients serve as an optimal estimate for the dependent data associated with the unknown data set.

One of the earliest applications of the chemometric approach to tissue spectroscopy was demonstrated by Ferrari et al.,[58] who developed a technique to estimate cerebral S_mO_2 in dogs from a measured attenuation spectrum. Previously, researchers had noted that the accurate measurement of overall tissue attenuation is not trivial, as scattering losses and instrumental effects (e.g., detection fiber geometry, numerical aperture, and fiber-to-tissue attachment) produce a baseline offset that is very hard to quantify accurately. To minimize the effects of baseline offset, therefore, the first derivative of the attenuation spectrum may be calculated and used as the independent data set. The corresponding dependent data (S_mO_2) must be separately determined; in Ferrari et al.'s method, this is achieved by drawing blood samples from the superior sagittal sinus for analysis by co-oximetry. Since the superior sagittal sinus drains 40% of all blood leaving the hemispheric tissues, and venous blood represents approximately 70% of the total blood volume in the illuminated region, we might expect a close correlation between S_vO_2 measured in sagittal sinus blood and S_mO_2. A calibration algorithm can then be generated by varying S_mO_2 over a large range—for example, by altering the inspired oxygen fraction (FiO_2) given to the dog. The resulting algorithm can then be validated by inducing hypoxia in other dogs and regressing the predicted S_mO_2 against S_vO_2 as measured by co-oximetry. The performance of Ferrari et al.'s algorithm is impressive, as can be seen in Fig 9.13. The major problem with this approach in humans, of course, is that it is difficult to alter S_mO_2 over a sufficiently large range to generate a useful calibration set without harming the patient.

The chemometric approach has been applied to other measurement types besides the simple attenuation spectrum. Dam et al. have proposed the method as a way of determining μ_a and μ_s' for a semi-infinite medium, given measurements of the spatial variation of backscattered intensity (see Section 9.5.1.1). In their approach, a calibration set is generated via Monte Carlo calculations and used to generate an empirical algorithm using a Sugeno Fuzzy Inference System (SFIS). This chemometric approach aims to partially overcome a major restriction of standard PLS and related techniques—namely, their implicit assumption that some linear mapping exists between the independent and dependent data. Since the mapping between μ_a, μ_s' (the dependent data), and, for example, attenuation (the independent data) is

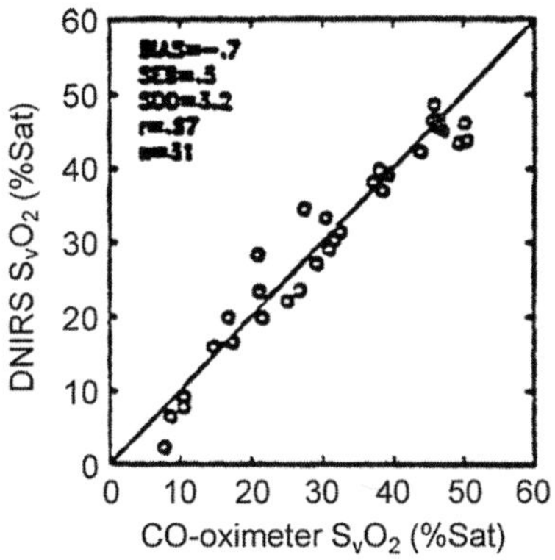

Figure 9.13 Absolute venous saturation measurements in dogs undergoing hypoxia. The plot shows the correlation between venous saturation determined chemometrically from derivative near-infrared spectra (DNIRS S_vO_2) and that measured using a co-oximeter to analyze blood draining from the superior sagittal sinus.[58]

fundamentally nonlinear (see Section 9.4), errors will occur if the overall span of the dependent data is too large. SFIS breaks the independent data into a number of distinct clusters, each spanning a small enough range of dependent data to possess a quasi-linear mapping. To assign optimal values of μ_a and μ_s' to a measured spatial reflectivity distribution, that distribution is assigned a "membership index" to each cluster based on its Euclidean distance to the center of that cluster. Each cluster's mapping is used to provide an estimate of μ_a and μ_s', and the final (μ_a, μ_s') is calculated as a weighted sum of these estimates. In this way, the estimate uses information from all clusters, rather than just one.

The work of Dam et al. follows an earlier suggestion of Farrell et al. to use a neural network classifier to extract optical properties from a measured spatial reflectivity distribution.[26]

Berger et al. have extended the approach to frequency-resolved measurements, using standard PLS to generate calibration algorithms for experimental Intralipid/dye phantoms of known μ_a and μ_s'.[59] In their experiments, modulation depth m_U and phase-shift $\Delta\Phi$ are measured for each phantom at a range of modulation frequencies from 50 to 600 MHz and at wavelengths of 674, 782, 803, 849, 947, and 956 nm, yielding data vectors m_U (ω, λ) and $\Delta\Phi(\omega, \lambda)$. By varying the Intralipid and dye concentrations, a calibration algorithm can be generated using PLS. An important data preprocessing step is to perform the PLS not on the data vectors themselves, but rather on transformed data vectors. This step is necessary because, as mentioned previously in the context of SFIS, chemometric methods implicitly assume a linear mapping between the independent and dependent data. Since we know from diffusion theory that $m_U(\omega)$ and $\Delta\Phi$ (ω) vary nonlinearly with μ_a and μ_s', comparatively poor performance can be expected if PLS is used to directly correlate these data sets. Berger et al. therefore employ a simple linearization step prior to PLS; that is, they solve the analytical diffusion theory relations and hence obtain initial estimates for μ_a and μ_s' at each frequency ω. These transformed data vectors $\mu_a(\omega)$ and $\mu_s'(\omega)$ will have a much more linear dependence on the dependent data pair (μ_a, μ_s') than do $A(\omega)$ and $\Delta\Phi(\omega)$. The ultimate goal of absolute quantification is to predict the concentration of a chromophore, and one can achieve this goal in two ways. The "two-step" method uses PLS to extract μ_a and μ_s' at each measurement wavelength and then uses a second PLS procedure to extract the dye concentration from the vector $\mu_a(\lambda)$. The one-step method generates a calibration algorithm that directly maps the dependent data onto the dye concentration. Both approaches seem promising, producing tighter correlations between measured and actual dye concentrations than are possible with diffusion theory coupled with standard multilinear regression (Fig 9.14).

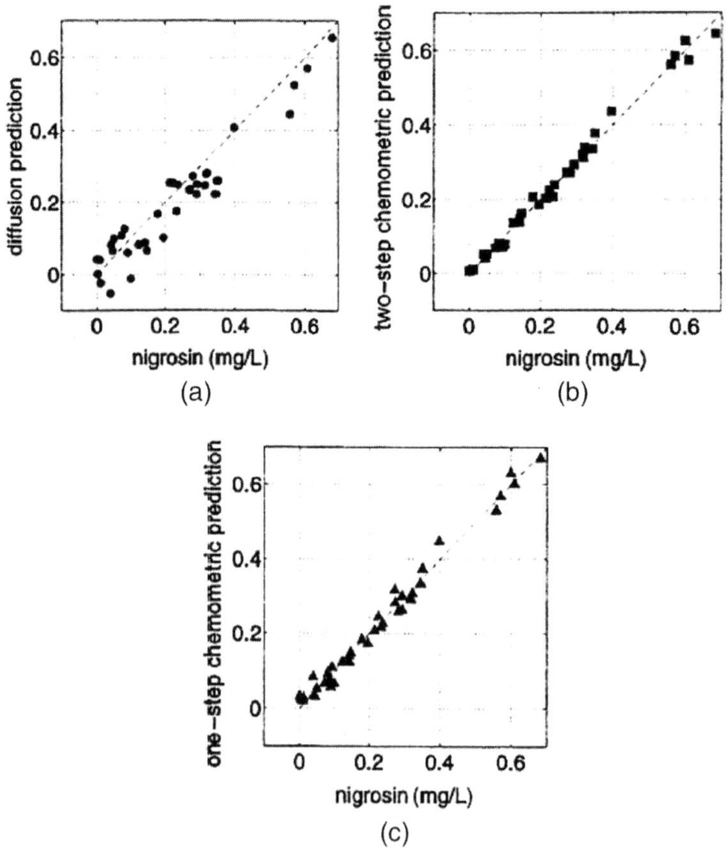

Figure 9.14 A comparison of PLS versus forward model fitting for determining dye concentrations in a scattering liquid from frequency-domain measurements.[59] Correlation plots show actual versus derived concentrations for the dye nigrosin: (a) results obtained by fitting a diffusion-theory forward model; (b) two-step chemometric procedure; and (c) one-step chemometric procedure.

9.6 Quantified Trend Measurements

While diffusion theory and other methods can clearly be used to predict absolute measurements of oxygenation, such applications are still in the minority compared with quantified trend measurements. These measurements do not attempt to predict the absolute oxygen level at any given time, but rather confine themselves to the simpler task of predicting how the oxygen level changes with time.

The physical basis of this method is the modified Beer–Lambert law:[40]

$$A \approx \mu_a \cdot r_{sd} \cdot \mathrm{DPF} + G, \tag{9.26}$$

which is a simple linearized form of the general $A(\mu_a, \mu_s')$ relationships predicted by diffusion theory. The meaning of the terms are

A Attenuation
μ_a Absorption coefficient of the medium
r_{sd} Physical source/detector spacing
DPF Differential path length factor
G Additive term due to fixed scattering losses.

The physical motivation for this simplified equation is that the effects of multiple scattering on attenuation can be crudely separated into effects due purely to absorption and effects due purely to scattering. Comparing this equation with the standard Beer–Lambert law for nonscattering samples,

$$A = \mu_a r_{\mathrm{sd}},$$

we see that

1. The attenuation produced by absorption is increased, due to the photon path lengthening effect of scattering. Hence, the source–detector spacing r_{sd} is multiplied by a factor termed the differential path length factor (DPF) to yield the true photon path length.
2. In addition to absorption, attenuation occurs due to scattering. Photons are not actually lost, but rather are scattered out of their normal straight-line path and consequently do not reach the detector. Losses due to scattering are included by introducing the additive term G into the equation.

The utility of this equation becomes apparent when we consider not the absolute attenuation A, but rather how A varies when μ_a varies by a small amount $\Delta\mu_a$. In the clinical context, such a change will occur via small changes in the tissue oxygenation or blood content. Provided these changes are small enough to prevent conformal changes in cell structure (as might occur, for example, if the cell energetics should become so compromised as to cause a failure of the membrane ion pumps), one can reasonably expect the scattering losses to remain approximately constant. The modified Beer–Lambert law then takes the form

$$\Delta A = \Delta\mu_a \cdot r_{sd} \cdot \mathrm{DPF}.$$

$\Delta\mu_a$ can, in turn, be expressed in terms of the change in concentration of all (oxygen-dependent) absorbers in the tissue. Given N such absorbers, the jth of which is present at concentration c_{ab_j} and possesses specific absorption coefficient ε_{ab_j}, then the equation becomes

$$\Delta A = \left(\sum_{\mathrm{j}=1}^{\mathrm{N}} \Delta c_{ab_j} \varepsilon_{ab_j} \right) \cdot r_{sd} \cdot \mathrm{DPF} \tag{9.27}$$

ΔA can be measured relatively easily. If at time $t = 0$ the detected light intensity is $I(0)$, while at time t it is $I(\mathrm{t})$, then the LHS can be determined via

$$\Delta A = \log_{10}[I(0)/I(t)];$$

In other words, it is not necessary to know either the absolute flux applied to the tissue or the absolute flux emerging from it. This is a major simplifying factor when developing instrumentation to make quantified trend measurements.

If a single absorber is present, then its change in concentration Δc_{ab} can be readily determined in absolute units, provided both and the DPF are known. If N absorbers are present ($N > 1$), then measurements of ΔA are required at M wavelengths, where $M \geq N$. In this case, the earlier equation becomes a matrix-vector equation:

$$\mathbf{\Delta A} = \mathbf{M} \cdot \mathbf{\Delta c_{ab}},$$

where $\mathbf{\Delta A}$ is an M-row vector containing the values of ΔA measured at M wavelengths. $\mathbf{\Delta c_{ab}}$ is now an N-row vector containing the changes in the N concentration values $c_{ab_1}, \ldots, c_{ab_N}$. $\mathbf{M}$ is an $M \times N$ matrix conventionally termed the "design set." Element M_{ij} is the specific absorption coefficient ε_{ab} of the jth absorber measured at wavelength i divided by the product $r_{sd} \times \mathrm{DPF}_i$. DPF_i is the DPF factor at the ith measurement wavelength. As we will see shortly, the DPF is determined by the absorption and scattering properties of the tissue, so it generally varies with wavelength.

If $M = N$, then the problem of determining $\mathbf{\Delta c_{ab}}$ from $\mathbf{\Delta A}$ is minimally constrained and can be solved by a direct matrix inversion:

$$\mathbf{\Delta c_{ab}} = \mathbf{M}^{-1} \cdot \mathbf{\Delta A},$$

provided M is nonsingular.[19] In practice, one must further ensure that M is "well conditioned" by paying careful attention to the selection of the measurement wavelengths. This step ensures that reliable values for $\Delta\mathbf{c_{ab}}$ can be extracted in the presence of measurement noise on $\Delta\mathbf{A}$.

A more robust estimate of $\Delta\mathbf{c_{ab}}$ can be achieved by over-constraining the problem—that is, by ensuring that $M > N$. In this case the vector $\Delta\mathbf{c_{ab}}$ is determined in a least-squares fashion, which can be implemented straightforwardly by multiplying $\Delta\mathbf{A}$ by the "pseudo-inverse" of $\mathbf{M}$ (also known as the "generalized" or "Moore–Penrose" inverse)[19]:

$$\Delta\mathbf{c_{ab}} = (\mathbf{M}^T\mathbf{M})^{-1}\mathbf{M}^T \cdot \Delta\mathbf{A}. \tag{9.28}$$

It is apparent that the concentrations can be obtained by computing appropriate linear combinations of the attenuation changes at each wavelength. The specific formula for mapping attenuation changes to concentration changes is generally referred to as an "algorithm" in the literature, and will be discussed more fully in Section 9.6.4.

9.6.1 Determination of the DPF at a given wavelength

Consider the modified Beer–Lambert law stated in the previous section. If the scattering-induced loss term G is considered to be independent of μ_a, then it is clear that the product $r_{sd} \times$ DPF [generally termed the "differential path" (DP)] can be expressed as

$$\mathrm{DP} = r_{sd} \times \mathrm{DPF} = \frac{\partial A}{\partial \mu_a}. \tag{9.29}$$

Thus, in principle, one could determine the DPF by merely increasing μ_a by a known small amount $\Delta\mu_a$ and then measuring the induced increase in attenuation ΔA. This "added absorber" method can be implemented by adding a known concentration of an absorbing compound of known specific absorption coefficient. In the clinical setting, this can be difficult to achieve, as it is very challenging to achieve a known concentration increase in the tissue volume of interest. In liquid phantoms, of course, it is extremely easy to achieve and has played a useful role in validating other methods.[40]

9.6.1.1 Time-resolved methods

Consider the schematic model of photon transport in tissue shown in Fig. 9.24 (later in this chapter). The total integrated intensity detected at the tissue surface results from contributions from randomly scattered photon paths that span a range of distances from r_{sd} to, in principle, infinity. This corresponds to a range of photon propagation times from r_{sd}/c to infinity. Consider first that the medium is absorption-free, so that the only source of attenuation is scattering. In a short interval of time t to $t + dt$, the flux of photons arriving at the detector will be denoted $S(t)dt$. Since $S(t) = 0$ for $t < r_{sd}/c$, the total detected flux in the absence of absorption can be written

$$I_0 = \int_0^\infty S(t)dt.$$

Now consider that this same medium is given a nonzero absorption coefficient μ_a, via the addition of an absorbing compound. The physical effect of this change is to cause photons to be lost and their energy converted to molecular rotational and vibrational energy. However, it will not cause a perturbation to the overall spatio-temporal distribution of photon paths. Those photons arriving at the detector in the interval t to $t + dt$ have clearly propagated a distance ct and so will be attenuated by a factor $\exp(-\mu_a ct)$ [this fact is termed the microscopic Beer–Lambert law (MBLL) in the literature; see Section 9.5.1.3]. Applying the previous expression, the total integrated intensity arising in the presence of absorption I can thus be written

$$I = \int_0^\infty S(t)\exp(-\mu_a ct)dt.$$

Defining the attenuation A as the natural logarithm of the reduction in light intensity (although it is more common to define attenuation in terms of the base 10 logarithm), we can write

$$A = \text{const} - \ln(I),$$

and hence

$$DP = \frac{\partial A}{\partial \mu_a} = -\frac{\partial(\ln(I))}{\partial \mu_a} = \frac{c\int_0^\infty tS(t)\exp(-\mu_a ct)dt}{\int_0^\infty S(t)\exp(-\mu_a ct)\mathrm{d}t}. \quad (9.25)$$

The final expression is recognizable as c multiplied by the mean propagation time for photons between source and detector. This simple derivation underpins the most flexible method for determining the DP (or DPF) to measure directly the mean time of flight of photons propagating from source to detector. This measurement can be performed directly in the time domain or, more simply, in the frequency domain. Both techniques will now be described.

9.6.1.2 Time-domain measurements

Delpy et al. pioneered the measurement of DPF in the time-domain.[40] Across 4 cm of biological tissue, the average propagation time for photons is approximately 1 ns, and the overall attenuation (e.g., at 800 nm) is approximately 5 OD. Delpy et al. showed that an ultrafast mode-locked dye laser emitting 10-nJ 10-ps pulses at 783 nm produced sufficient transmitted flux across this thickness of biological tissue to be detectable by a synchroscan streak camera. This electro-optic device employs a photoelectronic image intensifier to detect a photon-noise-limited light signal and amplify it onto an electroluminescent output phosphor screen, where it can be read out by a charge-coupled device (CCD) camera. The image intensifier contains electrostatic deflector plates similar to those in an oscilloscope. By applying a rapid, alternating potential to these plates, a time-varying input light signal is "streaked" onto the output phosphor so that time is encoded on it as distance. The overall temporal resolution of the device is determined by jitter in the triggering electronics; it can be as low as 10 ps for a device such as the Hamamatsu C1587. Figure 9.15 shows a schematic of a typical time-of-flight system.[40]

An optical pulse from the pump laser is used to generate a picosecond pulse of NIR light from the tunable dye laser. A portion of the pump pulse can also be focused onto a photodiode to generate a start pulse for the streak camera scan electronics (although in more recent systems this synchronization signal would probably be derived directly from the mode-locking electronics of the tunable laser). The NIR light is typically passed through a beam splitter, with the majority of the light being transmitted to the tissue. Power levels at the skin surface may reach several hundred milliwatts in a 1-mm^2 spot, when recording transmitted signals over more than 40 mm of tissue. These power levels exceed the maximum permissible exposure (MPE) for skin as defined, for example, in the European standard IEC60825; thus, such studies may require ethical approval (e.g., from an institutional review board) and should not to be performed routinely at the same site.

A fraction of the incident laser pulse is diverted into a reference delay line, before being applied to the streak camera along with the pulse transmitted through the tissue (the streak camera effectively comprises a 1D array of

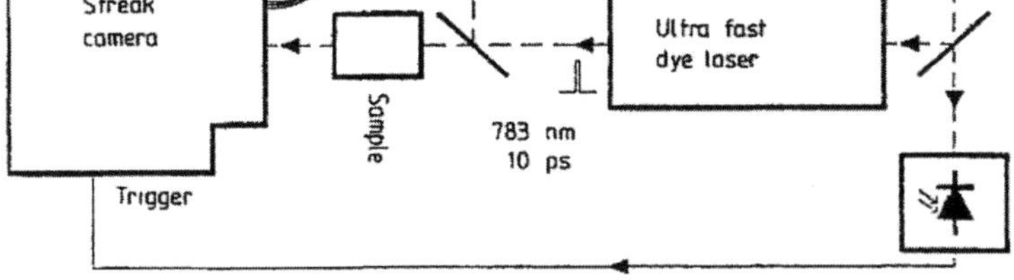

Figure 9.15 A typical time-of-flight system for measuring photon path length *in vivo*. In more recent systems, enhanced wavelength tunability in the NIR is obtained by using a Ti: sapphire laser instead of the dye laser.[40]

detector elements, which can record both these pulses simultaneously). The purpose of this reference delay is twofold. First, since the reference arm has a precisely known length, the reference pulse provides an absolute calibration of the streak-camera time axis. Second, the temporal dispersion of the reference pulse represents the instrumental response function, including streak camera jitter, multimode fiber dispersion, and so on. This can, in principle, be deconvolved from the measured temporal distribution of light transmitted through the tissue [generally referred to as a temporal response function (TPSF) in the literature[, although with state-of-the art lasers and streak cameras this step is not normally necessary. A typical TPSF has the form illustrated in Fig. 9.9; the mean time of such a distribution is readily computable using software.

9.6.1.3 Frequency-domain measurements

Rather than measure $I(t)$ directly, it is possible to measure the temporal Fourier transform of this quantity $\tilde{I}(\omega)$ by illuminating the tissue with a continuous-wave light beam whose intensity is modulated at angular frequency ω. The advantages of this approach and its practical implementation were described in Section 9.5.1.5, where its use to measure absolute tissue transport properties was discussed.

It is a standard result from Fourier theory[60] that if

$$\tilde{I}(\omega) \Leftrightarrow I(t),$$

then

$$\frac{d\tilde{I}(\omega)}{d\omega} \Leftrightarrow itI(t),$$

where $\Leftrightarrow$ denotes Fourier transformation. Since for any function $f(t)$

$$\int_{-\infty}^{\infty} f(t)dt = \tilde{f}(0),$$

it follows that

$$\langle t \rangle = \frac{\int_{-\infty}^{\infty} tf(t)dt}{\int_{-\infty}^{\infty} f(t)dt} = \frac{-i\left.\frac{d\tilde{I}(\omega)}{d\omega}\right|_{\omega=0}}{\tilde{I}(0)}.$$

Writing

$$\tilde{I}(\omega) = A(\omega)\exp[i\Delta\Phi(\omega)],$$

then

$$\langle t \rangle = -i \frac{\Delta\Phi'(\omega)A(\omega)\exp[i\Delta\Phi(\omega)] + A'(\omega)\exp[i\Delta\Phi(\omega)]}{A(\omega)\exp[i\Delta\Phi(\omega)]}\bigg|_{\omega=0}.$$

Since $I(t)$ is a real function, it follows that $\tilde{I}(\omega)$ is a Hermitian function,[60] and hence $A(\omega)$ is an even function. This in turn suggests that if $A'(\omega)$ remains continuous at the origin, then it must vanish there. Hence,

$$\langle t \rangle = \Delta\Phi'(\omega)\big|_{\omega=0}; \tag{9.31}$$

In other words, the mean time can be calculated by measuring the phase gradient of $I(\omega)$ at the origin.

In practice, the phase gradient is estimated by measuring the phase shift induced in the intensity-modulated beam at some finite frequency ω. To obtain an accurate estimate of $\Delta\Phi'(\omega)|_{\omega=0}$, this modulation frequency must not be too high. Arridge et al. have shown that modulation frequencies of 200 MHz or less yield measurements that are essentially identical to those obtained with direct time-of-flight measurements when measured across 5 cm of brain tissue.[24]

9.6.1.4 "Tracer" methods

It is possible, in principle, to obtain measurements of the path length *in vivo* using purely continuous-wave (CW) instrumentation.[10] Consider a multiple-scattering medium that contains, among others, a chromophore of known concentration c_{ab}. If this chromophore contains a distinct spectral feature at wavelength λ, which in the absence of scattering would have an amplitude A at unit concentration, then the path-lengthening effect of multiple scattering means that the feature will appear to have an amplitude $\mathrm{DPF}(\lambda) \times A$. By measuring the amplitude of this spectral feature in the multiple-scattering medium and comparing it with the amplitude in a nonscattering solution, DPF can then be determined. In effect, this method is an "added absorber" method in which $\Delta\mu_a$ (and hence ΔA) is measured over a step change in wavelength $\Delta\lambda$, rather than over a step change in chromophore concentration Δc_{ab}.

This method has the advantage that it uses comparatively simple instrumentation. The disadvantages of the method are that (1) one is forced to assume a value for c_{ab}, the intrinsic concentration of the chromophore, and (2) one obtains an estimate of DPF at only a single measurement wavelength. This latter disadvantage may not be so serious if the method is combined with techniques that estimate the wavelength dependence of path length using CW instrumentation as well (see Section 9.6.2).

To implement this idea on a practical level, Cope et al. suggested that tissue water, which has a distinct NIR feature at 975 nm[61] and is present in tissues with well-documented concentrations,[62–64] could serve as such a "tracer" chromophore and designed a CCD spectrometer to collect suitable

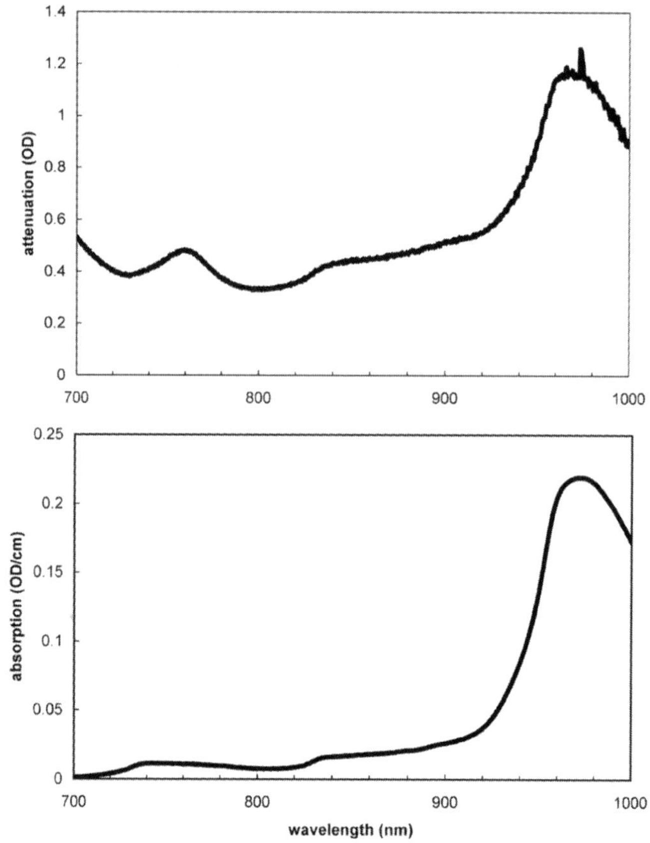

Figure 9.16 (top) A typical attenuation spectrum measured across 3.5 cm of human forearm muscle. (bottom) The spectrum of pure water measured across 1 cm. Note the substantial increase in the strength of the principle absorption band due to multiple scattering.

data.[65] Figure 9.16 shows a typical NIR attenuation spectrum measured over the 600- to 1000-nm wavelength range, and below it is the reference spectrum of pure water. It is clear to the eye that the amplitude of the 975-nm water peak is enhanced substantially due to multiple scattering in the *in vivo* spectrum.

The measurement of the amplitude of the absorption peak in the *in vivo* spectrum is complicated slightly, because the baseline of the spectrum is unknown. This ambiguity arises because (1) scattering-induced light loss produces an additive component G in the overall attenuation spectrum that is unknown unless μ_s' is known, and (2) most tissue spectrometers are not photometrically calibrated, so the measurement of absolute attenuation is difficult in practice. To overcome this problem, Cope et al. suggested using derivative spectroscopy, which is widely used in NIRS in such situations. Scattering will introduce a wavelength-dependent baseline offset, $G = G(\lambda)$. To a first-order approximation, this wavelength dependence will be linear, so it can be removed almost completely by forming the second differential of the attenuation spectrum.

9.6.2 Determination of the wavelength dependence of path length

It must always be borne in mind that the DPF is intrinsically a wavelength-dependent quantity, because it is determined by both the absorption and scattering properties of the tissue. To a first approximation, DPF can be approximated by the expression

$$DPF \approx \frac{3}{2}\sqrt{\frac{\mu_s{}'}{\mu_a}}, \tag{9.32}$$

That is, DPF falls as μ_s' falls, but rises as μ falls. μ_s' falls roughly linearly and by approximately 10% over the wavelength range 760 nm to 900 nm,[30] while μ_a shows substantial variations in the vicinity of the deoxygenated hemoglobin peak (760 nm), the lipid peak (930 nm), and the O—H water absorption overtone (975 nm). One thus expects DPF to fall with increasing wavelength but to display localized troughs in the vicinity of these wavelengths.

The importance of determining DPF(λ) depends on the application. If the goal is to measure changes in hemoglobin concentration, then the accuracy is not compromised too severely if DPF(λ) is assumed to be a constant. However, incorporating the correct DPF(λ) into the previously mentioned design set becomes crucial when attempting to detect small changes in the redox state of cytochrome oxidase. Such redox changes are generally brought about by changes in oxygen supply, and thus the small spectral changes due to CuA occur in the presence of much larger changes in the absorption spectrum of hemoglobin. Small inaccuracies in the design set can then produce "crosstalk" between the changes in hemoglobin concentration and cytochrome redox state.

The most direct solution to the problem is to measure DPF at each measurement wavelength used. This is the approach adopted in the UCL phase-resolved system "BAMBI,"[66] a four-wavelength frequency-domain system operating at 200 MHz. This system has been used to measure DPF on the adult head, forearm, and calf, and on the neonatal head.[67] In addition, valuable data on the variation of DPF with age on the adult head have been collected.[66] It appears that DPF as a function of age, over the range 0 to 50 years and at wavelengths 690, 744, 807, and 832 nm, can be estimated from the relations

$$\begin{aligned} \mathrm{DPF}_{690} &= 5.38 + 0.049A^{0.877}, \quad \mathrm{DPF}_{744} = 5.11 + 0.106A^{0.723}, \\ \mathrm{DPF}_{807} &= 4.99 + 0.067A^{0.814}, \quad \mathrm{DPF}_{832} = 4.67 + 0.062A^{0.819}, \end{aligned} \tag{9.33}$$

where A is the subject age in years.

Essenpreis et al.[73] have used a time-of-flight system that relies on a wavelength-tunable picosecond laser source to measure DPF *in vivo* in three different tissue types. At wavelengths between 740 and 840 nm, the DPF measured on adult head, forearm, and calf behaves broadly as described previously. These data have been incorporated into the design set used by several commercial NIR oximeters (NIRO500, Critikon 2020).

Delpy et al. have suggested a simple technique to determine DPF(λ) in tissues such as the arm and leg, where, in principle, both venous and arterial blood flow can be occluded.[68] The time course of hemoglobin changes in the muscle is then such that the total hemoglobin content of the tissue, Hb.T, remains approximately constant (assuming that pooling of blood due to gravity is negligible). One then knows that the change in $\mu_a(\lambda)$ is completely described by the absorption spectrum of Hb.D (the spectrum of HbO_2 minus the spectrum of Hb). Consider the initial period of the occlusion, where neither μ_a nor μ_s' will have changed significantly from their baseline values. In this case the DPF will not have departed significantly from its baseline value either. The change in attenuation at wavelength λ, $\Delta A(\lambda)$, is then given by

$$\Delta A(\lambda) \approx \mathrm{DPF}(\lambda) \cdot r_{sd} \cdot \varepsilon_{\mathrm{Hb.D}}(\lambda) \cdot \Delta c_{\mathrm{Hb.D}}.$$

Hence, if during the initial occlusion period the concentration of Hb.D rises approximately linearly with time t, such that $\Delta c_{\mathrm{Hb.D}} = const\ \ t$, then we have

$$\Delta A(\lambda,t) \propto \mathrm{DPF}(\lambda) \cdot \varepsilon_{\mathrm{Hb.D}}(\lambda) \cdot t; \tag{9.34}$$

and the wavelength dependence of path length can be determined by regressing ΔA against t for each wavelength and dividing the result by $\varepsilon_{\mathrm{Hb.D}}(\lambda)$.

In tissues such as those found in the head, it is clearly not possible to produce such a hemodynamic change. For this case, Kohl et al. have proposed a related technique to determine this wavelength dependence using CW instrumentation.[69] In their method, a CW spectrometer with very fast acquisition rate is used, so that several spectra can be obtained within one beat of the heart. In this case it is possible to measure directly the arrival of arterial blood into the tissue of interest (i.e., the brain) in a similar manner to that used by a pulse oximeter. Since the saturation of arterial blood can be measured independently, one finds a relationship similar to that mentioned earlier, but with $\varepsilon_{\mathrm{Hb.D}}(\lambda)$ replaced by $\varepsilon_{arterial}(\lambda)$, the specific absorption coefficient of the arterial blood. Kohl et al. have used this method to measure DPF(λ) over the occipital cortex, with a 3-cm source–detector spacing. This "pulse" measurement of DPF(λ) is clearly affected by the assumed value for the hemoglobin saturation of the blood arriving within each pulse, S_aO_2. Kohl et al. note that the results for DPF(λ) for wavelengths greater than 790 nm are, in fact, only weakly dependent on the assumed saturation. In general, the pulse method agrees well with time-of-flight spectroscopy, as shown in Fig. 9.17.

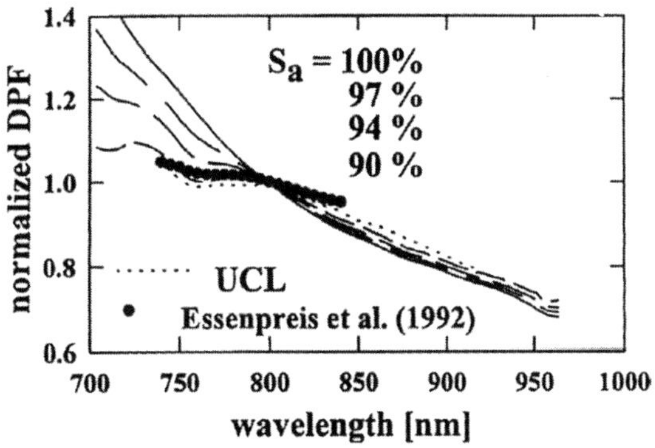

Figure 9.17 DPF(λ) as measured by the pulse method (lines) compared with direct time-of-flight measurements (filled circles). Several curves of DPF(λ) are shown for the pulse method, corresponding to different assumed values of S_aO_2.

At wavelengths less than 790 nm, the results show a strong dependence on the assumed S_aO_2, with the best agreement between methods arising when S_aO_2 is taken to be between 90% and 94%. This is somewhat lower than the expected saturation of purely arterial blood (97%).

It is worth noting that, while elegant, the pulse method relies on assumptions of tissue homogeneity that may not be realized in practice. For example, attenuation changes across the head are affected by the width of the cerebrospinal fluid (CSF) layer.[70] Since the brain surface visibly pulsates with the cardiac cycle, this may produce a confounding effect to distort DPF(λ). Also, the equality between $\Delta A/\Delta\mu_a$ and DP is only strictly valid if the perturbation in μ_a occurs uniformly throughout the medium. Since the CSF contains no blood, this will also produce a confounding effect. Detailed modeling using the theoretical models described in Section 9.8 is needed to determine the magnitude of these effects.

9.6.3 Instrumentation

By the standards required for absolute quantification, the instrumentation required for quantified trend measurements is comparatively simple. If previously measured values for the DPF are used, then all that is required is a source of CW light at the required wavelengths, a detector of adequate sensitivity, and a means of multiplexing the emission wavelengths onto the detector.

One of the earliest trend monitors was the system described by Cope and Delpy.[10] This system used laser diodes operating at 778, 813, 867, and 904 nm as the light source, guided to the tissue by fiber-optic cables. The detected light was collected by a similar fiber-optic cable and guided to a photomultiplier tube equipped with a cooled gallium arsenide (GaAs) photocathode. With a signal integration time of 20 s, this system could detect attenuation changes as small as 0.02 OD on a baseline attenuation of 10 OD. Such a system is capable of making transillumination measurements across the intact neonatal head. This device has been commercialized by Hamamatsu Photonics KK, first as the NIRO-1000 and subsequently as the NIRO-500.[4] Critikon produced a similar device in the mid-1990s, the Critikon 2020.

An example of a state-of-the-art trend monitor is the OXYMON device constructed by van der Sluijs et al.[71] By using a large-area avalanche photodiode as the detector, which offers detective quantum efficiency (DQE) exceeding 50% in the NIR spectrum and photon-counting performance, the acquisition rate can be improved to 50 Hz or more. This is fast enough to reveal pulsatile components in cerebral blood flow, as well as rapid changes in attenuation during functional brain activation.[72]

A recent development in trend-monitor instrumentation is to place the light detector directly onto the tissue surface rather than using fiber-optic bundles to relay the emerging light to a remote detector. This setup has the advantage that the detector collects exiting light over the full 2π steradians of solid angle, whereas fiber-optic bundles typically collect over a numerical aperture of 0.5 or less. A disadvantage is that electrical—rather than optical—signals must now be transmitted from patient to monitor, which makes these devices more susceptible to electrical interference. This can preclude the use of such devices in certain circumstances, such as in an operating theater when diathermy equipment is being used.

In addition to devices that employ light sources emitting at discrete wavelengths, trend monitors have been constructed using a white-light source and spectrometer.[66] Figure 9.18 shows the typical layout of such a device.

The tissue is illuminated with broadband light (typically from a quartz–tungsten–halogen bulb) via a fiber-optic bundle. The detected light is sent to the input slit of a grating spectrometer, where the light is dispersed and imaged onto a detector. The major advance that has made such an approach feasible for tissue oximetry is the advent of multielement array detectors such as the cooled low-noise charge-coupled device (CCD) detector, because with such devices all dispersed wavelengths are detected simultaneously by different elements in the array. Before the advent of such arrays, the light was imaged onto a single-element detector, which collected only a single wavelength at a time. The complete spectrum was obtained by scanning the detected wavelength across the entire spectral range of interest, typically by rotating the diffraction grating. Given N measurement wavelengths (where N is typically several hundred), such a procedure clearly generates $\sqrt{N}$ times more photon shot noise than using an N-element array detector.

For applications where a large dynamic range is required, the array detector of choice is the linear photodiode array. In applications where the lowest dark noise and highest sensitivity are required, however, the detector of choice is the CCD. Modern CCD detectors use multiphase pinning

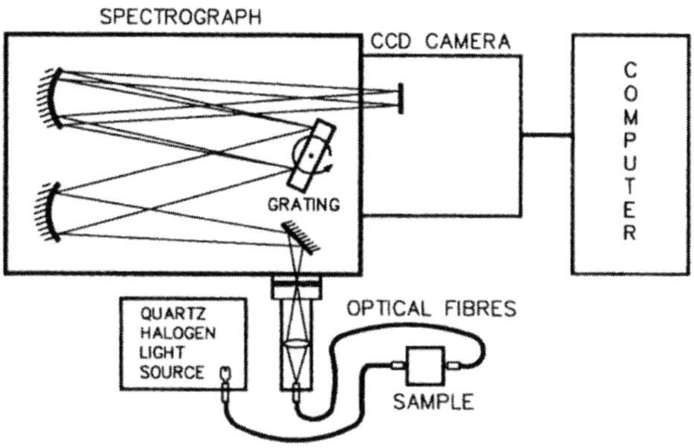

Figure 9.18 A schematic diagram of a multi-wavelength CCD-based tissue spectrometer.[11]

(MPP) to suppress thermal noise, and they can achieve single-pixel dark noise levels of 10 electrons per hour using simple Peltier cooling to, typically, −70°C.[73] DQE in the NIR can exceed 80%. An additional advantage of the CCD is that it is readily available as a 2D array detector. When combined with an imaging spectrometer, such a device can detect signals from several fibers in parallel, which simplifies instrument design in many situations.

One disadvantage of using a spectrometer is that light collection efficiency is usually markedly lower than for a other devices since (1) the input optics is generally of a rather low NA, and (2) to obtain high spectral resolution when detecting light from an extended source, an input slit must be used to reduce the area of the source, observed in a direction parallel to the dispersion direction of the grating. However, a high spectral resolution is not critical when performing tissue spectroscopy, as the spectral features of hemoglobin and other substances of instances are generally quite broad. Kohl et al., for example, have used an input slit width of 0.1 cm (equivalent to a spectral resolution of several tens of nanometers) to obtain measurements across the adult head using integration times as short as 20 ms.[69]

9.6.4 Algorithms

As mentioned earlier, multilinear regression provides a convenient tool for converting changes in light attenuation at several wavelengths into changes in chromophore concentration. Multilinear regression solves the problem by supplying a matrix $\mathbf{X}$, the "pseudo-inverse" matrix, relating the vector of chromophore concentration changes $\Delta\mathbf{c}$ to the vector of attenuation change $\Delta\mathbf{A}$.

Whether derived by multilinear regression or not, all current trend monitors use some mathematical formula to relate changes in chromophore concentration to changes in attenuation. Such a formula is termed an "algorithm" in the literature and generally varies between research groups and device manufacturers. There are several reasons for this variation:

1. Different manufacturers and researchers measure attenuation changes at different wavelengths.
2. The chromophores to be derived may vary.
3. There is often disagreement between groups on the precise form of the specific absorption coefficient spectra for the various chromophores.
4. Not all algorithms account for the wavelength-dependent dpf mentioned earlier.
5. Some algorithms are derived empirically, using data collected on a particular tissue *in vivo*.

To illustrate this, we will consider six algorithms that have been published in the literature, describing both their mathematical form and the method by which they were derived.

9.6.4.1 The UCL algorithm

Several algorithms have been derived for use in the oxygen trend monitors marketed by Hamamatsu Photonics (the four-wavelength NIRO-500 and the six-wavelength NIRO-1000). These algorithms are designed to monitor changes in oxyhemoglobin and deoxyhemoglobin concentrations and cytochrome oxidase redox state. The algorithms are obtained by performing multilinear regression on the spectra of purified oxyhemoglobin and deoxyhemoglobin[11] as well as the oxidized–reduced cytochrome oxidase difference spectrum measured *in vitro*. To account for wavelength-dependent DPF, the coefficients of the design set are scaled using the data obtained using time-of-flight spectroscopy.[74] The result of this procedure is the following two algorithms:

$$\begin{pmatrix} \Delta\mathrm{Hb} \\ \Delta\mathrm{HbO_2} \\ \Delta\mathrm{cyt}\text{-}aa_3 \end{pmatrix} = \begin{pmatrix} 1.58 & -1.35 & -0.57 & 0.68 \\ -0.66 & -0.85 & 0.56 & 1.5 \\ -0.26 & 1.17 & 0.12 & -0.92 \end{pmatrix} \times \begin{pmatrix} \Delta\mathrm{OD}_{775\,\mathrm{nm}} \\ \Delta\mathrm{OD}_{810\,\mathrm{nm}} \\ \Delta\mathrm{OD}_{870\,\mathrm{nm}} \\ \Delta\mathrm{OD}_{904\,\mathrm{nm}} \end{pmatrix}, \tag{9.35}$$

$$\begin{pmatrix} \Delta\mathrm{Hb} \\ \Delta\mathrm{HbO_2} \\ \Delta\mathrm{cyt}\text{-}aa_3 \end{pmatrix} = \begin{pmatrix} 1.56 & -0.51 & -0.78 & -0.57 & -0.023 & 0.65 \\ -0.55 & -0.72 & -0.39 & 0.12 & 0.74 & 0.31 \\ -0.3 & 0.65 & 0.6 & 0.26 & -0.26 & -0.83 \end{pmatrix} \times \begin{pmatrix} \Delta\mathrm{OD}_{775\,\mathrm{nm}} \\ \Delta\mathrm{OD}_{800\,\mathrm{nm}} \\ \Delta\mathrm{OD}_{825\,\mathrm{nm}} \\ \Delta\mathrm{OD}_{850\,\mathrm{nm}} \\ \Delta\mathrm{OD}_{875\,\mathrm{nm}} \\ \Delta\mathrm{OD}_{900\,\mathrm{nm}} \end{pmatrix}. \tag{9.36}$$

9.6.4.2 The SAPPORO algorithm

Tamura et al. have described a three-wavelength algorithm for deriving ΔHb, ΔHbO_2, and Δcyt aa_3 by solving simultaneous equations. A fourth measurement wavelength is used but functions as a reference wavelength, with the attenuation changes at the other three wavelengths then being measured relative to the change at this wavelength. The purpose of this step is to compensate for fluctuations in detected light intensity that arise for reasons other than fluctuations in chromophore concentration. Such fluctuations could occur, for example, in the tissue scattering coefficient, source brightness, or detector efficiency (e.g., due to poor coupling to the tissue surface). In matrix form, this algorithm has the form

$$\begin{pmatrix} \Delta\text{Hb} \\ \Delta\text{HbO}_2 \\ c_3 \times \Delta\text{cyt-}aa_3 \end{pmatrix} = \begin{pmatrix} 0.749 & -1.587 & 0.0 \\ -0.898 & -2.1 & 0.0 \\ 1.53 & -0.763 & 1.0 \end{pmatrix} \times \begin{pmatrix} \Delta\text{OD}_{700\text{ nm}} - \Delta\text{OD}_{750\text{ nm}} \\ \Delta\text{OD}_{730\text{ nm}} - \Delta\text{OD}_{750\text{ nm}} \\ \Delta\text{OD}_{805\text{ nm}} - \Delta\text{OD}_{750\text{ nm}} \end{pmatrix}. \quad (9.37)$$

Note that the algorithm does not attempt to quantify the changes in Δcyt aa_3; that is, the constant c_3 is undetermined. The algorithm is derived by solving a set of simultaneous equations involving the specific absorption coefficients for oxyhemoglobin and deoxyhemoglobin. Unlike the NIRO algorithms, these specific absorption coefficients are measured on a 1-cm cuvette containing whole rather than lysed blood cells.

9.6.4.3 The DUKE-P algorithm

Piantadosi has published a four-wavelength algorithm for determining ΔHb, ΔHbO_2, Δcyt aa_3 and a fourth, unidentified compound (this last signal being basically discarded).[75] Unlike the other algorithms, this one uses specific absorption coefficient spectra measured *in vivo* rather than *in vitro*. The procedure to determine the three spectra is to measure the attenuation spectrum of a cat head (1) with the blood fully oxygenated, (2) with the oxygenated blood then replaced with oxygenated perfluorocarbon, (3) with the perfluorocarbon fully deoxygenated, and (4) with the deoxygenated perfluorocarbon replaced by deoxyhemoglobin. Assuming that perfluorocarbon displays negligible absorption in the NIR yet can deliver oxygen at a rate sufficient to keep cytochrome oxidase fully oxidized, the difference spectra calculated during these maneuvers can then be used to estimate the absorption coefficient spectra as follows:

$$(1)\,\text{minus}\,(2) \Rightarrow \text{HbO}_2\,\text{spectrum}$$
$$(2)\,\text{minus}\,(3) \Rightarrow \text{cyt}\,aa_3\,\text{difference spectrum}$$
$$(4)\,\text{minus}\,(3) \Rightarrow \text{Hb}\,\text{spectrum}$$

In matrix form, Piantadosi quotes the following form for his algorithm:

$$\begin{pmatrix} \Delta\text{Hb} \\ \Delta\text{HbO}_2 \\ \Delta\text{cyt-}aa_3 \end{pmatrix} = \begin{pmatrix} 2.20 & -0.91 & 0.19 & -0.86 \\ -1.51 & 0.57 & -0.24 & 1.48 \\ -3.08 & 6.52 & -0.66 & -2.45 \end{pmatrix} \times \begin{pmatrix} \Delta\text{OD}_{775\,\text{nm}} \\ \Delta\text{OD}_{810\,\text{nm}} \\ \Delta\text{OD}_{870\,\text{nm}} \\ \Delta\text{OD}_{904\,\text{nm}} \end{pmatrix}. \tag{9.38}$$

9.6.4.4 The KEELE algorithm

Rolfe et al. have proposed a three-wavelength algorithm based on hemoglobin spectra measured on whole-blood samples and a cytochrome oxidase difference spectrum measured *in vitro*.[83] The wavelength-dependent DPF has been included in the same manner as the NIRO algorithms and using the same time-of-flight data. In matrix form, the algorithm is

$$\begin{pmatrix} \Delta\text{Hb} \\ \Delta\text{HbO}_2 \\ \Delta\text{cyt-}aa_3 \end{pmatrix} = \begin{pmatrix} 1.642 & -1.02 & -0.221 \\ -1.157 & 0.081 & 1.776 \\ -0.014 & 0.635 & -0.611 \end{pmatrix} \times \begin{pmatrix} \Delta\text{OD}_{775\,\text{nm}} \\ \Delta\text{OD}_{845\,\text{nm}} \\ \Delta\text{OD}_{904\,\text{nm}} \end{pmatrix}. \tag{9.39}$$

9.6.4.5 Algorithm comparison

Matcher et al. have published a critique of these published algorithms, by collecting data using a CCD spectrometer on several tissue samples during various maneuvers designed to change oxygenation.[35] Since changes in attenuation are measured at a continuum of wavelengths from 650 nm to 1000 nm using the CCD, all of the previously described algorithms can be applied to the common data set and the results compared directly. Matcher et al. concluded that while all four algorithms produced broadly similar results for the relative changes in Hb and HbO_2 concentration, substantial differences exist between algorithms for the derived change in cyt aa_3 redox state. This is not surprising given that the low concentration of cyt aa_3 *in vivo*, together with the distorting effect of multiple scattering on tissue spectra, makes the accurate determination of cyt aa_3 redox changes much more problematic than Hb and HbO_2 concentration changes.

9.7 Use of Quantified Trend Measurements to Infer Absolute Blood Flow, Blood Volume, Hemoglobin Saturation, and Tissue Oxygen Consumption

Although much interest currently surrounds those recent commercial NIR oximeters that offer some form of absolute quantification, it has actually been possible for some time to obtain several absolute hemodynamic variables by making quantified trend measurements during various controlled physiological maneuvers.

9.7.1 Venous saturation via venous occlusion plethysmography

A simple method for estimating venous hemoglobin saturation via venous occlusion plethysmography has been proposed by Yoxall et al.[76,77] Consider a tissue volume of interest, which contains arterial, capillary, and venous vascular beds. Under normal conditions, the saturation and blood volumes of these three compartments remain constant, with arterial inflow and venous outflow rates matching each other. Now consider that a mild pressure is applied, sufficient to prevent venous outflow from the tissue volume but insufficient to restrict arterial inflow. There will consequently be a pooling of venous blood within the field of view of the NIR oximeter. A quantified trend monitor will then register an increase in Hb and HbO_2 concentrations of ΔHb and ΔHbO_2, from which the venous hemoglobin saturation S_vO_2 can be obtained directly as

$$S_vO_2 = \frac{\Delta HbO_2}{\Delta Hb + \Delta HbO_2}. \tag{9.40}$$

Figure 9.19 shows some illustrative results obtained using this technique. The left graph shows changes in Hb and HbO_2 recorded on the adult head during a period of jugular venous occlusion produced using mild finger pressure. The right graph shows a linear regression of ΔHbO_2 versus $\Delta Hb + \Delta HbO_2$, the gradient of which yields S_vO_2.

9.7.2 Skeletal muscle blood flow

A similar principle can be applied to the measurement of muscle blood flow; that is, a quantified trend monitor is used as a plethysmograph.[78] To measure muscle blood flow, the rate of accumulation of $Hb + HbO_2$ in the muscle is measured during a venous occlusion. Let us denote the muscle blood flow rate as MBF (with typical units of milliliters of blood per 100 g of tissue per minute). Then if venous outflow is occluded, blood will accumulate in the muscle at this same rate, and this accumulation can be detected by a quantified trend monitor as an increase in total hemoglobin concentration Hb.T. Since Hb.T has conventional units of micromoles per liter (or μmolar, μM), it is necessary to introduce a scaling factor between these two units. This scaling factor depends on the tissue density ρ_t (units of grams per milliliter), on

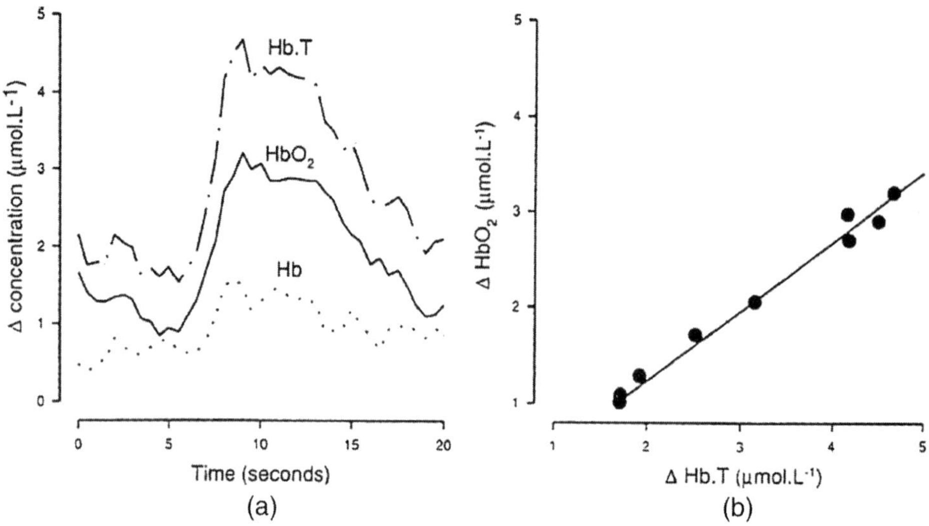

Figure 9.19 The venous occlusion technique applied to adult humans. The traces on the left show changes in Hb and HbO_2 in response to mild occlusion of the jugular vein. On the right is the resulting correlation plot of HbO_2 versus Hb.T.[32]

the molecular weight of hemoglobin MW_{Hb} (units of grams), and the concentration of hemoglobin within the blood *tHb* (units of grams per 100 mL). It is straightforward to show that these two units of concentration are related by the expression

$$c(\mu M) = c(\text{ml per } 100 \text{ g}) \times \frac{\rho_t \cdot tHb \cdot 10^5}{\text{MW}_{\text{Hb}}}.$$

Typically, ρ_t is taken to be 1.05 for muscle,[62] while $\text{MW}_{\text{Hb}} = 64{,}500$; *tHb* is typically 15 g per 100 mL in the large veins and arteries.[79]

To measure muscle blood flow, the initial rate of increase of Hb.T (measured immediately after the beginning of the occlusion) is determined using a quantified trend monitor. This rate of increase, $\dot{\text{H}}\text{b.T}$, can then be converted directly to an estimate of MBF via

$$MBF = \dot{\text{H}}\text{b.T} \times \frac{\text{MW}_{\text{Hb}}}{\rho_t \cdot tHb \cdot 10^5}. \tag{9.41}$$

9.7.3 Absolute muscle oxygen consumption

If an occlusion sufficient to prevent arterial inflow as well as venous outflow is applied to a limb, then, apart from a slight redistribution of stagnant blood due to pooling under gravity, Hb.T within the NIR field of view should remain constant. Under these conditions, the tissue will continue to consume oxygen at a rate of $\dot{O}_2$ until all oxygen within the capillaries has diffused into the surrounding tissue and been consumed. Hence, a trend monitor will detect an initially linear fall in HbO_2 that is accompanied by an almost symmetric rise in Hb. Since a single hemoglobin molecule binds four O_2 molecules, it is straightforward to convert the measured rate of fall of HbO_2, $\dot{H}bO_2$ (units of μM per minute), into the rate of oxygen consumption $\dot{O}_2$(conventional units of μmol per 100 g of tissue per minute),

$$\dot{O}_2 = \dot{H}bO_2 \times \frac{4}{10\rho_t}.$$

Since the time course of ΔHbO_2 and ΔHb are mirror images of each other, the signal-to-noise ratio can be improved by a factor $\sqrt{2}$ by calculating the change in Hb.D ($=HbO_2 - Hb$) and halving the value

$$\dot{O}_2 = \dot{H}b.D \times \frac{0.2}{\rho_t}. \tag{9.42}$$

9.7.4 Cerebral blood flow (CBF)

It is clearly impractical to completely occlude the venous return from the brain to measure brain blood flow. Edwards et al. suggested a novel variant of tracer clearance techniques such as radioactive-microsphere and xenon (Xe) washout, in which HbO_2 itself is used as a tracer.[80] Tracer washout techniques are all based on the Fick principle, which is a basic mathematical statement of material conservation.

Consider an organ into which blood flows at a rate F (with conventional units of milliliters per minute). If the inflowing arterial blood contains a measurable tracer at concentration c_a, and the outflowing venous blood contains a concentration c_v of this tracer, then Fick's law states that the amount of tracer in the organ Q increases according to

$$\frac{dQ}{dt} = F(c_a - c_v). \tag{9.43}$$

Now consider that a bolus of tracer of limited duration τ is administered. c_a and c_v then become functions of time. If τ is shorter than the transit time t_{transit} for the organ (the time taken for a packet of blood to traverse the organ from the arterial to venous compartment), then the venous concentration will effectively remain zero for the duration of the measurement. Under these conditions, Eq. (9.43) can be integrated with regard to time to yield the total amount of tracer accumulated in the organ:

$$Q = F \int_0^{\tau} c_a(t)dt.$$

Hence, the basis of all tracer methods for determining *F* is to measure the total accumulation of tracer *Q* over a given time τ and divide this by the known arterial inflow of tracer over this same time, $\int_0^\tau c_a(t)dt$.

In the NIR method for measuring CBF, HbO_2 itself is used as the tracer. To produce an effective "bolus" of arterial HbO_2, a volunteer's inspired oxygen fraction (FiO_2) is slowly reduced so that the resting arterial hemoglobin saturation (S_aO_2) falls from approximately 98% (the normal resting value in a healthy adult) to approximately 90%. Once S_aO_2 has stabilized at this value, FiO_2 is suddenly restored to its normal value so that, ideally, S_aO_2 is restored abruptly to 98%. $c_a(t)$ is thus approximated by a step function, which implies that $Q(t)$ increases linearly until the integration time-window τ approaches t_{transit}. $Q(t)$ is the increase in cerebral HbO_2 content, ΔHbO_2, whose change can be measured using a quantified trend monitor. $c_a(t)$ is the arterial hemoglobin saturation and can be measured using a pulse oximeter. As with the measurement of muscle blood flow, a conversion factor is required to express *F* in conventional clinical units (milliliters of blood per 100 g of tissue per minute), given the output of most trend monitors (μM) and pulse oximeters (% hemoglobin saturation). Since ΔHbO_2 is measured in units of μM, *F* can be obtained in units of μM per minute using

$$F = \frac{\Delta HbO_2}{\int_0^\tau S_aO_2(t)dt},$$

with *t* measured in minutes. *F* is then termed the "hemoglobin flow." To convert this into "cerebral blood flow" (CBF), we use the same conversion factor as before:

$$\text{CBF} = \frac{\Delta HbO_2}{\int_0^\tau S_aO_2(t)dt} \times \frac{MW_{Hb}}{\rho_t \cdot tHb \cdot 10^5}. \tag{9.44}$$

tHb is generally obtained from a blood sample drawn from a large vein. Since the hematocrit (and hence *tHb*) is generally lower in the small cerebral vessels than in the large vessels, it is necessary to apply the conversion

$$tHb_{cerebral} = tHb_{large\ vessel} \times \text{CLVHR},$$

where CLVHR is the "cerebral to large vessel hematocrit ratio" and typically has the value 0.7.[79] As with measurements of oxygen consumption, if Hb.T remains roughly constant (which should be the case if ΔSaO_2 is small), then SNR can be improved by considering changes in Hb.D rather than HbO_2.

The NIR method for measuring CBF has been validated against other methods such as Xe-133 clearance[81,82] and radioactive microspheres.[83] A good correlation between the different methods has been found both in neonatal humans and neonatal piglets.

9.7.5 Cerebral blood volume (CBV)

A slight variant of the basic idea behind CBF measurements has been proposed as a way of estimating a related hemodynamic quantity, the cerebral blood volume (CBV).[84] This quantity evidently closely correlates with the absolute concentration of hemoglobin in tissue Hb.T. As with the CBF method, the basic idea is to use HbO_2 as a tracer whose concentration is varied, although in the "oxygen-swing" CBV method, the change is induced over a much longer time scale (minutes rather than seconds).

Consider a simplified model of the circulation, in which blood flows at a uniform rate F_{Hb} (μM of hemoglobin per minute) through three distinct vascular compartments: arterial, capillary, and venous. If the arterial hemoglobin saturation (as measured by a pulse oximeter) is S_aO_2, then the rate of delivery of HbO_2 into the capillary compartment $\dot{H}bO_2)_d$ is given by

$$\dot{H}bO_2)_d = F \times S_aO_2.$$

If the capillary compartment converts HbO_2 to Hb at a rate (due to consumption) of $\dot{H}bO_2)_c$, then the venous saturation will be lower than S_aO_2 by an amount

$$S_aO_2 - S_vO_2 = \frac{\dot{H}bO_2)_c}{F};$$

Assuming that oxygen is exchanged uniformly along the length of the capillary compartment, then capillary oxygen saturation S_cO_2 will be the arithmetic mean of S_aO_2 and S_vO_2.

Now consider that S_aO_2 is lowered by an amount ΔS_aO_2—for example, by lowering the inspired oxygen fraction FiO_2 as in NIR CBF measurements. If this reduction is within the normal physiological range (e.g., from 98% to 90%), then, as with measurements of CBF, it is reasonable to assume that blood flow and oxygen consumption are unaffected. It is then evident that S_vO_2 and S_cO_2 must fall by this same amount and, therefore, the overall mixed arterial/capillary/venous hemoglobin saturation also falls by ΔS_aO_2. Since, by definition, the change in mixed hemoglobin saturation ΔS_mO_2 is given by

$$\Delta S_mO_2 = \frac{\Delta[HbO_2]}{[Hb.T]}, \tag{9.45}$$

we see that, in principle, regressing the measured change in S_aO_2 (which equals ΔS_mO_2) against ΔHbO_2 (as measured by a quantified trend monitor)

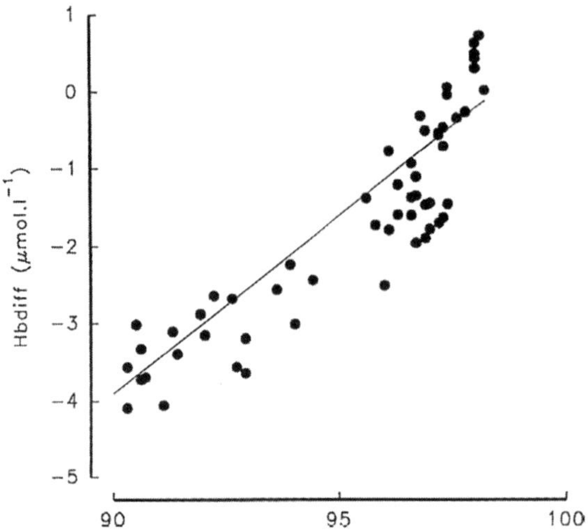

Figure 9.20 A typical swing in S_aO_2 (abscissa) induced by mild hypoxia and the resulting change in Hb.D, as measured by a quantified trend monitor.[85] The slope of the regression line yields CBV, as described in the text.

yields Hb.T. CBV (conventional units of milliliters of blood per 100 g of tissue) can then be obtained using the usual conversion factor:

$$\mathrm{CBV} = \mathrm{Hb.T} \times \frac{\mathrm{MW_{Hb}}}{\rho_t \cdot tHb \cdot 10^5}.$$

As with CBF measurements, the signal-to-noise ratio can be improved by regressing ΔHb.D against ΔS_mO_2, under the assumption that $\Delta\mathrm{Hb.T} \approx 0$. Figure 9.20 shows the results of a typical CBV maneuver. Note the reasonably tight and linear correlation between ΔHb.D and ΔS_aO_2, which verifies the underlying assumptions of the method.

Only a limited number of validation studies have been performed to test the NIR CBV technique. One early study in premature neonates[86] sought to correlate changes in CBV measured in this way against a direct measure of ΔHb.T obtained using quantified trend measurements. By increasing arterial pCO_2 by at least 0.5 kPa, an increase in CBV corresponding to 0.89 mL/100 g/kPa was measured by subtracting the absolute CBV values before and after the induction of hypercapnia. Surprisingly, the corresponding value obtained from ΔHb.T was almost four times smaller, 0.22 mL/100 g/kPa.

The precise cause of this discrepancy has so far not been adequately explained, but in an attempt to understand the effect, Firbank et al. have studied the CBV method both theoretically and experimentally.[87] In a series of experiments on anesthetized piglets, the correlation between changes in CBV (again induced by hypercapnia) as determined by subtracting absolute CBV measurements vs measurement of ΔHb.T was determined; Fig. 9.21

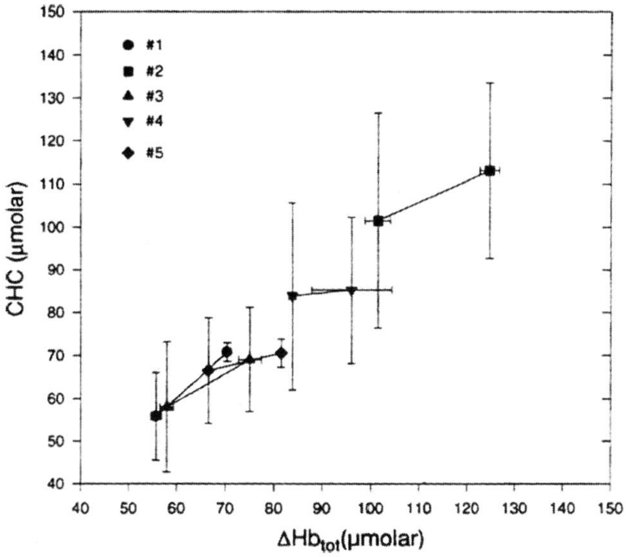

Figure 9.21 A study of the internal consistency of the CBV method. By inducing gross hypercapnia in piglets, CBV can be measured at two distinct values of Hb.T. The change in CBV is then compared with ΔHb.T, as measured by a quantified trend monitor.[87]

shows the correlation obtained. Note that each pair of connected symbols represents an individual animal, showing the absolute CBV (denoted as CHC here because only the absolute hemoglobin content, rather than blood content, is determined) before and after hypercapnia. Mainly for clarity, ΔHb. T (denoted as ΔHb_{tot} in Fig. 9.21) is converted into absolute units by assuming that its initial value equals CHC as determined by the oxygen-swing method. Little significance should thus be placed on the fact that the symbols are apparently clustered close to the unity line.

The important point of Fig. 9.21 is the marked tendency for the changes in CBV to under-read ΔHb.T, in contradiction to the findings of Brun and Greisen.[85] It is also apparent, however, that the measurement of absolute CBV is intrinsically less accurate than that of ΔHb.T, with absolute errors that obscure the small changes induced by hypercapnia. In part, this problem is attributable to random errors in pulse oximeter readings, which are often on the order of ±2% when the overall change in S_aO_2 is only on the order of 6–8%. However, other sources of error, including changes in path length due to absorption and (possibly) scattering changes, as well as probe movement due to swelling of the cranium, are also possible. By using analytical diffusion theory relations between μ_a, μ_s', and the resulting attenuation, Firbank et al. have attempted to quantify the magnitude of these effects. For a hypothetical increase in CBV of 15%, a 1-mm change in source–detector spacing between the two CBV measurements can cause ΔHb.T to over-read by as much as a factor of 2; however, the change in CBV appears to be much less affected by probe movement. The greatest source of systematic error affecting the apparent internal consistency of the CBV method appears to be changes in optical path length brought about by changes in μ_a. Recall that, to a rough approximation, diffusion theory predicts that the optical path length DP scales with the transport coefficients as (see Section 9.5.1.5)

$$\mathrm{DP} \propto \sqrt{\frac{\mu_s'}{\mu_a}}. \tag{9.46}$$

A hypercapnic maneuver increases CBV by approximately 15%. Since the overall volume fraction of blood in the brain is approximately 2%, this corresponds to an increase of blood volume fraction of roughly 0.3%. If the tissue μ_a arises solely from the μ_a of the blood hemoglobin, then this implies that the tissue μ_a will similarly rise by 0.3% during the hypercapnic maneuver. If the tissue μ_a also contains a contribution from fixed (i.e., non-oxygen-dependent) absorbers, then the corresponding percentage increase in μ_a will be somewhat lower. In either case, the increase in μ_a is sufficiently small that we can use Eq. (9.46) to infer that the DP will change by ΔDP where

$$\frac{\Delta \mathrm{DP}}{\mathrm{DP}} \approx -\frac{1}{2}\frac{\Delta\mu_a}{\mu_a};$$

In other words, we expect a fractional change in path length of 0.15% (or less, in the presence of background absorption). Recall that the absolute CBV method relies on regressing ΔHbO_2 against ΔS_aO_2, and that ΔHbO_2 is directly proportional to DP. Hence, if CBV is increased by 0.3%, we can expect the upper value of CBV to be underestimated by 0.15%. Since we are interested in the difference between these two CBV values, ΔCBV, it is clear that this will under-read the true ΔCBV by a factor of 2. The measurement of ΔHb.T is only slightly affected by a 0.15% fall in DP, however, so that we would expect, for a small change in CBV and zero background absorption, a factor of 2 discrepancy between a change in CBV and ΔHb.T, with ΔHb.T being the higher value. This result makes the results of Brun and Greisen even harder to explain, but helps to explain the results obtained by Firbank et al.[86] The large effect of even a modest decrease in path length points to the importance of obtaining real-time path length measurements using, for example, frequency-resolved instruments.

Recently, a direct validation of the NIR CBV method has been performed in immature lambs.[88] NIR CBV values were compared with CBV measurements obtained by injecting radio-labeled serum albumin and red blood cells. The NIR measurements (2.5 ± 0.2 mL/100 g) were identical to those obtained using radionuclides (2.5 ± 0.2 mL/100 g).

9.8 Effects of Tissue Geometry and Heterogeneity

So far, all of our discussions have tacitly assumed that the tissue being probed is homogeneous (i.e., possesses identical transport properties at all positions) and has a simple geometry. Neither of these assumptions is true in practice, of course, and it is the goal of this section to review (1) the tools available for modeling the true situation more realistically and (2) the quantitative effect that tissue heterogeneity has on absolute, quantified trend, and absolute flow/volume measurements.

We will consider four main complications produced by real tissues:

1. Real tissues have a complicated geometry; that is, the 3D volume occupied by the tissue is not a regular figure such as a slab, cylinder, or sphere of infinite half-space. Thus, analytical solutions to the diffusion equation are generally not available.
2. Real tissues do not possess identical transport properties throughout the volume, so light transport must be described by the spatially inhomogeneous diffusion equation. For example, the forearm consists of two bones ($\mu_a = 0.24$ cm^{-1}, $\mu_s' = 19$ cm^{-1} at 800 nm[89]) covered by muscle ($\mu_a =$ 0.23 cm^{-1}, $\mu_s' = 6.8$ cm^{-1} at 800 nm[30]), which in turn is covered by an overlying layer of adipose tissue and skin ($\mu_a = 0.083$ cm^{-1}, $\mu_s' = 11.3$ cm^{-1} at 800 nm).[90] Figure 9.22 shows an anatomical slice through the forearm of a male human cadaver, showing the spatial distribution of the tissue types mentioned here.
3. Regions of the tissue may exist where the diffusion equation itself becomes an unreliable model of light transport. This can occur in clear, low-scattering regions (known as "voids") where light transport is

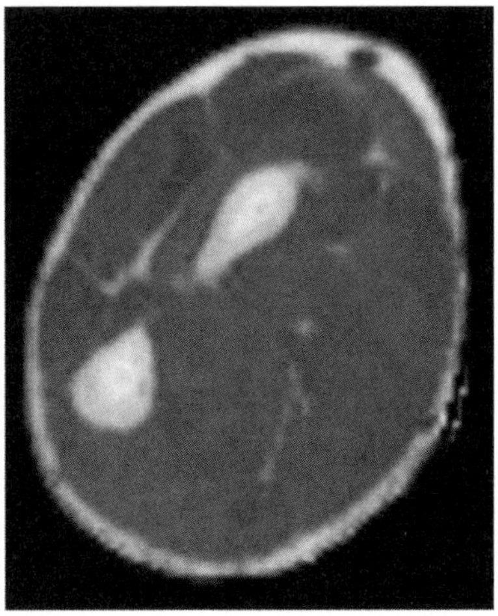

Figure 9.22 Anatomical slice through the forearm of the Visible Human male cadaver. Note the adipose surface layer and the radius and ulna bones, which make the forearm very heterogeneous for certain placements of the source and detector fibers.

approximated by the rectilinear propagation of geometrical optics. In real tissues, such voids include the cerebrospinal fluid layer surrounding the brain and the synovial fluid layer in the articular joints. In addition, whole blood in major veins and arteries displays a scattering coefficient not significantly greater than its absorption coefficient, which also strictly invalidates the use of the diffusion equation.[91]

4. Blood, which is the primary absorber in tissue, is not homogeneously distributed throughout the tissue. Instead, blood is present in microvessels, whose sizes range from some tens of microns to several millimeters. Hence, the absorber is distributed in "packets," which will also affect the details of light propagation.

9.8.1 Light transport models

To predict and ultimately correct for the effects of these tissue properties, one clearly needs a reliable mathematical model for light propagation through such tissues. The transport of neutral particles through a scattering medium can be described exactly by the Boltzmann transport equation. This equation thus provides the basis for studying the propagation of photons, if their wavelike properties are ignored. The Boltzmann equation is a complicated integro-differential equation that cannot be solved analytically except in very restricted special cases. Thus, a number of alternative methods have been developed and applied to tissue spectroscopy.

9.8.1.1 Two-layer diffusion models

A number of researchers have developed analytical models to deal with the simplest model of a heterogeneous medium, an infinite slab with optical properties $\mu_{a,1}$, $\mu_{s,1}'$ and thickness d above a semi-infinite half-space with optical properties $\mu_{a,2}$, $\mu_{s,2}'$, as shown schematically in Fig. 9.23.

In the late 1980s, a number of analytical and numerical models were developed based on diffusion and random walk theory.[93,94] The model of Takatani and Graham has been further developed by Schmitt et al.[95] to

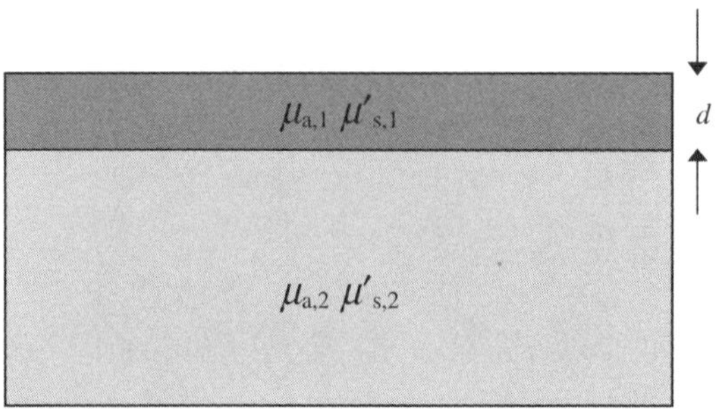

Figure 9.23 Schematic diagram of the simplest inhomogeneous geometry: an infinite slab of thickness *d* above a semi-infinite half-space.

include boundary reflections, yielding a somewhat cumbersome expression for the radial flux back-scattered from a three-layer medium, in which the upper layer is considered to act merely as a wavelength-dependent absorber. More recently, a frequency-domain model based on a spatial Fourier transform method has been proposed[96] and refined to include effects of boundary reflections by Kienle et al.[97] Consider the temporal form of the spatially homogeneous diffusion equation:

$$\left(D\nabla^2 - \mu_a c - \frac{\partial}{\partial t}\right)\Phi(\bar{r},t) = -q(\bar{r},t).$$

The appropriate differential equation to apply when considering a harmonically varying source at angular frequency ω is obtained from the temporal Fourier transform of this

$$(D\nabla^2 - (\mu_a c + i\omega))\Phi(\bar{r},\omega) = -q(\bar{r},\omega).$$

Now consider a two-layer medium as described previously. This equation is still valid in both layers, but with different transport properties D_1, $\mu_{a,1}$ and D_2, $\mu_{a,2}$, respectively, in each layer. Also consider the source term. When a multiple-scattering medium is illuminated by a thin collimated beam, it is a standard approximation to regard the medium as being illuminated by an equivalent isotropic source, located at a depth z_0 below the surface. In this case, we need to solve the following pair of differential equations, subject to certain boundary conditions:

$$\begin{aligned}&(D_1\nabla^2 - (\mu_{a,1}c + i\omega))\Phi_1(\bar{r}) = -q_0\delta(x,y,z - z_0) && 0 \le z < d,\\ &(D_2\nabla^2 - (\mu_{a,2} + i\omega))\Phi_2(\bar{r}) = 0 && d \le z.\end{aligned}$$

A convenient method of solution is then to use the Fourier transform approach to convert these equations into ordinary differential equations in z. To do this, consider the 2D spatial Fourier transform of Φ with regard to the x and y coordinates:

$$\phi(z,s_1,s_2) = \int_{-\infty}^{\infty}\int_{-\infty}^{\infty} \Phi(x,y,z) \cdot \exp[i(s_1 x + s_2 y)]dxdy.$$

Forming this 2D spatial Fourier transform and exploiting the azimuthal symmetry of the problem yields

$$\begin{aligned}&\tfrac{d^2}{dz^2}\phi_1(z,s) - \alpha_1^2\phi_1(z,s) = -\tfrac{q_0}{D_1}\delta(z - z_0) && 0 \le z < d,\\ &\tfrac{d^2}{dz^2}\phi_2(z,s) - \alpha_2^2\phi_2(z,s) = 0 && d \le z,\end{aligned} \tag{9.47}$$

where $\alpha_1^2 = (D_1 s^2 + \mu_{a,1}c + i\omega)/D_1$ (and the same for α_2) and $s^2 = s_1^2 + s_2^2$.

The boundary conditions that ϕ_1 and ϕ_2 must satisfy are as follows:

1. The presence of a refractive index mismatch at the upper boundary can be accurately included by setting $\phi_1 = 0$ on an "extrapolated" boundary at $z = -z_b$, rather than on the physical boundary at $z = 0$. The position of the extrapolated boundary is given by the relation

$$z_b = \frac{1 + R_{eff}}{1 - R_{eff}} 2D_1,$$

where R_{eff} depends on the refractive-index mismatch[93] and evaluates to 0.493 for a mismatch of 1.4:1 (typical of an air–tissue interface). Hence, a solution of these equations must satisfy

$$\phi(-z_b, s) = 0;$$

2. The requirement that the photon flux density and flux both be continuous across the internal boundary requires (assuming that this interface is refractive-index matched)[98]

$$\frac{\phi_1(d,s)}{\phi_2(d,s)} = \frac{n_1^2}{n_2^2} = 1, \quad D_1 \frac{\partial \phi_1(z,s)}{\partial z}\bigg|_{z=d} = D_2 \frac{\partial \phi_2(z,s)}{\partial z}\bigg|_{z=d}.$$

3. The requirement that the solution remain bounded in the semi-infinite lower region requires

$$\phi_2(\infty, s) = 0.$$

Applying these constraints to the general solution of

$$\begin{array}{ll}
\frac{d^2}{dz^2}\phi_1(z,s) - \alpha_{1\prime}^2\phi_1(z,s) = -\frac{q_0}{D_1}\delta(z - z_0) & 0 \le z < d \\
\frac{d^2}{dz^2}\phi_2(z,s) - \alpha_{2\prime}^2\phi_2(z,s) = 0 & d \le z \\
\frac{d^2}{dz^2}\phi_1(z,s) - {\alpha_1}^2\phi_1(z,s) = -\frac{1}{D_1}\delta(z - z_0) & 0 \le z < l \\
\frac{d^2}{dz^2}\phi_2(z,s) - {\alpha_2}^2\phi_2(z,s) = 0 & l \le z
\end{array}$$

yields the particular solutions

$$\phi_1(z,s) = \frac{\sinh[\alpha_1(z_b + z_0)]}{D_1\alpha_1} \frac{D_1\alpha_1 \cosh[\alpha_1(d - z)] + D_2\alpha_2 \sinh[\alpha_1(d - z)]}{D_1\alpha_1 \cosh[\alpha_1(d + z_b)] + D_2\alpha_2 \sinh[\alpha_1(d + z_b)]} - \frac{\sinh[\alpha_1(z_0 - z)]}{D_1\alpha_1}, \qquad 0 \le z < z_0,$$

$$\phi_1(z,s) = \frac{\sinh[\alpha_1(z_b + z_0)]}{D_1\alpha_1} \frac{D_1\alpha_1 \cosh[\alpha_1(d - z)] + D_2\alpha_2 \sinh[\alpha_1(d - z)]}{D_1\alpha_1 \cosh[\alpha_1(d + z_b)] + D_2\alpha_2 \sinh[\alpha_1(d + z_b)]}, \qquad z_0 \le z < d,$$

$$\phi_2(z,s) = \frac{\sinh[\alpha_1(z_b + z_0)] \exp[\alpha_2(d - z)]}{D_1\alpha_1 \cosh[\alpha_1(d + z_b)] + D_2\alpha_2 \sinh[\alpha_1(d + z_b)]}, \tag{9.48}$$

from which the photon density as a function of source–detector spacing r_{sd} and depth z can be obtained using an inverse Hankel transform:

$$\Phi(r_{sd},z) = \frac{1}{(2\pi)^2}\int_{-\infty}^{\infty}\int_{-\infty}^{\infty}\phi(z,s)\exp[-i(s_1x+s_2y)]ds_1ds_2$$
$$= \frac{1}{2\pi}\int_{0}^{\infty}\phi(z,s)sJ_0(sr)ds. \tag{9.49}$$

Finally, the remitted flux J can be calculated, either using the usual relation $J = -D_1\partial\Phi_1/\partial z|_{z=0}$, or, more accurately, in the case of a refractive-index mismatched boundary,[91,99]

$$J = -[0.118\Phi_1 + 0.306D_1\partial\Phi_1/\partial z]_{z=0}. \tag{9.50}$$

9.8.1.2 The Monte Carlo model

The Monte Carlo (MC) model is the most flexible and powerful model of light propagation in tissue, as it can in principle take account of all the effects described earlier.[100] Unfortunately, it is also the most computationally demanding. In this method, photons are simulated individually and their propagation through the tissue is modeled at the level of the individual scattering event. Figure 9.24 shows schematically a section of a single photon trajectory, as modeled by the MC technique.

After injection into the tissue, a uniformly distributed random deviate ξ in the range 0 to 1 is transformed, to yield an exponentially distributed random scattering length l (of mean value μ_s) using

$$l = -\frac{\ln(\xi)}{\mu_s}. \tag{9.51}$$

This length, together with the direction cosines describing the initial propagation direction of the photons, is then used to calculate the coordinates

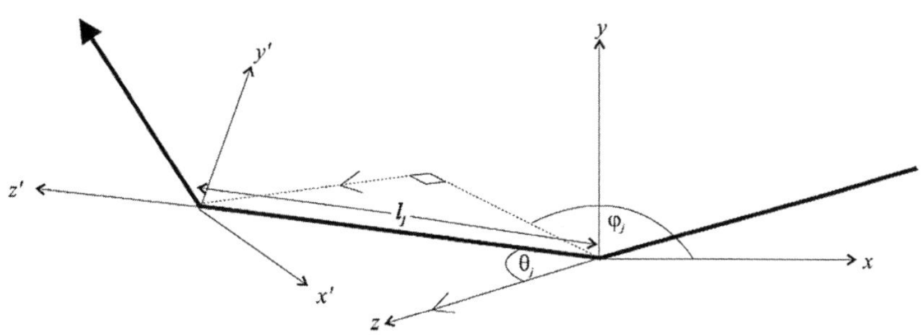

Figure 9.24 Photon random walk as simulated by the Monte Carlo technique. The distance between scattering events l_j and the scattering angles θ_j and φ_j at each scattering event are randomly generated, as described in the text.

of the first scattering event. At this point, two more random deviates similar to ξ are generated. The first, scaled to lie in the range 0 to 2π, describes the azimuthal scattering angle (it is almost always assumed that photon scattering is azimuthally symmetric about the incident propagation direction). The second is used to generate the altitudinal scattering angle, generally using a lookup table to produce angles distributed according to the required "scattering phase function" $p(\theta)$ (scattering is generally not isotropic with regard to θ).

The MC model can readily deal with spatially varying optical properties. When a photon trajectory traverses the boundary between two regions with different properties, the trajectory can be extrapolated back to the point of intersection with the boundary and appropriately reweighted.[100] The calculation of the trajectory can then be restarted from this new starting point, using the optical properties appropriate to the new medium. A difference in refractive index can be handled fully, by calculating the Fresnel reflection and transmission coefficients at the boundary. A random deviate ξ can then be compared with the intensity reflection coefficient R_F and the photon reflected, if $\xi < R_F$ or else is transmitted.[101]

9.8.1.3 The finite element method

The principal disadvantage of the MC technique is computational speed, since each photon trajectory may require several thousand scattering interactions to be simulated, and several tens of millions of photons may need to be calculated to obtain an acceptably low variance. For this reason, various groups have investigated numerical solutions to the Boltzmann transport equation (or, more often, its diffusion approximation). A number of numerical techniques exist for solving partial differential equations such as the diffusion equation, including the discrete ordinates method, the finite difference method, and the finite element method (FEM). FEM finds widespread application in structural engineering because it offers greater flexibility in modeling irregular geometries and faster computation speed than do the other techniques.

The essence of FEM is to subdivide the tissue volume of interest G with boundary ∂G into a large number of small non-overlapping "elements" of simple geometry. Popular element shapes include triangles (in 2D) and tetrahedrons (in 3D). Each element is then described by the location of its vertices, which are termed "nodes." The spatial arrangement of M elements and the associated N nodes is termed the finite element "mesh" and can be generated, given a border of arbitrary geometry, using tools such as the "adaptive mesh generator."[25] The example in Fig. 9.25 shows an anatomical slice through the abdomen of the Visible Human cadaver, with surface skin and adipose, bone, muscle, and internal organs clearly visible. Next to it is the output from an adaptive mesh generator, which identifies the boundaries

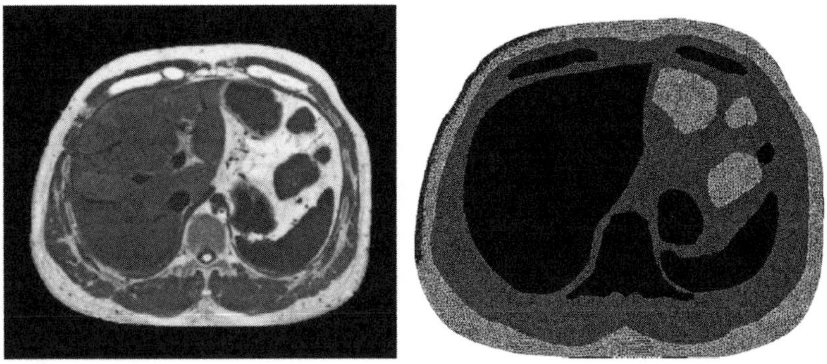

Figure 9.25 (left) An anatomical slice through the abdomen of the Visible Human male cadaver. (right) After segmentation and generation of "frontlines" defining different tissue regions, an adaptive mesh generator produces the 2D FEM mesh shown[25] (the color coding represents μ_a).

between various tissue types and generates a set of FEM elements that accurately represent the original image. Note the color coding, which represents μ_a and shows how the FEM method can ascribe different transport properties to different elements, thereby modeling inhomogeneous tissues.

In FEM, an "optimal" solution to the diffusion equation (for the sake of simplicity we will consider the time-independent case) $\Phi^h(\bar{r})$ is obtained, which is expressible as a linear combination of N basis functions ψ_i (also known as "element shape functions"). Element shape functions can have numerous forms, but in the simplest approach, the ith shape function is defined to be that piecewise linear function that has the value 1.0 at the ith node and that is zero at all other nodes. ψ_i thus has a support that is the union of all elements that share node i. Φ is then describable in terms of the basis ψ_i and a vector of weight coefficients ϕ_i (often termed "potentials"):

$$\Phi^h \approx \sum_{i=1}^{N} \phi_i \cdot \psi_i(\bar{r}). \tag{9.52}$$

Consider the spatially inhomogeneous time-independent diffusion equation:

$$[\nabla \cdot D(\bar{r})\nabla - \mu_a(\bar{r})c]\Phi(\bar{r}) = -q_0(\bar{r}). \tag{9.53}$$

If a solution $\Phi(\bar{r})$ exists, then it will also satisfy

$$\int_G \Psi(\bar{r})[\nabla \cdot D(\bar{r})\nabla - \mu_a(\bar{r})c]\Phi(\bar{r})dG = -\int_G \Psi(\bar{r})q_0(\bar{r})dG,$$

where $\Psi(\bar{r})$ is a suitable test function [technically, $\Psi(\bar{r})$ and all of its first derivatives must be integrable over G]. This equation can be integrated by parts to yield

$$\int_G D(\bar{r})\nabla\Phi(\bar{r}) \cdot \nabla\Psi(\bar{r}) + \mu_a(\bar{r})c\Psi(\bar{r})\Phi(\bar{r})dG$$
$$= \int_G \nabla \cdot [\Psi(\bar{r})D(\bar{r})\nabla\Phi(\bar{r})]dG + \int_G \Psi(\bar{r}) \cdot q_0(\bar{r})dG.$$

The first integral on the right-hand side can be evaluated using Gauss's theorem to yield

$$\int_G [D(\bar{r})\nabla\Phi(\bar{r}) \cdot \nabla\Psi(\bar{r}) + \mu_a(\bar{r})c\Psi(\bar{r})\Phi(\bar{r})]dG$$
$$= \int_{\partial G} \Psi(\bar{r})F(\bar{r})d(\partial G) + \int_G \Psi(\bar{r}) \cdot q_0(\bar{r})dG, \tag{9.54}$$

where we have used Fick's law to identify the term $-D(\bar{r})\nabla\Phi(\bar{r})\bar{N}$ as the irradiance emerging at the boundary, denoted $F(\bar{r})$. The problem of finding a function $\Phi(\bar{r})$ that satisfies Eq. (9.53) can thus be restated as the problem of finding $\Phi(\bar{r})$ such that Eq. (9.54) is true for any test function Ψ satisfying the integrability conditions. There is, of course, an infinite set of possible test functions to use. The basis of FEM is to use each of the basis functions ψ_i as test functions—that is, to find the function Φ^h that satisfies Eq. (9.54) for every ψ_i. To proceed further, we must establish an explicit relationship between Φ and F on the boundary. A variety of boundary conditions have been employed in the literature, but a popular choice for the case of a refractive-index mismatched boundary is the "Robin" condition:[102]

$$\xi\Phi(\bar{r}) - F(\bar{r}) = 0 \quad \forall\, \bar{r} \in \partial G, \tag{9.55}$$

where ξ is a constant whose value is determined by the refractive index mismatch (and which equals ½ for an index-matched boundary). Applying this boundary condition and substituting Eq. (9.52) into Eq. (9.54) leads to the requirement

$$\sum_{i=1}^{N} \phi_i \left\{ \int_G [D(\bar{r})\nabla\psi_i(\bar{r}) \cdot \nabla\psi_j(\bar{r}) + \mu_a c\psi_j(\bar{r})\psi_i(\bar{r})]dG + \xi \int_{\partial G} \psi_j(\bar{r})\psi_i(\bar{r})d(\partial G) \right\}$$
$$= \int_G \psi_j(\bar{r})q_0(\bar{r})dG,$$

for $j = 1$ to N.

This set of N linear equations for the N values of ϕ_i can conveniently be expressed in matrix form:

$$\mathbf{M} \cdot \Phi = \mathbf{Q}, \tag{9.56}$$

where

$$\Phi = [\phi_1, \phi_2, \phi_3, \cdots, \phi_N]^T,$$
$$M_{ij} = \int_G [D(\bar{r})\nabla\psi_i(\bar{r})\nabla\psi_j(\bar{r}) + \mu_a(\bar{r})c\psi_j(\bar{r})\psi_i(\bar{r})]dG + \xi\int_{\partial G}\psi_j(\bar{r})\psi_i(\bar{r})d(\partial G),$$
$$Q_j = \int_G \psi_j(\bar{r})q_0(\bar{r})dG; \tag{9.57}$$

The solution vector Φ can be obtained using a number of matrix inversion procedures such as Cholesky reduction or conjugate gradients.[19]

9.8.1.4 Hybrid diffusion–radiosity models

The aforementioned FEM can accurately model tissues of irregular geometry and spatially varying optical properties, but it is still limited to cases where the diffusion equation itself is valid. Formally the diffusion approximation assumes

1. $\mu_s' >> \mu_a$.
2. The radiance at all points can be approximated as a spherically symmetric distribution plus a small directional flux.

When the volume of interest contains regions of low scattering ("voids"), then these assumptions become invalid. In the limit that the scattering in the void falls to zero, the propagation of photons across the void becomes rectilinear and describable by geometrical optics rather than by the diffusion equation. The Monte Carlo technique can, of course, handle this situation easily, but its long computation times, especially for large tissue volumes, has motivated the development of faster techniques. In particular, Firbank et al.[103] and Arridge et al.[102] have developed a hybrid model, in which the diffusion equation and a "radiosity" model for rectilinear light propagation are combined.

The original diffusion–radiosity model of Firbank et al. was iterative; however, Arridge et al. have shown that, in fact, the calculation can be readily incorporated into the standard finite element diffusion model and hence solved directly. Consider a general void G_v with an arbitrary boundary ∂G_v, located within a diffusive medium, as shown in Fig. 9.26.

The medium within the void has a zero scattering coefficient and an absorption coefficient μ_a. Consider two arbitrary points on the surface, described by position vectors $\bar{m}$ and $\bar{m}'$. Within the framework of radiosity theory, all such points can be considered as acting as illumination sources for all other points on ∂G_v, and we wish then to calculate the illumination arising at $\bar{m}$ due to the outward photon current at $\bar{m}'$. To do this, we first introduce some terminology. The outward surface normals at $\bar{m}$ and $\bar{m}'$ are denoted as $\bar{N}(\bar{m})$ and $\bar{N}(\bar{m}')$, and the distance between $\bar{m}$ and $\bar{m}'$ is denoted as $d_{\bar{m},\bar{m}'}$. The

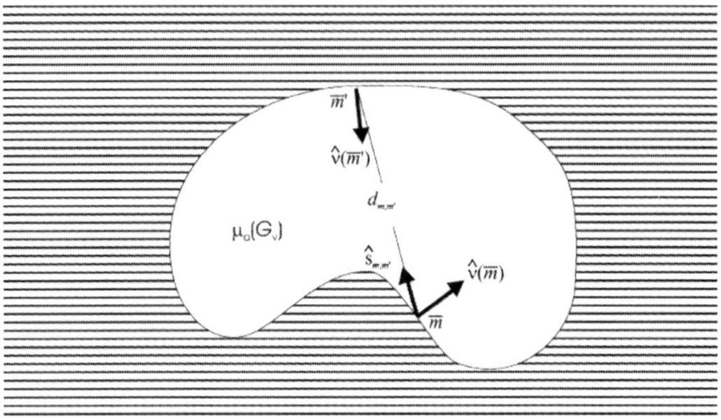

Figure 9.26 Schematic showing a void (nonscattering, but absorbing) region embedded within a diffusing medium. See the text for a description of the nomenclature.

unit vector separating $\bar{m}$ and $\bar{m}'$ is $\hat{s}(\bar{m},\bar{m}')$. At the point $\bar{m}'$, the radiant flux density from the surface is denoted $F(\bar{m}')$. If we assume that the surface emits diffusely so that Lambert's law describes the angular distribution of flux, then the radiance in the direction $\hat{s}(\bar{m},\bar{m}')$ is given by

$$I(\bar{m}',\hat{s}) = \frac{F(\bar{m}')}{\pi}.$$

It is worth noting that the assumption that the surface emits diffusely is not always completely correct,[103] since strictly speaking this requires that the scattered photon density is uniform throughout the scattering region; however, the approximation is generally adequate. Since an elemental area $d^2\bar{m}$ at $\bar{m}$ subtends a solid angle $d\Omega$ at $\bar{m}'$, where

$$d\Omega = \frac{d^2\bar{m} \cdot \bar{N}(\bar{m}) \cdot \hat{s}_{\bar{m},\bar{m}'}}{d^2_{\bar{m},\bar{m}'}},$$

and flux will be attenuated according to Beer's law as it traverses the void, an elemental area $d^2\bar{m}'$ at $\bar{m}'$ irradiates $d^2\bar{m}$ with an effective flux density $dF(\bar{m},\hat{s}_{\bar{m},\bar{m}'})$, where

$$dF(\bar{m},\hat{s}_{\bar{m},\bar{m}'})d^2\bar{m} = \frac{F(\bar{m}')}{\pi} \cdot \frac{d^2\bar{m}[\bar{N}(\bar{m}) \cdot \hat{s}_{\bar{m},\bar{m}'}]}{d^2_{\bar{m},\bar{m}'}} \cdot d^2\bar{m}'[\bar{N}\,(\bar{m}') \cdot \hat{s}_{\bar{m},\bar{m}'}]$$
$$\cdot \exp(-\mu_a d_{\bar{m},\bar{m}'}).$$

The integral of dF over the entire surface ∂G_v can then be considered as an additional source term in the FEM equations located at $\bar{m}$,

$$\int_{\partial G_v} h_{\bar{m},\bar{m}'} dF(\bar{m},\hat{s}_{\bar{m},\bar{m}'}) = q_0(\bar{m}),$$

where $h_{\bar{m},\bar{m}'}$ is a binary flag that indicates whether $d^2\bar{m}$ and $d^2\bar{m}'$ are mutually visible. Hence,

$$q_0(\bar{m}) = \int_{\partial G_v} h_{\bar{m},\bar{m}'} \frac{[\bar{N}(\bar{m}) \cdot \hat{s}_{\bar{m},\bar{m}'}][\bar{N}(\bar{m}') \cdot \hat{s}_{\bar{m},\bar{m}'}]}{\pi d^2_{\bar{m},\bar{m}'}} \exp(-\mu_a d_{\bar{m},\bar{m}'}) F(\bar{m}') d^2\bar{m}',$$

and applying the Robin boundary condition (Section 9.8.1.3),

$$q_0(\bar{m}) = \int_{\partial G_v} h_{\bar{m},\bar{m}'} \frac{[\bar{N}(\bar{m}) \cdot \hat{s}_{\bar{m},\bar{m}'}][\bar{N}(\bar{m}') \cdot \hat{s}_{\bar{m},\bar{m}'}]}{\pi d^2_{\bar{m},\bar{m}'}} \times \exp(-\mu_a d_{\bar{m},\bar{m}'}) \xi \Phi(\bar{m}') d^2\bar{m}'.$$

Such a spatial distribution of additional sources, which is expressible in terms of the photon density Φ, can readily be incorporated into the finite-element model described in the previous section. Recall that in that model, Φ was expressed as a linear sum of basis functions and was thus described by a vector Φ. This vector was obtained by solving the matrix equation

$$\mathbf{M} \cdot \Phi = \mathbf{Q},$$

where the source vector $\mathbf{Q}$ has entries

$$Q_j = \int_{G_v} \psi_j(\bar{r}) q_0(\bar{r}) d^3\bar{r}.$$

The direct diffusion–radiosity algorithm is thus implemented simply by augmenting the source vector with terms arising from $q_0(\bar{m})$ as defined previously. Note that the additional sources $q_0(\bar{m})$ exist only on the boundary ∂G_v. Also, since the FEM $\Phi(\bar{m}')$ is represented as

$$\Phi(\bar{m}') \approx \sum_{i=1}^{N} \phi_i \cdot \psi_i(\bar{m}'),$$

the additional terms in the source vector Q_j' become expressible as

$$Q'_j = \int\limits_{\partial G_v} \psi_j(\bar{m}) \int\limits_{\partial G_v} h_{\bar{m},\bar{m}'} \frac{[\bar{N}(\bar{m}) \cdot \hat{s}_{\bar{m},\bar{m}'}][\bar{N}(\bar{m}') \cdot \hat{s}_{\bar{m},\bar{m}'}]}{\pi d^2_{\bar{m},\bar{m}'}}$$

$$\times \exp(-\mu_a d_{\bar{m},\bar{m}'})\xi \sum_{i=1}^{N} \phi_i \cdot \psi_i(\bar{m}') d^2\bar{m} d^2\bar{m}'$$

$$= \sum_{i=1}^{N} \xi\phi_i \int\limits_{\partial G_v} \int\limits_{\partial G_v} h_{\bar{m},\bar{m}'} \psi_j(\bar{m}) \psi_i(\bar{m}') \frac{[\bar{N}(\bar{m}) \cdot \hat{s}_{\bar{m},\bar{m}'}][\bar{N}(\bar{m}') \cdot \hat{s}_{\bar{m},\bar{m}'}]}{\pi d^2_{\bar{m},\bar{m}'}}$$

$$\times \exp(-\mu_a d_{\bar{m},\bar{m}'}) d^2\bar{m} d^2\bar{m}' = \sum_{i=1}^{N} E_{ji}\phi_i;$$

That is,

$$\mathbf{Q}' = \mathbf{E} \cdot \Phi,$$

where $\mathbf{E}$ is now the "coupling" matrix, with entries

$$E_{ji} = \xi \int\limits_{\partial G_v} \int\limits_{\partial G_v} h_{\bar{m},\bar{m}'} \psi_j(\bar{m}) \psi_i(\bar{m}') \frac{[\bar{N}(\bar{m}) \cdot \hat{s}_{\bar{m},\bar{m}'}][\bar{N}(\bar{m}') \cdot \hat{s}_{\bar{m},\bar{m}'}]}{\pi d^2_{\bar{m},\bar{m}'}}$$

$$\exp(-\mu_a d_{\bar{m},\bar{m}'}) d^2\bar{m} d^2\bar{m}'. \tag{9.58}$$

The direct hybrid diffusion–radiosity model thus involves solving the modified matrix equation

$$\mathbf{M} \cdot \Phi = \mathbf{Q} + \mathbf{E} \cdot \Phi; \quad (\mathbf{M} - \mathbf{E})\Phi = \mathbf{Q}. \tag{9.59}$$

Hence, the algorithm is most simply implemented by augmenting the original system matrix entries M_{ij} with the new entries E_{ij} defined earlier. The solution can then be obtained as before.

Arridge et al. have compared this algorithm with both Monte Carlo calculations and with a "discrete ordinates" model called DANTSYS.[104] The agreement with the Monte Carlo results is excellent, with computation times being typically four orders of magnitude faster. Discrepancies with DANTSYS results appear to be due to the latter algorithm's failure to discretize propagation directions with sufficient accuracy to track all possible rectilinear trajectories through the nonscattering voids.

9.8.1.5 Discrete absorber models

One of the earliest attempts to analyze the situation in which the absorber is localized in discrete packets rather than distributed homogeneously was made by Duysens.[105] In this analysis, a simple nonscattering solution is considered in which cubic packets of absorber (cube side d) are randomly distributed on a cubic grid throughout a cuvette of thickness l so as to occupy a volume fraction c_p of the total volume. If the transmission across a single cubic packet is denoted T_p, then if the absorber could be treated as if it was homogeneously mixed with the host solution, the Beer–Lambert law predicts the relationship

$$A_{sol} = -c_p \frac{l}{d} \ln T_p$$

between these parameters and the apparent attenuation of the solution $A_{sol} = \ln(I_{out}/I_{in})$. By applying a simple probability theory, Duysens found that, in fact, the suspension displays a somewhat different attenuation A_{sus} given by

$$A_{sus} = -\frac{l}{d} \ln[1 - c_p(1 - T_p)]. \tag{9.60}$$

Consider the ratio of these attenuation values and characterize the deviation of the suspension from the ideal Beer–Lambert law by the quantity ξ:

$$\frac{A_{sus}}{A_{sol}} = 1 - \xi.$$

Expanding the expressions as power series, neglecting terms higher than first order, and rearranging allows us to establish a relationship between ξ, T_p, and c_p:[11]

$$\ln\left(\frac{1}{T_p}\right) = \ln\left(\frac{1 - \xi - c_p}{1 - 3\xi - c_p}\right);$$

If c_p and ξ are both small, then

$$\ln\left(\frac{1}{T_p}\right) = 2\xi. \tag{9.61}$$

This equation places a constraint on the maximum attenuation per packet if the deviation from the Beer–Lambert law is not to exceed ξ. Since, for nonscattering packets, $\ln(1/T_p) = \mu_a d$, where μ_a is the absorption coefficient of the packet, then if $\xi_{\max}$ is the maximum tolerable deviation from linearity, we have $\mu_a d < 2\xi_{\max}$. If we adopt a figure of 0.01 for $\xi_{\max}$, then we require

$$\mu_a d < 0.02 \tag{9.62}$$

for the Beer–Lambert law to remain valid. To put this into context, μ_a for whole blood at 800 nm (assuming 70% hemoglobin saturation) is approximately 4.5 cm^{-1}, while to a rough approximation d can be considered to be the diameter of a typical blood vessel, which can be anywhere from 0.002 to 0.1 cm. Thus, for the smallest vessels, Duysens's analysis does not predict a serious departure from the Beer–Lambert law. However, for larger vessels, the discrete localization of absorbers may produce noticeable effects.

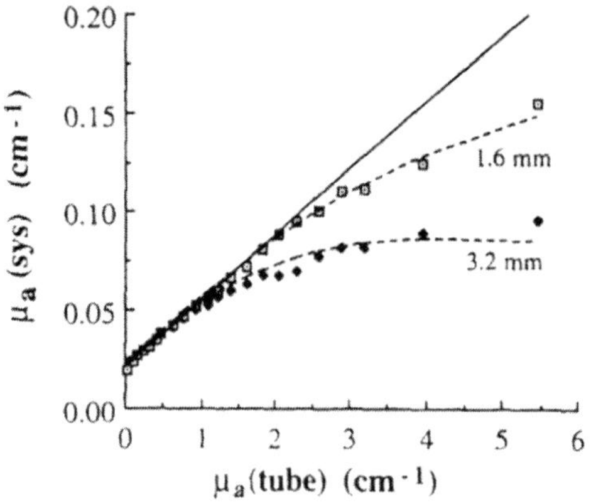

Figure 9.27 Experimental study of the effect of having an absorbing compound distributed discretely rather than homogeneously throughout a scattering medium.[106] As μ_a within the tubes is increased, the apparent μ_a (as determined by TRS) at first obeys Eq. (9.63), then systematically decreases to less than this value.

Liu et al. have reported an experimental and computational study that specifically included the effects of multiple scattering.[106] In their experiments, the authors fabricated a series of polyester-resin phantoms containing a network of fine channels (diameter d of order 0.1 cm) perfused with an Intralipid/dye mixture. If the channels occupy a volume fraction c_{tube} of the total phantom volume and possess an absorption coefficient $\mu_{a,tube}$, while the solid background phantom possesses an absorption coefficient $\mu_{a,back}$, then a naive analysis would suggest that the overall absorption coefficient $\mu_{a,sys}$ is given by

$$a_{a,sys} = c_{tube}\mu_{a,tube} + (1 - c_{tube})\mu_{a,back}. \tag{9.63}$$

By using TRS to estimate the apparent μ_a and μ_s' of these heterogeneous phantoms as $\mu_{a,\text{tube}}$ is varied, the authors have studied the accuracy of their approximations for various tube diameters and volume fractions. Figure 9.27 shows some typical results, in which $\mu_{a,\text{sys}}$ is plotted as a function of $\mu_{a,\text{tube}}$ for tube diameters of 0.16 cm and 0.32 cm and a volume fraction of 5.5%.

It is evident that a linear relationship, in agreement with Eq. (9.63), holds for small values of $\mu_{a,\text{tube}}$, but that the relationship becomes nonlinear at larger values of $\mu_{a,\text{tube}}$. $\mu_{a,\text{sys}}$ begins to decrease to less than the value predicted by Eq. (9.63), an effect that can be understood qualitatively in terms of the partial path length (see Section 9.8.2.1) experienced by photons in the solid phantom versus that in the tubes. As the absorption of the tubes reaches very high levels, photons traversing the tubes will almost certainly be absorbed and hence lost from the detected signal. The detected signal thus becomes biased increasingly heavily toward representing the optical properties of the background phantom rather than the tubes.

By analyzing the relationship between $\mu_{a,\mathrm{sys}}$ and $\mu_{a,\mathrm{tube}}$ in detail, the authors have proposed the following empirical relationship:

$$\mu_{a,sys} = \mu_{a,back} + c_{tube}\beta \exp(-\alpha d[\mu_{a,tube} - \mu_{a,back}])[\mu_{a,tube} - \mu_{a,back}], \quad (9.64)$$

where α and β are constants. α is weakly dependent on d and c_{tube}, generally lying between 0.4 and 0.7. β appears to be the same for all samples (0.7, in the present case), which suggests that its value may be determined by the materials used to construct the phantom (refractive index, g value). Substituting $\alpha = 0.5$ and $\beta = 1.0$ into Eq. (9.64) and expanding the exponential to the first order yields

$$\mu_{a,sys} = \mu_{a,back} + c_{tube}[1 - 0.5d(\mu_{a,tube} - \mu_{a,back})](\mu_{a,tube} - \mu_{a,back}).$$

If $0.5d[\mu_{\mathrm{a,tube}} - \mu_{\mathrm{a,back}}] \ll 1$, we revert to the linear relationship stated previously:

$$\mu_{a,sys} = c_{tube}\mu_{a,tube} + (1 - c_{tube})\mu_{a,back}.$$

To establish a criterion for this relationship to hold reasonably accurately, Liu et al. suggest $0.5d[\mu_{\mathrm{a,tube}} - \mu_{\mathrm{a,back}}] < 0.1$; that is,

$$[\mu_{a,tube} - \mu_{a,back}]d < 0.2, \quad (9.65)$$

which, in the case of tissue with low background (i.e., blood-free) absorption, is somewhat less stringent than the relationship suggested by Duysens's analysis. In particular, adopting the previous value of 4.6 cm^{-1} for the absorption of 70% saturated whole blood at 800 nm, we expect Eq. (9.63) to hold for vessel diameters $d < 0.2/4.6\ \mathrm{cm}^{-1}$, or approximately 400 μm.

In an attempt to understand the physical basis of Liu et al.'s empirical relationship, Firbank et al. have developed a simple "random hole" model for photon propagation through a regular array of tubes.[107] Consider such an array of tubes whose diameter is d and separation l, as shown in Fig. 9.28.

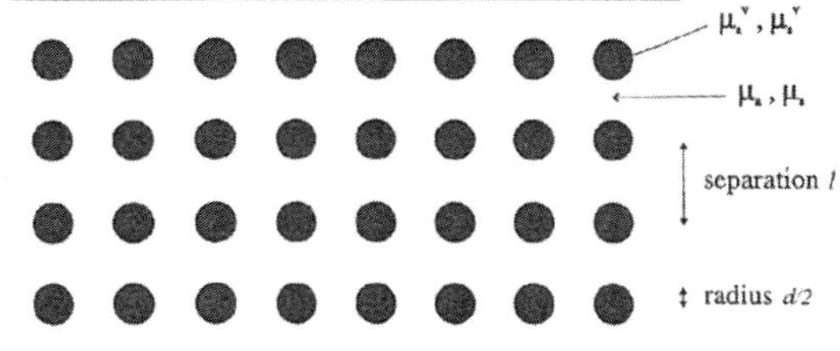

Figure 9.28 A simple model for tissue that contains absorber located in discrete packets rather than being distributed homogeneously. The absorber is considered to be located within tubes of diameter d, which are located on a regular grid with separation l.[107]

If first we assume that the absorption and scattering coefficients inside the tubes roughly match those outside, then to a good approximation we can use homogeneous diffusion theory to predict the TPSF for photons exiting at a distance r_{sd} from the source, $I(r_{sd},t)$. This then represents the probability that a photon detected at a distance r_{sd} from the source has traversed a total path length ct in the medium. Consider next that the absorption coefficient inside the tubes is increased from μ_a to μ_a^v. Photons traversing a physical distance ct in the medium will randomly encounter tubes. We can define the conditional probability $P(N,t)$ as the probability that a photon encounters N tubes in the course of traversing the path length ct. If we make the simplifying assumption that a photon does not scatter within a tube and experiences, on average, a distance d in each tube entered, then the total detected intensity at r_{sd}, $I(r_{sd})$can be expressed as

$$I(r_{sd}) = \int_0^\infty I(r_{sd},t) \sum_{\mathrm{N}=0}^{\infty} P(N,t) \exp[-(\mu_a^v - \mu_a)Nd]dt.$$

If we assume that the probability of any individual photon encountering a vessel is small, then $P(N,t)$ will be described by the Poisson distribution:

$$P(N,t) = \frac{\langle N \rangle^N \exp(-\langle N \rangle)}{N!},$$

where $\langle N \rangle$ is the average number of tubes encountered over the path length ct. Clearly, $\langle N \rangle$ is given by $\langle N \rangle = L'/l_{ph,v}$, where L' is the physical path length traversed in the tube-containing region, and $l_{ph,v}$ is the mean-free-path for a photon not to encounter a tube. $l_{ph,v}$ in turn is determined by the area packing fraction of tubes q and the effective interaction cross-section, which is d. Writing the number density of tubes as N_{tube}, we have

$$l_{ph,\mathrm{v}} \propto \frac{1}{N_{tube} d};$$

and since $N_{tube} = q/(\pi d^2/4)$, we have

$$l_{ph.\mathrm{v}} \propto \frac{\pi d}{4q}.$$

In addition, Firbank et al.'s random-hole model allows for the possibility that, due to a refractive-index mismatch between tube and surroundings, only a fraction ε of photons encountering a tube will actually enter it, the remainder being reflected. This increases the apparent interaction length. Finally, a constant k, close to unity (Firbank et al. use 1.236 to achieve optimal agreement with MC results), is introduced to account for other geometrical effects, yielding

$$l_{ph,v} = \frac{k\pi d}{4q\varepsilon}.$$

Not all of the physical path length ct is traversed in regions containing tubes. Specifically, after entering the medium at the surface, an incident photon must first traverse a linear distance d before reaching the tube-containing region. Using random-walk theory, if the scattering interaction length is $1/\mu_i$, then this requires a total path length $d^2\mu_s$ so that $L' = L - d^2\mu_s$. Then,

$$I(r_{sd}) = \int_0^\infty I(r_{sd},t) \sum_{N=0}^{\infty} \left(\frac{L - d^2\mu_s}{l_{ph,v}}\right)^N \frac{\exp[-(L - d^2\mu_s)/l_{ph,v}]}{N!} \exp[-(\mu_a^v - \mu_a)Nd]dt. \tag{9.66}$$

This model can be evaluated numerically, replacing the integral over transit times with a discrete sum extending up to, for example, three times the mean transit time and truncating the N-summation when $P(N, t)$ becomes sufficiently small. Firbank et al. have compared the predictions of this model with Monte Carlo simulations and with homogeneous diffusion theory (in which μ_a is calculated according to Liu et al.'s empirical relationship). They found that the random-hole model produces better agreement with MC results than Liu et al.'s formula, especially for r_{sd} less than 1 cm.

One consequence of this physical model is that it suggests an interpretation for β in Liu et al.'s formula. $\langle T \rangle$ can be calculated on the assumption that a tube of circular cross section experiences an isotropic radiance distribution of incident photons. Consider the flux of photons passing through an area element dA that are propagating at an angle θ to the surface normal. The total solid angle subtended in the interval θ to $\theta + d\theta$ is clearly $d\Omega = 2\pi \sin\theta d\theta$, and the resulting flux through dA is proportional to $dA \cos\theta$. Hence, the total intensity incident on dA at angle θ, $I(\theta)$ is given by $I(\theta) \propto \sin\theta \cos\theta$. The probability that such photons are reflected is given by the Fresnel reflection coefficient $R_F(\theta)$. Hence, the average transmission $\langle T \rangle$ can be calculated as

$$\langle T \rangle = 1 - \frac{\int_0^{\pi/2} R_F(\theta) \sin\theta \cos\theta d\theta}{\int_0^{\pi/2} \sin\theta \cos\theta d\theta}. \tag{9.67}$$

$R_F(\theta)$ is governed only by the relative refractive indices inside and outside the tube. For the polyester-resin/Intralipid phantom constructed by Liu et al., $\langle T \rangle$ evaluates to be 0.73, which agrees closely with Liu et al.'s empirical value of 0.74 ± 0.02. This agreement, together with the insensitivity of β to the scattering anisotropy coefficient g, led Firbank et al. to suggest that $\beta \equiv \langle T \rangle$.

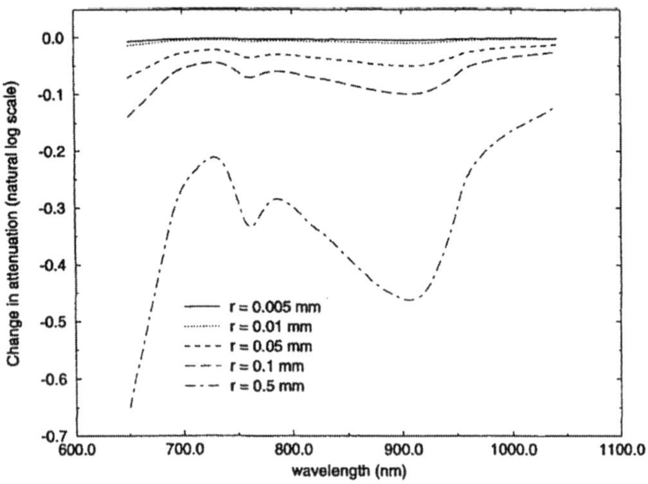

Figure 9.29 The deviation in attenuation away from that predicted by homogeneous diffusion theory as a function of tube radius. The deviation is plotted as a function of wavelength and for optical properties representative of blood-perfused tissue. Note how the deviation increases as μ_a increases.[107]

Firbank et al. conclude that homogeneous diffusion theory, with an average absorption coefficient given by $\mu_{a,sys} = q\mu_{a,tube} + (1 - q)\mu_{a,back}$ yields accurate results provided $[\mu_{a,tube} - \mu_{a,back}]d < 0.06$. Note that this criterion is more stringent than that suggested by Liu et al., and suggests that deviations from homogeneous diffusion theory may be apparent for vessels of diameter greater than 100 μm. Figure 9.29 shows the difference between the spectrum calculated by homogeneous diffusion theory and that calculated using the random-hole model, for vessels of different radii.

In conclusion, various analyses suggest that for vessels larger than a few hundred microns, discrete localization of hemoglobin absorption will produce noticeable effects on an NIR spectrum; in particular, the contribution of larger vessels to the overall signal will be reduced. This has implications for, among other things, measurements of CBV in which vasodilatation of larger veins and arteries may be difficult to detect.

9.8.2 Effects of tissue heterogeneity

Having described the various tools available to model tissue geometry and heterogeneity, we now consider the effects that various forms of heterogeneity have on absolute, quantified trend and absolute blood volume/flow measurements.

9.8.2.1 Quantified trend

The most important parameter used when making quantified trend measurements of hemoglobin concentration change is the differential path length. When the tissue is heterogeneous, the simple concept of an overall path length must be refined to account for the fact that photons traverse a different physical path length in different tissues. In the case of the adult head, only a small fraction of the overall path length will actually be traversed in the cerebral cortex. This has two effects:

1. A given hemoglobin concentration change in the cortex will produce a smaller change in attenuation as measured at the surface than if all the photon path length was traversed in brain. Correspondingly, if the concentration changes are quantified using the overall path length (as measured using any of the methods described in Section 9.6), then the actual concentration change will exceed the apparent concentration change.
2. Since the photon path length may now traverse blood-perfused tissues with different hemodynamic responses, the physiological significance of a given change can become difficult to interpret. This is a very basic criticism of the entire near-infrared tissue spectroscopy technique—namely, that at source–detector spacing used by most spectrometers (4–5 cm), the signals observed on the adult head are mainly representative of surface tissues such as skin and skull.

Hiraoka et al. have defined a rigorous extension to the concept of overall differential path length.[108] Consider a medium containing $j = 1,\ldots,n$ distinct subregions, each possessing its own transport properties. Whereas for a single, homogeneous medium the propagation of photons could be described by a single function, the TPSF, in this case an n-dimensional function is needed to describe the conditional probability that a photon spends a given amount of time t_j in each medium. In the case of a nonabsorbing medium, we denote this function $P_0(t_1, t_2 \ldots t_n)$. If we now consider that each medium possesses an absorption coefficient $\mu_{a,j}$, then in the same manner as in Section 9.6, we can use the microscopic Beer–Lambert law to express the overall number of detected photons N_{ph} as

$$N_{ph} = N_{ph0} \int_0^\infty \int_0^\infty \cdots \int_0^\infty P_0(t_1,t_2 \cdots t_n) \prod_{j=1}^{n} \exp(-\mu_{aj} c_j t_j) dt_1 dt_2 \cdots dt_n$$

$$= N_{ph0} \int_0^\infty \int_0^\infty \cdots \int_0^\infty P_0(t_1,t_2 \cdots t_n) \exp\left(-\sum_{j=1}^{n} \mu_{aj} c_j t_j\right) dt_1 dt_2 \cdots dt_n,$$

where N_{ph} is the total number of incident photons.

Given this relation, it is now possible to define a generalized version of the differential path length—namely, the partial differential path length in medium i PDP_i:

$$\mathrm{PDP}_i = \frac{\partial \ln(N_{ph0}/N_{ph})}{\partial \mu_{ai}},$$

which expresses the sensitivity of a surface measurement of attenuation to absorption coefficient changes occurring in region i of the medium. Following the procedure in Section 9.6, we can identify this with the mean time spent by photons in region i:

$$\begin{aligned}\mathrm{PDP}_i &= -\frac{1}{N_{ph}} \cdot \frac{\partial N_{ph}}{\partial \mu_{ai}} \\ &= \frac{\int_0^\infty \int_0^\infty \cdots \int_0^\infty P_0(t_1,t_2 \cdots t_n) c_i t_i \exp(-\sum_{j=1}^n \mu_{aj} c_j t_j) dt_1 dt_2 \cdots dt_n}{\int_0^\infty \int_0^\infty \cdots \int_0^\infty P_0(t_1,t_2 \cdots t_n) \exp(-\sum_{j=1}^n \mu_{aj} c_j t_j) dt_1 dt_2 \cdots dt_n} \\ &= c_i \cdot \langle t_i \rangle. \end{aligned} \tag{9.68}$$

The PDP is thus amenable to calculation using the Monte Carlo method, simply by recording the mean amount of time spent by all injected photons in the region of interest. This approach can be extended to the case that the medium is subdivided into individual pixels, in which case maps of the spatial distribution of PDP_i can be produced. Such maps have been variously termed the "photon hitting density" (PHD),[109] the "photon sampling volume" (PSV),[110] and the absorption-intensity "photon measurement density function" (PMDF).[111,112] Several authors have used this technique to calculate the total optical path length through the head and the corresponding partial optical path length in brain tissues, CSF, and surface tissues.[108,113,114]

Okada et al. have employed Monte Carlo and hybrid diffusion–radiosity models of varying sophistication to examine rigorously the penetration depth of light into the gray and white matter in the presence of overlying layers of skin, skull, and clear CSF.[114] In their study, four simple models of the adult head were considered; they are shown schematically in Fig. 9.30.

In Fig. 9.30, model (a) does not contain any nonscattering CSF, whereas model (d) includes renderings of real anatomical structures such as a convoluted gray/white matter interface with fluid-filled fissures (sulci). Figure 9.31 shows PMDFs calculated using the MC technique for each of these models, using a probe spacing of 4 cm.

Note how the clear CSF acts as a "light guide," preventing the penetration of photons deep into the gray and white matter layers. This effect is made more evident by plotting both the total DP and the PDP for each layer as a function of source–detector spacing. Figure 9.32 shows such plots, calculated by both Monte Carlo and diffusion–radiosity methods.

Note how the PDP in the surface layer increases almost linearly with source–detector spacing in the model without CSF, but displays a pronounced roll-off beyond $r_{sd} = 2$ cm in models with CSF. In the CSF models, this is

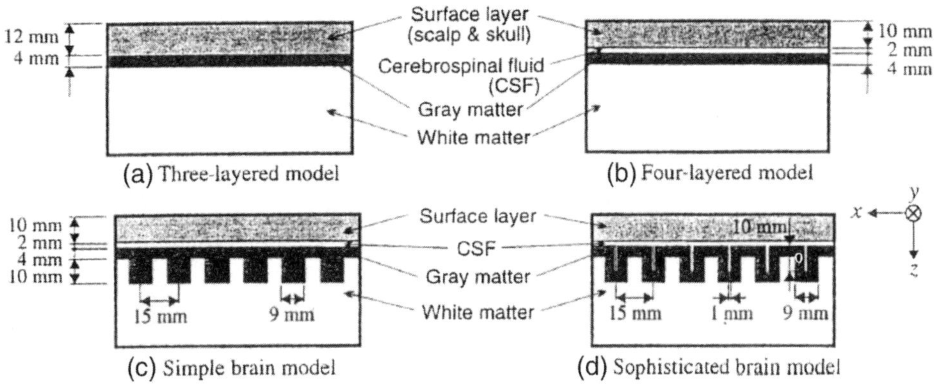

Figure 9.30 Schematic diagrams illustrating four phantoms of the human head of increasing sophistication. Model (a) contains no clear CSF layer, so diffusion theory applies throughout. Models (b) to (d) include a clear layer, progressing from a laminar to a corrugated geometry.[114]

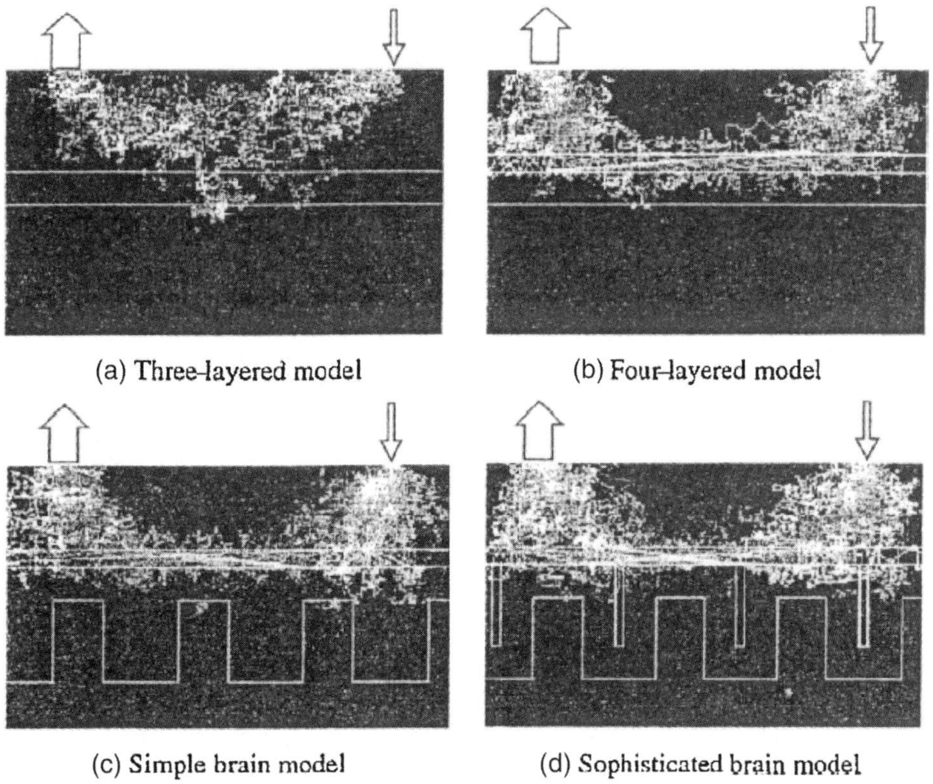

Figure 9.31 Absorption-intensity PMDFs calculated by Monte Carlo simulation for the four models shown in Fig. 9.30. Source/detector spacing is 4 cm.[111]

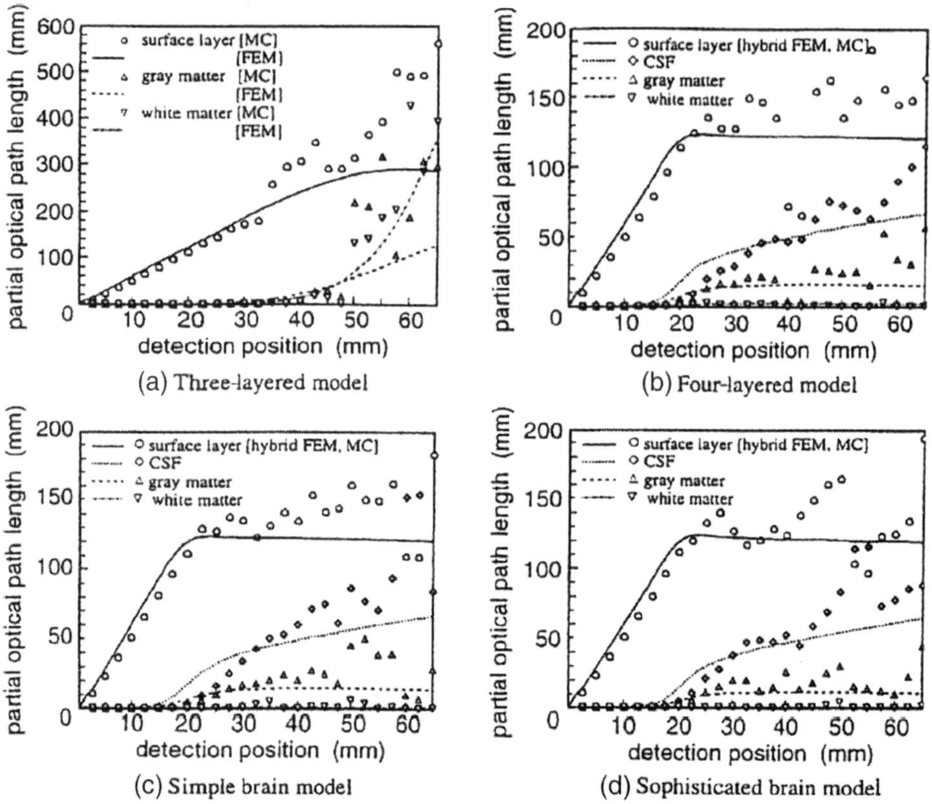

Figure 9.32 Partial optical path lengths as a function of source–detector spacing for the four models shown in Fig. 9.30. The path length through each layer is shown; open symbols are the results of MC calculation, and lines are the corresponding results obtained using the hybrid diffusion–radiosity model (see Section 9.8.1.4).

accompanied by an increased PDP in the gray matter for $r_{sd} > 2$ cm. Light penetration into white matter appears to be negligible at all usable values of r_{sd}.

Such techniques have also been applied to anatomically accurate models of the adult head (using the diffusion–radiosity model) and a simpler cylindrical geometry (using the Monte Carlo model) by Firbank et al.[115] These calculations allow one to directly visualize both intensity and mean-time PMDFs. Figures 9.33 and 9.34 show the FEM mesh used to represent the adult head in the region of the visual cortex and the corresponding intensity and mean-time PMDFs for source–detector spacing of 2, 3, and 4 cm. In the figures, it is evident that only the upper few millimeters of the cortex contributes to the detected signal at the tissue surface, with slightly better depth penetration being obtained by measuring mean time rather than intensity.

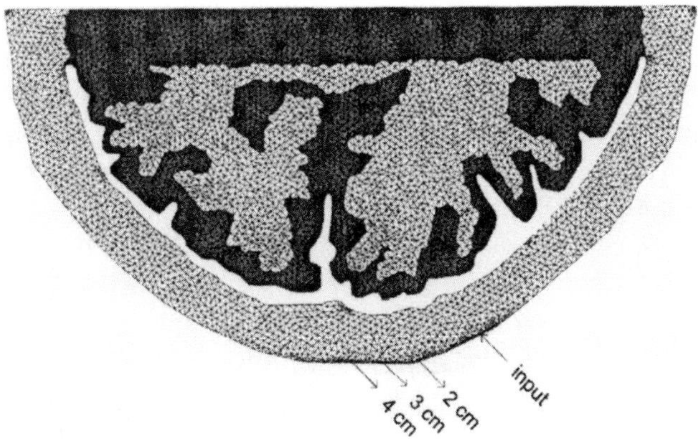

Figure 9.33 A finite-element mesh representing the skin/skull, CSF, gray matter of the visual cortex, and underlying white matter. This mesh is used together with the hybrid diffusion–radiosity model to predict the contribution of the gray matter to the NIR signal during visual evoked-response studies.[115]

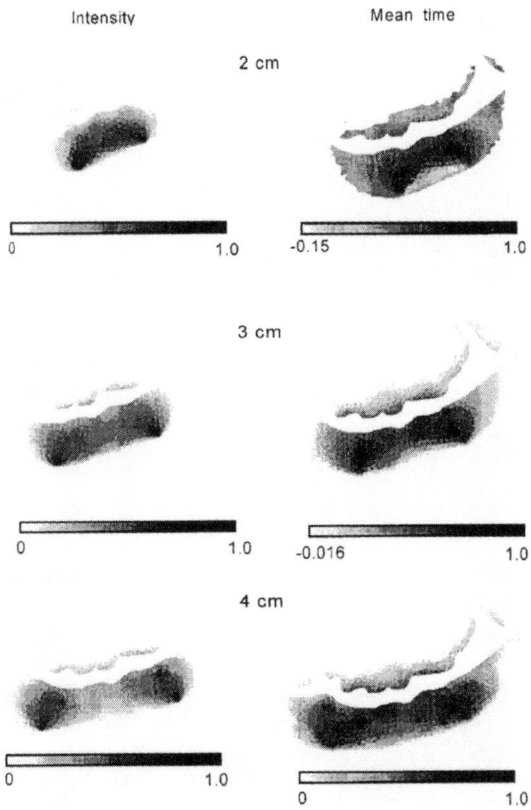

Figure 9.34 Intensity and mean-time PMDFs calculated using the hybrid diffusion–radiosity model together with the mesh shown in Fig. 9.32. Results are shown for source–detector spacing values of 2, 3 and 4 cm.[115]

Recently it has been suggested that these PMDFs should be recalculated, since the CSF is not clear but in fact contains a network of fine structures, called arachnoid trabecula, that connect the arachnoid to the pia mater. These structures tend to scatter photons and hence prevent their rectilinear propagation over large distances. One might then intuitively expect the resulting PMDFs to approach those obtained in the absence of the CSF layer. Okada and Delpy have performed such a study, in which the volume fraction of arachnoid trabeculae present in the CSF was varied between 0 and 50%. PMDFs calculated using a volume fraction of 50% were virtually identical to those calculated without a CSF layer. Interestingly, the presence of a mildly scattering CSF layer (e.g., 10–20% volume fraction of trabeculae) actually increased the PDP in gray matter relative to that without a CSF layer. Maps showing the spatial distribution of photons over the surface of the cortex also indicated an increased lateral spreading in the presence of mild CSF scattering. The increased PDP in gray matter thus seems to be associated with an increased number of surface interactions rather than an increase in the photon penetration depth.

The general consensus from various modeling studies is that at a source–detector spacing of 4 cm, the detected photons have traversed the majority of their total DP in surface tissues and CSF rather than gray matter. White matter is not believed to contribute significantly to the overall DP at such values of source–detector spacing. Nevertheless, this does not necessarily imply that the surface tissues and CSF primarily affect surface measurements of attenuation change. That is, the intrinsic concentration of absorber in the various tissues must also be taken into account. CSF is not perfused by any oxygen-dependent chromophore, so it will not contribute to any oxygen-dependent changes in attenuation as measured at the surface, regardless of the PDP traversed within it. Similarly, skin and skull are not expected to display oxygenation changes in response to functional activation of specific cortical areas, whereas gray matter will. Hence, in functional studies one can effectively ignore the substantial PDP traversed in the skin and skull—an observation that helps to explain the apparent success of functional NIR studies.[72]

9.8.2.2 Absolute measurements

Absolute measurements, at least in their simplest form, are based on measuring some property of the emerging light distribution (e.g., the temporal dispersion of an ultrashort laser pulse) and fitting this with an analytical model (see Section 9.5.1.2). Since the most popular models are based on solutions of the spatially homogeneous diffusion equation, the existence of regions with distinct optical properties will clearly cause a breakdown in the applicability of the underlying models and lead to erroneous values for the derived optical properties. In recent years, several authors have studied quantitatively the effects of layered tissue structures on absolute measurements. These studies have been carried out experimentally and using theoretical models such as those described previously.

Hielscher et al. have used gelatin phantoms and time-resolved MC calculations to investigate the effect of an overlying layer on absolute measurements of the transport properties of an underlying semi-infinite medium.[116] This scenario mimics important clinical measurements such as attempting to quantify muscle oxygenation through an overlying adipose layer, or measuring brain oxygenation through the overlying extracranial tissues. Importantly, Hielscher et al. note that in a large class of situations—namely, whenever the absorption of the upper layer dominates that of the lower medium—TRS measurements yield estimates of μ_a that are very close to μ_a of the lower medium. The reason for this behavior can be explained qualitatively as follows. Consider the relationship derived by Patterson et al. for the time-resolved reflectance $R_d(r_{sd},t)$ from a semi-infinite half-space:

$$R_d(r_{sd},t) = \frac{1}{(4\pi D)^{3/2}} z_0 t^{-5/2} \exp(-\mu_a ct) \exp\left(-\frac{r_{sd}^2 + z_0^2}{4Dt}\right).$$

Taking the logarithm of both sides and assuming $r_{sd} \gg z_0$ yields

$$\ln[R_d(r_{sd},t)] = \text{const} - \frac{5}{2}\ln(t) - \left(ct + \frac{a}{t}\right)\mu_a - \left(\frac{a}{t}\right)\mu_s',$$

where $a = 3r_{sd}^2/4c$.

One can see from this equation that for $t \gg a\mu$ it is primarily the value of μ_a that determines the shape of $\ln[R(r_{sd},t)]$, with μ_s' exerting only a weak effect. When the overlying layer possesses a very high μ_a, those detected photons that possess a very long transit time t will have traveled predominantly through the lower medium. This is because of the exponential nature of the microscopic Beer–Lambert law: Photons that spend large amounts of time in the high-μ_a overlayer are attenuated much more strongly than those that spend a similar amount of time in the lower-μ_a medium. The long-t tail of the TPSF is thus determined by the optical properties of the underlying medium; since this region of the TPSF primarily determines the μ_a value derived by curve-fitting this TPSF with a homogeneous diffusion model, the apparent μ_a is close to μ_a of the underlying medium. This fact is illustrated graphically in Fig. 9.35, in which the MC model is used to predict the TPSF from a two-layer medium (with a high-μ_a overlayer, curve III) and compare it with the TPSFs resulting from homogeneous media having the same transport properties as the two layers (curves I and II).

At large t, the TPSF from the two-layer system possesses the same decay constant as that from a homogeneous medium possessing the transport properties of the lower medium. Since this decay constant effectively yields an

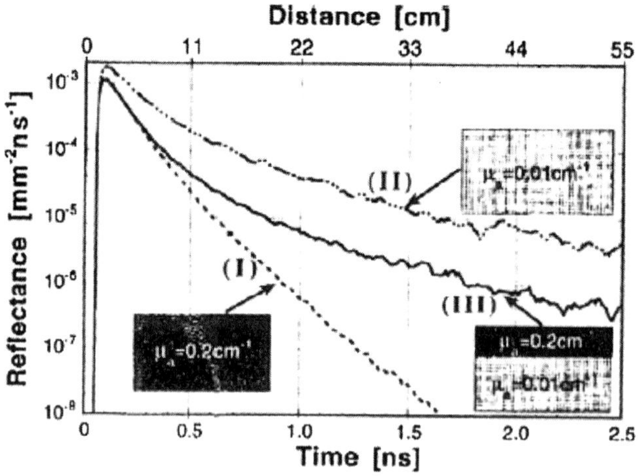

Figure 9.35 Monte Carlo simulations of the time-resolved reflectance from a two-layer medium (as depicted in Fig. 9.23) are shown as curve III. For comparison, curve I shows the time-resolved reflectance for a semi-infinite medium whose optical properties match those of the overlayer. Likewise, curve II corresponds to optical properties of the underlying medium.[116]

estimate for μ_a [see Eq. (9.14)], it is clear that the apparent μ_a will be close to μ_a of the underlying medium.

A purely experimental study into the relationship between the derived transport properties and the true properties of a two-layer system has been conducted using the OMNIA spatial and frequency-resolved system.[117] In this study, a variety of layers (thickness 0.08–1.6 cm) and underlying blocks were juxtaposed to produce 100 different two-layer systems. The transport properties of each region lay in the region $\mu_a = 0.07$ cm^{-1} to 0.15 cm^{-1} and $\mu_s' = 0.92$ cm^{-1} to 15.4 cm^{-1}. The OMNIA system was used to determine effective transport coefficients for each system using the data analysis procedure described by Eq. (9.24). Source–detector spacing was in the range 1.5–4.5 cm, while the modulation frequency was 110 Mhz. The main conclusion of this study was that for thin overlayers (typically thinner than 0.4 cm), the effective transport properties are essentially equal to the properties of the underlying medium; in contrast, for an overlayer thickness greater than 1.5 cm, the effective properties are essentially those of the overlayer.

Such studies reveal a certain, somewhat restricted set of circumstances in which models based on spatially homogeneous diffusion theory can return transport properties that bear a direct relationship to the true properties of one region of a two-layer system. However, they also show that often the derived transport properties are not representative of either region. In such cases it is necessary to employ a rigorous model of the inhomogeneous medium, such as that described in Section 9.8.1.1, to extract the transport properties of both regions independently. Kienle et al. have assessed the accuracy with which this model can be used to obtain the transport properties of both layers using spatially and temporally resolved measurements of reflectance.[97,118] These authors generated data for $R_d(r_{sd},t)$ using MC simulations and then used nonlinear regression of the model to recover the transport properties used in the simulation. Although the model can, in principle, be used to estimate the four transport properties and the thickness of the upper layer, in practice absolute measurements of $R_d(r_{sd},t)$ are required, which are difficult to obtain. Since the upper-layer thickness can be measured using other techniques in many cases, it is reasonable to concentrate on estimating the four transport properties, given knowledge of the layer thickness. Kienle et al. show that these can be derived from relative measurements of $R_d(r_{sd},t)$ at two distances, in which case μ_s' of the upper layer and μ_a of the lower medium can be recovered with errors of less than 10%.

In the frequency domain, the model yields good results if reflectance measurements are made at a minimum of three sites. Again, it appears that μ_a of the lower medium can be determined more accurately than that of the upper layer (a typical error of 10% arises when determining the former, compared to 30–40% for the latter). This is encouraging for applications such as brain and muscle oximetry, and it broadly agrees with the results obtained by Hielscher et al.[114]

It appears that some form of temporal or frequency-resolved measurements are needed to accurately extract optical properties for a two-layer model. Alexandrakis et al. have applied the two-layer model to MC-generated reflectance data to quantify the accuracy with which the four transport properties and the layer thickness can be estimated.[119] When only DC reflectance data are used, it appears that the five parameters cannot be recovered with any useful accuracy. Adding time or frequency-resolved data makes it possible to recover the layer thickness and the transport properties of the lower medium to within 10% of their true values.

A significant body of clinical experience has also been gained using instrumentation designed to produce absolute measurements of hemoglobin saturation using the spatially resolved technique (see Section 9.5.1.1). In particular, several studies have been undertaken to evaluate the performance of the INVOS 3100 (Somanetics Corporation) cerebral oximeter. This machine uses a source and two detectors (originally at distance 1 cm and 2.7 cm from the source, although subsequently increased to 3.0 and 4.0 cm). The basic strategy employed by the device is to assume that the signal detected by the near detector essentially reflects surface tissues only, and thus can be used to remove this contaminating signal from that detected by the distance detector. The machine is designed to produce a measurement of cerebral hemoglobin saturation that is essentially independent of the oxygenation of the surface tissues.

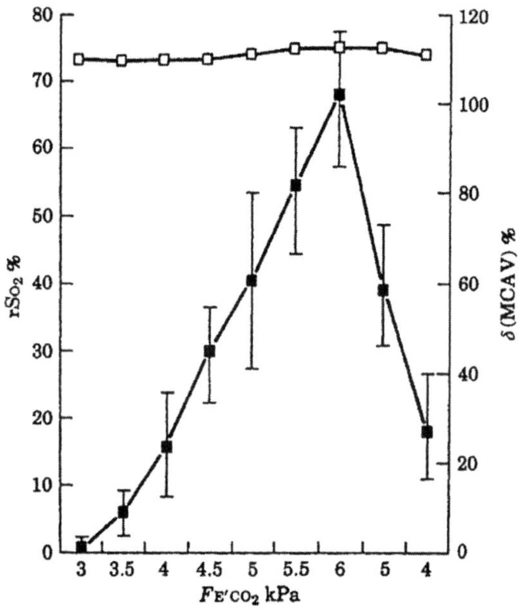

Figure 9.36 Measurements of mean cerebral hemoglobin saturation (here denoted rSo_2) measured by the INVOS 3100 during a period of induced hypercapnia (open symbols). Also shown is the corresponding change in middle-cerebral artery flow velocity as measured by transcranial Doppler ultrasound (filled symbols).[120]

Several authors have tested this claim, in a number of ways. Harris et al. have exploited the fact that the brain possesses a hemodynamic control mechanism known as "autoregulation" that is absent from the blood vessels supplying surface tissues.[120] One consequence of autoregulation is that the rate of blood flow to the brain (i.e., the CBF) is extremely sensitive to the partial pressure of dissolved carbon dioxide in the blood (pCO_2). The oxygen consumption of the brain, however, is not sensitive to pCO_2. Causing a subject to inspire elevated levels of carbon dioxide (thereby inducing hypercapnia) is expected to result in an oversupply of oxygen to the brain, subsequently leading to a relative fall in the fraction of Hb in the capillary and venous blood. Since all near-infrared oximeters (apart from pulse oximeters) measure mixed arterial/capillary/venous hemoglobin saturation, one might expect to see an increase in measured hemoglobin saturation after inducing hypercapnia. Figure 9.36 shows the actual saturation values recorded by the INVOS 3100 during such a maneuver.

The vasodilatory effect of CO_2 is confirmed by the large transient increase in blood flow velocity in the middle cerebral artery (measured by transcranial Doppler ultrasound), but the INVOS 3100 detects almost no change in hemoglobin saturation. Germon et al. have performed studies with a modified version of this device, in which the source–detector spacing has been increased

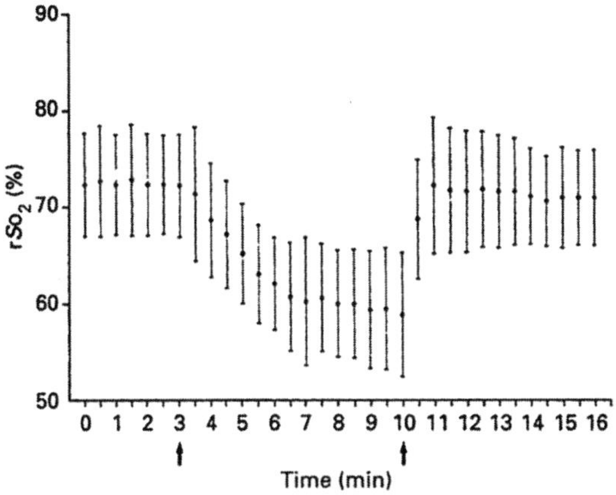

Figure 9.37 Changes in mean cerebral hemoglobin saturation, as measured by the INVOS 3100, during a period of induced scalp ischemia. Arrows indicate the times at which ischemia began and ended.[121]

to 3.0 and 4.0 cm for the near and far detectors, respectively.[121] Such an increase is expected to lead to deeper penetration by the detected photons, and consequently to less influence by the surface tissues. To assess the influence of surface tissues, experiments have been performed in which the blood supply to the scalp is occluded using a pneumatic cuff around the upper head. Figure 9.37 shows the resulting time course of hemoglobin saturation as measured by the INVOS 3100. A clear ischemia/hyperemia response is evident, again supporting the conclusion that the hemoglobin saturation reading is not independent of surface tissues.

9.8.3 Summary

A variety of theoretical tools are now available to model the effects of tissue geometry and spatial heterogeneity on quantified trend and absolute oxygenation measurements. The most flexible method is Monte Carlo simulation, while the fastest method that can deal with anatomically accurate features of the adult head is the finite element implementation of the hybrid diffusion–radiosity model. Both techniques have been successfully applied. As of this writing, a consensus is emerging that at source–detector spacing of 4.0 cm (typical of commercial instruments), NIR light probes only the upper few millimeters of gray matter. The area of the cortex surface over which the illuminating photons are distributed (and hence over which the detected signal is spatially averaged) extends approximately 1.0 cm to either side of a line connecting the source and the detector. It is thus apparent that functional NIR studies of the brain, while requiring relatively inexpensive, noninvasive, and portable instrumentation, are nevertheless comparatively primitive, in terms of spatial resolution, compared with competing techniques such as functional magnetic resonance (fMRI) and positron emission scanning (PET).

Similarly, the goal of absolute measurements of oxygenation in heterogeneous tissues has yet to be realized in clinical practice. The logical endpoint of such quantification is the production of tomographic maps of hemoglobin concentration and saturation. Much theoretical and experimental progress has been made in this field over the last 10 years, with iterative image reconstruction techniques utilizing time- or frequency-resolved measurements emerging as the most promising approach.

9.9 Chapter Summary

Quantification and localization of chromophores in biological tissues is a technologically challenging task. Multiple scatter and the spatial heterogeneity of tissue pose severe problems and render a straightforward analysis using the Beer–Lambert law invalid. Early NIR oximeters could determine only trend changes in tissue oxygenation and were of limited value. More recently, the introduction of a variety of instrumental, methodological, and technical advances has rendered such devices essentially obsolete.

Quantified trend monitors have become a reality since the advent of techniques to directly measure the photon path length through tissues. Originally such measurements were made using bulky time-of-flight systems, which allowed a useful database of DPF values to be constructed for a variety of tissues. These DPF values could be programmed into any trend monitor, to give an estimate of the change in hemoglobin oxygenation in useful units (e.g., μM). Since DPF is itself dependent on tissue oxygenation and is thus a dynamic variable, a major requirement is to measure photon path length continuously and in real time. This goal has essentially been achieved through the use of portable frequency-domain instruments, although the relatively high cost of the hardware, and the need for accurate initial calibration has meant that this technology has not yet migrated into commercial instruments. The use of "tracer" chromophores such as water provides a potentially low-cost way of measuring path length at a few wavelengths (760 nm, 820 nm, and 970 nm, where water displays distinct absorption overtones), but the accuracy of this technique is inferior to frequency-resolved measurements and it has not been applied commercially.

With the ability to quantify oxygenation changes comes the ability to measure certain hemodynamic variables such as muscle and brain blood flow and volume, muscle oxygen consumption, and muscle and brain venous oxygen saturation. This measurement requires certain physiological interventions to be performed—for example, changes in inspired oxygen concentration or arterio-venous occlusion—and it is not well suited to routine measurements or use in patients with respiratory or circulatory disorders.

The ultimate goal of absolute quantification of all chromophores can be realized if μ_a and μ_s' can be measured at a sufficient number of wavelengths. Many suggestions have been made for extracting these transport coefficients, generally by measuring the spatial and/or temporal variation of transmitted light intensity and modeling this behavior using the diffusion equation or Monte Carlo calculations. These methods have been verified on simple, homogeneous tissue-simulating phantoms, but it remains unclear how to interpret the output of these algorithms when they are used on real tissues possessing spatially varying transport properties, in all but the simplest nontrivial problem.

The "holy grail" of near-infrared spectroscopy is to map the spatial distribution of all chromophores in the tissue. The most promising approach to this problem currently appears to be near-infrared "tomography," in which the spatial distribution of μ_a and μ_s' is recovered from suitable light intensity measurements made on the tissue boundary. If these spatial distributions can be recovered accurately at a number of distinct wavelengths, then chromophore maps could be produced. The field of near-infrared tomography is currently attracting a large amount of attention, and several imaging instruments and a public domain image reconstruction software package are now available.[122] However, although reconstructed images have been demonstrated in which the spatial location of inhomogeneities has been successfully determined, it remains to be seen whether the algorithms will produce maps of μ_a and μ_s' that are sufficiently accurate to recover accurate spatial maps of chromophore concentration.

9.10 Recent Developments

This chapter was originally written in 2001. In the intervening years, substantial progress has been made, especially in the increased use of the technology for *in vivo* functional imaging. For a thorough update, the reader is referred to several excellent review articles describing recent advances in both NIR spectroscopy and imaging.[123,124]

Chief among the advances has been the increased uptake of spectrometers designed to produce an absolute measurement of tissue oxygenation, especially using the spatially resolved principle. A rough estimate, based on the number of scientific publications appearing each year, suggests that more than 2000 clinical oximeters are now in use in laboratories around the world. Despite the greatly increased uptake, however, the clinical impact of this technology remains a matter of debate.[125] Moreover, the pulse oximeter remains the only optical oximeter that had found a routine role in patient management at the time of writing.

The second major development has been the increased availability of near-infrared imagers and multisite near-infrared topography systems targeting the brain cortex. NIR imaging, as opposed to NIR spectroscopy, effectively increases the number of measurement channels from several to several dozen or several hundred, depending on the instrument. Although these instruments have not yet demonstrated an ability to image hemoglobin saturation with quantitative accuracy, the technique has found a role in functional studies of the brain *in vivo*, where relative changes in hemoglobin concentration are sufficient data to be useful.[126] The new discipline of functional NIR spectroscopy (fNIRS) is proving to be an especially powerful means of studying neonatal cognitive development.[127] In part, this is due to practical considerations. The neonatal skull and scalp are much thinner than their counterparts in the adult, which makes it considerably easier to collect high-quality data in neonates. In addition, the technique is much more usable in neonates than the obvious competing technique of functional MRI, which generally is not tolerated by neonates due to the excessive acoustic noise and stricter requirements on subject movement during scanning.

Although this chapter has concentrated on the use of near-infrared spectroscopy, it should be noted that the use of optical radiation, especially in the range from 450 to 600 nm, has been revitalized in recent years. Although obviously limited in terms of depth penetration due to intense scattering and absorption, optical oximetry has been rediscovered thanks to the introduction of high-performance hyperspectral imagers into the market over the last decade. By collecting 2D images at video rates over many different wavelength bands simultaneously (using scanning liquid crystal or acousto-optic filters), these devices form 2D maps of hemoglobin saturation in the surface microvasculature. To date, most applications have been in animal studies, where tumor microcirculation is of intense interest because many modern tumor-targeting drugs actually target the blood vessels rather than the malignant tumor cells themselves.[128]

A number of approaches have been investigated in an attempt to obtain localized hemoglobin saturation measurements with sub-millimeter spatial resolution and to a greater depth in tissues than is achievable by optical radiation. Optical coherence tomography (OCT) is an imaging tool that is widely used in ophthalmology and typically operates in the 800–1300 nm wavelength range. OCT is not well suited to measuring hemoglobin absorption because the combination of short path lengths (less than 1 mm) and the low absorption coefficient leads to very low contrast. In turn, a modification known as "pump probe" OCT has been proposed and demonstrated, in which pump radiation from a pulsed laser tuned to the intense visible bands of hemoglobin is used to transiently deplete the ground state of the hemoglobin, thereby transiently reducing the absorption. Differential detection between pump-on and pump-off states thus gives a signal specific to hemoglobin.[129] This technology has been demonstrated in simple models such as the zebrafish, but its clinical applicability is not obvious at this time. An alternative approach is to perform OCT at visible wavelengths, where the absorption coefficient is much higher. Advances in low-noise supercontinuum light sources make this a feasible proposition, and encouraging results have recently been reported in rodent brain and retina tissues.[130] Obviously, the imaging depth is reduced considerably with OCT variant compared with standard OCT.

A recent innovation that is attracting considerable interest is photoacoustic imaging and spectroscopy. This technique combines the advantages of ultrasound imaging (speed, spatial resolution, and tissue penetration) with those of optical spectroscopy (molecular sensitivity). The photoacoustic technique involves applying a pulse (typically of nanosecond duration) of optical radiation to tissue, where the optical wavelength is tuned to a strong absorption band of a molecular species of interest. In the case of hemoglobin, this typically covers the range 740 to 1000 nm. Because the pulse has a short duration, transient heating occurs at the site of strong molecular absorption, which results in rapid thermal expansion and the generation of a pulse of ultrasound energy. The pulse of ultrasound can be detected at the tissue surface and its amplitude is related to the local absorption coefficient. Since the speed of sound is known, the depth of the absorption site can be inferred from the ultrasound's arrival time. Thus, in principle, one has enough information to try to reconstruct an image of the absorption coefficient. Uniquely among the methods described in this chapter, photoacoustic imaging has a depth penetration and a depth resolution similar to high-frequency ultrasound (i.e., centimeters and approximately 0.1 mm, respectively), coupled with the molecular specificity of near-infrared spectroscopy.

Laufer et al. have demonstrated excellent results with this technology in tissue phantoms, but the data processing and image reconstruction problem shares many of the features of NIRI.[131] Since the photoacoustic signal amplitude depends not just on the local absorption coefficient, but also on the local fluence rate, numerical models such as those described in Section 9.8 are needed to achieve quantitative accuracy. With care and some assumptions, accuracy comparable to a lab-based co-oximeter has been demonstrated for determining localized hemoglobin saturation in vessels up to 1 cm deep into a phantom with realistic optical properties.

The field of photoacoustic imaging and spectroscopy has made substantial progress over the last few years,[132] but routine clinical application of quantitative photoacoustic spectroscopy has not yet arrived in the clinic. This technique has the unique ability to detect blood microvessels at distances significantly greater than the optical "attenuation limit" as determined by optical single scattering. It has been suggested that recent advances in adaptive optics could potentially allow photoacoustic imaging depths to exceed the "diffusion limit," as determined by optical multiple scattering and absorption, and become limited only by the "dissipation limit," which might potentially allow whole-body imaging.[133] Nevertheless, the feasibility of this approach remains to be demonstrated in a clinically applicable form.

References

1. A. Huch and R. Huch, "Transcutaneous, noninvasive monitoring of pO_2," *Hospital Pract.* **6**, 43–52 (1976).
2. A. Huch, D. W. Lübbers, and R. Huch, "Patientenuberwachung durch transcutane PCO_2 messung bei gleiechzeilger koutrolle der relatiuen Iokalen perfusion," *Anaesthetist* **22**, 379 (1973).
3. A. L. Stewart, R. J. Thorburn, P. L. Hope, M. Goldsmith, A. P. Lipscomb, and E. O. R. Reynolds, "Ultrasound appearance of the brain in the very preterm infants and neurodevelopmental outcome at 18 months of age," *Arch. Dis. Child.* **58**, 598–604 (1983).
4. C. E. Elwell, *A Practical Users Guide to Near Infrared Spectroscopy*, Hamamatsu Photonics KK, London (1995).
5. P. G. Al-Rawi, P. Smielewski, H. Hobbiger, S. Ghosh, and P. J. Kirkpatrick, "Assessment of spatially resolved spectroscopy during cardiopulmonary bypass," *J. Biomed. Opt.* **4**, 208–216 (1999).
6. G. A. Millikan, "The oximeter: an instrument for measuring continuously oxygen saturation of arterial blood in man," *Rev. Sci. Instrum.* **13**, 434–444 (1942).
7. S. Nakajima, Y. Hirai, H. Takase, A. Kuse, S. Aoyagi, M. Kishe, and K. Yamaguchi, "New pulsed type earpiece oximeter," *Kokyu To Junkan* **23**, 709–713 (1975).
8. F. F. Jöbsis, "Noninvasive, infrared monitoring of cerebral and myocardial oxygen sufficiency and circulatory parameters," *Science* **198**, 1264–1267 (1977).
9. M. Ferrari, I. Giannini, G. Sideri, and E. Zanette, "Continuous noninvasive monitoring of human-brain by near-infrared spectroscopy," *Adv. Exp. Med. Biol.* **191**, 873–882 (1985).
10. M. Cope and D. T. Delpy, "System for long-term measurement of cerebral blood and tissue oxygenation on newborn infants by near-infrared transillumination," *Med. Biol. Eng. Comput.* **26**, 289–294 (1988).
11. M. Cope, "The development of a near infrared spectroscopy system and its application for non invasive monitoring of cerebral blood and tissue oxygenation in the newborn infant," Ph.D. dissertation, University of London (1991).
12. J. Lin and C. W. Brown, "Spectroscopic measurement of NaCl and seawater salinity in the near-IR region of 680–1280 nm," *Appl. Spectrosc.* **47**, 239–241 (1993).
13. J. J. Kelly, K. A. Kelly, and C. H. Barlow, "Tissue temperature by near-infrared spectroscopy," *Proc. SPIE* **2389**, 818–828 (1995).

14. S. J. Matcher, M. Cope, and D. T. Delpy, "Use of the water absorption spectrum to quantify tissue chromophore concentration changes in near infrared spectroscopy," *Phys. Med. Biol.* **39**, 177–196 (1994).
15. J. M. Conway, K. H. Norris, and C. E. Bodwell, "A new approach for the estimation of body composition: infrared interactance," *Am. J. Clin. Nutr.* **40**, 1123–1130 (1984).
16. M. Kohl, M. Essenpreis, and M. Cope, "The influence of glucose-concentration upon the transport of light in tissue-simulating phantoms," *Phys. Med. Biol.* **40**, 1267–1287 (1995).
17. G. Gratton, P. M. Corballis, E. Cho, M. Fabiani, and D. C. Hood, "Shades of gray matter: noninvasive optical images of human brain responses during visual stimulation," *Psychophysiology* **32**, 505–509 (1995).
18. V. Quaresima, R. Sfareni, A. Pizzi, and M. Ferrari, "Measurement of the muscle optical properties on muscular dystrophy patients by a frequency-domain photometer," *Biomedical Optical Spectroscopy and Diagnostics.* E. Sevick-Muraca and D. Benaron, Eds., pp. 123–125, Optical Society of America, Washington D.C. (1996).
19. W. H. Press, S. A. Teukolsky, W. T. Vetterling, and B. P. Flannery, *Numerical Recipes in C*, Cambridge University Press, Cambridge, U.K. (1992).
20. G. I. Bell and S. Glasstone, *Nuclear Reactor Theory*, Van Nostrand Reinhold, New York (1970).
21. S. Chandrasekhar, *Radiative Transfer*, Dover, New York (1960).
22. W- F. Cheong, S. A. Prahl, and A. J. Welch, "A review of the optical properties of biological tissues," *IEEE J. Quantum Elec.* **26**, 2166–2185 (1990).
23. M. S. Patterson, B. Chance, and B. C. Wilson, "Time resolved reflectance and transmittance for the noninvasive measurement of tissue optical-properties," *Appl. Opt.* **28**, 2331–2336 (1989).
24. S. R. Arridge, M. Cope, and D. T. Delpy, "The theoretical basis for the determination of optical pathlengths in tissue: temporal and frequency analysis," *Phys. Med. Biol.* **37**, 1531–1560 (1992).
25. S. R. Arridge, M. Schweiger, M. Hiraoka, and D. T. Delpy, "A finite element approach for modelling photon transport in tissue," *Med. Phys.* **20**, 299–309 (1993).
26. T. J. Farrell, B. C. Wilson, and M. S. Patterson, "The use of a neural network to determine tissue optical properties from spatially resolved diffuse reflectance measurements," *Phys. Med. Biol.* **37**, 2281–2286 (1992).
27. R. Bays, G. Wagnières, D. Robert, D. Braichotte, J.-F. Savary, P. Monnier, and H. van den Bergh, "Clinical determination of tissue

optical properties by endoscopic spatially resolved reflectometry," *Appl. Opt.* **35**, 1756–1766 (1996).
28. P. W. McCormick, M. Stewart, P. Ray, G. Lewis, M. Dujovny, and J. I. Ausman, "Measurement of regional cerebrovascular haemoglobin oxygen saturation in cats using optical spectroscopy," *Neurol. Res.* **13**, 65–70 (1991).
29. S. J. Matcher, P. Kirkpatrick, K. Nahid, M. Cope, and D. T. Delpy, "Absolute quantification methods in tissue near infrared spectroscopy," *Proc. SPIE* **2389**, 486–494 (1995).
30. S. J. Matcher, D. T. Cope, and D. T. Delpy, "In vivo measurements of the wavelength dependence of tissue scattering coefficients between 760 and 900 nm measured with time-resolved spectroscopy," *Appl. Opt.* **36**, 386–396 (1997).
31. S. Suzuki, S. Takasaki, T. Ozaki, and Y. Kobayashi, "A tissue oxygenation monitor using NIR spatially resolved spectroscopy," *Proc. SPIE* **3597**, 582–592 (1999).
32. C. E. Elwell, S. J. Matcher, L. Tyszczuk, J. H. Meek, and D. T. Delpy, "Measurement of cerebral venous saturation in adults using near infrared spectroscopy," *Oxygen Transport to Tissue XVIII*, Nemoto and LaManna, Eds., pp. 453–460, Plenum Press, New York (1997).
33. V. Quaresima, S. Sacco, R. Totaro, and M. Ferrari, "Noninvasive measurement of cerebral hemoglobin oxygen saturation using two near infrared spectroscopy approaches," *J. Biomed. Opt.* **5**, 201–205 (2000).
34. B. Chance, S. Nioka, J. Kent, K. McCully, M. Fountain, R. Greenfeld, and G. Holtom, "Time resolved spectroscopy of hemoglobin and myoglobin in resting and ischemic muscles," *Anal. Biochem.* **174**, 698–707 (1988).
35. S. J. Matcher, "Closed-form expressions for obtaining the absorption and scattering coefficients of a turbid medium with time-resolved spectroscopy," *Appl. Opt.* **36**, 8298–8302 (1997).
36. J. M. Tualle, B. Gélébart, E. Tinet, S. Avrillier, and J. P. Ollivier, "Real time optical coefficients evaluation from time and space resolved measurements in biological tissues," *Opt. Commun.* **124**, 216–221 (1996).
37. S. J. Madsen, B. C. Wilson, M. S. Patterson, Y. D. Park, S. L. Jacques, and Y. Hefetz, "Experimental tests of a simple diffusion model for the estimation of scattering and absorption coefficients of turbid media from time-resolved diffuse reflectance measurements," *Appl. Opt.* **31**, 3509–3517 (1992).
38. H. Liu, B. Beauvoit, N. G. Wang, and B. Chance, "Characterization of absorption and scattering properties of small-volume biological samples using time-resolved spectroscopy," *Anal. Biochem.* **213**, 378–385 (1993).
39. H. Zhang, M. Miwa, Y. Yamashita, and Y. Tsuchiya, "Quantitation of absorbers in turbid media using time-resolved spectroscopy based on

microscopic Beer–Lambert law," *Jpn. J. Appl. Phys.* **37**, 2724–2727 (1998).

40. D. T. Delpy, M. Cope, P. van der Zee, S. R. Arridge, S. Wray, and J. Wyatt, "Estimation of optical pathlength through tissue from direct time of flight measurement," *Phys. Med. Biol.* **33**, 1433–1442 (1988).
41. H. Heussman, J. Kölzer, and G. Mitic, "Characterization of female breasts *in vivo* by time resolved and spectroscopic measurements in near infrared spectroscopy," *J. Biomed. Opt.* **1**, 425–434 (1996).
42. K. Suzuki, Y. Yamashita, K. Ohta, M. Kaneko, M. Yoshida, and B. Chance, "Quantitative measurement of optical parameters in normal breasts using time-resolved spectroscopy: *in vivo* results of 30 Japanese women," *J. Biomed. Opt.* **1**, 330–334 (1996).
43. M. Miwa, Y. Ueda, and B. Chance, "Development of time-resolved spectroscopy system for quantitative noninvasive tissue measurement," *Proc. SPIE* **2389**, 142–149 (1995).
44. L. L. Kalpaxis, L. M. Wang, P. Galland, X. Liang, P. P. Ho, and R. R. Alfano, "Three-dimensional temporal image reconstruction of an object hidden in highly scattering media by time-gated optical tomography," *Opt. Lett.* **18**, 1691–1693 (1993).
45. J. A. Moon, R. Mahon, M. D. Duncan, and J. Reintjes, "Three-dimensional reflective image reconstruction through a scattering medium based on time-gated Raman amplification," *Opt. Lett.* **19**, 1234–1236 (1994).
46. J. Watson, P. Georges, T. Lépine, B. Alonzi, and A. Brun, "Imaging in diffuse media with ultrafast degenerate optical parametric amplification," *Opt. Lett.* **20**, 231–233 (1995).
47. M. Barilli, G. Zaccanti, P. Bruscaglioni, A. Ismaelli, Q. N. Wei, and M. Ferrari, "Optical properties of *in vivo* human skeletal muscle from near infrared picosecond laser pulse," *Photodynamic Therapy and Biomedical Lasers*, P. Spinelli, M. Dal Fante, and R. Marchesini, Eds., pp. 930–934, Elsevier Science Publishers B.V., Amsterdam (1992).
48. M. S. Patterson, J. D. Moulton, B. C. Wilson, K. W. Berndt, and J. R. Lakowicz, "Frequency-domain reflectance for the determination of the scattering and absorption properties of tissue," *Appl. Opt.* **30**, 4474–4476 (1991).
49. J. R. Lakowicz, G. Laczko, I. Gryczynski, H. Szmacinki, and W. Wiczk, "Gigahertz frequency-domain fluorometry: resolution of complex decays, picosecond processes and future developments," *Photochem. Photobiol.* **2**, 295–311 (1988).
50. S. Fantini, M. A. Franceschini, J. B. Fishkin, B. Barbieri, and E. Gratton, "Quantitative determination of the absorption spectra of chromophores in strongly scattering media: a light-emitting-diode based technique," *Appl. Opt.* **33**, 5204–5213 (1994).

51. S. Fantini, M. A. Franchesini-Fantini, J. S. Maier, and S. A. Walker, "Frequency-domain multichannel optical detector for noninvasive tissue spectroscopy and oximetry," *Opt. Eng.* **34**, 32–42 (1995).
52. M. Gerken and G. W. Faris, "High-precision frequency-domain measurements of the optical properties of turbid media," *Opt. Lett.* **24**, 930–932 (1999).
53. B. J. Tromberg, O. Coquoz, J. B. Fishkin, T. Pham, E. R. Anderson, J. Butler, M. Cahn, J. D. Gross, V. Venugopalan, and D. Pham, "Noninvasive measurements of breast tissue optical properties using frequency-domain photon migration," *Phil. Trans. R. Soc. Lond. B* **352**, 661–668 (1997).
54. M. Firbank and D. T. Delpy, "A design for a stable and reproducible phantom for use in near infra-red imaging and spectroscopy," *Phys. Med. Biol.* **38**, 847–853 (1993).
55. M. Kohl, R. Watson, and M. Cope, "Determination of absorption coefficients in highly scattering media from changes in attenuation and phase," *Opt. Lett.* **21**, 1519–1521 (1996).
56. V. Quaresima, A. Pizzi, R. A. de Blasi, A. Ferrari, M. de Angelis, and M. Ferrari, "Quadriceps oxygenation changes during walking and running on a treadmill," *Proc. SPIE* **2383**, (1995).
57. J. W. Hall and A. Pollard, "Near-infrared spectrophotometry: a new dimension in clinical chemistry," *Clin. Chem.* **38**, 1623–1631 (1992).
58. M. Ferrari, D. A. Wilson, D. F. Hanley, J. F. Hartmann, M. C. Rogers, and R. J. Traystman, "Noninvasive determination of hemoglobin saturation in dogs by derivative near-infrared spectroscopy," *Heart Circ. Physiol.* **256**, 1493–1499 (1989).
59. A. J. Berger, V. Venugopalan, A. J. Durkin, T. Pham, and B. J. Tromberg, "Chemometric analysis of frequency-domain photon migration data: quantitative measurements of optical properties and chromophore concentrations in multicomponent turbid media," *Appl. Opt.* **39**, 1659–1667 (2000).
60. R. N. Bracewell, *The Fourier Transform and Its Application*, McGraw-Hill, Singapore (1986).
61. J. A. Curcio and C. C. Petty, "The near infrared absorption spectrum of liquid water," *J. Opt. Soc. Am.* **41**, 302–304 (1951).
62. H. Q. Woodard and D. R. White, "The composition of body tissues," *Br. J. Radiol.* **59**, 1209–1218 (1986).
63. C. B. McCardle, C. J. Richardson, D. A. Nicholas, M. Mirfakhraee, C. K. Hayden, and E. G. Amparo, "Developmental features of the neonatal brain: MR imaging," *Radiology* **162**, 223–229 (1987).
64. B. A. Holland, D. K. Haas, D. Norman, M. Brant-Zawadzki, and T. H. Newton, "MRI of normal brain maturation," *AJNR* **7**, 201–208 (1986).

65. M. Cope, D. T. Delpy, S. Wray, J. S. Wyatt, and E. O. R. Reynolds, "A CCD spectrometer to quantitate the concentration of chromophores in living tissue utilising the absorption peak of water at 975 nm," Eds., pp. 33–40, Plenum Press, New York (1989).
66. A. Duncan, J. H. Meek, M. Clemence, C. E. Elwell, P. Fallon, L. Tyszczuk, M. Cope, and D. T. Delpy, "Measurement of cranial optical path length as a function of age using phase resolved near infrared spectroscopy," *Pediatr. Res.* **39**, 889–894 (1996).
67. A. Duncan, J. H. Meek, M. Clemence, C. E. Elwell, L. Tyszczuk, M. Cope, and D. T. Delpy, "Optical pathlength measurements on adult head, calf and forearm and the head of the newborn infant using phase resolved spectroscopy," *Phys. Med. Biol.* **40**, 295–304 (1995).
68. D. T. Delpy, personal communication (1998).
69. M. Kohl, C. Nolte, H. R. Heekeren, S. Horst, U. Scholz, H. Obrig, and A. Villringer, "Determination of the wavelength dependence of the differential pathlength factor from near infrared pulse signals," *Phys. Med. Biol.* **43**, 1771–1782 (1998).
70. J. Plucinski, A. F. Frydrychowski, J. Kaczmarek, and W. Juzwa, "Theoretical foundations for noninvasive measurement of variations in the width of the subarachnoid space," *J. Biomed. Opt.* **5**, 291–299 (2000).
71. M. C. van der Sluijs, W.N.J.M. Collier, R. J. F. Houston, and B. Oeseburg, "A new and highly sensitive continuous wave near infrared spectrophotometer with multiple detectors," *Proc. SPIE* **3194**, 63–72 (1997).
72. V. Quaresima, M. Ferrari, B. Oeseburg, and W.N.J.M. Collier, "Left lateral frontal oxygenation changes upon word generation and control tasks by functional near infrared spectroscopy," abstract, presented at 6th meeting on Human Brain Mapping, San Antonio, Tex. (2000).
73. Wright Instruments, *CCD Camera Manual* (1997).
74. M. Essenpreis, C. E. Elwell, M. Cope, P. van der Zee, S. R. Arridge, and D. T. Delpy, "Spectral dependence of temporal point spread functions in human tissues," *Appl. Opt.* **32**, 418–425 (1993).
75. C. A. Piantadosi, "Absorption spectroscopy for assessment of mitochondrial function *in vivo*," *Methods in Toxicology, Vol. 2*, D. P. Jones and L. H. Lash, Eds., Academic Press, San Diego, CA pp. 107–125, (1993).
76. C. W. Yoxall and A. M. Weidling, "Measurement of venous saturation in the adult human forearm by near infrared spectroscopy with venous occlusion," *Med. Biol. Eng. Comput.* **35**, 331–336 (1997).
77. C. W. Yoxall, A. M. Weindling, N. H. Dawani, and I. Peart, "Measurement of cerebral venous oxyhaemoglobin saturation in children by near-infrared spectroscopy and partial jugular venous occlusion," *Pediatr. Res.* **38**, 319–323 (1995).

78. R. A. de Blasi, M. Ferrari, A. Natali, G. Conti, A. M. Mega, and A. Gasparetto, "Noninvasive measurement of forearm blood flow and oxygen consumption by near infrared spectroscopy," *J. Appl. Physiol.* **76**, 1388–1393 (1994).
79. C. E. Elwell, M. Cope, A. D. Edwards, J. S. Wyatt, D. T. Delpy, and E. O. R. Reynolds, "Quantification of adult cerebral hemodynamics by near-infrared spectroscopy," *J. Appl. Physiol.* **77**, 2753–2760 (1994).
80. A. D. Edwards, C. Richardson, P. van der Zee, C. E. Elwell, J. S. Wyatt, M. Cope, D. T. Delpy, and E. O. R. Reynolds, "Measurement of haemoglobin flow and blood flow by near infrared spectroscopy," *J. Appl. Physiol.* **75**, 1884–1889 (1993).
81. L. Skov, O. Pryds, and G. Greisen, "Estimating cerebral blood-flow in newborn infants: comparison of near-infrared spectroscopy and Xe-133 clearance," *Pediatr. Res.* **30**, 570–573 (1991).
82. H. U. Bucher, A. D. Edwards, A. E. Lipp, and G. Duc, "Comparison between near-infrared spectroscopy and Xe-133 clearance for estimation of cerebral blood-flow in critically ill preterm infants," *Pediatr. Res.* **33**, 56–60 (1993).
83. J. Goddard-Finegold, P. T. Louis, D. L. Rodriguez, Y. David, C. F. Contant, and P. Rolfe, "Correlation of near-infrared spectroscopy cerebral blood flow estimations and microsphere quantitations in newborn piglets," *Biol. Neonate* **74**, 376–384 (1998).
84. J. S. Wyatt, M. Cope, D. T. Delpy, C. E. Richardson, A. D. Edwards, S. Wray, and E. O. R. Reynolds, "Quantitation of cerebral blood volume in human infants by near-infrared spectroscopy," *J. Appl. Physiol.* **68**, 1086–1091 (1990).
85. C. E. Elwell, H. Owen-Reece, M. Cope, J. S. Wyatt, A. D. Edwards, D. T. Delpy, and E. O. R. Reynolds, "Measurement of adult cerebral haemodynamics using near infrared spectroscopy," *Acta Neurochir.* **59**, 74–80 (1993).
86. N. C. Brun and G. Greisen, "Cerebrovascular responses to carbon dioxide as detected by near-infrared spectrophotometry: comparison of three different measures," *Pediatr. Res.* **36**, 20–24 (1994).
87. M. Firbank, C. E. Elwell, C. E. Cooper, and D. T. Delpy, "Experimental and theoretical comparison of NIR spectroscopy measurements of cerebral hemoglobin changes," *J. Appl. Physiol.* **85**, 1915–1921 (1998).
88. C. P. Barnfield, V. Y. H. Yu, O. Noma, J. Kukita, L. J. Cussen, A. Oates, and A. M. Walker, "Cerebral blood volume measured using near-infrared spectroscopy and radiolabels in the immature lamb brain," *Pediatr. Res.* **46**, 50–56 (1999).
89. M. Firbank, M. Hiraoka, M. Essenpreis, and D. T. Delpy, "Measurement of the optical properties of the skull in the wavelength range 650–950 nm," *Phys. Med. Biol.* **38**, 503–510 (1993).

90. C. R. Simpson, M. Kohl, M. Essenpreis, and M. Cope, "Near-infrared optical properties of *ex vivo* human skin and subcutaneous tissues using the Monte Carlo inversion technique," *Phys. Med. Biol.* **43**, 2465–2478 (1998).
91. R. C. Haskell, L. O. Svaasand, T. T. Tsay, T. C. Feng, M. S. McAdams, and B. J. Tromberg, "Boundary conditions for the diffusion equation in radiative transfer," *J. Opt. Soc. Am.* **11**, 2727–2741 (1994).
92. S. Takatani and M. D. Graham, "Theoretical analysis of diffuse reflectance from a two-layer tissue model," *IEEE Trans. Biomed. Eng.* **26**, 656–664 (1979).
93. M. Keijzer, W. M. Star, and P. R. M. Storchi, "Optical diffusion in layered media," *Appl. Opt.* **27**, 1820–1824 (1988).
94. R. Nossal, J. Kiefer, G. H. Weiss, R. Bonner, H. Taitelbaum, and S. Halvin, "Photon migration in layered media," *Appl. Opt.* **27**, 3382–3391 (1988).
95. J. M. Schmitt, G. X. Zhou, E. C. Walker, and R. T. Wall, "Multilayer model of photon diffusion in skin," *J. Opt. Soc. Am.* **7**, 2141–2153 (1990).
96. I. Dayan, S. Havlin, and G. H. Weiss, "Photon migration in a two-layered turbid medium: a diffusion analysis," *J. Mod. Opt.* **39**, 1567–1582 (1992).
97. A. Kienle, M. S. Patterson, N. Dögnitz, R. Bays, G. Wagniéres, and H. van den Bergh, "Noninvasive determination of the optical properties of two-layered turbid media," *Appl. Opt.* **37**, 779–791 (1998).
98. R. L. Longini and R. Zdrojowski, "A note on the theory of backscattering of light by living tissue," *IEEE Trans. Biomed. Eng.* **15**, 4–10 (1968).
99. A. Kienle and M. S. Patterson, "Improved solutions of the steady-state and time-resolved diffusion equations for reflectance from a semi-infinite turbid medium," *J. Opt. Soc. Am.* **14**, 246–254 (1997).
100. P. van der Zee, "Measurement and modelling of the optical properties of human tissue in the near infrared," Ph.D. dissertation, University of London (1992).
101. S. L. Jacques and L. Wang, "Monte Carlo modeling of light transport in tissues," *Optical-Thermal Response of Laser-Irradiated Tissue*, A. J. Welch and M. J. C. van Gemert, Eds., pp. 73–100, Plenum Press, New York (1995).
102. S. R. Arridge, H. Dehghani, M. Schweiger, and E. Okada, "The finite element model for the propagation of light in scattering media: a direct method for domains with nonscattering media: a direct method for domains with nonscattering regions," *Med. Phys.* **27**, 252–264 (2000).

103. M. Firbank, S. R. Arridge, M. Schweiger, and D. T. Delpy, "An investigation of light transport through scattering bodies with non-scattering regions," submitted to PMB (1995).
104. R. E. Alcouffe, R. S. Baker, F. W. Brinkley, D. R. Marr, R. D. O'Dell, and W. F. O'Walters, "DANTSYS: a diffusion accelerated neutral particle transport code system," Los Alamos National Laboratory, report LA-12969-M (1995).
105. L. N. M. Duysens, "The flattening of the absorption spectrum of suspensions, as compared to that of solutions," *Biochim. Biophys. Acta* **19**, 1–12 (1956).
106. H. Liu, A. H. Hielscher, S. L. Jacques, F. K. Tittel, and B. Chance, "Influence of blood vessels on the measurement of hemoglobin oxygenation as determined by time-resolved reflectance spectroscopy," *Med. Phys.* **22**, 1209–1217 (1995).
107. M. Firbank, E. Okada, and D. T. Delpy, "Investigation of the effect of discrete absorbers upon the measurement of blood volume with near-infrared spectroscopy," *Phys. Med. Biol.* **42**, 465–477 (1997).
108. M. Hiraoka, M. Firbank, M. Essenpreis, M. Cope, S. R. Arridge, P. van der Zee, and D. T. Delpy, "A Monte Carlo investigation of optical pathlength in inhomogeneous tissue and its application to near-infrared spectroscopy," *Phys. Med. Biol.* **38**, 1859–1876 (1993).
109. J. C. Schotland, J. C. Haselgrove, and J. S. Leigh, "Photon hitting density," *Appl. Opt.* **32**, 448–453 (1993).
110. E. M. Sevick, C. L. Burch, J. K. Frisoli, M. L. Johnson, K. Nowaczyk, H. Szmacinski, and J. R. Lakowicz, "The physical basis of photon migration imaging using frequency-domain measurements," *Medical Optical Tomography: Functional Imaging and Monitoring*, G. Müller, Eds., pp. 485–512, SPIE Press, Bellingham, Wash. (1993).
111. S. R. Arridge, "Photon-measurement density functions. Part 1: analytical forms," *Appl. Opt.* **34**, 7395–7409 (1995).
112. S. R. Arridge and M. Schweiger, "Photon measurement density functions. Part 2: finite-element-method calculations," *Appl. Opt.* **34**, 8026–8037 (1995).
113. E. Okada, M. Firbank, and D. T. Delpy, "The effect of overlying tissue on the spatial sensitivity profile of near-infrared spectroscopy," *Phys. Med. Biol.* **40**, 2093–2108 (1995).
114. E. Okada, M. Firbank, M. Schweiger, S. R. Arridge, M. Cope, and D. T. Delpy, "Theoretical and experimental investigation of near-infrared light propagation in a model of the adult head," *Appl. Opt.* **36**, 21–31 (1997).
115. M. Firbank, E. Okada, and D. T. Delpy, "A theoretical study of the signal contribution of regions of the adult head to near-infrared spectroscopy studies of visual evoked responses," *Neuroimage* **8**, 69–78 (1998).

116. A. H. Hielscher, H. Liu, B. Chance, F. K. Tittel, and S. L. Jacques, "Time-resolved photon emission from layered turbid media," *Appl. Opt.* **35**, 719–728 (1996).
117. M. A. Franceschini, S. Fantini, L. A. Paunescu, J. S. Maier, and E. Gratton, "Influence of a superficial layer in the quantitative spectroscopic study of strongly scattering media," *Appl. Opt.* **37**, 7447–7458 (1998).
118. A. Kienle, T. Glanzmann, G. Wagniéres, and H. van den Bergh, "Investigation of two-layered turbid media with time-resolved reflectance," *Appl. Opt.* **37**, 6852–6862 (1998).
119. G. Alexandrakis, T. J. Farrell, and M. S. Patterson, "Accuracy of the diffusion approximation in determining the optical properties of a two-layer turbid medium," *Appl. Opt.* **37**, 7401–7409 (1998).
120. D. N. F. Harris and S. M. Bailey, "Near infrared spectroscopy in adults: does the Invos 3100 really measure intracerebral oxygenation," *Anaesthesia* **48**, 694–696 (1993).
121. T. J. Germon, N. M. Kane, A. R. Manara, and R. J. Nelson, "Near-infrared spectroscopy in adults: effects of extracranial ischaemia and intracranial hypoxia on estimation of cerebral oxygenation," *Br. J. Anaesth.* **73**, 503–506 (1994).
122. M. Schweiger and S. R. Arridge, "Comparison of two- and three-dimensional reconstruction methods in optical tomography," *Appl. Opt.* **37**, 7419–7428 (1998).
123. M. Wolf, M. Ferrari, and V. Quaresima, "Progress of near-infrared spectroscopy and topography for brain and muscle clinical applications," *J. Biomed. Opt.* **12**(6), 062104 (2007).
124. M. Ferrari, L. Mottola, and V. Quaresima, "Principles, techniques, and limitations of near infrared spectroscopy," *Can. J. Appl. Physiol.* **29**(4), 463–487 (2004).
125. G. Greisen, "Is near-infrared spectroscopy living up to its promises?," *Semin. Fetal Neonatal Med.* **11**(6), 498–502 (2006).
126. M. Ferrari and V. Quaresima, "A brief review on the history of human functional near-infrared spectroscopy (fNIRS) development and fields of application," *Neuroimage* **63**(2), 921–935 (2012).
127. S. Lloyd-Fox, A. Blasi, and C. E. Elwell, "Illuminating the developing brain: the past, present and future of functional near infrared spectroscopy," *Neurosci. Biobehav. Rev.* **34**(3), 269–284 (2010).
128. B. S. Sorg, B. J. Moeller, O. Donovan, Y. Cao, and M. W. Dewhirst, "Hyperspectral imaging of hemoglobin saturation in tumor microvasculature and tumor hypoxia development," *J. Biomed. Opt.* **10**(4), 044004 (2005).
129. B. E. Applegate and J. A. Izatt, "Molecular imaging of hemoglobin using ground state recovery pump-probe optical coherence tomography," *Proc. SPIE* **6429**, Y4291–Y4291 (2007).

130. S. P. Chong, C. W. Merkle, C. Leahy, H. Radhakrishnan, and V. J. Srinivasan, "Quantitative microvascular hemoglobin mapping using visible light spectroscopic optical coherence tomography," *Biomed. Opt. Express* **6**(4), 1429–1450 (2015).
131. J. Laufer, D. T. Delpy, C. Elwell, and P. Beard, "Quantitative spatially resolved measurement of tissue chromophore concentrations using photoacoustic spectroscopy: application to the measurement of blood oxygenation and haemoglobin concentration," *Phys. Med. Biol.* **52**, 141–168 (2007).
132. B. Cox, J. G. Laufer, A. R. Arridge, and P. C. Beard, "Quantitative spectroscopic photoacoustic imaging: a review," *J. Biomed. Opt.* **17**(6), 061202 (2012).
133. X. A. Xu, H. L. Liu, and L. V. Wang, "Time-reversed ultrasonically encoded optical focusing into scattering media," *Nat. Photonics* **5**(3), 154–157 (2011).

Stephen John Matcher is a professor of biophotonics at the University of Sheffield. He obtained his BSc and PhD degrees from Imperial College London. After working for the Max Planck Gesselschaft as a guest scientist, he joined University College London's Department of Medical Physics and Bioengineering, where he developed instrumentation and algorithms for tissue near-infrared spectroscopy. His current research interests cover near-infrared tissue imaging using optical coherence tomography, the use of OCT and nonlinear microscopy to characterise the structure of tissue engineered constructs, and biochemical characterization of biological samples using midinfrared lasers.

Chapter 10
Near-Infrared Spectroscopy in Multimodal Brain Research

Teemu Myllylä
University of Oulu, Oulu, Finland

Vladislav Toronov
Ryerson University, Toronto, Canada

Jurgen Claassen
Radboud University Medical Centre, Nijmegen, Netherlands

Vesa Kiviniemi
Medical Research Center of Oulu, Oulu, Finland

Valery V. Tuchin
Saratov National Research State University, Saratov, Russia

National Research Tomsk State University, Tomsk, Russia

Institute of Precision Mechanics and Control, Russian Academy of Sciences, Saratov, Russia

10.1 Introduction

The previous chapter discussed the application of near-infrared spectroscopy (NIRS) and near-infrared imaging (NIRI) to brain activity monitoring and compared them to other functional neuroimaging modalities. This chapter continues the theme by exploring the simultaneous use of functional near-infrared spectroscopy (fNIRS) and such functional neuroimaging modalities as electroencephalography (EEG), magnetoencephalography (MEG), and functional magnetic resonance imaging (fMRI). It also presents a review of some current studies and novel approaches, particularly those used to enhance penetration depth in NIRS imaging.

Multimodal measurements are increasingly being employed in the study of human physiology. Brain studies in particular can take advantage of the exciting possibility of combining data from different techniques, based on causality in brain activity—an increase in neuronal activity causes a metabolic demand for glucose and oxygen, which increases cerebral blood flow (CBF) to the active brain region. Studying the mechanisms and relationships between among activities, electrophysiology, and blood-flow-related parameters requires combining hemodynamic and electromagnetic-based imaging techniques. By simultaneously monitoring such physiological parameters as heart rate, blood pressure and blood flow, changes in respiration and end-tidal carbon dioxide levels, and shifts in blood flow distribution in the human body, we are able to study the functioning of organs, gaining a more comprehensive understanding of the interconnections between different physiological activities and their relation to brain function. Moreover, multimodal measurements of this type provide tools for identifying physiological components generated by the brain.

Performing multimodal measurements by combining EEG with MEG or fMRI is a common practice in modern-day neuroimaging. Since these three imaging modalities are increasingly being used with functional fNIRS, the following presentation provides a brief introduction to EEG, MEG, and fMRI; discusses their utilization in combination with fNIRS; and explores additional benefits offered by the simultaneous use of fNIRS and these imaging modalities. Furthermore, examples of recent multimodal studies involving the application fNIRS are presented.

10.1.1 Functional imaging of the brain

The initial studies of brain function were performed by Angelo Mosso, who measured brain pulsation via cranial holes and noticed changes in brain pulsations (and blood flow) during praying and mathematical tasks in 1881.[1] However, more widely used methods to study brain functions were developed with the discovery of electrical activity in the brain. These measurements started in 1924, when Hans Berger used an electrode to record electrical activity on the scalp.[2] Later, development of the EEG led to the use of topographic scalp maps, which record activity of the underlying brain structures through electrodes placed at certain points on the head. The early EEG amplifiers, which were coupled with ink printers, were unable to capture both microvolt-level neuronal changes and millivolt DC shifts. Consequently, for decades, high-pass filters set at more than 0.5 Hz were used as a norm, with DC changes being discarded as mostly noise.

In EEG measurements, scalp electric potential differences are determined by electric neuronal activity from the entire cortex and by the geometrical orientation of the cortex. However, to accurately determine the actual generators of current is by no means an easy task. This challenge arises because it is a mathematical impossibility to reconstruct current sources for a given EEG signal, as some currents produce potentials that cancel each other out—the so-called inverse problem.[3] In addition, EEG has a high temporal resolution, on the millisecond scale, which is adequate to follow brain activity changes. Its spatial resolution, however, is a relatively modest, being on the centimeter scale. This is a notable weakness, since brain studies place particular importance on precise localization of brain functions.

More advanced neuroimaging of the brain became a reality in the 1960s and 1970s, thanks to the development of MEG and MRI. MEG, which is closely related to EEG, measures the magnetic fields created by the electric currents in the body. The signals measured via both methods are generated by the same neuronal activities, which they are capable of measuring with a time resolution of less than one millisecond. Both methods can also utilize a large number of channels, ranging from 100 to 300 for MEG and up to 256 for EEG, often with a standardized 10–20 electrode placement system.[3] MEG has a spatial accuracy of few millimeters under favorable conditions.[4]

Work on radioactive tracer imaging by Lassen and Ingvar eventually led to the functional imaging of the brain. Indirect radioactive tracers reflect neuronal activation by detecting increases in regional blood flow and metabolism after neural activity increases. In the 1980s, Fox and Raichle discovered that with positron emission tomography (PET), blood flow actually increases more than what might be required based on glucose consumption changes after neuronal activation.[5]

The discrepancy in the PET results led the way for the blood oxygen level-dependent (BOLD) contrast discovered by Ogawa in 1990. Ogawa used MRI and was guided by the work of Faraday and Linus Pauling; the latter had discovered in 1936 that blood had different magnetic susceptibility based on whether it was oxygenated or deoxygenated. Paramagnetic deoxyhemoglobin (HHb) contains exposed Fe^{2+} ions that alter regional magnetic susceptibility; when oxygen binds the Fe^{2+}, the oxygenated blood is diamagnetic. Since Fox and Raichle showed that activated brain becomes over-oxygenated with PET, MRI can be used to detect this area after a 3–5-s delay when the blood increase in flow emerges. The activation hyperemia reduces regional magnetic susceptibility and T2* values, and the MRI signal increases.[5,6] The fNIRS signal can be used in imaging the same over-oxygenation phenomena locally after neuronal activity.

While between 1990 and 2005 researchers largely focused on detecting task activation-related hemodynamic responses with increasing accuracy, after 2005 the focus turned exponentially toward detecting spontaneous brain activity. Since the discovery of spontaneous resting state connectivity by Biswal and co-workers in 1995, there has been a significant shift in the attention of the neuroimaging community.[7] With the use of novel signal source separation methods such as independent component analysis (ICA), it has become possible to detect functional brain cortices and even entirely new functional networks.[8–12]

Currently, fMRI dominates brain mapping applications due to its high spatial accuracy and lack of radiation exposure. One major drawback, however, is its time resolution, which ranges from 1 to 2 Hz—a relatively poor rate. In recent years, new scanning techniques have enabled major improvements in this area. Magnetic resonance encephalography (MREG), for example, offers a time resolution of 10 to 16 Hz.[13,14]

One major weakness of fMRI is the fact that, during scanning, the subject must stay still in a lying position inside the scanner. As a result, the study of brain functions in real-world situations is beyond the method's current capability, allowing only fictive types of stimuli to be used. Another important consideration is the fact that the subject cannot move during the measurement. fNIRS, in contrast, does not suffer from these restrictions. It even allows measurements to be performed outside the hospital, in a daily life environment. Furthermore, a wearable fNIRS device can easily be combined with EEG measurements.[15] Complementing each other, the two methods have inspired a growing interest in the possibilities associated with a mobile application that enables study of brain activity in moving people.[16]

10.1.2 Toward multimodality

Multimodal measurements seek to combine the best features of various measurement methods and modalities. For example, a combination of EEG and fNIRS provides accurate temporal resolution for electrophysiology and blood flow-related parameters. The spatial accuracy of this combination is low, but can be improved by the simultaneous use of either MEG or fMRI.

The first simultaneous fMRI and fNIRS measurement was carried out in 1996, when researchers sought to measure cerebral blood oxygenation changes during human brain activation.[17] The combination of fMRI and fNIRS is still relatively rare, owing largely to MRI-compatibility requirements, and most publications related to the simultaneous use of fMRI and fNIRS focus on validating the correlation between BOLD signals and cortical hemodynamic changes.[18–29] Due to the development of magnetic resonance (MR)-compatible optical fiber for NIRS and their commercial availability, applications of combined fMRI-fNIRS are becoming more common. For instance, an fMRI/fNIRS-based system was recently built for diagnostic breast cancer imaging.[30] In brain imaging, fNIRS was recently used in the respiratory calibration of the BOLD signal in fMRI.[31]

Simultaneous measurements of MEG and fNIRS are still uncommon, due to requirements posed by the MEG device. For instance, a paper by Sander et al. reports on the simultaneous measurement of MEG and NIRS to study a motor activity paradigm.[32] Another recent paper by Seki et al. presents combined measurements using MEG, EEG, and fNIRS to study neuronal activity and cortical hemodynamics with neurovascular coupling.[33]

10.2 Realization of NIRS in Multimodal Setups

A large static magnetic field exists inside the MRI room, posing high demands on the materials and devices used in there. Ferromagnetic materials, for example, would be affected by these magnetic fields and are consequently

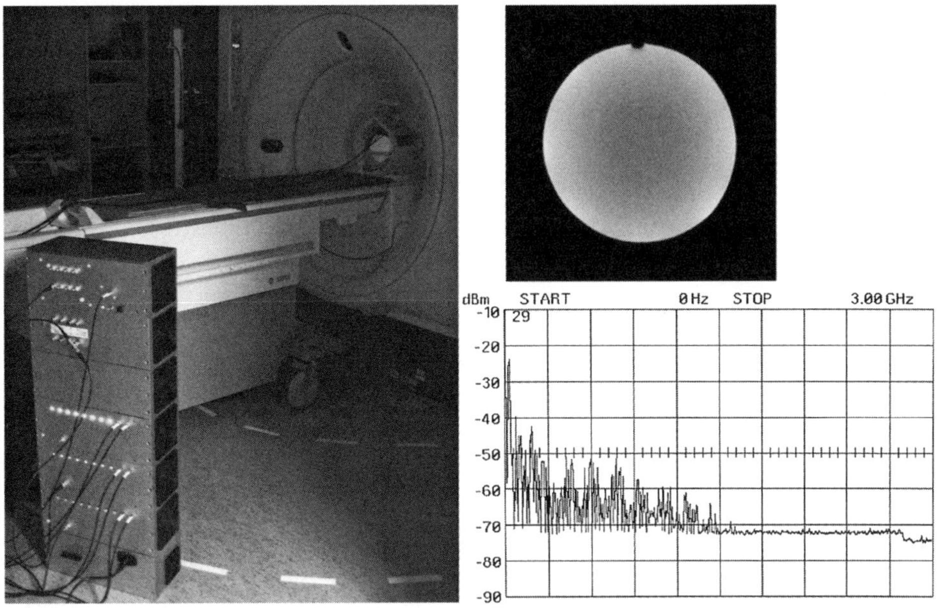

Figure 10.1 MR-compatible NIRS device in a 1.5-T MRI room during a calibration measurement. A phantom, visible in the figure as a yellow ball inside the MRI scanner, is illuminated with the NIRS device, while fiber optodes are attached to its surface.[36] In the upper-right corner is an undisturbed MR image of the phantom including the attached fiber optodes, which are invisible. To avoid disturbance, the device containing electronic parts is placed at some distance from the scanner. Different static magnetic field zones are marked on the floor. On the lower-right side is a typical electromagnetic noise spectrum between 0 and 3 GHz, measured inside an MRI room during scanning.

prohibited, while non-metallic materials, including metals such as aluminum and copper, are allowed. Based on safety considerations and the desire to avoid image distortion, materials and medical devices intended for MRI use can be divided into two classes. The first is MR safe, for objects that are completely non-magnetic, non-electrically conductive, and non-radio frequency (RF) reactive; the second is MR conditional, applied to objects that may contain magnetic, electrically conductive, or RF-reactive components, but are safe for operations in proximity to the MRI.[34] Moreover, restriction zones—normally marked by the manufacturer on the floor of the MRI room (observe the dash markings on the floor in Fig. 10.1)—guide the placement of MR-conditional devices and components.

In the case of a NIRS device, all optical fibers and optodes must be MR safe (i.e., free of ferromagnetic metals), since they will be attached to the subject's head inside the scanning volume of the MRI scanner. However, the NIRS device itself, which contains electrical parts, can be placed at a distance from the MRI scanner. It may be labeled as an MR-conditional device, provided that its electromagnetic shielding has been carefully designed. Figure 10.1 shows an MR-conditional NIRS device with effective radiation shielding at less than 1 GHz. According to our studies, electromagnetic radiation above the GHz range does not affect the quality of MR images.[35]

A common approach involves placing only optical fibers and optodes in the MRI room and interfacing the units in the control and the examination room through a waveguide, which is normally found in the RF-shielded walls of MRI rooms. In addition, some commercially available NIRS devices allow their integration into the MRI environment using MR-compatible fibers and optodes provided by the manufacturer. These fibers normally measure 7–10 m in length, which is sufficient for most MRI rooms. During scanning, a RF coil is typically fit closely around the subject's head. Thus, the tip of the fiber has to be bent to a small diameter of approximately 10 mm, or coupled to a prism, to turn the light path parallel to the skin.

Similar to MRI, a MEG device is placed inside an electromagnetically shielded room to prevent electromagnetic fields from disturbing the quality of the MEG recording. Materials used in instruments and devices near the MEG scanner must be non-magnetic. Moreover, devices containing electronic parts are forbidden inside the MEG room—only passive components are permitted. We have experimented with combining fNIRS measurements with MEG, allowing only plastic optical fibers in the MEG room. Another requirement was that the optodes (Fig. 10.2, right) attached on the subject's head had to be small in size because, as seen on the left in Fig. 10.2, the space between the subject's head and the MEG coil is limited. This MEG/NIRS setup, in combination with blood pressure measurements, was first tested in 2014 in the Clinic for Neurology at the University Hospital Magdeburg, which is home to a MEG lab equipped with a 248-channel whole-head device (BTI Magnes 3600 WH). No interference was observed in MEG data, indicating that the setup is fully MEG compatible.

10.2.1 NIRS head caps

A range of NIRS head caps and optodes for different brain monitoring purposes are now on the market due to a growing interest in commercializing clinical NIRS applications. For instance, one-channel single-use adhesive optodes are quite commonly used in intensive care for monitoring cerebral oxygenation ($SrcO_2$), to obtain early warning of a decreased oxygen level. Moreover, integrated NIRS/EEG head caps are being increasingly used in brain-computer interfaces (BCIs),[37] to improve the accuracy of decoding brain signals.[38] Hybrid NIRS/EEG caps with different configurations and sizes, providing comfortable fit and easy care, are now relatively widely available.[39]

Small optodes can be also fitted into the slit of the electrode joint, provided that the conductive gel between the electrode and the subject's skin is transparent. In terms of both electrodes and optodes, good contact between the sensor and the skin is the most important factor for achieving a stable

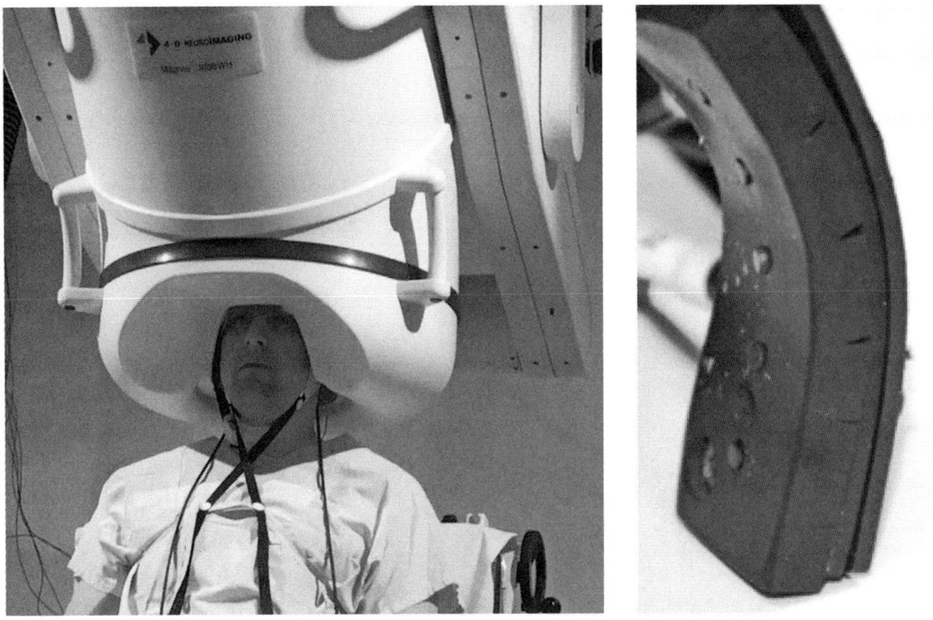

Figure 10.2 (left) The MEG-compatible optical fibers illuminate a subject's head inside the MEG coil. (right) An NIRS optode designed for MEG. The fiber clip, which is made of flexible rubber with a thickness of 10 mm, is sufficiently small for attachment on the head inside the MEG helmet without touching it. Made of hard plastic, the fiber tip emitting light into the head is bent to a 90-degree angle, allowing it to be placed inside the small space between the coil and the head. The diameter of the fiber source tip is 2.5 mm. The fiber detector has the same physical dimensions. Fiber detectors in the fiber clip are placed at various distances, 5 mm to 40 mm, from the source detector.

signal response. In the near future, this type of gel could potentially serve as an optical clearing agent (OCA) to improve the signal response of both modalities.

10.3 fNIRS Combined with Different Techniques: Possibilities and Challenges

10.3.1 fNIRS and neuroimaging

10.3.1.1 fMRI

A combination of fNIRS and fMRI can provide many advantages. First, accurate anatomical information for the brain, gathered by MRI, enables estimating NIR light propagation in the human head through different tissue layers.[40] As calculations of oxyhemoglobin (HbO_2) and HHb in fNIRS are based on simplified tissue geometries, absolute concentrations cannot be quantified reliably. However, a simultaneous measurement with fMRI provides spatial information that helps define where optodes are positioned with respect to the underlying brain. This allows investigation of spatial correspondences between the modalities.[19]

fNIRS is beneficial to the regression of low-frequency physiological noise from fMRI data, since fNIRS can measure physiological fluctuations at a higher temporal resolution.[41] Moreover, fNIRS can measure both HHb and HbO_2 as well as HbT changes, whereas fMRI alone is mainly sensitive to changes in HHb and blood flow. This information provides an estimate of the cerebral metabolic rate of oxygen and can improve the physiological interpretation of fMRI. Additionally, concurrent fNIRS offers a contrast mechanism for a direct characterization of CBF and other blood flow parameters,[26] such as the oxygen extraction fraction (OEF) and cerebral blood volume (CBV).[42]

An important brain activity measurement enabled by fNIRS is determining mitochondrial cytochrome aa_3 activity, which is responsible for more than 90% of cellular oxygen consumption. This enzyme directly reflects tissue metabolism, rather than the oxygen level of the surrounding blood vessels, which tends to fluctuate autonomously and masks activation patterns. As a result, measurement of cytochrome aa_3 activity has the potential to provide more direct information on neuronal activity than a simple Hb measurement.[43] Combined with fMRI, it presents a new method for studying the connection between metabolism and oxygenation. This approach would also be beneficial for exploring several brain disorders, such as autism, that have recently been shown to be linked with mitochondrial metabolism disorder.[44]

10.3.1.2 EEG

While both fNIRS and fMRI are hemodynamic techniques, EEG, which measures microvolt electric potential changes of millisecond duration on the scalp, is directly sensitive to integrated neuronal activity, mostly from the pyramidal neurons.[45] A fundamental disadvantage of EEG is its low spatial resolution due to the effects of the skull and tissue on the electric fields. Although recently some EEG groups have reported significant advances in EEG spatial resolution through the use of very many electrodes and novel reconstruction techniques, such novel EEG systems are still being validated. Given that fNIRS has a better spatial resolution than EEG and cerebral hemodynamic changes measured by fNIRS and neuronal activity are related through the neurovascular coupling,[46] simultaneous application of the EEG and fNIRS technologies potentially can provide a very powerful method to study the brain.[47]

EEG signals can be analyzed in both the frequency domain [alpha (8–14 Hz) and beta (15–30 Hz) rhythms] and the time domain [various characteristic event-related potential (ERP) markers, such as K-complexes, spindles, N100, P200, and N400].[48] An early simultaneous EEG and fNIRS study[49] during 15-minute rest found a correlation between the spontaneous changes in the HbO_2 concentration measured by fNIRS and the temporal changes in the EEG alpha-rhythm peak frequency. More recently, fNIRS and

EEG were used simultaneously to validate fNIRS as a tool to investigate visual, auditory, and language processing in human brain.[50] Significant correlations were found between visual and auditory ERPs measured by EEG and hemodynamic changes measured by fNIRS in the human visual and auditory cortices, respectively. The metabolic changes measured by fNIRS were compared with ERP to better understand the meaning of N400 and P200 ERPs as markers of the complex language processing.[51] Correlations were found between task-related changes in N400 amplitude and left Broca's area, and between task-related changes in P200 amplitude and left Wernicke's area. fNIRS and EEG were also used simultaneously in studies of schizophrenia,[52] cortical myoclonus,[53] epilepsy,[54] and amnesic mild cognitive impairment.[55]

10.3.1.3 MEG

Similar to EEG, MEG is directly sensitive to the activity in the pyramidal cortical neurons through the measurement of extra-weak 0.01–1.00-pT magnetic field variations near the surface of the head. Such variations can be measured using superconducting quantum interference devices (SQUIDs). The main difference between the sources of EEG and MEG signals is that MEG picks up only signals from the dendrites tangential to the head surface. The significant advantage of MEG over EEG is a much better spatial resolution.[56] Therefore the combined MEG/fNIRS recording potentially can be an even more powerful tool than combined EEG/fNIRS. Because the MEG machines are much more expensive than EEG,[57] the number of studies using combined MEG-fNIRS recordings is as yet extremely limited. A combined MEG/fNIRS system was used to record hemodynamic and neuronal responses to the auditory stimulation from the human auditory cortex.[58] Two SQUID MEG gradiometers were placed on both sides of the head near the auditory cortex, and a 10-channel non-metallic NIRS probe connected to an ETG-7100 NIRS system (Hitachi Medical Systems) was attached to the T4 areas (right auditory cortex) of subjects. In the study, 100-ms bursts of 1-kHz sound waves were applied to both of the subjects' ears repeatedly every second during 30 seconds. Clear and consistent N100 MEG responses of the opposite polarity were recorded on the opposite sides of the head in all 12 subjects. Also, some kinds of statistically significant hemodynamic responses were recorded on the auditory cortices in all subjects during 30 seconds following the onset of stimulations. However, the time-courses of the HbO_2, HHb, and HbT responses were significantly different in all subjects.

Since HHb is paramagnetic while the rest of the tissue is diamagnetic, when the human head is placed into the external magnetic field the BOLD changes in the total magnetic field can be also measured using SQUIDs. In one study, BOLD signals measured by SQUID in the human prefrontal area during Valsalva maneuvers were found highly correlated with Hb changes measured simultaneously by NIRS.[59] Thus, this MEG/fNIRS study showed that MEG can potentially measure both neuronal and hemodynamic functional brain signals.

10.3.2 Blood pressure and cerebral blood flow

10.3.2.1 Blood pressure

In recent decades, while we have started to better understand the functioning of the cardiovascular system, a growing emphasis has been placed on the mechanisms that link the cardiovascular and brain functions, particularly CBF. For example, vasomotor blood pressure fluctuations and their relationship to brain signals are of great interest.[60] One of the challenges in such studies involves measurement and separation of vasomotor wave (VMW) signals, which are physiologically related to neural activity fluctuations. Meeting this challenge requires multimodal methods that combine accurate blood pressure measurements with assessments of cerebral hemodynamics. Nonetheless, due to difficulties in obtaining blood pressure measurements during fMRI, to date relatively few studies have examined correlations between cerebral hemodynamics and the VMW. We have measured and correlated BOLD signals with VMW data in a resting state experiment,[61,62] using an MR-compatible noninvasive blood pressure (NIBP) measurement method.[63] One major challenge is to measure and separate blood pressure signals coming from large arteries and veins. To that end, additional measurements can be performed using various blood pressure measurement methods, including invasive ones outside the MRI room, while measuring cerebral hemodynamics with fNIRS (instead of fMRI).

The following graph shows an example of a blood pressure pulse propagation measurement conducted simultaneously with fNIRS to study the effect of hypercapnia on cerebral hemodynamics.

To induce mild hypercapnia, a breath hold (BH) task was used. One sensor, which was intended to detect the propagation of blood pressure pulses, was placed over the aortic valve on the sternum; another sensor was attached over the carotid artery. Figure 10.3 shows the calculated continuous blood pressure pulse transit time (PTT) between the sensors and corresponding changes in HbO_2 and HHb, measured by NIRS. The optode was placed on the subject's upper-left forehead to measure the ventromedial default mode network (DMNvmpf).

10.3.2.2 Cerebral blood flow

Since its development by Rune Aaslid in the 1980s, transcranial Doppler (TCD) has provided an opportunity to determine CBF velocity (CBFV) in intracranial arteries. In the majority of people, the Doppler ultrasound beam (2 MHz) is able to penetrate the temporal bone and reach the circle of Willis. The reflected ultrasound provides an accurate estimate of red blood cell velocity in these

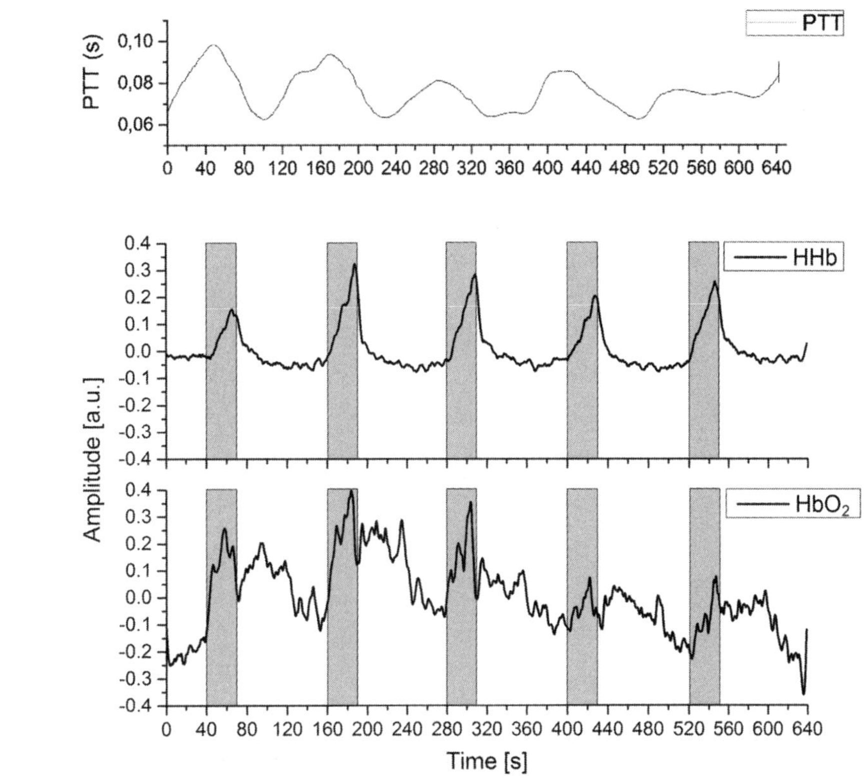

Figure 10.3 The breath hold task exhibits substantial changes in the PTT and NIRS signal. The BH paradigm consisted of blocks of 32 s of breath-holding followed by 88 s of rest, repeated five times. As seen, BH decreases PTT, indicating an increase in blood flow velocity, correlating with changes in HbO_2 and HHb.

vessels, from which the blood-flow velocity can be determined with high temporal resolution (greater than 10 Hz). Under the assumption that the vessel diameter remains constant, changes in flow velocity equal changes in CBF.

Three major limitations of this imaging modality must be noted. First, as many as 30% of older patients and female patients have a poorly penetrable temporal bone, resulting in a weak signal or even absence of signal. Second, there is limited research to confirm the stability of vessel diameter, such that possible confounding effects of diameter changes during measurements remain a "black box." Third, in many cases, blood vessels cannot be visualized during measurements (depending on equipment), which may lead to errors in the insonation angle (which affects velocity measurements) or even to incorrect blood vessel identification.

Most studies of cerebral hemodynamics have targeted the middle cerebral arteries (MCAs). These are the largest vessels to originate from the circle of Willis, and both MCAs carry approximately 70% of total CBF. Figure 10.4 shows an example of a transcranial Doppler recording of CBFV in MCA,

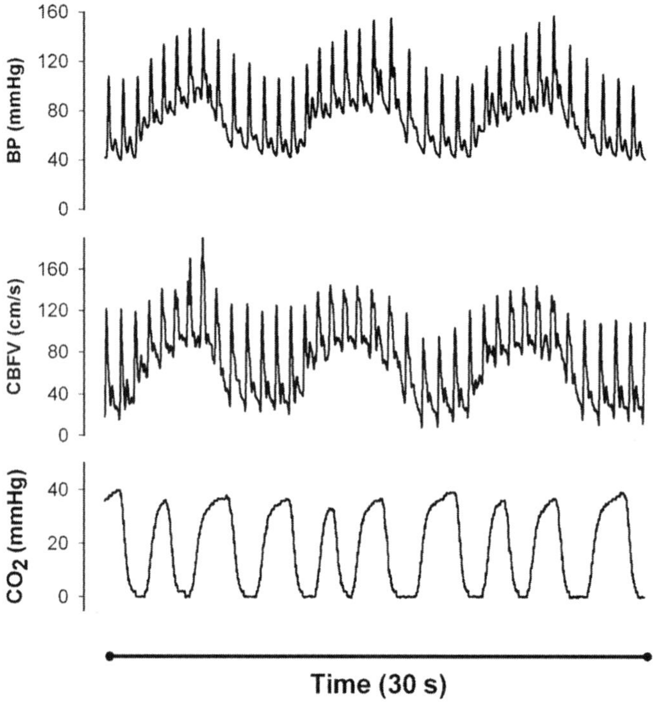

Figure 10.4 Transcranial Doppler recording of CBFV in MCA and recording of blood pressure (BP) and end-tidal CO_2 during repeated squat-stand maneuvers.[64]

together with a recording of blood pressure and end-tidal CO_2. In this example, blood pressure fluctuations were induced by repeated squatting and standing up. The TCD recording was able to capture both faster and slower changes in CBFV following changes in blood pressure (fast: systolic to diastolic pressure, slow: induced fluctuations).

The keen observer may note that the CBFV signal appears to be shifted slightly leftward compared to the blood pressure signal; that is, the upward and downward waves in CBFV appear to precede those in blood pressure. This is not a synchronization error but rather the manifestation of a physiological phenomenon known as phase shift, reflecting cerebral autoregulation.[65] The high temporal resolution of the blood pressure and CBFV registration allow us to quantify this phase shift and assess autoregulatory function.

10.4 Novel Approaches and Examples of Current Multimodal Studies

10.4.1 Combining TCD with fNIRS

fNIRS has the potential to complement TCD recordings of CBF. TCD has a high temporal resolution but lacks spatial resolution. It is used to record changes in blood flow in the MCA, a large conduit vessel. These changes result from downstream changes in the microvascular circulation (vasoconstriction and vasodilation), causing changes in resistance, which affect upstream blood flow in MCA. These upstream changes in blood flow can be accurately determined by TCD, but TCD cannot identify the location and extent of the downstream changes. In other words, a small increase in MCA flow measured by TCD could reflect a small global increase in CBF, or it could signal the presence of a larger but only regional increase in CBF. Moreover, whether such a regional change in CBF is located in the frontal lobe, parietal lobe, or temporal lobe cannot be distinguished.

fNIRS provides information on cerebral cortical tissue oxygenation in the brain region directly beneath the NIRS optodes. As such, fNIRS could help to improve the spatial resolution of TCD by revealing whether a change in flow observed in MCA (e.g., a reduction in flow) is followed by a reduction in tissue oxygenation in the whole brain or only in specific areas. Today, whole-brain coverage by fNIRS is still technically and practically difficult. fNIRS recordings over the frontal cortex have proven feasible, because the skull is relatively flat there with no hair. Figure 10.5 shows an example of a two-channel NIRS system (Oxymon) positioned over the frontal cortex.

We have modified a Spencer TCD headgear to include the NIRS optode holders, which are visible in Fig. 10.5. The headgear has TCD probe holders. This custom-made headgear allows stable fixation of two TCD probes and four NIRS optodes. Using this setup, we have been able to record TCD and NIRS signals during changes in body position, such as transitions from sitting to standing, to measure the effects of orthostatic hypotension on brain perfusion.

An example of a combined recording of TCD and NIRS is shown in Fig. 10.6. In this experiment, blood pressure fluctuations were induced, similar to those depicted Fig. 10.4, but we used transitions from sitting to standing, as this is more feasible in elderly subjects (the example is from a

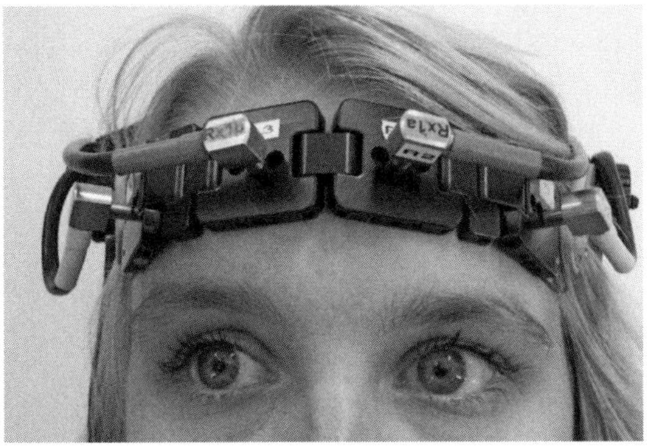

Figure 10.5 Measurement over the frontal cortex using a two-channel NIRS system (Oxymon). Two pairs of optodes are bilaterally attached on the forehead.[66]

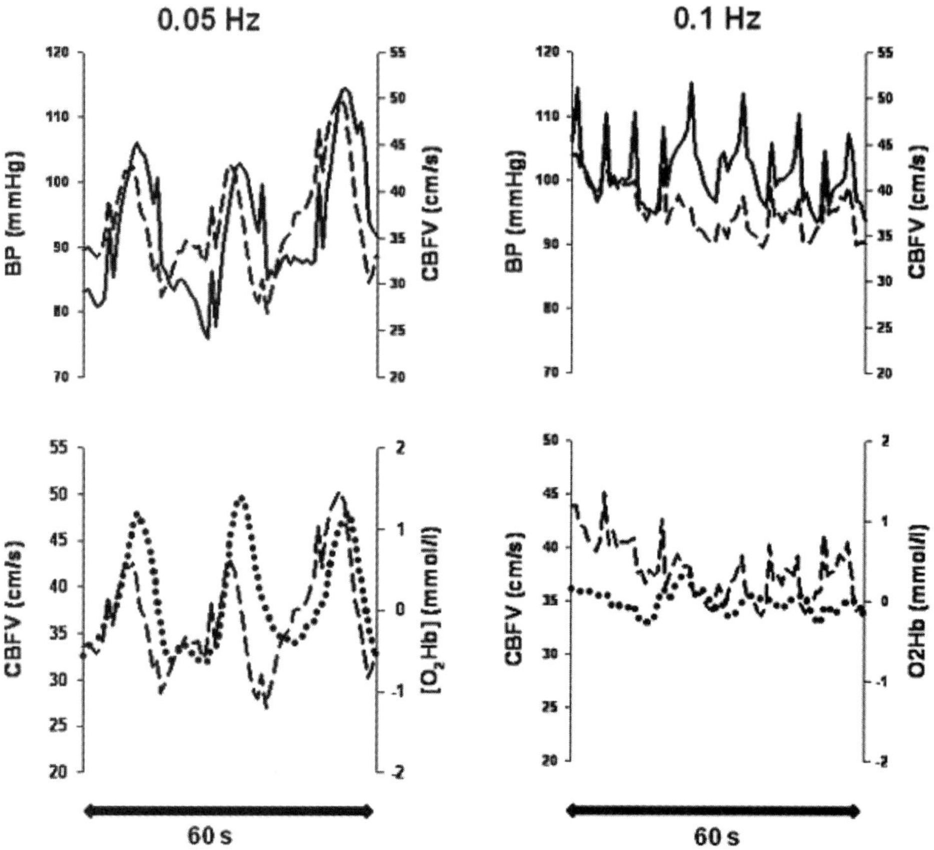

Figure 10.6 Simultaneous recording of BP (Finapres), CBFV (TCD), and NIRS (Oxymon) during induced blood pressure oscillations by repeated sit–stand maneuvers at 0.05 Hz and 0.1 Hz. Solid line: BP; dashed line: CBFV; dotted line: HbO_2.

patient with Alzheimer's disease). The purpose of inducing blood pressure oscillations is to study hemodynamic physiology. Fluctuations in blood pressure occur naturally and with different frequencies, and the brain vasculature, in turn, responds to these different frequencies in a different manner. Inducing a fluctuation, or oscillation, at a specific frequency allows us to study how brain blood flow responds to that particular frequency. In addition, induced oscillations are stronger than naturally occurring fluctuations, so they elicit a stronger response.

In Fig. 10.6, oscillations were induced at 0.05 Hz and at 0.1 Hz, by asking the subject to stand and sit at intervals of 10 s (0.05 Hz) and 5 s (0.1 Hz). The top graphs show the blood pressure and TCD signal; the bottom graphs show the TCD and NIRS signal (HbO_2 only). The induced blood pressure oscillations lead to oscillations in CBFV, which in turn lead to changes in HbO_2. Note that at 0.05 Hz, there is a phase shift due to cerebral autoregulation; this phase shift causes the CBFV signal to "lead" the blood pressure signal. There is also a phase shift between CBFV and HbO_2, where HbO_2 follows the CBFV signal with a short delay. We believe this delay is in part caused by the time delay between conduit flow in the MCA and cerebral tissue perfusion as measured by NIRS (transit time). Again, the high temporal resolution allows us to quantify these phase shifts. In this example, we noted a larger phase shift between CBFV and HbO_2 in patients with Alzheimer's disease versus controls, which could point toward an increase in transit time.

We have also used the combination of blood pressure, TCD, and fNIRS to assess changes in blood pressure and (global) CBF during cognitive activation. Such measurements should also include recordings of end-tidal CO_2. Performance of a cognitive task causes changes in autonomic activation (sympathetic activation), resulting in increases in heart rate and blood pressure. Respiration is also affected, through autonomic activation, or because subjects start to breathe in phase with the repetitions of the cognitive task. Respiration causes changes in end-tidal CO_2 that profoundly affect CBF. In addition, changes in intrathoracic pressure during breathing lead to oscillations in blood pressure. Collectively, these changes lead to fluctuations in CBF that occur outside of the neurovascular coupling effect of the cognitive activation per se. The resultant fluctuations in CBF are recognized in fNIRS experiments as "systemic noise." In part, these fluctuations can be dealt with by applying filtering, but this step requires that the neurovascular coupling response be identified and separated from this systemic noise. Achieving this feat can be difficult, however, if the systemic noise becomes phase locked with the repetitions of the cognitive task.

Experiments that combine blood pressure, TCD, fNIRS, and end-tidal CO_2 recordings can provide much information on systemic "noise" during cognitive activation experiments. In one experiment, we combined measurements of blood pressure and fNIRS during a working memory task (*N*-back) comparing young and older adults. During subjects' cognitive activation, we noted oscillations in blood pressure with different frequencies that correlated with oscillations in HbO_2. These oscillations changed when the task became more difficult; moreover, oscillations differed between young and older adults.[67]

Figure 10.7 shows recordings for young and old subjects, in whom oscillations in blood pressure and HbO_2 during the cognitive task are clearly visible. The example also illustrates that there can be different frequencies in oscillations. The young subject has prominent low-frequency oscillations, whereas the older subject has prominent high-frequency oscillations.

10.4.2 Development of hyperspectral fNIRS

Although many early fNIRS studies used hyperspectral (broadband) measurements, the need for multichannel systems and the high cost of sensitive spectrometers dictated the multispectral design used by all of the currently available commercial fNIRS systems. Recently, however, a number of vendors (e.g., Ocean Optics, Avantes, B&W Tek) have introduced highly

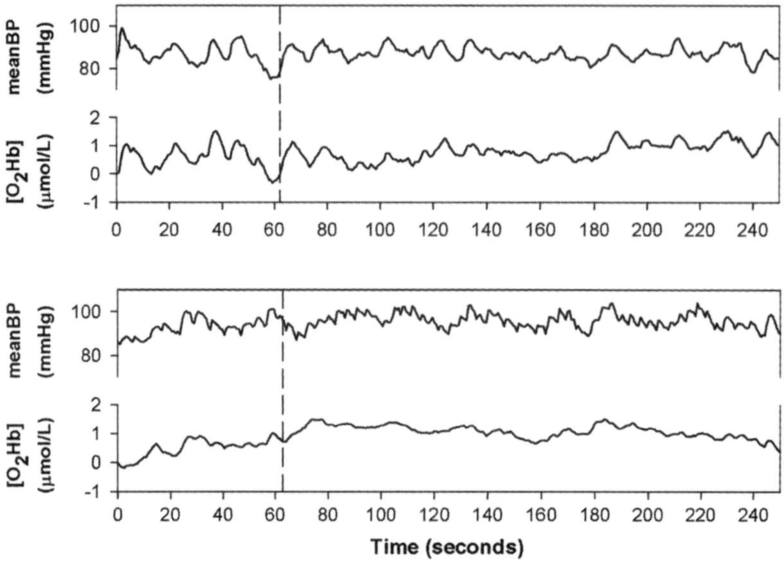

Figure 10.7 Beat-to-beat time series of mean blood pressure and HbO_2 (marked in figure as "O_2Hb") from a representative young participant (upper panel) and an older participant (lower panel) during baseline and 0-back performance. The dashed line marks the start of the task.[67]

sensitive portable NIR spectrometers. The suitability of such new products for the acquisition of fNIRS signals in adult humans was recently demonstrated.[68–70] In particular, the sensitivity of a hyperspectral fNIRS (hfNIRS) system can be greatly improved using novel multichannel signal processing techniques, such as independent component analysis.[70] Furthermore, hyperspectral NIRS can measure not only changes, but also the absolute properties of highly scattering media, including heterogeneous tissues.[68,71,72]

Another potential advantage of hfNIRS is its direct assessment of oxygen metabolism through measurements of changes between the oxidized and reduced states of cytochrome oxidase (Δox-CCO); CCO is the mitochondrial enzyme involved into the oxygen metabolism. The main challenge associated with measurements of Δox-CCO is the cross-talk with ΔHbO_2 and the much lower concentration of CCO compared to the concentration of hemoglobin species. Although hfNIRS has been used to measure cerebral Δox-CCO in multiple human and animal studies, due to the low signal-to-noise ratio and the complexity of the biochemical mechanism of O_2 metabolism even the direction of Δox-CCO remained controversial. A recent study on adult humans indicated that although an increase (decrease) in cerebral oxygen delivery should lead to the respective increase (decrease) in Δox-CCO, in some cases the opposite result may occur.[73]

To test the capability of our new hyperspectral system based on the QE65000 USB spectrometer (Ocean Optics, Florida) to measure Δox-CCO, we performed measurements on forearm muscles of healthy adult humans

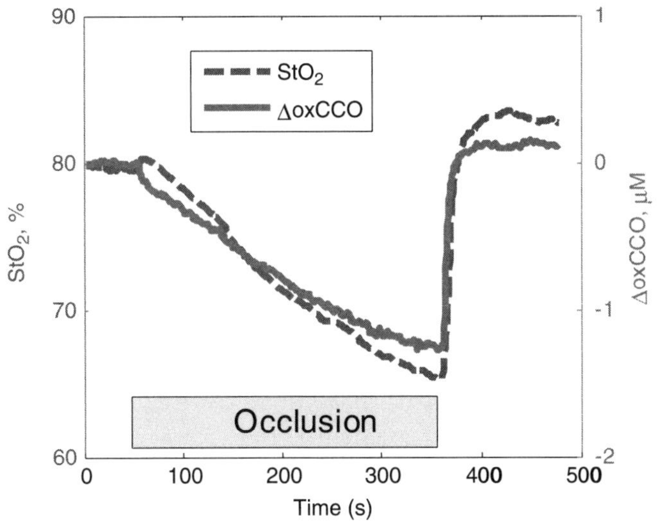

Figure 10.8 Changes of tissue oxygen saturation (StO_2) and Δox-CCO in the human forearm muscle during arterial occlusion.

during 5-minute arterial occlusions. The forearm occlusion was performed using the pressure cuff inflated at 150 mm Hg. The output of the halogen light source OSL2BIR (Thorlabs, United States) was delivered to the forearm using a 3-mm NIR fiber bundle (Edmund Optics, United States). The transmitted light was collected at 20 mm from the source by another 3-mm fiber bundle and projected onto the 200-μm slit of the spectrometer. Our custom Matlab algorithm used the nonlinear least squares curve fitting of the photon diffusion model to the hyperspectral data to perform the separation of ΔHbO_2, ΔHb, and Δox-CCO changes in two steps: (1) finding ΔHbO_2 and ΔHb using the 700- to 1000-nm waveband and (2) finding Δox-CCO using the 750- to 950-nm waveband. The results, which are shown in Fig. 10.8 can be easily interpreted: During the occlusion, Δox-CCO decreases, as does the tissue oxygen saturation, StO_2.

In a more recent study, we used hfNIRS to investigate how the healthy adult brain responds to complex events (including a secondary cognitive task) during driving. During complex motor and cognitive tasks, the prefrontal cortex (PFC) is highly activated.[74] In addition, this hairless area of the scalp minimizes the scattering and attenuation effect of the hairs. In our measurements, optodes were placed on the right and left PFC to measure ΔHbO_2, ΔHHb, and Δox-CCO during various simulated driving scenarios, and cranial activity of the PFC in complex driving tasks was assessed. The driving simulation setup consisted of a fully functional steering wheel, brake and accelerator pedals, an LCD monitor, and speakers.

For each subject, fNIRS was performed during two driving scenarios: one without distractions and another with distractions. Each of two blocks took approximately 10 minutes (depending on the speed of driving). Both driving scenarios included several routine driving tasks, such as right turns and left turns, and the turns combined with the presence of pedestrians or traffic at intersections. There was approximately 15 s of straight driving between all driving tasks. Each task was introduced by a recorded voice (similar to GPS commands) 5 seconds prior to the task (e.g., "Turn left at the traffic light"). In the second scenario, drivers were distracted during different driving tasks by general knowledge true/false questions (e.g., "A horse is greater than a puppy"), which they answered by pressing T or F buttons placed on the steering wheel. Driving tasks that were isolated and analyzed contained straight driving, straight driving with distraction, left turn in traffic, left turn in traffic with distraction, right turn, right turn with distraction, left turn on stop sign with distraction, and accident avoidance.

The positions of the optodes were determined using 10/20 EEG system (using an EEG standard cap). F7 and F8 locations were marked on each participant and optodes were placed on those locations at a 30-mm source-detector distance. Data analysis was performed using Matlab. The signal preprocessing included de-trending and de-noising the absorbance changes[70] using the Matlab independent component analysis algorithm FastICA[75] within the wavelength range of 715–900 nm. The number of independent components was reduced to five. These remaining five independent components were back-transformed to the original time-spectral domain using the inverse of the mixing matrix obtained using FastICA. All of these steps were taken to remove the baseline and to de-noise the absorbance changes. Subsequently, the fit of the absorbance changes was performed by modeling absorbance using the analytical solution to the diffusion equation for the semi-infinite medium. The nonlinear fit was performed using the Matlab function "lsqcurvefit," where the parameters of the fit were $\Delta[HbO_2]$, $\Delta[HHb]$, and $\Delta[\text{ox-CCO}]$. The baseline values of the absorption coefficient were computed assuming the baseline values $[HbO_2] = 60$ μM, $[HHb] =$ 20 μM, and water fraction of 85%. The reduced scattering coefficient as function of the wavelength was modeled using the power law.[71]

Seven driving tasks (events) were considered for analysis: straight with distraction, right turn, right turn with distraction, left turn in traffic, left turn in traffic with distraction, left turn on stop sign with distraction, and accident avoidance. For each participant and for each event, the changes in the chromophore concentrations $\Delta[HbO_2]$, $\Delta[HHb]$, and $\Delta[\text{ox-CCO}]$ were isolated within a 10-s time interval right after the onset of the event.

For each isolated time interval corresponding to the events, the initial value (at the very first moment of the event) was subtracted from the corresponding data trace. For each event, the event-related changes from the onset of each event were subject to the group statistical analysis. In addition, to assess the effect of distraction during driving on PFC activity, each distracted condition was compared to the same but non-distracted condition. For example, event-related changes in the PFC were compared for the right turn and distracted right turn events to assess how distraction, as a secondary cognitive task, affects PFC activity.

Statistical analysis including means, standard deviations, standard errors, paired-sample *t*-tests, and one-way analysis of variance (ANOVA) were performed in Matlab (R2013b version 8.2, Mathworks, United States).

After the isolation of the 10-second time intervals for each of the eight event types, all results were averaged over 16 subjects. The standard deviation and standard error were also calculated and are shown in the figures as error bars. To evaluate the statistical significance of the event-related changes, the paired-sample *t*-test was performed between the very first moment of the stimulation and the moment of the maximum changes in the group-averaged traces (5–10 s after the onset of the event). Statistical significance was assumed at $p < 0.05$.

To compare the event-related changes between distracted and non-distracted conditions, one-way ANOVA was conducted. All data for all of events (except accident avoidance) were put in two groups of "distracted" and "non-distracted," and then one-way ANOVA was performed between groups; mean changes in $\Delta[HbO_2]$, $\Delta[HHb]$, and $\Delta[\text{ox-CCO}]$ for each group and *p*-values for the difference between corresponding means were obtained.

We found that the right and left PFC were both significantly activated ($p < 0.05$) during all distracted driving conditions, similar to our fMRI findings, Table 10.1. All activations induced by distractions were identified by increasing HbO_2 concentrations, and all were accompanied by decreasing HHb concentrations and increasing ox-CCO concentrations. Among the various events, "left turn in traffic" resulted in least consistent changes between participants (Table 10.1).

These results dovetail nicely with our previous fMRI findings using similar driving scenarios.[76] However, to better understand the relationships during activation of different cortical regions, in the future we need to perform NIRS measurements on different positions simultaneously. Overall, our findings prove that fNIRS is a useful technique to measure brain activity during actual

Table 10.1 Event-related changes in concentrations of HbO_2, HHb, and ox-CCO. The trend of the changes that were statistically significant is shown by arrows and the changes that were not statistically significant are shown as ≈.

	HbO_2		HHb		ox-CCO	
Driving condition	**Right PFC**	**Left PFC**	**Right PFC**	**Left PFC**	**Right PFC**	**Left PFC**
Straight + distraction	↗	↗	↘	↘	≈	≈
Left turn + traffic + distraction	≈	↗	↘	↘	↗	↗
Right turn + distraction	↗	↗	↘	≈	≈	↗
Left turn + stop + distraction	↗	↗	≈	↘	↘	↗
Left turn + traffic	↗	≈	≈	≈	≈	≈
Right turn	↘	↘	↗	↗	↘	≈

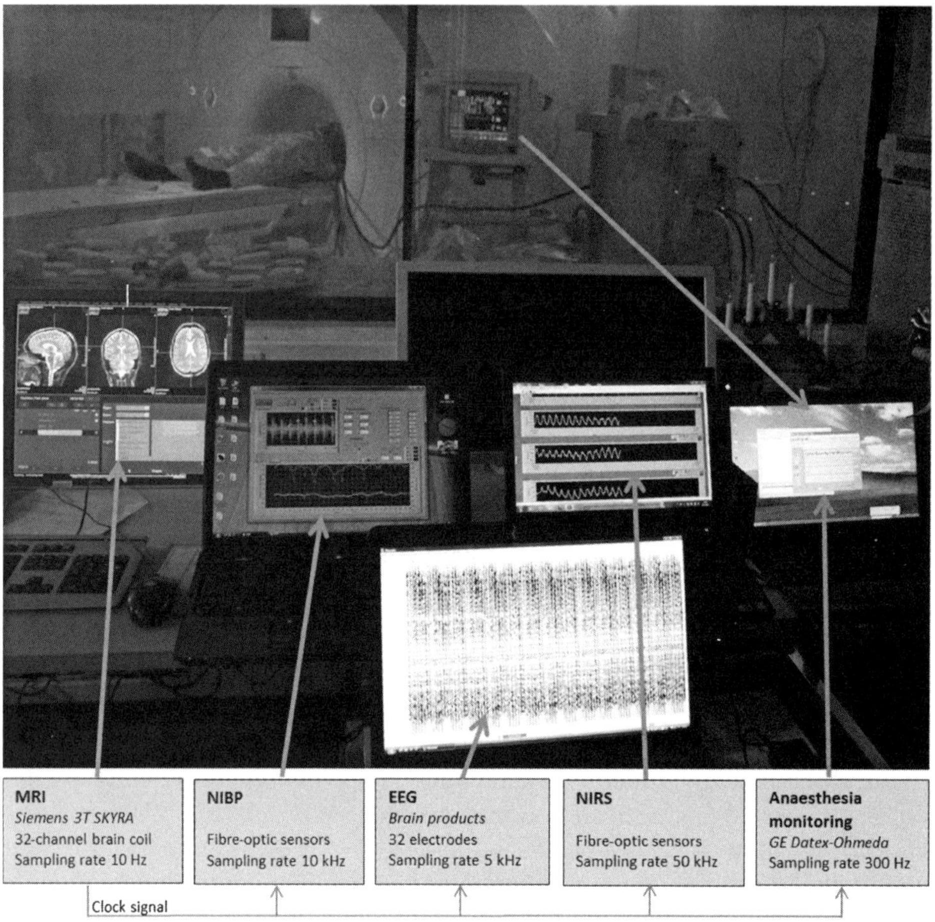

Figure 10.9 Multimodal measurement. The anesthesia monitoring device on the right is measuring ECG, CO_2 expiration, oxygen saturation (SpO_2), and blood pressure by cuff at the desired interval of 1 min or more.[78]

driving; fNIRS can also be developed to identify which cognitive or noncognitive tasks during driving (e.g., eating, drinking, speaking, texting) may result in distracted driving conditions (activation of PFC that may sacrifice visual attention).

10.4.3 Brain imaging utilizing fNIRS combined with seven modalities

The multimodal setup shown in Fig. 10.9, which is found in Medical Research Center of Oulu, Finland, is able to noninvasively, simultaneously, and continuously measure blood pressure fluctuation and pulse propagation, changes in respiration, brain and heart electrical activity, and brain hemodynamics. Hence, it provides for the possibility of obtaining precise assessments of the complex dynamics and interactions between these physiological variables.[62] For fMRI, the setup utilizes an ultrafast imaging technique, MREG, that is capable of providing a sampling rate approximately 20–25 times faster (10–16 Hz) than commonly used MRI sequences.[14,77] This makes unaliased sampling of BOLD signals possible.

Because so many modalities are part of this setup, several sensors are noninvasively attached to the subject. Figure 10.10 illustrates the sensor placement commonly used in the setup. For measuring blood flow pulsations and to continuously estimate blood pressure, one of the developed sensors is a specially designed fiber-optic acceleration sensor,[63] while the other is an optical fiber probe utilizing photoplethysmography (PPG). Simultaneous use of these two measurement methods enables extensive monitoring of the propagation of blood pressure waves in different areas of the human body.[79]

The use of combined, multimodal imaging can yield much higher precision of the acquired signal due to simultaneous data verification. For example, if one aims to develop robust biomonitoring equipment that generates data of the highest precision, acquiring data simultaneously enables the depiction of the needed markers with required specificity. Moreover, detection of new physiological phenomena can be made more robust by acquiring simultaneous data that can exclude the presence of mere artefactual data from just two sampled signals. Figure 10.11 illustrates synchronous oscillations of data coherence that currently have an unknown origin.[62] The use of multiple simultaneous measurements enables clear depiction of the phenomenon and helps to trace its origin.

10.4.3.1 Multimodal monitoring of blood-brain barrier disruption

The multimodal setup is currently being utilized in the study of such brain disorders as epilepsy, narcolepsy, and autism spectrum disorder (ASD), but it also provides a basis for the development of new monitoring methods for a range of clinical uses. Emerging applications include, for example, monitoring treatment of central nervous system (CNS) lymphoma that involves blood–brain barrier disruption (BBBD).

The so-called blood–brain barrier (BBB) protects the brain from substances in the blood circulation that might otherwise readily pass into the brain tissue. Many neuropsychiatric diseases, such as Alzheimer's disease,[80] as well as infections, edema, and tumors may potentially be affected by abnormalities of the BBB.[81] Hence, determining the condition of BBB is of great interest to researchers and clinicians alike. Although an important protection mechanism for the brain, the BBB also presents a major obstacle to delivery of drugs to the brain, which is especially problematic in the treatment of brain tumors.

Currently, only few clinical centers in the world are able to routinely disrupt the human BBB. One method involving BBBD, which was developed at the University of Portland, is being exploited at the Oulu University Hospital to treat CNS lymphoma. We have recently pioneered a direct current

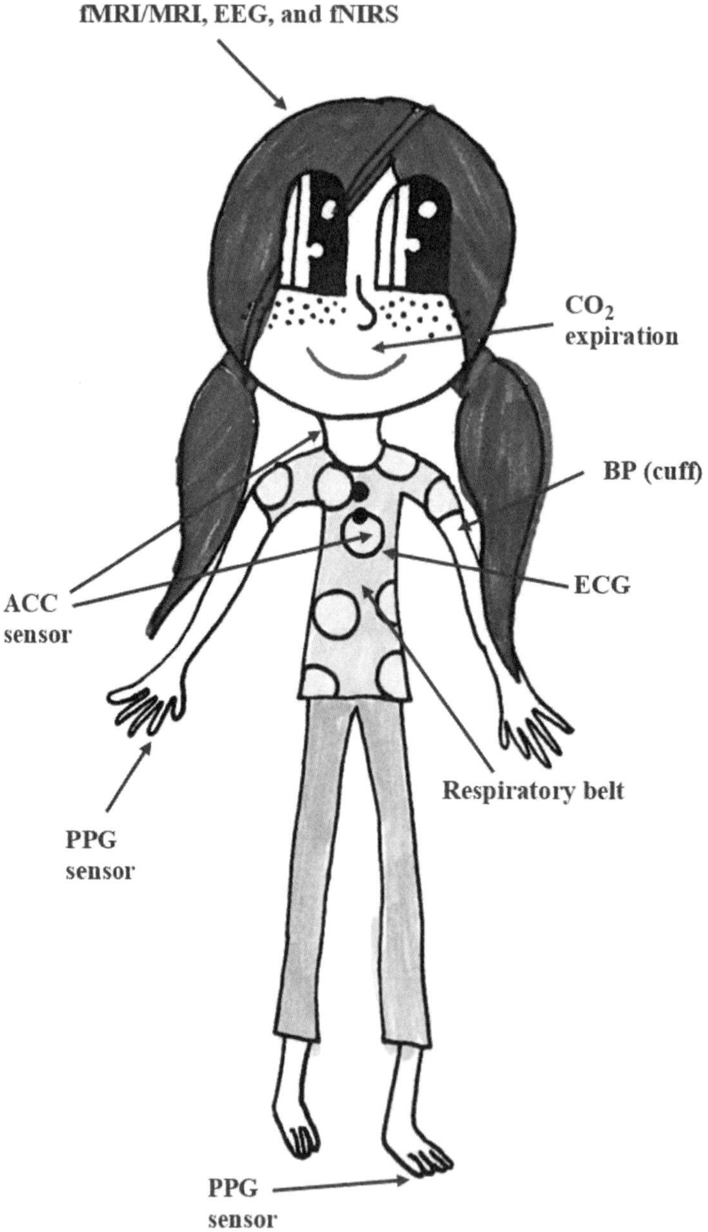

Figure 10.10 Placement of sensors. Simultaneous measurements include fNIRS, fMRI, NIBP, EEG, ECG, CO_2 expiration, and various sensors to measure blood pressure pulse propagation and pulse shape in small vessels. All sensors are MR compatible. Our experiences show that accelerometer (ACC) sensors are more sensitive to skin vibrations caused by large arteries, while PPG sensors mainly respond to blood flow pulsations in small vessels within a skin tissue. We usually place the acceleration sensors on the subject's chest and neck, and the PPG sensors on the peripherals—for example, on the fingertips and toes. These sensors also measure peripheral SpO_2. (Picture drawn by Auri Myllylä and colored by Anna Myllylä.)

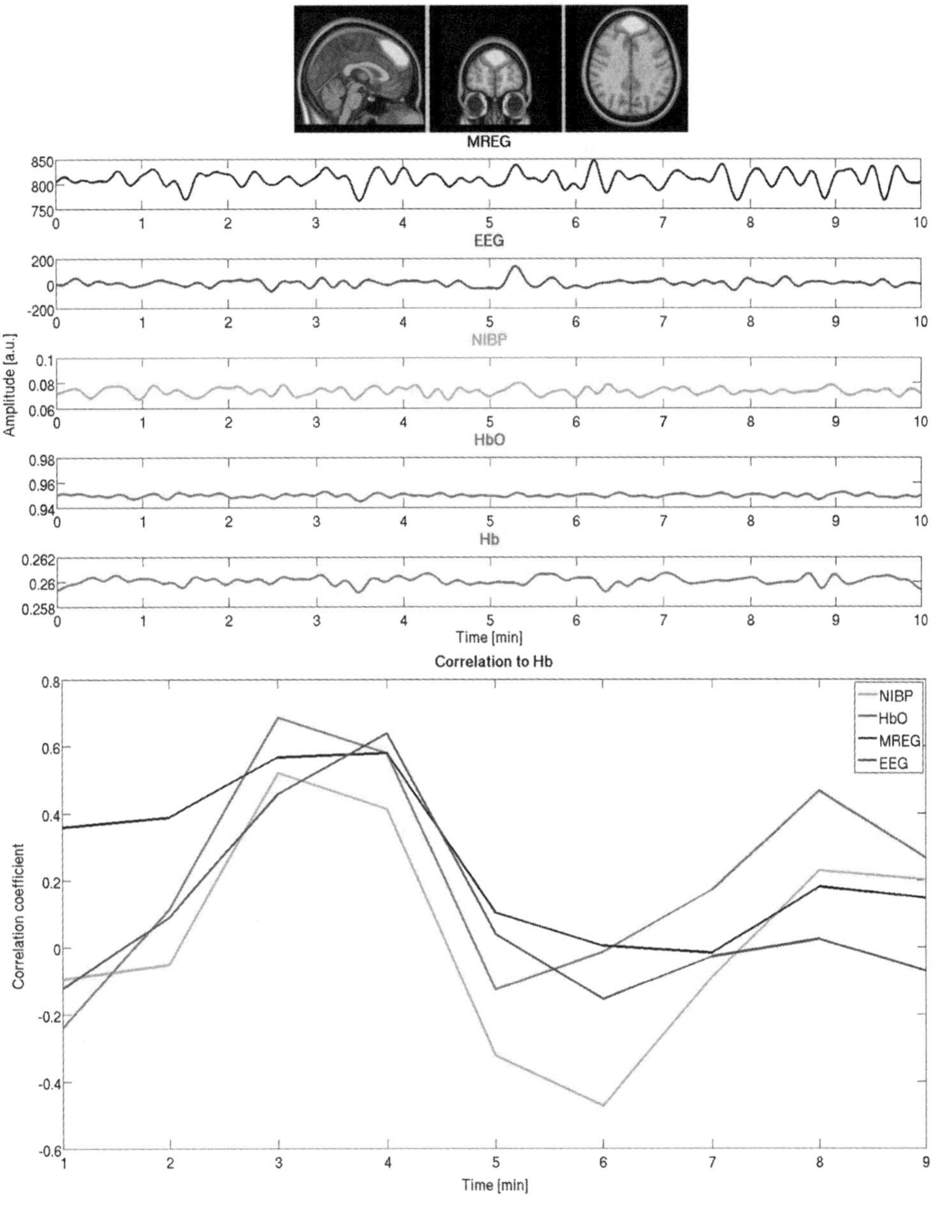

Figure 10.11 Synchronous human multimodal hepta-scan data measured during 3T MRI scanning. (top) Hot spots (bright colors) in a functional brain network belonging to the upmost anterior part of the default mode brain network (DMN) in MNI space. (middle) Time-domain activity of fMRI MREG data of the DMN in black sampled in 10 Hz. As seen below the MREG graph, the best-fit independent component is EEG, followed by NIBP, HbO_2, and HHb (here marked as Hb) measured over the DMN in the forehead using fNIRS. (bottom) Correlation of the HbO_2 signal to all other signals in sliding windows over 9 min, showing synchronous modulation of the measured signal correlation and indicating the strong additive power of multimodal imaging. (See color plates.)

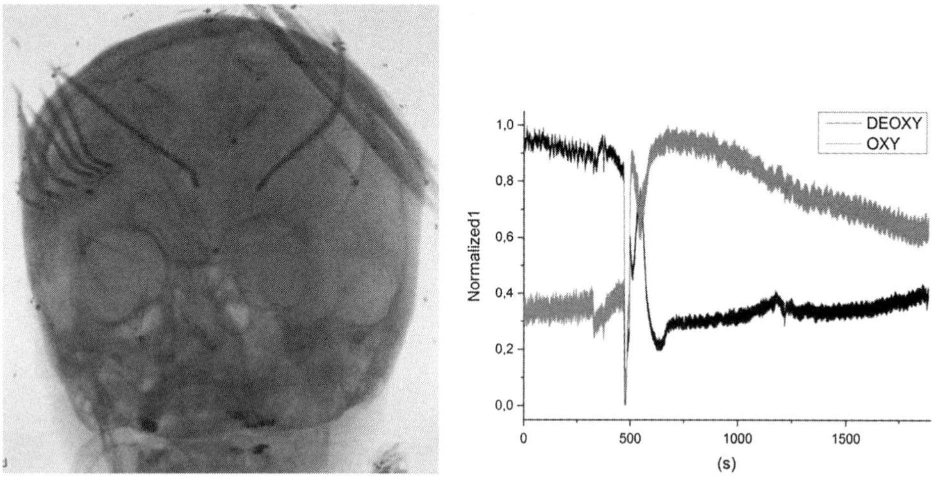

Figure 10.12 (left) A plain x-ray image from a digital subtraction angiography laboratory illustrating sensor positions during BBBD treatment. Clearly visible in black are optical fibers and EEG electrodes attached to the head, illustrating the positions at which the sensors are measuring. (right) The NIRS signal shows a robust change in amplitude after infusion of mannitol (time point of approximately 500 s) to disrupt the BBB. (See color plates.)

electroencephalographic (DC-EEG) and NIRS method for quantitative monitoring of the BBB opening *in vivo* for humans during CNS lymphoma treatment. For instance, robust changes in NIRS signals immediately after mannitol infusion that causes BBBD can be observed, as can be seen in Fig. 10.12.

The fNIRS/EEG multimodal method may potentially be used in monitoring of BBB status during the treatment. In the future, this method may also be combined with ultrafast MRI to monitor BBB integrity from the deep brain structures.

10.5 Enhancement of In-Depth NIRS Imaging

10.5.1 Transmittance of cranium tissues in the NIR

Figure 10.13 illustrates a multilayered head tissue structure—skin/scalp, skull, cerebral spinal fluid (CSF), gray and white matter—as well as photon trajectory maps (logarithmic color map) for the 905-nm wavelength 905 and a source–detector separation of 40 mm (b).[40]

As the NIR optical transmittance of human head tissues is of great importance for NIR spectroscopy, imaging, and treatment, such measurements were done in Ref. 83 for 20 human cadaver heads at 5 domains: forehead, left and right temples, occiput, and vertex. The researchers examines skulls with a shaved scalp and those free of scalp. The experimental setup is depicted in Fig. 10.14. A collimated, 1-W diode laser (810 nm) with a beam approximately 30 mm in diameter was used as a light source. A charge-coupled device (CCD) camera with a set of neutral optical filters was used to measure the spatial distribution of the transmitted radiation.

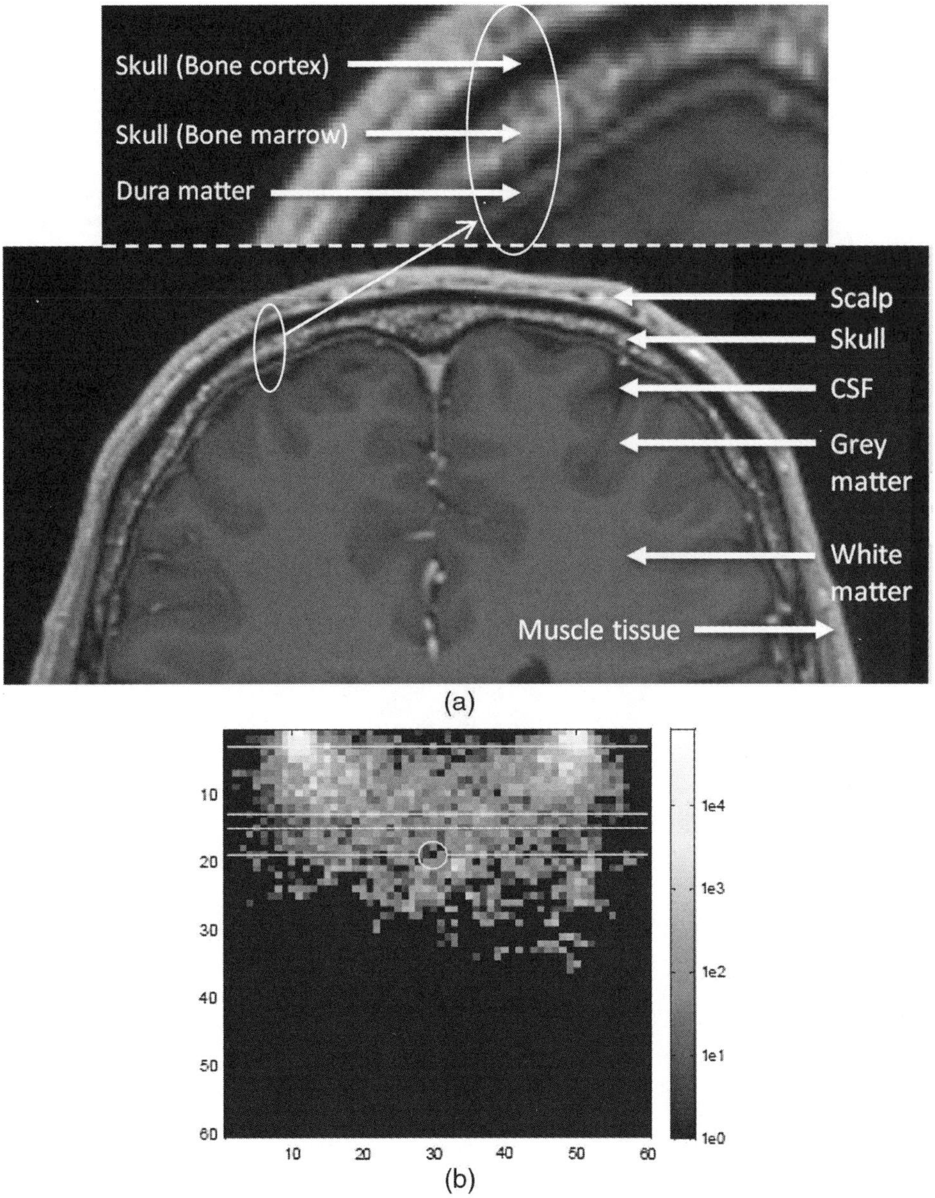

Figure 10.13 (a) Head tissue structure: skin/scalp (thickness 0.3 cm, $\mu_s' = 14.4\ \text{cm}^{-1}$), skull (thickness 1 cm, $\mu_s' = 15.0\ \text{cm}^{-1}$), cerebral fluid (CSF) (thickness 0.2 cm, $\mu_s' = 10^{-3}\ \text{cm}^{-1}$), gray matter (thickness 0.4 cm, $\mu_s' = 24.0\ \text{cm}^{-1}$), and white matter (thickness 2 cm, $\mu_s' = 78.0\ \text{cm}^{-1}$). (b) Photon trajectory maps (logarithmic color map) for the wavelength 905 nm and source-detector separation of 40 mm. White lines indicate the boundaries of tissue layers.[40]

The absolute value of power at the center of light distribution (maximum power) was measured using a 1-mm-diameter fiber with cosine corrector and a USB2000 (Ocean Optics) spectrometer. Thickness of a sample measured by a slide gauge was averaged for five sampling points. Sample temperature was

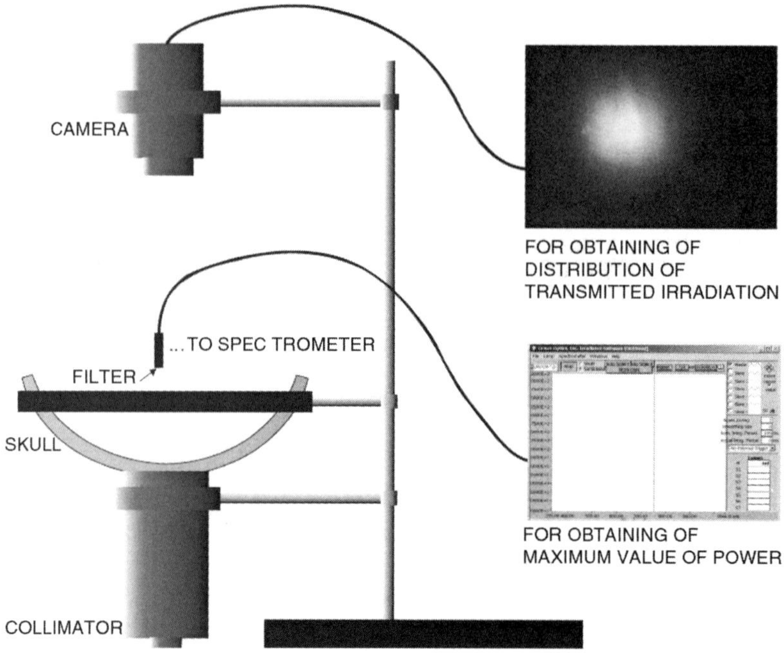

Figure 10.14 Experimental setup and method of measurements.[83]

measured by a contact temperature probe with a sensing area of 1 mm^2 at the surface of the skin and the skull within the center of irradiated field immediately after exposure. A sequence of CCD images (25 frames) within each domain of interest (temple, forehead, occiput, and vertex) was used for averaging. Subtraction of background (laser switched off) and image scaling were also done. The maximum power in the center of the transmitted light distribution was measured five times by the preliminary calibrated spectrometer, and the light distribution was normalized to the mean value. The resulting power density distributions were calculated for measurements carried out for 20 cadaver heads (forehead, left and right temples, occiput, and vertex). Data for transmittance via sample thickness were collected for 100 sampling points. Measurements were performed both for the skulls with shaved scalp and for those free of scalp.

Figure 10.15 shows the dependency of transmittance on sample thickness for both cases. The transmittance varied from 0.5% to 5% for skulls with scalp, and from 1% to 16% for skulls with removed scalp. For the incident power density of 68 mW/cm^2, the transmitted power density varied from 0.34 mW/cm^2 to 3.4 mW/cm^2, and from 0.68 mW/cm^2 to 9 mW/cm^2, respectively. It is evident that the much thinner skin reduces transmittance approximately similarly to how the much thicker skull does so. These experimental results well fit to Monte Carlo modeling.[83]

The scalp consists of both skin and underlying soft tissues, such as fat, connective, and muscular tissues. For a 60-year-old Caucasian female,

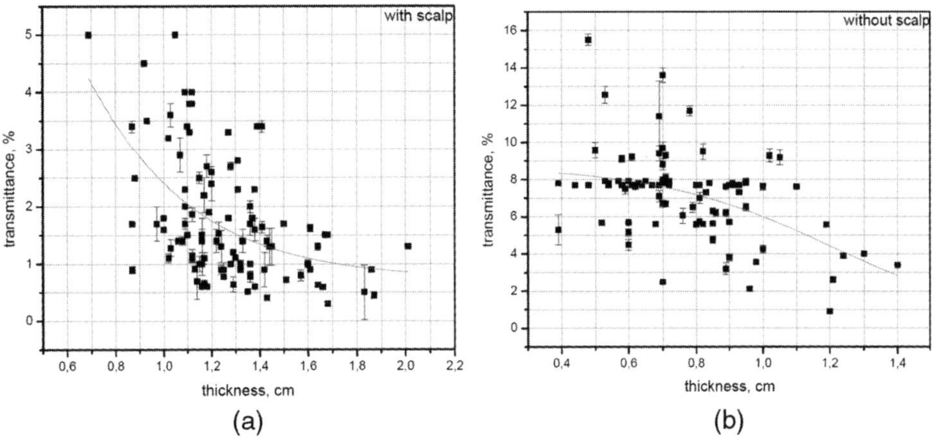

Figure 10.15 Dependence of optical transmittance on the sample thickness.[83]

thickness of the head skin was estimated as 2.5 ± 0.9 mm; it was thinner over the frontal and occipital lobes. The underlying soft tissues at the top of a head had an estimated thickness of 2.9 ± 1.1 mm. Skull thickness was estimated to be 6.5 ± 1.7 mm, and was 1–2 mm thinner at the parietal lobes.[84]

Transcranial light measurements of unfixed human cadaver brains allowed for quantification of light penetration into the CNS tissues.[85] Laser light at 808 nm penetrated the scalp, skull, meninges, and brain to a depth of approximately 40 mm, with an effective attenuation coefficient for the system of 2.22 cm^{-1}. No differences were found between transportation of the pulse and continuous-wave (CW) laser light. The 808-nm-wavelength light demonstrated superior CNS tissue penetration compared to the 660- or 940-nm wavelengths, owing to less absorption and scattering of the light.

Optical clearing of the skin and underlying soft tissues, as well as the skull and cerebral membrane, may have a further significant impact on NIR light transmittance, both for the incident and detected light, in the course of NIRS study. This issue is discussed in detail in the next subsection.

10.5.2 Optical clearing of tissues

The scattering coefficient of a tissue depends on the refractive index (RI) mismatch between the weighted-average RIs of interstitial fluid (ISF)/cell cytoplasm ($\bar{n}_0$) and scatterers ($\bar{n}_s$). For a tissue model that can be represented in the visible and NIR range as a system of noninteracting "optically soft" Mie scatterers with a mean diameter $2a$ ($g > 0.9$, $5 < (2\pi a/\lambda) < 50$, $1 < m < 1.1$), the reduced (or transport) scattering coefficient μ_s' is described as[86,87]

$$\mu_s' = \sigma_s' \rho_s = 3.28\pi a^2 \rho_s \left(\frac{2\pi \bar{n}_0 a}{\lambda_0}\right)^{0.37} (m-1)^{2.09}, \qquad (10.1)$$

where $\sigma_s' = \sigma_s(1 - g)$, σ_s is the scattering cross section, ρ_s is the volume density of the scattering centers, λ_0 is the wavelength of the incident light, and $m = \bar{n}_s/\bar{n}_0$ is the relative index of refraction (RI) of the scattering centers $\bar{n}_s$ and the background medium $\bar{n}_0$.

This tissue model is applicable to any soft tissue, including skin and muscle. To provide optical clearing, the RI of the OCA should be higher than the RI of the ISF and cell cytoplasm. At optical clearing, impregnation of a tissue by an OCA causes diffusion of OCA molecules inside tissue—that is, reduction of RI mismatch ($m \to 1$)—and correspondingly leads to a decrease in the scattering coefficient ($\mu_s \to 0$). The single-scattering directness is also sensitive to RI matching: It increases with the better matching conditions ($g \to 1$). As a result of these two processes, $\mu_s' = (1 - g)\mu_s$ becomes a very sensitive function of RI matching.[86–90] Therefore, the transport mean-free path of a photon, which is defined as

$$l_{\text{tr}} = \frac{1}{\mu_a + \mu_s'}, \tag{10.2}$$

and is a key parameter for NIRS probing depth, could be increased significantly at RI matching.

With optical clearing, not just index matching can be important; that is, other mechanisms of tissue clearing also may be essential. For instance, the action of hyperosmotic OCAs or strong mechanical compression may lead to a change in density and the packing ability of scatterers due to dehydration and corresponding tissue shrinkage.[86,88–90]

Skin has a complicated structure and performs a protective function by preventing pollutions and microorganisms from entering the body. The outermost cellular layer of skin is the epidermis, which consists of a layer of dead cells, known as the stratum corneum (SC), and a few layers of living cells. The excellent diffusional resistance of the SC makes the transdermal delivery of immersion agents and water lost by skin difficult. *In vivo*, the diffusion of water across the SC is a passive process that can be modified through the application of hyperosmotic OCAs. The established concentration gradient of water molecules is characterized by the parameter known as transepidermal water lost (TEWL).

The dermis is the next, thicker layer of the skin after the SC; its fibrous tissues can be easily impregnated by an OCA. The subcutaneous tissue of the head contains a small fat-cell layer, which is much less amenable to penetration by diffusing molecules than is the dermis. Experimental studies of optical clearing of the skin using glycerol, glycerol-water solutions, glucose, propylene glycol, oleic acid, DMSO, sunscreen creams, cosmetic lotions, gels. and pharmaceutical products are summarized in Refs. 86 and 88–90.

Figure 10.16 presents integrating sphere spectral measurements over the wavelength range of 400–1700 nm for porcine skin samples before and after

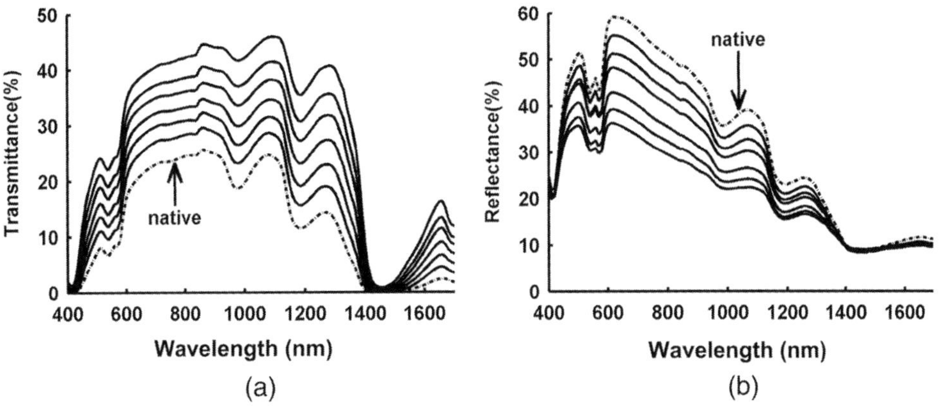

Figure 10.16 (a) Total transmittance and (b) diffuse reflectance of porcine skin sample measured with integrating sphere before and after treatment with the OCA [1,2-propylene glycol (1,2-propanediol)] at time intervals of 0, 10, 20, 30, 40, 50, and 60 min [from bottom to top in (a)] and [from top to bottom in (b)].[91]

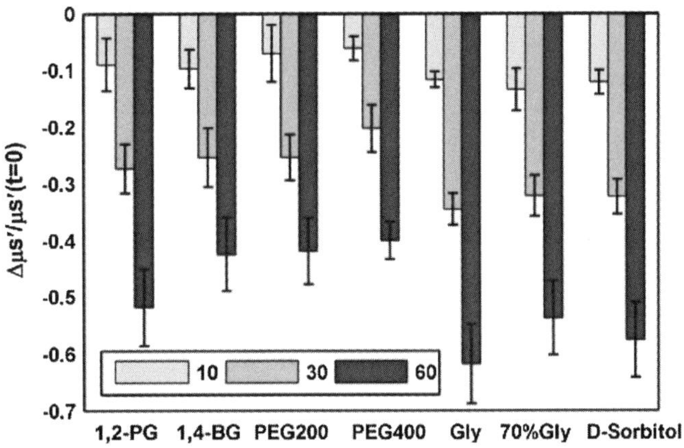

Figure 10.17 Relative decrease in the reduced scattering coefficient reconstructed from measurements of total transmittance and diffuse reflectance of porcine skin samples for the wavelength 760 nm after 10-, 30-, and 60-min treatment with different OCAs. 1,2-PG: 1,2-propylene glycol (1,2-propanediol); 1,4-BG: 1,4-butylene-glycol (1,4-butanediol); PEG-200: polyethylene glycol 200; PEG-400: polyethylene glycol 400; Gly: glycerol; and 70% Gly: 70% glycerol–water solution.[91]

treatment with 1,2-propanediol (1,2-propylene glycol).[91] After treatment by the agent during 60 min, transmittance was significantly increased and reflectance correspondingly decreased for the entire spectral range. The reduced scattering and absorption coefficients extracted from these measurements allow one to describe optical clearing mechanisms, one of which is connected with tissue dehydration. Experimental data for a few often-used OCAs, presented in Fig. 10.17, demonstrate a strong correlation between optical clearing efficiency and tissue dehydration estimated using the water absorption peak at 1450 nm. For small treatment times of approximately 10 min, the efficiency of a 70% glycerol solution is better than that of a 100% glycerol solution—a finding that reflects the lower viscosity of the less-concentrated solution. Topical application of glycerol, glucose, x-ray contrasts, propylene glycol, and cosmetic lotions, and gels to human skin *in vivo* also made the skin more transparent for NIR for a few minutes or hours.[86,88–90]

The internal structure of the muscle fiber (muscle cell) consists of a group of orientated myofibrils with many attached mitochondria that is surrounded by sarcoplasm (muscle cell cytoplasm); the sarcoplasm consists of mainly water and some dissolved salts and minerals. Due to the heterogeneous composition of skeletal muscle, its refractive index profile presents localized discontinuities at the boundaries between the myofibrils, mitochondria, and sarcoplasm.

For natural rat muscle, the global, the refractive index is 1.398, water content is 0.756, solid part content is 0.244, and index of refraction is 1.584; thus scatterers, due to some hydration of their material, have a lower index of refraction equal to 1.41.[86] Skeletal muscle tissue can be subjected to optical clearing to reduce such a refractive index mismatch.[92–94]

10.5.3 OCA diffusion

With the objective of studying OCA diffusion within skin or muscle, a simple method to estimate the diffusion time and diffusion coefficient of an OCA was proposed.[93–97] This method is based on measuring the temporal spectra for collimated transmittance of tissue samples under treatment. The analysis of the diffusion time as a function of OCA concentration in the solution allows for estimating the true diffusion coefficient of OCA in the tissue.[93–97]

Considering that OCA molecules undertake a free diffusion when entering into a tissue slab with thickness d, and that the agent diffuses through both surfaces of the slab at the same time, the time-dependent collimated transmittance of this slab is described as[86]

$$T_c(t,\lambda) = T_c(\lambda)\left\{1 - \exp\left(-\frac{t}{\tau_{\mathrm{OCA/W}}}\right)\right\}, \tag{10.3}$$

where $\tau_{\mathrm{OCA/W}}$ represents the diffusion time of the combined fluxes in each treatment [water (W) out and OCA into the tissue]:

$$\tau_{\mathrm{OCA/W}} = \frac{d^2}{\pi^2 D_{\mathrm{OCA/W}}}, \tag{10.4}$$

$D_{\mathrm{OCA/W}}$ is the diffusion coefficient of the combined fluxes.

To perform the study, the skeletal muscle from the abdominal wall of the rat species Whistar Han was selected.[94] After sacrificing one animal, the entire muscle block from the abdominal wall was retrieved. From that block and using a cryostat, samples with an approximated circular form of 10-mm diameter and 0.5-mm thickness were sliced. The collimated transmittance was measured at every second for whole spectra from 400 to 1000 nm (for each second, 37 spectra were averaged) for period of 30 min. Figure 10.18 presents the temporal dependencies of spectral components from the visible to near-infrared range of collimated transmittance of *ex vivo* rat muscle tissue of thickness 0.5 mm, modeling the muscle layer of the scalp, following application of glucose-water solutions of 20% (a), 40% (b), and 50% (c).[94] For the treatment containing the lowest concentration of glucose (20%), diffusion was fast and collimated transmittance was increased up to 5.7-fold during 3 min; for the 40% glucose solution, it increased up to 3.6-fold during 15 min. For more-concentrated solutions, diffusion again became faster, but optical clearing efficiency was lower. For the 50% glucose solution, transmittance increased only up to 2-fold during 6 min.

Such complex behavior can be explained by using the concept of multifluxes induced in a tissue. In the first case, two water fluxes are created. One is directed inside the tissue and caused by differences between the water concentrations inside and outside the tissue. The other water flux is directed from the tissue and caused by the osmotic action of glucose. Because of the small water molecules' strong involvement in the diffusion process, the total time response for the diffusion process is small (fast diffusion). After the saturation point is reached [after 3 min; Fig. 10.18(a)], the tissue needs to compensate for its partial dehydration caused by the glucose's effects, so it starts to receive water from the solution to create a balance. Tissue swelling then results, with some decay of collimated transmittance for longer times. As the water content in the solution becomes similar to the free water content in the tissue, the water flux intensity decreases (it becomes approximately zero when the glucose concentration is approximately 40%) [Fig. 10.18(b)]. Thus, only glucose diffuses freely in the system, and the diffusivity parameters depend only on diffusion parameters of the larger-sized glucose molecules; in turn, we see the maximal value of the measured diffusion time [Eq. (10.4), Fig. 10.19]. However, at a higher glucose concentrations, a decrease of diffusion time $\tau_{G/W}$ is seen again [Fig. 10.18(c)] because water is once again involved in the diffusion process, with the main flux being directed from the tissue to the surrounding solution due to glucose-induced hyperosmolarity.

For a particular glucose and ethylene glycol (EG) concentration, the mean and standard deviation for the diffusion time, measured at 11 wavelengths, was calculated.[94,95] Figure 10.19 shows the mean diffusion time of glucose and EG as a function of their concentrations in the water solutions. This dependence has a maximum at a concentration of 40.5%. The maximum indicates that an aqueous solution with that concentration of glucose would have the same amount of water as the free water in the tissue. In such a case, no or very weak water flux occurs with only glucose or EG flux into the tissue.

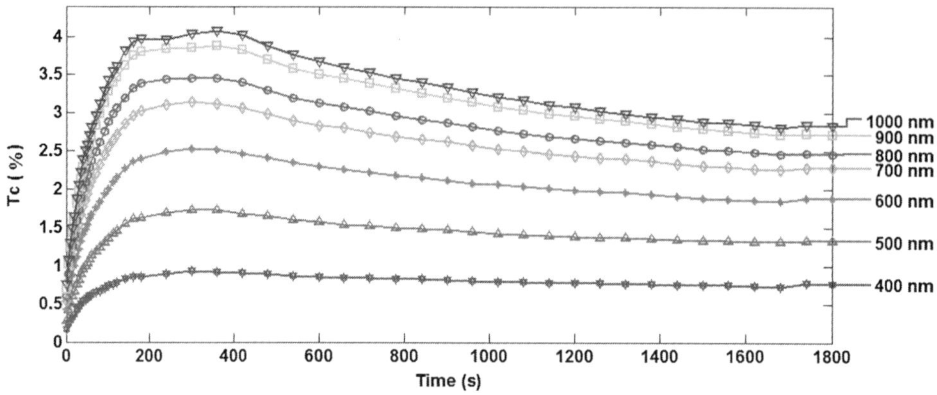

(a)

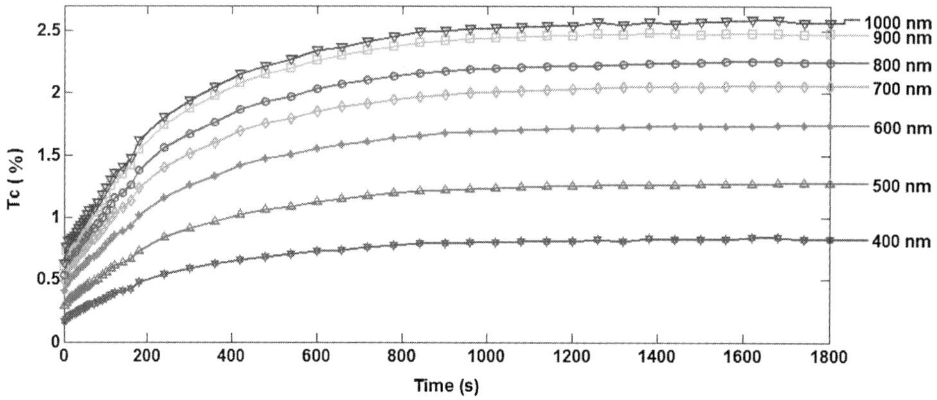

(b)

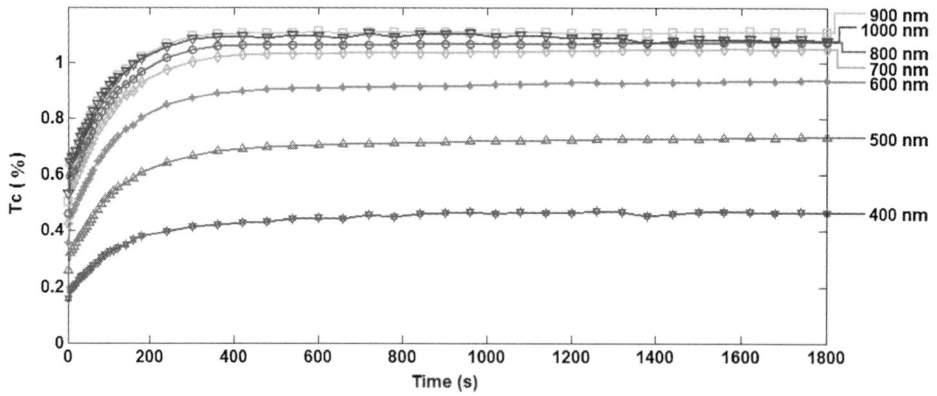

(c)

Figure 10.18 Temporal dependencies of spectral components from the visible to near-infrared range of collimated transmittance of *ex vivo* rat muscle tissue of thickness 0.5 mm modeling a muscle layer of the scalp, following application of three glucose-water solutions: (a) 20%, (b) 40%, and (c) 50%.[94]

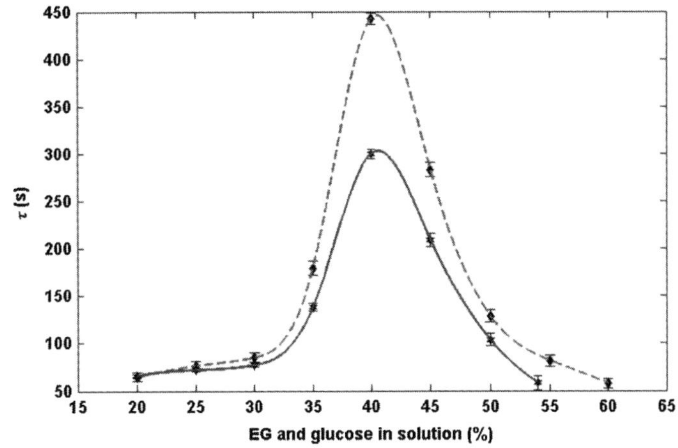

Figure 10.19 Mean diffusion time of glucose and ethylene glycol (EG) versus OCA concentration in solution. Glucose: solid line; EG: dashed line.[94,95]

Thus, the obtained diffusion time of 302.9 s for a 40.5% glucose solution is mostly due to glucose molecules' diffusion τ_G and represents the true diffusion time of glucose in the skeletal muscle. Using this value and accounting for the slab thickness of 0.5 mm, the diffusion coefficient of glucose in muscle was calculated with the help of Eq. (10.4):

$$D_G = \frac{d^2}{\pi^2 \tau_G} = \frac{0.05^2}{\pi^2 \times 302.9} = 8.36 \times 10^{-7}\ \mathrm{cm}^2 \cdot \mathrm{s}^{-1}. \tag{10.5}$$

From these measurements, the bound water content in muscle tissue can be evaluated as 75.6% (total water content in rat muscle tissue) −59.5% (amount of free water) = 16.1%. Thus, the remaining 16.1% of water is bound to the tissue proteins and does not participate in the dehydration mechanism of optical clearing—at least not on the time scale of 30 min.

Skull bones are typical representatives of hard tissues. Bone consists of both inorganic and organic matrices. The inorganic (mineral) matrix contains mostly hydroxyapatite [$Ca_{10}(PO_4)_6(OH)_2$], which is responsible for the compressive strength of bone. The main components of the organic matrix are collagen, other proteins, lipids, and blood cells. Normally bone is composed of 50–60% mineral components, 20–30% organic materials, 10–20% water, and less than 3% lipids.[86,98] It is important that porosity of the bones is on the order of 5–10%. At the microstructural level, cortical bone is organized into secondary osteons with a diameter of 200–300 μm; these osteons are composed of large vascular channels (50–90 μm diameter) surrounded by circumferential lamellar rings (3–7 μm thick), with the "cement lines" found at the outer boundary. At the nanostructural level, the lamellae are composed of organic type I mineralized collagen fibers (up to 15 μm in length, 50–70 nm in diameter) that are bound and impregnated with inorganic carbonated apatite nanocrystals (about 30 nm in length and width, 2–3 nm in thickness).

Water is of significant importance for the living bone and is one of its major components. Bone water is found at various locations and in different binding states—for example, bound to the mineral or organic components, or as free water (bulk water). The refractive index of the whole cranial bone at various stages of mineralization is in the range of 1.555 to 1.5.[86,98] Bone tissue components have the following refractive indices: apatite, greater than 1.623; hydrated collagen (type I), approximately 1.43; and lipids, approximately 1.45.[86]

The *in vitro* immersion optical clearing of human and porcine cranial bones under the action of propylene glycol and glycerol was for the first time demonstrated in the studies described in Refs. 98 and 99. The samples of human cranial bone were obtained from postmortem examinations. The thickness of each bone sample was measured with a micrometer in several points over the sample surface, and these measurements were then averaged. The thickness of the samples varied from 1.6 ± 0.1 to 5.0 ± 0.5 mm. The refractive index at 589 nm for propylene glycol and glycerol has been measured as 1.43 and 1.47, respectively.

Total transmittance and diffuse reflectance measurements have been performed in the 800–2000 nm wavelength range using a spectrophotometer CARY-2415 ("Varian," Australia) with an integrating sphere. For processing the experimental data to reconstruct reduced scattering and absorption coefficients with a fixed anisotropy factor ($g = 0.9$), the inverse adding doubling (IAD) method has been used.

Figure 10.20 demonstrates the change in the optical properties of human cranial bone in the spectral range 800–2000 nm before and after administration

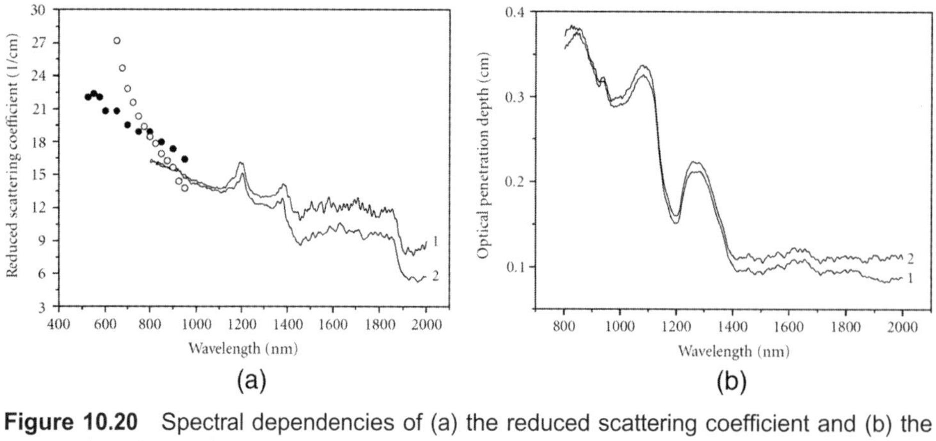

Figure 10.20 Spectral dependencies of (a) the reduced scattering coefficient and (b) the penetration depth of the human cranial bone sample measured (1) before and (2) after administration of glycerol during 1 hour.[98,99] The black circles represent data for untreated bone,[100] and the open circles represent untreated skull.[101]

of glycerol ($n = 1.47$). Reconstruction of absorption and reduced scattering coefficients was done with the IAD method, using the measured total transmittance and the diffuse reflectance of the bone samples. In the near-infrared range, the absorption bands of water at 978, 1192, 1464, and 1930 nm and the lipid band at 1745 nm were well observed. The administration of hyperosmotic solutions into cranial bone allows for effective reduction of its scattering properties and increases the penetration depth in the near-infrared range. The penetration depth was calculated by the diffusion approximation[86]

$$\delta = \frac{1}{\sqrt{3\mu_a(\mu_a + \mu_s')}}. \tag{10.6}$$

Optical clearing of the human dura mater—one of the membranes that separates the skull from the brain—also important because the dura mater is a highly scattering, rather thick membrane (0.4–1 mm thickness). The first results from *in vitro* experimental studies of the human and rabbit dura mater optical clearing under the action of mannitol, glucose, and glycerol solutions at various concentrations are presented in Refs. 102–104.

In vivo studies of glucose and glycerol action on the rabbit dura mater at the open cranium and epidural agent application also confirm the concept of effective optical clearing of fibrous tissue.[86,105] The total optical clearing was achieved very fast—50 s after tissue treatment by glycerol.

Figure 10.21 presents spectra for the transport (reduced) scattering coefficient μ_s' of a sample of human dura mater in the course of its

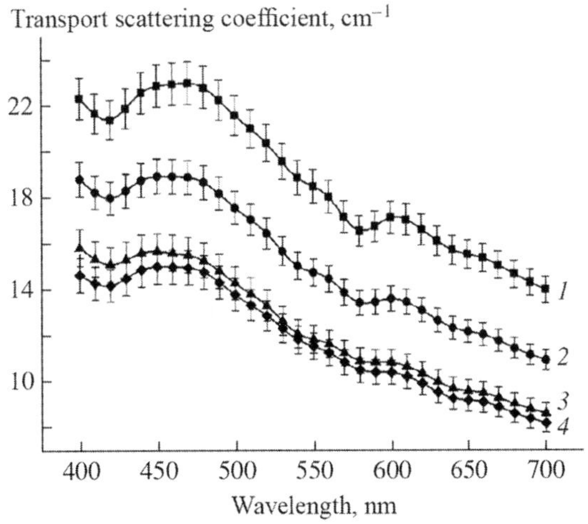

Figure 10.21 Spectra of the transport (reduced) scattering coefficient μ_s' of a sample of human dura mater in the course of its impregnation by a mannitol solution (0.16 g/mL), calculated by the IAD method from the measured diffuse reflectance and total transmittance spectra.[106] Vertical lines show the standard deviation. The time of action of the solution is (1) 0 min, (2) 2 min, (3) 5 min, and (4) 16 min.

impregnation by a mannitol solution (0.16 g/mL); the coefficient was calculated by the IAD method from the measured diffuse reflectance and total transmittance spectra.[106] Using such measurements for mannitol and glucose and the algorithm described in Ref. 104, the diffusion coefficients for a mannitol solution (0.16 g/mL) and a 40% glucose solution were found to be $D_M = (1.67 \pm 0.21) \times 10^{-6}\,cm^2\,s^{-1}$ and $D_G = (5.43 \pm 0.88) \times 10^{-6}\,cm^2\,s^{-1}$.[103,107]

10.5.4 *In vivo* optical clearing of skull

In vivo studies for mice demonstrating that the turbid mouse skull became transparent within 25 min after application of a skull optical clearing solution (SOCS) were done by Zhu et al. (Fig. 10.22).[108] The SOCS was composed from biocompatible agents including glucose, sorbitol, dimethyl sulfoxide, alcohol, laurinol, weak alkaline substances, and EDTA. As is evident in Fig. 10.22, the intact skull is turbid and the cortical blood vessels are hardly distinguishable. After treatment with SOCS, the skull becomes transparent and the cortical vessels can be observed clearly.

The minimum resolution diameter of a vessel was 14.4 ± 0.8 μm at imaging through the optically transparent skull after treatment with SOCS for 25 min [Fig. 10.22(e)] and 12.8 ± 0.9 μm for cranium window A with

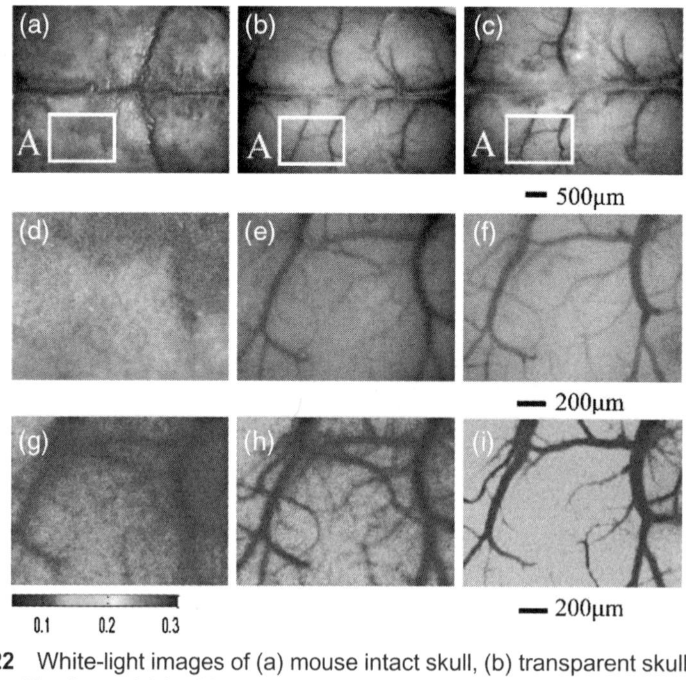

Figure 10.22 White-light images of (a) mouse intact skull, (b) transparent skull after SOCS treatment for 25 min, and (c) with removed rectangle area A. The corresponding magnified white images (d–f) and speckle contrast images (g–i) are within the rectangle area A shown in (a–c).[108] (See color plates.)

removed bone [Fig. 10.22(f)]. The speckle contrast images show that the blood flow of only a few large vessels can be measured hazily through the intact skull, while the blood flow distribution of the cortical microvessels can be distinguished clearly through the optically cleared skull, which is consistent with the view through the cranium window A. Therefore, it can be concluded that the IOC method enhances the contrast of both white-light and speckle images, which provides for nondestructive creation of a transparent cranial window for accessing high-resolution cortical structural and functional information.

10.6 Chapter Summary

Combination of fNIRS with neuroimaging techniques such as EEG, fMRI, and MEG provides many advantages. While fNIRS and fMRI are hemodynamic techniques, EEG and MEG are directly sensitive to the activity in the cortical neurons. Importantly, studying the mechanisms of neurovascular coupling requires combining electromagnetic- and hemodynamic-based imaging techniques. In this application, fNIRS can measure both HHb and HbO_2 as well as HbT changes, whereas fMRI is mainly sensitive to changes in HHb and blood flow. Moreover, fNIRS can be utilized in regression of low-frequency physiological noise from fMRI data. The combination of EEG with fNIRS provides accurate temporal resolution of electrophysiology and blood flow-related parameters and potentially represents a very powerful method for measuring these parameters, particularly in mobile applications. The significant advantage of MEG over EEG is the former's much better spatial resolution. However, because MEG is much more expensive than EEG, only a small number of studies have attempted combined MEG/fNIRS recordings.

Recently, a growing emphasis has been placed on the mechanisms that link cardiovascular and brain functions as well as the CBF. TCD provides a way to determine CBFV in intracranial arteries. fNIRS has the potential to complement TCD recordings of CBF by providing information about cerebral cortical tissue oxygenation in the brain region directly beneath the NIRS optodes. Hence, fNIRS could help, for instance, improve the spatial resolution of TCD by revealing whether a change in flow observed in the MCA is followed by a reduction in tissue oxygenation in the whole brain or only in specific areas.

Before NIRS can be fully exploited in the study of the brain, it is important to accurately determine the diffusive characteristics and penetration depth of near-infrared light in a multilayered head tissue. Importantly, this chapter presented transcranial light measurements showing that unfixed human cadaver brains allow for quantification of light penetration into the CNS tissues. Laser light at 808-nm has been found to penetrate the scalp, skull, meninges, and brain to a depth of approximately 40 mm. Moreover,

808-nm-wavelength light has demonstrated superior CNS tissue penetration compared to the 660- or 940-nm wavelengths, owing to its association with less absorption and scattering. Further, the presented results for OCAs show that optical clearing of skin and underlying soft tissues, as well as the skull and cerebral membrane, may have a further significant impact on near-infrared light transmittance both for the incident and detected light in the course of NIRS study.

Acknowledgments

Work done by VVT was supported by the Russia Science Foundation (Grant No. 14-15-0128). VVT is thankful to A.N. Bashkatov, E. A. Genina, V. I. Kochubey, D. Zhu, V. V. Lychagov, and L. Oliveira for their collaboration.

TM is grateful to Dr. Martin Walter and his colleagues for the opportunity to carry out multimodal imaging with magnetoencephalography at the Leibniz Institute for Neurobiology in Magdeburg.

References

1. S. Zago, R. Ferrucci, S. Marceglia, and A. Priori, "The Mosso method for recording brain pulsation: the forerunner of functional neuroimaging," *Neuroimage* **48**(4), 652–656 (2009).
2. L. F. Haas, "Hans Berger (1873–1941), Richard Caton (1842–1926), and electroencephalography," *J. Neurol. Neurosurg. Psychiatry* **74**(1), 9 (2003).
3. S. Tong and N. V. Thakor, *Quantitative EEG Analysis Methods and Clinical Applications*, Artech House, Norwood, MA (2009).
4. M. Hämäläinen, R. Hari, R. J. Ilmoniemi, J. Knuutila, and O. V. Lounasmaa, "Magnetoencephalography theory, instrumentation, and applications to noninvasive studies of the working human brain," *Rev. Mod. Phys.* **65**(2), 413–497 (1993).
5. P. T. Fox and M. E. Raichle, "Focal physiological uncoupling of cerebral blood flow and oxidative metabolism during somatosensory stimulation in human subjects," *Proc. Natl. Acad. Sci. U.S.A.* **83**(4), 1140–1144 (1986).
6. S. Ogawa, T. M. Lee, A. R. Kay, and D. W. Tank, "Brain magnetic resonance imaging with contrast dependent on blood oxygenation," *Proc. Natl. Acad. Sci. U.S.A.* **87**(24), 9868–9872 (1990).
7. B. Biswal, F. Zerrin Yetkin, V. M. Haughton, and J. S. Hyde, "Functional connectivity in the motor cortex of resting human brain using echo-planar MRI," *Magn. Resonance Med.* **34**(4), 537–541 (1995).
8. V. Kiviniemi, J. Kantola, J. Jauhiainen, A. Hyvärinen, and O. Tervonen, "Independent component analysis of nondeterministic fMRI signal sources," *Neuroimage* **19**(2), 253–260 (2003).

9. V. Kiviniemi, T. Starck, J. Remes, X. Long, J. Nikkinen, M. Haapea, J. Veijola, I. Moilanen, M. Isohanni, Y. Zang, and O. Tervonen, "Functional segmentation of the brain cortex using high model order group PICA," *Hum. Brain Mapp.* **30**(12), 3865–3886 (2009).
10. M. D. Greicius, G. Srivastava, A. L. Reiss, and V. Menon, "Default-mode network activity distinguishes Alzheimer's disease from healthy aging: evidence from functional MRI," *Proc. Natl. Acad. Sci. U.S.A.* **101**(13), 4637–4642 (2004).
11. C. F. Beckmann, M. DeLuca, J. T. Devlin, and S. M. Smith, "Investigations into resting-state connectivity using independent component analysis," *Phil. Trans. Roy. Soc. B Biol. Sci.* **360**(1457), 1001–1013 (2005).
12. S. M. Smith, P. T. Fox, K. L. Miller, D. C. Glahn, P. M. Fox, C. E. Mackay, N. Filippini, K. E. Watkins, R. Toro, A. R. Laird, and C. F. Beckmann, "Correspondence of the brain's functional architecture during activation and rest," *Proc. Natl. Acad. Sci. U.S.A.* **106**(31), 13040–13045 (2009).
13. T. Hugger, B. Zahneisen, P. LeVan, K. J. Lee, H. Lee, M. Zaitsev, and J. Hennig, "Fast undersampled functional magnetic resonance imaging using nonlinear regularized parallel image reconstruction," *PLoS One* **6**(12), e28822 (2011).
14. J. Assländer, B. Zahneisen, T. Hugger, M. Reisert, H. Lee, P. LeVan, and J. Hennig, "Single shot whole brain imaging using spherical stack of spirals trajectories," *Neuroimage* **73**(0), 59–70 (2013).
15. E. Lareau, F. Lesage, P. Pouliot, D. Nguyen, J. Le Lan, and M. Sawan, "Multichannel wearable system dedicated for simultaneous electroencephalography/near-infrared spectroscopy real-time data acquisitions," *BIOMEDO* **16**(9), 096014 (2011).
16. S. K. Piper, A. Krueger, S. P. Koch, J. Mehnert, C. Habermehl, J. Steinbrink, H. Obrig, and C. H. Schmitz, "A wearable multi-channel fNIRS system for brain imaging in freely moving subjects," *Neuroimage* **85**(Part I), 64–71 (2014).
17. A. Kleinschmidt, H. Obrig, M. Requardt, K. Merboldt, U. Dirnagl, A. Villringer, and J. Frahm, "Simultaneous recording of cerebral blood oxygenation changes during human brain activation by magnetic resonance imaging and near-infrared spectroscopy," *J. Cereb. Blood Flow Metab.* **16**(5), 817–826 (1996).
18. V. Toronov, A. Webb, J. H. Choi, M. Wolf, L. Safonova, U. Wolf, and E. Gratton, "Study of local cerebral hemodynamics by frequency-domain near-infrared spectroscopy and correlation with simultaneously acquired functional magnetic resonance imaging," *Opt. Express* **9**(8), 417–427 (2001).

19. G. Strangman, J. P. Culver, J. H. Thompson, and D. A. Boas, "A quantitative comparison of simultaneous BOLD fMRI and NIRS recordings during functional brain activation," *Neuroimage* **17**(2), 719–731 (2002).
20. D. J. Mehagnoul-Schipper, F. W. Bas van der Kallen, W. N. Colier, M. C. van der Sluijs, L. J. van Erning, O. Th, H. O. Thijssen, B. Oeseburg, W. H. Hoefnagels, and R. W. Jansen, "Simultaneous measurements of cerebral oxygenation changes during brain activation by near-infrared spectroscopy and functional magnetic resonance imaging in healthy young and elderly subjects," *Hum. Brain Mapp.* **16**(1), 14–23 (2002).
21. R. D. Hoge, M. A. Franceschini, R. J. M. Covolan, T. Huppert, J. F. Mandeville, and D. A. Boas, "Simultaneous recording of task-induced changes in blood oxygenation, volume, and flow using diffuse optical imaging and arterial spin-labeling MRI," *Neuroimage* **25**(3), 701–707 (2005).
22. T. J. Huppert, R. D. Hoge, S. G. Diamond, M. A. Franceschini, and D. A. Boas, "A temporal comparison of BOLD, ASL, and NIRS hemodynamic responses to motor stimuli in adult humans," *Neuroimage* **29**(2), 368–382 (2005).
23. J. Steinbrink, A. Villringer, F. Kempf, D. Haux, S. Boden, and H. Obrig, "Illuminating the BOLD signal: combined fMRI–fNIRS studies," *Magn. Reson. Imaging* **24**(4), 495–505 (2006).
24. X. Cui, S. Bray, D. M. Bryant, G. H. Glover, and A. L. Reiss, "A quantitative comparison of NIRS and fMRI across multiple cognitive tasks," *Neuroimage* **54**(4), 2808–2821 (2011).
25. L. Duan, Y. Zhang, and C. Zhu, "Quantitative comparison of resting-state functional connectivity derived from fNIRS and fMRI: a simultaneous recording study," *Neuroimage* **60**(4), 2008–2018 (2012).
26. Y. Tong and B. de Frederick, "Time lag dependent multimodal processing of concurrent fMRI and near-infrared spectroscopy (NIRS) data suggests a global circulatory origin for low-frequency oscillation signals in human brain," *Neuroimage* **53**(2), 553–564 (2010).
27. Y. Tong, P. R. Bergethon, and B. de Frederick, "An improved method for mapping cerebrovascular reserve using concurrent fMRI and near infrared spectroscopy with regressor interpolation at progressive time delays (RIPTiDe)," *Neuroimage* **56**(4), 2047–2057 (2011).
28. Y. Tong, L. M. Hocke, S. C. Licata, and B. de Frederick, "Low-frequency oscillations measured in the periphery with near-infrared spectroscopy are strongly correlated with blood oxygen level-dependent functional magnetic resonance imaging signals," *J. Biomed. Opt.* **17**(10), 106004 (2012).

29. Y. Tong, L. M. Hocke, L. D. Nickerson, S. C. Licata, and K. P. Lindsey, "Evaluating the effects of systemic low frequency oscillations measured in the periphery on the independent component analysis results of resting state networks," *Neuroimage* **76**(0), 202–215 (2013).
30. M. A. Mastanduno, F. El-Ghussein, S. Jiang, R. DiFlorio-Alexander, X. Junqing, Y. Hong, B. W. Pogue, and K. D. Paulsen, "Adaptable near-infrared spectroscopy fiber array for improved coupling to different breast sizes during clinical MRI," *Acad. Radiol.* **21**(2), 141–150 (2014).
31. T. Alderliesten, J. B. De Vis, P. M. A. Lemmers, F. van Bel, M. J. N. L. Benders, J. Hendrikse, and E. T. Petersen, "Simultaneous quantitative assessment of cerebral physiology using respiratory-calibrated MRI and near-infrared spectroscopy in healthy adults," *Neuroimage* **85**(0 Pt 1), 255–263 (2014).
32. T. H. Sander, A. Liebert, B. M. Mackert, H. Wabnitz, S. Leistner, G. Curio, M. Burghoff, R. Macdonald, and L. Trahms, "DC-magnetoencephalography and time-resolved near-infrared spectroscopy combined to study neuronal and vascular brain responses," *Physiol. Meas.* **28**(6), 651 (2007).
33. Y. Seki, T. Miyashita, A. Kandori, A. Maki, and H. Koizumi, "Simultaneous measurement of neuronal activity and cortical hemodynamics by unshielded magnetoencephalography and near-infrared spectroscopy," *J. Biomed. Opt.* **17**(10), 1070011–1070016 (2012).
34. American Society for Testing and Materials (ASTM) International, Designation: F2503–05. *Standard Practice for Marking Medical Devices and Other Items for Safety in the Magnetic Resonance Environment.* ASTM International, West Conshohocken, PA (2005).
35. T. Myllylä, J. Katisko, J. Harja, H. Sorvoja, and R. Myllylä, "EMI considerations for wireless data transfer in magnetic resonance imaging (MRI) operating room," *2nd International Symposium on Medical Information and Communication Technology (ISMICT'07)*, Oulu, Finland, 11–13 (2007).
36. H. S. S. Sorvoja, T. S. Myllylä, R. A. Myllylä, M. Kirillin, M. Yu, E. A. Sergeeva, A. A. Elseoud, J. Nikkinen, O. Tervonen, and V. Kiviniemi, "Non-invasive, MRI-compatible fibreoptic device for functional near-IR reflectometry of human brain," *Quantum Electronics* **40**(12), 1067–1073 (2011).
37. B. Z. Allison, S. Dunne, R. Leeb, J. D. R. Millán, and A. Nijholt, *Towards Practical Brain-Computer Interfaces: Bridging the Gap from Research to Real-World Applications*, Springer Science & Business Media, Berlin (2012).
38. R. Sitaram, H. Zhang, C. Guan, M. Thulasidas, Y. Hoshi, A. Ishikawa, K. Shimizu, and N. Birbaumer, "Temporal classification of multichannel

near-infrared spectroscopy signals of motor imagery for developing a brain–computer interface," *Neuroimage* **34**(4), 1416–1427 (2007).
39. A. Kassab, J. Lan, P. Vannasing, and M. Sawan, "Functional near-infrared spectroscopy caps for brain activity monitoring: a review," *Appl. Opt.* **54**(3), 576–586 (2015).
40. V. O. Korhonen, T. S. Myllyla, M. Y. Kirillin, A. P. Popov, A. V. Bykov, A. V. Gorshkov, E. A. Sergeeva, M. Kinnunen, and V. Kiviniemi, "Light propagation in NIR spectroscopy of the human brain," *IEEE J. Selected Top. Quantum Electron.* **20**(2), 289–298 (2014).
41. R. J. Cooper, L. Gagnon, D. M. Goldenholz, D. A. Boas, and D. N. Greve, "The utility of near-infrared spectroscopy in the regression of low-frequency physiological noise from functional magnetic resonance imaging data," *Neuroimage* **59(4)**, 3128–3138 (2012).
42. A. Bakker, B. Smith, P. Ainslie, and K. Smith, *Near-infrared spectroscopy, applied aspects of ultrasonography in humans*, P. Ainslie, Ed., InTech, Rijeka, Croatia, (2012).
43. H. R. Heekeren, M. Kohl, H. Obrig, R. Wenzel, W. von Pannwitz, S. J. Matcher, U. Dirnagl, C. E. Cooper, and A. Villringer, "Noninvasive assessment of changes in cytochrome-c oxidase oxidation in human subjects during visual stimulation," *J. Cereb. Blood Flow Metab.* **19**(6), 592–603 (1999).
44. K. Mori, Y. Toda, H. Ito, T. Mori, K. Mori, A. Goji, H. Hashimoto, H. Tani, M. Miyazaki, M. Harada, and S. Kagami, "Neuroimaging in autism spectrum disorders: 1H-MRS and NIRS study," *J. Med. Investig.* **62**(1), 29–36 (2015).
45. E. Niedermeyer, "Dipole theory and electroencephalography," *Clin. Electroencephalogr.* **27**(3), 121–131 (1996).
46. H. Shibasaki, "Human brain mapping: hemodynamic response and electrophysiology," *Clin. Neurophysiol.* **119**(4), 731–743 (2008).
47. F. Wallois, M. Mahmoudzadeh, A. Patil, and R. Grebe, "Usefulness of simultaneous EEG-NIRS recording in language studies," *Brain Lang.* **121**(2), 110–123 (2012).
48. B. A. Riedner, B. K. Hulse, M. J. Murphy, F. Ferrarelli, and G. Tononi, "Temporal dynamics of cortical sources underlying spontaneous and peripherally evoked slow waves," *Prog. Brain Res.* **193**, 201–218 (2011).
49. Y. Hoshi, S. Kosaka, Y. Xie, S. Kohri, and M. Tamura, "Relationship between fluctuations in the cerebral hemoglobin oxygenation state and neuronal activity under resting conditions in man," *Neurosci. Lett.* **245** (3), 147–150 (1998).
50. L. Chen, P. Sandmann, J. Thorne, C. Herrmann, and S. Debener, "Association of concurrent fNIRS and EEG signatures in response to auditory and visual stimuli," *Brain Topogr.*, 1–16 (2015).

51. S. Schneider, A. M. Rapp, F. B. Haeußinger, L. H. Ernst, F. Hamm, A. J. Fallgatter, and A. Ehlis, "Beyond the N400: complementary access to early neural correlates of novel metaphor comprehension using combined electrophysiological and haemodynamic measurements," *Cortex* **53**, 45–59 (2014).
52. S. Schneider, L. Wagels, F. B. Haeussinger, A. J. Fallgatter, A. Ehlis, and A. M. Rapp, "Haemodynamic and electrophysiological markers of pragmatic language comprehension in schizophrenia," *World J. Biol. Psychiatry*, 1–13 (2015).
53. E. Visani, L. Canafoglia, I. Gilioli, D. R. Sebastiano, V. Contarino, D. Duran, F. Panzica, R. Cubeddu, D. Contini, and L. Zucchelli, "Hemodynamic and EEG time-courses during unilateral hand movement in patients with cortical myoclonus: an EEG-fMRI and EEG-TD-fNIRS study," *Brain Topogr.*, 1–11 (2014).
54. P. Pouliot, T. P. Y. Tran, V. Birca, P. Vannasing, J. Tremblay, M. Lassonde, and D. K. Nguyen, "Hemodynamic changes during posterior epilepsies: an EEG-fNIRS study," *Epilepsy Res.* **108**(5), 883–890 (2014).
55. C. Babiloni, F. Vecchio, R. Altavilla, F. Tibuzzi, R. Lizio, C. Altamura, P. Palazzo, P. Maggio, F. Ursini, and M. Ercolani, "Hypercapnia affects the functional coupling of resting state electroencephalographic rhythms and cerebral haemodynamics in healthy elderly subjects and in patients with amnestic mild cognitive impairment," *Clin. Neurophysiol.* **125**(4), 685–693 (2014).
56. R. Hari and R. Salmelin, "Magnetoencephalography: from SQUIDs to neuroscience: Neuroimage 20th anniversary special edition," *Neuroimage* **61**(2), 386–396 (2012).
57. J. P. Wikswo, A. Gevins, and S. J. Williamson, "The future of the EEG and MEG," *Electroencephalogr. Clin. Neurophysiol.* **87**(1), 1–9 (1993).
58. Y. Seki, T. Miyashita, A. Kandori, A. Maki, and H. Koizumi, "Simultaneous measurement of neuronal activity and cortical hemodynamics by unshielded magnetoencephalography and near-infrared spectroscopy," *J. Biomed. Opt.* **17**(10), 1070011–1070016 (2012).
59. B. A. Burke and S. G. Diamond, "Measuring cerebral hemodynamics with a modified magnetoencephalography system," *Physiol. Meas.* **33**(12), 2079 (2012).
60. V. Kiviniemi, "Endogenous brain fluctuations and diagnostic imaging," *Hum. Brain Mapp.* **29**(7), 810–817 (2008).
61. V. Toronov, T. Myllylä, V. Kiviniemi, and V. Tuchin, "Dynamics of the brain: mathematical models and non-invasive experimental studies," *Eur. Phys. J. Spec. Top.* **222**(10), 2607–2622 (2013).
62. V. O. Korhonen, T. K. Hiltunen, T. S. Myllylä, S. Wang, J. Kantola, J. Nikkinen, Y. Zang, P. Levan, and V. Kiviniemi,

"Synchronous multi-scale neuroimaging environment for critically sampled physiological analysis of brain function: hepta-scan concept," *Brain Connectivity* **4**(9), 677–689 (2014).
63. T. S. Myllylä, A. A. Elseoud, H. S. S. Sorvoja, R. A. Myllylä, J. M. Harja, J. Nikkinen, O. Tervonen, and V. Kiviniemi, "Fibre optic sensor for non-invasive monitoring of blood pressure during MRI scanning," *J. Biophotonics* **4**(1–2), 98–107 (2011).
64. J. A. Claassen, B. D. Levine, and R. Zhang, "Dynamic cerebral autoregulation during repeated squat-stand maneuvers," *J. Appl. Physiol.* **106**(1), 153–160 (2009).
65. A. H. Van Beek, J. A. Claassen, M. G. O. Rikkert, and R. W. Jansen, "Cerebral autoregulation: an overview of current concepts and methodology with special focus on the elderly," *J. Cereb. Blood Flow Metab.* **28**(6), 1071–1085 (2008).
66. A. Vermeij, A. H. Van Beek, M. G. O. Rikkert, J. A. Claassen, and R. P. Kessels, "Effects of aging on cerebral oxygenation during working-memory performance: a functional near-infrared spectroscopy study," *PloS One* **7**(9), e46210 (2012).
67. A. Vermeij, Aisha S. S. Meel-van den Abeelen, R. P. Kessels, A. H. van Beek, and J. A. Claassen, "Very-low-frequency oscillations of cerebral hemodynamics and blood pressure are affected by aging and cognitive load," *Neuroimage* **85**, 608–615 (2014).
68. O. Pucci, V. Toronov, and K. St Lawrence, "Measurement of the optical properties of a two-layer model of the human head using broadband near-infrared spectroscopy," *Appl. Opt.* **49**(32), 6324–6332 (2010).
69. I. Schelkanova and V. Toronov, "Optimal quantitation of the cerebral hemodynamic response in functional near-infrared spectroscopy," *Opt. Express* **18**(18), 19386–19395 (2010).
70. I. Schelkanova and V. Toronov, "Independent component analysis of broadband near-infrared spectroscopy data acquired on adult human head," *Biomed Opt. Express* **3**(1), 64–74 (2012).
71. H. Z. Yeganeh, V. Toronov, J. T. Elliott, M. Diop, T. Lee, and K. St Lawrence, "Broadband continuous-wave technique to measure baseline values and changes in the tissue chromophore concentrations," *Biomed Opt. Express* **3**(11), 2761–2770 (2012).
72. M. Diop, E. Wright, V. Toronov, T. Lee, and K. S. Lawrence, "Improved light collection and wavelet de-noising enable quantification of cerebral blood flow and oxygen metabolism by a low-cost, off-the-shelf spectrometer," *J. Biomed. Opt.* **19**(5), 057007–057007 (2014).
73. C. Kolyva, I. Tachtsidis, A. Ghosh, T. Moroz, C. E. Cooper, M. Smith, and C. E. Elwell, "Systematic investigation of changes in oxidized cerebral cytochrome c oxidase concentration during frontal lobe

activation in healthy adults," *Biomed Opt. Express* **3**(10), 2550–2566 (2012).
74. E. K. Miller and J. D. Cohen, "An integrative theory of prefrontal cortex function," *Annu. Rev. Neurosci.* **24**(1), 167–202 (2001).
75. H. Gävert, J. Hurri, J. Särelä, and A. Hyvärinen, *The FastICA package for Matlab*, Laboratory of Computer and Information Science, Helsinki University of Technology (2005).
76. T. A. Schweizer, K. Kan, Y. Hung, F. Tam, G. Naglie, and S. J. Graham, "Brain activity during driving with distraction: an immersive fMRI study," *Frontiers Hum. Neurosci.* **7** (2013).
77. F. Testud, "Dynamic magnetic field estimation for magnetic resonance imaging," doctoral thesis, Fakultät für Mathematik und Physik der Albert-Ludwigs-Universität Freiburg im Breisgau Albert-Ludwigs-Universität Freiburg im Breisgau (2013).
78. T. Myllylä, "Multimodal biomedical measurement methods to study brain functions simultaneously with fMRI," Acta Universitatis Ouluensis, Series C, Technica 497 (2014).
79. T. Myllylä, V. Korhonen, E. Vihriälä, H. Sorvoja, T. Hiltunen, O. Tervonen, and V. Kiviniemi, "Human heart pulse wave responses measured simultaneously at several sensor placements by two MR-compatible fiber optic methods," *J. Sens.* **2012**, 1–8 (2012).
80. M. A. Erickson and W. A. Banks, "Blood–brain barrier dysfunction as a cause and consequence of Alzheimer's disease," *J. Cereb. Blood Flow Metab.* **33**(10), 1500–1513 (2013).
81. K. Schoknecht and H. Shalev, "Blood-brain barrier dysfunction in brain diseases: clinical experience," *Epilepsia* **53**(Suppl 6), 7-13, doi: 10.1111/j.1528-1167.2012.03697.x(2012).
82. V. O. Korhonen, T. S. Myllyla, M. Y. Kirillin, A. P. Popov, A. V. Bykov, A. V. Gorshkov, E. A. Sergeeva, M. Kinnunen, and V. Kiviniemi, "Light propagation in NIR spectroscopy of the human brain," *J. Selected Topics Quantum Electron.* **20**(2), 1–10 (2014).
83. V. V. Lychagov, V. V. Tuchin, M. A. Vilensky, B. N. Reznik, T. Ichim, and L. De Taboada, "Experimental study of NIR transmittance of the human skull," *Proc. SPIE* **6085**, 60850T (2006) [doi: 10.1117/12.650116].
84. K. Wendel, D. Stoliar, J. Malmivuo, and J. Hyttinen, "Measuring tissue thicknesses of the human head using centralized and normalized trajectories," *Proceedings of the Conference on Consciousness and Its Measures*, 29.11.-1.12. 2009, St. Raphael Hotel, Limassol, Cyprus (2009).
85. C. E. Tedford, S. DeLapp, S. Jacques, and J. Anders, "Quantitative analysis of transcranial and intraparenchymal light penetration in human cadaver brain tissue," *Lasers Surg. Med.* **47**(4), 312–322 (2015).
86. V. V. Tuchin, *Tissue Optics: Light Scattering Methods and Instruments for Medical Diagnosis*, 3rd ed., SPIE Press, Bellingham, Wash. (2015).

87. R. Graaff, J. Aarnoudse, J. R. Zijp, P. Sloot, F. De Mul, J. Greve, and M. Koelink, "Reduced light-scattering properties for mixtures of spherical particles: a simple approximation derived from Mie calculations," *Appl. Opt.* **31**(10), 1370–1376 (1992).
88. E. A. Genina, A. N. Bashkatov, and V. V. Tuchin, "Tissue optical immersion clearing," *Expert Rev. Med. Devices* **7**(6), 825–842 (2010).
89. K. V. Larin, M. G. Ghosn, A. N. Bashkatov, E. A. Genina, N. A. Trunina, and V. V. Tuchin, "Optical clearing for OCT image enhancement and in-depth monitoring of molecular diffusion," *IEEE J. Sel. Top. Quantum Electronics* **18**(3), 1244–1259 (2012).
90. D. Zhu, K. V. Larin, Q. Luo, and V. V. Tuchin, "Recent progress in tissue optical clearing," *Laser Photonics Rev.* **7**(5), 732–757 (2013).
91. T. Yu, X. Wen, V. V. Tuchin, Q. Luo, and D. Zhu, "Quantitative analysis of dehydration in porcine skin for assessing mechanism of optical clearing," *J. Biomed. Opt.* **16**(9), 095002–095002-9 (2011).
92. L. Oliveira, A. Lage, M. P. Clemente, and V. V. Tuchin, "Rat muscle opacity decrease due to the osmosis of a simple mixture," *J. Biomed. Opt.* **15**(5), 055004–055004-9 (2010).
93. L. Oliveira, M. I. Carvalho, E. Nogueira, and V. V. Tuchin, "Optical measurements of rat muscle samples under treatment with ethylene glycol and glucose," *J. Innov. Opt. Health Sci.* **6**(2), 1350012-1-15 (2013).
94. L. Oliveira, M. Carvalho, E. Nogueira, and V. Tuchin, "The characteristic time of glucose diffusion measured for muscle tissue at optical clearing," *Laser Physics* **23**(7), 075606 (2013).
95. L. M. Oliveira, M. I. Carvalho, E. M. Nogueira, and V. V. Tuchin, "Diffusion characteristics of ethylene glycol in skeletal muscle," *J. Biomed. Opt.* **20**(5), 051019–051019 (2015).
96. D. K. Tuchina, A. N. Bashkatov, E. A. Genina, and V. V. Tuchin, "Quantification of glucose and glycerol diffusion in myocardium," *J. Innov. Opt. Health Sci.* **8**(3), 1541006 (2015).
97. D. K. Tuchina, R. Shi, A. N. Bashkatov, E. A. Genina, D. Zhu, Q. Luo, and V. V. Tuchin, "Ex vivo optical measurements of glucose diffusion kinetics in native and diabetic mouse skin," *J. Biophotonics* **8**(4), 332–346 (2015).
98. E. A. Genina, A. N. Bashkatov, and V. V. Tuchin, "Optical clearing of cranial bone," *Adv. Opt. Technol.*, Article ID 267867, 8 pages (2008).
99. A. N. Bashkatov, E. A. Genina, V. I. Kochubey, and V. V. Tuchin, "Optical properties of human cranial bone in the spectral range from 800 to 2000 nm," Saratov Fall Meeting 2005: Optical Technologies in Biophysics and Medicine VII, 616310–616310-11 (2006).
100. N. Ugryumova, S. J. Matcher, and D. P. Attenburrow, "Measurement of bone mineral density via light scattering," *Phys. Med. Biol.* **49**(3), 469 (2004).

101. M. Firbank, M. Hiraoka, M. Essenpreis, and D. Delpy, "Measurement of the optical properties of the skull in the wavelength range 650–950 nm," *Phys. Med. Biol.* **38**(4), 503 (1993).
102. E. A. Genina, A. N. Bashkatov, N. A. Lakodina, S. A. Murikhina, Y. P. Sinichkin, and V. V. Tuchin, "Diffusion of glucose solution through fibrous tissues: in-vitro optical and weight measurements," Saratov Fall Meeting '99, pp. 255–261 (2000).
103. A. N. Bashkatov, E. A. Genina, Y. P. Sinichkin, V. I. Kochubey, N. A. Lakodina, and V. V. Tuchin, "Glucose and mannitol diffusion in human dura mater," *Biophys. J.* **85**(5), 3310–3318 (2003).
104. A. N. Bashkatov, E. A. Genina, and V. V. Tuchin, "Optical immersion as a tool for tissue scattering properties control," *Persp. Eng. Opt.* 313–334 (2002).
105. Q. Luo, C. Jiang, P. Li, H. Cheng, Z. Wang, Z. Wang, and V. V. Tuchin, "Laser speckle imaging of cerebral blood flow," *Coherent-Domain Optical Methods: Biomedical Diagnostics, Environmental Monitoring and Material Science 1*, 2nd ed., V. V. Tuchin, Ed., pp. 167–211, Springer-Verlag, Berlin (2013).
106. E. A. Genina, A. N. Bashkatov, V. I. Kochubey, and V. V. Tuchin, "Optical clearing of human dura mater," *Opt. Spectrosc.* **98**(3), 470–476 (2005).
107. V. V. Tuchin, *Optical Clearing of Tissues and Blood]*, SPIE Press Bellingham (2006).
108. J. Wang, Y. Zhang, T. Xu, Q. Luo, and D. Zhu, "An innovative transparent cranial window based on skull optical clearing," *Laser Phys. Lett.* **9**(6), 469 (2012).

Teemu Myllylä, DSc, Adjunct Professor of Biomedical Sensing and Instrumentation, studied at the University of Oulu, Technical University of Berlin, and Humboldt University of Berlin. He currently leads the Health and Wellness Measurements Research Group in the Optoelectronics and Measurement Techniques, University of Oulu, Finland. In addition, he actively participates in the Biophotonics Group, located in the same unit, as well as in the Oulu Functional Neuroimaging Research Group at the Medical Research Center of Oulu, Finland. His research focuses on development of biomedical measurement methods and instrumentations to study human physiology, such as blood pressure, blood flow, and metabolism, especially in the brain. Several developed methods have been utilised in medical research, particularly in magnetic resonance imaging. He collaborates closely with many university hospitals and medical research centers.

Vladislav Toronov received his Diploma in Engineering Physics from MEPhI, Moscow, USSR, 1986 and his PhD in Physics from Saratov State University, Saratov, Russian Federation, 1992. From 1992–1997, he served as a Research Professor at Saratov State University; from 1997–200, he conducted his postdoctoral work at the University of Illinois, Urbana, IL. He worked as a senior research scientist: at the Beckman Institute for Advanced Science and Technology from 2000–2006, after which he became an Assistant and Associate Professor of Physics at Ryerson University, Toronto, Canada.

Jurgen Claassen, MD PhD, is a geriatrician and researcher at the Radboud University Medical Center and the Donders Institute for Brain, Cognition, and Behaviour. His research focuses on late-onset Alzheimer's disease and cognitive aging, with specific attention on the role of cerebral hemodynamic changes. He uses TCD, NIRS, and continuous blood pressure monitoring to investigate cerebral autoregulation, vasomotor reactivity, and neurovascular coupling in cognitive aging and Alzheimer's disease. His research extends to blood pressure regulation in aging and to syncope in the elderly.

Vesa Kiviniemi, MD PhD, was the first European to detect spontaneous brain activity fluctuations independently in 1997 in a resting brain imaged with fMRI in Oulu. While still a PhD student, he was drafted by connectivity pioneer Bharat Biswal to MCW, USA where he learned about original resting state analytics. He pioneered the development of several methods for detecting spontaneous brain activity sources, most of which stem from collaborations with MCW and Oulu MIRG members. His most important work so far was the use of independent component analysis (ICA) in separating functional brain cortices from spontaneous brain activity, first announced in ISMRM 2001. This was made possible by joint ventures with Jussi Kantola and Aapo Hyvärinen, the developer of FastICA. After 2005, the resting state scanning and analysis became the fastest growing neuroimaging realm. Kiviniemi has had the privilege of giving several invited talks at HBM and ISMRM conferences on resting state. He is currently interested in detecting numerous physiological brain pulsations driving the glymphatic CSF system with an international research team.

Valery V. Tuchin is a professor and chairman of Optics and Biophotonics at National Research Saratov State University. He is also the head of laboratory at the Institute of Precision Mechanics and Control, RAS, and the supervisor of the Interdisciplinary Laboratory of Biophotonics at National Research Tomsk State University. His research interests include biophotonics, tissue optics, laser medicine, tissue optical clearing, and nanobiophotonics. He is a member of SPIE, OSA, and IEEE, guest professor of HUST (Wuhan) and Tianjin Universities of China, and adjunct professor of the Limerick University (Ireland) and National University of Ireland (Galway). He is a fellow of SPIE and OSA, and has been awarded the Honored Science Worker of the Russia, SPIE Educator Award, FiDiPro (Finland), Chime Bell Prize of Hubei Province (China), and Joseph W. Goodman Book Writing Award (OSA/SPIE).

Chapter 11

Measurement of Optical Fluence Distribution and Optical Properties of Tissues Using Time-Resolved Profiles of Optoacoustic Pressure

Ivan M. Pelivanov and Alexander A. Karabutov
Moscow State University, Moscow, Russia

Tatiana D. Khokhlova
University of Washington, Seattle, USA

Alexander A. Oraevsky
TomoWave Laboratories, Houston, USA
University of Houston, Houston, USA

11.1 Methods to Study Light Distribution in Tissue

Noninvasive determination of optical energy distribution in live biological tissues is an important aspect of any laser treatment or diagnostic procedure. It is particularly important for optical dosimetry in tissue,[1,2] tissue diagnostics based on optical properties,[3–5] quantitative molecular and functional imaging in the human body based on optical contrast,[6–9] laser therapeutic procedures (such as photodynamic therapy,[10–13] photothermal therapy,[14,15] and laser treatment of various skin lesions and hair removal[16–18]), and laser microsurgery.[19,20] Monitoring of changes in tissue optical properties can also be utilized for monitoring of laser–tissue interactions,[21] noninvasive optical biopsy,[22] the effects of laser ablation or denaturation of tissue,[23,24] and ablation by high-intensity focused ultrasound (HIFU).[25,26] The precision of these procedures would improve significantly if laser irradiation conditions could be adjusted in accordance with initial optical properties and with changes in optical properties during the course of the treatment. Therefore, there is a definite need for noninvasive measurement and visualization of the light distribution profile in the laser-irradiated tissue *in vivo*. Such feedback information helps predict the optimal laser wavelength, pulse duration, laser intensity, and light dose to be delivered so as to obtain diagnostic information or to cause a sufficient effect during treatment with minimal damage to adjacent tissue layers.

Various direct methods for monitoring the optical absorption in optically scattering media based on detection of scattered photons have been widely discussed in the literature.[27,28] While placement of a detector inside the tissue may cause significant distortion of the scattering and absorption patterns, noninvasive methods are much preferred, especially in the clinic.[23] In recent years, progress in noninvasive measurements of optical properties of biological tissues *in vivo* has been achieved using spectroscopic measurements of light diffusely back-scattered by the tissue structures (e.g., frequency-domain spectroscopy,[29–32] spatially resolved spectroscopy,[33,34] and time-resolved spectroscopy[35–38]). The time-resolved method is the most developed of these methods. Measurements are usually based on recording of the temporal profile of a picosecond or femtosecond laser pulse back-scattered by a medium. Its time delay with respect to the reference (unscattered) laser pulse determines the scattering coefficient of the medium, while the exponential decay index determines the absorption coefficient (see, for example, Ref. 35). Therefore, the contributions of absorption and scattering are taken into account separately. The diffusion approximation of the light transport equation is usually employed for reconstruction of tissue optical properties. Measurements can be conveniently performed by using radiation from optical parametric oscillators (lasers with the widest range of wavelength tunability), an ultrashort-pulse Ti:sapphire laser tunable between 650 and 1100 nm, or radiation from several diode lasers emitting in the same spectral range. These lasers are sufficiently reliable to be used in the clinic.[38] The absorption and scattering spectra of human tissues in the visible and near-infrared wavelength ranges of laser radiation were obtained in Refs. 37–39.

Despite the numerous advantages of these diffusion methods, they all have certain limitations. The main drawback is the mandatory assumption of homogeneity of the medium under study in terms of its optical properties. Even in a two-layer optically scattering medium, calculation of the optical properties presents great difficulties.[40] The profile of an optical signal back-scattered by the medium is not subject to drastic distortions caused by the presence of layers or inhomogeneous structures. The calculated optical coefficients, therefore, contain spatially integrated information from the medium, which reduces spatial resolution.[26] Thus, new methods of local and direct measurement of the optical energy distribution in biological tissues are of significant interest.

Optical coherence tomography (OCT) solves the problem of spatial resolution by limiting the detection to ballistic photons propagating along straight paths and detecting them only from a depth defined by the reference arm of an interferometer.[27] OCT is an advanced optical technology that offers a number of advantages when measuring depth profiles of optical energy scattered in tissue.[41,42] The main drawback of this approach is the shallow depth of monitoring, which is limited to a few millimeters or less in optically absorbing and scattering tissues.

One indirect method that can yield useful information on the absorbed energy distribution in the laser-irradiated tissue is pulsed photothermal radiometry (PPTR) with temporal resolution.[28,43,44] This method is useful primarily for imaging the temperature field in the surface plane. Such limitation arises because of the complexity of reconstructing the axial in-depth distribution of heat in PPTR, which is based on kinetic measurements of the surface temperature followed by computer simulations employing diffusion theory.

An alternative, accurate method of direct noninvasive characterization of optical properties in optically turbid media is the pulsed optoacoustic method, which is based on a thermal mechanism of pressure generation through absorption of pulsed laser radiation and time-resolved measurements of profiles of laser-induced pressure (LIP).[45–47] The main advantage of the time-resolved optoacoustic method is the proportionality of the amplitude of the laser-induced transient pressure to the product of optical absorption coefficient and the local optical fluence. The third factor of the optoacoustic signal is the thermo-acoustic efficiency (often called as Grüneisen parameter), which is constant for given type of tissue and a given temperature.[48,49]

Time-resolved optoacoustics can be employed without limitations in studies of optical absorption in optically homogeneous or heterogeneous media; in relatively transparent or opaque, uniformly absorbing or layered, strongly scattering or clear media; and for measurements of optical properties in optically scattering phantoms or biological tissues *in vivo*.[45–55]

Alternatively, optoacoustic measurements can be performed in the frequency domain by illuminating tissues with continuous optical waves chirp-modulated at a range of frequencies and then detecting optoacoustic signals with wide-band ultrasonic transducers.[56] The key advantage of this method is that it avoids the use of relatively expensive pulsed lasers. A significant disadvantage, however, is its low signal-to-noise ratio, which decreases linearly with optical fluence. As a result, frequency-domain measurement of the optoacoustic profiles in tissue is limited to a shallow depth due to the apparent need to detect a wide range of ultrasonic frequencies—from approximately 20 kHz (typical slope of the absorbed light distribution) to approximately 20 MHz (high-resolution requirement)—simultaneously. In this chapter, we will focus on the more sensitive method of optoacoustic measurements in the time domain.

11.2 Two Modes of Optoacoustic Detection

There are two modes of LIP detection, termed the optoacoustic forward and backward modes (Fig. 11.1).[57] In both cases, laser irradiation of the surface of an optically absorbing medium (such as tissue) takes place through a transparent medium (such as air, water, or fused silica). The thermoelastic pressure inside the heated layer results in generation of ultrasonic pressure transients that are launched into the absorbing (illuminated) medium and the transparent medium. The temporal profile of LIP depends on the ratio of acoustic impedances of the absorbing and transparent media.[45,49] In the forward mode, the LIP propagated forward along the laser beam into the depth of the absorbing medium (e.g., biological tissue) is detected at the rear surface of the irradiated medium [Fig. 11.1(a)].

The temporal profile of the LIP propagating forward into the absorbing medium is well established and has been quantitatively investigated both theoretically and experimentally.[44,49–52] Possible medical applications of optoacoustic detection in the forward mode are, however, limited to *in vitro* applications and applications in human organs with free access to the two opposite surfaces (such as the breast). For various medical applications with just one tissue surface access (such as skin or hollow organs), the backward detection mode is often the only choice. Simultaneously, the backward detection mode employing solid optically and acoustically transparent material [such as acrylic plastic or fused silica; see Fig. 11.1(b)] placed on the surface of tissue for optoacoustic coupling satisfies the need for a flat tissue

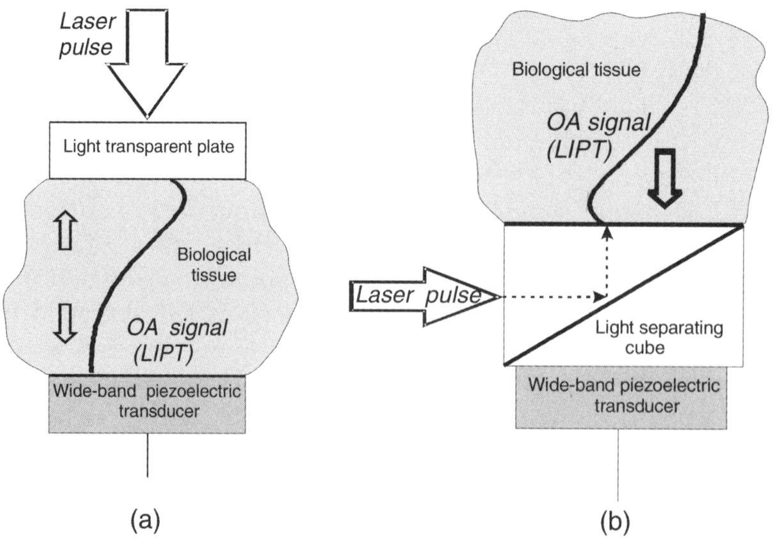

Figure 11.1 (a) Forward and (b) backward optoacoustic modes and their corresponding signal profiles.

surface, as required for generation of ultrawide ultrasonic frequency band in tissue.

The detection of laser-induced transient pressure at the front irradiated surface makes design of the piezoelectric detector more complex. The device used for backward detection is called a (optoacoustic) front surface transducer (FST).[55] An ultrawide-band acoustic transducer ($\Delta F_{ac} \approx$ 100 MHz) coupled to the top surface of the optoacoustic transparent medium can be used to detect LIP with temporal resolution sufficient for purposes of biomedical imaging and sensing. The transparent medium can be made from a proper material with sufficiently smooth and parallel surfaces, so that not only can bulk acoustic attenuation be minimized in the course of ultrasonic wave propagation to the piezoelectric detector, but ultrasonic scattering and absorption at the interfaces can also be avoided. That is why the backward mode permits ultrasonic detection in tissues in the range of up to 100 MHz, which enables measurements of optical absorption distribution for variety of tissues and in the entire optical spectral range.[48,49]

In the backward mode [see Fig. 11.1(b)], the LIP is detected at the side of the illuminated tissue surface. The optical source and piezoelectric detector can be arranged in several different ways. For example, laser illumination of a tissue surface can be performed by a light-separating prism, as illustrated in Fig. 11.1(b). In this case, a collimated laser beam is directed onto the tissue under study by a separating cube, which turns the light beam 90 deg and allows straight transmission of optoacoustic waves. The cube's top surface is in contact with the medium under study, and its bottom is in contact with a wideband piezoelectric transducer. The cube is made of two triangle glass or poly-methyl methacrylate (PMMA) prisms that are in optical contact through a thin optical mirror coated onto one of the surfaces. The thin layer of the optical mirror does not disturb acoustic waves with significantly longer wavelengths. The LIP generated in the medium under study propagates through this layer to the detector without any reflection losses. Various designs of FST permit tissue illumination either from the top or from the side of the transparent medium coupled to the absorbing tissue.[58–60]

In addition to piezoelectric transducers, various optical detectors can be used to detect LIP signals.[61–64] Optical methods of LIP detection have an advantage in sensitivity over point-like piezoelectric detectors when the total area is less than 0.5 mm^2, due to the fact that piezoelectric transducers with very low electric capacity shunt preamplifier circuits.[65] However, piezoelectric transducers with sufficient electrical capacity of approximately 20 pF possess sensitivity about 100- to 400-fold greater that those based on optical interferometry. Thus, optical detectors of acoustic waves represent useful tools for applications limited to the optoacoustic microscopy of blood vessels within or immediately underneath the skin and in clear media.[63] Tissue imaging at the depth of several centimeters is presently impossible with optical detectors due to their insufficient signal-to-noise ratio.

The profiles of LIP signals propagating forward into the depth of the optically absorbing tissue and backward into the optically transparent medium of the optoacoustic transducer differ significantly. The purpose of this chapter is to describe all stages of optoacoustic generation, propagation, and transformation in various media, including tissues and optoacoustic transducers. Due to a number of physical processes, the profiles of the laser-induced pressure signals gradually become altered as a function of distance from their origin. We describe these alterations and the algorithms currently used to recover the intrinsic depth profiles of the absorbed optical energy spatial distribution inside an optically scattering media (such as tissue). This, in turn, permits measurements of tissue optical properties using the amplitude and the profile of the optoacoustic signals.

11.3 Stages of the Optoacoustic Phenomena

Since LIP signal can be effectively generated only through a nonstationary deposition of thermal energy by laser radiation, optoacoustic monitoring is performed only using short (nanosecond) pulses of laser radiation. Pulsed optoacoustic phenomena in optically scattering medium can be treated in three stages:

1. Consideration of the light propagation and absorption inside the scattering medium (tissue) and determination of the distribution of the absorbed energy density.
2. Determination of the temperature field generated through optical absorption, taking into account the heat diffusion that occurs in the course of transient heat release.
3. Calculation of the acoustic wave profile generated in a rapidly heated volume by thermal energy (proportional to the coefficient of subsequent thermal expansion).

All three optoacoustic stages have no analytical solutions in the general case, and it is necessary to use certain assumptions to obtain analytical results.

A nanosecond duration of laser pulses ($\tau_L = 1$ to 15 ns) is the most effective for optoacoustic imaging and monitoring in optically turbid media, because both the thermal and the stress (pressure) confinement conditions are satisfied with this range of pulse durations, even for high-resolution imaging.[46] The thermal confinement condition requires that in the course of laser energy deposition, thermal diffusion can expand thermal sources, optically induced inside image voxels, to dimensions d_{th} many times smaller than the voxel size d_{voxel} to be resolved on the image:

$$d_{th} = \sqrt{\chi\tau_L} \ll d_{voxel}, \tag{11.1}$$

where χ is the thermal conductivity of the irradiated medium. In this case, the heat sources can be considered to be "frozen" within the laser pulse's duration and the distribution of heat release repeats the distribution of absorbed optical energy inside the medium (tissue). Therefore, under the heat confinement condition, the solution of stage 2 repeats that of stage 1 and can be omitted.

The stress confinement condition implies that the distance traveled by an acoustic wave within the time of the optical energy deposition (i.e., the laser pulse duration) is much shorter than the characteristic dimension of the image voxel:

$$d_{ac} = c_0\tau_L \ll d_{voxel} \tag{11.2}$$

Since the speed of sound, c_0, in tissue is much faster than the rate of heat diffusion, the condition of stress/pressure confinement is stronger than that of thermal confinement. Under the conditions in Eqs. (11.1) and (11.2), the initial pressure profile (the front of LIP) repeats the profile of heat release distribution—that is, the distribution of absorbed laser energy in the irradiated medium.[46–48] For purposes of measurement of tissue optical properties, it is often possible to weaken the pressure confinement condition by replacing the characteristic dimension of the voxel, d_{voxel}, in Eq. (11.2) with the effective optical penetration depth, $1/\mu_{eff}$.

Thus, at the first stage of the theoretical approach to the optoacoustic profile generation, the distribution of heat released within the medium should be considered. At the second stage, the problem of thermo-acoustic conversion should be solved. These stages will be considered later in this chapter. The main question concerns the possibility of solving the inverse optoacoustic problem—that is, reconstructing the distribution of heat sources in the medium under study (tissue) and determining the tissue optical properties from the measured profiles of LIPs.

11.4 Specific Features of Depth Distribution of the Absorbed Optical Energy in Optically Scattering Media

11.4.1 Monte Carlo method

Biological tissues are turbid media and, therefore, the study of light distribution within such media is an essential step for development of diagnostic tools. Multiple analytical approaches to the problem are based on the solution of the light transport equation.[66] This equation is rigorous, but can be solved only numerically. The analytical solutions can be obtained in the diffusion approximation,[67] in the P_3 and P_5 approximations,[68] and in a few high-order approximations. These solutions are valid only at some distance from the surface of the scattering medium equal to at least one mean-free path of a photon, whereas the characteristics of the optical energy distribution in the very subsurface layer are often the most essential for the effects of a laser treatment.

Monte Carlo (MC) simulation is currently the only tool that allows one to calculate the spatial distribution of absorbed optical energy in the subsurface layer of an optically turbid medium.[69] MC simulation accurately predicts light propagation through tissue even at the depths much smaller than photon transport mean-free path $l^* = 1/\mu_s'$, where μ_s' is the reduced optical scattering coefficient. The MC method is based on a statistical approach. That is, if an optically turbid medium is characterized by the optical absorption and scattering coefficients, the probability of a single act of either photon absorption or scattering can be defined by these coefficients. If more than 10^5 photons per 1 cm^3 are used in MC calculations, the absorbed optical energy distribution within the medium can be found quite accurately. Another advantage of the MC method is the ability to take into consideration the scattering directivity (anisotropy), irradiation conditions, and boundary conditions. Therefore, the MC model is a method of choice for computer modeling of the forward problem of light propagation in optoacoustic tomography.[70]

In this chapter, we use the Monte Carlo simulation technique developed in Ref. 69 to calculate spatial distributions of the optical energy in a scattering medium, such as tissue. The input parameters in the simulations are the optical absorption, μ_a, and the optical scattering, μ_s, coefficients; anisotropy factor $g = \langle\cos\theta\rangle$; and the laser beam diameter, d. Refractive indices n_1 and n_2 are also fixed and correspond to these of fused silica or PMMA and water. The optically scattering medium is modeled as semi-infinite. The number of photons used in each simulation is 10^5. The directivity pattern of a single photon scattering act is taken into account by the Henyey–Greenstein phase function. The cross-section spatial profile of the laser beam is considered Gaussian.

The results of the MC simulation show that the maximum of the axial depth distribution of the optical energy is located at $z_{\max} \approx l^*$ beneath the medium surface; in contrast, in a clear (non-scattering) absorbing medium, the maximum of the optical fluence is located at the boundary, $z = 0$ (Fig. 11.2).[52,53,67] The depth maximum fluence location $z_{\max}$ also depends on the effective diffuse optical reflection R_{eff} from the interface between the transparent medium and the optically turbid medium, which is determined, in turn, by the relationship between the refractive indexes n_1 and n_2 (see, for example, Refs. 52 and 53).

The empirical dependence of $z_{\max}$ on the reflection coefficient R_{eff}, which was studied numerically using the MC method in Refs. 53 and 67, can be presented in the form:

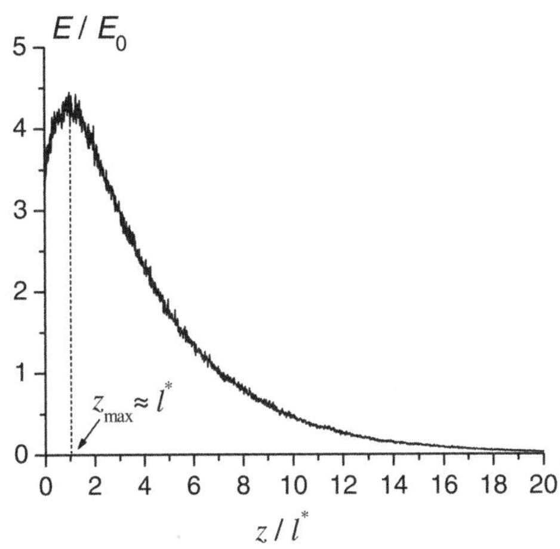

Figure 11.2 Typical depth profile of the absorbed optical energy distribution in an optically scattering medium in the case of plane wave geometry. Coordinate $z = 0$ corresponds to an interface between the transparent and the optically turbid medium.

$$z_{\max} \cong (1 - R_{eff})(1 - 0.4R_{eff}) \tag{11.3}$$

To avoid ambiguity in the simulations, the coefficient $R_{eff} = 0.0147$ (for the PMMA/water interface) was fixed in the study.

The absolute maximum value of the optical fluence inside the optically turbid medium can exceed the incident optical fluence 4 to 6 times in biological tissues and photons in the near-infrared spectral range (see Fig. 11.2). At the same time, the axial depth profile of the absorbed optical energy sufficiently far from the illuminated boundary is exponential: $\sim\exp(-\mu_{eff}z)$. The effective light attenuation coefficient, $\mu_{eff} = \sqrt{3\mu_a\mu_s'}$, depends on both optical absorption and scattering coefficients (see, for example, Refs. 53 and 71). In the case of $\mu_a \ll \mu_s$, the value of $Y = z_{\max}\mu_{eff} \sim \frac{1}{\mu_s'}\sqrt{3\mu_a\mu_s'} = 3\mu_a/\mu_{eff}$ depends only on the ratio of $X = \mu_a/\mu_{eff}$ and does not depend on the absolute value of optical properties or on anisotropy factor, g. However, if the medium is clear (i.e., uniformly absorbing and non-scattering), $\mu_a \gg \mu_s$, the maximum of laser fluence is located at the medium boundary. Thus, the laser fluence maximum location inside the turbid medium is shifted toward the surface of $z_{\max} = 0$ when the ratio of μ_a/μ_s' is increased.

MC simulation of the axial depth distribution of the absorbed optical energy in optically turbid media was performed for different values of μ_a, μ_s, and g.[68] The values of $z_{\max}$ and μ_{eff} were determined from the calculated optical distribution profiles, with relative errors of 4% for $z_{\max}$ and 2% for μ_{eff}, respectively (using 10^5 photons in the MC simulations).

The variation of the parameters of μ_a, μ_s, and g was performed so that the relationship $0.05 < \mu_a/\mu_{eff} < 0.35$ was fulfilled. Then the dependence of

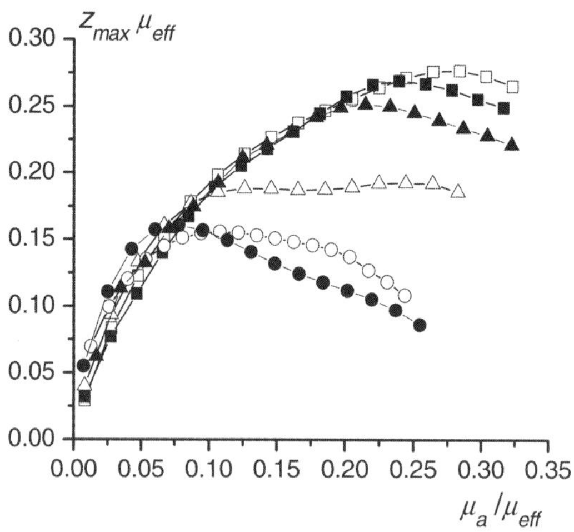

Figure 11.3 Numerically simulated dependence of $Y = z_{max}\,\mu_{eff}$ on the ratio $X = \mu_a/\mu_{eff}$ for different anisotropy factors g. Black circles: $g = 0$; open circles: $g = 0.3$; open triangles: $g = 0.5$; black triangles: $g = 0.8$; black squares: $g = 0.90$; open squares: $g = 0.95$.

$Y = z_{\max}\mu_{eff}$ on $X = \mu_a/\mu_{eff}$ was obtained (Fig. 11.3).[8] The relative errors for the values of X and Y were 5% and 8%, respectively.

As seen in Fig. 11.3, the value of the dependence is decreased with decreasing parameter g. It can be explained by the following reasons. When the ratio μ_a/μ_{eff} is fixed, the ratio $\mu_a/\mu_s' = 3(\mu_a/\mu_{eff})^2$ is also fixed, but the value of $\mu_a/\mu_s = \mu_a(1-g)/\mu_s'$ rises with increasing g. Therefore, the smaller the value of the anisotropy factor g, the larger the probability of a photon being absorbed in the subsurface layer of the turbid medium. At the same time, the curves corresponding to the case of $\mu_a/\mu_{eff} < 0.3$ converge within inaccuracy limits if the parameter $g \geq 0.8$. Under these conditions, the value of $Y_1 = z_{\max}\mu_{eff}$ depends on the ratio μ_a/μ_{eff} only and is not determined by the values of μ_a and μ_{eff} separately.

The study described here was performed for the plane wavefront of the incident laser beam. In many diagnostic applications, however, the use of a finite laser beam diameter is required. Two important parameters define the profiles of the optoacoustic signals in optically scattering media and, in turn, help to measure tissue optical properties using the time-resolved optoacoustic technique: (1) the location of the optical fluence maximum in an optically turbid medium and (2) the dependency of the optoacoustic signal amplitude (maximum of the absorbed optical energy) on the optical properties and the laser beam diameter. The qualitative analysis of these functions can be found in Refs. 53, 67, and 68.

Let us introduce the coefficient $k = F_{\max}/F_0$ to characterize the ratio of the subsurface maximum of the optical fluence inside the scattering medium and the incident fluence of the laser beam. To perform this analysis, the depth

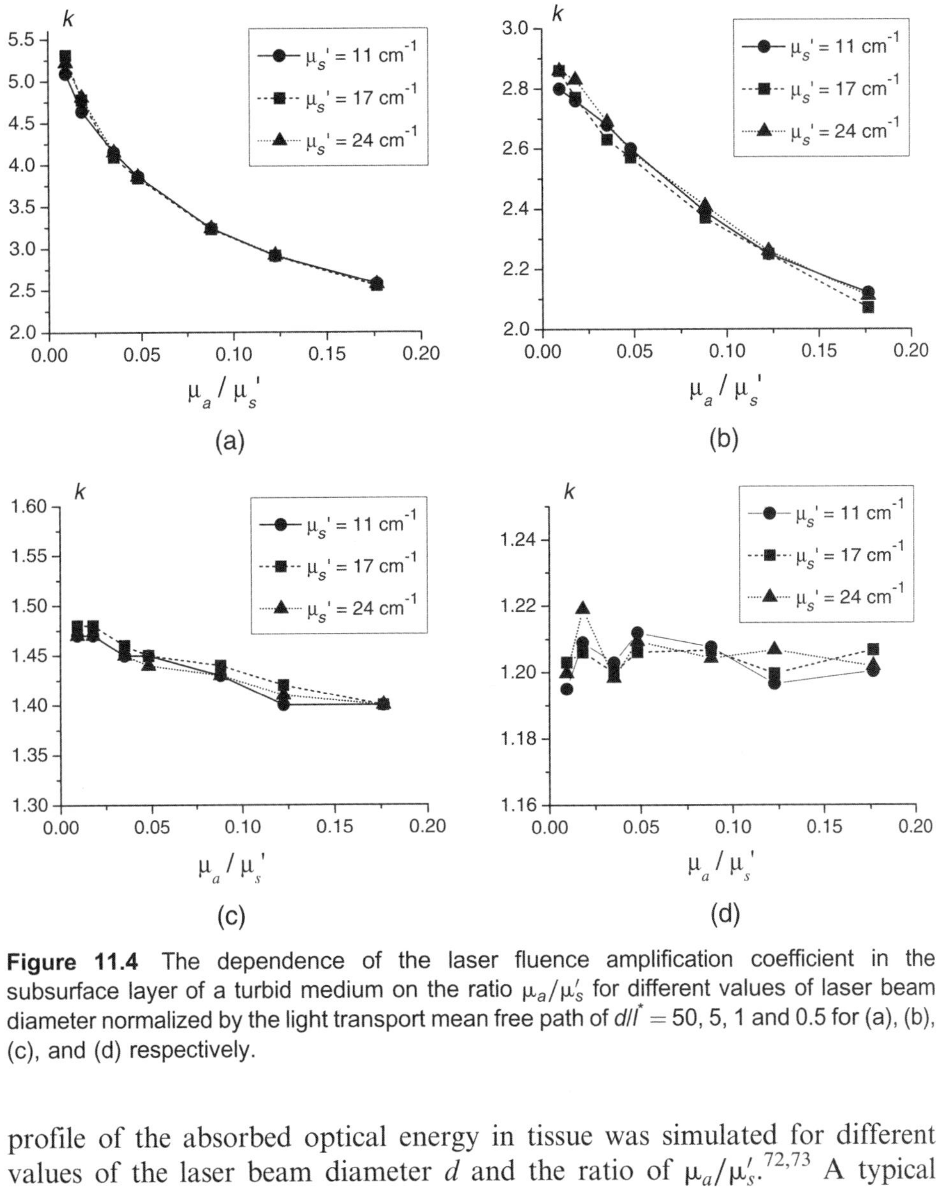

Figure 11.4 The dependence of the laser fluence amplification coefficient in the subsurface layer of a turbid medium on the ratio μ_a/μ_s' for different values of laser beam diameter normalized by the light transport mean free path of $d/l^* = 50$, 5, 1 and 0.5 for (a), (b), (c), and (d) respectively.

profile of the absorbed optical energy in tissue was simulated for different values of the laser beam diameter d and the ratio of μ_a/μ_s'.[72,73] A typical profile of such a distribution is shown in Fig. 11.2. The optical fluence amplification coefficient k represents a ratio of the maximum subsurface fluence to the incident fluence.

The set of curves of the optical fluence amplification coefficient k on the ratio of the optical coefficients, μ_a/μ_s', is illustrated in Fig. 11.4. Each individual curve corresponds to a fixed value of the effective scattering coefficient μ_s' (11 cm^{-1}, 17 cm^{-1}, and 24 cm^{-1}). Since media with different effective (reduced) scattering coefficients must be equivalent from the standpoint of illumination conditions, the laser beam diameter d was normalized by the light transport mean-free path, $l^* = 1/\mu_s'$. The normalized laser beam diameter, d/l^*, was the same for all curves corresponding to a single plot (i.e., *a*, *b*, *c*, and *d*), but was varied in the range from 0.5 to 50 from plot to plot.

If the laser beam is much wider than the effective light penetration depth, (e.g., $d/l^* = 50$), the wavefront of the incident laser field can be considered as planar, and the optical diffusion in the turbid medium can be described in terms of a 1D model along the beam axis. As seen in Fig. 11.4(a), the amplification of the laser fluence beneath the surface of the turbid medium reaches its maximum under two conditions: The laser beam is wide and $\mu_a/\mu_s' \ll 1$. However, even if the laser beam is wide, the coefficient decreases with an increasing ratio μ_a/μ_s'.

In the other limiting case of a narrow laser beam, the subsurface maximum of the optical fluence distribution is close to the incident fluence. For example, in case of $d/l^* = 0.5$ [Fig. 11.4(d)], the coefficient $k \approx 1.2$ and stays almost constant as the ratio μ_a/μ_s' changes more than 10 times. This finding can be explained by the fact that the light diffusion can no longer be described in terms of the 1D model. The optical flux is diffused in all directions with almost the same efficiency, thereby providing considerable light attenuation at depths smaller than l^* and leading to the absence of appreciable amplification of the laser fluence in subsurface regions of the optically scattering medium.

In addition to the dependencies of the laser fluence amplification on the laser beam diameter, Fig. 11.4 demonstrates another very important fact: The curves corresponding to the same normalized laser beam diameter, d/l^*, but different values of reduced light scattering coefficients, μ_s', coincide well within the wide range of μ_a/μ_s'. As a consequence, the distribution of laser fluence inside a turbid medium can be unambiguously determined by the two ratios of parameters, d/l^* and μ_a/μ_s', independently of the absolute values of these coefficients.

Figure 11.5 demonstrates the dependence of the laser fluence amplification coefficient k on the normalized laser beam d/l^* at different values of the μ_a/μ_s' ratio. As it can be expected, the maximal value of laser fluence within the turbid medium increases with an increase in d/l^* and reaches its limit at $d/l^* \gg 1$, where 1D light diffusion is realized. The increase in the ratio μ_a/μ_s' reduces the growth of k as a function of d/l^*. In case of $\mu_a/\mu_s' \ll 1$, the ratio reaches its absolute maximum of $k_{\max} \approx 5$–6. Finally, when the ratio of $\mu_a/\mu_s' > 1$, the laser fluence amplification coefficient $k \rightarrow 1$ (the medium is no longer strongly scattering), and the maximal value of laser fluence is located at the surface.

An important feature of the dependencies shown in Fig. 11.5 is that the curves corresponding to different values of μ_a/μ_s' in the range $0.01 < \mu_a/\mu_s' < 0.3$ coincide within a 10% error, when $d/l^* < 1$. In this case, the coefficient k is linearly proportional to d/l^*. This relationship can be explained by the fact that in case of a narrow laser beam, light diffusion across the beam

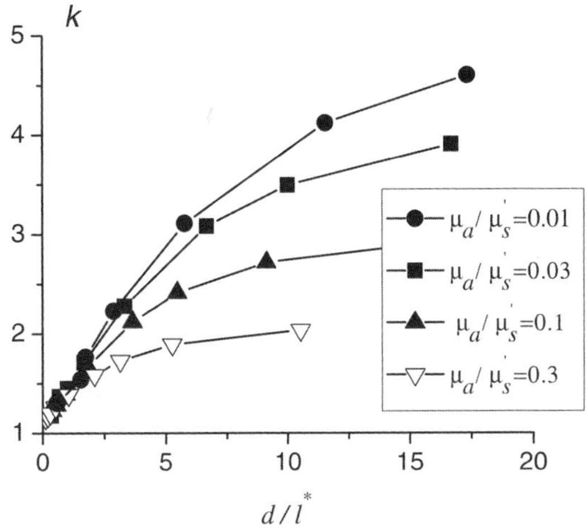

Figure 11.5 The dependence of the laser fluence amplification coefficient in the subsurface layer of a turbid medium on the laser beam diameter normalized by the light transport mean-free path for different ratios, μ_a/μ'_s.

affects the optical attenuation more strongly than it affects the optical absorption along the depth axis.

The set of dependencies of the maximal value of absorbed laser power density $\mu_a F_{max}$ on the optical absorption coefficient μ_a at different laser beam diameters is illustrated in Fig. 11.6. Similarly to Fig. 11.4, each individual curve is plotted for a fixed value of μ'_s (11 cm^{-1}, 17 cm^{-1}, and 24 cm^{-1}). The laser beam diameter d varies from plot to plot. The incident laser fluence is $F_0 = 1$ mJ/cm^2 for all plots presented in Fig. 11.4. Since the amplitude of the optoacoustic signal is proportional to the value of $\mu_a F_{max}$ (it will be demonstrated in Section 11.5.2), the curves presented in Fig. 11.6 also describe the dependencies of the optoacoustic signal amplitude on the optical absorption coefficient, μ_a.

The function $\mu_a F_{max}(\mu_a)$ is nonlinear [Fig. 11.6(a)] for all considered values of μ'_s under the condition of $d/l^* \gg 1$. This is due to the decrease in the coefficient $k = F_{max}/F_0$ with the increase in the optical absorption coefficient μ_a (see Fig. 11.5).

When the laser beam diameter is decreased to $d/l^* \sim 1$ [Fig. 11.6(b)], the diffusion of light across the laser beam affects the depth distribution of the optical fluence, which leads to the decrease in the coefficient k. Nevertheless, the subsurface amplification of the optical fluence is still present, and the curves corresponding to different values of μ'_s do not overlap.[72]

In the limiting case of $d/l^* \ll 1$, curves corresponding to different values of μ'_s must coincide. For example, in the case of $d = 0.5$ mm, the difference between the line slopes does not exceed 8%. The amplification of laser fluence in this case is on the order of 25%.

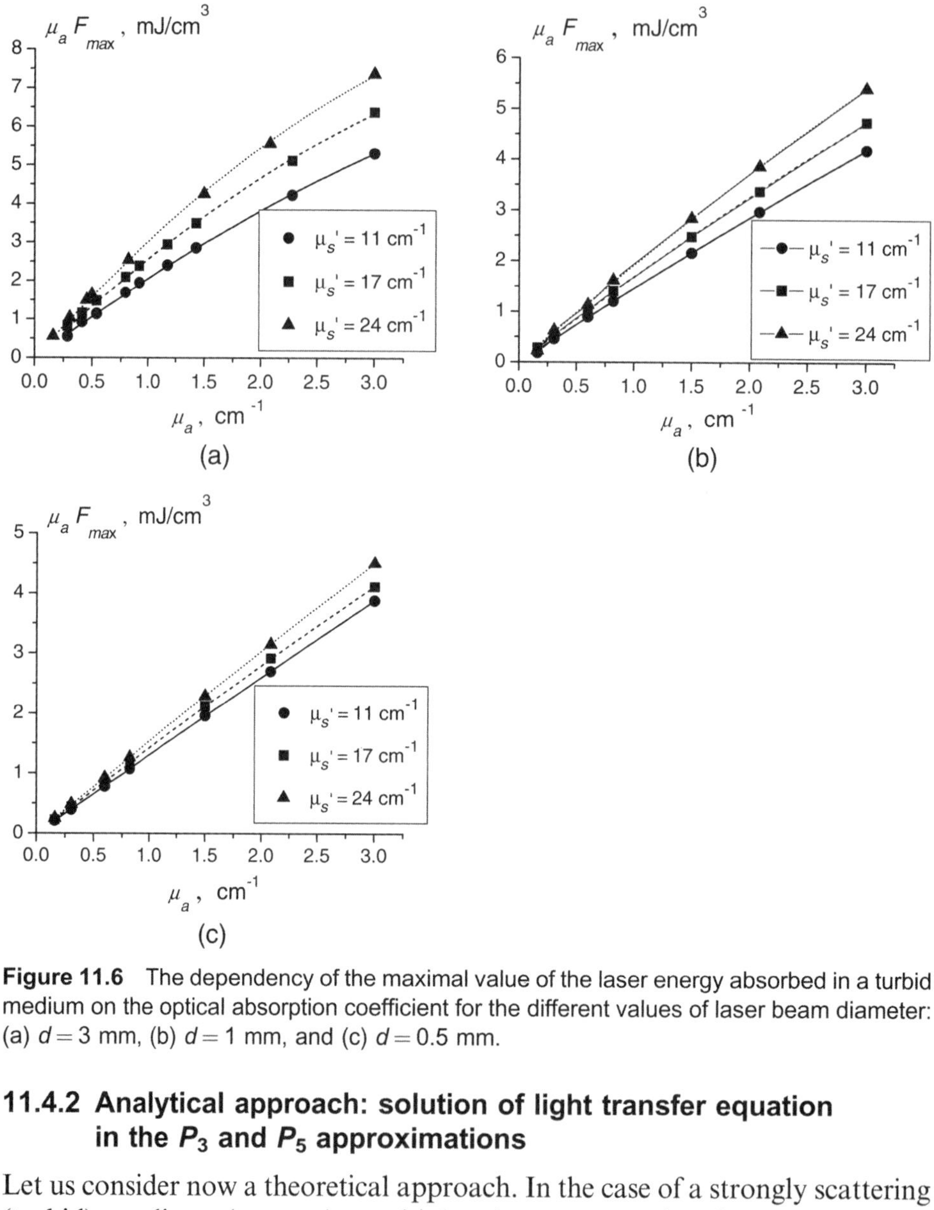

Figure 11.6 The dependency of the maximal value of the laser energy absorbed in a turbid medium on the optical absorption coefficient for the different values of laser beam diameter: (a) $d = 3$ mm, (b) $d = 1$ mm, and (c) $d = 0.5$ mm.

11.4.2 Analytical approach: solution of light transfer equation in the P_3 and P_5 approximations

Let us consider now a theoretical approach. In the case of a strongly scattering (turbid) medium ($\mu_a \ll \mu_s$), multiple photon scattering is achieved. The angular spectrum of the optical fluence within the medium can be found from the solution of light transport equation:[66]

$$(\vec{s} \cdot \vec{n}) \frac{\partial(L(z,\vec{s}))}{\partial z} = -\mu_t L(z,\vec{s}) + \mu_s \int_{4\pi} L(z,\vec{s}')p(\vec{s},\vec{s}')d\Omega', \qquad (11.4)$$

where $L(z,\vec{s})$ is the angular spectrum of the optical fluence inside of an optically scattering medium—that is, the part of the optical fluence at the point, z, within the unit solid angle, $d\Omega'$, propagated in the $\vec{s}$ direction; $P(\vec{s},\vec{s}')$ is the photon scattering anisotropy pattern; $\mu_t = \mu_a + \mu_s$ is the total optical attenuation coefficient; $\vec{s}$ and $\vec{s}'$ are the unit vectors determining the direction of the initial and the scattered photons, respectively; and $\vec{n}$ is the unit vector directed normally to the interface into the scattering medium.

In the case of strong scattering photons, the Henyey–Greenstein phase function is typically used to describe the photon scattering distribution $P(\vec{s},\vec{s}')$:[74]

$$p_{hg}(\vec{s},\vec{s}') = \frac{1}{4\pi}\frac{1-g^2}{[1+g^2-2g(\vec{s}\cdot\vec{s}')]^{3/2}}. \tag{11.5}$$

The optical fluence $F(z)$ at the point z can be obtained by integrating the function $L(z,s)$:[66]

$$F(z) = \frac{1}{4\pi}\int_{4\pi} L(z,\vec{s})d\Omega. \tag{11.6}$$

The angular spectrum of the optical fluence $L(z,\vec{s})$ inside the turbid medium can be expressed as the sum of coherent $L_{coh}(z)$ and diffuse $L_{\mathrm{dif}}(z,\vec{s})$ components.[53] The coherent component can be determined from the following solution:[75]

$$\frac{\partial(L_{coh}(z))}{\partial z} = -\mu_t L(z), \tag{11.7a}$$

The diffuse component obeys the photon transport equation:[73]

$$(\vec{s}\cdot\vec{n})\frac{\partial(L_{dif}(z,\vec{s}))}{\partial z} = -\mu_t L_{dif}(z,\vec{s}) + \mu_s\int_{4\pi} L_{dif}(z,\vec{s}')p(\vec{s},\vec{s}')d\Omega' + Q(z,\vec{s}), \tag{11.7b}$$

where $Q(z,\vec{s}) = \mu_s\int_{4\pi} L_{\mathrm{coh}}(z)P(\vec{s},\vec{s}')d\Omega'$ is the scattering optical source function. The boundary conditions for Eq. [11.7(b)] establish the relationship between the diffuse light propagated inside the optically turbid medium and the part of the diffuse light outgoing from the interface into the transparent medium (the diffusely back-scattered light):

$$\int_{\cos\theta>0} L(z=0,\vec{s})P_n(\vec{s}\cdot\vec{n})d\Omega = \int_{\cos\theta<0} R_{Fr}(-\vec{s}\cdot\vec{n})L(z=0,\vec{s})P_n(-\vec{s}\cdot\vec{n})d\Omega, \tag{11.8}$$

where $R_{Fr}(\vec{s}\cdot\vec{n})$ is the Fresnel reflection coefficient for unpolarized light.

The diffuse component of the angular spectrum of light fluence, $L_{dif}(z,\vec{s})$ in Eq. [11.7(b)], the phase function $P_{hg}(\vec{s},\vec{s}')$, and the optical source function $Q(z,\vec{s})$ can be presented as expansions in the Legendre polynomials:[75]

$$L(z,\vec{s}) = \sum_{k=0}^{N} \frac{2k+1}{4\pi} y_k(z) P_k(\vec{s}\cdot\vec{n}), \tag{11.9a}$$

$$p_{hg}(\vec{s},\vec{s}') = \sum_{k=0}^{N} \frac{2k+1}{4\pi} g^k P_k(\vec{s},\vec{s}'), \tag{11.9b}$$

$$Q(z,\vec{s}) = \sum_{k=0}^{N} \frac{2k+1}{4\pi} q_k(z) P_k(\vec{s}\cdot\vec{n}), \tag{11.9c}$$

where $P_k(\vec{s}\cdot\vec{n})$ is kth-order Legendre polynomials ($(\vec{s}\cdot\vec{n}) = \cos\theta$; θ is the angle between the z axis and the photon propagation direction inside the turbid medium); and $y_k(z)$ and $q_k(z)$ are the coefficients of the expansion of the functions $L_{dif}(z,\vec{s})$ and $Q(z,\vec{s})$ in Legendre polynomials, respectively.

The expansion coefficients of the source function $q_k(z)$ can be expressed in the form[75]

$$q_k(z) = \mu_s g^k F_0 \exp(-\mu_t z), \tag{11.10}$$

where F_0 is the incident optical fluence. The system of $(N+1)$ independent differential equations for expansion coefficients $y_k(z)$ can be obtained by substituting Eqs. (11.9a–c) in Eq. (11.7b), multiplying sequentially on $P_l(\vec{s}\cdot\vec{n})$, $l=(0, 1, \ldots, N)$, integrating over the whole solid angle 4π, and taking into account the orthogonality of the corresponding Legendre polynomials. In that case, the kth-order equation takes the form

$$\frac{k}{2k+1}\frac{dy_{k-1}(z)}{dz} + \mu_k y_k(z) + \frac{k+1}{2k+1}\frac{dy_{k+1}(z)}{dz} = q_k(z), \tag{11.11}$$

where $\mu_k = \mu_a + \mu_s(1-g^k)$, and coefficients $q_k(z)$ are determined from Eq. (11.10). The boundary conditions required for the solution of Eq. (11.11) can be derived in the form of Eq. (11.8). The function $L_{dif}(z,\vec{s})$ can be determined by solving Eq. (11.11), substituting the values of $y_k(z)$ into in Eq. [11.9(a)], and performing summation. The diffusion component of laser fluence $F_{dif}(z)$ inside the turbid medium can be found then with use of Eq. (11.6).

Since it is impossible to solve the system of infinite number of differential equations, the number of terms in the expansion of Eq. (11.9) should be restricted. The case $N=1$ corresponds to the diffusion approximation. In this approximation, the axial depth distribution of the optical fluence in a turbid medium $h(z) = F(z)/F_0$ can be described by[52,53]

$$h(z) = \exp(-\mu_t z) + \frac{3}{2\mu_{eff} l^*}\{\exp(\mu_{eff} l^*) - \exp[-\mu_{eff} l^*(2\Delta + 1)]\}\exp(-\mu_{eff} z), \tag{11.12}$$

where $\Delta = \frac{2}{3}\left(1 + R_{eff}/1 - R_{eff}\right)$.

The first term in Eq. (11.12) describes the coherent component of the optical fluence, which decreases rapidly with distance from the interface. At distances $z > (2\text{–}3)l^*$, the laser fluence is determined only by the diffused component of light, if $\mu_a \ll \mu_s$ and, consequently, $\mu_{eff} \ll \mu_s$. However, Eq. (11.12), which is obtained within frames of the diffusion model ($N = 1$), incorrectly describes the optical fluence beneath the surface of a turbid medium (in the region $0 \leq z \leq (2\text{–}3)l^*$; see, for example, Refs. 52 and 53). Therefore, to calculate the axial depth distribution of the optical fluence within the subsurface region of high expansion orders, N should be taken into account [Eq. (11.9)].[75]

For example, if $N = 3$ in Eq. (11.9), the system of four differential equations for the expansion coefficients $\{y_0(z)...y_3(z)\}$ of the function $L_{dif}(z,\vec{s})$ can be obtained from Eq. (11.11):

$$\mu_0 y_0(z) + \frac{dy_1(z)}{dz} = q_0(z), \tag{11.13a}$$

$$\frac{1}{3}\frac{dy_0(z)}{dz} + \mu_1 y_1(z) + \frac{2}{3}\frac{dy_2(z)}{dz} = q_1(z), \tag{11.13b}$$

$$\frac{2}{5}\frac{dy_1(z)}{dz} + \mu_2 y_2(z) + \frac{3}{5}\frac{dy_3(z)}{dz} = q_2(z), \tag{11.13c}$$

$$\frac{3}{7}\frac{dy_2(z)}{dz} + \mu_3 y_3(z) = q_3(z), \tag{11.13d}$$

where all values of $y_k(z)$ are taken for $z = 0$.

The boundary conditions in Eq. (11.8) for the system in Eq. (11.13) can be written as

$$4(-2R_1 + 1)y_0 + 8(3R_2 + 1)y_1 + 5(-12R_3 + 4R_1 + 1)y_2 + 28(5R_4 - 3R_2) = 0, \tag{11.14a}$$

$$(-20R_3 + 12R_1 + 1)y_0 + 12(5R_4 - 3R_2) + 5(-30R_5 + 28R_3 - 6R_1 + 1)y_2 + 2(175R_6 - 210R_4 + 63R_2 + 4)y_3 = 0, \tag{11.14b}$$

$$R_k = \int_0^{\pi/2} R_{Fr}(\theta)\cos^k\theta \cdot \sin\theta d\theta. \quad (11.15)$$

The expression for the axial depth distribution of the optical fluence in P_3 approximation can be determined by solving the system of Eq. (11.13) with the boundary conditions of Eq. (11.14), while taking into account that $y_k(z \to \infty) \to 0$:

$$h(z) = (F_{dif}(z) + F_{coh}(z))/F_0 = C_1 \exp(-\alpha_1 z) + C_2 \exp(-\alpha_2 z) + (m_0 + 1)\exp(-\mu_t z), \quad (11.16)$$

where the function $F_{coh}(z)$ can be found from Eq. [11.7(a)]; $C_{1,2}$ are the constants determined from Eq. (11.14); and m_0 and $\alpha_{1,2}$ are the complex functions of the turbid medium optical properties μ_a, μ_s, and g, which can be calculated numerically.

In the case of P_5 approximation to the light transport equation, the first six terms $y_k(z)$ in the expansion Eq. (11.9a) should be taken into account. In this case, the expression for the depth distribution of the optical fluence inside the scattering medium can be written in the form[68]

$$h(z) = (F_{dif}(z) + F_{coh}(z))/F_0 = \tilde{C}_1 \exp(-\tilde{\alpha}_1 z) + \tilde{C}_2 \exp(-\tilde{\alpha}_2 z) + \tilde{C}_3 \exp(-\tilde{\alpha}_3 z) + (m_k + 1)\exp(-\mu_t z). \quad (11.17)$$

The constants $\tilde{C}_{1,2,3}$ and m_k can be defined from the boundary conditions in Eq. (11.8) in the same manner as with P_3 approximation.

As it was mentioned earlier, the solution of the light transport equation in the diffusion approximation describes correctly the in-depth distribution of light fluence only far from the boundary, at distances where $z > (2\text{–}3)l^*$. Since only the first order in the expansion Eq. (11.9) is taken into account in the diffusion approximation, the angular spectrum of light fluence $L_{\text{dif}}(z,\vec{s})$ is formed only by the photons scattered in small angles. Near to the surface, however, the light fluence is defined by the whole angular spectrum of scattered light, including the back-scattered photons.

To find the validity range for the P_3 and P_5 approximations, the dependency of $Y = (z_{\max}\mu_{eff})$ on $X = (\mu_a/\mu_{eff})$ was plotted (Fig. 11.7) and the comparison of this dependency with Monte Carlo simulation was performed.

One can see a dramatic difference between three curves in Fig. 11.7 when $\mu_a/\mu_{eff} > 0.12$ and $\mu_a/\mu_{eff} > 0.15$ for three P_3 and P_5 approximations, respectively. Such discordance can be explained by the fact that with the increase of μ_a, the optical fluence maximum is shifted toward the surface of

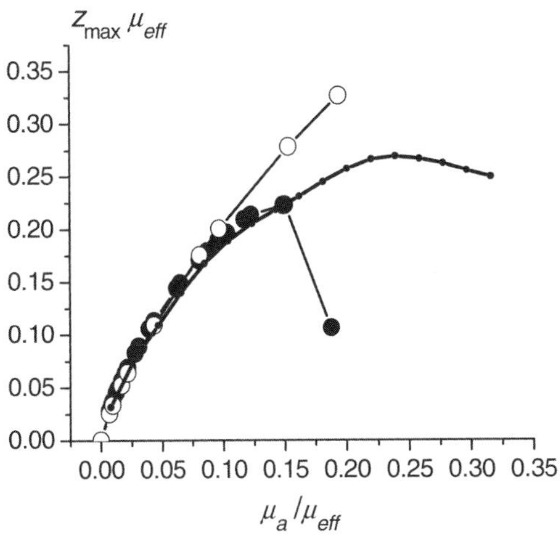

Figure 11.7 Comparison of the Monte-Carlo-simulated dependency of $Y=(z_{max}\ \mu_{eff})$ on $X=(\mu_a/\mu_{eff})$ (solid line) with the results obtained with the P_3 (open circles) and P_5 (dark circles) approximations. The anisotropy factor $g=0.95$.

the turbid medium. In this situation, six terms in the expansion Eq. (11.9) is not sufficient for accurate description of the optical fluence at the very subsurface layer of an optically scattering medium.

The results obtained allow us to evaluate the minimal distances from the turbid medium surface where the expressions defined in the P_3 and P_5 approximations describe light fluence correctly: $z>(0.6\text{–}0.7)l^*$ and $z>(0.4\text{–}0.5)l^*$, respectively.[68] For description of the optical fluence at smaller distances from the boundary, the higher orders in the expansion Eq. (11.9) are required. However, such a procedure is ambiguous, and Monte Carlo simulation represents a more suitable method for determining the optical distribution beneath the surface of an optically turbid medium.

11.5 Time-Resolved Optoacoustic Measurement of Depth Distribution of the Absorbed Optical Energy and Optical Properties in Scattering Media

11.5.1 Temporal profile of LIP

Let us assume that the medium under study is macrohomogeneous and can be described by some effective parameters: the specific heat C_p, the speed of sound c_0, the effective thermal expansion coefficient β, and the thermal conductivity χ. We consider a laser beam illuminating the boundary $z=0$ of an optically turbid medium from the side of a transparent medium. The laser fluence rate in the incident beam can be represented as

$$I_{ref}(r_\perp,\tau) = I_0 L(\tau) R(r_\perp), \tag{11.18}$$

where $R(r_\perp)$ and $L(\tau)$ are the dimensionless distributions of laser fluence rate over the transverse coordinate and time, respectively. The component $R(r_\perp)$ is usually taken in the Gaussian form:

$$R(r_\perp) = \exp\left(-\frac{4r_\perp^2}{d^2}\right). \tag{11.19}$$

The distribution of laser fluence rate along the beam axis inside the turbid medium can be represented as

$$I(z,r_\perp = 0,\tau) = I(z)L(\tau). \tag{11.20}$$

If the heat confinement condition (i.e., the relaxation time $\tau_{th} \sim 1/\mu_{eff}\chi$ of heat sources induced by laser impact over the heated region is much longer than the laser pulse duration, τ_L) is fulfilled and the irradiating laser pulse is infinitely short, $L(\tau) = \tau_L\delta(\tau)$, heating of a turbid medium can be considered as instantaneous and the thermal sources as "frozen." The profile of the LIP generated in the turbid medium and detected in forward mode (from the side of the scattering medium) can be presented as[46–48]

$$p(\tau = t - z/c_0,r_\perp = 0) = \Gamma\mu_a F_0\{H(c_0\tau)\vartheta(-\tau) + R_{ac}H(-c_0\tau)\vartheta(\tau)\}, \tag{11.21}$$

where $\vartheta(\tau)$ is the Heaviside function, $R_{ac} = (1 - N)/(1 + N)$ is the reflection coefficient of the ultrasonic wave from the interface of the turbid and transparent media, N is the ratio of acoustic impedances of these media, $\Gamma = \frac{\beta c_0^2}{2C_p}$ is the efficiency of the thermoacoustic transformation, and $F_0 = I_0\int_{-\infty}^{\infty} L(\tau)d\tau$ and $F(z) = I(z)\int_{-\infty}^{\infty} L(\tau)d\tau$ describe the relationship between laser fluence and its rate in the incident laser beam and in the turbid medium.

Moreover, the LIP also propagates in the opposite direction—that is, inside the transparent medium (backward detection mode; see Section 11.2). Its temporal profile has the following form:

$$p_{tr}(\tau_{tr} = t - z/c_{tr},r_\perp = 0) = (1 + R_{ac})\Gamma\mu_a F_0 H(-c_{tr}\tau_{tr})\vartheta(\tau_{tr}). \tag{11.22}$$

In the case of a short but finite laser pulse ($\mu_{eff}c_0\tau_L < 1$), it is necessary to take into account the expansion of the released heat sources in the course of the laser pulse, which results in spreading the front slope of the optoacoustic signal. Therefore, the profiles given by Eqs. (11.21) and (11.22) take the form

$$p(\tau,r_\perp=0)=\Gamma\mu_a I_0 * \int_{-\infty}^{\infty} L(t)\{H(c_0(t-\tau))\vartheta(t-\tau)+R_{ac}H(c_0(\tau-t))\vartheta(\tau-t)\}dt, \tag{11.23}$$

$$p_{tr}(\tau_{tr},r_\perp=0) = R_{ac}\Gamma\mu_a I_0 \int_0^{\infty} L(t)H(c_{tr}(t-\tau_{tr}))\vartheta(\tau_{tr}-t)dt. \tag{11.24}$$

11.5.2 Diffraction transformation of the LIP

The diameter of the laser-induced acoustic beam initially coincides with that of the laser beam. Depending on the relationship between the beam diameter d and the effective optical penetration depth μ_{eff}^{-1}, the phase front of the acoustic wave can be almost plane (at $\mu_{eff}d \gg 1$), cylindrical (at $\mu_{eff}d \sim 1$), or spherical (at $\mu_{eff}d \ll 1$). In the case of cylindrical and spherical phase fronts, the profile of the optoacoustic signal does not change in the course of propagation.[45] In contrast, if the wavefront is initially almost planar, the situation might be more complex. The acoustic beam cross section, $\pi d^2/4$, doubles after propagation of a distance called the effective diffraction length,

$$L_D = \pi d^2/4\lambda = \pi d^2 f/4c_o, \tag{11.25}$$

where λ is the wavelength of the acoustic wave with ultrasonic frequency f. The beam becomes almost spherical at distances of $z > 3L_D$.[45]

LIP profiles with sharp and smooth slopes contain a wide spectrum of ultrasonic frequencies, including low acoustic frequencies and high-frequency ultrasound. Since the effective length of the acoustic diffraction is directly proportional to the ultrasonic frequency, the influence of acoustic diffraction on the optoacoustic signal profile in the course of its propagation may vary depending on the laser pulse's duration and the optical properties of the irradiated turbid medium. The low-frequency components of the LIP spectrum diffract earlier (at shorter distances) than the high-frequency components. Nevertheless, to achieve a high spatial resolution and simultaneously image large objects in tissue, it is necessary to detect optoacoustic signals in as wide an ultrasonic frequency range as possible. In optoacoustic experiments, it is appropriate to minimize acoustic diffraction by employing laser beams as wide as is reasonable and by detecting the acoustic waves propagating normally to the boundary of the optically absorbing medium. The planar wave geometry of optoacoustic interaction enables direct studies of light intensity distribution and properties of optically turbid medium without complex signal processing.[67,68]

In practical situations, acoustic diffraction is unavoidable. As an example, consider detection of LIP in the backward mode using an optically transparent standoff (between the transducer and the tissue) made of fused silica or PMMA. The high ultrasound velocity of the standoff medium makes the effective diffraction length short. Therefore, acoustic diffraction is of great significance for optoacoustic measurements in tissue with only a single surface access that requires optoacoustic front-surface transducers operating in the backward mode.[48,60] The diffraction transformation of the optoacoustic signal profile in the paraxial approximation may be described by the equation[45]

$$\frac{\partial^2 p}{\partial\tau\partial z} = \frac{c_0}{2}\Delta_\perp p, \tag{11.26}$$

where $\Delta_\perp$ is the Laplace operator on the transverse coordinates $\{x, y\}$. In case of a wide beam ($\mu_{eff} d \geq 3$), the boundary condition can be specified as for the profile of planar LIP calculated earlier in Eqs. (11.21) through (11.24).

Assuming Gaussian transverse distribution of the optical fluence, as Eq. (11.19), in the laser beam, the boundary condition of the diffraction problem may be expressed as in Ref. 46:

$$p(z = 0,\tau,r_\perp) = p_0(\tau)\exp(-4r_\perp^2 d^2), \tag{11.27}$$

where $P_0(\tau)$ is the wave profile at the boundary of the light-absorbing medium. When the optical pulse propagates, the transverse light intensity distribution remains Gaussian for every harmonic component in Eq. (11.27), so that the solution may be written in the form

$$p(z,\tau,r_\perp) = \int_{-\infty}^{\infty} p_0(\tau)d\tau$$

$$\times \int_{-\infty}^{\infty} \exp\left(-i2\pi f(\tau - t) - \frac{4r_\perp^2}{d^2}\frac{f}{f + i\frac{4c_0 z}{\pi d^2}}\right)\left(f + i\frac{4c_0 z}{\pi d^2}\right)^{-1} f df. \tag{11.28}$$

On the beam axis ($r_\perp = 0$), the solution in Eq. (11.28) can be simplified as follows:

$$p(z,\tau,r_\perp = 0) = p_0(\tau) - \int_{-\infty}^{\tau} 2\pi f_D \exp(-2\pi f_D(\tau - t))p_0(t)dt, \tag{11.29}$$

where $f_D = \frac{4c_0 z}{\pi d^2}$ is the characteristic diffraction frequency.

The first term in Eq. (11.29) replicates the profile of the signal at the boundary and corresponds to the planar wave without diffraction. The second term describes the effect of a limited beam width. The characteristic diffraction frequency f_D increases with increasing propagation distance and decreasing cross-sectional area of the acoustic beam.

The diffraction transformations of the LIP profile can be corrected by the formula

$$p_0(\tau) = p(\tau) + 2\pi f_D \int_{-\infty}^{\tau} p(t)dt, \quad (11.30)$$

which can be derived from Eq. (11.29).

In the Fraunhofer (far-field) zone ($f_D t_a \gg 1$, $z \to \infty$), the solution of Eq. (11.29) becomes

$$p(z,\tau,r_\perp = 0) = \frac{d^2}{8c_0 z}\frac{dp_0(\tau)}{d\tau}. \quad (11.31)$$

As seen from Eq. (11.30), the profile of LIP in the Fraunhofer zone resembles the derivative of the profile at the boundary, and its amplitude decreases inversely proportional with the distance of LIP propagation to the detector.

For example, the LIP profile in Eq. (11.24) detected in the backward mode takes the following form in the Fraunhofer zone[73]:

$$p_d(z,\tau,r_\perp = 0) = \frac{d^2}{8c_{tr}z}\frac{\partial p}{\partial\tau} = \frac{d^2}{8c_{tr}L}(1+R_{ac})\Gamma\mu_a \times \left[\frac{I_0}{F_0}F(z=0)L(\tau) + \frac{\partial F(c_{tr}\tau)}{\partial\tau}\vartheta(\tau)\right], \quad (11.32)$$

where c_{tr} is the speed of sound in the transducer piezoelectric material.

11.5.3 Absorbed optical energy profiles measured in forward mode

The choice of the optoacoustic signal detection mode depends on the application and the diagnostic problem. First let us consider the forward mode. In this case, the medium under study is placed between a plate transparent to laser radiation, through which it is irradiated, and the surface of a wideband piezoelectric transducer [see Fig. 11.1(a)]. This provides the controllable thickness of the medium and the smoothness of its surfaces.

If the conditions of thermal and pressure confinement are fulfilled, the temporal profile of LIP can be described by Eq. (11.21). In forward mode under illumination with nanosecond laser pulses within the visible or infrared range of wavelengths, the influence of the laser pulse duration, described by Eq. (11.23), on the temporal profile of LIP generated in biological tissues is usually insignificant. A typical temporal profile of LIP detected in forward mode is shown in Fig. 11.8(a).

Such a bipolar form of LIP is explained by the influence of acoustic diffraction, as in Eq. (11.29). Low frequencies are subjected to stronger distortions associated with diffraction than are higher frequencies. This, in turn, results in shape distortions of the optoacoustic signal. The effect of the diffraction distortions can be eliminated analytically by using Eq. (11.30). The

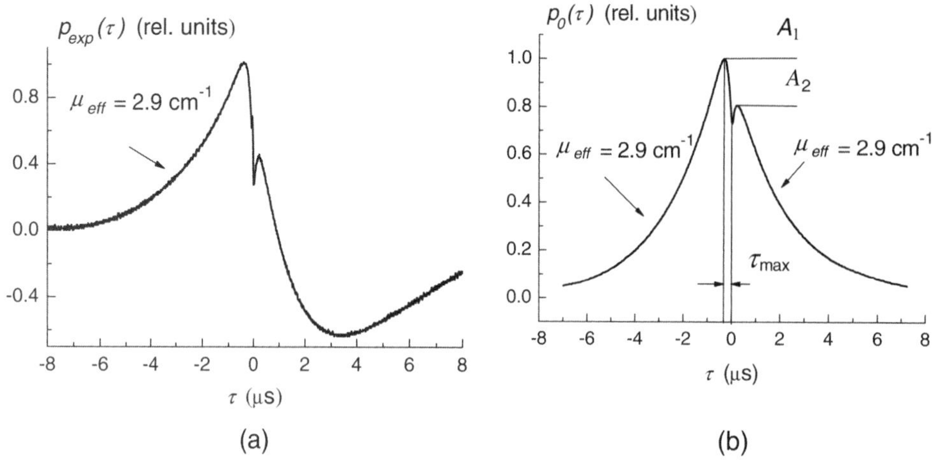

Figure 11.8 (a) Typical temporal profile of LIP experimentally measured in an optically turbid medium in forward mode. (b) The same profile of LIP after correction for acoustic diffraction.

temporal profile of LIP after diffraction correction is shown in Fig. 11.8(b). The procedure described by Eq. (11.30) is not always performed successfully, because the diffraction transformation can affect the detector, and the laser beam diameter can be determined inaccurately.

The leading edge of LIP, $P(\tau < 0)$, repeats the spatial distribution of absorbed optical energy and deposited heat. The LIP profile according to Eq. (11.21) is proportional to the optical fluence distribution $H(z)$ in the medium, and is influenced by the time domain variations in $P(\tau)$ and the spatial domain variations in $H(z)$. Time is related to space through the speed of sound in the medium as $z = -c_0\tau$. Therefore, if correction for diffraction is performed accurately, the leading edge of the LIP can be used for reconstructing the axial depth distribution of the optical fluence in a scattering medium.[46,48,67,68]

11.5.4 Determination of the effective optical attenuation, absorption, and reduced scattering coefficients

While knowledge of the depth distribution of the absorbed optical energy is important for many biomedical applications, the ultimate goal is to measure optical properties of the medium under study. Values of the optical properties can then be used to model the light distribution through MC simulation, if necessary.

Let us consider now the problem of optical properties determination using the optoacoustic method, while operating in the forward mode. As it was shown in Section 11.4, in the subsurface region of a turbid medium the function $H(z)$ can be calculated correctly only with MC simulation, while at distances $z > (2\text{–}3)l^*$ this function has the following analytic form within the framework of a simple diffusion model [see Eq. (11.12)]:[47,67]

$$p(z) = \frac{\beta c_0^2}{4c_p}\mu_{eff}F_0\frac{3}{2\mu_{eff}l^*}\{\exp(\mu_{eff}l^*) - \exp[-\mu_{eff}l^*(2\Delta+1)]\}\exp(-\mu_{eff}z). \tag{11.33}$$

Therefore, the first approach to measuring the optical properties of a scattering medium is to determine the absolute pressure value in the leading edge of the LIP and its subsequent approximation in the region of $|\tau| > (2\text{–}3)l^*/c_0$ by the theoretically derived expression given in Eq. (11.33). To realize this goal, the incident laser fluence F_0, the thermoacoustic efficiency (Gruneisen parameter, Γ), and the absolute sensitivity of the piezoelectric transducer must be known and the diffraction distortions of LIP profile compensated. Such a procedure was described in detail in Refs. 48 and 53.

The optical properties of a scattering medium can be also determined using a simpler approach[67,68] and without measurement of the absolute value of the signal pressure. Indeed, the effective optical attenuation coefficient μ_{eff} can be determined by fitting the leading edge of the LIP with an exponential function where compensation of the LIP profile for distortions associated with acoustic diffraction is not required. The position of the maximum $z_{\max}$ of the laser fluence can be also determined from the shape of the LIP. The dependence of $z_{\max}\mu_{eff}$ on the ratio of μ_a/μ_{eff} (which is similar to that in Fig. 11.7) was measured in Ref. 68 for media with different optical properties relevant to biological tissues in the visible and near-infrared ranges of the optical spectrum. This empirical dependence was shown to have universal character within the considered range of optical properties. In the range of the anisotropy factor of $0.8 < g < 0.95$, it depends only on the ratio of μ_a/μ_{eff}; it does not depend separately on the optical properties of tissue (μ_a and μ_{eff}):

$$Y_{fit} = 0.276\left[1 - \exp\left(-13.42\frac{\mu_a}{\mu_{eff}}\right)\right]. \tag{11.34}$$

Figure 11.9 compares the empirical and the Monte Carlo–simulated curves.

The optical absorption coefficient can be determined from the universal dependence, given by Eq. (11.34), if the values of μ_{eff} and $z_{\max}$ are measured:

$$\mu_a = -0.074\mu_{eff}\ln(1-3.62z_{\max}\mu_{eff}). \tag{11.35}$$

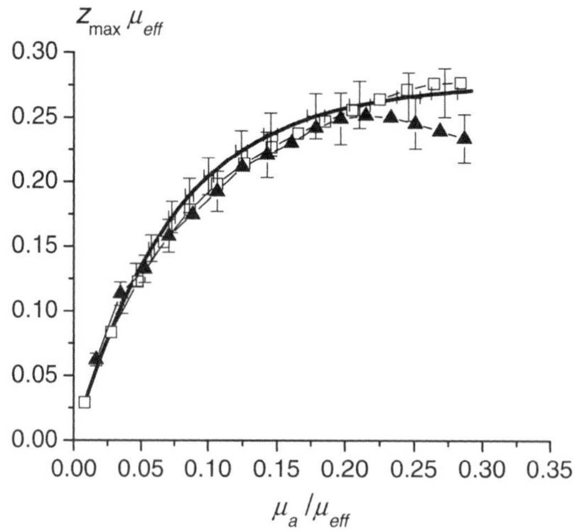

Figure 11.9 Comparison of the empirical dependency of $Y = (z_{max}\ \mu_{eff})$ on $X = (\mu_a/\mu_{eff})$ (solid line) with results obtained from Monte Carlo simulation for $g = 0.8$ (open squares) and $g = 0.95$ (solid dark triangles).

The reduced light scattering coefficient μ_s' can be determined with the diffusion relationship[67,71]

$$\mu_s' = \mu_{eff}^2/3\mu_a. \tag{11.36}$$

Thus, a complete set of optical properties for an optically turbid medium can be determined with a temporal profile of LIP, measured in relative units.

For example, using the procedure described here, the optical properties of quasi-homogeneous biological tissues (porcine and bovine liver and bovine muscle) and axial depth distributions of the optical fluence were measured *in vitro* at wavelengths of 1064 and 779 nm.[76] Unfortunately, the forward mode of LIP detection has some disadvantages. First, two-sided access to an object under study is required to perform such measurements. Second, the laser beam diameter should be wider than 10 mm to avoid dramatic diffraction distortions of the LIP profile. This requirement means that the measurements cannot be made locally with high spatial resolution. In addition, samples should be flat, plane-parallel, and sufficiently thick (approximately 2–3 cm). As a result, this method is basically suitable for *in vitro* measurement, while *in vivo* measurements can be performed only in special cases. The procedure used for *in vitro* measurements is usually associated with significant loss of blood and other effects that alter the tissue properties. Therefore, the measured optical properties will depend on the sample preparation procedure and differ from those found in the tissue of a living organism.

11.5.5 Possibility of *in vivo* measurements of tissue optical properties in backward mode

The ability to measure the optical absorption coefficient and the effective optical attenuation coefficient using the temporal profiles of the optoacoustic signals, detected in backward mode (both generation and detection of the optoacoustic signals are performed at the same surface of tissue), was originally demonstrated for the case of a uniformly absorbing and non-scattering medium.[57] The advantage of using the backward mode is that it opens up the opportunity to perform *in vivo* diagnostics having only one-sided access to the tissue surface. However, in case of an optically scattering medium, utilization of the backward mode for LIP detection and determination of the optical properties requires special attention to deconvolution of the medium and transducer transfer functions—specifically, reconstruction of the intrinsic profile of the absorbed optical energy through compensation of the optoacoustic signal distortions in the course of propagation and detection.[60]

Let us consider in detail the temporal profile of LIP detected in backward mode. The temporal shape of the intrinsic laser-induced pressure is described by Eq. (11.24) and shown in Fig. 11.10 (dashed line).

As one can see in Fig. 11.10, the temporal profile of LIP in the region of $\tau > 0$ corresponds to the axial depth distribution of heat. In the case of a wide laser beam, $d/l^* \gg 1$, the axial depth distribution of the optical fluence at the distances $z > (2\text{–}3)l^*$ is $F(z > (2\text{–}3)l^*) \sim \exp(-\mu_{eff}z)$. Therefore, if the laser beam is wide, the optical attenuation coefficient, μ_{eff}, can be determined by fitting the exponential function to the trailing edge of LIP within the range of

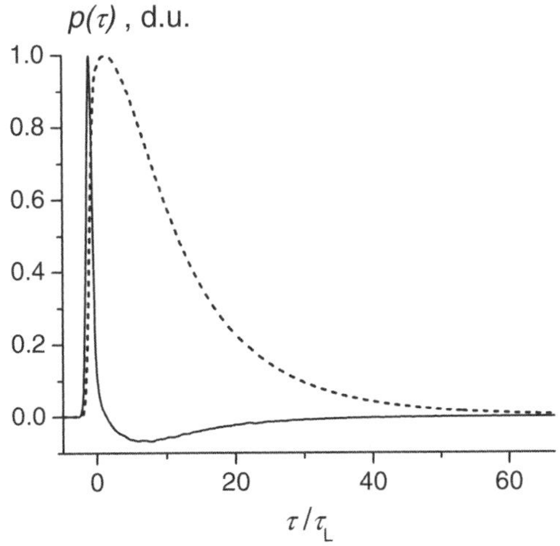

Figure 11.10 A typical temporal profile of LIP generated in an optically absorbing and scattering medium and detected in backward mode: near-field signal—that is, without taking into account the signal transformation due to acoustic diffraction (dashed line)—and far field signal strongly diffracted upon propagation through the transducer standoff (solid line).

$\tau > (2\text{–}3)l^*/c_0$. Such a procedure, however, put practical limitations on *in vivo* measurements. First, since the value of $\mu_s' \approx 10\ \text{cm}^{-1}$ is typical for most biological tissues within the near-infrared range,[77] the laser beam diameter has to be wider than 10 mm to provide a 1D regime of light diffusion into the medium. This leads to a substantial decrease in the lateral resolution of the measured optical properties. Second, if the laser beam is wide, the determination of the optical absorption coefficient from the LIP amplitude is also difficult, because the coefficient of subsurface amplification of the optical fluence, k, in turbid media depends on the ratio of μ_a/μ_s' and the value of d/l^* (see Figs. 11.4 and 11.5).

If the laser beam is narrow, $d/l^* < 1$, the coefficient $k \rightarrow 1$ and does not depend on the ratio of optical coefficients μ_a/μ_s' (see Fig. 11.4). In this case, a strong diffraction-driven transformation of the acoustic signals occurs within the transparent medium where the signals are detected.[46,48,60] This leads to suppression of the low-frequency components of the optoacoustic signal spectrum and, consequently, to formation of the rarefaction part in the LIP profile.[46,60,73] If the signal is strongly diffracted, $f_D \gg f_a$ ($f_D = 4Lc_{tr}/\pi d^2$ is the acoustic wave frequency at which the diffraction length is equal to the medium depth L; $f_a = \mu_{eff}c_0/2\pi$ is the characteristic frequency of the signal spectrum), then the LIP temporal profile $P_D(\tau)$ can be described by Eq. (11.32).

The normalized temporal profile of $P_D(\tau)$ is shown in Fig. 11.10 with a solid line. The first component of Eq. (11.32) repeats the laser pulse temporal profile and has a maximum at $\tau = 0$. Since the laser fluence $F(z)$ inside a turbid medium is a smoothly varying function, the second component of Eq. (11.32) is many times smaller than the first one. Therefore, if the laser beam is narrow, the LIP amplitude is proportional to the maximum value of the absorbed laser energy density $\mu_a F_{max}$ within the medium, which depends linearly on the optical absorption coefficient μ_a and practically does not depend on the reduced scattering coefficient μ_s' [see Fig. 11.6(c)].

Thus, the local optical absorption coefficient would be measured from the amplitude of LIP generated by a narrow laser beam, $d/l^* <\sim 1$, in a turbid medium. The *in vitro* determination of optical absorption for some tissues was performed in Refs. 76 and 77 without rigorous justification of the fact that the LIP amplitude is proportional to the optical absorption coefficient under the narrow laser beam condition. Experimental studies were performed to validate the Monte Carlo simulation results and the theoretical approach described in Section 11.4.[73] The dependencies of the LIP amplitude on the turbid media optical properties for different laser beam diameters were measured in the backward detection mode. A specially designed optoacoustic front-surface transducer (OAFST) was developed to perform these measurements. The operation principle and the photograph of the OAFST are depicted in Fig. 11.11. A diode-pumped Nd:YAG laser (pulse energy of $E_0 =$ 100 μJ, pulse duration, $\tau_L = 10$ ns, repetition rate, 500 Hz) operated at the

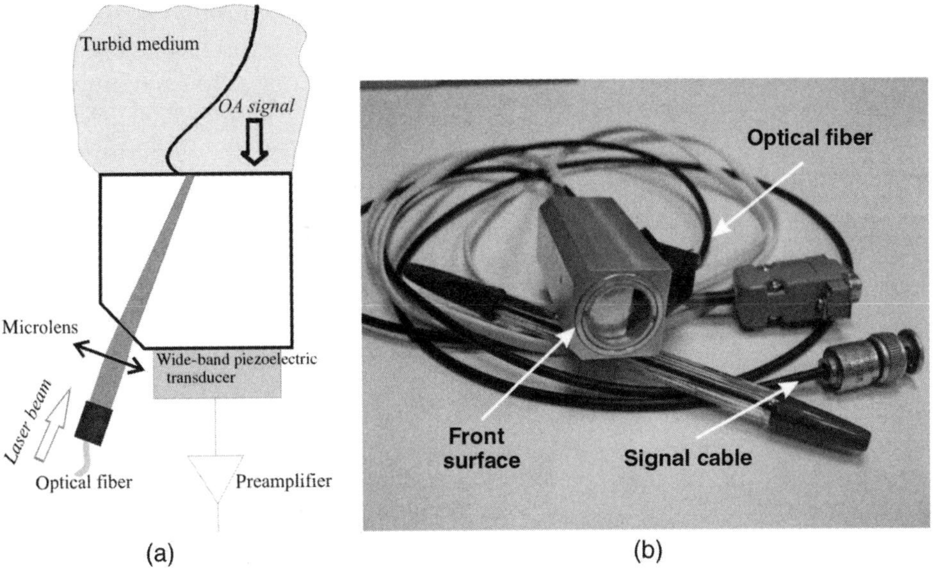

Figure 11.11 OAFST operating in backward mode: (a) schematic diagram of operating principles, and (b) photograph.

fundamental wavelength was used to illuminate the examined optically turbid media. Near-infrared light was delivered to OAFST with optical fiber and was incident onto the surface of the turbid medium through a standoff PMMA prism [Fig. 11.11(a)]. The incidence angle was approximately 18 deg and could not cause any significant differences in the light distribution compared with normal incidence owing to a strong optical scattering inherent to the media under study. The generated optoacoustic signals propagated back through the PMMA prism and were detected at its rear surface by a wide-band ultrasonic transducer made of 50-μm-thick polyvinylidene fluoride (PVDF). The processed electrical signal was amplified and digitized by an analog-to-digital converter (ADC). The final data processing was performed on a personal computer (PC).

The laser beam diameter on the illuminated surface could be varied by a set of microlenses embedded into the transducer. Measurements were performed for two different laser beam diameters, $d_1 = (3.0 \pm 0.2)$ mm and $d_2 = (0.6 \pm 0.2)$ mm, that were determined experimentally using a CCD camera.

The experimentally obtained dependencies of LIP amplitude as a function of the optical absorption coefficient, μ_a, for the laser beam diameters of d_1 and d_2 are shown in Figs. 11.12(a) and (b), respectively. Triangles, squares, and circles on the plots correspond to experimental data for the media with $\mu_s' = 11\,\text{cm}^{-1}$, $\mu_s' = 17\,\text{cm}^{-1}$, and $\mu_s' = 24\,\text{cm}^{-1}$, respectively, in black India ink diluted in milk with fat content varying from 1.5% to 3.5% to 6%. Note

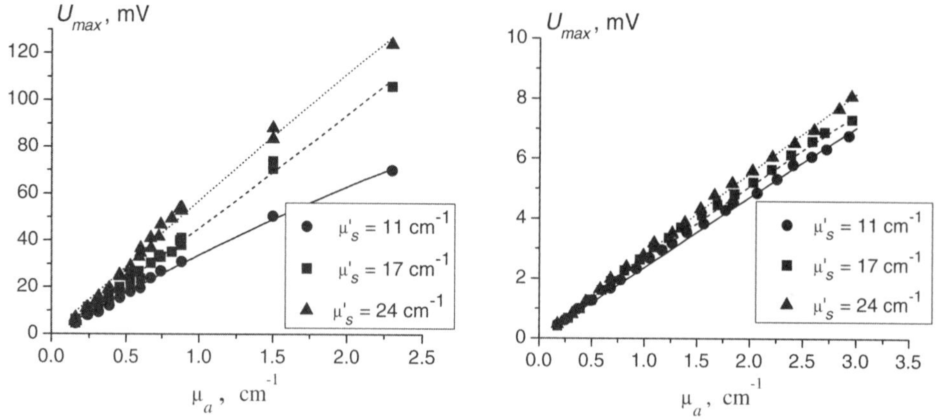

Figure 11.12 Curves showing LIP amplitude as a function of the optical absorption coefficient, μ_a, of a turbid medium for different reduced scattering coefficients, μ'_s (see circles, squares, and triangles), and for two of the laser beam diameters, (a) $d_1 = 3$ mm and (b) $d_2 = 0.6$ mm. Symbols correspond to the experimental data, whereas solid lines represent the results of Monte Carlo simulations.

that the same dependencies numerically calculated with the MC method were presented in Section 11.4 in Fig. 11.6. The experimental and numerically calculated curves were obtained in different units that were proportional to each other. The proportionality coefficient depends on a number of parameters, such as the transducer sensitivity, the thermoacoustic efficiency Γ, the laser beam diameter, and the PMMA prism length. Therefore, for direct comparison of the theoretical and experimental results, the normalization procedure was performed as follows. The angular coefficients of the experimental lines (Fig. 11.12, symbols) and the theoretical lines [Figs. 11.6(a) and (b)] for the reduced scattering coefficient of $\mu'_s = 17\,\mathrm{cm}^{-1}$ were calculated within the range of $\mu_a < 1$ cm^{-1}. Then the ratios of these coefficients for the laser beam diameters $d_1 = 3$ mm and $d_2 = 0.6$ mm were obtained: $\gamma_1 = 0.059$ (mJ/cm^3)/mV and $\gamma_2 = 0.49$ (mJ/cm^3)/mV, respectively. Finally, MC simulations performed for all reduced scattering coefficients were normalized by γ_1 [Fig. 11.3(a)] for the laser beam diameter $d_1 = 3$ mm, and by γ_2 [Fig. 11.12(b)] for the diameter, $d_2 = 0.6$ mm, respectively.

Figure 11.12(a) shows that with a sufficiently wide laser beam of $d_1 =$ 3 mm, the slopes of the dependencies of LIP amplitude on the optical absorption coefficient obtained for a twofold range of the reduced optical scattering coefficient, μ'_s, differ by 15%. As the laser beam diameter decreases down to $d_2 = 0.6$ mm, the difference becomes less noticeable—approximately 8% within the twofold range of μ'_s. Therefore, it is possible to plot a mean universal curve to determine the optical absorption coefficient from the LIP amplitude in any scattering media with unknown values of μ'_s.

In this section we have demonstrated the possibility of measuring the LIP amplitude and determining the optical absorption coefficient in optically turbid media with properties resembling biological tissues in the visible and near-infrared ranges. Since optoacoustic measurements were performed at the site of the optical illumination, this method is applicable to the clinical environment. Application of diode-pumped lasers makes it possible to perform noninvasive diagnostics and to achieve the high throughput of measurements that is necessary for biomedical applications. The specially designed OAFST is the essential component of this measurement technique. The mean universal curve that can be obtained from the dependencies illustrated in Fig. 11.12(b) and used for determining the optical absorption coefficient in turbid media is device specific, because the normalization coefficient γ depends on the transducer sensitivity, laser beam diameter, and PMMA prism length. It allows one to determine the local optical absorption coefficient of biological tissue *in vivo* from the LIP amplitude (if the thermoacoustic efficiency, Γ, is known). The smaller the incident laser beam, the better the accuracy and the spatial resolution of the measurements. For example, a relative error of 8% in measurements of the optical absorption coefficient can be achieved if the laser beam diameter is $d_2 = 0.6$ mm. The key advantages of this optoacoustic technique are the absence of complex data processing and the lack of *a priori* assumptions concerning the medium's micro and macro structures.

11.6 Technical Requirements for Time-Resolved Optoacoustic Detection

The main technical difficulty of the time-resolved measurements of optoacoustic profiles is associated with detection of the acoustic signals, which requires simultaneously high temporal resolution equal to or higher than the laser pulse duration and an ultrawide frequency-detection band extending to low ultrasonic frequencies. Ultrawide-band piezoelectric transducers, being low-noise detectors, have proved most suitable for this purpose.[58,65]

Wide-band piezoelectric transducers have two operation modes: short circuit and open (idler) circuit. The electrical circuits of the two modes of transducer operation are presented in Fig. 11.13.

A transducer operating in the short circuit mode has a substantial thickness that is larger than the spatial width of the detected ultrasonic transient [Fig. 11.13(a)]. The transducer thickness limits the lower limit of detected ultrasonic frequencies and the duration of the detection window. An excessively thick piezoelement would (1) lead to a more prominent acoustic diffraction at lower ultrasonic frequencies in the detected signal and (2) reduce the electric capacity to a value below the electric capacity of the electronic

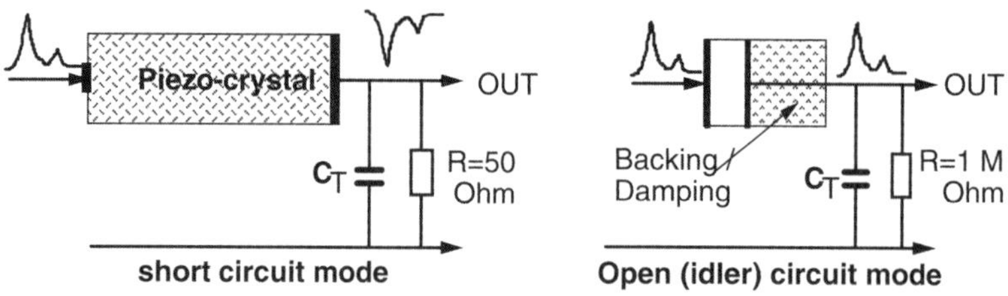

Figure 11.13 Electric circuitry for (a) short circuit and (b) open (idler) circuit operation modes of wide-band acoustic transducers.

circuitry. It is difficult to design acoustic transducers operating in the short circuit mode for detection of optoacoustic profiles longer than 1–2 μs. However, these transducers would be most optimal for detection of submicrosecond and nanosecond optoacoustic signals. The upper limit of ultrasonic frequency is defined by the discharge time of the transducer capacity, C_T, through the resistor, R. With low values of RC_T, the ultrasonic detection band could reach several hundred megahertz. The optimal value of the resistor R equals the wave resistance of the electrical cable (50–120 Ω). Transducers of this type do not require backing material to reduce resonant vibrations, which occur outside of the window for undisturbed measurements.

In case of an acoustic transducer operating in the open circuit mode [Fig. 11.13(b)], it is necessary that the thickness of the piezoelectric element be smaller than the acoustic wavelength detected in its upper ultrasonic frequency limit. The lower limit of detectable ultrasonic frequencies is defined by the discharge time of transducer electric capacity, C_T, through the input resistor of the electronic preamplifier, R. Therefore, it is desirable to design these acoustic transducers in one and the same housing with the charge preamplifier. To design acoustic transducers operating in the open circuit mode with a very wide ultrasonic detection band (≥100 MHz), one must employ very thin piezoelectric elements with a thickness of just a few microns.

The technical difficulties of such a design are associated with the nonflatness of the piezoelement and the necessity for effective damping of high-frequency resonances. The choice of appropriate backing material is crucial for widening the frequency band of these transducers and for simultaneous widening of the temporal window of undisturbed measurements. Design of the backing layer acoustically matched with the piezoelectric material for effecting damping of transducer resonances is one of the most important technical problems. These technical difficulties, however, are offset by the higher sensitivity of transducers operating in the open circuit mode. The sensitivity of open circuit transducers is greater than that of short circuit transducers due to the longer holding time of the electric charge at the piezoelement.[65]

One additional technical difficulty in providing high temporal resolution in the detection of optoacoustic profiles with thick transducers operating in the short circuit mode is the need for precise adjustment of the detector face with respect to the wavefront of the arriving optoacoustic signal. Obviously, the time difference between the instances of LIP arrival to the opposite edges of the piezoelectric element should be shorter than the temporal resolution of this transducer. Therefore, the angle between the direction of the LIP wavefront propagation and the normal to the piezoelectric detector must be as small as possible. For example, to achieve a temporal resolution of $\Delta\tau \approx 10$ ns, LIP has to be incident with an angle less than $\vartheta \approx 3'$ relative to the sensitive area aperture of $D \approx 3$ mm ($\vartheta << c_{tr}\ \Delta\tau/D$).

In summary, one may conclude that various biomedical applications will favor either open circuit or short circuit designs. While open circuits are more popular in the range of low-to-medium frequencies, the short circuit design can be beneficial in applications demanding very-high-frequency operation.

11.7 Summary and Biomedical Applications

The experimental data and theoretical models presented in this chapter show that the optoacoustic time-resolved detection of laser-induced pressure signals is a powerful method for directly visualizing axial depth profiles of the absorbed optical energy distribution in scattering and absorbing media, including biological tissues. The optical diffusion model can describe spatial distributions of the optical energy absorbed in an optically scattering medium only at depths greater than the effective optical diffusion length. The time-resolved optoacoustic detection technique permits accurate measurements of the absorbed optical energy distribution from the illuminated surface to the entire depth of tissue. By combining the optoacoustic measurement and diffusion theory, one can determine the effective optical attenuation coefficient in a turbid medium over a wide range of optical properties. The upper limit of this range is limited by the temporal resolution of the optoacoustic transducer and/or the laser pulse duration. For ultrawide-band acoustic transducers, the typical laser pulse duration of 6–8 ns limits the effective optical attenuation coefficient to approximately 1000 cm^{-1}. The lower limit of the measurement range is determined by the frequency range of the transducer and the diffraction of the ultrasonic waves. With a laser beam that is 2–3 cm in diameter, an effective optical attenuation coefficient of less than 1 cm^{-1} can be measured. The measurement accuracy for the effective optical attenuation coefficient over this range can be as high as 0.5%, but depends on the temporal resolution and the dynamic range of the optoacoustic signal detection.

Independent determination of the optical absorption and the optical scattering coefficients with the time-resolved optoacoustic method is possible if the thermomechanical properties of the medium and the sensitivity spectrum of the ultrasonic transducer are known. Due to a number of experimental parameters employed in calculating the optical absorption coefficient (i.e., the depth of the subsurface maximum of the optical fluence, the thermoacoustic efficiency, the absolute sensitivity of the optoacoustic transducer, the acoustic attenuation spectrum in the medium), the accuracy of optoacoustic measurements of optical properties in phantoms will not exceed 5%. With biological tissue samples, the uncertainty of parameters increases even further, so that it is reasonable to expect the accuracy of optical properties measurement to be at the level of 10–15%.

The optoacoustic method is based on optical absorption, so it is limited to media (tissues) and optical spectra where at least minimal optical absorption is present. The minimal required optical absorption is determined by the sensitivity of the acoustic detection system and the applied laser fluence. The demand for high sensitivity becomes even more challenging with understanding that optoacoustic must operate within a very wide band of ultrasonic frequencies, typically in the range of tens of kilohertz to tens of megahertz. The best transducers developed to date in our group allow detection of pressure amplitudes of approximately 1 Pa. With the high sensitivity of the piezoelectric transducers described previously, biological tissues with the optical absorption coefficient greater than 0.01 cm^{-1} can be successfully studied by the time-resolved optoacoustic method.

At present, time-intensive efforts are being made to develop various methods of optoacoustic tomography and noninvasive methods that could perform accurate and reliable measurements of concentrations of various physiologically important molecules in biological tissues. The optoacoustic spectroscopic imaging method offers not only selective excitation of biological chromophores by laser pulses at selected wavelengths, which in turn allows noninvasive measurement of molecular concentrations in tissues, but also high spatial resolution. Specific optical absorption by clinically relevant molecules such as hemoglobin and oxyhemoglobin, melanin, various lipids, and water enables safe and convenient *in vivo* measurement of these molecules by the optoacoustic method with resolution down to microns and depths in tissue not attainable by purely optical methods.

Another group of potential applications includes real-time monitoring of changes in tissue optical properties associated with therapeutic interventions. These interventions might be chemical, such as anti-angiogenesis cancer chemotherapy, or thermal, such as tissue ablation with high-intensity focused ultrasound (HIFU).[25,26,77] An interesting potential application enabled by the pulsed optoacoustic method's ability to accurately measure the effective optical attenuation is noninvasive continuous glucose monitoring in patients with diabetes.[78] Glucose, as a strong osmolyte, changes optical scattering in tissue, thereby affecting the profile of the optoacoustic signals. Therefore, optoacoustic measurements of the effective optical attenuation yield relative changes in glucose concentration in biological tissue. To correlate changes in the optoacoustic temporal profile with absolute values of glucose levels in the blood, the methods must be calibrated by relating the tissue optical scattering to the absolute concentration of glucose. The optoacoustic technique is more accurate than purely optical techniques that measure tissue scattering.

The merit of optoacoustic imaging and monitoring relates to its ability to provide information regarding functional parameters and molecular composition at the depth of tissues and with high spatial resolution. The sensitivity and resolution of the time-resolved optoacoustic method extend biomedical applications of optical spectroscopy to tissue depths of up to 5–6 cm. The most promising clinical applications include diagnostic imaging of cancer, detection of stroke and monitoring of the brain vasculature, detection of vulnerable-to-rupture atherosclerotic plaques and monitoring of coronary vessels of the heart, assessment of the functionality of deep veins and arteries, and blood oximetry with spatial resolution in critical care patients.[50,79,80]

References

1. W.M. Star, J.P.A. Marijnissen, and M.J.C. van Gemert, "Light dosimetry in optical phantoms and in tissues," *Phys. Med. Biol.* **33**, 437–454 (1988).
2. W.M. Star, "Light dosimetry *in vivo*," *Phys Med Biol.* **42**(5), 763–787 (1997).
3. V.V. Tuchin, "Laser light scattering in biomedical diagnostics and therapy," *J. Laser Appl.* **5**(2–3), 43–60 (1993).
4. B. Chance and R.R. Alfano, Eds., "Optical tomography: photon migration, and spectroscopy of tissue and model media: theory, human studies, and instrumentation," *Proc. SPIE* 2389 (1995).
5. V.V. Tuchin, Tissue Optics: Light Scattering Methods and Instruments for Medical Diagnosis, 3rd ed., PM 254, SPIE Press, Bellingham, Wash. (2015).
6. Deán-Ben X, Fehm TF, and Razansky D. Universal hand-held three-dimensional optoacoustic imaging probe for deep tissue human angiography and functional preclinical studies in real time. J Vis Exp. **4**(93), e51864 (2014).
7. J.G. Fujimoto, C. Pitris, S.A. Boppart, and M.E. Brezinski, "Optical coherence tomography: an emerging technology for biomedical imaging and optical biopsy," *Neoplasia* **2**(1–2), 9–25 (2000).
8. N. Shah, A. Cerussi, C. Eker, J. Espinoza, J. Butler, J. Fishkin, R. Hornung, and B. Tromberg, "Noninvasive functional optical spectroscopy of human breast tissue," *Proc. Natl. Acad. Sci. USA* **98**(8), 4420–4425 (2001).

9. G. J. Müller, B. Chance, and R. Alfano, et al., Eds., *Medical Optical Tomography: Functional Imaging and Monitoring*, SPIE Press, Bellingham, Wash. (1993).
10. J.A. Parrish and B.C. Wilson, "Current and future trends in laser medicine," *Photochem. Photobiol.* **53**, 731–738 (1991).
11. S.L. Jacques, "Simple theory, measurements and rules of thumb for dosimetry during photodynamic therapy," *Proc. SPIE* **1065**, 100–108, (1989).
12. L.I. Grossweiner, "PDT light dosimetry revisited," *J. Photochem Photobiol. B* **38**(2–3), 258–268 (1997).
13. L.I. Grossweiner, "Photodynamic therapy," *J. Laser Appl.* **7**(1), 51–57 (1995).
14. F. Fankhauser, H. Giger, P. Niederer, and T. Seiler, "Transpupillary laser phototherapy of tumors and vascular anomalies of retina and choroid: theoretical approach and clinical implications," *Technol. Health Care* **8**(2), 93–112 (2000).
15. O. Topaz, "Plaque removal and thrombus dissolution with the photoacoustic energy of pulsed-wave lasers–biotissue interactions and their clinical manifestations," *Cardiology* **87**(5), 384–391 (1996).
16. R. R. Anderson, Ed., "Lasers in surgery: advanced characterization, therapeutics, and systems: V," *Proc. SPIE* 2395 (1995).
17. L. Goldman, "Future of laser dermatology," *Lasers Surg. Med.* **22**(1), 3–8 (1998).
18. S.L. Kilmer and J.M. Garden, "Laser treatment of pigmented lesions and tattoos," *Semin. Cutan. Med. Surg.* **19**(4), 232–244 (2000).
19. P.R. Egbert, S. Fiadoyor, D.L. Budenz, P. Dadzie, and S. Byrd, "Diode laser transscleral cyclophotocoagulation as a primary surgical treatment for primary open-angle glaucoma," *Arch. Ophthalmol.* **119**(3), 345–350 (2001).
20. C.J. Chang and A. May-Kuen Wong, "Intraductal laser photocoagulation of the bilateral parotid ducts for reduction of drooling in patients with cerebral palsy," *Plast. Reconstr. Surg.* **107**(4), 907–913 (2001).
21. S.L. Jacques, "Laser–tissue interactions: photochemical, photothermal, and photomechanical," *Surg. Clin. North Am.* **72**(3), 531–558 (1992).
22. M.A. Suhr, C. Hopper, L. Jones, J.G. George, S.G. Bown, and A.J. MacRobert, "Optical biopsy systems for the diagnosis and monitoring of superficial cancer and precancer," *Int. J. Oral Maxillofac. Surg.* **29**(6), 453–457 (2000).
23. K. Nahen, W. Eisenbeis, and A. Vogel, "Acoustic on line monitoring of IR laser ablation of burnt skin," *Proc. SPIE* **3916**, 218–226 (2000).
24. R.O. Esenaliev, M. Motamedi, A.A. Karabutov, and A.A. Oraevsky, "Laser optoacoustic technique for real-time measurement of thermal damage in tissues," *Proc. SPIE* **3594**, 98–109 (1999).

25. F. Wu, Z-B. Wang, and P. Lu, Z-Li Xu, W-Z. Chen, H. Zhu, and C.-B. Jin, "Activated anti-tumor immunity in cancer patients after high intensity focused ultrasound ablation," *Ultrasound Med. Biol.* **30**(9), 1217–1222 (2004).
26. M.R. Bailey, V.A. Khokhlova, O.A. Sapozhnikov, S.G. Kargl, and L.A. Crum, "Physical mechanisms of the therapeutic effect of ultrasound," *Acoust. Phys.* **49**(4), 369–388 (2003).
27. D. Huang, E.A. Swanson, C.P. Lin, J.S. Schuman, W.G. Stinson, W. Chang, M.R. Hee, T. Flotte, K. Gregory, C. Puliafito, and J.G. Fujimoto, "Optical coherence tomography," *Science* **254**, 1178–1181 (1991).
28. S.A. Prahl, I.A. Vitkin, U. Bruggemann, B.S. Wilson, and R.R. Anderson, "Determination of optical properties of turbid media using pulsed photothermal radiometry," *Phys. Med. Biol.* **37**, 1203–1217 (1992).
29. T.H. Pham, O.C. Coquoz, J.B. Fishkin, E.A. Anderson, and B.J. Tromberg, "A broad bandwidth frequency domain instrument for quantitative tissue optical spectroscopy," *Rev. Sci. Instr.* **71**, 2500–2513 (2000).
30. F. Bevilacqua, A.J. Berger, A.E. Cerussi, D. Jakubowski, and B.J. Tromberg, "Broadband absorption spectroscopy in turbid media by combined frequency-domain and steady-state methods," *Appl. Opt.* **39** (34), 6498–6507 (2000).
31. Y. Yang, H. Liu, Xingde Li, and B. Chance, "Low-cost frequency-domain photon migration instrument for tissue spectroscopy, oxymetry and imaging," *Opt. Eng.* **36**, 1562–1569 (1997).
32. F. Bevilacqua, D. Piguet, P. Marquet, J.D. Gross, B.J. Tromberg, and C. Depeursinge, "In vivo local determination of tissue optical properties: applications to human brain," *Appl. Opt.* **38**(22), 4939–4950 (1999).
33. L. Kienle, M. Lilge, S. Patterson, R. Hibst, R. Steiner, and C. Wilson, "Spatially resolved absolute diffuse reflectance measurements for noninvasive determination of the optical scattering and absorption coefficients of biological tissue," *Appl. Opt.* **35**(13), 2304–2314 (1996).
34. R.K. Wang and Y.A. Wickramasinghe, "Fast algorithm to determine optical properties of a turbid medium from time-resolved measurements," *Appl. Opt.* **37**(31), 7342–7351 (1997).
35. A. Pifferi, J. Swartling, E. Chikoidze, A. Torricelli, P. Taroni, A. Bassi, S. Andersson-Engels, and R. Cubeddu, "Spectroscopic time-resolved diffuse reflectance and transmittance measurements of the female breast at different interfiber distances," *J. Biomed. Opt.* **9**, 1143–1151 (2004).
36. D. Comelli, A. Bassi, A. Pifferi, P. Taroni, A. Torricelli, R. Cubeddu, F. Martell, and G. Zaccanti, "In vivo time-resolved reflectance spectroscopy of the human forehead," *Appl. Opt.* **46**(10), 1717–1725 (2007).

37. C. Abrahamsson, T. Svensson, S. Svanberg, and S. Anderson-Engels, "Time and wavelength resolved spectroscopy of turbid media using light continuum generated in a crystal fiber," *Opt. Express* **12**(17), 4103–4117 (2004).
38. L. Spinelli, A. Torricelli, A. Pifferi, P. Taroni, G.M. Danesini, and R. Cubeddu, "Bulk optical properties and tissue components in the female breast from multiwavelength time-resolved optical mammography," *J. Biomed. Opt.* **9**, 1137–1142 (2004).
39. A. Kienle, M.S. Patterson, N. Dognitz, R. Bays, G. Wagnieres, and H. van de Bergh, "Noninvasive determination of the optical properties of two-layered turbid media," *Appl. Opt.* **37**(4), 779–791 (1998).
40. T.J. Farrell, M.S. Patterson, and M. Essenpreis, "Influence of layered tissue architecture on estimates of tissue optical properties obtained from spatially resolved diffuse reflectometry," *Appl. Opt.* **37**(10), 1958–1972 (1998).
41. A.I. Kholodnykh, I.Y. Petrova, K.VV. Larin, M. Motamedi, and R.O. Esenaliev, "Precision of measurement of tissue optical properties with optical coherence tomography," *Appl. Opt.* **42**(16), 3027–3037 (2003).
42. X.J. Wang, T.E. Milner, J.F. de Boer, Y. Zhang, D.H. Pashley, and J.S. Nelson, "Characterization of dentin and enamel by use of optical coherence tomography," *Appl. Opt.* **38**(10), 2092–2096 (1999).
43. I.A. Vitkin, B.C. Wilson, and R.R. Anderson, "Analysis of layered scattering materials by pulsed photothermal radiometry: application to photon propagation in tissue," *Appl. Opt.* **34**, 2973–2982 (1995).
44. S.L. Jacques, J.S. Nelson, and W.H. Wright, and T.E. Milner: "Pulsed photothermal radiometry of portwine-stain lesions," *Appl. Opt.* **32**, 2439–2446 (1993).
45. V.E. Gusev and A.A. Karabutov, *Laser Optoacoustics*, American Institute of Physics, New York (1993).
46. A.A. Oraevsky, S.L. Jacques, and F.K. Tittel, "Determination of tissue optical properties by time-resolved detection of laser-induced stress waves," *Proc. SPIE* **1882**, 86–101 (1993).
47. A.A. Oraevsky, S.L. Jacques, R.O. Esenaliev, and F.K. Tittel, "Direct measurement of laser fluence distribution and optoacoustic imaging in heterogeneous tissues," *Proc. SPIE* **2323**, 250–261 (1994).
48. A.A. Oraevsky, S.L. Jacques, and F.K. Tittel, "Measurement of tissue optical properties by time-resolved detection of laser-induced transient stress," *Appl. Opt.* **36**, 402–415 (1997).
49. R.O. Esenaliev, A.A. Karabutov, M. Motamedi, and A.A. Oraevsky, "Real-time optoacoustic monitoring of photothermal laser-tissue interactions," *Proc. SPIE* **3601**, 268–274 (1999).
50. A. A. Oraevsky, "Optoacoustic Tomography: From Fundamentals to Diagnostic Imaging of Breast Cancer," in *Biomedical Photonics Handbook, Second Edition: Fundamentals, Devices, and Techniques*,

T. Vo-Dinh, Ed., CRC Press, Boca Raton, Florida, **PM222**, 715–757 (2014).

51. A.A. Karabutov, N.B. Podymova, and V.S. Letokhov, "Time-resolved optoacoustic measurement of light absorption by inhomogeneous media," *Appl. Opt.* **34**, 1484–1487 (1995).
52. A.A. Oraevsky, R.O. Esenaliev, F.K. Tittel, M. Ostermeyer, L. Wang, and S.L. Jacques, "Laser opto-acoustic imaging of turbid media: determination of optical properties by comparison with diffusion theory and Monte Carlo simulation," *Proc. SPIE* **2681**, 277–284 (1996).
53. A.A. Karabutov, I.M. Pelivanov, N.B. Podymova, and S.E. Skipetrov, "Measurement of optical properties of turbid media by laser opto-acoustic method", *Quant. Electron.* **29**, 86–91 (1999).
54. A.A. Oraevsky, S.L. Jacques, R.O. Esenaliev, and F.K. Tittel, "Radial and z-axial resolution in laser optoacoustic imaging with ultrasonic transducers," *Proc. SPIE* **2389**, 209–217 (1995).
55. A. A. Oraevsky, S. L. Jacques, R. O. Esenaliev, and F. K. Tittel, "Time-resolved optoacoustic imaging in layered biological tissues," in *Advances in Optical Imaging and Photon Migration*, Robert R. Alfano, Ed., Academic Press, **21**, 161–165 (1994).
56. S.A. Telenkov and A. Mandelis, "Fourier-domain biophotoacoustic subsurface depth selective amplitude and phase imaging of turbid phantoms and biological tissue," *J. Biomed. Opt.* **11**(4), 044006-1–044006-10 (2006).
57. A.A. Oraevsky, R.O. Esenaliev, S.L. Jacques, and F.K. Tittel, "Laser opto-acoustic tomography for medical diagnostics: principles," *Proc. SPIE* **2676**, 22–31 (1996).
58. A.A. Oraevsky, "A nanosecond acoustic transducer with applications in laser medicine," *LEOS Newslett.* **8**(1), 6–8 (1994).
59. C. Kopp and R. Niessner, "Depth-resolved analysis of aqueous samples by optoacoustic spectroscopy," *Appl. Phys. B* **68**, 719–725 (1999).
60. A.A. Karabutov, E.V. Savateeva, N.B. Podymova, and A.A. Oraevsky, "Backward mode detection of laser-induced wide-band ultrasonic transients with optoacoustic transducer," *J. Appl. Phys.* **87**(4), 2003–2014 (2000).
61. E. Khachatryan, S. Maswadi, D. A. Tsyboulski, E. Barnes, D. Sardar, A. A. Oraevsky, K. Nash, and R. Glickman, "Optoacoustic microscopy using laser beam deflection technique," *Proc. SPIE* **v.8943**, 89432T (2014) doi: 10.1117/12.2040718
62. G. Paltauf, R. Nuster, M. Haltmeier, and P. Burgholzer, "Photoacoustic tomography using a Mach–Zehnder interferometer as an acoustic line detector," *Appl. Opt.* **46**(16), 3352–3358 (2007).

63. E. Zhang, J Laufer, and P. Beard, "Backward-mode multiwavelength photoacoustic scanner using a planar Fabry–Perot polymer film ultrasound sensor for high-resolution three-dimensional imaging of biological tissues," *Appl. Opt.* **47**, 561–577 (2008).
64. R. Pini, R. Salimbeni, M. Vannini, and G. Toci, "Probe-beam deflection diagnostics of shockwaves generated during laser drilling," *Appl. Phys. B* **61**, 505–510 (1995).
65. A.A. Karabutov and A.A. Oraevsky, "Ultimate sensitivity of wide-band detection for laser-induced ultrasonic transients," *Proc. SPIE* **3916**, 228–239 (2000).
66. A. Isimaru, Wave Propagation and Scattering in Random Media, Academic Press, New York (1978).
67. A. A. Karabutov, I. M. Pelivanov, N. B. Podymova, and S. E. Skipetrov, "Direct measurement of the spatial distribution of light intensity in a scattering medium," *J. Exp. Theor. Phys. Left. (JETP) Letters.* **70**(3), 183–188 (1999).
68. P.S. Grashin, A.A. Karabutov, I.M. Pelivanov, N.B. Podymova, A.A. Oraevsky, E.V. Savateeva, and V.S. Solomatin, "Distribution of the laser radiation intensity in turbid media: Monte-Carlo simulations, theoretical analysis and the results of optoacoustic measurements," *Quantum Electron.* **32**(10), 860–874 (2002).
69. L-H. Wang, S.L. Jacques, and L-Q. Zheng, "MCML: Monte Carlo modeling of photon transport in multi-layered tissues," *Comput. Meth. Progr. Biomed.* **47**, 131–146 (1995).
70. S. Ermilov, M. Fronheiser, H.-P. Brecht, R. Su, A. Conjusteau, K. Mehta, P. Otto, and A. Oraevsky, "Development of laser optoacoustic and ultrasonic imaging system for breast cancer utilizing handheld array probes," in *Photons Plus Ultrasound: Imaging and Sensing*, A. Oraevsky and L. Wang, Eds. *Proc. SPIE* **7177**, **717703**, (2009).
71. J. J. Duderstadt and L. J. Hamilton, *Nuclear Reactor Analysis*, Wiley, New York (1976).
72. I. M. Pelivanov, M. D. Barskaya, N. B. Podymova, T. D. Khokhlova, and A. A. Karabutov, "Optoacoustic measurement of the local optical absorption coefficient in turbid media: I. Monte-Carlo simulation of laser fluence distribution at the beam axis beneath the surface of a turbid medium," *Quant. Electron.* **39**(9), 831–834, (2009).
73. I. M. Pelivanov, M. D. Barskaya, N. B. Podymova, T. D. Khokhlova, and A. A. Karabutov, "Optoacoustic measurement of the local optical absorption coefficient in turbid media: II. on the possibility of optical absorption coefficient measurement in a turbid medium from the amplitude of the optoacoustic signal," *Quant. Electron.* **39**(9), 835–838, (2009).
74. H. C. van de Hulst, Multiple Light Scattering: Tables, Formulas and Applications, 2, Academic Press, New York (1980).

75. T. Spott and L. Svaasand, "Collimated light sources in the diffusion approximation," *Appl. Opt.* **39**(34), 6453–6465 (2000).
76. I.M. Pelivanov, S.A. Belov, V.S. Solomatin, T.D. Khokhlova, and A.A. Karabutov, "Direct opto-acoustic *in vitro* measurement of the spatial distribution of laser radiation in biological media," *Quantum Electron.* **36** (12), 1089–1096 (2006).
77. T.D. Khokhlova, I.M. Pelivanov, O.A. Sapozhnikov, V.S. Solomatin, and A.A. Karabutov, "Opto-acoustic diagnostics of the thermal action of high-intensity focused ultrasound on biological tissues: the possibility of its applications and model experiments," *Quantum Electron.* **36**(12), 1097–1102 (2006).
78. A.A. Bednov and A.A. Oraevsky, "Comparison of glucose effect in model media and biological tissue measured using time-resolved optoacoustic method," *Asian J. Phys.* **15**(1), 55–66 (2006).
79. A. Oraevsky and L. Wang, Eds., *Proceedings of Photons plus Ultrasound: Imaging and Sensing, SPIE/BIOS Conference held annually under auspices of Photonics West symposium*, SPIE Press, Bellingham, 2000–2015.
80. L. Wang, Ed., *Photoacoustic Imaging and Spectroscopy*, Taylor-Francis Group, Boca Raton, Fla. (2009).

Ivan M. Pelivanov is the assistant professor at the Physics Faculty of M.V. Lomonosov Moscow State University and visiting assistant professor at the University of Washington (Washington, USA). He graduated from the group of Professor A.A. Karabutov, which is a pioneering and renowned team in various physical and biological applications of optoacoustic spectroscopy, and received his PhD in 2000. His recent research focuses on designing sensitive wide-band detectors, application of optoacoustic method in NDT and material evaluation, analytic chemistry and in medicine. He is one of the most active members in optoacoustic research for more than a decade.

Tatiana D. Khokhlova received her PhD in physics in 2008 from Moscow State University, Moscow, Russia. After graduation from the PhD program she moved to the University of Washington (UW) in Seattle for postdoctoral training at the Applied Physics Laboratory, and then completed a research fellowship at the Department of Medicine, UW. In 2012, she received a Career Development Award from NIH and took a research faculty position at the Department of Medicine. Her research interests are in physical acoustics, biomedical applications of high intensity focused ultrasound, and photo-acoustic imaging.

Alexander A. Oraevsky has over 25 years of experience in managing research and development laboratories in academia and small businesses. He received his initial training in physics and mathematics from the Moscow Physical and Engineering Institute in Moscow, Russia, and obtained a doctorate in laser spectroscopy and laser biophysics from the USSR Academy of Sciences. He began his pioneering research in the field of optoacoustic imaging, sensing and monitoring in 1988. In 1992, as Whitaker Fellow, he joined the faculty at Rice University. In 1998, he founded the first business dedicated to optoacoustic imaging, LaserSonix Technologies. Prior to founding TomoWave Laboratories in 2010, he served as vice-president of Research and Development for Fairway Medical Technologies, was the director of the Optoacoustic Imaging and Spectroscopy Laboratory at the University of Texas Medical Branch in Galveston, Texas, USA, and an assistant professor at the Department of Ophthalmology and Visual Sciences. Currently, he leads TomoWave Laboratories as Chief Technology Officer and holds an adjunct Professor position at the Biomedical Engineering Department of the University of Houston. He is the recipient of multiple research awards advancing biomedical applications of the optoacoustic imaging sensing and monitoring, including the Berthold Leibinger Innovations Prize 2014. He is the primary inventor of 21 patents, and has published seven book chapters and over 200 highly cited scientific papers dealing with novel laser technologies applicable in biology and medicine.

Index

N

O

P

V

W